FOREWORD

This issue of *Industrial Structure Statistics* (ISS) is the sixteenth edition of the series. It comes in two volumes. Volume 1 - Core Data - provides detailed annual statistics on production, value added, employment, exports/imports, etc. for manufacturing and non-manufacturing sectors. This data facilitates an assessment of the evolution of the industrial structure of OECD Member countries. The publication was prepared by the Main Economic Indicators Division of the Statistics Directorate in the OECD Secretariat under the auspices of the Statistical Working Party on Industrial Statistics of the Industry Committee. The Secretariat wishes to thank the representatives of all the national authorities who have assisted in the preparation of this publication.

All series published in Volume 1, as well as data for earlier years, are held on the database, *Information System on Industrial* Structures (ISIS), and are available on diskette (ISS - Vol.1: Core Data, 1999). Moreover, the electronic version contains more detailed statistics relating to variables and to industrial and service sectors which are not presented in the paper publication (See inside cover page). These statistics are published on the responsibility of the Secretary-General of the OECD.

Volume 2 presents annual energy consumption data in manufacturing sectors. It was prepared by the Energy Statistics Division of the International Energy Agency (IEA). The energy data have also been collected under the auspices of the Statistical Working Party on Industrial Statistics of the Industry Committee. These data are also available on diskette (ISS - Vol. 2: Energy Consumption, 1999).

AVANT-PROPOS

Cette édition constitue la seizième parution des *Statistiques des structures industrielles* (SSI). Elle comprend deux volumes. Le Volume 1 - Données de base - présente des données structurelles détaillées relatives à la production, la valeur ajoutée, l'emploi, les exportations et les importations, etc. pour les secteurs manufacturier et non-manufacturier. Ces données permettent d'évaluer l'évolution de la structure industrielle des pays Membres de l'OCDE. Cette publication a été préparée par la Division des principaux indicateurs économiques de la Direction Statistique du Secrétariat de l'OCDE sous l'égide du Groupe de travail sur les Statistiques industrielles du Comité de l'industrie. Le Secrétariat tient à remercier les représentants de toutes les administrations nationales qui ont apporté leur concours à la préparation de cette publication.

L'ensemble des statistiques publiées dans le Volume 1 ainsi que celles portant sur des années antérieures sont contenues dans la base de données *Système d'Information sur les Structures Industrielles* (SISI) et sont disponibles sur disquette (SSI - Vol. 1 : Données de base, 1999). En outre, la version électronique contient davantage de statistiques détaillées relatives à des variables et à des secteurs de l'industrie et des services qui ne sont pas présentés dans la publication papier (voir en page de couverture). Ces statistiques sont publiées sous la responsabilité du Secrétaire général de l'OCDE.

Le Volume 2 présente des données annuelles de consommation d'énergie dans l'industrie manufacturière. Il a été préparé par la division des statistiques de l'énergie de l'Agence Internationale de l'Energie (AIE). Les données sur l'énergie ont elles aussi été collectées sous l'égide du Groupe de travail sur les Statistiques industrielles du Comité de l'industrie. Ces données sont également disponibles sur disquette (SSI - Vol. 2 : Consommation d'Energie, 1999).

Industrial Structure Statistics

Volume 2
ENERGY CONSUMPTION

Statistiques des structures industrielles

Volume 2
CONSOMMATION D'ÉNERGIE

1999 Edition

INTERNATIONAL ENERGY AGENCY

2, RUE ANDRÉ-PASCAL, 75775 PARIS CEDEX 16, FRANCE
9, RUE DE LA FÉDÉRATION, 75739 PARIS CEDEX 15, FRANCE

The International Energy Agency (IEA) is an autonomous body which was established in November 1974 within the framework of the Organisation for Economic Co-operation and Development (OECD) to implement an international energy programme.

It carries out a comprehensive programme of energy co-operation among twenty-four* of the OECD's twenty-nine Member countries. The basic aims of the IEA are:

i) co-operation among IEA participating countries to reduce excessive dependence on oil through energy conservation, development of alternative energy sources and energy research and development;

ii) an information system on the international oil market as well as consultation with oil companies;

iii) co-operation with oil producing and other oil consuming countries with a view to developing a stable international energy trade as well as the rational management and use of world energy resources in the interest of all countries;

iv) a plan to prepare participating countries against the risk of a major disruption of oil supplies and to share available oil in the event of an emergency.

IEA participating countries are: Australia, Austria, Belgium, Canada, Denmark, Finland, France, Germany, Greece, Hungary, Ireland, Italy, Japan, Luxembourg, the Netherlands, New Zealand, Norway (by special agreement), Portugal, Spain, Sweden, Switzerland, Turkey, the United Kingdom, the United States. The Commission of the European Communities takes part in the work of the IEA.

ORGANISATION FOR ECONOMIC CO-OPERATION AND DEVELOPMENT

Pursuant to Article 1 of the Convention signed in Paris on 14th December 1960, and which came into force on 30th September 1961, the Organisation for Economic Co-operation and Development (OECD) shall promote policies designed:

- to achieve the highest sustainable economic growth and employment and a rising standard of living in Member countries, while maintaining financial stability, and thus to contribute to the development of the world economy;

- to contribute to sound economic expansion in Member as well as non-member countries in the process of economic development; and

- to contribute to the expansion of world trade on a multilateral, non-discriminatory basis in accordance with international obligations.

The original Member countries of the OECD are Austria, Belgium, Canada, Denmark, France, Germany, Greece, Iceland, Ireland, Italy, Luxembourg, the Netherlands, Norway, Portugal, Spain, Sweden, Switzerland, Turkey, the United Kingdom and the United States. The following countries became Members subsequently through accession at the dates indicated hereafter: Japan (28th April 1964), Finland (28th January 1969), Australia (7th June 1971), New Zealand (29th May 1973), Mexico (18th May 1994), the Czech Republic (21st December 1995), Hungary (7th May 1996), Poland (22nd November 1996) and Korea (12th December 1996). The Commission of the European Communities takes part in the work of the OECD (Article 13 of the OECD Convention).

INTRODUCTORY REMARKS

The ISIS Energy Data Pilot Project was launched in 1995 in close collaboration between the Energy Statistics Division of the IEA and the Statistics Directorate of the OECD. Energy consumption data in manufacturing industry are collected as part of (Table 4) the annual ISIS questionnaire "Industrial Statistics". In the 48th meeting of the OECD Statistical Working Party of the Industry Committee in October 1997, the Pilot Phase of the Project came to an end and the collection of energy data has been made a permanent feature of the ISIS questionnaire.

The aims of the ISIS Energy Data Programme are:

- to establish a unified process for the collection of official manufacturing industry energy consumption data at a disaggregated level;

- to pursue review of energy efficiency indicators studies;

- to improve our understanding of where and how energy is used in the OECD member countries;

- to provide Member countries and researchers data to allow them to make inter-country comparisons;

- to support the Secretariat's programme on energy and environment;

- to provide the information required to track progress on energy efficiency by simplifying analysis of energy efficiency trends and by providing inter-country comparisons.

The energy data presented in this volume have been collected during the Pilot phase of the project. The data contains time series of annual energy consumption in manufacturing industry for most of the OECD Member countries (and one non-member country, the Slovak Republic) from 1990 to 1998 where available.

Although the consistency of the data with the methodology has been checked, discrepancies that still remain for some sectors and some countries (when known) together with explanatory notes on the collected data have been reported in the country notes section. However, there may still exist some problems which have not been identified. Consequently, we would be grateful if you could contact us about any anomaly you find in order to allow us to make corrections.

Since the data are submitted in either ISIC Revision 2[1] or Revision 3[2] format, databases are not complete. The latest details of these databases and the ISIS Energy Data Progamme Future Work Plan are described on the World Wide Web at http://iea.org/stats/files/isis.htm.

For the calculation of energy indicators we recommend that you use the corresponding socio-economic statistics collected by the OECD and especially those reported in the main ISIS publication.

---

[1] International Standard Industrial Classification of All Economic Activities. Statistical Papers, Series M, No. 4, Rev.2, United Nations, New York, 1968.
[2] International Standard Industrial Classification of All Economic Activities. Statistical Papers, Series M, No. 4, Rev.3, United Nations, New York, 1990.

TABLE OF CONTENTS

ISIS Energy Data Programme (IEA/OECD)

PART III

ENERGY CONSUMPTION IN MANUFACTURING INDUSTRY BY FUEL TYPES

PART IV

ENERGY CONSUMPTION IN MANUFACTURING INDUSTRY BY FUEL TYPES

METHODOLOGY

The type of energy data collected by the International Energy Agency (IEA) since it was established in 1974 has largely reflected the energy security concerns of its Member countries. IEA collects data on energy consumption in industry in four annual questionnaires (*Oil, Solid Fuels, Natural Gas* and *Electricity and Heat*) in a format designed to facilitate the construction of national energy balances. In these questionnaires, the use of fuels by industrial enterprises for transport, for the production of other fuels (i.e. for transformation), and for own consumption in energy producing industries, is not included in final consumption and allocated to the specific industry, but combined and reported separately as Transport, Transformation, Energy Sector etc. The ISIS energy data programme overcomes this shortcoming by requiring all uses of fuels to be reported by the actual consuming industry.

In addition, the ISIS Energy Data Programme has other advantages: By allowing for detailed analyses of energy demand in industry, it reveals opportunities for improving energy efficiency, as well as providing the information required to track progress on the energy efficiency front. Since it has a unified data collection methodology, the data are consistent and internationally comparable. Consistency of the ISIS energy data with OECD economic statistics provides a key tool to link economic and energy variables. Finally, its structure allows for energy efficiency studies in disaggregated manufacturing industry. IEA data are available in 2 digit ISIC level for manufacturing industry; ISIS data are available in 4 digit (in ISIC Revision 2 and/or Revision 3). It is therefore possible to calculate energy efficiency indicators for a number of manufacturing industry groups that can be classified according to different aggregation schemes, e.g., based on technology, wages, orientation, skills and environmental pollution.

A comparison of the IEA and ISIS approaches is illustrated in a schematic representation set out below.

ISIS Methodology:

Energy consumption in a manufacturing industry (in ISIS) = Energy consumption for

| the actual *production* activity |
| + |
| energy *transformation* activity |
| + |
| *own use* (of energy in transformation processes) |
| + |
| *transportation* activity |

in the industry

IEA Methodology:

Energy consumption in a manufacturing industry (in IEA) = Energy consumption for

the actual *production* activity   in the industry

Production activity comprises the use of purchased primary and secondary fuels that are not transformed (ie, disappear) in the production activity.

Transformation comprises the conversion of primary forms of energy to secondary and further transformation (e.g. coking coal to coke; crude oil to petroleum products; heavy fuel oil to electricity; PCI coal, coke oven to coke; natural gas and oil to blast furnace gas or coke oven gas; fuel inputs to electricity/heat etc.).

Own use refers to the primary and secondary energy consumed during transformation. It covers energy consumed for: heating; lighting; operation of all equipment used in the extraction process; traction; and distribution.

Transportation (on-site) relates to the movement of materials by pipeline, road, railway, air and internal navigation.

In other words, the energy data in the ISIS Energy Database covers the amount of primary and secondary fuels purchased to support the activity of the industry in question. Moreover, if the transformation output is sold to third parties (including electricity and steam) then the corresponding inputs are reduced accordingly if known. In addition, the quantity of electricity consumed and the quantity of electricity produced on-site for its own use are asked for separately. Therefore, "Electricity" refers to purchased electricity plus electricity that is generated and is consumed on-site, whereas "steam" refers to purchased steam only.

The ISIS Energy Data Programme covers 8 types of fuel classes: solid fuels, LPG, distillate oils, residual fuel oil, gas, biomass fuels, steam, and electricity.

- *Solid Fuels* (Solid) include anthracite, steam coal, coking coal, sub-bituminous coal, lignite, peat, gas coke, coke oven coke, patent fuel, BKB (Braunkohlenbrikettes), petroleum coke.

- *Liquefied Petroleum Gas* (LPG) includes ethane, propane, and butane.

- *Distillate Oils and Others* (Distiloil) include naphtha, gasolines (motor, aviation), kerosene, jet fuel (gasoline or kerosene type), gas oil/diesel oil, other petroleum products.

- *Gas* (Gas) includes natural gas, coke oven gas, blast furnace gas, refinery gases, gas works gas, oxygen steel furnace gas.

- B*iomass fuels* (Biomass) include wood and wood wastes, ethanol, black liquor, sludge/sewage gases, landfill gas, animal products and waste, industrial waste, and municipal waste.

- *Steam* (Steam) includes heat.

- *Electricity* (Electr) includes production from solar, hydro, wind and geothermal on-site)

- *of which generated on site for own use* (own use)

- *Total* = Solid + LPG + Distiloil + RFO + Gas + Biomass + Steam + Electr - Own use

ISIS Energy Data excludes the quantities of fuels used for non-energy purposes[3] and quantities of fuels purchased but resold. Non-energy use includes the use of energy products as raw materials (such as white and industrial spirits, lubricants, bitumen and petroleum waxes) in different sectors; that is, those not consumed as a fuel or transformed into another fuel.

---

[3]   Non-energy uses of fuels covers their use i) as raw material for the manufacture of, for example, plastics or fertilizers, ii) for their specific physical properties (such as white spirit, paraffin waxes, lubricants and bitumen) as lubricants or roofing materials, iii) for their chemical properties (petrochemical feedstocks).

The energy content of a fuel can be measured as the heat released on complete combustion. This energy content is referred to as a fuel's calorific value (or heat content), and it can be expressed as a gross (or higher) value, or a net (or lower) value. The burning of fossil fuels includes a loss of energy through the combination of hydrogen and oxygen and the vaporisation of water. The heat value of fossil fuels before vaporisation is the Gross Calorific Value (GCV). The Net Calorific Value (NCV) is the amount of heat which is actually available from the combustion process for capture and end use, after the evaporation of moisture. Except electricity, the data are expressed in terms of **Terajoules** (TJ) **using Net Calorific Values** (NCV) of individual fuel types. The unit of electricity is Megawatt hours (MWh). 1 MWh = 0.0036 TJ.

Due to market liberalisation, some data have become confidential. The qualifier "c" in the data tables indicates where these confidential data are.

The complete data shown in this publication are available on diskettes suitable for use on IBM-compatible personal computers. An order form has been provided inside of the book.

Enquiries, comments and suggestions are welcome and should be addressed to:

Mr. Lawrence Metzroth
Energy Statistics Division
International Energy Agency
9, rue de la Federation,
75739 Paris Cedex 15
France

Tel: +33 1 40 57 6631
Fax: +33 1 4057 6649

E-mail: lawrence.metzroth@ iea.org

Dr. Sohbet Karbuz
Energy Statistics Division
International Energy Agency
9, rue de la Federation,
75739 Paris Cedex 15
France

Tel: +33 1 40 57 6621
Fax: +33 1 4057 6649

E-mail: sohbet.karbuz@ iea.org

COUNTRY NOTES

ISIS Energy Data are collected in 4-digit ISIC Revision 2 and/or ISIC Revision 3.

The first character in the industry sector code specifies to which ISIC Revision that sector belongs. The industry sector codes beginning with the letter "S" in the database indicates ISIC Revision 2 and "C" indicates ISIC Revision 3.

AUSTRALIA

General notes on collected data
Ethane is classified as GAS.

No available data for transportation in manufacturing industry.

Biomass includes bagasse. ISIC Sector S3119, S3122, S3114 are included in S3118.

Australian data submission to the IEA refers to the fiscal year July to June. Therefore, July 1994 to June 1995, for example, is considered as 1995.

Source of ISIS energy data
Australian Bureau of Statistics using FES (fuel and electricity survey).

Publications
Australian energy consumption and production; historical trends and projections to 2009/10. Reports data collected in *FES*. Historical data set: 1973-1994, 1994-2010 (forecast). Data refer to fiscal year.

AUSTRIA

General notes on collected data
C15 includes C16.

C17 includes C18 from 1996 onward.

C20 includes C36 up to 1996.

C24 includes C25 up to 1996.

C29 includes C369.

C28 includes C273, C33.

C30 includes C31, C32.

C34 includes C35.

Except refining industry, transformation input is added to the final consumption. The assumption made is that establishments do not sell their transformed outputs to other establishments.

From 1996 onward, the industrial classification system was changed from Betriebsystematik 68 to NACE.

Source of ISIS energy data
Austrian Central Statistics Office (ÖSTAT)

Publications
Energieversorgung Österreich: Entgültige Energiebilanz 19XX, ÖSTAT

ISIS Energy Data Programme (IEA/OECD)

Concordance between Austrian Classification System (Betriebsystematik 68) and ISIC rev 3.

(10) manuf. of food, beverages and tobacco ....ISIC 15,16

(11) manuf. of textiles, textile products ....ISIC 17

(12) manuf. of wearing apparel and bedding ....ISIC 18

(13) manuf. of leather, leather substitutes and footwear...ISIC 19

(14) manuf. of wood and wooden sheets..ISIC 20

(15) manuf. and processing of paper and paper prod. ....ISIC 21

(16) printing and reproduction ....ISIC 22

(17) manuf. of chemicals, rubber and plastic products ....ISIC 24, 25

(18) manuf. of derivatives of oil and natural gas ....ISIC 23

(19) manuf. of glass and glass products ....ISIC 26(1)

(20) iron and non-iron basic industry, semi-final products.......................................ISIC 271, 272

(21) metal processing, steel and light metal construction ....ISIC 273

(22) manuf. of metallic products...ISIC 28

(23) manuf. of measurement and control equipment, medical and optical goods.......ISIC 33

(24) manuf. of machinery except electrical..ISIC 29,369

(25) manuf. of electrotechnik apparatus..ISIC 30-32

(26) manuf. of transport equipment...ISIC 34,35

BELGIUM

General notes on collected data
Refinery gas is included in Distillate oils.

Source of ISIS energy data
Bilans Annuels Detailles, Ministere des Affaires Economiques

Publications
Energie en Belgique 19XX, Ministere des Affaires Economiques

CANADA

General notes on collected data
Coke oven gas is included in Solid fuels.

Source of ISIS energy data
Natural Resources Canada

Publications
Quarterly Report on Energy Supply-Demand in Canada, Natural Resources Canada.

CZECH REPUBLIC

Source of ISIS energy data
Czech Statistical Office Prague

DENMARK

General notes on collected data
C1511 includes C1552
C1549 includes C1544
C2222 includes C2230
C2429 includes C2430 and 2421
C2519 includes C2511
C2691 includes C2692
C271 includes C2731
C272 includes C2732
C2912 includes C2911
C2929 includes C2923 and C2927
C3599 includes C3520
C2699 includes C2693

The data include sold electricity and steam which corresponds to less than five percent of total electricity and steam consumption.

Source of ISIS energy data
Danmarks Statistik

Publications
Statistiske Efterretninger, Industri og Energi, Statistics Denmark

FINLAND

General notes on collected data
The data in Energy and Emissions, Statistics Finland, have been converted to the ISIS format by making some adjustments (the fuel inputs have been adjusted to the electricity and steam sold to third parties).

Source of ISIS energy data
Statistics Finland

Publications
Yearbook of Industrial Statistics Volume 1, Statistics Finland.

Energy and Emissions, Statistics Finland.

FRANCE

General notes on collected data
Consumption of combustible fuels is not calculated from primary fuels.

Source of ISIS energy data
Ministère de l'Economie, des Finances et de l'Industrie, Secrétatriat d'Etat à l'Industrie.

Publications
Tableaux des consommations d'énergie en France

GERMANY

General notes on collected data
S3211 is included in S3215 for 1994
S3219 is included in S3214 for 1992, 1993, 1994
S3419 is included in S3412 for 1993, 1994
S3540 is included in S3530 for 1992, 1993, 1994
S3529 is included in S3521 for 1992, 1993, 1994
S3819 and S3845 are included in S3800 for 1992
S3829 is included in S3821 for 1993, 1994
S3901 is included in S3909 for 1993, 1994
Data for LPG, Biomass, Steam and Own use are not available.
Heavy Fuel Oil includes medium fuel oil.

Source of ISIS energy data
Monthly Report on Mining and Manufacturing, Central Statistical Office

Publications
Energie Daten, Bundesministerium fuer Wirtschaft

Produzierendes Gewerbe, Statistisches Bundesamt

Ausgewaelte Zahlen zur Energiewirtschaft, Statistisches Bundesamt

HUNGARY

General notes on collected data
TEOR (Hungarian industry classification system) has changed in 1992. The new system TEOR 92 is compatible with ISIC but the old system, TEOR 87, is not. Moreover, there is no direct concordance between TEOR 87 and TEOR 92. Peat is reported in LPG.

Source of ISIS energy data
Energy Information Agency

Publications
Statistical Yearbook

ICELAND

General notes on collected data
Steam is used in the context "energy of geothermal origin".

Source of ISIS energy data
National Energy Authority

Publications
Energy in Ireland:1980-1993, Statistical Bulletin

JAPAN

Source of ISIS energy data
MITI

Publications
Handbook of Energy & Economic Statistics in Japan, The Energy Conservation Centre.

Structural Survey of Energy Consumption in Commerce, Mining and Manufacturing

Overall Energy Statistics Yearbook, called as Red Book "AKAHON".

KOREA

General notes on collected data
Electricity figures are only given for 2 digit sectors.

The figure for total manufacturing (S3000) includes the figure for activities that can not be correctly allocated to the appropriate industry (they are put in ERR1).

Nonspecified industry for solid fuels is put in ERR1.

Source of ISIS energy data
Korea Energy Economics Institute (KEEI)

Publications
Report on Energy Census

LUXEMBOURG

Source of ISIS energy data
Fédération des Industries Luxembourgeoises

Publications
Rapport d'activité, Ministère de l'Energie

MEXICO

General notes on collected data
The figures for own use in sector S3909 prior to 1993 include mining & construction.

Own use figures are not available before 1993.

 Consumption figures reported for the years 1990-1993 in petroleum refineries S3530 should have to be reported in S3511.

Source of ISIS energy data
Table: Consumo de energia en el sector industrial por rama de actividad (net TJ), Ministry of Energy and Public Industry

Publications
Balance Nacional de Energia

NETHERLANDS

General notes on collected data

Data are taken from column 17 of "de nederlandse energihuishouding deel 1" of Central Bureau of Statistics. Own use electricity consumption is taken from column 13.

Table 3.1.1 Food, beverages and tobacco ....C15 (ISIC Rev3)

Table 3.1.2 Textile, clothes and leather industry C17

Table 3.1.3 Paper industry, printing and publishing C21

Table 3.1.4 Fertilizer industry...C2412

Table 3.1.5 Organic Chemicals industry C2411

Table 3.1.6 Anorganic Chemicals industry C2411

Table 3.1.7 Other basic chemicals industry C2420

Table 3.1.8 Chemical products industry..C2411

Table 3.1.9 Building materials industry C26

Table 3.1.10 Iron and steel industry...C271

Table 3.1.11 Non-ferro metals industry C272

Table 3.1.12 Metal products industry...C28

Table 3.1.13 Plastics, rubber C25

Table 3.1.19 Non specified manufacturing C3699

Table 2.2.2 Refineries C2320

Source of ISIS energy data

Survey on the production (conversion) and consumption of energy, Statistics Netherlands

Publications

The data of the first survey are published in *"Energy supply in the Netherlands"* part 1, chapters 2, 3 (energy balances) and part 2, chapter 5, tables 5.3 through 5.6. Part 1 gives energy consumption figures sector by sector in a very detailed form (including transformation and production activity, and identifies non-energy use) .

NEW ZEALAND

General notes on collected data

Electricity used in the manufacturing sector is for the year ended March 199X.

The updated "Own Use" data cover what is called "co-generation" or Combined Heat Power (CHP). More detailed database on this from 1995 onwards is now available. This provides comparable data for the period 1995-1998, but similar data is not available for previous years.

In C2320, Gas contains the loss of natural gas to synthetic petrol conversion. It is related to nil synthetic gasoline production since February 1997.

In C2411, Gas includes the loss of natural gas to AA methanol conversion.

The figures for activities that can not be correctly allocated to the appropriate industry were put in CERR1 in the database. These figures were submitted as "unallocated" by New Zealand Officials.

Several missing values for the sectors C2890 and C3310 are included in "CERR1" which is unallocated industry.

The data for 1990 to 1994, which were in ISIC Revision 2 in previous publication, are converted to ISIC Revision 3 by the Secretariat.

Source of ISIS energy data
Statistics New Zealand

Publications
Energy Data File, Ministry of Commerce: Energy consumption figures for 9 industries (including construction) are given in gross PJ in energy supply and demand balance tables.

NORWAY

General notes on collected data
Steam data are available after 1995.

Transformation is not included. Own use is included only for petroleum refineries.

Source of ISIS energy data
Statistics Norway. The consumption figures for the most of the products are taken from NOS (official statistics of Norway) Industry/Manufacturing and Electricity statistics.

Publications
Manufacturing Statistics, Statistics Norway.

Energy Statistics, Statistics Norway.

POLAND

General notes on collected data
Consumption refers to direct consumption and transformation input.

Source of ISIS energy data
Central Statistical Office

Publications
Gospodarka paliwowo-energetyczna

PORTUGAL

General notes on collected data
Biomass and electricity in 1994: S3220 includes S3240, S3510 includes S3520 and S3560

Source of ISIS energy data
Instituto Nacional de Estatistica

SLOVAK REPUBLIC

General notes on collected data
Disggregated sectors may not add up to aggregated sectors since some 3 or 4 digit sectors are included in 2 or 3 digit sector total.

Source of ISIS energy data
Annual Industrial Survey, Statistical Office of the Slovak Republic.

SWEDEN

General notes on collected data
LPG: Ethane is not included.

Distillate oils: Only kerosene, motor gasoline, and diesel/gas oil are included in manufacturing statistics.

Refinery gas is not included in gas.

All fuels used in transformation (and also for own use) were included for electricity but not for other energy producing industries where only purchased fuels are included. E.g. refinery gases and coke are not included for refinery industries.

Only fuels used for road transport are included.

For natural gas probably gross calorific values were used.

NACE 1592 and 1597 are included in ISIC 1552

No data are available for biomass in ISIC sectors 21 and 20 in disaggregated level

Source of ISIS energy data
Manufacturing 19XX, official statistics of Sweden, Statistics Sweden

Publications
Industri 19XX, Del 1, Branschdata fordelade enligt Svensk standard for naringsgrensindelning (*Manufacturing Part 1: Data by industry*), Official Statistics of Sweden (SOS), Statistics Sweden: Energy consumption data by industry and fuel type are presented in quantities and purchase values.

SWITZERLAND

General notes on collected data
In 1994 and 1995, S3800-S3810 are allocated to S3820-S3829.

Since the Swiss EKV survey covers up to 70% of the total energy consumption in industry, there may be some big differences between the figures reported in ISIS and *Schweizerische Gesamtenergiestatistik.*

Source of ISIS energy data
Union suisse des consommateurs d'energie de l'industrie et des autres branches economiques

Publications
Schweizerische Gesamtenergiestatistik (yearly), Bundesamt f. Energiewirtschaft

Energieverbrauch in der schweizerischen Industrie, EKV (yearly, from 1978 to 1990)

TURKEY

General notes on collected data
The survey for the year 1992 covers the manufacturing establishments with 25+ employees. The survey from 1995 onward, however covers only the establishments with energy consumption 500+

tonne oil equivalent which represents 90% of the total manufacturing industry.

Source of ISIS energy data
State Institute of Statistics

Publications
Energy Consumption in Manufacturing Industry.

UK

General notes on collected data
No breakdown of the biomass figures can yet be provided and the breakdown of electricity produced on-site is very limited.

The figure for total manufacturing includes the figure for activities that can not be correctly allocated to the appropriate industry (they are put in ERR1). The fuels used in transformation activity are put in ERR1 except for the sectors C1500, C2400 and C2700. Therefore in those sectors the sum of subsectoral data may not match the figures in main sectors.

Fuels used in transformation activity in sector C1600 are included in C1500.

Fuels used in transformation activity in sector C2200 are included in C2100.

Fuels used in transformation activity in sector C2800-C3400 are included in C3500.

Source of ISIS energy data
Department of Trade and Industry (DTI)

Publications
Digest of United Kingdom Energy Statistics, DTI.

USA

General notes on collected data
Estimates for some 4-digit ISIC categories are subject to error due to low sample coverage in some population subgroups.

The data are based on Table A4 (total inputs of energy for heat, power and electricity generation) of the US Manufacturing Energy Consumption Survey (MECS).

Sector S3300 includes SIC sector 2411.

Except for sectors S3411, S3530 and S3710 petroleum coke, blast furnace and coke oven gas are included in Biomass.

Biomass includes net steam as well.

Confidential qualifier "c" includes W (withheld to avoid disclosing data for individual establishments) and Q (withheld because relative standard error is greater than 50 percent) in MECS.

Source of ISIS energy data
Energy Information Administration

Publications
Manufacturing Energy Consumption Survey, Energy Information Administration.

Table 1. Availability of ISIS Energy Data by year and by ISIC Revision

	ISIC REVISION 2									ISIC REVISION 3								
	90	91	92	93	94	95	96	97	98	90	91	92	93	94	95	96	97	98
AUSTRALIA	..	✓	✓	✓	✓	✓	✓	✓	✓	..	..	..	..	..	..	..	..	..
AUSTRIA	..	..	..	..	..	..	..	..	..	✓	✓	✓	✓	✓	✓	✓	✓	✓
BELGIUM	..	..	..	✓	..	..	..	..	..	..	..	..	..	..	..	..	..	..
CANADA	..	..	..	..	..	..	..	..	..	✓	✓	✓	✓	✓	✓	✓	..	..
CZECH REP.	..	..	..	..	..	..	..	..	..	..	..	..	✓	✓	✓	✓	✓	..
DENMARK	..	..	..	..	..	..	..	..	..	✓	..	..	✓	..	✓	✓	✓	..
FINLAND	✓	✓	✓	✓	✓	✓	✓	..	..	..	..	..	..	..	..	..	..	..
FRANCE	..	..	..	..	..	..	..	..	..	..	..	..	..	✓	✓	✓	✓	✓
GERMANY	..	✓	✓	✓	✓	..	..	..	..	..	..	..	..	..	✓	✓	✓	..
GREECE	..	..	..	..	..	..	..	..	..	..	..	..	..	..	..	..	..	..
HUNGARY	✓	✓	✓	✓	✓	✓	✓	✓	..	..	..	..	..	..	..	..	..	..
ICELAND	..	..	..	..	✓	..	..	..	..	..	..	..	..	..	..	..	..	..
IRELAND	..	..	..	..	..	..	..	..	..	..	..	..	..	..	..	..	..	..
ITALY	..	..	..	..	..	..	..	..	..	..	..	..	..	..	..	..	..	..
JAPAN	..	..	..	..	..	..	..	..	..	✓	✓	✓	✓	✓	✓	✓	✓	..
KOREA	..	✓	✓	✓	✓	✓	✓	✓	✓	..	..	..	..	..	..	..	..	..
LUXEMBOURG	..	..	..	..	..	..	..	..	..	✓	✓	✓	✓	✓	✓	✓	✓	..
MEXICO	✓	✓	✓	✓	✓	✓	..	..	..	..	..	..	..	..	..	..	..	..
NETHERLANDS	..	..	..	..	..	..	..	..	..	..	..	..	..	✓	✓	✓	..	..
NEW ZEALAND	..	..	..	..	..	..	..	..	..	✓	✓	✓	✓	✓	✓	✓	✓	✓
NORWAY	..	..	✓	..	..	..	..	..	..	..	..	..	✓	✓	✓	✓	..	..
POLAND	..	..	..	..	..	..	..	..	..	..	..	..	..	✓	✓	✓	✓	..
PORTUGAL	..	..	..	..	✓	..	..	..	..	..	..	..	..	..	..	..	..	..
SLOVAK REP.	..	..	..	..	..	..	..	..	..	..	..	..	..	✓	✓	✓	✓	..
SPAIN	..	..	..	..	..	..	..	..	..	..	..	..	..	..	..	..	..	..
SWEDEN	..	✓	✓	✓	..	..	..	..	..	..	..	..	..	✓	✓	..	✓	..
SWITZERLAND	✓	✓	✓	✓	✓	✓	..	..	..	..	..	..	..	..	..	✓	..	..
TURKEY	..	..	✓	..	..	✓	✓	✓	..	..	..	..	..	..	..	..	..	..
UK	..	..	..	..	..	..	..	..	..	✓	✓	✓	✓	✓	✓	✓	..	..
USA	..	..	..	..	✓	..	..	..	..	..	..	..	..	..	..	..	..	..

Information System on Industrial Structures
(ISIS)

Energy Consumption Statistics

in

Manufacturing Industry

1990 – 1998

Total fuels

(ISIC Revision 2)

OECD
ORGANISATION FOR ECONOMIC
CO-OPERATION AND DEVELOPMENT

OCDE
ORGANISATION DE COOPERATION
ET DE DEVELOPPEMENT ECONOMIQUES

IEA
INTERNATIONAL ENERGY AGENCY

AIE
AGENCE INTERNATIONALE DE L'ENERGIE

9, RUE DE LA FÉDERATION, 75739 PARIS CEDEX 15
TEL. (33-1) 40.57.65.00 / TELEFAX (33-1) 40.57.66.49 / INTERNET: HTTP://WWW.IEA.ORG

Unit: TJ

AUSTRALIA

ISIC Revision 2 Industry Sector		1990	1991	1992	1993	1994	1995	1996	1997	1998
31	**FOOD, BEVERAGES AND TOBACCO**	..	133,098	120,457	134,625	143,304	153,332	161,909	168,680	153,733
311/2	FOOD	..	125,937	113,213	127,624	136,058	145,957	155,028	162,005	149,145
3111	Slaughtering, preparing and preserving meat	..	12,908	13,066	13,083	13,303	13,469	12,031	11,611	7,152
3112	Dairy products	..	11,149	11,655	11,888	12,566	13,175	13,035	13,180	10,657
3113	Canning, preserving of fruits and vegetables	..	4,728	4,756	5,037	5,345	5,511	5,581	5,729	4,319
3114	Canning, preserving and processing of fish	..	..			..				
3115	Vegetable and animal oils and fats	..	2,809	2,827	2,880	2,878	2,981	2,975	3,047	2,192
3116	Grain mill products	..	4,496	5,178	5,333	5,502	5,668	5,795	5,986	4,168
3117	Bakery products	..	4,078	4,045	4,143	4,295	4,610	4,578	4,650	3,160
3118	Sugar factories and refineries	..	85,769	71,688	85,261	92,170	100,543	111,033	117,802	117,497
3119	Cocoa, chocolate and sugar confectionery	..	..		..	..				
3121	Other food products	..	..		..	..				
3122	Prepared animal feeds	..								
313	BEVERAGES	..	6,297	6,404	6,172	6,464	6,582	6,083	5,889	4,100
3131	Distilling, rectifying and blending of spirits	..	..		..				..	
3132	Wine industries	..	..		..				..	
3133	Malt liquors and malts	..	..		..				..	
3134	Soft drinks	..	..		..				..	
314	TOBACCO	..	864	840	829	781	793	798	785	488
32	**TEXTILES, APPAREL AND LEATHER**	..	14,255	14,175	14,866	14,892	15,610	15,206	15,424	7,917
321	TEXTILES	..	..		..	..				
3211	Spinning weaving and finishing textiles	..	..		..					
3212	Made-up goods excluding wearing apparel	..	..		..					
3213	Knitting mills	..	..		..					
3214	Carpets and rugs	..	..		..					
3215	Cordage, rope and twine	..	..		..					
3219	Other textiles	..	..		..					
322	WEARING APPAREL, EXCEPT FOOTWEAR	..	..		..					
323	LEATHER AND FUR PRODUCTS	..	..		..					
3231	Tanneries and leather finishing	..	..		..					
3232	Fur dressing and dyeing industries	..	..		..					
3233	Leather prods. ex. footwear and wearing apparel	..	..		..					
324	FOOTWEAR, EX. RUBBER AND PLASTIC	..	..		..					
33	**WOOD PRODUCTS AND FURNITURE**	..	15,156	15,246	15,656	16,876	15,060	16,967	17,125	12,543
331	WOOD PRODUCTS, EXCEPT FURNITURE	..	..		..					
3311	Sawmills, planing and other wood mills	..	..		..					
3312	Wooden and cane containers	..	..		..					
3319	Other wood and cork products	..	..		..					
332	FURNITURE, FIXTURES, EXCL. METALLIC	..	..		..					
34	**PAPER, PUBLISHING AND PRINTING**	..	46,395	45,483	47,058	49,498	51,853	52,261	53,239	39,543
341	PAPER AND PRODUCTS	..	..		..					
3411	Pulp, paper and paperboard articles	..	..		..					
3412	Containers of paper and paperboard	..	..		..					
3419	Other pulp, paper and paperboard articles	..	..		..					
342	PRINTING AND PUBLISHING	..	..		..					
35	**CHEMICAL PRODUCTS**	..	147,832	141,548	143,896	153,747	156,567	154,148	149,309	130,496
351	INDUSTRIAL CHEMICALS	..	47,983	43,344	41,633	46,040	46,570	44,076	43,479	31,326
3511	Basic industrial chemicals excl. fertilizers	..	..		..				..	
3512	Fertilizers and pesticides	..	..		..				..	
3513	Synthetic resins and plastic materials	..	..		..				..	
352	OTHER CHEMICALS	..	4,098	4,061	4,152	4,529	4,773	4,828	5,135	2,352
3521	Paints, varnishes and lacquers	..	..		..				..	
3522	Drugs and medicines	..	..		..				..	
3523	Soap, cleaning preparations, perfumes, cosmetics	..	..		..				..	
3529	Other chemical products	..	..		..				..	
353	PETROLEUM REFINERIES	..	86,591	86,929	91,441	96,430	98,533	98,317	92,914	90,832
354	MISC. PETROLEUM AND COAL PRODUCTS	..	5,390	4,115	3,594	3,387	3,256	3,705	4,494	4,178
355	RUBBER PRODUCTS	..	1,715	1,628	1,579	1,507	1,442	1,549	1,604	996
3551	Tyres and tubes	..	..		..					
3559	Other rubber products	..	..		..					
356	PLASTIC PRODUCTS	..	2,054	1,471	1,498	1,854	1,993	1,673	1,682	813
36	**NON-METALLIC MINERAL PRODUCTS**	..	83,161	80,536	81,484	85,763	87,828	82,086	83,121	74,414
361	POTTERY, CHINA, EARTHENWARE	..	2,704	2,271	1,810	2,025	2,070	1,725	1,704	1,510
362	GLASS AND PRODUCTS	..	11,480	11,471	11,609	11,339	10,520	9,995	9,782	8,636
369	OTHER NON-METAL. MINERAL PRODUCTS	..	68,978	66,793	68,065	72,399	75,237	70,366	71,636	64,268
3691	Structural clay products	..	20,193	20,335	21,262	21,836	22,772	20,132	20,184	18,279
3692	Cement, lime and plaster	..	39,236	36,561	36,638	39,347	40,901	38,907	39,711	36,595
3699	Other non-metallic mineral products	..	9,549	9,897	10,165	11,216	11,564	11,327	11,741	9,394
37	**BASIC METAL INDUSTRIES**	..	627,865	623,587	623,285	652,600	654,168	642,788	655,083	572,030
371	IRON AND STEEL	..	368,268	358,808	351,972	372,552	379,578	368,407	374,731	384,200
372	NON-FERROUS METALS	..	259,597	264,778	271,314	280,048	274,590	274,381	280,352	187,830
38	**METAL PRODUCTS, MACHINERY, EQUIP.**	..	27,709	27,499	27,793	28,613	30,427	30,529	30,671	14,919
381	METAL PRODUCTS	..	..		..				..	
3811	Cutlery, hand tools and general hardware	..	..		..				..	
3812	Furniture and fixtures primarily of metal	..	..		..				..	
3813	Structural metal products	..	..		..				..	
3819	Other fabricated metal products	..	..		..				..	
382	NON-ELECTRICAL MACHINERY	..	..		..				..	
3821	Engines and turbines	..	..		..				..	
3822	Agricultural machinery and equipment	..	..		..				..	
3823	Metal and wood working machinery	..	..		..				..	
3824	Special industrial machinery	..	..		..				..	
3825	Office, computing and accounting machinery	..	..		..				..	
3829	Other non-electrical machinery and equipment	..	..		..				..	
383	ELECTRICAL MACHINERY	..	..		..				..	
3831	Electrical industrial machinery	..	..		..				..	
3832	Radio, TV and communications equipment	..	..		..				..	
3833	Electrical appliances and housewares	..	..		..				..	
3839	Other electrical apparatus and supplies	..	..		..				..	
384	TRANSPORT EQUIPMENT	..	..		..				..	
3841	Shipbuilding	..	..		..				..	
3842	Railroad equipment	..	..		..				..	
3843	Motor vehicles	..	..		..				..	
3844	Motorcycles and bicycles	..	..		..				..	
3845	Aircraft	..	..		..				..	
3849	Other transport equipment	..	..		..				..	
385	PROFESSIONAL AND SCIENTIFIC EQUIPMENT	..	..		..				..	
3851	Professional equipment	..	..		..				..	
3852	Photographic and optical goods	..	..		..				..	
3853	Watches and clocks	..	..		..				..	
39	**OTHER MANUFACTURING INDUSTRIES**	..	315	271	267	341	339	344	347	272
3901	Jewellery and related articles	..	..		..				..	
3902	Musical instruments	..	..		..				..	
3903	Sporting and athletic goods	..	..		..				..	
3909	Other manufactures	..	..		..				..	
SERR	**non-specified, unallocated industry**	..	..		..	..			..	
3	**TOTAL MANUFACTURING**	..	1,095,787	1,068,802	1,088,932	1,145,633	1,165,183	1,156,239	1,172,998	1,005,867

ISIS Energy Data Programme (IEA/OECD)

ISIC Revision 2 Industry Sector		1990	1991	1992	1993	1994	1995	1996	1997	1998
31	FOOD, BEVERAGES AND TOBACCO	..	..	..	27,482	..	..	..	..	..
311/2	FOOD	..	..	..	..	..	..	..	..	..
3111	Slaughtering, preparing and preserving meat	..	..	..	..	..	..	..	..	..
3112	Dairy products	..	..	..	..	..	..	..	..	..
3113	Canning, preserving of fruits and vegetables	..	..	..	..	..	..	..	..	..
3114	Canning, preserving and processing of fish	..	..	..	..	..	..	..	..	..
3115	Vegetable and animal oils and fats	..	..	..	..	..	..	..	..	..
3116	Grain mill products	..	..	..	..	..	..	..	..	..
3117	Bakery products	..	..	..	..	..	..	..	..	..
3118	Sugar factories and refineries	..	..	..	..	..	..	..	..	..
3119	Cocoa, chocolate and sugar confectionery	..	..	..	..	..	..	..	..	..
3121	Other food products	..	..	..	..	..	..	..	..	..
3122	Prepared animal feeds	..	..	..	..	..	..	..	..	..
313	BEVERAGES	..	..	..	..	..	..	..	..	..
3131	Distilling, rectifying and blending of spirits	..	..	..	..	..	..	..	..	..
3132	Wine industries	..	..	..	..	..	..	..	..	..
3133	Malt liquors and malts	..	..	..	..	..	..	..	..	..
3134	Soft drinks	..	..	..	..	..	..	..	..	..
314	TOBACCO	..	..	..	..	..	..	..	..	..
32	TEXTILES, APPAREL AND LEATHER	..	..	..	10,598	..	..	..	..	..
321	TEXTILES	..	..	..	..	..	..	..	..	..
3211	Spinning weaving and finishing textiles	..	..	..	..	..	..	..	..	..
3212	Made-up goods excluding wearing apparel	..	..	..	..	..	..	..	..	..
3213	Knitting mills	..	..	..	..	..	..	..	..	..
3214	Carpets and rugs	..	..	..	..	..	..	..	..	..
3215	Cordage, rope and twine	..	..	..	..	..	..	..	..	..
3219	Other textiles	..	..	..	..	..	..	..	..	..
322	WEARING APPAREL, EXCEPT FOOTWEAR	..	..	..	..	..	..	..	..	..
323	LEATHER AND FUR PRODUCTS	..	..	..	..	..	..	..	..	..
3231	Tanneries and leather finishing	..	..	..	..	..	..	..	..	..
3232	Fur dressing and dyeing industries	..	..	..	..	..	..	..	..	..
3233	Leather prods. ex. footwear and wearing apparel	..	..	..	..	..	..	..	..	..
324	FOOTWEAR, EX. RUBBER AND PLASTIC	..	..	..	..	..	..	..	..	..
33	WOOD PRODUCTS AND FURNITURE	..	..	..	1,883	..	..	..	..	..
331	WOOD PRODUCTS, EXCEPT FURNITURE	..	..	..	..	..	..	..	..	..
3311	Sawmills, planing and other wood mills	..	..	..	..	..	..	..	..	..
3312	Wooden and cane containers	..	..	..	..	..	..	..	..	..
3319	Other wood and cork products	..	..	..	..	..	..	..	..	..
332	FURNITURE, FIXTURES, EXCL. METALLIC	..	..	..	..	..	..	..	..	..
34	PAPER, PUBLISHING AND PRINTING	..	..	..	13,492	..	..	..	..	..
341	PAPER AND PRODUCTS	..	..	..	..	..	..	..	..	..
3411	Pulp, paper and paperboard articles	..	..	..	..	..	..	..	..	..
3412	Containers of paper and paperboard	..	..	..	..	..	..	..	..	..
3419	Other pulp, paper and paperboard articles	..	..	..	..	..	..	..	..	..
342	PRINTING AND PUBLISHING	..	..	..	..	..	..	..	..	..
35	CHEMICAL PRODUCTS	..	..	..	96,181	..	..	..	..	..
351	INDUSTRIAL CHEMICALS	..	..	..	..	..	..	..	..	..
3511	Basic industrial chemicals excl. fertilizers	..	..	..	..	..	..	..	..	..
3512	Fertilizers and pesticides	..	..	..	..	..	..	..	..	..
3513	Synthetic resins and plastic materials	..	..	..	..	..	..	..	..	..
352	OTHER CHEMICALS	..	..	..	..	..	..	..	..	..
3521	Paints, varnishes and lacquers	..	..	..	..	..	..	..	..	..
3522	Drugs and medicines	..	..	..	..	..	..	..	..	..
3523	Soap, cleaning preparations, perfumes, cosmetics	..	..	..	..	..	..	..	..	..
3529	Other chemical products	..	..	..	..	..	..	..	..	..
353	PETROLEUM REFINERIES	..	..	..	..	..	..	..	..	..
354	MISC. PETROLEUM AND COAL PRODUCTS	..	..	..	..	..	..	..	..	..
355	RUBBER PRODUCTS	..	..	..	..	..	..	..	..	..
3551	Tyres and tubes	..	..	..	..	..	..	..	..	..
3559	Other rubber products	..	..	..	..	..	..	..	..	..
356	PLASTIC PRODUCTS	..	..	..	..	..	..	..	..	..
36	NON-METALLIC MINERAL PRODUCTS	..	..	..	52,778	..	..	..	..	..
361	POTTERY, CHINA, EARTHENWARE	..	..	..	..	..	..	..	..	..
362	GLASS AND PRODUCTS	..	..	..	..	..	..	..	..	..
369	OTHER NON-METAL. MINERAL PRODUCTS	..	..	..	..	..	..	..	..	..
3691	Structural clay products	..	..	..	..	..	..	..	..	..
3692	Cement, lime and plaster	..	..	..	..	..	..	..	..	..
3699	Other non-metallic mineral products	..	..	..	..	..	..	..	..	..
37	BASIC METAL INDUSTRIES	..	..	..	214,111	..	..	..	..	..
371	IRON AND STEEL	..	..	..	201,198	..	..	..	..	..
372	NON-FERROUS METALS	..	..	..	12,913	..	..	..	..	..
38	METAL PRODUCTS, MACHINERY, EQUIP.	..	..	..	20,679	..	..	..	..	..
381	METAL PRODUCTS	..	..	..	..	..	..	..	..	..
3811	Cutlery, hand tools and general hardware	..	..	..	..	..	..	..	..	..
3812	Furniture and fixtures primarily of metal	..	..	..	..	..	..	..	..	..
3813	Structural metal products	..	..	..	..	..	..	..	..	..
3819	Other fabricated metal products	..	..	..	..	..	..	..	..	..
382	NON-ELECTRICAL MACHINERY	..	..	..	..	..	..	..	..	..
3821	Engines and turbines	..	..	..	..	..	..	..	..	..
3822	Agricultural machinery and equipment	..	..	..	..	..	..	..	..	..
3823	Metal and wood working machinery	..	..	..	..	..	..	..	..	..
3824	Special industrial machinery	..	..	..	..	..	..	..	..	..
3825	Office, computing and accounting machinery	..	..	..	..	..	..	..	..	..
3829	Other non-electrical machinery and equipment	..	..	..	..	..	..	..	..	..
383	ELECTRICAL MACHINERY	..	..	..	..	..	..	..	..	..
3831	Electrical industrial machinery	..	..	..	..	..	..	..	..	..
3832	Radio, TV and communications equipment	..	..	..	..	..	..	..	..	..
3833	Electrical appliances and housewares	..	..	..	..	..	..	..	..	..
3839	Other electrical apparatus and supplies	..	..	..	..	..	..	..	..	..
384	TRANSPORT EQUIPMENT	..	..	..	..	..	..	..	..	..
3841	Shipbuilding	..	..	..	..	..	..	..	..	..
3842	Railroad equipment	..	..	..	..	..	..	..	..	..
3843	Motor vehicles	..	..	..	..	..	..	..	..	..
3844	Motorcycles and bicycles	..	..	..	..	..	..	..	..	..
3845	Aircraft	..	..	..	..	..	..	..	..	..
3849	Other transport equipment	..	..	..	..	..	..	..	..	..
385	PROFESSIONAL AND SCIENTIFIC EQUIPMENT	..	..	..	..	..	..	..	..	..
3851	Professional equipment	..	..	..	..	..	..	..	..	..
3852	Photographic and optical goods	..	..	..	..	..	..	..	..	..
3853	Watches and clocks	..	..	..	..	..	..	..	..	..
39	OTHER MANUFACTURING INDUSTRIES	..	..	..	53,228	..	..	..	..	..
3901	Jewellery and related articles	..	..	..	..	..	..	..	..	..
3902	Musical instruments	..	..	..	..	..	..	..	..	..
3903	Sporting and athletic goods	..	..	..	..	..	..	..	..	..
3909	Other manufactures	..	..	..	..	..	..	..	..	..
SERR	non-specified, unallocated industry	..	..	..	..	..	..	..	..	..
3	TOTAL MANUFACTURING	..	..	..	490,441	..	..	..	..	..

ISIC Revision 2 Industry Sector		1990	1991	1992	1993	1994	1995	1996	1997	1998
31	FOOD, BEVERAGES AND TOBACCO	21,107	20,403	20,854	19,666	19,727	20,416	20,780	..	..
311/2	FOOD	18,176	17,654	17,690	16,677	16,961	17,273	17,634	..	..
3111	Slaughtering, preparing and preserving meat	..	..	..	..	..	..	..	..	..
3112	Dairy products	..	..	..	..	..	..	..	..	..
3113	Canning, preserving of fruits and vegetables	..	..	..	..	..	..	..	..	..
3114	Canning, preserving and processing of fish	..	..	..	..	..	..	..	..	..
3115	Vegetable and animal oils and fats	..	..	..	..	..	..	..	..	..
3116	Grain mill products	..	..	..	..	..	..	..	..	..
3117	Bakery products	..	..	..	..	..	..	..	..	..
3118	Sugar factories and refineries	..	..	..	..	..	..	..	..	..
3119	Cocoa, chocolate and sugar confectionery	..	..	..	..	..	..	..	..	..
3121	Other food products	..	..	..	..	..	..	..	..	..
3122	Prepared animal feeds	..	..	..	..	..	..	..	..	..
313	BEVERAGES	2,766	2,586	3,035	2,860	2,628	3,013	3,043	..	..
3131	Distilling, rectifying and blending of spirits	..	..	..	..	..	..	..	..	..
3132	Wine industries	..	..	..	..	..	..	..	..	..
3133	Malt liquors and malts	..	..	..	..	..	..	..	..	..
3134	Soft drinks	..	..	..	..	..	..	..	..	..
314	TOBACCO	165	163	129	130	138	130	104	..	..
32	TEXTILES, APPAREL AND LEATHER	3,264	3,154	3,063	3,118	3,020	3,175	3,527	..	..
321	TEXTILES	2,079	2,154	2,188	2,333	1,873	2,084	2,449	..	..
3211	Spinning weaving and finishing textiles	..	..	..	..	..	..	..	..	..
3212	Made-up goods excluding wearing apparel	..	..	..	..	..	..	..	..	..
3213	Knitting mills	..	..	..	..	..	..	..	..	..
3214	Carpets and rugs	..	..	..	..	..	..	..	..	..
3215	Cordage, rope and twine	..	..	..	..	..	..	..	..	..
3219	Other textiles	..	..	..	..	..	..	..	..	..
322	WEARING APPAREL, EXCEPT FOOTWEAR	678	583	468	425	752	711	688	..	..
323	LEATHER AND FUR PRODUCTS	307	258	256	218	167	108	86	..	..
3231	Tanneries and leather finishing	..	..	..	..	..	..	..	..	..
3232	Fur dressing and dyeing industries	..	..	..	..	..	..	..	..	..
3233	Leather prods. ex. footwear and wearing apparel	..	..	..	..	..	..	..	..	..
324	FOOTWEAR, EX. RUBBER AND PLASTIC	199	159	152	142	229	272	303	..	..
33	WOOD PRODUCTS AND FURNITURE	22,293	17,486	17,659	18,693	23,722	25,233	28,978	..	..
331	WOOD PRODUCTS, EXCEPT FURNITURE	20,222	15,820	16,053	17,333	22,066	23,708	27,273	..	..
3311	Sawmills, planing and other wood mills	..	..	..	..	..	..	..	..	..
3312	Wooden and cane containers	..	..	..	..	..	..	..	..	..
3319	Other wood and cork products	..	..	..	..	..	..	..	..	..
332	FURNITURE, FIXTURES, EXCL. METALLIC	2,071	1,666	1,606	1,360	1,657	1,526	1,706	..	..
34	PAPER, PUBLISHING AND PRINTING	223,543	215,177	214,972	232,889	250,975	252,003	260,225	..	..
341	PAPER AND PRODUCTS	220,527	212,201	212,169	230,102	248,251	249,213	257,069	..	..
3411	Pulp, paper and paperboard articles	..	..	..	..	..	..	..	..	..
3412	Containers of paper and paperboard	..	..	..	..	..	..	..	..	..
3419	Other pulp, paper and paperboard articles	..	..	..	..	..	..	..	..	..
342	PRINTING AND PUBLISHING	3,015	2,976	2,803	2,787	2,723	2,790	3,157	..	..
35	CHEMICAL PRODUCTS	56,839	51,829	55,543	59,171	70,882	71,645	75,228	..	..
351	INDUSTRIAL CHEMICALS	26,259	24,262	26,834	28,168	33,092	34,182	34,824	..	..
3511	Basic industrial chemicals excl. fertilizers	..	..	..	..	..	..	..	..	..
3512	Fertilizers and pesticides	..	..	..	..	..	..	..	..	..
3513	Synthetic resins and plastic materials	..	..	..	..	..	..	..	..	..
352	OTHER CHEMICALS	1,509	1,709	2,121	2,237	547	497	527	..	..
3521	Paints, varnishes and lacquers	..	..	..	..	..	..	..	..	..
3522	Drugs and medicines	..	..	..	..	..	..	..	..	..
3523	Soap, cleaning preparations, perfumes, cosmetics	..	..	..	..	..	..	..	..	..
3529	Other chemical products	..	..	..	..	..	..	..	..	..
353	PETROLEUM REFINERIES	26,277	22,740	22,398	23,450	33,074	32,657	34,876	..	..
354	MISC. PETROLEUM AND COAL PRODUCTS	429	942	2,091	3,114	263	351	239	..	..
355	RUBBER PRODUCTS	882	819	757	811	893	995	1,032	..	..
3551	Tyres and tubes	..	..	..	..	..	..	..	..	..
3559	Other rubber products	..	..	..	..	..	..	..	..	..
356	PLASTIC PRODUCTS	1,483	1,356	1,341	1,391	3,013	2,962	3,730	..	..
36	NON-METALLIC MINERAL PRODUCTS	19,923	16,789	14,732	12,697	11,910	12,997	13,615		..
361	POTTERY, CHINA, EARTHENWARE	465	400	488	280	441	560	562	..	..
362	GLASS AND PRODUCTS	2,392	2,365	2,548	2,704	2,985	3,712	3,368	..	..
369	OTHER NON-METAL. MINERAL PRODUCTS	17,065	14,024	11,696	9,713	8,484	8,726	9,685	..	..
3691	Structural clay products	..	..	..	..	..	..	..	..	..
3692	Cement, lime and plaster	..	..	..	..	..	..	..	..	..
3699	Other non-metallic mineral products	..	..	..	..	..	..	..	..	..
37	BASIC METAL INDUSTRIES	59,449	56,808	59,102	62,886	63,529	58,034	62,042	..	..
371	IRON AND STEEL	53,067	51,199	53,427	56,700	58,834	53,034	56,619	..	..
372	NON-FERROUS METALS	6,382	5,609	5,675	6,186	4,695	5,000	5,424	..	..
38	METAL PRODUCTS, MACHINERY, EQUIP.	16,991	15,842	15,627	15,584	15,922	18,194	21,015	..	..
381	METAL PRODUCTS	5,089	4,678	4,507	4,730	4,105	4,884	6,294	..	..
3811	Cutlery, hand tools and general hardware	..	..	..	..	..	..	..	..	..
3812	Furniture and fixtures primarily of metal	..	..	..	..	..	..	..	..	..
3813	Structural metal products	..	..	..	..	..	..	..	..	..
3819	Other fabricated metal products	..	..	..	..	..	..	..	..	..
382	NON-ELECTRICAL MACHINERY	5,344	4,995	4,810	4,820	5,520	6,170	7,070	..	..
3821	Engines and turbines	..	..	..	..	..	..	..	..	..
3822	Agricultural machinery and equipment	..	..	..	..	..	..	..	..	..
3823	Metal and wood working machinery	..	..	..	..	..	..	..	..	..
3824	Special industrial machinery	..	..	..	..	..	..	..	..	..
3825	Office, computing and accounting machinery	..	..	..	..	..	..	..	..	..
3829	Other non-electrical machinery and equipment	..	..	..	..	..	..	..	..	..
383	ELECTRICAL MACHINERY	2,400	2,292	2,358	2,226	2,536	2,711	3,662	..	..
3831	Electrical industrial machinery	..	..	..	..	..	..	..	..	..
3832	Radio, TV and communications equipment	..	..	..	..	..	..	..	..	..
3833	Electrical appliances and housewares	..	..	..	..	..	..	..	..	..
3839	Other electrical apparatus and supplies	..	..	..	..	..	..	..	..	..
384	TRANSPORT EQUIPMENT	3,726	3,458	3,497	3,366	3,305	3,960	3,503	..	..
3841	Shipbuilding	..	..	..	..	..	..	..	..	..
3842	Railroad equipment	..	..	..	..	..	..	..	..	..
3843	Motor vehicles	..	..	..	..	..	..	..	..	..
3844	Motorcycles and bicycles	..	..	..	..	..	..	..	..	..
3845	Aircraft	..	..	..	..	..	..	..	..	..
3849	Other transport equipment	..	..	..	..	..	..	..	..	..
385	PROFESSIONAL AND SCIENTIFIC EQUIPMENT	431	419	454	442	456	470	486	..	..
3851	Professional equipment	..	..	..	..	..	..	..	..	..
3852	Photographic and optical goods	..	..	..	..	..	..	..	..	..
3853	Watches and clocks	..	..	..	..	..	..	..	..	..
39	OTHER MANUFACTURING INDUSTRIES	386	370	380	397	440	524	530	..	..
3901	Jewellery and related articles	..	..	..	..	..	..	..	..	..
3902	Musical instruments	..	..	..	..	..	..	..	..	..
3903	Sporting and athletic goods	..	..	..	..	..	..	..	..	..
3909	Other manufactures	..	..	..	..	..	..	..	..	..
SERR	non-specified, unallocated industry	..	..	..	..	..			..	..
3	TOTAL MANUFACTURING	423,795	397,859	401,932	425,101	460,127	462,222	485,941	..	..

ISIS Energy Data Programme (IEA/OECD)

ISIC Revision 2 Industry Sector		1990	1991	1992	1993	1994	1995	1996	1997	1998
31	FOOD, BEVERAGES AND TOBACCO		42,545	215,856	202,487	204,821	..	..	..	..
311/2	FOOD	..	33,975	174,106	162,687	166,184				
3111	Slaughtering, preparing and preserving meat		5,222	17,211	18,129	19,257				
3112	Dairy products		5,706	29,630	28,180	27,632				
3113	Canning, preserving of fruits and vegetables		1,541	8,795	7,993	8,583				
3114	Canning, preserving and processing of fish		506	1,508	1,348	1,336				
3115	Vegetable and animal oils and fats		1,609	8,535	6,942	11,675				
3116	Grain mill products		1,693	2,489	2,423	2,317				
3117	Bakery products		3,222	17,076	15,624	15,032				
3118	Sugar factories and refineries		3,620	45,050	38,719	36,470				
3119	Cocoa, chocolate and sugar confectionery		2,916	7,971	7,243	7,190				
3121	Other food products		4,740	23,842	24,412	25,092				
3122	Prepared animal feeds		3,200	11,999	11,673	11,600				
313	BEVERAGES		7,268	38,940	37,296	35,977				
3131	Distilling, rectifying and blending of spirits		493	3,740	3,105	2,859				
3132	Wine industries		141	499	531	527				
3133	Malt liquors and malts		5,525	29,965	28,415	26,615				
3134	Soft drinks		1,108	4,736	5,246	5,976				
314	TOBACCO		1,303	2,811	2,503	2,660				
32	TEXTILES, APPAREL AND LEATHER		22,256	85,842	65,534	60,390	..	..	..	..
321	TEXTILES		19,920	75,671	59,059	54,644				
3211	Spinning weaving and finishing textiles		14,974	57,991	42,835	c				
3212	Made-up goods excluding wearing apparel		452	2,095	1,822	1,729				
3213	Knitting mills		1,361	4,132	3,552	3,210				
3214	Carpets and rugs		3,085	11,392	10,785	10,236				
3215	Cordage, rope and twine		48	61	63	39,468				
3219	Other textiles									
322	WEARING APPAREL, EXCEPT FOOTWEAR		1,273	6,005	3,379	2,899				
323	LEATHER AND FUR PRODUCTS		607	2,841	2,118	2,028				
3231	Tanneries and leather finishing		441	2,219	1,634	1,624				
3232	Fur dressing and dyeing industries		33	200	98	82				
3233	Leather prods. ex. footwear and wearing apparel		132	423	385	322				
324	FOOTWEAR, EX. RUBBER AND PLASTIC		456	1,325	978	819				
33	WOOD PRODUCTS AND FURNITURE		13,583	35,939	30,247	30,368	..	..	..	..
331	WOOD PRODUCTS, EXCEPT FURNITURE		9,416	18,978	19,159	21,190				
3311	Sawmills, planing and other wood mills		7,847	16,581	16,909	18,998				
3312	Wooden and cane containers		192	344	309	289				
3319	Other wood and cork products		1,376	2,053	1,941	1,903				
332	FURNITURE, FIXTURES, EXCL. METALLIC		4,167	16,961	11,088	9,179				
34	PAPER, PUBLISHING AND PRINTING		61,945	214,023	218,509	223,508	..	..	..	..
341	PAPER AND PRODUCTS		54,112	196,926	201,460	206,319				
3411	Pulp, paper and paperboard articles		45,575	170,406	175,511	179,091				
3412	Containers of paper and paperboard		8,536	26,520	25,949	27,229				
3419	Other pulp, paper and paperboard articles					c				
342	PRINTING AND PUBLISHING		7,833	17,097	17,049	17,189				
35	CHEMICAL PRODUCTS		c	1,045,397	953,688	933,364	..	..	..	..
351	INDUSTRIAL CHEMICALS		c	734,868	692,477	662,677				
3511	Basic industrial chemicals excl. fertilizers									
3512	Fertilizers and pesticides	..		..		..				
3513	Synthetic resins and plastic materials									
352	OTHER CHEMICALS		25,073	82,277	76,433	74,254				
3521	Paints, varnishes and lacquers		19,340	52,835	48,643	46,595				
3522	Drugs and medicines		3,505	14,398	15,128	15,169				
3523	Soap, cleaning preparations, perfumes, cosmetics		2,228	15,044	12,662	12,489				
3529	Other chemical products		c			c				
353	PETROLEUM REFINERIES		27,568	149,080	115,326	125,542				
354	MISC. PETROLEUM AND COAL PRODUCTS					c				
355	RUBBER PRODUCTS		7,572	25,403	21,024	20,625				
3551	Tyres and tubes									
3559	Other rubber products	..		..		..				
356	PLASTIC PRODUCTS		27,544	53,770	48,428	50,265				
36	NON-METALLIC MINERAL PRODUCTS	..	42,535	309,088	311,506	318,405	..	..	..	..
361	POTTERY, CHINA, EARTHENWARE		1,741	15,243	13,930	10,561				
362	GLASS AND PRODUCTS		13,850	71,016	72,053	73,360				
369	OTHER NON-METAL. MINERAL PRODUCTS		26,943	222,829	225,524	234,484				
3691	Structural clay products		4,741	46,627	50,409	52,569				
3692	Cement, lime and plaster		16,765	145,349	140,590	149,443				
3699	Other non-metallic mineral products		5,438	30,853	34,525	32,472				
37	BASIC METAL INDUSTRIES	..	c	847,053	787,855	836,846	..	..	..	..
371	IRON AND STEEL		c	739,538	690,807	744,782				
372	NON-FERROUS METALS		63,921	107,514	97,048	92,064				
38	METAL PRODUCTS, MACHINERY, EQUIP.	..	c	393,686	342,500	325,451	..	..	..	..
381	METAL PRODUCTS		c	31,207	80,062	78,379				
3811	Cutlery, hand tools and general hardware		4,877	11,103	10,398	9,961				
3812	Furniture and fixtures primarily of metal		1,138	3,889	3,794	3,538				
3813	Structural metal products		2,737	16,215	10,179	9,308				
3819	Other fabricated metal products		20,812	c	55,691	55,573				
382	NON-ELECTRICAL MACHINERY		35,757	109,684	84,847	75,256				
3821	Engines and turbines		13,943	40,540	33,643	30,080				
3822	Agricultural machinery and equipment		1,699	6,536	5,318	4,843				
3823	Metal and wood working machinery		4,497	16,504	8,563	8,143				
3824	Special industrial machinery		11,432	38,802	31,970	27,897				
3825	Office, computing and accounting machinery		4,187	7,302	5,353	4,293				
3829	Other non-electrical machinery and equipment					c				
383	ELECTRICAL MACHINERY		32,776	66,520	61,772	58,830				
3831	Electrical industrial machinery		14,828	31,555	30,976	29,964				
3832	Radio, TV and communications equipment		12,778	23,201	19,901	18,576				
3833	Electrical appliances and housewares		2,681	5,766	5,071	4,816				
3839	Other electrical apparatus and supplies		2,489	5,997	5,824	5,474				
384	TRANSPORT EQUIPMENT		c	112,758	109,052	106,143				
3841	Shipbuilding		1,640	4,907	5,050	4,080				
3842	Railroad equipment		1,330	9,087	8,414	6,865				
3843	Motor vehicles		44,254	97,418	88,503	88,639				
3844	Motorcycles and bicycles		531	1,084	1,072	886				
3845	Aircraft		c	c	5,813	5,485				
3849	Other transport equipment		88	262	200	188				
385	PROFESSIONAL AND SCIENTIFIC EQUIPMENT		3,372	7,716	6,768	6,843				
3851	Professional equipment		1,425	4,417	3,900	4,097				
3852	Photographic and optical goods		1,754	2,877	2,567	2,512				
3853	Watches and clocks		193	423	301	234				
39	OTHER MANUFACTURING INDUSTRIES	..	1,289	2,771	2,627	2,383	..	..	..	..
3901	Jewellery and related articles		120	257	c	c				
3902	Musical instruments		118	402	395	351				
3903	Sporting and athletic goods		105	317	381	364				
3909	Other manufactures		947	1,794	1,851	1,667				
SERR	non-specified, unallocated industry	..		..		..				
3	TOTAL MANUFACTURING	..	723,287	3,149,654	2,914,953	2,935,537	..	..	..	..

HUNGARY

ISIC Revision 2 Industry Sector		1990	1991	1992	1993	1994	1995	1996	1997	1998
31	FOOD, BEVERAGES AND TOBACCO	36,345	35,670	29,380	26,081	25,655	26,124	26,721	23,991	..
311/2	FOOD	29,979	29,655	24,132	21,010	21,519	21,685	23,226	20,828	..
3111	Slaughtering, preparing and preserving meat	6,953	6,605	5,422	5,299	4,767	4,898	5,152	5,128	..
3112	Dairy products	3,721	3,317	3,056	2,913	2,732	2,744	2,711	2,121	..
3113	Canning, preserving of fruits and vegetables	4,651	4,148	2,653	1,768	2,305	2,458	2,794	1,933	..
3114	Canning, preserving and processing of fish	..	..	..	..	..	..	..	..	..
3115	Vegetable and animal oils and fats	2,082	1,895	2,030	1,499	1,272	937	1,024	1,866	..
3116	Grain mill products	1,375	1,377	166	162	83	357	251	249	..
3117	Bakery products	3,237	2,782	2,256	1,712	1,174	1,591	1,863	1,636	..
3118	Sugar factories and refineries	6,176	7,818	5,722	4,841	5,992	5,949	6,428	5,397	..
3119	Cocoa, chocolate and sugar confectionery	768	771	585	519	520	277	278	249	..
3121	Other food products	1,017	943	1,615	1,909	2,411	2,191	2,436	1,958	..
3122	Prepared animal feeds	..	..	627	389	263	283	290	291	..
313	BEVERAGES	5,976	5,622	4,946	4,719	3,904	4,180	3,255	2,924	..
3131	Distilling, rectifying and blending of spirits	2,004	1,887	1,449	1,395	1,349	1,215	781	588	..
3132	Wine industries	478	331	271	197	163	226	82	71	..
3133	Malt liquors and malts	3,238	3,203	3,035	2,845	2,068	2,317	1,923	1,810	..
3134	Soft drinks	256	201	191	282	324	423	469	455	..
314	TOBACCO	389	393	302	352	231	259	239	239	..
32	TEXTILES, APPAREL AND LEATHER	10,764	8,641	6,478	5,860	4,989	4,027	3,867	3,628	..
321	TEXTILES	8,867	6,975	4,956	4,554	3,920	3,072	3,076	2,928	..
3211	Spinning weaving and finishing textiles	7,287	5,670	3,793	3,485	2,902	2,265	2,379	2,351	..
3212	Made-up goods excluding wearing apparel	..	..	101	64	36	57	35	27	..
3213	Knitting mills	1,085	866	637	610	601	471	374	313	..
3214	Carpets and rugs	..	..	..	..	..	..	..	..	..
3215	Cordage, rope and twine	495	439	425	395	381	279	288	237	..
3219	Other textiles	..	..	..	..	..	..	..	..	..
322	WEARING APPAREL, EXCEPT FOOTWEAR	809	671	571	502	502	371	357	307	..
323	LEATHER AND FUR PRODUCTS	631	580	630	540	364	410	292	297	..
3231	Tanneries and leather finishing	..	..	607	531	351	393	263	272	..
3232	Fur dressing and dyeing industries	316	289	..	..	..	..	25	21	..
3233	Leather prods. ex. footwear and wearing apparel	315	291	23	9	13	17	4	4	..
324	FOOTWEAR, EX. RUBBER AND PLASTIC	457	415	321	263	203	174	142	96	..
33	WOOD PRODUCTS AND FURNITURE	2,471	2,356	2,250	2,398	2,183	2,314	2,495	2,419	..
331	WOOD PRODUCTS, EXCEPT FURNITURE	1,225	1,423	1,822	2,036	1,945	1,967	2,055	2,009	..
3311	Sawmills, planing and other wood mills	1,225	1,423	1,692	1,913	1,836	1,865	1,943	1,934	..
3312	Wooden and cane containers	..	..	19	34	21	17	..	..	..
3319	Other wood and cork products	..	..	111	89	88	85	112	75	..
332	FURNITURE, FIXTURES, EXCL. METALLIC	1,246	933	428	362	238	347	440	410	..
34	PAPER, PUBLISHING AND PRINTING	7,436	6,385	6,130	4,982	4,727	5,275	6,010	6,090	..
341	PAPER AND PRODUCTS	6,945	5,926	5,821	4,583	4,285	4,944	5,609	5,759	..
3411	Pulp, paper and paperboard articles	6,907	5,926	2,605	2,381	2,163	2,549	3,132	3,251	..
3412	Containers of paper and paperboard	38	..	3,113	2,182	2,121	2,391	2,468	2,500	..
3419	Other pulp, paper and paperboard articles	..	..	103	20	1	4	9	8	..
342	PRINTING AND PUBLISHING	491	459	309	399	442	331	401	331	..
35	CHEMICAL PRODUCTS	61,882	56,322	54,499	49,091	44,964	43,045	48,123	44,260	..
351	INDUSTRIAL CHEMICALS	36,133	30,127	20,306	20,832	18,276	18,119	22,391	20,791	..
3511	Basic industrial chemicals excl. fertilizers	6,865	5,567	899	867	1,191	1,402	1,512	1,640	..
3512	Fertilizers and pesticides	14,321	11,775	8,632	9,511	6,259	6,431	10,489	8,728	..
3513	Synthetic resins and plastic materials	14,946	12,784	10,775	10,454	10,826	10,285	10,390	10,423	..
352	OTHER CHEMICALS	5,480	5,326	5,784	5,573	4,763	4,928	4,722	4,733	..
3521	Paints, varnishes and lacquers	..	..	498	538	325	306	240	254	..
3522	Drugs and medicines	5,158	5,019	4,436	4,218	3,853	3,823	3,744	3,778	..
3523	Soap, cleaning preparations, perfumes, cosmetics	322	307	574	570	436	594	534	517	..
3529	Other chemical products	..	..	275	247	149	205	204	184	..
353	PETROLEUM REFINERIES	17,076	18,097	25,749	18,866	18,02	16,421	16,742	14,867	..
354	MISC. PETROLEUM AND COAL PRODUCTS	..	..	47	1,608	1,580	1,156	1,792	1,579	..
355	RUBBER PRODUCTS	2,096	1,843	1,545	1,558	1,568	1,675	1,766	1,695	..
3551	Tyres and tubes	2,096	1,843	1,545	1,558	1,568	1,675	1,766	1,695	..
3559	Other rubber products	..	..	..	..	..	..	..	..	..
356	PLASTIC PRODUCTS	1,097	929	1,069	655	685	746	710	595	..
36	NON-METALLIC MINERAL PRODUCTS	44,136	33,545	26,249	27,223	28,708	25,999	25,455	27,024	..
361	POTTERY, CHINA, EARTHENWARE	1,139	1,140	1,184	1,219	1,164	1,149	1,171	1,219	..
362	GLASS AND PRODUCTS	10,527	9,495	7,872	7,461	8,054	5,044	5,320	7,144	..
369	OTHER NON-METAL. MINERAL PRODUCTS	32,470	22,910	17,193	18,543	19,490	19,806	18,964	18,661	..
3691	Structural clay products	12,441	9,651	5,708	6,700	6,747	7,061	6,927	6,300	..
3692	Cement, lime and plaster	17,567	11,782	10,309	10,712	11,983	12,072	11,395	11,713	..
3699	Other non-metallic mineral products	2,462	1,477	1,176	1,131	760	673	642	648	..
37	BASIC METAL INDUSTRIES	93,914	69,299	56,370	61,001	59,515	59,096	57,772	45,364	..
371	IRON AND STEEL	74,882	52,387	45,836	51,759	52,391	50,856	48,885	36,441	..
372	NON-FERROUS METALS	19,032	16,912	10,534	9,242	7,124	8,240	8,888	8,924	..
38	METAL PRODUCTS, MACHINERY, EQUIP.	24,954	21,559	19,174	16,865	14,319	13,187	13,395	13,184	..
381	METAL PRODUCTS	3,644	3,082	2,812	2,436	2,036	1,868	2,142	1,926	..
3811	Cutlery, hand tools and general hardware	1,328	1,105	1,026	913	786	533	678	592	..
3812	Furniture and fixtures primarily of metal	..	..	..	20	..	..	..	..	..
3813	Structural metal products	1,300	1,095	993	879	682	652	900	815	..
3819	Other fabricated metal products	1,016	882	793	624	568	683	564	519	..
382	NON-ELECTRICAL MACHINERY	5,171	4,465	3,920	3,741	2,359	2,016	2,114	1,991	..
3821	Engines and turbines	343	299	273	254	108	90	123	109	..
3822	Agricultural machinery and equipment	973	788	701	651	617	541	535	569	..
3823	Metal and wood working machinery	413	356	293	238	230	130	128	61	..
3824	Special industrial machinery	2,155	1,923	1,687	1,671	783	680	722	661	..
3825	Office, computing and accounting machinery	256	206	190	208	43	68	94	77	..
3829	Other non-electrical machinery and equipment	1,031	893	776	719	578	507	512	514	..
383	ELECTRICAL MACHINERY	7,099	6,200	5,564	5,251	4,870	4,577	4,702	4,660	..
3831	Electrical industrial machinery	1,308	1,153	990	922	636	553	623	529	..
3832	Radio, TV and communications equipment	1,494	1,334	1,107	990	947	745	679	705	..
3833	Electrical appliances and housewares	1,023	901	787	900	601	600	617	594	..
3839	Other electrical apparatus and supplies	3,274	2,812	2,680	2,439	2,686	2,679	2,783	2,832	..
384	TRANSPORT EQUIPMENT	8,176	7,071	6,302	5,079	4,663	4,347	4,086	4,302	..
3841	Shipbuilding	131	123	100	..	..	30	..	9	..
3842	Railroad equipment	1,938	1,735	1,433	1,145	872	862	983	963	..
3843	Motor vehicles	5,913	5,048	4,603	3,916	3,583	3,354	2,933	3,182	..
3844	Motorcycles and bicycles	106	74	84	18	4	12	..	..	..
3845	Aircraft	88	91	82	..	183	89	170	148	..
3849	Other transport equipment	..	..	..	..	21	..	..	..	..
385	PROFESSIONAL AND SCIENTIFIC EQUIPMENT	864	741	576	358	391	379	351	305	..
3851	Professional equipment	779	662	521	358	391	359	323	270	..
3852	Photographic and optical goods	85	79	55	..	..	20	28	35	..
3853	Watches and clocks	..	..	..	..	..	..	..	..	..
39	OTHER MANUFACTURING INDUSTRIES	442	385	365	364	235	147	9	8	..
3901	Jewellery and related articles	..	..	..	..	..	..	..	..	..
3902	Musical instruments	..	..	..	..	..	..	..	..	..
3903	Sporting and athletic goods	..	..	..	..	..	..	..	..	..
3909	Other manufactures	442	385	365	364	235	147	6	8	..
SERR	non-specified, unallocated industry	..	..	..	..	..	..	..	..	..
3	TOTAL MANUFACTURING	282,343	234,162	200,894	193,865	185,294	179,213	183,847	165,969	..

ISIC Revision 2 Industry Sector		1990	1991	1992	1993	1994	1995	1996	1997	1998
31	FOOD, BEVERAGES AND TOBACCO	..	..	..	..	3,652	..	..	..	..
311/2	FOOD					3,629				
3111	Slaughtering, preparing and preserving meat					107				
3112	Dairy products					268				
3113	Canning, preserving of fruits and vegetables									
3114	Canning, preserving and processing of fish					900				
3115	Vegetable and animal oils and fats					2,013				
3116	Grain mill products									
3117	Bakery products					65				
3118	Sugar factories and refineries									
3119	Cocoa, chocolate and sugar confectionery					13				
3121	Other food products					24				
3122	Prepared animal feeds					239				
313	BEVERAGES					23				
3131	Distilling, rectifying and blending of spirits									
3132	Wine industries									
3133	Malt liquors and malts									
3134	Soft drinks					23				
314	TOBACCO									
32	TEXTILES, APPAREL AND LEATHER	..	..	..	..	122	..	..	..	..
321	TEXTILES					108				
3211	Spinning weaving and finishing textiles					107				
3212	Made-up goods excluding wearing apparel									
3213	Knitting mills									
3214	Carpets and rugs									
3215	Cordage, rope and twine					1				
3219	Other textiles									
322	WEARING APPAREL, EXCEPT FOOTWEAR					6				
323	LEATHER AND FUR PRODUCTS					9				
3231	Tanneries and leather finishing									
3232	Fur dressing and dyeing industries									
3233	Leather prods. ex. footwear and wearing apparel									
324	FOOTWEAR, EX. RUBBER AND PLASTIC									
33	WOOD PRODUCTS AND FURNITURE	..	..	..	..	23	..	..	..	..
331	WOOD PRODUCTS, EXCEPT FURNITURE									
3311	Sawmills, planing and other wood mills									
3312	Wooden and cane containers									
3319	Other wood and cork products									
332	FURNITURE, FIXTURES, EXCL. METALLIC									
34	PAPER, PUBLISHING AND PRINTING	..	..	..	..	64	..	..	..	..
341	PAPER AND PRODUCTS					19				
3411	Pulp, paper and paperboard articles									
3412	Containers of paper and paperboard									
3419	Other pulp, paper and paperboard articles					19				
342	PRINTING AND PUBLISHING					44				
35	CHEMICAL PRODUCTS	..	..	..	..	527	..	..	..	..
351	INDUSTRIAL CHEMICALS					486				
3511	Basic industrial chemicals excl. fertilizers									
3512	Fertilizers and pesticides					486				
3513	Synthetic resins and plastic materials									
352	OTHER CHEMICALS					21				
3521	Paints, varnishes and lacquers					2				
3522	Drugs and medicines									
3523	Soap, cleaning preparations, perfumes, cosmetics									
3529	Other chemical products					20				
353	PETROLEUM REFINERIES									
354	MISC. PETROLEUM AND COAL PRODUCTS					2				
355	RUBBER PRODUCTS					2				
3551	Tyres and tubes									
3559	Other rubber products									
356	PLASTIC PRODUCTS					16				
36	NON-METALLIC MINERAL PRODUCTS	..	..	..	..	1,617	..	..	..	..
361	POTTERY, CHINA, EARTHENWARE					2				
362	GLASS AND PRODUCTS					2				
369	OTHER NON-METAL. MINERAL PRODUCTS					1,614				
3691	Structural clay products									
3692	Cement, lime and plaster					572				
3699	Other non-metallic mineral products					1,042				
37	BASIC METAL INDUSTRIES	..	..	..	..	12,498	..	..	..	..
371	IRON AND STEEL					4,668				
372	NON-FERROUS METALS					7,830				
38	METAL PRODUCTS, MACHINERY, EQUIP.	..	..	..	..	36	..	..	..	..
381	METAL PRODUCTS									
3811	Cutlery, hand tools and general hardware									
3812	Furniture and fixtures primarily of metal									
3813	Structural metal products									
3819	Other fabricated metal products									
382	NON-ELECTRICAL MACHINERY					11				
3821	Engines and turbines									
3822	Agricultural machinery and equipment									
3823	Metal and wood working machinery									
3824	Special industrial machinery									
3825	Office, computing and accounting machinery									
3829	Other non-electrical machinery and equipment									
383	ELECTRICAL MACHINERY					2				
3831	Electrical industrial machinery									
3832	Radio, TV and communications equipment									
3833	Electrical appliances and housewares									
3839	Other electrical apparatus and supplies									
384	TRANSPORT EQUIPMENT					23				
3841	Shipbuilding					23				
3842	Railroad equipment									
3843	Motor vehicles									
3844	Motorcycles and bicycles									
3845	Aircraft									
3849	Other transport equipment									
385	PROFESSIONAL AND SCIENTIFIC EQUIPMENT									
3851	Professional equipment									
3852	Photographic and optical goods									
3853	Watches and clocks									
39	OTHER MANUFACTURING INDUSTRIES	..	..	..	..	27	..	..	..	..
3901	Jewellery and related articles									
3902	Musical instruments									
3903	Sporting and athletic goods									
3909	Other manufactures									
SERR	non-specified, unallocated industry									
3	TOTAL MANUFACTURING	..	..	..	..	18,565	..	..	..	..

ISIC Revision 2 Industry Sector		1990	1991	1992	1993	1994	1995	1996	1997	1998
31	FOOD, BEVERAGES AND TOBACCO	..	56,916	73,854	62,851	67,861	73,854	75,676	72,183	63,761
311/2	FOOD	..	29,061	34,419	28,449	31,124	34,419	33,799	..	..
3111	Slaughtering, preparing and preserving meat	..	..	..	..	..	..	..	..	..
3112	Dairy products	..	..	..	..	..	..	..	..	..
3113	Canning, preserving of fruits and vegetables	..	..	..	..	..	..	..	..	..
3114	Canning, preserving and processing of fish	..	..	..	..	..	..	..	..	..
3115	Vegetable and animal oils and fats	..	..	..	..	..	..	..	..	..
3116	Grain mill products	..	..	..	..	..	..	..	..	..
3117	Bakery products	..	..	..	..	..	..	..	..	..
3118	Sugar factories and refineries	..	..	..	..	..	..	..	..	..
3119	Cocoa, chocolate and sugar confectionery	..	..	..	..	..	..	..	..	..
3121	Other food products	..	..	..	..	..	..	..	..	..
3122	Prepared animal feeds	..	..	..	..	..	..	..	..	..
313	BEVERAGES	..	12,389	17,957	16,682	17,994	17,957	18,415	..	..
3131	Distilling, rectifying and blending of spirits	..	..	..	..	..	..	..	..	..
3132	Wine industries	..	..	..	..	..	..	..	..	..
3133	Malt liquors and malts	..	..	..	..	..	..	..	..	..
3134	Soft drinks	..	..	..	..	..	..	..	..	..
314	TOBACCO	..	718	797	868	670	797	912	..	..
32	TEXTILES, APPAREL AND LEATHER	..	109,638	109,987	116,390	102,749	109,987	110,234	114,034	121,258
321	TEXTILES	..	57,702	57,303	54,163	54,975	57,303	54,361	..	..
3211	Spinning weaving and finishing textiles	..	..	..	..	..	..	..	..	..
3212	Made-up goods excluding wearing apparel	..	..	..	..	..	..	..	..	..
3213	Knitting mills	..	..	..	..	..	..	..	..	..
3214	Carpets and rugs	..	..	..	..	..	..	..	..	..
3215	Cordage, rope and twine	..	..	..	..	..	..	..	..	..
3219	Other textiles	..	..	..	..	..	..	..	..	..
322	WEARING APPAREL, EXCEPT FOOTWEAR	..	..	..	..	..	..	..	..	..
323	LEATHER AND FUR PRODUCTS	..	2,878	2,509	2,903	2,802	2,509	2,429	..	..
3231	Tanneries and leather finishing	..	..	..	..	..	..	..	..	..
3232	Fur dressing and dyeing industries	..	..	..	..	..	..	..	..	..
3233	Leather prods. ex. footwear and wearing apparel	..	..	..	..	..	..	..	..	..
324	FOOTWEAR, EX. RUBBER AND PLASTIC	..	..	..	..	..	..	..	..	..
33	WOOD PRODUCTS AND FURNITURE	..	5,538	6,338	5,896	7,810	6,338	6,928	7,305	5,596
331	WOOD PRODUCTS, EXCEPT FURNITURE	..	..	..	..	..	..	..	..	..
3311	Sawmills, planing and other wood mills	..	..	..	..	..	..	..	..	..
3312	Wooden and cane containers	..	..	..	..	..	..	..	..	..
3319	Other wood and cork products	..	..	..	..	..	..	..	..	..
332	FURNITURE, FIXTURES, EXCL. METALLIC	..	..	..	..	..	..	..	..	..
34	PAPER, PUBLISHING AND PRINTING	..	53,009	74,765	64,673	71,854	74,765	83,375	85,689	74,899
341	PAPER AND PRODUCTS	..	35,513	50,899	44,763	49,826	70,801	55,181	..	..
3411	Pulp, paper and paperboard articles	..	..	..	..	..	..	..	..	..
3412	Containers of paper and paperboard	..	..	..	..	..	..	..	..	..
3419	Other pulp, paper and paperboard articles	..	..	..	..	..	..	..	..	..
342	PRINTING AND PUBLISHING	..	575	369	352	322	2,917	352	..	..
35	CHEMICAL PRODUCTS	..	203,345	252,149	262,488	272,423	252,149	273,776	1,328,520	1,376,696
351	INDUSTRIAL CHEMICALS	..	99,507	129,042	148,704	158,451	129,042	138,030	..	..
3511	Basic industrial chemicals excl. fertilizers	..	71,560	81,388	96,665	102,423	81,388	89,402	..	..
3512	Fertilizers and pesticides	..	7,623	17,492	25,422	27,502	17,492	17,591	..	..
3513	Synthetic resins and plastic materials	..	20,324	30,162	26,617	28,526	30,162	31,037	..	..
352	OTHER CHEMICALS	..	32,221	32,150	27,911	28,610	32,150	34,064	..	..
3521	Paints, varnishes and lacquers	..	..	..	..	..	..	..	..	..
3522	Drugs and medicines	..	..	..	..	..	..	..	..	..
3523	Soap, cleaning preparations, perfumes, cosmetics	..	..	..	..	..	..	..	..	..
3529	Other chemical products	..	..	..	..	..	..	..	..	..
353	PETROLEUM REFINERIES	..	1,099	1,217	1,263	1,764	1,217	1,542	..	..
354	MISC. PETROLEUM AND COAL PRODUCTS	..	..	..	..	..	..	..	..	..
355	RUBBER PRODUCTS	..	6,571	6,560	5,969	6,202	6,560	6,624	..	..
3551	Tyres and tubes	..	..	..	..	..	..	..	..	..
3559	Other rubber products	..	..	..	..	..	..	..	..	..
356	PLASTIC PRODUCTS	..	532	828	827	731	828	1,572	..	..
36	NON-METALLIC MINERAL PRODUCTS	..	192,392	251,392	86,578	243,897	251,392	273,933	275,871	208,159
361	POTTERY, CHINA, EARTHENWARE	..	11,902	11,197	12,759	12,024	11,197	10,454	..	..
362	GLASS AND PRODUCTS	..	14,485	18,829	14,309	16,149	18,829	19,529	..	..
369	OTHER NON-METAL. MINERAL PRODUCTS	..	26,327	31,818	27,370	36,922	31,818	29,975	..	..
3691	Structural clay products	..	5,935	8,054	6,522	6,227	8,054	7,549	..	..
3692	Cement, lime and plaster	..	11,895	17,660	18,296	21,386	17,660	15,738	..	..
3699	Other non-metallic mineral products	..	8,497	6,104	2,552	9,309	6,104	6,688	..	..
37	BASIC METAL INDUSTRIES	..	492,215	622,020	723,095	591,314	622,020	645,808	665,071	666,130
371	IRON AND STEEL	..	40,928	61,995	53,848	62,292	61,995	56,957	..	..
372	NON-FERROUS METALS	..	6,034	8,863	5,500	7,106	8,863	6,149	..	..
38	METAL PRODUCTS, MACHINERY, EQUIP.	..	90,724	138,514	110,338	123,411	138,512	159,544	178,885	169,408
381	METAL PRODUCTS	..	5,238	7,469	5,997	5,992	7,469	7,194	..	..
3811	Cutlery, hand tools and general hardware	..	..	..	..	..	..	..	..	..
3812	Furniture and fixtures primarily of metal	..	..	..	..	..	..	..	..	..
3813	Structural metal products	..	..	..	..	..	..	..	..	..
3819	Other fabricated metal products	..	..	..	..	..	..	..	..	..
382	NON-ELECTRICAL MACHINERY	..	4,762	5,573	5,013	5,359	5,573	5,648	..	..
3821	Engines and turbines	..	..	..	..	..	..	..	..	..
3822	Agricultural machinery and equipment	..	..	..	..	..	..	..	..	..
3823	Metal and wood working machinery	..	..	..	..	..	..	..	..	..
3824	Special industrial machinery	..	..	..	..	..	..	..	..	..
3825	Office, computing and accounting machinery	..	..	..	..	..	..	..	..	..
3829	Other non-electrical machinery and equipment	..	..	..	..	..	..	..	..	..
383	ELECTRICAL MACHINERY	..	15,509	20,931	20,925	20,788	20,930	24,898	..	..
3831	Electrical industrial machinery	..	..	..	..	..	..	..	..	..
3832	Radio, TV and communications equipment	..	..	..	..	..	..	..	..	..
3833	Electrical appliances and housewares	..	..	..	..	..	..	..	..	..
3839	Other electrical apparatus and supplies	..	..	..	..	..	..	..	..	..
384	TRANSPORT EQUIPMENT	..	11,579	14,069	12,862	13,125	14,068	14,965	..	..
3841	Shipbuilding	..	3,491	2,814	3,547	3,278	2,814	2,481	..	..
3842	Railroad equipment	..	97	203	146	173	203	177	..	..
3843	Motor vehicles	..	7,655	10,786	8,911	9,375	10,785	12,142	..	..
3844	Motorcycles and bicycles	..	336	179	256	290	179	90	..	..
3845	Aircraft	..		87	2	9	87	75	..	..
3849	Other transport equipment	..	..	..	..	..	..	..	..	..
385	PROFESSIONAL AND SCIENTIFIC EQUIPMENT	..	1,389	1,328	1,740	1,906	1,328	1,047	..	..
3851	Professional equipment	..	..	..	..	..	..	..	..	..
3852	Photographic and optical goods	..	..	..	..	..	..	..	..	..
3853	Watches and clocks	..	..	..	..	..	..	..	..	..
39	OTHER MANUFACTURING INDUSTRIES	..	59,735	116,257	60,227	96,663	116,258	125,841	153,948	152,891
3901	Jewellery and related articles	..	..	..	..	..	..	..	..	..
3902	Musical instruments	..	..	..	..	..	..	..	..	..
3903	Sporting and athletic goods	..	..	..	..	..	..	..	..	..
3909	Other manufactures	..	..	..	..	..	..	..	..	..
SERR	non-specified, unallocated industry	..	..	..	..	..	..	..	..	..
3	TOTAL MANUFACTURING	..	1,263,512	1,645,277	1,492,535	1,577,982	1,645,276	1,755,116	2,881,506	2,838,798

ISIS Energy Data Programme (IEA/OECD)

ISIC Revision 2 Industry Sector		1990	1991	1992	1993	1994	1995	1996	1997	1998
31	**FOOD, BEVERAGES AND TOBACCO**	136,406	139,109	131,144	136,943	125,382	140,697	..	..	..
311/2	FOOD	118,517	119,988	110,848	116,345	103,018	122,400	..	..	..
3111	Slaughtering, preparing and preserving meat	..	..	..	..	..	..			..
3112	Dairy products	..	..	..	..	..	..			..
3113	Canning, preserving of fruits and vegetables	..	..	..	..	..	..			..
3114	Canning, preserving and processing of fish	..	..	..	..	..	..			..
3115	Vegetable and animal oils and fats	..	..	..	..	..	..			..
3116	Grain mill products	..	..	..	..	..	..			..
3117	Bakery products	..	..	..	..	..	..			..
3118	Sugar factories and refineries	118,517	119,988	110,848	116,345	103,018	122,400			..
3119	Cocoa, chocolate and sugar confectionery	..	..	..	..	..	..			..
3121	Other food products	..	..	..	..	..	..			..
3122	Prepared animal feeds	..	..	..	..	..	..			..
313	BEVERAGES	17,475	18,670	19,832	20,166	21,912	17,844			..
3131	Distilling, rectifying and blending of spirits	..	..	..	..	..	..			..
3132	Wine industries	..	..	..	..	..	..			..
3133	Malt liquors and malts	11,020	11,038	11,278	11,528	12,744	10,313			..
3134	Soft drinks	6,455	7,632	8,553	8,638	9,168	7,531			..
314	TOBACCO	414	452	464	432	452	453			..
32	**TEXTILES, APPAREL AND LEATHER**	..	..	..	..	..	..	..	..	..
321	TEXTILES	..	..	..	..	..	..			..
3211	Spinning weaving and finishing textiles	..	..	..	..	..	..			..
3212	Made-up goods excluding wearing apparel	..	..	..	..	..	..			..
3213	Knitting mills	..	..	..	..	..	..			..
3214	Carpets and rugs	..	..	..	..	..	..			..
3215	Cordage, rope and twine	..	..	..	..	..	..			..
3219	Other textiles	..	..	..	..	..	..			..
322	WEARING APPAREL, EXCEPT FOOTWEAR	..	..	..	..	..	..			..
323	LEATHER AND FUR PRODUCTS	..	..	..	..	..	..			..
3231	Tanneries and leather finishing	..	..	..	..	..	..			..
3232	Fur dressing and dyeing industries	..	..	..	..	..	..			..
3233	Leather prods. ex. footwear and wearing apparel	..	..	..	..	..	..			..
324	FOOTWEAR, EX. RUBBER AND PLASTIC	..	..	..	..	..	..			..
33	**WOOD PRODUCTS AND FURNITURE**	..	..	..	..	..	..	..	..	..
331	WOOD PRODUCTS, EXCEPT FURNITURE	..	..	..	..	..	..			..
3311	Sawmills, planing and other wood mills	..	..	..	..	..	..			..
3312	Wooden and cane containers	..	..	..	..	..	..			..
3319	Other wood and cork products	..	..	..	..	..	..			..
332	FURNITURE, FIXTURES, EXCL. METALLIC	..	..	..	..	..	..			..
34	**PAPER, PUBLISHING AND PRINTING**	55,378	52,009	48,558	47,500	49,161	44,639	..	..	..
341	PAPER AND PRODUCTS	55,378	52,009	48,558	47,500	49,161	44,639			..
3411	Pulp, paper and paperboard articles	55,378	52,009	48,558	47,500	49,161	44,639			..
3412	Containers of paper and paperboard	..	..	..	..	..	..			..
3419	Other pulp, paper and paperboard articles	..	..	..	..	..	..			..
342	PRINTING AND PUBLISHING	..	..	..	..	..	..			..
35	**CHEMICAL PRODUCTS**	293,725	313,065	297,662	284,925	331,179	326,396	..	..	..
351	INDUSTRIAL CHEMICALS	118,147	121,176	120,988	280,863	326,564	322,090			..
3511	Basic industrial chemicals excl. fertilizers	104,309	109,500	112,888	267,753	312,453	309,422			..
3512	Fertilizers and pesticides	13,839	11,676	8,100	13,110	14,111	12,668			..
3513	Synthetic resins and plastic materials	..	..	..	..	..	..			..
352	OTHER CHEMICALS	..	..	..	..	..	..			..
3521	Paints, varnishes and lacquers	..	..	..	..	..	..			..
3522	Drugs and medicines	..	..	..	..	..	..			..
3523	Soap, cleaning preparations, perfumes, cosmetics	..	..	..	..	..	..			..
3529	Other chemical products	..	..	..	..	..	..			..
353	PETROLEUM REFINERIES	170,476	186,556	171,453	..	..	..			..
354	MISC. PETROLEUM AND COAL PRODUCTS	..	..	..	..	..	..			..
355	RUBBER PRODUCTS	5,102	5,333	5,221	4,062	4,615	4,306			..
3551	Tyres and tubes	..	..	..	..	..	..			..
3559	Other rubber products	5,102	5,333	5,221	..	..	..			..
356	PLASTIC PRODUCTS	..	..	..	..	..	..			..
36	**NON-METALLIC MINERAL PRODUCTS**	126,747	130,979	139,629	138,603	137,284	128,383	..	..	..
361	POTTERY, CHINA, EARTHENWARE	..	..	..	..	..	..			..
362	GLASS AND PRODUCTS	31,311	31,221	32,717	33,282	30,870	27,885			..
369	OTHER NON-METAL. MINERAL PRODUCTS	95,435	99,757	106,912	105,321	106,414	100,498			..
3691	Structural clay products	..	..	..	..	..	..			..
3692	Cement, lime and plaster	95,435	99,757	106,912	105,321	106,414	100,498			..
3699	Other non-metallic mineral products	..	..	..	..	..	..			..
37	**BASIC METAL INDUSTRIES**	201,039	184,122	177,780	179,782	201,026	225,150	..	..	..
371	IRON AND STEEL	194,673	178,962	174,231	176,096	196,406	220,650			..
372	NON-FERROUS METALS	6,366	5,160	3,549	3,686	4,620	4,500			..
38	**METAL PRODUCTS, MACHINERY, EQUIP.**	6,120	5,423	5,883	6,927	6,951	5,172	..	..	..
381	METAL PRODUCTS	..	..	..	..	..	..			..
3811	Cutlery, hand tools and general hardware	..	..	..	..	..	..			..
3812	Furniture and fixtures primarily of metal	..	..	..	..	..	..			..
3813	Structural metal products	..	..	..	..	..	..			..
3819	Other fabricated metal products	..	..	..	..	..	..			..
382	NON-ELECTRICAL MACHINERY	..	..	..	..	..	..			..
3821	Engines and turbines	..	..	..	..	..	..			..
3822	Agricultural machinery and equipment	..	..	..	..	..	..			..
3823	Metal and wood working machinery	..	..	..	..	..	..			..
3824	Special industrial machinery	..	..	..	..	..	..			..
3825	Office, computing and accounting machinery	..	..	..	..	..	..			..
3829	Other non-electrical machinery and equipment	..	..	..	..	..	..			..
383	ELECTRICAL MACHINERY	..	..	..	..	..	..			..
3831	Electrical industrial machinery	..	..	..	..	..	..			..
3832	Radio, TV and communications equipment	..	..	..	..	..	..			..
3833	Electrical appliances and housewares	..	..	..	..	..	..			..
3839	Other electrical apparatus and supplies	..	..	..	..	..	..			..
384	TRANSPORT EQUIPMENT	6,120	5,423	5,883	6,927	6,951	5,172			..
3841	Shipbuilding	..	..	..	..	..	..			..
3842	Railroad equipment	..	..	..	..	..	..			..
3843	Motor vehicles	6,120	5,423	5,883	6,927	6,951	5,172			..
3844	Motorcycles and bicycles	..	..	..	..	..	..			..
3845	Aircraft	..	..	..	..	..	..			..
3849	Other transport equipment	..	..	..	..	..	..			..
385	PROFESSIONAL AND SCIENTIFIC EQUIPMENT	..	..	..	..	..	..			..
3851	Professional equipment	..	..	..	..	..	..			..
3852	Photographic and optical goods	..	..	..	..	..	..			..
3853	Watches and clocks	..	..	..	..	..	..			..
39	**OTHER MANUFACTURING INDUSTRIES**	259,050	285,703	296,230	327,817	308,055	351,557	..	..	..
3901	Jewellery and related articles	..	..	..	..	..	..			..
3902	Musical instruments	..	..	..	..	..	..			..
3903	Sporting and athletic goods	..	..	..	..	..	..			..
3909	Other manufactures	259,050	285,703	296,230	327,817	308,055	351,557			..
SERR	**non-specified, unallocated industry**	..	..	..	..	..	..			
3	**TOTAL MANUFACTURING**	1,078,464	1,110,409	1,096,886	1,122,497	1,159,038	1,221,994	..	..	..

ISIC Revision 2 Industry Sector	1990	1991	1992	1993	1994	1995	1996	1997	1998
31 FOOD, BEVERAGES AND TOBACCO	..	..	18,308	..	..	..	..	..	..
311/2 FOOD	..	..	16,899	..	..	..	..	..	..
3111 Slaughtering, preparing and preserving meat	..	..	2,780	..	..	..	..	..	..
3112 Dairy products	..	..	2,888	..	..	..	..	..	..
3113 Canning, preserving of fruits and vegetables	..	..	331	..	..	..	..	..	..
3114 Canning, preserving and processing of fish	..	..	2,033	..	..	..	..	..	..
3115 Vegetable and animal oils and fats	..	..	4,159	..	..	..	..	..	..
3116 Grain mill products	..	..	858	..	..	..	..	..	..
3117 Bakery products	..	..	1,340	..	..	..	..	..	..
3118 Sugar factories and refineries	..	..	..	..	..	..	..	..	..
3119 Cocoa, chocolate and sugar confectionery	..	..	372	..	..	..	..	..	..
3121 Other food products	..	..	1,018	..	..	..	..	..	..
3122 Prepared animal feeds	..	..	1,119	..	..	..	..	..	..
313 BEVERAGES	..	..	1,311	..	..	..	..	..	..
3131 Distilling, rectifying and blending of spirits	..	..	51	..	..	..	..	..	..
3132 Wine industries	..	..	..	..	..	..	..	..	..
3133 Malt liquors and malts	..	..	1,016	..	..	..	..	..	..
3134 Soft drinks	..	..	244	..	..	..	..	..	..
314 TOBACCO	..	..	98	..	..	..	..	..	..
32 TEXTILES, APPAREL AND LEATHER	..	..	1,097	..	..	..	..	..	..
321 TEXTILES	..	..	887	..	..	..	..	..	..
3211 Spinning weaving and finishing textiles	..	..	494	..	..	..	..	..	..
3212 Made-up goods excluding wearing apparel	..	..	49	..	..	..	..	..	..
3213 Knitting mills	..	..	124	..	..	..	..	..	..
3214 Carpets and rugs	..	..	3	..	..	..	..	..	..
3215 Cordage, rope and twine	..	..	125	..	..	..	..	..	..
3219 Other textiles	..	..	92	..	..	..	..	..	..
322 WEARING APPAREL, EXCEPT FOOTWEAR	..	..	95	..	..	..	..	..	..
323 LEATHER AND FUR PRODUCTS	..	..	101	..	..	..	..	..	..
3231 Tanneries and leather finishing	..	..	82	..	..	..	..	..	..
3232 Fur dressing and dyeing industries	..	..	8	..	..	..	..	..	..
3233 Leather prods. ex. footwear and wearing apparel	..	..	11	..	..	..	..	..	..
324 FOOTWEAR, EX. RUBBER AND PLASTIC	..	..	14	..	..	..	..	..	..
33 WOOD PRODUCTS AND FURNITURE	..	..	187,288	..	..	..	..	..	..
331 WOOD PRODUCTS, EXCEPT FURNITURE	..	..	7,109	..	..	..	..	..	..
3311 Sawmills, planing and other wood mills	..	..	6,444	..	..	..	..	..	..
3312 Wooden and cane containers	..	..	24	..	..	..	..	..	..
3319 Other wood and cork products	..	..	641	..	..	..	..	..	..
332 FURNITURE, FIXTURES, EXCL. METALLIC	..	..	180,178	..	..	..	..	..	..
34 PAPER, PUBLISHING AND PRINTING	..	..	43,173	..	..	..	..	..	..
341 PAPER AND PRODUCTS	..	..	41,402	..	..	..	..	..	..
3411 Pulp, paper and paperboard articles	..	..	40,310	..	..	..	..	..	..
3412 Containers of paper and paperboard	..	..	867	..	..	..	..	..	..
3419 Other pulp, paper and paperboard articles	..	..	225	..	..	..	..	..	..
342 PRINTING AND PUBLISHING	..	..	1,771	..	..	..	..	..	..
35 CHEMICAL PRODUCTS	..	..	69,212	..	..	..	..	..	..
351 INDUSTRIAL CHEMICALS	..	..	27,794	..	..	..	..	..	..
3511 Basic industrial chemicals excl. fertilizers	..	..	10,415	..	..	..	..	..	..
3512 Fertilizers and pesticides	..	..	15,754	..	..	..	..	..	..
3513 Synthetic resins and plastic materials	..	..	1,626	..	..	..	..	..	..
352 OTHER CHEMICALS	..	..	1,799	..	..	..	..	..	..
3521 Paints, varnishes and lacquers	..	..	238	..	..	..	..	..	..
3522 Drugs and medicines	..	..	785	..	..	..	..	..	..
3523 Soap, cleaning preparations, perfumes, cosmetics	..	..	135	..	..	..	..	..	..
3529 Other chemical products	..	..	641	..	..	..	..	..	..
353 PETROLEUM REFINERIES	..	..	35,505	..	..	..	..	..	..
354 MISC. PETROLEUM AND COAL PRODUCTS	..	..	2,331	..	..	..	..	..	..
355 RUBBER PRODUCTS	..	..	360	..	..	..	..	..	..
3551 Tyres and tubes	..	..	291	..	..	..	..	..	..
3559 Other rubber products	..	..	69	..	..	..	..	..	..
356 PLASTIC PRODUCTS	..	..	1,422	..	..	..	..	..	..
36 NON-METALLIC MINERAL PRODUCTS	..	..	10,551	..	..	..	..	..	..
361 POTTERY, CHINA, EARTHENWARE	..	..	164	..	..	..	..	..	..
362 GLASS AND PRODUCTS	..	..	1,591	..	..	..	..	..	..
369 OTHER NON-METAL. MINERAL PRODUCTS	..	..	8,796	..	..	..	..	..	..
3691 Structural clay products	..	..	311	..	..	..	..	..	..
3692 Cement, lime and plaster	..	..	4,685	..	..	..	..	..	..
3699 Other non-metallic mineral products	..	..	3,800	..	..	..	..	..	..
37 BASIC METAL INDUSTRIES	..	..	89,689	..	..	..	..	..	..
371 IRON AND STEEL	..	..	27,663	..	..	..	..	..	..
372 NON-FERROUS METALS	..	..	62,026	..	..	..	..	..	..
38 METAL PRODUCTS, MACHINERY, EQUIP.	..	..	11,698	..	..	..	..	..	..
381 METAL PRODUCTS	..	..	3,511	..	..	..	..	..	..
3811 Cutlery, hand tools and general hardware	..	..	254	..	..	..	..	..	..
3812 Furniture and fixtures primarily of metal	..	..	159	..	..	..	..	..	..
3813 Structural metal products	..	..	1,324	..	..	..	..	..	..
3819 Other fabricated metal products	..	..	1,774	..	..	..	..	..	..
382 NON-ELECTRICAL MACHINERY	..	..	3,584	..	..	..	..	..	..
3821 Engines and turbines	..	..	48	..	..	..	..	..	..
3822 Agricultural machinery and equipment	..	..	278	..	..	..	..	..	..
3823 Metal and wood working machinery	..	..	39	..	..	..	..	..	..
3824 Special industrial machinery	..	..	1,383	..	..	..	..	..	..
3825 Office, computing and accounting machinery	..	..	53	..	..	..	..	..	..
3829 Other non-electrical machinery and equipment	..	..	1,783	..	..	..	..	..	..
383 ELECTRICAL MACHINERY	..	..	1,328	..	..	..	..	..	..
3831 Electrical industrial machinery	..	..	477	..	..	..	..	..	..
3832 Radio, TV and communications equipment	..	..	298	..	..	..	..	..	..
3833 Electrical appliances and housewares	..	..	55	..	..	..	..	..	..
3839 Other electrical apparatus and supplies	..	..	498	..	..	..	..	..	..
384 TRANSPORT EQUIPMENT	..	..	3,090	..	..	..	..	..	..
3841 Shipbuilding	..	..	1,587	..	..	..	..	..	..
3842 Railroad equipment	..	..	201	..	..	..	..	..	..
3843 Motor vehicles	..	..	786	..	..	..	..	..	..
3844 Motorcycles and bicycles	..	..	34	..	..	..	..	..	..
3845 Aircraft	..	..	463	..	..	..	..	..	..
3849 Other transport equipment	..	..	19	..	..	..	..	..	..
385 PROFESSIONAL AND SCIENTIFIC EQUIPMENT	..	..	184	..	..	..	..	..	..
3851 Professional equipment	..	..	164	..	..	..	..	..	..
3852 Photographic and optical goods	..	..	20	..	..	..	..	..	..
3853 Watches and clocks	..	..	..	..	..	..	..	..	..
39 OTHER MANUFACTURING INDUSTRIES	..	..	263	..	..	..	..	..	..
3901 Jewellery and related articles	..	..	43	..	..	..	..	..	..
3902 Musical instruments	..	..	1	..	..	..	..	..	..
3903 Sporting and athletic goods	..	..	94	..	..	..	..	..	..
3909 Other manufactures	..	..	125	..	..	..	..	..	..
SERR non-specified, unallocated industry	..	..	..	..	..	..	..	..	..
3 TOTAL MANUFACTURING	..	..	431,278	..	..	..	..	..	..

ISIS Energy Data Programme (IEA/OECD)

ISIC Revision 2 Industry Sector		1990	1991	1992	1993	1994	1995	1996	1997	1998
31	FOOD, BEVERAGES AND TOBACCO	..	..	..	..	18,347	..	..	..	..
311/2	FOOD	..	..	..	..	15,823	..	..	..	..
3111	Slaughtering, preparing and preserving meat	..	..	..	..	813	..	..	..	..
3112	Dairy products	..	..	..	..	1,611	..	..	..	..
3113	Canning, preserving of fruits and vegetables	..	..	..	..	925	..	..	..	..
3114	Canning, preserving and processing of fish	..	..	..	..	290	..	..	..	..
3115	Vegetable and animal oils and fats	..	..	..	..	694	..	..	..	..
3116	Grain mill products	..	..	..	..	353	..	..	..	..
3117	Bakery products	..	..	..	..	777	..	..	..	..
3118	Sugar factories and refineries	..	..	..	..	1,280	..	..	..	..
3119	Cocoa, chocolate and sugar confectionery	..	..	..	..	120	..	..	..	..
3121	Other food products	..	..	..	..	1,250	..	..	..	..
3122	Prepared animal feeds	..	..	..	..	621	..	..	..	..
313	BEVERAGES	..	..	..	..	2,329	..	..	..	..
3131	Distilling, rectifying and blending of spirits	..	..	..	..	107	..	..	..	..
3132	Wine industries	..	..	..	..	144	..	..	..	..
3133	Malt liquors and malts	..	..	..	..	1,206	..	..	..	..
3134	Soft drinks	..	..	..	..	213	..	..	..	..
314	TOBACCO	..	..	..	..	196	..	..	..	..
32	TEXTILES, APPAREL AND LEATHER	..	..	..	..	18,397	..	..	..	..
321	TEXTILES	..	..	..	..	15,348	..	..	..	..
3211	Spinning weaving and finishing textiles	..	..	..	..	5,396	..	..	..	..
3212	Made-up goods excluding wearing apparel	..	..	..	..	..	..	..	..	..
3213	Knitting mills	..	..	..	..	..	..	..	..	..
3214	Carpets and rugs	..	..	..	..	..	..	..	..	..
3215	Cordage, rope and twine	..	..	..	..	..	..	..	..	..
3219	Other textiles	..	..	..	..	1,130	..	..	..	..
322	WEARING APPAREL, EXCEPT FOOTWEAR	..	..	..	..	2,517	..	..	..	..
323	LEATHER AND FUR PRODUCTS	..	..	..	..	490	..	..	..	..
3231	Tanneries and leather finishing	..	..	..	..	267	..	..	..	..
3232	Fur dressing and dyeing industries	..	..	..	..	..	..	..	..	..
3233	Leather prods. ex. footwear and wearing apparel	..	..	..	..	16	..	..	..	..
324	FOOTWEAR, EX. RUBBER AND PLASTIC	..	..	..	..	43	..	..	..	..
33	WOOD PRODUCTS AND FURNITURE	..	..	..	..	5,579	..	..	..	..
331	WOOD PRODUCTS, EXCEPT FURNITURE	..	..	..	..	1,689	..	..	..	..
3311	Sawmills, planing and other wood mills	..	..	..	..	2,049	..	..	..	..
3312	Wooden and cane containers	..	..	..	..	..	..	..	..	..
3319	Other wood and cork products	..	..	..	..	..	..	..	..	..
332	FURNITURE, FIXTURES, EXCL. METALLIC	..	..	..	..	106	..	..	..	..
34	PAPER, PUBLISHING AND PRINTING	..	..	..	..	15,968	..	..	..	..
341	PAPER AND PRODUCTS	..	..	..	..	15,606	..	..	..	..
3411	Pulp, paper and paperboard articles	..	..	..	..	7,001	..	..	..	..
3412	Containers of paper and paperboard	..	..	..	..	509	..	..	..	..
3419	Other pulp, paper and paperboard articles	..	..	..	..	1,257	..	..	..	..
342	PRINTING AND PUBLISHING	..	..	..	..	363	..	..	..	..
35	CHEMICAL PRODUCTS	..	..	..	..	15,060	..	..	..	..
351	INDUSTRIAL CHEMICALS	..	..	..	..	12,940	..	..	..	..
3511	Basic industrial chemicals excl. fertilizers	..	..	..	..	3,317	..	..	..	..
3512	Fertilizers and pesticides	..	..	..	..	1,274	..	..	..	..
3513	Synthetic resins and plastic materials	..	..	..	..	555	..	..	..	..
352	OTHER CHEMICALS	..	..	..	..	864	..	..	..	..
3521	Paints, varnishes and lacquers	..	..	..	..	81	..	..	..	..
3522	Drugs and medicines	..	..	..	..	208	..	..	..	..
3523	Soap, cleaning preparations, perfumes, cosmetics	..	..	..	..	428	..	..	..	..
3529	Other chemical products	..	..	..	..	147	..	..	..	..
353	PETROLEUM REFINERIES	..	..	..	..	14	..	..	..	..
354	MISC. PETROLEUM AND COAL PRODUCTS	..	..	..	..	295	..	..	..	..
355	RUBBER PRODUCTS	..	..	..	..	752	..	..	..	..
3551	Tyres and tubes	..	..	..	..	217	..	..	..	..
3559	Other rubber products	..	..	..	..	63	..	..	..	..
356	PLASTIC PRODUCTS	..	..	..	..	195	..	..	..	..
36	NON-METALLIC MINERAL PRODUCTS	..	..	..	..	58,831	..	..	..	..
361	POTTERY, CHINA, EARTHENWARE	..	..	..	..	20,752	..	..	..	..
362	GLASS AND PRODUCTS	..	..	..	..	6,942	..	..	..	..
369	OTHER NON-METAL. MINERAL PRODUCTS	..	..	..	..	31,136	..	..	..	..
3691	Structural clay products	..	..	..	..	3,945	..	..	..	..
3692	Cement, lime and plaster	..	..	..	..	22,351	..	..	..	..
3699	Other non-metallic mineral products	..	..	..	..	513	..	..	..	..
37	BASIC METAL INDUSTRIES	..	..	..	..	12,261	..	..	..	..
371	IRON AND STEEL	..	..	..	..	10,946	..	..	..	..
372	NON-FERROUS METALS	..	..	..	..	1,315	..	..	..	..
38	METAL PRODUCTS, MACHINERY, EQUIP.	..	..	..	..	6,367	..	..	..	..
381	METAL PRODUCTS	..	..	..	..	2,812	..	..	..	..
3811	Cutlery, hand tools and general hardware	..	..	..	..	83	..	..	..	..
3812	Furniture and fixtures primarily of metal	..	..	..	..	87	..	..	..	..
3813	Structural metal products	..	..	..	..	213	..	..	..	..
3819	Other fabricated metal products	..	..	..	..	575	..	..	..	..
382	NON-ELECTRICAL MACHINERY	..	..	..	..	497	..	..	..	..
3821	Engines and turbines	..	..	..	..	1	..	..	..	..
3822	Agricultural machinery and equipment	..	..	..	..	75	..	..	..	..
3823	Metal and wood working machinery	..	..	..	..	41	..	..	..	..
3824	Special industrial machinery	..	..	..	..	12	..	..	..	..
3825	Office, computing and accounting machinery	..	..	..	..	7	..	..	..	..
3829	Other non-electrical machinery and equipment	..	..	..	..	80	..	..	..	..
383	ELECTRICAL MACHINERY	..	..	..	..	1,422	..	..	..	..
3831	Electrical industrial machinery	..	..	..	..	180	..	..	..	..
3832	Radio, TV and communications equipment	..	..	..	..	45	..	..	..	..
3833	Electrical appliances and housewares	..	..	..	..	48	..	..	..	..
3839	Other electrical apparatus and supplies	..	..	..	..	194	..	..	..	..
384	TRANSPORT EQUIPMENT	..	..	..	..	1,588	..	..	..	..
3841	Shipbuilding	..	..	..	..	64	..	..	..	..
3842	Railroad equipment	..	..	..	..	11	..	..	..	..
3843	Motor vehicles	..	..	..	..	488	..	..	..	..
3844	Motorcycles and bicycles	..	..	..	..	26	..	..	..	..
3845	Aircraft	..	..	..	..	11	..	..	..	..
3849	Other transport equipment	..	..	..	..	131	..	..	..	..
385	PROFESSIONAL AND SCIENTIFIC EQUIPMENT	..	..	..	..	47	..	..	..	..
3851	Professional equipment	..	..	..	..	15	..	..	..	..
3852	Photographic and optical goods	..	..	..	..	..	..	..	..	..
3853	Watches and clocks	..	..	..	..	..	..	..	..	..
39	OTHER MANUFACTURING INDUSTRIES	..	..	..	..	3,518	..	..	..	..
3901	Jewellery and related articles	..	..	..	..	..	..	..	..	..
3902	Musical instruments	..	..	..	..	..	..	..	..	..
3903	Sporting and athletic goods	..	..	..	..	..	..	..	..	..
3909	Other manufactures	..	..	..	..	2,018	..	..	..	..
SERR	non-specified, unallocated industry	..	..	..	..	..	..	..	..	..
3	TOTAL MANUFACTURING	..	..	..	..	154,329	..	..	..	..

ISIC Revision 2 Industry Sector		1990	1991	1992	1993	1994	1995	1996	1997	1998
31	FOOD, BEVERAGES AND TOBACCO	..	23,902	23,129	23,353	..	..	..	..	..
311/2	FOOD	..	21,156	20,447	20,596	..	..	..	..	..
3111	Slaughtering, preparing and preserving meat	..	4,441	4,394	4,317	..	..	..	..	..
3112	Dairy products	..	4,126	3,660	3,503	..	..	..	..	..
3113	Canning, preserving of fruits and vegetables	..	1,749	1,687	1,718	..	..	..	..	..
3114	Canning, preserving and processing of fish	..	265	239	266	..	..	..	..	..
3115	Vegetable and animal oils and fats	..	1,692	1,649	1,619	..	..	..	..	..
3116	Grain mill products	..	419	356	347	..	..	..	..	..
3117	Bakery products	..	2,785	2,649	2,641	..	..	..	..	..
3118	Sugar factories and refineries	..	2,986	3,030	3,143	..	..	..	..	..
3119	Cocoa, chocolate and sugar confectionery	..	1,046	1,003	922	..	..	..	..	..
3121	Other food products	..	855	849	933	..	..	..	..	..
3122	Prepared animal feeds	..	789	931	1,188	..	..	..	..	..
313	BEVERAGES	..	2,567	2,502	2,588	..	..	..	..	..
3131	Distilling, rectifying and blending of spirits	..	733	736	823	..	..	..	..	..
3132	Wine industries	..	..	..	..	..	..	..	..	..
3133	Malt liquors and malts	..	1,684	1,581	825	..	..	..	..	..
3134	Soft drinks	..	151	186	941	..	..	..	..	..
314	TOBACCO	..	178	180	168	..	..	..	..	..
32	TEXTILES, APPAREL AND LEATHER	..	4,162	3,579	3,214	..	..	..	..	..
321	TEXTILES	..	3,647	3,211	2,836	..	..	..	..	..
3211	Spinning weaving and finishing textiles	..	2,089	1,937	1,705	..	..	..	..	..
3212	Made-up goods excluding wearing apparel	..	565	160	193	..	..	..	..	..
3213	Knitting mills	..	274	247	166	..	..	..	..	..
3214	Carpets and rugs	..	14	13	10	..	..	..	..	..
3215	Cordage, rope and twine	..	22	15	13	..	..	..	..	..
3219	Other textiles	..	682	839	749	..	..	..	..	..
322	WEARING APPAREL, EXCEPT FOOTWEAR	..	270	182	169	..	..	..	..	..
323	LEATHER AND FUR PRODUCTS	..	182	122	141	..	..	..	..	..
3231	Tanneries and leather finishing	..	156	102	116	..	..	..	..	..
3232	Fur dressing and dyeing industries	..	1	..	..	..	..	..	..	..
3233	Leather prods. ex. footwear and wearing apparel	..	26	20	25	..	..	..	..	..
324	FOOTWEAR, EX. RUBBER AND PLASTIC	..	63	65	68	..	..	..	..	..
33	WOOD PRODUCTS AND FURNITURE	..	36,908	36,986	38,099	..	..	..	..	..
331	WOOD PRODUCTS, EXCEPT FURNITURE	..	34,901	34,957	36,337	..	..	..	..	..
3311	Sawmills, planing and other wood mills	..	34,449	34,416	35,979	..	..	..	..	..
3312	Wooden and cane containers	..	80	182	59	..	..	..	..	..
3319	Other wood and cork products	..	372	359	299	..	..	..	..	..
332	FURNITURE, FIXTURES, EXCL. METALLIC	..	2,007	2,029	1,762	..	..	..	..	..
34	PAPER, PUBLISHING AND PRINTING	..	219,187	211,897	222,526	..	..	..	..	..
341	PAPER AND PRODUCTS	..	214,846	209,045	219,816	..	..	..	..	..
3411	Pulp, paper and paperboard articles	..	211,000	205,417	216,512	..	..	..	..	..
3412	Containers of paper and paperboard	..	1,189	1,260	1,298	..	..	..	..	..
3419	Other pulp, paper and paperboard articles	..	2,657	2,368	2,006	..	..	..	..	..
342	PRINTING AND PUBLISHING	..	4,341	2,852	2,710	..	..	..	..	..
35	CHEMICAL PRODUCTS	..	36,779	37,272	36,513	..	..	..	..	..
351	INDUSTRIAL CHEMICALS	..	24,466	24,502	23,825	..	..	..	..	..
3511	Basic industrial chemicals excl. fertilizers	..	14,045	12,897	12,484	..	..	..	..	..
3512	Fertilizers and pesticides	..	912	716	644	..	..	..	..	..
3513	Synthetic resins and plastic materials	..	9,509	10,888	10,697	..	..	..	..	..
352	OTHER CHEMICALS	..	5,180	5,124	4,782	..	..	..	..	..
3521	Paints, varnishes and lacquers	..	410	384	550	..	..	..	..	..
3522	Drugs and medicines	..	2,379	2,505	2,053	..	..	..	..	..
3523	Soap, cleaning preparations, perfumes, cosmetics	..	330	263	188	..	..	..	..	..
3529	Other chemical products	..	2,060	1,971	1,991	..	..	..	..	..
353	PETROLEUM REFINERIES	..	2,106	2,712	2,969	..	..	..	..	..
354	MISC. PETROLEUM AND COAL PRODUCTS	..	1,201	1,209	1,416	..	..	..	..	..
355	RUBBER PRODUCTS	..	1,134	1,212	1,097	..	..	..	..	..
3551	Tyres and tubes	..	359	383	310	..	..	..	..	..
3559	Other rubber products	..	775	829	787	..	..	..	..	..
356	PLASTIC PRODUCTS	..	2,693	2,514	2,425	..	..	..	..	..
36	NON-METALLIC MINERAL PRODUCTS	..	22,836	19,456	18,706	..	..	..	..	..
361	POTTERY, CHINA, EARTHENWARE	..	639	599	580	..	..	..	..	..
362	GLASS AND PRODUCTS	..	3,522	3,234	3,439	..	..	..	..	..
369	OTHER NON-METAL. MINERAL PRODUCTS	..	18,675	15,623	14,687	..	..	..	..	..
3691	Structural clay products	..	1,200	973	738	..	..	..	..	..
3692	Cement, lime and plaster	..	10,192	8,738	9,188	..	..	..	..	..
3699	Other non-metallic mineral products	..	7,283	5,912	4,761	..	..	..	..	..
37	BASIC METAL INDUSTRIES	..	80,228	80,341	83,260	..	..	..	..	..
371	IRON AND STEEL	..	66,810	68,366	70,342	..	..	..	..	..
372	NON-FERROUS METALS	..	13,418	11,975	12,918	..	..	..	..	..
38	METAL PRODUCTS, MACHINERY, EQUIP.	..	42,701	40,313	38,314	..	..	..	..	..
381	METAL PRODUCTS	..	10,583	9,603	8,723	..	..	..	..	..
3811	Cutlery, hand tools and general hardware	..	1,218	1,133	980	..	..	..	..	..
3812	Furniture and fixtures primarily of metal	..	685	512	374	..	..	..	..	..
3813	Structural metal products	..	2,042	1,765	1,687	..	..	..	..	..
3819	Other fabricated metal products	..	6,639	6,193	5,682	..	..	..	..	..
382	NON-ELECTRICAL MACHINERY	..	10,893	11,051	10,834	..	..	..	..	..
3821	Engines and turbines	..	355	368	337	..	..	..	..	..
3822	Agricultural machinery and equipment	..	240	512	215	..	..	..	..	..
3823	Metal and wood working machinery	..	772	728	696	..	..	..	..	..
3824	Special industrial machinery	..	2,535	2,219	2,156	..	..	..	..	..
3825	Office, computing and accounting machinery	..	628	550	623	..	..	..	..	..
3829	Other non-electrical machinery and equipment	..	6,363	6,674	6,808	..	..	..	..	..
383	ELECTRICAL MACHINERY	..	5,125	4,360	4,387	..	..	..	..	..
3831	Electrical industrial machinery	..	839	565	529	..	..	..	..	..
3832	Radio, TV and communications equipment	..	1,969	1,685	1,847	..	..	..	..	..
3833	Electrical appliances and housewares	..	252	229	229	..	..	..	..	..
3839	Other electrical apparatus and supplies	..	2,064	1,881	1,782	..	..	..	..	..
384	TRANSPORT EQUIPMENT	..	15,206	14,380	13,434	..	..	..	..	..
3841	Shipbuilding	..	901	946	864	..	..	..	..	..
3842	Railroad equipment	..	777	743	714	..	..	..	..	..
3843	Motor vehicles	..	11,679	10,909	10,292	..	..	..	..	..
3844	Motorcycles and bicycles	..	86	86	77	..	..	..	..	..
3845	Aircraft	..	1,687	1,611	1,420	..	..	..	..	..
3849	Other transport equipment	..	76	86	67	..	..	..	..	..
385	PROFESSIONAL AND SCIENTIFIC EQUIPMENT	..	894	920	935	..	..	..	..	..
3851	Professional equipment	..	833	840	854	..	..	..	..	..
3852	Photographic and optical goods	..	57	70	70	..	..	..	..	..
3853	Watches and clocks	..	4	10	11	..	..	..	..	..
39	OTHER MANUFACTURING INDUSTRIES	..	260	284	221	..	..	..	..	..
3901	Jewellery and related articles	..	45	44	32	..	..	..	..	..
3902	Musical instruments	..	5	4		..	..	..	..	..
3903	Sporting and athletic goods	..	83	80	53	..	..	..	..	..
3909	Other manufactures	..	127	156	135	..	..	..	..	..
SERR	non-specified, unallocated industry	..	..	..	..	..	..	..	..	..
3	TOTAL MANUFACTURING	..	466,963	453,257	464,205	..	..	..	..	..

ISIC Revision 2 Industry Sector	1990	1991	1992	1993	1994	1995	1996	1997	1998
31 **FOOD, BEVERAGES AND TOBACCO**	6,787	7,122	5,920	6,687	5,860	6,700	..	..	
311/2 FOOD	5,332	5,583	4,545	5,426	4,800	5,899	..	..	
3111 Slaughtering, preparing and preserving meat	..	..	..	..	..	..	..	..	
3112 Dairy products	..	..	..	..	..	..	..	..	
3113 Canning, preserving of fruits and vegetables	..	..	..	..	..	..	..	..	
3114 Canning, preserving and processing of fish	..	..	..	..	..	..	..	..	
3115 Vegetable and animal oils and fats	627	645	367	491	371	594	..	..	
3116 Grain mill products	157	181	190	195	158	144	..	..	
3117 Bakery products	..	..	..	..	..	..	..	..	
3118 Sugar factories and refineries	..	..	..	..	..	..	..	..	
3119 Cocoa, chocolate and sugar confectionery	683	694	697	690	714	710	..	..	
3121 Other food products	3,865	4,063	3,291	4,050	3,557	4,451	..	..	
3122 Prepared animal feeds	..	..	..	..	..	..	..	..	
313 BEVERAGES	946	983	897	848	802	801	..	..	
3131 Distilling, rectifying and blending of spirits	..	..	..	..	..	..	..	..	
3132 Wine industries	..	..	..	..	..	..	..	..	
3133 Malt liquors and malts	946	983	897	848	802	801	..	..	
3134 Soft drinks	..	..	..	..	..	..	..	..	
314 TOBACCO	509	556	478	413	258		..	..	
32 **TEXTILES, APPAREL AND LEATHER**	4,340	4,149	2,425	5,429	7,237	4,649	..	..	
321 TEXTILES	4,177	3,940	2,425	5,220	7,079	4,649	..	..	
3211 Spinning weaving and finishing textiles	..	..	..	..	..	..	..	..	
3212 Made-up goods excluding wearing apparel	..	..	..	..	..	..	..	..	
3213 Knitting mills	..	..	..	..	..	..	..	..	
3214 Carpets and rugs	..	..	..	..	..	..	..	..	
3215 Cordage, rope and twine	..	..	..	..	..	..	..	..	
3219 Other textiles	4,177	3,940	2,425	5,220	7,079	4,649	..	..	
322 WEARING APPAREL, EXCEPT FOOTWEAR	..	..	..	..	..	..	..	..	
323 LEATHER AND FUR PRODUCTS	..	..	..	..	..	..	..	..	
3231 Tanneries and leather finishing	..	..	..	..	..	..	..	..	
3232 Fur dressing and dyeing industries	..	..	..	..	..	..	..	..	
3233 Leather prods. ex. footwear and wearing apparel	..	..	..	..	..	..	..	..	
324 FOOTWEAR, EX. RUBBER AND PLASTIC	163	209		209	158		..	..	
33 **WOOD PRODUCTS AND FURNITURE**	..	..	..	..	..	..	..	..	
331 WOOD PRODUCTS, EXCEPT FURNITURE	..	..	..	..	..	..	..	..	
3311 Sawmills, planing and other wood mills	..	..	..	..	..	..	..	..	
3312 Wooden and cane containers	..	..	..	..	..	..	..	..	
3319 Other wood and cork products	..	..	..	..	..	..	..	..	
332 FURNITURE, FIXTURES, EXCL. METALLIC	..	..	..	..	..	..	..	..	
34 **PAPER, PUBLISHING AND PRINTING**	16,924	17,017	17,902	17,974	17,202	18,601	..	..	
341 PAPER AND PRODUCTS	16,924	17,017	17,902	17,974	17,202	18,601	..	..	
3411 Pulp, paper and paperboard articles	16,924	17,017	17,902	17,974	17,202	18,601	..	..	
3412 Containers of paper and paperboard	..	..	..	..	..	..	..	..	
3419 Other pulp, paper and paperboard articles	..	..	..	..	..	..	..	..	
342 PRINTING AND PUBLISHING	..	..	..	..	..	..	..	..	
35 **CHEMICAL PRODUCTS**	26,099	23,248	23,198	22,903	24,332	24,061	..	..	
351 INDUSTRIAL CHEMICALS	22,696	19,831	20,130	19,496	23,849		..	..	
3511 Basic industrial chemicals excl. fertilizers	22,696	19,831	20,130	19,496			..	..	
3512 Fertilizers and pesticides	..	..	..	..	..	..	..	..	
3513 Synthetic resins and plastic materials	..	..	..	..	..	..	..	..	
352 OTHER CHEMICALS	3,048	3,062	3,068	3,052	483	24,061	..	..	
3521 Paints, varnishes and lacquers	156	161	142	149	171	186	..	..	
3522 Drugs and medicines	..	..	..	..	..	..	..	..	
3523 Soap, cleaning preparations, perfumes, cosmetics	203	294	347	319	312	258	..	..	
3529 Other chemical products	2,689	2,607	2,579	2,584		23,617	..	..	
353 PETROLEUM REFINERIES	..	..	..	..	..	..	..	..	
354 MISC. PETROLEUM AND COAL PRODUCTS	..	..	..	..	..	..	..	..	
355 RUBBER PRODUCTS	..	..	..	..	..	..	..	..	
3551 Tyres and tubes	..	..	..	..	..	..	..	..	
3559 Other rubber products	..	..	..	..	..	..	..	..	
356 PLASTIC PRODUCTS	355	355		355			..	..	
36 **NON-METALLIC MINERAL PRODUCTS**	24,475	21,614	21,018	19,330	18,596	17,905	..	..	
361 POTTERY, CHINA, EARTHENWARE	1,069	780	746	761			..	..	
362 GLASS AND PRODUCTS	2,063	2,063	2,063	2,063			..	..	
369 OTHER NON-METAL. MINERAL PRODUCTS	21,343	18,771	18,209	16,506	18,596	17,905	..	..	
3691 Structural clay products	3,568	3,412	3,336	3,295	3,683	3,548	..	..	
3692 Cement, lime and plaster	17,775	15,359	14,873	13,211	14,913	14,357	..	..	
3699 Other non-metallic mineral products	..	..	..	..	..	..	..	..	
37 **BASIC METAL INDUSTRIES**	7,710	7,212	5,927	4,301	4,267	2,548	..	..	
371 IRON AND STEEL	..	..	..	..	..	..	..	..	
372 NON-FERROUS METALS	7,710	7,212	5,927	4,301	4,267	2,548	..	..	
38 **METAL PRODUCTS, MACHINERY, EQUIP.**	28,035	28,702	25,452	23,009	20,140	19,075	..	..	
381 METAL PRODUCTS	28,035	28,702	25,452	23,009	314	409	..	..	
3811 Cutlery, hand tools and general hardware	..	..	..	..	..	..	..	..	
3812 Furniture and fixtures primarily of metal	..	..	..	..	..	..	..	..	
3813 Structural metal products	107	108	108	108			..	..	
3819 Other fabricated metal products	27,928	28,594	25,344	22,901	314	409	..	..	
382 NON-ELECTRICAL MACHINERY	..	..	..	..	19,826	18,666	..	..	
3821 Engines and turbines	..	..	..	..	..	..	..	..	
3822 Agricultural machinery and equipment	..	..	..	..	..	..	..	..	
3823 Metal and wood working machinery	..	..	..	..	..	..	..	..	
3824 Special industrial machinery	..	..	..	..	..	..	..	..	
3825 Office, computing and accounting machinery	..	..	..	..	..	..	..	..	
3829 Other non-electrical machinery and equipment	..	..	..	..	19,826	18,666	..	..	
383 ELECTRICAL MACHINERY	..	..	..	..	..	..	..	..	
3831 Electrical industrial machinery	..	..	..	..	..	..	..	..	
3832 Radio, TV and communications equipment	..	..	..	..	..	..	..	..	
3833 Electrical appliances and housewares	..	..	..	..	..	..	..	..	
3839 Other electrical apparatus and supplies	..	..	..	..	..	..	..	..	
384 TRANSPORT EQUIPMENT	..	..	..	..	..	..	..	..	
3841 Shipbuilding	..	..	..	..	..	..	..	..	
3842 Railroad equipment	..	..	..	..	..	..	..	..	
3843 Motor vehicles	..	..	..	..	..	..	..	..	
3844 Motorcycles and bicycles	..	..	..	..	..	..	..	..	
3845 Aircraft	..	..	..	..	..	..	..	..	
3849 Other transport equipment	..	..	..	..	..	..	..	..	
385 PROFESSIONAL AND SCIENTIFIC EQUIPMENT	..	..	..	..	..	..	..	..	
3851 Professional equipment	..	..	..	..	..	..	..	..	
3852 Photographic and optical goods	..	..	..	..	..	..	..	..	
3853 Watches and clocks	..	..	..	..	..	..	..	..	
39 **OTHER MANUFACTURING INDUSTRIES**	..	..	..	..	..	..	..	..	
3901 Jewellery and related articles	..	..	..	..	..	..	..	..	
3902 Musical instruments	..	..	..	..	..	..	..	..	
3903 Sporting and athletic goods	..	..	..	..	..	..	..	..	
3909 Other manufactures	..	..	..	..	..	..	..	..	
SERR **non-specified, unallocated industry**	..	..	..	..	..	..	..	..	
3 **TOTAL MANUFACTURING**	114,370	109,064	101,842	99,633	97,634	93,539	..	..	

TURKEY

ISIC Revision 2 Industry Sector	1990	1991	1992	1993	1994	1995	1996	1997	1998
31 FOOD, BEVERAGES AND TOBACCO	..	..	54,648	..	..	59,221	48,199	52,415	..
311/2 FOOD	..	..	50,524	..	..	55,396	44,539	48,481	..
3111 Slaughtering, preparing and preserving meat	..	..	1,468	..	..	926	915	1,096	..
3112 Dairy products			1,273			873	1,207	1,092	
3113 Canning, preserving of fruits and vegetables			2,858			2,638	2,668	2,425	
3114 Canning, preserving and processing of fish			350			359	77	342	
3115 Vegetable and animal oils and fats			5,520			25,167	6,810	5,935	
3116 Grain mill products			1,169			341	497	95	
3117 Bakery products			1,707			1,228	1,938	1,945	
3118 Sugar factories and refineries			27,064			17,405	23,846	27,959	
3119 Cocoa, chocolate and sugar confectionery			783			840	400	685	
3121 Other food products			7,563			5,558	6,110	6,840	
3122 Prepared animal feeds			769			63	70	68	
313 BEVERAGES			3,242			2,855	2,798	3,007	
3131 Distilling, rectifying and blending of spirits			800			834	890	978	
3132 Wine industries			37			..	..	..	
3133 Malt liquors and malts			1,636			1,668	1,555	1,598	
3134 Soft drinks			769			353	353	431	
314 TOBACCO			883			970	862	927	
32 TEXTILES, APPAREL AND LEATHER	..		45,366	..		38,084	39,677	44,108	..
321 TEXTILES			41,111			36,363	37,099	42,000	
3211 Spinning weaving and finishing textiles			36,405			31,230	32,967	35,323	
3212 Made-up goods excluding wearing apparel			668			198	909	2,421	
3213 Knitting mills			2,882			3,204	2,188	3,067	
3214 Carpets and rugs			1,098			1,014	943	1,037	
3215 Cordage, rope and twine			10			..	..	..	
3219 Other textiles			49			716	92	152	
322 WEARING APPAREL, EXCEPT FOOTWEAR			3,505			1,357	2,176	1,705	
323 LEATHER AND FUR PRODUCTS			425			112	110	159	
3231 Tanneries and leather finishing			393			112	110	159	
3232 Fur dressing and dyeing industries			11			..	..	..	
3233 Leather prods. ex. footwear and wearing apparel			20			..	..	..	
324 FOOTWEAR, EX. RUBBER AND PLASTIC			326			253	292	245	
33 WOOD PRODUCTS AND FURNITURE			3,940			2,836	3,053	3,748	
331 WOOD PRODUCTS, EXCEPT FURNITURE			3,755			2,803	3,045	3,722	
3311 Sawmills, planing and other wood mills			3,738			2,803	3,045	3,633	
3312 Wooden and cane containers			3			..	..	89	
3319 Other wood and cork products			14			..	..	..	
332 FURNITURE, FIXTURES, EXCL. METALLIC			185			33	9	26	
34 PAPER, PUBLISHING AND PRINTING	..		19,294	..		19,687	24,102	25,145	..
341 PAPER AND PRODUCTS			18,457			19,492	23,741	18,243	
3411 Pulp, paper and paperboard articles			17,090			16,958	21,292	14,842	
3412 Containers of paper and paperboard			1,201			1,689	1,692	3,381	
3419 Other pulp, paper and paperboard articles			166			844	757	20	
342 PRINTING AND PUBLISHING			837			195	361	6,902	
35 CHEMICAL PRODUCTS			108,809			148,880	153,815	136,953	
351 INDUSTRIAL CHEMICALS			73,053			133,617	77,632	78,668	
3511 Basic industrial chemicals excl. fertilizers			46,023			12,538	13,895	15,169	
3512 Fertilizers and pesticides			21,111			29,843	34,165	32,603	
3513 Synthetic resins and plastic materials			5,919			91,236	29,571	30,896	
352 OTHER CHEMICALS			9,246			4,160	4,295	4,828	
3521 Paints, varnishes and lacquers			516			299	330	468	
3522 Drugs and medicines			5,876			721	666	692	
3523 Soap, cleaning preparations, perfumes, cosmetics			1,306			1,621	1,807	2,605	
3529 Other chemical products			1,548			1,519	1,492	1,063	
353 PETROLEUM REFINERIES			18,331			4,866	66,993	48,848	
354 MISC. PETROLEUM AND COAL PRODUCTS			1,758			2,027	503	190	
355 RUBBER PRODUCTS			4,625			3,979	4,098	3,805	
3551 Tyres and tubes			3,666			3,732	3,644	3,581	
3559 Other rubber products			959			247	454	224	
356 PLASTIC PRODUCTS			1,796			231	293	615	
36 NON-METALLIC MINERAL PRODUCTS			126,565			148,664	151,689	159,166	
361 POTTERY, CHINA, EARTHENWARE			8,062			11,507	11,773	5,627	
362 GLASS AND PRODUCTS			12,584			9,607	12,713	12,811	
369 OTHER NON-METAL. MINERAL PRODUCTS			105,919			127,551	127,203	140,728	
3691 Structural clay products			15,075			12,776	11,661	15,458	
3692 Cement, lime and plaster			87,660			113,598	114,310	118,839	
3699 Other non-metallic mineral products			3,184			1,177	1,232	6,432	
37 BASIC METAL INDUSTRIES			207,545			223,972	205,937	221,233	
371 IRON AND STEEL			194,576			210,887	190,535	204,191	
372 NON-FERROUS METALS			12,968			13,085	15,402	17,042	
38 METAL PRODUCTS, MACHINERY, EQUIP.			17,990			11,916	12,405	14,560	
381 METAL PRODUCTS			4,032			1,608	2,082	2,813	
3811 Cutlery, hand tools and general hardware			630			12	54	116	
3812 Furniture and fixtures primarily of metal			150			34	32	78	
3813 Structural metal products			644			231	441	383	
3819 Other fabricated metal products			2,608			1,330	1,555	2,236	
382 NON-ELECTRICAL MACHINERY			3,945			2,394	2,306	3,016	
3821 Engines and turbines			292			43	48	26	
3822 Agricultural machinery and equipment			992			546	567	1,010	
3823 Metal and wood working machinery			254			117	85	62	
3824 Special industrial machinery			256			..	..	..	
3825 Office, computing and accounting machinery			9			361	74	..	
3829 Other non-electrical machinery and equipment			2,142			1,329	1,532	1,918	
383 ELECTRICAL MACHINERY			3,130			2,132	1,977	2,284	
3831 Electrical industrial machinery			658			447	432	544	
3832 Radio, TV and communications equipment			668			499	524	523	
3833 Electrical appliances and housewares			165			11	13	16	
3839 Other electrical apparatus and supplies			1,640			1,175	1,008	1,202	
384 TRANSPORT EQUIPMENT			6,479			5,625	5,945	6,260	
3841 Shipbuilding			258			127	128	145	
3842 Railroad equipment			627			1,170	760	918	
3843 Motor vehicles			4,980			3,745	4,459	4,586	
3844 Motorcycles and bicycles			107			42	61	71	
3845 Aircraft			507			542	537	540	
3849 Other transport equipment			..			..	..	..	
385 PROFESSIONAL AND SCIENTIFIC EQUIPMENT			404			157	96	187	
3851 Professional equipment			232			157	96	50	
3852 Photographic and optical goods			6			..	..	..	
3853 Watches and clocks			166			..	..	138	
39 OTHER MANUFACTURING INDUSTRIES			250			5	7	10	
3901 Jewellery and related articles			26			..	7	..	
3902 Musical instruments			..			..	..	..	
3903 Sporting and athletic goods			..			..	..	..	
3909 Other manufactures			224			5	..	10	
SERR non-specified, unallocated industry			..			..	..	..	
3 TOTAL MANUFACTURING		..	584,409		..	653,265	638,891	657,340	..

ISIS Energy Data Programme (IEA/OECD)

UNITED STATES

ISIC Revision 2 Industry Sector		1990	1991	1992	1993	1994	1995	1996	1997	1998
31	FOOD, BEVERAGES AND TOBACCO	..	..	..	..	241,630	..	..	..	..
311/2	FOOD	..	..	..	..	546,879	..	..	..	..
3111	Slaughtering, preparing and preserving meat	..	..	..	..	126,931	..	..	..	..
3112	Dairy products	..	..	..	..	90,797	..	..	..	..
3113	Canning, preserving of fruits and vegetables	..	..	..	..	130,396	..	..	..	..
3114	Canning, preserving and processing of fish	..	..	..	..	8,483	..	..	..	..
3115	Vegetable and animal oils and fats	..	..	..	..	105,000	..	..	..	..
3116	Grain mill products	..	..	..	..	39,405	..	..	..	..
3117	Bakery products	..	..	..	..	55,967	..	..	..	..
3118	Sugar factories and refineries	..	..	..	..	122,031	..	..	..	..
3119	Cocoa, chocolate and sugar confectionery	..	..	..	..	26,735	..	..	..	..
3121	Other food products	..	..	..	..	159,112	..	..	..	..
3122	Prepared animal feeds	..	..	..	..	39,631	..	..	..	..
313	BEVERAGES	..	..	..	..	69,264	..	..	..	..
3131	Distilling, rectifying and blending of spirits	..	..	..	..	4,040	..	..	..	..
3132	Wine industries	..	..	..	..	11,923	..	..	..	..
3133	Malt liquors and malts	..	..	..	..	61,857	..	..	..	..
3134	Soft drinks	..	..	..	..	14,402	..	..	..	..
314	TOBACCO	..	..	..	..	3,864	..	..	..	..
32	TEXTILES, APPAREL AND LEATHER	..	..	..	..	364,268	..	..	..	..
321	TEXTILES	..	..	..	..	127,891	..	..	..	..
3211	Spinning weaving and finishing textiles	..	..	..	..	217,862	..	..	..	..
3212	Made-up goods excluding wearing apparel	..	..	..	..	20,862	..	..	..	..
3213	Knitting mills	..	..	..	..	55,468	..	..	..	..
3214	Carpets and rugs	..	..	..	..	18,572	..	..	..	..
3215	Cordage, rope and twine	..	..	..	..	612	..	..	..	..
3219	Other textiles	..	..	..	..	16,804	..	..	..	..
322	WEARING APPAREL, EXCEPT FOOTWEAR	..	..	..	..	32,872	..	..	..	..
323	LEATHER AND FUR PRODUCTS	..	..	..	..	6,915	..	..	..	..
3231	Tanneries and leather finishing	..	..	..	..	7,737	..	..	..	..
3232	Fur dressing and dyeing industries	..	..	..	..		..	..	..	..
3233	Leather prods. ex. footwear and wearing apparel	..	..	..	..	1,254	..	..	..	..
324	FOOTWEAR, EX. RUBBER AND PLASTIC	..	..	..	..	1,045	..	..	..	..
33	WOOD PRODUCTS AND FURNITURE	..	..	..	..	489,225	..	..	..	..
331	WOOD PRODUCTS, EXCEPT FURNITURE	..	..	..	..	352,168	..	..	..	..
3311	Sawmills, planing and other wood mills	..	..	..	..	342,200	..	..	..	..
3312	Wooden and cane containers	..	..	..	..	4,339	..	..	..	..
3319	Other wood and cork products	..	..	..	..	9,537	..	..	..	..
332	FURNITURE, FIXTURES, EXCL. METALLIC	..	..	..	..	44,697	..	..	..	..
34	PAPER, PUBLISHING AND PRINTING	..	..	..	..	2,188,916	..	..	..	..
341	PAPER AND PRODUCTS	..	..	..	..	2,075,937	..	..	..	..
3411	Pulp, paper and paperboard articles	..	..	..	..	2,239,484	..	..	..	..
3412	Containers of paper and paperboard	..	..	..	..	71,457	..	..	..	..
3419	Other pulp, paper and paperboard articles	..	..	..	..	73,233	..	..	..	..
342	PRINTING AND PUBLISHING	..	..	..	..	113,834	..	..	..	..
35	CHEMICAL PRODUCTS	..	..	..	..	5,528,378	..	..	..	..
351	INDUSTRIAL CHEMICALS	..	..	..	..	506,897	..	..	..	..
3511	Basic industrial chemicals excl. fertilizers	..	..	..	..	1,532,005	..	..	..	..
3512	Fertilizers and pesticides	..	..	..	..	332,311	..	..	..	..
3513	Synthetic resins and plastic materials	..	..	..	..	94,809	..	..	..	..
352	OTHER CHEMICALS	..	..	..	..	54,367	..	..	..	..
3521	Paints, varnishes and lacquers	..	..	..	..	4,561	..	..	..	..
3522	Drugs and medicines	..	..	..	..	84,494	..	..	..	..
3523	Soap, cleaning preparations, perfumes, cosmetics	..	..	..	..	32,786	..	..	..	..
3529	Other chemical products	..	..	..	..	105,544	..	..	..	..
353	PETROLEUM REFINERIES	..	..	..	..	998,109	..	..	..	..
354	MISC. PETROLEUM AND COAL PRODUCTS	..	..	..	..	94,359	..	..	..	..
355	RUBBER PRODUCTS	..	..	..	..	85,498	..	..	..	..
3551	Tyres and tubes	..	..	..	..	47,406	..	..	..	..
3559	Other rubber products	..	..	..	..	39,814	..	..	..	..
356	PLASTIC PRODUCTS	..	..	..	..	199,547	..	..	..	..
36	NON-METALLIC MINERAL PRODUCTS	..	..	..	..	943,948	..	..	..	..
361	POTTERY, CHINA, EARTHENWARE	..	..	..	..	22,655	..	..	..	..
362	GLASS AND PRODUCTS	..	..	..	..	206,360	..	..	..	..
369	OTHER NON-METAL. MINERAL PRODUCTS	..	..	..	..	343,951	..	..	..	..
3691	Structural clay products	..	..	..	..	61,028	..	..	..	..
3692	Cement, lime and plaster	..	..	..	..	413,570	..	..	..	..
3699	Other non-metallic mineral products	..	..	..	..	212,146	..	..	..	..
37	BASIC METAL INDUSTRIES	..	..	..	..	2,528,060	..	..	..	..
371	IRON AND STEEL	..	..	..	..	2,025,122	..	..	..	..
372	NON-FERROUS METALS	..	..	..	..	502,939	..	..	..	..
38	METAL PRODUCTS, MACHINERY, EQUIP.	..	..	..	..	1,238,191	..	..	..	..
381	METAL PRODUCTS	..	..	..	..	292,251	..	..	..	..
3811	Cutlery, hand tools and general hardware	..	..	..	..	27,240	..	..	..	..
3812	Furniture and fixtures primarily of metal	..	..	..	..	20,386	..	..	..	..
3813	Structural metal products	..	..	..	..	45,994	..	..	..	..
3819	Other fabricated metal products	..	..	..	..	206,788	..	..	..	..
382	NON-ELECTRICAL MACHINERY	..	..	..	..	121,109	..	..	..	..
3821	Engines and turbines	..	..	..	..	22,425	..	..	..	..
3822	Agricultural machinery and equipment	..	..	..	..	10,063	..	..	..	..
3823	Metal and wood working machinery	..	..	..	..	21,855	..	..	..	..
3824	Special industrial machinery	..	..	..	..	40,750	..	..	..	..
3825	Office, computing and accounting machinery	..	..	..	..	23,558	..	..	..	..
3829	Other non-electrical machinery and equipment	..	..	..	..	140,443	..	..	..	..
383	ELECTRICAL MACHINERY	..	..	..	..	81,056	..	..	..	..
3831	Electrical industrial machinery	..	..	..	..	32,156	..	..	..	..
3832	Radio, TV and communications equipment	..	..	..	..	91,588	..	..	..	..
3833	Electrical appliances and housewares	..	..	..	..	4,042	..	..	..	..
3839	Other electrical apparatus and supplies	..	..	..	..	77,853	..	..	..	..
384	TRANSPORT EQUIPMENT	..	..	..	..	c	..	..	..	..
3841	Shipbuilding	..	..	..	..	11,404	..	..	..	..
3842	Railroad equipment	..	..	..	..	3,844	..	..	..	..
3843	Motor vehicles	..	..	..	..	234,490	..	..	..	..
3844	Motorcycles and bicycles	..	..	..	..	793	..	..	..	..
3845	Aircraft	..	..	..	..	93,136	..	..	..	..
3849	Other transport equipment	..	..	..	..		..	..	..	..
385	PROFESSIONAL AND SCIENTIFIC EQUIPMENT	..	..	..	..	c	..	..	..	..
3851	Professional equipment	..	..	..	..	51,751	..	..	..	..
3852	Photographic and optical goods	..	..	..	..	21,438	..	..	..	..
3853	Watches and clocks	..	..	..	..	c	..	..	..	..
39	OTHER MANUFACTURING INDUSTRIES	..	..	..	..	18,349	..	..	..	..
3901	Jewellery and related articles	..	..	..	..	2,097	..	..	..	..
3902	Musical instruments	..	..	..	..	568	..	..	..	..
3903	Sporting and athletic goods	..	..	..	..	3,275	..	..	..	..
3909	Other manufactures	..	..	..	..	21,951	..	..	..	..
SERR	non-specified, unallocated industry	..	..	..	..		..	..	..	..
3	TOTAL MANUFACTURING	..	..	..	..	15,857,686	..	..	..	..

Information System on Industrial Structures (ISIS)

Energy Consumption Statistics

in

Manufacturing Industry

1990 – 1998

Total fuels

(ISIC Revision 3)

OECD
ORGANISATION FOR ECONOMIC
CO-OPERATION AND DEVELOPMENT

OCDE
ORGANISATION DE COOPERATION
ET DE DEVELOPPEMENT ECONOMIQUES

IEA
INTERNATIONAL ENERGY AGENCY

AIE
AGENCE INTERNATIONALE DE L'ENERGIE

9, RUE DE LA FÉDERATION, 75739 PARIS CEDEX 15
TEL. (33-1) 40.57.65.00 / TELEFAX (33-1) 40.57.66.49 / INTERNET: HTTP://WWW.IEA.ORG

ISIC Revision 3 Industry Sector		1990	1991	1992	1993	1994	1995	1996	1997	1998
15	FOOD PRODUCTS AND BEVERAGES	23,632	24,648	24,159	24,420	23,534	23,771	21,154	21,598	21,012
151	Production, processing and preserving (PPP)	..	..	..	..	..	..	..	..	..
1511	PPP of meat and meat products	..	..	..	..	..	..	..	..	..
1512	Processing and preserving of fish products	..	..	..	..	..	..	..	..	..
1513	Processing, preserving of fruit & vegetables	..	..	..	..	..	..	..	..	..
1514	Vegetable and animal oils and fats	..	..	..	..	..	..	..	..	..
152	Dairy products	..	..	..	..	..	..	..	..	..
153	Grain mill prod., starches & prepared animal feeds	..	..	..	..	..	..	..	..	..
1531	Grain mill products	..	..	..	..	..	..	..	..	..
1532	Starches and starch products	..	..	..	..	..	..	..	..	..
1533	Prepared animal feeds	..	..	..	..	..	..	..	..	..
154	Other food products	..	..	..	..	..	..	..	..	..
1541	Bakery products	..	..	..	..	..	..	..	..	..
1542	Sugar	..	..	..	..	..	..	..	..	..
1543	Cocoa, chocolate and sugar confectionery	..	..	..	..	..	..	..	..	..
1544	Macaroni, noodles, couscous & similar farinaceous prod.	..	..	..	..	..	..	..	..	..
1549	Other food products, nec	..	..	..	..	..	..	..	..	..
155	Beverages	..	..	..	..	..	..	..	..	..
1551	Distilling, rectifying and blending of spirits	..	..	..	..	..	..	..	..	..
1552	Wines	..	..	..	..	..	..	..	..	..
1553	Malt liquors and malt	..	..	..	..	..	..	..	..	..
1554	Soft drinks; production of mineral waters	..	..	..	..	..	..	..	..	..
16	TOBACCO PRODUCTS	..	..	..	..	..	..	..	..	..
17	TEXTILES	7,543	7,489	7,225	6,622	6,284	6,268	6,109	6,124	6,039
171	Spinning, weaving and finishing of textiles	..	..	..	..	..	..	..	..	..
1711	Preparation and spinning of textile fibres	..	..	..	..	..	..	..	..	..
1712	Finishing of textiles	..	..	..	..	..	..	..	..	..
172	Other textiles	..	..	..	..	..	..	..	..	..
1721	Made-up textile articles, except apparel	..	..	..	..	..	..	..	..	..
1722	Carpets and rugs	..	..	..	..	..	..	..	..	..
1723	Cordage, rope, twine and netting	..	..	..	..	..	..	..	..	..
1729	Other textiles, nec	..	..	..	..	..	..	..	..	..
173	Knitted and crocheted fabrics and articles	..	..	..	..	..	..	..	..	..
18	WEARING APPAREL, DRESSING & DYEING OF FUR	1,313	1,318	1,303	1,233	1,156	1,120	..	..	..
181	Wearing apparel, except fur apparel	..	..	..	..	..	..	..	..	..
182	Dressing and dyeing of fur; articles of fur	..	..	..	..	..	..	..	..	..
19	TANNING & DRESSING OF LEATHER, FOOTWEAR	748	748	775	730	658	688	529	541	556
191	Tanning and dressing of leather	..	..	..	..	..	..	..	..	..
1911	Tanning and dressing of leather	..	..	..	..	..	..	..	..	..
1912	Luggage, handbags, saddlery & harness	..	..	..	..	..	..	..	..	..
192	Footwear	..	..	..	..	..	..	..	..	..
20	WOOD AND WOOD PRODUCTS	14,002	15,182	16,213	14,878	15,743	16,929	13,263	12,117	15,200
201	Sawmilling and planing of wood	..	..	..	..	..	..	..	..	..
202	Products of wood, cork, straw & plaiting materials	..	..	..	..	..	..	..	..	..
2021	Veneer sheets	..	..	..	..	..	..	..	..	..
2022	Builders' carpentry and joinery	..	..	..	..	..	..	..	..	..
2023	Wooden containers	..	..	..	..	..	..	..	..	..
2029	Other products of wood	..	..	..	..	..	..	..	..	..
21	PAPER AND PAPER PRODUCTS	61,509	64,598	67,364	65,811	73,248	68,861	77,180	81,794	207,966
2101	Pulp, paper and paperboard	..	..	..	..	..	..	..	..	..
2102	Corrugated paper, paperboard and their containers	..	..	..	..	..	..	..	..	..
2109	Other articles of pulp and paperboard	..	..	..	..	..	..	..	..	..
22	PUBLISHING, PRINTING & REPRODUCTION	1,146	1,323	1,316	1,238	1,359	1,355	1,814	1,956	1,952
221	Publishing	..	..	..	..	..	..	..	..	..
2211	Publishing of books & brochures	..	..	..	..	..	..	..	..	..
2212	Publishing of newspapers and periodicals	..	..	..	..	..	..	..	..	..
2213	Publishing of recorded media	..	..	..	..	..	..	..	..	..
2219	Other publishing	..	..	..	..	..	..	..	..	..
222	Printing and related service activities	..	..	..	..	..	..	..	..	..
2221	Printing	..	..	..	..	..	..	..	..	..
2222	Service activities related to printing	..	..	..	..	..	..	..	..	..
223	Reproduction of recorded media	..	..	..	..	..	..	..	..	..
23	COKE, REFINED PETROLEUM PRODUCTS	33	38	39	39	36	29	127	123	195
231	Coke oven products	..	..	..	..	..	..	..	..	..
232	Refined petroleum products	33	38	39	39	36	29	127	123	195
233	Processing of nuclear fuel	..	..	..	..	..	..	..	..	..
24	CHEMICALS & CHEMICAL PRODUCTS	35,134	34,716	35,421	35,519	34,563	35,490	29,322	29,517	28,036
241	Basic chemicals	..	..	..	..	..	..	..	..	..
2411	Basic chemicals, exc. fertilizers & nitrogen compounds	..	..	..	..	..	..	..	..	..
2412	Fertilizers and nitrogen compounds	..	..	..	..	..	..	..	..	..
2413	Plastics in primary forms and synthetic rubber	..	..	..	..	..	..	..	..	..
242	Other chemical products	..	..	..	..	..	..	..	..	..
2421	Pesticides and other agro-chemical products	..	..	..	..	..	..	..	..	..
2422	Paints, varnishes and similar coatings	..	..	..	..	..	..	..	..	..
2423	Pharmaceuticals, medicinal chem. & botanical prod.	..	..	..	..	..	..	..	..	..
2424	Soap and detergents, perfumes etc.	..	..	..	..	..	..	..	..	..
2429	Other chemical products, nec	..	..	..	..	..	..	..	..	..
243	Man-made fibres	..	..	..	..	..	..	..	..	..
25	RUBBER AND PLASTICS PRODUCTS	..	..	..	..	..	..	4,947	5,131	5,381
251	Rubber products	..	..	..	..	..	..	..	..	..
2511	Rubber tyres and tubes	..	..	..	..	..	..	..	..	..
2519	Other rubber products	..	..	..	..	..	..	..	..	..
252	Plastic products	..	..	..	..	..	..	..	..	..
26	OTHER NON-METALLIC MINERAL PRODUCTS	5,048	5,119	4,395	4,688	4,782	4,871	31,249	31,792	30,010
261	Glass and glass products	..	..	..	..	..	..	..	..	..
269	Non-metallic mineral products, nec	..	..	..	..	..	..	..	..	..
2691	Pottery, china and earthenware	..	..	..	..	..	..	..	..	..
2692	Refractory ceramic products	..	..	..	..	..	..	..	..	..
2693	Structural non-refractory clay & ceramic prod.	..	..	..	..	..	..	..	..	..
2694	Cement, lime and plaster	..	..	..	..	..	..	..	..	..
2695	Articles of concrete, cement and plaster	..	..	..	..	..	..	..	..	..
2696	Cutting, shaping and finishing of stone	..	..	..	..	..	..	..	..	..
2699	Other non-metallic mineral products, nec	..	..	..	..	..	..	..	..	..

ISIC Revision 3 Industry Sector		1990	1991	1992	1993	1994	1995	1996	1997	1998
27	BASIC METALS	179,595	170,684	154,985	149,095	156,060	161,243	177,777	194,455	194,681
271	Basic iron and steel	..	..	..	..	..	..	170,500	187,906	187,520
272	Basic precious and non-ferrous metals	..	..	..	..	..	..	4,213	4,390	4,497
273	Casting of metals	..	..	..	..	..	..	3,064	2,159	2,664
2731	Casting of iron and steel	..	..	..	..	..	..	..	..	..
2732	Casting non-ferrous metals	..	..	..	..	..	..	..	..	..
28	FABRICATED METAL PRODUCTS	10,090	10,471	10,501	10,127	12,722	13,676	7,230	6,524	7,502
281	Str. metal prod., tanks, reservoirs, steam generators	..	..	..	..	..	..	..	..	..
2811	Structural metal products	..	..	..	..	..	..	..	..	..
2812	Tanks, reservoirs and containers of metal	..	..	..	..	..	..	..	..	..
2813	Steam generators, exc. central heating hot water boilers	..	..	..	..	..	..	..	..	..
289	Other fabricated metal products	..	..	..	..	..	..	..	..	..
2891	Forging, pressing, stamping & roll-forming of metal	..	..	..	..	..	..	..	..	..
2892	Treatment and coating of metals	..	..	..	..	..	..	..	..	..
2893	Cutlery, hand tools and general hardware	..	..	..	..	..	..	..	..	..
2899	Other fabricated metal products, nec	..	..	..	..	..	..	..	..	..
29	MACHINERY AND EQUIPMENT, NEC	6,425	6,727	6,588	6,445	6,295	6,403	6,613	5,108	5,096
291	General purpose machinery	..	..	..	..	..	..	..	..	..
2911	Engines and turbines	..	..	..	..	..	..	..	..	..
2912	Pumps, compressors, taps and valves	..	..	..	..	..	..	..	..	..
2913	Bearings, gears, gearing and driving elements	..	..	..	..	..	..	..	..	..
2914	Ovens, furnaces and furnace burners	..	..	..	..	..	..	..	..	..
2915	Lifting and handling equipment	..	..	..	..	..	..	..	..	..
2919	Other general purpose machinery	..	..	..	..	..	..	..	..	..
292	Special purpose machinery	..	..	..	..	..	..	..	..	..
2921	Agricultural and forestry machinery	..	..	..	..	..	..	..	..	..
2922	Machine-tools	..	..	..	..	..	..	..	..	..
2923	Machinery for metallurgy	..	..	..	..	..	..	..	..	..
2924	Machinery for mining, quarrying and construction	..	..	..	..	..	..	..	..	..
2925	Machinery for food, beverage & tobacco processing	..	..	..	..	..	..	..	..	..
2926	Machinery for textile, apparel & leather production	..	..	..	..	..	..	..	..	..
2927	Machinery for weapons and ammunition	..	..	..	..	..	..	..	..	..
2929	Other special purpose machinery	..	..	..	..	..	..	..	..	..
293	Domestic appliances, nec	..	..	..	..	..	..	..	..	..
30	OFFICE, ACCOUNTING & COMPUTING MACHINERY	6,508	7,644	7,143	6,876	6,952	6,811	5,966	5,565	5,682
31	ELECTRICAL MACHINERY & APPARATUS, NEC									
311	Electric motors, generators and transformers	..	..	..	..	..	..	..	..	..
312	Electricity distribution and control apparatus	..	..	..	..	..	..	..	..	..
313	Insulated wire and cable	..	..	..	..	..	..	..	..	..
314	Accumulators, primary cells & primary batteries	..	..	..	..	..	..	..	..	..
315	Electric lamps and lighting equipment	..	..	..	..	..	..	..	..	..
319	Other electrical equipment, nec	..	..	..	..	..	..	..	..	..
32	RADIO, TV & COMMUNICATION EQUIP. & APP.	..	..	..	..	..	..	..	..	..
321	Electronic valves, tubes, other electronic components	..	..	..	..	..	..	..	..	..
322	TV & radio transmitters, apparatus for line telephony	..	..	..	..	..	..	..	..	..
323	TV & radio receivers, recording apparatus	..	..	..	..	..	..	..	..	..
33	MEDICAL PRECISION &OPTICAL INSTRUMENTS	..	..	..	..	..	..	..	..	..
331	Medical appliances and instruments	..	..	..	..	..	..	..	..	..
3311	Medical, surgical equipment & orthopaedic app.	..	..	..	..	..	..	..	..	..
3312	Instruments & appliances for measuring, checking etc	..	..	..	..	..	..	..	..	..
3313	Industrial process control equipment	..	..	..	..	..	..	..	..	..
332	Optical instruments and photographic equipment	..	..	..	..	..	..	..	..	..
333	Watches and clocks	..	..	..	..	..	..	..	..	..
34	MOTOR VEHICLES, TRAILERS & SEMI-TRAILERS	6,038	6,445	6,506	6,570	6,595	7,123	5,143	4,544	4,820
341	Motor vehicles	..	..	..	..	..	..	..	..	..
342	Bodies (coachwork) for motor vehicles	..	..	..	..	..	..	..	..	..
343	Parts, accessories for motor vehicles & their engines	..	..	..	..	..	..	..	..	..
35	OTHER TRANSPORT EQUIPMENT	..	..	..	..	..	..	..	..	..
351	Building and repairing of ships and boats	..	..	..	..	..	..	..	..	..
3511	Building and repairing of ships	..	..	..	..	..	..	..	..	..
3512	Building, repairing of pleasure & sporting boats	..	..	..	..	..	..	..	..	..
352	Railway, tramway locomotives & rolling stock	..	..	..	..	..	..	..	..	..
353	Aircraft and spacecraft	..	..	..	..	..	..	..	..	..
359	Transport equipment, nec	..	..	..	..	..	..	..	..	..
3591	Motorcycles	..	..	..	..	..	..	..	..	..
3592	Bicycles and invalid carriages	..	..	..	..	..	..	..	..	..
3599	Other transport equipment, nec	..	..	..	..	..	..	..	..	..
36	FURNITURE; MANUFACTURING, NEC	..	..	..	..	..	..	3,522	3,341	3,166
361	Furniture	..	..	..	..	..	..	..	..	..
369	Manufacturing, nec	..	..	..	..	..	..	..	..	..
3691	Jewellery and related articles	..	..	..	..	..	..	..	..	..
3692	Musical instruments	..	..	..	..	..	..	..	..	..
3693	Sports goods	..	..	..	..	..	..	..	..	..
3694	Games and toys	..	..	..	..	..	..	..	..	..
3699	Other manufacturing, nec	..	..	..	..	..	..	..	..	..
37	RECYCLING	..	..	..	..	..	..	..	..	..
371	Recycling of metal waste and scrap	..	..	..	..	..	..	..	..	..
372	Recycling of non-metal waste and scrap	..	..	..	..	..	..	..	..	..
CERR	Non-specified industry	..	..	..	..	..	..	..	..	..
15-37	TOTAL MANUFACTURING	358,763	357,150	343,933	334,291	349,986	354,638	391,945	410,230	537,294

ISIC Revision 3 Industry Sector		1990	1991	1992	1993	1994	1995	1996	1997	1998
15	FOOD PRODUCTS AND BEVERAGES	93,562	88,280	81,140	86,497	100,594	89,107	90,070	..	..
151	Production, processing and preserving (PPP)	23,157	23,408	19,966	22,427	26,457	22,095	24,210	..	..
1511	PPP of meat and meat products	..	..	..	..	..	..	..	..	..
1512	Processing and preserving of fish products	..	..	..	..	..	..	..	..	..
1513	Processing, preserving of fruit & vegetables	..	..	..	..	..	..	..	..	..
1514	Vegetable and animal oils and fats	..	..	..	..	..	..	..	..	..
152	Dairy products	11,133	10,724	10,089	9,685	11,483	10,294	10,102	..	..
153	Grain mill prod., starches & prepared animal feeds	..	..	..	..	..	..	..	..	..
1531	Grain mill products	..	..	..	..	..	..	..	..	..
1532	Starches and starch products	..	..	..	..	..	..	..	..	..
1533	Prepared animal feeds	..	..	..	..	..	..	..	..	..
154	Other food products	44,622	41,579	39,943	42,129	50,092	45,990	44,581	..	..
1541	Bakery products	..	..	..	..	..	..	..	..	..
1542	Sugar	..	..	..	..	..	..	..	..	..
1543	Cocoa, chocolate and sugar confectionery	..	..	..	..	..	..	..	..	..
1544	Macaroni, noodles, couscous & similar farinaceous prod.	..	..	..	..	..	..	..	..	..
1549	Other food products, nec	..	..	..	..	..	..	..	..	..
155	Beverages	14,650	12,569	11,142	12,256	12,562	10,728	11,177	..	..
1551	Distilling, rectifying and blending of spirits	..	..	..	..	..	..	..	..	..
1552	Wines	..	..	..	..	..	..	..	..	..
1553	Malt liquors and malt	..	..	..	..	..	..	..	..	..
1554	Soft drinks; production of mineral waters	..	..	..	..	..	..	..	..	..
16	TOBACCO PRODUCTS	1,199		1,157	1,093	1,168	942	917	..	..
17	TEXTILES	19,415	22,200	22,553	21,875	29,575	25,637	25,091	..	..
171	Spinning, weaving and finishing of textiles	13,204	14,280	14,610	13,651	21,003	16,755	16,485	..	..
1711	Preparation and spinning of textile fibres	..	..	..	..	..	..	..	..	..
1712	Finishing of textiles	..	..	..	..	..	..	..	..	..
172	Other textiles	..	..	..	..	..	..	..	..	..
1721	Made-up textile articles, except apparel	..	..	..	..	..	..	..	..	..
1722	Carpets and rugs	..	..	..	..	..	..	..	..	..
1723	Cordage, rope, twine and netting	..	..	..	..	..	..	..	..	..
1729	Other textiles, nec	..	..	..	..	..	..	..	..	..
173	Knitted and crocheted fabrics and articles	6,211	7,920	7,943	8,224	8,572	8,882	8,606	..	..
18	WEARING APPAREL, DRESSING & DYEING OF FUR	4,836	4,506	5,234	5,104	5,270	5,119	5,125	..	..
181	Wearing apparel, except fur apparel	..	..	..	..	..	..	..	..	..
182	Dressing and dyeing of fur; articles of fur	..	..	..	..	..	..	..	..	..
19	TANNING & DRESSING OF LEATHER, FOOTWEAR	1,167	1,064	1,115	1,255	1,301	932	998	..	..
191	Tanning and dressing of leather	..	..	..	..	..	..	..	..	..
1911	Tanning and dressing of leather	..	..	..	..	..	..	..	..	..
1912	Luggage, handbags, saddlery & harness	..	..	..	..	..	..	..	..	..
192	Footwear	..	..	..	..	..	..	..	..	..
20	WOOD AND WOOD PRODUCTS	37,067	32,409	36,351	43,916	53,582	55,624	50,637	..	..
201	Sawmilling and planing of wood	15,112	12,515	14,152	17,584	21,389	21,221	22,070	..	..
202	Products of wood, cork, straw & plaiting materials	21,955	19,894	22,199	26,332	32,193	34,403	28,567	..	..
2021	Veneer sheets	..	..	..	..	..	..	..	..	..
2022	Builders' carpentry and joinery	..	..	..	..	..	..	..	..	..
2023	Wooden containers	..	..	..	..	..	..	..	..	..
2029	Other products of wood	..	..	..	..	..	..	..	..	..
21	PAPER AND PAPER PRODUCTS	628,245	648,540	643,673	648,147	696,005	738,502	712,078	..	..
2101	Pulp, paper and paperboard	..	..	..	..	..	..	..	..	..
2102	Corrugated paper, paperboard and their containers	..	..	..	..	..	..	..	..	..
2109	Other articles of pulp and paperboard	..	..	..	..	..	..	..	..	..
22	PUBLISHING, PRINTING & REPRODUCTION	8,147	8,013	9,616	9,921	10,708	10,275	9,181	..	..
221	Publishing	..	..	..	..	..	..	..	..	..
2211	Publishing of books & brochures	..	..	..	..	..	..	..	..	..
2212	Publishing of newspapers and periodicals	..	..	..	..	..	..	..	..	..
2213	Publishing of recorded media	..	..	..	..	..	..	..	..	..
2219	Other publishing	..	..	..	..	..	..	..	..	..
222	Printing and related service activities	..	..	..	..	..	..	..	..	..
2221	Printing	..	..	..	..	..	..	..	..	..
2222	Service activities related to printing	..	..	..	..	..	..	..	..	..
223	Reproduction of recorded media	..	..	..	..	..	..	..	..	..
23	COKE, REFINED PETROLEUM PRODUCTS	304,552	295,405	295,170	299,189	293,761	286,857	300,048	..	..
231	Coke oven products	..	..	..	..	..	..	..	..	..
232	Refined petroleum products	304,552	295,405	295,170	299,189	293,761	286,857	300,048	..	..
233	Processing of nuclear fuel	..	..	..	..	..	..	..	..	..
24	CHEMICALS & CHEMICAL PRODUCTS	210,210	212,174	199,653	226,104	243,728	258,899	242,478	..	..
241	Basic chemicals	114,235	113,812	108,220	135,573	126,522	178,579	161,033	..	..
2411	Basic chemicals, exc. fertilizers & nitrogen compounds	69,567	70,905	66,009	70,147	70,338	102,251	89,128	..	..
2412	Fertilizers and nitrogen compounds	28,837	28,632	28,966	41,049	41,277	48,319	44,132	..	..
2413	Plastics in primary forms and synthetic rubber	15,831	14,275	13,245	24,377	14,907	28,009	27,773	..	..
242	Other chemical products	95,975	98,362	91,433	90,531	117,206	80,320	81,445	..	..
2421	Pesticides and other agro-chemical products	..	..	..	..	..	..	..	..	..
2422	Paints, varnishes and similar coatings	..	..	..	..	..	..	..	..	..
2423	Pharmaceuticals, medicinal chem. & botanical prod.	..	..	..	..	..	..	..	..	..
2424	Soap and detergents, perfumes etc.	..	..	..	..	..	..	..	..	..
2429	Other chemical products, nec	..	..	..	..	..	..	..	..	..
243	Man-made fibres	..	..	..	..	..	..	..	..	..
25	RUBBER AND PLASTICS PRODUCTS	19,709	18,533	20,884	21,410	21,891	21,687	22,395	..	..
251	Rubber products	8,291	7,840	8,651	8,553	7,611	8,723	9,832	..	..
2511	Rubber tyres and tubes	..	..	..	..	..	..	..	..	..
2519	Other rubber products	..	..	..	..	..	..	..	..	..
252	Plastic products	11,418	10,693	12,233	12,857	14,280	12,964	12,563	..	..
26	OTHER NON-METALLIC MINERAL PRODUCTS	79,909	73,498	71,369	67,678	72,483	80,227	78,239	..	..
261	Glass and glass products	11,323	10,271	12,713	13,100	14,124	12,687	12,154	..	..
269	Non-metallic mineral products, nec	68,586	63,227	58,656	54,578	58,359	67,540	66,085	..	..
2691	Pottery, china and earthenware	..	..	..	..	..	..	..	..	..
2692	Refractory ceramic products	..	..	..	..	..	..	..	..	..
2693	Structural non-refractory clay & ceramic prod.	..	..	..	..	..	..	..	..	..
2694	Cement, lime and plaster	68,586	63,227	58,656	54,578	58,359	67,540	66,085	..	..
2695	Articles of concrete, cement and plaster	..	..	..	..	..	..	..	..	..
2696	Cutting, shaping and finishing of stone	..	..	..	..	..	..	..	..	..
2699	Other non-metallic mineral products, nec	..	..	..	..	..	..	..	..	..

ISIC Revision 3 Industry Sector		1990	1991	1992	1993	1994	1995	1996	1997	1998
27	**BASIC METALS**	386,039	410,321	431,727	444,437	433,178	435,296	461,302	..	..
271	Basic iron and steel	207,958	223,330	232,280	230,802	221,110	223,766	234,116	..	..
272	Basic precious and non-ferrous metals	178,081	186,991	199,447	213,635	212,068	211,530	227,186	..	..
273	Casting of metals	..	..	..	..	..	..	..	..	..
2731	Casting of iron and steel	..	..	..	..	..	..	..	..	..
2732	Casting non-ferrous metals	..	..	..	..	..	..	..	..	..
28	**FABRICATED METAL PRODUCTS**	25,575	23,233	27,173	29,389	33,516	35,159	35,347	..	..
281	Str. metal prod., tanks, reservoirs, steam generators	..	..	..	..	..	..	..	..	..
2811	Structural metal products	..	..	..	..	..	..	..	..	..
2812	Tanks, reservoirs and containers of metal	..	..	..	..	..	..	..	..	..
2813	Steam generators, exc. central heating hot water boilers	..	..	..	..	..	..	..	..	..
289	Other fabricated metal products	..	..	..	..	..	..	..	..	..
2891	Forging, pressing, stamping & roll-forming of metal	..	..	..	..	..	..	..	..	..
2892	Treatment and coating of metals	..	..	..	..	..	..	..	..	..
2893	Cutlery, hand tools and general hardware	..	..	..	..	..	..	..	..	..
2899	Other fabricated metal products, nec	..	..	..	..	..	..	..	..	..
29	**MACHINERY AND EQUIPMENT, NEC**	29,011	25,487	30,066	27,015	27,889	26,065	25,722	..	..
291	General purpose machinery	..	..	..	..	..	..	..	..	..
2911	Engines and turbines	..	..	..	..	..	..	..	..	..
2912	Pumps, compressors, taps and valves	..	..	..	..	..	..	..	..	..
2913	Bearings, gears, gearing and driving elements	..	..	..	..	..	..	..	..	..
2914	Ovens, furnaces and furnace burners	..	..	..	..	..	..	..	..	..
2915	Lifting and handling equipment	..	..	..	..	..	..	..	..	..
2919	Other general purpose machinery	..	..	..	..	..	..	..	..	..
292	Special purpose machinery	..	..	..	..	..	..	..	..	..
2921	Agricultural and forestry machinery	..	..	..	..	..	..	..	..	..
2922	Machine-tools	..	..	..	..	..	..	..	..	..
2923	Machinery for metallurgy	..	..	..	..	..	..	..	..	..
2924	Machinery for mining, quarrying and construction	..	..	..	..	..	..	..	..	..
2925	Machinery for food, beverage & tobacco processing	..	..	..	..	..	..	..	..	..
2926	Machinery for textile, apparel & leather production	..	..	..	..	..	..	..	..	..
2927	Machinery for weapons and ammunition	..	..	..	..	..	..	..	..	..
2929	Other special purpose machinery	..	..	..	..	..	..	..	..	..
293	Domestic appliances, nec	..	..	..	..	..	..	..	..	..
30	**OFFICE, ACCOUNTING & COMPUTING MACHINERY**	..	..	..	..	..	..	..	..	..
31	**ELECTRICAL MACHINERY & APPARATUS, NEC**	..	..	..	..	..	..	..	..	..
311	Electric motors, generators and transformers	..	..	..	..	..	..	..	..	..
312	Electricity distribution and control apparatus	..	..	..	..	..	..	..	..	..
313	Insulated wire and cable	..	..	..	..	..	..	..	..	..
314	Accumulators, primary cells & primary batteries	..	..	..	..	..	..	..	..	..
315	Electric lamps and lighting equipment	..	..	..	..	..	..	..	..	..
319	Other electrical equipment, nec	..	..	..	..	..	..	..	..	..
32	**RADIO, TV & COMMUNICATION EQUIP. & APP.**	..	..	..	..	..	..	..	..	..
321	Electronic valves, tubes, other electronic components	..	..	..	..	..	..	..	..	..
322	TV & radio transmitters, apparatus for line telephony	..	..	..	..	..	..	..	..	..
323	TV & radio receivers, recording apparatus	..	..	..	..	..	..	..	..	..
33	MEDICAL PRECISION &OPTICAL INSTRUMENTS	..	..	..	..	..	..	..	..	..
331	Medical appliances and instruments	..	..	..	..	..	..	..	..	..
3311	Medical, surgical equipment & orthopaedic app.	..	..	..	..	..	..	..	..	..
3312	Instruments & appliances for measuring, checking etc	..	..	..	..	..	..	..	..	..
3313	Industrial process control equipment	..	..	..	..	..	..	..	..	..
332	Optical instruments and photographic equipment	..	..	..	..	..	..	..	..	..
333	Watches and clocks	..	..	..	..	..	..	..	..	..
34	**MOTOR VEHICLES, TRAILERS & SEMI-TRAILERS**	49,004	47,033	48,838	53,911	55,510	55,235	58,375	..	..
341	Motor vehicles	21,847	22,034	24,707	28,221	26,854	28,217	29,636	..	..
342	Bodies (coachwork) for motor vehicles	..	..	..	..	..	..	..	..	..
343	Parts, accessories for motor vehicles & their engines	27,157	24,999	24,131	25,690	28,656	27,018	28,739	..	..
35	**OTHER TRANSPORT EQUIPMENT**	..	..	..	..	..	..	..	..	..
351	Building and repairing of ships and boats	..	..	..	..	..	..	..	..	..
3511	Building and repairing of ships	..	..	..	..	..	..	..	..	..
3512	Building, repairing of pleasure & sporting boats	..	..	..	..	..	..	..	..	..
352	Railway, tramway locomotives & rolling stock	..	..	..	..	..	..	..	..	..
353	Aircraft and spacecraft	..	..	..	..	..	..	..	..	..
359	Transport equipment, nec	..	..	..	..	..	..	..	..	..
3591	Motorcycles	..	..	..	..	..	..	..	..	..
3592	Bicycles and invalid carriages	..	..	..	..	..	..	..	..	..
3599	Other transport equipment, nec	..	..	..	..	..	..	..	..	..
36	**FURNITURE; MANUFACTURING, NEC**	185,171	162,054	132,352	105,602	76,973	141,890	123,052	..	..
361	Furniture	..	..	..	..	..	..	..	..	..
369	Manufacturing, nec	..	..	..	..	..	..	..	..	..
3691	Jewellery and related articles	..	..	..	..	..	..	..	..	..
3692	Musical instruments	..	..	..	..	..	..	..	..	..
3693	Sports goods	..	..	..	..	..	..	..	..	..
3694	Games and toys	..	..	..	..	..	..	..	..	..
3699	Other manufacturing, nec	..	..	..	..	..	..	..	..	..
37	**RECYCLING**	..	..	..	..	..	..	..	..	..
371	Recycling of metal waste and scrap	..	..	..	..	..	..	..	..	..
372	Recycling of non-metal waste and scrap	..	..	..	..	..	..	..	..	..
CERR	**Non-specified industry**	..	..	..	..	..	..	..	..	..
15-37	**TOTAL MANUFACTURING**	2,082,818	2,072,750	2,058,071	2,092,543	2,157,132	2,267,453	2,241,055	..	..

ISIC Revision 3 Industry Sector		1990	1991	1992	1993	1994	1995	1996	1997	1998
15	FOOD PRODUCTS AND BEVERAGES	..	..	..	59,411	52,530	53,574	58,233	53,751	..
151	Production, processing and preserving (PPP)	..	..	..	15,538	14,592	14,469	15,084	13,288	..
1511	PPP of meat and meat products	..	..	..	..	..	..	..	..	..
1512	Processing and preserving of fish products	..	..	..	..	..	..	..	..	..
1513	Processing, preserving of fruit & vegetables	..	..	..	..	..	..	..	..	..
1514	Vegetable and animal oils and fats	..	..	..	..	..	..	..	..	..
152	Dairy products	..	..	..	7,325	6,895	6,693	7,626	7,124	..
153	Grain mill prod., starches & prepared animal feeds	..	..	..	1,950	2,159	2,722	3,212	4,267	..
1531	Grain mill products	..	..	..	..	..	..	..	..	..
1532	Starches and starch products	..	..	..	..	..	..	..	..	..
1533	Prepared animal feeds	..	..	..	..	..	..	..	..	..
154	Other food products	..	..	..	23,622	16,947	18,109	20,962	18,674	..
1541	Bakery products	..	..	..	..	..	..	..	..	..
1542	Sugar	..	..	..	..	..	..	..	..	..
1543	Cocoa, chocolate and sugar confectionery	..	..	..	..	..	..	..	..	..
1544	Macaroni, noodles, couscous & similar farinaceous prod.	..	..	..	..	..	..	..	..	..
1549	Other food products, nec	..	..	..	..	..	..	..	..	..
155	Beverages	..	..	..	10,977	11,938	11,580	11,350	10,398	..
1551	Distilling, rectifying and blending of spirits	..	..	..	..	..	..	..	..	..
1552	Wines	..	..	..	..	..	..	..	..	..
1553	Malt liquors and malt	..	..	..	..	..	..	..	..	..
1554	Soft drinks; production of mineral waters	..	..	..	..	..	..	..	..	..
16	TOBACCO PRODUCTS	..	..	..	440	226	446	509	487	..
17	TEXTILES	..	..	..	33,075	31,349	28,338	28,653	23,769	..
171	Spinning, weaving and finishing of textiles	..	..	..	23,711	21,582	20,722	21,128	18,385	..
1711	Preparation and spinning of textile fibres	..	..	..	..	..	..	..	..	..
1712	Finishing of textiles	..	..	..	..	..	..	..	..	..
172	Other textiles	..	..	..	6,058	7,386	5,097	5,111	3,499	..
1721	Made-up textile articles, except apparel	..	..	..	..	..	..	..	..	..
1722	Carpets and rugs	..	..	..	..	..	..	..	..	..
1723	Cordage, rope, twine and netting	..	..	..	..	..	..	..	..	..
1729	Other textiles, nec	..	..	..	..	..	..	..	..	..
173	Knitted and crocheted fabrics and articles	..	..	..	3,306	2,381	2,520	2,414	1,885	..
18	WEARING APPAREL, DRESSING & DYEING OF FUR	..	..	..	2,116	2,552	2,961	2,707	2,285	..
181	Wearing apparel, except fur apparel	..	..	..	1,984	2,385	2,826	2,628	2,273	..
182	Dressing and dyeing of fur; articles of fur	..	..	..	132	167	135	79	12	..
19	TANNING & DRESSING OF LEATHER, FOOTWEAR	..	..	..	4,820	3,381	2,989	2,366	2,096	..
191	Tanning and dressing of leather	..	..	..	3,010	1,664	1,704	1,184	750	..
1911	Tanning and dressing of leather	..	..	..	..	..	..	..	..	..
1912	Luggage, handbags, saddlery & harness	..	..	..	..	..	..	..	..	..
192	Footwear	..	..	..	1,810	1,718	1,284	1,182	1,346	..
20	WOOD AND WOOD PRODUCTS	..	..	..	9,376	8,790	9,755	10,402	11,023	..
201	Sawmilling and planing of wood	..	..	..	2,978	3,648	3,840	3,910	2,871	..
202	Products of wood, cork, straw & plaiting materials	..	..	..	6,399	5,142	5,915	6,491	8,152	..
2021	Veneer sheets	..	..	..	..	..	..	..	..	..
2022	Builders' carpentry and joinery	..	..	..	..	..	..	..	..	..
2023	Wooden containers	..	..	..	..	..	..	..	..	..
2029	Other products of wood	..	..	..	..	..	..	..	..	..
21	PAPER AND PAPER PRODUCTS	..	..	..	58,200	51,777	49,195	47,672	50,347	..
2101	Pulp, paper and paperboard	..	..	..	..	..	..	..	..	..
2102	Corrugated paper, paperboard and their containers	..	..	..	..	..	..	..	..	..
2109	Other articles of pulp and paperboard	..	..	..	..	..	..	..	..	..
22	PUBLISHING, PRINTING & REPRODUCTION	..	..	..	3,521	1,146	1,176	1,230	1,468	..
221	Publishing	..	..	..	91	128	201	219	853	..
2211	Publishing of books & brochures	..	..	..	..	..	..	..	..	..
2212	Publishing of newspapers and periodicals	..	..	..	..	..	..	..	..	..
2213	Publishing of recorded media	..	..	..	..	..	..	..	..	..
2219	Other publishing	..	..	..	..	..	..	..	..	..
222	Printing and related service activities	..	..	..	3,291	907	859	888	494	..
2221	Printing	..	..	..	..	..	..	..	..	..
2222	Service activities related to printing	..	..	..	..	..	..	..	..	..
223	Reproduction of recorded media	..	..	..	139	112	117	123	121	..
23	COKE, REFINED PETROLEUM PRODUCTS	..	..	..	93,179	151,390	157,067	85,059	70,927	..
231	Coke oven products	..	..	..	6,188	67,549	70,662	69,388	51,060	..
232	Refined petroleum products	..	..	..	86,991	83,841	86,406	15,670	19,867	..
233	Processing of nuclear fuel	..	..	..	..	..	..	..	..	..
24	CHEMICALS & CHEMICAL PRODUCTS	..	..	..	49,506	50,629	52,956	96,214	178,219	..
241	Basic chemicals	..	..	..	41,198	43,309	44,133	88,562	171,940	..
2411	Basic chemicals, exc. fertilizers & nitrogen compounds	..	..	..	..	..	..	..	..	..
2412	Fertilizers and nitrogen compounds	..	..	..	..	..	..	..	..	..
2413	Plastics in primary forms and synthetic rubber	..	..	..	..	..	..	..	..	..
242	Other chemical products	..	..	..	4,702	3,900	5,972	4,807	5,753	..
2421	Pesticides and other agro-chemical products	..	..	..	..	..	..	..	..	..
2422	Paints, varnishes and similar coatings	..	..	..	..	..	..	..	..	..
2423	Pharmaceuticals, medicinal chem. & botanical prod.	..	..	..	..	..	..	..	..	..
2424	Soap and detergents, perfumes etc.	..	..	..	..	..	..	..	..	..
2429	Other chemical products, nec	..	..	..	..	..	..	..	..	..
243	Man-made fibres	..	..	..	3,606	3,420	2,851	2,845	526	..
25	RUBBER AND PLASTICS PRODUCTS	..	..	..	8,783	8,688	9,202	9,776	8,637	..
251	Rubber products	..	..	..	4,786	4,261	4,951	5,157	4,719	..
2511	Rubber tyres and tubes	..	..	..	..	..	..	..	..	..
2519	Other rubber products	..	..	..	..	..	..	..	..	..
252	Plastic products	..	..	..	3,997	4,427	4,251	4,618	3,918	..
26	OTHER NON-METALLIC MINERAL PRODUCTS	..	..	..	70,753	59,967	69,764	69,223	68,355	..
261	Glass and glass products	..	..	..	19,615	10,250	19,146	18,597	19,016	..
269	Non-metallic mineral products, nec	..	..	..	51,139	49,717	50,617	50,626	49,339	..
2691	Pottery, china and earthenware	..	..	..	..	..	..	..	..	..
2692	Refractory ceramic products	..	..	..	..	..	..	..	..	..
2693	Structural non-refractory clay & ceramic prod.	..	..	..	..	..	..	..	..	..
2694	Cement, lime and plaster	..	..	..	..	..	..	..	..	..
2695	Articles of concrete, cement and plaster	..	..	..	..	..	..	..	..	..
2696	Cutting, shaping and finishing of stone	..	..	..	..	..	..	..	..	..
2699	Other non-metallic mineral products, nec	..	..	..	..	..	..	..	..	..

ISIC Revision 3 Industry Sector		1990	1991	1992	1993	1994	1995	1996	1997	1998
27	**BASIC METALS**	..	..	..	269,195	293,960	296,202	280,391	282,184	..
271	Basic iron and steel	..	..	..	261,209	287,353	287,300	270,249	274,295	..
272	Basic precious and non-ferrous metals	..	..	..	3,492	2,569	2,990	3,731	2,924	..
273	Casting of metals	..	..	..	4,493	4,037	5,911	6,411	4,965	..
2731	Casting of iron and steel	..	..	..	..	..	..	..	..	..
2732	Casting non-ferrous metals	..	..	..	..	..	..	..	..	..
28	**FABRICATED METAL PRODUCTS**	..	..	..	13,303	13,119	14,210	15,531	11,058	..
281	Str. metal prod., tanks, reservoirs, steam generators	..	..	..	4,298	5,240	4,875	4,731	3,956	..
2811	Structural metal products	..	..	..	..	..	..	..	..	..
2812	Tanks, reservoirs and containers of metal	..	..	..	..	..	..	..	..	..
2813	Steam generators, exc. central heating hot water boilers	..	..	..	..	..	..	..	..	..
289	Other fabricated metal products	..	..	..	9,005	7,879	9,335	10,800	7,102	..
2891	Forging, pressing, stamping & roll-forming of metal	..	..	..	..	..	..	..	..	..
2892	Treatment and coating of metals	..	..	..	..	..	..	..	..	..
2893	Cutlery, hand tools and general hardware	..	..	..	..	..	..	..	..	..
2899	Other fabricated metal products, nec	..	..	..	..	..	..	..	..	..
29	**MACHINERY AND EQUIPMENT, NEC**	..	..	..	38,753	34,558	32,510	30,267	24,863	..
291	General purpose machinery	..	..	..	11,955	11,201	10,361	10,454	8,348	..
2911	Engines and turbines	..	..	..	..	..	..	..	..	..
2912	Pumps, compressors, taps and valves	..	..	..	..	..	..	..	..	..
2913	Bearings, gears, gearing and driving elements	..	..	..	..	..	..	..	..	..
2914	Ovens, furnaces and furnace burners	..	..	..	..	..	..	..	..	..
2915	Lifting and handling equipment	..	..	..	..	..	..	..	..	..
2919	Other general purpose machinery	..	..	..	..	..	..	..	..	..
292	Special purpose machinery	..	..	..	25,601	22,298	20,898	18,386	15,331	..
2921	Agricultural and forestry machinery	..	..	..	..	..	..	..	..	..
2922	Machine-tools	..	..	..	..	..	..	..	..	..
2923	Machinery for metallurgy	..	..	..	..	..	..	..	..	..
2924	Machinery for mining, quarrying and construction	..	..	..	..	..	..	..	..	..
2925	Machinery for food, beverage & tobacco processing	..	..	..	..	..	..	..	..	..
2926	Machinery for textile, apparel & leather production	..	..	..	..	..	..	..	..	..
2927	Machinery for weapons and ammunition	..	..	..	..	..	..	..	..	..
2929	Other special purpose machinery	..	..	..	..	..	..	..	..	..
293	Domestic appliances, nec	..	..	..	1,197	1,059	1,251	1,427	1,185	..
30	**OFFICE, ACCOUNTING & COMPUTING MACHINERY**	..	..	..	436	269	256	190	74	..
31	**ELECTRICAL MACHINERY & APPARATUS, NEC**	..	..	..	7,420	5,877	6,593	6,127	6,223	..
311	Electric motors, generators and transformers	..	..	..	3,056	2,326	2,504	2,002	1,600	..
312	Electricity distribution and control apparatus	..	..	..	1,312	1,287	1,295	1,067	1,225	..
313	Insulated wire and cable	..	..	..	985	830	925	923	813	..
314	Accumulators, primary cells & primary batteries	..	..	..	320	330	366	418	312	..
315	Electric lamps and lighting equipment	..	..	..	240	281	800	718	693	..
319	Other electrical equipment, nec	..	..	..	1,507	823	703	1,000	1,580	..
32	**RADIO, TV & COMMUNICATION EQUIP. & APP.**	..	..	..	1,722	1,690	1,698	1,798	1,539	..
321	Electronic valves, tubes, other electronic components	..	..	..	1,386	1,308	1,356	1,257	731	..
322	TV & radio transmitters, apparatus for line telephony	..	..	..	248	288	241	446	708	..
323	TV & radio receivers, recording apparatus	..	..	..	89	94	101	95	100	..
33	**MEDICAL PRECISION &OPTICAL INSTRUMENTS**	..	..	..	1,364	1,422	1,667	1,794	1,349	..
331	Medical appliances and instruments	..	..	..	1,030	978	1,193	1,277	1,186	..
3311	Medical, surgical equipment & orthopaedic app.	..	..	..	..	..	..	..	..	..
3312	Instruments & appliances for measuring, checking etc	..	..	..	..	..	..	..	..	..
3313	Industrial process control equipment	..	..	..	..	..	..	..	..	..
332	Optical instruments and photographic equipment	..	..	..	163	162	181	243	162	..
333	Watches and clocks	..	..	..	171	281	293	274	1	..
34	**MOTOR VEHICLES, TRAILERS & SEMI-TRAILERS**	..	..	..	22,602	16,861	13,934	10,957	10,634	..
341	Motor vehicles	..	..	..	17,301	12,809	8,666	6,835	6,725	..
342	Bodies (coachwork) for motor vehicles	..	..	..	1,002	834	884	844	381	..
343	Parts, accessories for motor vehicles & their engines	..	..	..	4,299	3,218	4,384	3,278	3,528	..
35	**OTHER TRANSPORT EQUIPMENT**	..	..	..	11,917	7,471	7,564	11,680	5,481	..
351	Building and repairing of ships and boats	..	..	..	122	94	453	74	74	..
3511	Building and repairing of ships	..	..	..	..	..	..	..	..	..
3512	Building, repairing of pleasure & sporting boats	..	..	..	..	..	..	..	..	..
352	Railway, tramway locomotives & rolling stock	..	..	..	8,209	4,084	4,309	7,690	3,490	..
353	Aircraft and spacecraft	..	..	..	2,516	2,040	1,569	2,015	1,440	..
359	Transport equipment, nec	..	..	..	1,071	1,252	1,232	1,901	479	..
3591	Motorcycles	..	..	..	..	..	..	..	..	..
3592	Bicycles and invalid carriages	..	..	..	..	..	..	..	..	..
3599	Other transport equipment, nec	..	..	..	..	..	..	..	..	..
36	**FURNITURE; MANUFACTURING, NEC**	..	..	..	10,277	8,243	9,128	8,879	6,123	..
361	Furniture	..	..	..	6,506	4,304	5,354	4,930	3,962	..
369	Manufacturing, nec	..	..	..	3,771	3,939	3,774	3,950	2,162	..
3691	Jewellery and related articles	..	..	..	..	..	..	..	..	..
3692	Musical instruments	..	..	..	..	..	..	..	..	..
3693	Sports goods	..	..	..	..	..	..	..	..	..
3694	Games and toys	..	..	..	..	..	..	..	..	..
3699	Other manufacturing, nec	..	..	..	..	..	..	..	..	..
37	**RECYCLING**	..	..	..	554	1,728	777	637	694	..
371	Recycling of metal waste and scrap	..	..	..	503	1,572	665	574	575	..
372	Recycling of non-metal waste and scrap	..	..	..	50	156	112	63	118	..
CERR	**Non-specified industry**	..	..	..	..	..	..	..	..	..
15-37	**TOTAL MANUFACTURING**	..	..	..	770,723	807,623	821,963	780,293	821,587	..

ISIC Revision 3 Industry Sector	1990	1991	1992	1993	1994	1995	1996	1997	1998
15 FOOD PRODUCTS AND BEVERAGES	32,325	..	..	32,694	..	35,708	34,102	35,066	..
151 Production, processing and preserving (PPP)	11,264	..	..	11,340	..	12,804	12,327	12,523	..
1511 PPP of meat and meat products	5,087	..	..	5,507	..	5,726	5,577	5,692	..
1512 Processing and preserving of fish products	3,295	..	..	3,588	..	4,782	3,824	4,343	..
1513 Processing, preserving of fruit & vegetables	645	..	..	760	..	845	899	884	..
1514 Vegetable and animal oils and fats	2,237	..	..	1,484	..	1,451	2,027	1,604	..
152 Dairy products	5,232	..	..	5,522	..	5,485	6,676	6,848	..
153 Grain mill prod., starches & prepared animal feeds	2,516	..	..	2,586	..	4,891	2,553	2,639	..
1531 Grain mill products	375	..	..	440	..	392	409	424	..
1532 Starches and starch products	207	..	..	429	..	139	330	341	..
1533 Prepared animal feeds	1,935	..	..	1,717	..	4,360	1,814	1,874	..
154 Other food products	9,049	..	..	8,847	..	8,909	8,854	9,043	..
1541 Bakery products	985	..	..	1,019	..	1,191	1,215	1,181	..
1542 Sugar	5,269	..	..	4,490	..	3,775	3,868	4,384	..
1543 Cocoa, chocolate and sugar confectionery	459	..	..	446	..	499	435	449	..
1544 Macaroni, noodles, couscous & similar farinaceous prod.	..	..	..	..	..	..	..	..	..
1549 Other food products, nec	2,336	..	..	2,892	..	3,445	3,336	3,029	..
155 Beverages	4,264	..	..	4,400	..	3,620	3,693	4,014	..
1551 Distilling, rectifying and blending of spirits	729	..	..	611	..	536	613	614	..
1552 Wines	..	..	..	..	..	..	..	..	..
1553 Malt liquors and malt	3,309	..	..	3,525	..	2,818	2,818	2,858	..
1554 Soft drinks; production of mineral waters	226	..	..	263	..	266	261	542	..
16 TOBACCO PRODUCTS	234	..	..	226	..	227	217	210	..
17 TEXTILES	2,338	..	..	2,120	..	1,891	1,904	1,819	..
171 Spinning, weaving and finishing of textiles	1,579	..	..	1,372	..	1,013	1,063	1,043	..
1711 Preparation and spinning of textile fibres	881	..	..	650	..	417	423	375	..
1712 Finishing of textiles	698	..	..	722	..	596	640	668	..
172 Other textiles	635	..	..	613	..	656	701	652	..
1721 Made-up textile articles, except apparel	135	..	..	152	..	205	236	173	..
1722 Carpets and rugs	304	..	..	246	..	197	214	219	..
1723 Cordage, rope, twine and netting	29	..	..	20	..	25	22	21	..
1729 Other textiles, nec	166	..	..	196	..	230	229	240	..
173 Knitted and crocheted fabrics and articles	124	..	..	135	..	222	141	123	..
18 WEARING APPAREL, DRESSING & DYEING OF FUR	216	..	..	221	..	220	229	213	..
181 Wearing apparel, except fur apparel	..	..	..	..	..	..	..	..	..
182 Dressing and dyeing of fur; articles of fur	..	..	..	..	..	..	..	..	..
19 TANNING & DRESSING OF LEATHER, FOOTWEAR	138	..	..	158	..	165	151	108	..
191 Tanning and dressing of leather	77	..	..	90	..	107	89	54	..
1911 Tanning and dressing of leather	69	..	..	80	..	94	77	42	..
1912 Luggage, handbags, saddlery & harness	8	..	..	9	..	13	12	11	..
192 Footwear	61	..	..	68	..	58	61	55	..
20 WOOD AND WOOD PRODUCTS	4,945	..	..	5,114	..	5,539	5,428	5,613	..
201 Sawmilling and planing of wood	642	..	..	562	..	691	531	625	..
202 Products of wood, cork, straw & plaiting materials	4,304	..	..	4,552	..	4,848	4,897	4,987	..
2021 Veneer sheets	1,331	..	..	1,182	..	1,017	1,199	1,208	..
2022 Builders' carpentry and joinery	2,611	..	..	3,086	..	3,438	3,217	3,380	..
2023 Wooden containers	75	..	..	66	..	152	222	236	..
2029 Other products of wood	288	..	..	218	..	242	259	163	..
21 PAPER AND PAPER PRODUCTS	6,221	..	..	5,229	..	5,198	6,075	6,268	..
2101 Pulp, paper and paperboard	4,548	..	..	3,526	..	3,334	3,416	3,565	..
2102 Corrugated paper, paperboard and their containers	698	..	..	766	..	839	922	954	..
2109 Other articles of pulp and paperboard	975	..	..	938	..	1,024	1,737	1,750	..
22 PUBLISHING, PRINTING & REPRODUCTION	1,185	..	..	1,168	..	1,438	1,431	1,282	..
221 Publishing	495	..	..	509	..	563	628	606	..
2211 Publishing of books & brochures	28	..	..	44	..	80	76	76	..
2212 Publishing of newspapers and periodicals	451	..	..	445	..	466	531	514	..
2213 Publishing of recorded media	..	..	..	..	..	..	..	..	..
2219 Other publishing	15	..	..	21	..	17	20	16	..
222 Printing and related service activities	690	..	..	658	..	875	803	676	..
2221 Printing	571	..	..	537	..	608	686	561	..
2222 Service activities related to printing	119	..	..	121	..	267	117	116	..
223 Reproduction of recorded media	..	..	..	..	..	..	..	..	..
23 COKE, REFINED PETROLEUM PRODUCTS	16,882	..	..	20,212	..	23,008	25,521	21,087	..
231 Coke oven products	..	..	..	..	..	..	..	..	..
232 Refined petroleum products	16,882	..	..	20,212	..	23,008	25,521	21,087	..
233 Processing of nuclear fuel	..	..	..	..	..	..	..	..	..
24 CHEMICALS & CHEMICAL PRODUCTS	8,818	..	..	10,220	..	10,483	11,583	11,535	..
241 Basic chemicals	4,281	..	..	5,543	..	5,605	5,766	6,364	..
2411 Basic chemicals, exc. fertilizers & nitrogen compounds	3,149	..	..	4,165	..	4,058	3,964	4,362	..
2412 Fertilizers and nitrogen compounds	973	..	..	1,222	..	1,452	1,631	1,819	..
2413 Plastics in primary forms and synthetic rubber	160	..	..	155	..	95	171	183	..
242 Other chemical products	4,537	..	..	4,678	..	4,878	5,816	5,170	..
2421 Pesticides and other agro-chemical products	..	..	..	..	..	..	..	..	..
2422 Paints, varnishes and similar coatings	282	..	..	284	..	296	324	323	..
2423 Pharmaceuticals, medicinal chem. & botanical prod.	2,382	..	..	2,470	..	2,486	3,234	2,569	..
2424 Soap and detergents, perfumes etc.	1,128	..	..	1,214	..	1,161	1,117	1,052	..
2429 Other chemical products, nec	744	..	..	709	..	934	1,142	1,227	..
243 Man-made fibres	..	..	..	..	..	..	..	..	..
25 RUBBER AND PLASTICS PRODUCTS	3,228	..	..	3,177	..	3,600	3,635	3,892	..
251 Rubber products	701	..	..	605	..	679	670	861	..
2511 Rubber tyres and tubes	..	..	..	..	..	..	..	..	..
2519 Other rubber products	701	..	..	605	..	679	670	861	..
252 Plastic products	2,527	..	..	2,571	..	2,921	2,966	3,030	..
26 OTHER NON-METALLIC MINERAL PRODUCTS	20,692	..	..	23,385	..	27,073	28,063	29,183	..
261 Glass and glass products	1,915	..	..	1,938	..	2,019	2,006	1,969	..
269 Non-metallic mineral products, nec	18,777	..	..	21,447	..	25,055	26,057	27,214	..
2691 Pottery, china and earthenware	308	..	..	234	..	240	239	252	..
2692 Refractory ceramic products	..	..	..	..	..	..	..	..	..
2693 Structural non-refractory clay & ceramic prod.	1,845	..	..	1,408	..	2,271	2,115	2,186	..
2694 Cement, lime and plaster	10,184	..	..	13,487	..	15,250	16,661	17,701	..
2695 Articles of concrete, cement and plaster	2,981	..	..	3,009	..	2,607	2,595	2,752	..
2696 Cutting, shaping and finishing of stone	12	..	..	12	..	1	14	15	..
2699 Other non-metallic mineral products, nec	3,447	..	..	3,297	..	4,686	4,433	4,309	..

ISIC Revision 3 Industry Sector	1990	1991	1992	1993	1994	1995	1996	1997	1998
27 **BASIC METALS**	4,341	..	..	4,143	..	4,442	4,548	4,706	..
271 Basic iron and steel	3,846	..	..	3,555	..	3,851	3,975	4,113	..
272 Basic precious and non-ferrous metals	496	..	..	587	..	590	573	593	..
273 Casting of metals	..	..	..	..	..	..	..	..	..
2731 Casting of iron and steel	..	..	..	..	..	..	..	..	..
2732 Casting non-ferrous metals	..	..	..	..	..	..	..	..	..
28 **FABRICATED METAL PRODUCTS**	3,519	..	..	3,691	..	3,928	4,256	4,204	..
281 Str. metal prod., tanks, reservoirs, steam generators	841	..	..	983	..	1,173	1,183	1,195	..
2811 Structural metal products	543	..	..	640	..	853	854	880	..
2812 Tanks, reservoirs and containers of metal	237	..	..	250	..	209	228	216	..
2813 Steam generators, exc. central heating hot water boilers	60	..	..	94	..	111	101	99	..
289 Other fabricated metal products	2,678	..	..	2,707	..	2,755	3,073	3,009	..
2891 Forging, pressing, stamping & roll-forming of metal	..	..	..	..	..	..	..	..	..
2892 Treatment and coating of metals	271	..	..	351	..	380	384	373	..
2893 Cutlery, hand tools and general hardware	952	..	..	1,108	..	883	1,187	1,177	..
2899 Other fabricated metal products, nec	1,454	..	..	1,248	..	1,492	1,502	1,458	..
29 **MACHINERY AND EQUIPMENT, NEC**	4,571	..	..	4,819	..	5,511	5,640	5,629	..
291 General purpose machinery	2,508	..	..	2,838	..	3,341	3,507	3,542	..
2911 Engines and turbines	..	..	..	..	..	..	..	..	..
2912 Pumps, compressors, taps and valves	1,418	..	..	1,640	..	2,062	2,217	2,136	..
2913 Bearings, gears, gearing and driving elements	38	..	..	81	..	129	105	109	..
2914 Ovens, furnaces and furnace burners	7	..	..	57	..	16	16	20	..
2915 Lifting and handling equipment	519	..	..	497	..	555	587	559	..
2919 Other general purpose machinery	525	..	..	563	..	579	582	719	..
292 Special purpose machinery	1,429	..	..	1,427	..	1,554	1,552	1,561	..
2921 Agricultural and forestry machinery	563	..	..	493	..	634	662	686	..
2922 Machine-tools	210	..	..	122	..	124	118	105	..
2923 Machinery for metallurgy	..	..	..	..	..	..	..	..	..
2924 Machinery for mining, quarrying and construction	254	..	..	273	..	204	187	207	..
2925 Machinery for food, beverage & tobacco processing	157	..	..	202	..	220	250	242	..
2926 Machinery for textile, apparel & leather production	73	..	..	40	..	50	36	27	..
2927 Machinery for weapons and ammunition	..	..	..	..	..	..	..	..	..
2929 Other special purpose machinery	173	..	..	297	..	323	300	294	..
293 Domestic appliances, nec	634	..	..	554	..	615	581	526	..
30 **OFFICE, ACCOUNTING & COMPUTING MACHINERY**	127	..	..	75	..	100	72	68	..
31 **ELECTRICAL MACHINERY & APPARATUS, NEC**	950	..	..	795	..	1,101	1,015	919	..
311 Electric motors, generators and transformers	156	..	..	171	..	248	286	276	..
312 Electricity distribution and control apparatus	166	..	..	181	..	330	201	187	..
313 Insulated wire and cable	399	..	..	186	..	235	243	205	..
314 Accumulators, primary cells & primary batteries	42	..	..	46	..	62	47	36	..
315 Electric lamps and lighting equipment	140	..	..	147	..	145	151	162	..
319 Other electrical equipment, nec	47	..	..	64	..	80	88	52	..
32 **RADIO, TV & COMMUNICATION EQUIP. & APP.**	539	..	..	539	..	603	715	699	..
321 Electronic valves, tubes, other electronic components	132	..	..	160	..	166	214	239	..
322 TV & radio transmitters, apparatus for line telephony	100	..	..	48	..	66	82	77	..
323 TV & radio receivers, recording apparatus	308	..	..	331	..	370	418	383	..
33 **MEDICAL PRECISION &OPTICAL INSTRUMENTS**	468	..	..	491	..	528	562	692	..
331 Medical appliances and instruments	399	..	..	431	..	459	484	587	..
3311 Medical, surgical equipment & orthopaedic app.	218	..	..	212	..	239	243	266	..
3312 Instruments & appliances for measuring, checking etc	169	..	..	210	..	218	238	319	..
3313 Industrial process control equipment	13	..	..	8	..	2	3	2	..
332 Optical instruments and photographic equipment	68	..	..	60	..	69	78	104	..
333 Watches and clocks	..	..	..	..	..	..	..	..	..
34 **MOTOR VEHICLES, TRAILERS & SEMI-TRAILERS**	653	..	..	799	..	723	805	756	..
341 Motor vehicles	..	..	..	..	..	..	..	..	..
342 Bodies (coachwork) for motor vehicles	220	..	..	280	..	324	391	386	..
343 Parts, accessories for motor vehicles & their engines	433	..	..	520	..	399	413	370	..
35 **OTHER TRANSPORT EQUIPMENT**	1,006	..	..	1,030	..	1,405	1,415	1,312	..
351 Building and repairing of ships and boats	833	..	..	856	..	1,169	1,194	1,123	..
3511 Building and repairing of ships	806	..	..	825	..	1,145	1,166	1,097	..
3512 Building, repairing of pleasure & sporting boats	27	..	..	30	..	23	27	27	..
352 Railway, tramway locomotives & rolling stock	..	..	..	..	..	..	..	..	..
353 Aircraft and spacecraft	43	..	..	34	..	56	69	72	..
359 Transport equipment, nec	130	..	..	140	..	180	152	117	..
3591 Motorcycles	..	..	..	..	..	..	..	..	..
3592 Bicycles and invalid carriages	54	..	..	65	..	40	59	36	..
3599 Other transport equipment, nec	76	..	..	75	..	140	93	81	..
36 **FURNITURE; MANUFACTURING, NEC**	2,856	..	..	3,500	..	3,574	3,480	3,372	..
361 Furniture	2,419	..	..	3,021	..	3,024	2,936	2,853	..
369 Manufacturing, nec	437	..	..	479	..	550	544	520	..
3691 Jewellery and related articles	13	..	..	9	..	14	35	8	..
3692 Musical instruments	5	..	..	5	..	12	7	6	..
3693 Sports goods	..	..	..	..	..	..	..	..	..
3694 Games and toys	299	..	..	343	..	398	383	382	..
3699 Other manufacturing, nec	120	..	..	121	..	127	119	123	..
37 **RECYCLING**	..	..	..	..	..	..	..	..	..
371 Recycling of metal waste and scrap	..	..	..	..	..	..	..	..	..
372 Recycling of non-metal waste and scrap	..	..	..	..	..	..	..	..	..
CERR **Non-specified industry**	..	..	..	..	..	..	..	..	..
15-37 **TOTAL MANUFACTURING**	116,253	..	..	123,807	..	136,464	140,847	138,632	..

FRANCE

ISIC Revision 3 Industry Sector	1990	1991	1992	1993	1994	1995	1996	1997	1998
15 **FOOD PRODUCTS AND BEVERAGES**	..	..	..	..	191,239	198,594	196,671	199,576	..
151 Production, processing and preserving (PPP)	..	..	..	..	37,545	39,159	40,286	39,461	..
1511 PPP of meat and meat products	..	..	..	..	..	..	..	..	..
1512 Processing and preserving of fish products	..	..	..	..	..	..	..	..	..
1513 Processing, preserving of fruit & vegetables	..	..	..	..	..	..	..	..	..
1514 Vegetable and animal oils and fats	..	..	..	..	..	..	..	..	..
152 Dairy products	..	..	..	..	35,554	38,868	39,291	38,713	..
153 Grain mill prod., starches & prepared animal feeds	..	..	..	..	39,947	40,527	38,911	41,351	..
1531 Grain mill products	..	..	..	..	..	..	..	..	..
1532 Starches and starch products	..	..	..	..	..	..	..	..	..
1533 Prepared animal feeds	..	..	..	..	..	..	..	..	..
154 Other food products	..	..	..	..	61,423	63,508	62,627	65,077	..
1541 Bakery products	..	..	..	..	..	..	..	..	..
1542 Sugar	..	..	..	..	..	..	..	..	..
1543 Cocoa, chocolate and sugar confectionery	..	..	..	..	..	..	..	..	..
1544 Macaroni, noodles, couscous & similar farinaceous prod.	..	..	..	..	..	..	..	..	..
1549 Other food products, nec	..	..	..	..	..	..	..	..	..
155 Beverages	..	..	..	..	16,770	16,532	15,556	14,974	..
1551 Distilling, rectifying and blending of spirits	..	..	..	..	..	..	..	..	..
1552 Wines	..	..	..	..	..	..	..	..	..
1553 Malt liquors and malt	..	..	..	..	..	..	..	..	..
1554 Soft drinks; production of mineral waters	..	..	..	..	..	..	..	..	..
16 **TOBACCO PRODUCTS**	..	..	..	..	..	..	..	..	..
17 **TEXTILES**	..	..	..	..	35,419	36,520	35,409	35,696	33,933
171 Spinning, weaving and finishing of textiles	..	..	..	..	26,606	27,346	25,900	26,277	24,691
1711 Preparation and spinning of textile fibres	..	..	..	..	..	..	..	..	..
1712 Finishing of textiles	..	..	..	..	..	..	..	..	..
172 Other textiles	..	..	..	..	6,326	6,714	6,919	7,045	6,808
1721 Made-up textile articles, except apparel	..	..	..	..	..	..	..	..	..
1722 Carpets and rugs	..	..	..	..	..	..	..	..	..
1723 Cordage, rope, twine and netting	..	..	..	..	..	..	..	..	..
1729 Other textiles, nec	..	..	..	..	..	..	..	..	..
173 Knitted and crocheted fabrics and articles	..	..	..	..	2,487	2,460	2,590	2,374	2,433
18 **WEARING APPAREL, DRESSING & DYEING OF FUR**	..	..	..	..	2,675	5,112	2,588	2,366	2,362
181 Wearing apparel, except fur apparel	..	..	..	..	2,575	5,014	2,481	2,270	2,279
182 Dressing and dyeing of fur; articles of fur	..	..	..	..	99	97	107	96	83
19 **TANNING & DRESSING OF LEATHER, FOOTWEAR**	..	..	..	..	2,353	2,331	2,312	2,250	1,949
191 Tanning and dressing of leather	..	..	..	..	1,242	1,278	1,182	1,254	996
1911 Tanning and dressing of leather	..	..	..	..	..	..	..	..	..
1912 Luggage, handbags, saddlery & harness	..	..	..	..	..	..	..	..	..
192 Footwear	..	..	..	..	1,111	1,053	1,130	996	954
20 **WOOD AND WOOD PRODUCTS**	..	..	..	..	10,453	11,320	11,922	12,940	10,196
201 Sawmilling and planing of wood	..	..	..	..	1,999	2,129	2,448	2,434	..
202 Products of wood, cork, straw & plaiting materials	..	..	..	..	8,453	9,191	9,475	10,506	10,196
2021 Veneer sheets	..	..	..	..	..	..	..	..	..
2022 Builders' carpentry and joinery	..	..	..	..	..	..	..	..	..
2023 Wooden containers	..	..	..	..	..	..	..	..	..
2029 Other products of wood	..	..	..	..	..	..	..	..	..
21 **PAPER AND PAPER PRODUCTS**	..	..	..	..	113,823	113,222	115,022	117,568	111,275
2101 Pulp, paper and paperboard	..	..	..	..	..	..	..	..	..
2102 Corrugated paper, paperboard and their containers	..	..	..	..	..	..	..	..	..
2109 Other articles of pulp and paperboard	..	..	..	..	..	..	..	..	..
22 **PUBLISHING, PRINTING & REPRODUCTION**	..	..	..	..	9,078	8,681	10,179	9,584	13,862
221 Publishing	..	..	..	..	1,077	858	916	943	1,080
2211 Publishing of books & brochures	..	..	..	..	..	..	..	..	..
2212 Publishing of newspapers and periodicals	..	..	..	..	..	..	..	..	..
2213 Publishing of recorded media	..	..	..	..	..	..	..	..	..
2219 Other publishing	..	..	..	..	..	..	..	..	..
222 Printing and related service activities	..	..	..	..	7,739	7,535	8,942	8,292	12,359
2221 Printing	..	..	..	..	..	..	..	..	..
2222 Service activities related to printing	..	..	..	..	..	..	..	..	..
223 Reproduction of recorded media	..	..	..	..	261	289	321	349	423
23 **COKE, REFINED PETROLEUM PRODUCTS**	..	..	..	..	..	..	..	..	..
231 Coke oven products	..	..	..	..	..	..	..	..	..
232 Refined petroleum products	..	..	..	..	..	..	..	..	..
233 Processing of nuclear fuel	..	..	..	..	..	..	..	..	..
24 **CHEMICALS & CHEMICAL PRODUCTS**	..	..	..	..	112,767	113,992	117,790	118,409	116,727
241 Basic chemicals	..	..	..	..	74,733	77,074	79,202	80,061	75,407
2411 Basic chemicals, exc. fertilizers & nitrogen compounds	..	..	..	..	..	..	..	..	..
2412 Fertilizers and nitrogen compounds	..	..	..	..	..	..	..	..	..
2413 Plastics in primary forms and synthetic rubber	..	..	..	..	..	..	..	..	..
242 Other chemical products	..	..	..	..	34,542	33,968	34,969	34,692	37,510
2421 Pesticides and other agro-chemical products	..	..	..	..	..	..	..	..	..
2422 Paints, varnishes and similar coatings	..	..	..	..	..	..	..	..	..
2423 Pharmaceuticals, medicinal chem. & botanical prod.	..	..	..	..	..	..	..	..	..
2424 Soap and detergents, perfumes etc.	..	..	..	..	..	..	..	..	..
2429 Other chemical products, nec	..	..	..	..	..	..	..	..	..
243 Man-made fibres	..	..	..	..	3,492	2,951	3,620	3,656	3,809
25 **RUBBER AND PLASTICS PRODUCTS**	..	..	..	..	39,003	41,115	42,655	45,297	49,124
251 Rubber products	..	..	..	..	15,335	16,897	17,845	19,003	20,535
2511 Rubber tyres and tubes	..	..	..	..	..	..	..	..	..
2519 Other rubber products	..	..	..	..	..	..	..	..	..
252 Plastic products	..	..	..	..	23,668	24,218	24,810	26,295	28,589
26 **OTHER NON-METALLIC MINERAL PRODUCTS**	..	..	..	..	169,867	172,086	169,509	171,462	173,226
261 Glass and glass products	..	..	..	..	54,554	56,847	56,700	59,079	60,265
269 Non-metallic mineral products, nec	..	..	..	..	115,313	115,239	112,810	112,383	112,961
2691 Pottery, china and earthenware	..	..	..	..	..	..	..	..	..
2692 Refractory ceramic products	..	..	..	..	..	..	..	..	..
2693 Structural non-refractory clay & ceramic prod.	..	..	..	..	..	..	..	..	..
2694 Cement, lime and plaster	..	..	..	..	..	..	..	..	..
2695 Articles of concrete, cement and plaster	..	..	..	..	..	..	..	..	..
2696 Cutting, shaping and finishing of stone	..	..	..	..	..	..	..	..	..
2699 Other non-metallic mineral products, nec	..	..	..	..	..	..	..	..	..

ISIC Revision 3 Industry Sector		1990	1991	1992	1993	1994	1995	1996	1997	1998
27	**BASIC METALS**	..	..	..	..	194,547	137,420	136,285	141,807	146,788
271	Basic iron and steel	..	..	..	..	148,949	91,522	89,877	94,548	96,731
272	Basic precious and non-ferrous metals	..	..	..	..	31,774	31,006	31,543	32,066	33,336
273	Casting of metals	..	..	..	..	13,823	14,892	14,865	15,194	16,721
2731	Casting of iron and steel	..	..	..	..	..	..	..	..	..
2732	Casting non-ferrous metals	..	..	..	..	..	..	..	..	..
28	**FABRICATED METAL PRODUCTS**	..	..	..	..	31,862	32,922	37,016	38,379	41,165
281	Str. metal prod., tanks, reservoirs, steam generators	..	..	..	..	6,219	6,185	7,250	7,048	7,475
2811	Structural metal products	..	..	..	..	..	..	..	..	..
2812	Tanks, reservoirs and containers of metal	..	..	..	..	..	..	..	..	..
2813	Steam generators, exc. central heating hot water boilers	..	..	..	..	..	..	..	..	..
289	Other fabricated metal products	..	..	..	..	25,643	26,737	29,766	31,331	33,690
2891	Forging, pressing, stamping & roll-forming of metal	..	..	..	..	..	..	..	..	..
2892	Treatment and coating of metals	..	..	..	..	..	..	..	..	..
2893	Cutlery, hand tools and general hardware	..	..	..	..	..	..	..	..	..
2899	Other fabricated metal products, nec	..	..	..	..	..	..	..	..	..
29	**MACHINERY AND EQUIPMENT, NEC**	..	..	..	..	20,522	21,400	22,386	22,072	22,651
291	General purpose machinery	..	..	..	..	9,666	10,134	10,420	10,320	10,988
2911	Engines and turbines	..	..	..	..	..	..	..	..	..
2912	Pumps, compressors, taps and valves	..	..	..	..	..	..	..	..	..
2913	Bearings, gears, gearing and driving elements	..	..	..	..	..	..	..	..	..
2914	Ovens, furnaces and furnace burners	..	..	..	..	..	..	..	..	..
2915	Lifting and handling equipment	..	..	..	..	..	..	..	..	..
2919	Other general purpose machinery	..	..	..	..	..	..	..	..	..
292	Special purpose machinery	..	..	..	..	7,687	7,923	8,511	8,646	8,452
2921	Agricultural and forestry machinery	..	..	..	..	..	..	..	..	..
2922	Machine-tools	..	..	..	..	..	..	..	..	..
2923	Machinery for metallurgy	..	..	..	..	..	..	..	..	..
2924	Machinery for mining, quarrying and construction	..	..	..	..	..	..	..	..	..
2925	Machinery for food, beverage & tobacco processing	..	..	..	..	..	..	..	..	..
2926	Machinery for textile, apparel & leather production	..	..	..	..	..	..	..	..	..
2927	Machinery for weapons and ammunition	..	..	..	..	..	..	..	..	..
2929	Other special purpose machinery	..	..	..	..	..	..	..	..	..
293	Domestic appliances, nec	..	..	..	..	3,169	3,343	3,455	3,107	3,211
30	**OFFICE, ACCOUNTING & COMPUTING MACHINERY**	..	..	..	..	1,302	1,456	1,558	1,415	1,204
31	**ELECTRICAL MACHINERY & APPARATUS, NEC**	..	..	..	..	14,973	15,947	17,257	16,987	18,142
311	Electric motors, generators and transformers	..	..	..	..	1,963	2,270	3,060	2,956	3,069
312	Electricity distribution and control apparatus	..	..	..	..	3,682	3,526	3,474	3,439	3,416
313	Insulated wire and cable	..	..	..	..	3,005	3,065	3,210	3,321	3,450
314	Accumulators, primary cells & primary batteries	..	..	..	..	1,789	1,770	1,775	1,749	1,742
315	Electric lamps and lighting equipment	..	..	..	..	1,047	1,184	1,215	1,133	1,276
319	Other electrical equipment, nec	..	..	..	..	3,486	4,132	4,522	4,390	5,189
32	**RADIO, TV & COMMUNICATION EQUIP. & APP.**	..	..	..	..	9,778	10,262	10,921	11,169	11,747
321	Electronic valves, tubes, other electronic components	..	..	..	..	6,970	7,450	8,000	8,355	8,713
322	TV & radio transmitters, apparatus for line telephony	..	..	..	..	2,139	2,074	2,172	2,082	2,311
323	TV & radio receivers, recording apparatus	..	..	..	..	669	738	748	732	723
33	**MEDICAL PRECISION &OPTICAL INSTRUMENTS**	..	..	..	..	4,779	5,328	5,421	5,449	5,449
331	Medical appliances and instruments	..	..	..	..	3,999	4,518	4,541	4,551	4,619
3311	Medical, surgical equipment & orthopaedic app.	..	..	..	..	..	..	..	..	..
3312	Instruments & appliances for measuring, checking etc	..	..	..	..	..	..	..	..	..
3313	Industrial process control equipment	..	..	..	..	..	..	..	..	..
332	Optical instruments and photographic equipment	..	..	..	..	522	547	604	693	612
333	Watches and clocks	..	..	..	..	258	262	276	205	219
34	**MOTOR VEHICLES, TRAILERS & SEMI-TRAILERS**	..	..	..	..	39,412	42,271	43,874	43,070	44,851
341	Motor vehicles	..	..	..	..	24,822	25,779	26,941	26,289	27,194
342	Bodies (coachwork) for motor vehicles	..	..	..	..	1,407	1,538	1,682	1,544	1,780
343	Parts, accessories for motor vehicles & their engines	..	..	..	..	13,182	14,954	15,251	15,238	15,877
35	**OTHER TRANSPORT EQUIPMENT**	..	..	..	..	10,526	11,572	12,316	12,716	12,390
351	Building and repairing of ships and boats	..	..	..	..	775	792	819	775	737
3511	Building and repairing of ships	..	..	..	..	..	..	..	..	..
3512	Building, repairing of pleasure & sporting boats	..	..	..	..	..	..	..	..	..
352	Railway, tramway locomotives & rolling stock	..	..	..	..	1,490	1,475	1,793	1,501	1,494
353	Aircraft and spacecraft	..	..	..	..	7,452	8,457	8,804	9,585	9,276
359	Transport equipment, nec	..	..	..	..	809	848	900	855	884
3591	Motorcycles	..	..	..	..	..	..	..	..	..
3592	Bicycles and invalid carriages	..	..	..	..	..	..	..	..	..
3599	Other transport equipment, nec	..	..	..	..	..	..	..	..	..
36	**FURNITURE; MANUFACTURING, NEC**	..	..	..	..	9,077	9,007	9,694	9,031	8,743
361	Furniture	..	..	..	..	6,492	6,551	7,108	6,258	5,992
369	Manufacturing, nec	..	..	..	..	2,585	2,456	2,586	2,773	2,752
3691	Jewellery and related articles	..	..	..	..	..	..	..	..	..
3692	Musical instruments	..	..	..	..	..	..	..	..	..
3693	Sports goods	..	..	..	..	..	..	..	..	..
3694	Games and toys	..	..	..	..	..	..	..	..	..
3699	Other manufacturing, nec	..	..	..	..	..	..	..	..	..
37	**RECYCLING**	..	..	..	..	..	..	..	..	..
371	Recycling of metal waste and scrap	..	..	..	..	..	..	..	..	..
372	Recycling of non-metal waste and scrap	..	..	..	..	..	..	..	..	..
CERR	**Non-specified industry**	..	..	..	..	..	..	..	..	..
15-37	**TOTAL MANUFACTURING**	..	..	..	..	1,208,708	1,187,831	1,196,359	1,205,879	1,016,434

ISIC Revision 3 Industry Sector		1990	1991	1992	1993	1994	1995	1996	1997	1998
15	**FOOD PRODUCTS AND BEVERAGES**	..	..	..	..	..	196,736	198,661	196,136	..
151	Production, processing and preserving (PPP)	..	..	..	..	..	..	..	..	..
1511	PPP of meat and meat products	..	..	..	..	..	..	..	..	..
1512	Processing and preserving of fish products	..	..	..	..	..	..	..	..	..
1513	Processing, preserving of fruit & vegetables	..	..	..	..	..	..	..	..	..
1514	Vegetable and animal oils and fats	..	..	..	..	..	..	..	..	..
152	Dairy products	..	..	..	..	..	..	..	..	..
153	Grain mill prod., starches & prepared animal feeds	..	..	..	..	..	..	..	..	..
1531	Grain mill products	..	..	..	..	..	..	..	..	..
1532	Starches and starch products	..	..	..	..	..	..	..	..	..
1533	Prepared animal feeds	..	..	..	..	..	..	..	..	..
154	Other food products	..	..	..	..	..	..	..	..	..
1541	Bakery products	..	..	..	..	..	..	..	..	..
1542	Sugar	..	..	..	..	..	..	..	..	..
1543	Cocoa, chocolate and sugar confectionery	..	..	..	..	..	..	..	..	..
1544	Macaroni, noodles, couscous & similar farinaceous prod.	..	..	..	..	..	..	..	..	..
1549	Other food products, nec	..	..	..	..	..	..	..	..	..
155	Beverages	..	..	..	..	..	..	..	..	..
1551	Distilling, rectifying and blending of spirits	..	..	..	..	..	..	..	..	..
1552	Wines	..	..	..	..	..	..	..	..	..
1553	Malt liquors and malt	..	..	..	..	..	..	..	..	..
1554	Soft drinks; production of mineral waters	..	..	..	..	..	..	..	..	..
16	**TOBACCO PRODUCTS**	..	..	..	..	..	3,033	2,572	2,378	..
17	**TEXTILES**	..	..	..	..	..	44,305	42,811	40,943	..
171	Spinning, weaving and finishing of textiles	..	..	..	..	..	..	..	..	..
1711	Preparation and spinning of textile fibres	..	..	..	..	..	..	..	..	..
1712	Finishing of textiles	..	..	..	..	..	..	..	..	..
172	Other textiles	..	..	..	..	..	..	..	..	..
1721	Made-up textile articles, except apparel	..	..	..	..	..	..	..	..	..
1722	Carpets and rugs	..	..	..	..	..	..	..	..	..
1723	Cordage, rope, twine and netting	..	..	..	..	..	..	..	..	..
1729	Other textiles, nec	..	..	..	..	..	..	..	..	..
173	Knitted and crocheted fabrics and articles	..	..	..	..	..	..	..	..	..
18	**WEARING APPAREL, DRESSING & DYEING OF FUR**	..	..	..	..	..	3,931	3,895	3,739	..
181	Wearing apparel, except fur apparel	..	..	..	..	..	..	..	..	..
182	Dressing and dyeing of fur; articles of fur	..	..	..	..	..	..	..	..	..
19	**TANNING & DRESSING OF LEATHER, FOOTWEAR**	..	..	..	..	..	2,558	2,489	2,090	..
191	Tanning and dressing of leather	..	..	..	..	..	..	..	..	..
1911	Tanning and dressing of leather	..	..	..	..	..	..	..	..	..
1912	Luggage, handbags, saddlery & harness	..	..	..	..	..	..	..	..	..
192	Footwear	..	..	..	..	..	..	..	..	..
20	**WOOD AND WOOD PRODUCTS**	..	..	..	..	..	22,893	23,301	23,556	..
201	Sawmilling and planing of wood	..	..	..	..	..	..	..	..	..
202	Products of wood, cork, straw & plaiting materials	..	..	..	..	..	..	..	..	..
2021	Veneer sheets	..	..	..	..	..	..	..	..	..
2022	Builders' carpentry and joinery	..	..	..	..	..	..	..	..	..
2023	Wooden containers	..	..	..	..	..	..	..	..	..
2029	Other products of wood	..	..	..	..	..	..	..	..	..
21	**PAPER AND PAPER PRODUCTS**	..	..	..	..	..	217,728	209,887	215,705	..
2101	Pulp, paper and paperboard	..	..	..	..	..	..	..	..	..
2102	Corrugated paper, paperboard and their containers	..	..	..	..	..	..	..	..	..
2109	Other articles of pulp and paperboard	..	..	..	..	..	..	..	..	..
22	**PUBLISHING, PRINTING & REPRODUCTION**	..	..	..	..	..	19,713	20,000	20,625	..
221	Publishing	..	..	..	..	..	..	..	..	..
2211	Publishing of books & brochures	..	..	..	..	..	..	..	..	..
2212	Publishing of newspapers and periodicals	..	..	..	..	..	..	..	..	..
2213	Publishing of recorded media	..	..	..	..	..	..	..	..	..
2219	Other publishing	..	..	..	..	..	..	..	..	..
222	Printing and related service activities	..	..	..	..	..	..	..	..	..
2221	Printing	..	..	..	..	..	..	..	..	..
2222	Service activities related to printing	..	..	..	..	..	..	..	..	..
223	Reproduction of recorded media	..	..	..	..	..	..	..	..	..
23	**COKE, REFINED PETROLEUM PRODUCTS**	..	..	..	..	..	105,520	105,882	102,807	..
231	Coke oven products	..	..	..	..	..	..	..	..	..
232	Refined petroleum products	..	..	..	..	..	..	..	..	..
233	Processing of nuclear fuel	..	..	..	..	..	..	..	..	..
24	**CHEMICALS & CHEMICAL PRODUCTS**	..	..	..	..	..	783,995	767,409	797,873	..
241	Basic chemicals	..	..	..	..	..	..	..	..	..
2411	Basic chemicals, exc. fertilizers & nitrogen compounds	..	..	..	..	..	..	..	..	..
2412	Fertilizers and nitrogen compounds	..	..	..	..	..	..	..	..	..
2413	Plastics in primary forms and synthetic rubber	..	..	..	..	..	..	..	..	..
242	Other chemical products	..	..	..	..	..	..	..	..	..
2421	Pesticides and other agro-chemical products	..	..	..	..	..	..	..	..	..
2422	Paints, varnishes and similar coatings	..	..	..	..	..	..	..	..	..
2423	Pharmaceuticals, medicinal chem. & botanical prod.	..	..	..	..	..	..	..	..	..
2424	Soap and detergents, perfumes etc.	..	..	..	..	..	..	..	..	..
2429	Other chemical products, nec	..	..	..	..	..	..	..	..	..
243	Man-made fibres	..	..	..	..	..	..	..	..	..
25	**RUBBER AND PLASTICS PRODUCTS**	..	..	..	..	..	71,996	70,450	72,271	..
251	Rubber products	..	..	..	..	..	..	..	..	..
2511	Rubber tyres and tubes	..	..	..	..	..	..	..	..	..
2519	Other rubber products	..	..	..	..	..	..	..	..	..
252	Plastic products	..	..	..	..	..	..	..	..	..
26	**OTHER NON-METALLIC MINERAL PRODUCTS**	..	..	..	..	..	322,290	311,773	309,459	..
261	Glass and glass products	..	..	..	..	..	..	..	..	..
269	Non-metallic mineral products, nec	..	..	..	..	..	..	..	..	..
2691	Pottery, china and earthenware	..	..	..	..	..	..	..	..	..
2692	Refractory ceramic products	..	..	..	..	..	..	..	..	..
2693	Structural non-refractory clay & ceramic prod.	..	..	..	..	..	..	..	..	..
2694	Cement, lime and plaster	..	..	..	..	..	..	..	..	..
2695	Articles of concrete, cement and plaster	..	..	..	..	..	..	..	..	..
2696	Cutting, shaping and finishing of stone	..	..	..	..	..	..	..	..	..
2699	Other non-metallic mineral products, nec	..	..	..	..	..	..	..	..	..

ISIC Revision 3 Industry Sector		1990	1991	1992	1993	1994	1995	1996	1997	1998
27	BASIC METALS	..	..	..	..	..	842,990	805,659	846,672	..
271	Basic iron and steel	..	..	..	..	..	..	..	..	..
272	Basic precious and non-ferrous metals	..	..	..	..	..	..	..	..	..
273	Casting of metals	..	..	..	..	..	..	..	..	..
2731	Casting of iron and steel	..	..	..	..	..	..	..	..	..
2732	Casting non-ferrous metals	..	..	..	..	..	..	..	..	..
28	FABRICATED METAL PRODUCTS	..	..	..	..	..	72,092	71,849	74,504	..
281	Str. metal prod., tanks, reservoirs, steam generators	..	..	..	..	..	..	..	..	..
2811	Structural metal products	..	..	..	..	..	..	..	..	..
2812	Tanks, reservoirs and containers of metal	..	..	..	..	..	..	..	..	..
2813	Steam generators, exc. central heating hot water boilers	..	..	..	..	..	..	..	..	..
289	Other fabricated metal products	..	..	..	..	..	..	..	..	..
2891	Forging, pressing, stamping & roll-forming of metal	..	..	..	..	..	..	..	..	..
2892	Treatment and coating of metals	..	..	..	..	..	..	..	..	..
2893	Cutlery, hand tools and general hardware	..	..	..	..	..	..	..	..	..
2899	Other fabricated metal products, nec	..	..	..	..	..	..	..	..	..
29	MACHINERY AND EQUIPMENT, NEC	..	..	..	..	..	80,517	81,181	76,601	..
291	General purpose machinery	..	..	..	..	..	..	..	..	..
2911	Engines and turbines	..	..	..	..	..	..	..	..	..
2912	Pumps, compressors, taps and valves	..	..	..	..	..	..	..	..	..
2913	Bearings, gears, gearing and driving elements	..	..	..	..	..	..	..	..	..
2914	Ovens, furnaces and furnace burners	..	..	..	..	..	..	..	..	..
2915	Lifting and handling equipment	..	..	..	..	..	..	..	..	..
2919	Other general purpose machinery	..	..	..	..	..	..	..	..	..
292	Special purpose machinery	..	..	..	..	..	..	..	..	..
2921	Agricultural and forestry machinery	..	..	..	..	..	..	..	..	..
2922	Machine-tools	..	..	..	..	..	..	..	..	..
2923	Machinery for metallurgy	..	..	..	..	..	..	..	..	..
2924	Machinery for mining, quarrying and construction	..	..	..	..	..	..	..	..	..
2925	Machinery for food, beverage & tobacco processing	..	..	..	..	..	..	..	..	..
2926	Machinery for textile, apparel & leather production	..	..	..	..	..	..	..	..	..
2927	Machinery for weapons and ammunition	..	..	..	..	..	..	..	..	..
2929	Other special purpose machinery	..	..	..	..	..	..	..	..	..
293	Domestic appliances, nec	..	..	..	..	..	..	..	..	..
30	OFFICE, ACCOUNTING & COMPUTING MACHINERY	..	..	..	..	..	3,123	2,901	2,906	..
31	ELECTRICAL MACHINERY & APPARATUS, NEC	..	..	..	..	..	36,230	35,159	33,033	..
311	Electric motors, generators and transformers	..	..	..	..	..	..	..	..	..
312	Electricity distribution and control apparatus	..	..	..	..	..	..	..	..	..
313	Insulated wire and cable	..	..	..	..	..	..	..	..	..
314	Accumulators, primary cells & primary batteries	..	..	..	..	..	..	..	..	..
315	Electric lamps and lighting equipment	..	..	..	..	..	..	..	..	..
319	Other electrical equipment, nec	..	..	..	..	..	..	..	..	..
32	RADIO, TV & COMMUNICATION EQUIP. & APP.	..	..	..	..	..	11,820	12,883	12,699	..
321	Electronic valves, tubes, other electronic components	..	..	..	..	..	..	..	..	..
322	TV & radio transmitters, apparatus for line telephony	..	..	..	..	..	..	..	..	..
323	TV & radio receivers, recording apparatus	..	..	..	..	..	..	..	..	..
33	MEDICAL PRECISION &OPTICAL INSTRUMENTS	..	..	..	..	..	11,439	11,072	10,160	..
331	Medical appliances and instruments	..	..	..	..	..	..	..	..	..
3311	Medical, surgical equipment & orthopaedic app.	..	..	..	..	..	..	..	..	..
3312	Instruments & appliances for measuring, checking etc	..	..	..	..	..	..	..	..	..
3313	Industrial process control equipment	..	..	..	..	..	..	..	..	..
332	Optical instruments and photographic equipment	..	..	..	..	..	..	..	..	..
333	Watches and clocks	..	..	..	..	..	..	..	..	..
34	MOTOR VEHICLES, TRAILERS & SEMI-TRAILERS	..	..	..	..	..	94,214	96,664	96,349	..
341	Motor vehicles	..	..	..	..	..	..	..	..	..
342	Bodies (coachwork) for motor vehicles	..	..	..	..	..	..	..	..	..
343	Parts, accessories for motor vehicles & their engines	..	..	..	..	..	..	..	..	..
35	OTHER TRANSPORT EQUIPMENT	..	..	..	..	..	17,382	17,635	16,202	..
351	Building and repairing of ships and boats	..	..	..	..	..	..	..	..	..
3511	Building and repairing of ships	..	..	..	..	..	..	..	..	..
3512	Building, repairing of pleasure & sporting boats	..	..	..	..	..	..	..	..	..
352	Railway, tramway locomotives & rolling stock	..	..	..	..	..	..	..	..	..
353	Aircraft and spacecraft	..	..	..	..	..	..	..	..	..
359	Transport equipment, nec	..	..	..	..	..	..	..	..	..
3591	Motorcycles	..	..	..	..	..	..	..	..	..
3592	Bicycles and invalid carriages	..	..	..	..	..	..	..	..	..
3599	Other transport equipment, nec	..	..	..	..	..	..	..	..	..
36	FURNITURE; MANUFACTURING, NEC	..	..	..	..	..	16,207	16,471	15,334	..
361	Furniture	..	..	..	..	..	..	..	..	..
369	Manufacturing, nec	..	..	..	..	..	..	..	..	..
3691	Jewellery and related articles	..	..	..	..	..	..	..	..	..
3692	Musical instruments	..	..	..	..	..	..	..	..	..
3693	Sports goods	..	..	..	..	..	..	..	..	..
3694	Games and toys	..	..	..	..	..	..	..	..	..
3699	Other manufacturing, nec	..	..	..	..	..	..	..	..	..
37	RECYCLING	..	..	..	..	..	1,200	1,477	1,799	..
371	Recycling of metal waste and scrap	..	..	..	..	..	..	..	..	..
372	Recycling of non-metal waste and scrap	..	..	..	..	..	..	..	..	..
CERR	Non-specified industry	..	..	..	..		..	..	..	
15-37	TOTAL MANUFACTURING	..	..	..	..	..	2,981,912	2,912,081	2,973,841	..

ISIC Revision 3 Industry Sector		1990	1991	1992	1993	1994	1995	1996	1997	1998
15	FOOD PRODUCTS AND BEVERAGES	212,392	222,187	224,451	228,314	232,079	243,860	248,904	189,756	..
151	Production, processing and preserving (PPP)	38,518	38,417	41,487	41,626	43,281	46,052	45,877	34,070	..
1511	PPP of meat and meat products	9,257	8,838	10,779	11,175	11,777	12,112	11,844	9,048	..
1512	Processing and preserving of fish products	13,019	13,110	13,399	13,344	15,422	16,825	16,783	10,802	..
1513	Processing, preserving of fruit & vegetables	5,607	5,705	5,809	5,959	5,609	5,767	5,602	4,262	..
1514	Vegetable and animal oils and fats	10,633	10,763	11,499	11,149	10,473	11,348	11,648	9,958	..
152	Dairy products	18,595	18,621	19,356	20,352	20,020	20,771	22,504	15,769	..
153	Grain mill prod., starches & prepared animal feeds	18,900	19,906	20,131	20,011	21,471	21,918	22,447	16,469	..
1531	Grain mill products	2,842	2,912	2,884	3,151	3,136	3,241	3,123	906	..
1532	Starches and starch products	13,886	14,790	15,007	14,672	14,789	15,271	15,547	13,160	..
1533	Prepared animal feeds	2,172	2,204	2,240	2,188	3,547	3,406	3,777	2,403	..
154	Other food products	95,266	99,780	101,608	103,914	105,420	112,212	114,979	88,591	..
1541	Bakery products	22,724	23,727	23,748	24,308	24,817	26,480	27,747	20,742	..
1542	Sugar	16,583	17,084	17,368	15,216	14,994	15,686	15,506	13,178	..
1543	Cocoa, chocolate and sugar confectionery	6,787	7,567	7,697	8,184	7,681	7,605	7,784	5,535	..
1544	Macaroni, noodles, couscous & similar farinaceous prod.	6,324	6,940	7,040	8,056	7,639	8,538	8,773	7,589	..
1549	Other food products, nec	42,848	44,462	45,755	48,149	50,290	53,904	55,168	41,547	..
155	Beverages	41,114	45,464	41,869	42,411	41,887	42,907	43,097	34,858	..
1551	Distilling, rectifying and blending of spirits	8,444	8,185	7,878	8,340	7,464	7,507	7,781	6,610	..
1552	Wines	3,407	3,279	3,344	3,441	3,263	3,377	3,332	2,187	..
1553	Malt liquors and malt	15,799	16,939	17,395	17,648	17,001	16,311	16,211	12,541	..
1554	Soft drinks; production of mineral waters	13,464	17,061	13,252	12,982	14,159	15,713	15,773	13,519	..
16	TOBACCO PRODUCTS	3,544	3,736	3,763	3,681	2,398	2,375	2,340	932	..
17	TEXTILES	102,824	104,833	101,785	94,038	95,394	90,483	89,689	66,140	..
171	Spinning, weaving and finishing of textiles	92,143	93,806	91,015	83,638	81,411	76,200	75,291	57,440	..
1711	Preparation and spinning of textile fibres	33,181	34,099	33,162	28,721	27,478	24,686	22,840	12,914	..
1712	Finishing of textiles	58,962	59,707	57,853	54,917	53,933	51,513	52,451	44,527	..
172	Other textiles	9,624	9,956	9,601	9,324	8,666	9,374	9,511	6,051	..
1721	Made-up textile articles, except apparel	1,523	1,759	1,759	1,928	1,553	1,729	1,851	807	..
1722	Carpets and rugs	1,364	1,368	1,246	1,303	1,114	1,180	1,212	790	..
1723	Cordage, rope, twine and netting	590	607	515	464	401	435	408	255	..
1729	Other textiles, nec	6,148	6,223	6,081	5,629	5,598	6,030	6,040	4,198	..
173	Knitted and crocheted fabrics and articles	1,057	1,071	1,169	1,076	5,317	4,909	4,888	2,649	..
18	WEARING APPAREL, DRESSING & DYEING OF FUR	4,675	5,002	4,931	4,723	5,290	4,944	4,552	2,952	..
181	Wearing apparel, except fur apparel	4,597	4,925	4,867	4,660	5,237	4,929	4,494	2,915	..
182	Dressing and dyeing of fur; articles of fur	78	77	64	63	54	15	57	37	..
19	TANNING & DRESSING OF LEATHER, FOOTWEAR	6,029	5,634	5,520	5,281	4,507	4,654	7,891	5,688	..
191	Tanning and dressing of leather	4,381	3,921	3,748	3,587	3,016	3,172	6,417	4,832	..
1911	Tanning and dressing of leather	4,273	3,810	3,589	3,470	2,918	3,065	6,314	4,806	..
1912	Luggage, handbags, saddlery & harness	108	112	159	117	98	107	103	26	..
192	Footwear	1,648	1,713	1,772	1,694	1,491	1,482	1,474	856	..
20	WOOD AND WOOD PRODUCTS	11,722	12,382	12,770	13,129	13,276	13,390	13,822	6,544	..
201	Sawmilling and planing of wood	2,609	2,487	2,521	2,351	2,380	2,487	2,649	1,178	..
202	Products of wood, cork, straw & plaiting materials	9,113	9,895	10,250	10,778	10,896	10,903	11,173	5,366	..
2021	Veneer sheets	7,807	8,410	8,689	9,224	9,151	8,791	9,085	4,452	..
2022	Builders' carpentry and joinery	822	972	1,027	1,061	1,189	1,648	1,604	673	..
2023	Wooden containers	137	134	136	128	157	121	137	49	..
2029	Other products of wood	347	380	398	364	399	343	347	192	..
21	PAPER AND PAPER PRODUCTS	431,976	435,142	437,019	443,887	455,532	478,173	491,208	361,288	..
2101	Pulp, paper and paperboard	411,352	414,035	415,393	422,820	434,645	456,958	468,939	347,243	..
2102	Corrugated paper, paperboard and their containers	11,699	12,942	13,106	12,733	12,921	13,394	13,231	8,849	..
2109	Other articles of pulp and paperboard	8,926	8,164	8,520	8,335	7,965	7,821	9,038	5,195	..
22	PUBLISHING, PRINTING & REPRODUCTION	20,429	21,618	23,080	23,919	25,940	29,001	29,259	17,028	..
221	Publishing	3,322	3,002	3,252	3,549	3,772	4,533	4,233	1,446	..
2211	Publishing of books & brochures	379	403	446	380	474	499	668	198	..
2212	Publishing of newspapers and periodicals	2,944	2,599	2,806	3,169	3,298	4,034	3,565	1,249	..
2213	Publishing of recorded media	..	..	..	..	..	..	..	..	..
2219	Other publishing	..	..	..	..	..	..	..	..	..
222	Printing and related service activities	17,068	18,568	19,704	20,251	21,046	23,430	24,011	15,168	..
2221	Printing	15,955	17,383	18,505	18,890	19,615	21,899	22,281	14,764	..
2222	Service activities related to printing	1,112	1,185	1,199	1,361	1,431	1,531	1,729	404	..
223	Reproduction of recorded media	39	48	124	119	1,122	1,039	1,015	414	..
23	COKE, REFINED PETROLEUM PRODUCTS	930,636	972,074	998,444	1,032,960	1,046,567	1,081,863	1,081,352	1,077,841	..
231	Coke oven products	589,608	596,278	573,851	568,389	547,514	565,685	555,432	543,926	..
232	Refined petroleum products	339,696	374,491	423,055	463,037	497,443	514,470	524,058	533,167	..
233	Processing of nuclear fuel	1,333	1,306	1,537	1,533	1,609	1,707	1,861	748	..
24	CHEMICALS & CHEMICAL PRODUCTS	1,154,242	1,187,086	1,193,729	1,195,939	1,121,127	1,170,142	1,190,216	988,072	..
241	Basic chemicals	950,961	980,646	971,678	974,941	889,103	930,489	956,902	801,279	..
2411	Basic chemicals, exc. fertilizers & nitrogen compounds	584,661	605,683	591,520	609,031	539,050	567,979	641,351	534,451	..
2412	Fertilizers and nitrogen compounds	45,795	47,589	47,115	44,197	41,915	40,594	40,407	34,655	..
2413	Plastics in primary forms and synthetic rubber	320,505	327,374	333,043	321,712	308,138	321,916	275,144	232,174	..
242	Other chemical products	109,881	113,701	119,655	117,082	132,451	140,936	140,169	108,884	..
2421	Pesticides and other agro-chemical products	4,312	5,050	4,110	5,265	6,130	6,285	6,528	4,674	..
2422	Paints, varnishes and similar coatings	3,328	3,615	3,557	3,533	3,491	3,734	4,044	2,082	..
2423	Pharmaceuticals, medicinal chem. & botanical prod.	30,949	31,509	32,809	33,876	33,113	34,181	34,853	25,613	..
2424	Soap and detergents, perfumes etc.	12,568	13,345	12,454	12,721	14,599	14,846	14,770	11,084	..
2429	Other chemical products, nec	58,723	60,182	66,724	61,688	75,118	81,889	79,974	65,430	..
243	Man-made fibres	93,400	92,739	102,396	103,916	99,574	98,716	93,145	77,909	..
25	RUBBER AND PLASTICS PRODUCTS	100,323	105,396	105,127	103,051	113,456	122,193	121,047	65,707	..
251	Rubber products	32,376	33,056	33,120	31,332	30,544	32,785	34,708	20,725	..
2511	Rubber tyres and tubes	17,464	17,899	18,026	16,186	16,604	18,127	18,665	12,059	..
2519	Other rubber products	14,912	15,158	15,094	15,146	13,939	14,658	16,043	8,666	..
252	Plastic products	67,947	72,340	72,006	71,719	82,912	89,408	86,339	44,981	..
26	OTHER NON-METALLIC MINERAL PRODUCTS	554,920	571,644	590,779	563,066	588,427	572,260	576,164	507,291	..
261	Glass and glass products	79,927	75,706	75,757	72,089	76,889	73,106	71,915	53,198	..
269	Non-metallic mineral products, nec	474,993	495,938	515,022	490,977	511,539	499,154	504,250	454,093	..
2691	Pottery, china and earthenware	14,848	14,315	14,902	14,196	14,160	14,470	14,760	10,808	..
2692	Refractory ceramic products	26,397	26,913	24,728	22,507	21,318	21,615	22,055	17,608	..
2693	Structural non-refractory clay & ceramic prod.	12,830	12,211	12,270	11,907	11,890	11,698	11,384	9,884	..
2694	Cement, lime and plaster	332,384	354,870	371,424	350,895	376,665	360,022	363,466	331,682	..
2695	Articles of concrete, cement and plaster	34,195	35,623	36,959	37,369	36,944	38,374	39,769	34,209	..
2696	Cutting, shaping and finishing of stone	2,662	2,737	3,053	3,021	2,634	2,678	2,673	2,397	..
2699	Other non-metallic mineral products, nec	51,677	49,269	51,687	51,082	47,929	50,297	50,142	47,504	..

ISIS Energy Data Programme (IEA/OECD)

ISIC Revision 3 Industry Sector		1990	1991	1992	1993	1994	1995	1996	1997	1998
27	**BASIC METALS**	3,746,866	3,768,410	3,563,671	3,554,140	3,483,349	3,574,322	3,568,842	3,396,090	..
271	Basic iron and steel	3,554,200	3,569,820	3,367,133	3,361,604	3,343,510	3,428,661	3,423,240	3,307,179	..
272	Basic precious and non-ferrous metals	107,408	108,822	110,449	106,284	107,269	111,294	111,246	69,061	..
273	Casting of metals	85,258	89,768	86,089	86,253	32,570	34,366	34,356	19,849	..
2731	Casting of iron and steel	16,387	16,108	14,819	13,508	22,979	24,575	24,847	12,819	..
2732	Casting non-ferrous metals	68,871	73,660	71,270	72,745	9,592	9,791	9,509	7,030	..
28	**FABRICATED METAL PRODUCTS**	173,184	182,353	177,118	170,048	90,892	95,267	98,589	54,630	..
281	Str. metal prod., tanks, reservoirs, steam generators	51,883	56,320	56,221	54,599	24,835	25,903	27,616	14,746	..
2811	Structural metal products	2,194	2,614	2,356	2,487	18,371	18,632	20,124	10,675	..
2812	Tanks, reservoirs and containers of metal	2,197	2,914	3,079	2,937	2,545	2,527	2,633	1,400	..
2813	Steam generators, exc. central heating hot water boilers	47,492	50,791	50,786	49,175	3,918	4,743	4,859	2,670	..
289	Other fabricated metal products	121,302	126,033	120,897	115,449	66,058	69,364	70,973	39,884	..
2891	Forging, pressing, stamping & roll-forming of metal	13,557	13,914	14,117	13,937	22,609	24,858	26,065	14,830	..
2892	Treatment and coating of metals	4,102	4,078	3,764	3,643	17,113	18,306	18,369	11,501	..
2893	Cutlery, hand tools and general hardware	22,993	24,211	23,971	22,386	3,671	3,540	3,674	1,303	..
2899	Other fabricated metal products, nec	80,651	83,831	79,045	75,483	22,664	22,659	22,865	12,250	..
29	**MACHINERY AND EQUIPMENT, NEC**	109,143	118,882	107,523	102,173	78,064	80,463	84,194	40,013	..
291	General purpose machinery	62,001	68,659	60,931	58,080	41,940	42,226	43,494	23,595	..
2911	Engines and turbines	6,263	5,632	3,683	6,465	8,684	5,260	5,512	3,292	..
2912	Pumps, compressors, taps and valves	12,485	12,428	11,924	12,396	6,500	6,523	5,411	3,680	..
2913	Bearings, gears, gearing and driving elements	554	274	540	520	12,151	13,307	13,542	6,237	..
2914	Ovens, furnaces and furnace burners	4,497	4,809	3,714	3,794	446	520	520	240	..
2915	Lifting and handling equipment	7,946	13,683	12,040	7,760	3,537	3,095	3,320	1,603	..
2919	Other general purpose machinery	30,256	31,834	29,031	27,144	10,623	13,521	15,189	8,543	..
292	Special purpose machinery	37,608	39,998	36,809	34,652	26,596	28,704	30,999	12,442	..
2921	Agricultural and forestry machinery	8,846	9,572	8,063	7,323	1,754	1,915	1,901	1,481	..
2922	Machine-tools	704	704	613	619	6,768	8,055	8,076	3,138	..
2923	Machinery for metallurgy	7,238	7,542	6,541	6,190	608	106	278	46	..
2924	Machinery for mining, quarrying and construction	270	313	292	299	6,425	6,464	6,880	2,826	..
2925	Machinery for food, beverage & tobacco processing	2,151	2,226	2,304	1,872	301	309	308	148	..
2926	Machinery for textile, apparel & leather production	628	684	818	797	1,434	1,336	1,354	458	..
2927	Machinery for weapons and ammunition	8,236	8,736	8,394	8,113	651	651	680	299	..
2929	Other special purpose machinery	9,534	10,222	9,785	9,439	8,655	9,868	11,523	4,046	..
293	Domestic appliances, nec	9,534	10,225	9,783	9,442	9,528	9,532	9,701	3,976	..
30	**OFFICE, ACCOUNTING & COMPUTING MACHINERY**	15,164	16,682	16,433	16,253	16,241	16,313	15,810	5,080	..
31	**ELECTRICAL MACHINERY & APPARATUS, NEC**	62,565	66,445	64,533	66,043	54,558	56,951	55,480	22,160	..
311	Electric motors, generators and transformers	6,677	7,531	6,987	6,604	6,006	6,177	6,089	2,492	..
312	Electricity distribution and control apparatus	6,902	7,645	7,632	8,399	7,459	7,201	7,557	3,299	..
313	Insulated wire and cable	22,207	23,008	21,895	21,798	12,289	12,513	13,089	5,985	..
314	Accumulators, primary cells & primary batteries	4,127	4,378	4,440	4,283	4,861	5,362	5,223	2,337	..
315	Electric lamps and lighting equipment	3,506	3,745	3,619	3,688	3,733	3,904	4,113	1,665	..
319	Other electrical equipment, nec	19,147	20,138	19,960	21,270	20,210	21,795	19,409	6,382	..
32	**RADIO, TV & COMMUNICATION EQUIP. & APP.**	90,689	100,557	97,964	101,461	109,742	117,614	129,102	55,660	..
321	Electronic valves, tubes, other electronic components	44,894	51,098	51,591	53,668	72,835	77,316	82,497	33,534	..
322	TV & radio transmitters, apparatus for line telephony	4,214	3,704	3,630	4,736	4,275	4,309	4,640	1,115	..
323	TV & radio receivers, recording apparatus	41,581	45,755	42,743	43,057	32,632	35,988	41,965	21,012	..
33	**MEDICAL PRECISION &OPTICAL INSTRUMENTS**	14,178	15,292	14,376	13,992	13,172	12,916	12,505	5,388	..
331	Medical appliances and instruments	7,528	8,267	7,690	7,421	7,003	7,236	7,077	3,678	..
3311	Medical, surgical equipment & orthopaedic app.	3,362	3,641	3,293	3,214	3,271	3,456	3,593	2,117	..
3312	Instruments & appliances for measuring, checking etc	454	453	424	394	3,334	3,419	3,066	1,473	..
3313	Industrial process control equipment	3,712	4,173	3,973	3,812	399	361	418	88	..
332	Optical instruments and photographic equipment	3,710	4,173	3,972	3,810	3,899	3,491	3,329	1,124	..
333	Watches and clocks	2,940	2,851	2,715	2,761	2,270	2,189	2,099	586	..
34	**MOTOR VEHICLES, TRAILERS & SEMI-TRAILERS**	159,804	172,237	170,940	163,445	165,048	171,511	166,781	92,563	..
341	Motor vehicles	61,742	69,391	68,685	62,693	61,933	61,307	63,441	39,303	..
342	Bodies (coachwork) for motor vehicles	9,716	10,543	10,553	9,655	9,264	9,223	9,504	5,789	..
343	Parts, accessories for motor vehicles & their engines	88,347	92,303	91,702	91,097	93,851	100,982	93,836	47,470	..
35	**OTHER TRANSPORT EQUIPMENT**	15,411	17,120	20,310	17,190	9,678	9,604	10,329	6,507	..
351	Building and repairing of ships and boats	738	1,205	1,770	820	5,536	5,416	5,713	3,844	..
3511	Building and repairing of ships	146	542	185	129	5,416	5,280	5,557	3,760	..
3512	Building, repairing of pleasure & sporting boats	592	662	1,585	692	120	135	157	83	..
352	Railway, tramway locomotives & rolling stock	592	662	1,584	691	615	720	625	325	..
353	Aircraft and spacecraft	2,288	2,295	2,439	2,339	2,404	2,449	3,056	1,802	..
359	Transport equipment, nec	11,793	12,958	14,517	13,340	1,123	1,019	934	536	..
3591	Motorcycles	..	..	..	..		..	..	..	..
3592	Bicycles and invalid carriages	107	128	114	96	974	909	820	483	..
3599	Other transport equipment, nec	11,686	12,829	14,403	13,244	149	110	114	53	..
36	**FURNITURE; MANUFACTURING, NEC**	12,458	13,735	15,008	14,086	14,137	14,613	16,211	8,578	..
361	Furniture	5,309	5,536	5,380	5,385	5,476	5,770	6,487	2,925	..
369	Manufacturing, nec	7,149	8,200	9,628	8,701	8,661	8,843	9,724	5,653	..
3691	Jewellery and related articles	1,112	1,065	980	1,328	244	272	242	59	..
3692	Musical instruments	709	818	822	850	1,495	1,098	1,073	522	..
3693	Sports goods	354	355	469	310	798	815	913	349	..
3694	Games and toys	3,945	4,801	6,500	5,126	256	294	291	58	..
3699	Other manufacturing, nec	1,029	1,161	857	1,088	5,868	6,364	7,205	4,664	..
37	**RECYCLING**	1,029	1,161	856	1,088	989	1,126	1,066	627	..
371	Recycling of metal waste and scrap	1,029	1,161	856	1,088	989	1,126	1,066	627	..
372	Recycling of non-metal waste and scrap	..	..	..	..		..	..	..	..
CERR	**Non-specified industry**	..	..	..	..		..	..	..	..
15-37	**TOTAL MANUFACTURING**	7,934,203	8,119,610	7,950,131	7,931,908	7,739,864	7,964,041	8,015,351	6,976,536	..

LUXEMBOURG

ISIC Revision 3 Industry Sector		1990	1991	1992	1993	1994	1995	1996	1997	1998
15	**FOOD PRODUCTS AND BEVERAGES**	61,320	58,532	54,735	57,373	50,538	37,302	34,980	31,068	..
151	Production, processing and preserving (PPP)	..	..	..	..	..	..	..	..	..
1511	PPP of meat and meat products	..	..	..	..	..	..	..	..	..
1512	Processing and preserving of fish products	..	..	..	..	..	..	..	..	..
1513	Processing, preserving of fruit & vegetables	..	..	..	..	..	..	..	..	..
1514	Vegetable and animal oils and fats	..	..	..	..	..	..	..	..	..
152	Dairy products	..	..	..	..	..	..	..	..	..
153	Grain mill prod., starches & prepared animal feeds	..	..	..	..	..	..	..	..	..
1531	Grain mill products	..	..	..	..	..	..	..	..	..
1532	Starches and starch products	..	..	..	..	..	..	..	..	..
1533	Prepared animal feeds	..	..	..	..	..	..	..	..	..
154	Other food products	..	..	..	..	..	..	..	..	..
1541	Bakery products	..	..	..	..	..	..	..	..	..
1542	Sugar	..	..	..	..	..	..	..	..	..
1543	Cocoa, chocolate and sugar confectionery	..	..	..	..	..	..	..	..	..
1544	Macaroni, noodles, couscous & similar farinaceous prod.	..	..	..	..	..	..	..	..	..
1549	Other food products, nec	..	..	..	..	..	..	..	..	..
155	Beverages	..	..	..	..	..	..	..	..	..
1551	Distilling, rectifying and blending of spirits	..	..	..	..	..	..	..	..	..
1552	Wines	..	..	..	..	..	..	..	..	..
1553	Malt liquors and malt	..	..	..	..	..	..	..	..	..
1554	Soft drinks; production of mineral waters	..	..	..	..	..	..	..	..	..
16	**TOBACCO PRODUCTS**	..	..	..	..	..	..	..	..	..
17	**TEXTILES**	..	..	..	..	..	..	..	..	..
171	Spinning, weaving and finishing of textiles	..	..	..	..	..	..	..	..	..
1711	Preparation and spinning of textile fibres	..	..	..	..	..	..	..	..	..
1712	Finishing of textiles	..	..	..	..	..	..	..	..	..
172	Other textiles	..	..	..	..	..	..	..	..	..
1721	Made-up textile articles, except apparel	..	..	..	..	..	..	..	..	..
1722	Carpets and rugs	..	..	..	..	..	..	..	..	..
1723	Cordage, rope, twine and netting	..	..	..	..	..	..	..	..	..
1729	Other textiles, nec	..	..	..	..	..	..	..	..	..
173	Knitted and crocheted fabrics and articles	..	..	..	..	..	..	..	..	..
18	**WEARING APPAREL, DRESSING & DYEING OF FUR**	..	..	..	..	..	..	..	..	..
181	Wearing apparel, except fur apparel	..	..	..	..	..	..	..	..	..
182	Dressing and dyeing of fur; articles of fur	..	..	..	..	..	..	..	..	..
19	**TANNING & DRESSING OF LEATHER, FOOTWEAR**	..	..	..	..	..	..	..	..	..
191	Tanning and dressing of leather	..	..	..	..	..	..	..	..	..
1911	Tanning and dressing of leather	..	..	..	..	..	..	..	..	..
1912	Luggage, handbags, saddlery & harness	..	..	..	..	..	..	..	..	..
192	Footwear	..	..	..	..	..	..	..	..	..
20	**WOOD AND WOOD PRODUCTS**	..	..	..	..	..	..	..	..	..
201	Sawmilling and planing of wood	..	..	..	..	..	..	..	..	..
202	Products of wood, cork, straw & plaiting materials	..	..	..	..	..	..	..	..	..
2021	Veneer sheets	..	..	..	..	..	..	..	..	..
2022	Builders' carpentry and joinery	..	..	..	..	..	..	..	..	..
2023	Wooden containers	..	..	..	..	..	..	..	..	..
2029	Other products of wood	..	..	..	..	..	..	..	..	..
21	**PAPER AND PAPER PRODUCTS**	..	..	..	..	..	..	..	..	..
2101	Pulp, paper and paperboard	..	..	..	..	..	..	..	..	..
2102	Corrugated paper, paperboard and their containers	..	..	..	..	..	..	..	..	..
2109	Other articles of pulp and paperboard	..	..	..	..	..	..	..	..	..
22	**PUBLISHING, PRINTING & REPRODUCTION**	..	..	..	..	..	..	..	..	..
221	Publishing	..	..	..	..	..	..	..	..	..
2211	Publishing of books & brochures	..	..	..	..	..	..	..	..	..
2212	Publishing of newspapers and periodicals	..	..	..	..	..	..	..	..	..
2213	Publishing of recorded media	..	..	..	..	..	..	..	..	..
2219	Other publishing	..	..	..	..	..	..	..	..	..
222	Printing and related service activities	..	..	..	..	..	..	..	..	..
2221	Printing	..	..	..	..	..	..	..	..	..
2222	Service activities related to printing	..	..	..	..	..	..	..	..	..
223	Reproduction of recorded media	..	..	..	..	..	..	..	..	..
23	**COKE, REFINED PETROLEUM PRODUCTS**	..	..	..	..	..	..	..	..	..
231	Coke oven products	..	..	..	..	..	..	..	..	..
232	Refined petroleum products	..	..	..	..	..	..	..	..	..
233	Processing of nuclear fuel	..	..	..	..	..	..	..	..	..
24	**CHEMICALS & CHEMICAL PRODUCTS**	2,317	2,521	2,736	2,737	2,903	2,968	2,885	3,046	..
241	Basic chemicals	..	..	..	..	..	..	..	..	..
2411	Basic chemicals, exc. fertilizers & nitrogen compounds	..	..	..	..	..	..	..	..	..
2412	Fertilizers and nitrogen compounds	..	..	..	..	..	..	..	..	..
2413	Plastics in primary forms and synthetic rubber	..	..	..	..	..	..	..	..	..
242	Other chemical products	..	..	..	..	..	..	..	..	..
2421	Pesticides and other agro-chemical products	..	..	..	..	..	..	..	..	..
2422	Paints, varnishes and similar coatings	..	..	..	..	..	..	..	..	..
2423	Pharmaceuticals, medicinal chem. & botanical prod.	..	..	..	..	..	..	..	..	..
2424	Soap and detergents, perfumes etc.	..	..	..	..	..	..	..	..	..
2429	Other chemical products, nec	..	..	..	..	..	..	..	..	..
243	Man-made fibres	..	..	..	..	..	..	..	..	..
25	**RUBBER AND PLASTICS PRODUCTS**	..	..	..	..	..	..	..	..	..
251	Rubber products	..	..	..	..	..	..	..	..	..
2511	Rubber tyres and tubes	..	..	..	..	..	..	..	..	..
2519	Other rubber products	..	..	..	..	..	..	..	..	..
252	Plastic products	..	..	..	..	..	..	..	..	..
26	**OTHER NON-METALLIC MINERAL PRODUCTS**	5,074	5,334	5,802	5,745	6,117	5,843	5,730	6,106	..
261	Glass and glass products	..	..	..	..	..	..	..	..	..
269	Non-metallic mineral products, nec	..	..	..	..	..	..	..	..	..
2691	Pottery, china and earthenware	..	..	..	..	..	..	..	..	..
2692	Refractory ceramic products	..	..	..	..	..	..	..	..	..
2693	Structural non-refractory clay & ceramic prod.	..	..	..	..	..	..	..	..	..
2694	Cement, lime and plaster	..	..	..	..	..	..	..	..	..
2695	Articles of concrete, cement and plaster	..	..	..	..	..	..	..	..	..
2696	Cutting, shaping and finishing of stone	..	..	..	..	..	..	..	..	..
2699	Other non-metallic mineral products, nec	..	..	..	..	..	..	..	..	..

ISIC Revision 3 Industry Sector		1990	1991	1992	1993	1994	1995	1996	1997	1998
27	BASIC METALS	7,858	7,297	7,744	7,254	7,963	7,766	7,699	7,517	..
271	Basic iron and steel	..	..	..	..	..	..	..	..	..
272	Basic precious and non-ferrous metals	..	..	..	..	..	..	..	..	..
273	Casting of metals	..	..	..	..	..	..	..	..	..
2731	Casting of iron and steel	..	..	..	..	..	..	..	..	..
2732	Casting non-ferrous metals	..	..	..	..	..	..	..	..	..
28	FABRICATED METAL PRODUCTS	286	281	271	272	268	276	257	263	..
281	Str. metal prod., tanks, reservoirs, steam generators	..	..	..	..	..	..	..	..	..
2811	Structural metal products	..	..	..	..	..	..	..	..	..
2812	Tanks, reservoirs and containers of metal	..	..	..	..	..	..	..	..	..
2813	Steam generators, exc. central heating hot water boilers	..	..	..	..	..	..	..	..	..
289	Other fabricated metal products	..	..	..	..	..	..	..	..	..
2891	Forging, pressing, stamping & roll-forming of metal	..	..	..	..	..	..	..	..	..
2892	Treatment and coating of metals	..	..	..	..	..	..	..	..	..
2893	Cutlery, hand tools and general hardware	..	..	..	..	..	..	..	..	..
2899	Other fabricated metal products, nec	..	..	..	..	..	..	..	..	..
29	MACHINERY AND EQUIPMENT, NEC	..	..	..	..	..	..	..	..	..
291	General purpose machinery	..	..	..	..	..	..	..	..	..
2911	Engines and turbines	..	..	..	..	..	..	..	..	..
2912	Pumps, compressors, taps and valves	..	..	..	..	..	..	..	..	..
2913	Bearings, gears, gearing and driving elements	..	..	..	..	..	..	..	..	..
2914	Ovens, furnaces and furnace burners	..	..	..	..	..	..	..	..	..
2915	Lifting and handling equipment	..	..	..	..	..	..	..	..	..
2919	Other general purpose machinery	..	..	..	..	..	..	..	..	..
292	Special purpose machinery	..	..	..	..	..	..	..	..	..
2921	Agricultural and forestry machinery	..	..	..	..	..	..	..	..	..
2922	Machine-tools	..	..	..	..	..	..	..	..	..
2923	Machinery for metallurgy	..	..	..	..	..	..	..	..	..
2924	Machinery for mining, quarrying and construction	..	..	..	..	..	..	..	..	..
2925	Machinery for food, beverage & tobacco processing	..	..	..	..	..	..	..	..	..
2926	Machinery for textile, apparel & leather production	..	..	..	..	..	..	..	..	..
2927	Machinery for weapons and ammunition	..	..	..	..	..	..	..	..	..
2929	Other special purpose machinery	..	..	..	..	..	..	..	..	..
293	Domestic appliances, nec	..	..	..	..	..	..	..	..	..
30	OFFICE, ACCOUNTING & COMPUTING MACHINERY	..	..	..	..	..	..	..	..	..
31	ELECTRICAL MACHINERY & APPARATUS, NEC	..	..	..	..	..	..	..	..	..
311	Electric motors, generators and transformers	..	..	..	..	..	..	..	..	..
312	Electricity distribution and control apparatus	..	..	..	..	..	..	..	..	..
313	Insulated wire and cable	..	..	..	..	..	..	..	..	..
314	Accumulators, primary cells & primary batteries	..	..	..	..	..	..	..	..	..
315	Electric lamps and lighting equipment	..	..	..	..	..	..	..	..	..
319	Other electrical equipment, nec	..	..	..	..	..	..	..	..	..
32	RADIO, TV & COMMUNICATION EQUIP. & APP.	..	..	..	..	..	..	..	..	..
321	Electronic valves, tubes, other electronic components	..	..	..	..	..	..	..	..	..
322	TV & radio transmitters, apparatus for line telephony	..	..	..	..	..	..	..	..	..
323	TV & radio receivers, recording apparatus	..	..	..	..	..	..	..	..	..
33	MEDICAL PRECISION &OPTICAL INSTRUMENTS	..	..	..	..	..	..	..	..	..
331	Medical appliances and instruments	..	..	..	..	..	..	..	..	..
3311	Medical, surgical equipment & orthopaedic app.	..	..	..	..	..	..	..	..	..
3312	Instruments & appliances for measuring, checking etc	..	..	..	..	..	..	..	..	..
3313	Industrial process control equipment	..	..	..	..	..	..	..	..	..
332	Optical instruments and photographic equipment	..	..	..	..	..	..	..	..	..
333	Watches and clocks	..	..	..	..	..	..	..	..	..
34	MOTOR VEHICLES, TRAILERS & SEMI-TRAILERS	..	..	..	..	..	..	..	..	..
341	Motor vehicles	..	..	..	..	..	..	..	..	..
342	Bodies (coachwork) for motor vehicles	..	..	..	..	..	..	..	..	..
343	Parts, accessories for motor vehicles & their engines	..	..	..	..	..	..	..	..	..
35	OTHER TRANSPORT EQUIPMENT	..	..	..	..	..	..	..	..	..
351	Building and repairing of ships and boats	..	..	..	..	..	..	..	..	..
3511	Building and repairing of ships	..	..	..	..	..	..	..	..	..
3512	Building, repairing of pleasure & sporting boats	..	..	..	..	..	..	..	..	..
352	Railway, tramway locomotives & rolling stock	..	..	..	..	..	..	..	..	..
353	Aircraft and spacecraft	..	..	..	..	..	..	..	..	..
359	Transport equipment, nec	..	..	..	..	..	..	..	..	..
3591	Motorcycles	..	..	..	..	..	..	..	..	..
3592	Bicycles and invalid carriages	..	..	..	..	..	..	..	..	..
3599	Other transport equipment, nec	..	..	..	..	..	..	..	..	..
36	FURNITURE; MANUFACTURING, NEC	..	..	..	..	..	..	..	..	..
361	Furniture	..	..	..	..	..	..	..	..	..
369	Manufacturing, nec	..	..	..	..	..	..	..	..	..
3691	Jewellery and related articles	..	..	..	..	..	..	..	..	..
3692	Musical instruments	..	..	..	..	..	..	..	..	..
3693	Sports goods	..	..	..	..	..	..	..	..	..
3694	Games and toys	..	..	..	..	..	..	..	..	..
3699	Other manufacturing, nec	..	..	..	..	..	..	..	..	..
37	RECYCLING	..	..	..	..	..	..	..	..	..
371	Recycling of metal waste and scrap	..	..	..	..	..	..	..	..	..
372	Recycling of non-metal waste and scrap	..	..	..	..	..	..	..	..	..
CERR	Non-specified industry	..	..	..	..	..	..	..	..	..
15-37	TOTAL MANUFACTURING	76,855	73,965	71,288	73,381	67,789	54,155	51,551	48,000	..

ISIC Revision 3 Industry Sector	1990	1991	1992	1993	1994	1995	1996	1997	1998
15 FOOD PRODUCTS AND BEVERAGES	..	..	..	87,580	83,520	85,070	..	..	..
151 Production, processing and preserving (PPP)	..	..	..	..	..	..	..	..	..
1511 PPP of meat and meat products	..	..	..	..	..	..	..	..	..
1512 Processing and preserving of fish products	..	..	..	..	..	..	..	..	..
1513 Processing, preserving of fruit & vegetables	..	..	..	..	..	..	..	..	..
1514 Vegetable and animal oils and fats	..	..	..	..	..	..	..	..	..
152 Dairy products	..	..	..	..	..	..	..	..	..
153 Grain mill prod., starches & prepared animal feeds	..	..	..	..	..	..	..	..	..
1531 Grain mill products	..	..	..	..	..	..	..	..	..
1532 Starches and starch products	..	..	..	..	..	..	..	..	..
1533 Prepared animal feeds	..	..	..	..	..	..	..	..	..
154 Other food products	..	..	..	..	..	..	..	..	..
1541 Bakery products	..	..	..	..	..	..	..	..	..
1542 Sugar	..	..	..	..	..	..	..	..	..
1543 Cocoa, chocolate and sugar confectionery	..	..	..	..	..	..	..	..	..
1544 Macaroni, noodles, couscous & similar farinaceous prod.	..	..	..	..	..	..	..	..	..
1549 Other food products, nec	..	..	..	..	..	..	..	..	..
155 Beverages	..	..	..	..	..	..	..	..	..
1551 Distilling, rectifying and blending of spirits	..	..	..	..	..	..	..	..	..
1552 Wines	..	..	..	..	..	..	..	..	..
1553 Malt liquors and malt	..	..	..	..	..	..	..	..	..
1554 Soft drinks; production of mineral waters	..	..	..	..	..	..	..	..	..
16 TOBACCO PRODUCTS	..	..	..	..	..	..	..	..	..
17 TEXTILES	..	..	..	8,170	8,110	8,710	..	..	..
171 Spinning, weaving and finishing of textiles	..	..	..	..	..	..	..	..	..
1711 Preparation and spinning of textile fibres	..	..	..	..	..	..	..	..	..
1712 Finishing of textiles	..	..	..	..	..	..	..	..	..
172 Other textiles	..	..	..	..	..	..	..	..	..
1721 Made-up textile articles, except apparel	..	..	..	..	..	..	..	..	..
1722 Carpets and rugs	..	..	..	..	..	..	..	..	..
1723 Cordage, rope, twine and netting	..	..	..	..	..	..	..	..	..
1729 Other textiles, nec	..	..	..	..	..	..	..	..	..
173 Knitted and crocheted fabrics and articles	..	..	..	..	..	..	..	..	..
18 WEARING APPAREL, DRESSING & DYEING OF FUR	..	..	..	..	..	..	..	..	..
181 Wearing apparel, except fur apparel	..	..	..	..	..	..	..	..	..
182 Dressing and dyeing of fur; articles of fur	..	..	..	..	..	..	..	..	..
19 TANNING & DRESSING OF LEATHER, FOOTWEAR	..	..	..	..	..	..	..	..	..
191 Tanning and dressing of leather	..	..	..	..	..	..	..	..	..
1911 Tanning and dressing of leather	..	..	..	..	..	..	..	..	..
1912 Luggage, handbags, saddlery & harness	..	..	..	..	..	..	..	..	..
192 Footwear	..	..	..	..	..	..	..	..	..
20 WOOD AND WOOD PRODUCTS	..	..	..	..	..	..	..	..	..
201 Sawmilling and planing of wood	..	..	..	..	..	..	..	..	..
202 Products of wood, cork, straw & plaiting materials	..	..	..	..	..	..	..	..	..
2021 Veneer sheets	..	..	..	..	..	..	..	..	..
2022 Builders' carpentry and joinery	..	..	..	..	..	..	..	..	..
2023 Wooden containers	..	..	..	..	..	..	..	..	..
2029 Other products of wood	..	..	..	..	..	..	..	..	..
21 PAPER AND PAPER PRODUCTS	..	..	..	31,450	30,540	32,830	..	..	..
2101 Pulp, paper and paperboard	..	..	..	..	..	..	..	..	..
2102 Corrugated paper, paperboard and their containers	..	..	..	..	..	..	..	..	..
2109 Other articles of pulp and paperboard	..	..	..	..	..	..	..	..	..
22 PUBLISHING, PRINTING & REPRODUCTION	..	..	..	..	..	..	..	..	..
221 Publishing	..	..	..	..	..	..	..	..	..
2211 Publishing of books & brochures	..	..	..	..	..	..	..	..	..
2212 Publishing of newspapers and periodicals	..	..	..	..	..	..	..	..	..
2213 Publishing of recorded media	..	..	..	..	..	..	..	..	..
2219 Other publishing	..	..	..	..	..	..	..	..	..
222 Printing and related service activities	..	..	..	..	..	..	..	..	..
2221 Printing	..	..	..	..	..	..	..	..	..
2222 Service activities related to printing	..	..	..	..	..	..	..	..	..
223 Reproduction of recorded media	..	..	..	..	..	..	..	..	..
23 COKE, REFINED PETROLEUM PRODUCTS	..	..	..	139,360	143,170	129,870	..	..	..
231 Coke oven products	..	..	..	..	..	..	..	..	..
232 Refined petroleum products	..	..	..	139,360	143,170	129,870	..	..	..
233 Processing of nuclear fuel	..	..	..	..	..	..	..	..	..
24 CHEMICALS & CHEMICAL PRODUCTS	..	..	..	242,670	246,680	274,740	..	..	..
241 Basic chemicals	..	..	..	221,140	221,200	247,580	..	..	..
2411 Basic chemicals, exc. fertilizers & nitrogen compounds	..	..	..	183,940	188,690	215,280	..	..	..
2412 Fertilizers and nitrogen compounds	..	..	..	37,200	32,510	32,300	..	..	..
2413 Plastics in primary forms and synthetic rubber	..	..	..	..	..	..	..	..	..
242 Other chemical products	..	..	..	21,530	25,480	27,160	..	..	..
2421 Pesticides and other agro-chemical products	..	..	..	..	..	..	..	..	..
2422 Paints, varnishes and similar coatings	..	..	..	..	..	..	..	..	..
2423 Pharmaceuticals, medicinal chem. & botanical prod.	..	..	..	..	..	..	..	..	..
2424 Soap and detergents, perfumes etc.	..	..	..	..	..	..	..	..	..
2429 Other chemical products, nec	..	..	..	..	..	..	..	..	..
243 Man-made fibres	..	..	..	..	..	..	..	..	..
25 RUBBER AND PLASTICS PRODUCTS	..	..	..	13,540	13,510	13,310	..	..	..
251 Rubber products	..	..	..	..	..	..	..	..	..
2511 Rubber tyres and tubes	..	..	..	..	..	..	..	..	..
2519 Other rubber products	..	..	..	..	..	..	..	..	..
252 Plastic products	..	..	..	..	..	..	..	..	..
26 OTHER NON-METALLIC MINERAL PRODUCTS	..	..	..	38,320	37,680	33,360	..	..	..
261 Glass and glass products	..	..	..	..	..	..	..	..	..
269 Non-metallic mineral products, nec	..	..	..	..	..	..	..	..	..
2691 Pottery, china and earthenware	..	..	..	..	..	..	..	..	..
2692 Refractory ceramic products	..	..	..	..	..	..	..	..	..
2693 Structural non-refractory clay & ceramic prod.	..	..	..	..	..	..	..	..	..
2694 Cement, lime and plaster	..	..	..	..	..	..	..	..	..
2695 Articles of concrete, cement and plaster	..	..	..	..	..	..	..	..	..
2696 Cutting, shaping and finishing of stone	..	..	..	..	..	..	..	..	..
2699 Other non-metallic mineral products, nec	..	..	..	..	..	..	..	..	..

ISIC Revision 3 Industry Sector		1990	1991	1992	1993	1994	1995	1996	1997	1998
27	**BASIC METALS**	..	..	..	101,780	91,250	105,420	..	..	..
271	Basic iron and steel	..	..	..	92,060	80,630	97,600	..	..	..
272	Basic precious and non-ferrous metals	..	..	..	9,720	10,620	7,820	..	..	..
273	Casting of metals	..	..	..	..	..	..	..	..	..
2731	Casting of iron and steel	..	..	..	..	..	..	..	..	..
2732	Casting non-ferrous metals	..	..	..	..	..	..	..	..	..
28	**FABRICATED METAL PRODUCTS**	..	..	..	35,870	35,670	39,130	..	..	..
281	Str. metal prod., tanks, reservoirs, steam generators	..	..	..	..	..	..	..	..	..
2811	Structural metal products	..	..	..	..	..	..	..	..	..
2812	Tanks, reservoirs and containers of metal	..	..	..	..	..	..	..	..	..
2813	Steam generators, exc. central heating hot water boilers	..	..	..	..	..	..	..	..	..
289	Other fabricated metal products	..	..	..	..	..	..	..	..	..
2891	Forging, pressing, stamping & roll-forming of metal	..	..	..	..	..	..	..	..	..
2892	Treatment and coating of metals	..	..	..	..	..	..	..	..	..
2893	Cutlery, hand tools and general hardware	..	..	..	..	..	..	..	..	..
2899	Other fabricated metal products, nec	..	..	..	..	..	..	..	..	..
29	**MACHINERY AND EQUIPMENT, NEC**	..	..	..	..	..	..	..	..	..
291	General purpose machinery	..	..	..	..	..	..	..	..	..
2911	Engines and turbines	..	..	..	..	..	..	..	..	..
2912	Pumps, compressors, taps and valves	..	..	..	..	..	..	..	..	..
2913	Bearings, gears, gearing and driving elements	..	..	..	..	..	..	..	..	..
2914	Ovens, furnaces and furnace burners	..	..	..	..	..	..	..	..	..
2915	Lifting and handling equipment	..	..	..	..	..	..	..	..	..
2919	Other general purpose machinery	..	..	..	..	..	..	..	..	..
292	Special purpose machinery	..	..	..	..	..	..	..	..	..
2921	Agricultural and forestry machinery	..	..	..	..	..	..	..	..	..
2922	Machine-tools	..	..	..	..	..	..	..	..	..
2923	Machinery for metallurgy	..	..	..	..	..	..	..	..	..
2924	Machinery for mining, quarrying and construction	..	..	..	..	..	..	..	..	..
2925	Machinery for food, beverage & tobacco processing	..	..	..	..	..	..	..	..	..
2926	Machinery for textile, apparel & leather production	..	..	..	..	..	..	..	..	..
2927	Machinery for weapons and ammunition	..	..	..	..	..	..	..	..	..
2929	Other special purpose machinery	..	..	..	..	..	..	..	..	..
293	Domestic appliances, nec	..	..	..	..	..	..	..	..	..
30	**OFFICE, ACCOUNTING & COMPUTING MACHINERY**	..	..	..	..	..	..	..	..	..
31	**ELECTRICAL MACHINERY & APPARATUS, NEC**	..	..	..	..	..	..	..	..	..
311	Electric motors, generators and transformers	..	..	..	..	..	..	..	..	..
312	Electricity distribution and control apparatus	..	..	..	..	..	..	..	..	..
313	Insulated wire and cable	..	..	..	..	..	..	..	..	..
314	Accumulators, primary cells & primary batteries	..	..	..	..	..	..	..	..	..
315	Electric lamps and lighting equipment	..	..	..	..	..	..	..	..	..
319	Other electrical equipment, nec	..	..	..	..	..	..	..	..	..
32	**RADIO, TV & COMMUNICATION EQUIP. & APP.**	..	..	..	..	..	..	..	..	..
321	Electronic valves, tubes, other electronic components	..	..	..	..	..	..	..	..	..
322	TV & radio transmitters, apparatus for line telephony	..	..	..	..	..	..	..	..	..
323	TV & radio receivers, recording apparatus	..	..	..	..	..	..	..	..	..
33	**MEDICAL PRECISION &OPTICAL INSTRUMENTS**	..	..	..	..	..	..	..	..	..
331	Medical appliances and instruments	..	..	..	..	..	..	..	..	..
3311	Medical, surgical equipment & orthopaedic app.	..	..	..	..	..	..	..	..	..
3312	Instruments & appliances for measuring, checking etc	..	..	..	..	..	..	..	..	..
3313	Industrial process control equipment	..	..	..	..	..	..	..	..	..
332	Optical instruments and photographic equipment	..	..	..	..	..	..	..	..	..
333	Watches and clocks	..	..	..	..	..	..	..	..	..
34	**MOTOR VEHICLES, TRAILERS & SEMI-TRAILERS**	..	..	..	..	..	..	..	..	..
341	Motor vehicles	..	..	..	..	..	..	..	..	..
342	Bodies (coachwork) for motor vehicles	..	..	..	..	..	..	..	..	..
343	Parts, accessories for motor vehicles & their engines	..	..	..	..	..	..	..	..	..
35	**OTHER TRANSPORT EQUIPMENT**	..	..	..	..	..	..	..	..	..
351	Building and repairing of ships and boats	..	..	..	..	..	..	..	..	..
3511	Building and repairing of ships	..	..	..	..	..	..	..	..	..
3512	Building, repairing of pleasure & sporting boats	..	..	..	..	..	..	..	..	..
352	Railway, tramway locomotives & rolling stock	..	..	..	..	..	..	..	..	..
353	Aircraft and spacecraft	..	..	..	..	..	..	..	..	..
359	Transport equipment, nec	..	..	..	..	..	..	..	..	..
3591	Motorcycles	..	..	..	..	..	..	..	..	..
3592	Bicycles and invalid carriages	..	..	..	..	..	..	..	..	..
3599	Other transport equipment, nec	..	..	..	..	..	..	..	..	..
36	**FURNITURE; MANUFACTURING, NEC**	..	..	..	..	220	260	..	..	..
361	Furniture	..	..	..	..	..	..	..	..	..
369	Manufacturing, nec	..	..	..	..	220	260	..	..	..
3691	Jewellery and related articles	..	..	..	..	..	..	..	..	..
3692	Musical instruments	..	..	..	..	..	..	..	..	..
3693	Sports goods	..	..	..	..	..	..	..	..	..
3694	Games and toys	..	..	..	..	..	..	..	..	..
3699	Other manufacturing, nec	..	..	..	..	220	260	..	..	..
37	**RECYCLING**	..	..	..	..	..	..	..	..	..
371	Recycling of metal waste and scrap	..	..	..	..	..	..	..	..	..
372	Recycling of non-metal waste and scrap	..	..	..	..	..	..	..	..	..
CERR	**Non-specified industry**	..	..	..	..	..	..	..	..	..
15-37	**TOTAL MANUFACTURING**	..	..	..	698,740	690,350	722,700	..	..	..

ISIC Revision 3 Industry Sector	1990	1991	1992	1993	1994	1995	1996	1997	1998
15 FOOD PRODUCTS AND BEVERAGES	8,057	8,088	7,750	7,633	7,673	6,442	6,118	7,184	4,539
151 Production, processing and preserving (PPP)	..	..	..	..	..	..	..	..	2,209
1511 PPP of meat and meat products	..	..	..	..	..	..	..	..	..
1512 Processing and preserving of fish products	..	..	..	..	..	..	..	..	..
1513 Processing, preserving of fruit & vegetables	..	..	..	..	..	..	..	..	..
1514 Vegetable and animal oils and fats	..	..	..	..	..	..	..	..	..
152 Dairy products	..	..	..	..	..	..	..	..	1,716
153 Grain mill prod., starches & prepared animal feeds	..	..	..	..	..	..	..	..	..
1531 Grain mill products	..	..	..	..	..	..	..	..	..
1532 Starches and starch products	..	..	..	..	..	..	..	..	..
1533 Prepared animal feeds	..	..	..	..	..	..	..	..	..
154 Other food products	..	..	..	..	..	..	..	..	2,255
1541 Bakery products	..	..	..	..	..	..	..	..	..
1542 Sugar	..	..	..	..	..	..	..	..	..
1543 Cocoa, chocolate and sugar confectionery	..	..	..	..	..	..	..	..	..
1544 Macaroni, noodles, couscous & similar farinaceous prod.	..	..	..	..	..	..	..	..	..
1549 Other food products, nec	..	..	..	..	..	..	..	..	..
155 Beverages	..	..	..	..	..	..	..	..	..
1551 Distilling, rectifying and blending of spirits	..	..	..	..	..	..	..	..	..
1552 Wines	..	..	..	..	..	..	..	..	..
1553 Malt liquors and malt	..	..	..	..	..	..	..	..	..
1554 Soft drinks; production of mineral waters	..	..	..	..	..	..	..	..	..
16 TOBACCO PRODUCTS	..	..	..	..	..	..	..	..	..
17 TEXTILES	775	775	775	775	775	800	775	720	831
171 Spinning, weaving and finishing of textiles	..	..	..	..	..	..	..	..	..
1711 Preparation and spinning of textile fibres	..	..	..	..	..	..	..	..	..
1712 Finishing of textiles	..	..	..	..	..	..	..	..	..
172 Other textiles	..	..	..	..	..	..	..	..	..
1721 Made-up textile articles, except apparel	..	..	..	..	..	..	..	..	..
1722 Carpets and rugs	..	..	..	..	..	..	..	..	..
1723 Cordage, rope, twine and netting	..	..	..	..	..	..	..	..	..
1729 Other textiles, nec	..	..	..	..	..	..	..	..	..
173 Knitted and crocheted fabrics and articles	..	..	..	..	..	..	..	..	..
18 WEARING APPAREL, DRESSING & DYEING OF FUR	..	..	..	..	..	..	..	..	..
181 Wearing apparel, except fur apparel	..	..	..	..	..	..	..	..	..
182 Dressing and dyeing of fur; articles of fur	..	..	..	..	..	..	..	..	..
19 TANNING & DRESSING OF LEATHER, FOOTWEAR	..	..	..	..	..	..	..	..	..
191 Tanning and dressing of leather	..	..	..	..	..	..	..	..	..
1911 Tanning and dressing of leather	..	..	..	..	..	..	..	..	..
1912 Luggage, handbags, saddlery & harness	..	..	..	..	..	..	..	..	..
192 Footwear	..	..	..	..	..	..	..	..	..
20 WOOD AND WOOD PRODUCTS	6,867	6,867	6,867	6,867	6,867	7,006	6,760	6,308	3,064
201 Sawmilling and planing of wood	..	..	..	..	..	..	..	..	..
202 Products of wood, cork, straw & plaiting materials	..	..	..	..	..	..	..	..	..
2021 Veneer sheets	..	..	..	..	..	..	..	..	..
2022 Builders' carpentry and joinery	..	..	..	..	..	..	..	..	..
2023 Wooden containers	..	..	..	..	..	..	..	..	..
2029 Other products of wood	..	..	..	..	..	..	..	..	..
21 PAPER AND PAPER PRODUCTS	4,878	4,878	4,878	4,878	4,878	3,419	3,918	4,017	2,038
2101 Pulp, paper and paperboard	..	..	..	..	..	..	..	..	..
2102 Corrugated paper, paperboard and their containers	..	..	..	..	..	..	..	..	..
2109 Other articles of pulp and paperboard	..	..	..	..	..	..	..	..	..
22 PUBLISHING, PRINTING & REPRODUCTION	..	..	..	..	..	..	..	..	631
221 Publishing	..	..	..	..	..	..	..	..	..
2211 Publishing of books & brochures	..	..	..	..	..	..	..	..	..
2212 Publishing of newspapers and periodicals	..	..	..	..	..	..	..	..	..
2213 Publishing of recorded media	..	..	..	..	..	..	..	..	..
2219 Other publishing	..	..	..	..	..	..	..	..	..
222 Printing and related service activities	..	..	..	..	..	..	..	..	..
2221 Printing	..	..	..	..	..	..	..	..	..
2222 Service activities related to printing	..	..	..	..	..	..	..	..	..
223 Reproduction of recorded media	..	..	..	..	..	..	..	..	..
23 COKE, REFINED PETROLEUM PRODUCTS	43,524	39,424	43,177	42,476	40,125	30,114	25,562	19,875	20,630
231 Coke oven products	..	..	..	..	..	..	..	..	27
232 Refined petroleum products	43,524	39,424	43,177	42,476	40,125	30,114	25,562	19,875	20,603
233 Processing of nuclear fuel	..	..	..	..	..	..	..	..	..
24 CHEMICALS & CHEMICAL PRODUCTS	13,347	10,780	12,241	11,812	17,483	23,096	30,169	35,598	30,825
241 Basic chemicals	9,552	6,985	8,446	8,017	13,688	19,763	26,850	32,508	30,299
2411 Basic chemicals, exc. fertilizers & nitrogen compounds	..	..	..	..	..	..	..	32,508	30,299
2412 Fertilizers and nitrogen compounds	..	..	..	..	..	..	..	..	..
2413 Plastics in primary forms and synthetic rubber	..	..	..	..	..	..	..	..	..
242 Other chemical products	..	..	..	..	..	..	..	..	..
2421 Pesticides and other agro-chemical products	..	..	..	..	..	..	..	..	..
2422 Paints, varnishes and similar coatings	..	..	..	..	..	..	..	..	..
2423 Pharmaceuticals, medicinal chem. & botanical prod.	..	..	..	..	..	..	..	..	..
2424 Soap and detergents, perfumes etc.	..	..	..	..	..	..	..	..	..
2429 Other chemical products, nec	..	..	..	..	..	..	..	..	..
243 Man-made fibres	..	..	..	..	..	..	..	..	..
25 RUBBER AND PLASTICS PRODUCTS	..	..	..	..	..	..	..	..	924
251 Rubber products	..	..	..	..	..	..	..	..	..
2511 Rubber tyres and tubes	..	..	..	..	..	..	..	..	..
2519 Other rubber products	..	..	..	..	..	..	..	..	..
252 Plastic products	..	..	..	..	..	..	..	..	..
26 OTHER NON-METALLIC MINERAL PRODUCTS	859	859	859	859	859	868	859	663	866
261 Glass and glass products	..	..	..	..	..	..	..	..	..
269 Non-metallic mineral products, nec	..	..	..	..	..	..	..	..	..
2691 Pottery, china and earthenware	..	..	..	..	..	..	..	..	..
2692 Refractory ceramic products	..	..	..	..	..	..	..	..	..
2693 Structural non-refractory clay & ceramic prod.	..	..	..	..	..	..	..	..	..
2694 Cement, lime and plaster	..	..	..	..	..	..	..	..	..
2695 Articles of concrete, cement and plaster	..	..	..	..	..	..	..	..	..
2696 Cutting, shaping and finishing of stone	..	..	..	..	..	..	..	..	..
2699 Other non-metallic mineral products, nec	..	..	..	..	..	..	..	..	..

ISIC Revision 3 Industry Sector	1990	1991	1992	1993	1994	1995	1996	1997	1998
27 BASIC METALS	33,854	35,035	35,151	35,043	33,894	34,032	33,707	30,750	31,747
271 Basic iron and steel	14,115	15,296	15,412	15,304	14,155	15,115	14,563	13,720	15,208
272 Basic precious and non-ferrous metals	19,739	19,739	19,739	19,739	19,739	18,917	19,144	17,030	17,632
273 Casting of metals	..	..	..	..	..	..	..	..	..
2731 Casting of iron and steel	..	..	..	..	..	..	..	..	..
2732 Casting non-ferrous metals	..	..	..	..	..	..	..	..	..
28 FABRICATED METAL PRODUCTS	1,481	1,481	1,481	1,481	1,481	1,491	1,481	1,444	511
281 Str. metal prod., tanks, reservoirs, steam generators	..	..	..	..	..	..	..	..	..
2811 Structural metal products	..	..	..	..	..	..	..	..	..
2812 Tanks, reservoirs and containers of metal	..	..	..	..	..	..	..	..	..
2813 Steam generators, exc. central heating hot water boilers	..	..	..	..	..	..	..	..	..
289 Other fabricated metal products	..	..	..	..	..	..	..	..	511
2891 Forging, pressing, stamping & roll-forming of metal	..	..	..	..	..	..	..	..	..
2892 Treatment and coating of metals	..	..	..	..	..	..	..	..	..
2893 Cutlery, hand tools and general hardware	..	..	..	..	..	..	..	..	..
2899 Other fabricated metal products, nec	..	..	..	..	..	..	..	..	..
29 MACHINERY AND EQUIPMENT, NEC	..	..	..	..	..	..	..	..	..
291 General purpose machinery	..	..	..	..	..	..	..	..	..
2911 Engines and turbines	..	..	..	..	..	..	..	..	..
2912 Pumps, compressors, taps and valves	..	..	..	..	..	..	..	..	..
2913 Bearings, gears, gearing and driving elements	..	..	..	..	..	..	..	..	..
2914 Ovens, furnaces and furnace burners	..	..	..	..	..	..	..	..	..
2915 Lifting and handling equipment	..	..	..	..	..	..	..	..	..
2919 Other general purpose machinery	..	..	..	..	..	..	..	..	..
292 Special purpose machinery	..	..	..	..	..	..	..	..	..
2921 Agricultural and forestry machinery	..	..	..	..	..	..	..	..	..
2922 Machine-tools	..	..	..	..	..	..	..	..	..
2923 Machinery for metallurgy	..	..	..	..	..	..	..	..	..
2924 Machinery for mining, quarrying and construction	..	..	..	..	..	..	..	..	..
2925 Machinery for food, beverage & tobacco processing	..	..	..	..	..	..	..	..	..
2926 Machinery for textile, apparel & leather production	..	..	..	..	..	..	..	..	..
2927 Machinery for weapons and ammunition	..	..	..	..	..	..	..	..	..
2929 Other special purpose machinery	..	..	..	..	..	..	..	..	..
293 Domestic appliances, nec	..	..	..	..	..	..	..	..	..
30 OFFICE, ACCOUNTING & COMPUTING MACHINERY	..	..	..	..	..	..	..	..	..
31 ELECTRICAL MACHINERY & APPARATUS, NEC	..	..	..	..	..	..	..	..	..
311 Electric motors, generators and transformers	..	..	..	..	..	..	..	..	..
312 Electricity distribution and control apparatus	..	..	..	..	..	..	..	..	..
313 Insulated wire and cable	..	..	..	..	..	..	..	..	..
314 Accumulators, primary cells & primary batteries	..	..	..	..	..	..	..	..	..
315 Electric lamps and lighting equipment	..	..	..	..	..	..	..	..	..
319 Other electrical equipment, nec	..	..	..	..	..	..	..	..	..
32 RADIO, TV & COMMUNICATION EQUIP. & APP.	..	..	..	..	..	..	..	..	..
321 Electronic valves, tubes, other electronic components	..	..	..	..	..	..	..	..	..
322 TV & radio transmitters, apparatus for line telephony	..	..	..	..	..	..	..	..	..
323 TV & radio receivers, recording apparatus	..	..	..	..	..	..	..	..	..
33 MEDICAL PRECISION &OPTICAL INSTRUMENTS	..	..	..	..	..	..	..	..	825
331 Medical appliances and instruments	..	..	..	..	..	..	..	..	825
3311 Medical, surgical equipment & orthopaedic app.	..	..	..	..	..	..	..	..	11
3312 Instruments & appliances for measuring, checking etc	..	..	..	..	..	..	..	..	..
3313 Industrial process control equipment	..	..	..	..	..	..	..	..	815
332 Optical instruments and photographic equipment	..	..	..	..	..	..	..	..	..
333 Watches and clocks	..	..	..	..	..	..	..	..	..
34 MOTOR VEHICLES, TRAILERS & SEMI-TRAILERS	..	..	..	..	..	..	..	..	..
341 Motor vehicles	..	..	..	..	..	..	..	..	..
342 Bodies (coachwork) for motor vehicles	..	..	..	..	..	..	..	..	..
343 Parts, accessories for motor vehicles & their engines	..	..	..	..	..	..	..	..	..
35 OTHER TRANSPORT EQUIPMENT	..	..	..	..	..	..	..	..	350
351 Building and repairing of ships and boats	..	..	..	..	..	..	..	..	..
3511 Building and repairing of ships	..	..	..	..	..	..	..	..	..
3512 Building, repairing of pleasure & sporting boats	..	..	..	..	..	..	..	..	..
352 Railway, tramway locomotives & rolling stock	..	..	..	..	..	..	..	..	..
353 Aircraft and spacecraft	..	..	..	..	..	..	..	..	..
359 Transport equipment, nec	..	..	..	..	..	..	..	..	..
3591 Motorcycles	..	..	..	..	..	..	..	..	..
3592 Bicycles and invalid carriages	..	..	..	..	..	..	..	..	..
3599 Other transport equipment, nec	..	..	..	..	..	..	..	..	..
36 FURNITURE; MANUFACTURING, NEC	167	167	167	167	167	285	167	604	299
361 Furniture	..	..	..	..	..	..	..	..	..
369 Manufacturing, nec	..	..	..	..	..	..	..	604	299
3691 Jewellery and related articles	..	..	..	..	..	..	..	..	..
3692 Musical instruments	..	..	..	..	..	..	..	..	..
3693 Sports goods	..	..	..	..	..	..	..	..	..
3694 Games and toys	..	..	..	..	..	..	..	..	..
3699 Other manufacturing, nec	..	..	..	..	..	..	..	..	..
37 RECYCLING	..	..	..	..	..	..	..	..	..
371 Recycling of metal waste and scrap	..	..	..	..	..	..	..	..	..
372 Recycling of non-metal waste and scrap	..	..	..	..	..	..	..	..	..
CERR Non-specified industry	47,180	47,761	46,860	48,787	46,965	42,647	38,855	66,882	72,044
15-37 TOTAL MANUFACTURING	160,988	156,115	160,206	160,778	161,167	150,200	148,371	174,045	170,124

Unit: TJ

ISIC Revision 3 Industry Sector		1990	1991	1992	1993	1994	1995	1996	1997	1998
15	**FOOD PRODUCTS AND BEVERAGES**	..	..	..	18,647	18,440	18,352	17,723	..	..
151	Production, processing and preserving (PPP)				9,658	9,447	9,761	9,218		
1511	PPP of meat and meat products				2,895	2,830	2,770	2,548		
1512	Processing and preserving of fish products				5,426	5,011	5,108	4,888		
1513	Processing, preserving of fruit & vegetables				731	789	780	697		
1514	Vegetable and animal oils and fats				606	818	1,103	1,086		
152	Dairy products				2,694	3,163	2,599	2,434		
153	Grain mill prod., starches & prepared animal feeds				2,401	1,923	2,002	2,184		
1531	Grain mill products				854	615	510	485		
1532	Starches and starch products				76	98	110	87		
1533	Prepared animal feeds				1,471	1,209	1,382	1,612		
154	Other food products				2,423	2,361	2,651	2,412		
1541	Bakery products				1,480	1,360	1,729	1,523		
1542	Sugar				..	..	..	..		
1543	Cocoa, chocolate and sugar confectionery				369	428	389	407		
1544	Macaroni, noodles, couscous & similar farinaceous prod.				52	49	51	48		
1549	Other food products, nec				522	525	482	434		
155	Beverages				1,471	1,546	1,340	1,474		
1551	Distilling, rectifying and blending of spirits				40	37	44	128		
1552	Wines				3	2	3	4		
1553	Malt liquors and malt				1,186	923	955	808		
1554	Soft drinks; production of mineral waters				242	584	337	534		
16	**TOBACCO PRODUCTS**	..	..	..	121	110	148	97	..	..
17	**TEXTILES**	..	..	..	949	1,099	1,014	1,142	..	..
171	Spinning, weaving and finishing of textiles				473	647	576	588		
1711	Preparation and spinning of textile fibres				346	517	445	442		
1712	Finishing of textiles				127	130	131	146		
172	Other textiles				325	315	316	438		
1721	Made-up textile articles, except apparel				63	68	85	189		
1722	Carpets and rugs				7	3	3	8		
1723	Cordage, rope, twine and netting				131	135	148	165		
1729	Other textiles, nec				124	109	80	74		
173	Knitted and crocheted fabrics and articles				151	137	122	116		
18	**WEARING APPAREL, DRESSING & DYEING OF FUR**	..	..	..	86	106	121	152	..	..
181	Wearing apparel, except fur apparel				72	74	102	132		
182	Dressing and dyeing of fur; articles of fur				14	32	19	20		
19	**TANNING & DRESSING OF LEATHER, FOOTWEAR**	..	..	..	93	106	116	106	..	..
191	Tanning and dressing of leather				79	93	88	83		
1911	Tanning and dressing of leather				71	87	83	77		
1912	Luggage, handbags, saddlery & harness				8	6	5	6		
192	Footwear				14	13	28	23		
20	**WOOD AND WOOD PRODUCTS**	..	..	..	8,530	8,978	9,093	8,741	..	..
201	Sawmilling and planing of wood				5,992	6,223	6,288	5,994		
202	Products of wood, cork, straw & plaiting materials				2,538	2,755	2,805	2,747		
2021	Veneer sheets				1,804	2,043	1,875	1,861		
2022	Builders' carpentry and joinery				623	622	802	767		
2023	Wooden containers				75	61	99	80		
2029	Other products of wood				36	29	28	39		
21	**PAPER AND PAPER PRODUCTS**	..	..	..	41,230	44,767	44,797	44,995	..	..
2101	Pulp, paper and paperboard				40,184	43,859	43,496	43,865		
2102	Corrugated paper, paperboard and their containers				656	489	529	352		
2109	Other articles of pulp and paperboard				391	419	772	778		
22	**PUBLISHING, PRINTING & REPRODUCTION**	..	..	..	2,092	1,914	2,007	1,966	..	..
221	Publishing				799	831	888	912		
2211	Publishing of books & brochures				127	120	130	112		
2212	Publishing of newspapers and periodicals				593	628	656	691		
2213	Publishing of recorded media				44	41	54	44		
2219	Other publishing				35	41	48	64		
222	Printing and related service activities				1,244	1,037	1,080	1,021		
2221	Printing				1,028	839	899	850		
2222	Service activities related to printing				216	198	181	172		
223	Reproduction of recorded media				48	46	39	33		
23	**COKE, REFINED PETROLEUM PRODUCTS**	..	..	..	37,645	37,105	31,651	35,449	..	..
231	Coke oven products				..	..	..	..		
232	Refined petroleum products				37,645	37,105	31,651	35,449		
233	Processing of nuclear fuel				..	..	..	..		
24	**CHEMICALS & CHEMICAL PRODUCTS**	..	..	..	38,766	37,899	46,876	36,857	..	..
241	Basic chemicals				37,189	35,889	45,313	35,498		
2411	Basic chemicals, exc. fertilizers & nitrogen compounds				28,819	28,336	38,837	29,185		
2412	Fertilizers and nitrogen compounds				4,407	3,693	3,651	3,849		
2413	Plastics in primary forms and synthetic rubber				3,964	3,859	2,825	2,464		
242	Other chemical products				1,576	2,010	1,563	1,358		
2421	Pesticides and other agro-chemical products				..	536	..	..		
2422	Paints, varnishes and similar coatings				272	263	262	234		
2423	Pharmaceuticals, medicinal chem. & botanical prod.				726	702	727	663		
2424	Soap and detergents, perfumes etc.				324	217	278	260		
2429	Other chemical products, nec				254	292	296	202		
243	Man-made fibres				..	..	..	1		
25	**RUBBER AND PLASTICS PRODUCTS**	..	..	..	1,814	1,945	1,955	1,687	..	..
251	Rubber products				392	412	246	254		
2511	Rubber tyres and tubes				290	88	68	85		
2519	Other rubber products				102	324	178	169		
252	Plastic products				1,421	1,533	1,709	1,433		
26	**OTHER NON-METALLIC MINERAL PRODUCTS**	..	..	..	11,995	12,960	16,704	14,231	..	..
261	Glass and glass products				1,554	1,792	1,806	1,709		
269	Non-metallic mineral products, nec				10,441	11,167	14,898	12,522		
2691	Pottery, china and earthenware				206	232	237	232		
2692	Refractory ceramic products				69	71	171	166		
2693	Structural non-refractory clay & ceramic prod.				224	277	173	223		
2694	Cement, lime and plaster				6,777	6,908	7,364	7,191		
2695	Articles of concrete, cement and plaster				1,933	2,027	2,471	2,238		
2696	Cutting, shaping and finishing of stone				290	581	214	284		
2699	Other non-metallic mineral products, nec				942	1,071	4,268	2,186		

ISIC Revision 3 Industry Sector		1990	1991	1992	1993	1994	1995	1996	1997	1998
27	**BASIC METALS**	..	..	..	86,655	91,668	125,838	87,387	..	..
271	Basic iron and steel	..	..	..	22,584	25,543	51,664	25,507	..	..
272	Basic precious and non-ferrous metals	..	..	..	64,030	65,264	73,294	61,049	..	..
273	Casting of metals	..	..	..	41	861	880	831	..	..
2731	Casting of iron and steel	..	..	..	..	726	712	673	..	..
2732	Casting non-ferrous metals	..	..	..	41	135	168	158	..	..
28	**FABRICATED METAL PRODUCTS**	..	..	..	3,461	2,639	3,224	2,605	..	..
281	Str. metal prod., tanks, reservoirs, steam generators	..	..	..	1,347	1,281	1,566	988	..	..
2811	Structural metal products	..	..	..	1,270	1,228	1,492	904	..	..
2812	Tanks, reservoirs and containers of metal	..	..	..	66	39	66	68	..	..
2813	Steam generators, exc. central heating hot water boilers	..	..	..	11	15	8	17	..	..
289	Other fabricated metal products	..	..	..	2,114	1,358	1,659	1,616	..	..
2891	Forging, pressing, stamping & roll-forming of metal	..	..	..	..	18	23	22	..	..
2892	Treatment and coating of metals	..	..	..	668	345	569	659	..	..
2893	Cutlery, hand tools and general hardware	..	..	..	225	245	209	185	..	..
2899	Other fabricated metal products, nec	..	..	..	1,221	750	858	751	..	..
29	**MACHINERY AND EQUIPMENT, NEC**	..	..	..	2,840	2,825	3,037	2,906	..	..
291	General purpose machinery	..	..	..	1,327	1,270	1,403	1,465	..	..
2911	Engines and turbines	..	..	..	246	335	227	251	..	..
2912	Pumps, compressors, taps and valves	..	..	..	240	228	425	381	..	..
2913	Bearings, gears, gearing and driving elements	..	..	..	82	58	106	72	..	..
2914	Ovens, furnaces and furnace burners	..	..	..	39	50	47	53	..	..
2915	Lifting and handling equipment	..	..	..	411	350	366	371	..	..
2919	Other general purpose machinery	..	..	..	309	249	233	337	..	..
292	Special purpose machinery	..	..	..	1,261	1,304	1,364	1,192	..	..
2921	Agricultural and forestry machinery	..	..	..	482	420	527	531	..	..
2922	Machine-tools	..	..	..	43	46	55	63	..	..
2923	Machinery for metallurgy	..	..	..	21	21	24	7	..	..
2924	Machinery for mining, quarrying and construction	..	..	..	139	183	166	171	..	..
2925	Machinery for food, beverage & tobacco processing	..	..	..	28	47	119	151	..	..
2926	Machinery for textile, apparel & leather production	..	..	..	5	6	6	6	..	..
2927	Machinery for weapons and ammunition	..	..	..	406	423	362	162	..	..
2929	Other special purpose machinery	..	..	..	139	159	105	101	..	..
293	Domestic appliances, nec	..	..	..	251	250	270	249	..	..
30	**OFFICE, ACCOUNTING & COMPUTING MACHINERY**	..	..	..	40	54	59	30	..	..
31	**ELECTRICAL MACHINERY & APPARATUS, NEC**	..	..	..	1,302	1,974	5,456	1,553	..	..
311	Electric motors, generators and transformers	..	..	..	261	592	364	281	..	..
312	Electricity distribution and control apparatus	..	..	..	166	182	206	162	..	..
313	Insulated wire and cable	..	..	..	243	236	370	258	..	..
314	Accumulators, primary cells & primary batteries	..	..	..	24	25	26	34	..	..
315	Electric lamps and lighting equipment	..	..	..	141	151	116	156	..	..
319	Other electrical equipment, nec	..	..	..	466	788	4,375	662	..	..
32	**RADIO, TV & COMMUNICATION EQUIP. & APP.**	..	..	..	280	242	236	210	..	..
321	Electronic valves, tubes, other electronic components	..	..	..	43	52	52	57	..	..
322	TV & radio transmitters, apparatus for line telephony	..	..	..	153	140	159	128	..	..
323	TV & radio receivers, recording apparatus	..	..	..	85	50	25	24	..	..
33	**MEDICAL PRECISION &OPTICAL INSTRUMENTS**	..	..	..	277	336	326	306	..	..
331	Medical appliances and instruments	..	..	..	265	319	314	289	..	..
3311	Medical, surgical equipment & orthopaedic app.	..	..	..	70	81	69	91	..	..
3312	Instruments & appliances for measuring, checking etc	..	..	..	151	183	212	156	..	..
3313	Industrial process control equipment	..	..	..	44	55	33	42	..	..
332	Optical instruments and photographic equipment	..	..	..	11	18	12	17	..	..
333	Watches and clocks	..	..	..	..	..	..	..	..	..
34	**MOTOR VEHICLES, TRAILERS & SEMI-TRAILERS**	..	..	..	937	809	933	1,139	..	..
341	Motor vehicles	..	..	..	13	19	25	52	..	..
342	Bodies (coachwork) for motor vehicles	..	..	..	164	159	169	225	..	..
343	Parts, accessories for motor vehicles & their engines	..	..	..	760	631	738	862	..	..
35	**OTHER TRANSPORT EQUIPMENT**	..	..	..	2,871	3,349	3,263	3,102	..	..
351	Building and repairing of ships and boats	..	..	..	2,064	2,556	2,565	2,369	..	..
3511	Building and repairing of ships	..	..	..	1,970	2,396	2,489	2,269	..	..
3512	Building, repairing of pleasure & sporting boats	..	..	..	94	159	75	100	..	..
352	Railway, tramway locomotives & rolling stock	..	..	..	319	291	231	208	..	..
353	Aircraft and spacecraft	..	..	..	392	430	404	448	..	..
359	Transport equipment, nec	..	..	..	96	72	64	77	..	..
3591	Motorcycles	..	..	..	..	..	..	..	..	..
3592	Bicycles and invalid carriages	..	..	..	83	58	57	70	..	..
3599	Other transport equipment, nec	..	..	..	13	14	7	7	..	..
36	**FURNITURE; MANUFACTURING, NEC**	..	..	..	1,235	1,310	1,342	1,746	..	..
361	Furniture	..	..	..	941	1,035	1,013	1,367	..	..
369	Manufacturing, nec	..	..	..	295	275	328	379	..	..
3691	Jewellery and related articles	..	..	..	53	46	57	46	..	..
3692	Musical instruments	..	..	..	1	1	1	2	..	..
3693	Sports goods	..	..	..	71	108	127	164	..	..
3694	Games and toys	..	..	..	14	2	9	6	..	..
3699	Other manufacturing, nec	..	..	..	155	117	135	160	..	..
37	**RECYCLING**	..	..	..	54	253	206	226	..	..
371	Recycling of metal waste and scrap	..	..	..	14	213	135	188	..	..
372	Recycling of non-metal waste and scrap	..	..	..	40	41	70	38	..	..
CERR	**Non-specified industry**	..	..	..	..	..	..	..	..	..
15-37	**TOTAL MANUFACTURING**	..	..	..	261,919	270,886	316,752	264,355	..	..

POLAND

ISIC Revision 3 Industry Sector		1990	1991	1992	1993	1994	1995	1996	1997	1998
15	**FOOD PRODUCTS AND BEVERAGES**	..	..	..	..	132,024	121,734	160,312	142,767	..
151	Production, processing and preserving (PPP)	..	..	..	..	30,573	34,735	44,931	38,055	..
1511	PPP of meat and meat products	..	..	..	..	..	..	..	..	..
1512	Processing and preserving of fish products	..	..	..	..	..	..	..	..	..
1513	Processing, preserving of fruit & vegetables	..	..	..	..	..	..	..	..	..
1514	Vegetable and animal oils and fats	..	..	..	..	..	..	..	..	..
152	Dairy products	..	..	..	..	21,623	20,962	28,566	24,009	..
153	Grain mill prod., starches & prepared animal feeds	..	..	..	..	2,554	5,322	5,702	4,526	..
1531	Grain mill products	..	..	..	..	..	..	..	..	..
1532	Starches and starch products	..	..	..	..	..	..	..	..	..
1533	Prepared animal feeds	..	..	..	..	..	..	..	..	..
154	Other food products	..	..	..	..	61,861	46,733	61,604	58,861	..
1541	Bakery products	..	..	..	..	..	..	..	..	..
1542	Sugar	..	..	..	..	..	..	..	..	..
1543	Cocoa, chocolate and sugar confectionery	..	..	..	..	..	..	..	..	..
1544	Macaroni, noodles, couscous & similar farinaceous prod.	..	..	..	..	..	..	..	..	..
1549	Other food products, nec	..	..	..	..	..	..	..	..	..
155	Beverages	..	..	..	..	15,413	13,982	19,509	17,315	..
1551	Distilling, rectifying and blending of spirits	..	..	..	..	..	..	..	..	..
1552	Wines	..	..	..	..	..	..	..	..	..
1553	Malt liquors and malt	..	..	..	..	..	..	..	..	..
1554	Soft drinks; production of mineral waters	..	..	..	..	..	..	..	..	..
16	**TOBACCO PRODUCTS**	..	..	..	..	1,367	1,667	1,576	1,485	..
17	**TEXTILES**	..	..	..	..	36,599	31,383	35,644	28,650	..
171	Spinning, weaving and finishing of textiles	..	..	..	..	30,292	23,641	23,518	19,371	..
1711	Preparation and spinning of textile fibres	..	..	..	..	..	..	..	..	..
1712	Finishing of textiles	..	..	..	..	..	..	..	..	..
172	Other textiles	..	..	..	..	2,844	4,063	8,358	5,698	..
1721	Made-up textile articles, except apparel	..	..	..	..	..	..	..	..	..
1722	Carpets and rugs	..	..	..	..	..	..	..	..	..
1723	Cordage, rope, twine and netting	..	..	..	..	..	..	..	..	..
1729	Other textiles, nec	..	..	..	..	..	..	..	..	..
173	Knitted and crocheted fabrics and articles	..	..	..	..	3,463	3,679	3,767	3,582	..
18	**WEARING APPAREL, DRESSING & DYEING OF FUR**	..	..	..	..	5,352	4,900	5,797	5,120	..
181	Wearing apparel, except fur apparel	..	..	..	..	5,352	4,715	5,593	4,859	..
182	Dressing and dyeing of fur; articles of fur	..	..	..	..	..	185	204	262	..
19	**TANNING & DRESSING OF LEATHER, FOOTWEAR**	..	..	..	..	4,136	5,722	5,421	5,242	..
191	Tanning and dressing of leather	..	..	..	..	2,604	2,597	3,711	3,264	..
1911	Tanning and dressing of leather	..	..	..	..	..	..	..	..	..
1912	Luggage, handbags, saddlery & harness	..	..	..	..	..	..	..	..	..
192	Footwear	..	..	..	..	1,531	3,125	1,710	1,978	..
20	**WOOD AND WOOD PRODUCTS**	..	..	..	..	24,520	23,145	31,849	26,942	..
201	Sawmilling and planing of wood	..	..	..	..	4,141	4,718	4,061	4,070	..
202	Products of wood, cork, straw & plaiting materials	..	..	..	..	20,379	18,427	27,788	22,872	..
2021	Veneer sheets	..	..	..	..	..	..	..	..	..
2022	Builders' carpentry and joinery	..	..	..	..	..	..	..	..	..
2023	Wooden containers	..	..	..	..	..	..	..	..	..
2029	Other products of wood	..	..	..	..	..	..	..	..	..
21	**PAPER AND PAPER PRODUCTS**	..	..	..	..	73,436	55,037	53,945	57,677	..
2101	Pulp, paper and paperboard	..	..	..	..	..	..	..	..	..
2102	Corrugated paper, paperboard and their containers	..	..	..	..	..	..	..	..	..
2109	Other articles of pulp and paperboard	..	..	..	..	..	..	..	..	..
22	**PUBLISHING, PRINTING & REPRODUCTION**	..	..	..	..	821	1,186	3,195	3,260	..
221	Publishing	..	..	..	..	..	141	142	1,845	..
2211	Publishing of books & brochures	..	..	..	..	..	..	..	..	..
2212	Publishing of newspapers and periodicals	..	..	..	..	..	..	..	..	..
2213	Publishing of recorded media	..	..	..	..	..	..	..	..	..
2219	Other publishing	..	..	..	..	..	..	..	..	..
222	Printing and related service activities	..	..	..	..	821	1,044	3,053	1,415	..
2221	Printing	..	..	..	..	..	..	..	..	..
2222	Service activities related to printing	..	..	..	..	..	..	..	..	..
223	Reproduction of recorded media	..	..	..	..	..	..	..	..	..
23	**COKE, REFINED PETROLEUM PRODUCTS**	..	..	..	..	1,114,866	1,137,918	1,228,987	1,530,744	..
231	Coke oven products	..	..	..	..	390,471	413,358	365,031	390,568	..
232	Refined petroleum products	..	..	..	..	724,395	724,560	863,956	1,140,177	..
233	Processing of nuclear fuel	..	..	..	..	..	..	..	..	..
24	**CHEMICALS & CHEMICAL PRODUCTS**	..	..	..	..	213,432	162,879	157,113	153,908	..
241	Basic chemicals	..	..	..	..	170,852	129,791	126,355	127,481	..
2411	Basic chemicals, exc. fertilizers & nitrogen compounds	..	..	..	..	..	..	..	..	..
2412	Fertilizers and nitrogen compounds	..	..	..	..	..	..	..	..	..
2413	Plastics in primary forms and synthetic rubber	..	..	..	..	..	..	..	..	..
242	Other chemical products	..	..	..	..	19,395	17,829	16,263	13,794	..
2421	Pesticides and other agro-chemical products	..	..	..	..	..	..	..	..	..
2422	Paints, varnishes and similar coatings	..	..	..	..	..	..	..	..	..
2423	Pharmaceuticals, medicinal chem. & botanical prod.	..	..	..	..	..	..	..	..	..
2424	Soap and detergents, perfumes etc.	..	..	..	..	..	..	..	..	..
2429	Other chemical products, nec	..	..	..	..	..	..	..	..	..
243	Man-made fibres	..	..	..	..	23,184	15,259	14,495	12,632	..
25	**RUBBER AND PLASTICS PRODUCTS**	..	..	..	..	20,556	21,180	19,139	19,055	..
251	Rubber products	..	..	..	..	11,233	11,377	11,526	10,688	..
2511	Rubber tyres and tubes	..	..	..	..	..	..	..	..	..
2519	Other rubber products	..	..	..	..	..	..	..	..	..
252	Plastic products	..	..	..	..	9,323	9,803	7,613	8,367	..
26	**OTHER NON-METALLIC MINERAL PRODUCTS**	..	..	..	..	145,867	151,521	160,104	148,628	..
261	Glass and glass products	..	..	..	..	24,829	26,602	26,900	26,276	..
269	Non-metallic mineral products, nec	..	..	..	..	121,038	124,918	133,204	122,352	..
2691	Pottery, china and earthenware	..	..	..	..	..	..	..	..	..
2692	Refractory ceramic products	..	..	..	..	..	..	..	..	..
2693	Structural non-refractory clay & ceramic prod.	..	..	..	..	..	..	..	..	..
2694	Cement, lime and plaster	..	..	..	..	..	..	..	..	..
2695	Articles of concrete, cement and plaster	..	..	..	..	..	..	..	..	..
2696	Cutting, shaping and finishing of stone	..	..	..	..	..	..	..	..	..
2699	Other non-metallic mineral products, nec	..	..	..	..	..	..	..	..	..

ISIS Energy Data Programme (IEA/OECD)

ISIC Revision 3 Industry Sector		1990	1991	1992	1993	1994	1995	1996	1997	1998
27	**BASIC METALS**	..	..	..	..	495,868	453,713	420,628	433,062	..
271	Basic iron and steel		..	..	..	448,566	409,345	374,301	384,090	..
272	Basic precious and non-ferrous metals		..	..	..	35,971	33,720	34,864	35,438	..
273	Casting of metals		..	..	..	11,331	10,647	11,463	13,533	..
2731	Casting of iron and steel		..	..	..					
2732	Casting non-ferrous metals									
28	**FABRICATED METAL PRODUCTS**	..	..	..	..	15,411	14,602	16,334	19,148	..
281	Str. metal prod., tanks, reservoirs, steam generators				..	2,801	4,191	6,628	7,097	
2811	Structural metal products									
2812	Tanks, reservoirs and containers of metal					..	..	..	..	
2813	Steam generators, exc. central heating hot water boilers									
289	Other fabricated metal products				..	12,611	10,411	9,705	12,051	
2891	Forging, pressing, stamping & roll-forming of metal									
2892	Treatment and coating of metals					..	..	..	..	
2893	Cutlery, hand tools and general hardware					..	..	..	..	
2899	Other fabricated metal products, nec					..	..	..	..	
29	**MACHINERY AND EQUIPMENT, NEC**	..	..	..	..	43,594	42,226	46,240	40,953	..
291	General purpose machinery				..	9,119	12,086	13,171	14,068	..
2911	Engines and turbines									
2912	Pumps, compressors, taps and valves					..	..	..	..	
2913	Bearings, gears, gearing and driving elements					..	..	..	..	
2914	Ovens, furnaces and furnace burners					..	..	..	..	
2915	Lifting and handling equipment					..	..	..	..	
2919	Other general purpose machinery					..	..	..	..	
292	Special purpose machinery				..	26,937	25,882	29,273	22,806	..
2921	Agricultural and forestry machinery					..	..	..	..	
2922	Machine-tools					..	..	..	..	
2923	Machinery for metallurgy					..	..	..	..	
2924	Machinery for mining, quarrying and construction					..	..	..	..	
2925	Machinery for food, beverage & tobacco processing					..	..	..	..	
2926	Machinery for textile, apparel & leather production					..	..	..	..	
2927	Machinery for weapons and ammunition					..	..	..	..	
2929	Other special purpose machinery					..	..	..	..	
293	Domestic appliances, nec		..	..	..	7,538	4,258	3,796	4,078	..
30	**OFFICE, ACCOUNTING & COMPUTING MACHINERY**	..	..	..	..	..	181	210	161	..
31	**ELECTRICAL MACHINERY & APPARATUS, NEC**	..	..	..	..	13,632	11,044	12,343	11,813	..
311	Electric motors, generators and transformers		..	..	..	2,135	2,177	2,660	2,334	..
312	Electricity distribution and control apparatus					4,387	2,090	2,349	2,105	
313	Insulated wire and cable					1,238	1,772	1,919	2,418	
314	Accumulators, primary cells & primary batteries					469	368	416	438	
315	Electric lamps and lighting equipment					1,317	1,359	1,507	1,438	
319	Other electrical equipment, nec					4,086	3,278	3,491	3,081	
32	**RADIO, TV & COMMUNICATION EQUIP. & APP.**	..	..	..	..	5,480	4,273	3,772	3,013	..
321	Electronic valves, tubes, other electronic components					357	789	663	505	
322	TV & radio transmitters, apparatus for line telephony					1,215	725	854	601	
323	TV & radio receivers, recording apparatus					3,908	2,759	2,255	1,906	
33	**MEDICAL PRECISION &OPTICAL INSTRUMENTS**	..	..	..	..	1,569	1,958	2,426	2,626	..
331	Medical appliances and instruments					1,359	1,875	2,188	2,447	
3311	Medical, surgical equipment & orthopaedic app.					..	..	..	..	
3312	Instruments & appliances for measuring, checking etc					..	..	..	..	
3313	Industrial process control equipment					..	..	..	..	
332	Optical instruments and photographic equipment					184	52	210	159	
333	Watches and clocks					27	31	28	20	
34	**MOTOR VEHICLES, TRAILERS & SEMI-TRAILERS**	..	..	..	..	17,474	17,046	18,406	17,125	..
341	Motor vehicles				..	12,863	12,083	12,789	11,584	..
342	Bodies (coachwork) for motor vehicles					347	674	337	431	
343	Parts, accessories for motor vehicles & their engines					4,264	4,289	5,280	5,109	
35	**OTHER TRANSPORT EQUIPMENT**	..	..	..	..	17,564	15,410	15,573	13,460	..
351	Building and repairing of ships and boats				..	6,120	5,631	5,756	4,850	..
3511	Building and repairing of ships					..	..	..	..	
3512	Building, repairing of pleasure & sporting boats					..	..	..	..	
352	Railway, tramway locomotives & rolling stock					6,382	5,533	6,084	5,095	
353	Aircraft and spacecraft					4,329	3,559	3,066	2,931	
359	Transport equipment, nec					732	687	667	584	
3591	Motorcycles					..	..	..	..	
3592	Bicycles and invalid carriages					..	..	..	..	
3599	Other transport equipment, nec					..	..	..	..	
36	**FURNITURE; MANUFACTURING, NEC**	..	..	..	..	11,501	12,289	14,138	20,539	..
361	Furniture				..	10,796	10,593	12,297	10,301	..
369	Manufacturing, nec					704	1,696	1,841	10,237	
3691	Jewellery and related articles					..	..	..	..	
3692	Musical instruments					..	..	..	..	
3693	Sports goods					..	..	..	..	
3694	Games and toys					..	..	..	..	
3699	Other manufacturing, nec					..	..	..	..	
37	**RECYCLING**	..	..	..	..	244	515	897	646	..
371	Recycling of metal waste and scrap				..	152	423	571	309	..
372	Recycling of non-metal waste and scrap					93	92	327	336	
CERR	**Non-specified industry**	..	..	..	..	..	..	..	..	
15-37	**TOTAL MANUFACTURING**	..	..	..	..	2,395,314	2,291,526	2,414,049	2,686,024	..

ISIC Revision 3 Industry Sector		1990	1991	1992	1993	1994	1995	1996	1997	1998
15	FOOD PRODUCTS AND BEVERAGES	..	..	..	..	23,428	21,917	27,572	18,444	..
151	Production, processing and preserving (PPP)	..	..	..	..	3,136	2,943	2,258	4,852	..
1511	PPP of meat and meat products	..	..	..	..	1,963	..	1,585	1,616	..
1512	Processing and preserving of fish products	..	..	..	..	..	..	41	43	..
1513	Processing, preserving of fruit & vegetables	..	..	..	..	310	..	332	2,396	..
1514	Vegetable and animal oils and fats	..	..	..	..	..	..	..	..	..
152	Dairy products	..	..	..	..	2,013	1,837	1,894	1,869	..
153	Grain mill prod., starches & prepared animal feeds	..	..	..	..	1,678	1,486	4,017	1,711	..
1531	Grain mill products	..	..	..	..	353	347	356	315	..
1532	Starches and starch products	..	..	..	..	560	468	561	567	..
1533	Prepared animal feeds	..	..	..	..	738	640	3,096	826	..
154	Other food products	..	..	..	..	5,761	5,847	6,864	6,488	..
1541	Bakery products	..	..	..	..	1,680	1,835	1,854	1,742	..
1542	Sugar	..	..	..	..	3,327	3,689	4,642	4,365	..
1543	Cocoa, chocolate and sugar confectionery	..	..	..	..	..	..	120	122	..
1544	Macaroni, noodles, couscous & similar farinaceous prod.	..	..	..	..	..	..	..	..	..
1549	Other food products, nec	..	..	..	..	323	219	246	205	..
155	Beverages	..	..	..	..	10,838	9,803	11,629	3,413	..
1551	Distilling, rectifying and blending of spirits	..	..	..	..	913	834	917	982	..
1552	Wines	..	..	..	..	7,380	6,278	8,169	127	..
1553	Malt liquors and malt	..	..	..	..	2,324	2,493	2,310	2,015	..
1554	Soft drinks; production of mineral waters	..	..	..	..	..	164	152	142	..
16	TOBACCO PRODUCTS	..	..	..	..	..	..	..	..	..
17	TEXTILES	..	..	..	..	7,062	6,566	6,550	5,307	..
171	Spinning, weaving and finishing of textiles	..	..	..	..	5,480	5,470	4,119	4,785	..
1711	Preparation and spinning of textile fibres	..	..	..	..	4,142	..	2,682	4,764	..
1712	Finishing of textiles	..	..	..	..	..	..	..	..	..
172	Other textiles	..	..	..	..	383	114	39	39	..
1721	Made-up textile articles, except apparel	..	..	..	..	28	32	36	31	..
1722	Carpets and rugs	..	..	..	..	..	..	..	..	..
1723	Cordage, rope, twine and netting	..	..	..	..	..	..	..	..	..
1729	Other textiles, nec	..	..	..	..	57	41	..	2	..
173	Knitted and crocheted fabrics and articles	..	..	..	..	1,197	980	894	444	..
18	WEARING APPAREL, DRESSING & DYEING OF FUR	..	..	..	..	1,141	1,314	1,247	1,209	..
181	Wearing apparel, except fur apparel	..	..	..	..	1,130	..	1,233	1,190	..
182	Dressing and dyeing of fur; articles of fur	..	..	..	..	..	..	..	14	..
19	TANNING & DRESSING OF LEATHER, FOOTWEAR	..	..	..	..	2,329	2,263	2,439	1,522	..
191	Tanning and dressing of leather	..	..	..	..	910	902	797	577	..
1911	Tanning and dressing of leather	..	..	..	..	879	857	757	541	..
1912	Luggage, handbags, saddlery & harness	..	..	..	..	31	44	40	36	..
192	Footwear	..	..	..	..	1,418	1,361	1,641	945	..
20	WOOD AND WOOD PRODUCTS	..	..	..	..	3,888	3,980	3,316	2,876	..
201	Sawmilling and planing of wood	..	..	..	..	1,634	1,854	1,302	1,134	..
202	Products of wood, cork, straw & plaiting materials	..	..	..	..	2,249	2,124	1,997	1,720	..
2021	Veneer sheets	..	..	..	..	1,423	..	1,077	922	..
2022	Builders' carpentry and joinery	..	..	..	..	779	..	616	511	..
2023	Wooden containers	..	..	..	..	..	..	13	49	..
2029	Other products of wood	..	..	..	..	45	..	289	238	..
21	PAPER AND PAPER PRODUCTS	..	..	..	..	16,561	16,280	19,992	20,076	..
2101	Pulp, paper and paperboard	..	..	..	..	15,328	..	18,596	18,737	..
2102	Corrugated paper, paperboard and their containers	..	..	..	..	..	..	62	115	..
2109	Other articles of pulp and paperboard	..	..	..	..	1,174	..	1,331	1,220	..
22	PUBLISHING, PRINTING & REPRODUCTION	..	..	..	..	359	380	406	422	..
221	Publishing	..	..	..	..	36	..	37	29	..
2211	Publishing of books & brochures	..	..	..	..	13	..	19	10	..
2212	Publishing of newspapers and periodicals	..	..	..	..	11	..	9	15	..
2213	Publishing of recorded media	..	..	..	..	..	..	..	..	..
2219	Other publishing	..	..	..	..	10	..	..	..	..
222	Printing and related service activities	..	..	..	..	315	..	362	389	..
2221	Printing	..	..	..	..	236	..	277	323	..
2222	Service activities related to printing	..	..	..	..	53	..	50	54	..
223	Reproduction of recorded media	..	..	..	..	..	..	..	..	..
23	COKE, REFINED PETROLEUM PRODUCTS	..	..	..	..	40,295	38,041	40,570	39,562	..
231	Coke oven products	..	..	..	..	..	..	..	..	..
232	Refined petroleum products	..	..	..	..	..	..	..	..	..
233	Processing of nuclear fuel	..	..	..	..	..	..	..	..	..
24	CHEMICALS & CHEMICAL PRODUCTS	..	..	..	..	51,663	36,814	38,224	37,144	..
241	Basic chemicals	..	..	..	..	28,712	..	12,386	11,833	..
2411	Basic chemicals, exc. fertilizers & nitrogen compounds	..	..	..	..	708	..	600	670	..
2412	Fertilizers and nitrogen compounds	..	..	..	..	..	..	..	15	..
2413	Plastics in primary forms and synthetic rubber	..	..	..	..	..	..	..	..	..
242	Other chemical products	..	..	..	..	2,504	..	2,936	3,104	..
2421	Pesticides and other agro-chemical products	..	..	..	..	..	..	..	..	..
2422	Paints, varnishes and similar coatings	..	..	..	..	..	..	213	208	..
2423	Pharmaceuticals, medicinal chem. & botanical prod.	..	..	..	..	2,011	..	2,552	2,489	..
2424	Soap and detergents, perfumes etc.	..	..	..	..	181	..	171	203	..
2429	Other chemical products, nec	..	..	..	..	..	..	..	203	..
243	Man-made fibres	..	..	..	..	..	..	..	..	..
25	RUBBER AND PLASTICS PRODUCTS	..	..	..	..	6,220	6,488	6,580	4,171	..
251	Rubber products	..	..	..	..	3,359	3,441	3,413	3,375	..
2511	Rubber tyres and tubes	..	..	..	..	2,352	2,485	2,443	2,219	..
2519	Other rubber products	..	..	..	..	981	912	929	915	..
252	Plastic products	..	..	..	..	2,860	3,046	3,164	796	..
26	OTHER NON-METALLIC MINERAL PRODUCTS	..	..	..	..	25,154	27,868	25,269	27,391	..
261	Glass and glass products	..	..	..	..	4,881	4,913	4,710	3,895	..
269	Non-metallic mineral products, nec	..	..	..	..	20,270	22,953	20,559	23,495	..
2691	Pottery, china and earthenware	..	..	..	..	184	..	227	264	..
2692	Refractory ceramic products	..	..	..	..	2,780	..	3,330	4,846	..
2693	Structural non-refractory clay & ceramic prod.	..	..	..	..	2,592	..	2,447	2,453	..
2694	Cement, lime and plaster	..	..	..	..	14,015	..	13,735	14,547	..
2695	Articles of concrete, cement and plaster	..	..	..	..	674	..	805	698	..
2696	Cutting, shaping and finishing of stone	..	..	..	..	23	..	11	27	..
2699	Other non-metallic mineral products, nec	..	..	..	..	..	..	..	..	..

ISIS Energy Data Programme (IEA/OECD)

ISIC Revision 3 Industry Sector		1990	1991	1992	1993	1994	1995	1996	1997	1998
27	BASIC METALS	..	..	..	..	199,435	199,086	186,683	190,151	
271	Basic iron and steel	..	..	..	..	189,416	189,407	174,387	178,269	..
272	Basic precious and non-ferrous metals					9,483	8,917	11,445	11,260	..
273	Casting of metals					534	763	842	565	
2731	Casting of iron and steel			..		..	..	425	163	
2732	Casting non-ferrous metals									..
28	FABRICATED METAL PRODUCTS				..	2,010	2,107	2,853	1,792	
281	Str. metal prod., tanks, reservoirs, steam generators				..	1,198	1,145	1,049	972	
2811	Structural metal products					386		407	786	..
2812	Tanks, reservoirs and containers of metal					244	..	236	181	..
2813	Steam generators, exc. central heating hot water boilers								3	
289	Other fabricated metal products					801	941	937	765	
2891	Forging, pressing, stamping & roll-forming of metal							..	..	
2892	Treatment and coating of metals					41	..	61	67	
2893	Cutlery, hand tools and general hardware					183	..	172	194	
2899	Other fabricated metal products, nec				..	570	..	508	481	
29	MACHINERY AND EQUIPMENT, NEC	..			..	11,576	8,257	8,914	11,790	..
291	General purpose machinery				..	5,410	2,648	2,976	5,950	
2911	Engines and turbines				..	185	193	189	117	..
2912	Pumps, compressors, taps and valves				..	598	526	627	583	..
2913	Bearings, gears, gearing and driving elements				..	4,154	1,522	1,633	1,476	..
2914	Ovens, furnaces and furnace burners				..	..	..	..	..	
2915	Lifting and handling equipment					327	236	227	130	
2919	Other general purpose machinery					99	113	210	3,532	
292	Special purpose machinery				..	5,380	4,880	5,260	5,208	
2921	Agricultural and forestry machinery					632	622	570	612	
2922	Machine-tools					736	559	1,192	1,701	
2923	Machinery for metallurgy					..	..	..	..	
2924	Machinery for mining, quarrying and construction				..	1,423	821	659	446	..
2925	Machinery for food, beverage & tobacco processing				..	66	61	57	61	..
2926	Machinery for textile, apparel & leather production				..	191	203	144	208	..
2927	Machinery for weapons and ammunition					..	..	..	..	
2929	Other special purpose machinery				..	2,183	2,524	2,619	2,124	
293	Domestic appliances, nec				..	679	642	600	555	
30	OFFICE, ACCOUNTING & COMPUTING MACHINERY	..	..	..	..	153	141	40	48	..
31	ELECTRICAL MACHINERY & APPARATUS, NEC	..			..	18,369	1,795	1,710	1,253	..
311	Electric motors, generators and transformers					337	..	249	336	
312	Electricity distribution and control apparatus					178	..	174	145	
313	Insulated wire and cable				..	346	..	357	115	
314	Accumulators, primary cells & primary batteries					..		..	..	
315	Electric lamps and lighting equipment				..	17,225	..	448	225	
319	Other electrical equipment, nec					243	..	336	405	
32	RADIO, TV & COMMUNICATION EQUIP. & APP.				..	2,010	967	820	796	
321	Electronic valves, tubes, other electronic components					1,264	..	163	191	
322	TV & radio transmitters, apparatus for line telephony					219	..	127	98	
323	TV & radio receivers, recording apparatus				..	526	..	..	483	
33	MEDICAL PRECISION &OPTICAL INSTRUMENTS				..	748	764	720	585	..
331	Medical appliances and instruments					700	686	660	531	
3311	Medical, surgical equipment & orthopaedic app.				..	402	388	384	353	
3312	Instruments & appliances for measuring, checking etc				..	286	280	241	145	
3313	Industrial process control equipment					12	19	35	31	
332	Optical instruments and photographic equipment					..	..	..	..	
333	Watches and clocks					..		..	..	
34	MOTOR VEHICLES, TRAILERS & SEMI-TRAILERS	..			..	1,720	1,976	2,231	2,189	
341	Motor vehicles					1,098	..	1,464	687	
342	Bodies (coachwork) for motor vehicles					18	..	54	78	
343	Parts, accessories for motor vehicles & their engines				..	605	..	713	624	
35	OTHER TRANSPORT EQUIPMENT	..			..	3,283	3,680	4,039	3,591	
351	Building and repairing of ships and boats					..	..	..	229	
3511	Building and repairing of ships					..	..	..	229	
3512	Building, repairing of pleasure & sporting boats					..	..	..	..	
352	Railway, tramway locomotives & rolling stock				..	856	876	1,241	1,123	
353	Aircraft and spacecraft					..	..	..	..	
359	Transport equipment, nec				..	166	337	383	309	
3591	Motorcycles					65	..	..	..	
3592	Bicycles and invalid carriages						..	..	..	
3599	Other transport equipment, nec					..	..	..	..	
36	FURNITURE; MANUFACTURING, NEC	..			..	1,576	1,605	1,131	2,350	
361	Furniture				..	1,418	1,412	958	2,141	
369	Manufacturing, nec					154	188	158	198	
3691	Jewellery and related articles					69	75	86	64	
3692	Musical instruments					..	..	..	..	
3693	Sports goods					..		..	..	
3694	Games and toys					..	8	11	13	
3699	Other manufacturing, nec					83	105	61	57	
37	RECYCLING				..	140	146	140	167	
371	Recycling of metal waste and scrap					75	72	75	109	
372	Recycling of non-metal waste and scrap				..	65	73	65	49	
CERR	Non-specified industry					..	..	..	..	
15-37	TOTAL MANUFACTURING	..	..	..	..	419,119	382,435	381,446	372,835	

ISIC Revision 3 Industry Sector		1990	1991	1992	1993	1994	1995	1996	1997	1998
15	FOOD PRODUCTS AND BEVERAGES	..	..	..	..	23,582	24,458	..	23,461	..
151	Production, processing and preserving (PPP)	..	..	..	..	6,909	7,370	..	8,188	..
1511	PPP of meat and meat products	..	..	..	..	3,480	3,471	..	4,492	..
1512	Processing and preserving of fish products	..	..	..	..	409	548	..	457	..
1513	Processing, preserving of fruit & vegetables	..	..	..	..	1,365	1,631	..	1,680	..
1514	Vegetable and animal oils and fats	..	..	..	..	1,655	1,721	..	1,558	..
152	Dairy products	..	..	..	..	3,837	3,858	..	3,565	..
153	Grain mill prod., starches & prepared animal feeds	..	..	..	..	1,620	1,735	..	2,371	..
1531	Grain mill products	..	..	..	..	427	420	..	581	..
1532	Starches and starch products	..	..	..	..	305	415	..	911	..
1533	Prepared animal feeds	..	..	..	..	888	900	..	880	..
154	Other food products	..	..	..	..	8,822	8,938	..	6,959	..
1541	Bakery products	..	..	..	..	2,632	2,659	..	2,646	..
1542	Sugar	..	..	..	..	3,618	3,788	..	1,807	..
1543	Cocoa, chocolate and sugar confectionery	..	..	..	..	989	1,021	..	1,151	..
1544	Macaroni, noodles, couscous & similar farinaceous prod.	..	..	..	..	130	145	..	2	..
1549	Other food products, nec	..	..	..	..	1,453	1,325	..	1,353	..
155	Beverages	..	..	..	..	2,394	2,558	..	2,377	..
1551	Distilling, rectifying and blending of spirits	..	..	..	..	433	394	..	384	..
1552	Wines	..	..	..	..	518	519	..	101	..
1553	Malt liquors and malt	..	..	..	..	574	806	..	388	..
1554	Soft drinks; production of mineral waters	..	..	..	..	868	838	..	1,503	..
16	TOBACCO PRODUCTS	..	..	..	..	182	191	..	157	..
17	TEXTILES	..	..	..	..	2,449	2,315	..	2,561	..
171	Spinning, weaving and finishing of textiles	..	..	..	..	1,204	1,159	..	1,286	..
1711	Preparation and spinning of textile fibres	..	..	..	..	840	865	..	685	..
1712	Finishing of textiles	..	..	..	..	364	294	..	601	..
172	Other textiles	..	..	..	..	1,116	1,035	..	1,197	..
1721	Made-up textile articles, except apparel	..	..	..	..	201	132	..	254	..
1722	Carpets and rugs	..	..	..	..	6	11	..	9	..
1723	Cordage, rope, twine and netting	..	..	..	..	13	13	..	22	..
1729	Other textiles, nec	..	..	..	..	896	879	..	912	..
173	Knitted and crocheted fabrics and articles	..	..	..	..	129	120	..	78	..
18	WEARING APPAREL, DRESSING & DYEING OF FUR	..	..	..	..	144	132	..	401	..
181	Wearing apparel, except fur apparel	..	..	..	..	144	132	..	401	..
182	Dressing and dyeing of fur; articles of fur	..	..	..	..	..	..	..	..	..
19	TANNING & DRESSING OF LEATHER, FOOTWEAR	..	..	..	..	221	215	..	214	..
191	Tanning and dressing of leather	..	..	..	..	177	177	..	165	..
1911	Tanning and dressing of leather	..	..	..	..	153	154	..	153	..
1912	Luggage, handbags, saddlery & harness	..	..	..	..	24	22	..	12	..
192	Footwear	..	..	..	..	45	37	..	49	..
20	WOOD AND WOOD PRODUCTS	..	..	..	..	40,331	42,026	..	41,692	..
201	Sawmilling and planing of wood	..	..	..	..	7,201	7,688	..	32,275	..
202	Products of wood, cork, straw & plaiting materials	..	..	..	..	3,188	4,019	..	9,417	..
2021	Veneer sheets	..	..	..	..	1,453	2,197	..	6,390	..
2022	Builders' carpentry and joinery	..	..	..	..	1,462	1,490	..	2,584	..
2023	Wooden containers	..	..	..	..	169	165	..	142	..
2029	Other products of wood	..	..	..	..	104	167	..	302	..
21	PAPER AND PAPER PRODUCTS	..	..	..	..	236,947	257,228	..	215,631	..
2101	Pulp, paper and paperboard	..	..	..	..	106,814	108,228	..	209,189	..
2102	Corrugated paper, paperboard and their containers	..	..	..	..	1,186	1,295	..	4,084	..
2109	Other articles of pulp and paperboard	..	..	..	..	769	946	..	2,358	..
22	PUBLISHING, PRINTING & REPRODUCTION	..	..	..	..	2,785	2,964	..	3,164	..
221	Publishing	..	..	..	..	846	872	..	978	..
2211	Publishing of books & brochures	..	..	..	..	90	87	..	137	..
2212	Publishing of newspapers and periodicals	..	..	..	..	660	695	..	801	..
2213	Publishing of recorded media	..	..	..	..	83	69	..	37	..
2219	Other publishing	..	..	..	..	12	22	..	3	..
222	Printing and related service activities	..	..	..	..	1,934	2,088	..	2,182	..
2221	Printing	..	..	..	..	1,732	1,878	..	1,832	..
2222	Service activities related to printing	..	..	..	..	202	210	..	350	..
223	Reproduction of recorded media	..	..	..	..	5	3	..	4	..
23	COKE, REFINED PETROLEUM PRODUCTS	..	..	..	..	2,713	2,998	..	7,002	..
231	Coke oven products	..	..	..	..	..	..	..	..	..
232	Refined petroleum products	..	..	..	..	2,599	2,861	..	6,853	..
233	Processing of nuclear fuel	..	..	..	..	115	137	..	149	..
24	CHEMICALS & CHEMICAL PRODUCTS	..	..	..	..	25,738	24,619	..	25,916	..
241	Basic chemicals	..	..	..	..	19,828	18,849	..	20,250	..
2411	Basic chemicals, exc. fertilizers & nitrogen compounds	..	..	..	..	14,327	12,217	..	15,398	..
2412	Fertilizers and nitrogen compounds	..	..	..	..	609	572	..	598	..
2413	Plastics in primary forms and synthetic rubber	..	..	..	..	4,892	6,060	..	4,253	..
242	Other chemical products	..	..	..	..	4,935	4,926	..	4,830	..
2421	Pesticides and other agro-chemical products	..	..	..	..	24	27	..	65	..
2422	Paints, varnishes and similar coatings	..	..	..	..	504	579	..	602	..
2423	Pharmaceuticals, medicinal chem. & botanical prod.	..	..	..	..	2,631	2,637	..	2,604	..
2424	Soap and detergents, perfumes etc.	..	..	..	..	334	365	..	260	..
2429	Other chemical products, nec	..	..	..	..	1,442	1,318	..	1,299	..
243	Man-made fibres	..	..	..	..	975	843	..	836	..
25	RUBBER AND PLASTICS PRODUCTS	..	..	..	..	5,514	6,048	..	5,323	..
251	Rubber products	..	..	..	..	1,187	1,205	..	1,365	..
2511	Rubber tyres and tubes	..	..	..	..	344	358	..	215	..
2519	Other rubber products	..	..	..	..	842	848	..	1,150	..
252	Plastic products	..	..	..	..	4,328	4,843	..	3,958	..
26	OTHER NON-METALLIC MINERAL PRODUCTS	..	..	..	..	19,705	21,847	..	22,690	..
261	Glass and glass products	..	..	..	..	4,772	4,856	..	5,001	..
269	Non-metallic mineral products, nec	..	..	..	..	14,933	16,991	..	17,689	..
2691	Pottery, china and earthenware	..	..	..	..	554	563	..	820	..
2692	Refractory ceramic products	..	..	..	..	136	74	..	139	..
2693	Structural non-refractory clay & ceramic prod.	..	..	..	..	469	438	..	515	..
2694	Cement, lime and plaster	..	..	..	..	8,297	9,750	..	10,584	..
2695	Articles of concrete, cement and plaster	..	..	..	..	2,461	2,665	..	2,000	..
2696	Cutting, shaping and finishing of stone	..	..	..	..	231	397	..	112	..
2699	Other non-metallic mineral products, nec	..	..	..	..	2,786	3,105	..	3,518	..

ISIC Revision 3 Industry Sector	1990	1991	1992	1993	1994	1995	1996	1997	1998
27 **BASIC METALS**	..	..	..	..	86,580	92,304	..	55,601	..
271 Basic iron and steel	..	..	..	..	74,601	79,578	..	41,983	..
272 Basic precious and non-ferrous metals	..	..	..	..	11,931	12,668	..	13,521	..
273 Casting of metals	..	..	..	..	48	57	..	97	..
2731 Casting of iron and steel	..	..	..	..	7	30	..	29	..
2732 Casting non-ferrous metals	..	..	..	..	40	27	..	68	..
28 **FABRICATED METAL PRODUCTS**	..	..	..	..	8,095	9,074	..	10,308	..
281 Str. metal prod., tanks, reservoirs, steam generators	..	..	..	..	1,883	1,904	..	1,874	..
2811 Structural metal products	..	..	..	..	1,584	1,577	..	1,606	..
2812 Tanks, reservoirs and containers of metal	..	..	..	..	267	299	..	210	..
2813 Steam generators, exc. central heating hot water boilers	..	..	..	..	32	29	..	57	..
289 Other fabricated metal products	..	..	..	..	6,212	7,170	..	8,435	..
2891 Forging, pressing, stamping & roll-forming of metal	..	..	..	..	30	46	..	1,039	..
2892 Treatment and coating of metals	..	..	..	..	1,069	1,491	..	2,136	..
2893 Cutlery, hand tools and general hardware	..	..	..	..	1,345	1,445	..	1,371	..
2899 Other fabricated metal products, nec	..	..	..	..	3,768	4,188	..	3,889	..
29 **MACHINERY AND EQUIPMENT, NEC**	..	..	..	..	10,811	11,483	..	12,801	..
291 General purpose machinery	..	..	..	..	5,759	6,546	..	6,091	..
2911 Engines and turbines	..	..	..	..	427	416	..	416	..
2912 Pumps, compressors, taps and valves	..	..	..	..	1,010	1,039	..	974	..
2913 Bearings, gears, gearing and driving elements	..	..	..	..	1,128	1,310	..	1,217	..
2914 Ovens, furnaces and furnace burners	..	..	..	..	31	29	..	20	..
2915 Lifting and handling equipment	..	..	..	..	1,148	1,269	..	1,312	..
2919 Other general purpose machinery	..	..	..	..	2,015	2,482	..	2,153	..
292 Special purpose machinery	..	..	..	..	4,161	4,050	..	5,742	..
2921 Agricultural and forestry machinery	..	..	..	..	278	389	..	394	..
2922 Machine-tools	..	..	..	..	762	740	..	1,349	..
2923 Machinery for metallurgy	..	..	..	..	243	341	..	331	..
2924 Machinery for mining, quarrying and construction	..	..	..	..	1,072	1,049	..	1,154	..
2925 Machinery for food, beverage & tobacco processing	..	..	..	..	117	117	..	227	..
2926 Machinery for textile, apparel & leather production	..	..	..	..	99	87	..	97	..
2927 Machinery for weapons and ammunition	..	..	..	..	629	359	..	667	..
2929 Other special purpose machinery	..	..	..	..	961	968	..	1,524	..
293 Domestic appliances, nec	..	..	..	..	891	887	..	969	..
30 **OFFICE, ACCOUNTING & COMPUTING MACHINERY**	..	..	..	..	377	215	..	217	..
31 **ELECTRICAL MACHINERY & APPARATUS, NEC**	..	..	..	..	2,459	2,491	..	3,339	..
311 Electric motors, generators and transformers	..	..	..	..	274	339	..	561	..
312 Electricity distribution and control apparatus	..	..	..	..	508	516	..	1,021	..
313 Insulated wire and cable	..	..	..	..	642	696	..	771	..
314 Accumulators, primary cells & primary batteries	..	..	..	..	448	506	..	543	..
315 Electric lamps and lighting equipment	..	..	..	..	282	279	..	322	..
319 Other electrical equipment, nec	..	..	..	..	304	155	..	121	..
32 **RADIO, TV & COMMUNICATION EQUIP. & APP.**	..	..	..	..	1,833	2,014	..	2,225	..
321 Electronic valves, tubes, other electronic components	..	..	..	..	410	507	..	596	..
322 TV & radio transmitters, apparatus for line telephony	..	..	..	..	1,390	1,463	..	1,457	..
323 TV & radio receivers, recording apparatus	..	..	..	..	33	44	..	172	..
33 **MEDICAL PRECISION &OPTICAL INSTRUMENTS**	..	..	..	..	1,248	1,092	..	1,150	..
331 Medical appliances and instruments	..	..	..	..	1,173	1,019	..	1,027	..
3311 Medical, surgical equipment & orthopaedic app.	..	..	..	..	353	486	..	371	..
3312 Instruments & appliances for measuring, checking etc	..	..	..	..	751	450	..	613	..
3313 Industrial process control equipment	..	..	..	..	69	84	..	44	..
332 Optical instruments and photographic equipment	..	..	..	..	67	65	..	107	..
333 Watches and clocks	..	..	..	..	8	7	..	16	..
34 **MOTOR VEHICLES, TRAILERS & SEMI-TRAILERS**	..	..	..	..	11,382	12,294	..	12,333	..
341 Motor vehicles	..	..	..	..	7,697	8,194	..	7,929	..
342 Bodies (coachwork) for motor vehicles	..	..	..	..	1,151	1,312	..	1,330	..
343 Parts, accessories for motor vehicles & their engines	..	..	..	..	2,534	2,788	..	3,074	..
35 **OTHER TRANSPORT EQUIPMENT**	..	..	..	..	2,808	2,705	..	3,075	..
351 Building and repairing of ships and boats	..	..	..	..	535	544	..	355	..
3511 Building and repairing of ships	..	..	..	..	417	424	..	265	..
3512 Building, repairing of pleasure & sporting boats	..	..	..	..	118	120	..	89	..
352 Railway, tramway locomotives & rolling stock	..	..	..	..	662	548	..	1,005	..
353 Aircraft and spacecraft	..	..	..	..	1,510	1,502	..	1,607	..
359 Transport equipment, nec	..	..	..	..	101	111	..	108	..
3591 Motorcycles	..	..	..	..	6	10	..	5	..
3592 Bicycles and invalid carriages	..	..	..	..	60	67	..	93	..
3599 Other transport equipment, nec	..	..	..	..	35	33	..	11	..
36 **FURNITURE; MANUFACTURING, NEC**	..	..	..	..	2,257	2,488	..	2,711	..
361 Furniture	..	..	..	..	1,886	1,987	..	2,079	..
369 Manufacturing, nec	..	..	..	..	371	501	..	631	..
3691 Jewellery and related articles	..	..	..	..	36	35	..	288	..
3692 Musical instruments	..	..	..	..	1	1	..	15	..
3693 Sports goods	..	..	..	..	52	58	..	26	..
3694 Games and toys	..	..	..	..	36	152	..	39	..
3699 Other manufacturing, nec	..	..	..	..	246	254	..	263	..
37 **RECYCLING**	..	..	..	..	829	951	..	869	..
371 Recycling of metal waste and scrap	..	..	..	..	810	942	..	856	..
372 Recycling of non-metal waste and scrap	..	..	..	..	19	10	..	13	..
CERR **Non-specified industry**	..	..	..	..	..	..	..	..	..
15-37 **TOTAL MANUFACTURING**	..	..	..	..	488,992	522,152	..	452,843	..

ISIC Revision 3 Industry Sector		1990	1991	1992	1993	1994	1995	1996	1997	1998
15	FOOD PRODUCTS AND BEVERAGES	..	..	..	..	..	..	5,534	..	..
151	Production, processing and preserving (PPP)	..	..	..	..	..	..	551	..	..
1511	PPP of meat and meat products	..	..	..	..	..	..		..	..
1512	Processing and preserving of fish products	..	..	..	..	..	..		..	..
1513	Processing, preserving of fruit & vegetables	..	..	..	..	..	..		..	..
1514	Vegetable and animal oils and fats	..	..	..	..	..	..	551	..	..
152	Dairy products	..	..	..	..	..	..		..	..
153	Grain mill prod., starches & prepared animal feeds	..	..	..	..	..	..	78	..	..
1531	Grain mill products	..	..	..	..	..	..	78	..	..
1532	Starches and starch products	..	..	..	..	..	..		..	..
1533	Prepared animal feeds	..	..	..	..	..	..		..	..
154	Other food products	..	..	..	..	..	..	4,324	..	..
1541	Bakery products	..	..	..	..	..	..		..	..
1542	Sugar	..	..	..	..	..	..		..	..
1543	Cocoa, chocolate and sugar confectionery	..	..	..	..	..	..	399	..	..
1544	Macaroni, noodles, couscous & similar farinaceous prod.	..	..	..	..	..	..		..	..
1549	Other food products, nec	..	..	..	..	..	..	3,925	..	..
155	Beverages	..	..	..	..	..	..	580	..	..
1551	Distilling, rectifying and blending of spirits	..	..	..	..	..	..		..	..
1552	Wines	..	..	..	..	..	..		..	..
1553	Malt liquors and malt	..	..	..	..	..	..	580	..	..
1554	Soft drinks; production of mineral waters	..	..	..	..	..	..		..	..
16	TOBACCO PRODUCTS	..	..	..	..	..	..		..	..
17	TEXTILES	..	..	..	..	..	..	2,383	..	..
171	Spinning, weaving and finishing of textiles	..	..	..	..	..	..		..	..
1711	Preparation and spinning of textile fibres	..	..	..	..	..	..		..	..
1712	Finishing of textiles	..	..	..	..	..	..		..	..
172	Other textiles	..	..	..	..	..	..	2,383	..	..
1721	Made-up textile articles, except apparel	..	..	..	..	..	..		..	..
1722	Carpets and rugs	..	..	..	..	..	..		..	..
1723	Cordage, rope, twine and netting	..	..	..	..	..	..		..	..
1729	Other textiles, nec	..	..	..	..	..	..	2,383	..	..
173	Knitted and crocheted fabrics and articles	..	..	..	..	..	..		..	..
18	WEARING APPAREL, DRESSING & DYEING OF FUR	..	..	..	..	..	..		..	..
181	Wearing apparel, except fur apparel	..	..	..	..	..	..		..	..
182	Dressing and dyeing of fur; articles of fur	..	..	..	..	..	..		..	..
19	TANNING & DRESSING OF LEATHER, FOOTWEAR	..	..	..	..	..	..		..	..
191	Tanning and dressing of leather	..	..	..	..	..	..		..	..
1911	Tanning and dressing of leather	..	..	..	..	..	..		..	..
1912	Luggage, handbags, saddlery & harness	..	..	..	..	..	..		..	..
192	Footwear	..	..	..	..	..	..		..	..
20	WOOD AND WOOD PRODUCTS	..	..	..	..	..	..		..	..
201	Sawmilling and planing of wood	..	..	..	..	..	..		..	..
202	Products of wood, cork, straw & plaiting materials	..	..	..	..	..	..		..	..
2021	Veneer sheets	..	..	..	..	..	..		..	..
2022	Builders' carpentry and joinery	..	..	..	..	..	..		..	..
2023	Wooden containers	..	..	..	..	..	..		..	..
2029	Other products of wood	..	..	..	..	..	..		..	..
21	PAPER AND PAPER PRODUCTS	..	..	..	..	..	..	15,894	..	..
2101	Pulp, paper and paperboard	..	..	..	..	..	..	15,894	..	..
2102	Corrugated paper, paperboard and their containers	..	..	..	..	..	..		..	..
2109	Other articles of pulp and paperboard	..	..	..	..	..	..		..	..
22	PUBLISHING, PRINTING & REPRODUCTION	..	..	..	..	..	..		..	..
221	Publishing	..	..	..	..	..	..		..	..
2211	Publishing of books & brochures	..	..	..	..	..	..		..	..
2212	Publishing of newspapers and periodicals	..	..	..	..	..	..		..	..
2213	Publishing of recorded media	..	..	..	..	..	..		..	..
2219	Other publishing	..	..	..	..	..	..		..	..
222	Printing and related service activities	..	..	..	..	..	..		..	..
2221	Printing	..	..	..	..	..	..		..	..
2222	Service activities related to printing	..	..	..	..	..	..		..	..
223	Reproduction of recorded media	..	..	..	..	..	..		..	..
23	COKE, REFINED PETROLEUM PRODUCTS	..	..	..	..	..	..		..	..
231	Coke oven products	..	..	..	..	..	..		..	..
232	Refined petroleum products	..	..	..	..	..	..		..	..
233	Processing of nuclear fuel	..	..	..	..	..	..		..	..
24	CHEMICALS & CHEMICAL PRODUCTS	..	..	..	..	..	..	19,909	..	..
241	Basic chemicals	..	..	..	..	..	..		..	..
2411	Basic chemicals, exc. fertilizers & nitrogen compounds	..	..	..	..	..	..		..	..
2412	Fertilizers and nitrogen compounds	..	..	..	..	..	..		..	..
2413	Plastics in primary forms and synthetic rubber	..	..	..	..	..	..		..	..
242	Other chemical products	..	..	..	..	..	..	19,909	..	..
2421	Pesticides and other agro-chemical products	..	..	..	..	..	..		..	..
2422	Paints, varnishes and similar coatings	..	..	..	..	..	..	186	..	..
2423	Pharmaceuticals, medicinal chem. & botanical prod.	..	..	..	..	..	..		..	..
2424	Soap and detergents, perfumes etc.	..	..	..	..	..	..	168	..	..
2429	Other chemical products, nec	..	..	..	..	..	..	19,556	..	..
243	Man-made fibres	..	..	..	..	..	..		..	..
25	RUBBER AND PLASTICS PRODUCTS	..	..	..	..	..	..		..	..
251	Rubber products	..	..	..	..	..	..		..	..
2511	Rubber tyres and tubes	..	..	..	..	..	..		..	..
2519	Other rubber products	..	..	..	..	..	..		..	..
252	Plastic products	..	..	..	..	..	..		..	..
26	OTHER NON-METALLIC MINERAL PRODUCTS	..	..	..	..	..	..	14,561	..	..
261	Glass and glass products	..	..	..	..	..	..		..	..
269	Non-metallic mineral products, nec	..	..	..	..	..	..	14,561	..	..
2691	Pottery, china and earthenware	..	..	..	..	..	..		..	..
2692	Refractory ceramic products	..	..	..	..	..	..		..	..
2693	Structural non-refractory clay & ceramic prod.	..	..	..	..	..	..	3,121	..	..
2694	Cement, lime and plaster	..	..	..	..	..	..	11,440	..	..
2695	Articles of concrete, cement and plaster	..	..	..	..	..	..		..	..
2696	Cutting, shaping and finishing of stone	..	..	..	..	..	..		..	..
2699	Other non-metallic mineral products, nec	..	..	..	..	..	..		..	..

ISIC Revision 3 Industry Sector		1990	1991	1992	1993	1994	1995	1996	1997	1998
27	BASIC METALS	..	..	..	..	..	..	1,618	..	..
271	Basic iron and steel	..	..				..		..	..
272	Basic precious and non-ferrous metals	..	..				..		..	..
273	Casting of metals	..	..				..	1,618	..	..
2731	Casting of iron and steel	..	..				..		..	..
2732	Casting non-ferrous metals	..	..				..	1,618	..	..
28	FABRICATED METAL PRODUCTS	..	..	..	..	..	..	234	..	..
281	Str. metal prod., tanks, reservoirs, steam generators	..	..				..		..	..
2811	Structural metal products	..	..				..		..	..
2812	Tanks, reservoirs and containers of metal	..	..				..	..	..	..
2813	Steam generators, exc. central heating hot water boilers	..	..				..		..	..
289	Other fabricated metal products	..	..				..	234	..	..
2891	Forging, pressing, stamping & roll-forming of metal	..	..				..		..	..
2892	Treatment and coating of metals	..	..				..		..	..
2893	Cutlery, hand tools and general hardware	..	..				..		..	..
2899	Other fabricated metal products, nec	..	..				..	234	..	..
29	MACHINERY AND EQUIPMENT, NEC	..	..	..	..	..	..	11,265	..	..
291	General purpose machinery	..	..				..	11,265	..	..
2911	Engines and turbines	..	..				..	..	..	..
2912	Pumps, compressors, taps and valves	..	..				..	..	..	..
2913	Bearings, gears, gearing and driving elements	..	..				..	..	..	..
2914	Ovens, furnaces and furnace burners	..	..				..	..	..	..
2915	Lifting and handling equipment	..	..				..	..	..	..
2919	Other general purpose machinery	..	..				..	11,265	..	..
292	Special purpose machinery	..	..				..	..	..	..
2921	Agricultural and forestry machinery	..	..				..	..	..	..
2922	Machine-tools	..	..				..	..	..	..
2923	Machinery for metallurgy	..	..				..	..	..	..
2924	Machinery for mining, quarrying and construction	..	..				..	..	..	..
2925	Machinery for food, beverage & tobacco processing	..	..				..	..	..	..
2926	Machinery for textile, apparel & leather production	..	..				..	..	..	..
2927	Machinery for weapons and ammunition	..	..				..	..	..	..
2929	Other special purpose machinery	..	..				..	..	..	..
293	Domestic appliances, nec	..	..				..	..	..	..
30	OFFICE, ACCOUNTING & COMPUTING MACHINERY	..	..	..	..	..	..	..	..	..
31	ELECTRICAL MACHINERY & APPARATUS, NEC	..	..	..	..	..	..	..	..	..
311	Electric motors, generators and transformers	..	..				..	..	..	..
312	Electricity distribution and control apparatus	..	..				..	..	..	..
313	Insulated wire and cable	..	..				..	..	..	..
314	Accumulators, primary cells & primary batteries	..	..				..	..	..	..
315	Electric lamps and lighting equipment	..	..				..	..	..	..
319	Other electrical equipment, nec	..	..				..	..	..	..
32	RADIO, TV & COMMUNICATION EQUIP. & APP.	..	..	..	..	..	..	..	..	..
321	Electronic valves, tubes, other electronic components	..	..				..	..	..	..
322	TV & radio transmitters, apparatus for line telephony	..	..				..	..	..	..
323	TV & radio receivers, recording apparatus	..	..				..	..	..	..
33	MEDICAL PRECISION &OPTICAL INSTRUMENTS	..	..	..	..	..	..	..	..	..
331	Medical appliances and instruments	..	..				..	..	..	..
3311	Medical, surgical equipment & orthopaedic app.	..	..				..	..	..	..
3312	Instruments & appliances for measuring, checking etc	..	..				..	..	..	..
3313	Industrial process control equipment	..	..				..	..	..	..
332	Optical instruments and photographic equipment	..	..				..	..	..	..
333	Watches and clocks	..	..				..	..	..	..
34	MOTOR VEHICLES, TRAILERS & SEMI-TRAILERS	..	..	..	..	..	..	..	..	..
341	Motor vehicles	..	..				..	..	..	..
342	Bodies (coachwork) for motor vehicles	..	..				..	..	..	..
343	Parts, accessories for motor vehicles & their engines	..	..				..	..	..	..
35	OTHER TRANSPORT EQUIPMENT	..	..	..	..	..	..	..	..	..
351	Building and repairing of ships and boats	..	..				..	..	..	..
3511	Building and repairing of ships	..	..				..	..	..	..
3512	Building, repairing of pleasure & sporting boats	..	..				..	..	..	..
352	Railway, tramway locomotives & rolling stock	..	..				..	..	..	..
353	Aircraft and spacecraft	..	..				..	..	..	..
359	Transport equipment, nec	..	..				..	..	..	..
3591	Motorcycles	..	..				..	..	..	..
3592	Bicycles and invalid carriages	..	..				..	..	..	..
3599	Other transport equipment, nec	..	..				..	..	..	..
36	FURNITURE; MANUFACTURING, NEC	..	..	..	..	..	..	..	..	..
361	Furniture	..	..				..	..	..	..
369	Manufacturing, nec	..	..				..	..	..	..
3691	Jewellery and related articles	..	..				..	..	..	..
3692	Musical instruments	..	..				..	..	..	..
3693	Sports goods	..	..				..	..	..	..
3694	Games and toys	..	..				..	..	..	..
3699	Other manufacturing, nec	..	..				..	..	..	..
37	RECYCLING	..	..	..	..	..	..	..	..	
371	Recycling of metal waste and scrap	..	..				..	..	..	
372	Recycling of non-metal waste and scrap	..	..				..	..	..	
CERR	Non-specified industry	..	..	..	..	..	..	..		
15-37	TOTAL MANUFACTURING	..	..	..	..	..	..	71,398	..	..

ISIC Revision 3 Industry Sector	1990	1991	1992	1993	1994	1995	1996	1997	1998
15 **FOOD PRODUCTS AND BEVERAGES**	182,326	177,230	179,700	169,573	177,614	182,913	188,214	..	..
151 Production, processing and preserving (PPP)	..	..	..	..	..	..	..	..	..
1511 PPP of meat and meat products	..	..	..	..	..	..	..	..	..
1512 Processing and preserving of fish products	..	..	..	..	..	..	..	..	..
1513 Processing, preserving of fruit & vegetables	..	..	..	..	..	..	..	..	..
1514 Vegetable and animal oils and fats	..	..	..	..	..	..	..	..	..
152 Dairy products	..	..	..	..	..	..	..	..	..
153 Grain mill prod., starches & prepared animal feeds	..	..	..	..	..	..	..	..	..
1531 Grain mill products	..	..	..	..	..	..	..	..	..
1532 Starches and starch products	..	..	..	..	..	..	..	..	..
1533 Prepared animal feeds	..	..	..	..	..	..	..	..	..
154 Other food products	..	..	..	..	..	..	..	..	..
1541 Bakery products	..	..	..	..	..	..	..	..	..
1542 Sugar	..	..	..	..	..	..	..	..	..
1543 Cocoa, chocolate and sugar confectionery	..	..	..	..	..	..	..	..	..
1544 Macaroni, noodles, couscous & similar farinaceous prod.	..	..	..	..	..	..	..	..	..
1549 Other food products, nec	..	..	..	..	..	..	..	..	..
155 Beverages	..	..	..	..	..	..	..	..	..
1551 Distilling, rectifying and blending of spirits	..	..	..	..	..	..	..	..	..
1552 Wines	..	..	..	..	..	..	..	..	..
1553 Malt liquors and malt	..	..	..	..	..	..	..	..	..
1554 Soft drinks; production of mineral waters	..	..	..	..	..	..	..	..	..
16 **TOBACCO PRODUCTS**	1,881	1,984	1,967	1,858	2,111	2,173	2,356	..	..
17 **TEXTILES**	37,909	37,448	35,615	42,729	39,092	37,267	35,984	..	..
171 Spinning, weaving and finishing of textiles	..	..	..	..	..	..	..	..	..
1711 Preparation and spinning of textile fibres	..	..	..	..	..	..	..	..	..
1712 Finishing of textiles	..	..	..	..	..	..	..	..	..
172 Other textiles	..	..	..	..	..	..	..	..	..
1721 Made-up textile articles, except apparel	..	..	..	..	..	..	..	..	..
1722 Carpets and rugs	..	..	..	..	..	..	..	..	..
1723 Cordage, rope, twine and netting	..	..	..	..	..	..	..	..	..
1729 Other textiles, nec	..	..	..	..	..	..	..	..	..
173 Knitted and crocheted fabrics and articles	..	..	..	..	..	..	..	..	..
18 **WEARING APPAREL, DRESSING & DYEING OF FUR**	8,834	8,700	8,247	8,871	8,401	7,237	7,400	..	..
181 Wearing apparel, except fur apparel	..	..	..	..	..	..	..	..	..
182 Dressing and dyeing of fur; articles of fur	..	..	..	..	..	..	..	..	..
19 **TANNING & DRESSING OF LEATHER, FOOTWEAR**	2,762	2,861	2,645	3,096	2,925	3,016	2,936	..	..
191 Tanning and dressing of leather	..	..	..	..	..	..	..	..	..
1911 Tanning and dressing of leather	..	..	..	..	..	..	..	..	..
1912 Luggage, handbags, saddlery & harness	..	..	..	..	..	..	..	..	..
192 Footwear	..	..	..	..	..	..	..	..	..
20 **WOOD AND WOOD PRODUCTS**	14,139	15,718	11,661	13,908	17,327	18,299	19,768	..	..
201 Sawmilling and planing of wood	..	..	..	..	..	..	..	..	..
202 Products of wood, cork, straw & plaiting materials	..	..	..	..	..	..	..	..	..
2021 Veneer sheets	..	..	..	..	..	..	..	..	..
2022 Builders' carpentry and joinery	..	..	..	..	..	..	..	..	..
2023 Wooden containers	..	..	..	..	..	..	..	..	..
2029 Other products of wood	..	..	..	..	..	..	..	..	..
21 **PAPER AND PAPER PRODUCTS**	77,619	82,637	84,238	80,803	87,496	90,545	78,789	..	..
2101 Pulp, paper and paperboard	..	..	..	..	..	..	..	..	..
2102 Corrugated paper, paperboard and their containers	..	..	..	..	..	..	..	..	..
2109 Other articles of pulp and paperboard	..	..	..	..	..	..	..	..	..
22 **PUBLISHING, PRINTING & REPRODUCTION**	18,196	19,190	19,163	21,080	24,229	26,490	27,259	..	..
221 Publishing	..	..	..	..	..	..	..	..	..
2211 Publishing of books & brochures	..	..	..	..	..	..	..	..	..
2212 Publishing of newspapers and periodicals	..	..	..	..	..	..	..	..	..
2213 Publishing of recorded media	..	..	..	..	..	..	..	..	..
2219 Other publishing	..	..	..	..	..	..	..	..	..
222 Printing and related service activities	..	..	..	..	..	..	..	..	..
2221 Printing	..	..	..	..	..	..	..	..	..
2222 Service activities related to printing	..	..	..	..	..	..	..	..	..
223 Reproduction of recorded media	..	..	..	..	..	..	..	..	..
23 **COKE, REFINED PETROLEUM PRODUCTS**	323,449	328,195	284,455	348,320	340,110	353,515	362,163	..	..
231 Coke oven products	..	..	..	..	..	..	..	..	..
232 Refined petroleum products	..	..	..	..	..	..	..	..	..
233 Processing of nuclear fuel	..	..	..	..	..	..	..	..	..
24 **CHEMICALS & CHEMICAL PRODUCTS**	223,179	200,011	204,618	211,595	249,569	215,291	286,446	..	..
241 Basic chemicals	172,089	152,652	156,121	160,683	174,565	149,367	200,129	..	..
2411 Basic chemicals, exc. fertilizers & nitrogen compounds	139,664	128,687	134,082	138,904	150,462	125,261	173,738	..	..
2412 Fertilizers and nitrogen compounds	10,496	6,864	5,257	4,640	4,932	5,836	5,848	..	..
2413 Plastics in primary forms and synthetic rubber	21,929	17,101	16,782	17,139	19,172	18,270	20,543	..	..
242 Other chemical products	45,301	41,822	44,999	50,422	57,675	52,491	72,351	..	..
2421 Pesticides and other agro-chemical products	2,728	3,018	3,457	4,562	5,013	3,221	6,478	..	..
2422 Paints, varnishes and similar coatings	5,217	4,747	4,754	5,608	6,841	6,321	6,724	..	..
2423 Pharmaceuticals, medicinal chem. & botanical prod.	19,373	17,541	18,276	19,370	21,536	21,109	27,206	..	..
2424 Soap and detergents, perfumes etc.	7,092	5,687	5,874	6,521	7,191	6,648	9,465	..	..
2429 Other chemical products, nec	10,891	10,830	12,637	14,361	17,094	15,192	22,478	..	..
243 Man-made fibres	18,885	18,637	16,001	13,447	13,616	10,609	11,840	..	..
25 **RUBBER AND PLASTICS PRODUCTS**	89,416	92,279	78,036	84,289	92,298	94,290	94,943	..	..
251 Rubber products	20,644	20,380	15,521	17,425	19,706	20,531	18,437	..	..
2511 Rubber tyres and tubes	..	..	..	..	..	..	..	..	..
2519 Other rubber products	..	..	..	..	..	..	..	..	..
252 Plastic products	68,772	71,898	62,515	66,863	72,592	73,759	76,506	..	..
26 **OTHER NON-METALLIC MINERAL PRODUCTS**	154,968	133,483	125,386	101,098	107,547	105,762	106,194	..	..
261 Glass and glass products	22,623	23,286	22,005	21,585	25,919	23,392	20,796	..	..
269 Non-metallic mineral products, nec	132,346	110,197	103,381	79,512	81,628	82,370	85,398	..	..
2691 Pottery, china and earthenware	13,422	12,376	11,705	11,079	11,230	12,259	13,166	..	..
2692 Refractory ceramic products	17,909	14,818	13,677	12,009	13,841	14,935	14,675	..	..
2693 Structural non-refractory clay & ceramic prod.	..	..	..	..	..	..	..	..	..
2694 Cement, lime and plaster	79,534	64,378	60,650	40,250	42,493	40,727	45,310	..	..
2695 Articles of concrete, cement and plaster	21,481	18,626	17,350	16,174	14,063	14,449	12,246	..	..
2696 Cutting, shaping and finishing of stone	..	..	..	..	..	..	..	..	..
2699 Other non-metallic mineral products, nec	..	..	..	..	..	..	..	..	..

ISIS Energy Data Programme (IEA/OECD)

ISIC Revision 3 Industry Sector	1990	1991	1992	1993	1994	1995	1996	1997	1998
27	**BASIC METALS**	323,980	313,971	307,018	320,915	370,639	372,410	391,570	..	..
271	Basic iron and steel	258,955	251,792	245,428	259,552	275,333	279,948	291,428	..	..
272	Basic precious and non-ferrous metals	39,806	39,783	40,474	41,529	48,618	44,575	43,781	..	..
273	Casting of metals	25,219	22,396	21,117	19,834	23,044	20,162	20,342	..	..
2731	Casting of iron and steel	..	..	..	..	..	..	..	..	..
2732	Casting non-ferrous metals	..	..	..	..	..	..	..	..	..
28	**FABRICATED METAL PRODUCTS**	35,755	33,357	29,938	28,899	24,068	25,472	27,383	..	..
281	Str. metal prod., tanks, reservoirs, steam generators	..	..	..	..	..	..	..	..	..
2811	Structural metal products	..	..	..	..	..	..	..	..	..
2812	Tanks, reservoirs and containers of metal	..	..	..	..	..	..	..	..	..
2813	Steam generators, exc. central heating hot water boilers	..	..	..	..	..	..	..	..	..
289	Other fabricated metal products	..	..	..	..	..	..	..	..	..
2891	Forging, pressing, stamping & roll-forming of metal	..	..	..	..	..	..	..	..	..
2892	Treatment and coating of metals	..	..	..	..	..	..	..	..	..
2893	Cutlery, hand tools and general hardware	..	..	..	..	..	..	..	..	..
2899	Other fabricated metal products, nec	..	..	..	..	..	..	..	..	..
29	**MACHINERY AND EQUIPMENT, NEC**	64,110	58,831	56,689	56,986	50,346	50,313	51,786	..	..
291	General purpose machinery	..	..	..	..	..	..	..	..	..
2911	Engines and turbines	..	..	..	..	..	..	..	..	..
2912	Pumps, compressors, taps and valves	..	..	..	..	..	..	..	..	..
2913	Bearings, gears, gearing and driving elements	..	..	..	..	..	..	..	..	..
2914	Ovens, furnaces and furnace burners	..	..	..	..	..	..	..	..	..
2915	Lifting and handling equipment	..	..	..	..	..	..	..	..	..
2919	Other general purpose machinery	..	..	..	..	..	..	..	..	..
292	Special purpose machinery	..	..	..	..	..	..	..	..	..
2921	Agricultural and forestry machinery	..	..	..	..	..	..	..	..	..
2922	Machine-tools	..	..	..	..	..	..	..	..	..
2923	Machinery for metallurgy	..	..	..	..	..	..	..	..	..
2924	Machinery for mining, quarrying and construction	..	..	..	..	..	..	..	..	..
2925	Machinery for food, beverage & tobacco processing	..	..	..	..	..	..	..	..	..
2926	Machinery for textile, apparel & leather production	..	..	..	..	..	..	..	..	..
2927	Machinery for weapons and ammunition	..	..	..	..	..	..	..	..	..
2929	Other special purpose machinery	..	..	..	..	..	..	..	..	..
293	Domestic appliances, nec	..	..	..	..	..	..	..	..	..
30	**OFFICE, ACCOUNTING & COMPUTING MACHINERY**	3,520	3,952	3,829	3,400	2,556	2,787	3,077	..	..
31	**ELECTRICAL MACHINERY & APPARATUS, NEC**	18,847	20,272	19,303	19,058	16,386	15,852	16,280	..	..
311	Electric motors, generators and transformers	..	..	..	..	..	..	..	..	..
312	Electricity distribution and control apparatus	..	..	..	..	..	..	..	..	..
313	Insulated wire and cable	..	..	..	..	..	..	..	..	..
314	Accumulators, primary cells & primary batteries	..	..	..	..	..	..	..	..	..
315	Electric lamps and lighting equipment	..	..	..	..	..	..	..	..	..
319	Other electrical equipment, nec	..	..	..	..	..	..	..	..	..
32	**RADIO, TV & COMMUNICATION EQUIP. & APP.**	18,977	18,690	16,825	15,971	13,556	13,863	15,049	..	..
321	Electronic valves, tubes, other electronic components	..	..	..	..	..	..	..	..	..
322	TV & radio transmitters, apparatus for line telephony	..	..	..	..	..	..	..	..	..
323	TV & radio receivers, recording apparatus	..	..	..	..	..	..	..	..	..
33	**MEDICAL PRECISION &OPTICAL INSTRUMENTS**	6,754	6,552	6,284	5,941	4,935	4,794	5,098	..	..
331	Medical appliances and instruments	..	..	..	..	..	..	..	..	..
3311	Medical, surgical equipment & orthopaedic app.	..	..	..	..	..	..	..	..	..
3312	Instruments & appliances for measuring, checking etc	..	..	..	..	..	..	..	..	..
3313	Industrial process control equipment	..	..	..	..	..	..	..	..	..
332	Optical instruments and photographic equipment	..	..	..	..	..	..	..	..	..
333	Watches and clocks	..	..	..	..	..	..	..	..	..
34	**MOTOR VEHICLES, TRAILERS & SEMI-TRAILERS**	39,273	38,175	42,200	40,723	37,546	40,907	42,495	..	..
341	Motor vehicles	..	..	..	..	..	..	..	..	..
342	Bodies (coachwork) for motor vehicles	..	..	..	..	..	..	..	..	..
343	Parts, accessories for motor vehicles & their engines	..	..	..	..	..	..	..	..	..
35	**OTHER TRANSPORT EQUIPMENT**	32,151	32,114	32,178	29,880	26,124	25,267	26,630	..	..
351	Building and repairing of ships and boats	..	..	..	..	..	..	..	..	..
3511	Building and repairing of ships	..	..	..	..	..	..	..	..	..
3512	Building, repairing of pleasure & sporting boats	..	..	..	..	..	..	..	..	..
352	Railway, tramway locomotives & rolling stock	..	..	..	..	..	..	..	..	..
353	Aircraft and spacecraft	..	..	..	..	..	..	..	..	..
359	Transport equipment, nec	..	..	..	..	..	..	..	..	..
3591	Motorcycles	..	..	..	..	..	..	..	..	..
3592	Bicycles and invalid carriages	..	..	..	..	..	..	..	..	..
3599	Other transport equipment, nec	..	..	..	..	..	..	..	..	..
36	**FURNITURE; MANUFACTURING, NEC**	35,778	38,807	30,071	31,531	33,039	30,888	31,157	..	..
361	Furniture	..	..	..	..	..	..	..	..	..
369	Manufacturing, nec	..	..	..	..	..	..	..	..	..
3691	Jewellery and related articles	..	..	..	..	..	..	..	..	..
3692	Musical instruments	..	..	..	..	..	..	..	..	..
3693	Sports goods	..	..	..	..	..	..	..	..	..
3694	Games and toys	..	..	..	..	..	..	..	..	..
3699	Other manufacturing, nec	..	..	..	..	..	..	..	..	..
37	**RECYCLING**	..	..	..	..	..	..	..	..	..
371	Recycling of metal waste and scrap	..	..	..	..	..	..	..	..	..
372	Recycling of non-metal waste and scrap	..	..	..	..	..	..	..	..	..
CERR	**Non-specified industry**	-12,528	-12,715	-12,136	-8,010	20,780	27,484	19,871	..	..
15-37	**TOTAL MANUFACTURING**	1,716,097	1,666,791	1,583,985	1,647,994	1,748,695	1,746,835	1,842,853	..	..

Information System on Industrial Structures (ISIS)

Energy Consumption Statistics

in

Manufacturing Industry

1990 – 1998

Separately by fuel types

(ISIC Revision 2)

OECD
OCDE
OECD
ORGANISATION FOR ECONOMIC
CO-OPERATION AND DEVELOPMENT
OCDE
ORGANISATION DE COOPERATION
ET DE DEVELOPPEMENT ECONOMIQUES

IEA
INTERNATIONAL ENERGY AGENCY
AIE
AGENCE INTERNATIONALE DE L'ENERGIE

9, RUE DE LA FÉDERATION, 75739 PARIS CEDEX 15
TEL. (33-1) 40.57.65.00 / TELEFAX (33-1) 40.57.66.49 / INTERNET: HTTP://WWW.IEA.ORG

ISIC Revision 2 Industry Sector	Solid TJ	LPG TJ	Distiloil TJ	RFO TJ	Gas TJ	Biomass TJ	Steam TJ	Electr MWh	Own Use MWh	TOTAL TJ
31 FOOD, BEVERAGES AND TOBACCO	13,991	3,010	997	2,160	21,309	76,835	..	4,525,000	415,000	133,098
311/2 FOOD	13,204	2,866	977	2,002	17,255	76,835	..	3,970,000	415,000	125,937
3111 Slaughtering, preparing and preserving meat	2,818	687	196	1,073	3,648	627	..	1,072,000	0	12,908
3112 Dairy products	3,240	656	128	345	3,595	1,018	..	602,000	0	11,149
3113 Canning, preserving of fruits and vegetables	1,318	471	2	190	1,284	376	..	302,000	0	4,728
3114 Canning, preserving and processing of fish	..	..	..	..	..	..	..	..	..	..
3115 Vegetable and animal oils and fats	947	6	8	18	998	0	..	231,000	0	2,809
3116 Grain mill products	1,099	234	136	33	1,471	0	..	423,000	0	4,496
3117 Bakery products	0	355	99	91	2,255	0	..	355,000	0	4,078
3118 Sugar factories and refineries	3,782	457	408	252	4,004	74,814	..	985,000	415,000	85,769
3119 Cocoa, chocolate and sugar confectionery	..	..	..	..	..	..	..	..	..	..
3121 Other food products	..	..	..	..	..	..	..	..	..	..
3122 Prepared animal feeds	..	..	..	..	..	..	..	..	..	..
313 BEVERAGES	694	141	20	158	3,527	0	..	488,000	0	6,297
3131 Distilling, rectifying and blending of spirits	..	..	..	..	..	..	..	..	..	..
3132 Wine industries	..	..	..	..	..	..	..	..	..	..
3133 Malt liquors and malts	..	..	..	..	..	..	..	..	..	..
3134 Soft drinks	..	..	..	..	..	..	..	..	..	..
314 TOBACCO	93	3	0	0	527	0	..	67,000	0	864
32 TEXTILES, APPAREL AND LEATHER	1,212	424	51	215	5,632	0	..	1,867,000	0	14,255
321 TEXTILES	..	..	..	..	..	..	..	..	..	..
3211 Spinning weaving and finishing textiles	..	..	..	..	..	..	..	..	..	..
3212 Made-up goods excluding wearing apparel	..	..	..	..	..	..	..	..	..	..
3213 Knitting mills	..	..	..	..	..	..	..	..	..	..
3214 Carpets and rugs	..	..	..	..	..	..	..	..	..	..
3215 Cordage, rope and twine	..	..	..	..	..	..	..	..	..	..
3219 Other textiles	..	..	..	..	..	..	..	..	..	..
322 WEARING APPAREL, EXCEPT FOOTWEAR	..	..	..	..	..	..	..	..	..	..
323 LEATHER AND FUR PRODUCTS	..	..	..	..	..	..	..	..	..	..
3231 Tanneries and leather finishing	..	..	..	..	..	..	..	..	..	..
3232 Fur dressing and dyeing industries	..	..	..	..	..	..	..	..	..	..
3233 Leather prods. ex. footwear and wearing apparel	..	..	..	..	..	..	..	..	..	..
324 FOOTWEAR, EX. RUBBER AND PLASTIC	..	..	..	..	..	..	..	..	..	..
33 WOOD PRODUCTS AND FURNITURE	814	194	865	653	1,117	7,503	..	1,133,000	19,000	15,156
331 WOOD PRODUCTS, EXCEPT FURNITURE	..	..	..	..	..	..	..	..	..	..
3311 Sawmills, planing and other wood mills	..	..	..	..	..	..	..	..	..	..
3312 Wooden and cane containers	..	..	..	..	..	..	..	..	..	..
3319 Other wood and cork products	..	..	..	..	..	..	..	..	..	..
332 FURNITURE, FIXTURES, EXCL. METALLIC	..	..	..	..	..	..	..	..	..	..
34 PAPER, PUBLISHING AND PRINTING	8,716	748	50	871	15,117	7,850	..	4,093,000	470,000	46,395
341 PAPER AND PRODUCTS	..	..	..	..	..	..	..	..	..	..
3411 Pulp, paper and paperboard articles	..	..	..	..	..	..	..	..	..	..
3412 Containers of paper and paperboard	..	..	..	..	..	..	..	..	..	..
3419 Other pulp, paper and paperboard articles	..	..	..	..	..	..	..	..	..	..
342 PRINTING AND PUBLISHING	..	..	..	..	..	..	..	..	..	..
35 CHEMICAL PRODUCTS	7,745	887	22,174	494	99,461	0	..	5,104,000	362,000	147,832
351 INDUSTRIAL CHEMICALS	4,056	641	86	63	33,547	0	..	2,820,000	156,000	47,983
3511 Basic industrial chemicals excl. fertilizers	..	..	..	..	..	..	..	..	..	..
3512 Fertilizers and pesticides	..	..	..	..	..	..	..	..	..	..
3513 Synthetic resins and plastic materials	..	..	..	..	..	..	..	..	..	..
352 OTHER CHEMICALS	312	48	22	29	1,484	0	..	612,000	0	4,098
3521 Paints, varnishes and lacquers	..	..	..	..	..	..	..	..	..	..
3522 Drugs and medicines	..	..	..	..	..	..	..	..	..	..
3523 Soap, cleaning preparations, perfumes, cosmetics	..	..	..	..	..	..	..	..	..	..
3529 Other chemical products	..	..	..	..	..	..	..	..	..	..
353 PETROLEUM REFINERIES	0	0	21,999	0	60,884	0	..	1,236,000	206,000	86,591
354 MISC. PETROLEUM AND COAL PRODUCTS	3,278	133	30	402	1,230	0	..	88,000	0	5,390
355 RUBBER PRODUCTS	43	16	0	0	1,102	0	..	154,000	0	1,715
3551 Tyres and tubes	..	..	..	..	..	..	..	..	..	..
3559 Other rubber products	..	..	..	..	..	..	..	..	..	..
356 PLASTIC PRODUCTS	56	49	37	0	1,214	0	..	194,000	0	2,054
36 NON-METALLIC MINERAL PRODUCTS	22,071	1,553	1,172	1,047	45,879	981	..	2,905,000	0	83,161
361 POTTERY, CHINA, EARTHENWARE	48	198	27	0	2,193	0	..	66,000	0	2,704
362 GLASS AND PRODUCTS	0	195	21	219	9,745	0	..	361,000	0	11,480
369 OTHER NON-METAL. MINERAL PRODUCTS	22,023	1,160	1,124	828	33,941	981	..	2,478,000	0	68,978
3691 Structural clay products	1,438	282	414	358	14,327	973	..	667,000	0	20,193
3692 Cement, lime and plaster	16,671	512	577	360	16,446	8	..	1,295,000	0	39,236
3699 Other non-metallic mineral products	3,914	366	133	110	3,168	0	..	516,000	0	9,549
37 BASIC METAL INDUSTRIES	329,494	2,443	3,353	30,914	160,431	970	..	31,343,000	3,493,000	627,865
371 IRON AND STEEL	277,031	622	152	1,315	74,892	198	..	4,764,000	859,000	368,268
372 NON-FERROUS METALS	52,463	1,821	3,201	29,599	85,539	772	..	26,579,000	2,634,000	259,597
38 METAL PRODUCTS, MACHINERY, EQUIP.	64	1,377	176	19	11,925	0	..	3,931,000	1,000	27,709
381 METAL PRODUCTS	..	..	..	..	..	..	..	..	..	..
3811 Cutlery, hand tools and general hardware	..	..	..	..	..	..	..	..	..	..
3812 Furniture and fixtures primarily of metal	..	..	..	..	..	..	..	..	..	..
3813 Structural metal products	..	..	..	..	..	..	..	..	..	..
3819 Other fabricated metal products	..	..	..	..	..	..	..	..	..	..
382 NON-ELECTRICAL MACHINERY	..	..	..	..	..	..	..	..	..	..
3821 Engines and turbines	..	..	..	..	..	..	..	..	..	..
3822 Agricultural machinery and equipment	..	..	..	..	..	..	..	..	..	..
3823 Metal and wood working machinery	..	..	..	..	..	..	..	..	..	..
3824 Special industrial machinery	..	..	..	..	..	..	..	..	..	..
3825 Office, computing and accounting machinery	..	..	..	..	..	..	..	..	..	..
3829 Other non-electrical machinery and equipment	..	..	..	..	..	..	..	..	..	..
383 ELECTRICAL MACHINERY	..	..	..	..	..	..	..	..	..	..
3831 Electrical industrial machinery	..	..	..	..	..	..	..	..	..	..
3832 Radio, TV and communications equipment	..	..	..	..	..	..	..	..	..	..
3833 Electrical appliances and housewares	..	..	..	..	..	..	..	..	..	..
3839 Other electrical apparatus and supplies	..	..	..	..	..	..	..	..	..	..
384 TRANSPORT EQUIPMENT	..	..	..	..	..	..	..	..	..	..
3841 Shipbuilding	..	..	..	..	..	..	..	..	..	..
3842 Railroad equipment	..	..	..	..	..	..	..	..	..	..
3843 Motor vehicles	..	..	..	..	..	..	..	..	..	..
3844 Motorcycles and bicycles	..	..	..	..	..	..	..	..	..	..
3845 Aircraft	..	..	..	..	..	..	..	..	..	..
3849 Other transport equipment	..	..	..	..	..	..	..	..	..	..
385 PROFESSIONAL AND SCIENTIFIC EQUIPMENT	..	..	..	..	..	..	..	..	..	..
3851 Professional equipment	..	..	..	..	..	..	..	..	..	..
3852 Photographic and optical goods	..	..	..	..	..	..	..	..	..	..
3853 Watches and clocks	..	..	..	..	..	..	..	..	..	..
39 OTHER MANUFACTURING INDUSTRIES	0	3	0	0	233	0	..	22,000	0	315
3901 Jewellery and related articles	..	..	..	..	..	..	..	..	..	..
3902 Musical instruments	..	..	..	..	..	..	..	..	..	..
3903 Sporting and athletic goods	..	..	..	..	..	..	..	..	..	..
3909 Other manufactures	..	..	..	..	..	..	..	..	..	..
SERR non-specified, unallocated industry	..	..	..	..	..	..	..	..	..	..
3 TOTAL MANUFACTURING	384,107	10,639	28,838	36,373	361,104	94,139	..	54,923,000	4,760,000	1,095,787

ISIS Energy Data Programme (IEA/OECD)

Year: **1992** **AUSTRALIA**

ISIC Revision 2 Industry Sector	Solid TJ	LPG TJ	Distiloil TJ	RFO TJ	Gas TJ	Biomass TJ	Steam TJ	Electr MWh	Own Use MWh	TOTAL TJ
31 **FOOD, BEVERAGES AND TOBACCO**	13,508	2,884	941	2,013	22,619	63,588	..	4,550,000	410,000	120,457
311/2 FOOD	12,849	2,753	913	1,852	18,352	63,588		3,995,000	410,000	113,213
3111 Slaughtering, preparing and preserving meat	2,922	677	197	963	3,803	634		1,075,000	0	13,066
3112 Dairy products	3,113	665	128	391	4,104	1,090		601,000	0	11,655
3113 Canning, preserving of fruits and vegetables	1,268	469	3	200	1,345	351		311,000	0	4,756
3114 Canning, preserving and processing of fish	..	..	..	..	..	..		..	..	..
3115 Vegetable and animal oils and fats	977	6	11	19	982	0		231,000	0	2,827
3116 Grain mill products	1,683	225	23	33	1,619	0		443,000	0	5,178
3117 Bakery products	0	266	83	96	2,336	0		351,000	0	4,045
3118 Sugar factories and refineries	2,886	445	468	150	4,163	61,513		983,000	410,000	71,688
3119 Cocoa, chocolate and sugar confectionery	..	..	..	..	..	..		..	..	..
3121 Other food products	..	..	..	..	..	..		..	..	..
3122 Prepared animal feeds	..	..	..	..	..	..		..	..	..
313 BEVERAGES	621	129	19	161	3,739	0		482,000	0	6,404
3131 Distilling, rectifying and blending of spirits	..	..	..	..	..	..		..	..	..
3132 Wine industries	..	..	..	..	..	..		..	..	..
3133 Malt liquors and malts	..	..	..	..	..	..		..	..	..
3134 Soft drinks	..	..	..	..	..	..		..	..	..
314 TOBACCO	38	2	9	0	528	0		73,000	0	840
32 **TEXTILES, APPAREL AND LEATHER**	1,091	432	43	223	5,658	0	..	1,869,000	0	14,175
321 TEXTILES	..	..	..	..	..	..		..	..	..
3211 Spinning weaving and finishing textiles	..	..	..	..	..	..		..	..	..
3212 Made-up goods excluding wearing apparel	..	..	..	..	..	..		..	..	..
3213 Knitting mills	..	..	..	..	..	..		..	..	..
3214 Carpets and rugs	..	..	..	..	..	..		..	..	..
3215 Cordage, rope and twine	..	..	..	..	..	..		..	..	..
3219 Other textiles	..	..	..	..	..	..		..	..	..
322 WEARING APPAREL, EXCEPT FOOTWEAR	..	..	..	..	..	..		..	..	..
323 LEATHER AND FUR PRODUCTS	..	..	..	..	..	..		..	..	..
3231 Tanneries and leather finishing	..	..	..	..	..	..		..	..	..
3232 Fur dressing and dyeing industries	..	..	..	..	..	..		..	..	..
3233 Leather prods. ex. footwear and wearing apparel	..	..	..	..	..	..		..	..	..
324 FOOTWEAR, EX. RUBBER AND PLASTIC	..	..	..	..	..	..		..	..	..
33 **WOOD PRODUCTS AND FURNITURE**	947	190	873	640	986	7,657	..	1,117,000	19,000	15,246
331 WOOD PRODUCTS, EXCEPT FURNITURE	..	..	..	..	..	..		..	..	..
3311 Sawmills, planing and other wood mills	..	..	..	..	..	..		..	..	..
3312 Wooden and cane containers	..	..	..	..	..	..		..	..	..
3319 Other wood and cork products	..	..	..	..	..	..		..	..	..
332 FURNITURE, FIXTURES, EXCL. METALLIC	..	..	..	..	..	..		..	..	..
34 **PAPER, PUBLISHING AND PRINTING**	7,957	445	46	728	15,498	7,763	..	4,114,000	490,000	45,483
341 PAPER AND PRODUCTS	..	..	..	..	..	..		..	..	..
3411 Pulp, paper and paperboard articles	..	..	..	..	..	..		..	..	..
3412 Containers of paper and paperboard	..	..	..	..	..	..		..	..	..
3419 Other pulp, paper and paperboard articles	..	..	..	..	..	..		..	..	..
342 PRINTING AND PUBLISHING	..	..	..	..	..	..		..	..	..
35 **CHEMICAL PRODUCTS**	7,466	761	24,981	493	90,787	0	..	5,115,000	376,000	141,548
351 INDUSTRIAL CHEMICALS	4,645	534	117	22	28,623	0		2,771,000	159,000	43,344
3511 Basic industrial chemicals excl. fertilizers	..	..	..	..	..	..		..	..	..
3512 Fertilizers and pesticides	..	..	..	..	..	..		..	..	..
3513 Synthetic resins and plastic materials	..	..	..	..	..	..		..	..	..
352 OTHER CHEMICALS	202	50	15	0	1,580	0		615,000	0	4,061
3521 Paints, varnishes and lacquers	..	..	..	..	..	..		..	..	..
3522 Drugs and medicines	..	..	..	..	..	..		..	..	..
3523 Soap, cleaning preparations, perfumes, cosmetics	..	..	..	..	..	..		..	..	..
3529 Other chemical products	..	..	..	..	..	..		..	..	..
353 PETROLEUM REFINERIES	0	0	24,786	0	58,219	0		1,307,000	217,000	86,929
354 MISC. PETROLEUM AND COAL PRODUCTS	2,532	141	31	471	652	0		80,000	0	4,115
355 RUBBER PRODUCTS	34	16	0	0	1,027	0		153,000	0	1,628
3551 Tyres and tubes	..	..	..	..	..	..		..	..	..
3559 Other rubber products	..	..	..	..	..	..		..	..	..
356 PLASTIC PRODUCTS	53	20	32	0	686	0		189,000	0	1,471
36 **NON-METALLIC MINERAL PRODUCTS**	20,298	1,385	1,435	782	45,560	737	..	2,872,000	0	80,536
361 POTTERY, CHINA, EARTHENWARE	48	13	22	0	1,976	0		59,000	0	2,271
362 GLASS AND PRODUCTS	0	199	22	215	9,743	0		359,000	0	11,471
369 OTHER NON-METAL. MINERAL PRODUCTS	20,250	1,173	1,391	567	33,841	737		2,454,000	0	66,793
3691 Structural clay products	1,148	291	387	118	15,232	729		675,000	0	20,335
3692 Cement, lime and plaster	14,954	510	851	392	15,274	8		1,270,000	0	36,561
3699 Other non-metallic mineral products	4,148	372	153	57	3,335	0		509,000	0	9,897
37 **BASIC METAL INDUSTRIES**	317,897	2,742	3,292	29,671	169,148	152	..	31,800,000	3,832,000	623,587
371 IRON AND STEEL	262,285	729	144	1,416	79,907	147		4,926,000	987,000	358,808
372 NON-FERROUS METALS	55,612	2,013	3,148	28,255	89,241	5		26,874,000	2,845,000	264,778
38 **METAL PRODUCTS, MACHINERY, EQUIP.**	70	1,379	170	16	11,442	0	..	4,006,000	0	27,499
381 METAL PRODUCTS	..	..	..	..	..	..		..	..	..
3811 Cutlery, hand tools and general hardware	..	..	..	..	..	..		..	..	..
3812 Furniture and fixtures primarily of metal	..	..	..	..	..	..		..	..	..
3813 Structural metal products	..	..	..	..	..	..		..	..	..
3819 Other fabricated metal products	..	..	..	..	..	..		..	..	..
382 NON-ELECTRICAL MACHINERY	..	..	..	..	..	..		..	..	..
3821 Engines and turbines	..	..	..	..	..	..		..	..	..
3822 Agricultural machinery and equipment	..	..	..	..	..	..		..	..	..
3823 Metal and wood working machinery	..	..	..	..	..	..		..	..	..
3824 Special industrial machinery	..	..	..	..	..	..		..	..	..
3825 Office, computing and accounting machinery	..	..	..	..	..	..		..	..	..
3829 Other non-electrical machinery and equipment	..	..	..	..	..	..		..	..	..
383 ELECTRICAL MACHINERY	..	..	..	..	..	..		..	..	..
3831 Electrical industrial machinery	..	..	..	..	..	..		..	..	..
3832 Radio, TV and communications equipment	..	..	..	..	..	..		..	..	..
3833 Electrical appliances and housewares	..	..	..	..	..	..		..	..	..
3839 Other electrical apparatus and supplies	..	..	..	..	..	..		..	..	..
384 TRANSPORT EQUIPMENT	..	..	..	..	..	..		..	..	..
3841 Shipbuilding	..	..	..	..	..	..		..	..	..
3842 Railroad equipment	..	..	..	..	..	..		..	..	..
3843 Motor vehicles	..	..	..	..	..	..		..	..	..
3844 Motorcycles and bicycles	..	..	..	..	..	..		..	..	..
3845 Aircraft	..	..	..	..	..	..		..	..	..
3849 Other transport equipment	..	..	..	..	..	..		..	..	..
385 PROFESSIONAL AND SCIENTIFIC EQUIPMENT	..	..	..	..	..	..		..	..	..
3851 Professional equipment	..	..	..	..	..	..		..	..	..
3852 Photographic and optical goods	..	..	..	..	..	..		..	..	..
3853 Watches and clocks	..	..	..	..	..	..		..	..	..
39 **OTHER MANUFACTURING INDUSTRIES**	0	3	0	0	203	0	..	18,000	0	271
3901 Jewellery and related articles	..	..	..	..	..	..		..	..	..
3902 Musical instruments	..	..	..	..	..	..		..	..	..
3903 Sporting and athletic goods	..	..	..	..	..	..		..	..	..
3909 Other manufactures	..	..	..	..	..	..		..	..	..
SERR **non-specified, unallocated industry**	..	..	..							
3 **TOTAL MANUFACTURING**	369,234	10,221	31,781	34,566	361,901	79,897	..	55,461,000	5,127,000	1,068,802

Year: **1993** **AUSTRALIA**

ISIC Revision 2 Industry Sector	Solid TJ	LPG TJ	Distiloil TJ	RFO TJ	Gas TJ	Biomass TJ	Steam TJ	Electr MWh	Own Use MWh	TOTAL TJ
31 **FOOD, BEVERAGES AND TOBACCO**	12,746	3,267	937	2,030	23,002	77,275	..	4,723,000	454,000	134,625
311/2 FOOD	12,084	3,126	898	1,876	19,020	77,275	..	4,161,000	454,000	127,624
3111 Slaughtering, preparing and preserving meat	2,780	725	160	909	3,925	613	..	1,103,000	0	13,083
3112 Dairy products	3,111	819	157	383	4,308	885	..	618,000	0	11,888
3113 Canning, preserving of fruits and vegetables	1,341	528	3	204	1,393	405	..	323,000	0	5,037
3114 Canning, preserving and processing of fish	..	..	..	..	..	..	..	..	..	..
3115 Vegetable and animal oils and fats	976	6	11	19	1,000	0	..	241,000	0	2,880
3116 Grain mill products	1,769	248	21	33	1,624	0	..	455,000	0	5,333
3117 Bakery products	0	297	86	95	2,362	0	..	362,000	0	4,143
3118 Sugar factories and refineries	2,107	503	460	233	4,408	75,372	..	1,059,000	454,000	85,261
3119 Cocoa, chocolate and sugar confectionery	..	..	..	..	..	..	..	..	..	..
3121 Other food products	..	..	..	..	..	..	..	..	..	..
3122 Prepared animal feeds	..	..	..	..	..	..	..	..	..	..
313 BEVERAGES	624	139	23	154	3,461	0	..	492,000	0	6,172
3131 Distilling, rectifying and blending of spirits	..	..	..	..	..	..	..	..	..	..
3132 Wine industries	..	..	..	..	..	..	..	..	..	..
3133 Malt liquors and malts	..	..	..	..	..	..	..	..	..	..
3134 Soft drinks	..	..	..	..	..	..	..	..	..	..
314 TOBACCO	38	2	16	0	521	0	..	70,000	0	829
32 **TEXTILES, APPAREL AND LEATHER**	1,125	474	42	242	6,089	0	..	1,915,000	0	14,866
321 TEXTILES	..	..	..	..	..	..	..	..	..	..
3211 Spinning weaving and finishing textiles	..	..	..	..	..	..	..	..	..	..
3212 Made-up goods excluding wearing apparel	..	..	..	..	..	..	..	..	..	..
3213 Knitting mills	..	..	..	..	..	..	..	..	..	..
3214 Carpets and rugs	..	..	..	..	..	..	..	..	..	..
3215 Cordage, rope and twine	..	..	..	..	..	..	..	..	..	..
3219 Other textiles	..	..	..	..	..	..	..	..	..	..
322 WEARING APPAREL, EXCEPT FOOTWEAR	..	..	..	..	..	..	..	..	..	..
323 LEATHER AND FUR PRODUCTS	..	..	..	..	..	..	..	..	..	..
3231 Tanneries and leather finishing	..	..	..	..	..	..	..	..	..	..
3232 Fur dressing and dyeing industries	..	..	..	..	..	..	..	..	..	..
3233 Leather prods. ex. footwear and wearing apparel	..	..	..	..	..	..	..	..	..	..
324 FOOTWEAR, EX. RUBBER AND PLASTIC	..	..	..	..	..	..	..	..	..	..
33 **WOOD PRODUCTS AND FURNITURE**	1,046	220	893	707	861	7,850	..	1,153,000	20,000	15,656
331 WOOD PRODUCTS, EXCEPT FURNITURE	..	..	..	..	..	..	..	..	..	..
3311 Sawmills, planing and other wood mills	..	..	..	..	..	..	..	..	..	..
3312 Wooden and cane containers	..	..	..	..	..	..	..	..	..	..
3319 Other wood and cork products	..	..	..	..	..	..	..	..	..	..
332 FURNITURE, FIXTURES, EXCL. METALLIC	..	..	..	..	..	..	..	..	..	..
34 **PAPER, PUBLISHING AND PRINTING**	7,551	473	45	764	16,225	8,583	..	4,322,000	595,000	47,058
341 PAPER AND PRODUCTS	..	..	..	..	..	..	..	..	..	..
3411 Pulp, paper and paperboard articles	..	..	..	..	..	..	..	..	..	..
3412 Containers of paper and paperboard	..	..	..	..	..	..	..	..	..	..
3419 Other pulp, paper and paperboard articles	..	..	..	..	..	..	..	..	..	..
342 PRINTING AND PUBLISHING	..	..	..	..	..	..	..	..	..	..
35 **CHEMICAL PRODUCTS**	6,781	776	25,523	409	93,041	0	..	5,241,000	417,000	143,896
351 INDUSTRIAL CHEMICALS	4,385	534	82	12	26,947	0	..	2,850,000	163,000	41,633
3511 Basic industrial chemicals excl. fertilizers	..	..	..	..	..	..	..	..	..	..
3512 Fertilizers and pesticides	..	..	..	..	..	..	..	..	..	..
3513 Synthetic resins and plastic materials	..	..	..	..	..	..	..	..	..	..
352 OTHER CHEMICALS	219	53	15	0	1,622	0	..	623,000	0	4,152
3521 Paints, varnishes and lacquers	..	..	..	..	..	..	..	..	..	..
3522 Drugs and medicines	..	..	..	..	..	..	..	..	..	..
3523 Soap, cleaning preparations, perfumes, cosmetics	..	..	..	..	..	..	..	..	..	..
3529 Other chemical products	..	..	..	..	..	..	..	..	..	..
353 PETROLEUM REFINERIES	0	0	25,362	0	62,187	0	..	1,335,000	254,000	91,441
354 MISC. PETROLEUM AND COAL PRODUCTS	2,081	152	32	397	644	0	..	80,000	0	3,594
355 RUBBER PRODUCTS	35	16	0	0	959	0	..	158,000	0	1,579
3551 Tyres and tubes	..	..	..	..	..	..	..	..	..	..
3559 Other rubber products	..	..	..	..	..	..	..	..	..	..
356 PLASTIC PRODUCTS	61	21	32	0	682	0	..	195,000	0	1,498
36 **NON-METALLIC MINERAL PRODUCTS**	20,264	1,530	1,730	774	45,877	671	..	2,955,000	0	81,484
361 POTTERY, CHINA, EARTHENWARE	49	37	21	0	1,501	0	..	56,000	0	1,810
362 GLASS AND PRODUCTS	0	203	22	211	9,888	0	..	357,000	0	11,609
369 OTHER NON-METAL. MINERAL PRODUCTS	20,215	1,290	1,687	563	34,488	671	..	2,542,000	0	68,065
3691 Structural clay products	999	345	413	120	16,191	663	..	703,000	0	21,262
3692 Cement, lime and plaster	14,762	573	1,112	393	15,038	8	..	1,320,000	0	36,638
3699 Other non-metallic mineral products	4,454	372	162	50	3,259	0	..	519,000	0	10,165
37 **BASIC METAL INDUSTRIES**	311,404	2,940	3,572	28,220	172,924	63	..	32,899,000	3,965,000	623,285
371 IRON AND STEEL	254,698	804	146	1,214	79,957	58	..	5,143,000	950,000	351,972
372 NON-FERROUS METALS	56,706	2,136	3,426	27,006	92,967	5	..	27,756,000	3,015,000	271,314
38 **METAL PRODUCTS, MACHINERY, EQUIP.**	48	1,582	157	13	11,316	0	..	4,077,000	0	27,793
381 METAL PRODUCTS	..	..	..	..	..	..	..	..	..	..
3811 Cutlery, hand tools and general hardware	..	..	..	..	..	..	..	..	..	..
3812 Furniture and fixtures primarily of metal	..	..	..	..	..	..	..	..	..	..
3813 Structural metal products	..	..	..	..	..	..	..	..	..	..
3819 Other fabricated metal products	..	..	..	..	..	..	..	..	..	..
382 NON-ELECTRICAL MACHINERY	..	..	..	..	..	..	..	..	..	..
3821 Engines and turbines	..	..	..	..	..	..	..	..	..	..
3822 Agricultural machinery and equipment	..	..	..	..	..	..	..	..	..	..
3823 Metal and wood working machinery	..	..	..	..	..	..	..	..	..	..
3824 Special industrial machinery	..	..	..	..	..	..	..	..	..	..
3825 Office, computing and accounting machinery	..	..	..	..	..	..	..	..	..	..
3829 Other non-electrical machinery and equipment	..	..	..	..	..	..	..	..	..	..
383 ELECTRICAL MACHINERY	..	..	..	..	..	..	..	..	..	..
3831 Electrical industrial machinery	..	..	..	..	..	..	..	..	..	..
3832 Radio, TV and communications equipment	..	..	..	..	..	..	..	..	..	..
3833 Electrical appliances and housewares	..	..	..	..	..	..	..	..	..	..
3839 Other electrical apparatus and supplies	..	..	..	..	..	..	..	..	..	..
384 TRANSPORT EQUIPMENT	..	..	..	..	..	..	..	..	..	..
3841 Shipbuilding	..	..	..	..	..	..	..	..	..	..
3842 Railroad equipment	..	..	..	..	..	..	..	..	..	..
3843 Motor vehicles	..	..	..	..	..	..	..	..	..	..
3844 Motorcycles and bicycles	..	..	..	..	..	..	..	..	..	..
3845 Aircraft	..	..	..	..	..	..	..	..	..	..
3849 Other transport equipment	..	..	..	..	..	..	..	..	..	..
385 PROFESSIONAL AND SCIENTIFIC EQUIPMENT	..	..	..	..	..	..	..	..	..	..
3851 Professional equipment	..	..	..	..	..	..	..	..	..	..
3852 Photographic and optical goods	..	..	..	..	..	..	..	..	..	..
3853 Watches and clocks	..	..	..	..	..	..	..	..	..	..
39 **OTHER MANUFACTURING INDUSTRIES**	0	2	0	0	204	..	..	17,000	0	267
3901 Jewellery and related articles	..	..	..	..	..	..	..	..	..	..
3902 Musical instruments	..	..	..	..	..	..	..	..	..	..
3903 Sporting and athletic goods	..	..	..	..	..	..	..	..	..	..
3909 Other manufactures	..	..	..	..	..	..	..	..	..	..
SERR non-specified, unallocated industry	..	..	..	..	..	..	..	..	..	..
3 **TOTAL MANUFACTURING**	360,965	11,264	32,899	33,159	369,539	94,442	..	57,302,000	5,451,000	1,088,932

77

ISIC Revision 2 Industry Sector	Solid TJ	LPG TJ	Distiloil TJ	RFO TJ	Gas TJ	Biomass TJ	Steam TJ	Electr MWh	Own Use MWh	TOTAL TJ
31 **FOOD, BEVERAGES AND TOBACCO**	13,221	3,431	946	1,976	24,312	83,430		4,863,000	422,000	143,304
311/2 FOOD	12,502	3,284	917	1,864	20,187	83,430		4,276,000	422,000	136,058
3111 Slaughtering, preparing and preserving meat	2,961	735	165	832	4,001	483		1,146,000	0	13,303
3112 Dairy products	2,778	861	161	381	4,808	1,291		635,000	0	12,566
3113 Canning, preserving of fruits and vegetables	1,370	526	3	252	1,556	439		333,000	0	5,345
3114 Canning, preserving and processing of fish	..						..			..
3115 Vegetable and animal oils and fats	961	6	10	20	1,021	0		239,000	0	2,878
3116 Grain mill products	1,768	295	22	32	1,664	0		478,000	0	5,502
3117 Bakery products	0	289	87	77	2,470	0		381,000	0	4,295
3118 Sugar factories and refineries	2,664	572	469	270	4,667	81,217		1,064,000	422,000	92,170
3119 Cocoa, chocolate and sugar confectionery	..		..				..			..
3121 Other food products	..		..				..			..
3122 Prepared animal feeds	..			..			..			..
313 BEVERAGES	681	145	24	112	3,659	0		512,000	0	6,464
3131 Distilling, rectifying and blending of spirits	..		..				..			..
3132 Wine industries	..			..			..			..
3133 Malt liquors and malts	..			..			..			..
3134 Soft drinks	..			..			..			..
314 TOBACCO	38	2	5	0	466	0		75,000	0	781
32 **TEXTILES, APPAREL AND LEATHER**	1,133	479	51	295	6,180	0		1,876,000	0	14,892
321 TEXTILES	..	..	..				..			..
3211 Spinning weaving and finishing textiles	..		..				..			..
3212 Made-up goods excluding wearing apparel	..		..				..			..
3213 Knitting mills	..		..				..			..
3214 Carpets and rugs	..		..				..			..
3215 Cordage, rope and twine	..		..				..			..
3219 Other textiles	..		..				..			..
322 WEARING APPAREL, EXCEPT FOOTWEAR	..		..				..			..
323 LEATHER AND FUR PRODUCTS	..						..			..
3231 Tanneries and leather finishing	..		..				..			..
3232 Fur dressing and dyeing industries	..		..				..			..
3233 Leather prods. ex. footwear and wearing apparel	..		..				..			..
324 FOOTWEAR, EX. RUBBER AND PLASTIC	..		..				..			..
33 **WOOD PRODUCTS AND FURNITURE**	1,246	241	902	650	1,352	8,183		1,216,000	21,000	16,876
331 WOOD PRODUCTS, EXCEPT FURNITURE	..						..			..
3311 Sawmills, planing and other wood mills	..		..				..			..
3312 Wooden and cane containers	..		..				..			..
3319 Other wood and cork products	..		..				..			..
332 FURNITURE, FIXTURES, EXCL. METALLIC	..		..				..			..
34 **PAPER, PUBLISHING AND PRINTING**	7,143	455	71	966	17,287	10,249		4,252,000	550,000	49,498
341 PAPER AND PRODUCTS	..						..			..
3411 Pulp, paper and paperboard articles	..		..				..			..
3412 Containers of paper and paperboard	..		..				..			..
3419 Other pulp, paper and paperboard articles	..		..				..			..
342 PRINTING AND PUBLISHING	..		..				..			..
35 **CHEMICAL PRODUCTS**	6,800	813	27,108	417	99,259	0		5,803,000	428,000	153,747
351 INDUSTRIAL CHEMICALS	4,297	542	98	22	30,051	0		3,243,000	179,000	46,040
3511 Basic industrial chemicals excl. fertilizers	..		..				..			..
3512 Fertilizers and pesticides	..		..				..			..
3513 Synthetic resins and plastic materials	..		..				..			..
352 OTHER CHEMICALS	398	67	27	0	1,733	0		640,000	0	4,529
3521 Paints, varnishes and lacquers	..		..				..			..
3522 Drugs and medicines	..		..				..			..
3523 Soap, cleaning preparations, perfumes, cosmetics	..		..				..			..
3529 Other chemical products	..		..				..			..
353 PETROLEUM REFINERIES	0	0	26,873	0	65,309	0		1,429,000	249,000	96,430
354 MISC. PETROLEUM AND COAL PRODUCTS	2,017	156	35	395	482	0		84,000	0	3,387
355 RUBBER PRODUCTS	33	25	0	0	869	0		161,000	0	1,507
3551 Tyres and tubes	..		..				..			..
3559 Other rubber products	..		..				..			..
356 PLASTIC PRODUCTS	55	23	75	0	815	0		246,000	0	1,854
36 **NON-METALLIC MINERAL PRODUCTS**	21,738	1,709	1,891	760	47,754	740		3,103,000	0	85,763
361 POTTERY, CHINA, EARTHENWARE	50	16	23	0	1,716	0		61,000	0	2,025
362 GLASS AND PRODUCTS	0	208	22	206	9,557	0		374,000	0	11,339
369 OTHER NON-METAL. MINERAL PRODUCTS	21,688	1,485	1,846	554	36,481	740		2,668,000	0	72,399
3691 Structural clay products	947	372	442	146	16,522	732		743,000	0	21,836
3692 Cement, lime and plaster	17,086	616	1,242	377	15,162	8		1,349,000	0	39,347
3699 Other non-metallic mineral products	3,655	497	162	31	4,797	0		576,000	0	11,216
37 **BASIC METAL INDUSTRIES**	323,637	2,939	3,200	28,066	185,344	3		34,437,000	4,045,000	652,600
371 IRON AND STEEL	266,693	853	375	895	87,219	0		5,479,000	891,000	372,552
372 NON-FERROUS METALS	56,944	2,086	2,825	27,171	98,125	3		28,958,000	3,154,000	280,048
38 **METAL PRODUCTS, MACHINERY, EQUIP.**	46	1,761	168	12	11,391	0		4,232,000	0	28,613
381 METAL PRODUCTS	..	..	..				..			..
3811 Cutlery, hand tools and general hardware	..		..				..			..
3812 Furniture and fixtures primarily of metal	..		..				..			..
3813 Structural metal products	..		..				..			..
3819 Other fabricated metal products	..		..				..			..
382 NON-ELECTRICAL MACHINERY	..		..				..			..
3821 Engines and turbines	..		..				..			..
3822 Agricultural machinery and equipment	..		..				..			..
3823 Metal and wood working machinery	..		..				..			..
3824 Special industrial machinery	..		..				..			..
3825 Office, computing and accounting machinery	..		..				..			..
3829 Other non-electrical machinery and equipment	..		..				..			..
383 ELECTRICAL MACHINERY	..		..				..			..
3831 Electrical industrial machinery	..		..				..			..
3832 Radio, TV and communications equipment	..		..				..			..
3833 Electrical appliances and housewares	..		..				..			..
3839 Other electrical apparatus and supplies	..		..				..			..
384 TRANSPORT EQUIPMENT	..		..				..			..
3841 Shipbuilding	..		..				..			..
3842 Railroad equipment	..		..				..			..
3843 Motor vehicles	..		..				..			..
3844 Motorcycles and bicycles	..		..				..			..
3845 Aircraft	..		..				..			..
3849 Other transport equipment	..		..				..			..
385 PROFESSIONAL AND SCIENTIFIC EQUIPMENT	..		..				..			..
3851 Professional equipment	..		..				..			..
3852 Photographic and optical goods	..		..				..			..
3853 Watches and clocks	..		..				..			..
39 **OTHER MANUFACTURING INDUSTRIES**	0	0	0	0	258	0		23,000	0	341
3901 Jewellery and related articles	..		..				..			..
3902 Musical instruments	..		..				..			..
3903 Sporting and athletic goods	..		..				..			..
3909 Other manufactures	..		..				..			..
SERR **non-specified, unallocated industry**										
3 **TOTAL MANUFACTURING**	374,964	11,828	34,337	33,142	393,137	102,605		59,805,000	5,466,000	1,145,633

ISIC Revision 2 Industry Sector		Solid TJ	LPG TJ	Distiloil TJ	RFO TJ	Gas TJ	Biomass TJ	Steam TJ	Electr MWh	Own Use MWh	TOTAL TJ
31	FOOD, BEVERAGES AND TOBACCO	13,315	3,689	928	2,081	25,741	90,629	..	5,107,000	399,000	153,332
311/2	FOOD	12,613	3,539	901	1,983	21,507	90,629	..	4,506,000	399,000	145,957
3111	Slaughtering, preparing and preserving meat	2,683	851	169	827	4,196	495	..	1,180,000	0	13,469
3112	Dairy products	2,779	883	116	410	5,236	1,321	..	675,000	0	13,175
3113	Canning, preserving of fruits and vegetables	1,341	543	3	249	1,611	497	..	352,000	0	5,511
3114	Canning, preserving and processing of fish	..	..	..	..	..	..	..	..	..	..
3115	Vegetable and animal oils and fats	973	6	11	21	1,048	0	..	256,000	0	2,981
3116	Grain mill products	1,778	340	22	32	1,718	0	..	495,000	1,000	5,668
3117	Bakery products	0	295	94	76	2,683	0	..	406,000	0	4,610
3118	Sugar factories and refineries	3,059	621	486	368	5,015	88,316	..	1,142,000	398,000	100,543
3119	Cocoa, chocolate and sugar confectionery	..	..	..	..	..	..	..	..	..	..
3121	Other food products	..	..	..	..	..	..	..	..	..	..
3122	Prepared animal feeds	..	..	..	..	..	..	..	..	..	..
313	BEVERAGES	664	148	24	98	3,758	0	..	525,000	0	6,582
3131	Distilling, rectifying and blending of spirits	..	..	..	..	..	..	..	..	..	..
3132	Wine industries	..	..	..	..	..	..	..	..	..	..
3133	Malt liquors and malts	..	..	..	..	..	..	..	..	..	..
3134	Soft drinks	..	..	..	..	..	..	..	..	..	..
314	TOBACCO	38	2	3	0	476	0	..	76,000	0	793
32	TEXTILES, APPAREL AND LEATHER	1,250	554	55	249	6,532	0	..	1,936,000	0	15,610
321	TEXTILES	..	..	..	..	..	..	..	..	..	..
3211	Spinning weaving and finishing textiles	..	..	..	..	..	..	..	..	..	..
3212	Made-up goods excluding wearing apparel	..	..	..	..	..	..	..	..	..	..
3213	Knitting mills	..	..	..	..	..	..	..	..	..	..
3214	Carpets and rugs	..	..	..	..	..	..	..	..	..	..
3215	Cordage, rope and twine	..	..	..	..	..	..	..	..	..	..
3219	Other textiles	..	..	..	..	..	..	..	..	..	..
322	WEARING APPAREL, EXCEPT FOOTWEAR	..	..	..	..	..	..	..	..	..	..
323	LEATHER AND FUR PRODUCTS	..	..	..	..	..	..	..	..	..	..
3231	Tanneries and leather finishing	..	..	..	..	..	..	..	..	..	..
3232	Fur dressing and dyeing industries	..	..	..	..	..	..	..	..	..	..
3233	Leather prods. ex. footwear and wearing apparel	..	..	..	..	..	..	..	..	..	..
324	FOOTWEAR, EX. RUBBER AND PLASTIC	..	..	..	..	..	..	..	..	..	..
33	WOOD PRODUCTS AND FURNITURE	1,399	255	922	557	1,313	8,080	..	1,286,000	582,000	15,060
331	WOOD PRODUCTS, EXCEPT FURNITURE	..	..	..	..	..	..	..	..	..	..
3311	Sawmills, planing and other wood mills	..	..	..	..	..	..	..	..	..	..
3312	Wooden and cane containers	..	..	..	..	..	..	..	..	..	..
3319	Other wood and cork products	..	..	..	..	..	..	..	..	..	..
332	FURNITURE, FIXTURES, EXCL. METALLIC	..	..	..	..	..	..	..	..	..	..
34	PAPER, PUBLISHING AND PRINTING	6,914	477	29	995	17,674	10,403	..	4,267,000	0	51,853
341	PAPER AND PRODUCTS	..	..	..	..	..	..	..	..	..	..
3411	Pulp, paper and paperboard articles	..	..	..	..	..	..	..	..	..	..
3412	Containers of paper and paperboard	..	..	..	..	..	..	..	..	..	..
3419	Other pulp, paper and paperboard articles	..	..	..	..	..	..	..	..	..	..
342	PRINTING AND PUBLISHING	..	..	..	..	..	..	..	..	..	..
35	CHEMICAL PRODUCTS	6,603	881	29,926	543	98,526	0	..	6,008,000	428,000	156,567
351	INDUSTRIAL CHEMICALS	4,349	604	97	108	29,957	0	..	3,346,000	164,000	46,570
3511	Basic industrial chemicals excl. fertilizers	..	..	..	..	..	..	..	..	..	..
3512	Fertilizers and pesticides	..	..	..	..	..	..	..	..	..	..
3513	Synthetic resins and plastic materials	..	..	..	..	..	..	..	..	..	..
352	OTHER CHEMICALS	334	66	21	0	1,911	0	..	678,000	0	4,773
3521	Paints, varnishes and lacquers	..	..	..	..	..	..	..	..	..	..
3522	Drugs and medicines	..	..	..	..	..	..	..	..	..	..
3523	Soap, cleaning preparations, perfumes, cosmetics	..	..	..	..	..	..	..	..	..	..
3529	Other chemical products	..	..	..	..	..	..	..	..	..	..
353	PETROLEUM REFINERIES	0	0	29,699	0	64,478	0	..	1,456,000	246,000	98,533
354	MISC. PETROLEUM AND COAL PRODUCTS	1,827	162	33	435	482	0	..	88,000	0	3,256
355	RUBBER PRODUCTS	34	25	0	0	865	0	..	162,000	18,000	1,442
3551	Tyres and tubes	..	..	..	..	..	..	..	..	..	..
3559	Other rubber products	..	..	..	..	..	..	..	..	..	..
356	PLASTIC PRODUCTS	59	24	76	0	833	0	..	278,000	0	1,993
36	NON-METALLIC MINERAL PRODUCTS	22,218	1,801	2,018	813	48,837	718	..	3,174,000	1,000	87,828
361	POTTERY, CHINA, EARTHENWARE	52	16	24	0	1,748	0	..	64,000	0	2,070
362	GLASS AND PRODUCTS	0	213	11	202	8,852	0	..	346,000	1,000	10,520
369	OTHER NON-METAL. MINERAL PRODUCTS	22,166	1,572	1,983	611	38,237	718	..	2,764,000	0	75,237
3691	Structural clay products	1,143	401	420	165	17,125	710	..	780,000	0	22,772
3692	Cement, lime and plaster	17,348	673	1,423	415	16,070	8	..	1,379,000	0	40,901
3699	Other non-metallic mineral products	3,675	498	140	31	5,042	0	..	605,000	0	11,564
37	BASIC METAL INDUSTRIES	325,037	3,840	2,615	28,469	191,780	3	..	33,590,000	5,139,000	654,168
371	IRON AND STEEL	267,171	899	399	1,277	91,810	0	..	5,723,000	717,000	379,578
372	NON-FERROUS METALS	57,866	2,941	2,216	27,192	99,970	3	..	27,867,000	4,422,000	274,590
38	METAL PRODUCTS, MACHINERY, EQUIP.	46	1,938	173	13	12,233	0	..	4,451,000	0	30,427
381	METAL PRODUCTS	..	..	..	..	..	..	..	..	..	..
3811	Cutlery, hand tools and general hardware	..	..	..	..	..	..	..	..	..	..
3812	Furniture and fixtures primarily of metal	..	..	..	..	..	..	..	..	..	..
3813	Structural metal products	..	..	..	..	..	..	..	..	..	..
3819	Other fabricated metal products	..	..	..	..	..	..	..	..	..	..
382	NON-ELECTRICAL MACHINERY	..	..	..	..	..	..	..	..	..	..
3821	Engines and turbines	..	..	..	..	..	..	..	..	..	..
3822	Agricultural machinery and equipment	..	..	..	..	..	..	..	..	..	..
3823	Metal and wood working machinery	..	..	..	..	..	..	..	..	..	..
3824	Special industrial machinery	..	..	..	..	..	..	..	..	..	..
3825	Office, computing and accounting machinery	..	..	..	..	..	..	..	..	..	..
3829	Other non-electrical machinery and equipment	..	..	..	..	..	..	..	..	..	..
383	ELECTRICAL MACHINERY	..	..	..	..	..	..	..	..	..	..
3831	Electrical industrial machinery	..	..	..	..	..	..	..	..	..	..
3832	Radio, TV and communications equipment	..	..	..	..	..	..	..	..	..	..
3833	Electrical appliances and housewares	..	..	..	..	..	..	..	..	..	..
3839	Other electrical apparatus and supplies	..	..	..	..	..	..	..	..	..	..
384	TRANSPORT EQUIPMENT	..	..	..	..	..	..	..	..	..	..
3841	Shipbuilding	..	..	..	..	..	..	..	..	..	..
3842	Railroad equipment	..	..	..	..	..	..	..	..	..	..
3843	Motor vehicles	..	..	..	..	..	..	..	..	..	..
3844	Motorcycles and bicycles	..	..	..	..	..	..	..	..	..	..
3845	Aircraft	..	..	..	..	..	..	..	..	..	..
3849	Other transport equipment	..	..	..	..	..	..	..	..	..	..
385	PROFESSIONAL AND SCIENTIFIC EQUIPMENT	..	..	..	..	..	..	..	..	..	..
3851	Professional equipment	..	..	..	..	..	..	..	..	..	..
3852	Photographic and optical goods	..	..	..	..	..	..	..	..	..	..
3853	Watches and clocks	..	..	..	..	..	..	..	..	..	..
39	OTHER MANUFACTURING INDUSTRIES	0	0	0	0	263	0	..	21,000	0	339
3901	Jewellery and related articles	..	..	..	..	..	..	..	..	..	..
3902	Musical instruments	..	..	..	..	..	..	..	..	..	..
3903	Sporting and athletic goods	..	..	..	..	..	..	..	..	..	..
3909	Other manufactures	..	..	..	..	..	..	..	..	..	..
SERR	non-specified, unallocated industry	..	..	..	..	..	..	..	..	..	..
3	TOTAL MANUFACTURING	376,782	13,435	36,666	33,720	402,899	109,833	..	59,840,000	6,549,000	1,165,183

Year: 1996   **AUSTRALIA**

ISIC Revision 2 Industry Sector	Solid TJ	LPG TJ	Distiloil TJ	RFO TJ	Gas TJ	Biomass TJ	Steam TJ	Electr MWh	Own Use MWh	TOTAL TJ
31 FOOD, BEVERAGES AND TOBACCO	14,336	3,442	884	1,643	24,759	99,947		5,117,000	423,000	161,909
311/2 FOOD	13,620	3,297	869	1,528	21,036	99,947		4,515,000	423,000	155,028
3111 Slaughtering, preparing and preserving meat	2,762	769	134	240	3,738	450		1,094,000	0	12,031
3112 Dairy products	2,456	804	131	130	5,702	1,368		679,000	0	13,035
3113 Canning, preserving of fruits and vegetables	1,357	580	3	219	1,594	464		379,000	0	5,581
3114 Canning, preserving and processing of fish	..		..		..					..
3115 Vegetable and animal oils and fats	727	7	8	23	1,263	0		263,000	0	2,975
3116 Grain mill products	1,873	367	22	0	1,693	0		512,000	1,000	5,795
3117 Bakery products	0	314	82	76	2,619	0		413,000	0	4,578
3118 Sugar factories and refineries	4,445	456	489	840	4,427	97,665		1,175,000	422,000	111,033
3119 Cocoa, chocolate and sugar confectionery	..		..		..					..
3121 Other food products	..		..		..					..
3122 Prepared animal feeds	..		..		..					..
313 BEVERAGES	678	143	11	115	3,250	0		524,000	0	6,083
3131 Distilling, rectifying and blending of spirits	..		..		..					..
3132 Wine industries	..		..		..					..
3133 Malt liquors and malts	..		..		..					..
3134 Soft drinks	..		..		..					..
314 TOBACCO	38	2	4	0	473	0		78,000	0	798
32 TEXTILES, APPAREL AND LEATHER	1,099	620	56	192	6,277	0		1,934,000	0	15,206
321 TEXTILES	..		..		..					..
3211 Spinning weaving and finishing textiles	..		..		..					..
3212 Made-up goods excluding wearing apparel	..		..		..					..
3213 Knitting mills	..		..		..					..
3214 Carpets and rugs	..		..		..					..
3215 Cordage, rope and twine	..		..		..					..
3219 Other textiles	..		..		..					..
322 WEARING APPAREL, EXCEPT FOOTWEAR	..		..		..					..
323 LEATHER AND FUR PRODUCTS	..		..		..					..
3231 Tanneries and leather finishing	..		..		..					..
3232 Fur dressing and dyeing industries	..		..		..					..
3233 Leather prods. ex. footwear and wearing apparel	..		..		..					..
324 FOOTWEAR, EX. RUBBER AND PLASTIC	..		..		..					..
33 WOOD PRODUCTS AND FURNITURE	1,156	297	930	294	1,203	8,533		1,281,000	16,000	16,967
331 WOOD PRODUCTS, EXCEPT FURNITURE	..		..		..					..
3311 Sawmills, planing and other wood mills	..		..		..					..
3312 Wooden and cane containers	..		..		..					..
3319 Other wood and cork products	..		..		..					..
332 FURNITURE, FIXTURES, EXCL. METALLIC	..		..		..					..
34 PAPER, PUBLISHING AND PRINTING	7,330	476	34	753	20,192	10,170		4,281,000	585,000	52,261
341 PAPER AND PRODUCTS	..		..		..					..
3411 Pulp, paper and paperboard articles	..		..		..					..
3412 Containers of paper and paperboard	..		..		..					..
3419 Other pulp, paper and paperboard articles	..		..		..					..
342 PRINTING AND PUBLISHING	..		..		..					..
35 CHEMICAL PRODUCTS	6,646	685	30,315	410	96,753	0		5,623,000	251,000	154,148
351 INDUSTRIAL CHEMICALS	3,934	404	609	33	28,404	0		3,068,000	98,000	44,076
3511 Basic industrial chemicals excl. fertilizers	..		..		..					..
3512 Fertilizers and pesticides	..		..		..					..
3513 Synthetic resins and plastic materials	..		..		..					..
352 OTHER CHEMICALS	309	68	14	0	1,960	0		688,000	0	4,828
3521 Paints, varnishes and lacquers	..		..		..					..
3522 Drugs and medicines	..		..		..					..
3523 Soap, cleaning preparations, perfumes, cosmetics	..		..		..					..
3529 Other chemical products	..		..		..					..
353 PETROLEUM REFINERIES	0	0	29,586	0	64,303	0		1,368,000	138,000	98,317
354 MISC. PETROLEUM AND COAL PRODUCTS	2,316	170	32	377	490	0		89,000	0	3,705
355 RUBBER PRODUCTS	33	23	0	0	921	0		174,000	15,000	1,549
3551 Tyres and tubes	..		..		..					..
3559 Other rubber products	..		..		..					..
356 PLASTIC PRODUCTS	54	20	74	0	675	0		236,000	0	1,673
36 NON-METALLIC MINERAL PRODUCTS	19,779	1,879	2,197	510	45,768	696		3,128,000	1,000	82,086
361 POTTERY, CHINA, EARTHENWARE	53	13	24	0	1,423	0		59,000	0	1,725
362 GLASS AND PRODUCTS	0	194	11	199	8,486	0		308,000	1,000	9,995
369 OTHER NON-METAL. MINERAL PRODUCTS	19,726	1,672	2,162	311	35,859	696		2,761,000	0	70,366
3691 Structural clay products	964	424	395	127	14,884	688		736,000	0	20,132
3692 Cement, lime and plaster	15,295	738	1,635	165	15,904	8		1,434,000	0	38,907
3699 Other non-metallic mineral products	3,467	510	132	19	5,071	0		591,000	0	11,327
37 BASIC METAL INDUSTRIES	312,449	3,960	3,737	28,528	190,556	0		33,870,000	5,104,000	642,788
371 IRON AND STEEL	256,251	1,054	391	1,150	91,215	0		5,706,000	610,000	368,407
372 NON-FERROUS METALS	56,198	2,906	3,346	27,378	99,341	0		28,164,000	4,494,000	274,381
38 METAL PRODUCTS, MACHINERY, EQUIP.	46	1,975	95	12	12,543	0		4,405,000	0	30,529
381 METAL PRODUCTS	..		..		..					..
3811 Cutlery, hand tools and general hardware	..		..		..					..
3812 Furniture and fixtures primarily of metal	..		..		..					..
3813 Structural metal products	..		..		..					..
3819 Other fabricated metal products	..		..		..					..
382 NON-ELECTRICAL MACHINERY	..		..		..					..
3821 Engines and turbines	..		..		..					..
3822 Agricultural machinery and equipment	..		..		..					..
3823 Metal and wood working machinery	..		..		..					..
3824 Special industrial machinery	..		..		..					..
3825 Office, computing and accounting machinery	..		..		..					..
3829 Other non-electrical machinery and equipment	..		..		..					..
383 ELECTRICAL MACHINERY	..		..		..					..
3831 Electrical industrial machinery	..		..		..					..
3832 Radio, TV and communications equipment	..		..		..					..
3833 Electrical appliances and housewares	..		..		..					..
3839 Other electrical apparatus and supplies	..		..		..					..
384 TRANSPORT EQUIPMENT	..		..		..					..
3841 Shipbuilding	..		..		..					..
3842 Railroad equipment	..		..		..					..
3843 Motor vehicles	..		..		..					..
3844 Motorcycles and bicycles	..		..		..					..
3845 Aircraft	..		..		..					..
3849 Other transport equipment	..		..		..					..
385 PROFESSIONAL AND SCIENTIFIC EQUIPMENT	..		..		..					..
3851 Professional equipment	..		..		..					..
3852 Photographic and optical goods	..		..		..					..
3853 Watches and clocks	..		..		..					..
39 OTHER MANUFACTURING INDUSTRIES	0	0	0	0	265	0		22,000	0	344
3901 Jewellery and related articles	..		..		..					..
3902 Musical instruments	..		..		..					..
3903 Sporting and athletic goods	..		..		..					..
3909 Other manufactures	..		..		..					..
SERR non-specified, unallocated industry										
3 TOTAL MANUFACTURING	362,841	13,334	38,248	32,342	398,316	119,346		59,661,000	6,380,000	1,156,239

Year: 1997

AUSTRALIA

ISIC Revision 2 Industry Sector		Solid TJ	LPG TJ	Distiloil TJ	RFO TJ	Gas TJ	Biomass TJ	Steam TJ	Electr MWh	Own Use MWh	TOTAL TJ
31	FOOD, BEVERAGES AND TOBACCO	14,100	3,573	957	1,132	24,690	107,286	..	5,138,000	432,000	168,680
311/2	FOOD	13,377	3,426	936	1,019	21,082	107,286	..	4,565,000	432,000	162,005
3111	Slaughtering, preparing and preserving meat	2,536	805	138	161	3,569	438	..	1,101,000	0	11,611
3112	Dairy products	2,473	840	142	48	5,809	1,406	..	684,000	0	13,180
3113	Canning, preserving of fruits and vegetables	1,411	605	3	178	1,647	481	..	390,000	0	5,729
3114	Canning, preserving and processing of fish	..	..	..	..	..	..	..	..	..	..
3115	Vegetable and animal oils and fats	740	6	7	0	1,322	0	..	270,000	0	3,047
3116	Grain mill products	1,955	375	26	0	1,765	0	..	519,000	1,000	5,986
3117	Bakery products	0	324	82	76	2,634	0	..	426,000	0	4,650
3118	Sugar factories and refineries	4,262	471	538	556	4,336	104,961	..	1,175,000	431,000	117,802
3119	Cocoa, chocolate and sugar confectionery	..	..	..	..	..	..	..	..	..	..
3121	Other food products	..	..	..	..	..	..	..	..	..	..
3122	Prepared animal feeds	..	..	..	..	..	..	..	..	..	..
313	BEVERAGES	685	145	17	113	3,151	0	..	494,000	0	5,889
3131	Distilling, rectifying and blending of spirits	..	..	..	..	..	..	..	..	..	..
3132	Wine industries	..	..	..	..	..	..	..	..	..	..
3133	Malt liquors and malts	..	..	..	..	..	..	..	..	..	..
3134	Soft drinks	..	..	..	..	..	..	..	..	..	..
314	TOBACCO	38	2	4	0	457	0	..	79,000	0	785
32	TEXTILES, APPAREL AND LEATHER	1,172	658	57	131	6,289	0	..	1,977,000	0	15,424
321	TEXTILES	..	..	..	..	..	..	..	..	..	..
3211	Spinning weaving and finishing textiles	..	..	..	..	..	..	..	..	..	..
3212	Made-up goods excluding wearing apparel	..	..	..	..	..	..	..	..	..	..
3213	Knitting mills	..	..	..	..	..	..	..	..	..	..
3214	Carpets and rugs	..	..	..	..	..	..	..	..	..	..
3215	Cordage, rope and twine	..	..	..	..	..	..	..	..	..	..
3219	Other textiles	..	..	..	..	..	..	..	..	..	..
322	WEARING APPAREL, EXCEPT FOOTWEAR	..	..	..	..	..	..	..	..	..	..
323	LEATHER AND FUR PRODUCTS	..	..	..	..	..	..	..	..	..	..
3231	Tanneries and leather finishing	..	..	..	..	..	..	..	..	..	..
3232	Fur dressing and dyeing industries	..	..	..	..	..	..	..	..	..	..
3233	Leather prods. ex. footwear and wearing apparel	..	..	..	..	..	..	..	..	..	..
324	FOOTWEAR, EX. RUBBER AND PLASTIC	..	..	..	..	..	..	..	..	..	..
33	WOOD PRODUCTS AND FURNITURE	1,096	307	966	148	1,261	8,728	..	1,300,000	17,000	17,125
331	WOOD PRODUCTS, EXCEPT FURNITURE	..	..	..	..	..	..	..	..	..	..
3311	Sawmills, planing and other wood mills	..	..	..	..	..	..	..	..	..	..
3312	Wooden and cane containers	..	..	..	..	..	..	..	..	..	..
3319	Other wood and cork products	..	..	..	..	..	..	..	..	..	..
332	FURNITURE, FIXTURES, EXCL. METALLIC	..	..	..	..	..	..	..	..	..	..
34	PAPER, PUBLISHING AND PRINTING	8,095	512	36	842	20,259	10,063	..	4,355,000	624,000	53,239
341	PAPER AND PRODUCTS	..	..	..	..	..	..	..	..	..	..
3411	Pulp, paper and paperboard articles	..	..	..	..	..	..	..	..	..	..
3412	Containers of paper and paperboard	..	..	..	..	..	..	..	..	..	..
3419	Other pulp, paper and paperboard articles	..	..	..	..	..	..	..	..	..	..
342	PRINTING AND PUBLISHING	..	..	..	..	..	..	..	..	..	..
35	CHEMICAL PRODUCTS	7,345	488	26,532	403	95,360	0	..	5,758,000	430,000	149,309
351	INDUSTRIAL CHEMICALS	3,876	185	646	0	27,947	0	..	3,097,000	90,000	43,479
3511	Basic industrial chemicals excl. fertilizers	..	..	..	..	..	..	..	..	..	..
3512	Fertilizers and pesticides	..	..	..	..	..	..	..	..	..	..
3513	Synthetic resins and plastic materials	..	..	..	..	..	..	..	..	..	..
352	OTHER CHEMICALS	309	79	13	0	2,200	0	..	704,000	0	5,135
3521	Paints, varnishes and lacquers	..	..	..	..	..	..	..	..	..	..
3522	Drugs and medicines	..	..	..	..	..	..	..	..	..	..
3523	Soap, cleaning preparations, perfumes, cosmetics	..	..	..	..	..	..	..	..	..	..
3529	Other chemical products	..	..	..	..	..	..	..	..	..	..
353	PETROLEUM REFINERIES	0	0	25,766	0	63,127	0	..	1,441,000	324,000	92,914
354	MISC. PETROLEUM AND COAL PRODUCTS	3,072	179	33	403	476	0	..	92,000	0	4,494
355	RUBBER PRODUCTS	34	24	0	0	952	0	..	181,000	16,000	1,604
3551	Tyres and tubes	..	..	..	..	..	..	..	..	..	..
3559	Other rubber products	..	..	..	..	..	..	..	..	..	..
356	PLASTIC PRODUCTS	54	21	74	0	658	0	..	243,000	0	1,682
36	NON-METALLIC MINERAL PRODUCTS	19,962	2,025	2,277	308	46,196	909	..	3,180,000	1,000	83,121
361	POTTERY, CHINA, EARTHENWARE	54	13	26	0	1,391	0	..	61,000	0	1,704
362	GLASS AND PRODUCTS	0	219	11	0	8,465	0	..	303,000	1,000	9,782
369	OTHER NON-METAL. MINERAL PRODUCTS	19,908	1,793	2,240	308	36,340	909	..	2,816,000	0	71,636
3691	Structural clay products	709	457	397	127	14,923	903	..	741,000	0	20,184
3692	Cement, lime and plaster	15,518	788	1,699	178	16,183	6	..	1,483,000	0	39,711
3699	Other non-metallic mineral products	3,681	548	144	3	5,234	0	..	592,000	0	11,741
37	BASIC METAL INDUSTRIES	319,482	3,019	2,617	29,422	192,147	0	..	35,282,000	5,172,000	655,083
371	IRON AND STEEL	261,975	1,083	365	983	91,663	0	..	5,825,000	641,000	374,731
372	NON-FERROUS METALS	57,507	1,936	2,252	28,439	100,484	0	..	29,457,000	4,531,000	280,352
38	METAL PRODUCTS, MACHINERY, EQUIP.	47	2,083	94	12	12,422	0	..	4,448,000	0	30,671
381	METAL PRODUCTS	..	..	..	..	..	..	..	..	..	..
3811	Cutlery, hand tools and general hardware	..	..	..	..	..	..	..	..	..	..
3812	Furniture and fixtures primarily of metal	..	..	..	..	..	..	..	..	..	..
3813	Structural metal products	..	..	..	..	..	..	..	..	..	..
3819	Other fabricated metal products	..	..	..	..	..	..	..	..	..	..
382	NON-ELECTRICAL MACHINERY	..	..	..	..	..	..	..	..	..	..
3821	Engines and turbines	..	..	..	..	..	..	..	..	..	..
3822	Agricultural machinery and equipment	..	..	..	..	..	..	..	..	..	..
3823	Metal and wood working machinery	..	..	..	..	..	..	..	..	..	..
3824	Special industrial machinery	..	..	..	..	..	..	..	..	..	..
3825	Office, computing and accounting machinery	..	..	..	..	..	..	..	..	..	..
3829	Other non-electrical machinery and equipment	..	..	..	..	..	..	..	..	..	..
383	ELECTRICAL MACHINERY	..	..	..	..	..	..	..	..	..	..
3831	Electrical industrial machinery	..	..	..	..	..	..	..	..	..	..
3832	Radio, TV and communications equipment	..	..	..	..	..	..	..	..	..	..
3833	Electrical appliances and housewares	..	..	..	..	..	..	..	..	..	..
3839	Other electrical apparatus and supplies	..	..	..	..	..	..	..	..	..	..
384	TRANSPORT EQUIPMENT	..	..	..	..	..	..	..	..	..	..
3841	Shipbuilding	..	..	..	..	..	..	..	..	..	..
3842	Railroad equipment	..	..	..	..	..	..	..	..	..	..
3843	Motor vehicles	..	..	..	..	..	..	..	..	..	..
3844	Motorcycles and bicycles	..	..	..	..	..	..	..	..	..	..
3845	Aircraft	..	..	..	..	..	..	..	..	..	..
3849	Other transport equipment	..	..	..	..	..	..	..	..	..	..
385	PROFESSIONAL AND SCIENTIFIC EQUIPMENT	..	..	..	..	..	..	..	..	..	..
3851	Professional equipment	..	..	..	..	..	..	..	..	..	..
3852	Photographic and optical goods	..	..	..	..	..	..	..	..	..	..
3853	Watches and clocks	..	..	..	..	..	..	..	..	..	..
39	OTHER MANUFACTURING INDUSTRIES	0	0	0	0	268	0	..	22,000	0	347
3901	Jewellery and related articles	..	..	..	..	..	..	..	..	..	..
3902	Musical instruments	..	..	..	..	..	..	..	..	..	..
3903	Sporting and athletic goods	..	..	..	..	..	..	..	..	..	..
3909	Other manufactures	..	..	..	..	..	..	..	..	..	..
SERR	non-specified, unallocated industry	..	..	..	..	..	..	..	..	..	..
3	TOTAL MANUFACTURING	371,299	12,665	33,536	32,398	398,892	126,986	..	61,460,000	6,676,000	1,172,998

ISIS Energy Data Programme (IEA/OECD)

Year: **1998** AUSTRALIA

ISIC Revision 2 Industry Sector	Solid TJ	LPG TJ	Distiloil TJ	RFO TJ	Gas TJ	Biomass TJ	Steam TJ	Electr MWh	Own Use MWh	TOTAL TJ
31 FOOD, BEVERAGES AND TOBACCO	13,739	3,559	885	1,063	24,988	109,482	..	5,351	695	153,733
311/2 FOOD	13,016	3,414	865	949	21,404	109,482	..	4,769	695	149,145
3111 Slaughtering, preparing and preserving meat	2,219	814	121	161	3,555	278	..	1,129	0	7,152
3112 Dairy products	2,471	810	126	34	5,842	1,371	..	715	0	10,657
3113 Canning, preserving of fruits and vegetables	1,640	617	3	183	1,626	249	..	406	0	4,319
3114 Canning, preserving and processing of fish	..	..	..	..	..	..	..	..	..	..
3115 Vegetable and animal oils and fats	749	6	7	0	1,429	0	..	284	0	2,192
3116 Grain mill products	1,962	381	26	0	1,797	0	..	541	1	4,168
3117 Bakery products	0	329	67	76	2,686	0	..	444	0	3,160
3118 Sugar factories and refineries	3,975	457	515	495	4,469	107,584	..	1,250	694	117,497
3119 Cocoa, chocolate and sugar confectionery	..	..	..	..	..	..	..	..	..	..
3121 Other food products	..	..	..	..	..	..	..	..	..	..
3122 Prepared animal feeds	..	..	..	..	..	..	..	..	..	..
313 BEVERAGES	685	143	16	114	3,140	0	..	504	0	4,100
3131 Distilling, rectifying and blending of spirits	..	..	..	..	..	..	..	..	..	..
3132 Wine industries	..	..	..	..	..	..	..	..	..	..
3133 Malt liquors and malts	..	..	..	..	..	..	..	..	..	..
3134 Soft drinks	..	..	..	..	..	..	..	..	..	..
314 TOBACCO	38	2	4	0	444	0	..	78	0	488
32 TEXTILES, APPAREL AND LEATHER	1,150	663	54	131	5,912	0	..	2,018	0	7,917
321 TEXTILES	..	..	..	..	..	..	..	..	..	..
3211 Spinning weaving and finishing textiles	..	..	..	..	..	..	..	..	..	..
3212 Made-up goods excluding wearing apparel	..	..	..	..	..	..	..	..	..	..
3213 Knitting mills	..	..	..	..	..	..	..	..	..	..
3214 Carpets and rugs	..	..	..	..	..	..	..	..	..	..
3215 Cordage, rope and twine	..	..	..	..	..	..	..	..	..	..
3219 Other textiles	..	..	..	..	..	..	..	..	..	..
322 WEARING APPAREL, EXCEPT FOOTWEAR	..	..	..	..	..	..	..	..	..	..
323 LEATHER AND FUR PRODUCTS	..	..	..	..	..	..	..	..	..	..
3231 Tanneries and leather finishing	..	..	..	..	..	..	..	..	..	..
3232 Fur dressing and dyeing industries	..	..	..	..	..	..	..	..	..	..
3233 Leather prods. ex. footwear and wearing apparel	..	..	..	..	..	..	..	..	..	..
324 FOOTWEAR, EX. RUBBER AND PLASTIC	..	..	..	..	..	..	..	..	..	..
33 WOOD PRODUCTS AND FURNITURE	710	312	984	62	1,349	9,121	..	1,354	22	12,543
331 WOOD PRODUCTS, EXCEPT FURNITURE	..	..	..	..	..	..	..	..	..	..
3311 Sawmills, planing and other wood mills	..	..	..	..	..	..	..	..	..	..
3312 Wooden and cane containers	..	..	..	..	..	..	..	..	..	..
3319 Other wood and cork products	..	..	..	..	..	..	..	..	..	..
332 FURNITURE, FIXTURES, EXCL. METALLIC	..	..	..	..	..	..	..	..	..	..
34 PAPER, PUBLISHING AND PRINTING	7,475	522	37	792	20,489	10,214	..	4,529	602	39,543
341 PAPER AND PRODUCTS	..	..	..	..	..	..	..	..	..	..
3411 Pulp, paper and paperboard articles	..	..	..	..	..	..	..	..	..	..
3412 Containers of paper and paperboard	..	..	..	..	..	..	..	..	..	..
3419 Other pulp, paper and paperboard articles	..	..	..	..	..	..	..	..	..	..
342 PRINTING AND PUBLISHING	..	..	..	..	..	..	..	..	..	..
35 CHEMICAL PRODUCTS	7,241	499	30,720	397	91,619	0	..	5,860	277	130,496
351 INDUSTRIAL CHEMICALS	3,734	187	302	0	27,092	0	..	3,143	159	31,326
3511 Basic industrial chemicals excl. fertilizers	..	..	..	..	..	..	..	..	..	..
3512 Fertilizers and pesticides	..	..	..	..	..	..	..	..	..	..
3513 Synthetic resins and plastic materials	..	..	..	..	..	..	..	..	..	..
352 OTHER CHEMICALS	309	84	12	0	1,944	0	..	758	0	2,352
3521 Paints, varnishes and lacquers	..	..	..	..	..	..	..	..	..	..
3522 Drugs and medicines	..	..	..	..	..	..	..	..	..	..
3523 Soap, cleaning preparations, perfumes, cosmetics	..	..	..	..	..	..	..	..	..	..
3529 Other chemical products	..	..	..	..	..	..	..	..	..	..
353 PETROLEUM REFINERIES	0	0	30,299	0	60,528	0	..	1,435	118	90,832
354 MISC. PETROLEUM AND COAL PRODUCTS	3,108	182	33	397	458	0	..	93	0	4,178
355 RUBBER PRODUCTS	34	24	0	0	937	0	..	183	0	996
3551 Tyres and tubes	..	..	..	..	..	..	..	..	..	..
3559 Other rubber products	..	..	..	..	..	..	..	..	..	..
356 PLASTIC PRODUCTS	56	22	74	0	660	0	..	248	0	813
36 NON-METALLIC MINERAL PRODUCTS	20,578	2,032	2,799	346	47,743	904	..	3,355	0	74,414
361 POTTERY, CHINA, EARTHENWARE	55	13	26	0	1,416	0	..	61	0	1,510
362 GLASS AND PRODUCTS	0	219	11	0	8,405	0	..	296	0	8,636
369 OTHER NON-METAL. MINERAL PRODUCTS	20,523	1,800	2,762	346	37,922	904	..	2,998	0	64,268
3691 Structural clay products	747	465	404	127	15,635	898	..	788	0	18,279
3692 Cement, lime and plaster	16,731	759	2,136	178	16,779	6	..	1,578	0	36,595
3699 Other non-metallic mineral products	3,045	576	222	41	5,508	0	..	632	0	9,394
37 BASIC METAL INDUSTRIES	345,504	2,129	2,108	27,924	194,238	0	..	39,164	3,977	572,030
371 IRON AND STEEL	289,627	1,150	112	471	92,820	0	..	6,072	500	384,200
372 NON-FERROUS METALS	55,877	979	1,996	27,453	101,418	0	..	33,092	3,477	187,830
38 METAL PRODUCTS, MACHINERY, EQUIP.	47	2,129	157	12	12,557	0	..	4,676	0	14,919
381 METAL PRODUCTS	..	..	..	..	..	..	..	..	..	..
3811 Cutlery, hand tools and general hardware	..	..	..	..	..	..	..	..	..	..
3812 Furniture and fixtures primarily of metal	..	..	..	..	..	..	..	..	..	..
3813 Structural metal products	..	..	..	..	..	..	..	..	..	..
3819 Other fabricated metal products	..	..	..	..	..	..	..	..	..	..
382 NON-ELECTRICAL MACHINERY	..	..	..	..	..	..	..	..	..	..
3821 Engines and turbines	..	..	..	..	..	..	..	..	..	..
3822 Agricultural machinery and equipment	..	..	..	..	..	..	..	..	..	..
3823 Metal and wood working machinery	..	..	..	..	..	..	..	..	..	..
3824 Special industrial machinery	..	..	..	..	..	..	..	..	..	..
3825 Office, computing and accounting machinery	..	..	..	..	..	..	..	..	..	..
3829 Other non-electrical machinery and equipment	..	..	..	..	..	..	..	..	..	..
383 ELECTRICAL MACHINERY	..	..	..	..	..	..	..	..	..	..
3831 Electrical industrial machinery	..	..	..	..	..	..	..	..	..	..
3832 Radio, TV and communications equipment	..	..	..	..	..	..	..	..	..	..
3833 Electrical appliances and housewares	..	..	..	..	..	..	..	..	..	..
3839 Other electrical apparatus and supplies	..	..	..	..	..	..	..	..	..	..
384 TRANSPORT EQUIPMENT	..	..	..	..	..	..	..	..	..	..
3841 Shipbuilding	..	..	..	..	..	..	..	..	..	..
3842 Railroad equipment	..	..	..	..	..	..	..	..	..	..
3843 Motor vehicles	..	..	..	..	..	..	..	..	..	..
3844 Motorcycles and bicycles	..	..	..	..	..	..	..	..	..	..
3845 Aircraft	..	..	..	..	..	..	..	..	..	..
3849 Other transport equipment	..	..	..	..	..	..	..	..	..	..
385 PROFESSIONAL AND SCIENTIFIC EQUIPMENT	..	..	..	..	..	..	..	..	..	..
3851 Professional equipment	..	..	..	..	..	..	..	..	..	..
3852 Photographic and optical goods	..	..	..	..	..	..	..	..	..	..
3853 Watches and clocks	..	..	..	..	..	..	..	..	..	..
39 OTHER MANUFACTURING INDUSTRIES	0	0	0	0	272	0	..	22	0	272
3901 Jewellery and related articles	..	..	..	..	..	..	..	..	..	..
3902 Musical instruments	..	..	..	..	..	..	..	..	..	..
3903 Sporting and athletic goods	..	..	..	..	..	..	..	..	..	..
3909 Other manufactures	..	..	..	..	..	..	..	..	..	..
SERR non-specified, unallocated industry	..	..	..	..	..	..	..	..	..	..
3 TOTAL MANUFACTURING	396,444	11,845	37,744	30,727	399,167	129,721	..	66,329	5,573	1,005,867

ISIC Revision 2 Industry Sector		Solid TJ	LPG TJ	Distiloil TJ	RFO TJ	Gas TJ	Biomass TJ	Steam TJ	Electr MWh	Own Use MWh	TOTAL TJ
31	**FOOD, BEVERAGES AND TOBACCO**	3,657	92	1,622	7,056	4,413	..	..	2,956,000	..	27,482
311/2	FOOD	..	..	..	..	..	..	..	..	..	..
3111	Slaughtering, preparing and preserving meat	..	..	..	..	..	..	..	..	..	..
3112	Dairy products	..	..	..	..	..	..	..	..	..	..
3113	Canning, preserving of fruits and vegetables	..	..	..	..	..	..	..	..	..	..
3114	Canning, preserving and processing of fish	..	..	..	..	..	..	..	..	..	..
3115	Vegetable and animal oils and fats	..	..	..	..	..	..	..	..	..	..
3116	Grain mill products	..	..	..	..	..	..	..	..	..	..
3117	Bakery products	..	..	..	..	..	..	..	..	..	..
3118	Sugar factories and refineries	..	..	..	..	..	..	..	..	..	..
3119	Cocoa, chocolate and sugar confectionery	..	..	..	..	..	..	..	..	..	..
3121	Other food products	..	..	..	..	..	..	..	..	..	..
3122	Prepared animal feeds	..	..	..	..	..	..	..	..	..	..
313	BEVERAGES	..	..	..	..	..	..	..	..	..	..
3131	Distilling, rectifying and blending of spirits	..	..	..	..	..	..	..	..	..	..
3132	Wine industries	..	..	..	..	..	..	..	..	..	..
3133	Malt liquors and malts	..	..	..	..	..	..	..	..	..	..
3134	Soft drinks	..	..	..	..	..	..	..	..	..	..
314	TOBACCO	..	..	..	..	..	..	..	..	..	..
32	**TEXTILES, APPAREL AND LEATHER**	147	0	555	1,908	2,232	..	..	1,599,000	..	10,598
321	TEXTILES	..	..	..	..	..	..	..	..	..	..
3211	Spinning weaving and finishing textiles	..	..	..	..	..	..	..	..	..	..
3212	Made-up goods excluding wearing apparel	..	..	..	..	..	..	..	..	..	..
3213	Knitting mills	..	..	..	..	..	..	..	..	..	..
3214	Carpets and rugs	..	..	..	..	..	..	..	..	..	..
3215	Cordage, rope and twine	..	..	..	..	..	..	..	..	..	..
3219	Other textiles	..	..	..	..	..	..	..	..	..	..
322	WEARING APPAREL, EXCEPT FOOTWEAR	..	..	..	..	..	..	..	..	..	..
323	LEATHER AND FUR PRODUCTS	..	..	..	..	..	..	..	..	..	..
3231	Tanneries and leather finishing	..	..	..	..	..	..	..	..	..	..
3232	Fur dressing and dyeing industries	..	..	..	..	..	..	..	..	..	..
3233	Leather prods. ex. footwear and wearing apparel	..	..	..	..	..	..	..	..	..	..
324	FOOTWEAR, EX. RUBBER AND PLASTIC	..	..	..	..	..	..	..	..	..	..
33	**WOOD PRODUCTS AND FURNITURE**	0	0	0	0	0	..	..	523,000	..	1,883
331	WOOD PRODUCTS, EXCEPT FURNITURE	..	..	..	..	..	..	..	..	..	..
3311	Sawmills, planing and other wood mills	..	..	..	..	..	..	..	..	..	..
3312	Wooden and cane containers	..	..	..	..	..	..	..	..	..	..
3319	Other wood and cork products	..	..	..	..	..	..	..	..	..	..
332	FURNITURE, FIXTURES, EXCL. METALLIC	..	..	..	..	..	..	..	..	..	..
34	**PAPER, PUBLISHING AND PRINTING**	1,290	0	128	1,827	2,885	..	..	2,045,000	..	13,492
341	PAPER AND PRODUCTS	..	..	..	..	..	..	..	..	..	..
3411	Pulp, paper and paperboard articles	..	..	..	..	..	..	..	..	..	..
3412	Containers of paper and paperboard	..	..	..	..	..	..	..	..	..	..
3419	Other pulp, paper and paperboard articles	..	..	..	..	..	..	..	..	..	..
342	PRINTING AND PUBLISHING	..	..	..	..	..	..	..	..	..	..
35	**CHEMICAL PRODUCTS**	2,022	1,106	599	20,789	34,290	..	..	10,382,000	..	96,181
351	INDUSTRIAL CHEMICALS	..	..	..	..	..	..	..	..	..	..
3511	Basic industrial chemicals excl. fertilizers	..	..	..	..	..	..	..	..	..	..
3512	Fertilizers and pesticides	..	..	..	..	..	..	..	..	..	..
3513	Synthetic resins and plastic materials	..	..	..	..	..	..	..	..	..	..
352	OTHER CHEMICALS	..	..	..	..	..	..	..	..	..	..
3521	Paints, varnishes and lacquers	..	..	..	..	..	..	..	..	..	..
3522	Drugs and medicines	..	..	..	..	..	..	..	..	..	..
3523	Soap, cleaning preparations, perfumes, cosmetics	..	..	..	..	..	..	..	..	..	..
3529	Other chemical products	..	..	..	..	..	..	..	..	..	..
353	PETROLEUM REFINERIES	..	..	..	..	..	..	..	..	..	..
354	MISC. PETROLEUM AND COAL PRODUCTS	..	..	..	..	..	..	..	..	..	..
355	RUBBER PRODUCTS	..	..	..	..	..	..	..	..	..	..
3551	Tyres and tubes	..	..	..	..	..	..	..	..	..	..
3559	Other rubber products	..	..	..	..	..	..	..	..	..	..
356	PLASTIC PRODUCTS	..	..	..	..	..	..	..	..	..	..
36	**NON-METALLIC MINERAL PRODUCTS**	19,602	0	512	6,903	17,629	..	..	2,259,000	..	52,778
361	POTTERY, CHINA, EARTHENWARE	..	..	..	..	..	..	..	..	..	..
362	GLASS AND PRODUCTS	..	..	..	..	..	..	..	..	..	..
369	OTHER NON-METAL. MINERAL PRODUCTS	..	..	..	..	..	..	..	..	..	..
3691	Structural clay products	..	..	..	..	..	..	..	..	..	..
3692	Cement, lime and plaster	..	..	..	..	..	..	..	..	..	..
3699	Other non-metallic mineral products	..	..	..	..	..	..	..	..	..	..
37	**BASIC METAL INDUSTRIES**	132,883	46	683	5,359	49,724	..	..	7,060,000	..	214,111
371	IRON AND STEEL	132,736	0	299	2,720	46,277	..	..	5,324,000	..	201,198
372	NON-FERROUS METALS	147	46	384	2,639	3,447	..	..	1,736,000	..	12,913
38	**METAL PRODUCTS, MACHINERY, EQUIP.**	850	46	684	3,654	6,211	..	..	2,565,000	..	20,679
381	METAL PRODUCTS	..	..	..	..	..	..	..	..	..	..
3811	Cutlery, hand tools and general hardware	..	..	..	..	..	..	..	..	..	..
3812	Furniture and fixtures primarily of metal	..	..	..	..	..	..	..	..	..	..
3813	Structural metal products	..	..	..	..	..	..	..	..	..	..
3819	Other fabricated metal products	..	..	..	..	..	..	..	..	..	..
382	NON-ELECTRICAL MACHINERY	..	..	..	..	..	..	..	..	..	..
3821	Engines and turbines	..	..	..	..	..	..	..	..	..	..
3822	Agricultural machinery and equipment	..	..	..	..	..	..	..	..	..	..
3823	Metal and wood working machinery	..	..	..	..	..	..	..	..	..	..
3824	Special industrial machinery	..	..	..	..	..	..	..	..	..	..
3825	Office, computing and accounting machinery	..	..	..	..	..	..	..	..	..	..
3829	Other non-electrical machinery and equipment	..	..	..	..	..	..	..	..	..	..
383	ELECTRICAL MACHINERY	..	..	..	..	..	..	..	..	..	..
3831	Electrical industrial machinery	..	..	..	..	..	..	..	..	..	..
3832	Radio, TV and communications equipment	..	..	..	..	..	..	..	..	..	..
3833	Electrical appliances and housewares	..	..	..	..	..	..	..	..	..	..
3839	Other electrical apparatus and supplies	..	..	..	..	..	..	..	..	..	..
384	TRANSPORT EQUIPMENT	..	..	..	..	..	..	..	..	..	..
3841	Shipbuilding	..	..	..	..	..	..	..	..	..	..
3842	Railroad equipment	..	..	..	..	..	..	..	..	..	..
3843	Motor vehicles	..	..	..	..	..	..	..	..	..	..
3844	Motorcycles and bicycles	..	..	..	..	..	..	..	..	..	..
3845	Aircraft	..	..	..	..	..	..	..	..	..	..
3849	Other transport equipment	..	..	..	..	..	..	..	..	..	..
385	PROFESSIONAL AND SCIENTIFIC EQUIPMENT	..	..	..	..	..	..	..	..	..	..
3851	Professional equipment	..	..	..	..	..	..	..	..	..	..
3852	Photographic and optical goods	..	..	..	..	..	..	..	..	..	..
3853	Watches and clocks	..	..	..	..	..	..	..	..	..	..
39	**OTHER MANUFACTURING INDUSTRIES**	850	1,194	11,231	9,826	25,443	..	..	1,301,000	..	53,228
3901	Jewellery and related articles	..	..	..	..	..	..	..	..	..	..
3902	Musical instruments	..	..	..	..	..	..	..	..	..	..
3903	Sporting and athletic goods	..	..	..	..	..	..	..	..	..	..
3909	Other manufactures	..	..	..	..	..	..	..	..	..	..
SERR	non-specified, unallocated industry	..	..	..	..	..	..	..	..	..	..
3	**TOTAL MANUFACTURING**	161,301	2,484	16,014	57,331	142,827	..	..	30,690,000	..	490,441

ISIS Energy Data Programme (IEA/OECD)

Year: **1990**   **FINLAND**

ISIC Revision 2 Industry Sector	Solid TJ	LPG TJ	Distiloil TJ	RFO TJ	Gas TJ	Biomass TJ	Steam TJ	Electr MWh	Own Use MWh	TOTAL TJ
31 FOOD, BEVERAGES AND TOBACCO	3,120	..	2,222	6,372	1,964	180	2,552	1,371,944	67,142	21,107
311/2 FOOD	2,540	..	2,037	5,503	1,641	179	2,119	1,217,228	62,476	18,176
3111 Slaughtering, preparing and preserving meat	..	..	..	..	..	..	..	..	..	..
3112 Dairy products	..	..	..	..	..	..	..	..	..	..
3113 Canning, preserving of fruits and vegetables	..	..	..	..	..	..	..	..	..	..
3114 Canning, preserving and processing of fish	..	..	..	..	..	..	..	..	..	..
3115 Vegetable and animal oils and fats	..	..	..	..	..	..	..	..	..	..
3116 Grain mill products	..	..	..	..	..	..	..	..	..	..
3117 Bakery products	..	..	..	..	..	..	..	..	..	..
3118 Sugar factories and refineries	..	..	..	..	..	..	..	..	..	..
3119 Cocoa, chocolate and sugar confectionery	..	..	..	..	..	..	..	..	..	..
3121 Other food products	..	..	..	..	..	..	..	..	..	..
3122 Prepared animal feeds	..	..	..	..	..	..	..	..	..	..
313 BEVERAGES	580	..	170	827	323	1	394	135,474	4,666	2,766
3131 Distilling, rectifying and blending of spirits	..	..	..	..	..	..	..	..	..	..
3132 Wine industries	..	..	..	..	..	..	..	..	..	..
3133 Malt liquors and malts	..	..	..	..	..	..	..	..	..	..
3134 Soft drinks	..	..	..	..	..	..	..	..	..	..
314 TOBACCO	0	..	15	42	0	0	39	19,242	0	165
32 TEXTILES, APPAREL AND LEATHER	1	..	676	334	366	1	706	331,558	3,883	3,264
321 TEXTILES	0	..	307	154	344	0	513	211,518	0	2,079
3211 Spinning weaving and finishing textiles	..	..	..	..	..	..	..	..	..	..
3212 Made-up goods excluding wearing apparel	..	..	..	..	..	..	..	..	..	..
3213 Knitting mills	..	..	..	..	..	..	..	..	..	..
3214 Carpets and rugs	..	..	..	..	..	..	..	..	..	..
3215 Cordage, rope and twine	..	..	..	..	..	..	..	..	..	..
3219 Other textiles	..	..	..	..	..	..	..	..	..	..
322 WEARING APPAREL, EXCEPT FOOTWEAR	0	..	237	61	22	0	112	68,464	0	678
323 LEATHER AND FUR PRODUCTS	1	..	57	117	0	0	45	24,214	0	307
3231 Tanneries and leather finishing	..	..	..	..	..	..	..	..	..	..
3232 Fur dressing and dyeing industries	..	..	..	..	..	..	..	..	..	..
3233 Leather prods. ex. footwear and wearing apparel	..	..	..	..	..	..	..	..	..	..
324 FOOTWEAR, EX. RUBBER AND PLASTIC	0	..	75	2	0	1	36	27,362	3,883	199
33 WOOD PRODUCTS AND FURNITURE	223	..	1,364	2,045	507	9,850	3,517	1,360,873	31,019	22,293
331 WOOD PRODUCTS, EXCEPT FURNITURE	220	..	995	1,753	428	9,167	3,517	1,181,574	31,019	20,222
3311 Sawmills, planing and other wood mills	..	..	..	..	..	..	..	..	..	..
3312 Wooden and cane containers	..	..	..	..	..	..	..	..	..	..
3319 Other wood and cork products	..	..	..	..	..	..	..	..	..	..
332 FURNITURE, FIXTURES, EXCL. METALLIC	3	..	369	292	79	683	0	179,299	0	2,071
34 PAPER, PUBLISHING AND PRINTING	26,174	..	970	14,104	31,708	113,278	867	18,644,469	8,521,758	223,543
341 PAPER AND PRODUCTS	26,174	..	527	13,936	31,654	113,278	0	18,231,812	8,521,166	220,527
3411 Pulp, paper and paperboard articles	..	..	..	..	..	..	..	..	..	..
3412 Containers of paper and paperboard	..	..	..	..	..	..	..	..	..	..
3419 Other pulp, paper and paperboard articles	..	..	..	..	..	..	..	..	..	..
342 PRINTING AND PUBLISHING	0	..	443	168	54	0	867	412,657	592	3,015
35 CHEMICAL PRODUCTS	5,276	..	901	6,279	31,319	3,472	952	3,599,516	1,199,403	56,839
351 INDUSTRIAL CHEMICALS	4,860	..	271	3,581	6,933	898	0	3,128,891	430,016	26,259
3511 Basic industrial chemicals excl. fertilizers	..	..	..	..	..	..	..	..	..	..
3512 Fertilizers and pesticides	..	..	..	..	..	..	..	..	..	..
3513 Synthetic resins and plastic materials	..	..	..	..	..	..	..	..	..	..
352 OTHER CHEMICALS	0	..	241	390	19	0	406	125,897	0	1,509
3521 Paints, varnishes and lacquers	..	..	..	..	..	..	..	..	..	..
3522 Drugs and medicines	..	..	..	..	..	..	..	..	..	..
3523 Soap, cleaning preparations, perfumes, cosmetics	..	..	..	..	..	..	..	..	..	..
3529 Other chemical products	..	..	..	..	..	..	..	..	..	..
353 PETROLEUM REFINERIES	416	..	44	1,793	24,220	2,574	0	0	769,387	26,277
354 MISC. PETROLEUM AND COAL PRODUCTS	0	..	79	191	62	0	40	15,796	0	429
355 RUBBER PRODUCTS	0	..	32	103	56	0	420	75,382	0	882
3551 Tyres and tubes	..	..	..	..	..	..	..	..	..	..
3559 Other rubber products	..	..	..	..	..	..	..	..	..	..
356 PLASTIC PRODUCTS	0	..	234	221	29	0	86	253,550	0	1,483
36 NON-METALLIC MINERAL PRODUCTS	7,953	..	1,963	2,766	3,649	0	349	905,971	5,254	19,923
361 POTTERY, CHINA, EARTHENWARE	0	..	38	88	222	0	23	26,192	0	465
362 GLASS AND PRODUCTS	0	..	106	67	1,667	0	0	153,369	0	2,392
369 OTHER NON-METAL. MINERAL PRODUCTS	7,953	..	1,819	2,611	1,760	0	326	726,410	5,254	17,065
3691 Structural clay products	..	..	..	..	..	..	..	..	..	..
3692 Cement, lime and plaster	..	..	..	..	..	..	..	..	..	..
3699 Other non-metallic mineral products	..	..	..	..	..	..	..	..	..	..
37 BASIC METAL INDUSTRIES	17,061	..	857	9,720	21,873	74	0	3,327,270	587,321	59,449
371 IRON AND STEEL	15,879	..	737	8,590	21,653	65	0	2,154,304	447,893	53,067
372 NON-FERROUS METALS	1,182	..	120	1,130	220	9	0	1,172,966	139,428	6,382
38 METAL PRODUCTS, MACHINERY, EQUIP.	4	..	3,202	1,778	811	6	3,774	2,064,080	4,078	16,991
381 METAL PRODUCTS	1	..	1,341	303	494	1	748	615,469	4,078	5,089
3811 Cutlery, hand tools and general hardware	..	..	..	..	..	..	..	..	..	..
3812 Furniture and fixtures primarily of metal	..	..	..	..	..	..	..	..	..	..
3813 Structural metal products	..	..	..	..	..	..	..	..	..	..
3819 Other fabricated metal products	..	..	..	..	..	..	..	..	..	..
382 NON-ELECTRICAL MACHINERY	1	..	890	357	145	5	1,561	662,471	0	5,344
3821 Engines and turbines	..	..	..	..	..	..	..	..	..	..
3822 Agricultural machinery and equipment	..	..	..	..	..	..	..	..	..	..
3823 Metal and wood working machinery	..	..	..	..	..	..	..	..	..	..
3824 Special industrial machinery	..	..	..	..	..	..	..	..	..	..
3825 Office, computing and accounting machinery	..	..	..	..	..	..	..	..	..	..
3829 Other non-electrical machinery and equipment	..	..	..	..	..	..	..	..	..	..
383 ELECTRICAL MACHINERY	0	..	270	99	138	0	562	369,827	0	2,400
3831 Electrical industrial machinery	..	..	..	..	..	..	..	..	..	..
3832 Radio, TV and communications equipment	..	..	..	..	..	..	..	..	..	..
3833 Electrical appliances and housewares	..	..	..	..	..	..	..	..	..	..
3839 Other electrical apparatus and supplies	..	..	..	..	..	..	..	..	..	..
384 TRANSPORT EQUIPMENT	2	..	624	949	32	0	805	365,084	0	3,726
3841 Shipbuilding	..	..	..	..	..	..	..	..	..	..
3842 Railroad equipment	..	..	..	..	..	..	..	..	..	..
3843 Motor vehicles	..	..	..	..	..	..	..	..	..	..
3844 Motorcycles and bicycles	..	..	..	..	..	..	..	..	..	..
3845 Aircraft	..	..	..	..	..	..	..	..	..	..
3849 Other transport equipment	..	..	..	..	..	..	..	..	..	..
385 PROFESSIONAL AND SCIENTIFIC EQUIPMENT	0	..	77	70	2	0	98	51,229	0	431
3851 Professional equipment	..	..	..	..	..	..	..	..	..	..
3852 Photographic and optical goods	..	..	..	..	..	..	..	..	..	..
3853 Watches and clocks	..	..	..	..	..	..	..	..	..	..
39 OTHER MANUFACTURING INDUSTRIES	0	..	104	52	18	7	52	42,587	0	386
3901 Jewellery and related articles	..	..	..	..	..	..	..	..	..	..
3902 Musical instruments	..	..	..	..	..	..	..	..	..	..
3903 Sporting and athletic goods	..	..	..	..	..	..	..	..	..	..
3909 Other manufactures	..	..	..	..	..	..	..	..	..	..
SERR non-specified, unallocated industry	..	..	..	..	..	..	..	..	..	..
3 TOTAL MANUFACTURING	59,812	..	12,259	43,450	92,215	126,868	12,769	31,648,268	10,419,858	423,795

Year: **1991**   **FINLAND**

ISIC Revision 2 Industry Sector	Solid TJ	LPG TJ	Distiloil TJ	RFO TJ	Gas TJ	Biomass TJ	Steam TJ	Electr MWh	Own Use MWh	TOTAL TJ
31 FOOD, BEVERAGES AND TOBACCO	3,093	..	2,083	6,018	1,896	186	2,465	1,320,423	25,422	20,403
311/2 FOOD	2,526	..	1,936	5,210	1,567	186	2,053	1,185,299	25,422	17,654
3111 Slaughtering, preparing and preserving meat	..	..	..	..	..	..	..	..	..	..
3112 Dairy products	..	..	..	..	..	..	..	..	..	..
3113 Canning, preserving of fruits and vegetables	..	..	..	..	..	..	..	..	..	..
3114 Canning, preserving and processing of fish	..	..	..	..	..	..	..	..	..	..
3115 Vegetable and animal oils and fats	..	..	..	..	..	..	..	..	..	..
3116 Grain mill products	..	..	..	..	..	..	..	..	..	..
3117 Bakery products	..	..	..	..	..	..	..	..	..	..
3118 Sugar factories and refineries	..	..	..	..	..	..	..	..	..	..
3119 Cocoa, chocolate and sugar confectionery	..	..	..	..	..	..	..	..	..	..
3121 Other food products	..	..	..	..	..	..	..	..	..	..
3122 Prepared animal feeds	..	..	..	..	..	..	..	..	..	..
313 BEVERAGES	567	..	131	760	329	0	368	119,845	0	2,586
3131 Distilling, rectifying and blending of spirits	..	..	..	..	..	..	..	..	..	..
3132 Wine industries	..	..	..	..	..	..	..	..	..	..
3133 Malt liquors and malts	..	..	..	..	..	..	..	..	..	..
3134 Soft drinks	..	..	..	..	..	..	..	..	..	..
314 TOBACCO	0	..	16	48	0	0	44	15,279	0	163
32 TEXTILES, APPAREL AND LEATHER	1	..	601	454	538	3	534	291,602	7,361	3,154
321 TEXTILES	0	..	286	292	519	0	373	190,015	0	2,154
3211 Spinning weaving and finishing textiles	..	..	..	..	..	..	..	..	..	..
3212 Made-up goods excluding wearing apparel	..	..	..	..	..	..	..	..	..	..
3213 Knitting mills	..	..	..	..	..	..	..	..	..	..
3214 Carpets and rugs	..	..	..	..	..	..	..	..	..	..
3215 Cordage, rope and twine	..	..	..	..	..	..	..	..	..	..
3219 Other textiles	..	..	..	..	..	..	..	..	..	..
322 WEARING APPAREL, EXCEPT FOOTWEAR	0	..	201	58	19	0	108	54,782	0	583
323 LEATHER AND FUR PRODUCTS	1	..	48	102	0	1	36	21,974	2,455	258
3231 Tanneries and leather finishing	..	..	..	..	..	..	..	..	..	..
3232 Fur dressing and dyeing industries	..	..	..	..	..	..	..	..	..	..
3233 Leather prods. ex. footwear and wearing apparel	..	..	..	..	..	..	..	..	..	..
324 FOOTWEAR, EX. RUBBER AND PLASTIC	0	..	66	2	0	2	17	24,831	4,906	159
33 WOOD PRODUCTS AND FURNITURE	187	..	1,188	1,402	318	8,305	2,169	1,088,077	0	17,486
331 WOOD PRODUCTS, EXCEPT FURNITURE	184	..	846	1,236	220	7,773	2,169	942,298	0	15,820
3311 Sawmills, planing and other wood mills	..	..	..	..	..	..	..	..	..	..
3312 Wooden and cane containers	..	..	..	..	..	..	..	..	..	..
3319 Other wood and cork products	..	..	..	..	..	..	..	..	..	..
332 FURNITURE, FIXTURES, EXCL. METALLIC	3	..	342	166	98	532	0	145,779	0	1,666
34 PAPER, PUBLISHING AND PRINTING	23,526	..	924	13,094	31,055	107,765	839	18,853,666	8,305,291	215,177
341 PAPER AND PRODUCTS	23,526	..	500	12,922	31,014	107,765	0	18,436,920	8,305,291	212,201
3411 Pulp, paper and paperboard articles	..	..	..	..	..	..	..	..	..	..
3412 Containers of paper and paperboard	..	..	..	..	..	..	..	..	..	..
3419 Other pulp, paper and paperboard articles	..	..	..	..	..	..	..	..	..	..
342 PRINTING AND PUBLISHING	0	..	424	172	41	0	839	416,746	0	2,976
35 CHEMICAL PRODUCTS	4,087	..	978	5,900	28,754	3,148	801	3,058,321	791,460	51,829
351 INDUSTRIAL CHEMICALS	3,866	..	256	3,292	6,979	634	0	2,592,543	27,251	24,262
3511 Basic industrial chemicals excl. fertilizers	..	..	..	..	..	..	..	..	..	..
3512 Fertilizers and pesticides	..	..	..	..	..	..	..	..	..	..
3513 Synthetic resins and plastic materials	..	..	..	..	..	..	..	..	..	..
352 OTHER CHEMICALS	0	..	242	668	41	0	308	125,010	0	1,709
3521 Paints, varnishes and lacquers	..	..	..	..	..	..	..	..	..	..
3522 Drugs and medicines	..	..	..	..	..	..	..	..	..	..
3523 Soap, cleaning preparations, perfumes, cosmetics	..	..	..	..	..	..	..	..	..	..
3529 Other chemical products	..	..	..	..	..	..	..	..	..	..
353 PETROLEUM REFINERIES	221	..	187	1,556	21,001	2,514	0	0	760,846	22,740
354 MISC. PETROLEUM AND COAL PRODUCTS	0	..	45	108	657	0	0	36,715	0	942
355 RUBBER PRODUCTS	0	..	35	104	51	0	379	72,923	3,363	819
3551 Tyres and tubes	..	..	..	..	..	..	..	..	..	..
3559 Other rubber products	..	..	..	..	..	..	..	..	..	..
356 PLASTIC PRODUCTS	0	..	213	172	25	0	114	231,130	0	1,356
36 NON-METALLIC MINERAL PRODUCTS	6,413	..	1,910	1,895	3,457	0	243	797,572	0	16,789
361 POTTERY, CHINA, EARTHENWARE	0	..	36	80	225	0	2	15,898	0	400
362 GLASS AND PRODUCTS	0	..	83	51	1,696	0	0	148,623	0	2,365
369 OTHER NON-METAL. MINERAL PRODUCTS	6,413	..	1,791	1,764	1,536	0	241	633,051	0	14,024
3691 Structural clay products	..	..	..	..	..	..	..	..	..	..
3692 Cement, lime and plaster	..	..	..	..	..	..	..	..	..	..
3699 Other non-metallic mineral products	..	..	..	..	..	..	..	..	..	..
37 BASIC METAL INDUSTRIES	15,553	..	803	10,275	20,202	90	0	3,279,969	534,034	56,808
371 IRON AND STEEL	14,957	..	723	9,147	20,028	61	0	2,159,425	414,077	51,199
372 NON-FERROUS METALS	596	..	80	1,128	174	29	0	1,120,544	119,957	5,609
38 METAL PRODUCTS, MACHINERY, EQUIP.	10	..	2,981	1,462	771	5	3,839	1,889,154	7,446	15,842
381 METAL PRODUCTS	5	..	1,252	241	475	2	734	546,849	0	4,678
3811 Cutlery, hand tools and general hardware	..	..	..	..	..	..	..	..	..	..
3812 Furniture and fixtures primarily of metal	..	..	..	..	..	..	..	..	..	..
3813 Structural metal products	..	..	..	..	..	..	..	..	..	..
3819 Other fabricated metal products	..	..	..	..	..	..	..	..	..	..
382 NON-ELECTRICAL MACHINERY	5	..	850	299	127	3	1,578	600,048	7,446	4,995
3821 Engines and turbines	..	..	..	..	..	..	..	..	..	..
3822 Agricultural machinery and equipment	..	..	..	..	..	..	..	..	..	..
3823 Metal and wood working machinery	..	..	..	..	..	..	..	..	..	..
3824 Special industrial machinery	..	..	..	..	..	..	..	..	..	..
3825 Office, computing and accounting machinery	..	..	..	..	..	..	..	..	..	..
3829 Other non-electrical machinery and equipment	..	..	..	..	..	..	..	..	..	..
383 ELECTRICAL MACHINERY	0	..	246	97	126	0	587	343,361	0	2,292
3831 Electrical industrial machinery	..	..	..	..	..	..	..	..	..	..
3832 Radio, TV and communications equipment	..	..	..	..	..	..	..	..	..	..
3833 Electrical appliances and housewares	..	..	..	..	..	..	..	..	..	..
3839 Other electrical apparatus and supplies	..	..	..	..	..	..	..	..	..	..
384 TRANSPORT EQUIPMENT	0	..	552	782	42	0	834	346,694	0	3,458
3841 Shipbuilding	..	..	..	..	..	..	..	..	..	..
3842 Railroad equipment	..	..	..	..	..	..	..	..	..	..
3843 Motor vehicles	..	..	..	..	..	..	..	..	..	..
3844 Motorcycles and bicycles	..	..	..	..	..	..	..	..	..	..
3845 Aircraft	..	..	..	..	..	..	..	..	..	..
3849 Other transport equipment	..	..	..	..	..	..	..	..	..	..
385 PROFESSIONAL AND SCIENTIFIC EQUIPMENT	0	..	81	43	1	0	106	52,202	0	419
3851 Professional equipment	..	..	..	..	..	..	..	..	..	..
3852 Photographic and optical goods	..	..	..	..	..	..	..	..	..	..
3853 Watches and clocks	..	..	..	..	..	..	..	..	..	..
39 OTHER MANUFACTURING INDUSTRIES	0	..	102	45	17	5	50	41,938	0	370
3901 Jewellery and related articles	..	..	..	..	..	..	..	..	..	..
3902 Musical instruments	..	..	..	..	..	..	..	..	..	..
3903 Sporting and athletic goods	..	..	..	..	..	..	..	..	..	..
3909 Other manufactures	..	..	..	..	..	..	..	..	..	..
SERR non-specified, unallocated industry	..	..	..	..	..	..	..	..	..	..
3 TOTAL MANUFACTURING	52,870	..	11,570	40,545	87,008	119,507	10,940	30,620,722	9,671,014	397,859

ISIS Energy Data Programme (IEA/OECD)

ISIC Revision 2 Industry Sector	Solid TJ	LPG TJ	Distiloil TJ	RFO TJ	Gas TJ	Biomass TJ	Steam TJ	Electr MWh	Own Use MWh	TOTAL TJ
31 FOOD, BEVERAGES AND TOBACCO	3,200	..	2,008	4,760	2,469	171	3,229	1,456,603	63,033	20,854
311/2 FOOD	2,559		1,901	4,144	1,862	171	2,648	1,283,690	60,166	17,690
3111 Slaughtering, preparing and preserving meat	..		..	..	..		..	..	..	..
3112 Dairy products	..		..	..	..		..	..	..	..
3113 Canning, preserving of fruits and vegetables	..		..	..	..		..	..	..	..
3114 Canning, preserving and processing of fish	..		..	..	..		..	..	..	..
3115 Vegetable and animal oils and fats	..		..	..	..		..	..	..	..
3116 Grain mill products	..		..	..	..		..	..	..	..
3117 Bakery products	..		..	..	..		..	..	..	..
3118 Sugar factories and refineries	..		..	..	..		..	..	..	..
3119 Cocoa, chocolate and sugar confectionery	..		..	..	..		..	..	..	..
3121 Other food products	..		..	..	..		..	..	..	..
3122 Prepared animal feeds	..		..	..	..		..	..	..	..
313 BEVERAGES	641	..	93	594	607	0	542	157,896	2,867	3,035
3131 Distilling, rectifying and blending of spirits	..		..	..	..		..	..	..	..
3132 Wine industries	..		..	..	..		..	..	..	..
3133 Malt liquors and malts	..		..	..	..		..	..	..	..
3134 Soft drinks	..		..	..	..		..	..	..	..
314 TOBACCO	0		14	22	0	0	39	15,017	0	129
32 TEXTILES, APPAREL AND LEATHER	1	..	566	656	359	2	457	289,781	5,760	3,063
321 TEXTILES	0		307	518	344	0	323	193,227	0	2,188
3211 Spinning weaving and finishing textiles	..		..	..	..		..	..	..	..
3212 Made-up goods excluding wearing apparel	..		..	..	..		..	..	..	..
3213 Knitting mills	..		..	..	..		..	..	..	..
3214 Carpets and rugs	..		..	..	..		..	..	..	..
3215 Cordage, rope and twine	..		..	..	..		..	..	..	..
3219 Other textiles	..		..	..	..		..	..	..	..
322 WEARING APPAREL, EXCEPT FOOTWEAR	0		148	31	15	0	88	52,805	1,034	468
323 LEATHER AND FUR PRODUCTS	1		44	105	0	1	32	20,161	0	256
3231 Tanneries and leather finishing	..		..	..	..		..	..	..	..
3232 Fur dressing and dyeing industries	..		..	..	..		..	..	..	..
3233 Leather prods. ex. footwear and wearing apparel	..		..	..	..		..	..	..	..
324 FOOTWEAR, EX. RUBBER AND PLASTIC	0		67	2	0	1	14	23,588	4,726	152
33 WOOD PRODUCTS AND FURNITURE	318	..	1,084	1,512	408	7,740	2,387	1,193,704	24,263	17,659
331 WOOD PRODUCTS, EXCEPT FURNITURE	315		764	1,473	302	7,158	2,387	1,039,221	24,263	16,053
3311 Sawmills, planing and other wood mills	..		..	..	..		..	..	..	..
3312 Wooden and cane containers	..		..	..	..		..	..	..	..
3319 Other wood and cork products	..		..	..	..		..	..	..	..
332 FURNITURE, FIXTURES, EXCL. METALLIC	3		320	39	106	582	0	154,483	0	1,606
34 PAPER, PUBLISHING AND PRINTING	23,519	..	886	11,867	31,433	107,960	1,778	18,542,563	8,117,790	214,972
341 PAPER AND PRODUCTS	23,519		497	11,765	31,378	107,960	964	18,141,649	8,117,790	212,169
3411 Pulp, paper and paperboard articles	..		..	..	..		..	..	..	..
3412 Containers of paper and paperboard	..		..	..	..		..	..	..	..
3419 Other pulp, paper and paperboard articles	..		..	..	..		..	..	..	..
342 PRINTING AND PUBLISHING	0		389	102	55	0	814	400,914	0	2,803
35 CHEMICAL PRODUCTS	4,594	..	846	6,917	30,393	4,022	1,009	3,653,548	1,497,492	55,543
351 INDUSTRIAL CHEMICALS	4,301		247	4,158	7,629	1,300	0	3,111,626	556,350	26,834
3511 Basic industrial chemicals excl. fertilizers	..		..	..	..		..	..	..	..
3512 Fertilizers and pesticides	..		..	..	..		..	..	..	..
3513 Synthetic resins and plastic materials	..		..	..	..		..	..	..	..
352 OTHER CHEMICALS	0		221	650	39	0	648	160,716	4,310	2,121
3521 Paints, varnishes and lacquers	..		..	..	..		..	..	..	..
3522 Drugs and medicines	..		..	..	..		..	..	..	..
3523 Soap, cleaning preparations, perfumes, cosmetics	..		..	..	..		..	..	..	..
3529 Other chemical products	..		..	..	..		..	..	..	..
353 PETROLEUM REFINERIES	293		106	1,687	20,963	2,722	0	0	936,832	22,398
354 MISC. PETROLEUM AND COAL PRODUCTS	0		71	127	1,684	0	0	58,157	0	2,091
355 RUBBER PRODUCTS	0		27	127	53	0	282	74,550	0	757
3551 Tyres and tubes	..		..	..	..		..	..	..	..
3559 Other rubber products	..		..	..	..		..	..	..	..
356 PLASTIC PRODUCTS	0		174	168	25	0	79	248,499	0	1,341
36 NON-METALLIC MINERAL PRODUCTS	5,690	..	1,478	1,348	3,368	0	111	766,612	6,210	14,732
361 POTTERY, CHINA, EARTHENWARE	0		34	78	292	0	3	22,562	0	488
362 GLASS AND PRODUCTS	0		100	13	1,796	0	0	177,507	0	2,548
369 OTHER NON-METAL. MINERAL PRODUCTS	5,690		1,344	1,257	1,280	0	108	566,543	6,210	11,696
3691 Structural clay products	..		..	..	..		..	..	..	..
3692 Cement, lime and plaster	..		..	..	..		..	..	..	..
3699 Other non-metallic mineral products	..		..	..	..		..	..	..	..
37 BASIC METAL INDUSTRIES	16,653	..	840	10,421	20,253	131	0	3,464,040	462,932	59,102
371 IRON AND STEEL	16,009		736	9,940	19,984	0	0	2,322,451	445,121	53,427
372 NON-FERROUS METALS	644		104	481	269	131	0	1,141,589	17,811	5,675
38 METAL PRODUCTS, MACHINERY, EQUIP.	1	..	2,709	1,370	818	5	3,647	1,966,066	299	15,627
381 METAL PRODUCTS	0		1,049	188	452	1	750	574,547	299	4,507
3811 Cutlery, hand tools and general hardware	..		..	..	..		..	..	..	..
3812 Furniture and fixtures primarily of metal	..		..	..	..		..	..	..	..
3813 Structural metal products	..		..	..	..		..	..	..	..
3819 Other fabricated metal products	..		..	..	..		..	..	..	..
382 NON-ELECTRICAL MACHINERY	1		815	282	211	4	1,332	601,525	0	4,810
3821 Engines and turbines	..		..	..	..		..	..	..	..
3822 Agricultural machinery and equipment	..		..	..	..		..	..	..	..
3823 Metal and wood working machinery	..		..	..	..		..	..	..	..
3824 Special industrial machinery	..		..	..	..		..	..	..	..
3825 Office, computing and accounting machinery	..		..	..	..		..	..	..	..
3829 Other non-electrical machinery and equipment	..		..	..	..		..	..	..	..
383 ELECTRICAL MACHINERY	0		232	99	138	0	568	366,843	0	2,358
3831 Electrical industrial machinery	..		..	..	..		..	..	..	..
3832 Radio, TV and communications equipment	..		..	..	..		..	..	..	..
3833 Electrical appliances and housewares	..		..	..	..		..	..	..	..
3839 Other electrical apparatus and supplies	..		..	..	..		..	..	..	..
384 TRANSPORT EQUIPMENT	0		541	766	16	0	861	364,703	0	3,497
3841 Shipbuilding	..		..	..	..		..	..	..	..
3842 Railroad equipment	..		..	..	..		..	..	..	..
3843 Motor vehicles	..		..	..	..		..	..	..	..
3844 Motorcycles and bicycles	..		..	..	..		..	..	..	..
3845 Aircraft	..		..	..	..		..	..	..	..
3849 Other transport equipment	..		..	..	..		..	..	..	..
385 PROFESSIONAL AND SCIENTIFIC EQUIPMENT	0		72	35	1	0	136	58,448	0	454
3851 Professional equipment	..		..	..	..		..	..	..	..
3852 Photographic and optical goods	..		..	..	..		..	..	..	..
3853 Watches and clocks	..		..	..	..		..	..	..	..
39 OTHER MANUFACTURING INDUSTRIES	0	..	86	40	31	2	64	43,517	0	380
3901 Jewellery and related articles	..		..	..	..		..	..	..	..
3902 Musical instruments	..		..	..	..		..	..	..	..
3903 Sporting and athletic goods	..		..	..	..		..	..	..	..
3909 Other manufactures	..		..	..	..		..	..	..	..
SERR non-specified, unallocated industry										
3 TOTAL MANUFACTURING	53,976	..	10,503	38,891	89,532	120,033	12,682	31,376,434	10,177,779	401,932

ISIC Revision 2 Industry Sector		Solid TJ	LPG TJ	Distiloil TJ	RFO TJ	Gas TJ	Biomass TJ	Steam TJ	Electr MWh	Own Use MWh	TOTAL TJ
31	FOOD, BEVERAGES AND TOBACCO	3,404	..	1,902	4,435	2,340	154	2,553	1,409,709	54,596	19,666
311/2	FOOD	2,577	..	1,779	3,812	2,060	154	1,999	1,239,553	46,195	16,677
3111	Slaughtering, preparing and preserving meat	..	..	..	..	..	..	..	..	..	..
3112	Dairy products	..	..	..	..	..	..	..	..	..	..
3113	Canning, preserving of fruits and vegetables	..	..	..	..	..	..	..	..	..	..
3114	Canning, preserving and processing of fish	..	..	..	..	..	..	..	..	..	..
3115	Vegetable and animal oils and fats	..	..	..	..	..	..	..	..	..	..
3116	Grain mill products	..	..	..	..	..	..	..	..	..	..
3117	Bakery products	..	..	..	..	..	..	..	..	..	..
3118	Sugar factories and refineries	..	..	..	..	..	..	..	..	..	..
3119	Cocoa, chocolate and sugar confectionery	..	..	..	..	..	..	..	..	..	..
3121	Other food products	..	..	..	..	..	..	..	..	..	..
3122	Prepared animal feeds	..	..	..	..	..	..	..	..	..	..
313	BEVERAGES	827	..	112	606	280	0	503	156,072	8,401	2,860
3131	Distilling, rectifying and blending of spirits	..	..	..	..	..	..	..	..	..	..
3132	Wine industries	..	..	..	..	..	..	..	..	..	..
3133	Malt liquors and malts	..	..	..	..	..	..	..	..	..	..
3134	Soft drinks	..	..	..	..	..	..	..	..	..	..
314	TOBACCO	0	..	11	17	0	0	51	14,084	0	130
32	TEXTILES, APPAREL AND LEATHER	0	..	522	410	651	3	465	298,173	1,881	3,118
321	TEXTILES	0	..	286	284	629	0	358	215,601	0	2,333
3211	Spinning weaving and finishing textiles	..	..	..	..	..	..	..	..	..	..
3212	Made-up goods excluding wearing apparel	..	..	..	..	..	..	..	..	..	..
3213	Knitting mills	..	..	..	..	..	..	..	..	..	..
3214	Carpets and rugs	..	..	..	..	..	..	..	..	..	..
3215	Cordage, rope and twine	..	..	..	..	..	..	..	..	..	..
3219	Other textiles	..	..	..	..	..	..	..	..	..	..
322	WEARING APPAREL, EXCEPT FOOTWEAR	0	..	130	15	22	0	95	45,312	0	425
323	LEATHER AND FUR PRODUCTS	0	..	43	109	0	2	1	17,411	0	218
3231	Tanneries and leather finishing	..	..	..	..	..	..	..	..	..	..
3232	Fur dressing and dyeing industries	..	..	..	..	..	..	..	..	..	..
3233	Leather prods. ex. footwear and wearing apparel	..	..	..	..	..	..	..	..	..	..
324	FOOTWEAR, EX. RUBBER AND PLASTIC	0	..	63	2	0	1	11	19,849	1,881	142
33	WOOD PRODUCTS AND FURNITURE	352	..	1,061	1,612	460	7,871	2,941	1,221,032	0	18,693
331	WOOD PRODUCTS, EXCEPT FURNITURE	352	..	767	1,510	349	7,540	2,941	1,076,134	0	17,333
3311	Sawmills, planing and other wood mills	..	..	..	..	..	..	..	..	..	..
3312	Wooden and cane containers	..	..	..	..	..	..	..	..	..	..
3319	Other wood and cork products	..	..	..	..	..	..	..	..	..	..
332	FURNITURE, FIXTURES, EXCL. METALLIC	0	..	294	102	111	331	0	144,898	0	1,360
34	PAPER, PUBLISHING AND PRINTING	22,074	..	834	11,927	28,972	124,355	4,139	20,033,014	8,758,594	232,889
341	PAPER AND PRODUCTS	22,074	..	470	11,819	28,910	124,355	3,292	19,642,462	8,758,594	230,102
3411	Pulp, paper and paperboard articles	..	..	..	..	..	..	..	..	..	..
3412	Containers of paper and paperboard	..	..	..	..	..	..	..	..	..	..
3419	Other pulp, paper and paperboard articles	..	..	..	..	..	..	..	..	..	..
342	PRINTING AND PUBLISHING	0	..	364	108	62	0	847	390,552	0	2,787
35	CHEMICAL PRODUCTS	5,278	..	1,009	7,473	31,520	3,520	1,003	3,893,793	1,291,493	59,171
351	INDUSTRIAL CHEMICALS	4,613	..	232	4,259	8,236	1,032	0	3,308,598	587,448	28,168
3511	Basic industrial chemicals excl. fertilizers	..	..	..	..	..	..	..	..	..	..
3512	Fertilizers and pesticides	..	..	..	..	..	..	..	..	..	..
3513	Synthetic resins and plastic materials	..	..	..	..	..	..	..	..	..	..
352	OTHER CHEMICALS	0	..	242	771	36	0	614	163,672	4,361	2,237
3521	Paints, varnishes and lacquers	..	..	..	..	..	..	..	..	..	..
3522	Drugs and medicines	..	..	..	..	..	..	..	..	..	..
3523	Soap, cleaning preparations, perfumes, cosmetics	..	..	..	..	..	..	..	..	..	..
3529	Other chemical products	..	..	..	..	..	..	..	..	..	..
353	PETROLEUM REFINERIES	665	..	268	2,032	20,516	2,488	0	0	699,649	23,450
354	MISC. PETROLEUM AND COAL PRODUCTS	0	..	68	138	2,640	0	0	74,512	35	3,114
355	RUBBER PRODUCTS	0	..	28	118	61	0	316	80,053	0	811
3551	Tyres and tubes	..	..	..	..	..	..	..	..	..	..
3559	Other rubber products	..	..	..	..	..	..	..	..	..	..
356	PLASTIC PRODUCTS	0	..	171	155	31	0	73	266,958	0	1,391
36	NON-METALLIC MINERAL PRODUCTS	4,490	..	1,261	1,110	3,269	0	57	697,137	0	12,697
361	POTTERY, CHINA, EARTHENWARE	0	..	19	99	122	0	2	10,448	0	280
362	GLASS AND PRODUCTS	0	..	82	11	1,971	0	0	177,762	0	2,704
369	OTHER NON-METAL. MINERAL PRODUCTS	4,490	..	1,160	1,000	1,176	0	55	508,927	0	9,713
3691	Structural clay products	..	..	..	..	..	..	..	..	..	..
3692	Cement, lime and plaster	..	..	..	..	..	..	..	..	..	..
3699	Other non-metallic mineral products	..	..	..	..	..	..	..	..	..	..
37	BASIC METAL INDUSTRIES	17,305	..	573	10,275	23,574	0	70	3,673,904	593,521	62,886
371	IRON AND STEEL	16,305	..	451	9,657	23,289	0	0	2,514,042	570,181	56,700
372	NON-FERROUS METALS	1,000	..	122	618	285	0	70	1,159,862	23,340	6,186
38	METAL PRODUCTS, MACHINERY, EQUIP.	2	..	2,719	1,347	1,095	3	3,726	1,858,785	0	15,584
381	METAL PRODUCTS	0	..	1,033	174	733	1	726	573,096	0	4,730
3811	Cutlery, hand tools and general hardware	..	..	..	..	..	..	..	..	..	..
3812	Furniture and fixtures primarily of metal	..	..	..	..	..	..	..	..	..	..
3813	Structural metal products	..	..	..	..	..	..	..	..	..	..
3819	Other fabricated metal products	..	..	..	..	..	..	..	..	..	..
382	NON-ELECTRICAL MACHINERY	2	..	790	301	213	2	1,397	587,395	0	4,820
3821	Engines and turbines	..	..	..	..	..	..	..	..	..	..
3822	Agricultural machinery and equipment	..	..	..	..	..	..	..	..	..	..
3823	Metal and wood working machinery	..	..	..	..	..	..	..	..	..	..
3824	Special industrial machinery	..	..	..	..	..	..	..	..	..	..
3825	Office, computing and accounting machinery	..	..	..	..	..	..	..	..	..	..
3829	Other non-electrical machinery and equipment	..	..	..	..	..	..	..	..	..	..
383	ELECTRICAL MACHINERY	0	..	231	78	119	0	601	332,469	0	2,226
3831	Electrical industrial machinery	..	..	..	..	..	..	..	..	..	..
3832	Radio, TV and communications equipment	..	..	..	..	..	..	..	..	..	..
3833	Electrical appliances and housewares	..	..	..	..	..	..	..	..	..	..
3839	Other electrical apparatus and supplies	..	..	..	..	..	..	..	..	..	..
384	TRANSPORT EQUIPMENT	0	..	602	763	29	0	854	310,461	0	3,366
3841	Shipbuilding	..	..	..	..	..	..	..	..	..	..
3842	Railroad equipment	..	..	..	..	..	..	..	..	..	..
3843	Motor vehicles	..	..	..	..	..	..	..	..	..	..
3844	Motorcycles and bicycles	..	..	..	..	..	..	..	..	..	..
3845	Aircraft	..	..	..	..	..	..	..	..	..	..
3849	Other transport equipment	..	..	..	..	..	..	..	..	..	..
385	PROFESSIONAL AND SCIENTIFIC EQUIPMENT	0	..	63	31	1	0	148	55,364	0	442
3851	Professional equipment	..	..	..	..	..	..	..	..	..	..
3852	Photographic and optical goods	..	..	..	..	..	..	..	..	..	..
3853	Watches and clocks	..	..	..	..	..	..	..	..	..	..
39	OTHER MANUFACTURING INDUSTRIES	0	..	81	48	38	3	52	48,640	0	397
3901	Jewellery and related articles	..	..	..	..	..	..	..	..	..	..
3902	Musical instruments	..	..	..	..	..	..	..	..	..	..
3903	Sporting and athletic goods	..	..	..	..	..	..	..	..	..	..
3909	Other manufactures	..	..	..	..	..	..	..	..	..	..
SERR	non-specified, unallocated industry	..	..	..	..	..	..	..	..	..	..
3	TOTAL MANUFACTURING	52,905	..	9,962	38,637	91,919	135,909	15,006	33,134,187	10,700,085	425,101

ISIS Energy Data Programme (IEA/OECD)

ISIC Revision 2 Industry Sector	Solid TJ	LPG TJ	Distiloil TJ	RFO TJ	Gas TJ	Biomass TJ	Steam TJ	Electr MWh	Own Use MWh	TOTAL TJ
31 FOOD, BEVERAGES AND TOBACCO	3,388	..	1,737	4,330	2,367	143	2,759	1,461,987	72,228	19,727
311/2 FOOD	2,632		1,654	3,897	2,077	143	2,161	1,291,062	69,550	16,961
3111 Slaughtering, preparing and preserving meat	..	..	..	..	..		..	..	..	..
3112 Dairy products	..	..	..	..	..		..	..	..	..
3113 Canning, preserving of fruits and vegetables	..	..	..	..	..		..	..	..	..
3114 Canning, preserving and processing of fish	..	..	..	..	..		..	..	..	..
3115 Vegetable and animal oils and fats	..	..	..	..	..		..	..	..	..
3116 Grain mill products	..	..	..	..	..		..	..	..	..
3117 Bakery products	..	..	..	..	..		..	..	..	..
3118 Sugar factories and refineries	..	..	..	..	..		..	..	..	..
3119 Cocoa, chocolate and sugar confectionery	..	..	..	..	..		..	..	..	..
3121 Other food products	..	..	..	..	..		..	..	..	..
3122 Prepared animal feeds	..	..	..	..	..		..	..	..	..
313 BEVERAGES	756		62	419	290	0	555	154,286	2,678	2,628
3131 Distilling, rectifying and blending of spirits	..	..	..	..	..		..	..	..	..
3132 Wine industries	..	..	..	..	..		..	..	..	..
3133 Malt liquors and malts	..	..	..	..	..		..	..	..	..
3134 Soft drinks	..	..	..	..	..		..	..	..	..
314 TOBACCO	0		21	14	0	0	43	16,639	0	138
32 TEXTILES, APPAREL AND LEATHER	0	..	571	309	657	2	465	283,022	736	3,020
321 TEXTILES	0		304	190	436	0	292	180,843	0	1,873
3211 Spinning weaving and finishing textiles	..	..	..	..	..		..	..	..	..
3212 Made-up goods excluding wearing apparel	..	..	..	..	..		..	..	..	..
3213 Knitting mills	..	..	..	..	..		..	..	..	..
3214 Carpets and rugs	..	..	..	..	..		..	..	..	..
3215 Cordage, rope and twine	..	..	..	..	..		..	..	..	..
3219 Other textiles	..	..	..	..	..		..	..	..	..
322 WEARING APPAREL, EXCEPT FOOTWEAR	0		175	34	220	0	95	63,973	736	752
323 LEATHER AND FUR PRODUCTS	0		33	83	0	0	2	13,537	0	167
3231 Tanneries and leather finishing	..	..	..	..	..		..	..	..	..
3232 Fur dressing and dyeing industries	..	..	..	..	..		..	..	..	..
3233 Leather prods. ex. footwear and wearing apparel	..	..	..	..	..		..	..	..	..
324 FOOTWEAR, EX. RUBBER AND PLASTIC	0		59	2	1	2	76	24,669	0	229
33 WOOD PRODUCTS AND FURNITURE	600		1,174	2,072	597	9,763	4,204	1,528,529	52,918	23,722
331 WOOD PRODUCTS, EXCEPT FURNITURE	599		879	1,855	506	9,415	4,204	1,332,821	52,918	22,066
3311 Sawmills, planing and other wood mills	..	..	..	..	..		..	..	..	..
3312 Wooden and cane containers	..	..	..	..	..		..	..	..	..
3319 Other wood and cork products	..	..	..	..	..		..	..	..	..
332 FURNITURE, FIXTURES, EXCL. METALLIC	1		295	217	91	348	0	195,708	0	1,657
34 PAPER, PUBLISHING AND PRINTING	23,556	..	773	12,668	31,968	135,810	4,808	19,734,814	8,237,163	250,975
341 PAPER AND PRODUCTS	23,556		471	12,580	31,893	135,810	4,032	19,323,109	8,237,163	248,251
3411 Pulp, paper and paperboard articles	..	..	..	..	..		..	..	..	..
3412 Containers of paper and paperboard	..	..	..	..	..		..	..	..	..
3419 Other pulp, paper and paperboard articles	..	..	..	..	..		..	..	..	..
342 PRINTING AND PUBLISHING	0		302	88	75	0	776	411,705	0	2,723
35 CHEMICAL PRODUCTS	7,001	..	847	11,283	36,031	3,546	1,932	4,262,468	1,417,406	70,882
351 INDUSTRIAL CHEMICALS	6,142		319	4,542	9,641	886	1,067	3,579,650	664,380	33,092
3511 Basic industrial chemicals excl. fertilizers	..	..	..	..	..		..	..	..	..
3512 Fertilizers and pesticides	..	..	..	..	..		..	..	..	..
3513 Synthetic resins and plastic materials	..	..	..	..	..		..	..	..	..
352 OTHER CHEMICALS	0		68	300	1	0	28	41,704	0	547
3521 Paints, varnishes and lacquers	..	..	..	..	..		..	..	..	..
3522 Drugs and medicines	..	..	..	..	..		..	..	..	..
3523 Soap, cleaning preparations, perfumes, cosmetics	..	..	..	..	..		..	..	..	..
3529 Other chemical products	..	..	..	..	..		..	..	..	..
353 PETROLEUM REFINERIES	859		137	5,949	26,181	2,659	0	0	753,026	33,074
354 MISC. PETROLEUM AND COAL PRODUCTS	0		1	0	22	0	0	66,729	0	263
355 RUBBER PRODUCTS	0		60	127	63	0	324	88,546	0	893
3551 Tyres and tubes	..	..	..	..	..		..	..	..	..
3559 Other rubber products	..	..	..	..	..		..	..	..	..
356 PLASTIC PRODUCTS	0		262	365	123	1	513	485,839	0	3,013
36 NON-METALLIC MINERAL PRODUCTS	4,345	..	871	1,240	3,166	0	34	709,592	83,501	11,910
361 POTTERY, CHINA, EARTHENWARE	0		9	83	254	0	18	21,278	0	441
362 GLASS AND PRODUCTS	0		70	12	2,192	0	16	276,363	83,198	2,985
369 OTHER NON-METAL. MINERAL PRODUCTS	4,345		792	1,145	720	0	0	411,951	303	8,484
3691 Structural clay products	..	..	..	..	..		..	..	..	..
3692 Cement, lime and plaster	..	..	..	..	..		..	..	..	..
3699 Other non-metallic mineral products	..	..	..	..	..		..	..	..	..
37 BASIC METAL INDUSTRIES	23,172	..	487	10,276	18,309	0	0	3,714,810	580,101	63,529
371 IRON AND STEEL	22,856		409	9,948	18,144	0	0	2,652,068	575,142	58,834
372 NON-FERROUS METALS	316		78	328	165	0	0	1,062,742	4,959	4,695
38 METAL PRODUCTS, MACHINERY, EQUIP.	0	..	2,566	1,250	1,148	3	3,632	2,034,195	0	15,922
381 METAL PRODUCTS	0		883	152	748	1	569	486,550	0	4,105
3811 Cutlery, hand tools and general hardware	..	..	..	..	..		..	..	..	..
3812 Furniture and fixtures primarily of metal	..	..	..	..	..		..	..	..	..
3813 Structural metal products	..	..	..	..	..		..	..	..	..
3819 Other fabricated metal products	..	..	..	..	..		..	..	..	..
382 NON-ELECTRICAL MACHINERY	0		905	386	222	2	1,474	703,076	0	5,520
3821 Engines and turbines	..	..	..	..	..		..	..	..	..
3822 Agricultural machinery and equipment	..	..	..	..	..		..	..	..	..
3823 Metal and wood working machinery	..	..	..	..	..		..	..	..	..
3824 Special industrial machinery	..	..	..	..	..		..	..	..	..
3825 Office, computing and accounting machinery	..	..	..	..	..		..	..	..	..
3829 Other non-electrical machinery and equipment	..	..	..	..	..		..	..	..	..
383 ELECTRICAL MACHINERY	0		280	65	147	0	580	406,652	0	2,536
3831 Electrical industrial machinery	..	..	..	..	..		..	..	..	..
3832 Radio, TV and communications equipment	..	..	..	..	..		..	..	..	..
3833 Electrical appliances and housewares	..	..	..	..	..		..	..	..	..
3839 Other electrical apparatus and supplies	..	..	..	..	..		..	..	..	..
384 TRANSPORT EQUIPMENT	0		435	643	29	0	877	366,987	0	3,305
3841 Shipbuilding	..	..	..	..	..		..	..	..	..
3842 Railroad equipment	..	..	..	..	..		..	..	..	..
3843 Motor vehicles	..	..	..	..	..		..	..	..	..
3844 Motorcycles and bicycles	..	..	..	..	..		..	..	..	..
3845 Aircraft	..	..	..	..	..		..	..	..	..
3849 Other transport equipment	..	..	..	..	..		..	..	..	..
385 PROFESSIONAL AND SCIENTIFIC EQUIPMENT	0		63	4	2	0	132	70,930	0	456
3851 Professional equipment	..	..	..	..	..		..	..	..	..
3852 Photographic and optical goods	..	..	..	..	..		..	..	..	..
3853 Watches and clocks	..	..	..	..	..		..	..	..	..
39 OTHER MANUFACTURING INDUSTRIES	0		90	45	35	2	70	54,948	0	440
3901 Jewellery and related articles	..	..	..	..	..		..	..	..	..
3902 Musical instruments	..	..	..	..	..		..	..	..	..
3903 Sporting and athletic goods	..	..	..	..	..		..	..	..	..
3909 Other manufactures	..	..	..	..	..		..	..	..	..
SERR non-specified, unallocated industry	..	..	..	..	..		..	..	..	..
3 TOTAL MANUFACTURING	62,062		9,116	43,473	94,278	149,269	17,904	33,784,365	10,444,053	460,127

ISIC Revision 2 Industry Sector	Solid TJ	LPG TJ	Distiloil TJ	RFO TJ	Gas TJ	Biomass TJ	Steam TJ	Electr MWh	Own Use MWh	TOTAL TJ
31 FOOD, BEVERAGES AND TOBACCO	3,572	..	1,528	4,017	2,184	220	3,621	1,544,499	79,522	20,416
311/2 FOOD	2,771	..	1,464	3,620	1,856	141	2,815	1,358,990	79,522	17,273
3111 Slaughtering, preparing and preserving meat	..	..	..	..	..	..	..	..	..	..
3112 Dairy products	..	..	..	..	..	..	..	..	..	..
3113 Canning, preserving of fruits and vegetables	..	..	..	..	..	..	..	..	..	..
3114 Canning, preserving and processing of fish	..	..	..	..	..	..	..	..	..	..
3115 Vegetable and animal oils and fats	..	..	..	..	..	..	..	..	..	..
3116 Grain mill products	..	..	..	..	..	..	..	..	..	..
3117 Bakery products	..	..	..	..	..	..	..	..	..	..
3118 Sugar factories and refineries	..	..	..	..	..	..	..	..	..	..
3119 Cocoa, chocolate and sugar confectionery	..	..	..	..	..	..	..	..	..	..
3121 Other food products	..	..	..	..	..	..	..	..	..	..
3122 Prepared animal feeds	..	..	..	..	..	..	..	..	..	..
313 BEVERAGES	801	..	52	382	328	79	765	168,354	0	3,013
3131 Distilling, rectifying and blending of spirits	..	..	..	..	..	..	..	..	..	..
3132 Wine industries	..	..	..	..	..	..	..	..	..	..
3133 Malt liquors and malts	..	..	..	..	..	..	..	..	..	..
3134 Soft drinks	..	..	..	..	..	..	..	..	..	..
314 TOBACCO	0	..	12	15	0	0	41	17,155	0	130
32 TEXTILES, APPAREL AND LEATHER	0	..	565	385	680	8	468	297,879	833	3,175
321 TEXTILES	0	..	318	307	468	0	280	197,448	0	2,084
3211 Spinning weaving and finishing textiles	..	..	..	..	..	..	..	..	..	..
3212 Made-up goods excluding wearing apparel	..	..	..	..	..	..	..	..	..	..
3213 Knitting mills	..	..	..	..	..	..	..	..	..	..
3214 Carpets and rugs	..	..	..	..	..	..	..	..	..	..
3215 Cordage, rope and twine	..	..	..	..	..	..	..	..	..	..
3219 Other textiles	..	..	..	..	..	..	..	..	..	..
322 WEARING APPAREL, EXCEPT FOOTWEAR	0	..	165	37	212	0	75	62,482	833	711
323 LEATHER AND FUR PRODUCTS	0	..	31	40	0	3	2	9,017	0	108
3231 Tanneries and leather finishing	..	..	..	..	..	..	..	..	..	..
3232 Fur dressing and dyeing industries	..	..	..	..	..	..	..	..	..	..
3233 Leather prods. ex. footwear and wearing apparel	..	..	..	..	..	..	..	..	..	..
324 FOOTWEAR, EX. RUBBER AND PLASTIC	0	..	51	1	0	5	111	28,932	0	272
33 WOOD PRODUCTS AND FURNITURE	870	..	1,069	1,815	900	10,218	5,543	1,406,920	68,462	25,233
331 WOOD PRODUCTS, EXCEPT FURNITURE	870	..	820	1,760	795	9,936	5,429	1,206,086	67,878	23,708
3311 Sawmills, planing and other wood mills	..	..	..	..	..	..	..	..	..	..
3312 Wooden and cane containers	..	..	..	..	..	..	..	..	..	..
3319 Other wood and cork products	..	..	..	..	..	..	..	..	..	..
332 FURNITURE, FIXTURES, EXCL. METALLIC	0	..	249	55	105	282	114	200,834	584	1,526
34 PAPER, PUBLISHING AND PRINTING	19,997	..	846	13,025	30,223	129,085	7,122	21,829,559	7,467,060	252,003
341 PAPER AND PRODUCTS	19,997	..	575	12,946	30,137	129,085	6,336	21,394,141	7,467,060	249,213
3411 Pulp, paper and paperboard articles	..	..	..	..	..	..	..	..	..	..
3412 Containers of paper and paperboard	..	..	..	..	..	..	..	..	..	..
3419 Other pulp, paper and paperboard articles	..	..	..	..	..	..	..	..	..	..
342 PRINTING AND PUBLISHING	0	..	271	79	86	0	786	435,418	0	2,790
35 CHEMICAL PRODUCTS	7,020	..	876	10,122	35,640	3,464	2,801	3,998,861	742,864	71,645
351 INDUSTRIAL CHEMICALS	6,227	..	419	3,371	9,507	743	2,132	3,273,139	0	34,182
3511 Basic industrial chemicals excl. fertilizers	..	..	..	..	..	..	..	..	..	..
3512 Fertilizers and pesticides	..	..	..	..	..	..	..	..	..	..
3513 Synthetic resins and plastic materials	..	..	..	..	..	..	..	..	..	..
352 OTHER CHEMICALS	0	..	55	285	0	0	10	40,886	0	497
3521 Paints, varnishes and lacquers	..	..	..	..	..	..	..	..	..	..
3522 Drugs and medicines	..	..	..	..	..	..	..	..	..	..
3523 Soap, cleaning preparations, perfumes, cosmetics	..	..	..	..	..	..	..	..	..	..
3529 Other chemical products	..	..	..	..	..	..	..	..	..	..
353 PETROLEUM REFINERIES	793	..	98	5,877	25,942	2,621	0	0	742,864	32,657
354 MISC. PETROLEUM AND COAL PRODUCTS	0	..	0	0	0	0	0	97,568	0	351
355 RUBBER PRODUCTS	0	..	62	142	63	0	369	99,759	0	995
3551 Tyres and tubes	..	..	..	..	..	..	..	..	..	..
3559 Other rubber products	..	..	..	..	..	..	..	..	..	..
356 PLASTIC PRODUCTS	0	..	242	447	128	100	290	487,509	0	2,962
36 NON-METALLIC MINERAL PRODUCTS	4,676	..	1,064	1,215	3,734	0	223	680,558	101,281	12,997
361 POTTERY, CHINA, EARTHENWARE	0	..	9	94	353	0	9	26,306	0	560
362 GLASS AND PRODUCTS	0	..	81	165	2,622	0	172	274,764	88,058	3,712
369 OTHER NON-METAL. MINERAL PRODUCTS	4,676	..	974	956	759	0	42	379,488	13,223	8,726
3691 Structural clay products	..	..	..	..	..	..	..	..	..	..
3692 Cement, lime and plaster	..	..	..	..	..	..	..	..	..	..
3699 Other non-metallic mineral products	..	..	..	..	..	..	..	..	..	..
37 BASIC METAL INDUSTRIES	21,602	..	529	8,491	15,897	0	12	3,724,406	529,135	58,034
371 IRON AND STEEL	21,464	..	383	8,140	15,587	0	0	2,601,334	529,135	53,034
372 NON-FERROUS METALS	138	..	146	351	310	0	12	1,123,072	0	5,000
38 METAL PRODUCTS, MACHINERY, EQUIP.	0	..	2,661	1,168	1,269	2	4,542	2,375,550	0	18,194
381 METAL PRODUCTS	0	..	924	152	867	1	858	578,285	0	4,884
3811 Cutlery, hand tools and general hardware	..	..	..	..	..	..	..	..	..	..
3812 Furniture and fixtures primarily of metal	..	..	..	..	..	..	..	..	..	..
3813 Structural metal products	..	..	..	..	..	..	..	..	..	..
3819 Other fabricated metal products	..	..	..	..	..	..	..	..	..	..
382 NON-ELECTRICAL MACHINERY	0	..	817	274	215	1	1,976	801,873	0	6,170
3821 Engines and turbines	..	..	..	..	..	..	..	..	..	..
3822 Agricultural machinery and equipment	..	..	..	..	..	..	..	..	..	..
3823 Metal and wood working machinery	..	..	..	..	..	..	..	..	..	..
3824 Special industrial machinery	..	..	..	..	..	..	..	..	..	..
3825 Office, computing and accounting machinery	..	..	..	..	..	..	..	..	..	..
3829 Other non-electrical machinery and equipment	..	..	..	..	..	..	..	..	..	..
383 ELECTRICAL MACHINERY	0	..	269	56	148	0	628	447,134	0	2,711
3831 Electrical industrial machinery	..	..	..	..	..	..	..	..	..	..
3832 Radio, TV and communications equipment	..	..	..	..	..	..	..	..	..	..
3833 Electrical appliances and housewares	..	..	..	..	..	..	..	..	..	..
3839 Other electrical apparatus and supplies	..	..	..	..	..	..	..	..	..	..
384 TRANSPORT EQUIPMENT	0	..	593	686	37	0	945	471,849	0	3,960
3841 Shipbuilding	..	..	..	..	..	..	..	..	..	..
3842 Railroad equipment	..	..	..	..	..	..	..	..	..	..
3843 Motor vehicles	..	..	..	..	..	..	..	..	..	..
3844 Motorcycles and bicycles	..	..	..	..	..	..	..	..	..	..
3845 Aircraft	..	..	..	..	..	..	..	..	..	..
3849 Other transport equipment	..	..	..	..	..	..	..	..	..	..
385 PROFESSIONAL AND SCIENTIFIC EQUIPMENT	0	..	58	0	2	0	135	76,409	0	470
3851 Professional equipment	..	..	..	..	..	..	..	..	..	..
3852 Photographic and optical goods	..	..	..	..	..	..	..	..	..	..
3853 Watches and clocks	..	..	..	..	..	..	..	..	..	..
39 OTHER MANUFACTURING INDUSTRIES	0	..	103	46	40	4	91	66,680	0	524
3901 Jewellery and related articles	..	..	..	..	..	..	..	..	..	..
3902 Musical instruments	..	..	..	..	..	..	..	..	..	..
3903 Sporting and athletic goods	..	..	..	..	..	..	..	..	..	..
3909 Other manufactures	..	..	..	..	..	..	..	..	..	..
SERR non-specified, unallocated industry	..	..	..	..	..	..	..	..	..	..
3 TOTAL MANUFACTURING	57,737	..	9,241	40,284	90,567	143,001	24,423	35,924,912	8,989,157	462,222

ISIS Energy Data Programme (IEA/OECD)

ISIC Revision 2 Industry Sector	Solid TJ	LPG TJ	Distiloil TJ	RFO TJ	Gas TJ	Biomass TJ	Steam TJ	Electr MWh	Own Use MWh	TOTAL TJ
31 FOOD, BEVERAGES AND TOBACCO	3,565	..	1,517	3,675	2,285	262	3,946	1,622,863	86,688	20,780
311/2 FOOD	2,793	..	1,439	3,277	1,968	181	3,083	1,445,738	86,688	17,634
3111 Slaughtering, preparing and preserving meat	..	..	..	..	..	..	..	..	..	..
3112 Dairy products	..	..	..	..	..	..	..	..	..	..
3113 Canning, preserving of fruits and vegetables	..	..	..	..	..	..	..	..	..	..
3114 Canning, preserving and processing of fish	..	..	..	..	..	..	..	..	..	..
3115 Vegetable and animal oils and fats	..	..	..	..	..	..	..	..	..	..
3116 Grain mill products	..	..	..	..	..	..	..	..	..	..
3117 Bakery products	..	..	..	..	..	..	..	..	..	..
3118 Sugar factories and refineries	..	..	..	..	..	..	..	..	..	..
3119 Cocoa, chocolate and sugar confectionery	..	..	..	..	..	..	..	..	..	..
3121 Other food products	..	..	..	..	..	..	..	..	..	..
3122 Prepared animal feeds	..	..	..	..	..	..	..	..	..	..
313 BEVERAGES	772	..	73	376	317	81	822	167,196	0	3,043
3131 Distilling, rectifying and blending of spirits	..	..	..	..	..	..	..	..	..	..
3132 Wine industries	..	..	..	..	..	..	..	..	..	..
3133 Malt liquors and malts	..	..	..	..	..	..	..	..	..	..
3134 Soft drinks	..	..	..	..	..	..	..	..	..	..
314 TOBACCO	0	..	5	22	0	0	41	9,929	0	104
32 TEXTILES, APPAREL AND LEATHER	0	..	637	301	658	6	625	360,982	0	3,527
321 TEXTILES	0	..	389	266	456	0	399	260,730	0	2,449
3211 Spinning weaving and finishing textiles	..	..	..	..	..	..	..	..	..	..
3212 Made-up goods excluding wearing apparel	..	..	..	..	..	..	..	..	..	..
3213 Knitting mills	..	..	..	..	..	..	..	..	..	..
3214 Carpets and rugs	..	..	..	..	..	..	..	..	..	..
3215 Cordage, rope and twine	..	..	..	..	..	..	..	..	..	..
3219 Other textiles	..	..	..	..	..	..	..	..	..	..
322 WEARING APPAREL, EXCEPT FOOTWEAR	0	..	143	23	201	0	101	61,218	0	688
323 LEATHER AND FUR PRODUCTS	0	..	53	10	0	0	1	6,134	0	86
3231 Tanneries and leather finishing	..	..	..	..	..	..	..	..	..	..
3232 Fur dressing and dyeing industries	..	..	..	..	..	..	..	..	..	..
3233 Leather prods. ex. footwear and wearing apparel	..	..	..	..	..	..	..	..	..	..
324 FOOTWEAR, EX. RUBBER AND PLASTIC	0	..	52	2	1	6	124	32,900	0	303
33 WOOD PRODUCTS AND FURNITURE	670	..	1,025	1,946	582	9,715	9,059	1,743,810	82,290	28,978
331 WOOD PRODUCTS, EXCEPT FURNITURE	670	..	762	1,876	472	9,443	8,883	1,516,836	81,664	27,273
3311 Sawmills, planing and other wood mills	..	..	..	..	..	..	..	..	..	..
3312 Wooden and cane containers	..	..	..	..	..	..	..	..	..	..
3319 Other wood and cork products	..	..	..	..	..	..	..	..	..	..
332 FURNITURE, FIXTURES, EXCL. METALLIC	0	..	263	70	110	272	176	226,974	626	1,706
34 PAPER, PUBLISHING AND PRINTING	17,676	..	770	11,643	28,051	122,591	22,331	21,849,036	5,970,348	260,225
341 PAPER AND PRODUCTS	17,676	..	475	11,567	27,940	122,591	21,262	21,403,015	5,970,348	257,069
3411 Pulp, paper and paperboard articles	..	..	..	..	..	..	..	..	..	..
3412 Containers of paper and paperboard	..	..	..	..	..	..	..	..	..	..
3419 Other pulp, paper and paperboard articles	..	..	..	..	..	..	..	..	..	..
342 PRINTING AND PUBLISHING	0	..	295	76	111	0	1,069	446,021	0	3,157
35 CHEMICAL PRODUCTS	6,143	..	872	11,454	36,526	3,635	4,229	4,707,096	1,271,228	75,228
351 INDUSTRIAL CHEMICALS	5,380	..	309	3,673	9,221	449	3,227	3,959,726	469,423	34,824
3511 Basic industrial chemicals excl. fertilizers	..	..	..	..	..	..	..	..	..	..
3512 Fertilizers and pesticides	..	..	..	..	..	..	..	..	..	..
3513 Synthetic resins and plastic materials	..	..	..	..	..	..	..	..	..	..
352 OTHER CHEMICALS	0	..	59	308	0	0	14	40,480	0	527
3521 Paints, varnishes and lacquers	..	..	..	..	..	..	..	..	..	..
3522 Drugs and medicines	..	..	..	..	..	..	..	..	..	..
3523 Soap, cleaning preparations, perfumes, cosmetics	..	..	..	..	..	..	..	..	..	..
3529 Other chemical products	..	..	..	..	..	..	..	..	..	..
353 PETROLEUM REFINERIES	763	..	200	6,600	27,058	3,141	0	0	801,805	34,876
354 MISC. PETROLEUM AND COAL PRODUCTS	0	..	0	0	0	0	0	66,523	0	239
355 RUBBER PRODUCTS	0	..	42	156	59	0	404	103,100	0	1,032
3551 Tyres and tubes	..	..	..	..	..	..	..	..	..	..
3559 Other rubber products	..	..	..	..	..	..	..	..	..	..
356 PLASTIC PRODUCTS	0	..	262	717	188	45	584	537,267	0	3,730
36 NON-METALLIC MINERAL PRODUCTS	4,831	..	1,076	1,153	3,282	0	154	866,397	0	13,615
361 POTTERY, CHINA, EARTHENWARE	0	..	11	118	266	0	38	35,834	0	562
362 GLASS AND PRODUCTS	0	..	66	176	2,078	0	14	287,356	0	3,368
369 OTHER NON-METAL. MINERAL PRODUCTS	4,831	..	999	859	938	0	102	543,207	0	9,685
3691 Structural clay products	..	..	..	..	..	..	..	..	..	..
3692 Cement, lime and plaster	..	..	..	..	..	..	..	..	..	..
3699 Other non-metallic mineral products	..	..	..	..	..	..	..	..	..	..
37 BASIC METAL INDUSTRIES	22,081	..	672	8,969	17,770	0	378	3,873,370	492,146	62,042
371 IRON AND STEEL	21,909	..	541	8,633	17,554	0	0	2,709,289	492,146	56,619
372 NON-FERROUS METALS	172	..	131	336	216	0	378	1,164,081	0	5,424
38 METAL PRODUCTS, MACHINERY, EQUIP.	0	..	2,723	1,135	1,250	1	6,573	2,592,377	0	21,015
381 METAL PRODUCTS	0	..	1,117	169	923	0	1,382	750,947	0	6,294
3811 Cutlery, hand tools and general hardware	..	..	..	..	..	..	..	..	..	..
3812 Furniture and fixtures primarily of metal	..	..	..	..	..	..	..	..	..	..
3813 Structural metal products	..	..	..	..	..	..	..	..	..	..
3819 Other fabricated metal products	..	..	..	..	..	..	..	..	..	..
382 NON-ELECTRICAL MACHINERY	0	..	858	340	130	1	2,646	859,618	0	7,070
3821 Engines and turbines	..	..	..	..	..	..	..	..	..	..
3822 Agricultural machinery and equipment	..	..	..	..	..	..	..	..	..	..
3823 Metal and wood working machinery	..	..	..	..	..	..	..	..	..	..
3824 Special industrial machinery	..	..	..	..	..	..	..	..	..	..
3825 Office, computing and accounting machinery	..	..	..	..	..	..	..	..	..	..
3829 Other non-electrical machinery and equipment	..	..	..	..	..	..	..	..	..	..
383 ELECTRICAL MACHINERY	0	..	235	68	144	0	1,214	555,813	0	3,662
3831 Electrical industrial machinery	..	..	..	..	..	..	..	..	..	..
3832 Radio, TV and communications equipment	..	..	..	..	..	..	..	..	..	..
3833 Electrical appliances and housewares	..	..	..	..	..	..	..	..	..	..
3839 Other electrical apparatus and supplies	..	..	..	..	..	..	..	..	..	..
384 TRANSPORT EQUIPMENT	0	..	461	558	52	0	1,171	350,255	0	3,503
3841 Shipbuilding	..	..	..	..	..	..	..	..	..	..
3842 Railroad equipment	..	..	..	..	..	..	..	..	..	..
3843 Motor vehicles	..	..	..	..	..	..	..	..	..	..
3844 Motorcycles and bicycles	..	..	..	..	..	..	..	..	..	..
3845 Aircraft	..	..	..	..	..	..	..	..	..	..
3849 Other transport equipment	..	..	..	..	..	..	..	..	..	..
385 PROFESSIONAL AND SCIENTIFIC EQUIPMENT	0	..	52	0	1	0	160	75,744	0	486
3851 Professional equipment	..	..	..	..	..	..	..	..	..	..
3852 Photographic and optical goods	..	..	..	..	..	..	..	..	..	..
3853 Watches and clocks	..	..	..	..	..	..	..	..	..	..
39 OTHER MANUFACTURING INDUSTRIES	0	..	106	38	44	34	90	60,536	0	530
3901 Jewellery and related articles	..	..	..	..	..	..	..	..	..	..
3902 Musical instruments	..	..	..	..	..	..	..	..	..	..
3903 Sporting and athletic goods	..	..	..	..	..	..	..	..	..	..
3909 Other manufactures	..	..	..	..	..	..	..	..	..	..
SERR non-specified, unallocated industry	..	..	..	..	..	..	..	..	..	..
3 TOTAL MANUFACTURING	54,966	..	9,398	40,314	90,448	136,244	47,385	37,676,467	7,902,700	485,941

ISIC Revision 2 Industry Sector		Solid TJ	LPG TJ	Distiloil TJ	RFO TJ	Gas TJ	Biomass TJ	Steam TJ	Electr MWh	Own Use MWh	TOTAL TJ
31	**FOOD, BEVERAGES AND TOBACCO**	..	..	..	..	..	..	..	11,817,944	..	42,545
311/2	FOOD	..	..	..	..	..	..	..	9,437,389	..	33,975
3111	Slaughtering, preparing and preserving meat								1,450,639	..	5,222
3112	Dairy products								1,584,944	..	5,706
3113	Canning, preserving of fruits and vegetables								428,111	..	1,541
3114	Canning, preserving and processing of fish								140,444	..	506
3115	Vegetable and animal oils and fats								447,000	..	1,609
3116	Grain mill products								470,222	..	1,693
3117	Bakery products								894,944	..	3,222
3118	Sugar factories and refineries								1,005,556	..	3,620
3119	Cocoa, chocolate and sugar confectionery								809,917	..	2,916
3121	Other food products								1,316,722	..	4,740
3122	Prepared animal feeds								888,861	..	3,200
313	BEVERAGES								2,018,750	..	7,268
3131	Distilling, rectifying and blending of spirits								136,889	..	493
3132	Wine industries								39,167	..	141
3133	Malt liquors and malts								1,534,833	..	5,525
3134	Soft drinks								307,889	..	1,108
314	TOBACCO								361,806	..	1,303
32	**TEXTILES, APPAREL AND LEATHER**	..	..	..	..	..	..	..	6,182,250	..	22,256
321	TEXTILES								5,533,444	..	19,920
3211	Spinning weaving and finishing textiles								4,159,306	..	14,974
3212	Made-up goods excluding wearing apparel								125,694	..	452
3213	Knitting mills								378,056	..	1,361
3214	Carpets and rugs								857,028	..	3,085
3215	Cordage, rope and twine								13,361	..	48
3219	Other textiles										
322	WEARING APPAREL, EXCEPT FOOTWEAR								353,667	..	1,273
323	LEATHER AND FUR PRODUCTS								168,528	..	607
3231	Tanneries and leather finishing								122,556	..	441
3232	Fur dressing and dyeing industries								9,194	..	33
3233	Leather prods. ex. footwear and wearing apparel								36,750	..	132
324	FOOTWEAR, EX. RUBBER AND PLASTIC								126,611	..	456
33	**WOOD PRODUCTS AND FURNITURE**	..	..	..	..	..	..	..	3,773,083	..	13,583
331	WOOD PRODUCTS, EXCEPT FURNITURE								2,615,528	..	9,416
3311	Sawmills, planing and other wood mills								2,179,806	..	7,847
3312	Wooden and cane containers								53,472	..	192
3319	Other wood and cork products								382,250	..	1,376
332	FURNITURE, FIXTURES, EXCL. METALLIC								1,157,556	..	4,167
34	**PAPER, PUBLISHING AND PRINTING**	..	..	..	..	..	..	..	17,206,889	..	61,945
341	PAPER AND PRODUCTS								15,031,000	..	54,112
3411	Pulp, paper and paperboard articles								12,659,778	..	45,575
3412	Containers of paper and paperboard								2,371,222	..	8,536
3419	Other pulp, paper and paperboard articles										
342	PRINTING AND PUBLISHING								2,175,861	..	7,833
35	**CHEMICAL PRODUCTS**	c	..	c	c	c	..	..	c	..	c
351	INDUSTRIAL CHEMICALS	c	..	c	c	c	..	..	c	..	c
3511	Basic industrial chemicals excl. fertilizers										..
3512	Fertilizers and pesticides										..
3513	Synthetic resins and plastic materials										
352	OTHER CHEMICALS								6,964,639	..	25,073
3521	Paints, varnishes and lacquers								5,372,139	..	19,340
3522	Drugs and medicines								973,611	..	3,505
3523	Soap, cleaning preparations, perfumes, cosmetics								618,889	..	2,228
3529	Other chemical products	c	..	c	c	c			c	..	c
353	PETROLEUM REFINERIES								7,657,778	..	27,568
354	MISC. PETROLEUM AND COAL PRODUCTS										..
355	RUBBER PRODUCTS								2,103,306	..	7,572
3551	Tyres and tubes										
3559	Other rubber products										
356	PLASTIC PRODUCTS								7,651,222	..	27,544
36	**NON-METALLIC MINERAL PRODUCTS**	..	..	..	..	..	..	..	11,815,167	..	42,535
361	POTTERY, CHINA, EARTHENWARE								483,722	..	1,741
362	GLASS AND PRODUCTS								3,847,278	..	13,850
369	OTHER NON-METAL. MINERAL PRODUCTS								7,484,139	..	26,943
3691	Structural clay products								1,316,806	..	4,741
3692	Cement, lime and plaster								4,656,944	..	16,765
3699	Other non-metallic mineral products								1,510,417	..	5,438
37	**BASIC METAL INDUSTRIES**	c	..	c	c	c	..	..	c	..	c
371	IRON AND STEEL	c	..	c	c	c	..	..	c	..	c
372	NON-FERROUS METALS								17,755,778	..	63,921
38	**METAL PRODUCTS, MACHINERY, EQUIP.**	c	..	c	c	c	..	..	c	..	c
381	METAL PRODUCTS	c	..	c	c	c	..	..	c	..	c
3811	Cutlery, hand tools and general hardware								1,354,861	..	4,877
3812	Furniture and fixtures primarily of metal								316,056	..	1,138
3813	Structural metal products								760,222	..	2,737
3819	Other fabricated metal products								5,781,167	..	20,812
382	NON-ELECTRICAL MACHINERY								9,932,472	..	35,757
3821	Engines and turbines								3,872,917	..	13,943
3822	Agricultural machinery and equipment								472,028	..	1,699
3823	Metal and wood working machinery								1,249,083	..	4,497
3824	Special industrial machinery								3,175,556	..	11,432
3825	Office, computing and accounting machinery								1,162,917	..	4,187
3829	Other non-electrical machinery and equipment										
383	ELECTRICAL MACHINERY								9,104,333	..	32,776
3831	Electrical industrial machinery								4,119,000	..	14,828
3832	Radio, TV and communications equipment								3,549,361	..	12,778
3833	Electrical appliances and housewares								744,722	..	2,681
3839	Other electrical apparatus and supplies								691,250	..	2,489
384	TRANSPORT EQUIPMENT	c	..	c	c	c	..	..	c	..	c
3841	Shipbuilding								455,417	..	1,640
3842	Railroad equipment								369,556	..	1,330
3843	Motor vehicles								12,292,639	..	44,254
3844	Motorcycles and bicycles								147,472	..	531
3845	Aircraft	c	..	c	c	c	..	..	c	..	c
3849	Other transport equipment								24,556	..	88
385	PROFESSIONAL AND SCIENTIFIC EQUIPMENT								936,722	..	3,372
3851	Professional equipment								395,833	..	1,425
3852	Photographic and optical goods								487,361	..	1,754
3853	Watches and clocks								53,528	..	193
39	**OTHER MANUFACTURING INDUSTRIES**	..	..	..	..	..	..	..	358,139	..	1,289
3901	Jewellery and related articles								33,222	..	120
3902	Musical instruments								32,889	..	118
3903	Sporting and athletic goods								29,056	..	105
3909	Other manufactures								262,972	..	947
SERR	**non-specified, unallocated industry**	..	..	..	..	..	..	..	..	..	..
3	**TOTAL MANUFACTURING**	..	..	..	..	..	..	..	200,913,028	..	723,287

ISIS Energy Data Programme (IEA/OECD)

ISIC Revision 2 Industry Sector	Solid TJ	LPG TJ	Distiloil TJ	RFO TJ	Gas TJ	Biomass TJ	Steam TJ	Electr MWh	Own Use MWh	TOTAL TJ
31 **FOOD, BEVERAGES AND TOBACCO**	29,608	..	34,367	25,065	83,834	..	..	11,939,445	..	215,856
311/2 FOOD	26,136	..	24,681	23,815	64,922	..	..	9,597,639	..	174,106
3111 Slaughtering, preparing and preserving meat	1,564	..	3,215	808	6,348	..	..	1,465,528	..	17,211
3112 Dairy products	2,990	..	3,944	2,983	13,752	..	..	1,655,861	..	29,630
3113 Canning, preserving of fruits and vegetables	201	..	2,682	462	3,860	..	..	441,778	..	8,795
3114 Canning, preserving and processing of fish	37	..	260	2	710	..	..	138,611	..	1,508
3115 Vegetable and animal oils and fats	464	..	300	754	5,360	..	..	460,222	..	8,535
3116 Grain mill products	21	..	306	13	553	..	..	443,389	..	2,489
3117 Bakery products	330	..	5,993	131	7,373	..	..	902,444	..	17,076
3118 Sugar factories and refineries	15,950	..	646	15,593	9,082	..	..	1,049,639	..	45,050
3119 Cocoa, chocolate and sugar confectionery	366	..	1,610	661	2,369	..	..	823,528	..	7,971
3121 Other food products	2,745	..	3,278	2,003	10,973	..	..	1,345,333	..	23,842
3122 Prepared animal feeds	1,468	..	2,447	405	4,542	..	..	871,306	..	11,999
313 BEVERAGES	3,210	..	8,833	1,250	18,238	..	..	2,057,973	..	38,940
3131 Distilling, rectifying and blending of spirits	1,222	..	733	154	1,181	..	..	124,917	..	3,740
3132 Wine industries	7	..	156	18	175	..	..	39,722	..	499
3133 Malt liquors and malts	1,924	..	6,267	1,050	15,063	..	..	1,572,528	..	29,965
3134 Soft drinks	57	..	1,677	28	1,819	..	..	320,806	..	4,736
314 TOBACCO	262	..	853	0	674	..	..	283,833	..	2,811
32 **TEXTILES, APPAREL AND LEATHER**	19,473	..	10,709	5,779	29,288	..	..	5,720,249	..	85,842
321 TEXTILES	17,995	..	6,325	5,226	27,541	..	..	5,162,305	..	75,671
3211 Spinning weaving and finishing textiles	16,472	..	3,528	3,390	20,790	..	..	3,836,361	..	57,991
3212 Made-up goods excluding wearing apparel	524	..	292	125	702	..	..	125,639	..	2,095
3213 Knitting mills	292	..	1,238	256	1,155	..	..	330,861	..	4,132
3214 Carpets and rugs	706	..	1,255	1,455	4,885	..	..	858,583	..	11,392
3215 Cordage, rope and twine	1	..	12	0	9	..	..	10,861	..	61
3219 Other textiles	..	..	..	..	..	..	..	..	..	..
322 WEARING APPAREL, EXCEPT FOOTWEAR	629	..	3,351	36	907	..	..	300,417	..	6,005
323 LEATHER AND FUR PRODUCTS	564	..	671	408	671	..	..	146,527	..	2,841
3231 Tanneries and leather finishing	518	..	403	407	504	..	..	107,444	..	2,219
3232 Fur dressing and dyeing industries	1	..	68	0	104	..	..	7,500	..	200
3233 Leather prods. ex. footwear and wearing apparel	45	..	200	1	63	..	..	31,583	..	423
324 FOOTWEAR, EX. RUBBER AND PLASTIC	285	..	362	109	169	..	..	111,000	..	1,325
33 **WOOD PRODUCTS AND FURNITURE**	1,764	..	12,553	3,310	4,963	..	..	3,708,028	..	35,939
331 WOOD PRODUCTS, EXCEPT FURNITURE	988	..	2,520	3,054	3,149	..	..	2,574,222	..	18,978
3311 Sawmills, planing and other wood mills	869	..	2,117	3,019	2,747	..	..	2,174,778	..	16,581
3312 Wooden and cane containers	38	..	116	0	9	..	..	50,361	..	344
3319 Other wood and cork products	81	..	287	35	393	..	..	349,083	..	2,053
332 FURNITURE, FIXTURES, EXCL. METALLIC	776	..	10,033	256	1,814	..	..	1,133,806	..	16,961
34 **PAPER, PUBLISHING AND PRINTING**	38,939	..	7,955	22,083	82,537	..	..	17,363,555	..	214,023
341 PAPER AND PRODUCTS	38,709	..	5,986	22,013	75,734	..	..	15,134,444	..	196,926
3411 Pulp, paper and paperboard articles	37,557	..	2,956	21,431	62,424	..	..	12,788,333	..	170,406
3412 Containers of paper and paperboard	1,152	..	3,030	582	13,310	..	..	2,346,111	..	26,520
3419 Other pulp, paper and paperboard articles	..	..	..	..	..	..	..	..	..	..
342 PRINTING AND PUBLISHING	230	..	1,969	70	6,803	..	..	2,229,111	..	17,097
35 **CHEMICAL PRODUCTS**	154,770	..	56,842	157,976	446,344	..	..	63,740,277	..	1,045,397
351 INDUSTRIAL CHEMICALS	132,872	..	17,377	79,998	359,686	..	..	40,259,611	..	734,868
3511 Basic industrial chemicals excl. fertilizers	..	..	..	..	..	..	..	..	..	..
3512 Fertilizers and pesticides	..	..	..	..	..	..	..	..	..	..
3513 Synthetic resins and plastic materials	..	..	..	..	..	..	..	..	..	..
352 OTHER CHEMICALS	13,592	..	8,648	4,875	30,108	..	..	6,959,444	..	82,277
3521 Paints, varnishes and lacquers	9,036	..	5,013	2,324	17,209	..	..	5,348,083	..	52,835
3522 Drugs and medicines	2,236	..	2,503	1,549	4,527	..	..	995,222	..	14,398
3523 Soap, cleaning preparations, perfumes, cosmetics	2,320	..	1,132	1,002	8,372	..	..	616,139	..	15,044
3529 Other chemical products	..	..	..	..	..	..	..	..	..	..
353 PETROLEUM REFINERIES	2,026	..	23,247	68,517	31,011	..	..	6,744,083	..	149,080
354 MISC. PETROLEUM AND COAL PRODUCTS	..	..	..	..	..	..	..	..	..	..
355 RUBBER PRODUCTS	3,351	..	1,517	1,774	11,170	..	..	2,108,611	..	25,403
3551 Tyres and tubes	..	..	..	..	..	..	..	..	..	..
3559 Other rubber products	..	..	..	..	..	..	..	..	..	..
356 PLASTIC PRODUCTS	2,929	..	6,053	2,812	14,369	..	..	7,668,528	..	53,770
36 **NON-METALLIC MINERAL PRODUCTS**	109,504	..	15,222	31,222	109,648	..	..	12,081,223	..	309,088
361 POTTERY, CHINA, EARTHENWARE	1,241	..	1,889	18	10,242	..	..	514,722	..	15,243
362 GLASS AND PRODUCTS	531	..	1,742	11,646	43,471	..	..	3,785,028	..	71,016
369 OTHER NON-METAL. MINERAL PRODUCTS	107,732	..	11,591	19,558	55,935	..	..	7,781,473	..	222,829
3691 Structural clay products	3,048	..	3,284	2,781	32,896	..	..	1,282,806	..	46,627
3692 Cement, lime and plaster	101,511	..	2,642	11,253	12,251	..	..	4,914,389	..	145,349
3699 Other non-metallic mineral products	3,173	..	5,665	5,524	10,788	..	..	1,584,278	..	30,853
37 **BASIC METAL INDUSTRIES**	445,106	..	7,246	50,126	203,265	..	..	39,252,639	..	847,053
371 IRON AND STEEL	432,524	..	3,073	48,441	173,152	..	..	22,874,583	..	739,538
372 NON-FERROUS METALS	12,582	..	4,173	1,685	30,113	..	..	16,378,056	..	107,514
38 **METAL PRODUCTS, MACHINERY, EQUIP.**	35,403	..	59,218	6,051	143,809	..	..	41,445,861	..	393,686
381 METAL PRODUCTS	1,597	..	10,443	194	10,240	..	..	2,425,722	..	31,207
3811 Cutlery, hand tools and general hardware	364	..	2,387	69	3,520	..	..	1,323,111	..	11,103
3812 Furniture and fixtures primarily of metal	98	..	1,106	51	1,431	..	..	334,083	..	3,889
3813 Structural metal products	1,135	..	6,950	74	5,289	..	..	768,528	..	16,215
3819 Other fabricated metal products	c	..	c	c	c	..	..	c	..	c
382 NON-ELECTRICAL MACHINERY	13,403	..	21,978	2,413	38,171	..	..	9,366,305	..	109,684
3821 Engines and turbines	5,509	..	5,623	766	15,101	..	..	3,761,444	..	40,540
3822 Agricultural machinery and equipment	1,908	..	1,049	248	1,759	..	..	436,722	..	6,536
3823 Metal and wood working machinery	1,064	..	8,834	78	2,424	..	..	1,139,917	..	16,504
3824 Special industrial machinery	3,740	..	5,959	1,196	17,215	..	..	2,969,972	..	38,802
3825 Office, computing and accounting machinery	1,182	..	513	125	1,672	..	..	1,058,250	..	7,302
3829 Other non-electrical machinery and equipment	..	..	..	..	..	..	..	..	..	..
383 ELECTRICAL MACHINERY	4,956	..	9,091	1,942	18,476	..	..	8,904,083	..	66,520
3831 Electrical industrial machinery	2,343	..	4,388	404	9,678	..	..	4,095,111	..	31,555
3832 Radio, TV and communications equipment	1,812	..	3,399	942	4,928	..	..	3,366,694	..	23,201
3833 Electrical appliances and housewares	608	..	404	172	1,956	..	..	729,472	..	5,766
3839 Other electrical apparatus and supplies	193	..	900	424	1,914	..	..	712,806	..	5,997
384 TRANSPORT EQUIPMENT	12,803	..	8,396	886	42,800	..	..	13,298,167	..	112,758
3841 Shipbuilding	1,233	..	348	116	1,635	..	..	437,417	..	4,907
3842 Railroad equipment	4,718	..	934	58	2,036	..	..	372,556	..	9,087
3843 Motor vehicles	6,710	..	6,724	685	38,844	..	..	12,348,500	..	97,418
3844 Motorcycles and bicycles	80	..	309	12	251	..	..	120,111	..	1,084
3845 Aircraft	c	..	c	c	c	..	..	c	..	c
3849 Other transport equipment	62	..	81	15	34	..	..	19,583	..	262
385 PROFESSIONAL AND SCIENTIFIC EQUIPMENT	996	..	1,400	18	2,110	..	..	886,722	..	7,716
3851 Professional equipment	565	..	800	10	1,489	..	..	431,250	..	4,417
3852 Photographic and optical goods	297	..	508	8	587	..	..	410,222	..	2,877
3853 Watches and clocks	134	..	92	0	34	..	..	45,250	..	423
39 **OTHER MANUFACTURING INDUSTRIES**	180	..	687	46	582	..	..	354,389	..	2,771
3901 Jewellery and related articles	9	..	84	2	42	..	..	33,472	..	257
3902 Musical instruments	43	..	167	41	48	..	..	28,639	..	402
3903 Sporting and athletic goods	48	..	48	0	86	..	..	37,611	..	317
3909 Other manufactures	80	..	388	3	406	..	..	254,667	..	1,794
SERR non-specified, unallocated industry	..	..	..	..	..	..	..	..	..	..
3 **TOTAL MANUFACTURING**	834,747	..	204,799	301,658	1,104,270	..	..	195,605,666	..	3,149,654

		Solid	LPG	Distiloil	RFO	Gas	Biomass	Steam	Electr	Own Use	TOTAL
	ISIC Revision 2 Industry Sector	TJ	TJ	TJ	TJ	TJ	TJ	TJ	MWh	MWh	TJ
31	**FOOD, BEVERAGES AND TOBACCO**	19,772	..	33,045	25,275	81,487	..	..	11,918,833	..	202,487
311/2	FOOD	17,481	..	24,902	24,290	61,278	..	..	9,648,917	..	162,687
3111	Slaughtering, preparing and preserving meat	1,089	..	3,450	858	7,240	..	..	1,525,500	..	18,129
3112	Dairy products	544	..	4,937	2,726	14,222	..	..	1,597,528	..	28,180
3113	Canning, preserving of fruits and vegetables	153	..	2,088	382	3,817	..	..	431,472	..	7,993
3114	Canning, preserving and processing of fish	16	..	239	2	632	..	..	127,500	..	1,348
3115	Vegetable and animal oils and fats	348	..	378	774	3,811	..	..	453,194	..	6,942
3116	Grain mill products	10	..	311	12	511	..	..	438,500	..	2,423
3117	Bakery products	98	..	4,382	51	7,763	..	..	925,056	..	15,624
3118	Sugar factories and refineries	12,523	..	582	17,304	4,572	..	..	1,038,417	..	38,719
3119	Cocoa, chocolate and sugar confectionery	38	..	1,764	134	2,331	..	..	826,750	..	7,243
3121	Other food products	2,278	..	3,451	1,820	11,767	..	..	1,415,667	..	24,412
3122	Prepared animal feeds	384	..	3,320	227	4,612	..	..	869,333	..	11,673
313	BEVERAGES	2,158	..	7,270	985	19,621	..	..	2,017,305	..	37,296
3131	Distilling, rectifying and blending of spirits	511	..	804	117	1,255	..	..	116,028	..	3,105
3132	Wine industries	8	..	146	21	210	..	..	40,444	..	531
3133	Malt liquors and malts	1,599	..	4,595	814	15,975	..	..	1,508,861	..	28,415
3134	Soft drinks	40	..	1,725	33	2,181	..	..	351,972	..	5,246
314	TOBACCO	133	..	873	0	588	..	..	252,611	..	2,503
32	**TEXTILES, APPAREL AND LEATHER**	6,778	..	7,930	4,134	28,323	..	..	5,102,499	..	65,534
321	TEXTILES	6,218	..	5,676	3,730	26,826	..	..	4,613,527	..	59,059
3211	Spinning weaving and finishing textiles	5,373	..	3,011	2,287	20,072	..	..	3,359,000	..	42,835
3212	Made-up goods excluding wearing apparel	356	..	295	123	595	..	..	125,944	..	1,822
3213	Knitting mills	74	..	1,117	207	1,104	..	..	291,694	..	3,552
3214	Carpets and rugs	414	..	1,235	1,113	5,045	..	..	827,361	..	10,785
3215	Cordage, rope and twine	1	..	18	0	10	..	..	9,528	..	63
3219	Other textiles	..	..	..	..	..	..	..	..	..	..
322	WEARING APPAREL, EXCEPT FOOTWEAR	218	..	1,373	32	799	..	..	265,944	..	3,379
323	LEATHER AND FUR PRODUCTS	212	..	599	305	546	..	..	126,556	..	2,118
3231	Tanneries and leather finishing	189	..	341	304	475	..	..	90,306	..	1,634
3232	Fur dressing and dyeing industries	0	..	69	0	8	..	..	5,889	..	98
3233	Leather prods. ex. footwear and wearing apparel	23	..	189	1	63	..	..	30,361	..	385
324	FOOTWEAR, EX. RUBBER AND PLASTIC	130	..	282	67	152	..	..	96,472	..	978
33	**WOOD PRODUCTS AND FURNITURE**	2,931	..	6,276	3,589	3,963	..	..	3,746,528	..	30,247
331	WOOD PRODUCTS, EXCEPT FURNITURE	690	..	2,576	3,322	3,165	..	..	2,612,639	..	19,159
3311	Sawmills, planing and other wood mills	632	..	2,159	3,301	2,772	..	..	2,234,806	..	16,909
3312	Wooden and cane containers	10	..	118	0	16	..	..	45,722	..	309
3319	Other wood and cork products	48	..	299	21	377	..	..	332,111	..	1,941
332	FURNITURE, FIXTURES, EXCL. METALLIC	2,241	..	3,700	267	798	..	..	1,133,889	..	11,088
34	**PAPER, PUBLISHING AND PRINTING**	35,312	..	7,920	17,909	94,781	..	..	17,385,222	..	218,509
341	PAPER AND PRODUCTS	35,258	..	6,055	17,830	88,057	..	..	15,072,305	..	201,460
3411	Pulp, paper and paperboard articles	34,148	..	3,093	16,857	75,758	..	..	12,682,083	..	175,511
3412	Containers of paper and paperboard	1,110	..	2,962	973	12,299	..	..	2,390,222	..	25,949
3419	Other pulp, paper and paperboard articles	..	..	..	..	..	..	..	..	..	..
342	PRINTING AND PUBLISHING	54	..	1,865	79	6,724	..	..	2,312,917	..	17,049
35	**CHEMICAL PRODUCTS**	121,145	..	49,166	153,664	409,766	..	..	61,096,361	..	953,688
351	INDUSTRIAL CHEMICALS	109,597	..	23,559	81,124	340,137	..	..	38,350,028	..	692,477
3511	Basic industrial chemicals excl. fertilizers	..	..	..	..	..	..	..	..	..	..
3512	Fertilizers and pesticides	..	..	..	..	..	..	..	..	..	..
3513	Synthetic resins and plastic materials	..	..	..	..	..	..	..	..	..	..
352	OTHER CHEMICALS	9,296	..	8,051	3,859	30,629	..	..	6,832,750	..	76,433
3521	Paints, varnishes and lacquers	5,219	..	4,209	1,967	18,558	..	..	5,191,722	..	48,643
3522	Drugs and medicines	2,810	..	2,582	1,097	4,892	..	..	1,040,722	..	15,128
3523	Soap, cleaning preparations, perfumes, cosmetics	1,267	..	1,260	795	7,179	..	..	600,306	..	12,662
3529	Other chemical products	..	..	..	..	..	..	..	..	..	..
353	PETROLEUM REFINERIES	28	..	10,679	65,960	14,747	..	..	6,642,139	..	115,326
354	MISC. PETROLEUM AND COAL PRODUCTS	..	..	..	..	..	..	..	..	..	..
355	RUBBER PRODUCTS	863	..	1,535	1,415	10,271	..	..	1,927,750	..	21,024
3551	Tyres and tubes	..	..	..	..	..	..	..	..	..	..
3559	Other rubber products	..	..	..	..	..	..	..	..	..	..
356	PLASTIC PRODUCTS	1,361	..	5,342	1,306	13,982	..	..	7,343,694	..	48,428
36	**NON-METALLIC MINERAL PRODUCTS**	105,185	..	16,184	28,244	118,366	..	..	12,090,945	..	311,506
361	POTTERY, CHINA, EARTHENWARE	908	..	2,486	5	9,011	..	..	422,139	..	13,930
362	GLASS AND PRODUCTS	103	..	1,671	9,833	47,247	..	..	3,666,389	..	72,053
369	OTHER NON-METAL. MINERAL PRODUCTS	104,174	..	12,027	18,406	62,108	..	..	8,002,417	..	225,524
3691	Structural clay products	2,116	..	3,711	2,642	36,904	..	..	1,398,778	..	50,409
3692	Cement, lime and plaster	95,415	..	2,600	11,862	12,826	..	..	4,968,639	..	140,590
3699	Other non-metallic mineral products	6,643	..	5,716	3,902	12,378	..	..	1,635,000	..	34,525
37	**BASIC METAL INDUSTRIES**	397,679	..	7,166	53,117	196,779	..	..	36,976,139	..	787,855
371	IRON AND STEEL	388,806	..	2,763	51,583	169,511	..	..	21,706,639	..	690,807
372	NON-FERROUS METALS	8,873	..	4,403	1,534	27,268	..	..	15,269,500	..	97,048
38	**METAL PRODUCTS, MACHINERY, EQUIP.**	20,120	..	46,986	3,803	134,522	..	..	38,074,861	..	342,500
381	METAL PRODUCTS	2,254	..	13,071	525	35,781	..	..	7,897,417	..	80,062
3811	Cutlery, hand tools and general hardware	104	..	2,292	81	3,454	..	..	1,240,833	..	10,398
3812	Furniture and fixtures primarily of metal	20	..	1,071	47	1,503	..	..	320,194	..	3,794
3813	Structural metal products	1,366	..	2,671	78	3,261	..	..	778,611	..	10,179
3819	Other fabricated metal products	764	..	7,037	319	27,563	..	..	5,557,779	..	55,691
382	NON-ELECTRICAL MACHINERY	5,750	..	14,735	1,875	32,733	..	..	8,265,055	..	84,847
3821	Engines and turbines	2,118	..	5,513	563	13,484	..	..	3,323,639	..	33,643
3822	Agricultural machinery and equipment	1,205	..	864	124	1,731	..	..	387,333	..	5,318
3823	Metal and wood working machinery	452	..	2,159	80	2,317	..	..	987,583	..	8,563
3824	Special industrial machinery	1,379	..	5,824	1,008	13,804	..	..	2,765,194	..	31,970
3825	Office, computing and accounting machinery	596	..	375	100	1,397	..	..	801,306	..	5,353
3829	Other non-electrical machinery and equipment	..	..	..	..	..	..	..	..	..	..
383	ELECTRICAL MACHINERY	2,376	..	9,154	820	18,933	..	..	8,469,056	..	61,772
3831	Electrical industrial machinery	1,273	..	5,303	352	9,931	..	..	3,921,306	..	30,976
3832	Radio, TV and communications equipment	756	..	2,660	27	4,951	..	..	3,196,333	..	19,901
3833	Electrical appliances and housewares	310	..	405	46	1,854	..	..	682,278	..	5,071
3839	Other electrical apparatus and supplies	37	..	786	395	2,197	..	..	669,139	..	5,824
384	TRANSPORT EQUIPMENT	9,525	..	8,697	566	44,946	..	..	12,588,305	..	109,052
3841	Shipbuilding	998	..	483	36	2,023	..	..	419,444	..	5,050
3842	Railroad equipment	3,186	..	1,210	0	2,716	..	..	361,667	..	8,414
3843	Motor vehicles	5,152	..	6,228	452	37,015	..	..	11,015,583	..	88,503
3844	Motorcycles and bicycles	58	..	275	4	311	..	..	117,694	..	1,072
3845	Aircraft	131	..	388	74	2,854	..	..	657,250	..	5,813
3849	Other transport equipment	0	..	113	0	27	..	..	16,667	..	200
385	PROFESSIONAL AND SCIENTIFIC EQUIPMENT	215	..	1,329	17	2,129	..	..	855,028	..	6,768
3851	Professional equipment	46	..	761	10	1,555	..	..	424,361	..	3,900
3852	Photographic and optical goods	109	..	484	7	550	..	..	393,667	..	2,567
3853	Watches and clocks	60	..	84	0	24	..	..	37,000	..	301
39	**OTHER MANUFACTURING INDUSTRIES**	108	..	670	40	625	..	..	328,861	..	2,627
3901	Jewellery and related articles	c	..	c	c	c	..	..	c	..	c
3902	Musical instruments	31	..	176	39	57	..	..	25,417	..	395
3903	Sporting and athletic goods	1	..	42	0	184	..	..	42,889	..	381
3909	Other manufactures	76	..	452	1	384	..	..	260,555	..	1,851
SERR	**non-specified, unallocated industry**	..	..	..	..	..	..	..	..	..	..
3	**TOTAL MANUFACTURING**	709,030	..	175,343	289,775	1,068,612	..	..	186,720,249	..	2,914,953

ISIS Energy Data Programme (IEA/OECD)

ISIC Revision 2 Industry Sector		Solid TJ	LPG TJ	Distiloil TJ	RFO TJ	Gas TJ	Biomass TJ	Steam TJ	Electr MWh	Own Use MWh	TOTAL TJ
31	**FOOD, BEVERAGES AND TOBACCO**	17,858	..	31,434	22,706	88,806	..	..	12,226,944	..	204,821
311/2	FOOD	16,378	..	23,831	21,831	68,381	..	..	9,934,249	..	166,184
3111	Slaughtering, preparing and preserving meat	772	..	3,413	616	8,570	..	..	1,634,972	..	19,257
3112	Dairy products	260	..	4,451	2,434	14,561	..	..	1,646,110	..	27,632
3113	Canning, preserving of fruits and vegetables	63	..	2,418	345	4,152	..	..	445,972	..	8,583
3114	Canning, preserving and processing of fish	6	..	194	1	698	..	..	121,361	..	1,336
3115	Vegetable and animal oils and fats	304	..	500	866	8,309	..	..	471,028	..	11,675
3116	Grain mill products	2	..	278	42	475	..	..	422,333	..	2,317
3117	Bakery products	65	..	4,114	51	7,344	..	..	960,528	..	15,032
3118	Sugar factories and refineries	12,538	..	900	15,492	3,950	..	..	997,222	..	36,470
3119	Cocoa, chocolate and sugar confectionery	16	..	1,431	10	2,551	..	..	883,889	..	7,190
3121	Other food products	2,088	..	2,908	1,779	12,942	..	..	1,493,167	..	25,092
3122	Prepared animal feeds	264	..	3,224	195	4,829	..	..	857,667	..	11,600
313	BEVERAGES	1,334	..	6,669	875	19,858	..	..	2,011,417	..	35,977
3131	Distilling, rectifying and blending of spirits	242	..	827	128	1,287	..	..	104,194	..	2,859
3132	Wine industries	13	..	137	20	203	..	..	42,722	..	527
3133	Malt liquors and malts	1,059	..	4,116	713	15,357	..	..	1,491,751	..	26,615
3134	Soft drinks	20	..	1,589	14	3,011	..	..	372,750	..	5,976
314	TOBACCO	146	..	934	0	567	..	..	281,278	..	2,660
32	**TEXTILES, APPAREL AND LEATHER**	4,909	..	7,216	3,061	27,378	..	..	4,951,667	..	60,390
321	TEXTILES	4,493	..	5,270	2,835	25,896	..	..	4,486,000	..	54,644
3211	Spinning weaving and finishing textiles	c	..	c	c	c	..	..	c	..	
3212	Made-up goods excluding wearing apparel	41	..	385	165	629	..	..	141,500	..	1,729
3213	Knitting mills	13	..	977	97	1,161	..	..	267,139	..	3,210
3214	Carpets and rugs	217	..	1,116	1,089	4,815	..	..	833,194	..	10,236
3215	Cordage, rope and twine	4,222	..	2,792	1,484	19,291	..	..	3,244,167	..	39,468
3219	Other textiles	0	..	0	0	0	..	..	0	..	
322	WEARING APPAREL, EXCEPT FOOTWEAR	102	..	1,124	28	780	..	..	240,278	..	2,899
323	LEATHER AND FUR PRODUCTS	276	..	551	161	558	..	..	133,972	..	2,028
3231	Tanneries and leather finishing	274	..	329	159	498	..	..	101,111	..	1,624
3232	Fur dressing and dyeing industries	0	..	50	2	13	..	..	4,667	..	82
3233	Leather prods. ex. footwear and wearing apparel	2	..	172	0	47	..	..	28,194	..	322
324	FOOTWEAR, EX. RUBBER AND PLASTIC	38	..	271	37	144	..	..	91,417	..	819
33	**WOOD PRODUCTS AND FURNITURE**	847	..	7,074	3,985	4,256	..	..	3,946,194	..	30,368
331	WOOD PRODUCTS, EXCEPT FURNITURE	704	..	3,092	3,715	3,449	..	..	2,841,555	..	21,190
3311	Sawmills, planing and other wood mills	669	..	2,734	3,702	3,062	..	..	2,453,111	..	18,998
3312	Wooden and cane containers	6	..	123	0	14	..	..	40,444	..	289
3319	Other wood and cork products	29	..	235	13	373	..	..	348,000	..	1,903
332	FURNITURE, FIXTURES, EXCL. METALLIC	143	..	3,982	270	807	..	..	1,104,639	..	9,179
34	**PAPER, PUBLISHING AND PRINTING**	34,556	..	7,501	16,261	99,855	..	..	18,148,722	..	223,508
341	PAPER AND PRODUCTS	34,532	..	5,842	16,187	92,987	..	..	15,769,805	..	206,319
3411	Pulp, paper and paperboard articles	33,647	..	3,093	15,635	78,846	..	..	13,297,111	..	179,091
3412	Containers of paper and paperboard	885	..	2,749	552	14,141	..	..	2,472,694	..	27,229
3419	Other pulp, paper and paperboard articles	c	..	c	c	c	..	..	c	..	
342	PRINTING AND PUBLISHING	24	..	1,659	74	6,868	..	..	2,378,917	..	17,189
35	**CHEMICAL PRODUCTS**	111,376	..	57,039	145,419	396,912	..	..	61,838,388	..	933,364
351	INDUSTRIAL CHEMICALS	100,869	..	23,405	75,030	325,981	..	..	38,164,556	..	662,677
3511	Basic industrial chemicals excl. fertilizers	0	..	0	0	0	..	..	0	..	
3512	Fertilizers and pesticides	0	..	0	0	0	..	..	0	..	
3513	Synthetic resins and plastic materials	0	..	0	0	0	..	..	0	..	
352	OTHER CHEMICALS	9,030	..	7,822	2,671	29,178	..	..	7,098,027	..	74,254
3521	Paints, varnishes and lacquers	5,074	..	4,027	1,398	16,546	..	..	5,430,694	..	46,595
3522	Drugs and medicines	2,676	..	2,669	766	5,215	..	..	1,067,583	..	15,169
3523	Soap, cleaning preparations, perfumes, cosmetics	1,280	..	1,126	507	7,417	..	..	599,750	..	12,489
3529	Other chemical products	c	..	c	c	c	..	..	c	..	
353	PETROLEUM REFINERIES	24	..	19,425	65,767	16,072	..	..	6,737,278	..	125,542
354	MISC. PETROLEUM AND COAL PRODUCTS	c	..	c	c	c	..	..	c	..	
355	RUBBER PRODUCTS	495	..	1,223	978	10,801	..	..	1,980,083	..	20,625
3551	Tyres and tubes	0	..	0	0	0	..	..	0	..	
3559	Other rubber products	0	..	0	0	0	..	..	0	..	
356	PLASTIC PRODUCTS	958	..	5,164	973	14,880	..	..	7,858,444	..	50,265
36	**NON-METALLIC MINERAL PRODUCTS**	104,771	..	14,148	29,126	124,650	..	..	12,697,278	..	318,405
361	POTTERY, CHINA, EARTHENWARE	259	..	287	3	8,560	..	..	403,333	..	10,561
362	GLASS AND PRODUCTS	7	..	1,402	9,941	48,159	..	..	3,847,556	..	73,360
369	OTHER NON-METAL. MINERAL PRODUCTS	104,505	..	12,459	19,182	67,931	..	..	8,446,389	..	234,484
3691	Structural clay products	2,016	..	3,582	2,389	39,204	..	..	1,493,972	..	52,569
3692	Cement, lime and plaster	100,649	..	2,162	13,298	14,676	..	..	5,182,695	..	149,443
3699	Other non-metallic mineral products	1,840	..	6,715	3,495	14,051	..	..	1,769,722	..	32,472
37	**BASIC METAL INDUSTRIES**	436,943	..	5,059	58,429	201,445	..	..	37,491,723	..	836,846
371	IRON AND STEEL	429,763	..	1,674	56,952	174,219	..	..	22,826,056	..	744,782
372	NON-FERROUS METALS	7,180	..	3,385	1,477	27,226	..	..	14,665,667	..	92,064
38	**METAL PRODUCTS, MACHINERY, EQUIP.**	10,902	..	40,490	2,518	131,065	..	..	39,021,029	..	325,451
381	METAL PRODUCTS	866	..	11,228	480	36,390	..	..	8,170,722	..	78,379
3811	Cutlery, hand tools and general hardware	31	..	2,061	74	3,218	..	..	1,271,250	..	9,961
3812	Furniture and fixtures primarily of metal	4	..	938	64	1,360	..	..	325,444	..	3,538
3813	Structural metal products	567	..	2,156	97	3,579	..	..	807,917	..	9,308
3819	Other fabricated metal products	264	..	6,073	245	28,233	..	..	5,766,111	..	55,573
382	NON-ELECTRICAL MACHINERY	2,290	..	12,743	1,186	29,751	..	..	8,135,000	..	75,256
3821	Engines and turbines	350	..	4,879	490	12,411	..	..	3,319,472	..	30,080
3822	Agricultural machinery and equipment	1,057	..	706	160	1,562	..	..	377,278	..	4,843
3823	Metal and wood working machinery	251	..	1,863	65	2,430	..	..	981,528	..	8,143
3824	Special industrial machinery	632	..	5,033	445	12,059	..	..	2,702,222	..	27,897
3825	Office, computing and accounting machinery	0	..	262	26	1,289	..	..	754,500	..	4,293
3829	Other non-electrical machinery and equipment	c	..	c	c	c	..	..	c	..	
383	ELECTRICAL MACHINERY	1,248	..	7,678	360	18,696	..	..	8,568,779	..	58,830
3831	Electrical industrial machinery	574	..	4,449	180	10,386	..	..	3,993,056	..	29,964
3832	Radio, TV and communications equipment	355	..	2,206	2	4,363	..	..	3,236,056	..	18,576
3833	Electrical appliances and housewares	303	..	345	67	1,723	..	..	660,528	..	4,816
3839	Other electrical apparatus and supplies	16	..	678	111	2,224	..	..	679,139	..	5,474
384	TRANSPORT EQUIPMENT	6,366	..	7,715	475	43,680	..	..	13,307,528	..	106,143
3841	Shipbuilding	407	..	454	-14	1,775	..	..	397,139	..	4,080
3842	Railroad equipment	1,615	..	1,298	0	2,791	..	..	322,611	..	6,865
3843	Motor vehicles	4,304	..	5,208	386	36,197	..	..	11,817,750	..	88,639
3844	Motorcycles and bicycles	1	..	287	0	188	..	..	113,917	..	886
3845	Aircraft	38	..	364	75	2,702	..	..	640,611	..	5,485
3849	Other transport equipment	1	..	104	0	27	..	..	15,500	..	188
385	PROFESSIONAL AND SCIENTIFIC EQUIPMENT	132	..	1,126	17	2,548	..	..	839,000	..	6,843
3851	Professional equipment	7	..	654	9	1,862	..	..	434,639	..	4,097
3852	Photographic and optical goods	119	..	395	8	666	..	..	367,861	..	2,512
3853	Watches and clocks	6	..	77	0	20	..	..	36,500	..	234
39	**OTHER MANUFACTURING INDUSTRIES**	49	..	583	38	527	..	..	329,361	..	2,383
3901	Jewellery and related articles	c	..	c	c	c	..	..	c	..	c
3902	Musical instruments	15	..	142	37	69	..	..	24,528	..	351
3903	Sporting and athletic goods	0	..	35	0	168	..	..	44,806	..	364
3909	Other manufactures	34	..	406	1	290	..	..	260,027	..	1,667
SERR	**non-specified, unallocated industry**										
3	**TOTAL MANUFACTURING**	722,211	..	170,544	281,543	1,074,894	..	..	190,651,306	..	2,935,537

Year: **1990**   **HUNGARY**

ISIC Revision 2 Industry Sector		Solid TJ	LPG TJ	Distiloil TJ	RFO TJ	Gas TJ	Biomass TJ	Steam TJ	Electr MWh	Own Use MWh	TOTAL TJ
31	**FOOD, BEVERAGES AND TOBACCO**	782	47	4,452	522	3,038	..	22,717	1,520,000	190,394	36,345
311/2	FOOD	727	36	4,027	458	2,559		18,200	1,270,834	167,389	29,979
3111	Slaughtering, preparing and preserving meat	76	5	946	0	189		4,304	398,056	0	6,953
3112	Dairy products	5	0	797	0	288		2,098	148,056	0	3,721
3113	Canning, preserving of fruits and vegetables	9	15	205	6	139		3,532	206,944	0	4,651
3114	Canning, preserving and processing of fish	..	..	..	..	..		..	..	..	..
3115	Vegetable and animal oils and fats	0	0	89	13	72		1,694	67,500	8,127	2,082
3116	Grain mill products	59	1	403	0	89		169	181,667	0	1,375
3117	Bakery products	14	14	1,338	0	1,486		182	56,389	0	3,237
3118	Sugar factories and refineries	540	0	177	406	69		5,017	150,000	159,262	6,176
3119	Cocoa, chocolate and sugar confectionery	3	1	45	30	117		454	32,778	0	768
3121	Other food products	21	0	27	3	110		750	29,444	0	1,017
3122	Prepared animal feeds	..	..	..	..	..		..	..	..	..
313	BEVERAGES	53	11	406	64	460		4,248	226,944	23,005	5,976
3131	Distilling, rectifying and blending of spirits	0	0	53	6	221		1,512	58,889	0	2,004
3132	Wine industries	5	0	123	0	23		249	21,667	0	478
3133	Malt liquors and malts	48	9	165	58	204		2,338	138,610	23,005	3,238
3134	Soft drinks	0	2	65	0	12		149	7,778	0	256
314	TOBACCO	2	0	19	0	19		269	22,222	0	389
32	**TEXTILES, APPAREL AND LEATHER**	39	2	476	53	595	..	6,965	731,667	0	10,764
321	TEXTILES	28	1	297	53	543		5,730	615,278	0	8,867
3211	Spinning weaving and finishing textiles	16	1	182	24	441		4,697	535,000	0	7,287
3212	Made-up goods excluding wearing apparel	..	..	..	..	..		..	..	..	..
3213	Knitting mills	4	0	60	19	71		752	49,722	0	1,085
3214	Carpets and rugs	..	..	..	..	..		..	..	..	..
3215	Cordage, rope and twine	8	0	55	10	31		281	30,556	0	495
3219	Other textiles	..	..	..	..	..		..	..	..	..
322	WEARING APPAREL, EXCEPT FOOTWEAR	4	1	87	0	44		480	53,611	0	809
323	LEATHER AND FUR PRODUCTS	0	0	28	0	0		512	25,278	0	631
3231	Tanneries and leather finishing	..	..	..	..	..		..	..	..	..
3232	Fur dressing and dyeing industries	0	0	14	0	0		256	12,778	0	316
3233	Leather prods. ex. footwear and wearing apparel	0	0	14	0	0		256	12,500	0	315
324	FOOTWEAR, EX. RUBBER AND PLASTIC	7	0	64	0	8		243	37,500	0	457
33	**WOOD PRODUCTS AND FURNITURE**	11	0	165	0	111	..	1,771	114,722	0	2,471
331	WOOD PRODUCTS, EXCEPT FURNITURE	5	0	60	0	67		884	58,056	0	1,225
3311	Sawmills, planing and other wood mills	5	0	60	0	67		884	58,056	0	1,225
3312	Wooden and cane containers	..	..	..	..	..		..	..	..	..
3319	Other wood and cork products	..	..	..	..	..		..	..	..	..
332	FURNITURE, FIXTURES, EXCL. METALLIC	6	0	105	0	44		887	56,666	0	1,246
34	**PAPER, PUBLISHING AND PRINTING**	1	6	142	0	123	..	5,447	537,222	60,407	7,436
341	PAPER AND PRODUCTS	0	5	83	0	79		5,300	470,833	60,407	6,945
3411	Pulp, paper and paperboard articles	0	5	45	0	79		5,300	470,833	60,407	6,907
3412	Containers of paper and paperboard	0	0	38	0	0		0	0	0	38
3419	Other pulp, paper and paperboard articles	..	..	..	..	..		..	..	..	..
342	PRINTING AND PUBLISHING	1	1	59	0	44		147	66,389	0	491
35	**CHEMICAL PRODUCTS**	16	130	2,653	29	17,149	..	30,540	3,329,167	172,231	61,882
351	INDUSTRIAL CHEMICALS	16	13	1,045	28	6,932		19,895	2,395,274	116,465	36,133
3511	Basic industrial chemicals excl. fertilizers	14	9	795	0	4,742		1,079	71,389	8,563	6,865
3512	Fertilizers and pesticides	2	0	157	20	1,468		11,051	505,833	54,914	14,321
3513	Synthetic resins and plastic materials	0	4	93	8	722		7,765	1,818,056	52,988	14,946
352	OTHER CHEMICALS	0	21	180	0	33		4,346	267,222	17,248	5,480
3521	Paints, varnishes and lacquers	..	..	..	..	..		..	..	..	..
3522	Drugs and medicines	0	2	165	0	32		4,094	257,500	17,248	5,158
3523	Soap, cleaning preparations, perfumes, cosmetics	0	19	15	0	1		252	9,722	0	322
3529	Other chemical products	..	..	..	..	..		..	..	..	..
353	PETROLEUM REFINERIES	0	96	1,311	1	10,148		4,062	443,612	38,518	17,076
354	MISC. PETROLEUM AND COAL PRODUCTS	..	..	..	..	..		..	..	..	..
355	RUBBER PRODUCTS	0	0	64	0	7		1,652	103,611	0	2,096
3551	Tyres and tubes	0	0	64	0	7		1,652	103,611	0	2,096
3559	Other rubber products	..	..	..	..	..		..	..	..	..
356	PLASTIC PRODUCTS	0	0	53	0	29		585	119,444	0	1,097
36	**NON-METALLIC MINERAL PRODUCTS**	3,382	67	1,241	4,868	27,686	..	2,892	1,111,111	0	44,136
361	POTTERY, CHINA, EARTHENWARE	0	32	17	0	680		300	30,556	0	1,139
362	GLASS AND PRODUCTS	3	3	93	309	8,734		429	265,555	0	10,527
369	OTHER NON-METAL. MINERAL PRODUCTS	3,379	32	1,131	4,559	18,272		2,163	815,000	0	32,470
3691	Structural clay products	2,475	32	403	80	7,875		643	259,167	0	12,441
3692	Cement, lime and plaster	671	0	148	4,445	10,070		501	481,111	0	17,567
3699	Other non-metallic mineral products	233	0	580	34	327		1,019	74,722	0	2,462
37	**BASIC METAL INDUSTRIES**	33,413	2	822	4,494	26,619	..	17,032	3,522,222	318,896	93,914
371	IRON AND STEEL	33,405	1	550	3,073	22,836		9,590	1,745,833	238,270	74,882
372	NON-FERROUS METALS	8	1	272	1,421	3,783		7,442	1,776,389	80,626	19,032
38	**METAL PRODUCTS, MACHINERY, EQUIP.**	610	29	2,231	52	4,993	..	11,637	1,500,555	0	24,954
381	METAL PRODUCTS	232	3	423	27	699		1,396	240,000	0	3,644
3811	Cutlery, hand tools and general hardware	25	2	136	18	343		410	109,444	0	1,328
3812	Furniture and fixtures primarily of metal	..	..	..	..	..		..	..	..	..
3813	Structural metal products	137	1	210	9	191		507	68,056	0	1,300
3819	Other fabricated metal products	70	0	77	0	165		479	62,500	0	1,016
382	NON-ELECTRICAL MACHINERY	177	9	625	25	749		2,560	285,000	0	5,171
3821	Engines and turbines	7	0	12	0	102		162	16,667	0	343
3822	Agricultural machinery and equipment	30	0	282	16	163		311	47,500	0	973
3823	Metal and wood working machinery	0	0	109	0	5		230	19,167	0	413
3824	Special industrial machinery	130	9	104	7	264		1,304	93,611	0	2,155
3825	Office, computing and accounting machinery	10	0	37	0	54		49	29,444	0	256
3829	Other non-electrical machinery and equipment	0	0	81	2	161		504	78,611	0	1,031
383	ELECTRICAL MACHINERY	48	13	379	0	1,548		3,491	450,000	0	7,099
3831	Electrical industrial machinery	32	2	132	0	121		764	71,389	0	1,308
3832	Radio, TV and communications equipment	5	0	105	0	47		975	100,556	0	1,494
3833	Electrical appliances and housewares	11	1	60	0	158		565	63,333	0	1,023
3839	Other electrical apparatus and supplies	0	10	82	0	1,222		1,187	214,722	0	3,274
384	TRANSPORT EQUIPMENT	153	4	759	0	1,797		3,841	450,555	0	8,176
3841	Shipbuilding	0	0	0	0	4		111	4,444	0	131
3842	Railroad equipment	17	0	237	0	118		1,333	64,722	0	1,938
3843	Motor vehicles	136	4	515	0	1,562		2,338	377,222	0	5,913
3844	Motorcycles and bicycles	0	0	7	0	84		0	4,167	0	106
3845	Aircraft	0	0	0	0	29		59	0	0	88
3849	Other transport equipment	..	..	..	..	..		..	..	..	..
385	PROFESSIONAL AND SCIENTIFIC EQUIPMENT	0	0	45	0	200		349	75,000	0	864
3851	Professional equipment	0	0	43	0	200		280	71,111	0	779
3852	Photographic and optical goods	0	0	2	0	0		69	3,889	0	85
3853	Watches and clocks	..	..	..	..	..		..	..	..	..
39	**OTHER MANUFACTURING INDUSTRIES**	46	0	25	0	180	..	90	28,056	0	442
3901	Jewellery and related articles	..	..	..	..	..		..	..	..	..
3902	Musical instruments	..	..	..	..	..		..	..	..	..
3903	Sporting and athletic goods	..	..	..	..	..		..	..	..	..
3909	Other manufactures	46	0	25	0	180		90	28,056	0	442
SERR	**non-specified, unallocated industry**	..	..	..	..	..		..	..	..	..
3	**TOTAL MANUFACTURING**	38,300	283	12,207	10,018	80,494	..	99,091	12,394,722	741,928	282,343

95

ISIS Energy Data Programme (IEA/OECD)

Year: **1991**   **HUNGARY**

ISIC Revision 2 Industry Sector	Solid TJ	LPG TJ	Distiloil TJ	RFO TJ	Gas TJ	Biomass TJ	Steam TJ	Electr MWh	Own Use MWh	TOTAL TJ
31 FOOD, BEVERAGES AND TOBACCO	821	51	3,642	800	2,881	..	22,986	1,465,278	218,370	35,670
311/2 FOOD	759	32	3,341	780	2,464	..	18,578	1,223,056	194,971	29,655
3111 Slaughtering, preparing and preserving meat	75	4	704	0	209	..	4,230	384,167	0	6,605
3112 Dairy products	3	0	671	6	194	..	1,964	133,056	0	3,317
3113 Canning, preserving of fruits and vegetables	9	14	164	6	155	..	3,155	179,167	0	4,148
3114 Canning, preserving and processing of fish	..	..	..	..	..	..	..	..	..	..
3115 Vegetable and animal oils and fats	0	0	68	11	75	..	1,560	59,722	9,579	1,895
3116 Grain mill products	53	0	442	0	120	..	138	173,333	0	1,377
3117 Bakery products	6	12	1,064	0	1,372	..	149	49,722	0	2,782
3118 Sugar factories and refineries	595	0	179	707	130	..	6,220	181,667	185,392	7,818
3119 Cocoa, chocolate and sugar confectionery	16	1	40	47	123	..	424	33,333	0	771
3121 Other food products	2	1	9	3	86	..	738	28,889	0	943
3122 Prepared animal feeds	..	..	..	..	..	..	..	..	..	..
313 BEVERAGES	61	19	282	20	394	..	4,137	220,278	23,399	5,622
3131 Distilling, rectifying and blending of spirits	5	1	18	4	173	..	1,477	58,056	0	1,887
3132 Wine industries	3	0	63	0	17	..	193	15,278	0	331
3133 Malt liquors and malts	43	15	145	16	197	..	2,368	139,722	23,399	3,203
3134 Soft drinks	10	3	56	0	7	..	99	7,222	0	201
314 TOBACCO	1	0	19	0	23	..	271	21,944	0	393
32 TEXTILES, APPAREL AND LEATHER	31	0	295	51	415	..	5,962	524,167	0	8,641
321 TEXTILES	22	0	177	51	372	..	4,784	435,833	0	6,975
3211 Spinning weaving and finishing textiles	12	0	120	24	289	..	3,867	377,222	0	5,670
3212 Made-up goods excluding wearing apparel	..	..	..	..	..	..	..	..	..	..
3213 Knitting mills	3	0	33	15	55	..	637	34,167	0	866
3214 Carpets and rugs	..	..	..	..	..	..	..	..	..	..
3215 Cordage, rope and twine	7	0	24	12	28	..	280	24,444	0	439
3219 Other textiles	..	..	..	..	..	..	..	..	..	..
322 WEARING APPAREL, EXCEPT FOOTWEAR	0	0	61	0	28	..	445	38,056	0	671
323 LEATHER AND FUR PRODUCTS	2	0	18	0	0	..	485	20,834	0	580
3231 Tanneries and leather finishing	..	..	..	..	..	..	..	..	..	..
3232 Fur dressing and dyeing industries	1	0	9	0	0	..	242	10,278	0	289
3233 Leather prods. ex. footwear and wearing apparel	1	0	9	0	0	..	243	10,556	0	291
324 FOOTWEAR, EX. RUBBER AND PLASTIC	7	0	39	0	15	..	248	29,444	0	415
33 WOOD PRODUCTS AND FURNITURE	13	0	123	0	120	..	1,709	108,611	0	2,356
331 WOOD PRODUCTS, EXCEPT FURNITURE	9	0	48	0	68	..	1,094	56,667	0	1,423
3311 Sawmills, planing and other wood mills	9	0	48	0	68	..	1,094	56,667	0	1,423
3312 Wooden and cane containers	..	..	..	..	..	..	..	..	..	..
3319 Other wood and cork products	..	..	..	..	..	..	..	..	..	..
332 FURNITURE, FIXTURES, EXCL. METALLIC	4	0	75	0	52	..	615	51,944	0	933
34 PAPER, PUBLISHING AND PRINTING	0	6	132	0	118	..	4,655	460,556	51,044	6,385
341 PAPER AND PRODUCTS	0	5	76	0	76	..	4,507	401,667	51,044	5,926
3411 Pulp, paper and paperboard articles	0	5	76	0	76	..	4,507	401,667	51,044	5,926
3412 Containers of paper and paperboard	..	..	..	..	..	..	..	..	..	..
3419 Other pulp, paper and paperboard articles	..	..	..	..	..	..	..	..	..	..
342 PRINTING AND PUBLISHING	0	1	56	0	42	..	148	58,889	0	459
35 CHEMICAL PRODUCTS	4	109	2,376	19	15,488	..	28,894	2,783,055	163,150	56,322
351 INDUSTRIAL CHEMICALS	4	14	752	19	4,492	..	18,298	1,928,056	109,290	30,127
3511 Basic industrial chemicals excl. fertilizers	3	7	564	0	2,688	..	1,853	136,667	11,176	5,567
3512 Fertilizers and pesticides	1	0	113	9	1,154	..	9,205	401,667	42,390	11,775
3513 Synthetic resins and plastic materials	0	7	75	10	650	..	7,240	1,389,722	55,724	12,784
352 OTHER CHEMICALS	0	22	186	0	23	..	4,291	238,333	15,035	5,326
3521 Paints, varnishes and lacquers	..	..	..	..	..	..	..	..	..	..
3522 Drugs and medicines	0	3	171	0	22	..	4,049	230,000	15,035	5,019
3523 Soap, cleaning preparations, perfumes, cosmetics	0	19	15	0	1	..	242	8,333	0	307
3529 Other chemical products	..	..	..	..	..	..	..	..	..	..
353 PETROLEUM REFINERIES	0	73	1,364	0	10,945	..	4,262	442,500	38,825	18,097
354 MISC. PETROLEUM AND COAL PRODUCTS	..	..	..	..	..	..	..	..	..	..
355 RUBBER PRODUCTS	0	0	37	0	15	..	1,487	84,444	0	1,843
3551 Tyres and tubes	0	0	37	0	15	..	1,487	84,444	0	1,843
3559 Other rubber products	..	..	..	..	..	..	..	..	..	..
356 PLASTIC PRODUCTS	0	0	37	0	13	..	556	89,722	0	929
36 NON-METALLIC MINERAL PRODUCTS	2,568	87	660	4,007	20,375	..	2,678	880,555	0	33,545
361 POTTERY, CHINA, EARTHENWARE	0	43	15	0	670	..	310	28,333	0	1,140
362 GLASS AND PRODUCTS	1	0	61	244	7,864	..	421	251,111	0	9,495
369 OTHER NON-METAL. MINERAL PRODUCTS	2,567	44	584	3,763	11,841	..	1,947	601,111	0	22,910
3691 Structural clay products	1,410	44	254	56	6,493	..	584	225,000	0	9,651
3692 Cement, lime and plaster	917	0	126	3,691	5,254	..	556	343,889	0	11,782
3699 Other non-metallic mineral products	240	0	204	16	94	..	807	32,222	0	1,477
37 BASIC METAL INDUSTRIES	26,730	3	533	1,481	17,147	..	14,866	2,671,667	299,657	69,299
371 IRON AND STEEL	26,720	1	331	553	13,667	..	8,084	1,077,500	235,595	52,387
372 NON-FERROUS METALS	10	2	202	928	3,480	..	6,782	1,594,167	64,062	16,912
38 METAL PRODUCTS, MACHINERY, EQUIP.	499	27	1,494	29	4,089	..	11,227	1,165,000	0	21,559
381 METAL PRODUCTS	190	3	284	15	572	..	1,347	186,388	0	3,082
3811 Cutlery, hand tools and general hardware	19	2	91	10	281	..	396	85,000	0	1,105
3812 Furniture and fixtures primarily of metal	..	..	..	..	..	..	..	..	..	..
3813 Structural metal products	113	1	141	5	156	..	489	52,777	0	1,095
3819 Other fabricated metal products	58	0	52	0	135	..	462	48,611	0	882
382 NON-ELECTRICAL MACHINERY	145	8	418	14	613	..	2,470	221,389	0	4,465
3821 Engines and turbines	4	0	8	0	84	..	156	13,056	0	299
3822 Agricultural machinery and equipment	24	0	189	9	133	..	300	36,944	0	788
3823 Metal and wood working machinery	0	0	73	3	4	..	222	15,000	0	356
3824 Special industrial machinery	109	8	69	1	216	..	1,258	72,778	0	1,923
3825 Office, computing and accounting machinery	8	0	25	0	44	..	47	22,778	0	206
3829 Other non-electrical machinery and equipment	0	0	54	1	132	..	487	60,833	0	893
383 ELECTRICAL MACHINERY	40	12	254	0	1,268	..	3,368	349,444	0	6,200
3831 Electrical industrial machinery	27	2	89	0	99	..	737	55,278	0	1,153
3832 Radio, TV and communications equipment	4	0	70	0	38	..	941	78,056	0	1,334
3833 Electrical appliances and housewares	9	1	40	0	129	..	545	49,167	0	901
3839 Other electrical apparatus and supplies	0	9	55	0	1,002	..	1,145	166,943	0	2,812
384 TRANSPORT EQUIPMENT	124	4	508	0	1,472	..	3,704	349,723	0	7,071
3841 Shipbuilding	0	0	0	0	3	..	107	3,611	0	123
3842 Railroad equipment	13	0	158	0	97	..	1,286	50,278	0	1,735
3843 Motor vehicles	111	4	345	0	1,280	..	2,254	292,778	0	5,048
3844 Motorcycles and bicycles	0	0	5	0	69	..	0	0	0	74
3845 Aircraft	0	0	0	0	23	..	57	3,056	0	91
3849 Other transport equipment	..	..	..	..	..	..	..	..	..	..
385 PROFESSIONAL AND SCIENTIFIC EQUIPMENT	0	0	30	0	164	..	338	58,056	0	741
3851 Professional equipment	0	0	28	0	164	..	272	55,000	0	662
3852 Photographic and optical goods	0	0	2	0	0	..	66	3,056	0	79
3853 Watches and clocks	..	..	..	..	..	..	..	..	..	..
39 OTHER MANUFACTURING INDUSTRIES	38	0	23	0	165	..	67	25,556	0	385
3901 Jewellery and related articles	..	..	..	..	..	..	..	..	..	..
3902 Musical instruments	..	..	..	..	..	..	..	..	..	..
3903 Sporting and athletic goods	..	..	..	..	..	..	..	..	..	..
3909 Other manufactures	38	0	23	0	165	..	67	25,556	0	385
SERR non-specified, unallocated industry	..	..	..	..	..	..	..	..	..	..
3 TOTAL MANUFACTURING	30,704	283	9,278	6,387	60,798	..	93,044	10,084,445	732,221	234,162

Year: **1992**   **HUNGARY**

ISIC Revision 2 Industry Sector	Solid TJ	LPG TJ	Distiloil TJ	RFO TJ	Gas TJ	Biomass TJ	Steam TJ	Electr MWh	Own Use MWh	TOTAL TJ
31 FOOD, BEVERAGES AND TOBACCO	646	61	2,928	512	2,912	..	18,310	1,270,000	155,884	29,380
311/2 FOOD	594	29	2,590	472	2,563	..	14,645	1,033,611	133,875	24,132
3111 Slaughtering, preparing and preserving meat	128	7	577	0	217	..	3,353	316,667	0	5,422
3112 Dairy products	4	0	697	27	194	..	1,670	128,889	0	3,056
3113 Canning, preserving of fruits and vegetables	8	13	101	3	102	..	1,923	139,722	0	2,653
3114 Canning, preserving and processing of fish	..	..	..	..	..	..	..	..	..	..
3115 Vegetable and animal oils and fats	0	0	52	15	84	..	1,680	67,500	12,116	2,030
3116 Grain mill products	3	0	29	0	6	..	37	25,278	0	166
3117 Bakery products	0	5	731	50	1,132	..	191	40,833	0	2,256
3118 Sugar factories and refineries	431	1	129	375	280	..	4,462	133,889	121,759	5,722
3119 Cocoa, chocolate and sugar confectionery	1	2	24	0	151	..	314	25,833	0	585
3121 Other food products	2	0	76	2	330	..	960	68,056	0	1,615
3122 Prepared animal feeds	17	1	174	0	67	..	55	86,944	0	627
313 BEVERAGES	46	32	313	12	305	..	3,532	218,056	22,009	4,946
3131 Distilling, rectifying and blending of spirits	3	4	112	0	61	..	1,113	43,333	0	1,449
3132 Wine industries	2	0	46	0	33	..	143	13,056	0	271
3133 Malt liquors and malts	39	20	112	12	183	..	2,195	153,611	22,009	3,035
3134 Soft drinks	2	8	43	0	28	..	81	8,056	0	191
314 TOBACCO	6	0	25	28	44	..	133	18,333	0	302
32 TEXTILES, APPAREL AND LEATHER	60	3	266	103	459	..	4,114	409,167	0	6,478
321 TEXTILES	47	3	183	63	328	..	3,138	331,667	0	4,956
3211 Spinning weaving and finishing textiles	32	3	115	26	178	..	2,489	263,889	0	3,793
3212 Made-up goods excluding wearing apparel	14	0	17	0	0	..	28	11,667	0	101
3213 Knitting mills	0	0	24	37	98	..	384	26,111	0	637
3214 Carpets and rugs	..	..	..	..	..	..	..	..	..	..
3215 Cordage, rope and twine	1	0	27	0	52	..	237	30,000	0	425
3219 Other textiles	..	..	..	..	..	..	..	..	..	..
322 WEARING APPAREL, EXCEPT FOOTWEAR	13	0	29	40	95	..	278	32,222	0	571
323 LEATHER AND FUR PRODUCTS	0	0	17	0	32	..	495	23,889	0	630
3231 Tanneries and leather finishing	0	0	17	0	21	..	495	20,556	0	607
3232 Fur dressing and dyeing industries	..	..	..	..	..	..	..	..	..	..
3233 Leather prods. ex. footwear and wearing apparel	0	0	0	0	11	..	0	3,333	0	23
324 FOOTWEAR, EX. RUBBER AND PLASTIC	0	0	37	0	4	..	203	21,389	0	321
33 WOOD PRODUCTS AND FURNITURE	33	0	116	0	251	..	1,479	103,056	0	2,250
331 WOOD PRODUCTS, EXCEPT FURNITURE	32	0	80	0	210	..	1,219	78,056	0	1,822
3311 Sawmills, planing and other wood mills	30	0	74	0	210	..	1,117	72,500	0	1,692
3312 Wooden and cane containers	2	0	3	0	0	..	3	3,056	0	19
3319 Other wood and cork products	0	0	3	0	0	..	99	2,500	0	111
332 FURNITURE, FIXTURES, EXCL. METALLIC	1	0	36	0	41	..	260	25,000	0	428
34 PAPER, PUBLISHING AND PRINTING	0	7	107	0	157	..	4,423	450,278	51,503	6,130
341 PAPER AND PRODUCTS	0	6	80	0	122	..	4,333	406,945	51,503	5,821
3411 Pulp, paper and paperboard articles	0	2	35	0	89	..	1,890	182,222	18,646	2,605
3412 Containers of paper and paperboard	0	4	35	0	33	..	2,374	218,056	32,857	3,113
3419 Other pulp, paper and paperboard articles	0	0	10	0	0	..	69	6,667	0	103
342 PRINTING AND PUBLISHING	0	1	27	0	35	..	90	43,333	0	309
35 CHEMICAL PRODUCTS	27	128	1,815	16	20,402	..	24,127	2,369,444	151,648	54,499
351 INDUSTRIAL CHEMICALS	2	9	211	16	1,592	..	13,402	1,485,000	75,509	20,306
3511 Basic industrial chemicals excl. fertilizers	2	1	59	0	245	..	459	39,444	2,609	899
3512 Fertilizers and pesticides	0	0	73	8	650	..	6,994	285,278	33,294	8,632
3513 Synthetic resins and plastic materials	0	8	79	8	697	..	5,949	1,160,278	39,606	10,775
352 OTHER CHEMICALS	7	19	503	0	97	..	4,448	228,889	31,798	5,784
3521 Paints, varnishes and lacquers	0	1	296	0	43	..	136	6,111	0	498
3522 Drugs and medicines	0	1	184	0	44	..	3,541	200,278	15,169	4,436
3523 Soap, cleaning preparations, perfumes, cosmetics	0	17	22	0	1	..	545	13,611	16,629	574
3529 Other chemical products	7	0	1	0	9	..	226	8,889	0	275
353 PETROLEUM REFINERIES	0	100	1,025	0	18,591	..	4,401	493,054	39,790	25,749
354 MISC. PETROLEUM AND COAL PRODUCTS	0	0	0	0	0	..	0	13,056	0	47
355 RUBBER PRODUCTS	18	0	46	0	22	..	1,206	70,278	0	1,545
3551 Tyres and tubes	18	0	46	0	22	..	1,206	70,278	0	1,545
3559 Other rubber products	..	..	..	..	..	..	..	..	..	..
356 PLASTIC PRODUCTS	0	0	30	0	100	..	670	79,167	4,551	1,069
36 NON-METALLIC MINERAL PRODUCTS	2,229	84	403	3,622	15,516	..	1,958	676,944	0	26,249
361 POTTERY, CHINA, EARTHENWARE	0	49	18	0	665	..	328	34,444	0	1,184
362 GLASS AND PRODUCTS	0	5	34	26	6,806	..	310	191,944	0	7,872
369 OTHER NON-METAL. MINERAL PRODUCTS	2,229	30	351	3,596	8,045	..	1,320	450,556	0	17,193
3691 Structural clay products	744	29	164	48	4,100	..	134	135,833	0	5,708
3692 Cement, lime and plaster	1,305	0	106	3,534	3,811	..	530	284,167	0	10,309
3699 Other non-metallic mineral products	180	1	81	14	134	..	656	30,556	0	1,176
37 BASIC METAL INDUSTRIES	24,266	4	439	1,065	12,556	..	12,713	1,782,222	302,558	56,370
371 IRON AND STEEL	24,260	3	329	151	10,817	..	7,653	971,389	242,770	45,836
372 NON-FERROUS METALS	6	1	110	914	1,739	..	5,060	810,833	59,788	10,534
38 METAL PRODUCTS, MACHINERY, EQUIP.	588	61	1,229	23	4,727	..	8,930	1,004,444	0	19,174
381 METAL PRODUCTS	222	7	227	12	682	..	1,083	160,833	0	2,812
3811 Cutlery, hand tools and general hardware	22	6	73	8	335	..	318	73,333	0	1,026
3812 Furniture and fixtures primarily of metal	..	..	..	..	..	..	..	..	..	..
3813 Structural metal products	132	1	113	4	186	..	393	45,556	0	993
3819 Other fabricated metal products	68	0	41	0	161	..	372	41,944	0	793
382 NON-ELECTRICAL MACHINERY	172	18	350	11	695	..	1,977	193,611	0	3,920
3821 Engines and turbines	5	0	7	0	95	..	125	11,389	0	273
3822 Agricultural machinery and equipment	29	0	158	7	151	..	240	32,222	0	701
3823 Metal and wood working machinery	0	0	61	2	5	..	178	13,056	0	293
3824 Special industrial machinery	129	18	58	1	245	..	1,007	63,611	0	1,687
3825 Office, computing and accounting machinery	9	0	21	0	50	..	38	20,000	0	190
3829 Other non-electrical machinery and equipment	0	0	45	1	149	..	389	53,333	0	776
383 ELECTRICAL MACHINERY	46	27	209	0	1,490	..	2,713	299,722	0	5,564
3831 Electrical industrial machinery	31	5	73	0	116	..	594	47,500	0	990
3832 Radio, TV and communications equipment	5	0	58	0	45	..	758	66,944	0	1,107
3833 Electrical appliances and housewares	10	1	33	0	152	..	439	42,222	0	787
3839 Other electrical apparatus and supplies	0	21	45	0	1,177	..	922	143,056	0	2,680
384 TRANSPORT EQUIPMENT	148	8	420	0	1,707	..	2,933	301,667	0	6,302
3841 Shipbuilding	0	0	0	0	4	..	85	3,056	0	100
3842 Railroad equipment	16	0	131	0	112	..	1,018	43,333	0	1,433
3843 Motor vehicles	132	8	285	0	1,484	..	1,785	252,500	0	4,603
3844 Motorcycles and bicycles	0	0	4	0	80	..	0	0	0	84
3845 Aircraft	0	0	0	0	27	..	45	2,778	0	82
3849 Other transport equipment	..	..	..	..	..	..	..	..	..	..
385 PROFESSIONAL AND SCIENTIFIC EQUIPMENT	0	1	23	0	153	..	224	48,611	0	576
3851 Professional equipment	0	0	22	0	153	..	180	46,111	0	521
3852 Photographic and optical goods	0	1	1	0	0	..	44	2,500	0	55
3853 Watches and clocks	..	..	..	..	..	..	..	..	..	..
39 OTHER MANUFACTURING INDUSTRIES	32	0	22	0	157	..	63	25,278	0	365
3901 Jewellery and related articles	..	..	..	..	..	..	..	..	..	..
3902 Musical instruments	..	..	..	..	..	..	..	..	..	..
3903 Sporting and athletic goods	..	..	..	..	..	..	..	..	..	..
3909 Other manufactures	32	0	22	0	157	..	63	25,278	0	365
SERR non-specified, unallocated industry	..	..	..	..	..	..	..	..	..	..
3 TOTAL MANUFACTURING	27,881	348	7,325	5,341	57,137	..	76,117	8,090,833	661,593	200,894

ISIS Energy Data Programme (IEA/OECD)

ISIC Revision 2 Industry Sector	Solid TJ	LPG TJ	Distiloil TJ	RFO TJ	Gas TJ	Biomass TJ	Steam TJ	Electr MWh	Own Use MWh	TOTAL TJ
31	**FOOD, BEVERAGES AND TOBACCO**	**468**	**61**	**2,493**	**331**	**2,991**	..	**16,288**	**1,096,111**	**137,952**	**26,081**
311/2	FOOD	389	20	2,133	294	2,458	..	12,964	878,611	114,034	21,010
3111	Slaughtering, preparing and preserving meat	67	6	553	0	279	..	3,231	323,056	0	5,299
3112	Dairy products	12	1	635	49	216	..	1,572	118,889	0	2,913
3113	Canning, preserving of fruits and vegetables	1	2	35	0	97	..	1,357	76,667	0	1,768
3114	Canning, preserving and processing of fish	..	..	..		..	..		..		..
3115	Vegetable and animal oils and fats	0	0	30	12	81	..	1,206	54,167	6,960	1,499
3116	Grain mill products	5	0	22	0	11	..	40	23,333	0	162
3117	Bakery products	15	0	523	34	838	..	196	29,444	0	1,712
3118	Sugar factories and refineries	282	1	120	199	301	..	3,888	120,833	107,074	4,841
3119	Cocoa, chocolate and sugar confectionery	0	10	8	0	187	..	240	20,556	0	519
3121	Other food products	0	0	108	0	342	..	1,182	76,944	0	1,909
3122	Prepared animal feeds	7	0	99	0	106	..	52	34,722	0	389
313	BEVERAGES	46	39	290	0	433	..	3,264	203,611	23,918	4,719
3131	Distilling, rectifying and blending of spirits	20	3	126	0	97	..	1,018	38,056	1,621	1,395
3132	Wine industries	0	0	32	0	54	..	85	7,222	0	197
3133	Malt liquors and malts	23	22	77	0	220	..	2,050	148,055	22,297	2,845
3134	Soft drinks	3	14	55	0	62	..	111	10,278	0	282
314	TOBACCO	33	2	70	37	100	..	60	13,889	0	352
32	**TEXTILES, APPAREL AND LEATHER**	**80**	**0**	**261**	**89**	**644**	..	**3,512**	**358,056**	**4,230**	**5,860**
321	TEXTILES	48	0	140	68	439	..	2,816	293,333	3,476	4,554
3211	Spinning weaving and finishing textiles	37	0	99	20	220	..	2,276	235,000	3,476	3,485
3212	Made-up goods excluding wearing apparel	10	0	14	20	0	..	13	1,944	0	64
3213	Knitting mills	1	0	14	28	130	..	347	25,000	0	610
3214	Carpets and rugs	..	..	..		..	..		..		..
3215	Cordage, rope and twine	0	0	13	0	89	..	180	31,389	0	395
3219	Other textiles	..	..	..		..	..	..		..	..
322	WEARING APPAREL, EXCEPT FOOTWEAR	2	0	70	0	181	..	149	27,779	0	502
323	LEATHER AND FUR PRODUCTS	20	0	33	21	22	..	375	20,000	754	540
3231	Tanneries and leather finishing	20	0	33	21	22	..	375	17,500	754	531
3232	Fur dressing and dyeing industries	..	..	..		..	..		..		..
3233	Leather prods. ex. footwear and wearing apparel	0	0	0	0	0	..		2,500	0	9
324	FOOTWEAR, EX. RUBBER AND PLASTIC	10	0	18	0	2	..	172	16,944	0	263
33	**WOOD PRODUCTS AND FURNITURE**	**28**	**20**	**87**	**0**	**527**	..	**1,321**	**115,277**	**0**	**2,398**
331	WOOD PRODUCTS, EXCEPT FURNITURE	28	20	63	0	486	..	1,094	95,833	0	2,036
3311	Sawmills, planing and other wood mills	21	10	57	0	486	..	1,016	89,722	0	1,913
3312	Wooden and cane containers	7	10	2	0	0	..	2	3,611	0	34
3319	Other wood and cork products	0	0	4	0	0	..	76	2,500	0	89
332	FURNITURE, FIXTURES, EXCL. METALLIC	0	0	24	0	41	..	227	19,444	0	362
34	**PAPER, PUBLISHING AND PRINTING**	**5**	**15**	**108**	**29**	**162**	..	**3,437**	**374,444**	**33,939**	**4,982**
341	PAPER AND PRODUCTS	0	4	71	29	86	..	3,320	331,944	33,939	4,583
3411	Pulp, paper and paperboard articles	0	2	38	29	61	..	1,677	165,000	5,589	2,381
3412	Containers of paper and paperboard	0	2	27	0	25	..	1,638	164,444	28,350	2,182
3419	Other pulp, paper and paperboard articles	0	0	6	0	0	..	5	2,500	0	20
342	PRINTING AND PUBLISHING	5	11	37	0	76	..	117	42,500	0	399
35	**CHEMICAL PRODUCTS**	**20**	**130**	**1,345**	**15**	**14,095**	..	**25,395**	**2,389,167**	**141,581**	**49,091**
351	INDUSTRIAL CHEMICALS	3	11	201	14	1,684	..	13,918	1,458,333	69,228	20,832
3511	Basic industrial chemicals excl. fertilizers	3	0	67	0	149	..	523	36,111	1,476	867
3512	Fertilizers and pesticides	0	0	56	7	774	..	7,796	274,722	30,804	9,511
3513	Synthetic resins and plastic materials	0	11	78	7	761	..	5,599	1,147,500	36,948	10,454
352	OTHER CHEMICALS	1	23	405	0	125	..	4,317	220,279	25,325	5,573
3521	Paints, varnishes and lacquers	0	2	264	0	80	..	172	5,556	0	538
3522	Drugs and medicines	1	1	108	0	45	..	3,404	193,056	10,063	4,218
3523	Soap, cleaning preparations, perfumes, cosmetics	0	20	23	0	0	..	532	13,889	15,262	570
3529	Other chemical products	0	0	10	0	0	..	209	7,778	0	247
353	PETROLEUM REFINERIES	0	70	678	0	12,109	..	4,339	504,722	40,960	18,866
354	MISC. PETROLEUM AND COAL PRODUCTS	0	0	0	0	38	..	1,363	57,500	0	1,608
355	RUBBER PRODUCTS	16	25	48	0	47	..	1,117	84,722	0	1,558
3551	Tyres and tubes	16	25	48	0	47	..	1,117	84,722	0	1,558
3559	Other rubber products	..	..	..		..	..	..		..	..
356	PLASTIC PRODUCTS	0	1	13	1	92	..	341	63,611	6,068	655
36	**NON-METALLIC MINERAL PRODUCTS**	**2,297**	**49**	**367**	**4,130**	**15,790**	..	**2,050**	**705,556**	**0**	**27,223**
361	POTTERY, CHINA, EARTHENWARE	0	34	16	0	696	..	347	35,000	0	1,219
362	GLASS AND PRODUCTS	0	0	44	0	6,406	..	363	180,000	0	7,461
369	OTHER NON-METAL. MINERAL PRODUCTS	2,297	15	307	4,130	8,688	..	1,340	490,556	0	18,543
3691	Structural clay products	706	15	123	73	5,092	..	148	150,833	0	6,700
3692	Cement, lime and plaster	1,414	0	107	4,050	3,461	..	558	311,667	0	10,712
3699	Other non-metallic mineral products	177	0	77	7	135	..	634	28,056	0	1,131
37	**BASIC METAL INDUSTRIES**	**30,444**	**15**	**383**	**828**	**12,918**	..	**11,261**	**1,721,667**	**290,524**	**61,001**
371	IRON AND STEEL	30,436	14	303	0	11,654	..	6,809	933,611	227,151	51,759
372	NON-FERROUS METALS	8	1	80	828	1,264	..	4,452	788,056	63,373	9,242
38	**METAL PRODUCTS, MACHINERY, EQUIP.**	**582**	**51**	**867**	**23**	**4,014**	..	**8,169**	**877,778**	**380**	**16,865**
381	METAL PRODUCTS	198	27	191	10	551	..	966	137,223	380	2,436
3811	Cutlery, hand tools and general hardware	22	4	72	5	265	..	300	68,056	0	913
3812	Furniture and fixtures primarily of metal	0	0	0	0	0	..	0	5,556	0	20
3813	Structural metal products	99	22	89	5	203	..	286	48,889	380	879
3819	Other fabricated metal products	77	1	30	0	83	..	380	14,722	0	624
382	NON-ELECTRICAL MACHINERY	173	1	258	12	771	..	1,887	177,500	0	3,741
3821	Engines and turbines	12	1	7	0	40	..	153	11,389	0	254
3822	Agricultural machinery and equipment	46	0	117	10	132	..	222	34,444	0	651
3823	Metal and wood working machinery	10	0	42	2	4	..	134	12,778	0	238
3824	Special industrial machinery	84	0	47	0	375	..	936	63,611	0	1,671
3825	Office, computing and accounting machinery	21	0	14	0	65	..	91	4,722	0	208
3829	Other non-electrical machinery and equipment	0	0	31	0	155	..	351	50,556	0	719
383	ELECTRICAL MACHINERY	56	22	141	1	1,534	..	2,479	282,778	0	5,251
3831	Electrical industrial machinery	29	5	46	0	166	..	524	42,222	0	922
3832	Radio, TV and communications equipment	10	0	47	1	91	..	632	58,056	0	990
3833	Electrical appliances and housewares	17	2	19	0	199	..	474	52,500	0	900
3839	Other electrical apparatus and supplies	0	15	29	0	1,078	..	849	130,000	0	2,439
384	TRANSPORT EQUIPMENT	145	0	247	0	1,089	..	2,695	250,833	0	5,079
3841	Shipbuilding	..	..	..		..	..	..		..	..
3842	Railroad equipment	4	0	56	0	104	..	910	19,722	0	1,145
3843	Motor vehicles	141	0	186	0	985	..	1,785	227,500	0	3,916
3844	Motorcycles and bicycles	0	0	5	0	0	..	0	3,611	0	18
3845	Aircraft	..	..	..		..	..	..		..	..
3849	Other transport equipment	..	..	..		..	..	..		..	..
385	PROFESSIONAL AND SCIENTIFIC EQUIPMENT	10	1	30	0	69	..	142	29,444	0	358
3851	Professional equipment	10	1	30	0	69	..	142	29,444	0	358
3852	Photographic and optical goods	..	..	..		..	..	..		..	..
3853	Watches and clocks	..	..	..		..	..	..		..	..
39	**OTHER MANUFACTURING INDUSTRIES**	**27**	**1**	**31**	**0**	**159**	..	**66**	**22,222**	**0**	**364**
3901	Jewellery and related articles	..	..	..		..	..		..		..
3902	Musical instruments	..	..	..		..	..		..		..
3903	Sporting and athletic goods	..	..	..		..	..		..		..
3909	Other manufactures	27	1	31	0	159	..	66	22,222	0	364
SERR	non-specified, unallocated industry	..	..	..		..	..		..		..
3	**TOTAL MANUFACTURING**	**33,951**	**342**	**5,942**	**5,445**	**51,300**	..	**71,499**	**7,660,278**	**608,606**	**193,865**

ISIC Rev 2 Industry Sector		Solid TJ	LPG TJ	Distiloil TJ	RFO TJ	Gas TJ	Biomass TJ	Steam TJ	Electr MWh	Own Use MWh	TOTAL TJ
31	FOOD, BEVERAGES AND TOBACCO	510	196	2,198	480	3,291	..	15,402	1,150,278	156,524	25,655
311/2	FOOD	473	120	1,864	453	2,806	..	12,965	922,222	133,775	21,519
3111	Slaughtering, preparing and preserving meat	51	52	530	0	329	..	2,694	308,611	0	4,767
3112	Dairy products	3	0	569	63	218	..	1,458	116,944	0	2,732
3113	Canning, preserving of fruits and vegetables	6	35	62	0	114	..	1,622	129,444	0	2,305
3114	Canning, preserving and processing of fish	..	..	..	..	..	..	..	..	..	..
3115	Vegetable and animal oils and fats	0	0	31	0	92	..	985	53,889	8,347	1,272
3116	Grain mill products	0	0	12	0	2	..	21	13,333	0	83
3117	Bakery products	0	0	361	13	634	..	96	19,444	0	1,174
3118	Sugar factories and refineries	407	32	111	359	479	..	4,555	139,167	125,428	5,992
3119	Cocoa, chocolate and sugar confectionery	0	0	10	0	229	..	213	18,889	0	520
3121	Other food products	1	1	116	18	639	..	1,274	100,556	0	2,411
3122	Prepared animal feeds	5	0	62	0	70	..	47	21,945	0	263
313	BEVERAGES	31	75	304	0	400	..	2,413	211,945	22,749	3,904
3131	Distilling, rectifying and blending of spirits	1	31	129	0	100	..	940	43,611	2,587	1,349
3132	Wine industries	0	0	13	0	54	..	73	6,389	0	163
3133	Malt liquors and malts	30	23	76	0	184	..	1,291	149,167	20,162	2,068
3134	Soft drinks	0	21	86	0	62	..	109	12,778	0	324
314	TOBACCO	6	1	30	27	85	..	24	16,111	0	231
32	TEXTILES, APPAREL AND LEATHER	51	1	218	71	590	..	2,848	339,167	3,103	4,989
321	TEXTILES	34	1	118	30	378	..	2,353	281,944	2,621	3,920
3211	Spinning weaving and finishing textiles	23	1	86	0	169	..	1,839	220,277	2,621	2,902
3212	Made-up goods excluding wearing apparel	10	0	12	0	0	..	9	1,389	0	36
3213	Knitting mills	0	0	5	30	126	..	340	27,778	0	601
3214	Carpets and rugs	..	..	..	..	..	..	..	..	..	..
3215	Cordage, rope and twine	1	0	15	0	83	..	165	32,500	0	381
3219	Other textiles	..	..	..	..	..	..	..	..	..	..
322	WEARING APPAREL, EXCEPT FOOTWEAR	13	0	75	41	164	..	111	27,223	0	502
323	LEATHER AND FUR PRODUCTS	0	0	13	0	47	..	245	16,944	482	364
3231	Tanneries and leather finishing	0	0	13	0	37	..	245	16,111	482	351
3232	Fur dressing and dyeing industries	..	..	..	..	..	..	..	..	..	..
3233	Leather prods. ex. footwear and wearing apparel	0	0	0	0	10	..	0	833	0	13
324	FOOTWEAR, EX. RUBBER AND PLASTIC	4	0	12	0	1	..	139	13,056	0	203
33	WOOD PRODUCTS AND FURNITURE	92	1	76	0	514	..	1,089	114,167	0	2,183
331	WOOD PRODUCTS, EXCEPT FURNITURE	92	1	50	0	490	..	968	95,556	0	1,945
3311	Sawmills, planing and other wood mills	92	1	44	0	490	..	888	89,167	0	1,836
3312	Wooden and cane containers	0	0	3	0	0	..	2	4,445	0	21
3319	Other wood and cork products	0	0	3	0	0	..	78	1,944	0	88
332	FURNITURE, FIXTURES, EXCL. METALLIC	0	0	26	0	24	..	121	18,611	0	238
34	PAPER, PUBLISHING AND PRINTING	5	7	93	20	313	..	3,010	389,166	34,012	4,727
341	PAPER AND PRODUCTS	0	6	65	20	168	..	2,911	343,611	34,012	4,285
3411	Pulp, paper and paperboard articles	0	3	31	20	73	..	1,442	169,722	4,748	2,163
3412	Containers of paper and paperboard	0	3	34	0	95	..	1,469	173,611	29,264	2,121
3419	Other pulp, paper and paperboard articles	0	0	0	0	0	..	0	278	0	1
342	PRINTING AND PUBLISHING	5	1	28	0	145	..	99	45,555	0	442
35	CHEMICAL PRODUCTS	16	145	991	24	14,141	..	21,128	2,507,222	140,847	44,964
351	INDUSTRIAL CHEMICALS	2	12	215	22	2,194	..	10,197	1,629,722	64,789	18,276
3511	Basic industrial chemicals excl. fertilizers	2	0	89	0	102	..	484	144,444	1,553	1,191
3512	Fertilizers and pesticides	0	0	47	7	1,297	..	3,993	283,333	29,211	6,259
3513	Synthetic resins and plastic materials	0	12	79	15	795	..	5,720	1,201,945	34,025	10,826
352	OTHER CHEMICALS	0	22	276	1	118	..	3,700	210,277	30,774	4,763
3521	Paints, varnishes and lacquers	0	2	157	0	42	..	107	4,722	0	325
3522	Drugs and medicines	0	2	90	1	70	..	3,061	189,722	15,065	3,853
3523	Soap, cleaning preparations, perfumes, cosmetics	0	18	22	0	1	..	416	10,000	15,709	436
3529	Other chemical products	0	0	7	0	5	..	116	5,833	0	149
353	PETROLEUM REFINERIES	0	70	442	1	11,662	..	4,348	476,389	40,558	18,092
354	MISC. PETROLEUM AND COAL PRODUCTS	0	0	0	0	39	..	1,376	45,834	0	1,580
355	RUBBER PRODUCTS	14	41	51	0	24	..	1,114	90,000	0	1,568
3551	Tyres and tubes	14	41	51	0	24	..	1,114	90,000	0	1,568
3559	Other rubber products	..	..	..	..	..	..	..	..	..	..
356	PLASTIC PRODUCTS	0	0	7	0	104	..	393	55,000	4,726	685
36	NON-METALLIC MINERAL PRODUCTS	2,738	529	373	4,173	16,568	..	1,646	744,722	0	28,708
361	POTTERY, CHINA, EARTHENWARE	0	32	12	0	625	..	375	33,334	0	1,164
362	GLASS AND PRODUCTS	0	497	43	0	6,490	..	291	203,611	0	8,054
369	OTHER NON-METAL. MINERAL PRODUCTS	2,738	0	318	4,173	9,453	..	980	507,777	0	19,490
3691	Structural clay products	856	0	136	0	5,165	..	64	146,111	0	6,747
3692	Cement, lime and plaster	1,746	0	124	4,164	4,190	..	537	339,444	0	11,983
3699	Other non-metallic mineral products	136	0	58	9	98	..	379	22,222	0	760
37	BASIC METAL INDUSTRIES	30,475	16	363	430	13,193	..	9,855	1,684,167	244,450	59,515
371	IRON AND STEEL	30,475	14	307	0	12,411	..	6,657	890,000	188,070	52,391
372	NON-FERROUS METALS	0	2	56	430	782	..	3,198	794,167	56,380	7,124
38	METAL PRODUCTS, MACHINERY, EQUIP.	517	42	688	81	3,680	..	6,391	811,111	0	14,319
381	METAL PRODUCTS	135	25	143	30	499	..	800	112,222	0	2,036
3811	Cutlery, hand tools and general hardware	10	1	58	10	274	..	249	51,111	0	786
3812	Furniture and fixtures primarily of metal	..	..	..	..	..	..	..	..	..	..
3813	Structural metal products	80	23	63	20	154	..	229	31,389	0	682
3819	Other fabricated metal products	45	1	22	0	71	..	322	29,722	0	568
382	NON-ELECTRICAL MACHINERY	221	7	211	18	407	..	1,017	132,778	0	2,359
3821	Engines and turbines	11	6	3	0	24	..	28	10,000	0	108
3822	Agricultural machinery and equipment	53	0	121	18	110	..	201	31,667	0	617
3823	Metal and wood working machinery	0	0	25	0	1	..	170	9,445	0	230
3824	Special industrial machinery	139	1	25	0	96	..	401	33,611	0	783
3825	Office, computing and accounting machinery	17	0	12	0	0	..	6	2,222	0	43
3829	Other non-electrical machinery and equipment	1	0	25	0	176	..	211	45,833	0	578
383	ELECTRICAL MACHINERY	13	9	114	18	1,519	..	2,175	283,889	0	4,870
3831	Electrical industrial machinery	1	0	41	0	100	..	390	28,889	0	636
3832	Radio, TV and communications equipment	4	0	37	18	42	..	609	65,833	0	947
3833	Electrical appliances and housewares	8	2	12	0	122	..	316	39,167	0	601
3839	Other electrical apparatus and supplies	0	7	24	0	1,255	..	860	150,000	0	2,686
384	TRANSPORT EQUIPMENT	148	0	220	0	1,105	..	2,299	247,500	0	4,663
3841	Shipbuilding	..	..	..	..	..	..	..	..	..	..
3842	Railroad equipment	0	0	40	0	71	..	705	15,556	0	872
3843	Motor vehicles	148	0	143	0	1,007	..	1,473	225,556	0	3,583
3844	Motorcycles and bicycles	0	0	1	0	0	..	0	833	0	4
3845	Aircraft	0	0	15	0	27	..	121	5,555	0	183
3849	Other transport equipment	0	0	21	0	0	..	0	0	0	21
385	PROFESSIONAL AND SCIENTIFIC EQUIPMENT	0	1	0	15	150	..	100	34,722	0	391
3851	Professional equipment	0	1	0	15	150	..	100	34,722	0	391
3852	Photographic and optical goods	..	..	..	..	..	..	..	..	..	..
3853	Watches and clocks	..	..	..	..	..	..	..	..	..	..
39	OTHER MANUFACTURING INDUSTRIES	17	0	23	0	90	..	52	14,722	0	235
3901	Jewellery and related articles	..	..	..	..	..	..	..	..	..	..
3902	Musical instruments	..	..	..	..	..	..	..	..	..	..
3903	Sporting and athletic goods	..	..	..	..	..	..	..	..	..	..
3909	Other manufactures	17	0	23	0	90	..	52	14,722	0	235
SERR	non-specified, unallocated industry	..	..	..	..	..	..	..	..	..	..
3	TOTAL MANUFACTURING	34,421	937	5,023	5,279	52,380	..	61,421	7,754,722	578,936	185.294

ISIS Energy Data Programme (IEA/OECD)

ISIC Revision 2 Industry Sector	Solid TJ	LPG TJ	Distiloil TJ	RFO TJ	Gas TJ	Biomass TJ	Steam TJ	Electr MWh	Own Use MWh	TOTAL TJ
31 **FOOD, BEVERAGES AND TOBACCO**	500	145	2,087	532	3,310	..	15,960	1,155,000	157,821	26,124
311/2 FOOD	482	85	1,762	509	2,865	..	13,011	967,500	142,354	21,685
3111 Slaughtering, preparing and preserving meat	52	59	516	0	199	..	2,930	317,222	0	4,898
3112 Dairy products	1	1	537	61	372	..	1,363	113,611	0	2,744
3113 Canning, preserving of fruits and vegetables	4	15	51	0	316	..	1,606	129,444	0	2,458
3114 Canning, preserving and processing of fish	..	..	..	..	..	..	..	..	..	..
3115 Vegetable and animal oils and fats	0	0	23	0	23	..	779	38,611	7,632	937
3116 Grain mill products	0	0	5	0	128	..	21	56,389	0	357
3117 Bakery products	0	0	300	28	1,051	..	94	32,778	0	1,591
3118 Sugar factories and refineries	418	8	151	383	247	..	4,742	134,722	134,722	5,949
3119 Cocoa, chocolate and sugar confectionery	1	0	9	0	33	..	170	17,778	0	277
3121 Other food products	0	1	111	37	436	..	1,286	88,889	0	2,191
3122 Prepared animal feeds	6	1	59	0	60	..	20	38,056	0	283
313 BEVERAGES	10	60	268	1	355	..	2,924	171,667	15,467	4,180
3131 Distilling, rectifying and blending of spirits	1	13	29	0	48	..	1,020	31,667	2,831	1,215
3132 Wine industries	0	0	7	0	121	..	66	8,889	0	226
3133 Malt liquors and malts	9	24	62	1	137	..	1,720	113,611	12,636	2,317
3134 Soft drinks	0	23	170	0	49	..	118	17,500	0	423
314 TOBACCO	8	0	57	22	90	..	25	15,833	0	259
32 **TEXTILES, APPAREL AND LEATHER**	42	5	156	94	454	..	2,226	291,944	366	4,027
321 TEXTILES	23	5	82	51	310	..	1,723	243,889	0	3,072
3211 Spinning weaving and finishing textiles	11	5	58	0	123	..	1,349	199,723	0	2,265
3212 Made-up goods excluding wearing apparel	8	0	3	0	0	..	9	10,278	0	57
3213 Knitting mills	4	0	5	51	109	..	241	16,944	0	471
3214 Carpets and rugs	..	..	..	..	..	..	..	..	..	..
3215 Cordage, rope and twine	0	0	16	0	78	..	124	16,944	0	279
3219 Other textiles	..	..	..	..	..	..	..	..	..	..
322 WEARING APPAREL, EXCEPT FOOTWEAR	15	0	51	43	111	..	72	21,944	0	371
323 LEATHER AND FUR PRODUCTS	0	0	12	0	33	..	312	15,000	366	410
3231 Tanneries and leather finishing	0	0	12	0	23	..	312	13,056	366	393
3232 Fur dressing and dyeing industries	..	..	..	..	..	..	..	..	..	..
3233 Leather prods. ex. footwear and wearing apparel	0	0	0	0	10	..	..	1,944	0	17
324 FOOTWEAR, EX. RUBBER AND PLASTIC	4	0	11	0	0	..	119	11,111	0	174
33 **WOOD PRODUCTS AND FURNITURE**	25	0	64	0	603	..	1,162	127,778	0	2,314
331 WOOD PRODUCTS, EXCEPT FURNITURE	25	0	55	0	469	..	1,052	101,667	0	1,967
3311 Sawmills, planing and other wood mills	25	0	49	0	469	..	967	98,611	0	1,865
3312 Wooden and cane containers	0	0	3	0	0	..	12	556	0	17
3319 Other wood and cork products	0	0	3	0	0	..	73	2,500	0	85
332 FURNITURE, FIXTURES, EXCL. METALLIC	0	0	9	0	134	..	110	26,111	0	347
34 **PAPER, PUBLISHING AND PRINTING**	0	9	87	0	333	..	3,590	383,889	34,903	5,275
341 PAPER AND PRODUCTS	0	9	68	0	199	..	3,530	351,111	34,903	4,944
3411 Pulp, paper and paperboard articles	0	7	31	0	79	..	1,834	170,833	4,676	2,549
3412 Containers of paper and paperboard	0	2	36	0	120	..	1,696	179,445	30,227	2,391
3419 Other pulp, paper and paperboard articles	0	0	1	0	0	..	0	833	0	4
342 PRINTING AND PUBLISHING	0	0	19	0	134	..	60	32,778	0	331
35 **CHEMICAL PRODUCTS**	15	168	846	13	12,250	..	21,123	2,534,167	136,992	43,045
351 INDUSTRIAL CHEMICALS	0	16	170	13	1,945	..	10,189	1,677,778	70,464	18,119
3511 Basic industrial chemicals excl. fertilizers	0	0	65	0	136	..	603	170,556	4,316	1,402
3512 Fertilizers and pesticides	0	0	39	0	1,059	..	4,443	277,500	30,161	6,431
3513 Synthetic resins and plastic materials	0	16	66	13	750	..	5,143	1,229,722	35,987	10,285
352 OTHER CHEMICALS	0	25	267	0	204	..	3,787	207,778	28,658	4,928
3521 Paints, varnishes and lacquers	0	2	131	0	56	..	98	5,278	0	306
3522 Drugs and medicines	0	1	90	0	49	..	3,083	180,278	13,658	3,823
3523 Soap, cleaning preparations, perfumes, cosmetics	0	22	22	0	81	..	469	15,000	15,000	594
3529 Other chemical products	0	0	24	0	18	..	137	7,222	0	205
353 PETROLEUM REFINERIES	0	61	361	0	9,889	..	4,517	480,277	37,870	16,421
354 MISC. PETROLEUM AND COAL PRODUCTS	0	0	0	0	40	..	1,034	22,778	0	1,156
355 RUBBER PRODUCTS	15	66	42	0	62	..	1,223	74,167	0	1,675
3551 Tyres and tubes	15	66	42	0	62	..	1,223	74,167	0	1,675
3559 Other rubber products	..	0	..	..	..	..	..	..	..	..
356 PLASTIC PRODUCTS	0	0	6	0	110	..	373	71,389	0	746
36 **NON-METALLIC MINERAL PRODUCTS**	2,468	483	325	4,844	13,853	..	1,557	685,833	0	25,999
361 POTTERY, CHINA, EARTHENWARE	0	34	10	0	642	..	342	33,611	0	1,149
362 GLASS AND PRODUCTS	0	447	42	0	3,890	..	170	137,500	0	5,044
369 OTHER NON-METAL. MINERAL PRODUCTS	2,468	2	273	4,844	9,321	..	1,045	514,722	0	19,806
3691 Structural clay products	666	1	130	173	5,321	..	198	158,889	0	7,061
3692 Cement, lime and plaster	1,735	0	112	4,668	3,851	..	520	329,444	0	12,072
3699 Other non-metallic mineral products	67	1	31	3	149	..	327	26,389	0	673
37 **BASIC METAL INDUSTRIES**	28,993	17	359	416	13,601	..	10,475	1,718,056	263,985	59,096
371 IRON AND STEEL	28,992	17	296	0	12,544	..	6,652	874,445	220,311	50,856
372 NON-FERROUS METALS	1	0	63	416	1,057	..	3,823	843,611	43,674	8,240
38 **METAL PRODUCTS, MACHINERY, EQUIP.**	366	73	543	56	3,768	..	5,543	788,333	0	13,187
381 METAL PRODUCTS	69	56	84	16	591	..	640	114,444	0	1,868
3811 Cutlery, hand tools and general hardware	2	1	37	10	179	..	181	34,167	0	533
3812 Furniture and fixtures primarily of metal	..	..	..	..	..	..	..	..	..	..
3813 Structural metal products	14	55	29	6	290	..	114	40,000	0	652
3819 Other fabricated metal products	53	0	18	0	122	..	345	40,277	0	683
382 NON-ELECTRICAL MACHINERY	129	6	181	8	435	..	803	126,111	0	2,016
3821 Engines and turbines	2	5	3	0	6	..	26	13,333	0	90
3822 Agricultural machinery and equipment	33	0	86	8	136	..	190	24,444	0	541
3823 Metal and wood working machinery	0	0	22	0	19	..	65	6,667	0	130
3824 Special industrial machinery	94	1	31	0	119	..	319	32,222	0	680
3825 Office, computing and accounting machinery	0	0	24	0	21	..	17	1,667	0	68
3829 Other non-electrical machinery and equipment	0	0	15	0	134	..	186	47,778	0	507
383 ELECTRICAL MACHINERY	11	10	90	31	1,613	..	1,856	268,333	0	4,577
3831 Electrical industrial machinery	0	0	37	0	64	..	360	25,556	0	553
3832 Radio, TV and communications equipment	0	0	23	31	73	..	455	45,278	0	745
3833 Electrical appliances and housewares	11	3	13	0	123	..	304	40,556	0	600
3839 Other electrical apparatus and supplies	0	7	17	0	1,353	..	737	156,943	0	2,679
384 TRANSPORT EQUIPMENT	157	1	173	1	965	..	2,158	247,778	0	4,347
3841 Shipbuilding	0	0	0	0	13	..	10	1,944	0	30
3842 Railroad equipment	0	0	36	1	69	..	702	15,000	0	862
3843 Motor vehicles	157	1	130	0	876	..	1,374	226,667	0	3,354
3844 Motorcycles and bicycles	0	0	0	0	7	..	0	1,389	0	12
3845 Aircraft	0	0	7	0	0	..	72	2,778	0	89
3849 Other transport equipment	..	..	..	..	..	..	..	..	..	..
385 PROFESSIONAL AND SCIENTIFIC EQUIPMENT	0	0	15	0	164	..	86	31,667	0	379
3851 Professional equipment	0	0	14	0	161	..	85	27,500	0	359
3852 Photographic and optical goods	0	0	1	0	3	..	1	4,167	0	20
3853 Watches and clocks	..	..	..	..	..	..	..	..	..	..
39 **OTHER MANUFACTURING INDUSTRIES**	82	0	24	0	0	..	38	833	0	147
3901 Jewellery and related articles	..	..	..	..	..	..	..	..	..	..
3902 Musical instruments	..	..	..	..	..	..	..	..	..	..
3903 Sporting and athletic goods	..	..	..	..	..	..	..	..	..	..
3909 Other manufactures	82	0	24	0	0	..	38	833	0	147
SERR non-specified, unallocated industry										
3 **TOTAL MANUFACTURING**	32,491	900	4,491	5,955	48,172	..	61,674	7,685,833	594,067	179,213

ISIC Revision 2 Industry Sector	Solid TJ	LPG TJ	Distiloil TJ	RFO TJ	Gas TJ	Biomass TJ	Steam TJ	Electr MWh	Own Use MWh	TOTAL TJ
31 FOOD, BEVERAGES AND TOBACCO	536	355	1,923	727	3,561	..	16,135	1,176,389	208,730	26,721
311/2 FOOD	534	294	1,626	698	3,169	..	13,989	1,011,111	201,068	23,226
3111 Slaughtering, preparing and preserving meat	61	151	489	0	282	..	3,021	318,887	0	5,152
3112 Dairy products	1	1	482	72	380	..	1,373	111,667	0	2,711
3113 Canning, preserving of fruits and vegetables	1	87	85	0	415	..	1,682	145,556	0	2,794
3114 Canning, preserving and processing of fish	..	0	20	0	24	..	854	41,667	7,314	1,024
3115 Vegetable and animal oils and fats	2	0	43	0	65	..	29	30,556	0	251
3116 Grain mill products	2	2	295	10	1,258	..	121	40,000	0	1,863
3117 Bakery products	461	35	49	561	286	..	5,081	144,444	148,695	6,428
3118 Sugar factories and refineries	0	5	11	0	25	..	145	26,667	0	278
3119 Cocoa, chocolate and sugar confectionery	0	1	69	55	369	..	1,658	120,278	45,059	2,436
3121 Other food products	2	12	83	0	65	..	25	31,389	0	290
3122 Prepared animal feeds	4	0	..	0	..	..	..	..	0	..
313 BEVERAGES	2	60	266	1	297	..	2,122	148,611	7,662	3,255
3131 Distilling, rectifying and blending of spirits	2	2	11	0	70	..	641	15,278	0	781
3132 Wine industries	0	0	9	1	2	..	49	5,833	0	82
3133 Malt liquors and malts	0	27	41	0	171	..	1,349	100,833	7,662	1,923
3134 Soft drinks	0	31	205	0	54	..	83	26,667	0	469
314 TOBACCO	0	1	31	28	95	..	24	16,667	0	239
32 TEXTILES, APPAREL AND LEATHER	33	6	137	77	438	..	2,123	292,500	0	3,867
321 TEXTILES	18	0	71	34	311	..	1,749	248,056	0	3,076
3211 Spinning weaving and finishing textiles	10	0	35	3	171	..	1,457	195,278	0	2,379
3212 Made-up goods excluding wearing apparel	0	0	10	0	9	..	11	1,389	0	35
3213 Knitting mills	8	0	12	31	83	..	150	25,000	0	374
3214 Carpets and rugs	..	..	..	..	..	..	131	26,389	0	288
3215 Cordage, rope and twine	0	0	14	0	48	..	..	..	0	..
3219 Other textiles	..	..	..	..	..	..	80	20,278	0	357
322 WEARING APPAREL, EXCEPT FOOTWEAR	11	6	48	43	96	..	204	13,056	0	292
323 LEATHER AND FUR PRODUCTS	1	0	9	0	31	..	204	12,223	0	263
3231 Tanneries and leather finishing	1	0	8	0	6	..	0	0	0	25
3232 Fur dressing and dyeing industries	0	0	0	0	25	..	0	833	0	4
3233 Leather prods. ex. footwear and wearing apparel	0	0	1	0	0	..	90	11,110	0	142
324 FOOTWEAR, EX. RUBBER AND PLASTIC	3	0	9	0	0	..	1,141	136,111	0	2,495
33 WOOD PRODUCTS AND FURNITURE	98	4	82	0	680	..	1,141	136,111	0	2,495
331 WOOD PRODUCTS, EXCEPT FURNITURE	18	3	58	0	518	..	1,065	109,167	0	2,055
3311 Sawmills, planing and other wood mills	18	3	54	0	516	..	981	103,056	0	1,943
3312 Wooden and cane containers	..	..	4	0	2	..	84	6,111	0	112
3319 Other wood and cork products	0	0	24	0	162	..	76	26,944	0	440
332 FURNITURE, FIXTURES, EXCL. METALLIC	80	1	24	0	162	..	76	26,944	0	440
34 PAPER, PUBLISHING AND PRINTING	0	44	91	113	316	..	4,033	423,611	30,992	6,010
341 PAPER AND PRODUCTS	0	11	70	113	195	..	3,940	386,667	30,992	5,609
3411 Pulp, paper and paperboard articles	0	7	29	113	56	..	2,222	198,056	2,192	3,132
3412 Containers of paper and paperboard	0	4	35	0	137	..	1,718	188,333	28,800	2,468
3419 Other pulp, paper and paperboard articles	0	0	6	0	2	..	0	278	0	9
342 PRINTING AND PUBLISHING	0	33	21	0	121	..	93	36,944	0	401
35 CHEMICAL PRODUCTS	16	181	747	21	13,320	..	25,066	2,588,611	151,908	48,123
351 INDUSTRIAL CHEMICALS	0	18	148	15	2,623	..	13,759	1,696,111	77,317	22,391
3511 Basic industrial chemicals excl. fertilizers	0	0	34	0	145	..	742	167,222	3,171	1,512
3512 Fertilizers and pesticides	0	0	30	0	1,747	..	7,661	324,444	32,538	10,489
3513 Synthetic resins and plastic materials	0	18	84	15	731	..	5,356	1,204,445	41,608	10,390
352 OTHER CHEMICALS	0	49	182	0	229	..	3,641	202,778	30,274	4,722
3521 Paints, varnishes and lacquers	0	2	77	0	75	..	76	2,778	0	240
3522 Drugs and medicines	0	1	70	0	25	..	3,034	184,444	13,806	3,744
3523 Soap, cleaning preparations, perfumes, cosmetics	0	46	14	0	100	..	397	10,000	16,468	534
3529 Other chemical products	0	0	21	0	29	..	134	5,556	0	204
353 PETROLEUM REFINERIES	0	55	367	6	10,231	..	4,459	490,556	39,382	16,742
354 MISC. PETROLEUM AND COAL PRODUCTS	0	0	0	0	45	..	1,539	57,778	0	1,792
355 RUBBER PRODUCTS	16	50	34	0	92	..	1,303	75,278	0	1,766
3551 Tyres and tubes	16	50	34	0	92	..	1,303	75,278	0	1,766
3559 Other rubber products	..	..	..	..	..	..	..	..	..	..
356 PLASTIC PRODUCTS	..	9	16	0	100	..	365	66,110	4,935	710
36 NON-METALLIC MINERAL PRODUCTS	2,492	535	328	3,481	14,639	..	1,514	685,000	0	25,455
361 POTTERY, CHINA, EARTHENWARE	0	56	17	0	625	..	338	37,500	0	1,171
362 GLASS AND PRODUCTS	0	395	33	0	4,184	..	178	147,222	0	5,320
369 OTHER NON-METAL. MINERAL PRODUCTS	2,492	84	278	3,481	9,830	..	998	500,278	0	18,964
3691 Structural clay products	656	81	136	137	5,173	..	196	152,222	0	6,927
3692 Cement, lime and plaster	1,764	0	110	3,340	4,496	..	533	320,000	0	11,395
3699 Other non-metallic mineral products	72	3	32	4	161	..	269	28,056	0	642
37 BASIC METAL INDUSTRIES	28,419	35	346	787	12,129	..	10,915	1,708,056	279,936	57,772
371 IRON AND STEEL	28,417	34	287	0	11,325	..	6,617	819,167	206,787	48,885
372 NON-FERROUS METALS	2	1	59	787	804	..	4,298	888,889	73,149	8,888
38 METAL PRODUCTS, MACHINERY, EQUIP.	307	75	467	45	3,930	..	5,731	788,889	0	13,395
381 METAL PRODUCTS	51	57	92	17	653	..	857	115,278	0	2,142
3811 Cutlery, hand tools and general hardware	0	11	30	10	210	..	265	42,222	0	678
3812 Furniture and fixtures primarily of metal	..	..	..	..	..	..	..	..	..	..
3813 Structural metal products	0	44	46	7	349	..	302	42,222	0	900
3819 Other fabricated metal products	51	2	16	0	94	..	290	30,834	0	564
382 NON-ELECTRICAL MACHINERY	149	6	148	23	496	..	861	119,722	0	2,114
3821 Engines and turbines	12	5	39	5	10	..	12	11,111	0	123
3822 Agricultural machinery and equipment	18	0	27	5	184	..	198	28,611	0	535
3823 Metal and wood working machinery	0	0	34	0	26	..	50	5,000	0	128
3824 Special industrial machinery	106	0	31	13	92	..	380	27,778	0	722
3825 Office, computing and accounting machinery	13	1	6	0	42	..	10	6,111	0	94
3829 Other non-electrical machinery and equipment	0	0	11	0	142	..	211	41,111	0	512
383 ELECTRICAL MACHINERY	14	11	92	4	1,671	..	1,860	291,667	0	4,702
3831 Electrical industrial machinery	0	0	38	0	100	..	384	28,056	0	623
3832 Radio, TV and communications equipment	2	0	24	4	45	..	423	50,278	0	679
3833 Electrical appliances and housewares	11	4	9	0	132	..	318	39,722	0	617
3839 Other electrical apparatus and supplies	1	7	21	0	1,394	..	735	173,611	0	2,783
384 TRANSPORT EQUIPMENT	93	1	122	1	940	..	2,097	231,111	0	4,086
3841 Shipbuilding	..	..	..	..	..	..	..	..	..	..
3842 Railroad equipment	0	0	33	1	93	..	796	16,667	0	983
3843 Motor vehicles	93	1	88	0	827	..	1,173	208,611	0	2,933
3844 Motorcycles and bicycles	..	..	..	..	..	..	..	..	..	..
3845 Aircraft	0	0	1	0	20	..	128	5,833	0	170
3849 Other transport equipment	..	..	..	..	..	..	..	..	..	..
385 PROFESSIONAL AND SCIENTIFIC EQUIPMENT	0	0	13	0	170	..	56	31,111	0	351
3851 Professional equipment	0	0	12	0	165	..	56	25,000	0	323
3852 Photographic and optical goods	0	0	1	0	5	..	0	6,111	0	28
3853 Watches and clocks	..	..	..	..	..	..	..	833	0	9
39 OTHER MANUFACTURING INDUSTRIES	0	0	3	0	3	..	0	833	0	9
3901 Jewellery and related articles	..	..	..	..	..	..	..	..	..	..
3902 Musical instruments	..	..	..	..	..	..	..	..	..	..
3903 Sporting and athletic goods	..	..	..	..	..	..	..	..	..	..
3909 Other manufactures	0	0	3	0	3	..	0	0	0	6
SERR non-specified, unallocated industry	..	..	..	..	..	..	..	..	..	..
3 TOTAL MANUFACTURING	31,901	1,235	4,124	5,251	49,016	..	66,658	7,800,000	671,566	183,847

ISIS Energy Data Programme (IEA/OECD)

Year: **1997**   # HUNGARY

ISIC Revision 2 Industry Sector	Solid TJ	LPG TJ	Distiloil TJ	RFO TJ	Gas TJ	Biomass TJ	Steam TJ	Electr MWh	Own Use MWh	TOTAL TJ
31 FOOD, BEVERAGES AND TOBACCO	472	359	1,721	521	3,217	..	14,444	1,129,444	224,816	23,991
311/2 FOOD	470	301	1,415	487	2,855	..	12,590	973,333	220,552	20,828
3111 Slaughtering, preparing and preserving meat	64	179	463	0	330	..	2,943	319,167	0	5,128
3112 Dairy products	1	1	361	39	358	..	1,033	91,111	0	2,121
3113 Canning, preserving of fruits and vegetables	1	78	77	0	417	..	1,006	98,331	0	1,933
3114 Canning, preserving and processing of fish	..	..	..	..		..				
3115 Vegetable and animal oils and fats	2	0	7	0	44	..	1,487	100,556	10,059	1,866
3116 Grain mill products	2	1	43	0	65	..	24	31,667	0	249
3117 Bakery products	0	28	271	0	1,130	..	80	35,278	0	1,636
3118 Sugar factories and refineries	395	6	34	394	123	..	4,445	123,056	123,056	5,397
3119 Cocoa, chocolate and sugar confectionery	0	0	12	0	27	..	120	25,000	0	249
3121 Other food products	2	7	65	54	282	..	1,439	117,778	87,437	1,958
3122 Prepared animal feeds	3	1	82	0	79	..	13	31,389	0	291
313 BEVERAGES	2	57	280	3	263	..	1,835	138,611	4,264	2,924
3131 Distilling, rectifying and blending of spirits	2	2	15	0	52	..	468	13,611	0	588
3132 Wine industries	0	0	4	3	1	..	42	5,833	0	71
3133 Malt liquors and malts	0	30	32	0	165	..	1,254	95,556	4,264	1,810
3134 Soft drinks	0	25	229	0	45	..	71	23,611	0	455
314 TOBACCO	0	1	26	31	99	..	19	17,500	0	239
32 TEXTILES, APPAREL AND LEATHER	66	6	103	61	414	..	2,039	260,834	0	3,628
321 TEXTILES	45	0	54	26	300	..	1,696	224,167	0	2,928
3211 Spinning weaving and finishing textiles	30	0	30	1	172	..	1,450	185,556	0	2,351
3212 Made-up goods excluding wearing apparel	0	0	6	0	8	..	9	1,111	0	27
3213 Knitting mills	15	0	8	25	74	..	128	17,500	0	313
3214 Carpets and rugs	..	..	..	..	..	..	..	..		
3215 Cordage, rope and twine	0	0	10	0	46	..	109	20,000	0	237
3219 Other textiles	..	..	..	..	..	..	..	..		
322 WEARING APPAREL, EXCEPT FOOTWEAR	18	6	34	35	88	..	72	15,000	0	307
323 LEATHER AND FUR PRODUCTS	3	0	9	0	26	..	214	12,500	0	297
3231 Tanneries and leather finishing	3	0	8	0	5	..	214	11,667	0	272
3232 Fur dressing and dyeing industries	0	0	0	0	21	..	0	0	0	21
3233 Leather prods. ex. footwear and wearing apparel	0	0	1	0	0	..	0	833	0	4
324 FOOTWEAR, EX. RUBBER AND PLASTIC	0	0	6	0	0	..	57	9,167	0	96
33 WOOD PRODUCTS AND FURNITURE	84	6	82	0	645	..	1,103	138,612	0	2,419
331 WOOD PRODUCTS, EXCEPT FURNITURE	6	5	58	0	495	..	1,038	113,056	0	2,009
3311 Sawmills, planing and other wood mills	6	5	55	0	495	..	985	107,778	0	1,934
3312 Wooden and cane containers	..	..	..	..	..	..	..			
3319 Other wood and cork products	0	0	3	0	0	..	53	5,278	0	75
332 FURNITURE, FIXTURES, EXCL. METALLIC	78	1	24	0	150	..	65	25,556	0	410
34 PAPER, PUBLISHING AND PRINTING	9	11	81	121	302	..	4,124	442,778	42,147	6,090
341 PAPER AND PRODUCTS	9	11	61	121	194	..	4,040	409,722	42,147	5,759
3411 Pulp, paper and paperboard articles	9	8	27	121	45	..	2,304	219,444	14,730	3,251
3412 Containers of paper and paperboard	0	3	29	0	147	..	1,736	190,000	27,417	2,500
3419 Other pulp, paper and paperboard articles	0	0	5	0	2	..	0	278	0	8
342 PRINTING AND PUBLISHING	0	0	20	0	108	..	84	33,056	0	331
35 CHEMICAL PRODUCTS	9	173	796	16	11,672	..	22,605	2,647,500	150,473	44,260
351 INDUSTRIAL CHEMICALS	0	20	152	1	2,593	..	12,004	1,752,500	79,972	20,791
3511 Basic industrial chemicals excl. fertilizers	0	1	51	0	125	..	688	216,944	1,598	1,640
3512 Fertilizers and pesticides	0	0	30	1	1,734	..	6,015	295,000	31,650	8,728
3513 Synthetic resins and plastic materials	0	19	71	0	734	..	5,301	1,240,556	46,724	10,423
352 OTHER CHEMICALS	0	50	215	15	206	..	3,570	209,167	21,082	4,733
3521 Paints, varnishes and lacquers	0	2	83	15	67	..	78	2,500	0	254
3522 Drugs and medicines	0	0	99	0	34	..	2,996	192,778	12,471	3,778
3523 Soap, cleaning preparations, perfumes, cosmetics	0	48	12	0	80	..	377	8,611	8,611	517
3529 Other chemical products	0	0	21	0	25	..	119	5,278	0	184
353 PETROLEUM REFINERIES	0	62	386	0	8,657	..	4,140	486,667	36,097	14,867
354 MISC. PETROLEUM AND COAL PRODUCTS	0	0	0	0	49	..	1,353	56,944	7,726	1,579
355 RUBBER PRODUCTS	9	40	34	0	81	..	1,251	77,778	0	1,695
3551 Tyres and tubes	9	40	34	0	81	..	1,251	77,778	0	1,695
3559 Other rubber products	..	..	..	..	..	..	..	..		
356 PLASTIC PRODUCTS	0	1	9	0	86	..	287	64,444	5,596	595
36 NON-METALLIC MINERAL PRODUCTS	2,623	495	308	3,580	15,838	..	1,516	740,000	0	27,024
361 POTTERY, CHINA, EARTHENWARE	0	35	15	0	675	..	345	41,389	0	1,219
362 GLASS AND PRODUCTS	0	371	33	0	5,741	..	252	207,500	0	7,144
369 OTHER NON-METAL. MINERAL PRODUCTS	2,623	89	260	3,580	9,422	..	919	491,111	0	18,661
3691 Structural clay products	647	82	126	44	4,730	..	162	141,389	0	6,300
3692 Cement, lime and plaster	1,904	1	108	3,531	4,511	..	503	320,833	0	11,713
3699 Other non-metallic mineral products	72	6	26	5	181	..	254	28,889	0	648
37 BASIC METAL INDUSTRIES	21,812	47	251	88	10,700	..	6,823	1,693,889	126,290	45,364
371 IRON AND STEEL	21,803	46	184	0	8,158	..	3,879	752,778	94,208	36,441
372 NON-FERROUS METALS	9	1	67	88	2,542	..	2,944	941,111	32,082	8,924
38 METAL PRODUCTS, MACHINERY, EQUIP.	315	101	439	28	4,023	..	5,204	853,888	0	13,184
381 METAL PRODUCTS	60	67	81	0	687	..	608	117,500	0	1,926
3811 Cutlery, hand tools and general hardware	0	15	37	0	198	..	207	37,500	0	592
3812 Furniture and fixtures primarily of metal	..	..	..	..	..	..	..	..		
3813 Structural metal products	6	47	30	0	399	..	166	46,389	0	815
3819 Other fabricated metal products	54	5	14	0	90	..	235	33,611	0	519
382 NON-ELECTRICAL MACHINERY	140	20	122	25	478	..	769	121,388	0	1,991
3821 Engines and turbines	13	2	32	2	17	..	9	9,444	0	109
3822 Agricultural machinery and equipment	20	17	43	10	185	..	174	33,333	0	569
3823 Metal and wood working machinery	0	0	7	0	23	..	18	3,611	0	61
3824 Special industrial machinery	98	1	26	13	94	..	333	26,667	0	661
3825 Office, computing and accounting machinery	9	0	3	0	34	..	14	4,722	0	77
3829 Other non-electrical machinery and equipment	0	0	11	0	125	..	221	43,611	0	514
383 ELECTRICAL MACHINERY	3	14	94	3	1,662	..	1,728	321,111	0	4,660
3831 Electrical industrial machinery	0	0	31	0	107	..	293	27,222	0	529
3832 Radio, TV and communications equipment	2	0	30	3	45	..	416	58,056	0	705
3833 Electrical appliances and housewares	0	2	8	0	105	..	317	45,000	0	594
3839 Other electrical apparatus and supplies	1	12	25	0	1,405	..	702	190,833	0	2,832
384 TRANSPORT EQUIPMENT	112	0	131	0	1,040	..	2,065	265,000	0	4,302
3841 Shipbuilding	0	0	0	0	0	..	8	278	0	9
3842 Railroad equipment	0	0	36	0	87	..	778	17,222	0	963
3843 Motor vehicles	112	0	90	0	935	..	1,173	242,222	0	3,182
3844 Motorcycles and bicycles	..	..	..	..	..	..	..	..		
3845 Aircraft	0	0	5	0	18	..	106	5,278	0	148
3849 Other transport equipment	..	..	..	..	..	..	..	..		
385 PROFESSIONAL AND SCIENTIFIC EQUIPMENT	0	0	11	0	156	..	34	28,889	0	305
3851 Professional equipment	0	0	9	0	148	..	33	22,222	0	270
3852 Photographic and optical goods	0	0	2	0	8	..	1	6,667	0	35
3853 Watches and clocks	..	..	..	..	..	..	..	..		
39 OTHER MANUFACTURING INDUSTRIES	0	0	2	0	3	..	0	833	0	8
3901 Jewellery and related articles	..	..	..	..	..	..	..	..		
3902 Musical instruments	..	..	..	..	..	..	..	..		
3903 Sporting and athletic goods	..	..	..	..	..	..	..	..		
3909 Other manufactures	0	0	2	0	3	..	0	833	0	8
SERR non-specified, unallocated industry										
3 TOTAL MANUFACTURING	25,390	1,198	3,783	4,415	46,814	..	57,858	7,907,778	543,726	165,969

Year: **1994**

ICELAND

ISIC Revision 2 Industry Sector		Solid TJ	LPG TJ	Distiloil TJ	RFO TJ	Gas TJ	Biomass TJ	Steam TJ	Electr MWh	Own Use MWh	TOTAL TJ
31	FOOD, BEVERAGES AND TOBACCO	0	0	171	2,026	..	0	465	274,944	..	3,652
311/2	FOOD	0	0	171	2,026	..	0	465	268,495	..	3,629
3111	Slaughtering, preparing and preserving meat	0	0	0	0	..	0	45	17,211	..	107
3112	Dairy products	0	0	85	0	..	0	68	31,943	..	268
3113	Canning, preserving of fruits and vegetables	..	..	86	202	..	..	145	129,695	..	900
3114	Canning, preserving and processing of fish	0	0	0	1,824	..	0	0	52,568	..	2,013
3115	Vegetable and animal oils and fats	..	..	..	..	..	..	..	..	..	
3116	Grain mill products	0	0	0	0	..	0	0	18,016	..	65
3117	Bakery products	..	..	..	..	..	..	..	..	..	
3118	Sugar factories and refineries	..	..	..	..	..	0	0	3,690	..	13
3119	Cocoa, chocolate and sugar confectionery	0	0	0	0	..	0	0	6,603	..	24
3121	Other food products	0	0	0	0	..	0	207	8,769	..	239
3122	Prepared animal feeds	0	0	0	0	..	0	0	6,449	..	23
313	BEVERAGES	..	..	..	..	..	..	..	..	..	
3131	Distilling, rectifying and blending of spirits	..	..	..	..	..	..	..	..	..	
3132	Wine industries	..	..	..	..	..	..	..	..	..	
3133	Malt liquors and malts	..	..	..	..	..	..	..	..	..	
3134	Soft drinks	0	0	0	0	..	0	0	6,449	..	23
314	TOBACCO	..	..	..	..	..	..	57	18,123	..	122
32	TEXTILES, APPAREL AND LEATHER	0	0	0	0	..	0	57	14,073	..	108
321	TEXTILES	0	0	0	0	..	0	57	13,929	..	107
3211	Spinning weaving and finishing textiles	..	..	..	..	..	..	..	..	..	
3212	Made-up goods excluding wearing apparel	..	..	..	..	..	..	..	..	..	
3213	Knitting mills	..	..	..	..	..	..	..	..	..	
3214	Carpets and rugs	..	..	..	..	..	..	..	..	..	
3215	Cordage, rope and twine	0	0	0	0	..	0	0	144	..	1
3219	Other textiles	..	..	..	..	..	0	0	1,554	..	6
322	WEARING APPAREL, EXCEPT FOOTWEAR	0	0	0	0	..	0	0	2,496	..	9
323	LEATHER AND FUR PRODUCTS	..	..	..	..	..	..	..	..	..	
3231	Tanneries and leather finishing	..	..	..	..	..	..	..	..	..	
3232	Fur dressing and dyeing industries	..	..	..	..	..	..	..	..	..	
3233	Leather prods. ex. footwear and wearing apparel	..	..	..	..	..	..	..	..	..	
324	FOOTWEAR, EX. RUBBER AND PLASTIC	0	0	0	0	..	0	0	6,264	..	23
33	WOOD PRODUCTS AND FURNITURE	0	0	0	0	..	0	0	17,641	..	64
331	WOOD PRODUCTS, EXCEPT FURNITURE	..	..	..	..	..	..	..	..	..	
3311	Sawmills, planing and other wood mills	..	..	..	..	..	..	..	..	..	
3312	Wooden and cane containers	..	..	..	..	..	..	..	..	..	
3319	Other wood and cork products	..	..	..	..	..	..	..	..	..	
332	FURNITURE, FIXTURES, EXCL. METALLIC	0	0	0	0	..	0	0	17,641	..	64
34	PAPER, PUBLISHING AND PRINTING	0	0	0	0	..	0	0	146,454	..	527
341	PAPER AND PRODUCTS	0	0	0	0	..	0	0	5,293	..	19
3411	Pulp, paper and paperboard articles	..	..	..	..	..	..	..	..	..	
3412	Containers of paper and paperboard	0	0	0	0	..	0	0	5,293	..	19
3419	Other pulp, paper and paperboard articles	0	0	0	0	..	0	0	12,348	..	44
342	PRINTING AND PUBLISHING	0	0	0	0	..	0	0	146,454	..	527
35	CHEMICAL PRODUCTS	0	0	0	0	..	0	0	135,119	..	486
351	INDUSTRIAL CHEMICALS	..	..	..	..	..	..	..	135,119	..	486
3511	Basic industrial chemicals excl. fertilizers	0	0	0	0	..	0	0	135,119	..	486
3512	Fertilizers and pesticides	..	..	..	..	..	..	..	..	..	
3513	Synthetic resins and plastic materials	0	0	0	0	..	0	0	5,934	..	21
352	OTHER CHEMICALS	0	0	0	0	..	0	0	513	..	2
3521	Paints, varnishes and lacquers	..	..	..	..	..	..	..	..	..	
3522	Drugs and medicines	..	..	..	..	..	..	..	..	..	
3523	Soap, cleaning preparations, perfumes, cosmetics	0	0	0	0	..	0	0	5,421	..	20
3529	Other chemical products	..	..	..	..	..	..	..	..	..	
353	PETROLEUM REFINERIES	0	0	0	0	..	0	0	604	..	2
354	MISC. PETROLEUM AND COAL PRODUCTS	0	0	0	0	..	0	0	448	..	2
355	RUBBER PRODUCTS	..	..	..	..	..	..	..	..	..	
3551	Tyres and tubes	..	..	..	..	..	..	..	..	..	
3559	Other rubber products	0	0	0	0	..	0	0	4,349	..	16
356	PLASTIC PRODUCTS	..	..	..	..	..	..	..	..	..	
36	NON-METALLIC MINERAL PRODUCTS	529	0	0	0	..	0	927	44,793	..	1,617
361	POTTERY, CHINA, EARTHENWARE	0	0	0	0	..	0	0	538	..	2
362	GLASS AND PRODUCTS	0	0	0	0	..	0	0	426	..	2
369	OTHER NON-METAL. MINERAL PRODUCTS	529	0	0	0	..	0	927	43,829	..	1,614
3691	Structural clay products	529	0	0	0	..	0	0	11,984	..	572
3692	Cement, lime and plaster	0	0	0	0	..	0	927	31,845	..	1,042
3699	Other non-metallic mineral products	0	0	0	0	..	0	0		..	
37	BASIC METAL INDUSTRIES	4,228	19	7	202	..	223	0	2,171,928	..	12,498
371	IRON AND STEEL	2,286	0	7	0	..	223	0	597,680	..	4,668
372	NON-FERROUS METALS	1,942	19	0	202	..	0	0	1,574,248	..	7,830
38	METAL PRODUCTS, MACHINERY, EQUIP.	0	0	0	0	..	0	0	10,005	..	36
381	METAL PRODUCTS	..	..	..	..	..	..	..	..	..	
3811	Cutlery, hand tools and general hardware	..	..	..	..	..	..	..	..	..	
3812	Furniture and fixtures primarily of metal	..	..	..	..	..	..	..	..	..	
3813	Structural metal products	..	..	..	..	..	..	..	..	..	
3819	Other fabricated metal products	0	0	0	0	..	0	0	3,053	..	11
382	NON-ELECTRICAL MACHINERY	..	..	..	..	..	..	..	..	..	
3821	Engines and turbines	..	..	..	..	..	..	..	..	..	
3822	Agricultural machinery and equipment	..	..	..	..	..	..	..	..	..	
3823	Metal and wood working machinery	..	..	..	..	..	..	..	..	..	
3824	Special industrial machinery	..	..	..	..	..	..	..	..	..	
3825	Office, computing and accounting machinery	..	..	..	..	..	..	..	..	..	
3829	Other non-electrical machinery and equipment	0	0	0	0	..	0	0	448	..	2
383	ELECTRICAL MACHINERY	..	..	..	..	..	..	..	..	..	
3831	Electrical industrial machinery	..	..	..	..	..	..	..	..	..	
3832	Radio, TV and communications equipment	..	..	..	..	..	..	..	..	..	
3833	Electrical appliances and housewares	..	..	..	..	..	..	..	..	..	
3839	Other electrical apparatus and supplies	0	0	0	0	..	0	0	6,504	..	23
384	TRANSPORT EQUIPMENT	0	0	0	0	..	0	0	6,460	..	23
3841	Shipbuilding	..	..	..	..	..	..	..	..	..	
3842	Railroad equipment	..	..	..	..	..	..	..	44	..	
3843	Motor vehicles	..	..	..	..	..	..	..	..	..	
3844	Motorcycles and bicycles	..	..	..	..	..	..	..	..	..	
3845	Aircraft	..	..	..	..	..	..	..	..	..	
3849	Other transport equipment	..	..	..	..	..	..	..	..	..	
385	PROFESSIONAL AND SCIENTIFIC EQUIPMENT	..	..	..	..	..	..	..	..	..	
3851	Professional equipment	..	..	..	..	..	..	..	..	..	
3852	Photographic and optical goods	..	..	..	..	..	..	..	..	..	
3853	Watches and clocks	..	..	..	..	..	..	..	..	..	
39	OTHER MANUFACTURING INDUSTRIES	0	0	0	0	..	0	0	7,393	..	27
3901	Jewellery and related articles	..	..	..	..	..	..	..	..	..	
3902	Musical instruments	..	..	..	..	..	..	..	..	..	
3903	Sporting and athletic goods	..	..	..	..	..	..	..	..	..	
3909	Other manufactures	..	..	..	..	..	..	..	..	..	
SERR	non-specified, unallocated industry	..	..	..	..	..	..	..	..	..	
3	TOTAL MANUFACTURING	4,757	19	178	2,228	..	223	1,449	2,697,545	..	18,565

ISIS Energy Data Programme (IEA/OECD)

ISIC Revision 2 Industry Sector		Solid TJ	LPG TJ	Distiloil TJ	RFO TJ	Gas TJ	Biomass TJ	Steam TJ	Electr MWh	Own Use MWh	TOTAL TJ
31	**FOOD, BEVERAGES AND TOBACCO**	633	547	5,760	35,861	489	..	..	3,785,000		56,916
311/2	FOOD	0	484	3,286	25,291	0			0	..	29,061
3111	Slaughtering, preparing and preserving meat	..	..	..	..						..
3112	Dairy products	..	..	..	..						..
3113	Canning, preserving of fruits and vegetables	..	..	..	..						..
3114	Canning, preserving and processing of fish	..	..	..	..						..
3115	Vegetable and animal oils and fats	..	..	..	..						..
3116	Grain mill products	..	..	..	..						..
3117	Bakery products	..	..	..	..						..
3118	Sugar factories and refineries	..	..	..	..						..
3119	Cocoa, chocolate and sugar confectionery	..	..	..	..						..
3121	Other food products	..	..	..	..						..
3122	Prepared animal feeds	..	..	..	..						..
313	BEVERAGES	0	63	2,456	9,870	0			0		12,389
3131	Distilling, rectifying and blending of spirits	..	..	..	..						..
3132	Wine industries	..	..	..	..						..
3133	Malt liquors and malts	..	..	..	..						..
3134	Soft drinks	..	..	..	..						..
314	TOBACCO	0	0	18	700	0			0		718
32	**TEXTILES, APPAREL AND LEATHER**	12,824	274	1,973	58,333	572	..	..	9,906,000		109,638
321	TEXTILES	0	263	1,748	55,691	0			0		57,702
3211	Spinning weaving and finishing textiles	..	..	..	..						..
3212	Made-up goods excluding wearing apparel	..	..	..	..						..
3213	Knitting mills	..	..	..	..						..
3214	Carpets and rugs	..	..	..	..						..
3215	Cordage, rope and twine	..	..	..	..						..
3219	Other textiles	..	..	..	..						..
322	WEARING APPAREL, EXCEPT FOOTWEAR	..	..	..	..						..
323	LEATHER AND FUR PRODUCTS	0	11	225	2,642	0			0		2,878
3231	Tanneries and leather finishing	..	..	..	..						..
3232	Fur dressing and dyeing industries	..	..	..	..						..
3233	Leather prods. ex. footwear and wearing apparel	..	..	..	..						..
324	FOOTWEAR, EX. RUBBER AND PLASTIC	..	..	..	..						..
33	**WOOD PRODUCTS AND FURNITURE**	5	0	148	2,911	1	..	..	687,000		5,538
331	WOOD PRODUCTS, EXCEPT FURNITURE	..	..	..	..						..
3311	Sawmills, planing and other wood mills	..	..	..	..						..
3312	Wooden and cane containers	..	..	..	..						..
3319	Other wood and cork products	..	..	..	..						..
332	FURNITURE, FIXTURES, EXCL. METALLIC	..	..	..	..						..
34	**PAPER, PUBLISHING AND PRINTING**	1,579	125	447	35,516	110	..	..	4,231,000		53,009
341	PAPER AND PRODUCTS	0	104	287	35,122	0			0		35,513
3411	Pulp, paper and paperboard articles	..	..	..	..						..
3412	Containers of paper and paperboard	..	..	..	..						..
3419	Other pulp, paper and paperboard articles	..	..	..	..						..
342	PRINTING AND PUBLISHING	0	21	160	394	0			0		575
35	**CHEMICAL PRODUCTS**	16,642	5,922	14,671	119,337	592	..	..	12,828,000		203,345
351	INDUSTRIAL CHEMICALS	0	5,332	4,819	89,356	0			0		99,507
3511	Basic industrial chemicals excl. fertilizers	0	2,328	3,180	66,052	0			0		71,560
3512	Fertilizers and pesticides	0	1,090	146	6,387	0			0		7,623
3513	Synthetic resins and plastic materials	0	1,914	1,493	16,917	0			0		20,324
352	OTHER CHEMICALS	0	568	8,772	22,881	0			0		32,221
3521	Paints, varnishes and lacquers	..	..	..	..						..
3522	Drugs and medicines	..	..	..	..						..
3523	Soap, cleaning preparations, perfumes, cosmetics	..	..	..	..						..
3529	Other chemical products	..	..	..	..						..
353	PETROLEUM REFINERIES	0	0	349	750	0			0		1,099
354	MISC. PETROLEUM AND COAL PRODUCTS	..	..	..	..						..
355	RUBBER PRODUCTS	0	6	575	5,990	0			0		6,571
3551	Tyres and tubes	..	..	..	..						..
3559	Other rubber products	..	..	..	..						..
356	PLASTIC PRODUCTS	0	16	156	360	0			0		532
36	**NON-METALLIC MINERAL PRODUCTS**	114,422	2,763	13,438	36,513	1,698	..	..	6,544,000		192,392
361	POTTERY, CHINA, EARTHENWARE	0	1,352	4,619	5,931	0			0		11,902
362	GLASS AND PRODUCTS	0	847	1,217	12,421	0			0		14,485
369	OTHER NON-METAL. MINERAL PRODUCTS	0	564	7,602	18,161	0			0		26,327
3691	Structural clay products	0	84	1,376	4,475	0			0		5,935
3692	Cement, lime and plaster	0	66	4,029	7,800	0			0		11,895
3699	Other non-metallic mineral products	0	414	2,197	5,886	0			0		8,497
37	**BASIC METAL INDUSTRIES**	404,736	1,567	6,483	38,912	2,933	..	..	10,440,000		492,215
371	IRON AND STEEL	0	880	5,126	34,922	0			0		40,928
372	NON-FERROUS METALS	0	687	1,357	3,990	0			0		6,034
38	**METAL PRODUCTS, MACHINERY, EQUIP.**	134	2,902	9,060	26,515	3,905	..	..	13,391,000		90,724
381	METAL PRODUCTS	0	615	1,299	3,324	0			0		5,238
3811	Cutlery, hand tools and general hardware	..	..	..	..						..
3812	Furniture and fixtures primarily of metal	..	..	..	..						..
3813	Structural metal products	..	..	..	..						..
3819	Other fabricated metal products	..	..	..	..						..
382	NON-ELECTRICAL MACHINERY	0	267	1,642	2,853	0			0		4,762
3821	Engines and turbines	..	..	..	..						..
3822	Agricultural machinery and equipment	..	..	..	..						..
3823	Metal and wood working machinery	..	..	..	..						..
3824	Special industrial machinery	..	..	..	..						..
3825	Office, computing and accounting machinery	..	..	..	..						..
3829	Other non-electrical machinery and equipment	..	..	..	..						..
383	ELECTRICAL MACHINERY	0	1,384	1,717	12,408	0			0		15,509
3831	Electrical industrial machinery	..	..	..	..						..
3832	Radio, TV and communications equipment	..	..	..	..						..
3833	Electrical appliances and housewares	..	..	..	..						..
3839	Other electrical apparatus and supplies	..	..	..	..						..
384	TRANSPORT EQUIPMENT	0	611	4,028	6,940	0			0		11,579
3841	Shipbuilding	0	313	1,469	1,709	0			0		3,491
3842	Railroad equipment	0	17	40	40	0			0		97
3843	Motor vehicles	0	263	2,366	5,026	0			0		7,655
3844	Motorcycles and bicycles	0	18	153	165	0			0		336
3845	Aircraft	..	..	..	..						..
3849	Other transport equipment	..	..	..	..						..
385	PROFESSIONAL AND SCIENTIFIC EQUIPMENT	0	25	374	990	0			0		1,389
3851	Professional equipment	..	..	..	..						..
3852	Photographic and optical goods	..	..	..	..						..
3853	Watches and clocks	..	..	..	..						..
39	**OTHER MANUFACTURING INDUSTRIES**	56	2,638	39,096	15,396	429	..	..	589,000		59,735
3901	Jewellery and related articles	..	..	..	..						..
3902	Musical instruments	..	..	..	..						..
3903	Sporting and athletic goods	..	..	..	..						..
3909	Other manufactures	..	..	..	..						..
SERR	non-specified, unallocated industry										
3	**TOTAL MANUFACTURING**	551,031	16,738	91,076	369,294	10,729	..	..	62,401,000		1,263,512

Year: **1992**　　　　　　　　　　　　**KOREA**

ISIC Revision 2 Industry Sector	Solid TJ	LPG TJ	Distiloil TJ	RFO TJ	Gas TJ	Biomass TJ	Steam TJ	Electr MWh	Own Use MWh	TOTAL TJ
31 FOOD, BEVERAGES AND TOBACCO	1,337	1,590	5,376	46,207	1,058	..	..	5,079,427	..	73,854
311/2 FOOD	0	1,478	4,261	28,680	0	..	..	0	..	34,419
3111 Slaughtering, preparing and preserving meat	..	..	..	..	..	..	..	..	..	..
3112 Dairy products	..	..	..	..	..	..	..	..	..	..
3113 Canning, preserving of fruits and vegetables	..	..	..	..	..	..	..	..	..	..
3114 Canning, preserving and processing of fish	..	..	..	..	..	..	..	..	..	..
3115 Vegetable and animal oils and fats	..	..	..	..	..	..	..	..	..	..
3116 Grain mill products	..	..	..	..	..	..	..	..	..	..
3117 Bakery products	..	..	..	..	..	..	..	..	..	..
3118 Sugar factories and refineries	..	..	..	..	..	..	..	..	..	..
3119 Cocoa, chocolate and sugar confectionery	..	..	..	..	..	..	..	..	..	..
3121 Other food products	..	..	..	..	..	..	..	..	..	..
3122 Prepared animal feeds	..	..	..	..	..	..	..	..	..	..
313 BEVERAGES	0	112	1,107	16,738	0	..	..	0	..	17,957
3131 Distilling, rectifying and blending of spirits	..	..	..	..	..	..	..	..	..	..
3132 Wine industries	..	..	..	..	..	..	..	..	..	..
3133 Malt liquors and malts	..	..	..	..	..	..	..	..	..	..
3134 Soft drinks	..	..	..	..	..	..	..	..	..	..
314 TOBACCO	0	0	8	789	0	..	..	0	..	797
32 TEXTILES, APPAREL AND LEATHER	3,696	210	1,831	57,771	2,250	..	..	12,285,805	..	109,987
321 TEXTILES	0	198	1,602	55,503	0	..	..	0	..	57,303
3211 Spinning weaving and finishing textiles	..	..	..	..	..	..	..	..	..	..
3212 Made-up goods excluding wearing apparel	..	..	..	..	..	..	..	..	..	..
3213 Knitting mills	..	..	..	..	..	..	..	..	..	..
3214 Carpets and rugs	..	..	..	..	..	..	..	..	..	..
3215 Cordage, rope and twine	..	..	..	..	..	..	..	..	..	..
3219 Other textiles	..	..	..	..	..	..	..	..	..	..
322 WEARING APPAREL, EXCEPT FOOTWEAR	..	..	..	..	..	..	..	..	..	..
323 LEATHER AND FUR PRODUCTS	0	12	229	2,268	0	..	..	0	..	2,509
3231 Tanneries and leather finishing	..	..	..	..	..	..	..	..	..	..
3232 Fur dressing and dyeing industries	..	..	..	..	..	..	..	..	..	..
3233 Leather prods. ex. footwear and wearing apparel	..	..	..	..	..	..	..	..	..	..
324 FOOTWEAR, EX. RUBBER AND PLASTIC	..	..	..	..	..	..	..	..	..	..
33 WOOD PRODUCTS AND FURNITURE	49	123	172	3,070	94	..	..	786,156	..	6,338
331 WOOD PRODUCTS, EXCEPT FURNITURE	..	..	..	..	..	..	..	..	..	..
3311 Sawmills, planing and other wood mills	..	..	..	..	..	..	..	..	..	..
3312 Wooden and cane containers	..	..	..	..	..	..	..	..	..	..
3319 Other wood and cork products	..	..	..	..	..	..	..	..	..	..
332 FURNITURE, FIXTURES, EXCL. METALLIC	..	..	..	..	..	..	..	..	..	..
34 PAPER, PUBLISHING AND PRINTING	308	236	904	50,128	739	..	..	6,236,114	..	74,765
341 PAPER AND PRODUCTS	0	215	815	49,869	0	..	..	0	..	50,899
3411 Pulp, paper and paperboard articles	..	..	..	..	..	..	..	..	..	..
3412 Containers of paper and paperboard	..	..	..	..	..	..	..	..	..	..
3419 Other pulp, paper and paperboard articles	..	..	..	..	..	..	..	..	..	..
342 PRINTING AND PUBLISHING	0	21	89	259	0	..	..	0	..	369
35 CHEMICAL PRODUCTS	12,437	21,242	15,896	132,659	1,860	..	..	18,904,236	..	252,149
351 INDUSTRIAL CHEMICALS	0	18,441	4,385	106,216	0	..	..	0	..	129,042
3511 Basic industrial chemicals excl. fertilizers	0	773	1,647	78,968	0	..	..	0	..	81,388
3512 Fertilizers and pesticides	0	7,846	143	9,503	0	..	..	0	..	17,492
3513 Synthetic resins and plastic materials	0	9,822	2,595	17,745	0	..	..	0	..	30,162
352 OTHER CHEMICALS	0	2,673	10,847	18,630	0	..	..	0	..	32,150
3521 Paints, varnishes and lacquers	..	..	..	..	..	..	..	..	..	..
3522 Drugs and medicines	..	..	..	..	..	..	..	..	..	..
3523 Soap, cleaning preparations, perfumes, cosmetics	..	..	..	..	..	..	..	..	..	..
3529 Other chemical products	..	..	..	..	..	..	..	..	..	..
353 PETROLEUM REFINERIES	0	22	343	852	0	..	..	0	..	1,217
354 MISC. PETROLEUM AND COAL PRODUCTS	..	..	..	..	..	..	..	..	..	..
355 RUBBER PRODUCTS	0	78	273	6,209	0	..	..	0	..	6,560
3551 Tyres and tubes	..	..	..	..	..	..	..	..	..	..
3559 Other rubber products	..	..	..	..	..	..	..	..	..	..
356 PLASTIC PRODUCTS	0	28	48	752	0	..	..	0	..	828
36 NON-METALLIC MINERAL PRODUCTS	152,260	5,282	11,440	45,122	3,718	..	..	9,324,901	..	251,392
361 POTTERY, CHINA, EARTHENWARE	0	3,242	2,939	5,016	0	..	..	0	..	11,197
362 GLASS AND PRODUCTS	0	1,310	858	16,661	0	..	..	0	..	18,829
369 OTHER NON-METAL. MINERAL PRODUCTS	0	730	7,643	23,445	0	..	..	0	..	31,818
3691 Structural clay products	0	403	2,284	5,367	0	..	..	0	..	8,054
3692 Cement, lime and plaster	0	190	3,398	14,072	0	..	..	0	..	17,660
3699 Other non-metallic mineral products	0	137	1,961	4,006	0	..	..	0	..	6,104
37 BASIC METAL INDUSTRIES	486,025	2,714	7,822	60,322	7,173	..	..	16,101,183	..	622,020
371 IRON AND STEEL	0	1,515	5,958	54,522	0	..	..	0	..	61,995
372 NON-FERROUS METALS	0	1,199	1,864	5,800	0	..	..	0	..	8,863
38 METAL PRODUCTS, MACHINERY, EQUIP.	0	4,680	12,945	31,745	9,524	..	..	22,116,700	..	138,514
381 METAL PRODUCTS	0	1,271	1,916	4,282	0	..	..	..	..	7,469
3811 Cutlery, hand tools and general hardware	..	..	..	..	..	..	..	..	..	..
3812 Furniture and fixtures primarily of metal	..	..	..	..	..	..	..	..	..	..
3813 Structural metal products	..	..	..	..	..	..	..	..	..	..
3819 Other fabricated metal products	..	..	..	..	..	..	..	0	..	..
382 NON-ELECTRICAL MACHINERY	0	456	1,967	3,150	0	..	..	0	..	5,573
3821 Engines and turbines	..	..	..	..	..	..	..	..	..	..
3822 Agricultural machinery and equipment	..	..	..	..	..	..	..	..	..	..
3823 Metal and wood working machinery	..	..	..	..	..	..	..	..	..	..
3824 Special industrial machinery	..	..	..	..	..	..	..	..	..	..
3825 Office, computing and accounting machinery	..	..	..	..	..	..	..	..	..	..
3829 Other non-electrical machinery and equipment	..	..	..	..	..	..	..	..	..	..
383 ELECTRICAL MACHINERY	0	1,739	3,627	15,565	0	..	..	0	..	20,931
3831 Electrical industrial machinery	..	..	..	..	..	..	..	..	..	..
3832 Radio, TV and communications equipment	..	..	..	..	..	..	..	..	..	..
3833 Electrical appliances and housewares	..	..	..	..	..	..	..	..	..	..
3839 Other electrical apparatus and supplies	..	..	..	..	..	..	..	..	..	..
384 TRANSPORT EQUIPMENT	0	1,191	4,903	7,975	0	..	..	0	..	14,069
3841 Shipbuilding	0	348	1,735	731	0	..	..	0	..	2,814
3842 Railroad equipment	0	181	13	9	0	..	..	0	..	203
3843 Motor vehicles	0	645	3,011	7,130	0	..	..	0	..	10,786
3844 Motorcycles and bicycles	0	15	78	86	0	..	..	0	..	179
3845 Aircraft	0	2	66	19	0	..	..	0	..	87
3849 Other transport equipment	..	..	..	..	..	..	..	..	..	..
385 PROFESSIONAL AND SCIENTIFIC EQUIPMENT	0	23	532	773	0	..	..	0	..	1,328
3851 Professional equipment	..	..	..	..	..	..	..	..	..	..
3852 Photographic and optical goods	..	..	..	..	..	..	..	..	..	..
3853 Watches and clocks	..	..	..	..	..	..	..	..	..	..
39 OTHER MANUFACTURING INDUSTRIES	41,374	3,158	50,745	13,477	3,142	..	..	1,211,500	..	116,257
3901 Jewellery and related articles	..	..	..	..	..	..	..	..	..	..
3902 Musical instruments	..	..	..	..	..	..	..	..	..	..
3903 Sporting and athletic goods	..	..	..	..	..	..	..	..	..	..
3909 Other manufactures	..	..	..	..	..	..	..	..	..	..
SERR non-specified, unallocated industry										
3 TOTAL MANUFACTURING	697,486	39,235	107,131	440,501	29,558	..	..	92,046,022	..	1,645,277

ISIS Energy Data Programme (IEA/OECD)

ISIC Revision 2 Industry Sector	Solid TJ	LPG TJ	Distiloil TJ	RFO TJ	Gas TJ	Biomass TJ	Steam TJ	Electr MWh	Own Use MWh	TOTAL TJ
31 FOOD, BEVERAGES AND TOBACCO	1,541	694	4,030	41,275	551	..	..	4,100,000	..	62,851
311/2 FOOD	0	560	2,808	25,081	0	..	..	0	..	28,449
3111 Slaughtering, preparing and preserving meat	..	..	..	..	..	..	..	..	..	..
3112 Dairy products	..	..	..	..	..	..	..	..	..	..
3113 Canning, preserving of fruits and vegetables	..	..	..	..	..	..	..	..	..	..
3114 Canning, preserving and processing of fish	..	..	..	..	..	..	..	..	..	..
3115 Vegetable and animal oils and fats	..	..	..	..	..	..	..	..	..	..
3116 Grain mill products	..	..	..	..	..	..	..	..	..	..
3117 Bakery products	..	..	..	..	..	..	..	..	..	..
3118 Sugar factories and refineries	..	..	..	..	..	..	..	..	..	..
3119 Cocoa, chocolate and sugar confectionery	..	..	..	..	..	..	..	..	..	..
3121 Other food products	..	..	..	..	..	..	..	..	..	..
3122 Prepared animal feeds	..	..	..	..	..	..	..	..	..	..
313 BEVERAGES	0	134	1,202	15,346	0	..	..	0	..	16,682
3131 Distilling, rectifying and blending of spirits	..	..	..	..	..	..	..	..	..	..
3132 Wine industries	..	..	..	..	..	..	..	..	..	..
3133 Malt liquors and malts	..	..	..	..	..	..	..	..	..	..
3134 Soft drinks	..	..	..	..	..	..	..	..	..	..
314 TOBACCO	0	0	20	848	0	..	..	0	..	868
32 TEXTILES, APPAREL AND LEATHER	22,331	206	2,125	54,735	932	..	..	10,017,000	..	116,390
321 TEXTILES	0	191	1,856	52,116	0	..	..	0	..	54,163
3211 Spinning weaving and finishing textiles	..	..	..	..	..	..	..	..	..	..
3212 Made-up goods excluding wearing apparel	..	..	..	..	..	..	..	..	..	..
3213 Knitting mills	..	..	..	..	..	..	..	..	..	..
3214 Carpets and rugs	..	..	..	..	..	..	..	..	..	..
3215 Cordage, rope and twine	..	..	..	..	..	..	..	..	..	..
3219 Other textiles	..	..	..	..	..	..	..	..	..	..
322 WEARING APPAREL, EXCEPT FOOTWEAR	..	..	..	..	..	..	..	..	..	..
323 LEATHER AND FUR PRODUCTS	0	15	269	2,619	0	..	..	0	..	2,903
3231 Tanneries and leather finishing	..	..	..	..	..	..	..	..	..	..
3232 Fur dressing and dyeing industries	..	..	..	..	..	..	..	..	..	..
3233 Leather prods. ex. footwear and wearing apparel	..	..	..	..	..	..	..	..	..	..
324 FOOTWEAR, EX. RUBBER AND PLASTIC	..	..	..	..	..	..	..	..	..	..
33 WOOD PRODUCTS AND FURNITURE	44	0	160	2,772	4	..	..	810,000	..	5,896
331 WOOD PRODUCTS, EXCEPT FURNITURE	..	..	..	..	..	..	..	..	..	..
3311 Sawmills, planing and other wood mills	..	..	..	..	..	..	..	..	..	..
3312 Wooden and cane containers	..	..	..	..	..	..	..	..	..	..
3319 Other wood and cork products	..	..	..	..	..	..	..	..	..	..
332 FURNITURE, FIXTURES, EXCL. METALLIC	..	..	..	..	..	..	..	..	..	..
34 PAPER, PUBLISHING AND PRINTING	930	58	628	44,429	243	..	..	5,107,000	..	64,673
341 PAPER AND PRODUCTS	0	24	537	44,202	0	..	..	0	..	44,763
3411 Pulp, paper and paperboard articles	..	..	..	..	..	..	..	..	..	..
3412 Containers of paper and paperboard	..	..	..	..	..	..	..	..	..	..
3419 Other pulp, paper and paperboard articles	..	..	..	..	..	..	..	..	..	..
342 PRINTING AND PUBLISHING	0	34	91	227	0	..	..	0	..	352
35 CHEMICAL PRODUCTS	18,362	28,149	14,325	142,200	959	..	..	16,248,000	..	262,488
351 INDUSTRIAL CHEMICALS	0	26,190	5,801	116,713	0	..	..	0	..	148,704
3511 Basic industrial chemicals excl. fertilizers	0	321	3,364	92,980	0	..	..	0	..	96,665
3512 Fertilizers and pesticides	0	17,995	164	7,263	0	..	..	0	..	25,422
3513 Synthetic resins and plastic materials	0	7,874	2,273	16,470	0	..	..	0	..	26,617
352 OTHER CHEMICALS	0	1,866	7,542	18,503	0	..	..	0	..	27,911
3521 Paints, varnishes and lacquers	..	..	..	..	..	..	..	..	..	..
3522 Drugs and medicines	..	..	..	..	..	..	..	..	..	..
3523 Soap, cleaning preparations, perfumes, cosmetics	..	..	..	..	..	..	..	..	..	..
3529 Other chemical products	..	..	..	..	..	..	..	..	..	..
353 PETROLEUM REFINERIES	0	32	536	695	0	..	..	0	..	1,263
354 MISC. PETROLEUM AND COAL PRODUCTS	..	..	..	..	..	..	..	..	..	..
355 RUBBER PRODUCTS	0	53	326	5,590	0	..	..	0	..	5,969
3551 Tyres and tubes	..	..	..	..	..	..	..	..	..	..
3559 Other rubber products	..	..	..	..	..	..	..	..	..	..
356 PLASTIC PRODUCTS	0	8	120	699	0	..	..	0	..	827
36 NON-METALLIC MINERAL PRODUCTS	1,121	3,748	12,793	37,897	2,449	..	..	7,936,000	..	86,578
361 POTTERY, CHINA, EARTHENWARE	0	1,815	3,884	7,060	0	..	..	0	..	12,759
362 GLASS AND PRODUCTS	0	1,172	969	12,168	0	..	..	0	..	14,309
369 OTHER NON-METAL. MINERAL PRODUCTS	0	761	7,940	18,669	0	..	..	0	..	27,370
3691 Structural clay products	0	405	1,710	4,407	0	..	..	0	..	6,522
3692 Cement, lime and plaster	0	69	3,965	14,262	0	..	..	0	..	18,296
3699 Other non-metallic mineral products	0	287	2,265	699	0	..	..	0	..	2,552
37 BASIC METAL INDUSTRIES	614,095	1,259	7,827	50,262	4,306	..	..	12,596,000	..	723,095
371 IRON AND STEEL	0	649	6,332	46,867	0	..	..	0	..	53,848
372 NON-FERROUS METALS	0	610	1,495	3,395	0	..	..	0	..	5,500
38 METAL PRODUCTS, MACHINERY, EQUIP.	756	3,287	11,895	31,355	5,463	..	..	15,995,000	..	110,338
381 METAL PRODUCTS	0	997	1,511	3,489	0	..	..	0	..	5,997
3811 Cutlery, hand tools and general hardware	..	..	..	..	..	..	..	..	..	..
3812 Furniture and fixtures primarily of metal	..	..	..	..	..	..	..	..	..	..
3813 Structural metal products	..	..	..	..	..	..	..	..	..	..
3819 Other fabricated metal products	..	..	..	..	..	..	..	..	..	..
382 NON-ELECTRICAL MACHINERY	0	187	1,894	2,932	0	..	..	0	..	5,013
3821 Engines and turbines	..	..	..	..	..	..	..	..	..	..
3822 Agricultural machinery and equipment	..	..	..	..	..	..	..	..	..	..
3823 Metal and wood working machinery	..	..	..	..	..	..	..	..	..	..
3824 Special industrial machinery	..	..	..	..	..	..	..	..	..	..
3825 Office, computing and accounting machinery	..	..	..	..	..	..	..	..	..	..
3829 Other non-electrical machinery and equipment	..	..	..	..	..	..	..	..	..	..
383 ELECTRICAL MACHINERY	0	1,264	2,992	16,669	0	..	..	0	..	20,925
3831 Electrical industrial machinery	..	..	..	..	..	..	..	..	..	..
3832 Radio, TV and communications equipment	..	..	..	..	..	..	..	..	..	..
3833 Electrical appliances and housewares	..	..	..	..	..	..	..	..	..	..
3839 Other electrical apparatus and supplies	..	..	..	..	..	..	..	..	..	..
384 TRANSPORT EQUIPMENT	0	830	4,683	7,349	0	..	..	0	..	12,862
3841 Shipbuilding	0	339	1,793	1,415	0	..	..	0	..	3,547
3842 Railroad equipment	0	86	25	35	0	..	..	0	..	146
3843 Motor vehicles	0	387	2,757	5,767	0	..	..	0	..	8,911
3844 Motorcycles and bicycles	0	18	106	132	0	..	..	0	..	256
3845 Aircraft	0	0	2	0	0	..	..	0	..	2
3849 Other transport equipment	..	..	..	..	..	..	..	..	..	..
385 PROFESSIONAL AND SCIENTIFIC EQUIPMENT	0	9	815	916	0	..	..	0	..	1,740
3851 Professional equipment	..	..	..	..	..	..	..	..	..	..
3852 Photographic and optical goods	..	..	..	..	..	..	..	..	..	..
3853 Watches and clocks	..	..	..	..	..	..	..	..	..	..
39 OTHER MANUFACTURING INDUSTRIES	130	1,070	41,449	14,468	849	..	..	628,000	..	60,227
3901 Jewellery and related articles	..	..	..	..	..	..	..	..	..	..
3902 Musical instruments	..	..	..	..	..	..	..	..	..	..
3903 Sporting and athletic goods	..	..	..	..	..	..	..	..	..	..
3909 Other manufactures	..	..	..	..	..	..	..	..	..	..
SERR non-specified, unallocated industry	..	..	..	..	..	..	..	..	..	..
3 TOTAL MANUFACTURING	659,310	38,471	95,232	419,393	15,756	..	..	73,437,000	..	1,492,535

ISIC Revision 2 Industry Sector	Solid TJ	LPG TJ	Distiloil TJ	RFO TJ	Gas TJ	Biomass TJ	Steam TJ	Electr MWh	Own Use MWh	TOTAL TJ
31 FOOD, BEVERAGES AND TOBACCO	1,010	1,146	4,032	44,610	669	..	..	4,553,895	..	67,861
311/2 FOOD	0	994	2,901	27,229	0			0		31,124
3111 Slaughtering, preparing and preserving meat	..	..	..	..	..					..
3112 Dairy products	..	..	..	..	..					..
3113 Canning, preserving of fruits and vegetables	..	..	..	..	..					..
3114 Canning, preserving and processing of fish	..	..	..	..	..					..
3115 Vegetable and animal oils and fats	..	..	..	..	..					..
3116 Grain mill products	..	..	..	..	..					..
3117 Bakery products	..	..	..	..	..					..
3118 Sugar factories and refineries	..	..	..	..	..					..
3119 Cocoa, chocolate and sugar confectionery	..	..	..	..	..					..
3121 Other food products	..	..	..	..	..					..
3122 Prepared animal feeds	..	..	..	..	..					..
313 BEVERAGES	0	152	1,117	16,725	0			0		17,994
3131 Distilling, rectifying and blending of spirits	..	..	..	..	..					..
3132 Wine industries	..	..	..	..	..					..
3133 Malt liquors and malts	..	..	..	..	..					..
3134 Soft drinks	..	..	..	..	..					..
314 TOBACCO	0	0	14	656	0			0		670
32 TEXTILES, APPAREL AND LEATHER	3,540	240	1,691	55,846	1,508	..	..	11,090,138	..	102,749
321 TEXTILES	0	228	1,451	53,296	0			0		54,975
3211 Spinning weaving and finishing textiles	..	..	..	..	..					..
3212 Made-up goods excluding wearing apparel	..	..	..	..	..					..
3213 Knitting mills	..	..	..	..	..					..
3214 Carpets and rugs	..	..	..	..	..					..
3215 Cordage, rope and twine	..	..	..	..	..					..
3219 Other textiles	..	..	..	..	..					..
322 WEARING APPAREL, EXCEPT FOOTWEAR	..	..	..	..	..					..
323 LEATHER AND FUR PRODUCTS	0	12	240	2,550	0			0		2,802
3231 Tanneries and leather finishing	..	..	..	..	..					..
3232 Fur dressing and dyeing industries	..	..	..	..	..					..
3233 Leather prods. ex. footwear and wearing apparel	..	..	..	..	..					..
324 FOOTWEAR, EX. RUBBER AND PLASTIC	..	..	..	..	..					..
33 WOOD PRODUCTS AND FURNITURE	39	28	182	4,060	45	..	..	960,006	..	7,810
331 WOOD PRODUCTS, EXCEPT FURNITURE	..	..	..	..	..					..
3311 Sawmills, planing and other wood mills	..	..	..	..	..					..
3312 Wooden and cane containers	..	..	..	..	..					..
3319 Other wood and cork products	..	..	..	..	..					..
332 FURNITURE, FIXTURES, EXCL. METALLIC	..	..	..	..	..					..
34 PAPER, PUBLISHING AND PRINTING	292	70	611	49,467	561	..	..	5,792,409	..	71,854
341 PAPER AND PRODUCTS	0	53	514	49,259	0			0		49,826
3411 Pulp, paper and paperboard articles	..	..	..	..	..					..
3412 Containers of paper and paperboard	..	..	..	..	..					..
3419 Other pulp, paper and paperboard articles	..	..	..	..	..					..
342 PRINTING AND PUBLISHING	0	17	97	208	0			0		322
35 CHEMICAL PRODUCTS	12,037	31,847	14,957	148,954	1,213	..	..	17,615,372	..	272,423
351 INDUSTRIAL CHEMICALS	0	29,536	4,511	124,404	0			0		158,451
3511 Basic industrial chemicals excl. fertilizers	0	784	2,381	99,258	0			0		102,423
3512 Fertilizers and pesticides	0	18,748	134	8,620	0			0		27,502
3513 Synthetic resins and plastic materials	0	10,004	1,996	16,526	0			0		28,526
352 OTHER CHEMICALS	0	2,193	8,976	17,441	0			0		28,610
3521 Paints, varnishes and lacquers	..	..	..	..	..					..
3522 Drugs and medicines	..	..	..	..	..					..
3523 Soap, cleaning preparations, perfumes, cosmetics	..	..	..	..	..					..
3529 Other chemical products	..	..	..	..	..					..
353 PETROLEUM REFINERIES	0	26	1,046	692	0			0		1,764
354 MISC. PETROLEUM AND COAL PRODUCTS	..	..	..	..	..					..
355 RUBBER PRODUCTS	0	82	300	5,820	0			0		6,202
3551 Tyres and tubes	..	..	..	..	..					..
3559 Other rubber products	..	..	..	..	..					..
356 PLASTIC PRODUCTS	0	10	124	597	0			0		731
36 NON-METALLIC MINERAL PRODUCTS	144,365	4,919	11,574	48,602	2,931	..	..	8,751,699	..	243,897
361 POTTERY, CHINA, EARTHENWARE	0	2,249	3,191	6,584	0			0		12,024
362 GLASS AND PRODUCTS	0	1,389	762	13,998	0			0		16,149
369 OTHER NON-METAL. MINERAL PRODUCTS	0	1,281	7,621	28,020	0			0		36,922
3691 Structural clay products	0	344	1,797	4,086	0			0		6,227
3692 Cement, lime and plaster	0	206	3,728	17,452	0			0		21,386
3699 Other non-metallic mineral products	0	731	2,096	6,482	0			0		9,309
37 BASIC METAL INDUSTRIES	467,348	1,707	7,465	60,226	5,781	..	..	13,551,920	..	591,314
371 IRON AND STEEL	0	729	5,742	55,821	0			0		62,292
372 NON-FERROUS METALS	0	978	1,723	4,405	0			0		7,106
38 METAL PRODUCTS, MACHINERY, EQUIP.	0	3,868	11,161	32,141	6,678	..	..	19,322,940	..	123,411
381 METAL PRODUCTS	0	1,184	1,303	3,505	0			0		5,992
3811 Cutlery, hand tools and general hardware	..	..	..	..	..					..
3812 Furniture and fixtures primarily of metal	..	..	..	..	..					..
3813 Structural metal products	..	..	..	..	..					..
3819 Other fabricated metal products	..	..	..	..	..					..
382 NON-ELECTRICAL MACHINERY	0	270	1,867	3,222	0			0		5,359
3821 Engines and turbines	..	..	..	..	..					..
3822 Agricultural machinery and equipment	..	..	..	..	..					..
3823 Metal and wood working machinery	..	..	..	..	..					..
3824 Special industrial machinery	..	..	..	..	..					..
3825 Office, computing and accounting machinery	..	..	..	..	..					..
3829 Other non-electrical machinery and equipment	..	..	..	..	..					..
383 ELECTRICAL MACHINERY	0	1,433	2,520	16,835	0			0		20,788
3831 Electrical industrial machinery	..	..	..	..	..					..
3832 Radio, TV and communications equipment	..	..	..	..	..					..
3833 Electrical appliances and housewares	..	..	..	..	..					..
3839 Other electrical apparatus and supplies	..	..	..	..	..					..
384 TRANSPORT EQUIPMENT	0	953	4,773	7,399	0			0		13,125
3841 Shipbuilding	0	349	1,614	1,315	0			0		3,278
3842 Railroad equipment	0	118	27	28	0			0		173
3843 Motor vehicles	0	467	2,986	5,922	0			0		9,375
3844 Motorcycles and bicycles	0	17	139	134	0			0		290
3845 Aircraft	0	2	7	0	0			0		9
3849 Other transport equipment	..	..	..	..	..					..
385 PROFESSIONAL AND SCIENTIFIC EQUIPMENT	0	28	698	1,180	0			0		1,906
3851 Professional equipment	..	..	..	..	..					..
3852 Photographic and optical goods	..	..	..	..	..					..
3853 Watches and clocks	..	..	..	..	..					..
39 OTHER MANUFACTURING INDUSTRIES	33,315	1,781	43,108	15,133	1,184	..	..	594,870	..	96,663
3901 Jewellery and related articles	..	..	..	..	..					..
3902 Musical instruments	..	..	..	..	..					..
3903 Sporting and athletic goods	..	..	..	..	..					..
3909 Other manufactures	..	..	..	..	..					..
SERR non-specified, unallocated industry										
3 TOTAL MANUFACTURING	661,946	45,606	94,781	459,039	20,570	..	..	82,233,249	..	1,577,982

ISIC Revision 2 Industry Sector	Solid TJ	LPG TJ	Distiloil TJ	RFO TJ	Gas TJ	Biomass TJ	Steam TJ	Electr MWh	Own Use MWh	TOTAL TJ
31 FOOD, BEVERAGES AND TOBACCO	1,337	1,590	5,376	46,207	1,058	..	..	5,079,427	..	73,854
311/2 FOOD	0	1,478	4,261	28,680	0			0		34,419
3111 Slaughtering, preparing and preserving meat	..	..	..	..	..			..		..
3112 Dairy products	..	..	..	..	..			..		..
3113 Canning, preserving of fruits and vegetables	..	..	..	..	..			..		..
3114 Canning, preserving and processing of fish	..	..	..	..	..			..		..
3115 Vegetable and animal oils and fats	..	..	..	..	..			..		..
3116 Grain mill products	..	..	..	..	..			..		..
3117 Bakery products	..	..	..	..	..			..		..
3118 Sugar factories and refineries	..	..	..	..	..			..		..
3119 Cocoa, chocolate and sugar confectionery	..	..	..	..	..			..		..
3121 Other food products	..	..	..	..	..			..		..
3122 Prepared animal feeds	..	..	..	..	..			..		..
313 BEVERAGES	0	112	1,107	16,738	0			0		17,957
3131 Distilling, rectifying and blending of spirits	..	..	..	..	..			..		..
3132 Wine industries	..	..	..	..	..			..		..
3133 Malt liquors and malts	..	..	..	..	..			..		..
3134 Soft drinks	..	..	..	..	..			..		..
314 TOBACCO	0	0	8	789	0			0		797
32 TEXTILES, APPAREL AND LEATHER	3,696	210	1,831	57,771	2,250			12,285,804		109,987
321 TEXTILES	0	198	1,602	55,503	0			0		57,303
3211 Spinning weaving and finishing textiles	..	..	..	..	..			..		..
3212 Made-up goods excluding wearing apparel	..	..	..	..	..			..		..
3213 Knitting mills	..	..	..	..	..			..		..
3214 Carpets and rugs	..	..	..	..	..			..		..
3215 Cordage, rope and twine	..	..	..	..	..			..		..
3219 Other textiles	..	..	..	..	..			..		..
322 WEARING APPAREL, EXCEPT FOOTWEAR	..	..	..	..	..			..		..
323 LEATHER AND FUR PRODUCTS	0	12	229	2,268	0			0		2,509
3231 Tanneries and leather finishing	..	..	..	..	..			..		..
3232 Fur dressing and dyeing industries	..	..	..	..	..			..		..
3233 Leather prods. ex. footwear and wearing apparel	..	..	..	..	..			..		..
324 FOOTWEAR, EX. RUBBER AND PLASTIC	..	..	..	..	..			..		..
33 WOOD PRODUCTS AND FURNITURE	49	123	172	3,070	94			786,156		6,338
331 WOOD PRODUCTS, EXCEPT FURNITURE	..	..	..	..	..			..		..
3311 Sawmills, planing and other wood mills	..	..	..	..	..			..		..
3312 Wooden and cane containers	..	..	..	..	..			..		..
3319 Other wood and cork products	..	..	..	..	..			..		..
332 FURNITURE, FIXTURES, EXCL. METALLIC	..	..	..	..	..			..		..
34 PAPER, PUBLISHING AND PRINTING	308	236	904	50,128	739			6,236,112		74,765
341 PAPER AND PRODUCTS	0	215	815	49,869	0			5,528,215		70,801
3411 Pulp, paper and paperboard articles	..	..	..	..	..			..		..
3412 Containers of paper and paperboard	..	..	..	..	..			..		..
3419 Other pulp, paper and paperboard articles	..	..	..	..	..			..		..
342 PRINTING AND PUBLISHING	0	21	89	259	0			707,897		2,917
35 CHEMICAL PRODUCTS	12,437	21,242	15,896	132,659	1,860			18,904,240		252,149
351 INDUSTRIAL CHEMICALS	0	18,441	4,385	106,216	0			0		129,042
3511 Basic industrial chemicals excl. fertilizers	0	773	1,647	78,968	0			0		81,388
3512 Fertilizers and pesticides	0	7,846	143	9,503	0			0		17,492
3513 Synthetic resins and plastic materials	0	9,822	2,595	17,745	0			0		30,162
352 OTHER CHEMICALS	0	2,673	10,847	18,630	0			0		32,150
3521 Paints, varnishes and lacquers	..	..	..	..	..			..		..
3522 Drugs and medicines	..	..	..	..	..			..		..
3523 Soap, cleaning preparations, perfumes, cosmetics	..	..	..	..	..			..		..
3529 Other chemical products	..	..	..	..	..			..		..
353 PETROLEUM REFINERIES	..	..	..	..	..			..		..
354 MISC. PETROLEUM AND COAL PRODUCTS	0	22	343	852	0			0		1,217
355 RUBBER PRODUCTS	0	78	273	6,209	0			0		6,560
3551 Tyres and tubes	..	..	..	..	..			..		..
3559 Other rubber products	..	..	..	..	..			..		..
356 PLASTIC PRODUCTS	0	28	48	752	0			0		828
36 NON-METALLIC MINERAL PRODUCTS	152,260	5,282	11,440	45,122	3,718			9,324,900		251,392
361 POTTERY, CHINA, EARTHENWARE	0	3,242	2,939	5,016	0			0		11,197
362 GLASS AND PRODUCTS	0	1,310	858	16,661	0			0		18,829
369 OTHER NON-METAL. MINERAL PRODUCTS	0	730	7,643	23,445	0			0		31,818
3691 Structural clay products	0	403	2,284	5,367	0			0		8,054
3692 Cement, lime and plaster	0	190	3,398	14,072	0			0		17,660
3699 Other non-metallic mineral products	0	137	1,961	4,006	0			0		6,104
37 BASIC METAL INDUSTRIES	486,025	2,714	7,822	60,322	7,173			16,101,181		622,020
371 IRON AND STEEL	0	1,515	5,958	54,522	0			0		61,995
372 NON-FERROUS METALS	0	1,199	1,864	5,800	0			0		8,863
38 METAL PRODUCTS, MACHINERY, EQUIP.	0	4,680	12,945	31,743	9,524			22,116,707		138,512
381 METAL PRODUCTS	0	1,271	1,916	4,282	0			0		7,469
3811 Cutlery, hand tools and general hardware	..	..	..	..	..			..		..
3812 Furniture and fixtures primarily of metal	..	..	..	..	..			..		..
3813 Structural metal products	..	..	..	..	..			..		..
3819 Other fabricated metal products	..	..	..	..	..			..		..
382 NON-ELECTRICAL MACHINERY	0	456	1,967	3,150	0			0		5,573
3821 Engines and turbines	..	..	..	..	..			..		..
3822 Agricultural machinery and equipment	..	..	..	..	..			..		..
3823 Metal and wood working machinery	..	..	..	..	..			..		..
3824 Special industrial machinery	..	..	..	..	..			..		..
3825 Office, computing and accounting machinery	..	..	..	..	..			..		..
3829 Other non-electrical machinery and equipment	..	..	..	..	..			..		..
383 ELECTRICAL MACHINERY	0	1,739	3,627	15,564	0			0		20,930
3831 Electrical industrial machinery	..	..	..	..	..			..		..
3832 Radio, TV and communications equipment	..	..	..	..	..			..		..
3833 Electrical appliances and housewares	..	..	..	..	..			..		..
3839 Other electrical apparatus and supplies	..	..	..	..	..			..		..
384 TRANSPORT EQUIPMENT	0	1,191	4,903	7,974	0			0		14,068
3841 Shipbuilding	0	348	1,735	731	0			0		2,814
3842 Railroad equipment	0	181	13	9	0			0		203
3843 Motor vehicles	0	645	3,011	7,129	0			0		10,785
3844 Motorcycles and bicycles	0	15	78	86	0			0		179
3845 Aircraft	0	2	66	19	0			0		87
3849 Other transport equipment	..	..	..	..	..			..		..
385 PROFESSIONAL AND SCIENTIFIC EQUIPMENT	0	23	532	773	0			0		1,328
3851 Professional equipment	..	..	..	..	..			..		..
3852 Photographic and optical goods	..	..	..	..	..			..		..
3853 Watches and clocks	..	..	..	..	..			..		..
39 OTHER MANUFACTURING INDUSTRIES	41,375	3,158	50,745	13,477	3,142			1,211,463		116,258
3901 Jewellery and related articles	..	..	..	..	..			..		..
3902 Musical instruments	..	..	..	..	..			..		..
3903 Sporting and athletic goods	..	..	..	..	..			..		..
3909 Other manufactures	..	..	..	..	..			..		..
SERR non-specified, unallocated industry	..	..	..	..	..			..		..
3 TOTAL MANUFACTURING	697,487	39,235	107,131	440,499	29,558			92,045,990		1,645,276

ISIC Revision 2 Industry Sector		Solid TJ	LPG TJ	Distiloil TJ	RFO TJ	Gas TJ	Biomass TJ	Steam TJ	Electr MWh	Own Use MWh	TOTAL TJ
31	FOOD, BEVERAGES AND TOBACCO	1,426	1,661	4,415	47,050	1,365	..	..	5,488,693	..	75,676
311/2	FOOD	0	1,548	3,366	28,885	0	..	..	0	..	33,799
3111	Slaughtering, preparing and preserving meat	..	..	..	..	..	..	..	..	..	..
3112	Dairy products	..	..	..	..	..	..	..	..	..	..
3113	Canning, preserving of fruits and vegetables	..	..	..	..	..	..	..	..	..	..
3114	Canning, preserving and processing of fish	..	..	..	..	..	..	..	..	..	..
3115	Vegetable and animal oils and fats	..	..	..	..	..	..	..	..	..	..
3116	Grain mill products	..	..	..	..	..	..	..	..	..	..
3117	Bakery products	..	..	..	..	..	..	..	..	..	..
3118	Sugar factories and refineries	..	..	..	..	..	..	..	..	..	..
3119	Cocoa, chocolate and sugar confectionery	..	..	..	..	..	..	..	..	..	..
3121	Other food products	..	..	..	..	..	..	..	..	..	..
3122	Prepared animal feeds	..	..	..	..	..	..	..	..	..	..
313	BEVERAGES	0	113	1,039	17,263	0	..	..	0	..	18,415
3131	Distilling, rectifying and blending of spirits	..	..	..	..	..	..	..	..	..	..
3132	Wine industries	..	..	..	..	..	..	..	..	..	..
3133	Malt liquors and malts	..	..	..	..	..	..	..	..	..	..
3134	Soft drinks	..	..	..	..	..	..	..	..	..	..
314	TOBACCO	0	0	10	902	0	..	..	0	..	912
32	TEXTILES, APPAREL AND LEATHER	3,996	153	2,051	54,586	3,201	..	..	12,846,272	..	110,234
321	TEXTILES	0	141	1,864	52,356	0	..	..	0	..	54,361
3211	Spinning weaving and finishing textiles	..	..	..	..	..	..	..	..	..	..
3212	Made-up goods excluding wearing apparel	..	..	..	..	..	..	..	..	..	..
3213	Knitting mills	..	..	..	..	..	..	..	..	..	..
3214	Carpets and rugs	..	..	..	..	..	..	..	..	..	..
3215	Cordage, rope and twine	..	..	..	..	..	..	..	..	..	..
3219	Other textiles	..	..	..	..	..	..	..	..	..	..
322	WEARING APPAREL, EXCEPT FOOTWEAR	..	..	..	..	..	..	..	..	..	..
323	LEATHER AND FUR PRODUCTS	0	12	187	2,230	0	..	..	0	..	2,429
3231	Tanneries and leather finishing	..	..	..	..	..	..	..	..	..	..
3232	Fur dressing and dyeing industries	..	..	..	..	..	..	..	..	..	..
3233	Leather prods. ex. footwear and wearing apparel	..	..	..	..	..	..	..	..	..	..
324	FOOTWEAR, EX. RUBBER AND PLASTIC	..	..	..	..	..	..	..	..	..	..
33	WOOD PRODUCTS AND FURNITURE	52	84	186	3,336	90	..	..	883,400	..	6,928
331	WOOD PRODUCTS, EXCEPT FURNITURE	..	..	..	..	..	..	..	..	..	..
3311	Sawmills, planing and other wood mills	..	..	..	..	..	..	..	..	..	..
3312	Wooden and cane containers	..	..	..	..	..	..	..	..	..	..
3319	Other wood and cork products	..	..	..	..	..	..	..	..	..	..
332	FURNITURE, FIXTURES, EXCL. METALLIC	..	..	..	..	..	..	..	..	..	..
34	PAPER, PUBLISHING AND PRINTING	259	349	731	54,453	853	..	..	7,425,100	..	83,375
341	PAPER AND PRODUCTS	0	325	640	54,216	0	..	..	0	..	55,181
3411	Pulp, paper and paperboard articles	..	..	..	..	..	..	..	..	..	..
3412	Containers of paper and paperboard	..	..	..	..	..	..	..	..	..	..
3419	Other pulp, paper and paperboard articles	..	..	..	..	..	..	..	..	..	..
342	PRINTING AND PUBLISHING	0	24	91	237	0	..	..	0	..	352
35	CHEMICAL PRODUCTS	13,801	19,413	15,843	146,576	2,752	..	..	20,941,999	..	273,776
351	INDUSTRIAL CHEMICALS	0	16,489	4,738	116,803	0	..	..	0	..	138,030
3511	Basic industrial chemicals excl. fertilizers	0	299	1,713	87,390	0	..	..	0	..	89,402
3512	Fertilizers and pesticides	0	5,709	114	11,768	0	..	..	0	..	17,591
3513	Synthetic resins and plastic materials	0	10,481	2,911	17,645	0	..	..	0	..	31,037
352	OTHER CHEMICALS	0	2,758	10,436	20,870	0	..	..	0	..	34,064
3521	Paints, varnishes and lacquers	..	..	..	..	..	..	..	..	..	..
3522	Drugs and medicines	..	..	..	..	..	..	..	..	..	..
3523	Soap, cleaning preparations, perfumes, cosmetics	..	..	..	..	..	..	..	..	..	..
3529	Other chemical products	..	..	..	..	..	..	..	..	..	..
353	PETROLEUM REFINERIES	0	34	350	1,158	0	..	..	0	..	1,542
354	MISC. PETROLEUM AND COAL PRODUCTS	..	..	..	..	..	..	..	..	..	..
355	RUBBER PRODUCTS	0	81	252	6,291	0	..	..	0	..	6,624
3551	Tyres and tubes	..	..	..	..	..	..	..	..	..	..
3559	Other rubber products	..	..	..	..	..	..	..	..	..	..
356	PLASTIC PRODUCTS	0	51	67	1,454	0	..	..	0	..	1,572
36	NON-METALLIC MINERAL PRODUCTS	175,419	5,720	10,014	44,224	4,036	..	..	9,588,781	..	273,933
361	POTTERY, CHINA, EARTHENWARE	0	3,437	2,288	4,729	0	..	..	0	..	10,454
362	GLASS AND PRODUCTS	0	1,523	766	17,240	0	..	..	0	..	19,529
369	OTHER NON-METAL. MINERAL PRODUCTS	0	760	6,960	22,255	0	..	..	0	..	29,975
3691	Structural clay products	0	436	1,812	5,301	0	..	..	0	..	7,549
3692	Cement, lime and plaster	0	172	3,097	12,469	0	..	..	0	..	15,738
3699	Other non-metallic mineral products	0	152	2,051	4,485	0	..	..	0	..	6,688
37	BASIC METAL INDUSTRIES	517,587	3,559	7,236	52,311	0	..	..	18,087,553	..	645,808
371	IRON AND STEEL	0	2,683	5,675	48,599	0	..	..	0	..	56,957
372	NON-FERROUS METALS	0	876	1,561	3,712	0	..	..	0	..	6,149
38	METAL PRODUCTS, MACHINERY, EQUIP.	0	5,447	14,187	34,118	14,828	..	..	25,267,900	..	159,544
381	METAL PRODUCTS	0	1,368	2,066	3,760	0	..	..	0	..	7,194
3811	Cutlery, hand tools and general hardware	..	..	..	..	..	..	..	..	..	..
3812	Furniture and fixtures primarily of metal	..	..	..	..	..	..	..	..	..	..
3813	Structural metal products	..	..	..	..	..	..	..	..	..	..
3819	Other fabricated metal products	..	..	..	..	..	..	..	..	..	..
382	NON-ELECTRICAL MACHINERY	0	435	2,068	3,145	0	..	..	0	..	5,648
3821	Engines and turbines	..	..	..	..	..	..	..	..	..	..
3822	Agricultural machinery and equipment	..	..	..	..	..	..	..	..	..	..
3823	Metal and wood working machinery	..	..	..	..	..	..	..	..	..	..
3824	Special industrial machinery	..	..	..	..	..	..	..	..	..	..
3825	Office, computing and accounting machinery	..	..	..	..	..	..	..	..	..	..
3829	Other non-electrical machinery and equipment	..	..	..	..	..	..	..	..	..	..
383	ELECTRICAL MACHINERY	0	2,471	3,759	18,668	0	..	..	0	..	24,898
3831	Electrical industrial machinery	..	..	..	..	..	..	..	..	..	..
3832	Radio, TV and communications equipment	..	..	..	..	..	..	..	..	..	..
3833	Electrical appliances and housewares	..	..	..	..	..	..	..	..	..	..
3839	Other electrical apparatus and supplies	..	..	..	..	..	..	..	..	..	..
384	TRANSPORT EQUIPMENT	0	1,153	5,670	8,142	0	..	..	0	..	14,965
3841	Shipbuilding	0	291	1,621	569	0	..	..	0	..	2,481
3842	Railroad equipment	0	166	10	1	0	..	..	0	..	177
3843	Motor vehicles	0	679	3,934	7,529	0	..	..	0	..	12,142
3844	Motorcycles and bicycles	0	15	32	43	0	..	..	0	..	90
3845	Aircraft	0	2	73	0	0	..	..	0	..	75
3849	Other transport equipment	0	20	624	403	0	..	..	0	..	1,047
385	PROFESSIONAL AND SCIENTIFIC EQUIPMENT	..	..	..	..	..	..	..	..	..	..
3851	Professional equipment	..	..	..	..	..	..	..	..	..	..
3852	Photographic and optical goods	..	..	..	..	..	..	..	..	..	..
3853	Watches and clocks	..	..	..	..	..	..	..	..	..	..
39	OTHER MANUFACTURING INDUSTRIES	45,190	3,931	52,194	15,649	4,192	..	..	1,301,400	..	125,841
3901	Jewellery and related articles	..	..	..	..	..	..	..	..	..	..
3902	Musical instruments	..	..	..	..	..	..	..	..	..	..
3903	Sporting and athletic goods	..	..	..	..	..	..	..	..	..	..
3909	Other manufactures	..	..	..	..	..	..	..	..	..	..
SERR	non-specified, unallocated industry	..	..	..	..	..	..	..	..	..	..
3	TOTAL MANUFACTURING	757,730	40,317	106,857	452,303	31,317	..	..	101,831,098	..	1,755,116

ISIS Energy Data Programme (IEA/OECD)

Year: **1997**   **KOREA**

ISIC Revision 2 Industry Sector		Solid TJ	LPG TJ	Distiloil TJ	RFO TJ	Gas TJ	Biomass TJ	Steam TJ	Electr MWh	Own Use MWh	TOTAL TJ
31	**FOOD, BEVERAGES AND TOBACCO**	1,482	1,742	4,342	42,630	1,428	..	..	5,710,833	..	72,183
311/2	FOOD	..	..	..	..	..	..	..	..	..	..
3111	Slaughtering, preparing and preserving meat	..	..	..	..	..	..	..	..	..	..
3112	Dairy products	..	..	..	..	..	..	..	..	..	..
3113	Canning, preserving of fruits and vegetables	..	..	..	..	..	..	..	..	..	..
3114	Canning, preserving and processing of fish	..	..	..	..	..	..	..	..	..	..
3115	Vegetable and animal oils and fats	..	..	..	..	..	..	..	..	..	..
3116	Grain mill products	..	..	..	..	..	..	..	..	..	..
3117	Bakery products	..	..	..	..	..	..	..	..	..	..
3118	Sugar factories and refineries	..	..	..	..	..	..	..	..	..	..
3119	Cocoa, chocolate and sugar confectionery	..	..	..	..	..	..	..	..	..	..
3121	Other food products	..	..	..	..	..	..	..	..	..	..
3122	Prepared animal feeds	..	..	..	..	..	..	..	..	..	..
313	BEVERAGES	..	..	..	..	..	..	..	..	..	..
3131	Distilling, rectifying and blending of spirits	..	..	..	..	..	..	..	..	..	..
3132	Wine industries	..	..	..	..	..	..	..	..	..	..
3133	Malt liquors and malts	..	..	..	..	..	..	..	..	..	..
3134	Soft drinks	..	..	..	..	..	..	..	..	..	..
314	TOBACCO	..	..	..	..	..	..	..	..	..	..
32	**TEXTILES, APPAREL AND LEATHER**	4,023	115	2,084	55,225	4,672	..	..	13,309,722	..	114,034
321	TEXTILES	..	..	..	..	..	..	..	..	..	..
3211	Spinning weaving and finishing textiles	..	..	..	..	..	..	..	..	..	..
3212	Made-up goods excluding wearing apparel	..	..	..	..	..	..	..	..	..	..
3213	Knitting mills	..	..	..	..	..	..	..	..	..	..
3214	Carpets and rugs	..	..	..	..	..	..	..	..	..	..
3215	Cordage, rope and twine	..	..	..	..	..	..	..	..	..	..
3219	Other textiles	..	..	..	..	..	..	..	..	..	..
322	WEARING APPAREL, EXCEPT FOOTWEAR	..	..	..	..	..	..	..	..	..	..
323	LEATHER AND FUR PRODUCTS	..	..	..	..	..	..	..	..	..	..
3231	Tanneries and leather finishing	..	..	..	..	..	..	..	..	..	..
3232	Fur dressing and dyeing industries	..	..	..	..	..	..	..	..	..	..
3233	Leather prods. ex. footwear and wearing apparel	..	..	..	..	..	..	..	..	..	..
324	FOOTWEAR, EX. RUBBER AND PLASTIC	..	..	..	..	..	..	..	..	..	..
33	**WOOD PRODUCTS AND FURNITURE**	0	78	157	3,353	92	..	..	1,006,944	..	7,305
331	WOOD PRODUCTS, EXCEPT FURNITURE	..	..	..	..	..	..	..	..	..	..
3311	Sawmills, planing and other wood mills	..	..	..	..	..	..	..	..	..	..
3312	Wooden and cane containers	..	..	..	..	..	..	..	..	..	..
3319	Other wood and cork products	..	..	..	..	..	..	..	..	..	..
332	FURNITURE, FIXTURES, EXCL. METALLIC	..	..	..	..	..	..	..	..	..	..
34	**PAPER, PUBLISHING AND PRINTING**	289	408	762	54,500	1,098	..	..	7,953,333	..	85,689
341	PAPER AND PRODUCTS	..	..	..	..	..	..	..	..	..	..
3411	Pulp, paper and paperboard articles	..	..	..	..	..	..	..	..	..	..
3412	Containers of paper and paperboard	..	..	..	..	..	..	..	..	..	..
3419	Other pulp, paper and paperboard articles	..	..	..	..	..	..	..	..	..	..
342	PRINTING AND PUBLISHING	..	..	..	..	..	..	..	..	..	..
35	**CHEMICAL PRODUCTS**	7,279	38,173	1,085,068	110,451	3,917	..	..	23,231,111	..	1,328,520
351	INDUSTRIAL CHEMICALS	..	..	..	..	..	..	..	..	..	..
3511	Basic industrial chemicals excl. fertilizers	..	..	..	..	..	..	..	..	..	..
3512	Fertilizers and pesticides	..	..	..	..	..	..	..	..	..	..
3513	Synthetic resins and plastic materials	..	..	..	..	..	..	..	..	..	..
352	OTHER CHEMICALS	..	..	..	..	..	..	..	..	..	..
3521	Paints, varnishes and lacquers	..	..	..	..	..	..	..	..	..	..
3522	Drugs and medicines	..	..	..	..	..	..	..	..	..	..
3523	Soap, cleaning preparations, perfumes, cosmetics	..	..	..	..	..	..	..	..	..	..
3529	Other chemical products	..	..	..	..	..	..	..	..	..	..
353	PETROLEUM REFINERIES	..	..	..	..	..	..	..	..	..	..
354	MISC. PETROLEUM AND COAL PRODUCTS	..	..	..	..	..	..	..	..	..	..
355	RUBBER PRODUCTS	..	..	..	..	..	..	..	..	..	..
3551	Tyres and tubes	..	..	..	..	..	..	..	..	..	..
3559	Other rubber products	..	..	..	..	..	..	..	..	..	..
356	PLASTIC PRODUCTS	..	..	..	..	..	..	..	..	..	..
36	**NON-METALLIC MINERAL PRODUCTS**	178,875	6,520	9,177	40,645	4,969	..	..	9,912,500	..	275,871
361	POTTERY, CHINA, EARTHENWARE	..	..	..	..	..	..	..	..	..	..
362	GLASS AND PRODUCTS	..	..	..	..	..	..	..	..	..	..
369	OTHER NON-METAL. MINERAL PRODUCTS	..	..	..	..	..	..	..	..	..	..
3691	Structural clay products	..	..	..	..	..	..	..	..	..	..
3692	Cement, lime and plaster	..	..	..	..	..	..	..	..	..	..
3699	Other non-metallic mineral products	..	..	..	..	..	..	..	..	..	..
37	**BASIC METAL INDUSTRIES**	537,805	3,222	6,418	30,391	11,866	..	..	20,935,833	..	665,071
371	IRON AND STEEL	..	..	..	..	..	..	..	..	..	..
372	NON-FERROUS METALS	..	..	..	..	..	..	..	..	..	..
38	**METAL PRODUCTS, MACHINERY, EQUIP.**	1,371	6,114	13,428	41,013	16,986	..	..	27,770,278	..	178,885
381	METAL PRODUCTS	..	..	..	..	..	..	..	..	..	..
3811	Cutlery, hand tools and general hardware	..	..	..	..	..	..	..	..	..	..
3812	Furniture and fixtures primarily of metal	..	..	..	..	..	..	..	..	..	..
3813	Structural metal products	..	..	..	..	..	..	..	..	..	..
3819	Other fabricated metal products	..	..	..	..	..	..	..	..	..	..
382	NON-ELECTRICAL MACHINERY	..	..	..	..	..	..	..	..	..	..
3821	Engines and turbines	..	..	..	..	..	..	..	..	..	..
3822	Agricultural machinery and equipment	..	..	..	..	..	..	..	..	..	..
3823	Metal and wood working machinery	..	..	..	..	..	..	..	..	..	..
3824	Special industrial machinery	..	..	..	..	..	..	..	..	..	..
3825	Office, computing and accounting machinery	..	..	..	..	..	..	..	..	..	..
3829	Other non-electrical machinery and equipment	..	..	..	..	..	..	..	..	..	..
383	ELECTRICAL MACHINERY	..	..	..	..	..	..	..	..	..	..
3831	Electrical industrial machinery	..	..	..	..	..	..	..	..	..	..
3832	Radio, TV and communications equipment	..	..	..	..	..	..	..	..	..	..
3833	Electrical appliances and housewares	..	..	..	..	..	..	..	..	..	..
3839	Other electrical apparatus and supplies	..	..	..	..	..	..	..	..	..	..
384	TRANSPORT EQUIPMENT	..	..	..	..	..	..	..	..	..	..
3841	Shipbuilding	..	..	..	..	..	..	..	..	..	..
3842	Railroad equipment	..	..	..	..	..	..	..	..	..	..
3843	Motor vehicles	..	..	..	..	..	..	..	..	..	..
3844	Motorcycles and bicycles	..	..	..	..	..	..	..	..	..	..
3845	Aircraft	..	..	..	..	..	..	..	..	..	..
3849	Other transport equipment	..	..	..	..	..	..	..	..	..	..
385	PROFESSIONAL AND SCIENTIFIC EQUIPMENT	..	..	..	..	..	..	..	..	..	..
3851	Professional equipment	..	..	..	..	..	..	..	..	..	..
3852	Photographic and optical goods	..	..	..	..	..	..	..	..	..	..
3853	Watches and clocks	..	..	..	..	..	..	..	..	..	..
39	**OTHER MANUFACTURING INDUSTRIES**	54,606	3,607	61,490	23,085	6,204	..	..	1,376,668	..	153,948
3901	Jewellery and related articles	..	..	..	..	..	..	..	..	..	..
3902	Musical instruments	..	..	..	..	..	..	..	..	..	..
3903	Sporting and athletic goods	..	..	..	..	..	..	..	..	..	..
3909	Other manufactures	..	..	..	..	..	..	..	..	..	..
SERR	non-specified, unallocated industry	..	..	..	..	..	..	..	..	..	..
3	**TOTAL MANUFACTURING**	785,730	59,979	1,182,926	401,293	51,232	..	..	111,207,222	..	2,881,506

KOREA

ISIC Revision 2 Industry Sector	Solid TJ	LPG TJ	Distiloil TJ	RFO TJ	Gas TJ	Biomass TJ	Steam TJ	Electr MWh	Own Use MWh	TOTAL TJ
31 FOOD, BEVERAGES AND TOBACCO	1,304	1,514	4,843	34,254	2,317	..	..	5,424,722	..	63,761
311/2 FOOD	..	..	..	..	..	..	..	..	..	..
3111 Slaughtering, preparing and preserving meat	..	..	..	..	..	..	..	..	..	..
3112 Dairy products	..	..	..	..	..	..	..	..	..	..
3113 Canning, preserving of fruits and vegetables	..	..	..	..	..	..	..	..	..	..
3114 Canning, preserving and processing of fish	..	..	..	..	..	..	..	..	..	..
3115 Vegetable and animal oils and fats	..	..	..	..	..	..	..	..	..	..
3116 Grain mill products	..	..	..	..	..	..	..	..	..	..
3117 Bakery products	..	..	..	..	..	..	..	..	..	..
3118 Sugar factories and refineries	..	..	..	..	..	..	..	..	..	..
3119 Cocoa, chocolate and sugar confectionery	..	..	..	..	..	..	..	..	..	..
3121 Other food products	..	..	..	..	..	..	..	..	..	..
3122 Prepared animal feeds	..	..	..	..	..	..	..	..	..	..
313 BEVERAGES	..	..	..	..	..	..	..	..	..	..
3131 Distilling, rectifying and blending of spirits	..	..	..	..	..	..	..	..	..	..
3132 Wine industries	..	..	..	..	..	..	..	..	..	..
3133 Malt liquors and malts	..	..	..	..	..	..	..	..	..	..
3134 Soft drinks	..	..	..	..	..	..	..	..	..	..
314 TOBACCO	..	..	..	..	..	..	..	13,128,889	..	121,258
32 TEXTILES, APPAREL AND LEATHER	3,974	255	4,337	57,555	7,873	..	..	..	..	..
321 TEXTILES	..	..	..	..	..	..	..	..	..	..
3211 Spinning weaving and finishing textiles	..	..	..	..	..	..	..	..	..	..
3212 Made-up goods excluding wearing apparel	..	..	..	..	..	..	..	..	..	..
3213 Knitting mills	..	..	..	..	..	..	..	..	..	..
3214 Carpets and rugs	..	..	..	..	..	..	..	..	..	..
3215 Cordage, rope and twine	..	..	..	..	..	..	..	..	..	..
3219 Other textiles	..	..	..	..	..	..	..	..	..	..
322 WEARING APPAREL, EXCEPT FOOTWEAR	..	..	..	..	..	..	..	..	..	..
323 LEATHER AND FUR PRODUCTS	..	..	..	..	..	..	..	..	..	..
3231 Tanneries and leather finishing	..	..	..	..	..	..	..	..	..	..
3232 Fur dressing and dyeing industries	..	..	..	..	..	..	..	..	..	..
3233 Leather prods. ex. footwear and wearing apparel	..	..	..	..	..	..	..	..	..	..
324 FOOTWEAR, EX. RUBBER AND PLASTIC	..	..	..	..	..	..	..	818,611	..	5,596
33 WOOD PRODUCTS AND FURNITURE	0	20	69	2,467	93	..	..	..	..	..
331 WOOD PRODUCTS, EXCEPT FURNITURE	..	..	..	..	..	..	..	..	..	..
3311 Sawmills, planing and other wood mills	..	..	..	..	..	..	..	..	..	..
3312 Wooden and cane containers	..	..	..	..	..	..	..	..	..	..
3319 Other wood and cork products	..	..	..	..	..	..	..	..	..	..
332 FURNITURE, FIXTURES, EXCL. METALLIC	..	..	..	..	..	..	..	7,286,111	..	74,899
34 PAPER, PUBLISHING AND PRINTING	327	363	1,372	45,295	1,312	..	..	..	..	..
341 PAPER AND PRODUCTS	..	..	..	..	..	..	..	..	..	..
3411 Pulp, paper and paperboard articles	..	..	..	..	..	..	..	..	..	..
3412 Containers of paper and paperboard	..	..	..	..	..	..	..	..	..	..
3419 Other pulp, paper and paperboard articles	..	..	..	..	..	..	..	..	..	..
342 PRINTING AND PUBLISHING	..	..	..	..	..	..	..	23,033,889	..	1,376,696
35 CHEMICAL PRODUCTS	6,988	51,842	1,154,845	74,480	5,619	..	..	..	..	..
351 INDUSTRIAL CHEMICALS	..	..	..	..	..	..	..	..	..	..
3511 Basic industrial chemicals excl. fertilizers	..	..	..	..	..	..	..	..	..	..
3512 Fertilizers and pesticides	..	..	..	..	..	..	..	..	..	..
3513 Synthetic resins and plastic materials	..	..	..	..	..	..	..	..	..	..
352 OTHER CHEMICALS	..	..	..	..	..	..	..	..	..	..
3521 Paints, varnishes and lacquers	..	..	..	..	..	..	..	..	..	..
3522 Drugs and medicines	..	..	..	..	..	..	..	..	..	..
3523 Soap, cleaning preparations, perfumes, cosmetics	..	..	..	..	..	..	..	..	..	..
3529 Other chemical products	..	..	..	..	..	..	..	..	..	..
353 PETROLEUM REFINERIES	..	..	..	..	..	..	..	..	..	..
354 MISC. PETROLEUM AND COAL PRODUCTS	..	..	..	..	..	..	..	..	..	..
355 RUBBER PRODUCTS	..	..	..	..	..	..	..	..	..	..
3551 Tyres and tubes	..	..	..	..	..	..	..	..	..	..
3559 Other rubber products	..	..	..	..	..	..	..	..	..	..
356 PLASTIC PRODUCTS	..	..	..	..	..	..	..	7,738,333	..	208,159
36 NON-METALLIC MINERAL PRODUCTS	139,219	4,765	5,712	25,240	5,365	..	..	..	..	..
361 POTTERY, CHINA, EARTHENWARE	..	..	..	..	..	..	..	..	..	..
362 GLASS AND PRODUCTS	..	..	..	..	..	..	..	..	..	..
369 OTHER NON-METAL. MINERAL PRODUCTS	..	..	..	..	..	..	..	..	..	..
3691 Structural clay products	..	..	..	..	..	..	..	..	..	..
3692 Cement, lime and plaster	..	..	..	..	..	..	..	..	..	..
3699 Other non-metallic mineral products	..	..	..	..	..	..	..	20,146,667	..	666,130
37 BASIC METAL INDUSTRIES	548,106	3,401	4,815	24,543	12,737	..	..	..	..	..
371 IRON AND STEEL	..	..	..	..	..	..	..	..	..	..
372 NON-FERROUS METALS	..	..	..	..	..	..	..	25,044,444	..	169,408
38 METAL PRODUCTS, MACHINERY, EQUIP.	360	3,257	7,836	51,957	15,838	..	..	..	..	..
381 METAL PRODUCTS	..	..	..	..	..	..	..	..	..	..
3811 Cutlery, hand tools and general hardware	..	..	..	..	..	..	..	..	..	..
3812 Furniture and fixtures primarily of metal	..	..	..	..	..	..	..	..	..	..
3813 Structural metal products	..	..	..	..	..	..	..	..	..	..
3819 Other fabricated metal products	..	..	..	..	..	..	..	..	..	..
382 NON-ELECTRICAL MACHINERY	..	..	..	..	..	..	..	..	..	..
3821 Engines and turbines	..	..	..	..	..	..	..	..	..	..
3822 Agricultural machinery and equipment	..	..	..	..	..	..	..	..	..	..
3823 Metal and wood working machinery	..	..	..	..	..	..	..	..	..	..
3824 Special industrial machinery	..	..	..	..	..	..	..	..	..	..
3825 Office, computing and accounting machinery	..	..	..	..	..	..	..	..	..	..
3829 Other non-electrical machinery and equipment	..	..	..	..	..	..	..	..	..	..
383 ELECTRICAL MACHINERY	..	..	..	..	..	..	..	..	..	..
3831 Electrical industrial machinery	..	..	..	..	..	..	..	..	..	..
3832 Radio, TV and communications equipment	..	..	..	..	..	..	..	..	..	..
3833 Electrical appliances and housewares	..	..	..	..	..	..	..	..	..	..
3839 Other electrical apparatus and supplies	..	..	..	..	..	..	..	..	..	..
384 TRANSPORT EQUIPMENT	..	..	..	..	..	..	..	..	..	..
3841 Shipbuilding	..	..	..	..	..	..	..	..	..	..
3842 Railroad equipment	..	..	..	..	..	..	..	..	..	..
3843 Motor vehicles	..	..	..	..	..	..	..	..	..	..
3844 Motorcycles and bicycles	..	..	..	..	..	..	..	..	..	..
3845 Aircraft	..	..	..	..	..	..	..	..	..	..
3849 Other transport equipment	..	..	..	..	..	..	..	..	..	..
385 PROFESSIONAL AND SCIENTIFIC EQUIPMENT	..	..	..	..	..	..	..	..	..	..
3851 Professional equipment	..	..	..	..	..	..	..	..	..	..
3852 Photographic and optical goods	..	..	..	..	..	..	..	..	..	..
3853 Watches and clocks	..	..	..	..	..	..	..	1,209,167	..	152,891
39 OTHER MANUFACTURING INDUSTRIES	62,403	6,368	49,647	19,204	10,916	..	..	..	..	..
3901 Jewellery and related articles	..	..	..	..	..	..	..	..	..	..
3902 Musical instruments	..	..	..	..	..	..	..	..	..	..
3903 Sporting and athletic goods	..	..	..	..	..	..	..	..	..	..
3909 Other manufactures	..	..	..	..	..	..	..	..	..	..
SERR non-specified, unallocated industry	..	..	..	..	..	..	..	103,830,833	..	2,838,798
3 TOTAL MANUFACTURING	762,681	71,785	1,233,476	334,995	62,070	..	..	..	..	..

ISIS Energy Data Programme (IEA/OECD)

ISIC Revision 2 Industry Sector		Solid TJ	LPG TJ	Distiloil TJ	RFO TJ	Gas TJ	Biomass TJ	Steam TJ	Electr MWh	Own Use MWh	TOTAL TJ
31	**FOOD, BEVERAGES AND TOBACCO**	0	731	10,222	44,703	5,199	72,813		1,365,116	604,651	136,406
311/2	FOOD	0	0	8,460	36,943	0	72,813		525,581	441,860	118,517
3111	Slaughtering, preparing and preserving meat										
3112	Dairy products	..	..	..	..	..			..	..	..
3113	Canning, preserving of fruits and vegetables										
3114	Canning, preserving and processing of fish										
3115	Vegetable and animal oils and fats										
3116	Grain mill products	..	..	..	..	..					
3117	Bakery products										
3118	Sugar factories and refineries										
3119	Cocoa, chocolate and sugar confectionery	0	0	8,460	36,943	0	72,813		525,581	441,860	118,517
3121	Other food products										
3122	Prepared animal feeds	..	..	..	..	..					
313	BEVERAGES	0	731	1,762	7,728	4,947	0		803,488	162,791	17,475
3131	Distilling, rectifying and blending of spirits										
3132	Wine industries	..		..	..	..					
3133	Malt liquors and malts	0	103	382	6,360	3,459	0		361,628	162,791	11,020
3134	Soft drinks	0	628	1,380	1,368	1,488	0		441,860	0	6,455
314	TOBACCO	0	0	0	32	252	0		36,047	0	414
32	**TEXTILES, APPAREL AND LEATHER**	..		..	..	..			..		..
321	TEXTILES	..	..	..	..	..			..		..
3211	Spinning weaving and finishing textiles										
3212	Made-up goods excluding wearing apparel	..	..	..	..	..					
3213	Knitting mills										
3214	Carpets and rugs	..	..	..	..	..					
3215	Cordage, rope and twine										
3219	Other textiles	..	..	..	..	..			..		..
322	WEARING APPAREL, EXCEPT FOOTWEAR										
323	LEATHER AND FUR PRODUCTS	..	..	..	..	..			..		..
3231	Tanneries and leather finishing										
3232	Fur dressing and dyeing industries	..	..	..	..	..					
3233	Leather prods. ex. footwear and wearing apparel										
324	FOOTWEAR, EX. RUBBER AND PLASTIC	..	..	..	..	..			..		..
33	**WOOD PRODUCTS AND FURNITURE**	..	..	..	..	..			..		..
331	WOOD PRODUCTS, EXCEPT FURNITURE	..	..	..	..	..			..		..
3311	Sawmills, planing and other wood mills										
3312	Wooden and cane containers	..	..	..	..	..					
3319	Other wood and cork products										
332	FURNITURE, FIXTURES, EXCL. METALLIC	..	..	..	..	..			..		..
34	**PAPER, PUBLISHING AND PRINTING**	0	247	517	31,080	14,266	0		3,202,326	627,907	55,378
341	PAPER AND PRODUCTS	0	247	517	31,080	14,266	0		3,202,326	627,907	55,378
3411	Pulp, paper and paperboard articles	0	247	517	31,080	14,266	0		3,202,326	627,907	55,378
3412	Containers of paper and paperboard										
3419	Other pulp, paper and paperboard articles										
342	PRINTING AND PUBLISHING	..	..	..	..	..			..		..
35	**CHEMICAL PRODUCTS**	0	187	1,862	49,495	223,189	0		7,287,209	2,011,628	293,725
351	INDUSTRIAL CHEMICALS	0	183	1,623	37,571	60,787	0		5,448,836	453,488	118,147
3511	Basic industrial chemicals excl. fertilizers	0	183	1,488	35,324	51,235	0		4,919,766	453,488	104,309
3512	Fertilizers and pesticides	0	0	135	2,247	9,552	0		529,070	0	13,839
3513	Synthetic resins and plastic materials										
352	OTHER CHEMICALS	..	..	..	..	..			..		..
3521	Paints, varnishes and lacquers										
3522	Drugs and medicines	..	..	..	..	..					
3523	Soap, cleaning preparations, perfumes, cosmetics										
3529	Other chemical products	..	..	..	..	..					
353	PETROLEUM REFINERIES	..	..	..	..	..			..		..
354	MISC. PETROLEUM AND COAL PRODUCTS	0	0	0	11,224	159,252	0		1,558,140	1,558,140	170,476
355	RUBBER PRODUCTS	0	4	239	700	3,150	0		280,233	0	5,102
3551	Tyres and tubes										
3559	Other rubber products	0	4	239	700	3,150	0		280,233	0	5,102
356	PLASTIC PRODUCTS	..	..	..	..	..			..		..
36	**NON-METALLIC MINERAL PRODUCTS**	1,014	20	1,536	80,906	30,989	0		3,420,930	9,302	126,747
361	POTTERY, CHINA, EARTHENWARE										
362	GLASS AND PRODUCTS	1,014	20	585	6,205	20,574	0		809,302	0	31,311
369	OTHER NON-METAL. MINERAL PRODUCTS	0	0	951	74,701	10,415	0		2,611,628	9,302	95,435
3691	Structural clay products										
3692	Cement, lime and plaster	0	0	951	74,701	10,415	0		2,611,628	9,302	95,435
3699	Other non-metallic mineral products										
37	**BASIC METAL INDUSTRIES**	56,468	1,289	1,269	26,414	81,885	0		10,260,465	895,349	201,039
371	IRON AND STEEL	56,468	1,257	780	26,414	80,280	0		9,082,558	895,349	194,673
372	NON-FERROUS METALS	0	32	489	0	1,605	0		1,177,907	0	6,366
38	**METAL PRODUCTS, MACHINERY, EQUIP.**	0	1,146	255	56	1,745	0		810,465	0	6,120
381	METAL PRODUCTS	..	..	..	..	..			..		..
3811	Cutlery, hand tools and general hardware										
3812	Furniture and fixtures primarily of metal	..	..	..	..	..					
3813	Structural metal products										
3819	Other fabricated metal products	..	..	..	..	..					
382	NON-ELECTRICAL MACHINERY	..	..	..	..	..			..		..
3821	Engines and turbines										
3822	Agricultural machinery and equipment	..	..	..	..	..					
3823	Metal and wood working machinery										
3824	Special industrial machinery	..	..	..	..	..					
3825	Office, computing and accounting machinery										
3829	Other non-electrical machinery and equipment	..	..	..	..	..					
383	ELECTRICAL MACHINERY	..	..	..	..	..			..		..
3831	Electrical industrial machinery										
3832	Radio, TV and communications equipment	..	..	..	..	..					
3833	Electrical appliances and housewares										
3839	Other electrical apparatus and supplies	..	..	..	..	..					
384	TRANSPORT EQUIPMENT	0	1,146	255	56	1,745	0		810,465	0	6,120
3841	Shipbuilding	..		..	..	..					
3842	Railroad equipment										
3843	Motor vehicles	0	1,146	255	56	1,745	0		810,465	0	6,120
3844	Motorcycles and bicycles										
3845	Aircraft	..	..	..	..	..					
3849	Other transport equipment										
385	PROFESSIONAL AND SCIENTIFIC EQUIPMENT	..	..	..	..	..			..		..
3851	Professional equipment										
3852	Photographic and optical goods	..	..	..	..	..					
3853	Watches and clocks										
39	**OTHER MANUFACTURING INDUSTRIES**	0	10,930	20,138	57,900	88,555	0		24,192,778	1,546,512	259,050
3901	Jewellery and related articles										
3902	Musical instruments	..	..	..	..	..					
3903	Sporting and athletic goods										
3909	Other manufactures	0	10,930	20,138	57,900	88,555	0		24,192,778	1,546,512	259,050
SERR	**non-specified, unallocated industry**	..		..	..	..					
3	**TOTAL MANUFACTURING**	57,482	14,550	35,799	290,554	445,828	72,813		50,539,289	5,695,349	1,078,464

ISIC Revision 2 Industry Sector	Solid TJ	LPG TJ	Distiloil TJ	RFO TJ	Gas TJ	Biomass TJ	Steam TJ	Electr MWh	Own Use MWh	TOTAL TJ
31 FOOD, BEVERAGES AND TOBACCO	0	875	1,992	46,815	6,575	79,746	..	1,512,791	650,000	139,109
311/2 FOOD	0	0	56	39,918	0	79,746	..	562,791	488,372	119,988
3111 Slaughtering, preparing and preserving meat	..	..	..	..	..	..	..	..	..	..
3112 Dairy products	..	..	..	..	..	..	..	..	..	..
3113 Canning, preserving of fruits and vegetables	..	..	..	..	..	..	..	..	..	..
3114 Canning, preserving and processing of fish	..	..	..	..	..	..	..	..	..	..
3115 Vegetable and animal oils and fats	..	..	..	..	..	..	..	..	..	..
3116 Grain mill products	..	..	..	..	..	..	..	..	..	..
3117 Bakery products	..	..	..	..	..	..	..	..	..	..
3118 Sugar factories and refineries	0	0	56	39,918	0	79,746	..	562,791	488,372	119,988
3119 Cocoa, chocolate and sugar confectionery	..	..	..	..	..	..	..	..	..	..
3121 Other food products	..	..	..	..	..	..	..	..	..	..
3122 Prepared animal feeds	..	..	..	..	..	..	..	..	..	..
313 BEVERAGES	0	875	1,932	6,837	6,334	0	..	909,302	161,628	18,670
3131 Distilling, rectifying and blending of spirits	..	..	..	..	..	..	..	..	..	..
3132 Wine industries	..	..	..	..	..	..	..	..	..	..
3133 Malt liquors and malts	0	32	103	5,612	4,420	0	..	403,488	161,628	11,038
3134 Soft drinks	0	843	1,829	1,225	1,914	0	..	505,814	0	7,632
314 TOBACCO	0	0	4	60	241	0	..	40,698	..	452
32 TEXTILES, APPAREL AND LEATHER	..	..	..	..	..	..	..	..	..	..
321 TEXTILES	..	..	..	..	..	..	..	..	..	..
3211 Spinning weaving and finishing textiles	..	..	..	..	..	..	..	..	..	..
3212 Made-up goods excluding wearing apparel	..	..	..	..	..	..	..	..	..	..
3213 Knitting mills	..	..	..	..	..	..	..	..	..	..
3214 Carpets and rugs	..	..	..	..	..	..	..	..	..	..
3215 Cordage, rope and twine	..	..	..	..	..	..	..	..	..	..
3219 Other textiles	..	..	..	..	..	..	..	..	..	..
322 WEARING APPAREL, EXCEPT FOOTWEAR	..	..	..	..	..	..	..	..	..	..
323 LEATHER AND FUR PRODUCTS	..	..	..	..	..	..	..	..	..	..
3231 Tanneries and leather finishing	..	..	..	..	..	..	..	..	..	..
3232 Fur dressing and dyeing industries	..	..	..	..	..	..	..	..	..	..
3233 Leather prods. ex. footwear and wearing apparel	..	..	..	..	..	..	..	..	..	..
324 FOOTWEAR, EX. RUBBER AND PLASTIC	..	..	..	..	..	..	..	..	..	..
33 WOOD PRODUCTS AND FURNITURE	..	..	..	..	..	..	..	..	..	..
331 WOOD PRODUCTS, EXCEPT FURNITURE	..	..	..	..	..	..	..	..	..	..
3311 Sawmills, planing and other wood mills	..	..	..	..	..	..	..	..	..	..
3312 Wooden and cane containers	..	..	..	..	..	..	..	..	..	..
3319 Other wood and cork products	..	..	..	..	..	..	..	..	..	..
332 FURNITURE, FIXTURES, EXCL. METALLIC	..	..	..	..	..	..	..	..	..	..
34 PAPER, PUBLISHING AND PRINTING	0	171	994	18,630	24,248	0	..	2,732,558	519,767	52,009
341 PAPER AND PRODUCTS	0	171	994	18,630	24,248	0	..	2,732,558	519,767	52,009
3411 Pulp, paper and paperboard articles	0	171	994	18,630	24,248	0	..	2,732,558	519,767	52,009
3412 Containers of paper and paperboard	..	..	..	..	..	..	..	..	..	..
3419 Other pulp, paper and paperboard articles	..	..	..	..	..	..	..	..	..	..
342 PRINTING AND PUBLISHING	..	..	..	..	..	..	..	..	..	..
35 CHEMICAL PRODUCTS	0	497	4,681	52,009	239,033	0	..	7,043,023	2,363,953	313,065
351 INDUSTRIAL CHEMICALS	0	493	4,128	41,803	58,937	0	..	4,837,209	444,186	121,176
3511 Basic industrial chemicals excl. fertilizers	0	493	3,981	39,043	51,160	0	..	4,561,628	444,186	109,500
3512 Fertilizers and pesticides	0	0	147	2,760	7,777	0	..	275,581	0	11,676
3513 Synthetic resins and plastic materials	..	..	..	..	..	..	..	..	..	..
352 OTHER CHEMICALS	..	..	..	..	..	..	..	..	..	..
3521 Paints, varnishes and lacquers	..	..	..	..	..	..	..	..	..	..
3522 Drugs and medicines	..	..	..	..	..	..	..	..	..	..
3523 Soap, cleaning preparations, perfumes, cosmetics	..	..	..	..	..	..	..	..	..	..
3529 Other chemical products	..	..	..	..	..	..	..	..	..	..
353 PETROLEUM REFINERIES	0	0	0	9,637	176,919	0	..	1,919,767	1,919,767	186,556
354 MISC. PETROLEUM AND COAL PRODUCTS	..	..	..	..	..	..	..	..	..	..
355 RUBBER PRODUCTS	0	4	553	569	3,177	0	..	286,047	0	5,333
3551 Tyres and tubes	..	..	..	..	..	..	..	..	..	..
3559 Other rubber products	0	4	553	569	3,177	0	..	286,047	0	5,333
356 PLASTIC PRODUCTS	..	..	..	..	..	..	..	..	..	..
36 NON-METALLIC MINERAL PRODUCTS	875	72	609	80,449	34,222	0	..	4,097,675	0	130,979
361 POTTERY, CHINA, EARTHENWARE	..	..	..	..	..	..	..	..	..	..
362 GLASS AND PRODUCTS	875	72	609	1,317	25,284	0	..	851,163	0	31,221
369 OTHER NON-METAL. MINERAL PRODUCTS	0	0	0	79,132	8,938	0	..	3,246,512	0	99,757
3691 Structural clay products	..	..	..	..	..	..	..	..	..	..
3692 Cement, lime and plaster	0	0	0	79,132	8,938	0	..	3,246,512	0	99,757
3699 Other non-metallic mineral products	..	..	..	..	..	..	..	..	..	..
37 BASIC METAL INDUSTRIES	50,880	1,022	688	24,668	79,805	0	..	8,463,953	947,674	184,122
371 IRON AND STEEL	50,880	931	680	24,668	77,959	0	..	7,570,930	947,674	178,962
372 NON-FERROUS METALS	0	91	8	0	1,846	0	..	893,023	0	5,160
38 METAL PRODUCTS, MACHINERY, EQUIP.	0	871	414	0	1,748	0	..	663,953	0	5,423
381 METAL PRODUCTS	..	..	..	..	..	..	..	..	..	..
3811 Cutlery, hand tools and general hardware	..	..	..	..	..	..	..	..	..	..
3812 Furniture and fixtures primarily of metal	..	..	..	..	..	..	..	..	..	..
3813 Structural metal products	..	..	..	..	..	..	..	..	..	..
3819 Other fabricated metal products	..	..	..	..	..	..	..	..	..	..
382 NON-ELECTRICAL MACHINERY	..	..	..	..	..	..	..	..	..	..
3821 Engines and turbines	..	..	..	..	..	..	..	..	..	..
3822 Agricultural machinery and equipment	..	..	..	..	..	..	..	..	..	..
3823 Metal and wood working machinery	..	..	..	..	..	..	..	..	..	..
3824 Special industrial machinery	..	..	..	..	..	..	..	..	..	..
3825 Office, computing and accounting machinery	..	..	..	..	..	..	..	..	..	..
3829 Other non-electrical machinery and equipment	..	..	..	..	..	..	..	..	..	..
383 ELECTRICAL MACHINERY	..	..	..	..	..	..	..	..	..	..
3831 Electrical industrial machinery	..	..	..	..	..	..	..	..	..	..
3832 Radio, TV and communications equipment	..	..	..	..	..	..	..	..	..	..
3833 Electrical appliances and housewares	..	..	..	..	..	..	..	..	..	..
3839 Other electrical apparatus and supplies	..	..	..	..	..	..	..	..	..	..
384 TRANSPORT EQUIPMENT	0	871	414	0	1,748	0	..	663,953	0	5,423
3841 Shipbuilding	..	..	..	..	..	..	..	..	..	..
3842 Railroad equipment	..	..	..	..	..	..	..	..	..	..
3843 Motor vehicles	0	871	414	0	1,748	0	..	663,953	0	5,423
3844 Motorcycles and bicycles	..	..	..	..	..	..	..	..	..	..
3845 Aircraft	..	..	..	..	..	..	..	..	..	..
3849 Other transport equipment	..	..	..	..	..	..	..	..	..	..
385 PROFESSIONAL AND SCIENTIFIC EQUIPMENT	..	..	..	..	..	..	..	..	..	..
3851 Professional equipment	..	..	..	..	..	..	..	..	..	..
3852 Photographic and optical goods	..	..	..	..	..	..	..	..	..	..
3853 Watches and clocks	..	..	..	..	..	..	..	..	..	..
39 OTHER MANUFACTURING INDUSTRIES	0	11,730	31,927	42,686	99,735	0	..	28,391,053	717,442	285,703
3901 Jewellery and related articles	..	..	..	..	..	..	..	..	..	..
3902 Musical instruments	..	..	..	..	..	..	..	..	..	..
3903 Sporting and athletic goods	..	..	..	..	..	..	..	..	..	..
3909 Other manufactures	0	11,730	31,927	42,686	99,735	0	..	28,391,053	717,442	285,703
SERR non-specified, unallocated industry	..	..	..	..	..	..	..	..	..	..
3 TOTAL MANUFACTURING	51,755	15,238	41,305	265,257	485,366	79,746	..	52,905,006	5,198,836	1,110,409

113

ISIC Revision 2 Industry Sector	Solid TJ	LPG TJ	Distiloil TJ	RFO TJ	Gas TJ	Biomass TJ	Steam TJ	Electr MWh	Own Use MWh	TOTAL TJ
31 **FOOD, BEVERAGES AND TOBACCO**	0	1,488	4,785	45,912	6,222	69,890		1,529,070	738,372	131,144
311/2 FOOD	0	0	12	40,737	0	69,890		612,791	554,651	110,848
3111 Slaughtering, preparing and preserving meat	..	..	..	..	..	..		..	..	..
3112 Dairy products	..	..	..	..	..	..		..	..	..
3113 Canning, preserving of fruits and vegetables	..	..	..	..	..	..		..	..	..
3114 Canning, preserving and processing of fish	..	..	..	..	..	..		..	..	..
3115 Vegetable and animal oils and fats	..	..	..	..	..	..		..	..	..
3116 Grain mill products	..	..	..	..	..	..		..	..	..
3117 Bakery products	..	..	..	..	..	..		..	..	..
3118 Sugar factories and refineries	0	0	12	40,737		69,890		612,791	554,651	110,848
3119 Cocoa, chocolate and sugar confectionery	..	..	..	..	..	..		..	..	..
3121 Other food products	..	..	..	..	..	..		..	..	..
3122 Prepared animal feeds	..	..	..	..	..	..		..	..	..
313 BEVERAGES	0	1,488	4,769	5,111	5,977	0		874,419	183,721	19,832
3131 Distilling, rectifying and blending of spirits	..	..	..	..	..	..		..	..	..
3132 Wine industries	..	..	..	..	..	..		..	..	..
3133 Malt liquors and malts	0	32	410	4,268	5,434	0		498,838	183,721	11,278
3134 Soft drinks	0	1,456	4,359	843	543	0		375,581	0	8,553
314 TOBACCO	0	0	4	64	245	0		41,860	0	464
32 **TEXTILES, APPAREL AND LEATHER**	..	..	..	..	..	..		..	..	..
321 TEXTILES	..	..	..	..	..	..		..	..	..
3211 Spinning weaving and finishing textiles	..	..	..	..	..	..		..	..	..
3212 Made-up goods excluding wearing apparel	..	..	..	..	..	..		..	..	..
3213 Knitting mills	..	..	..	..	..	..		..	..	..
3214 Carpets and rugs	..	..	..	..	..	..		..	..	..
3215 Cordage, rope and twine	..	..	..	..	..	..		..	..	..
3219 Other textiles	..	..	..	..	..	..		..	..	..
322 WEARING APPAREL, EXCEPT FOOTWEAR	..	..	..	..	..	..		..	..	..
323 LEATHER AND FUR PRODUCTS	..	..	..	..	..	..		..	..	..
3231 Tanneries and leather finishing	..	..	..	..	..	..		..	..	..
3232 Fur dressing and dyeing industries	..	..	..	..	..	..		..	..	..
3233 Leather prods. ex. footwear and wearing apparel	..	..	..	..	..	..		..	..	..
324 FOOTWEAR, EX. RUBBER AND PLASTIC	..	..	..	..	..	..		..	..	..
33 **WOOD PRODUCTS AND FURNITURE**	..	..	..	..	..	..		..	..	..
331 WOOD PRODUCTS, EXCEPT FURNITURE	..	..	..	..	..	..		..	..	..
3311 Sawmills, planing and other wood mills	..	..	..	..	..	..		..	..	..
3312 Wooden and cane containers	..	..	..	..	..	..		..	..	..
3319 Other wood and cork products	..	..	..	..	..	..		..	..	..
332 FURNITURE, FIXTURES, EXCL. METALLIC	..	..	..	..	..	..		..	..	..
34 **PAPER, PUBLISHING AND PRINTING**	0	231	4,160	14,915	19,432	0		3,344,186	616,279	48,558
341 PAPER AND PRODUCTS	0	231	4,160	14,915	19,432	0		3,344,186	616,279	48,558
3411 Pulp, paper and paperboard articles	0	231	4,160	14,915	19,432	0		3,344,186	616,279	48,558
3412 Containers of paper and paperboard	..	..	..	..	..	..		..	..	..
3419 Other pulp, paper and paperboard articles	..	..	..	..	..	..		..	..	..
342 PRINTING AND PUBLISHING	..	..	..	..	..	..		..	..	..
35 **CHEMICAL PRODUCTS**	0	505	4,765	51,024	223,213	0		7,712,791	2,669,768	297,662
351 INDUSTRIAL CHEMICALS	0	501	4,272	40,133	58,919	0		5,368,605	601,163	120,988
3511 Basic industrial chemicals excl. fertilizers	0	501	4,188	38,991	52,769	0		5,132,558	566,279	112,888
3512 Fertilizers and pesticides	0	0	84	1,142	6,150	0		236,047	34,884	8,100
3513 Synthetic resins and plastic materials	..	..	..	..	..	..		..	..	..
352 OTHER CHEMICALS	..	..	..	..	..	..		..	..	..
3521 Paints, varnishes and lacquers	..	..	..	..	..	..		..	..	..
3522 Drugs and medicines	..	..	..	..	..	..		..	..	..
3523 Soap, cleaning preparations, perfumes, cosmetics	..	..	..	..	..	..		..	..	..
3529 Other chemical products	..	..	..	..	..	..		..	..	..
353 PETROLEUM REFINERIES	0	0	0	10,302	161,151	0		2,068,605	2,068,605	171,453
354 MISC. PETROLEUM AND COAL PRODUCTS	..	..	..	..	..	..		..	..	..
355 RUBBER PRODUCTS	0	4	493	589	3,143	0		275,581	0	5,221
3551 Tyres and tubes	..	..	..	..	..	..		..	..	..
3559 Other rubber products	0	4	493	589	3,143	0		275,581	0	5,221
356 PLASTIC PRODUCTS	..	..	..	..	..	..		..	..	..
36 **NON-METALLIC MINERAL PRODUCTS**	887	80	1,285	82,986	39,271	0		4,200,000	0	139,629
361 POTTERY, CHINA, EARTHENWARE	..	..	..	..	..	..		..	..	..
362 GLASS AND PRODUCTS	887	80	1,285	1,372	25,928	0		879,070	0	32,717
369 OTHER NON-METAL. MINERAL PRODUCTS	0	0	0	81,614	13,343	0		3,320,930	0	106,912
3691 Structural clay products	..	..	..	..	..	..		..	..	..
3692 Cement, lime and plaster	0	0	0	81,614	13,343	0		3,320,930	0	106,912
3699 Other non-metallic mineral products	..	..	..	..	..	..		..	..	..
37 **BASIC METAL INDUSTRIES**	58,461	465	883	13,969	79,225	0		7,795,349	912,791	177,780
371 IRON AND STEEL	58,461	338	879	13,969	76,602	0		7,574,419	912,791	174,231
372 NON-FERROUS METALS	0	127	4	0	2,623	0		220,930	0	3,549
38 **METAL PRODUCTS, MACHINERY, EQUIP.**	0	951	314	0	1,470	0		874,419	0	5,883
381 METAL PRODUCTS	..	..	..	..	..	..		..	..	..
3811 Cutlery, hand tools and general hardware	..	..	..	..	..	..		..	..	..
3812 Furniture and fixtures primarily of metal	..	..	..	..	..	..		..	..	..
3813 Structural metal products	..	..	..	..	..	..		..	..	..
3819 Other fabricated metal products	..	..	..	..	..	..		..	..	..
382 NON-ELECTRICAL MACHINERY	..	..	..	..	..	..		..	..	..
3821 Engines and turbines	..	..	..	..	..	..		..	..	..
3822 Agricultural machinery and equipment	..	..	..	..	..	..		..	..	..
3823 Metal and wood working machinery	..	..	..	..	..	..		..	..	..
3824 Special industrial machinery	..	..	..	..	..	..		..	..	..
3825 Office, computing and accounting machinery	..	..	..	..	..	..		..	..	..
3829 Other non-electrical machinery and equipment	..	..	..	..	..	..		..	..	..
383 ELECTRICAL MACHINERY	..	..	..	..	..	..		..	..	..
3831 Electrical industrial machinery	..	..	..	..	..	..		..	..	..
3832 Radio, TV and communications equipment	..	..	..	..	..	..		..	..	..
3833 Electrical appliances and housewares	..	..	..	..	..	..		..	..	..
3839 Other electrical apparatus and supplies	..	..	..	..	..	..		..	..	..
384 TRANSPORT EQUIPMENT	0	951	314	0	1,470	0		874,419	0	5,883
3841 Shipbuilding	..	..	..	..	..	..		..	..	..
3842 Railroad equipment	..	..	..	..	..	..		..	..	..
3843 Motor vehicles	0	951	314	0	1,470	0		874,419	0	5,883
3844 Motorcycles and bicycles	..	..	..	..	..	..		..	..	..
3845 Aircraft	..	..	..	..	..	..		..	..	..
3849 Other transport equipment	..	..	..	..	..	..		..	..	..
385 PROFESSIONAL AND SCIENTIFIC EQUIPMENT	..	..	..	..	..	..		..	..	..
3851 Professional equipment	..	..	..	..	..	..		..	..	..
3852 Photographic and optical goods	..	..	..	..	..	..		..	..	..
3853 Watches and clocks	..	..	..	..	..	..		..	..	..
39 **OTHER MANUFACTURING INDUSTRIES**	0	12,716	47,172	42,237	98,114	0		27,726,944	1,062,791	296,230
3901 Jewellery and related articles	..	..	..	..	..	..		..	..	..
3902 Musical instruments	..	..	..	..	..	..		..	..	..
3903 Sporting and athletic goods	..	..	..	..	..	..		..	..	..
3909 Other manufactures	0	12,716	47,172	42,237	98,114	0		27,726,944	1,062,791	296,230
SERR non-specified, unallocated industry										
3 **TOTAL MANUFACTURING**	59,348	16,436	63,364	251,043	466,947	69,890		53,182,759	6,000,001	1,096,886

Year: **1993**   MEXICO

ISIC Revision 2 Industry Sector	Solid TJ	LPG TJ	Distiloil TJ	RFO TJ	Gas TJ	Biomass TJ	Steam TJ	Electr MWh	Own Use MWh	TOTAL TJ
31 **FOOD, BEVERAGES AND TOBACCO**	0	656	3,107	41,000	7,661	80,634	..	1,726,944	647,778	136,943
311/2 FOOD	0	0	32	35,248	0	80,634	..	580,278	460,556	116,345
3111 Slaughtering, preparing and preserving meat	..	..	..	..	..	..	..	..	..	..
3112 Dairy products	..	..	..	..	..	..	..	..	..	..
3113 Canning, preserving of fruits and vegetables	..	..	..	..	..	..	..	..	..	..
3114 Canning, preserving and processing of fish	..	..	..	..	..	..	..	..	..	..
3115 Vegetable and animal oils and fats	..	..	..	..	..	..	..	..	..	..
3116 Grain mill products	..	..	..	..	..	..	..	..	..	..
3117 Bakery products	..	..	..	..	..	..	..	..	..	..
3118 Sugar factories and refineries	0	0	32	35,248	0	80,634	..	580,278	460,556	116,345
3119 Cocoa, chocolate and sugar confectionery	..	..	..	..	..	..	..	..	..	..
3121 Other food products	..	..	..	..	..	..	..	..	..	..
3122 Prepared animal feeds	..	..	..	..	..	..	..	..	..	..
313 BEVERAGES	0	656	3,071	5,692	7,431	0	..	1,108,333	187,222	20,166
3131 Distilling, rectifying and blending of spirits	..	..	..	..	..	..	..	..	..	..
3132 Wine industries	..	..	..	..	..	..	..	..	..	..
3133 Malt liquors and malts	0	32	418	4,364	5,554	0	..	509,444	187,222	11,528
3134 Soft drinks	0	624	2,653	1,328	1,877	0	..	598,889	0	8,638
314 TOBACCO	0	0	4	60	230	0	..	38,333	0	432
32 **TEXTILES, APPAREL AND LEATHER**	..	..	..	..	..	..	..	..	..	..
321 TEXTILES	..	..	..	..	..	..	..	..	..	..
3211 Spinning weaving and finishing textiles	..	..	..	..	..	..	..	..	..	..
3212 Made-up goods excluding wearing apparel	..	..	..	..	..	..	..	..	..	..
3213 Knitting mills	..	..	..	..	..	..	..	..	..	..
3214 Carpets and rugs	..	..	..	..	..	..	..	..	..	..
3215 Cordage, rope and twine	..	..	..	..	..	..	..	..	..	..
3219 Other textiles	..	..	..	..	..	..	..	..	..	..
322 WEARING APPAREL, EXCEPT FOOTWEAR	..	..	..	..	..	..	..	..	..	..
323 LEATHER AND FUR PRODUCTS	..	..	..	..	..	..	..	..	..	..
3231 Tanneries and leather finishing	..	..	..	..	..	..	..	..	..	..
3232 Fur dressing and dyeing industries	..	..	..	..	..	..	..	..	..	..
3233 Leather prods. ex. footwear and wearing apparel	..	..	..	..	..	..	..	..	..	..
324 FOOTWEAR, EX. RUBBER AND PLASTIC	..	..	..	..	..	..	..	..	..	..
33 **WOOD PRODUCTS AND FURNITURE**	..	..	..	..	..	..	..	..	..	..
331 WOOD PRODUCTS, EXCEPT FURNITURE	..	..	..	..	..	..	..	..	..	..
3311 Sawmills, planing and other wood mills	..	..	..	..	..	..	..	..	..	..
3312 Wooden and cane containers	..	..	..	..	..	..	..	..	..	..
3319 Other wood and cork products	..	..	..	..	..	..	..	..	..	..
332 FURNITURE, FIXTURES, EXCL. METALLIC	..	..	..	..	..	..	..	..	..	..
34 **PAPER, PUBLISHING AND PRINTING**	0	227	4,069	14,589	19,006	0	..	3,322,778	653,611	47,500
341 PAPER AND PRODUCTS	0	227	4,069	14,589	19,006	0	..	3,322,778	653,611	47,500
3411 Pulp, paper and paperboard articles	0	227	4,069	14,589	19,006	0	..	3,322,778	653,611	47,500
3412 Containers of paper and paperboard	..	..	..	..	..	..	..	..	..	..
3419 Other pulp, paper and paperboard articles	..	..	..	..	..	..	..	..	..	..
342 PRINTING AND PUBLISHING	..	..	..	..	..	..	..	..	..	..
35 **CHEMICAL PRODUCTS**	0	513	5,059	51,922	208,339	0	..	7,830,555	2,527,222	284,925
351 INDUSTRIAL CHEMICALS	0	509	4,403	51,389	206,395	0	..	7,573,611	2,527,222	280,863
3511 Basic industrial chemicals excl. fertilizers	0	509	4,252	48,740	197,563	0	..	7,039,722	2,403,889	267,753
3512 Fertilizers and pesticides	0	0	151	2,649	8,832	0	..	533,889	123,333	13,110
3513 Synthetic resins and plastic materials	..	..	..	..	..	..	..	..	..	..
352 OTHER CHEMICALS	..	..	..	..	..	..	..	..	..	..
3521 Paints, varnishes and lacquers	..	..	..	..	..	..	..	..	..	..
3522 Drugs and medicines	..	..	..	..	..	..	..	..	..	..
3523 Soap, cleaning preparations, perfumes, cosmetics	..	..	..	..	..	..	..	..	..	..
3529 Other chemical products	..	..	..	..	..	..	..	..	..	..
353 PETROLEUM REFINERIES	..	..	..	..	..	..	..	..	..	..
354 MISC. PETROLEUM AND COAL PRODUCTS	..	..	..	..	..	..	..	..	..	..
355 RUBBER PRODUCTS	0	4	656	533	1,944	0	..	256,944	0	4,062
3551 Tyres and tubes	..	..	..	..	..	..	..	..	..	..
3559 Other rubber products	..	..	..	..	..	..	..	..	..	..
356 PLASTIC PRODUCTS	..	..	..	..	..	..	..	..	..	..
36 **NON-METALLIC MINERAL PRODUCTS**	895	84	1,193	82,628	38,216	0	..	4,329,722	0	138,603
361 POTTERY, CHINA, EARTHENWARE	..	..	..	..	..	..	..	..	..	..
362 GLASS AND PRODUCTS	895	84	1,193	1,400	26,448	0	..	906,111	0	33,282
369 OTHER NON-METAL. MINERAL PRODUCTS	0	0	0	81,228	11,768	0	..	3,423,611	0	105,321
3691 Structural clay products	..	..	..	..	..	..	..	..	..	..
3692 Cement, lime and plaster	0	0	0	81,228	11,768	0	..	3,423,611	0	105,321
3699 Other non-metallic mineral products	..	..	..	..	..	..	..	..	..	..
37 **BASIC METAL INDUSTRIES**	58,874	406	871	17,246	82,744	0	..	6,560,556	1,104,722	179,782
371 IRON AND STEEL	58,874	299	871	17,246	79,986	0	..	6,332,500	1,104,722	176,096
372 NON-FERROUS METALS	0	107	0	0	2,758	0	..	228,056	0	3,686
38 **METAL PRODUCTS, MACHINERY, EQUIP.**	0	1,305	310	0	2,042	0	..	908,333	0	6,927
381 METAL PRODUCTS	..	..	..	..	..	..	..	..	..	..
3811 Cutlery, hand tools and general hardware	..	..	..	..	..	..	..	..	..	..
3812 Furniture and fixtures primarily of metal	..	..	..	..	..	..	..	..	..	..
3813 Structural metal products	..	..	..	..	..	..	..	..	..	..
3819 Other fabricated metal products	..	..	..	..	..	..	..	..	..	..
382 NON-ELECTRICAL MACHINERY	..	..	..	..	..	..	..	..	..	..
3821 Engines and turbines	..	..	..	..	..	..	..	..	..	..
3822 Agricultural machinery and equipment	..	..	..	..	..	..	..	..	..	..
3823 Metal and wood working machinery	..	..	..	..	..	..	..	..	..	..
3824 Special industrial machinery	..	..	..	..	..	..	..	..	..	..
3825 Office, computing and accounting machinery	..	..	..	..	..	..	..	..	..	..
3829 Other non-electrical machinery and equipment	..	..	..	..	..	..	..	..	..	..
383 ELECTRICAL MACHINERY	..	..	..	..	..	..	..	..	..	..
3831 Electrical industrial machinery	..	..	..	..	..	..	..	..	..	..
3832 Radio, TV and communications equipment	..	..	..	..	..	..	..	..	..	..
3833 Electrical appliances and housewares	..	..	..	..	..	..	..	..	..	..
3839 Other electrical apparatus and supplies	..	..	..	..	..	..	..	..	..	..
384 TRANSPORT EQUIPMENT	0	1,305	310	0	2,042	0	..	908,333	0	6,927
3841 Shipbuilding	..	..	..	..	..	..	..	..	..	..
3842 Railroad equipment	..	..	..	..	..	..	..	..	..	..
3843 Motor vehicles	0	1,305	310	0	2,042	0	..	908,333	0	6,927
3844 Motorcycles and bicycles	..	..	..	..	..	..	..	..	..	..
3845 Aircraft	..	..	..	..	..	..	..	..	..	..
3849 Other transport equipment	..	..	..	..	..	..	..	..	..	..
385 PROFESSIONAL AND SCIENTIFIC EQUIPMENT	..	..	..	..	..	..	..	..	..	..
3851 Professional equipment	..	..	..	..	..	..	..	..	..	..
3852 Photographic and optical goods	..	..	..	..	..	..	..	..	..	..
3853 Watches and clocks	..	..	..	..	..	..	..	..	..	..
39 **OTHER MANUFACTURING INDUSTRIES**	0	13,587	45,534	40,602	121,352	0	..	30,406,389	755,833	327,817
3901 Jewellery and related articles	..	..	..	..	..	..	..	..	..	..
3902 Musical instruments	..	..	..	..	..	..	..	..	..	..
3903 Sporting and athletic goods	..	..	..	..	..	..	..	..	..	..
3909 Other manufactures	0	13,587	45,534	40,602	121,352	0	..	30,406,389	755,833	327,817
SERR **non-specified, unallocated industry**	..	..	..	..	..	..	..	..	..	..
3 **TOTAL MANUFACTURING**	59,769	16,778	60,143	247,987	479,360	80,634	..	55,085,277	5,689,166	1,122,497

ISIS Energy Data Programme (IEA/OECD)

ISIC Revision 2 Industry Sector	Solid TJ	LPG TJ	Distiloil TJ	RFO TJ	Gas TJ	Biomass TJ	Steam TJ	Electr MWh	Own Use MWh	TOTAL TJ
31 FOOD, BEVERAGES AND TOBACCO	0	696	3,218	35,821	8,463	72,855		1,972,500	770,000	125,382
311/2 FOOD	0	0	24	29,767	0	72,855		687,222	583,889	103,018
3111 Slaughtering, preparing and preserving meat	..	..	..	..	..	..		..	..	..
3112 Dairy products	..	..	..	..	..	..		..	..	..
3113 Canning, preserving of fruits and vegetables	..	..	..	..	..	..		..	..	..
3114 Canning, preserving and processing of fish	..	..	..	..	..	..		..	..	..
3115 Vegetable and animal oils and fats	..	..	..	..	..	..		..	..	..
3116 Grain mill products	..	..	..	..	..	..		..	..	..
3117 Bakery products	..	..	..	..	..	..		..	..	..
3118 Sugar factories and refineries	0	0	24	29,767	0	72,855		687,222	583,889	103,018
3119 Cocoa, chocolate and sugar confectionery	..	..	..	..	..	..		..	..	..
3121 Other food products	..	..	..	..	..	..		..	..	..
3122 Prepared animal feeds	..	..	..	..	..	..		..	..	..
313 BEVERAGES	0	696	3,190	5,994	8,222	0		1,244,445	186,111	21,912
3131 Distilling, rectifying and blending of spirits	..	..	..	..	..	..		..	..	..
3132 Wine industries	..	..	..	..	..	..		..	..	..
3133 Malt liquors and malts	0	36	346	4,590	6,240	0		611,667	186,111	12,744
3134 Soft drinks	0	660	2,844	1,404	1,982	0		632,778	0	9,168
314 TOBACCO	0	0	4	60	241	0		40,833	0	452
32 TEXTILES, APPAREL AND LEATHER	..	..	..	..	..	..		..	..	..
321 TEXTILES	..	..	..	..	..	..		..	..	..
3211 Spinning weaving and finishing textiles	..	..	..	..	..	..		..	..	..
3212 Made-up goods excluding wearing apparel	..	..	..	..	..	..		..	..	..
3213 Knitting mills	..	..	..	..	..	..		..	..	..
3214 Carpets and rugs	..	..	..	..	..	..		..	..	..
3215 Cordage, rope and twine	..	..	..	..	..	..		..	..	..
3219 Other textiles	..	..	..	..	..	..		..	..	..
322 WEARING APPAREL, EXCEPT FOOTWEAR	..	..	..	..	..	..		..	..	..
323 LEATHER AND FUR PRODUCTS	..	..	..	..	..	..		..	..	..
3231 Tanneries and leather finishing	..	..	..	..	..	..		..	..	..
3232 Fur dressing and dyeing industries	..	..	..	..	..	..		..	..	..
3233 Leather prods. ex. footwear and wearing apparel	..	..	..	..	..	..		..	..	..
324 FOOTWEAR, EX. RUBBER AND PLASTIC	..	..	..	..	..	..		..	..	..
33 WOOD PRODUCTS AND FURNITURE	..	..	..	..	..	..		..	..	..
331 WOOD PRODUCTS, EXCEPT FURNITURE	..	..	..	..	..	..		..	..	..
3311 Sawmills, planing and other wood mills	..	..	..	..	..	..		..	..	..
3312 Wooden and cane containers	..	..	..	..	..	..		..	..	..
3319 Other wood and cork products	..	..	..	..	..	..		..	..	..
332 FURNITURE, FIXTURES, EXCL. METALLIC	..	..	..	..	..	..		..	..	..
34 PAPER, PUBLISHING AND PRINTING	0	235	4,212	15,098	19,673	0		3,419,167	657,222	49,161
341 PAPER AND PRODUCTS	0	235	4,212	15,098	19,673	0		3,419,167	657,222	49,161
3411 Pulp, paper and paperboard articles	0	235	4,212	15,098	19,673	0		3,419,167	657,222	49,161
3412 Containers of paper and paperboard	..	..	..	..	..	..		..	..	..
3419 Other pulp, paper and paperboard articles	..	..	..	..	..	..		..	..	..
342 PRINTING AND PUBLISHING	..	..	..	..	..	..		..	..	..
35 CHEMICAL PRODUCTS	0	573	5,656	51,258	252,582	0		8,324,722	2,460,833	331,179
351 INDUSTRIAL CHEMICALS	0	569	4,880	50,749	250,265	0		8,044,444	2,460,833	326,564
3511 Basic industrial chemicals excl. fertilizers	0	569	4,769	46,736	241,655	0		7,520,000	2,318,889	312,453
3512 Fertilizers and pesticides	0	0	111	4,013	8,610	0		524,444	141,944	14,111
3513 Synthetic resins and plastic materials	..	..	..	..	..	..		..	..	..
352 OTHER CHEMICALS	..	..	..	..	..	..		..	..	..
3521 Paints, varnishes and lacquers	..	..	..	..	..	..		..	..	..
3522 Drugs and medicines	..	..	..	..	..	..		..	..	..
3523 Soap, cleaning preparations, perfumes, cosmetics	..	..	..	..	..	..		..	..	..
3529 Other chemical products	..	..	..	..	..	..		..	..	..
353 PETROLEUM REFINERIES	..	..	..	..	..	..		..	..	..
354 MISC. PETROLEUM AND COAL PRODUCTS	..	..	..	..	..	..		..	..	..
355 RUBBER PRODUCTS	0	4	776	509	2,317	0		280,278	0	4,615
3551 Tyres and tubes	..	..	..	..	..	..		..	..	..
3559 Other rubber products	..	..	..	..	..	..		..	..	..
356 PLASTIC PRODUCTS	..	..	..	..	..	..		..	..	..
36 NON-METALLIC MINERAL PRODUCTS	903	84	1,555	83,459	33,774	0		4,863,611	0	137,284
361 POTTERY, CHINA, EARTHENWARE	..	..	..	..	..	..		..	..	..
362 GLASS AND PRODUCTS	903	84	1,555	1,432	23,253	0		1,011,944	0	30,870
369 OTHER NON-METAL. MINERAL PRODUCTS	0	0	0	82,027	10,521	0		3,851,667	0	106,414
3691 Structural clay products	..	..	..	..	..	..		..	..	..
3692 Cement, lime and plaster	0	0	0	82,027	10,521	0		3,851,667	0	106,414
3699 Other non-metallic mineral products	..	..	..	..	..	..		..	..	..
37 BASIC METAL INDUSTRIES	65,664	469	974	19,235	92,665	0		7,381,667	1,265,278	201,026
371 IRON AND STEEL	65,664	334	974	19,235	89,210	0		7,095,556	1,265,278	196,406
372 NON-FERROUS METALS	0	135	0	0	3,455	0		286,111	0	4,620
38 METAL PRODUCTS, MACHINERY, EQUIP.	0	1,309	310	0	2,050	0		911,667	0	6,951
381 METAL PRODUCTS	..	..	..	..	..	..		..	..	..
3811 Cutlery, hand tools and general hardware	..	..	..	..	..	..		..	..	..
3812 Furniture and fixtures primarily of metal	..	..	..	..	..	..		..	..	..
3813 Structural metal products	..	..	..	..	..	..		..	..	..
3819 Other fabricated metal products	..	..	..	..	..	..		..	..	..
382 NON-ELECTRICAL MACHINERY	..	..	..	..	..	..		..	..	..
3821 Engines and turbines	..	..	..	..	..	..		..	..	..
3822 Agricultural machinery and equipment	..	..	..	..	..	..		..	..	..
3823 Metal and wood working machinery	..	..	..	..	..	..		..	..	..
3824 Special industrial machinery	..	..	..	..	..	..		..	..	..
3825 Office, computing and accounting machinery	..	..	..	..	..	..		..	..	..
3829 Other non-electrical machinery and equipment	..	..	..	..	..	..		..	..	..
383 ELECTRICAL MACHINERY	..	..	..	..	..	..		..	..	..
3831 Electrical industrial machinery	..	..	..	..	..	..		..	..	..
3832 Radio, TV and communications equipment	..	..	..	..	..	..		..	..	..
3833 Electrical appliances and housewares	..	..	..	..	..	..		..	..	..
3839 Other electrical apparatus and supplies	..	..	..	..	..	..		..	..	..
384 TRANSPORT EQUIPMENT	0	1,309	310	0	2,050	0		911,667	0	6,951
3841 Shipbuilding	..	..	..	..	..	..		..	..	..
3842 Railroad equipment	..	..	..	..	..	..		..	..	..
3843 Motor vehicles	0	1,309	310	0	2,050	0		911,667	0	6,951
3844 Motorcycles and bicycles	..	..	..	..	..	..		..	..	..
3845 Aircraft	..	..	..	..	..	..		..	..	..
3849 Other transport equipment	..	..	..	..	..	..		..	..	..
385 PROFESSIONAL AND SCIENTIFIC EQUIPMENT	..	..	..	..	..	..		..	..	..
3851 Professional equipment	..	..	..	..	..	..		..	..	..
3852 Photographic and optical goods	..	..	..	..	..	..		..	..	..
3853 Watches and clocks	..	..	..	..	..	..		..	..	..
39 OTHER MANUFACTURING INDUSTRIES	0	14,013	33,960	32,086	111,657	0		33,076,944	760,556	308,055
3901 Jewellery and related articles	..	..	..	..	..	..		..	..	..
3902 Musical instruments	..	..	..	..	..	..		..	..	..
3903 Sporting and athletic goods	..	..	..	..	..	..		..	..	..
3909 Other manufactures	0	14,013	33,960	32,086	111,657	0		33,076,944	760,556	308,055
SERR non-specified, unallocated industry										
3 TOTAL MANUFACTURING	66,567	17,379	49,885	236,957	520,864	72,855		59,950,278	5,913,889	1,159,038

ISIC Revision 2 Industry Sector		Solid TJ	LPG TJ	Distiloil TJ	RFO TJ	Gas TJ	Biomass TJ	Steam TJ	Electr MWh	Own Use MWh	TOTAL TJ
		0	1,062	2,029	42,969	5,735	84,925	..	1,957,223	852,500	140,697
31	**FOOD, BEVERAGES AND TOBACCO**	0	1,062	2,029	42,969	5,735	84,925	..	1,957,223	852,500	140,697
311/2	FOOD	0	0	32	36,991	0	84,925		820,000	694,444	122,400
3111	Slaughtering, preparing and preserving meat	..	..	..	..	..	..		..	..	..
3112	Dairy products	..	..	..	..	..	..		..	..	..
3113	Canning, preserving of fruits and vegetables	..	..	..	..	..	..		..	..	..
3114	Canning, preserving and processing of fish	..	..	..	..	..	..		..	..	..
3115	Vegetable and animal oils and fats	..	..	..	..	..	..		..	..	..
3116	Grain mill products	..	..	..	..	..	..		..	..	..
3117	Bakery products	..	..	..	..	..	..		..	..	..
3118	Sugar factories and refineries	0	0	32	36,991	0	84,925		820,000	694,444	122,400
3119	Cocoa, chocolate and sugar confectionery	..	..	..	..	..	..		..	..	..
3121	Other food products	..	..	..	..	..	..		..	..	..
3122	Prepared animal feeds	..	..	..	..	..	..		..	..	..
313	BEVERAGES	0	1,062	1,993	5,922	5,501	0		1,093,056	158,056	17,844
3131	Distilling, rectifying and blending of spirits	..	..	..	..	..	..		..	..	..
3132	Wine industries	..	..	..	..	..	..		..	..	..
3133	Malt liquors and malts	0	91	36	4,311	4,623	0		505,833	158,056	10,313
3134	Soft drinks	0	971	1,957	1,611	878	0		587,223	0	7,531
314	TOBACCO	0	0	4	56	234	0		44,167	0	453
32	**TEXTILES, APPAREL AND LEATHER**	..	..	..	..	..	..		..	..	..
321	TEXTILES	..	..	..	..	..	..		..	..	..
3211	Spinning weaving and finishing textiles	..	..	..	..	..	..		..	..	..
3212	Made-up goods excluding wearing apparel	..	..	..	..	..	..		..	..	..
3213	Knitting mills	..	..	..	..	..	..		..	..	..
3214	Carpets and rugs	..	..	..	..	..	..		..	..	..
3215	Cordage, rope and twine	..	..	..	..	..	..		..	..	..
3219	Other textiles	..	..	..	..	..	..		..	..	..
322	WEARING APPAREL, EXCEPT FOOTWEAR	..	..	..	..	..	..		..	..	..
323	LEATHER AND FUR PRODUCTS	..	..	..	..	..	..		..	..	..
3231	Tanneries and leather finishing	..	..	..	..	..	..		..	..	..
3232	Fur dressing and dyeing industries	..	..	..	..	..	..		..	..	..
3233	Leather prods. ex. footwear and wearing apparel	..	..	..	..	..	..		..	..	..
324	FOOTWEAR, EX. RUBBER AND PLASTIC	..	..	..	..	..	..		..	..	..
33	**WOOD PRODUCTS AND FURNITURE**	..	..	..	..	..	..		..	..	..
331	WOOD PRODUCTS, EXCEPT FURNITURE	..	..	..	..	..	..		..	..	..
3311	Sawmills, planing and other wood mills	..	..	..	..	..	..		..	..	..
3312	Wooden and cane containers	..	..	..	..	..	..		..	..	..
3319	Other wood and cork products	..	..	..	..	..	..		..	..	..
332	FURNITURE, FIXTURES, EXCL. METALLIC	..	..	..	..	..	..		..	..	..
34	**PAPER, PUBLISHING AND PRINTING**	0	211	3,822	13,710	17,865	0		3,106,389	597,778	44,639
341	PAPER AND PRODUCTS	0	211	3,822	13,710	17,865	0		3,106,389	597,778	44,639
3411	Pulp, paper and paperboard articles	0	211	3,822	13,710	17,865	0		3,106,389	597,778	44,639
3412	Containers of paper and paperboard	..	..	..	..	..	..		..	..	..
3419	Other pulp, paper and paperboard articles	..	..	..	..	..	..		..	..	..
342	PRINTING AND PUBLISHING	..	..	..	..	..	..		..	..	..
35	**CHEMICAL PRODUCTS**	0	533	5,274	47,041	253,828	0		8,543,334	3,065,556	326,396
351	INDUSTRIAL CHEMICALS	0	529	4,550	46,568	251,665	0		8,281,667	3,065,556	322,090
3511	Basic industrial chemicals excl. fertilizers	0	529	4,506	43,450	243,507	0		7,779,167	2,937,500	309,422
3512	Fertilizers and pesticides	0	0	44	3,118	8,158	0		502,500	128,056	12,668
3513	Synthetic resins and plastic materials	..	..	..	..	..	..		..	..	..
352	OTHER CHEMICALS	..	..	..	..	..	..		..	..	..
3521	Paints, varnishes and lacquers	..	..	..	..	..	..		..	..	..
3522	Drugs and medicines	..	..	..	..	..	..		..	..	..
3523	Soap, cleaning preparations, perfumes, cosmetics	..	..	..	..	..	..		..	..	..
3529	Other chemical products	..	..	..	..	..	..		..	..	..
353	PETROLEUM REFINERIES	..	..	..	..	..	..		..	..	..
354	MISC. PETROLEUM AND COAL PRODUCTS	..	..	..	..	..	..		..	..	..
355	RUBBER PRODUCTS	0	4	724	473	2,163	0		261,667	0	4,306
3551	Tyres and tubes	..	..	..	..	..	..		..	..	..
3559	Other rubber products	..	..	..	..	..	..		..	..	..
356	PLASTIC PRODUCTS	..	..	..	..	..	..		..	..	..
36	**NON-METALLIC MINERAL PRODUCTS**	843	4	346	80,058	30,590	0		4,595,000	0	128,383
361	POTTERY, CHINA, EARTHENWARE	..	..	..	..	..	..		..	..	..
362	GLASS AND PRODUCTS	843	4	346	2,593	20,653	0		957,222	0	27,885
369	OTHER NON-METAL. MINERAL PRODUCTS	0	0	0	77,465	9,937	0		3,637,778	0	100,498
3691	Structural clay products	..	..	..	..	..	..		..	..	..
3692	Cement, lime and plaster	0	0	0	77,465	9,937	0		3,637,778	0	100,498
3699	Other non-metallic mineral products	..	..	..	..	..	..		..	..	..
37	**BASIC METAL INDUSTRIES**	73,643	501	1,149	22,735	101,490	0		8,549,445	1,429,444	225,150
371	IRON AND STEEL	73,643	394	1,149	22,735	97,918	0		8,321,389	1,429,444	220,650
372	NON-FERROUS METALS	0	107	0	0	3,572	0		228,056	0	4,500
38	**METAL PRODUCTS, MACHINERY, EQUIP.**	0	489	103	0	1,880	0		750,000	0	5,172
381	METAL PRODUCTS	..	..	..	..	..	..		..	..	..
3811	Cutlery, hand tools and general hardware	..	..	..	..	..	..		..	..	..
3812	Furniture and fixtures primarily of metal	..	..	..	..	..	..		..	..	..
3813	Structural metal products	..	..	..	..	..	..		..	..	..
3819	Other fabricated metal products	..	..	..	..	..	..		..	..	..
382	NON-ELECTRICAL MACHINERY	..	..	..	..	..	..		..	..	..
3821	Engines and turbines	..	..	..	..	..	..		..	..	..
3822	Agricultural machinery and equipment	..	..	..	..	..	..		..	..	..
3823	Metal and wood working machinery	..	..	..	..	..	..		..	..	..
3824	Special industrial machinery	..	..	..	..	..	..		..	..	..
3825	Office, computing and accounting machinery	..	..	..	..	..	..		..	..	..
3829	Other non-electrical machinery and equipment	..	..	..	..	..	..		..	..	..
383	ELECTRICAL MACHINERY	..	..	..	..	..	..		..	..	..
3831	Electrical industrial machinery	..	..	..	..	..	..		..	..	..
3832	Radio, TV and communications equipment	..	..	..	..	..	..		..	..	..
3833	Electrical appliances and housewares	..	..	..	..	..	..		..	..	..
3839	Other electrical apparatus and supplies	..	..	..	..	..	..		..	..	..
384	TRANSPORT EQUIPMENT	0	489	103	0	1,880	0		750,000	0	5,172
3841	Shipbuilding	..	..	..	..	..	..		..	..	..
3842	Railroad equipment	..	..	..	..	..	..		..	..	..
3843	Motor vehicles	0	489	103	0	1,880	0		750,000	0	5,172
3844	Motorcycles and bicycles	..	..	..	..	..	..		..	..	..
3845	Aircraft	..	..	..	..	..	..		..	..	..
3849	Other transport equipment	..	..	..	..	..	..		..	..	..
385	PROFESSIONAL AND SCIENTIFIC EQUIPMENT	..	..	..	..	..	..		..	..	..
3851	Professional equipment	..	..	..	..	..	..		..	..	..
3852	Photographic and optical goods	..	..	..	..	..	..		..	..	..
3853	Watches and clocks	..	..	..	..	..	..		..	..	..
39	**OTHER MANUFACTURING INDUSTRIES**	0	12,998	45,279	6,376	159,433	0		36,175,000	766,389	351,557
3901	Jewellery and related articles	..	..	..	..	..	..		..	..	..
3902	Musical instruments	..	..	..	..	..	..		..	..	..
3903	Sporting and athletic goods	..	..	..	..	..	..		..	..	..
3909	Other manufactures	0	12,998	45,279	6,376	159,433	0		36,175,000	766,389	351,557
SERR	**non-specified, unallocated industry**	..	..	..	..	..	..		..	..	..
3	**TOTAL MANUFACTURING**	74,486	15,798	58,002	212,889	570,821	84,925	..	63,676,391	6,711,667	1,221,994

117

ISIC Revision 2 Industry Sector	Solid TJ	LPG TJ	Distiloil TJ	RFO TJ	Gas TJ	Biomass TJ	Steam TJ	Electr MWh	Own Use MWh	TOTAL TJ
31 FOOD, BEVERAGES AND TOBACCO	16	244	4,183	3,410	0	4	..	2,903,010	..	18,308
311/2 FOOD	16	243	3,696	3,226	0	4	..	2,698,271	..	16,899
3111 Slaughtering, preparing and preserving meat	0	149	685	63	0	4	..	521,986	..	2,780
3112 Dairy products	0	11	781	94	0	0	..	556,156	..	2,888
3113 Canning, preserving of fruits and vegetables	0	1	79	11	0	0	..	66,672	..	331
3114 Canning, preserving and processing of fish	0	17	453	44	0	0	..	421,978	..	2,033
3115 Vegetable and animal oils and fats	16	5	432	2,822	0	0	..	245,575	..	4,159
3116 Grain mill products	0	17	180	0	0	0	..	183,626	..	858
3117 Bakery products	0	27	294	11	0	0	..	280,022	..	1,340
3118 Sugar factories and refineries	..	..	..	..	0	..	..		..	
3119 Cocoa, chocolate and sugar confectionery	0	7	45	47	0	0	..	75,839	..	372
3121 Other food products	0	7	230	134	0	0	..	179,737	..	1,018
3122 Prepared animal feeds	0	2	517	0	0	0	..	166,680	..	1,119
313 BEVERAGES	0	1	425	184	0	0	..	194,738	..	1,311
3131 Distilling, rectifying and blending of spirits	0	0	0	0	0	0	..	14,168	..	51
3132 Wine industries	..	..	..	..	0	..	..		..	
3133 Malt liquors and malts	0	1	311	177	0	0	..	146,401	..	1,016
3134 Soft drinks	0	0	114	7	0	0	..	34,169	..	244
314 TOBACCO	0	0	62	0	0	0	..	10,001	..	98
32 TEXTILES, APPAREL AND LEATHER	0	1	239	99	0	0	..	210,572	..	1,097
321 TEXTILES	0	1	189	99	0	0	..	166,124	..	887
3211 Spinning weaving and finishing textiles	0	1	99	99	0	0	..	81,951	..	494
3212 Made-up goods excluding wearing apparel	0	0	6	0	0	0	..	11,945	..	49
3213 Knitting mills	0	0	36	0	0	0	..	24,446	..	124
3214 Carpets and rugs	0	0	1	0	0	0	..	556	..	3
3215 Cordage, rope and twine	0	0	36	0	0	0	..	24,724	..	125
3219 Other textiles	0	0	11	0	0	0	..	22,502	..	92
322 WEARING APPAREL, EXCEPT FOOTWEAR	0	0	17	0	0	0	..	21,668	..	95
323 LEATHER AND FUR PRODUCTS	0	0	30	0	0	0	..	19,724	..	101
3231 Tanneries and leather finishing	0	0	24	0	0	0	..	16,113	..	82
3232 Fur dressing and dyeing industries	0	0	3	0	0	0	..	1,389	..	8
3233 Leather prods. ex. footwear and wearing apparel	0	0	3	0	0	0	..	2,222	..	11
324 FOOTWEAR, EX. RUBBER AND PLASTIC	0	0	3	0	0	0	..	3,056	..	14
33 WOOD PRODUCTS AND FURNITURE	0	5	670	154	0	4,357	..	50,583,769	..	187,288
331 WOOD PRODUCTS, EXCEPT FURNITURE	0	3	518	153	0	4,320	..	587,547	..	7,109
3311 Sawmills, planing and other wood mills	0	3	454	153	0	3,814	..	561,156	..	6,444
3312 Wooden and cane containers	0	0	15	0	0	0	..	2,500	..	24
3319 Other wood and cork products	0	0	49	0	0	506	..	23,891	..	641
332 FURNITURE, FIXTURES, EXCL. METALLIC	0	2	152	1	0	37	..	49,996,222	..	180,178
34 PAPER, PUBLISHING AND PRINTING	242	115	372	2,385	0	15,472	..	6,829,713	..	43,173
341 PAPER AND PRODUCTS	242	12	243	2,383	0	15,472	..	6,402,734	..	41,402
3411 Pulp, paper and paperboard articles	242	8	152	2,353	0	15,472	..	6,134,102	..	40,310
3412 Containers of paper and paperboard	0	0	90	19	0	0	..	210,572	..	867
3419 Other pulp, paper and paperboard articles	0	4	1	11	0	0	..	58,060	..	225
342 PRINTING AND PUBLISHING	0	103	129	2	0	0	..	426,979	..	1,771
35 CHEMICAL PRODUCTS	0	9,215	1,675	2,194	33,562	0	..	6,268,279	..	69,212
351 INDUSTRIAL CHEMICALS	0	9,101	290	1,321	0	0	..	4,745,102	..	27,794
3511 Basic industrial chemicals excl. fertilizers	0	39	181	906	0	0	..	2,580,206	..	10,415
3512 Fertilizers and pesticides	0	8,965	6	320	0	0	..	1,795,144	..	15,754
3513 Synthetic resins and plastic materials	0	97	103	95	0	0	..	369,752	..	1,626
352 OTHER CHEMICALS	0	58	235	341	0	0	..	323,637	..	1,799
3521 Paints, varnishes and lacquers	0	1	107	0	0	0	..	36,114	..	238
3522 Drugs and medicines	0	0	84	104	0	0	..	165,846	..	785
3523 Soap, cleaning preparations, perfumes, cosmetics	0	56	13	0	0	0	..	18,335	..	135
3529 Other chemical products	0	1	31	237	0	0	..	103,342	..	641
353 PETROLEUM REFINERIES	0	0	96	149	33,562	0	..	471,704	..	35,505
354 MISC. PETROLEUM AND COAL PRODUCTS	0	22	877	359	0	0	..	298,079	..	2,331
355 RUBBER PRODUCTS	0	0	13	1	0	0	..	96,119	..	360
3551 Tyres and tubes	0	0	7	1	0	0	..	78,618	..	291
3559 Other rubber products	0	0	6	0	0	0	..	17,501	..	69
356 PLASTIC PRODUCTS	0	34	164	23	0	0	..	333,638	..	1,422
36 NON-METALLIC MINERAL PRODUCTS	4,255	1,005	918	965	0	0	..	946,742	..	10,551
361 POTTERY, CHINA, EARTHENWARE	0	38	27	0	0	0	..	27,502	..	164
362 GLASS AND PRODUCTS	0	198	79	741	0	0	..	159,179	..	1,591
369 OTHER NON-METAL. MINERAL PRODUCTS	4,255	769	812	224	0	0	..	760,061	..	8,796
3691 Structural clay products	0	66	54	115	0	0	..	21,113	..	311
3692 Cement, lime and plaster	3,621	12	147	91	0	0	..	226,129	..	4,685
3699 Other non-metallic mineral products	634	691	611	18	0	0	..	512,819	..	3,800
37 BASIC METAL INDUSTRIES	14	299	2,143	1,195	371	1	..	23,796,070	..	89,689
371 IRON AND STEEL	14	8	339	501	371	1	..	7,341,420	..	27,663
372 NON-FERROUS METALS	0	291	1,804	694	0	0	..	16,454,650	..	62,026
38 METAL PRODUCTS, MACHINERY, EQUIP.	3	503	1,995	143	0	1	..	2,514,646	..	11,698
381 METAL PRODUCTS	3	257	560	7	0	1	..	745,337	..	3,511
3811 Cutlery, hand tools and general hardware	0	6	17	0	0	0	..	64,172	..	254
3812 Furniture and fixtures primarily of metal	0	0	44	0	0	0	..	31,947	..	159
3813 Structural metal products	3	193	263	6	0	1	..	238,352	..	1,324
3819 Other fabricated metal products	0	58	236	1	0	0	..	410,866	..	1,774
382 NON-ELECTRICAL MACHINERY	0	66	660	15	0	0	..	789,786	..	3,584
3821 Engines and turbines	0	3	17	0	0	0	..	7,778	..	48
3822 Agricultural machinery and equipment	0	18	90	0	0	0	..	47,226	..	278
3823 Metal and wood working machinery	0	0	6	0	0	0	..	9,167	..	39
3824 Special industrial machinery	0	30	352	15	0	0	..	273,911	..	1,383
3825 Office, computing and accounting machinery	0	0	8	0	0	0	..	12,501	..	53
3829 Other non-electrical machinery and equipment	0	15	187	0	0	0	..	439,203	..	1,783
383 ELECTRICAL MACHINERY	0	11	161	2	0	0	..	320,581	..	1,328
3831 Electrical industrial machinery	0	2	78	2	0	0	..	109,731	..	477
3832 Radio, TV and communications equipment	0	0	28	0	0	0	..	75,006	..	298
3833 Electrical appliances and housewares	0	0	14	0	0	0	..	11,390	..	55
3839 Other electrical apparatus and supplies	0	9	41	0	0	0	..	124,454	..	498
384 TRANSPORT EQUIPMENT	0	169	610	119	0	0	..	608,938	..	3,090
3841 Shipbuilding	0	77	390	23	0	0	..	304,747	..	1,587
3842 Railroad equipment	0	11	55	25	0	0	..	30,558	..	201
3843 Motor vehicles	0	39	113	13	0	0	..	172,514	..	786
3844 Motorcycles and bicycles	0	0	15	0	0	0	..	5,278	..	34
3845 Aircraft	0	40	33	58	0	0	..	92,230	..	463
3849 Other transport equipment	0	2	4	0	0	0	..	3,611	..	19
385 PROFESSIONAL AND SCIENTIFIC EQUIPMENT	0	0	4	0	0	0	..	50,004	..	184
3851 Professional equipment	0	0	2	0	0	0	..	45,004	..	164
3852 Photographic and optical goods	0	0	2	0	0	0	..	5,000	..	20
3853 Watches and clocks	..	..	..	..	..	..	..		..	
39 OTHER MANUFACTURING INDUSTRIES	0	4	44	0	0	8	..	57,505	..	263
3901 Jewellery and related articles	0	4	9	0	0	0	..	8,334	..	43
3902 Musical instruments	0	0	0	0	0	0	..	278	..	1
3903 Sporting and athletic goods	0	0	22	0	0	0	..	20,002	..	94
3909 Other manufactures	0	0	13	0	0	8	..	28,891	..	125
SERR non-specified, unallocated industry	..	..	..	..	..	..	..		..	
3 TOTAL MANUFACTURING	4,530	11,391	12,239	10,545	33,933	19,843	..	94,110,306	..	431,278

Year: 1994 — PORTUGAL

ISIC	ISIC Revision 2 Industry Sector	Solid (TJ)	LPG (TJ)	Distiloil (TJ)	RFO (TJ)	Gas (TJ)	Biomass (TJ)	Steam (TJ)	Electr (MWh)	Own Use (MWh)	TOTAL (TJ)
31	FOOD, BEVERAGES AND TOBACCO	0	1,363	1,566	7,688	..	3,777	..	1,123,000	24,833	18,347
311/2	FOOD	0	1,232	1,397	6,168	..	3,777	..	920,000	17,543	15,823
3111	Slaughtering, preparing and preserving meat	0	170	258	385	..	0	..	0	0	813
3112	Dairy products	0	37	460	1,114	..	0	..	0	0	1,611
3113	Canning, preserving of fruits and vegetables	0	5	37	883	..	0	..	0	0	925
3114	Canning, preserving and processing of fish	0	16	48	226	..	0	..	0	0	290
3115	Vegetable and animal oils and fats	0	14	38	642	..	0	..	0	0	694
3116	Grain mill products	0	46	74	233	..	0	..	0	0	353
3117	Bakery products	0	652	67	58	..	0	..	0	0	777
3118	Sugar factories and refineries	0	3	7	1,333	..	0	..	0	0	1,280
3119	Cocoa, chocolate and sugar confectionery	0	1	8	111	..	0	..	0	0	120
3121	Other food products	0	262	224	764	..	0	..	0	0	1,250
3122	Prepared animal feeds	0	26	176	419	..	0	..	0	0	621
313	BEVERAGES	0	102	158	1,436	..	0	..	183,000	7,290	2,329
3131	Distilling, rectifying and blending of spirits	0	0	27	80	..	0	..	0	0	107
3132	Wine industries	0	74	31	39	..	0	..	0	0	144
3133	Malt liquors and malts	0	11	66	1,155	..	0	..	0	7,290	1,206
3134	Soft drinks	0	17	34	162	..	0	..	20,000	0	213
314	TOBACCO	0	29	11	84	..	0	..	0	0	196
32	TEXTILES, APPAREL AND LEATHER	0	548	154	7,848	..	1,344	..	2,534,000	171,959	18,397
321	TEXTILES	0	327	94	6,690	..	1,078	..	2,151,000	162,526	15,348
3211	Spinning weaving and finishing textiles	0	173	60	5,748	..	0	..	0	162,526	5,396
3212	Made-up goods excluding wearing apparel	..	..	..	..	..	..	..	..	..	..
3213	Knitting mills	..	..	..	..	..	..	..	..	..	..
3214	Carpets and rugs	..	..	..	..	..	..	..	..	..	..
3215	Cordage, rope and twine	..	..	..	..	..	..	..	..	..	..
3219	Other textiles	0	154	34	942	..	0	..	316,000	0	1,130
322	WEARING APPAREL, EXCEPT FOOTWEAR	0	189	25	899	..	266	..	67,000	9,433	2,517
323	LEATHER AND FUR PRODUCTS	0	16	20	247	..	0	..	0	0	490
3231	Tanneries and leather finishing	0	0	20	247	..	0	..	0	0	267
3232	Fur dressing and dyeing industries	..	..	..	..	..	..	..	..	..	..
3233	Leather prods. ex. footwear and wearing apparel	0	16	0	0	..	0	..	0	0	16
324	FOOTWEAR, EX. RUBBER AND PLASTIC	0	16	15	12	..	0	..	706,000	99,957	43
33	WOOD PRODUCTS AND FURNITURE	0	115	379	1,661	..	1,242	..	706,000	99,957	5,579
331	WOOD PRODUCTS, EXCEPT FURNITURE	0	100	307	1,642	..	0	..	0	0	2,049
3311	Sawmills, planing and other wood mills	0	100	307	1,642	..	0	..	0	0	1,689
3312	Wooden and cane containers	..	..	..	..	..	..	..	..	..	..
3319	Other wood and cork products	0	15	72	19	..	0	..	0	0	106
332	FURNITURE, FIXTURES, EXCL. METALLIC	0	15	72	19	..	1,242	..	0	0	1,348
34	PAPER, PUBLISHING AND PRINTING	0	359	182	12,583	..	855	..	1,723,000	1,170,450	15,968
341	PAPER AND PRODUCTS	0	272	144	12,565	..	855	..	1,662,000	1,170,450	15,606
3411	Pulp, paper and paperboard articles	0	244	99	10,872	..	0	..	0	1,170,450	7,001
3412	Containers of paper and paperboard	0	26	27	456	..	0	..	0	0	509
3419	Other pulp, paper and paperboard articles	0	2	18	1,237	..	0	..	0	0	1,257
342	PRINTING AND PUBLISHING	0	87	38	18	..	0	..	61,000	0	363
35	CHEMICAL PRODUCTS	461	1,158	442	5,479	..	1,041	..	2,465,000	665,178	15,060
351	INDUSTRIAL CHEMICALS	461	1,049	110	4,344	..	997	..	1,888,000	227,060	12,940
3511	Basic industrial chemicals excl. fertilizers	461	984	70	2,614	..	0	..	0	225,469	3,317
3512	Fertilizers and pesticides	0	16	28	1,236	..	0	..	0	1,591	1,274
3513	Synthetic resins and plastic materials	0	49	12	494	..	0	..	0	0	555
352	OTHER CHEMICALS	0	42	138	684	..	0	..	0	0	864
3521	Paints, varnishes and lacquers	0	3	52	26	..	0	..	0	0	81
3522	Drugs and medicines	0	16	41	151	..	0	..	0	0	208
3523	Soap, cleaning preparations, perfumes, cosmetics	0	12	18	398	..	0	..	0	0	428
3529	Other chemical products	0	11	27	109	..	0	..	442,000	438,118	147
353	PETROLEUM REFINERIES	0	0	0	0	..	0	..	16,000	0	14
354	MISC. PETROLEUM AND COAL PRODUCTS	0	13	128	96	..	0	..	119,000	0	295
355	RUBBER PRODUCTS	0	30	32	218	..	44	..	0	0	752
3551	Tyres and tubes	0	16	15	186	..	0	..	0	0	217
3559	Other rubber products	0	14	17	32	..	0	..	0	0	63
356	PLASTIC PRODUCTS	0	24	34	137	..	0	..	1,760,000	38,634	195
36	NON-METALLIC MINERAL PRODUCTS	20,896	9,551	1,069	9,036	..	12,082	..	439,000	0	58,831
361	POTTERY, CHINA, EARTHENWARE	0	6,876	36	423	..	11,837	..	119,000	38,634	20,752
362	GLASS AND PRODUCTS	0	1,265	34	5,353	..	1	..	1,202,000	0	6,942
369	OTHER NON-METAL. MINERAL PRODUCTS	20,896	1,410	999	3,260	..	244	..	0	0	31,136
3691	Structural clay products	0	1,348	285	2,312	..	244	..	0	0	3,945
3692	Cement, lime and plaster	20,896	28	359	824	..	0	..	0	0	22,351
3699	Other non-metallic mineral products	0	34	355	124	..	0	..	770,000	77,228	513
37	BASIC METAL INDUSTRIES	7,179	946	88	1,419	..	135	..	770,000	77,228	12,261
371	IRON AND STEEL	7,179	735	63	983	..	0	..	629,000	77,228	10,946
372	NON-FERROUS METALS	0	211	25	436	..	135	..	141,000	0	1,315
38	METAL PRODUCTS, MACHINERY, EQUIP.	0	1,619	405	401	..	27	..	1,098,000	10,577	6,367
381	METAL PRODUCTS	0	756	120	101	..	0	..	515,000	5,183	2,812
3811	Cutlery, hand tools and general hardware	0	78	5	0	..	0	..	0	0	83
3812	Furniture and fixtures primarily of metal	0	72	13	2	..	0	..	0	0	87
3813	Structural metal products	0	173	25	15	..	0	..	0	5,183	213
3819	Other fabricated metal products	0	433	77	84	..	0	..	78,000	0	575
382	NON-ELECTRICAL MACHINERY	0	162	51	3	..	0	..	0	0	497
3821	Engines and turbines	0	0	1	1	..	0	..	0	0	1
3822	Agricultural machinery and equipment	0	61	13	0	..	0	..	0	0	75
3823	Metal and wood working machinery	0	39	2	0	..	0	..	0	0	41
3824	Special industrial machinery	0	6	6	0	..	0	..	0	0	12
3825	Office, computing and accounting machinery	0	0	5	2	..	0	..	0	0	7
3829	Other non-electrical machinery and equipment	0	56	24	0	..	0	..	258,000	5,394	80
383	ELECTRICAL MACHINERY	0	203	112	171	..	27	..	0	0	1,422
3831	Electrical industrial machinery	0	56	41	102	..	0	..	0	0	180
3832	Radio, TV and communications equipment	0	31	13	1	..	0	..	0	0	45
3833	Electrical appliances and housewares	0	21	14	13	..	0	..	0	0	48
3839	Other electrical apparatus and supplies	0	95	44	55	..	0	..	238,000	0	194
384	TRANSPORT EQUIPMENT	0	497	117	117	..	0	..	0	0	1,588
3841	Shipbuilding	0	35	29	0	..	0	..	0	0	64
3842	Railroad equipment	0	0	11	0	..	0	..	0	0	11
3843	Motor vehicles	0	299	73	116	..	0	..	0	0	488
3844	Motorcycles and bicycles	0	24	1	1	..	0	..	0	0	26
3845	Aircraft	0	11	0	0	..	0	..	0	0	11
3849	Other transport equipment	0	128	3	0	..	0	..	9,000	0	131
385	PROFESSIONAL AND SCIENTIFIC EQUIPMENT	0	1	5	9	..	0	..	0	0	47
3851	Professional equipment	0	1	5	9	..	0	..	0	0	15
3852	Photographic and optical goods	..	..	..	..	..	..	..	..	..	..
3853	Watches and clocks	..	..	..	..	..	..	..	415,000	0	3,518
39	OTHER MANUFACTURING INDUSTRIES	0	309	1,051	658	..	6	..	..	..	..
3901	Jewellery and related articles	..	..	..	..	..	..	..	..	..	..
3902	Musical instruments	..	..	..	..	..	..	..	..	..	..
3903	Sporting and athletic goods	..	..	..	..	..	0	..	0	0	2,018
3909	Other manufactures	..	309	1,051	658	..	..	..	..	..	..
SERR	non-specified, unallocated industry	..	..	..	..	..	..	..	..	..	..
3	TOTAL MANUFACTURING	28,536	15,968	5,336	46,773	..	20,509	..	12,594,000	2,258,816	154,329

ISIS Energy Data Programme (IEA/OECD)

Year: 1991 SWEDEN

ISIC Revision 2 Industry Sector	Solid TJ	LPG TJ	Distiloil TJ	RFO TJ	Gas TJ	Biomass TJ	Steam TJ	Electr MWh	Own Use MWh	TOTAL TJ
31 FOOD, BEVERAGES AND TOBACCO	600	1,191	3,185	4,201	3,701	954	920	2,600,304	58,751	23,902
311/2 FOOD	600	1,026	2,872	3,606	3,099	954	875	2,315,397	58,751	21,156
3111 Slaughtering, preparing and preserving meat	0	110	718	421	221	802	270	527,529	0	4,441
3112 Dairy products	0	82	461	943	352	78	52	599,553	0	4,126
3113 Canning, preserving of fruits and vegetables	0	64	82	314	745	1	22	144,837	0	1,749
3114 Canning, preserving and processing of fish	0	3	124	3	0	9	0	35,057	0	265
3115 Vegetable and animal oils and fats	0	446	26	254	169	0	336	128,573	398	1,692
3116 Grain mill products	0	2	54	29	8	6	17	84,294	0	419
3117 Bakery products	4	107	937	79	232	39	83	362,232	0	2,785
3118 Sugar factories and refineries	593	0	91	936	1,210	0	0	101,741	58,353	2,986
3119 Cocoa, chocolate and sugar confectionery	0	0	71	379	67	1	24	140,058	0	1,046
3121 Other food products	3	97	202	168	39	3	26	88,159	0	855
3122 Prepared animal feeds	0	115	106	80	56	15	45	103,364	0	789
313 BEVERAGES	0	165	313	567	570	0	10	261,715	0	2,567
3131 Distilling, rectifying and blending of spirits	0	163	33	189	37	0	3	85,470	0	733
3132 Wine industries	..	..	..	..	..	..	..	..	..	..
3133 Malt liquors and malts	0	0	212	344	525	0	7	165,494	0	1,684
3134 Soft drinks	0	2	68	34	8	0	0	10,751	0	151
314 TOBACCO	0	0	0	28	32	0	35	23,192	0	178
32 TEXTILES, APPAREL AND LEATHER	0	794	354	668	116	82	232	532,319	0	4,162
321 TEXTILES	0	785	260	617	116	8	140	478,009	0	3,647
3211 Spinning weaving and finishing textiles	0	413	110	542	109	7	69	233,040	0	2,089
3212 Made-up goods excluding wearing apparel	0	101	65	37	0	1	50	86,332	0	565
3213 Knitting mills	0	76	51	17	5	0	15	30,565	0	274
3214 Carpets and rugs	0	0	7	0	0	0	1	1,788	0	14
3215 Cordage, rope and twine	0	0	6	1	0	0	2	3,707	0	22
3219 Other textiles	0	195	21	20	2	0	3	122,577	0	682
322 WEARING APPAREL, EXCEPT FOOTWEAR	0	9	77	44	0	7	20	31,388	0	270
323 LEATHER AND FUR PRODUCTS	0	0	8	2	0	58	65	13,687	0	182
3231 Tanneries and leather finishing	0	0	2	0	0	58	61	9,597	0	156
3232 Fur dressing and dyeing industries	0	0	0	0	0	0	0	179	0	1
3233 Leather prods. ex. footwear and wearing apparel	0	0	6	2	0	0	4	3,911	0	26
324 FOOTWEAR, EX. RUBBER AND PLASTIC	0	0	9	5	0	9	7	9,235	0	63
33 WOOD PRODUCTS AND FURNITURE	11	16	1,944	1,138	2	26,265	741	1,896,189	9,800	36,908
331 WOOD PRODUCTS, EXCEPT FURNITURE	11	11	1,653	1,111	1	25,367	709	1,686,890	9,800	34,901
3311 Sawmills, planing and other wood mills	11	11	1,537	1,110	1	25,246	700	1,630,028	9,800	34,449
3312 Wooden and cane containers	0	0	21	0	0	13	2	12,127	0	80
3319 Other wood and cork products	0	0	95	1	0	108	7	44,735	0	372
332 FURNITURE, FIXTURES, EXCL. METALLIC	0	5	291	27	1	898	32	209,299	0	2,007
34 PAPER, PUBLISHING AND PRINTING	2,695	1,231	2,338	12,890	936	132,596	3,684	20,020,998	2,571,874	219,187
341 PAPER AND PRODUCTS	2,694	937	1,623	12,739	908	132,468	3,188	19,318,908	2,571,874	214,846
3411 Pulp, paper and paperboard articles	2,630	847	1,352	12,234	732	132,177	3,061	18,673,857	2,571,874	211,000
3412 Containers of paper and paperboard	0	9	111	336	173	10	75	131,987	0	1,189
3419 Other pulp, paper and paperboard articles	64	81	160	169	3	281	52	513,064	0	2,657
342 PRINTING AND PUBLISHING	1	294	715	151	28	128	496	702,090	0	4,341
35 CHEMICAL PRODUCTS	1,262	663	1,886	2,903	2,168	2,195	3,190	6,394,785	141,485	36,779
351 INDUSTRIAL CHEMICALS	1,260	471	604	1,463	1,595	1,800	1,564	4,408,323	44,815	24,466
3511 Basic industrial chemicals excl. fertilizers	0	91	335	548	1,065	27	1,243	3,019,634	37,491	14,045
3512 Fertilizers and pesticides	0	0	45	7	231	2	1	173,900	0	912
3513 Synthetic resins and plastic materials	1,260	380	224	908	299	1,771	320	1,214,789	7,324	9,509
352 OTHER CHEMICALS	0	108	553	592	358	121	1,174	631,615	0	5,180
3521 Paints, varnishes and lacquers	0	0	119	10	51	0	40	52,854	0	410
3522 Drugs and medicines	0	36	82	322	5	0	801	314,824	0	2,379
3523 Soap, cleaning preparations, perfumes, cosmetics	0	0	56	46	10	0	13	57,009	0	330
3529 Other chemical products	0	72	296	214	292	121	320	206,928	0	2,060
353 PETROLEUM REFINERIES	2	0	44	231	0	1	153	561,845	96,670	2,106
354 MISC. PETROLEUM AND COAL PRODUCTS	0	38	306	348	187	1	43	77,150	0	1,201
355 RUBBER PRODUCTS	0	1	113	80	12	4	117	224,125	0	1,134
3551 Tyres and tubes	0	1	45	31	0	3	6	75,740	0	359
3559 Other rubber products	0	0	68	49	12	1	111	148,385	0	775
356 PLASTIC PRODUCTS	0	45	266	189	16	268	139	491,727	0	2,693
36 NON-METALLIC MINERAL PRODUCTS	8,825	3,944	2,015	2,006	684	264	118	1,383,348	0	22,836
361 POTTERY, CHINA, EARTHENWARE	4	260	17	13	0	0	22	89,618	0	639
362 GLASS AND PRODUCTS	3	1,436	341	804	0	21	3	253,951	0	3,522
369 OTHER NON-METAL. MINERAL PRODUCTS	8,818	2,248	1,657	1,189	684	243	93	1,039,779	0	18,675
3691 Structural clay products	0	289	165	38	324	226	9	41,491	0	1,200
3692 Cement, lime and plaster	7,959	514	63	528	8	0	1	310,905	0	10,192
3699 Other non-metallic mineral products	859	1,445	1,429	623	352	17	83	687,383	0	7,283
37 BASIC METAL INDUSTRIES	35,936	6,079	1,583	5,210	4,440	74	480	7,377,783	37,155	80,228
371 IRON AND STEEL	34,179	5,396	1,060	4,674	4,305	26	363	4,705,874	37,155	66,810
372 NON-FERROUS METALS	1,757	683	523	536	135	48	117	2,671,909	0	13,418
38 METAL PRODUCTS, MACHINERY, EQUIP.	39	1,759	7,789	2,654	1,040	818	4,047	6,820,733	0	42,701
381 METAL PRODUCTS	2	436	2,283	335	187	381	691	1,741,161	0	10,583
3811 Cutlery, hand tools and general hardware	0	4	248	36	4	2	160	212,251	0	1,218
3812 Furniture and fixtures primarily of metal	0	26	117	16	1	286	25	59,354	0	685
3813 Structural metal products	0	48	634	21	9	76	85	324,611	0	2,042
3819 Other fabricated metal products	2	358	1,284	262	173	17	421	1,144,945	0	6,639
382 NON-ELECTRICAL MACHINERY	34	251	2,155	558	169	345	1,292	1,691,431	0	10,893
3821 Engines and turbines	0	1	82	34	0	0	7	64,199	0	355
3822 Agricultural machinery and equipment	0	1	57	1	2	24	4	42,056	0	240
3823 Metal and wood working machinery	0	11	148	68	0	1	86	127,188	0	772
3824 Special industrial machinery	19	56	470	178	112	17	287	387,777	0	2,535
3825 Office, computing and accounting machinery	0	18	82	9	0	189	14	87,814	0	628
3829 Other non-electrical machinery and equipment	15	164	1,316	268	55	114	894	982,397	0	6,363
383 ELECTRICAL MACHINERY	3	123	641	415	1	2	508	953,269	0	5,125
3831 Electrical industrial machinery	0	50	170	28	0	0	80	142,077	0	839
3832 Radio, TV and communications equipment	0	2	160	171	0	0	269	379,767	0	1,969
3833 Electrical appliances and housewares	0	13	32	28	0	0	13	46,085	0	252
3839 Other electrical apparatus and supplies	3	58	279	188	1	2	146	385,340	0	2,064
384 TRANSPORT EQUIPMENT	0	948	2,515	1,316	683	90	1,428	2,284,990	0	15,206
3841 Shipbuilding	0	0	185	100	87	17	12	138,966	0	901
3842 Railroad equipment	0	65	218	5	0	0	154	93,110	0	777
3843 Motor vehicles	0	879	1,726	1,007	569	73	1,074	1,764,057	0	11,679
3844 Motorcycles and bicycles	0	0	13	0	27	0	0	12,776	0	86
3845 Aircraft	0	2	344	204	0	0	183	264,931	0	1,687
3849 Other transport equipment	0	2	29	0	0	0	5	11,150	0	76
385 PROFESSIONAL AND SCIENTIFIC EQUIPMENT	0	1	195	30	0	0	128	149,882	0	894
3851 Professional equipment	0	1	176	30	0	0	119	140,759	0	833
3852 Photographic and optical goods	0	0	18	0	0	0	9	8,310	0	57
3853 Watches and clocks	0	0	1	0	0	0	0	813	0	4
39 OTHER MANUFACTURING INDUSTRIES	0	3	73	24	0	12	13	37,620	0	260
3901 Jewellery and related articles	0	0	10	4	0	0	9	6,116	0	45
3902 Musical instruments	0	0	2	2	0	0	0	318	0	5
3903 Sporting and athletic goods	0	3	15	15	0	12	0	10,641	0	83
3909 Other manufactures	0	0	46	3	0	0	4	20,545	0	127
SERR non-specified, unallocated industry	..	..	..	..	..	..	..	..	..	..
3 TOTAL MANUFACTURING	49,368	15,680	21,167	31,694	13,087	163,260	13,425	47,064,079	2,819,065	466,963

Year: **1992**   SWEDEN

ISIC Revision 2 Industry Sector		Solid TJ	LPG TJ	Distiloil TJ	RFO TJ	Gas TJ	Biomass TJ	Steam TJ	Electr MWh	Own Use MWh	TOTAL TJ
31	FOOD, BEVERAGES AND TOBACCO	441	1,455	3,051	3,635	3,947	799	1,094	2,493,693	75,217	23,129
311/2	FOOD	441	1,216	2,728	3,146	3,275	798	1,043	2,241,758	75,217	20,447
3111	Slaughtering, preparing and preserving meat	0	120	736	372	227	747	340	514,362	0	4,394
3112	Dairy products	0	96	392	780	375	20	50	540,929	0	3,660
3113	Canning, preserving of fruits and vegetables	0	60	74	274	744	0	24	141,855	0	1,687
3114	Canning, preserving and processing of fish	0	3	104	2	0	4	3	34,030	0	239
3115	Vegetable and animal oils and fats	0	550	20	75	188	0	380	124,706	3,598	1,649
3116	Grain mill products	0	2	61	0	3	0	20	75,023	0	356
3117	Bakery products	0	121	909	18	230	15	82	353,845	0	2,649
3118	Sugar factories and refineries	441	0	64	1,063	1,315	0	0	112,406	71,619	3,030
3119	Cocoa, chocolate and sugar confectionery	0	4	67	272	48	0	71	150,402	0	1,003
3121	Other food products	0	108	205	176	36	3	23	82,748	0	849
3122	Prepared animal feeds	0	152	96	114	109	9	50	111,452	0	931
313	BEVERAGES	0	239	323	459	642	1	16	228,455	0	2,502
3131	Distilling, rectifying and blending of spirits	0	237	41	185	38	0	3	64,432	0	736
3132	Wine industries	..	..	..	..	..	..	..	..	..	..
3133	Malt liquors and malts	0	0	222	221	601	0	2	148,514	0	1,581
3134	Soft drinks	0	2	60	53	3	1	11	15,509	0	186
314	TOBACCO	0	0	0	30	30	0	35	23,480	0	180
32	TEXTILES, APPAREL AND LEATHER	0	750	312	572	134	17	172	450,610	0	3,579
321	TEXTILES	0	745	226	552	134	1	93	405,502	0	3,211
3211	Spinning weaving and finishing textiles	0	417	103	489	122	0	46	211,083	0	1,937
3212	Made-up goods excluding wearing apparel	0	2	42	2	4	0	27	22,952	0	160
3213	Knitting mills	0	37	53	12	0	1	15	35,801	0	247
3214	Carpets and rugs	0	0	6	1	0	0	1	1,388	0	13
3215	Cordage, rope and twine	0	1	1	1	0	0	2	2,776	0	15
3219	Other textiles	0	288	21	47	8	0	2	131,502	0	839
322	WEARING APPAREL, EXCEPT FOOTWEAR	0	4	65	13	0	1	12	24,028	0	182
323	LEATHER AND FUR PRODUCTS	0	1	12	3	0	0	62	12,172	0	122
3231	Tanneries and leather finishing	0	1	5	2	0	0	60	9,481	0	102
3232	Fur dressing and dyeing industries	..	..	..	..	..	..	..	..	..	..
3233	Leather prods. ex. footwear and wearing apparel	0	0	7	1	0	0	2	2,691	0	20
324	FOOTWEAR, EX. RUBBER AND PLASTIC	0	0	9	4	0	15	5	8,908	0	65
33	WOOD PRODUCTS AND FURNITURE	27	24	1,860	962	2	26,494	792	1,907,927	11,984	36,986
331	WOOD PRODUCTS, EXCEPT FURNITURE	27	22	1,561	900	0	25,601	766	1,700,921	11,984	34,957
3311	Sawmills, planing and other wood mills	23	22	1,466	890	0	25,413	760	1,634,668	11,984	34,416
3312	Wooden and cane containers	4	0	21	10	0	57	3	24,172	0	182
3319	Other wood and cork products	0	0	74	0	0	131	3	42,081	0	359
332	FURNITURE, FIXTURES, EXCL. METALLIC	0	2	299	62	2	893	26	207,006	0	2,029
34	PAPER, PUBLISHING AND PRINTING	2,206	1,035	2,179	12,685	1,365	131,825	3,005	18,945,121	2,945,971	211,897
341	PAPER AND PRODUCTS	2,206	888	1,629	12,597	1,340	131,768	2,720	18,473,010	2,945,971	209,045
3411	Pulp, paper and paperboard articles	2,177	826	1,363	12,124	1,173	131,381	2,675	17,861,989	2,945,971	205,417
3412	Containers of paper and paperboard	7	23	103	368	165	100	5	135,907	0	1,260
3419	Other pulp, paper and paperboard articles	22	39	163	105	2	287	40	475,114	0	2,368
342	PRINTING AND PUBLISHING	0	147	550	88	25	57	285	472,111	0	2,852
35	CHEMICAL PRODUCTS	470	296	2,019	3,271	1,060	4,926	3,181	6,308,824	184,223	37,272
351	INDUSTRIAL CHEMICALS	470	130	717	1,072	436	4,515	1,697	4,363,292	67,526	24,502
3511	Basic industrial chemicals excl. fertilizers	0	46	433	398	75	27	1,347	2,977,808	41,334	12,897
3512	Fertilizers and pesticides	0	0	38	15	76	2	2	162,002	0	716
3513	Synthetic resins and plastic materials	470	84	246	659	285	4,486	348	1,223,482	26,192	10,888
352	OTHER CHEMICALS	0	59	618	736	377	135	1,170	563,527	0	5,124
3521	Paints, varnishes and lacquers	0	0	124	10	51	0	37	45,123	0	384
3522	Drugs and medicines	0	0	164	482	4	2	825	285,566	0	2,505
3523	Soap, cleaning preparations, perfumes, cosmetics	0	0	68	17	8	0	17	42,622	0	263
3529	Other chemical products	0	59	262	227	314	133	291	190,216	0	1,971
353	PETROLEUM REFINERIES	0	0	19	920	0	0	0	609,184	116,697	2,712
354	MISC. PETROLEUM AND COAL PRODUCTS	0	43	264	349	217	13	40	78,515	0	1,209
355	RUBBER PRODUCTS	0	1	111	62	14	0	163	239,073	0	1,212
3551	Tyres and tubes	0	1	50	28	0	0	5	82,976	0	383
3559	Other rubber products	0	0	61	34	14	0	158	156,097	0	829
356	PLASTIC PRODUCTS	0	63	290	132	16	263	111	455,233	0	2,514
36	NON-METALLIC MINERAL PRODUCTS	7,296	3,612	1,561	1,656	542	182	97	1,252,732	0	19,456
361	POTTERY, CHINA, EARTHENWARE	7	257	17	33	0	0	16	74,819	0	599
362	GLASS AND PRODUCTS	1	1,477	231	702	0	16	4	223,040	0	3,234
369	OTHER NON-METAL. MINERAL PRODUCTS	7,288	1,878	1,313	921	542	166	77	954,873	0	15,623
3691	Structural clay products	0	233	148	35	312	100	9	37,781	0	973
3692	Cement, lime and plaster	6,868	472	39	299	7	0	0	292,386	0	8,738
3699	Other non-metallic mineral products	420	1,173	1,126	587	223	66	68	624,706	0	5,912
37	BASIC METAL INDUSTRIES	37,322	6,262	2,214	4,147	4,572	66	624	7,009,377	27,758	80,341
371	IRON AND STEEL	35,660	5,636	1,710	3,750	4,440	5	555	4,641,532	27,758	68,366
372	NON-FERROUS METALS	1,662	626	504	397	132	61	69	2,367,845	0	11,975
38	METAL PRODUCTS, MACHINERY, EQUIP.	5	1,896	7,176	2,157	961	912	4,225	6,383,733	0	40,313
381	METAL PRODUCTS	5	450	1,973	301	162	237	643	1,620,054	0	9,603
3811	Cutlery, hand tools and general hardware	0	17	228	25	3	1	136	200,817	0	1,133
3812	Furniture and fixtures primarily of metal	0	23	90	44	1	107	27	61,147	0	512
3813	Structural metal products	0	36	534	12	19	69	74	283,636	0	1,765
3819	Other fabricated metal products	5	374	1,121	220	139	60	406	1,074,454	0	6,193
382	NON-ELECTRICAL MACHINERY	0	303	2,003	435	182	611	1,700	1,615,704	0	11,051
3821	Engines and turbines	0	1	71	46	0	0	14	65,427	0	368
3822	Agricultural machinery and equipment	0	1	59	0	1	306	2	39,780	0	512
3823	Metal and wood working machinery	0	13	143	37	0	1	93	122,419	0	728
3824	Special industrial machinery	0	89	438	125	125	5	317	311,199	0	2,219
3825	Office, computing and accounting machinery	0	9	60	7	0	188	5	77,986	0	550
3829	Other non-electrical machinery and equipment	0	190	1,232	220	56	111	1,269	998,893	0	6,674
383	ELECTRICAL MACHINERY	0	133	561	302	2	2	414	818,253	0	4,360
3831	Electrical industrial machinery	0	12	115	25	0	0	30	106,323	0	565
3832	Radio, TV and communications equipment	0	2	172	139	0	0	220	319,983	0	1,685
3833	Electrical appliances and housewares	0	22	29	11	0	0	13	42,720	0	229
3839	Other electrical apparatus and supplies	0	97	245	127	2	2	151	349,227	0	1,881
384	TRANSPORT EQUIPMENT	0	1,006	2,441	1,096	615	59	1,328	2,176,474	0	14,380
3841	Shipbuilding	0	1	159	96	70	58	36	145,984	0	946
3842	Railroad equipment	0	74	218	3	0	0	137	86,384	0	743
3843	Motor vehicles	0	928	1,694	869	523	1	873	1,672,499	0	10,909
3844	Motorcycles and bicycles	0	0	15	0	22	0	4	12,632	0	86
3845	Aircraft	0	0	333	128	0	0	272	243,797	0	1,611
3849	Other transport equipment	0	3	22	0	0	0	6	15,178	0	86
385	PROFESSIONAL AND SCIENTIFIC EQUIPMENT	0	4	198	23	0	3	140	153,248	0	920
3851	Professional equipment	0	4	176	23	0	3	125	141,296	0	840
3852	Photographic and optical goods	0	0	18	0	0	0	15	10,382	0	70
3853	Watches and clocks	0	0	4	0	0	0	0	1,570	0	10
39	OTHER MANUFACTURING INDUSTRIES	0	3	75	25	0	12	12	43,696	0	284
3901	Jewellery and related articles	0	1	9	4	0	0	8	6,051	0	44
3902	Musical instruments	0	0	2	1	0	0	0	231	0	4
3903	Sporting and athletic goods	0	2	21	16	0	12	0	8,172	0	80
3909	Other manufactures	0	0	43	4	0	0	4	29,242	0	156
SERR	non-specified, unallocated industry	..	..	..	..	..	..	..	..	..	..
3	TOTAL MANUFACTURING	47,767	15,333	20,447	29,110	12,583	165,233	13,202	44,795,713	3,245,153	453,257

121

Year: **1993**   SWEDEN

ISIC Revision 2 Industry Sector	Solid TJ	LPG TJ	Distiloil TJ	RFO TJ	Gas TJ	Biomass TJ	Steam TJ	Electr MWh	Own Use MWh	TOTAL TJ
31 **FOOD, BEVERAGES AND TOBACCO**	700	1,780	3,039	3,633	4,013	484	1,117	2,462,765	77,576	23,353
311/2 FOOD	700	1,496	2,693	3,046	3,375	484	1,070	2,225,467	77,576	20,596
3111 Slaughtering, preparing and preserving meat	0	166	731	369	290	467	335	544,039	0	4,317
3112 Dairy products	0	78	393	935	334	0	77	468,308	0	3,503
3113 Canning, preserving of fruits and vegetables	0	59	130	218	740	0	36	148,638	0	1,718
3114 Canning, preserving and processing of fish	0	13	101	6	0	0	3	39,779	0	266
3115 Vegetable and animal oils and fats	0	625	19	122	139	0	275	126,628	4,697	1,619
3116 Grain mill products	0	2	55	1	4	0	11	75,976	0	347
3117 Bakery products	0	121	855	32	254	3	92	356,687	0	2,641
3118 Sugar factories and refineries	700	0	61	842	1,367	0	0	120,850	72,879	3,143
3119 Cocoa, chocolate and sugar confectionery	0	6	58	236	52	0	110	127,885	0	922
3121 Other food products	0	131	196	170	38	2	39	99,153	0	933
3122 Prepared animal feeds	0	295	94	115	157	12	92	117,524	0	1,188
313 BEVERAGES	0	284	346	559	612	0	12	215,345	0	2,588
3131 Distilling, rectifying and blending of spirits	0	279	36	223	49	0	3	64,731	0	823
3132 Wine industries	..	..	..	..	..	..	..	..	..	..
3133 Malt liquors and malts	0	0	245	206	109	0	..	..	..	..
3134 Soft drinks	0	5	65	130	454	0	9	73,490	0	825
314 TOBACCO	0	0	0	28	26	0	35	77,124	0	941
								21,953	0	168
32 **TEXTILES, APPAREL AND LEATHER**	0	685	387	389	142	12	180	394,134	0	3,214
321 TEXTILES	0	665	309	364	142	1	90	351,462	0	2,836
3211 Spinning weaving and finishing textiles	0	436	204	275	125	0	61	167,643	0	1,705
3212 Made-up goods excluding wearing apparel	0	5	42	28	5	0	22	25,412	0	193
3213 Knitting mills	0	5	49	16	0	0	3	25,704	0	166
3214 Carpets and rugs	0	0	4	1	0	0	1	1,040	0	10
3215 Cordage, rope and twine	0	0	1	1	0	0	2	2,496	0	13
3219 Other textiles	0	219	9	43	12	0	1	129,167	0	749
322 WEARING APPAREL, EXCEPT FOOTWEAR	0	17	48	15	0	0	10	21,869	0	169
323 LEATHER AND FUR PRODUCTS	0	2	17	3	0	0	73	12,676	0	141
3231 Tanneries and leather finishing	0	1	8	1	0	0	71	9,589	0	116
3232 Fur dressing and dyeing industries	..	..	..	..	..	..	..	..	..	..
3233 Leather prods. ex. footwear and wearing apparel	0	1	9	2	0	0	2	3,087	0	25
324 FOOTWEAR, EX. RUBBER AND PLASTIC	0	1	13	7	0	11	7	8,127	0	68
33 **WOOD PRODUCTS AND FURNITURE**	68	59	1,821	1,104	1	27,098	814	1,994,583	12,906	38,099
331 WOOD PRODUCTS, EXCEPT FURNITURE	68	53	1,572	1,060	1	26,329	777	1,811,981	12,906	36,337
3311 Sawmills, planing and other wood mills	68	53	1,492	1,059	1	26,209	774	1,769,199	12,906	35,979
3312 Wooden and cane containers	0	0	16	0	0	12	0	8,692	0	59
3319 Other wood and cork products	0	0	64	1	0	108	3	34,090	0	299
332 FURNITURE, FIXTURES, EXCL. METALLIC	0	6	249	44	0	769	37	182,602	0	1,762
34 **PAPER, PUBLISHING AND PRINTING**	2,208	1,183	2,324	17,215	2,062	136,491	2,727	19,379,942	3,180,994	222,526
341 PAPER AND PRODUCTS	2,208	1,022	1,827	17,152	2,035	136,489	2,413	18,922,621	3,180,994	219,816
3411 Pulp, paper and paperboard articles	2,188	921	1,565	16,803	1,856	136,197	2,365	18,352,294	3,180,994	216,512
3412 Containers of paper and paperboard	8	53	105	292	175	102	12	153,043	0	1,298
3419 Other pulp, paper and paperboard articles	12	48	157	57	4	190	36	417,284	0	2,006
342 PRINTING AND PUBLISHING	0	161	497	63	27	2	314	457,321	0	2,710
35 **CHEMICAL PRODUCTS**	222	369	1,994	3,451	1,618	4,412	2,849	6,198,593	199,113	36,513
351 INDUSTRIAL CHEMICALS	221	201	780	1,596	955	3,972	1,179	4,241,596	96,976	23,825
3511 Basic industrial chemicals excl. fertilizers	0	187	472	543	657	0	816	2,785,459	60,790	12,484
3512 Fertilizers and pesticides	0	0	36	57	44	2	2	139,657	0	644
3513 Synthetic resins and plastic materials	221	14	272	996	254	3,970	361	1,316,480	36,186	10,697
352 OTHER CHEMICALS	0	72	467	328	407	145	1,348	559,724	0	4,782
3521 Paints, varnishes and lacquers	0	3	117	19	43	4	89	76,447	0	550
3522 Drugs and medicines	0	1	60	82	3	0	949	266,121	0	2,053
3523 Soap, cleaning preparations, perfumes, cosmetics	0	0	60	7	10	0	9	28,331	0	188
3529 Other chemical products	0	68	230	220	351	141	301	188,825	0	1,991
353 PETROLEUM REFINERIES	0	0	42	1,064	0	0	77	598,205	102,137	2,969
354 MISC. PETROLEUM AND COAL PRODUCTS	1	44	340	293	229	13	47	124,711	0	1,416
355 RUBBER PRODUCTS	0	0	104	52	16	0	105	227,803	0	1,097
3551 Tyres and tubes	0	0	40	19	0	0	2	69,058	0	310
3559 Other rubber products	0	0	64	33	16	0	103	158,745	0	787
356 PLASTIC PRODUCTS	0	52	261	118	11	282	93	446,554	0	2,425
36 **NON-METALLIC MINERAL PRODUCTS**	8,113	3,328	1,512	1,363	485	10	65	1,063,917	0	18,706
361 POTTERY, CHINA, EARTHENWARE	8	248	16	36	0	0	19	70,173	0	580
362 GLASS AND PRODUCTS	97	1,636	317	634	7	2	1	206,961	0	3,439
369 OTHER NON-METAL. MINERAL PRODUCTS	8,008	1,444	1,179	693	478	8	45	786,783	0	14,687
3691 Structural clay products	0	205	118	37	247	0	6	34,859	0	738
3692 Cement, lime and plaster	7,416	286	32	402	8	0	0	290,033	0	9,188
3699 Other non-metallic mineral products	592	953	1,029	254	223	8	39	461,891	0	4,761
37 **BASIC METAL INDUSTRIES**	38,669	6,382	2,308	4,774	4,725	64	700	7,141,661	20,099	83,260
371 IRON AND STEEL	36,590	5,763	1,762	4,379	4,608	3	536	4,659,146	20,099	70,342
372 NON-FERROUS METALS	2,079	619	546	395	117	61	164	2,482,515	0	12,918
38 **METAL PRODUCTS, MACHINERY, EQUIP.**	12	1,550	6,953	2,328	829	639	4,158	6,067,928	0	38,314
381 METAL PRODUCTS	2	491	1,839	261	157	75	577	1,478,120	0	8,723
3811 Cutlery, hand tools and general hardware	0	11	175	26	4	0	139	173,605	0	980
3812 Furniture and fixtures primarily of metal	0	24	71	51	1	5	19	56,471	0	374
3813 Structural metal products	0	31	513	29	21	48	63	272,801	0	1,687
3819 Other fabricated metal products	2	425	1,080	155	131	22	356	975,243	0	5,682
382 NON-ELECTRICAL MACHINERY	10	267	2,027	467	181	495	1,612	1,604,269	0	10,834
3821 Engines and turbines	0	1	46	41	0	0	23	62,852	0	337
3822 Agricultural machinery and equipment	0	1	61	24	1	24	0	28,854	0	215
3823 Metal and wood working machinery	0	16	140	34	0	0	106	111,163	0	696
3824 Special industrial machinery	0	88	491	100	116	1	294	296,006	0	2,156
3825 Office, computing and accounting machinery	0	4	72	11	0	214	13	85,784	0	623
3829 Other non-electrical machinery and equipment	10	157	1,217	257	64	256	1,176	1,019,610	0	6,808
383 ELECTRICAL MACHINERY	0	143	518	291	4	3	421	835,248	0	4,387
3831 Electrical industrial machinery	0	15	102	5	1	0	22	106,623	0	529
3832 Radio, TV and communications equipment	0	15	141	158	0	0	227	362,869	0	1,847
3833 Electrical appliances and housewares	0	23	34	12	0	0	14	40,586	0	229
3839 Other electrical apparatus and supplies	0	90	241	116	3	3	158	325,170	0	1,782
384 TRANSPORT EQUIPMENT	0	643	2,364	1,290	487	66	1,394	1,997,220	0	13,434
3841 Shipbuilding	0	2	183	91	71	64	34	116,496	0	864
3842 Railroad equipment	0	63	234	1	5	1	102	85,662	0	714
3843 Motor vehicles	0	574	1,587	1,106	395	1	1,034	1,554,110	0	10,292
3844 Motorcycles and bicycles	0	0	16	0	16	0	4	11,260	0	77
3845 Aircraft	0	3	327	92	0	0	217	216,869	0	1,420
3849 Other transport equipment	0	1	17	0	0	0	3	12,823	0	67
385 PROFESSIONAL AND SCIENTIFIC EQUIPMENT	0	6	205	19	0	0	154	153,071	0	935
3851 Professional equipment	0	6	184	19	0	0	138	140,880	0	854
3852 Photographic and optical goods	0	0	16	0	0	0	16	10,568	0	70
3853 Watches and clocks	0	0	5	0	0	0	0	1,623	0	11
39 **OTHER MANUFACTURING INDUSTRIES**	0	1	73	3	0	13	13	32,702	0	221
3901 Jewellery and related articles	0	0	5	0	0	0	9	5,052	0	32
3902 Musical instruments	..	..	..	..	..	..	..	..	..	..
3903 Sporting and athletic goods	0	0	14	0	0	13	..	5,782	0	53
3909 Other manufactures	0	0	54	0	0	0	2	21,868	0	135
SERR non-specified, unallocated industry	..	..	..	..	..	..	..	..	..	..
3 **TOTAL MANUFACTURING**	49,992	15,337	20,411	34,260	13,875	169,223	12,623	44,736,225	3,490,688	464,205

ISIC Revision 2 Industry Sector		Solid TJ	LPG TJ	Distiloil TJ	RFO TJ	Gas TJ	Biomass TJ	Steam TJ	Electr MWh	Own Use MWh	TOTAL TJ
31	**FOOD, BEVERAGES AND TOBACCO**	5	..	1,603	1,096	2,153	211	73	465,833	8,611	6,787
311/2	FOOD	5	..	1,315	1,037	1,449	178	73	357,777	3,611	5,332
3111	Slaughtering, preparing and preserving meat	..	..	..	..	..	..	..	..	..	..
3112	Dairy products	..	..	..	..	..	..	..	..	..	..
3113	Canning, preserving of fruits and vegetables	..	..	..	..	..	..	..	..	..	..
3114	Canning, preserving and processing of fish	..	..	..	..	..	..	..	..	..	..
3115	Vegetable and animal oils and fats	0	..	27	472	6	20	0	28,333	0	627
3116	Grain mill products	0	..	56	0	16	0	0	23,611	0	157
3117	Bakery products	..	..	..	..	..	..	..	..	..	..
3118	Sugar factories and refineries	..	..	..	..	..	..	..	..	..	..
3119	Cocoa, chocolate and sugar confectionery	0	..	120	30	129	19	73	90,000	3,333	683
3121	Other food products	5	..	1,112	535	1,298	139	0	215,833	278	3,865
3122	Prepared animal feeds	..	..	..	..	..	..	..	..	..	..
313	BEVERAGES	0	..	152	59	496	5	0	65,000	0	946
3131	Distilling, rectifying and blending of spirits	..	..	..	..	..	..	..	..	..	..
3132	Wine industries	..	..	..	..	..	..	..	..	..	..
3133	Malt liquors and malts	0	..	152	59	496	5	0	65,000	0	946
3134	Soft drinks	..	..	..	..	..	..	..	..	..	..
314	TOBACCO	0	..	136	0	208	28	0	43,056	5,000	509
32	**TEXTILES, APPAREL AND LEATHER**	9	..	415	700	725	115	73	643,056	3,333	4,340
321	TEXTILES	9	..	360	648	725	110	73	628,889	3,333	4,177
3211	Spinning weaving and finishing textiles	..	..	..	..	..	..	..	..	..	..
3212	Made-up goods excluding wearing apparel	..	..	..	..	..	..	..	..	..	..
3213	Knitting mills	..	..	..	..	..	..	..	..	..	..
3214	Carpets and rugs	..	..	..	..	..	..	..	..	..	..
3215	Cordage, rope and twine	..	..	..	..	..	..	..	..	..	..
3219	Other textiles	9	..	360	648	725	110	73	628,889	3,333	4,177
322	WEARING APPAREL, EXCEPT FOOTWEAR	..	..	..	..	..	..	..	..	..	..
323	LEATHER AND FUR PRODUCTS	..	..	..	..	..	..	..	..	..	..
3231	Tanneries and leather finishing	..	..	..	..	..	..	..	..	..	..
3232	Fur dressing and dyeing industries	..	..	..	..	..	..	..	..	..	..
3233	Leather prods. ex. footwear and wearing apparel	..	..	..	..	..	..	..	..	..	..
324	FOOTWEAR, EX. RUBBER AND PLASTIC	0	..	55	52	0	5	0	14,167	0	163
33	**WOOD PRODUCTS AND FURNITURE**	..	..	..	..	..	..	..	..	..	..
331	WOOD PRODUCTS, EXCEPT FURNITURE	..	..	..	..	..	..	..	..	..	..
3311	Sawmills, planing and other wood mills	..	..	..	..	..	..	..	..	..	..
3312	Wooden and cane containers	..	..	..	..	..	..	..	..	..	..
3319	Other wood and cork products	..	..	..	..	..	..	..	..	..	..
332	FURNITURE, FIXTURES, EXCL. METALLIC	..	..	..	..	..	..	..	..	..	..
34	**PAPER, PUBLISHING AND PRINTING**	1,085	..	373	5,258	2,466	2,006	1,232	1,513,611	262,500	16,924
341	PAPER AND PRODUCTS	1,085	..	373	5,258	2,466	2,006	1,232	1,513,611	262,500	16,924
3411	Pulp, paper and paperboard articles	1,085	..	373	5,258	2,466	2,006	1,232	1,513,611	262,500	16,924
3412	Containers of paper and paperboard	..	..	..	..	..	..	..	..	..	..
3419	Other pulp, paper and paperboard articles	..	..	..	..	..	..	..	..	..	..
342	PRINTING AND PUBLISHING	..	..	..	..	..	..	..	..	..	..
35	**CHEMICAL PRODUCTS**	118	..	3,223	2,032	9,034	3,752	275	2,883,611	754,318	26,099
351	INDUSTRIAL CHEMICALS	0	..	2,841	1,667	8,504	3,691	275	2,341,111	752,778	22,696
3511	Basic industrial chemicals excl. fertilizers	0	..	2,841	1,667	8,504	3,691	275	2,341,111	752,778	22,696
3512	Fertilizers and pesticides	..	..	..	..	..	..	..	..	..	..
3513	Synthetic resins and plastic materials	..	..	..	..	..	..	..	..	..	..
352	OTHER CHEMICALS	118	..	300	365	529	32	0	475,000	1,540	3,048
3521	Paints, varnishes and lacquers	0	..	108	7	5	0	0	10,278	180	156
3522	Drugs and medicines	..	..	..	..	..	..	..	..	..	..
3523	Soap, cleaning preparations, perfumes, cosmetics	0	..	113	33	11	0	0	12,778	0	203
3529	Other chemical products	118	..	79	325	513	32	0	451,944	1,360	2,689
353	PETROLEUM REFINERIES	..	..	..	..	..	..	..	..	..	..
354	MISC. PETROLEUM AND COAL PRODUCTS	..	..	..	..	..	..	..	..	..	..
355	RUBBER PRODUCTS	..	..	..	..	..	..	..	..	..	..
3551	Tyres and tubes	..	..	..	..	..	..	..	..	..	..
3559	Other rubber products	..	..	..	..	..	..	..	..	..	..
356	PLASTIC PRODUCTS	0	..	82	0	1	29	0	67,500	0	355
36	**NON-METALLIC MINERAL PRODUCTS**	11,622	..	973	5,158	2,274	2,079	0	710,000	51,944	24,475
361	POTTERY, CHINA, EARTHENWARE	0	..	393	0	559	1	0	32,222	0	1,069
362	GLASS AND PRODUCTS	0	..	27	1,536	150	0	0	97,223	0	2,063
369	OTHER NON-METAL. MINERAL PRODUCTS	11,622	..	553	3,622	1,565	2,078	0	580,555	51,944	21,343
3691	Structural clay products	0	..	327	1,770	1,194	8	0	74,722	0	3,568
3692	Cement, lime and plaster	11,622	..	226	1,852	371	2,070	0	505,833	51,944	17,775
3699	Other non-metallic mineral products	..	..	..	..	..	..	..	..	..	..
37	**BASIC METAL INDUSTRIES**	0	..	504	2	1,484	69	0	1,569,722	0	7,710
371	IRON AND STEEL	..	..	..	..	..	..	..	..	..	..
372	NON-FERROUS METALS	0	..	504	2	1,484	69	0	1,569,722	0	7,710
38	**METAL PRODUCTS, MACHINERY, EQUIP.**	827	..	5,683	2,074	6,070	386	345	3,625,833	111,944	28,035
381	METAL PRODUCTS	827	..	5,683	2,074	6,070	386	345	3,625,833	111,944	28,035
3811	Cutlery, hand tools and general hardware	..	..	..	..	..	..	..	..	..	..
3812	Furniture and fixtures primarily of metal	..	..	..	..	..	..	..	..	..	..
3813	Structural metal products	0	..	47	3	4	0	0	14,722	0	107
3819	Other fabricated metal products	827	..	5,636	2,071	6,066	386	345	3,611,111	111,944	27,928
382	NON-ELECTRICAL MACHINERY	..	..	..	..	..	..	..	..	..	..
3821	Engines and turbines	..	..	..	..	..	..	..	..	..	..
3822	Agricultural machinery and equipment	..	..	..	..	..	..	..	..	..	..
3823	Metal and wood working machinery	..	..	..	..	..	..	..	..	..	..
3824	Special industrial machinery	..	..	..	..	..	..	..	..	..	..
3825	Office, computing and accounting machinery	..	..	..	..	..	..	..	..	..	..
3829	Other non-electrical machinery and equipment	..	..	..	..	..	..	..	..	..	..
383	ELECTRICAL MACHINERY	..	..	..	..	..	..	..	..	..	..
3831	Electrical industrial machinery	..	..	..	..	..	..	..	..	..	..
3832	Radio, TV and communications equipment	..	..	..	..	..	..	..	..	..	..
3833	Electrical appliances and housewares	..	..	..	..	..	..	..	..	..	..
3839	Other electrical apparatus and supplies	..	..	..	..	..	..	..	..	..	..
384	TRANSPORT EQUIPMENT	..	..	..	..	..	..	..	..	..	..
3841	Shipbuilding	..	..	..	..	..	..	..	..	..	..
3842	Railroad equipment	..	..	..	..	..	..	..	..	..	..
3843	Motor vehicles	..	..	..	..	..	..	..	..	..	..
3844	Motorcycles and bicycles	..	..	..	..	..	..	..	..	..	..
3845	Aircraft	..	..	..	..	..	..	..	..	..	..
3849	Other transport equipment	..	..	..	..	..	..	..	..	..	..
385	PROFESSIONAL AND SCIENTIFIC EQUIPMENT	..	..	..	..	..	..	..	..	..	..
3851	Professional equipment	..	..	..	..	..	..	..	..	..	..
3852	Photographic and optical goods	..	..	..	..	..	..	..	..	..	..
3853	Watches and clocks	..	..	..	..	..	..	..	..	..	..
39	**OTHER MANUFACTURING INDUSTRIES**	..	..	..	..	..	..	..	..	..	..
3901	Jewellery and related articles	..	..	..	..	..	..	..	..	..	..
3902	Musical instruments	..	..	..	..	..	..	..	..	..	..
3903	Sporting and athletic goods	..	..	..	..	..	..	..	..	..	..
3909	Other manufactures	..	..	..	..	..	..	..	..	..	..
SERR	non-specified, unallocated industry	..	..	..	..	..	..	..	..	..	..
3	**TOTAL MANUFACTURING**	13,666	..	12,774	16,320	24,206	8,618	1,998	11,411,666	1,192,650	114,370

123

SWITZERLAND

ISIC Rev. 2 Industry Sector		Solid TJ	LPG TJ	Distiloil TJ	RFO TJ	Gas TJ	Biomass TJ	Steam TJ	Electr MWh	Own Use MWh	TOTAL TJ
31	FOOD, BEVERAGES AND TOBACCO	5		1,806	1,139	2,174	192	97	481,944	7,222	7,122
311/2	FOOD	0		1,496	1,093	1,436	158	97	369,167	7,222	5,583
3111	Slaughtering, preparing and preserving meat										
3112	Dairy products										
3113	Canning, preserving of fruits and vegetables										
3114	Canning, preserving and processing of fish										
3115	Vegetable and animal oils and fats	0		30	485	6	17	0	29,722	0	645
3116	Grain mill products	0		79	1	13	0	0	31,389	6,944	181
3117	Bakery products										
3118	Sugar factories and refineries										
3119	Cocoa, chocolate and sugar confectionery	0		150	0	126	18	74	90,556	0	694
3121	Other food products			1,237	607	1,291	123	23	217,500	278	4,063
3122	Prepared animal feeds										
313	BEVERAGES	5		159	46	518	5	0	69,444	0	983
3131	Distilling, rectifying and blending of spirits										
3132	Wine industries										
3133	Malt liquors and malts										
3134	Soft drinks	5		159	46	518	5	0	69,444	0	983
314	TOBACCO	0		151	0	220	29	0	43,333	0	556
32	TEXTILES, APPAREL AND LEATHER	0		553	521	597	104	62	643,056	833	4,149
321	TEXTILES	0		422	521	593	104	62	622,500	833	3,940
3211	Spinning weaving and finishing textiles										
3212	Made-up goods excluding wearing apparel										
3213	Knitting mills										
3214	Carpets and rugs										
3215	Cordage, rope and twine										
3219	Other textiles	0									
322	WEARING APPAREL, EXCEPT FOOTWEAR			422	521	593	104	62	622,500	833	3,940
323	LEATHER AND FUR PRODUCTS										
3231	Tanneries and leather finishing										
3232	Fur dressing and dyeing industries										
3233	Leather prods. ex. footwear and wearing apparel										
324	FOOTWEAR, EX. RUBBER AND PLASTIC	0		131	0	4	0	0	20,556	0	209
33	WOOD PRODUCTS AND FURNITURE										
331	WOOD PRODUCTS, EXCEPT FURNITURE										
3311	Sawmills, planing and other wood mills										
3312	Wooden and cane containers										
3319	Other wood and cork products										
332	FURNITURE, FIXTURES, EXCL. METALLIC										
34	PAPER, PUBLISHING AND PRINTING	662		615	4,819	3,096	2,025	1,303	1,520,556	271,389	17,017
341	PAPER AND PRODUCTS	662		615	4,819	3,096	2,025	1,303	1,520,556	271,389	17,017
3411	Pulp, paper and paperboard articles	662		615	4,819	3,096	2,025	1,303	1,520,556	271,389	17,017
3412	Containers of paper and paperboard										
3419	Other pulp, paper and paperboard articles										
342	PRINTING AND PUBLISHING										
35	CHEMICAL PRODUCTS	0		3,335	1,253	8,908	3,605	433	2,527,778	940,556	23,248
351	INDUSTRIAL CHEMICALS	0		2,841	1,104	7,905	3,566	424	2,048,056	939,444	19,831
3511	Basic industrial chemicals excl. fertilizers	0		2,841	1,104	7,905	3,566	424	2,048,056	939,444	19,831
3512	Fertilizers and pesticides										
3513	Synthetic resins and plastic materials										
352	OTHER CHEMICALS	0		412	149	1,002	10	9	412,222	1,112	3,062
3521	Paints, varnishes and lacquers	0		109	6	8	0	0	10,556	0	161
3522	Drugs and medicines										
3523	Soap, cleaning preparations, perfumes, cosmetics										
3529	Other chemical products	0		133	32	46	0	9	20,556	0	294
353	PETROLEUM REFINERIES	0		170	111	948	10	0	381,110	1,112	2,607
354	MISC. PETROLEUM AND COAL PRODUCTS										
355	RUBBER PRODUCTS										
3551	Tyres and tubes										
3559	Other rubber products										
356	PLASTIC PRODUCTS	0		82	0	1	29	0	67,500	0	355
36	NON-METALLIC MINERAL PRODUCTS	8,675		828	5,996	1,834	2,109	0	654,444	51,111	21,614
361	POTTERY, CHINA, EARTHENWARE	0		321		356	1	0	28,333	0	780
362	GLASS AND PRODUCTS	0		27	1,536	150	0	0	97,222	0	2,063
369	OTHER NON-METAL. MINERAL PRODUCTS	8,675		480	4,460	1,328	2,108	0	528,889	51,111	18,771
3691	Structural clay products	0		284	1,541	1,310	8	0	74,722	0	3,412
3692	Cement, lime and plaster	8,675		196	2,919	18	2,100	0	454,167	51,111	15,359
3699	Other non-metallic mineral products										
37	BASIC METAL INDUSTRIES	0		528	0	1,475	90	0	1,422,778	833	7,212
371	IRON AND STEEL										
372	NON-FERROUS METALS	0		528	0	1,475	90	0	1,422,778	833	7,212
38	METAL PRODUCTS, MACHINERY, EQUIP.	698		6,399	1,328	6,806	284	515	3,546,111	26,111	28,702
381	METAL PRODUCTS	698		6,399	1,328	6,806	284	515	3,546,111	26,111	28,702
3811	Cutlery, hand tools and general hardware										
3812	Furniture and fixtures primarily of metal										
3813	Structural metal products	0		48	3	4	0	0	14,722	0	108
3819	Other fabricated metal products	698									
382	NON-ELECTRICAL MACHINERY			6,351	1,325	6,802	284	515	3,531,389	26,111	28,594
3821	Engines and turbines										
3822	Agricultural machinery and equipment										
3823	Metal and wood working machinery										
3824	Special industrial machinery										
3825	Office, computing and accounting machinery										
3829	Other non-electrical machinery and equipment										
383	ELECTRICAL MACHINERY										
3831	Electrical industrial machinery										
3832	Radio, TV and communications equipment										
3833	Electrical appliances and housewares										
3839	Other electrical apparatus and supplies										
384	TRANSPORT EQUIPMENT										
3841	Shipbuilding										
3842	Railroad equipment										
3843	Motor vehicles										
3844	Motorcycles and bicycles										
3845	Aircraft										
3849	Other transport equipment										
385	PROFESSIONAL AND SCIENTIFIC EQUIPMENT										
3851	Professional equipment										
3852	Photographic and optical goods										
3853	Watches and clocks										
39	OTHER MANUFACTURING INDUSTRIES										
3901	Jewellery and related articles										
3902	Musical instruments										
3903	Sporting and athletic goods										
3909	Other manufactures										
SERR	non-specified, unallocated industry										
3	TOTAL MANUFACTURING	10,040		14,064	15,056	24,890	8,409	2,410	10,796,667	1,298,055	109,064

ISIC Revision 2 Industry Sector	Solid TJ	LPG TJ	Distiloil TJ	RFO TJ	Gas TJ	Biomass TJ	Steam TJ	Electr MWh	Own Use MWh	TOTAL TJ
31 FOOD, BEVERAGES AND TOBACCO	5	..	1,473	845	1,837	177	93	422,222	8,333	5,920
311/2 FOOD	0	..	1,149	836	1,204	128	93	323,611	8,333	4,545
3111 Slaughtering, preparing and preserving meat	..	..	..	..	..	..	..	..	..	..
3112 Dairy products	..	..	..	..	..	..	..	..	..	..
3113 Canning, preserving of fruits and vegetables	..	..	..	..	..	..	..	..	..	..
3114 Canning, preserving and processing of fish	..	..	..	..	..	..	..	..	..	..
3115 Vegetable and animal oils and fats	0	..	22	252	18	7	0	18,889	0	367
3116 Grain mill products	0	..	79	11	0	0	0	35,833	8,055	190
3117 Bakery products	..	..	..	..	..	..	..	..	..	..
3118 Sugar factories and refineries	..	..	..	..	..	..	..	..	..	..
3119 Cocoa, chocolate and sugar confectionery	0	..	150	0	126	18	75	91,111	0	697
3121 Other food products	0	..	898	573	1,060	103	18	177,778	278	3,291
3122 Prepared animal feeds	..	..	..	..	..	..	..	..	..	..
313 BEVERAGES	5	..	196	9	455	11	0	61,389	0	897
3131 Distilling, rectifying and blending of spirits	..	..	..	..	..	..	..	..	..	..
3132 Wine industries	..	..	..	..	..	..	..	..	..	..
3133 Malt liquors and malts	5	..	196	9	455	11	0	61,389	0	897
3134 Soft drinks	..	..	..	..	..	..	..	..	..	..
314 TOBACCO	0	..	128	0	178	38	0	37,222	0	478
32 TEXTILES, APPAREL AND LEATHER	0	..	520	159	679	20	59	337,500	63,056	2,425
321 TEXTILES	0	..	520	159	679	20	59	337,500	63,056	2,425
3211 Spinning weaving and finishing textiles	..	..	..	..	..	..	..	..	..	..
3212 Made-up goods excluding wearing apparel	..	..	..	..	..	..	..	..	..	..
3213 Knitting mills	..	..	..	..	..	..	..	..	..	..
3214 Carpets and rugs	..	..	..	..	..	..	..	..	..	..
3215 Cordage, rope and twine	..	..	..	..	..	..	..	..	..	..
3219 Other textiles	0	..	520	159	679	20	59	337,500	63,056	2,425
322 WEARING APPAREL, EXCEPT FOOTWEAR	..	..	..	..	..	..	..	..	..	..
323 LEATHER AND FUR PRODUCTS	..	..	..	..	..	..	..	..	..	..
3231 Tanneries and leather finishing	..	..	..	..	..	..	..	..	..	..
3232 Fur dressing and dyeing industries	..	..	..	..	..	..	..	..	..	..
3233 Leather prods. ex. footwear and wearing apparel	..	..	..	..	..	..	..	..	..	..
324 FOOTWEAR, EX. RUBBER AND PLASTIC	..	..	..	..	..	..	..	..	..	..
33 WOOD PRODUCTS AND FURNITURE	..	..	..	..	..	..	..	..	..	..
331 WOOD PRODUCTS, EXCEPT FURNITURE	..	..	..	..	..	..	..	..	..	..
3311 Sawmills, planing and other wood mills	..	..	..	..	..	..	..	..	..	..
3312 Wooden and cane containers	..	..	..	..	..	..	..	..	..	..
3319 Other wood and cork products	..	..	..	..	..	..	..	..	..	..
332 FURNITURE, FIXTURES, EXCL. METALLIC	..	..	..	..	..	..	..	..	..	..
34 PAPER, PUBLISHING AND PRINTING	120	..	833	3,908	5,334	2,145	1,270	1,556,667	364,444	17,902
341 PAPER AND PRODUCTS	120	..	833	3,908	5,334	2,145	1,270	1,556,667	364,444	17,902
3411 Pulp, paper and paperboard articles	120	..	833	3,908	5,334	2,145	1,270	1,556,667	364,444	17,902
3412 Containers of paper and paperboard	..	..	..	..	..	..	..	..	..	..
3419 Other pulp, paper and paperboard articles	..	..	..	..	..	..	..	..	..	..
342 PRINTING AND PUBLISHING	0	..	2,784	759	9,850	3,856	253	2,554,167	971,944	23,198
35 CHEMICAL PRODUCTS	0	..	2,386	725	8,747	3,843	244	2,133,333	970,833	20,130
351 INDUSTRIAL CHEMICALS	0	..	2,386	725	8,747	3,843	244	2,133,333	970,833	20,130
3511 Basic industrial chemicals excl. fertilizers	..	..	..	..	..	..	..	..	..	..
3512 Fertilizers and pesticides	..	..	..	..	..	..	..	..	..	..
3513 Synthetic resins and plastic materials	0	..	398	34	1,103	13	9	420,834	1,111	3,068
352 OTHER CHEMICALS	0	..	94	5	7	0	0	10,000	0	142
3521 Paints, varnishes and lacquers	..	..	..	..	..	..	..	..	..	..
3522 Drugs and medicines	..	..	171	21	41	0	9	29,167	0	347
3523 Soap, cleaning preparations, perfumes, cosmetics	0	..	133	8	1,055	13	0	381,667	1,111	2,579
3529 Other chemical products	0	..	..	..	..	..	..	..	..	..
353 PETROLEUM REFINERIES	..	..	..	..	..	..	..	..	..	..
354 MISC. PETROLEUM AND COAL PRODUCTS	..	..	..	..	..	..	..	..	..	..
355 RUBBER PRODUCTS	..	..	..	..	..	..	..	..	..	..
3551 Tyres and tubes	..	..	..	..	..	..	..	..	..	..
3559 Other rubber products	..	..	..	..	..	..	..	..	..	..
356 PLASTIC PRODUCTS	..	..	698	7,255	2,100	2,098	0	616,111	4,361	21,018
36 NON-METALLIC MINERAL PRODUCTS	6,665	..	..	..	..	..	..	..	..	..
361 POTTERY, CHINA, EARTHENWARE	0	..	204	0	453	1	0	24,444	0	746
362 GLASS AND PRODUCTS	0	..	27	1,536	150	0	0	97,223	0	2,063
369 OTHER NON-METAL. MINERAL PRODUCTS	6,665	..	467	5,719	1,497	2,097	0	494,444	4,361	18,209
3691 Structural clay products	1	..	284	1,349	1,432	6	0	73,333	0	3,336
3692 Cement, lime and plaster	6,664	..	183	4,370	65	2,091	0	421,111	4,361	14,873
3699 Other non-metallic mineral products	..	..	..	..	..	..	..	..	..	..
37 BASIC METAL INDUSTRIES	0	..	331	0	1,405	79	0	1,142,500	278	5,927
371 IRON AND STEEL	0	..	331	0	1,405	79	0	1,142,500	278	5,927
372 NON-FERROUS METALS	..	..	..	..	..	..	..	..	..	..
38 METAL PRODUCTS, MACHINERY, EQUIP.	731	..	5,338	937	6,328	270	520	3,238,611	91,944	25,452
381 METAL PRODUCTS	731	..	5,338	937	6,328	270	520	3,238,611	91,944	25,452
3811 Cutlery, hand tools and general hardware	..	..	..	..	..	..	..	..	..	..
3812 Furniture and fixtures primarily of metal	0	..	48	3	4	0	0	14,722	0	108
3813 Structural metal products	731	..	5,290	934	6,324	270	520	3,223,889	91,944	25,344
3819 Other fabricated metal products	..	..	..	..	..	..	..	..	..	..
382 NON-ELECTRICAL MACHINERY	..	..	..	..	..	..	..	..	..	..
3821 Engines and turbines	..	..	..	..	..	..	..	..	..	..
3822 Agricultural machinery and equipment	..	..	..	..	..	..	..	..	..	..
3823 Metal and wood working machinery	..	..	..	..	..	..	..	..	..	..
3824 Special industrial machinery	..	..	..	..	..	..	..	..	..	..
3825 Office, computing and accounting machinery	..	..	..	..	..	..	..	..	..	..
3829 Other non-electrical machinery and equipment	..	..	..	..	..	..	..	..	..	..
383 ELECTRICAL MACHINERY	..	..	..	..	..	..	..	..	..	..
3831 Electrical industrial machinery	..	..	..	..	..	..	..	..	..	..
3832 Radio, TV and communications equipment	..	..	..	..	..	..	..	..	..	..
3833 Electrical appliances and housewares	..	..	..	..	..	..	..	..	..	..
3839 Other electrical apparatus and supplies	..	..	..	..	..	..	..	..	..	..
384 TRANSPORT EQUIPMENT	..	..	..	..	..	..	..	..	..	..
3841 Shipbuilding	..	..	..	..	..	..	..	..	..	..
3842 Railroad equipment	..	..	..	..	..	..	..	..	..	..
3843 Motor vehicles	..	..	..	..	..	..	..	..	..	..
3844 Motorcycles and bicycles	..	..	..	..	..	..	..	..	..	..
3845 Aircraft	..	..	..	..	..	..	..	..	..	..
3849 Other transport equipment	..	..	..	..	..	..	..	..	..	..
385 PROFESSIONAL AND SCIENTIFIC EQUIPMENT	..	..	..	..	..	..	..	..	..	..
3851 Professional equipment	..	..	..	..	..	..	..	..	..	..
3852 Photographic and optical goods	..	..	..	..	..	..	..	..	..	..
3853 Watches and clocks	..	..	..	..	..	..	..	..	..	..
39 OTHER MANUFACTURING INDUSTRIES	..	..	..	..	..	..	..	..	..	..
3901 Jewellery and related articles	..	..	..	..	..	..	..	..	..	..
3902 Musical instruments	..	..	..	..	..	..	..	..	..	..
3903 Sporting and athletic goods	..	..	..	..	..	..	..	..	..	..
3909 Other manufactures	..	..	..	..	..	..	..	..	..	..
SERR non-specified, unallocated industry	..	..	..	..	..	..	..	..	..	..
3 TOTAL MANUFACTURING	7,521	..	11,977	13,863	27,533	8,645	2,195	9,867,778	1,504,360	101,842

ISIS Energy Data Programme (IEA/OECD)

ISIC Revision 2 Industry Sector	Solid TJ	LPG TJ	Distiloil TJ	RFO TJ	Gas TJ	Biomass TJ	Steam TJ	Electr MWh	Own Use MWh	TOTAL TJ
31 **FOOD, BEVERAGES AND TOBACCO**	4	..	2,021	338	2,464	257	101	424,945	7,778	6,687
311/2 FOOD	0		1,739	338	1,874	203	101	333,000	7,778	5,426
3111 Slaughtering, preparing and preserving meat	..		..		..			..		..
3112 Dairy products	..		..		..			..		..
3113 Canning, preserving of fruits and vegetables	..		..		..			..		..
3114 Canning, preserving and processing of fish	..		..		..			..		..
3115 Vegetable and animal oils and fats	0		172	4	304	0	0	3,000	0	491
3116 Grain mill products	0		96	0	0	0	0	35,000	7,500	195
3117 Bakery products	..		..		..			..		..
3118 Sugar factories and refineries	..		..		..			..		..
3119 Cocoa, chocolate and sugar confectionery	0		149	0	108	17	85	91,944	0	690
3121 Other food products	0		1,322	334	1,462	186	16	203,056	278	4,050
3122 Prepared animal feeds	..		..		..			..		..
313 BEVERAGES	4		171	0	430	22	0	61,389	0	848
3131 Distilling, rectifying and blending of spirits	..		..		..			..		..
3132 Wine industries	..		..		..			..		..
3133 Malt liquors and malts	4		171	0	430	22	0	61,389	0	848
3134 Soft drinks	..		..		..			..		..
314 TOBACCO	0		111	0	160	32	0	30,556	0	413
32 **TEXTILES, APPAREL AND LEATHER**	0	..	922	289	1,626	75	62	751,389	69,444	5,429
321 TEXTILES	0		791	289	1,622	75	62	730,833	69,444	5,220
3211 Spinning weaving and finishing textiles	..		..		..			..		..
3212 Made-up goods excluding wearing apparel	..		..		..			..		..
3213 Knitting mills	..		..		..			..		..
3214 Carpets and rugs	..		..		..			..		..
3215 Cordage, rope and twine	..		..		..			..		..
3219 Other textiles	0		791	289	1,622	75	62	730,833	69,444	5,220
322 WEARING APPAREL, EXCEPT FOOTWEAR	..		..		..			..		..
323 LEATHER AND FUR PRODUCTS	..		..		..			..		..
3231 Tanneries and leather finishing	..		..		..			..		..
3232 Fur dressing and dyeing industries	..		..		..			..		..
3233 Leather prods. ex. footwear and wearing apparel	..		..		..			..		..
324 FOOTWEAR, EX. RUBBER AND PLASTIC	0		131	0	4	0	0	20,556	0	209
33 **WOOD PRODUCTS AND FURNITURE**	..	..	..		..			..		..
331 WOOD PRODUCTS, EXCEPT FURNITURE	..		..		..			..		..
3311 Sawmills, planing and other wood mills	..		..		..			..		..
3312 Wooden and cane containers	..		..		..			..		..
3319 Other wood and cork products	..		..		..			..		..
332 FURNITURE, FIXTURES, EXCL. METALLIC	..		..		..			..		..
34 **PAPER, PUBLISHING AND PRINTING**	0	..	757	3,595	6,094	1,929	1,550	1,518,056	393,333	17,974
341 PAPER AND PRODUCTS	0		757	3,595	6,094	1,929	1,550	1,518,056	393,333	17,974
3411 Pulp, paper and paperboard articles	0		757	3,595	6,094	1,929	1,550	1,518,056	393,333	17,974
3412 Containers of paper and paperboard	..		..		..			..		..
3419 Other pulp, paper and paperboard articles	..		..		..			..		..
342 PRINTING AND PUBLISHING	..		..		..			..		..
35 **CHEMICAL PRODUCTS**	0	..	2,753	1,036	8,802	4,331	340	2,471,667	904,722	22,903
351 INDUSTRIAL CHEMICALS	0		2,287	1,007	7,695	4,289	331	1,984,446	904,722	19,496
3511 Basic industrial chemicals excl. fertilizers	0		2,287	1,007	7,695	4,289	331	1,984,446	904,722	19,496
3512 Fertilizers and pesticides	..		..		..			..		..
3513 Synthetic resins and plastic materials	..		..		..			..		..
352 OTHER CHEMICALS	0		384	29	1,106	13	9	419,721	0	3,052
3521 Paints, varnishes and lacquers	0		102	1	7	0	0	10,833	0	149
3522 Drugs and medicines	..		..		..			..		..
3523 Soap, cleaning preparations, perfumes, cosmetics	0		149	20	44	0	9	26,944	0	319
3529 Other chemical products	0		133	8	1,055	13	0	381,944	0	2,584
353 PETROLEUM REFINERIES	..		..		..			..		..
354 MISC. PETROLEUM AND COAL PRODUCTS	..		..		..			..		..
355 RUBBER PRODUCTS	..		..		..			..		..
3551 Tyres and tubes	..		..		..			..		..
3559 Other rubber products	..		..		..			..		..
356 PLASTIC PRODUCTS	0		82	0	1	29	0	67,500	0	355
36 **NON-METALLIC MINERAL PRODUCTS**	5,809	..	686	6,048	2,161	2,657	0	586,389	39,444	19,330
361 POTTERY, CHINA, EARTHENWARE	0		157	0	517	0	0	24,167	0	761
362 GLASS AND PRODUCTS	0		27	1,536	150	0	0	97,222	0	2,063
369 OTHER NON-METAL. MINERAL PRODUCTS	5,809		502	4,512	1,494	2,657	0	465,000	39,444	16,506
3691 Structural clay products	1		264	1,294	1,440	37	0	71,944	0	3,295
3692 Cement, lime and plaster	5,808		238	3,218	54	2,620	0	393,056	39,444	13,211
3699 Other non-metallic mineral products										
37 **BASIC METAL INDUSTRIES**	0	..	313	0	1,111	61	0	864,444	82,222	4,301
371 IRON AND STEEL	..		..		..			..		..
372 NON-FERROUS METALS	0		313	0	1,111	61	0	864,444	82,222	4,301
38 **METAL PRODUCTS, MACHINERY, EQUIP.**	488	..	4,922	808	5,943	260	456	2,888,056	73,611	23,009
381 METAL PRODUCTS	488		4,922	808	5,943	260	456	2,888,056	73,611	23,009
3811 Cutlery, hand tools and general hardware	..		..		..			..		..
3812 Furniture and fixtures primarily of metal	..		..		..			..		..
3813 Structural metal products	0		48	3	4	0	0	14,722	0	108
3819 Other fabricated metal products	488		4,874	805	5,939	260	456	2,873,334	73,611	22,901
382 NON-ELECTRICAL MACHINERY	..		..		..			..		..
3821 Engines and turbines	..		..		..			..		..
3822 Agricultural machinery and equipment	..		..		..			..		..
3823 Metal and wood working machinery	..		..		..			..		..
3824 Special industrial machinery	..		..		..			..		..
3825 Office, computing and accounting machinery	..		..		..			..		..
3829 Other non-electrical machinery and equipment	..		..		..			..		..
383 ELECTRICAL MACHINERY	..		..		..			..		..
3831 Electrical industrial machinery	..		..		..			..		..
3832 Radio, TV and communications equipment	..		..		..			..		..
3833 Electrical appliances and housewares	..		..		..			..		..
3839 Other electrical apparatus and supplies	..		..		..			..		..
384 TRANSPORT EQUIPMENT	..		..		..			..		..
3841 Shipbuilding	..		..		..			..		..
3842 Railroad equipment	..		..		..			..		..
3843 Motor vehicles	..		..		..			..		..
3844 Motorcycles and bicycles	..		..		..			..		..
3845 Aircraft	..		..		..			..		..
3849 Other transport equipment	..		..		..			..		..
385 PROFESSIONAL AND SCIENTIFIC EQUIPMENT	..		..		..			..		..
3851 Professional equipment	..		..		..			..		..
3852 Photographic and optical goods	..		..		..			..		..
3853 Watches and clocks	..		..		..			..		..
39 **OTHER MANUFACTURING INDUSTRIES**	..	..	..		..			..		..
3901 Jewellery and related articles	..		..		..			..		..
3902 Musical instruments	..		..		..			..		..
3903 Sporting and athletic goods	..		..		..			..		..
3909 Other manufactures	..		..		..			..		..
SERR non-specified, unallocated industry	..		..		..			..		..
3 **TOTAL MANUFACTURING**	6,301	..	12,374	12,114	28,201	9,570	2,509	9,504,946	1,570,554	99,633

Year: 1994

SWITZERLAND

ISIC Revision 2 Industry Sector	Solid TJ	LPG TJ	Distiloil TJ	RFO TJ	Gas TJ	Biomass TJ	Steam TJ	Electr MWh	Own Use MWh	TOTAL TJ
31 FOOD, BEVERAGES AND TOBACCO	4	8	1,599	339	2,083	118	150	440,555	7,500	5,860
311/2 FOOD	0	8	1,286	339	1,706	104	150	342,778	7,500	4,800
3111 Slaughtering, preparing and preserving meat	..	..	..	..	..	..	..	..	..	..
3112 Dairy products	..	..	..	..	..	..	..	..	..	..
3113 Canning, preserving of fruits and vegetables	..	..	..	..	..	..	..	..	..	..
3114 Canning, preserving and processing of fish	..	..	..	..	..	..	..	..	..	..
3115 Vegetable and animal oils and fats	0	0	20	0	285	0	0	18,333	0	371
3116 Grain mill products	0	0	42	20	0	0	0	34,167	7,500	158
3117 Bakery products	..	..	..	..	..	..	..	..	..	..
3118 Sugar factories and refineries	0	1	148	0	114	17	86	96,667	0	714
3119 Cocoa, chocolate and sugar confectionery	0	7	1,076	319	1,307	87	64	193,611	0	3,557
3121 Other food products	..	..	..	..	..	..	..	..	..	..
3122 Prepared animal feeds	4	0	189	0	377	13	0	60,833	0	802
313 BEVERAGES	4	0	189	0	377	13	0	60,833	0	802
3131 Distilling, rectifying and blending of spirits	..	..	..	..	..	..	..	..	..	..
3132 Wine industries	..	..	..	..	..	..	..	..	..	..
3133 Malt liquors and malts	4	0	189	0	377	13	0	60,833	0	802
3134 Soft drinks	..	..	..	..	..	..	..	..	..	..
314 TOBACCO	0	0	124	0	0	1	0	36,944	0	258
32 TEXTILES, APPAREL AND LEATHER	0	0	803	255	2,413	36	69	1,085,278	68,333	7,237
321 TEXTILES	0	0	700	255	2,408	36	69	1,071,389	68,333	7,079
3211 Spinning weaving and finishing textiles	..	..	..	..	..	..	..	..	..	..
3212 Made-up goods excluding wearing apparel	..	..	..	..	..	..	..	..	..	..
3213 Knitting mills	..	..	..	..	..	..	..	..	..	..
3214 Carpets and rugs	..	..	..	..	..	..	..	..	..	..
3215 Cordage, rope and twine	..	..	..	..	..	..	..	..	..	..
3219 Other textiles	0	0	700	255	2,408	36	69	1,071,389	68,333	7,079
322 WEARING APPAREL, EXCEPT FOOTWEAR	..	..	..	..	..	..	..	..	..	..
323 LEATHER AND FUR PRODUCTS	..	..	..	..	..	..	..	..	..	..
3231 Tanneries and leather finishing	..	..	..	..	..	..	..	..	..	..
3232 Fur dressing and dyeing industries	..	..	..	..	..	..	..	..	..	..
3233 Leather prods. ex. footwear and wearing apparel	..	..	..	..	..	..	..	..	..	..
324 FOOTWEAR, EX. RUBBER AND PLASTIC	0	0	103	0	5	0	0	13,889	0	158
33 WOOD PRODUCTS AND FURNITURE	..	..	..	..	..	..	..	..	..	..
331 WOOD PRODUCTS, EXCEPT FURNITURE	..	..	..	..	..	..	..	..	..	..
3311 Sawmills, planing and other wood mills	..	..	..	..	..	..	..	..	..	..
3312 Wooden and cane containers	..	..	..	..	..	..	..	..	..	..
3319 Other wood and cork products	..	..	..	..	..	..	..	..	..	..
332 FURNITURE, FIXTURES, EXCL. METALLIC	..	..	..	..	..	..	..	..	..	..
34 PAPER, PUBLISHING AND PRINTING	0	32	706	3,267	7,494	0	1,669	1,189,722	69,167	17,202
341 PAPER AND PRODUCTS	0	32	706	3,267	7,494	0	1,669	1,189,722	69,167	17,202
3411 Pulp, paper and paperboard articles	0	32	706	3,267	7,494	0	1,669	1,189,722	69,167	17,202
3412 Containers of paper and paperboard	..	..	..	..	..	..	..	..	..	..
3419 Other pulp, paper and paperboard articles	..	..	..	..	..	..	..	..	..	..
342 PRINTING AND PUBLISHING	..	..	..	..	..	..	..	..	..	..
35 CHEMICAL PRODUCTS	0	20	2,508	778	8,715	4,143	711	2,697,738	626,389	24,332
351 INDUSTRIAL CHEMICALS	0	20	2,238	758	8,666	4,143	711	2,657,738	626,389	23,849
3511 Basic industrial chemicals excl. fertilizers	..	..	..	..	..	..	..	..	..	..
3512 Fertilizers and pesticides	..	..	..	..	..	..	..	..	..	..
3513 Synthetic resins and plastic materials	0	0	270	20	49	0	0	40,000	0	483
352 OTHER CHEMICALS	0	0	113	0	6	0	0	14,444	0	171
3521 Paints, varnishes and lacquers	..	..	..	..	..	..	..	..	..	..
3522 Drugs and medicines	0	0	157	20	43	0	0	25,556	0	312
3523 Soap, cleaning preparations, perfumes, cosmetics	..	..	..	..	..	..	..	..	..	..
3529 Other chemical products	..	..	..	..	..	..	..	..	..	..
353 PETROLEUM REFINERIES	..	..	..	..	..	..	..	..	..	..
354 MISC. PETROLEUM AND COAL PRODUCTS	..	..	..	..	..	..	..	..	..	..
355 RUBBER PRODUCTS	..	..	..	..	..	..	..	..	..	..
3551 Tyres and tubes	..	..	..	..	..	..	..	..	..	..
3559 Other rubber products	..	..	..	..	..	..	..	..	..	..
356 PLASTIC PRODUCTS	..	..	..	..	..	..	..	..	..	..
36 NON-METALLIC MINERAL PRODUCTS	6,239	0	597	5,830	1,682	2,417	0	551,388	42,778	18,596
361 POTTERY, CHINA, EARTHENWARE	..	..	..	..	..	..	..	..	..	..
362 GLASS AND PRODUCTS	..	..	..	..	..	..	..	..	..	..
369 OTHER NON-METAL. MINERAL PRODUCTS	6,239	0	597	5,830	1,682	2,417	0	551,388	42,778	18,596
3691 Structural clay products	1	0	304	1,396	1,658	38	0	79,444	0	3,683
3692 Cement, lime and plaster	6,238	0	293	4,434	24	2,379	0	471,944	42,778	14,913
3699 Other non-metallic mineral products	..	..	..	..	..	..	..	..	..	..
37 BASIC METAL INDUSTRIES	0	0	313	0	1,111	27	0	864,444	82,222	4,267
371 IRON AND STEEL	0	0	313	0	1,111	27	0	864,444	82,222	4,267
372 NON-FERROUS METALS	..	..	..	..	..	..	..	..	..	..
38 METAL PRODUCTS, MACHINERY, EQUIP.	607	124	4,116	449	4,933	85	389	2,687,223	65,833	20,140
381 METAL PRODUCTS	0	0	67	0	137	0	0	30,556	0	314
3811 Cutlery, hand tools and general hardware	..	..	..	..	..	..	..	..	..	..
3812 Furniture and fixtures primarily of metal	..	..	..	..	..	..	..	..	..	..
3813 Structural metal products	..	..	..	..	..	..	..	..	..	..
3819 Other fabricated metal products	0	0	67	0	137	0	0	30,556	0	314
382 NON-ELECTRICAL MACHINERY	607	124	4,049	449	4,796	85	389	2,656,667	65,833	19,826
3821 Engines and turbines	..	..	..	..	..	..	..	..	..	..
3822 Agricultural machinery and equipment	..	..	..	..	..	..	..	..	..	..
3823 Metal and wood working machinery	..	..	..	..	..	..	..	..	..	..
3824 Special industrial machinery	..	..	..	..	..	..	..	..	..	..
3825 Office, computing and accounting machinery	..	..	..	..	..	..	..	..	..	..
3829 Other non-electrical machinery and equipment	607	124	4,049	449	4,796	85	389	2,656,667	65,833	19,826
383 ELECTRICAL MACHINERY	..	..	..	..	..	..	..	..	..	..
3831 Electrical industrial machinery	..	..	..	..	..	..	..	..	..	..
3832 Radio, TV and communications equipment	..	..	..	..	..	..	..	..	..	..
3833 Electrical appliances and housewares	..	..	..	..	..	..	..	..	..	..
3839 Other electrical apparatus and supplies	..	..	..	..	..	..	..	..	..	..
384 TRANSPORT EQUIPMENT	..	..	..	..	..	..	..	..	..	..
3841 Shipbuilding	..	..	..	..	..	..	..	..	..	..
3842 Railroad equipment	..	..	..	..	..	..	..	..	..	..
3843 Motor vehicles	..	..	..	..	..	..	..	..	..	..
3844 Motorcycles and bicycles	..	..	..	..	..	..	..	..	..	..
3845 Aircraft	..	..	..	..	..	..	..	..	..	..
3849 Other transport equipment	..	..	..	..	..	..	..	..	..	..
385 PROFESSIONAL AND SCIENTIFIC EQUIPMENT	..	..	..	..	..	..	..	..	..	..
3851 Professional equipment	..	..	..	..	..	..	..	..	..	..
3852 Photographic and optical goods	..	..	..	..	..	..	..	..	..	..
3853 Watches and clocks	..	..	..	..	..	..	..	..	..	..
39 OTHER MANUFACTURING INDUSTRIES	..	..	..	..	..	..	..	..	..	..
3901 Jewellery and related articles	..	..	..	..	..	..	..	..	..	..
3902 Musical instruments	..	..	..	..	..	..	..	..	..	..
3903 Sporting and athletic goods	..	..	..	..	..	..	..	..	..	..
3909 Other manufactures	..	..	..	..	..	..	..	..	..	..
SERR non-specified, unallocated industry	..	..	..	..	..	..	..	..	..	..
3 TOTAL MANUFACTURING	6,850	184	10,642	10,918	28,431	6,826	2,988	9,516,348	962,222	97,634

ISIS Energy Data Programme (IEA/OECD)

ISIC Revision 2 Industry Sector	Solid TJ	LPG TJ	Distiloil TJ	RFO TJ	Gas TJ	Biomass TJ	Steam TJ	Electr MWh	Own Use MWh	TOTAL TJ
31 FOOD, BEVERAGES AND TOBACCO	2	55	1,785	215	2,619	160	165	472,222	278	6,700
311/2 FOOD	0	55	1,615	215	2,217	149	165	412,222	278	5,899
3111 Slaughtering, preparing and preserving meat	..									..
3112 Dairy products	..									..
3113 Canning, preserving of fruits and vegetables	..									..
3114 Canning, preserving and processing of fish	..									..
3115 Vegetable and animal oils and fats	0	0	41	36	411	0	0	29,444	0	594
3116 Grain mill products	0	0	65	0	2	0	0	21,389	0	144
3117 Bakery products	..									..
3118 Sugar factories and refineries	..									..
3119 Cocoa, chocolate and sugar confectionery	0	1	139	0	123	11	78	99,445	0	710
3121 Other food products	0	54	1,370	179	1,681	138	87	261,944	278	4,451
3122 Prepared animal feeds	..									..
313 BEVERAGES	2	0	170	0	402	11	0	60,000	0	801
3131 Distilling, rectifying and blending of spirits	..									..
3132 Wine industries	..									..
3133 Malt liquors and malts	2	0	170	0	402	11	0	60,000	0	801
3134 Soft drinks	..									..
314 TOBACCO	..									..
32 TEXTILES, APPAREL AND LEATHER	0	0	955	0	1,353	36	69	621,111	0	4,649
321 TEXTILES	0	0	955	0	1,353	36	69	621,111	0	4,649
3211 Spinning weaving and finishing textiles	..									..
3212 Made-up goods excluding wearing apparel	..									..
3213 Knitting mills	..									..
3214 Carpets and rugs	..									..
3215 Cordage, rope and twine	..									..
3219 Other textiles	0	0	955	0	1,353	36	69	621,111	0	4,649
322 WEARING APPAREL, EXCEPT FOOTWEAR	..									..
323 LEATHER AND FUR PRODUCTS	..									..
3231 Tanneries and leather finishing	..									..
3232 Fur dressing and dyeing industries	..									..
3233 Leather prods. ex. footwear and wearing apparel	..									..
324 FOOTWEAR, EX. RUBBER AND PLASTIC	..									..
33 WOOD PRODUCTS AND FURNITURE	..									..
331 WOOD PRODUCTS, EXCEPT FURNITURE	..									..
3311 Sawmills, planing and other wood mills	..									..
3312 Wooden and cane containers	..									..
3319 Other wood and cork products	..									..
332 FURNITURE, FIXTURES, EXCL. METALLIC	..									..
34 PAPER, PUBLISHING AND PRINTING	0	49	812	3,101	6,204	2,375	1,734	1,302,778	101,111	18,601
341 PAPER AND PRODUCTS	0	49	812	3,101	6,204	2,375	1,734	1,302,778	101,111	18,601
3411 Pulp, paper and paperboard articles	0	49	812	3,101	6,204	2,375	1,734	1,302,778	101,111	18,601
3412 Containers of paper and paperboard	..									..
3419 Other pulp, paper and paperboard articles	..									..
342 PRINTING AND PUBLISHING	..									..
35 CHEMICAL PRODUCTS	0	1	3,645	402	9,266	4,873	12	2,218,056	589,722	24,061
351 INDUSTRIAL CHEMICALS	..									..
3511 Basic industrial chemicals excl. fertilizers	..									..
3512 Fertilizers and pesticides	..									..
3513 Synthetic resins and plastic materials	..									..
352 OTHER CHEMICALS	0	1	3,645	402	9,266	4,873	12	2,218,056	589,722	24,061
3521 Paints, varnishes and lacquers	0	0	123	0	7	0	0	15,556	0	186
3522 Drugs and medicines	..									..
3523 Soap, cleaning preparations, perfumes, cosmetics	0	0	129	0	42	0	12	20,833	0	258
3529 Other chemical products	0	1	3,393	402	9,217	4,873	0	2,181,667	589,722	23,617
353 PETROLEUM REFINERIES	..									..
354 MISC. PETROLEUM AND COAL PRODUCTS	..									..
355 RUBBER PRODUCTS	..									..
3551 Tyres and tubes	..									..
3559 Other rubber products	..									..
356 PLASTIC PRODUCTS	..									..
36 NON-METALLIC MINERAL PRODUCTS	6,810	136	545	4,004	1,613	3,185	2	486,389	39,167	17,905
361 POTTERY, CHINA, EARTHENWARE	..									..
362 GLASS AND PRODUCTS	..									..
369 OTHER NON-METAL. MINERAL PRODUCTS	6,810	136	545	4,004	1,613	3,185	2	486,389	39,167	17,905
3691 Structural clay products	0	0	274	1,323	1,577	88	0	79,444	0	3,548
3692 Cement, lime and plaster	6,810	136	271	2,681	36	3,097	2	406,945	39,167	14,357
3699 Other non-metallic mineral products	..									..
37 BASIC METAL INDUSTRIES	0	110	329	0	1,359	58	0	698,889	506,667	2,548
371 IRON AND STEEL	..									..
372 NON-FERROUS METALS	0	110	329	0	1,359	58	0	698,889	506,667	2,548
38 METAL PRODUCTS, MACHINERY, EQUIP.	560	252	3,873	320	4,560	86	541	2,526,944	59,444	19,075
381 METAL PRODUCTS	0	75	69	0	86	0	0	49,722	0	409
3811 Cutlery, hand tools and general hardware	..									..
3812 Furniture and fixtures primarily of metal	..									..
3813 Structural metal products	..									..
3819 Other fabricated metal products	0	75	69	0	86	0	0	49,722	0	409
382 NON-ELECTRICAL MACHINERY	560	177	3,804	320	4,474	86	541	2,477,222	59,444	18,666
3821 Engines and turbines	..									..
3822 Agricultural machinery and equipment	..									..
3823 Metal and wood working machinery	..									..
3824 Special industrial machinery	..									..
3825 Office, computing and accounting machinery	..									..
3829 Other non-electrical machinery and equipment	560	177	3,804	320	4,474	86	541	2,477,222	59,444	18,666
383 ELECTRICAL MACHINERY	..									..
3831 Electrical industrial machinery	..									..
3832 Radio, TV and communications equipment	..									..
3833 Electrical appliances and housewares	..									..
3839 Other electrical apparatus and supplies	..									..
384 TRANSPORT EQUIPMENT	..									..
3841 Shipbuilding	..									..
3842 Railroad equipment	..									..
3843 Motor vehicles	..									..
3844 Motorcycles and bicycles	..									..
3845 Aircraft	..									..
3849 Other transport equipment	..									..
385 PROFESSIONAL AND SCIENTIFIC EQUIPMENT	..									..
3851 Professional equipment	..									..
3852 Photographic and optical goods	..									..
3853 Watches and clocks	..									..
39 OTHER MANUFACTURING INDUSTRIES	..									..
3901 Jewellery and related articles	..									..
3902 Musical instruments	..									..
3903 Sporting and athletic goods	..									..
3909 Other manufactures	..									..
SERR non-specified, unallocated industry	..									..
3 TOTAL MANUFACTURING	7,372	603	11,944	8,042	26,974	10,773	2,523	8,326,389	1,296,389	93,539

ISIC Revision 2 Industry Sector	Solid TJ	LPG TJ	Distiloil TJ	RFO TJ	Gas TJ	Biomass TJ	Steam TJ	Electr MWh	Own Use MWh	TOTAL TJ
31 FOOD, BEVERAGES AND TOBACCO	**27,679**	**662**	**2,210**	**17,995**	**1,135**	**446**	..	**1,725,541**	**469,631**	**54,648**
311/2 FOOD	27,354	627	1,962	15,143	1,105	442	..	1,546,957	466,200	50,524
3111 Slaughtering, preparing and preserving meat	253	15	344	587	0	1	..	74,385	27	1,468
3112 Dairy products	96	15	135	878	1	9	..	38,627	2	1,273
3113 Canning, preserving of fruits and vegetables	25	119	136	2,282	0	2	..	81,780	0	2,858
3114 Canning, preserving and processing of fish	8	41	20	174	39	0	..	18,968	0	350
3115 Vegetable and animal oils and fats	2,510	15	281	1,412	107	391	..	229,087	5,834	5,520
3116 Grain mill products	163	7	179	127	0	1	..	194,010	1,907	1,169
3117 Bakery products	85	241	247	641	128	12	..	98,414	250	1,707
3118 Sugar factories and refineries	20,149	6	178	5,764	830	19	..	490,219	457,429	27,064
3119 Cocoa, chocolate and sugar confectionery	45	139	48	397	0	2	..	42,128	4	783
3121 Other food products	3,705	22	274	2,764	0	3	..	221,042	326	7,563
3122 Prepared animal feeds	315	7	120	117	0	2	..	58,297	421	769
313 BEVERAGES	190	22	184	2,400	25	1	..	120,069	3,429	3,242
3131 Distilling, rectifying and blending of spirits	39	22	32	644	25	0	..	10,444	0	800
3132 Wine industries	3	0	3	23	0	1	..	1,958	0	37
3133 Malt liquors and malts	69	0	37	1,287	0	0	..	70,211	2,592	1,636
3134 Soft drinks	79	0	112	446	0	0	..	37,456	837	769
314 TOBACCO	135	13	64	452	5	3	..	58,515	2	883
32 TEXTILES, APPAREL AND LEATHER	**7,419**	**243**	**1,185**	**19,700**	**3,113**	**8**	..	**3,826,496**	**21,449**	**45,366**
321 TEXTILES	6,623	206	923	17,592	2,795	6	..	3,622,924	21,267	41,111
3211 Spinning weaving and finishing textiles	5,846	201	722	15,328	2,468	2	..	3,309,426	21,230	36,405
3212 Made-up goods excluding wearing apparel	130	1	21	271	0	0	..	67,964	9	668
3213 Knitting mills	587	2	136	1,569	31	1	..	154,390	9	2,882
3214 Carpets and rugs	57	2	29	408	287	3	..	86,627	28	1,098
3215 Cordage, rope and twine	3	0	0	1	0	0	..	1,661	0	10
3219 Other textiles	0	0	15	15	9	0	..	2,856	0	49
322 WEARING APPAREL, EXCEPT FOOTWEAR	735	15	226	1,622	317	1	..	163,681	124	3,505
323 LEATHER AND FUR PRODUCTS	43	0	29	267	0	0	..	23,812	0	425
3231 Tanneries and leather finishing	41	0	24	253	0	0	..	20,951	0	393
3232 Fur dressing and dyeing industries	0	0	0	9	0	0	..	604	0	11
3233 Leather prods. ex. footwear and wearing apparel	2	0	5	5	0	0	..	2,257	0	20
324 FOOTWEAR, EX. RUBBER AND PLASTIC	18	22	7	219	1	1	..	16,079	58	326
33 WOOD PRODUCTS AND FURNITURE	**1,558**	**124**	**173**	**878**	**1**	**313**	..	**248,174**	**1**	**3,940**
331 WOOD PRODUCTS, EXCEPT FURNITURE	1,553	124	135	804	1	312	..	229,567	1	3,755
3311 Sawmills, planing and other wood mills	1,553	124	133	796	1	311	..	227,907	1	3,738
3312 Wooden and cane containers	0	0	1	0	0	1	..	195	0	3
3319 Other wood and cork products	0	0	1	8	0	0	..	1,465	0	14
332 FURNITURE, FIXTURES, EXCL. METALLIC	5	0	38	74	0	1	..	18,607	0	185
34 PAPER, PUBLISHING AND PRINTING	**1,151**	**62**	**524**	**13,346**	**610**	**97**	..	**1,384,976**	**411,618**	**19,294**
341 PAPER AND PRODUCTS	1,117	34	405	13,016	565	96	..	1,307,202	411,595	18,457
3411 Pulp, paper and paperboard articles	599	5	297	12,582	514	96	..	1,244,001	411,501	17,090
3412 Containers of paper and paperboard	508	27	86	344	48	0	..	52,412	94	1,201
3419 Other pulp, paper and paperboard articles	10	2	22	90	3	0	..	10,789	0	166
342 PRINTING AND PUBLISHING	34	28	119	330	45	1	..	77,774	23	837
35 CHEMICAL PRODUCTS	**12,030**	**990**	**1,758**	**60,058**	**21,737**	**73**	..	**3,824,384**	**445,661**	**108,809**
351 INDUSTRIAL CHEMICALS	10,429	925	430	31,035	21,625	55	..	2,590,964	214,731	73,053
3511 Basic industrial chemicals excl. fertilizers	10,384	56	208	25,425	3,538	55	..	1,836,768	70,844	46,023
3512 Fertilizers and pesticides	0	1	154	2,692	17,633	0	..	319,126	143,887	21,111
3513 Synthetic resins and plastic materials	45	868	68	2,918	454	0	..	435,070	0	5,919
352 OTHER CHEMICALS	364	13	207	7,866	82	15	..	194,082	12	9,246
3521 Paints, varnishes and lacquers	20	0	30	349	15	0	..	28,241	0	516
3522 Drugs and medicines	1	1	65	5,598	3	1	..	57,446	10	5,876
3523 Soap, cleaning preparations, perfumes, cosmetics	76	0	24	933	57	14	..	56,193	0	1,306
3529 Other chemical products	267	12	88	986	7	0	..	52,202	2	1,548
353 PETROLEUM REFINERIES	0	17	304	17,249	12	0	..	438,538	230,604	18,331
354 MISC. PETROLEUM AND COAL PRODUCTS	10	25	477	1,143	0	0	..	28,752	112	1,758
355 RUBBER PRODUCTS	1,209	4	96	2,325	0	0	..	275,331	0	4,625
3551 Tyres and tubes	1,169	0	10	1,772	0	0	..	198,636	0	3,666
3559 Other rubber products	40	4	86	553	0	0	..	76,695	0	959
356 PLASTIC PRODUCTS	18	6	244	440	18	3	..	296,717	202	1,796
36 NON-METALLIC MINERAL PRODUCTS	**75,392**	**5,764**	**1,782**	**11,120**	**18,251**	**273**	..	**3,885,227**	**991**	**126,565**
361 POTTERY, CHINA, EARTHENWARE	279	3,908	86	240	2,733	4	..	225,739	57	8,062
362 GLASS AND PRODUCTS	125	1,222	122	3,980	5,826	0	..	363,787	108	12,584
369 OTHER NON-METAL. MINERAL PRODUCTS	74,988	634	1,574	6,900	9,692	269	..	3,295,701	826	105,919
3691 Structural clay products	12,050	1	381	1,741	0	47	..	238,161	768	15,075
3692 Cement, lime and plaster	62,587	512	531	4,682	8,554	220	..	2,937,317	0	87,660
3699 Other non-metallic mineral products	351	121	662	477	1,138	2	..	120,223	58	3,184
37 BASIC METAL INDUSTRIES	**146,930**	**1,256**	**1,674**	**38,161**	**1,968**	**72**	..	**5,357,538**	**500,971**	**207,545**
371 IRON AND STEEL	145,025	904	1,404	31,095	1,967	23	..	4,395,710	462,839	194,576
372 NON-FERROUS METALS	1,905	352	270	7,066	1	49	..	961,828	38,132	12,968
38 METAL PRODUCTS, MACHINERY, EQUIP.	**2,016**	**1,002**	**1,865**	**5,648**	**1,510**	**71**	..	**1,634,057**	**1,161**	**17,990**
381 METAL PRODUCTS	507	350	269	1,446	115	45	..	361,573	391	4,032
3811 Cutlery, hand tools and general hardware	27	21	32	271	0	1	..	77,547	334	630
3812 Furniture and fixtures primarily of metal	1	0	14	114	0	0	..	5,835	0	150
3813 Structural metal products	77	54	74	307	0	1	..	36,484	57	644
3819 Other fabricated metal products	402	275	149	754	115	43	..	241,707	0	2,608
382 NON-ELECTRICAL MACHINERY	522	26	381	1,214	618	21	..	323,331	371	3,945
3821 Engines and turbines	138	4	10	120	0	0	..	5,622	0	292
3822 Agricultural machinery and equipment	80	4	178	460	3	20	..	68,796	237	992
3823 Metal and wood working machinery	84	3	10	83	0	0	..	20,439	0	254
3824 Special industrial machinery	18	4	40	84	0	0	..	30,594	44	256
3825 Office, computing and accounting machinery	0	0	4	2	0	0	..	931	8	9
3829 Other non-electrical machinery and equipment	202	11	139	465	615	1	..	196,949	82	2,142
383 ELECTRICAL MACHINERY	91	378	272	1,231	26	2	..	314,128	125	3,130
3831 Electrical industrial machinery	17	38	57	320	0	0	..	62,912	20	658
3832 Radio, TV and communications equipment	8	5	67	299	0	0	..	80,308	80	668
3833 Electrical appliances and housewares	6	4	13	47	0	0	..	26,275	0	165
3839 Other electrical apparatus and supplies	60	331	135	565	26	2	..	144,633	25	1,640
384 TRANSPORT EQUIPMENT	807	236	906	1,601	751	3	..	604,379	269	6,479
3841 Shipbuilding	9	49	36	86	0	0	..	21,590	0	258
3842 Railroad equipment	332	19	141	21	0	0	..	31,770	32	627
3843 Motor vehicles	171	168	707	1,331	751	3	..	513,914	237	4,980
3844 Motorcycles and bicycles	0	0	9	58	0	0	..	11,085	0	107
3845 Aircraft	295	0	13	105	0	0	..	26,020	0	507
3849 Other transport equipment	..	..	..	..	..	..	..	..	..	..
385 PROFESSIONAL AND SCIENTIFIC EQUIPMENT	89	12	37	156	0	0	..	30,646	5	404
3851 Professional equipment	7	10	27	82	0	0	..	29,512	5	232
3852 Photographic and optical goods	0	0	1	1	0	0	..	1,134	0	6
3853 Watches and clocks	82	2	9	73	0	0	..	15,897	0	166
39 OTHER MANUFACTURING INDUSTRIES	**5**	**6**	**10**	**172**	**0**	**0**	..	**1,942**	**0**	**250**
3901 Jewellery and related articles	2	1	3	13	..	..	..	..	..	26
3902 Musical instruments	..	..	..	..	..	..	..	42	..	..
3903 Sporting and athletic goods	..	..	..	..	..	..	..	..	..	..
3909 Other manufactures	3	5	7	159	0	0	..	13,913	0	224
SERR non-specified, unallocated industry	..	..	..	..	..	..	..	..	..	..
3 TOTAL MANUFACTURING	**274,180**	**10,109**	**11,181**	**167,078**	**48,325**	**1,353**	..	**21,902,290**	**1,851,483**	**584,409**

ISIS Energy Data Programme (IEA/OECD)

TURKEY

ISIC Revision 2 Industry Sector	Solid TJ	LPG TJ	Distiloil TJ	RFO TJ	Gas TJ	Biomass TJ	Steam TJ	Electr MWh	Own Use MWh	TOTAL TJ
31 **FOOD, BEVERAGES AND TOBACCO**	17,579	828	782	13,484	3,002	20,227	..	1,193,918	272,011	59,221
311/2 FOOD	17,357	784	716	11,139	2,607	20,226	..	985,166	272,001	55,396
3111 Slaughtering, preparing and preserving meat	125	47	91	336	154	0	..	48,142	114	926
3112 Dairy products	105	2	39	600	4	0	..	34,187	5	873
3113 Canning, preserving of fruits and vegetables	6	33	37	2,287	1	0	..	76,082	33	2,638
3114 Canning, preserving and processing of fish	0	283	27	0	0	0	..	13,644	0	359
3115 Vegetable and animal oils and fats	2,118	3	139	1,287	749	20,113	..	213,307	2,835	25,167
3116 Grain mill products	247	0	10	67	0	0	..	4,590	0	341
3117 Bakery products	45	222	98	482	93	0	..	80,134	100	1,228
3118 Sugar factories and refineries	12,518	22	84	3,961	537	113	..	315,960	268,788	17,405
3119 Cocoa, chocolate and sugar confectionery	0	154	21	137	406	0	..	33,753	0	840
3121 Other food products	2,148	18	160	1,982	663	0	..	163,209	126	5,558
3122 Prepared animal feeds	45	0	10	0	0	0	..	2,158	0	63
313 BEVERAGES	0	3	53	2,050	309	0	..	122,170	10	2,855
3131 Distilling, rectifying and blending of spirits	0	1	20	767	0	0	..	12,868	10	834
3132 Wine industries	..	..	..	..	..	..	..	..	..	..
3133 Malt liquors and malts	0	1	22	1,077	274	0	..	81,557	0	1,668
3134 Soft drinks	0	1	11	206	35	0	..	27,745	0	353
314 TOBACCO	222	41	13	295	86	1	..	86,582	0	970
32 **TEXTILES, APPAREL AND LEATHER**	3,803	1,186	598	13,228	10,452	894	..	2,260,782	59,882	38,084
321 TEXTILES	3,574	1,182	575	12,190	10,292	894	..	2,186,380	59,822	36,363
3211 Spinning weaving and finishing textiles	3,347	1,182	497	11,353	6,953	894	..	2,005,308	59,662	31,230
3212 Made-up goods excluding wearing apparel	41	0	1	33	100	0	..	6,378	55	198
3213 Knitting mills	152	0	68	624	2,022	0	..	94,024	100	3,204
3214 Carpets and rugs	34	0	9	180	501	0	..	80,670	5	1,014
3215 Cordage, rope and twine	..	..	..	..	..	..	..	..	..	..
3219 Other textiles	0	0	0	0	716	0	..	0	0	716
322 WEARING APPAREL, EXCEPT FOOTWEAR	229	3	19	731	160	0	..	59,761	57	1,357
323 LEATHER AND FUR PRODUCTS	0	0	2	81	0	0	..	8,026	0	112
3231 Tanneries and leather finishing	0	0	2	81	0	0	..	8,026	0	112
3232 Fur dressing and dyeing industries	..	..	..	..	..	..	..	..	..	..
3233 Leather prods. ex. footwear and wearing apparel	..	..	..	..	..	..	..	..	..	..
324 FOOTWEAR, EX. RUBBER AND PLASTIC	0	1	2	226	0	0	..	6,615	3	253
33 **WOOD PRODUCTS AND FURNITURE**	1,084	17	51	538	235	158	..	233,247	24,002	2,836
331 WOOD PRODUCTS, EXCEPT FURNITURE	1,084	17	51	527	235	158	..	227,016	24,002	2,803
3311 Sawmills, planing and other wood mills	1,084	17	51	527	235	158	..	227,016	24,002	2,803
3312 Wooden and cane containers	..	..	..	..	..	..	..	..	..	..
3319 Other wood and cork products	..	..	..	..	..	..	..	..	..	..
332 FURNITURE, FIXTURES, EXCL. METALLIC	0	0	0	11	0	0	..	6,231	0	33
34 **PAPER, PUBLISHING AND PRINTING**	535	188	110	11,886	2,873	744	..	1,415,030	484,192	19,687
341 PAPER AND PRODUCTS	535	188	101	11,872	2,858	744	..	1,371,323	484,189	19,492
3411 Pulp, paper and paperboard articles	281	86	65	10,839	2,298	744	..	1,218,890	484,189	16,958
3412 Containers of paper and paperboard	254	100	8	955	51	0	..	89,246	0	1,689
3419 Other pulp, paper and paperboard articles	0	2	28	78	509	0	..	63,187	0	844
342 PRINTING AND PUBLISHING	0	0	9	14	15	0	..	43,707	3	195
35 **CHEMICAL PRODUCTS**	5,630	1,505	66,350	41,702	26,997	111	..	3,098,455	1,269,285	148,880
351 INDUSTRIAL CHEMICALS	4,201	141	66,051	33,289	25,287	17	..	2,260,697	974,343	133,617
3511 Basic industrial chemicals excl. fertilizers	3,421	45	53	8,195	0	17	..	260,686	36,534	12,538
3512 Fertilizers and pesticides	780	96	122	3,168	24,828	0	..	383,549	147,626	29,843
3513 Synthetic resins and plastic materials	0	0	65,876	21,926	459	0	..	1,616,462	790,183	91,236
352 OTHER CHEMICALS	604	39	171	1,213	1,489	94	..	153,363	678	4,160
3521 Paints, varnishes and lacquers	0	31	17	204	0	0	..	13,094	0	299
3522 Drugs and medicines	0	1	26	339	170	0	..	51,895	601	721
3523 Soap, cleaning preparations, perfumes, cosmetics	248	0	6	242	846	94	..	51,494	70	1,621
3529 Other chemical products	356	7	122	428	473	0	..	36,880	7	1,519
353 PETROLEUM REFINERIES	2	2	46	4,514	11	0	..	375,053	294,264	4,866
354 MISC. PETROLEUM AND COAL PRODUCTS	0	1,323	66	565	0	0	..	20,198	0	2,027
355 RUBBER PRODUCTS	823	0	11	1,971	210	0	..	267,905	0	3,979
3551 Tyres and tubes	823	0	10	1,870	210	0	..	227,494	0	3,732
3559 Other rubber products	0	0	1	101	0	0	..	40,411	0	247
356 PLASTIC PRODUCTS	0	0	5	150	0	0	..	21,239	0	231
36 **NON-METALLIC MINERAL PRODUCTS**	105,992	7,153	954	8,778	8,846	21	..	4,700,860	726	148,664
361 POTTERY, CHINA, EARTHENWARE	339	5,602	85	74	4,160	0	..	346,634	149	11,507
362 GLASS AND PRODUCTS	4	1,133	41	3,336	3,728	18	..	374,243	212	9,607
369 OTHER NON-METAL. MINERAL PRODUCTS	105,649	418	828	5,368	958	3	..	3,979,983	365	127,551
3691 Structural clay products	9,898	64	281	1,481	228	1	..	228,689	170	12,776
3692 Cement, lime and plaster	95,500	161	374	3,489	730	2	..	3,706,108	15	113,598
3699 Other non-metallic mineral products	251	193	173	398	0	0	..	45,186	180	1,177
37 **BASIC METAL INDUSTRIES**	152,456	1,375	1,132	33,342	10,539	48	..	8,445,950	1,479,311	223,972
371 IRON AND STEEL	151,458	986	767	27,964	10,199	3	..	6,866,318	1,446,873	210,887
372 NON-FERROUS METALS	998	389	365	5,378	340	45	..	1,579,632	32,438	13,085
38 **METAL PRODUCTS, MACHINERY, EQUIP.**	778	1,130	634	2,698	2,460	6	..	1,172,243	2,903	11,916
381 METAL PRODUCTS	67	285	36	329	225	2	..	184,335	0	1,608
3811 Cutlery, hand tools and general hardware	0	1	0	0	0	0	..	3,184	0	12
3812 Furniture and fixtures primarily of metal	0	0	0	0	27	0	..	2,000	0	34
3813 Structural metal products	41	13	5	79	0	1	..	25,635	0	231
3819 Other fabricated metal products	26	271	31	250	198	1	..	153,516	0	1,330
382 NON-ELECTRICAL MACHINERY	224	38	88	890	391	3	..	212,763	1,531	2,394
3821 Engines and turbines	0	1	13	17	0	0	..	3,256	0	43
3822 Agricultural machinery and equipment	47	1	21	119	232	0	..	34,957	0	546
3823 Metal and wood working machinery	72	1	2	8	0	0	..	9,325	0	117
3824 Special industrial machinery	..	..	..	..	..	..	..	..	..	..
3825 Office, computing and accounting machinery	0	0	4	347	0	0	..	2,715	0	361
3829 Other non-electrical machinery and equipment	105	35	48	399	159	3	..	162,510	1,531	1,329
383 ELECTRICAL MACHINERY	8	316	91	655	294	0	..	214,544	1,329	2,132
3831 Electrical industrial machinery	0	3	6	134	132	0	..	47,687	13	447
3832 Radio, TV and communications equipment	0	1	21	212	37	0	..	63,331	12	499
3833 Electrical appliances and housewares	0	0	0	1	0	0	..	2,883	0	11
3839 Other electrical apparatus and supplies	8	312	64	308	125	0	..	100,643	1,304	1,175
384 TRANSPORT EQUIPMENT	479	491	416	759	1,521	1	..	543,891	36	5,625
3841 Shipbuilding	1	9	15	37	2	0	..	17,364	0	127
3842 Railroad equipment	218	5	82	105	519	1	..	66,618	6	1,170
3843 Motor vehicles	34	477	301	429	1,000	0	..	417,730	30	3,745
3844 Motorcycles and bicycles	0	0	3	14	0	0	..	6,840	0	42
3845 Aircraft	226	0	15	174	0	0	..	35,339	0	542
3849 Other transport equipment	..	..	..	..	..	..	..	..	..	..
385 PROFESSIONAL AND SCIENTIFIC EQUIPMENT	0	0	3	65	29	0	..	16,710	7	157
3851 Professional equipment	0	0	3	65	29	0	..	16,710	7	157
3852 Photographic and optical goods	..	..	..	..	..	..	..	..	..	..
3853 Watches and clocks	..	..	..	..	..	..	..	..	..	..
39 **OTHER MANUFACTURING INDUSTRIES**	0	0	0	2	0	0	..	800	0	5
3901 Jewellery and related articles	..	..	..	..	..	..	..	..	..	..
3902 Musical instruments	..	..	..	..	..	..	..	..	..	..
3903 Sporting and athletic goods	..	..	..	..	..	..	..	..	..	..
3909 Other manufactures	0	0	0	2	0	0	..	800	0	5
SERR non-specified, unallocated industry	..	..	..	..	..	..	..	..	..	..
3 **TOTAL MANUFACTURING**	287,857	13,382	70,611	125,658	65,404	22,209	..	22,521,285	3,592,312	653,265

ISIC Revision 2 Industry Sector	Solid TJ	LPG TJ	Distiloil TJ	RFO TJ	Gas TJ	Biomass TJ	Steam TJ	Electr MWh	Own Use MWh	TOTAL TJ
31	**FOOD, BEVERAGES AND TOBACCO**	22,523	776	925	15,503	4,144	555	0	1,421,486	373,404	48,199
311/2	FOOD	22,386	755	839	13,284	3,692	553	0	1,214,898	373,349	44,539
3111	Slaughtering, preparing and preserving meat	87	176	115	307	11	0	0	60,855	58	915
3112	Dairy products	95	1	67	854	0	0	0	52,705	0	1,207
3113	Canning, preserving of fruits and vegetables	1	21	111	2,298	0	0	0	65,916	70	2,668
3114	Canning, preserving and processing of fish	0	0	0	0	0	0	0	21,359	0	77
3115	Vegetable and animal oils and fats	2,815	18	217	1,344	970	552	0	251,075	2,802	6,810
3116	Grain mill products	387	0	18	74	0	0	0	6,969	1,834	497
3117	Bakery products	59	511	134	581	231	0	0	117,351	78	1,938
3118	Sugar factories and refineries	16,333	13	102	5,218	2,029	1	0	410,023	368,343	23,846
3119	Cocoa, chocolate and sugar confectionery	0	12	24	189	68	0	0	29,912	68	400
3121	Other food products	2,561	3	43	2,413	383	0	0	196,363	79	6,110
3122	Prepared animal feeds	48	0	8	6	0	0	0	2,370	17	70
313	BEVERAGES	49	1	63	1,919	318	0	0	124,501	43	2,798
3131	Distilling, rectifying and blending of spirits	0	1	23	772	45	0	0	13,692	0	890
3132	Wine industries	..	..	..	..	..	..	..	..	..	..
3133	Malt liquors and malts	0	0	24	999	236	0	0	82,272	43	1,555
3134	Soft drinks	49	0	16	148	37	0	0	28,537	0	353
314	TOBACCO	88	20	23	300	134	2	0	82,087	12	862
32	**TEXTILES, APPAREL AND LEATHER**	5,113	601	415	14,074	9,240	113	106	2,844,164	62,246	39,677
321	TEXTILES	4,712	598	365	12,630	8,926	113	106	2,742,466	62,194	37,099
3211	Spinning weaving and finishing textiles	4,341	597	323	11,455	7,877	3	0	2,387,080	61,763	32,967
3212	Made-up goods excluding wearing apparel	313	0	3	319	116	0	0	44,041	23	909
3213	Knitting mills	20	0	24	547	639	110	106	206,555	403	2,188
3214	Carpets and rugs	38	1	15	309	212	0	0	102,099	5	943
3215	Cordage, rope and twine	..	..	..	..	..	..	..	..	..	..
3219	Other textiles	0	0	0	0	82	0	0	2,691	0	92
322	WEARING APPAREL, EXCEPT FOOTWEAR	401	2	46	1,138	275	0	0	87,090	3	2,176
323	LEATHER AND FUR PRODUCTS	0	0	2	40	39	0	0	8,075	0	110
3231	Tanneries and leather finishing	0	0	2	40	39	0	0	8,075	0	110
3232	Fur dressing and dyeing industries	..	..	..	..	..	..	..	..	..	..
3233	Leather prods. ex. footwear and wearing apparel	..	..	..	..	..	..	..	..	..	..
324	FOOTWEAR, EX. RUBBER AND PLASTIC	0	1	2	266	0	0	0	6,533	49	292
33	**WOOD PRODUCTS AND FURNITURE**	1,069	57	57	504	399	52	0	254,300	2	3,053
331	WOOD PRODUCTS, EXCEPT FURNITURE	1,069	57	54	504	399	52	0	252,708	2	3,045
3311	Sawmills, planing and other wood mills	1,069	57	54	504	399	52	0	252,708	2	3,045
3312	Wooden and cane containers	..	..	..	..	..	..	..	..	..	..
3319	Other wood and cork products	..	..	..	..	..	..	..	..	..	..
332	FURNITURE, FIXTURES, EXCL. METALLIC	0	0	3	0	0	0	0	1,592	0	9
34	**PAPER, PUBLISHING AND PRINTING**	960	294	103	10,310	2,689	6,568	226	1,204,888	384,949	24,102
341	PAPER AND PRODUCTS	960	287	80	10,157	2,657	6,568	226	1,164,386	384,929	23,741
3411	Pulp, paper and paperboard articles	614	159	58	9,156	2,225	6,568	226	1,015,655	380,656	21,292
3412	Containers of paper and paperboard	346	126	17	859	48	0	0	86,597	4,273	1,692
3419	Other pulp, paper and paperboard articles	0	2	5	142	384	0	0	62,134	0	757
342	PRINTING AND PUBLISHING	0	7	23	153	32	0	0	40,502	20	361
35	**CHEMICAL PRODUCTS**	10,033	426	1,482	98,926	30,832	3	4,511	3,730,987	1,619,305	153,815
351	INDUSTRIAL CHEMICALS	5,298	195	1,234	33,170	28,553	3	4,511	2,353,527	1,056,857	77,632
3511	Basic industrial chemicals excl. fertilizers	4,469	146	78	8,312	71	3	0	318,013	91,239	13,895
3512	Fertilizers and pesticides	829	49	1,127	3,355	28,057	0	0	386,738	178,920	34,165
3513	Synthetic resins and plastic materials	0	0	29	21,503	425	0	4,511	1,648,776	786,698	29,571
352	OTHER CHEMICALS	616	133	121	1,170	1,619	0	0	186,991	10,254	4,295
3521	Paints, varnishes and lacquers	0	2	22	238	0	0	0	19,070	84	330
3522	Drugs and medicines	0	1	22	280	175	0	0	52,326	109	666
3523	Soap, cleaning preparations, perfumes, cosmetics	226	123	8	199	1,028	0	0	66,896	5,053	1,807
3529	Other chemical products	390	7	69	453	416	0	0	48,699	5,008	1,492
353	PETROLEUM REFINERIES	3,304	0	34	62,653	0	0	0	830,491	552,027	66,993
354	MISC. PETROLEUM AND COAL PRODUCTS	0	98	68	277	0	0	0	16,812	55	503
355	RUBBER PRODUCTS	815	0	18	1,521	660	0	0	300,978	0	4,098
3551	Tyres and tubes	815	0	12	1,278	660	0	0	244,154	0	3,644
3559	Other rubber products	0	0	6	243	0	0	0	56,824	0	454
356	PLASTIC PRODUCTS	0	0	7	135	0	0	0	42,188	112	293
36	**NON-METALLIC MINERAL PRODUCTS**	105,001	5,546	1,544	7,628	14,620	25	0	4,821,334	8,824	151,689
361	POTTERY, CHINA, EARTHENWARE	145	4,485	124	76	5,692	0	0	355,447	8,061	11,773
362	GLASS AND PRODUCTS	0	810	48	2,819	7,543	0	0	415,194	351	12,713
369	OTHER NON-METAL. MINERAL PRODUCTS	104,856	251	1,372	4,733	1,385	25	0	4,050,693	412	127,203
3691	Structural clay products	8,788	40	475	1,180	312	2	0	240,205	284	11,661
3692	Cement, lime and plaster	95,941	67	633	3,031	1,073	21	0	3,762,301	118	114,310
3699	Other non-metallic mineral products	127	144	264	522	0	2	0	48,187	10	1,232
37	**BASIC METAL INDUSTRIES**	123,688	1,603	1,197	37,670	12,900	16	0	10,020,421	2,002,899	205,937
371	IRON AND STEEL	121,137	949	807	31,889	12,523	12	0	8,429,104	1,979,706	190,535
372	NON-FERROUS METALS	2,551	654	390	5,781	377	4	0	1,591,317	23,193	15,402
38	**METAL PRODUCTS, MACHINERY, EQUIP.**	926	1,147	625	2,067	3,284	8	73	1,242,343	54,796	12,405
381	METAL PRODUCTS	221	373	46	255	349	5	73	211,059	0	2,082
3811	Cutlery, hand tools and general hardware	0	38	2	0	25	0	0	3,884	0	54
3812	Furniture and fixtures primarily of metal	0	0	0	0	25	0	0	1,956	0	32
3813	Structural metal products	175	63	15	61	18	4	0	29,109	0	441
3819	Other fabricated metal products	46	272	29	194	306	1	73	176,110	0	1,555
382	NON-ELECTRICAL MACHINERY	185	31	116	517	801	3	0	221,601	40,232	2,306
3821	Engines and turbines	0	2	15	19	0	0	0	3,317	2	48
3822	Agricultural machinery and equipment	29	1	36	93	266	0	0	39,556	3	567
3823	Metal and wood working machinery	32	1	2	14	0	0	0	10,128	0	85
3824	Special industrial machinery	..	..	..	..	..	..	..	..	..	74
3825	Office, computing and accounting machinery	9	0	5	48	0	0	0	3,203	0	74
3829	Other non-electrical machinery and equipment	115	27	58	343	535	3	0	165,397	40,227	1,532
383	ELECTRICAL MACHINERY	0	187	108	510	419	0	0	222,644	13,493	1,977
3831	Electrical industrial machinery	0	1	7	103	136	0	0	51,301	20	432
3832	Radio, TV and communications equipment	0	0	20	133	156	0	0	67,815	8,361	524
3833	Electrical appliances and housewares	0	0	0	2	0	0	0	3,134	0	13
3839	Other electrical apparatus and supplies	0	185	81	272	127	0	0	100,394	5,112	1,008
384	TRANSPORT EQUIPMENT	520	556	352	755	1,685	0	0	577,976	1,064	5,945
3841	Shipbuilding	0	9	10	41	1	0	0	18,659	0	128
3842	Railroad equipment	306	8	51	78	234	0	0	23,514	359	760
3843	Motor vehicles	41	516	280	437	1,431	0	0	487,870	705	4,459
3844	Motorcycles and bicycles	0	23	0	5	0	0	0	9,085	0	61
3845	Aircraft	173	0	11	194	19	0	0	38,848	0	537
3849	Other transport equipment	..	..	..	..	..	..	..	..	..	..
385	PROFESSIONAL AND SCIENTIFIC EQUIPMENT	0	0	3	30	30	0	0	9,063	7	96
3851	Professional equipment	0	0	3	30	30	0	0	9,063	7	96
3852	Photographic and optical goods	..	..	..	..	..	..	..	..	..	..
3853	Watches and clocks	..	..	..	..	..	..	..	..	..	..
39	**OTHER MANUFACTURING INDUSTRIES**	0	0	0	6	0	0	0	184	0	7
3901	Jewellery and related articles	0	0	0	6	0	0	0	184	0	7
3902	Musical instruments	..	..	..	..	..	..	..	..	..	..
3903	Sporting and athletic goods	..	..	..	..	..	..	..	..	..	..
3909	Other manufactures	..	..	..	..	..	..	..	..	..	..
SERR	**non-specified, unallocated industry**										
3	**TOTAL MANUFACTURING**	269,313	10,450	6,348	186,694	78,108	7,340	4,916	25,540,291	4,506,425	638,891

ISIC Rev. 2 Industry Sector		Solid TJ	LPG TJ	Distiloil TJ	RFO TJ	Gas TJ	Biomass TJ	Steam TJ	Electr MWh	Own Use MWh	TOTAL TJ
31	FOOD, BEVERAGES AND TOBACCO	24,987	1,207	1,213	16,387	4,306	424	0	1,540,634	459,665	52,415
311/2	FOOD	24,734	1,110	1,117	14,085	3,841	423	0	1,340,297	459,463	48,481
3111	Slaughtering, preparing and preserving meat	264	158	153	249	16	0	0	71,325	306	1,096
3112	Dairy products	59	15	74	770	0	0	0	48,556	161	1,092
3113	Canning, preserving of fruits and vegetables	1	29	120	2,012	12	0	0	69,679	0	2,425
3114	Canning, preserving and processing of fish	0	142	9	118	0	0	0	20,339	0	342
3115	Vegetable and animal oils and fats	1,988	8	254	1,315	1,051	421	0	251,228	1,892	5,935
3116	Grain mill products	0	0	7	71	0	0	0	4,696	0	95
3117	Bakery products	53	523	199	523	264	0	0	106,338	12	1,945
3118	Sugar factories and refineries	19,787	14	100	5,923	1,944	2	0	495,095	442,520	27,959
3119	Cocoa, chocolate and sugar confectionery	0	190	53	172	61	0	0	57,945	0	685
3121	Other food products	2,537	31	141	2,924	493	0	0	212,779	14,554	6,840
3122	Prepared animal feeds	45	0	7	8	0	0	0	2,317	18	68
313	BEVERAGES	135	46	63	1,994	322	0	0	124,398	189	3,007
3131	Distilling, rectifying and blending of spirits	1	1	33	852	44	0	0	13,116	164	978
3132	Wine industries	..	..	..	..	..	..	..	..	..	..
3133	Malt liquors and malts	0	0	22	1,032	249	0	0	82,091	24	1,598
3134	Soft drinks	134	45	8	110	29	0	0	29,191	1	431
314	TOBACCO	118	51	33	308	143	1	0	75,939	13	927
32	TEXTILES, APPAREL AND LEATHER	4,221	1,530	669	14,565	11,733	109	197	3,196,878	117,868	44,108
321	TEXTILES	3,854	1,397	600	13,761	11,260	109	197	3,124,062	117,832	42,000
3211	Spinning weaving and finishing textiles	3,016	1,317	524	11,888	9,573	47	197	2,606,015	117,596	35,323
3212	Made-up goods excluding wearing apparel	656	11	30	1,185	54	0	0	134,811	0	2,421
3213	Knitting mills	141	68	29	289	1,308	62	0	270,559	230	3,067
3214	Carpets and rugs	41	1	13	358	233	0	0	108,554	0	1,037
3215	Cordage, rope and twine	..	..	4	41	92	0	0	4,123	6	152
3219	Other textiles	..	..	..	..	..	..	..	..	..	..
322	WEARING APPAREL, EXCEPT FOOTWEAR	340	133	63	555	410	0	0	56,529	0	1,705
323	LEATHER AND FUR PRODUCTS	27	0	3	29	63	0	0	10,247	0	159
3231	Tanneries and leather finishing	27	0	3	29	63	0	0	10,247	0	159
3232	Fur dressing and dyeing industries	..	..	..	..	..	..	..	..	..	..
3233	Leather prods. ex. footwear and wearing apparel	..	..	..	..	..	..	..	..	..	..
324	FOOTWEAR, EX. RUBBER AND PLASTIC	0	0	3	220	0	0	0	6,040	36	245
33	WOOD PRODUCTS AND FURNITURE	1,018	211	82	588	909	44	0	273,258	24,389	3,748
331	WOOD PRODUCTS, EXCEPT FURNITURE	1,018	211	80	588	909	43	0	266,904	24,389	3,722
3311	Sawmills, planing and other wood mills	983	211	78	551	909	43	0	262,699	24,389	3,633
3312	Wooden and cane containers	35	0	2	37	0	0	0	4,205	0	89
3319	Other wood and cork products	..	..	..	..	..	..	..	..	..	..
332	FURNITURE, FIXTURES, EXCL. METALLIC	0	0	2	0	0	1	0	6,354	..	26
34	PAPER, PUBLISHING AND PRINTING	407	883	106	11,355	3,114	6,051	236	1,302,834	471,498	25,145
341	PAPER AND PRODUCTS	407	791	73	7,723	2,653	4,757	0	860,741	349,826	18,243
3411	Pulp, paper and paperboard articles	254	211	60	5,507	2,612	4,757	0	678,808	278,456	14,842
3412	Containers of paper and paperboard	153	580	12	2,201	41	0	0	180,932	71,368	3,381
3419	Other pulp, paper and paperboard articles	0	0	1	15	0	0	0	1,001	2	20
342	PRINTING AND PUBLISHING	0	92	33	3,632	461	1,294	236	442,093	121,672	6,902
35	CHEMICAL PRODUCTS	8,872	730	2,231	78,761	33,950	0	4,778	3,870,765	1,750,929	136,953
351	INDUSTRIAL CHEMICALS	5,678	253	1,853	34,496	26,444	0	4,778	2,530,665	1,095,720	78,668
3511	Basic industrial chemicals excl. fertilizers	4,962	171	216	8,723	262	0	0	343,696	111,797	15,169
3512	Fertilizers and pesticides	716	74	1,590	3,556	25,854	0	0	396,881	171,074	32,603
3513	Synthetic resins and plastic materials	0	8	47	22,217	328	0	4,778	1,790,088	812,849	30,896
352	OTHER CHEMICALS	851	278	149	1,290	1,583	0	0	207,070	19,054	4,828
3521	Paints, varnishes and lacquers	0	1	37	323	0	0	0	29,927	108	468
3522	Drugs and medicines	0	1	29	260	203	0	0	55,289	107	692
3523	Soap, cleaning preparations, perfumes, cosmetics	552	261	10	402	1,122	0	0	84,692	12,990	2,605
3529	Other chemical products	299	15	73	305	258	0	0	37,162	5,849	1,063
353	PETROLEUM REFINERIES	1,674	143	117	41,144	5,256	0	0	778,794	635,961	48,848
354	MISC. PETROLEUM AND COAL PRODUCTS	0	2	75	85	0	0	0	7,855	0	190
355	RUBBER PRODUCTS	669	43	18	1,449	667	0	0	266,261	0	3,805
3551	Tyres and tubes	669	43	12	1,307	667	0	0	245,228	0	3,581
3559	Other rubber products	0	0	6	142	0	0	0	21,033	0	224
356	PLASTIC PRODUCTS	0	11	19	297	0	0	0	80,120	194	615
36	NON-METALLIC MINERAL PRODUCTS	108,842	5,704	2,145	8,118	16,264	2	0	5,055,312	29,993	159,166
361	POTTERY, CHINA, EARTHENWARE	86	2,084	48	112	2,644	0	0	183,460	2,062	5,627
362	GLASS AND PRODUCTS	0	932	55	3,051	7,222	0	0	431,175	269	12,811
369	OTHER NON-METAL. MINERAL PRODUCTS	108,756	2,688	2,042	4,955	6,398	2	0	4,440,677	27,662	140,728
3691	Structural clay products	10,567	2,114	452	1,243	127	1	0	291,364	26,475	15,458
3692	Cement, lime and plaster	97,978	110	1,261	3,356	1,900	1	0	3,953,689	196	118,839
3699	Other non-metallic mineral products	211	464	329	356	4,371	0	0	195,624	991	6,432
37	BASIC METAL INDUSTRIES	131,578	1,471	1,230	35,988	20,087	1,075	0	10,891,967	2,613,012	221,233
371	IRON AND STEEL	127,711	1,220	953	29,622	19,788	1,022	0	1,671,224	24,209	186,245
372	NON-FERROUS METALS	3,867	251	277	6,366	299	53	0	9,220,743	2,588,803	34,988
38	METAL PRODUCTS, MACHINERY, EQUIP.	1,028	1,345	659	2,283	4,221	233	74	1,396,888	86,497	14,560
381	METAL PRODUCTS	366	540	30	345	342	233	0	265,972	180	2,813
3811	Cutlery, hand tools and general hardware	0	84	5	10	0	0	0	4,791	0	116
3812	Furniture and fixtures primarily of metal	0	1	1	22	31	0	0	6,525	165	78
3813	Structural metal products	163	55	8	24	18	3	0	31,155	0	383
3819	Other fabricated metal products	203	400	16	289	293	230	0	223,501	15	2,236
382	NON-ELECTRICAL MACHINERY	37	43	98	599	1,250	0	74	323,266	69,087	3,016
3821	Engines and turbines	0	1	4	8	0	0	0	3,578	0	26
3822	Agricultural machinery and equipment	11	7	50	144	418	0	0	105,621	2	1,010
3823	Metal and wood working machinery	25	0	0	10	0	0	0	7,545	0	62
3824	Special industrial machinery	..	..	..	..	..	..	..	..	..	..
3825	Office, computing and accounting machinery	..	..	..	..	..	..	..	..	..	..
3829	Other non-electrical machinery and equipment	1	35	44	437	832	0	74	206,522	69,085	1,918
383	ELECTRICAL MACHINERY	14	254	97	551	430	0	0	277,556	16,925	2,284
3831	Electrical industrial machinery	0	3	5	112	202	0	0	71,850	10,295	544
3832	Radio, TV and communications equipment	0	2	26	152	94	0	0	70,373	1,262	523
3833	Electrical appliances and housewares	0	0	0	2	0	0	0	3,889	0	16
3839	Other electrical apparatus and supplies	14	249	66	285	134	0	0	131,444	5,368	1,202
384	TRANSPORT EQUIPMENT	610	507	423	699	2,176	0	0	512,858	297	6,260
3841	Shipbuilding	0	12	10	41	1	0	0	22,565	0	145
3842	Railroad equipment	395	10	56	76	297	0	0	23,351	12	918
3843	Motor vehicles	1	463	344	428	1,843	0	0	418,885	285	4,586
3844	Motorcycles and bicycles	0	22	2	6	0	0	0	11,357	0	71
3845	Aircraft	214	0	11	148	35	0	0	36,700	0	540
3849	Other transport equipment	..	..	..	..	..	..	..	..	..	..
385	PROFESSIONAL AND SCIENTIFIC EQUIPMENT	1	1	11	89	23	0	0	17,236	8	187
3851	Professional equipment	1	1	4	4	23	0	0	4,595	8	50
3852	Photographic and optical goods	..	..	..	..	..	..	..	..	..	..
3853	Watches and clocks	0	0	7	85	0	0	0	12,641	0	138
39	OTHER MANUFACTURING INDUSTRIES	0	0	0	5	0	0	0	1,399	0	10
3901	Jewellery and related articles	..	..	..	..	..	..	..	..	..	..
3902	Musical instruments	..	..	..	..	..	..	..	..	..	..
3903	Sporting and athletic goods	..	..	..	..	..	..	..	..	..	..
3909	Other manufactures	0	0	0	5	0	0	0	1,399	0	10
SERR	non-specified, unallocated industry	..	..	..	..	..	..	..	..	..	..
3	TOTAL MANUFACTURING	280,953	13,081	8,335	168,050	94,584	7,938	5,285	27,529,935	5,553,851	657,340

Year: **1994** | **UNITED STATES**

ISIC Revision 2 Industry Sector		Solid TJ	LPG TJ	Distilloil TJ	RFO TJ	Gas TJ	Biomass TJ	Steam TJ	Electr MWh	Own Use MWh	TOTAL TJ
31	**FOOD, BEVERAGES AND TOBACCO**	c	c	c	30,027	c	c	..	58,778,486	..	241,630
311/2	FOOD	c	c	c	c	546,879	c	..	c	..	546,879
3111	Slaughtering, preparing and preserving meat	738	1,132	2,012	4,482	74,119	2,285	..	11,711,896	..	126,931
3112	Dairy products	c	c	2,594	2,040	59,591	876	..	7,137,847	..	90,797
3113	Canning, preserving of fruits and vegetables	3,813	932	1,304	4,095	98,433	4,438	..	4,828,134	..	130,396
3114	Canning, preserving and processing of fish	0	c	3,543	c	2,165	c	..	770,928	..	8,483
3115	Vegetable and animal oils and fats	17,629	c	1,690	3,094	64,468	4,609	..	3,752,759	..	105,000
3116	Grain mill products	c	c	c	c	22,876	1,422	..	4,196,444	..	39,405
3117	Bakery products	0	740	1,113	c	39,361	c	..	4,097,963	..	55,967
3118	Sugar factories and refineries	c	c	c	c	34,343	87,688	..	c	..	122,031
3119	Cocoa, chocolate and sugar confectionery	0	c	c	c	16,375	2,784	..	2,104,394	..	26,735
3121	Other food products	c	c	787	2,340	110,306	5,848	..	11,064,221	..	159,112
3122	Prepared animal feeds	c	c	2,070	1,425	24,842	752	..	2,928,282	..	39,631
313	BEVERAGES	19,391	c	c	c	49,873	c	..	c	..	69,264
3131	Distilling, rectifying and blending of spirits	1,945	c	c	c	2,095	c	..	c	..	4,040
3132	Wine industries	0	c	c	c	10,393	c	..	425,138	..	11,923
3133	Malt liquors and malts	17,446	c	c	3,661	30,020	1,573	..	2,543,484	..	61,857
3134	Soft drinks	c	c	c	755	7,365	c	..	1,745,117	..	14,402
314	TOBACCO	c	c	0	832	c	c	..	842,094	..	3,864
32	**TEXTILES, APPAREL AND LEATHER**	43,141	c	8,517	19,253	142,626	c	..	41,869,790	..	364,268
321	TEXTILES	c	c	c	c	c	c	..	35,525,241	..	127,891
3211	Spinning weaving and finishing textiles	32,653	1,255	3,379	9,888	72,030	6,700	..	25,543,484	..	217,862
3212	Made-up goods excluding wearing apparel	c	c	c	c	12,346	c	..	2,365,501	..	20,862
3213	Knitting mills	3,384	1,674	3,715	5,240	21,415	4,766	..	4,242,834	..	55,468
3214	Carpets and rugs	3,519	c	c	1,431	8,988	c	..	1,287,147	..	18,572
3215	Cordage, rope and twine	0	c	c	0	c	c	..	169,951	..	612
3219	Other textiles	c	c	c	c	9,905	c	..	1,916,324	..	16,804
322	WEARING APPAREL, EXCEPT FOOTWEAR	c	c	c	c	12,736	536	..	5,444,396	..	32,872
323	LEATHER AND FUR PRODUCTS	0	c	c	c	4,719	c	..	609,880	..	6,915
3231	Tanneries and leather finishing	0	c	c	2,076	4,172	c	..	413,576	..	7,737
3232	Fur dressing and dyeing industries	..	..	..	..	..	..	..	..	..	..
3233	Leather prods. ex. footwear and wearing apparel	0	c	c	c	547	c	..	196,304	..	1,254
324	FOOTWEAR, EX. RUBBER AND PLASTIC	0	c	c	c	c	c	..	290,273	..	1,045
33	**WOOD PRODUCTS AND FURNITURE**	c	4,437	22,644	c	54,037	320,594	..	24,309,285	..	489,225
331	WOOD PRODUCTS, EXCEPT FURNITURE	c	c	c	c	43,109	241,114	..	18,873,695	..	352,168
3311	Sawmills, planing and other wood mills	c	3,359	c	0	38,843	237,545	..	17,348,157	..	342,200
3312	Wooden and cane containers	0	c	c	0	1,269	1,198	..	520,105	..	4,339
3319	Other wood and cork products	0	c	549	c	2,997	2,371	..	1,005,433	..	9,537
332	FURNITURE, FIXTURES, EXCL. METALLIC	1,401	551	c	c	10,823	12,934	..	5,274,354	..	44,697
34	**PAPER, PUBLISHING AND PRINTING**	c	c	10,499	171,236	590,620	1,124,268	..	81,192,430	..	2,188,916
341	PAPER AND PRODUCTS	c	c	8,993	170,925	544,349	1,122,172	..	63,749,329	..	2,075,937
3411	Pulp, paper and paperboard articles	304,010	2,491	7,119	164,852	468,726	1,111,782	..	50,139,920	..	2,239,484
3412	Containers of paper and paperboard	0	1,736	749	2,051	41,958	1,152	..	6,614,111	..	71,457
3419	Other pulp, paper and paperboard articles	c	c	1,125	4,022	33,665	9,238	..	6,995,298	..	73,233
342	PRINTING AND PUBLISHING	0	1,166	1,506	c	46,271	2,096	..	17,443,101	..	113,834
35	**CHEMICAL PRODUCTS**	c	53,589	c	140,657	3,946,331	547,785	..	233,337,814	..	5,528,378
351	INDUSTRIAL CHEMICALS	c	c	8,766	c	c	c	..	138,369,629	..	506,897
3511	Basic industrial chemicals excl. fertilizers	c	2,393	5,566	26,887	1,114,924	c	..	106,176,279	..	1,532,005
3512	Fertilizers and pesticides	2,366	c	1,324	1,698	297,268	6,095	..	6,544,487	..	332,311
3513	Synthetic resins and plastic materials	c	597	1,876	c	c	c	..	25,648,863	..	94,809
352	OTHER CHEMICALS	c	c	c	c	c	c	..	15,101,865	..	54,367
3521	Paints, varnishes and lacquers	c	c	c	c	c	c	..	1,266,909	..	4,561
3522	Drugs and medicines	10,364	c	2,140	4,466	39,506	3,320	..	6,860,458	..	84,494
3523	Soap, cleaning preparations, perfumes, cosmetics	c	c	1,051	c	23,886	c	..	2,180,335	..	32,786
3529	Other chemical products	18,658	c	755	12,559	53,875	2,438	..	4,794,163	..	105,544
353	PETROLEUM REFINERIES	c	c	6,693	67,906	719,472	84,034	..	33,334,560	..	998,109
354	MISC. PETROLEUM AND COAL PRODUCTS	c	c	14,513	3,036	52,013	17,259	..	2,093,764	..	94,359
355	RUBBER PRODUCTS	2,524	c	1,521	6,803	42,086	c	..	9,045,554	..	85,498
3551	Tyres and tubes	1,800	c	663	4,491	21,941	1,722	..	4,663,559	..	47,406
3559	Other rubber products	724	c	858	2,312	20,145	c	..	4,381,995	..	39,814
356	PLASTIC PRODUCTS	2,161	2,354	1,529	3,286	60,777	2,027	..	35,392,442	..	199,547
36	**NON-METALLIC MINERAL PRODUCTS**	284,152	3,806	22,562	7,496	413,459	79,757	..	36,865,500	..	943,948
361	POTTERY, CHINA, EARTHENWARE	0	c	c	0	18,690	c	..	1,101,270	..	22,655
362	GLASS AND PRODUCTS	c	c	831	4,269	160,495	c	..	11,323,669	..	206,360
369	OTHER NON-METAL. MINERAL PRODUCTS	c	c	21,691	c	234,274	c	..	24,440,561	..	343,951
3691	Structural clay products	c	c	878	c	47,959	6,639	..	1,542,255	..	61,028
3692	Cement, lime and plaster	274,413	c	6,148	1,635	35,589	53,317	..	11,796,729	..	413,570
3699	Other non-metallic mineral products	6,789	c	14,665	c	150,726	c	..	11,101,577	..	212,146
37	**BASIC METAL INDUSTRIES**	743,827	4,981	12,230	42,909	1,199,282	16,037	..	141,331,784	..	2,528,060
371	IRON AND STEEL	724,423	2,257	6,126	42,316	1,001,297	9,741	..	66,378,302	..	2,025,122
372	NON-FERROUS METALS	19,404	2,724	6,104	593	197,985	6,296	..	74,953,482	..	502,939
38	**METAL PRODUCTS, MACHINERY, EQUIP.**	c	c	17,907	22,932	572,044	65,318	..	155,552,701	..	1,238,191
381	METAL PRODUCTS	c	c	c	c	186,161	c	..	29,469,306	..	292,251
3811	Cutlery, hand tools and general hardware	c	c	c	1,090	14,828	c	..	3,144,931	..	27,240
3812	Furniture and fixtures primarily of metal	c	c	c	c	13,370	c	..	1,949,012	..	20,386
3813	Structural metal products	c	1,640	c	c	26,012	c	..	5,095,027	..	45,994
3819	Other fabricated metal products	c	2,334	2,184	c	131,951	910	..	19,280,336	..	206,788
382	NON-ELECTRICAL MACHINERY	c	c	c	c	121,109	c	..	c	..	121,109
3821	Engines and turbines	c	c	1,587	c	10,214	c	..	2,951,049	..	22,425
3822	Agricultural machinery and equipment	c	533	c	c	9,530	c	..	c	..	10,063
3823	Metal and wood working machinery	0	c	c	c	8,749	c	..	3,640,563	..	21,855
3824	Special industrial machinery	c	c	690	c	20,112	1,004	..	5,262,143	..	40,750
3825	Office, computing and accounting machinery	0	c	c	c	5,899	c	..	4,905,264	..	23,558
3829	Other non-electrical machinery and equipment	c	2,240	1,394	c	66,605	2,279	..	18,867,973	..	140,443
383	ELECTRICAL MACHINERY	c	c	c	c	81,056	c	..	c	..	81,056
3831	Electrical industrial machinery	c	c	c	c	14,215	1,563	..	4,549,523	..	32,156
3832	Radio, TV and communications equipment	c	c	805	1,472	27,677	c	..	17,120,591	..	91,588
3833	Electrical appliances and housewares	c	c	c	c	4,042	c	..	c	..	4,042
3839	Other electrical apparatus and supplies	c	1,455	c	1,230	35,121	c	..	11,124,124	..	77,853
384	TRANSPORT EQUIPMENT	c	c	c	c	c	c	..	c	..	c
3841	Shipbuilding	0	c	1,105	3,085	c	c	..	2,004,012	..	11,404
3842	Railroad equipment	c	c	c	c	c	c	..	c	..	3,844
3843	Motor vehicles	c	3,099	c	3,353	117,828	10,072	..	27,816,056	..	234,490
3844	Motorcycles and bicycles	0	c	c	c	793	c	..	c	..	793
3845	Aircraft	c	c	2,975	5,854	32,381	8,092	..	12,176,215	..	93,136
3849	Other transport equipment	..	..	..	..	c	c	..	c	..	c
385	PROFESSIONAL AND SCIENTIFIC EQUIPMENT	c	c	c	c	c	c	..	c	..	c
3851	Professional equipment	c	c	839	659	16,567	c	..	9,357,228	..	51,751
3852	Photographic and optical goods	c	c	c	c	10,292	c	..	3,096,050	..	21,438
3853	Watches and clocks	0	c	c	0	c	c	..	c	..	c
39	**OTHER MANUFACTURING INDUSTRIES**	c	c	c	c	c	c	..	5,096,848	..	18,349
3901	Jewellery and related articles	0	c	c	c	c	c	..	582,422	..	2,097
3902	Musical instruments	0	0	c	c	c	c	..	157,796	..	568
3903	Sporting and athletic goods	0	c	c	c	c	c	..	909,731	..	3,275
3909	Other manufactures	c	c	c	c	9,542	c	..	3,446,899	..	21,951
SERR	**non-specified, unallocated industry**	..	..	..	..	..	..	..	..	..	..
3	**TOTAL MANUFACTURING**	2,643,888	98,243	150,890	437,154	7,531,834	2,193,672	..	778,334,638	..	15,857,686

ISIS Energy Data Programme (IEA/OECD)

Information System on Industrial Structures (ISIS)

Energy Consumption Statistics

in

Manufacturing Industry

1990 – 1998

Separately by fuel types

(ISIC Revision 3)

OECD
ORGANISATION FOR ECONOMIC
CO-OPERATION AND DEVELOPMENT

OCDE
ORGANISATION DE COOPERATION
ET DE DEVELOPPEMENT ECONOMIQUES

IEA
INTERNATIONAL ENERGY AGENCY

AIE
AGENCE INTERNATIONALE DE L'ENERGIE

9, RUE DE LA FÉDERATION, 75739 PARIS CEDEX 15
TEL. (33-1) 40.57.65.00 / TELEFAX (33-1) 40.57.66.49 / INTERNET: HTTP://WWW.IEA.ORG

ISIC Revision 3 Industry Sector	Solid TJ	LPG TJ	Distiloil TJ	RFO TJ	Gas TJ	Biomass TJ	Steam TJ	Electr MWh	Own Use MWh	TOTAL TJ
15　FOOD PRODUCTS AND BEVERAGES	161	157	2,640	6,289	8,649	341	652	1,317,375	..	23,632
151　Production, processing and preserving (PPP)	..	..	..	..	..	..	..			..
1511　PPP of meat and meat products	..	..	..	..	..	..	..			..
1512　Processing and preserving of fish products	..	..	..	..	..	..	..			..
1513　Processing, preserving of fruit & vegetables	..	..	..	..	..	..	..			..
1514　Vegetable and animal oils and fats	..	..	..	..	..	..	..			..
152　Dairy products	..	..	..	..	..	..	..			..
153　Grain mill prod., starches & prepared animal feeds	..	..	..	..	..	..	..			..
1531　Grain mill products	..	..	..	..	..	..	..			..
1532　Starches and starch products	..	..	..	..	..	..	..			..
1533　Prepared animal feeds	..	..	..	..	..	..	..			..
154　Other food products	..	..	..	..	..	..	..			..
1541　Bakery products	..	..	..	..	..	..	..			..
1542　Sugar	..	..	..	..	..	..	..			..
1543　Cocoa, chocolate and sugar confectionery	..	..	..	..	..	..	..			..
1544　Macaroni, noodles, couscous & similar farinaceous prod.	..	..	..	..	..	..	..			..
1549　Other food products, nec	..	..	..	..	..	..	..			..
155　Beverages	..	..	..	..	..	..	..			..
1551　Distilling, rectifying and blending of spirits	..	..	..	..	..	..	..			..
1552　Wines	..	..	..	..	..	..	..			..
1553　Malt liquors and malt	..	..	..	..	..	..	..			..
1554　Soft drinks; production of mineral waters	..	..	..	..	..	..	..			..
16　TOBACCO PRODUCTS										
17　TEXTILES	4	22	159	1,827	2,499	374	36	728,222	..	7,543
171　Spinning, weaving and finishing of textiles	..	..	..	..	..	..	..			..
1711　Preparation and spinning of textile fibres	..	..	..	..	..	..	..			..
1712　Finishing of textiles	..	..	..	..	..	..	..			..
172　Other textiles	..	..	..	..	..	..	..			..
1721　Made-up textile articles, except apparel	..	..	..	..	..	..	..			..
1722　Carpets and rugs	..	..	..	..	..	..	..			..
1723　Cordage, rope, twine and netting	..	..	..	..	..	..	..			..
1729　Other textiles, nec	..	..	..	..	..	..	..			..
173　Knitted and crocheted fabrics and articles	..	..	..	..	..	..	..			..
18　WEARING APPAREL, DRESSING & DYEING OF FUR	11	8	142	734	93	15	5	84,731	..	1,313
181　Wearing apparel, except fur apparel	..	..	..	..	..	..	..			..
182　Dressing and dyeing of fur; articles of fur	..	..	..	..	..	..	..			..
19　TANNING & DRESSING OF LEATHER, FOOTWEAR	6	1	79	294	95	21	11	66,972	..	748
191　Tanning and dressing of leather	..	..	..	..	..	..	..			..
1911　Tanning and dressing of leather	..	..	..	..	..	..	..			..
1912　Luggage, handbags, saddlery & harness	..	..	..	..	..	..	..			..
192　Footwear	..	..	..	..	..	..	..			..
20　WOOD AND WOOD PRODUCTS	74	11	2,692	1,809	1,398	4,354	90	992,867	..	14,002
201　Sawmilling and planing of wood	..	..	..	..	..	..	..			..
202　Products of wood, cork, straw & plaiting materials	..	..	..	..	..	..	..			..
2021　Veneer sheets	..	..	..	..	..	..	..			..
2022　Builders' carpentry and joinery	..	..	..	..	..	..	..			..
2023　Wooden containers	..	..	..	..	..	..	..			..
2029　Other products of wood	..	..	..	..	..	..	..			..
21　PAPER AND PAPER PRODUCTS	3,835	38	243	5,456	17,378	21,192	76	3,691,839	..	61,509
2101　Pulp, paper and paperboard	..	..	..	..	..	..	..			..
2102　Corrugated paper, paperboard and their containers	..	..	..	..	..	..	..			..
2109　Other articles of pulp and paperboard	..	..	..	..	..	..	..			..
22　PUBLISHING, PRINTING & REPRODUCTION	3	29	178	159	241	6	69	128,132	..	1,146
221　Publishing	..	..	..	..	..	..	..			..
2211　Publishing of books & brochures	..	..	..	..	..	..	..			..
2212　Publishing of newspapers and periodicals	..	..	..	..	..	..	..			..
2213　Publishing of recorded media	..	..	..	..	..	..	..			..
2219　Other publishing	..	..	..	..	..	..	..			..
222　Printing and related service activities	..	..	..	..	..	..	..			..
2221　Printing	..	..	..	..	..	..	..			..
2222　Service activities related to printing	..	..	..	..	..	..	..			..
223　Reproduction of recorded media	..	..	..	..	..	..	..			..
23　COKE, REFINED PETROLEUM PRODUCTS	0	0	15	0	0	0	0	4,885	..	33
231　Coke oven products	..	..	..	..	..	..	..			..
232　Refined petroleum products	0	0	15	0	0	0	0	4,885	..	33
233　Processing of nuclear fuel	..	..	..	..	..	..	..			..
24　CHEMICALS & CHEMICAL PRODUCTS	775	46	800	4,145	10,653	4,996	906	3,559,221	..	35,134
241　Basic chemicals	..	..	..	..	..	..	..			..
2411　Basic chemicals, exc. fertilizers & nitrogen compounds	..	..	..	..	..	..	..			..
2412　Fertilizers and nitrogen compounds	..	..	..	..	..	..	..			..
2413　Plastics in primary forms and synthetic rubber	..	..	..	..	..	..	..			..
242　Other chemical products	..	..	..	..	..	..	..			..
2421　Pesticides and other agro-chemical products	..	..	..	..	..	..	..			..
2422　Paints, varnishes and similar coatings	..	..	..	..	..	..	..			..
2423　Pharmaceuticals, medicinal chem. & botanical prod.	..	..	..	..	..	..	..			..
2424　Soap and detergents, perfumes etc.	..	..	..	..	..	..	..			..
2429　Other chemical products, nec	..	..	..	..	..	..	..			..
243　Man-made fibres	..	..	..	..	..	..	..			..
25　RUBBER AND PLASTICS PRODUCTS										
251　Rubber products	..	..	..	..	..	..	..			..
2511　Rubber tyres and tubes	..	..	..	..	..	..	..			..
2519　Other rubber products	..	..	..	..	..	..	..			..
252　Plastic products	..	..	..	..	..	..	..			..
26　OTHER NON-METALLIC MINERAL PRODUCTS	6	150	169	326	3,061	238	7	302,961	..	5,048
261　Glass and glass products	..	..	..	..	..	..	..			..
269　Non-metallic mineral products, nec	..	..	..	..	..	..	..			..
2691　Pottery, china and earthenware	..	..	..	..	..	..	..			..
2692　Refractory ceramic products	..	..	..	..	..	..	..			..
2693　Structural non-refractory clay & ceramic prod.	..	..	..	..	..	..	..			..
2694　Cement, lime and plaster	..	..	..	..	..	..	..			..
2695　Articles of concrete, cement and plaster	..	..	..	..	..	..	..			..
2696　Cutting, shaping and finishing of stone	..	..	..	..	..	..	..			..
2699　Other non-metallic mineral products, nec	..	..	..	..	..	..	..			..

ISIC Revision 3 Industry Sector		Solid TJ	LPG TJ	Distiloil TJ	RFO TJ	Gas TJ	Biomass TJ	Steam TJ	Electr MWh	Own Use MWh	TOTAL TJ
27	**BASIC METALS**	118,115	595	380	5,674	37,625	1,721	0	4,301,450	..	179,595
271	Basic iron and steel	..	..	..	..	..	..	..	..	..	..
272	Basic precious and non-ferrous metals	..	..	..	..	..	..	..	..	..	..
273	Casting of metals	..	..	..	..	..	..	..	..	..	..
2731	Casting of iron and steel	..	..	..	..	..	..	..	..	..	..
2732	Casting non-ferrous metals	..	..	..	..	..	..	..	..	..	..
28	**FABRICATED METAL PRODUCTS**	24	238	1,620	1,491	2,690	872	156	832,952	..	10,090
281	Str. metal prod., tanks, reservoirs, steam generators	..	..	..	..	..	..	..	..	..	..
2811	Structural metal products	..	..	..	..	..	..	..	..	..	..
2812	Tanks, reservoirs and containers of metal	..	..	..	..	..	..	..	..	..	..
2813	Steam generators, exc. central heating hot water boilers	..	..	..	..	..	..	..	..	..	..
289	Other fabricated metal products	..	..	..	..	..	..	..	..	..	..
2891	Forging, pressing, stamping & roll-forming of metal	..	..	..	..	..	..	..	..	..	..
2892	Treatment and coating of metals	..	..	..	..	..	..	..	..	..	..
2893	Cutlery, hand tools and general hardware	..	..	..	..	..	..	..	..	..	..
2899	Other fabricated metal products, nec	..	..	..	..	..	..	..	..	..	..
29	**MACHINERY AND EQUIPMENT, NEC**	12	133	1,042	1,481	1,442	45	348	533,843	..	6,425
291	General purpose machinery	..	..	..	..	..	..	..	..	..	..
2911	Engines and turbines	..	..	..	..	..	..	..	..	..	..
2912	Pumps, compressors, taps and valves	..	..	..	..	..	..	..	..	..	..
2913	Bearings, gears, gearing and driving elements	..	..	..	..	..	..	..	..	..	..
2914	Ovens, furnaces and furnace burners	..	..	..	..	..	..	..	..	..	..
2915	Lifting and handling equipment	..	..	..	..	..	..	..	..	..	..
2919	Other general purpose machinery	..	..	..	..	..	..	..	..	..	..
292	Special purpose machinery	..	..	..	..	..	..	..	..	..	..
2921	Agricultural and forestry machinery	..	..	..	..	..	..	..	..	..	..
2922	Machine-tools	..	..	..	..	..	..	..	..	..	..
2923	Machinery for metallurgy	..	..	..	..	..	..	..	..	..	..
2924	Machinery for mining, quarrying and construction	..	..	..	..	..	..	..	..	..	..
2925	Machinery for food, beverage & tobacco processing	..	..	..	..	..	..	..	..	..	..
2926	Machinery for textile, apparel & leather production	..	..	..	..	..	..	..	..	..	..
2927	Machinery for weapons and ammunition	..	..	..	..	..	..	..	..	..	..
2929	Other special purpose machinery	..	..	..	..	..	..	..	..	..	..
293	Domestic appliances, nec	..	..	..	..	..	..	..	..	..	..
30	**OFFICE, ACCOUNTING & COMPUTING MACHINERY**	30	75	999	1,019	1,158	107	380	761,190	..	6,508
31	**ELECTRICAL MACHINERY & APPARATUS, NEC**	..	..	..	..	..	..	..	..	..	..
311	Electric motors, generators and transformers	..	..	..	..	..	..	..	..	..	..
312	Electricity distribution and control apparatus	..	..	..	..	..	..	..	..	..	..
313	Insulated wire and cable	..	..	..	..	..	..	..	..	..	..
314	Accumulators, primary cells & primary batteries	..	..	..	..	..	..	..	..	..	..
315	Electric lamps and lighting equipment	..	..	..	..	..	..	..	..	..	..
319	Other electrical equipment, nec	..	..	..	..	..	..	..	..	..	..
32	**RADIO, TV & COMMUNICATION EQUIP. & APP.**	..	..	..	..	..	..	..	..	..	..
321	Electronic valves, tubes, other electronic components	..	..	..	..	..	..	..	..	..	..
322	TV & radio transmitters, apparatus for line telephony	..	..	..	..	..	..	..	..	..	..
323	TV & radio receivers, recording apparatus	..	..	..	..	..	..	..	..	..	..
33	**MEDICAL PRECISION &OPTICAL INSTRUMENTS**	..	..	..	..	..	..	..	..	..	..
331	Medical appliances and instruments	..	..	..	..	..	..	..	..	..	..
3311	Medical, surgical equipment & orthopaedic app.	..	..	..	..	..	..	..	..	..	..
3312	Instruments & appliances for measuring, checking etc	..	..	..	..	..	..	..	..	..	..
3313	Industrial process control equipment	..	..	..	..	..	..	..	..	..	..
332	Optical instruments and photographic equipment	..	..	..	..	..	..	..	..	..	..
333	Watches and clocks	..	..	..	..	..	..	..	..	..	..
34	**MOTOR VEHICLES, TRAILERS & SEMI-TRAILERS**	14	117	949	734	1,743	54	530	526,849	..	6,038
341	Motor vehicles	..	..	..	..	..	..	..	..	..	..
342	Bodies (coachwork) for motor vehicles	..	..	..	..	..	..	..	..	..	..
343	Parts, accessories for motor vehicles & their engines	..	..	..	..	..	..	..	..	..	..
35	**OTHER TRANSPORT EQUIPMENT**	..	..	..	..	..	..	..	..	..	..
351	Building and repairing of ships and boats	..	..	..	..	..	..	..	..	..	..
3511	Building and repairing of ships	..	..	..	..	..	..	..	..	..	..
3512	Building, repairing of pleasure & sporting boats	..	..	..	..	..	..	..	..	..	..
352	Railway, tramway locomotives & rolling stock	..	..	..	..	..	..	..	..	..	..
353	Aircraft and spacecraft	..	..	..	..	..	..	..	..	..	..
359	Transport equipment, nec	..	..	..	..	..	..	..	..	..	..
3591	Motorcycles	..	..	..	..	..	..	..	..	..	..
3592	Bicycles and invalid carriages	..	..	..	..	..	..	..	..	..	..
3599	Other transport equipment, nec	..	..	..	..	..	..	..	..	..	..
36	**FURNITURE; MANUFACTURING, NEC**	..	..	..	..	..	..	..	..	..	..
361	Furniture	..	..	..	..	..	..	..	..	..	..
369	Manufacturing, nec	..	..	..	..	..	..	..	..	..	..
3691	Jewellery and related articles	..	..	..	..	..	..	..	..	..	..
3692	Musical instruments	..	..	..	..	..	..	..	..	..	..
3693	Sports goods	..	..	..	..	..	..	..	..	..	..
3694	Games and toys	..	..	..	..	..	..	..	..	..	..
3699	Other manufacturing, nec	..	..	..	..	..	..	..	..	..	..
37	**RECYCLING**	..	..	..	..	..	..	..	..	..	..
371	Recycling of metal waste and scrap	..	..	..	..	..	..	..	..	..	..
372	Recycling of non-metal waste and scrap	..	..	..	..	..	..	..	..	..	..
CERR	Non-specified industry	..	..	..	..	..	..	..	..	..	..
15-37	**TOTAL MANUFACTURING**	123,070	1,620	12,107	31,438	88,725	34,336	3,266	17,833,489	..	358,763

ISIC Revision 3 Industry Sector	Solid TJ	LPG TJ	Distiloil TJ	RFO TJ	Gas TJ	Biomass TJ	Steam TJ	Electr MWh	Own Use MWh	TOTAL TJ
15 **FOOD PRODUCTS AND BEVERAGES**	196	160	2,723	6,716	8,926	337	644	1,373,984	..	24,648
151 Production, processing and preserving (PPP)	..	..	..	..	..	..	..	..	..	..
1511 PPP of meat and meat products	..	..	..	..	..	..	..	..	..	..
1512 Processing and preserving of fish products	..	..	..	..	..	..	..	..	..	..
1513 Processing, preserving of fruit & vegetables	..	..	..	..	..	..	..	..	..	..
1514 Vegetable and animal oils and fats	..	..	..	..	..	..	..	..	..	..
152 Dairy products	..	..	..	..	..	..	..	..	..	..
153 Grain mill prod., starches & prepared animal feeds	..	..	..	..	..	..	..	..	..	..
1531 Grain mill products	..	..	..	..	..	..	..	..	..	..
1532 Starches and starch products	..	..	..	..	..	..	..	..	..	..
1533 Prepared animal feeds	..	..	..	..	..	..	..	..	..	..
154 Other food products	..	..	..	..	..	..	..	..	..	..
1541 Bakery products	..	..	..	..	..	..	..	..	..	..
1542 Sugar	..	..	..	..	..	..	..	..	..	..
1543 Cocoa, chocolate and sugar confectionery	..	..	..	..	..	..	..	..	..	..
1544 Macaroni, noodles, couscous & similar farinaceous prod.	..	..	..	..	..	..	..	..	..	..
1549 Other food products, nec	..	..	..	..	..	..	..	..	..	..
155 Beverages	..	..	..	..	..	..	..	..	..	..
1551 Distilling, rectifying and blending of spirits	..	..	..	..	..	..	..	..	..	..
1552 Wines	..	..	..	..	..	..	..	..	..	..
1553 Malt liquors and malt	..	..	..	..	..	..	..	..	..	..
1554 Soft drinks; production of mineral waters	..	..	..	..	..	..	..	..	..	..
16 **TOBACCO PRODUCTS**	..	..	..	..	..	..	..	..	..	..
17 **TEXTILES**	9	24	158	2,054	2,206	323	47	741,128	..	7,489
171 Spinning, weaving and finishing of textiles	..	..	..	..	..	..	..	..	..	..
1711 Preparation and spinning of textile fibres	..	..	..	..	..	..	..	..	..	..
1712 Finishing of textiles	..	..	..	..	..	..	..	..	..	..
172 Other textiles	..	..	..	..	..	..	..	..	..	..
1721 Made-up textile articles, except apparel	..	..	..	..	..	..	..	..	..	..
1722 Carpets and rugs	..	..	..	..	..	..	..	..	..	..
1723 Cordage, rope, twine and netting	..	..	..	..	..	..	..	..	..	..
1729 Other textiles, nec	..	..	..	..	..	..	..	..	..	..
173 Knitted and crocheted fabrics and articles	..	..	..	..	..	..	..	..	..	..
18 **WEARING APPAREL, DRESSING & DYEING OF FUR**	10	9	124	693	146	23	3	86,242	..	1,318
181 Wearing apparel, except fur apparel	..	..	..	..	..	..	..	..	..	..
182 Dressing and dyeing of fur; articles of fur	..	..	..	..	..	..	..	..	..	..
19 **TANNING & DRESSING OF LEATHER, FOOTWEAR**	5	0	73	298	98	27	15	64,523	..	748
191 Tanning and dressing of leather	..	..	..	..	..	..	..	..	..	..
1911 Tanning and dressing of leather	..	..	..	..	..	..	..	..	..	..
1912 Luggage, handbags, saddlery & harness	..	..	..	..	..	..	..	..	..	..
192 Footwear	..	..	..	..	..	..	..	..	..	..
20 **WOOD AND WOOD PRODUCTS**	74	12	2,577	1,756	1,408	5,669	91	998,474	..	15,182
201 Sawmilling and planing of wood	..	..	..	..	..	..	..	..	..	..
202 Products of wood, cork, straw & plaiting materials	..	..	..	..	..	..	..	..	..	..
2021 Veneer sheets	..	..	..	..	..	..	..	..	..	..
2022 Builders' carpentry and joinery	..	..	..	..	..	..	..	..	..	..
2023 Wooden containers	..	..	..	..	..	..	..	..	..	..
2029 Other products of wood	..	..	..	..	..	..	..	..	..	..
21 **PAPER AND PAPER PRODUCTS**	4,495	40	246	6,002	18,676	20,925	83	3,925,175	..	64,598
2101 Pulp, paper and paperboard	..	..	..	..	..	..	..	..	..	..
2102 Corrugated paper, paperboard and their containers	..	..	..	..	..	..	..	..	..	..
2109 Other articles of pulp and paperboard	..	..	..	..	..	..	..	..	..	..
22 **PUBLISHING, PRINTING & REPRODUCTION**	2	30	182	178	341	16	70	139,871	..	1,323
221 Publishing	..	..	..	..	..	..	..	..	..	..
2211 Publishing of books & brochures	..	..	..	..	..	..	..	..	..	..
2212 Publishing of newspapers and periodicals	..	..	..	..	..	..	..	..	..	..
2213 Publishing of recorded media	..	..	..	..	..	..	..	..	..	..
2219 Other publishing	..	..	..	..	..	..	..	..	..	..
222 Printing and related service activities	..	..	..	..	..	..	..	..	..	..
2221 Printing	..	..	..	..	..	..	..	..	..	..
2222 Service activities related to printing	..	..	..	..	..	..	..	..	..	..
223 Reproduction of recorded media	..	..	..	..	..	..	..	..	..	..
23 **COKE, REFINED PETROLEUM PRODUCTS**	0	0	20	0	0	0	0	5,080	..	38
231 Coke oven products	..	..	..	..	..	..	..	..	..	..
232 Refined petroleum products	0	0	20	0	0	0	0	5,080	..	38
233 Processing of nuclear fuel	..	..	..	..	..	..	..	..	..	..
24 **CHEMICALS & CHEMICAL PRODUCTS**	885	33	794	4,067	10,260	4,710	1,160	3,557,436	..	34,716
241 Basic chemicals	..	..	..	..	..	..	..	..	..	..
2411 Basic chemicals, exc. fertilizers & nitrogen compounds	..	..	..	..	..	..	..	..	..	..
2412 Fertilizers and nitrogen compounds	..	..	..	..	..	..	..	..	..	..
2413 Plastics in primary forms and synthetic rubber	..	..	..	..	..	..	..	..	..	..
242 Other chemical products	..	..	..	..	..	..	..	..	..	..
2421 Pesticides and other agro-chemical products	..	..	..	..	..	..	..	..	..	..
2422 Paints, varnishes and similar coatings	..	..	..	..	..	..	..	..	..	..
2423 Pharmaceuticals, medicinal chem. & botanical prod.	..	..	..	..	..	..	..	..	..	..
2424 Soap and detergents, perfumes etc.	..	..	..	..	..	..	..	..	..	..
2429 Other chemical products, nec	..	..	..	..	..	..	..	..	..	..
243 Man-made fibres	..	..	..	..	..	..	..	..	..	..
25 **RUBBER AND PLASTICS PRODUCTS**	..	..	..	..	..	..	..	..	..	..
251 Rubber products	..	..	..	..	..	..	..	..	..	..
2511 Rubber tyres and tubes	..	..	..	..	..	..	..	..	..	..
2519 Other rubber products	..	..	..	..	..	..	..	..	..	..
252 Plastic products	..	..	..	..	..	..	..	..	..	..
26 **OTHER NON-METALLIC MINERAL PRODUCTS**	2	136	115	184	3,309	216	8	319,128	..	5,119
261 Glass and glass products	..	..	..	..	..	..	..	..	..	..
269 Non-metallic mineral products, nec	..	..	..	..	..	..	..	..	..	..
2691 Pottery, china and earthenware	..	..	..	..	..	..	..	..	..	..
2692 Refractory ceramic products	..	..	..	..	..	..	..	..	..	..
2693 Structural non-refractory clay & ceramic prod.	..	..	..	..	..	..	..	..	..	..
2694 Cement, lime and plaster	..	..	..	..	..	..	..	..	..	..
2695 Articles of concrete, cement and plaster	..	..	..	..	..	..	..	..	..	..
2696 Cutting, shaping and finishing of stone	..	..	..	..	..	..	..	..	..	..
2699 Other non-metallic mineral products, nec	..	..	..	..	..	..	..	..	..	..

ISIC Revision 3 Industry Sector	Solid TJ	LPG TJ	Distiloil TJ	RFO TJ	Gas TJ	Biomass TJ	Steam TJ	Electr MWh	Own Use MWh	TOTAL TJ
27 **BASIC METALS**	110,651	532	352	5,571	37,582	1,685	0	3,975,214	..	170,684
271 Basic iron and steel	..	..	..	..	..	..	..	..	..	..
272 Basic precious and non-ferrous metals	..	..	..	..	..	..	..	..	..	..
273 Casting of metals	..	..	..	..	..	..	..	..	..	..
2731 Casting of iron and steel	..	..	..	..	..	..	..	..	..	..
2732 Casting non-ferrous metals	..	..	..	..	..	..	..	..	..	..
28 **FABRICATED METAL PRODUCTS**	20	241	1,706	1,588	2,842	868	152	848,417	..	10,471
281 Str. metal prod., tanks, reservoirs, steam generators	..	..	..	..	..	..	..	..	..	..
2811 Structural metal products	..	..	..	..	..	..	..	..	..	..
2812 Tanks, reservoirs and containers of metal	..	..	..	..	..	..	..	..	..	..
2813 Steam generators, exc. central heating hot water boilers	..	..	..	..	..	..	..	..	..	..
289 Other fabricated metal products	..	..	..	..	..	..	..	..	..	..
2891 Forging, pressing, stamping & roll-forming of metal	..	..	..	..	..	..	..	..	..	..
2892 Treatment and coating of metals	..	..	..	..	..	..	..	..	..	..
2893 Cutlery, hand tools and general hardware	..	..	..	..	..	..	..	..	..	..
2899 Other fabricated metal products, nec	..	..	..	..	..	..	..	..	..	..
29 **MACHINERY AND EQUIPMENT, NEC**	12	163	1,119	1,389	1,587	67	480	530,439	..	6,727
291 General purpose machinery	..	..	..	..	..	..	..	..	..	..
2911 Engines and turbines	..	..	..	..	..	..	..	..	..	..
2912 Pumps, compressors, taps and valves	..	..	..	..	..	..	..	..	..	..
2913 Bearings, gears, gearing and driving elements	..	..	..	..	..	..	..	..	..	..
2914 Ovens, furnaces and furnace burners	..	..	..	..	..	..	..	..	..	..
2915 Lifting and handling equipment	..	..	..	..	..	..	..	..	..	..
2919 Other general purpose machinery	..	..	..	..	..	..	..	..	..	..
292 Special purpose machinery	..	..	..	..	..	..	..	..	..	..
2921 Agricultural and forestry machinery	..	..	..	..	..	..	..	..	..	..
2922 Machine-tools	..	..	..	..	..	..	..	..	..	..
2923 Machinery for metallurgy	..	..	..	..	..	..	..	..	..	..
2924 Machinery for mining, quarrying and construction	..	..	..	..	..	..	..	..	..	..
2925 Machinery for food, beverage & tobacco processing	..	..	..	..	..	..	..	..	..	..
2926 Machinery for textile, apparel & leather production	..	..	..	..	..	..	..	..	..	..
2927 Machinery for weapons and ammunition	..	..	..	..	..	..	..	..	..	..
2929 Other special purpose machinery	..	..	..	..	..	..	..	..	..	..
293 Domestic appliances, nec	..	..	..	..	..	..	..	..	..	..
30 **OFFICE, ACCOUNTING & COMPUTING MACHINERY**	18	92	1,077	1,204	1,283	131	463	937,711	..	7,644
31 **ELECTRICAL MACHINERY & APPARATUS, NEC**	..	..	..	..	..	..	..	..	..	..
311 Electric motors, generators and transformers	..	..	..	..	..	..	..	..	..	..
312 Electricity distribution and control apparatus	..	..	..	..	..	..	..	..	..	..
313 Insulated wire and cable	..	..	..	..	..	..	..	..	..	..
314 Accumulators, primary cells & primary batteries	..	..	..	..	..	..	..	..	..	..
315 Electric lamps and lighting equipment	..	..	..	..	..	..	..	..	..	..
319 Other electrical equipment, nec	..	..	..	..	..	..	..	..	..	..
32 **RADIO, TV & COMMUNICATION EQUIP. & APP.**	..	..	..	..	..	..	..	..	..	..
321 Electronic valves, tubes, other electronic components	..	..	..	..	..	..	..	..	..	..
322 TV & radio transmitters, apparatus for line telephony	..	..	..	..	..	..	..	..	..	..
323 TV & radio receivers, recording apparatus	..	..	..	..	..	..	..	..	..	..
33 **MEDICAL PRECISION &OPTICAL INSTRUMENTS**	..	..	..	..	..	..	..	..	..	..
331 Medical appliances and instruments	..	..	..	..	..	..	..	..	..	..
3311 Medical, surgical equipment & orthopaedic app.	..	..	..	..	..	..	..	..	..	..
3312 Instruments & appliances for measuring, checking etc	..	..	..	..	..	..	..	..	..	..
3313 Industrial process control equipment	..	..	..	..	..	..	..	..	..	..
332 Optical instruments and photographic equipment	..	..	..	..	..	..	..	..	..	..
333 Watches and clocks	..	..	..	..	..	..	..	..	..	..
34 **MOTOR VEHICLES, TRAILERS & SEMI-TRAILERS**	12	201	1,033	807	1,882	65	492	542,633	..	6,445
341 Motor vehicles	..	..	..	..	..	..	..	..	..	..
342 Bodies (coachwork) for motor vehicles	..	..	..	..	..	..	..	..	..	..
343 Parts, accessories for motor vehicles & their engines	..	..	..	..	..	..	..	..	..	..
35 **OTHER TRANSPORT EQUIPMENT**	..	..	..	..	..	..	..	..	..	..
351 Building and repairing of ships and boats	..	..	..	..	..	..	..	..	..	..
3511 Building and repairing of ships	..	..	..	..	..	..	..	..	..	..
3512 Building, repairing of pleasure & sporting boats	..	..	..	..	..	..	..	..	..	..
352 Railway, tramway locomotives & rolling stock	..	..	..	..	..	..	..	..	..	..
353 Aircraft and spacecraft	..	..	..	..	..	..	..	..	..	..
359 Transport equipment, nec	..	..	..	..	..	..	..	..	..	..
3591 Motorcycles	..	..	..	..	..	..	..	..	..	..
3592 Bicycles and invalid carriages	..	..	..	..	..	..	..	..	..	..
3599 Other transport equipment, nec	..	..	..	..	..	..	..	..	..	..
36 **FURNITURE; MANUFACTURING, NEC**	..	..	..	..	..	..	..	..	..	..
361 Furniture	..	..	..	..	..	..	..	..	..	..
369 Manufacturing, nec	..	..	..	..	..	..	..	..	..	..
3691 Jewellery and related articles	..	..	..	..	..	..	..	..	..	..
3692 Musical instruments	..	..	..	..	..	..	..	..	..	..
3693 Sports goods	..	..	..	..	..	..	..	..	..	..
3694 Games and toys	..	..	..	..	..	..	..	..	..	..
3699 Other manufacturing, nec	..	..	..	..	..	..	..	..	..	..
37 **RECYCLING**	..	..	..	..	..	..	..	..	..	..
371 Recycling of metal waste and scrap	..	..	..	..	..	..	..	..	..	..
372 Recycling of non-metal waste and scrap	..	..	..	..	..	..	..	..	..	..
CERR **Non-specified industry**	..	..	..	..	..	..	..	..	..	..
15-37 **TOTAL MANUFACTURING**	116,391	1,673	12,299	32,507	90,546	35,062	3,708	18,045,455	..	357,150

ISIC Revision 3 Industry Sector	Solid TJ	LPG TJ	Distiloil TJ	RFO TJ	Gas TJ	Biomass TJ	Steam TJ	Electr MWh	Own Use MWh	TOTAL TJ
15 **FOOD PRODUCTS AND BEVERAGES**	218	149	2,863	6,294	8,556	294	618	1,435,269	..	24,159
151 Production, processing and preserving (PPP)	..	..	..	..	..	..	..	..	..	..
1511 PPP of meat and meat products	..	..	..	..	..	..	..	..	..	..
1512 Processing and preserving of fish products	..	..	..	..	..	..	..	..	..	..
1513 Processing, preserving of fruit & vegetables	..	..	..	..	..	..	..	..	..	..
1514 Vegetable and animal oils and fats	..	..	..	..	..	..	..	..	..	..
152 Dairy products	..	..	..	..	..	..	..	..	..	..
153 Grain mill prod., starches & prepared animal feeds	..	..	..	..	..	..	..	..	..	..
1531 Grain mill products	..	..	..	..	..	..	..	..	..	..
1532 Starches and starch products	..	..	..	..	..	..	..	..	..	..
1533 Prepared animal feeds	..	..	..	..	..	..	..	..	..	..
154 Other food products	..	..	..	..	..	..	..	..	..	..
1541 Bakery products	..	..	..	..	..	..	..	..	..	..
1542 Sugar	..	..	..	..	..	..	..	..	..	..
1543 Cocoa, chocolate and sugar confectionery	..	..	..	..	..	..	..	..	..	..
1544 Macaroni, noodles, couscous & similar farinaceous prod.	..	..	..	..	..	..	..	..	..	..
1549 Other food products, nec	..	..	..	..	..	..	..	..	..	..
155 Beverages	..	..	..	..	..	..	..	..	..	..
1551 Distilling, rectifying and blending of spirits	..	..	..	..	..	..	..	..	..	..
1552 Wines	..	..	..	..	..	..	..	..	..	..
1553 Malt liquors and malt	..	..	..	..	..	..	..	..	..	..
1554 Soft drinks; production of mineral waters	..	..	..	..	..	..	..	..	..	..
16 **TOBACCO PRODUCTS**										
17 **TEXTILES**	7	17	168	1,483	2,507	333	58	736,637	..	7,225
171 Spinning, weaving and finishing of textiles	..	..	..	..	..	..	..	..	..	..
1711 Preparation and spinning of textile fibres	..	..	..	..	..	..	..	..	..	..
1712 Finishing of textiles	..	..	..	..	..	..	..	..	..	..
172 Other textiles	..	..	..	..	..	..	..	..	..	..
1721 Made-up textile articles, except apparel	..	..	..	..	..	..	..	..	..	..
1722 Carpets and rugs	..	..	..	..	..	..	..	..	..	..
1723 Cordage, rope, twine and netting	..	..	..	..	..	..	..	..	..	..
1729 Other textiles, nec	..	..	..	..	..	..	..	..	..	..
173 Knitted and crocheted fabrics and articles	..	..	..	..	..	..	..	..	..	..
18 **WEARING APPAREL, DRESSING & DYEING OF FUR**	13	9	141	662	154	13	6	84,663	..	1,303
181 Wearing apparel, except fur apparel	..	..	..	..	..	..	..	..	..	..
182 Dressing and dyeing of fur; articles of fur	..	..	..	..	..	..	..	..	..	..
19 **TANNING & DRESSING OF LEATHER, FOOTWEAR**	3	1	76	299	119	17	16	67,886	..	775
191 Tanning and dressing of leather	..	..	..	..	..	..	..	..	..	..
1911 Tanning and dressing of leather	..	..	..	..	..	..	..	..	..	..
1912 Luggage, handbags, saddlery & harness	..	..	..	..	..	..	..	..	..	..
192 Footwear	..	..	..	..	..	..	..	..	..	..
20 **WOOD AND WOOD PRODUCTS**	92	11	2,759	1,618	1,490	6,284	94	1,073,646	..	16,213
201 Sawmilling and planing of wood	..	..	..	..	..	..	..	..	..	..
202 Products of wood, cork, straw & plaiting materials	..	..	..	..	..	..	..	..	..	..
2021 Veneer sheets	..	..	..	..	..	..	..	..	..	..
2022 Builders' carpentry and joinery	..	..	..	..	..	..	..	..	..	..
2023 Wooden containers	..	..	..	..	..	..	..	..	..	..
2029 Other products of wood	..	..	..	..	..	..	..	..	..	..
21 **PAPER AND PAPER PRODUCTS**	4,398	48	241	5,562	20,309	22,046	72	4,080,030	..	67,364
2101 Pulp, paper and paperboard	..	..	..	..	..	..	..	..	..	..
2102 Corrugated paper, paperboard and their containers	..	..	..	..	..	..	..	..	..	..
2109 Other articles of pulp and paperboard	..	..	..	..	..	..	..	..	..	..
22 **PUBLISHING, PRINTING & REPRODUCTION**	1	31	187	142	342	3	44	157,110	..	1,316
221 Publishing	..	..	..	..	..	..	..	..	..	..
2211 Publishing of books & brochures	..	..	..	..	..	..	..	..	..	..
2212 Publishing of newspapers and periodicals	..	..	..	..	..	..	..	..	..	..
2213 Publishing of recorded media	..	..	..	..	..	..	..	..	..	..
2219 Other publishing	..	..	..	..	..	..	..	..	..	..
222 Printing and related service activities	..	..	..	..	..	..	..	..	..	..
2221 Printing	..	..	..	..	..	..	..	..	..	..
2222 Service activities related to printing	..	..	..	..	..	..	..	..	..	..
223 Reproduction of recorded media	..	..	..	..	..	..	..	..	..	..
23 **COKE, REFINED PETROLEUM PRODUCTS**	0	0	20	0	0	0	0	5,352	..	39
231 Coke oven products	..	..	..	..	..	..	..	..	..	..
232 Refined petroleum products	0	0	20	0	0	0	0	5,352	..	39
233 Processing of nuclear fuel	..	..	..	..	..	..	..	..	..	..
24 **CHEMICALS & CHEMICAL PRODUCTS**	1,734	24	759	3,125	10,830	4,910	1,096	3,595,365	..	35,421
241 Basic chemicals	..	..	..	..	..	..	..	..	..	..
2411 Basic chemicals, exc. fertilizers & nitrogen compounds	..	..	..	..	..	..	..	..	..	..
2412 Fertilizers and nitrogen compounds	..	..	..	..	..	..	..	..	..	..
2413 Plastics in primary forms and synthetic rubber	..	..	..	..	..	..	..	..	..	..
242 Other chemical products	..	..	..	..	..	..	..	..	..	..
2421 Pesticides and other agro-chemical products	..	..	..	..	..	..	..	..	..	..
2422 Paints, varnishes and similar coatings	..	..	..	..	..	..	..	..	..	..
2423 Pharmaceuticals, medicinal chem. & botanical prod.	..	..	..	..	..	..	..	..	..	..
2424 Soap and detergents, perfumes etc.	..	..	..	..	..	..	..	..	..	..
2429 Other chemical products, nec	..	..	..	..	..	..	..	..	..	..
243 Man-made fibres	..	..	..	..	..	..	..	..	..	..
25 **RUBBER AND PLASTICS PRODUCTS**	..	..	..	..	..	..	..	..	..	..
251 Rubber products	..	..	..	..	..	..	..	..	..	..
2511 Rubber tyres and tubes	..	..	..	..	..	..	..	..	..	..
2519 Other rubber products	..	..	..	..	..	..	..	..	..	..
252 Plastic products	..	..	..	..	..	..	..	..	..	..
26 **OTHER NON-METALLIC MINERAL PRODUCTS**	5	124	153	136	2,569	230	6	325,637	..	4,395
261 Glass and glass products	..	..	..	..	..	..	..	..	..	..
269 Non-metallic mineral products, nec	..	..	..	..	..	..	..	..	..	..
2691 Pottery, china and earthenware	..	..	..	..	..	..	..	..	..	..
2692 Refractory ceramic products	..	..	..	..	..	..	..	..	..	..
2693 Structural non-refractory clay & ceramic prod.	..	..	..	..	..	..	..	..	..	..
2694 Cement, lime and plaster	..	..	..	..	..	..	..	..	..	..
2695 Articles of concrete, cement and plaster	..	..	..	..	..	..	..	..	..	..
2696 Cutting, shaping and finishing of stone	..	..	..	..	..	..	..	..	..	..
2699 Other non-metallic mineral products, nec	..	..	..	..	..	..	..	..	..	..

ISIC Revision 3 Industry Sector	Solid TJ	LPG TJ	Distiloil TJ	RFO TJ	Gas TJ	Biomass TJ	Steam TJ	Electr MWh	Own Use MWh	TOTAL TJ
27 BASIC METALS	101,566	468	310	5,374	34,922	1,153	0	3,108,969	..	154,985
271 Basic iron and steel	..	..	..	..	..	..	..	..	..	..
272 Basic precious and non-ferrous metals	..	..	..	..	..	..	..	..	..	..
273 Casting of metals	..	..	..	..	..	..	..	..	..	..
2731 Casting of iron and steel	..	..	..	..	..	..	..	..	..	..
2732 Casting non-ferrous metals	..	..	..	..	..	..	..	..	..	..
28 FABRICATED METAL PRODUCTS	22	200	1,763	1,430	2,848	910	215	864,611	..	10,501
281 Str. metal prod., tanks, reservoirs, steam generators	..	..	..	..	..	..	..	..	..	..
2811 Structural metal products	..	..	..	..	..	..	..	..	..	..
2812 Tanks, reservoirs and containers of metal	..	..	..	..	..	..	..	..	..	..
2813 Steam generators, exc. central heating hot water boilers	..	..	..	..	..	..	..	..	..	..
289 Other fabricated metal products	..	..	..	..	..	..	..	..	..	..
2891 Forging, pressing, stamping & roll-forming of metal	..	..	..	..	..	..	..	..	..	..
2892 Treatment and coating of metals	..	..	..	..	..	..	..	..	..	..
2893 Cutlery, hand tools and general hardware	..	..	..	..	..	..	..	..	..	..
2899 Other fabricated metal products, nec	..	..	..	..	..	..	..	..	..	..
29 MACHINERY AND EQUIPMENT, NEC	13	146	1,212	1,205	1,606	98	379	535,919	..	6,588
291 General purpose machinery	..	..	..	..	..	..	..	..	..	..
2911 Engines and turbines	..	..	..	..	..	..	..	..	..	..
2912 Pumps, compressors, taps and valves	..	..	..	..	..	..	..	..	..	..
2913 Bearings, gears, gearing and driving elements	..	..	..	..	..	..	..	..	..	..
2914 Ovens, furnaces and furnace burners	..	..	..	..	..	..	..	..	..	..
2915 Lifting and handling equipment	..	..	..	..	..	..	..	..	..	..
2919 Other general purpose machinery	..	..	..	..	..	..	..	..	..	..
292 Special purpose machinery	..	..	..	..	..	..	..	..	..	..
2921 Agricultural and forestry machinery	..	..	..	..	..	..	..	..	..	..
2922 Machine-tools	..	..	..	..	..	..	..	..	..	..
2923 Machinery for metallurgy	..	..	..	..	..	..	..	..	..	..
2924 Machinery for mining, quarrying and construction	..	..	..	..	..	..	..	..	..	..
2925 Machinery for food, beverage & tobacco processing	..	..	..	..	..	..	..	..	..	..
2926 Machinery for textile, apparel & leather production	..	..	..	..	..	..	..	..	..	..
2927 Machinery for weapons and ammunition	..	..	..	..	..	..	..	..	..	..
2929 Other special purpose machinery	..	..	..	..	..	..	..	..	..	..
293 Domestic appliances, nec	..	..	..	..	..	..	..	..	..	..
30 OFFICE, ACCOUNTING & COMPUTING MACHINERY	19	82	1,111	1,066	1,121	117	471	876,535	..	7,143
31 ELECTRICAL MACHINERY & APPARATUS, NEC	..	..	..	..	..	..	..	..	..	..
311 Electric motors, generators and transformers	..	..	..	..	..	..	..	..	..	..
312 Electricity distribution and control apparatus	..	..	..	..	..	..	..	..	..	..
313 Insulated wire and cable	..	..	..	..	..	..	..	..	..	..
314 Accumulators, primary cells & primary batteries	..	..	..	..	..	..	..	..	..	..
315 Electric lamps and lighting equipment	..	..	..	..	..	..	..	..	..	..
319 Other electrical equipment, nec	..	..	..	..	..	..	..	..	..	..
32 RADIO, TV & COMMUNICATION EQUIP. & APP.	..	..	..	..	..	..	..	..	..	..
321 Electronic valves, tubes, other electronic components	..	..	..	..	..	..	..	..	..	..
322 TV & radio transmitters, apparatus for line telephony	..	..	..	..	..	..	..	..	..	..
323 TV & radio receivers, recording apparatus	..	..	..	..	..	..	..	..	..	..
33 MEDICAL PRECISION & OPTICAL INSTRUMENTS	..	..	..	..	..	..	..	..	..	..
331 Medical appliances and instruments	..	..	..	..	..	..	..	..	..	..
3311 Medical, surgical equipment & orthopaedic app.	..	..	..	..	..	..	..	..	..	..
3312 Instruments & appliances for measuring, checking etc	..	..	..	..	..	..	..	..	..	..
3313 Industrial process control equipment	..	..	..	..	..	..	..	..	..	..
332 Optical instruments and photographic equipment	..	..	..	..	..	..	..	..	..	..
333 Watches and clocks	..	..	..	..	..	..	..	..	..	..
34 MOTOR VEHICLES, TRAILERS & SEMI-TRAILERS	15	108	1,094	720	1,937	65	520	568,503	..	6,506
341 Motor vehicles	..	..	..	..	..	..	..	..	..	..
342 Bodies (coachwork) for motor vehicles	..	..	..	..	..	..	..	..	..	..
343 Parts, accessories for motor vehicles & their engines	..	..	..	..	..	..	..	..	..	..
35 OTHER TRANSPORT EQUIPMENT	..	..	..	..	..	..	..	..	..	..
351 Building and repairing of ships and boats	..	..	..	..	..	..	..	..	..	..
3511 Building and repairing of ships	..	..	..	..	..	..	..	..	..	..
3512 Building, repairing of pleasure & sporting boats	..	..	..	..	..	..	..	..	..	..
352 Railway, tramway locomotives & rolling stock	..	..	..	..	..	..	..	..	..	..
353 Aircraft and spacecraft	..	..	..	..	..	..	..	..	..	..
359 Transport equipment, nec	..	..	..	..	..	..	..	..	..	..
3591 Motorcycles	..	..	..	..	..	..	..	..	..	..
3592 Bicycles and invalid carriages	..	..	..	..	..	..	..	..	..	..
3599 Other transport equipment, nec	..	..	..	..	..	..	..	..	..	..
36 FURNITURE; MANUFACTURING, NEC	..	..	..	..	..	..	..	..	..	..
361 Furniture	..	..	..	..	..	..	..	..	..	..
369 Manufacturing, nec	..	..	..	..	..	..	..	..	..	..
3691 Jewellery and related articles	..	..	..	..	..	..	..	..	..	..
3692 Musical instruments	..	..	..	..	..	..	..	..	..	..
3693 Sports goods	..	..	..	..	..	..	..	..	..	..
3694 Games and toys	..	..	..	..	..	..	..	..	..	..
3699 Other manufacturing, nec	..	..	..	..	..	..	..	..	..	..
37 RECYCLING	..	..	..	..	..	..	..	..	..	..
371 Recycling of metal waste and scrap	..	..	..	..	..	..	..	..	..	..
372 Recycling of non-metal waste and scrap	..	..	..	..	..	..	..	..	..	..
CERR Non-specified industry	..	..	..	..	..	..	..	..	..	..
15-37 TOTAL MANUFACTURING	108,106	1,418	12,857	29,116	89,310	36,473	3,595	17,516,132	..	343,933

ISIC Revision 3 Industry Sector	Solid TJ	LPG TJ	Distiloil TJ	RFO TJ	Gas TJ	Biomass TJ	Steam TJ	Electr MWh	Own Use MWh	TOTAL TJ
15 FOOD PRODUCTS AND BEVERAGES	190	141	2,621	5,987	9,250	346	676	1,447,042	..	24,420
151 Production, processing and preserving (PPP)	..	..	..	..	..	..	..	..	..	..
1511 PPP of meat and meat products	..	..	..	..	..	..	..	..	..	..
1512 Processing and preserving of fish products	..	..	..	..	..	..	..	..	..	..
1513 Processing, preserving of fruit & vegetables	..	..	..	..	..	..	..	..	..	..
1514 Vegetable and animal oils and fats	..	..	..	..	..	..	..	..	..	..
152 Dairy products	..	..	..	..	..	..	..	..	..	..
153 Grain mill prod., starches & prepared animal feeds	..	..	..	..	..	..	..	..	..	..
1531 Grain mill products	..	..	..	..	..	..	..	..	..	..
1532 Starches and starch products	..	..	..	..	..	..	..	..	..	..
1533 Prepared animal feeds	..	..	..	..	..	..	..	..	..	..
154 Other food products	..	..	..	..	..	..	..	..	..	..
1541 Bakery products	..	..	..	..	..	..	..	..	..	..
1542 Sugar	..	..	..	..	..	..	..	..	..	..
1543 Cocoa, chocolate and sugar confectionery	..	..	..	..	..	..	..	..	..	..
1544 Macaroni, noodles, couscous & similar farinaceous prod.	..	..	..	..	..	..	..	..	..	..
1549 Other food products, nec	..	..	..	..	..	..	..	..	..	..
155 Beverages	..	..	..	..	..	..	..	..	..	..
1551 Distilling, rectifying and blending of spirits	..	..	..	..	..	..	..	..	..	..
1552 Wines	..	..	..	..	..	..	..	..	..	..
1553 Malt liquors and malt	..	..	..	..	..	..	..	..	..	..
1554 Soft drinks; production of mineral waters	..	..	..	..	..	..	..	..	..	..
16 TOBACCO PRODUCTS	..	..	..	..	..	..	..	..	..	..
17 TEXTILES	3	14	144	1,505	2,133	340	54	674,706	..	6,622
171 Spinning, weaving and finishing of textiles	..	..	..	..	..	..	..	..	..	..
1711 Preparation and spinning of textile fibres	..	..	..	..	..	..	..	..	..	..
1712 Finishing of textiles	..	..	..	..	..	..	..	..	..	..
172 Other textiles	..	..	..	..	..	..	..	..	..	..
1721 Made-up textile articles, except apparel	..	..	..	..	..	..	..	..	..	..
1722 Carpets and rugs	..	..	..	..	..	..	..	..	..	..
1723 Cordage, rope, twine and netting	..	..	..	..	..	..	..	..	..	..
1729 Other textiles, nec	..	..	..	..	..	..	..	..	..	..
173 Knitted and crocheted fabrics and articles	..	..	..	..	..	..	..	..	..	..
18 WEARING APPAREL, DRESSING & DYEING OF FUR	7	11	125	617	171	10	12	78,027	..	1,233
181 Wearing apparel, except fur apparel	..	..	..	..	..	..	..	..	..	..
182 Dressing and dyeing of fur; articles of fur	..	..	..	..	..	..	..	..	..	..
19 TANNING & DRESSING OF LEATHER, FOOTWEAR	5	1	69	262	109	33	14	66,267	..	730
191 Tanning and dressing of leather	..	..	..	..	..	..	..	..	..	..
1911 Tanning and dressing of leather	..	..	..	..	..	..	..	..	..	..
1912 Luggage, handbags, saddlery & harness	..	..	..	..	..	..	..	..	..	..
192 Footwear	..	..	..	..	..	..	..	..	..	..
20 WOOD AND WOOD PRODUCTS	55	31	2,542	1,501	1,453	5,431	76	1,052,271	..	14,878
201 Sawmilling and planing of wood	..	..	..	..	..	..	..	..	..	..
202 Products of wood, cork, straw & plaiting materials	..	..	..	..	..	..	..	..	..	..
2021 Veneer sheets	..	..	..	..	..	..	..	..	..	..
2022 Builders' carpentry and joinery	..	..	..	..	..	..	..	..	..	..
2023 Wooden containers	..	..	..	..	..	..	..	..	..	..
2029 Other products of wood	..	..	..	..	..	..	..	..	..	..
21 PAPER AND PAPER PRODUCTS	4,316	68	228	6,070	20,603	19,628	68	4,119,142	..	65,811
2101 Pulp, paper and paperboard	..	..	..	..	..	..	..	..	..	..
2102 Corrugated paper, paperboard and their containers	..	..	..	..	..	..	..	..	..	..
2109 Other articles of pulp and paperboard	..	..	..	..	..	..	..	..	..	..
22 PUBLISHING, PRINTING & REPRODUCTION	2	29	165	143	308	2	46	150,413	..	1,238
221 Publishing	..	..	..	..	..	..	..	..	..	..
2211 Publishing of books & brochures	..	..	..	..	..	..	..	..	..	..
2212 Publishing of newspapers and periodicals	..	..	..	..	..	..	..	..	..	..
2213 Publishing of recorded media	..	..	..	..	..	..	..	..	..	..
2219 Other publishing	..	..	..	..	..	..	..	..	..	..
222 Printing and related service activities	..	..	..	..	..	..	..	..	..	..
2221 Printing	..	..	..	..	..	..	..	..	..	..
2222 Service activities related to printing	..	..	..	..	..	..	..	..	..	..
223 Reproduction of recorded media	..	..	..	..	..	..	..	..	..	..
23 COKE, REFINED PETROLEUM PRODUCTS	0	0	20	0	0	0	0	5,224	..	39
231 Coke oven products	..	..	..	..	..	..	..	..	..	..
232 Refined petroleum products	0	0	20	0	0	0	0	5,224	..	39
233 Processing of nuclear fuel	..	..	..	..	..	..	..	..	..	..
24 CHEMICALS & CHEMICAL PRODUCTS	1,671	42	734	2,994	11,121	5,094	1,149	3,531,752	..	35,519
241 Basic chemicals	..	..	..	..	..	..	..	..	..	..
2411 Basic chemicals, exc. fertilizers & nitrogen compounds	..	..	..	..	..	..	..	..	..	..
2412 Fertilizers and nitrogen compounds	..	..	..	..	..	..	..	..	..	..
2413 Plastics in primary forms and synthetic rubber	..	..	..	..	..	..	..	..	..	..
242 Other chemical products	..	..	..	..	..	..	..	..	..	..
2421 Pesticides and other agro-chemical products	..	..	..	..	..	..	..	..	..	..
2422 Paints, varnishes and similar coatings	..	..	..	..	..	..	..	..	..	..
2423 Pharmaceuticals, medicinal chem. & botanical prod.	..	..	..	..	..	..	..	..	..	..
2424 Soap and detergents, perfumes etc.	..	..	..	..	..	..	..	..	..	..
2429 Other chemical products, nec	..	..	..	..	..	..	..	..	..	..
243 Man-made fibres	..	..	..	..	..	..	..	..	..	..
25 RUBBER AND PLASTICS PRODUCTS	..	..	..	..	..	..	..	..	..	..
251 Rubber products	..	..	..	..	..	..	..	..	..	..
2511 Rubber tyres and tubes	..	..	..	..	..	..	..	..	..	..
2519 Other rubber products	..	..	..	..	..	..	..	..	..	..
252 Plastic products	..	..	..	..	..	..	..	..	..	..
26 OTHER NON-METALLIC MINERAL PRODUCTS	1	164	162	137	2,744	269	9	334,015	..	4,688
261 Glass and glass products	..	..	..	..	..	..	..	..	..	..
269 Non-metallic mineral products, nec	..	..	..	..	..	..	..	..	..	..
2691 Pottery, china and earthenware	..	..	..	..	..	..	..	..	..	..
2692 Refractory ceramic products	..	..	..	..	..	..	..	..	..	..
2693 Structural non-refractory clay & ceramic prod.	..	..	..	..	..	..	..	..	..	..
2694 Cement, lime and plaster	..	..	..	..	..	..	..	..	..	..
2695 Articles of concrete, cement and plaster	..	..	..	..	..	..	..	..	..	..
2696 Cutting, shaping and finishing of stone	..	..	..	..	..	..	..	..	..	..
2699 Other non-metallic mineral products, nec	..	..	..	..	..	..	..	..	..	..

ISIC Revision 3 Industry Sector	Solid TJ	LPG TJ	Distiloil TJ	RFO TJ	Gas TJ	Biomass TJ	Steam TJ	Electr MWh	Own Use MWh	TOTAL TJ
27 **BASIC METALS**	96,997	415	298	5,524	35,540	1,223	0	2,526,710	..	149,095
271 Basic iron and steel	..	..	..	..	..	..	..	..	..	..
272 Basic precious and non-ferrous metals	..	..	..	..	..	..	..	..	..	..
273 Casting of metals	..	..	..	..	..	..	..	..	..	..
2731 Casting of iron and steel	..	..	..	..	..	..	..	..	..	..
2732 Casting non-ferrous metals	..	..	..	..	..	..	..	..	..	..
28 **FABRICATED METAL PRODUCTS**	22	242	1,730	1,407	2,640	914	132	844,740	..	10,127
281 Str. metal prod., tanks, reservoirs, steam generators	..	..	..	..	..	..	..	..	..	..
2811 Structural metal products	..	..	..	..	..	..	..	..	..	..
2812 Tanks, reservoirs and containers of metal	..	..	..	..	..	..	..	..	..	..
2813 Steam generators, exc. central heating hot water boilers	..	..	..	..	..	..	..	..	..	..
289 Other fabricated metal products	..	..	..	..	..	..	..	..	..	..
2891 Forging, pressing, stamping & roll-forming of metal	..	..	..	..	..	..	..	..	..	..
2892 Treatment and coating of metals	..	..	..	..	..	..	..	..	..	..
2893 Cutlery, hand tools and general hardware	..	..	..	..	..	..	..	..	..	..
2899 Other fabricated metal products, nec	..	..	..	..	..	..	..	..	..	..
29 **MACHINERY AND EQUIPMENT, NEC**	12	158	1,175	1,240	1,494	95	434	509,875	..	6,445
291 General purpose machinery	..	..	..	..	..	..	..	..	..	..
2911 Engines and turbines	..	..	..	..	..	..	..	..	..	..
2912 Pumps, compressors, taps and valves	..	..	..	..	..	..	..	..	..	..
2913 Bearings, gears, gearing and driving elements	..	..	..	..	..	..	..	..	..	..
2914 Ovens, furnaces and furnace burners	..	..	..	..	..	..	..	..	..	..
2915 Lifting and handling equipment	..	..	..	..	..	..	..	..	..	..
2919 Other general purpose machinery	..	..	..	..	..	..	..	..	..	..
292 Special purpose machinery	..	..	..	..	..	..	..	..	..	..
2921 Agricultural and forestry machinery	..	..	..	..	..	..	..	..	..	..
2922 Machine-tools	..	..	..	..	..	..	..	..	..	..
2923 Machinery for metallurgy	..	..	..	..	..	..	..	..	..	..
2924 Machinery for mining, quarrying and construction	..	..	..	..	..	..	..	..	..	..
2925 Machinery for food, beverage & tobacco processing	..	..	..	..	..	..	..	..	..	..
2926 Machinery for textile, apparel & leather production	..	..	..	..	..	..	..	..	..	..
2927 Machinery for weapons and ammunition	..	..	..	..	..	..	..	..	..	..
2929 Other special purpose machinery	..	..	..	..	..	..	..	..	..	..
293 Domestic appliances, nec	..	..	..	..	..	..	..	..	..	..
30 **OFFICE, ACCOUNTING & COMPUTING MACHINERY**	17	144	1,098	1,064	1,235	47	453	782,941	..	6,876
31 **ELECTRICAL MACHINERY & APPARATUS, NEC**	..	..	..	..	..	..	..	..	..	..
311 Electric motors, generators and transformers	..	..	..	..	..	..	..	..	..	..
312 Electricity distribution and control apparatus	..	..	..	..	..	..	..	..	..	..
313 Insulated wire and cable	..	..	..	..	..	..	..	..	..	..
314 Accumulators, primary cells & primary batteries	..	..	..	..	..	..	..	..	..	..
315 Electric lamps and lighting equipment	..	..	..	..	..	..	..	..	..	..
319 Other electrical equipment, nec	..	..	..	..	..	..	..	..	..	..
32 **RADIO, TV & COMMUNICATION EQUIP. & APP.**	..	..	..	..	..	..	..	..	..	..
321 Electronic valves, tubes, other electronic components	..	..	..	..	..	..	..	..	..	..
322 TV & radio transmitters, apparatus for line telephony	..	..	..	..	..	..	..	..	..	..
323 TV & radio receivers, recording apparatus	..	..	..	..	..	..	..	..	..	..
33 **MEDICAL PRECISION &OPTICAL INSTRUMENTS**	..	..	..	..	..	..	..	..	..	..
331 Medical appliances and instruments	..	..	..	..	..	..	..	..	..	..
3311 Medical, surgical equipment & orthopaedic app.	..	..	..	..	..	..	..	..	..	..
3312 Instruments & appliances for measuring, checking etc	..	..	..	..	..	..	..	..	..	..
3313 Industrial process control equipment	..	..	..	..	..	..	..	..	..	..
332 Optical instruments and photographic equipment	..	..	..	..	..	..	..	..	..	..
333 Watches and clocks	..	..	..	..	..	..	..	..	..	..
34 **MOTOR VEHICLES, TRAILERS & SEMI-TRAILERS**	15	81	1,067	712	2,076	49	594	548,924	..	6,570
341 Motor vehicles	..	..	..	..	..	..	..	..	..	..
342 Bodies (coachwork) for motor vehicles	..	..	..	..	..	..	..	..	..	..
343 Parts, accessories for motor vehicles & their engines	..	..	..	..	..	..	..	..	..	..
35 **OTHER TRANSPORT EQUIPMENT**	..	..	..	..	..	..	..	..	..	..
351 Building and repairing of ships and boats	..	..	..	..	..	..	..	..	..	..
3511 Building and repairing of ships	..	..	..	..	..	..	..	..	..	..
3512 Building, repairing of pleasure & sporting boats	..	..	..	..	..	..	..	..	..	..
352 Railway, tramway locomotives & rolling stock	..	..	..	..	..	..	..	..	..	..
353 Aircraft and spacecraft	..	..	..	..	..	..	..	..	..	..
359 Transport equipment, nec	..	..	..	..	..	..	..	..	..	..
3591 Motorcycles	..	..	..	..	..	..	..	..	..	..
3592 Bicycles and invalid carriages	..	..	..	..	..	..	..	..	..	..
3599 Other transport equipment, nec	..	..	..	..	..	..	..	..	..	..
36 **FURNITURE; MANUFACTURING, NEC**	..	..	..	..	..	..	..	..	..	..
361 Furniture	..	..	..	..	..	..	..	..	..	..
369 Manufacturing, nec	..	..	..	..	..	..	..	..	..	..
3691 Jewellery and related articles	..	..	..	..	..	..	..	..	..	..
3692 Musical instruments	..	..	..	..	..	..	..	..	..	..
3693 Sports goods	..	..	..	..	..	..	..	..	..	..
3694 Games and toys	..	..	..	..	..	..	..	..	..	..
3699 Other manufacturing, nec	..	..	..	..	..	..	..	..	..	..
37 **RECYCLING**	..	..	..	..	..	..	..	..	..	..
371 Recycling of metal waste and scrap	..	..	..	..	..	..	..	..	..	..
372 Recycling of non-metal waste and scrap	..	..	..	..	..	..	..	..	..	..
CERR Non-specified industry										
15-37 **TOTAL MANUFACTURING**	103,312	1,541	12,180	29,164	90,878	33,480	3,717	16,672,049	..	334,291

ISIC Revision 3 Industry Sector	Solid TJ	LPG TJ	Distiloil TJ	RFO TJ	Gas TJ	Biomass TJ	Steam TJ	Electr MWh	Own Use MWh	TOTAL TJ
15　FOOD PRODUCTS AND BEVERAGES	219	119	2,519	4,773	9,773	255	626	1,458,196	..	23,534
151　Production, processing and preserving (PPP)	..	..	..	..	..	..	..	..	..	..
1511　PPP of meat and meat products	..	..	..	..	..	..	..	..	..	..
1512　Processing and preserving of fish products	..	..	..	..	..	..	..	..	..	..
1513　Processing, preserving of fruit & vegetables	..	..	..	..	..	..	..	..	..	..
1514　Vegetable and animal oils and fats	..	..	..	..	..	..	..	..	..	..
152　Dairy products	..	..	..	..	..	..	..	..	..	..
153　Grain mill prod., starches & prepared animal feeds	..	..	..	..	..	..	..	..	..	..
1531　Grain mill products	..	..	..	..	..	..	..	..	..	..
1532　Starches and starch products	..	..	..	..	..	..	..	..	..	..
1533　Prepared animal feeds	..	..	..	..	..	..	..	..	..	..
154　Other food products	..	..	..	..	..	..	..	..	..	..
1541　Bakery products	..	..	..	..	..	..	..	..	..	..
1542　Sugar	..	..	..	..	..	..	..	..	..	..
1543　Cocoa, chocolate and sugar confectionery	..	..	..	..	..	..	..	..	..	..
1544　Macaroni, noodles, couscous & similar farinaceous prod.	..	..	..	..	..	..	..	..	..	..
1549　Other food products, nec	..	..	..	..	..	..	..	..	..	..
155　Beverages	..	..	..	..	..	..	..	..	..	..
1551　Distilling, rectifying and blending of spirits	..	..	..	..	..	..	..	..	..	..
1552　Wines	..	..	..	..	..	..	..	..	..	..
1553　Malt liquors and malt	..	..	..	..	..	..	..	..	..	..
1554　Soft drinks; production of mineral waters	..	..	..	..	..	..	..	..	..	..
16　TOBACCO PRODUCTS	..	..	..	..	..	..	..	..	..	..
17　TEXTILES	3	15	141	1,386	2,011	305	47	659,866	..	6,284
171　Spinning, weaving and finishing of textiles	..	..	..	..	..	..	..	..	..	..
1711　Preparation and spinning of textile fibres	..	..	..	..	..	..	..	..	..	..
1712　Finishing of textiles	..	..	..	..	..	..	..	..	..	..
172　Other textiles	..	..	..	..	..	..	..	..	..	..
1721　Made-up textile articles, except apparel	..	..	..	..	..	..	..	..	..	..
1722　Carpets and rugs	..	..	..	..	..	..	..	..	..	..
1723　Cordage, rope, twine and netting	..	..	..	..	..	..	..	..	..	..
1729　Other textiles, nec	..	..	..	..	..	..	..	..	..	..
173　Knitted and crocheted fabrics and articles	..	..	..	..	..	..	..	..	..	..
18　WEARING APPAREL, DRESSING & DYEING OF FUR	5	12	119	591	163	10	9	68,506	..	1,156
181　Wearing apparel, except fur apparel	..	..	..	..	..	..	..	..	..	..
182　Dressing and dyeing of fur; articles of fur	..	..	..	..	..	..	..	..	..	..
19　TANNING & DRESSING OF LEATHER, FOOTWEAR	3	1	63	185	152	19	13	61,665	..	658
191　Tanning and dressing of leather	..	..	..	..	..	..	..	..	..	..
1911　Tanning and dressing of leather	..	..	..	..	..	..	..	..	..	..
1912　Luggage, handbags, saddlery & harness	..	..	..	..	..	..	..	..	..	..
192　Footwear	..	..	..	..	..	..	..	..	..	..
20　WOOD AND WOOD PRODUCTS	28	23	2,622	1,382	1,441	6,157	113	1,104,726	..	15,743
201　Sawmilling and planing of wood	..	..	..	..	..	..	..	..	..	..
202　Products of wood, cork, straw & plaiting materials	..	..	..	..	..	..	..	..	..	..
2021　Veneer sheets	..	..	..	..	..	..	..	..	..	..
2022　Builders' carpentry and joinery	..	..	..	..	..	..	..	..	..	..
2023　Wooden containers	..	..	..	..	..	..	..	..	..	..
2029　Other products of wood	..	..	..	..	..	..	..	..	..	..
21　PAPER AND PAPER PRODUCTS	4,324	56	229	5,331	23,455	23,792	866	4,220,843	..	73,248
2101　Pulp, paper and paperboard	..	..	..	..	..	..	..	..	..	..
2102　Corrugated paper, paperboard and their containers	..	..	..	..	..	..	..	..	..	..
2109　Other articles of pulp and paperboard	..	..	..	..	..	..	..	..	..	..
22　PUBLISHING, PRINTING & REPRODUCTION	0	27	163	140	395	3	53	160,547	..	1,359
221　Publishing	..	..	..	..	..	..	..	..	..	..
2211　Publishing of books & brochures	..	..	..	..	..	..	..	..	..	..
2212　Publishing of newspapers and periodicals	..	..	..	..	..	..	..	..	..	..
2213　Publishing of recorded media	..	..	..	..	..	..	..	..	..	..
2219　Other publishing	..	..	..	..	..	..	..	..	..	..
222　Printing and related service activities	..	..	..	..	..	..	..	..	..	..
2221　Printing	..	..	..	..	..	..	..	..	..	..
2222　Service activities related to printing	..	..	..	..	..	..	..	..	..	..
223　Reproduction of recorded media	..	..	..	..	..	..	..	..	..	..
23　COKE, REFINED PETROLEUM PRODUCTS	0	4	12	0	0	0	0	5,458	..	36
231　Coke oven products	..	..	..	..	..	..	..	..	..	..
232　Refined petroleum products	0	4	12	0	0	0	0	5,458	..	36
233　Processing of nuclear fuel	..	..	..	..	..	..	..	..	..	..
24　CHEMICALS & CHEMICAL PRODUCTS	1,838	40	787	3,228	10,660	4,526	650	3,565,095	..	34,563
241　Basic chemicals	..	..	..	..	..	..	..	..	..	..
2411　Basic chemicals, exc. fertilizers & nitrogen compounds	..	..	..	..	..	..	..	..	..	..
2412　Fertilizers and nitrogen compounds	..	..	..	..	..	..	..	..	..	..
2413　Plastics in primary forms and synthetic rubber	..	..	..	..	..	..	..	..	..	..
242　Other chemical products	..	..	..	..	..	..	..	..	..	..
2421　Pesticides and other agro-chemical products	..	..	..	..	..	..	..	..	..	..
2422　Paints, varnishes and similar coatings	..	..	..	..	..	..	..	..	..	..
2423　Pharmaceuticals, medicinal chem. & botanical prod.	..	..	..	..	..	..	..	..	..	..
2424　Soap and detergents, perfumes etc.	..	..	..	..	..	..	..	..	..	..
2429　Other chemical products, nec	..	..	..	..	..	..	..	..	..	..
243　Man-made fibres	..	..	..	..	..	..	..	..	..	..
25　RUBBER AND PLASTICS PRODUCTS	..	..	..	..	..	..	..	..	..	..
251　Rubber products	..	..	..	..	..	..	..	..	..	..
2511　Rubber tyres and tubes	..	..	..	..	..	..	..	..	..	..
2519　Other rubber products	..	..	..	..	..	..	..	..	..	..
252　Plastic products	..	..	..	..	..	..	..	..	..	..
26　OTHER NON-METALLIC MINERAL PRODUCTS	3	88	170	97	2,967	236	11	336,001	..	4,782
261　Glass and glass products	..	..	..	..	..	..	..	..	..	..
269　Non-metallic mineral products, nec	..	..	..	..	..	..	..	..	..	..
2691　Pottery, china and earthenware	..	..	..	..	..	..	..	..	..	..
2692　Refractory ceramic products	..	..	..	..	..	..	..	..	..	..
2693　Structural non-refractory clay & ceramic prod.	..	..	..	..	..	..	..	..	..	..
2694　Cement, lime and plaster	..	..	..	..	..	..	..	..	..	..
2695　Articles of concrete, cement and plaster	..	..	..	..	..	..	..	..	..	..
2696　Cutting, shaping and finishing of stone	..	..	..	..	..	..	..	..	..	..
2699　Other non-metallic mineral products, nec	..	..	..	..	..	..	..	..	..	..

ISIC Revision 3 Industry Sector		Solid TJ	LPG TJ	Distiloil TJ	RFO TJ	Gas TJ	Biomass TJ	Steam TJ	Electr MWh	Own Use MWh	TOTAL TJ
27	**BASIC METALS**	101,842	450	299	5,658	37,664	1,133	0	2,503,852	..	156,060
271	Basic iron and steel	..	..	..	..	..	..	..	..	..	..
272	Basic precious and non-ferrous metals	..	..	..	..	..	..	..	..	..	..
273	Casting of metals	..	..	..	..	..	..	..	..	..	..
2731	Casting of iron and steel	..	..	..	..	..	..	..	..	..	..
2732	Casting non-ferrous metals	..	..	..	..	..	..	..	..	..	..
28	**FABRICATED METAL PRODUCTS**	18	234	2,045	1,327	3,157	1,028	138	1,326,448	..	12,722
281	Str. metal prod., tanks, reservoirs, steam generators	..	..	..	..	..	..	..	..	..	..
2811	Structural metal products	..	..	..	..	..	..	..	..	..	..
2812	Tanks, reservoirs and containers of metal	..	..	..	..	..	..	..	..	..	..
2813	Steam generators, exc. central heating hot water boilers	..	..	..	..	..	..	..	..	..	..
289	Other fabricated metal products	..	..	..	..	..	..	..	..	..	..
2891	Forging, pressing, stamping & roll-forming of metal	..	..	..	..	..	..	..	..	..	..
2892	Treatment and coating of metals	..	..	..	..	..	..	..	..	..	..
2893	Cutlery, hand tools and general hardware	..	..	..	..	..	..	..	..	..	..
2899	Other fabricated metal products, nec	..	..	..	..	..	..	..	..	..	..
29	**MACHINERY AND EQUIPMENT, NEC**	10	159	1,190	1,093	1,520	107	373	512,039	..	6,295
291	General purpose machinery	..	..	..	..	..	..	..	..	..	..
2911	Engines and turbines	..	..	..	..	..	..	..	..	..	..
2912	Pumps, compressors, taps and valves	..	..	..	..	..	..	..	..	..	..
2913	Bearings, gears, gearing and driving elements	..	..	..	..	..	..	..	..	..	..
2914	Ovens, furnaces and furnace burners	..	..	..	..	..	..	..	..	..	..
2915	Lifting and handling equipment	..	..	..	..	..	..	..	..	..	..
2919	Other general purpose machinery	..	..	..	..	..	..	..	..	..	..
292	Special purpose machinery	..	..	..	..	..	..	..	..	..	..
2921	Agricultural and forestry machinery	..	..	..	..	..	..	..	..	..	..
2922	Machine-tools	..	..	..	..	..	..	..	..	..	..
2923	Machinery for metallurgy	..	..	..	..	..	..	..	..	..	..
2924	Machinery for mining, quarrying and construction	..	..	..	..	..	..	..	..	..	..
2925	Machinery for food, beverage & tobacco processing	..	..	..	..	..	..	..	..	..	..
2926	Machinery for textile, apparel & leather production	..	..	..	..	..	..	..	..	..	..
2927	Machinery for weapons and ammunition	..	..	..	..	..	..	..	..	..	..
2929	Other special purpose machinery	..	..	..	..	..	..	..	..	..	..
293	Domestic appliances, nec	..	..	..	..	..	..	..	..	..	..
30	**OFFICE, ACCOUNTING & COMPUTING MACHINERY**	15	126	1,134	995	1,157	49	422	848,345	..	6,952
31	**ELECTRICAL MACHINERY & APPARATUS, NEC**	..	..	..	..	..	..	..	..	..	..
311	Electric motors, generators and transformers	..	..	..	..	..	..	..	..	..	..
312	Electricity distribution and control apparatus	..	..	..	..	..	..	..	..	..	..
313	Insulated wire and cable	..	..	..	..	..	..	..	..	..	..
314	Accumulators, primary cells & primary batteries	..	..	..	..	..	..	..	..	..	..
315	Electric lamps and lighting equipment	..	..	..	..	..	..	..	..	..	..
319	Other electrical equipment, nec	..	..	..	..	..	..	..	..	..	..
32	**RADIO, TV & COMMUNICATION EQUIP. & APP.**	..	..	..	..	..	..	..	..	..	..
321	Electronic valves, tubes, other electronic components	..	..	..	..	..	..	..	..	..	..
322	TV & radio transmitters, apparatus for line telephony	..	..	..	..	..	..	..	..	..	..
323	TV & radio receivers, recording apparatus	..	..	..	..	..	..	..	..	..	..
33	**MEDICAL PRECISION &OPTICAL INSTRUMENTS**	..	..	..	..	..	..	..	..	..	..
331	Medical appliances and instruments	..	..	..	..	..	..	..	..	..	..
3311	Medical, surgical equipment & orthopaedic app.	..	..	..	..	..	..	..	..	..	..
3312	Instruments & appliances for measuring, checking etc	..	..	..	..	..	..	..	..	..	..
3313	Industrial process control equipment	..	..	..	..	..	..	..	..	..	..
332	Optical instruments and photographic equipment	..	..	..	..	..	..	..	..	..	..
333	Watches and clocks	..	..	..	..	..	..	..	..	..	..
34	**MOTOR VEHICLES, TRAILERS & SEMI-TRAILERS**	12	90	1,110	658	1,996	58	549	589,418	..	6,595
341	Motor vehicles	..	..	..	..	..	..	..	..	..	..
342	Bodies (coachwork) for motor vehicles	..	..	..	..	..	..	..	..	..	..
343	Parts, accessories for motor vehicles & their engines	..	..	..	..	..	..	..	..	..	..
35	**OTHER TRANSPORT EQUIPMENT**	..	..	..	..	..	..	..	..	..	..
351	Building and repairing of ships and boats	..	..	..	..	..	..	..	..	..	..
3511	Building and repairing of ships	..	..	..	..	..	..	..	..	..	..
3512	Building, repairing of pleasure & sporting boats	..	..	..	..	..	..	..	..	..	..
352	Railway, tramway locomotives & rolling stock	..	..	..	..	..	..	..	..	..	..
353	Aircraft and spacecraft	..	..	..	..	..	..	..	..	..	..
359	Transport equipment, nec	..	..	..	..	..	..	..	..	..	..
3591	Motorcycles	..	..	..	..	..	..	..	..	..	..
3592	Bicycles and invalid carriages	..	..	..	..	..	..	..	..	..	..
3599	Other transport equipment, nec	..	..	..	..	..	..	..	..	..	..
36	**FURNITURE; MANUFACTURING, NEC**	..	..	..	..	..	..	..	..	..	..
361	Furniture	..	..	..	..	..	..	..	..	..	..
369	Manufacturing, nec	..	..	..	..	..	..	..	..	..	..
3691	Jewellery and related articles	..	..	..	..	..	..	..	..	..	..
3692	Musical instruments	..	..	..	..	..	..	..	..	..	..
3693	Sports goods	..	..	..	..	..	..	..	..	..	..
3694	Games and toys	..	..	..	..	..	..	..	..	..	..
3699	Other manufacturing, nec	..	..	..	..	..	..	..	..	..	..
37	**RECYCLING**	..	..	..	..	..	..	..	..	..	..
371	Recycling of metal waste and scrap	..	..	..	..	..	..	..	..	..	..
372	Recycling of non-metal waste and scrap	..	..	..	..	..	..	..	..	..	..
CERR	**Non-specified industry**	..	..	..	..	..	..	..	..	..	..
15-37	**TOTAL MANUFACTURING**	108,320	1,444	12,603	26,844	96,511	37,678	3,870	17,421,005	..	349,986

ISIC Revision 3 Industry Sector	Solid TJ	LPG TJ	Distiloil TJ	RFO TJ	Gas TJ	Biomass TJ	Steam TJ	Electr MWh	Own Use MWh	TOTAL TJ
15 FOOD PRODUCTS AND BEVERAGES	134	107	2,359	4,722	10,143	261	590	1,515,187		23,771
151 Production, processing and preserving (PPP)	..	..	..	..	..	..	..	..		..
1511 PPP of meat and meat products	..	..	..	..	..	..	..	..		..
1512 Processing and preserving of fish products	..	..	..	..	..	..	..	..		..
1513 Processing, preserving of fruit & vegetables	..	..	..	..	..	..	..	..		..
1514 Vegetable and animal oils and fats	..	..	..	..	..	..	..	..		..
152 Dairy products	..	..	..	..	..	..	..	..		..
153 Grain mill prod., starches & prepared animal feeds	..	..	..	..	..	..	..	..		..
1531 Grain mill products	..	..	..	..	..	..	..	..		..
1532 Starches and starch products	..	..	..	..	..	..	..	..		..
1533 Prepared animal feeds	..	..	..	..	..	..	..	..		..
154 Other food products	..	..	..	..	..	..	..	..		..
1541 Bakery products	..	..	..	..	..	..	..	..		..
1542 Sugar	..	..	..	..	..	..	..	..		..
1543 Cocoa, chocolate and sugar confectionery	..	..	..	..	..	..	..	..		..
1544 Macaroni, noodles, couscous & similar farinaceous prod.	..	..	..	..	..	..	..	..		..
1549 Other food products, nec	..	..	..	..	..	..	..	..		..
155 Beverages	..	..	..	..	..	..	..	..		..
1551 Distilling, rectifying and blending of spirits	..	..	..	..	..	..	..	..		..
1552 Wines	..	..	..	..	..	..	..	..		..
1553 Malt liquors and malt	..	..	..	..	..	..	..	..		..
1554 Soft drinks; production of mineral waters	..	..	..	..	..	..	..	..		..
16 TOBACCO PRODUCTS	..	..	..	..	..	..	..	..		..
17 TEXTILES	0	14	129	1,083	2,313	306	34	663,747	..	6,268
171 Spinning, weaving and finishing of textiles	..	..	..	..	..	..	..	..		..
1711 Preparation and spinning of textile fibres	..	..	..	..	..	..	..	..		..
1712 Finishing of textiles	..	..	..	..	..	..	..	..		..
172 Other textiles	..	..	..	..	..	..	..	..		..
1721 Made-up textile articles, except apparel	..	..	..	..	..	..	..	..		..
1722 Carpets and rugs	..	..	..	..	..	..	..	..		..
1723 Cordage, rope, twine and netting	..	..	..	..	..	..	..	..		..
1729 Other textiles, nec	..	..	..	..	..	..	..	..		..
173 Knitted and crocheted fabrics and articles	..	..	..	..	..	..	..	..		..
18 WEARING APPAREL, DRESSING & DYEING OF FUR	6	9	107	582	163	11	13	63,506	..	1,120
181 Wearing apparel, except fur apparel	..	..	..	..	..	..	..	..		..
182 Dressing and dyeing of fur; articles of fur	..	..	..	..	..	..	..	..		..
19 TANNING & DRESSING OF LEATHER, FOOTWEAR	3	1	64	215	144	20	13	63,434	..	688
191 Tanning and dressing of leather	..	..	..	..	..	..	..	..		..
1911 Tanning and dressing of leather	..	..	..	..	..	..	..	..		..
1912 Luggage, handbags, saddlery & harness	..	..	..	..	..	..	..	..		..
192 Footwear	..	..	..	..	..	..	..	..		..
20 WOOD AND WOOD PRODUCTS	21	25	2,679	1,756	1,494	6,815	123	1,115,586	..	16,929
201 Sawmilling and planing of wood	..	..	..	..	..	..	..	..		..
202 Products of wood, cork, straw & plaiting materials	..	..	..	..	..	..	..	..		..
2021 Veneer sheets	..	..	..	..	..	..	..	..		..
2022 Builders' carpentry and joinery	..	..	..	..	..	..	..	..		..
2023 Wooden containers	..	..	..	..	..	..	..	..		..
2029 Other products of wood	..	..	..	..	..	..	..	..		..
21 PAPER AND PAPER PRODUCTS	4,135	61	258	4,287	23,989	19,006	1,469	4,348,884	..	68,861
2101 Pulp, paper and paperboard	..	..	..	..	..	..	..	..		..
2102 Corrugated paper, paperboard and their containers	..	..	..	..	..	..	..	..		..
2109 Other articles of pulp and paperboard	..	..	..	..	..	..	..	..		..
22 PUBLISHING, PRINTING & REPRODUCTION	0	27	163	137	394	3	53	160,547	..	1,355
221 Publishing	..	..	..	..	..	..	..	..		..
2211 Publishing of books & brochures	..	..	..	..	..	..	..	..		..
2212 Publishing of newspapers and periodicals	..	..	..	..	..	..	..	..		..
2213 Publishing of recorded media	..	..	..	..	..	..	..	..		..
2219 Other publishing	..	..	..	..	..	..	..	..		..
222 Printing and related service activities	..	..	..	..	..	..	..	..		..
2221 Printing	..	..	..	..	..	..	..	..		..
2222 Service activities related to printing	..	..	..	..	..	..	..	..		..
223 Reproduction of recorded media	..	..	..	..	..	..	..	..		..
23 COKE, REFINED PETROLEUM PRODUCTS	0	1	9	0	0	0	0	5,286	..	29
231 Coke oven products	..	..	..	..	..	..	..	..		..
232 Refined petroleum products	0	1	9	0	0	0	0	5,286	..	29
233 Processing of nuclear fuel	..	..	..	..	..	..	..	..		..
24 CHEMICALS & CHEMICAL PRODUCTS	1,750	36	879	3,154	11,129	4,656	813	3,631,348	..	35,490
241 Basic chemicals	..	..	..	..	..	..	..	..		..
2411 Basic chemicals, exc. fertilizers & nitrogen compounds	..	..	..	..	..	..	..	..		..
2412 Fertilizers and nitrogen compounds	..	..	..	..	..	..	..	..		..
2413 Plastics in primary forms and synthetic rubber	..	..	..	..	..	..	..	..		..
242 Other chemical products	..	..	..	..	..	..	..	..		..
2421 Pesticides and other agro-chemical products	..	..	..	..	..	..	..	..		..
2422 Paints, varnishes and similar coatings	..	..	..	..	..	..	..	..		..
2423 Pharmaceuticals, medicinal chem. & botanical prod.	..	..	..	..	..	..	..	..		..
2424 Soap and detergents, perfumes etc.	..	..	..	..	..	..	..	..		..
2429 Other chemical products, nec	..	..	..	..	..	..	..	..		..
243 Man-made fibres	..	..	..	..	..	..	..	..		..
25 RUBBER AND PLASTICS PRODUCTS	..	..	..	..	..	..	..	..		..
251 Rubber products	..	..	..	..	..	..	..	..		..
2511 Rubber tyres and tubes	..	..	..	..	..	..	..	..		..
2519 Other rubber products	..	..	..	..	..	..	..	..		..
252 Plastic products	..	..	..	..	..	..	..	..		..
26 OTHER NON-METALLIC MINERAL PRODUCTS	3	83	175	105	3,005	237	10	347,949	..	4,871
261 Glass and glass products	..	..	..	..	..	..	..	..		..
269 Non-metallic mineral products, nec	..	..	..	..	..	..	..	..		..
2691 Pottery, china and earthenware	..	..	..	..	..	..	..	..		..
2692 Refractory ceramic products	..	..	..	..	..	..	..	..		..
2693 Structural non-refractory clay & ceramic prod.	..	..	..	..	..	..	..	..		..
2694 Cement, lime and plaster	..	..	..	..	..	..	..	..		..
2695 Articles of concrete, cement and plaster	..	..	..	..	..	..	..	..		..
2696 Cutting, shaping and finishing of stone	..	..	..	..	..	..	..	..		..
2699 Other non-metallic mineral products, nec	..	..	..	..	..	..	..	..		..

ISIC Revision 3 Industry Sector		Solid TJ	LPG TJ	Distiloil TJ	RFO TJ	Gas TJ	Biomass TJ	Steam TJ	Electr MWh	Own Use MWh	TOTAL TJ
27	**BASIC METALS**	106,088	442	310	6,539	37,474	1,133	0	2,571,519	..	161,243
271	Basic iron and steel	..	..	..	..	..	..	..	..	..	..
272	Basic precious and non-ferrous metals	..	..	..	..	..	..	..	..	..	..
273	Casting of metals	..	..	..	..	..	..	..	..	..	..
2731	Casting of iron and steel	..	..	..	..	..	..	..	..	..	..
2732	Casting non-ferrous metals	..	..	..	..	..	..	..	..	..	..
28	**FABRICATED METAL PRODUCTS**	11	253	2,236	1,273	3,527	1,032	173	1,436,290	..	13,676
281	Str. metal prod., tanks, reservoirs, steam generators	..	..	..	..	..	..	..	..	..	..
2811	Structural metal products	..	..	..	..	..	..	..	..	..	..
2812	Tanks, reservoirs and containers of metal	..	..	..	..	..	..	..	..	..	..
2813	Steam generators, exc. central heating hot water boilers	..	..	..	..	..	..	..	..	..	..
289	Other fabricated metal products	..	..	..	..	..	..	..	..	..	..
2891	Forging, pressing, stamping & roll-forming of metal	..	..	..	..	..	..	..	..	..	..
2892	Treatment and coating of metals	..	..	..	..	..	..	..	..	..	..
2893	Cutlery, hand tools and general hardware	..	..	..	..	..	..	..	..	..	..
2899	Other fabricated metal products, nec	..	..	..	..	..	..	..	..	..	..
29	**MACHINERY AND EQUIPMENT, NEC**	10	189	1,261	854	1,592	102	448	540,834	..	6,403
291	General purpose machinery	..	..	..	..	..	..	..	..	..	..
2911	Engines and turbines	..	..	..	..	..	..	..	..	..	..
2912	Pumps, compressors, taps and valves	..	..	..	..	..	..	..	..	..	..
2913	Bearings, gears, gearing and driving elements	..	..	..	..	..	..	..	..	..	..
2914	Ovens, furnaces and furnace burners	..	..	..	..	..	..	..	..	..	..
2915	Lifting and handling equipment	..	..	..	..	..	..	..	..	..	..
2919	Other general purpose machinery	..	..	..	..	..	..	..	..	..	..
292	Special purpose machinery	..	..	..	..	..	..	..	..	..	..
2921	Agricultural and forestry machinery	..	..	..	..	..	..	..	..	..	..
2922	Machine-tools	..	..	..	..	..	..	..	..	..	..
2923	Machinery for metallurgy	..	..	..	..	..	..	..	..	..	..
2924	Machinery for mining, quarrying and construction	..	..	..	..	..	..	..	..	..	..
2925	Machinery for food, beverage & tobacco processing	..	..	..	..	..	..	..	..	..	..
2926	Machinery for textile, apparel & leather production	..	..	..	..	..	..	..	..	..	..
2927	Machinery for weapons and ammunition	..	..	..	..	..	..	..	..	..	..
2929	Other special purpose machinery	..	..	..	..	..	..	..	..	..	..
293	Domestic appliances, nec	..	..	..	..	..	..	..	..	..	..
30	**OFFICE, ACCOUNTING & COMPUTING MACHINERY**	4	119	1,135	853	1,288	52	371	830,165	..	6,811
31	**ELECTRICAL MACHINERY & APPARATUS, NEC**	..	..	..	..	..	..	..	..	..	..
311	Electric motors, generators and transformers	..	..	..	..	..	..	..	..	..	..
312	Electricity distribution and control apparatus	..	..	..	..	..	..	..	..	..	..
313	Insulated wire and cable	..	..	..	..	..	..	..	..	..	..
314	Accumulators, primary cells & primary batteries	..	..	..	..	..	..	..	..	..	..
315	Electric lamps and lighting equipment	..	..	..	..	..	..	..	..	..	..
319	Other electrical equipment, nec	..	..	..	..	..	..	..	..	..	..
32	**RADIO, TV & COMMUNICATION EQUIP. & APP.**	..	..	..	..	..	..	..	..	..	..
321	Electronic valves, tubes, other electronic components	..	..	..	..	..	..	..	..	..	..
322	TV & radio transmitters, apparatus for line telephony	..	..	..	..	..	..	..	..	..	..
323	TV & radio receivers, recording apparatus	..	..	..	..	..	..	..	..	..	..
33	**MEDICAL PRECISION &OPTICAL INSTRUMENTS**	..	..	..	..	..	..	..	..	..	..
331	Medical appliances and instruments	..	..	..	..	..	..	..	..	..	..
3311	Medical, surgical equipment & orthopaedic app.	..	..	..	..	..	..	..	..	..	..
3312	Instruments & appliances for measuring, checking etc	..	..	..	..	..	..	..	..	..	..
3313	Industrial process control equipment	..	..	..	..	..	..	..	..	..	..
332	Optical instruments and photographic equipment	..	..	..	..	..	..	..	..	..	..
333	Watches and clocks	..	..	..	..	..	..	..	..	..	..
34	**MOTOR VEHICLES, TRAILERS & SEMI-TRAILERS**	12	227	1,151	606	2,093	60	710	628,978	..	7,123
341	Motor vehicles	..	..	..	..	..	..	..	..	..	..
342	Bodies (coachwork) for motor vehicles	..	..	..	..	..	..	..	..	..	..
343	Parts, accessories for motor vehicles & their engines	..	..	..	..	..	..	..	..	..	..
35	**OTHER TRANSPORT EQUIPMENT**	..	..	..	..	..	..	..	..	..	..
351	Building and repairing of ships and boats	..	..	..	..	..	..	..	..	..	..
3511	Building and repairing of ships	..	..	..	..	..	..	..	..	..	..
3512	Building, repairing of pleasure & sporting boats	..	..	..	..	..	..	..	..	..	..
352	Railway, tramway locomotives & rolling stock	..	..	..	..	..	..	..	..	..	..
353	Aircraft and spacecraft	..	..	..	..	..	..	..	..	..	..
359	Transport equipment, nec	..	..	..	..	..	..	..	..	..	..
3591	Motorcycles	..	..	..	..	..	..	..	..	..	..
3592	Bicycles and invalid carriages	..	..	..	..	..	..	..	..	..	..
3599	Other transport equipment, nec	..	..	..	..	..	..	..	..	..	..
36	**FURNITURE; MANUFACTURING, NEC**	..	..	..	..	..	..	..	..	..	..
361	Furniture	..	..	..	..	..	..	..	..	..	..
369	Manufacturing, nec	..	..	..	..	..	..	..	..	..	..
3691	Jewellery and related articles	..	..	..	..	..	..	..	..	..	..
3692	Musical instruments	..	..	..	..	..	..	..	..	..	..
3693	Sports goods	..	..	..	..	..	..	..	..	..	..
3694	Games and toys	..	..	..	..	..	..	..	..	..	..
3699	Other manufacturing, nec	..	..	..	..	..	..	..	..	..	..
37	**RECYCLING**	..	..	..	..	..	..	..	..	..	..
371	Recycling of metal waste and scrap	..	..	..	..	..	..	..	..	..	..
372	Recycling of non-metal waste and scrap	..	..	..	..	..	..	..	..	..	..
CERR	**Non-specified industry**	..	..	..	..	..	..	..	..	..	..
15-37	**TOTAL MANUFACTURING**	112,177	1,594	12,915	26,166	98,748	33,694	4,820	17,923,260	..	354,638

ISIC Revision 3 Industry Sector	Solid TJ	LPG TJ	Distiloil TJ	RFO TJ	Gas TJ	Biomass TJ	Steam TJ	Electr MWh	Own Use MWh	TOTAL TJ
15 FOOD PRODUCTS AND BEVERAGES	148	91	2,162	3,027	10,719	113	452	1,234,244	..	21,154
151 Production, processing and preserving (PPP)	..	..	..	..	..	..	..	..	..	..
1511 PPP of meat and meat products	..	..	..	..	..	..	..	..	..	..
1512 Processing and preserving of fish products	..	..	..	..	..	..	..	..	..	..
1513 Processing, preserving of fruit & vegetables	..	..	..	..	..	..	..	..	..	..
1514 Vegetable and animal oils and fats	..	..	..	..	..	..	..	..	..	..
152 Dairy products	..	..	..	..	..	..	..	..	..	..
153 Grain mill prod., starches & prepared animal feeds	..	..	..	..	..	..	..	..	..	..
1531 Grain mill products	..	..	..	..	..	..	..	..	..	..
1532 Starches and starch products	..	..	..	..	..	..	..	..	..	..
1533 Prepared animal feeds	..	..	..	..	..	..	..	..	..	..
154 Other food products	..	..	..	..	..	..	..	..	..	..
1541 Bakery products	..	..	..	..	..	..	..	..	..	..
1542 Sugar	..	..	..	..	..	..	..	..	..	..
1543 Cocoa, chocolate and sugar confectionery	..	..	..	..	..	..	..	..	..	..
1544 Macaroni, noodles, couscous & similar farinaceous prod.	..	..	..	..	..	..	..	..	..	..
1549 Other food products, nec	..	..	..	..	..	..	..	..	..	..
155 Beverages	..	..	..	..	..	..	..	..	..	..
1551 Distilling, rectifying and blending of spirits	..	..	..	..	..	..	..	..	..	..
1552 Wines	..	..	..	..	..	..	..	..	..	..
1553 Malt liquors and malt	..	..	..	..	..	..	..	..	..	..
1554 Soft drinks; production of mineral waters	..	..	..	..	..	..	..	..	..	..
16 TOBACCO PRODUCTS	0	13	113	1,082	2,300	234	104	628,675	..	6,109
17 TEXTILES	..	..	..	..	..	..	..	..	..	..
171 Spinning, weaving and finishing of textiles	..	..	..	..	..	..	..	..	..	..
1711 Preparation and spinning of textile fibres	..	..	..	..	..	..	..	..	..	..
1712 Finishing of textiles	..	..	..	..	..	..	..	..	..	..
172 Other textiles	..	..	..	..	..	..	..	..	..	..
1721 Made-up textile articles, except apparel	..	..	..	..	..	..	..	..	..	..
1722 Carpets and rugs	..	..	..	..	..	..	..	..	..	..
1723 Cordage, rope, twine and netting	..	..	..	..	..	..	..	..	..	..
1729 Other textiles, nec	..	..	..	..	..	..	..	..	..	..
173 Knitted and crocheted fabrics and articles	..	..	..	..	..	..	..	..	..	..
18 WEARING APPAREL, DRESSING & DYEING OF FUR	..	..	..	..	..	..	..	..	..	..
181 Wearing apparel, except fur apparel	..	..	..	..	..	..	..	..	..	..
182 Dressing and dyeing of fur; articles of fur	..	..	..	..	..	..	..	..	..	..
19 TANNING & DRESSING OF LEATHER, FOOTWEAR	0	0	24	101	192	7	8	54,349	..	529
191 Tanning and dressing of leather	..	..	..	..	..	..	..	..	..	..
1911 Tanning and dressing of leather	..	..	..	..	..	..	..	..	..	..
1912 Luggage, handbags, saddlery & harness	..	..	..	..	..	..	..	..	..	..
192 Footwear	..	..	..	..	..	..	..	..	..	..
20 WOOD AND WOOD PRODUCTS	3	28	1,275	1,036	1,699	6,316	155	764,317	..	13,263
201 Sawmilling and planing of wood	..	..	..	..	..	..	..	..	..	..
202 Products of wood, cork, straw & plaiting materials	..	..	..	..	..	..	..	..	..	..
2021 Veneer sheets	..	..	..	..	..	..	..	..	..	..
2022 Builders' carpentry and joinery	..	..	..	..	..	..	..	..	..	..
2023 Wooden containers	..	..	..	..	..	..	..	..	..	..
2029 Other products of wood	..	..	..	..	..	..	..	..	..	..
21 PAPER AND PAPER PRODUCTS	4,359	51	196	4,528	23,962	27,455	1,465	4,212,224	..	77,180
2101 Pulp, paper and paperboard	..	..	..	..	..	..	..	..	..	..
2102 Corrugated paper, paperboard and their containers	..	..	..	..	..	..	..	..	..	..
2109 Other articles of pulp and paperboard	..	..	..	..	..	..	..	..	..	..
22 PUBLISHING, PRINTING & REPRODUCTION	1	47	97	103	607	11	69	244,244	..	1,814
221 Publishing	..	..	..	..	..	..	..	..	..	..
2211 Publishing of books & brochures	..	..	..	..	..	..	..	..	..	..
2212 Publishing of newspapers and periodicals	..	..	..	..	..	..	..	..	..	..
2213 Publishing of recorded media	..	..	..	..	..	..	..	..	..	..
2219 Other publishing	..	..	..	..	..	..	..	..	..	..
222 Printing and related service activities	..	..	..	..	..	..	..	..	..	..
2221 Printing	..	..	..	..	..	..	..	..	..	..
2222 Service activities related to printing	..	..	..	..	..	..	..	..	..	..
223 Reproduction of recorded media	..	..	..	..	..	..	..	..	..	..
23 COKE, REFINED PETROLEUM PRODUCTS	0	0	31	0	0	0	0	26,757	..	127
231 Coke oven products	..	..	..	..	..	..	..	..	..	..
232 Refined petroleum products	0	0	31	0	0	0	0	26,757	..	127
233 Processing of nuclear fuel	..	..	..	..	..	..	..	..	..	..
24 CHEMICALS & CHEMICAL PRODUCTS	1,892	14	219	1,074	9,223	6,060	894	2,762,505	..	29,322
241 Basic chemicals	..	..	..	..	..	..	..	..	..	..
2411 Basic chemicals, exc. fertilizers & nitrogen compounds	..	..	..	..	..	..	..	..	..	..
2412 Fertilizers and nitrogen compounds	..	..	..	..	..	..	..	..	..	..
2413 Plastics in primary forms and synthetic rubber	..	..	..	..	..	..	..	..	..	..
242 Other chemical products	..	..	..	..	..	..	..	..	..	..
2421 Pesticides and other agro-chemical products	..	..	..	..	..	..	..	..	..	..
2422 Paints, varnishes and similar coatings	..	..	..	..	..	..	..	..	..	..
2423 Pharmaceuticals, medicinal chem. & botanical prod.	..	..	..	..	..	..	..	..	..	..
2424 Soap and detergents, perfumes etc.	..	..	..	..	..	..	..	..	..	..
2429 Other chemical products, nec	..	..	..	..	..	..	..	..	..	..
243 Man-made fibres	..	..	..	..	..	..	..	..	..	..
25 RUBBER AND PLASTICS PRODUCTS	0	17	248	635	1,315	61	120	708,749	..	4,947
251 Rubber products	..	..	..	..	..	..	..	..	..	..
2511 Rubber tyres and tubes	..	..	..	..	..	..	..	..	..	..
2519 Other rubber products	..	..	..	..	..	..	..	..	..	..
252 Plastic products	..	..	..	..	..	..	..	..	..	..
26 OTHER NON-METALLIC MINERAL PRODUCTS	5,306	1,235	2,573	4,881	11,503	827	29	1,359,776	..	31,249
261 Glass and glass products	..	..	..	..	..	..	..	..	..	..
269 Non-metallic mineral products, nec	..	..	..	..	..	..	..	..	..	..
2691 Pottery, china and earthenware	..	..	..	..	..	..	..	..	..	..
2692 Refractory ceramic products	..	..	..	..	..	..	..	..	..	..
2693 Structural non-refractory clay & ceramic prod.	..	..	..	..	..	..	..	..	..	..
2694 Cement, lime and plaster	..	..	..	..	..	..	..	..	..	..
2695 Articles of concrete, cement and plaster	..	..	..	..	..	..	..	..	..	..
2696 Cutting, shaping and finishing of stone	..	..	..	..	..	..	..	..	..	..
2699 Other non-metallic mineral products, nec	..	..	..	..	..	..	..	..	..	..

ISIC Revision 3 Industry Sector	Solid TJ	LPG TJ	Distiloil TJ	RFO TJ	Gas TJ	Biomass TJ	Steam TJ	Electr MWh	Own Use MWh	TOTAL TJ
27 **BASIC METALS**	111,499	1,075	333	5,841	42,403	1,391	11	4,227,813	..	177,777
271 Basic iron and steel	111,282	706	260	5,393	40,336	513	3	3,334,762	..	170,500
272 Basic precious and non-ferrous metals	193	328	61	353	1,590	64	8	448,646		4,213
273 Casting of metals	24	41	12	95	477	814	0	444,405		3,064
2731 Casting of iron and steel	..	..	..	..	..	..	..	..		..
2732 Casting non-ferrous metals	..	..	..	..	..	..	..	..		..
28 **FABRICATED METAL PRODUCTS**	69	396	593	699	2,469	320	57	729,682	..	7,230
281 Str. metal prod., tanks, reservoirs, steam generators	..	..	..	..	..	..	..	..	..	..
2811 Structural metal products	..	..	..	..	..	..	..	..		..
2812 Tanks, reservoirs and containers of metal	..	..	..	..	..	..	..	..		..
2813 Steam generators, exc. central heating hot water boilers	..	..	..	..	..	..	..	..		..
289 Other fabricated metal products	..	..	..	..	..	..	..	..		..
2891 Forging, pressing, stamping & roll-forming of metal	..	..	..	..	..	..	..	..		..
2892 Treatment and coating of metals	..	..	..	..	..	..	..	..		..
2893 Cutlery, hand tools and general hardware	..	..	..	..	..	..	..	..		..
2899 Other fabricated metal products, nec	..	..	..	..	..	..	..	..		..
29 **MACHINERY AND EQUIPMENT, NEC**	12	301	500	640	1,442	26	288	946,199	..	6,613
291 General purpose machinery	..	..	..	..	..	..	..	..		..
2911 Engines and turbines	..	..	..	..	..	..	..	..		..
2912 Pumps, compressors, taps and valves	..	..	..	..	..	..	..	..		..
2913 Bearings, gears, gearing and driving elements	..	..	..	..	..	..	..	..		..
2914 Ovens, furnaces and furnace burners	..	..	..	..	..	..	..	..		..
2915 Lifting and handling equipment	..	..	..	..	..	..	..	..		..
2919 Other general purpose machinery	..	..	..	..	..	..	..	..		..
292 Special purpose machinery	..	..	..	..	..	..	..	..		..
2921 Agricultural and forestry machinery	..	..	..	..	..	..	..	..		..
2922 Machine-tools	..	..	..	..	..	..	..	..		..
2923 Machinery for metallurgy	..	..	..	..	..	..	..	..		..
2924 Machinery for mining, quarrying and construction	..	..	..	..	..	..	..	..		..
2925 Machinery for food, beverage & tobacco processing	..	..	..	..	..	..	..	..		..
2926 Machinery for textile, apparel & leather production	..	..	..	..	..	..	..	..		..
2927 Machinery for weapons and ammunition	..	..	..	..	..	..	..	..		..
2929 Other special purpose machinery	..	..	..	..	..	..	..	..		..
293 Domestic appliances, nec	..	..	..	..	..	..	..	..		..
30 **OFFICE, ACCOUNTING & COMPUTING MACHINERY**	1	38	381	473	1,485	1	455	869,884	..	5,966
31 **ELECTRICAL MACHINERY & APPARATUS, NEC**	..	..	..	..	..	..	..	..	..	..
311 Electric motors, generators and transformers	..	..	..	..	..	..	..	..		..
312 Electricity distribution and control apparatus	..	..	..	..	..	..	..	..		..
313 Insulated wire and cable	..	..	..	..	..	..	..	..		..
314 Accumulators, primary cells & primary batteries	..	..	..	..	..	..	..	..		..
315 Electric lamps and lighting equipment	..	..	..	..	..	..	..	..		..
319 Other electrical equipment, nec	..	..	..	..	..	..	..	..		..
32 **RADIO, TV & COMMUNICATION EQUIP. & APP.**	..	..	..	..	..	..	..	..	..	..
321 Electronic valves, tubes, other electronic components	..	..	..	..	..	..	..	..		..
322 TV & radio transmitters, apparatus for line telephony	..	..	..	..	..	..	..	..		..
323 TV & radio receivers, recording apparatus	..	..	..	..	..	..	..	..		..
33 **MEDICAL PRECISION &OPTICAL INSTRUMENTS**	..	..	..	..	..	..	..	..	..	..
331 Medical appliances and instruments	..	..	..	..	..	..	..	..		..
3311 Medical, surgical equipment & orthopaedic app.	..	..	..	..	..	..	..	..		..
3312 Instruments & appliances for measuring, checking etc	..	..	..	..	..	..	..	..		..
3313 Industrial process control equipment	..	..	..	..	..	..	..	..		..
332 Optical instruments and photographic equipment	..	..	..	..	..	..	..	..		..
333 Watches and clocks	..	..	..	..	..	..	..	..		..
34 **MOTOR VEHICLES, TRAILERS & SEMI-TRAILERS**	0	128	173	147	2,104	6	698	523,719	..	5,143
341 Motor vehicles	..	..	..	..	..	..	..	..		..
342 Bodies (coachwork) for motor vehicles	..	..	..	..	..	..	..	..		..
343 Parts, accessories for motor vehicles & their engines	..	..	..	..	..	..	..	..		..
35 **OTHER TRANSPORT EQUIPMENT**	..	..	..	..	..	..	..	..	..	..
351 Building and repairing of ships and boats	..	..	..	..	..	..	..	..		..
3511 Building and repairing of ships	..	..	..	..	..	..	..	..		..
3512 Building, repairing of pleasure & sporting boats	..	..	..	..	..	..	..	..		..
352 Railway, tramway locomotives & rolling stock	..	..	..	..	..	..	..	..		..
353 Aircraft and spacecraft	..	..	..	..	..	..	..	..		..
359 Transport equipment, nec	..	..	..	..	..	..	..	..		..
3591 Motorcycles	..	..	..	..	..	..	..	..		..
3592 Bicycles and invalid carriages	..	..	..	..	..	..	..	..		..
3599 Other transport equipment, nec	..	..	..	..	..	..	..	..		..
36 **FURNITURE; MANUFACTURING, NEC**	12	12	616	633	225	726	44	348,613	..	3,522
361 Furniture	..	..	..	..	..	..	..	..		..
369 Manufacturing, nec	..	..	..	..	..	..	..	..		..
3691 Jewellery and related articles	..	..	..	..	..	..	..	..		..
3692 Musical instruments	..	..	..	..	..	..	..	..		..
3693 Sports goods	..	..	..	..	..	..	..	..		..
3694 Games and toys	..	..	..	..	..	..	..	..		..
3699 Other manufacturing, nec	..	..	..	..	..	..	..	..		..
37 **RECYCLING**	..	..	..	..	..	..	..	..	..	..
371 Recycling of metal waste and scrap	..	..	..	..	..	..	..	..		..
372 Recycling of non-metal waste and scrap	..	..	..	..	..	..	..	..		..
CERR Non-specified industry	..	..	..	..	..	..	..	..	..	..
15-37 **TOTAL MANUFACTURING**	123,302	3,446	9,534	24,900	111,648	43,554	4,849	19,641,750	..	391,945

ISIC Revision 3 Industry Sector		Solid TJ	LPG TJ	Distiloil TJ	RFO TJ	Gas TJ	Biomass TJ	Steam TJ	Electr MWh	Own Use MWh	TOTAL TJ
15	**FOOD PRODUCTS AND BEVERAGES**	148	108	2,062	2,535	11,208	135	387	1,392,742	..	21,598
151	Production, processing and preserving (PPP)	..	..	..	..	..	..	..	..	..	..
1511	PPP of meat and meat products	..	..	..	..	..	..	..	..	..	..
1512	Processing and preserving of fish products	..	..	..	..	..	..	..	..	..	..
1513	Processing, preserving of fruit & vegetables	..	..	..	..	..	..	..	..	..	..
1514	Vegetable and animal oils and fats	..	..	..	..	..	..	..	..	..	..
152	Dairy products	..	..	..	..	..	..	..	..	..	..
153	Grain mill prod., starches & prepared animal feeds	..	..	..	..	..	..	..	..	..	..
1531	Grain mill products	..	..	..	..	..	..	..	..	..	..
1532	Starches and starch products	..	..	..	..	..	..	..	..	..	..
1533	Prepared animal feeds	..	..	..	..	..	..	..	..	..	..
154	Other food products	..	..	..	..	..	..	..	..	..	..
1541	Bakery products	..	..	..	..	..	..	..	..	..	..
1542	Sugar	..	..	..	..	..	..	..	..	..	..
1543	Cocoa, chocolate and sugar confectionery	..	..	..	..	..	..	..	..	..	..
1544	Macaroni, noodles, couscous & similar farinaceous prod.	..	..	..	..	..	..	..	..	..	..
1549	Other food products, nec	..	..	..	..	..	..	..	..	..	..
155	Beverages	..	..	..	..	..	..	..	..	..	..
1551	Distilling, rectifying and blending of spirits	..	..	..	..	..	..	..	..	..	..
1552	Wines	..	..	..	..	..	..	..	..	..	..
1553	Malt liquors and malt	..	..	..	..	..	..	..	..	..	..
1554	Soft drinks; production of mineral waters	..	..	..	..	..	..	..	..	..	..
16	**TOBACCO PRODUCTS**	..	..	..	..	..	..	..	..	..	..
17	**TEXTILES**	1	13	97	1,050	2,360	224	63	643,741	..	6,124
171	Spinning, weaving and finishing of textiles	..	..	..	..	..	..	..	..	..	..
1711	Preparation and spinning of textile fibres	..	..	..	..	..	..	..	..	..	..
1712	Finishing of textiles	..	..	..	..	..	..	..	..	..	..
172	Other textiles	..	..	..	..	..	..	..	..	..	..
1721	Made-up textile articles, except apparel	..	..	..	..	..	..	..	..	..	..
1722	Carpets and rugs	..	..	..	..	..	..	..	..	..	..
1723	Cordage, rope, twine and netting	..	..	..	..	..	..	..	..	..	..
1729	Other textiles, nec	..	..	..	..	..	..	..	..	..	..
173	Knitted and crocheted fabrics and articles	..	..	..	..	..	..	..	..	..	..
18	**WEARING APPAREL, DRESSING & DYEING OF FUR**	..	..	..	..	..	..	..	..	..	..
181	Wearing apparel, except fur apparel	..	..	..	..	..	..	..	..	..	..
182	Dressing and dyeing of fur; articles of fur	..	..	..	..	..	..	..	..	..	..
19	**TANNING & DRESSING OF LEATHER, FOOTWEAR**	0	0	28	69	238	6	7	53,413	..	541
191	Tanning and dressing of leather	..	..	..	..	..	..	..	..	..	..
1911	Tanning and dressing of leather	..	..	..	..	..	..	..	..	..	..
1912	Luggage, handbags, saddlery & harness	..	..	..	..	..	..	..	..	..	..
192	Footwear	..	..	..	..	..	..	..	..	..	..
20	**WOOD AND WOOD PRODUCTS**	5	30	1,211	997	1,857	5,011	156	791,289	..	12,117
201	Sawmilling and planing of wood	..	..	..	..	..	..	..	..	..	..
202	Products of wood, cork, straw & plaiting materials	..	..	..	..	..	..	..	..	..	..
2021	Veneer sheets	..	..	..	..	..	..	..	..	..	..
2022	Builders' carpentry and joinery	..	..	..	..	..	..	..	..	..	..
2023	Wooden containers	..	..	..	..	..	..	..	..	..	..
2029	Other products of wood	..	..	..	..	..	..	..	..	..	..
21	**PAPER AND PAPER PRODUCTS**	4,180	54	200	4,236	24,423	31,342	1,624	4,370,658	..	81,794
2101	Pulp, paper and paperboard	..	..	..	..	..	..	..	..	..	..
2102	Corrugated paper, paperboard and their containers	..	..	..	..	..	..	..	..	..	..
2109	Other articles of pulp and paperboard	..	..	..	..	..	..	..	..	..	..
22	**PUBLISHING, PRINTING & REPRODUCTION**	1	38	98	109	682	12	76	261,340	..	1,956
221	Publishing	..	..	..	..	..	..	..	..	..	..
2211	Publishing of books & brochures	..	..	..	..	..	..	..	..	..	..
2212	Publishing of newspapers and periodicals	..	..	..	..	..	..	..	..	..	..
2213	Publishing of recorded media	..	..	..	..	..	..	..	..	..	..
2219	Other publishing	..	..	..	..	..	..	..	..	..	..
222	Printing and related service activities	..	..	..	..	..	..	..	..	..	..
2221	Printing	..	..	..	..	..	..	..	..	..	..
2222	Service activities related to printing	..	..	..	..	..	..	..	..	..	..
223	Reproduction of recorded media	..	..	..	..	..	..	..	..	..	..
23	**COKE, REFINED PETROLEUM PRODUCTS**	0	0	29	0	0	0	0	26,228	..	123
231	Coke oven products	..	..	..	..	..	..	..	..	..	..
232	Refined petroleum products	0	0	29	0	0	0	0	26,228	..	123
233	Processing of nuclear fuel	..	..	..	..	..	..	..	..	..	..
24	**CHEMICALS & CHEMICAL PRODUCTS**	1,990	27	235	806	9,720	6,209	451	2,799,419	..	29,517
241	Basic chemicals	..	..	..	..	..	..	..	..	..	..
2411	Basic chemicals, exc. fertilizers & nitrogen compounds	..	..	..	..	..	..	..	..	..	..
2412	Fertilizers and nitrogen compounds	..	..	..	..	..	..	..	..	..	..
2413	Plastics in primary forms and synthetic rubber	..	..	..	..	..	..	..	..	..	..
242	Other chemical products	..	..	..	..	..	..	..	..	..	..
2421	Pesticides and other agro-chemical products	..	..	..	..	..	..	..	..	..	..
2422	Paints, varnishes and similar coatings	..	..	..	..	..	..	..	..	..	..
2423	Pharmaceuticals, medicinal chem. & botanical prod.	..	..	..	..	..	..	..	..	..	..
2424	Soap and detergents, perfumes etc.	..	..	..	..	..	..	..	..	..	..
2429	Other chemical products, nec	..	..	..	..	..	..	..	..	..	..
243	Man-made fibres	..	..	..	..	..	..	..	..	..	..
25	**RUBBER AND PLASTICS PRODUCTS**	0	15	217	457	1,657	46	103	732,435	..	5,131
251	Rubber products	..	..	..	..	..	..	..	..	..	..
2511	Rubber tyres and tubes	..	..	..	..	..	..	..	..	..	..
2519	Other rubber products	..	..	..	..	..	..	..	..	..	..
252	Plastic products	..	..	..	..	..	..	..	..	..	..
26	**OTHER NON-METALLIC MINERAL PRODUCTS**	6,099	1,270	2,670	4,371	11,854	510	45	1,381,191	..	31,792
261	Glass and glass products	..	..	..	..	..	..	..	..	..	..
269	Non-metallic mineral products, nec	..	..	..	..	..	..	..	..	..	..
2691	Pottery, china and earthenware	..	..	..	..	..	..	..	..	..	..
2692	Refractory ceramic products	..	..	..	..	..	..	..	..	..	..
2693	Structural non-refractory clay & ceramic prod.	..	..	..	..	..	..	..	..	..	..
2694	Cement, lime and plaster	..	..	..	..	..	..	..	..	..	..
2695	Articles of concrete, cement and plaster	..	..	..	..	..	..	..	..	..	..
2696	Cutting, shaping and finishing of stone	..	..	..	..	..	..	..	..	..	..
2699	Other non-metallic mineral products, nec	..	..	..	..	..	..	..	..	..	..

ISIC Revision 3 Industry Sector		Solid TJ	LPG TJ	Distiloil TJ	RFO TJ	Gas TJ	Biomass TJ	Steam TJ	Electr MWh	Own Use MWh	TOTAL TJ
27	**BASIC METALS**	116,706	1,216	420	7,366	55,909	1,205	8	3,229,236	..	194,455
271	Basic iron and steel	116,504	821	350	6,962	53,742	378	1	2,541,032	..	187,906
272	Basic precious and non-ferrous metals	191	355	58	302	1,727	52	7	471,524	..	4,390
273	Casting of metals	11	40	12	102	440	775	0	216,680	..	2,159
2731	Casting of iron and steel	..	..	..	..	..	..	..	..	..	..
2732	Casting non-ferrous metals	..	..	..	..	..	..	..	..	..	..
28	**FABRICATED METAL PRODUCTS**	78	480	627	619	1,861	318	62	688,655	..	6,524
281	Str. metal prod., tanks, reservoirs, steam generators	..	..	..	..	..	..	..	..	..	..
2811	Structural metal products	..	..	..	..	..	..	..	..	..	..
2812	Tanks, reservoirs and containers of metal	..	..	..	..	..	..	..	..	..	..
2813	Steam generators, exc. central heating hot water boilers	..	..	..	..	..	..	..	..	..	..
289	Other fabricated metal products	..	..	..	..	..	..	..	..	..	..
2891	Forging, pressing, stamping & roll-forming of metal	..	..	..	..	..	..	..	..	..	..
2892	Treatment and coating of metals	..	..	..	..	..	..	..	..	..	..
2893	Cutlery, hand tools and general hardware	..	..	..	..	..	..	..	..	..	..
2899	Other fabricated metal products, nec	..	..	..	..	..	..	..	..	..	..
29	**MACHINERY AND EQUIPMENT, NEC**	9	326	513	534	1,362	29	203	592,823	..	5,108
291	General purpose machinery	..	..	..	..	..	..	..	..	..	..
2911	Engines and turbines	..	..	..	..	..	..	..	..	..	..
2912	Pumps, compressors, taps and valves	..	..	..	..	..	..	..	..	..	..
2913	Bearings, gears, gearing and driving elements	..	..	..	..	..	..	..	..	..	..
2914	Ovens, furnaces and furnace burners	..	..	..	..	..	..	..	..	..	..
2915	Lifting and handling equipment	..	..	..	..	..	..	..	..	..	..
2919	Other general purpose machinery	..	..	..	..	..	..	..	..	..	..
292	Special purpose machinery	..	..	..	..	..	..	..	..	..	..
2921	Agricultural and forestry machinery	..	..	..	..	..	..	..	..	..	..
2922	Machine-tools	..	..	..	..	..	..	..	..	..	..
2923	Machinery for metallurgy	..	..	..	..	..	..	..	..	..	..
2924	Machinery for mining, quarrying and construction	..	..	..	..	..	..	..	..	..	..
2925	Machinery for food, beverage & tobacco processing	..	..	..	..	..	..	..	..	..	..
2926	Machinery for textile, apparel & leather production	..	..	..	..	..	..	..	..	..	..
2927	Machinery for weapons and ammunition	..	..	..	..	..	..	..	..	..	..
2929	Other special purpose machinery	..	..	..	..	..	..	..	..	..	..
293	Domestic appliances, nec	..	..	..	..	..	..	..	..	..	..
30	**OFFICE, ACCOUNTING & COMPUTING MACHINERY**	0	17	297	418	1,406	0	282	873,689	..	5,565
31	**ELECTRICAL MACHINERY & APPARATUS, NEC**	..	..	..	..	..	..	..	..	..	..
311	Electric motors, generators and transformers	..	..	..	..	..	..	..	..	..	..
312	Electricity distribution and control apparatus	..	..	..	..	..	..	..	..	..	..
313	Insulated wire and cable	..	..	..	..	..	..	..	..	..	..
314	Accumulators, primary cells & primary batteries	..	..	..	..	..	..	..	..	..	..
315	Electric lamps and lighting equipment	..	..	..	..	..	..	..	..	..	..
319	Other electrical equipment, nec	..	..	..	..	..	..	..	..	..	..
32	**RADIO, TV & COMMUNICATION EQUIP. & APP.**	..	..	..	..	..	..	..	..	..	..
321	Electronic valves, tubes, other electronic components	..	..	..	..	..	..	..	..	..	..
322	TV & radio transmitters, apparatus for line telephony	..	..	..	..	..	..	..	..	..	..
323	TV & radio receivers, recording apparatus	..	..	..	..	..	..	..	..	..	..
33	**MEDICAL PRECISION &OPTICAL INSTRUMENTS**	..	..	..	..	..	..	..	..	..	..
331	Medical appliances and instruments	..	..	..	..	..	..	..	..	..	..
3311	Medical, surgical equipment & orthopaedic app.	..	..	..	..	..	..	..	..	..	..
3312	Instruments & appliances for measuring, checking etc	..	..	..	..	..	..	..	..	..	..
3313	Industrial process control equipment	..	..	..	..	..	..	..	..	..	..
332	Optical instruments and photographic equipment	..	..	..	..	..	..	..	..	..	..
333	Watches and clocks	..	..	..	..	..	..	..	..	..	..
34	**MOTOR VEHICLES, TRAILERS & SEMI-TRAILERS**	0	84	191	117	1,651	7	507	551,910	..	4,544
341	Motor vehicles	..	..	..	..	..	..	..	..	..	..
342	Bodies (coachwork) for motor vehicles	..	..	..	..	..	..	..	..	..	..
343	Parts, accessories for motor vehicles & their engines	..	..	..	..	..	..	..	..	..	..
35	**OTHER TRANSPORT EQUIPMENT**	..	..	..	..	..	..	..	..	..	..
351	Building and repairing of ships and boats	..	..	..	..	..	..	..	..	..	..
3511	Building and repairing of ships	..	..	..	..	..	..	..	..	..	..
3512	Building, repairing of pleasure & sporting boats	..	..	..	..	..	..	..	..	..	..
352	Railway, tramway locomotives & rolling stock	..	..	..	..	..	..	..	..	..	..
353	Aircraft and spacecraft	..	..	..	..	..	..	..	..	..	..
359	Transport equipment, nec	..	..	..	..	..	..	..	..	..	..
3591	Motorcycles	..	..	..	..	..	..	..	..	..	..
3592	Bicycles and invalid carriages	..	..	..	..	..	..	..	..	..	..
3599	Other transport equipment, nec	..	..	..	..	..	..	..	..	..	..
36	**FURNITURE; MANUFACTURING, NEC**	9	9	594	524	205	775	46	328,039	..	3,341
361	Furniture	..	..	..	..	..	..	..	..	..	..
369	Manufacturing, nec	..	..	..	..	..	..	..	..	..	..
3691	Jewellery and related articles	..	..	..	..	..	..	..	..	..	..
3692	Musical instruments	..	..	..	..	..	..	..	..	..	..
3693	Sports goods	..	..	..	..	..	..	..	..	..	..
3694	Games and toys	..	..	..	..	..	..	..	..	..	..
3699	Other manufacturing, nec	..	..	..	..	..	..	..	..	..	..
37	**RECYCLING**	..	..	..	..	..	..	..	..	..	..
371	Recycling of metal waste and scrap	..	..	..	..	..	..	..	..	..	..
372	Recycling of non-metal waste and scrap	..	..	..	..	..	..	..	..	..	..
CERR	**Non-specified industry**	..	..	..	..	..	..	..	..	..	..
15-37	**TOTAL MANUFACTURING**	129,226	3,687	9,489	24,208	126,393	45,829	4,020	18,716,808	..	410,230

ISIC Revision 3 Industry Sector		Solid TJ	LPG TJ	Distiloil TJ	RFO TJ	Gas TJ	Biomass TJ	Steam TJ	Electr MWh	Own Use MWh	TOTAL TJ
15	**FOOD PRODUCTS AND BEVERAGES**	98	97	1,979	2,236	11,008	175	539	1,355,919	..	21,012
151	Production, processing and preserving (PPP)	..	..	..	..	..	..	..	..	..	..
1511	PPP of meat and meat products	..	..	..	..	..	..	..	..	..	..
1512	Processing and preserving of fish products	..	..	..	..	..	..	..	..	..	..
1513	Processing, preserving of fruit & vegetables	..	..	..	..	..	..	..	..	..	..
1514	Vegetable and animal oils and fats	..	..	..	..	..	..	..	..	..	..
152	Dairy products	..	..	..	..	..	..	..	..	..	..
153	Grain mill prod., starches & prepared animal feeds	..	..	..	..	..	..	..	..	..	..
1531	Grain mill products	..	..	..	..	..	..	..	..	..	..
1532	Starches and starch products	..	..	..	..	..	..	..	..	..	..
1533	Prepared animal feeds	..	..	..	..	..	..	..	..	..	..
154	Other food products	..	..	..	..	..	..	..	..	..	..
1541	Bakery products	..	..	..	..	..	..	..	..	..	..
1542	Sugar	..	..	..	..	..	..	..	..	..	..
1543	Cocoa, chocolate and sugar confectionery	..	..	..	..	..	..	..	..	..	..
1544	Macaroni, noodles, couscous & similar farinaceous prod.	..	..	..	..	..	..	..	..	..	..
1549	Other food products, nec	..	..	..	..	..	..	..	..	..	..
155	Beverages	..	..	..	..	..	..	..	..	..	..
1551	Distilling, rectifying and blending of spirits	..	..	..	..	..	..	..	..	..	..
1552	Wines	..	..	..	..	..	..	..	..	..	..
1553	Malt liquors and malt	..	..	..	..	..	..	..	..	..	..
1554	Soft drinks; production of mineral waters	..	..	..	..	..	..	..	..	..	..
16	**TOBACCO PRODUCTS**	0	9	114	979	2,300	218	72	652,570	..	6,039
17	**TEXTILES**	..	..	..	..	..	..	..	..	..	..
171	Spinning, weaving and finishing of textiles	..	..	..	..	..	..	..	..	..	..
1711	Preparation and spinning of textile fibres	..	..	..	..	..	..	..	..	..	..
1712	Finishing of textiles	..	..	..	..	..	..	..	..	..	..
172	Other textiles	..	..	..	..	..	..	..	..	..	..
1721	Made-up textile articles, except apparel	..	..	..	..	..	..	..	..	..	..
1722	Carpets and rugs	..	..	..	..	..	..	..	..	..	..
1723	Cordage, rope, twine and netting	..	..	..	..	..	..	..	..	..	..
1729	Other textiles, nec	..	..	..	..	..	..	..	..	..	..
173	Knitted and crocheted fabrics and articles	..	..	..	..	..	..	..	..	..	..
18	**WEARING APPAREL, DRESSING & DYEING OF FUR**	..	..	..	..	..	..	..	..	..	..
181	Wearing apparel, except fur apparel	..	..	..	..	..	..	..	..	..	..
182	Dressing and dyeing of fur; articles of fur	..	..	..	..	..	..	..	..	..	..
19	**TANNING & DRESSING OF LEATHER, FOOTWEAR**	0	0	25	65	260	2	8	54,429	..	556
191	Tanning and dressing of leather	..	..	..	..	..	..	..	..	..	..
1911	Tanning and dressing of leather	..	..	..	..	..	..	..	..	..	..
1912	Luggage, handbags, saddlery & harness	..	..	..	..	..	..	..	..	..	..
192	Footwear	..	..	..	..	..	..	..	..	..	..
20	**WOOD AND WOOD PRODUCTS**	0	28	1,397	875	1,978	7,848	149	812,448	..	15,200
201	Sawmilling and planing of wood	..	..	..	..	..	..	..	..	..	..
202	Products of wood, cork, straw & plaiting materials	..	..	..	..	..	..	..	..	..	..
2021	Veneer sheets	..	..	..	..	..	..	..	..	..	..
2022	Builders' carpentry and joinery	..	..	..	..	..	..	..	..	..	..
2023	Wooden containers	..	..	..	..	..	..	..	..	..	..
2029	Other products of wood	..	..	..	..	..	..	..	..	..	..
21	**PAPER AND PAPER PRODUCTS**	4,513	40	200	3,509	25,825	156,112	1,671	4,471,223	..	207,966
2101	Pulp, paper and paperboard	..	..	..	..	..	..	..	..	..	..
2102	Corrugated paper, paperboard and their containers	..	..	..	..	..	..	..	..	..	..
2109	Other articles of pulp and paperboard	..	..	..	..	..	..	..	..	..	..
22	**PUBLISHING, PRINTING & REPRODUCTION**	0	36	95	79	683	6	73	272,035	..	1,952
221	Publishing	..	..	..	..	..	..	..	..	..	..
2211	Publishing of books & brochures	..	..	..	..	..	..	..	..	..	..
2212	Publishing of newspapers and periodicals	..	..	..	..	..	..	..	..	..	..
2213	Publishing of recorded media	..	..	..	..	..	..	..	..	..	..
2219	Other publishing	..	..	..	..	..	..	..	..	..	..
222	Printing and related service activities	..	..	..	..	..	..	..	..	..	..
2221	Printing	..	..	..	..	..	..	..	..	..	..
2222	Service activities related to printing	..	..	..	..	..	..	..	..	..	..
223	Reproduction of recorded media	..	..	..	..	..	..	..	..	..	..
23	**COKE, REFINED PETROLEUM PRODUCTS**	0	0	30	0	0	0	0	45,916	..	195
231	Coke oven products	..	..	..	..	..	..	..	..	..	..
232	Refined petroleum products	0	0	30	0	0	0	0	45,916	..	195
233	Processing of nuclear fuel	..	..	..	..	..	..	..	..	..	..
24	**CHEMICALS & CHEMICAL PRODUCTS**	2,276	7	216	748	9,979	3,370	932	2,918,817	..	28,036
241	Basic chemicals	..	..	..	..	..	..	..	..	..	..
2411	Basic chemicals, exc. fertilizers & nitrogen compounds	..	..	..	..	..	..	..	..	..	..
2412	Fertilizers and nitrogen compounds	..	..	..	..	..	..	..	..	..	..
2413	Plastics in primary forms and synthetic rubber	..	..	..	..	..	..	..	..	..	..
242	Other chemical products	..	..	..	..	..	..	..	..	..	..
2421	Pesticides and other agro-chemical products	..	..	..	..	..	..	..	..	..	..
2422	Paints, varnishes and similar coatings	..	..	..	..	..	..	..	..	..	..
2423	Pharmaceuticals, medicinal chem. & botanical prod.	..	..	..	..	..	..	..	..	..	..
2424	Soap and detergents, perfumes etc.	..	..	..	..	..	..	..	..	..	..
2429	Other chemical products, nec	..	..	..	..	..	..	..	..	..	..
243	Man-made fibres	..	..	..	..	..	..	..	..	..	..
25	**RUBBER AND PLASTICS PRODUCTS**	0	31	242	436	1,580	45	100	818,497	..	5,381
251	Rubber products	..	..	..	..	..	..	..	..	..	..
2511	Rubber tyres and tubes	..	..	..	..	..	..	..	..	..	..
2519	Other rubber products	..	..	..	..	..	..	..	..	..	..
252	Plastic products	..	..	..	..	..	..	..	..	..	..
26	**OTHER NON-METALLIC MINERAL PRODUCTS**	4,938	1,026	2,646	3,990	12,088	474	8	1,344,627	..	30,010
261	Glass and glass products	..	..	..	..	..	..	..	..	..	..
269	Non-metallic mineral products, nec	..	..	..	..	..	..	..	..	..	..
2691	Pottery, china and earthenware	..	..	..	..	..	..	..	..	..	..
2692	Refractory ceramic products	..	..	..	..	..	..	..	..	..	..
2693	Structural non-refractory clay & ceramic prod.	..	..	..	..	..	..	..	..	..	..
2694	Cement, lime and plaster	..	..	..	..	..	..	..	..	..	..
2695	Articles of concrete, cement and plaster	..	..	..	..	..	..	..	..	..	..
2696	Cutting, shaping and finishing of stone	..	..	..	..	..	..	..	..	..	..
2699	Other non-metallic mineral products, nec	..	..	..	..	..	..	..	..	..	..

ISIS Energy Data Programme (IEA/OECD)

ISIC Revision 3 Industry Sector		Solid TJ	LPG TJ	Distiloil TJ	RFO TJ	Gas TJ	Biomass TJ	Steam TJ	Electr MWh	Own Use MWh	TOTAL TJ
27	**BASIC METALS**	111,531	1,282	406	7,802	58,780	1,184	7	3,802,407	..	194,681
271	Basic iron and steel	111,318	933	333	7,412	56,376	357	1	2,997,209	..	187,520
272	Basic precious and non-ferrous metals	203	299	62	296	1,901	52	6	466,034		4,497
273	Casting of metals	10	50	11	94	503	775	0	339,164	..	2,664
2731	Casting of iron and steel	..	..	..	..	..	..	..	..	..	..
2732	Casting non-ferrous metals	..	..	..	..	..	..	..	..	..	..
28	**FABRICATED METAL PRODUCTS**	78	451	610	565	2,263	328	51	876,917		7,502
281	Str. metal prod., tanks, reservoirs, steam generators	..	..	..	..	..	..	..	..	..	..
2811	Structural metal products	..	..	..	..	..	..	..	..	..	..
2812	Tanks, reservoirs and containers of metal	..	..	..	..	..	..	..	..	..	..
2813	Steam generators, exc. central heating hot water boilers	..	..	..	..	..	..	..	..	..	..
289	Other fabricated metal products	..	..	..	..	..	..	..	..	..	..
2891	Forging, pressing, stamping & roll-forming of metal	..	..	..	..	..	..	..	..	..	..
2892	Treatment and coating of metals	..	..	..	..	..	..	..	..	..	..
2893	Cutlery, hand tools and general hardware	..	..	..	..	..	..	..	..	..	..
2899	Other fabricated metal products, nec	..	..	..	..	..	..	..	..	..	..
29	**MACHINERY AND EQUIPMENT, NEC**	3	307	504	482	1,319	37	198	624,355	..	5,096
291	General purpose machinery	..	..	..	..	..	..	..	..	..	..
2911	Engines and turbines	..	..	..	..	..	..	..	..	..	..
2912	Pumps, compressors, taps and valves	..	..	..	..	..	..	..	..	..	..
2913	Bearings, gears, gearing and driving elements	..	..	..	..	..	..	..	..	..	..
2914	Ovens, furnaces and furnace burners	..	..	..	..	..	..	..	..	..	..
2915	Lifting and handling equipment	..	..	..	..	..	..	..	..	..	..
2919	Other general purpose machinery	..	..	..	..	..	..	..	..	..	..
292	Special purpose machinery	..	..	..	..	..	..	..	..	..	..
2921	Agricultural and forestry machinery	..	..	..	..	..	..	..	..	..	..
2922	Machine-tools	..	..	..	..	..	..	..	..	..	..
2923	Machinery for metallurgy	..	..	..	..	..	..	..	..	..	..
2924	Machinery for mining, quarrying and construction	..	..	..	..	..	..	..	..	..	..
2925	Machinery for food, beverage & tobacco processing	..	..	..	..	..	..	..	..	..	..
2926	Machinery for textile, apparel & leather production	..	..	..	..	..	..	..	..	..	..
2927	Machinery for weapons and ammunition	..	..	..	..	..	..	..	..	..	..
2929	Other special purpose machinery	..	..	..	..	..	..	..	..	..	..
293	Domestic appliances, nec	..	..	..	..	..	..	..	..	..	..
30	**OFFICE, ACCOUNTING & COMPUTING MACHINERY**	0	16	301	409	1,431	0	317	891,034	..	5,682
31	**ELECTRICAL MACHINERY & APPARATUS, NEC**	..	..	..	..	..	..	..	..	..	..
311	Electric motors, generators and transformers	..	..	..	..	..	..	..	..	..	..
312	Electricity distribution and control apparatus	..	..	..	..	..	..	..	..	..	..
313	Insulated wire and cable	..	..	..	..	..	..	..	..	..	..
314	Accumulators, primary cells & primary batteries	..	..	..	..	..	..	..	..	..	..
315	Electric lamps and lighting equipment	..	..	..	..	..	..	..	..	..	..
319	Other electrical equipment, nec	..	..	..	..	..	..	..	..	..	..
32	**RADIO, TV & COMMUNICATION EQUIP. & APP.**	..	..	..	..	..	..	..	..	..	..
321	Electronic valves, tubes, other electronic components	..	..	..	..	..	..	..	..	..	..
322	TV & radio transmitters, apparatus for line telephony	..	..	..	..	..	..	..	..	..	..
323	TV & radio receivers, recording apparatus	..	..	..	..	..	..	..	..	..	..
33	**MEDICAL PRECISION &OPTICAL INSTRUMENTS**	..	..	..	..	..	..	..	..	..	..
331	Medical appliances and instruments	..	..	..	..	..	..	..	..	..	..
3311	Medical, surgical equipment & orthopaedic app.	..	..	..	..	..	..	..	..	..	..
3312	Instruments & appliances for measuring, checking etc	..	..	..	..	..	..	..	..	..	..
3313	Industrial process control equipment	..	..	..	..	..	..	..	..	..	..
332	Optical instruments and photographic equipment	..	..	..	..	..	..	..	..	..	..
333	Watches and clocks	..	..	..	..	..	..	..	..	..	..
34	**MOTOR VEHICLES, TRAILERS & SEMI-TRAILERS**	0	12	191	114	1,829	0	486	607,884		4,820
341	Motor vehicles	..	..	..	..	..	..	..	..	..	..
342	Bodies (coachwork) for motor vehicles	..	..	..	..	..	..	..	..	..	..
343	Parts, accessories for motor vehicles & their engines	..	..	..	..	..	..	..	..	..	..
35	**OTHER TRANSPORT EQUIPMENT**	..	..	..	..	..	..	..	..	..	..
351	Building and repairing of ships and boats	..	..	..	..	..	..	..	..	..	..
3511	Building and repairing of ships	..	..	..	..	..	..	..	..	..	..
3512	Building, repairing of pleasure & sporting boats	..	..	..	..	..	..	..	..	..	..
352	Railway, tramway locomotives & rolling stock	..	..	..	..	..	..	..	..	..	..
353	Aircraft and spacecraft	..	..	..	..	..	..	..	..	..	..
359	Transport equipment, nec	..	..	..	..	..	..	..	..	..	..
3591	Motorcycles	..	..	..	..	..	..	..	..	..	..
3592	Bicycles and invalid carriages	..	..	..	..	..	..	..	..	..	..
3599	Other transport equipment, nec	..	..	..	..	..	..	..	..	..	..
36	**FURNITURE; MANUFACTURING, NEC**	2	9	535	457	222	679	52	335,853	..	3,166
361	Furniture	..	..	..	..	..	..	..	..	..	..
369	Manufacturing, nec	..	..	..	..	..	..	..	..	..	..
3691	Jewellery and related articles	..	..	..	..	..	..	..	..	..	..
3692	Musical instruments	..	..	..	..	..	..	..	..	..	..
3693	Sports goods	..	..	..	..	..	..	..	..	..	..
3694	Games and toys	..	..	..	..	..	..	..	..	..	..
3699	Other manufacturing, nec	..	..	..	..	..	..	..	..	..	..
37	**RECYCLING**										
371	Recycling of metal waste and scrap	..	..	..	..	..	..	..	..	..	..
372	Recycling of non-metal waste and scrap	..	..	..	..	..	..	..	..	..	..
CERR	**Non-specified industry**										
15-37	**TOTAL MANUFACTURING**	123,439	3,351	9,491	22,746	131,545	170,478	4,663	19,884,931		537,294

ISIC Revision 3 Industry Sector	Solid TJ	LPG TJ	Distiloil TJ	RFO TJ	Gas TJ	Biomass TJ	Steam TJ	Electr MWh	Own Use MWh	TOTAL TJ
15 FOOD PRODUCTS AND BEVERAGES	95	1,881	2,071	5,605	57,240	0	0	7,408,333	..	93,562
151 Production, processing and preserving (PPP)	0	475	408	2,859	13,725	0	0	1,580,556	..	23,157
1511 PPP of meat and meat products									..	
1512 Processing and preserving of fish products									..	
1513 Processing, preserving of fruit & vegetables									..	
1514 Vegetable and animal oils and fats									..	
152 Dairy products	76	380	304	789	6,444	0	0	872,222	..	11,133
153 Grain mill prod., starches & prepared animal feeds									..	
1531 Grain mill products									..	
1532 Starches and starch products									..	
1533 Prepared animal feeds									..	
154 Other food products	19	893	1,036	1,577	25,947	0	0	4,208,333	..	44,622
1541 Bakery products									..	
1542 Sugar									..	
1543 Cocoa, chocolate and sugar confectionery									..	
1544 Macaroni, noodles, couscous & similar farinaceous prod.									..	
1549 Other food products, nec									..	
155 Beverages	0	133	323	380	11,124	0	0	747,222	..	14,650
1551 Distilling, rectifying and blending of spirits									..	
1552 Wines									..	
1553 Malt liquors and malt									..	
1554 Soft drinks; production of mineral waters									..	
16 TOBACCO PRODUCTS	0	10	0	0	729	0	0	127,778	..	1,199
17 TEXTILES	0	219	257	969	12,150	0	0	1,616,667	..	19,415
171 Spinning, weaving and finishing of textiles	0	86	190	636	8,262	0	0	1,119,445	..	13,204
1711 Preparation and spinning of textile fibres									..	
1712 Finishing of textiles									..	
172 Other textiles									..	
1721 Made-up textile articles, except apparel									..	
1722 Carpets and rugs									..	
1723 Cordage, rope, twine and netting									..	
1729 Other textiles, nec	0	133	67	333	3,888	0	0	497,222	..	6,211
173 Knitted and crocheted fabrics and articles	0	76	114	152	2,394	0	0	583,333	..	4,836
18 WEARING APPAREL, DRESSING & DYEING OF FUR									..	
181 Wearing apparel, except fur apparel									..	
182 Dressing and dyeing of fur; articles of fur	0	19	76	0	612	0	0	127,778	..	1,167
19 TANNING & DRESSING OF LEATHER, FOOTWEAR									..	
191 Tanning and dressing of leather									..	
1911 Tanning and dressing of leather									..	
1912 Luggage, handbags, saddlery & harness									..	
192 Footwear	10	789	912	1,131	16,965	0	0	4,794,444	..	37,067
20 WOOD AND WOOD PRODUCTS	10	352	475	770	8,325	0	0	1,438,888	..	15,112
201 Sawmilling and planing of wood	0	437	437	361	8,640	0	0	3,355,556	..	21,955
202 Products of wood, cork, straw & plaiting materials									..	
2021 Veneer sheets									..	
2022 Builders' carpentry and joinery									..	
2023 Wooden containers									..	
2029 Other products of wood									..	
21 PAPER AND PAPER PRODUCTS	3,924	0	9,985	78,252	91,494	267,440	1,450	48,805,556	..	628,245
2101 Pulp, paper and paperboard									..	
2102 Corrugated paper, paperboard and their containers									..	
2109 Other articles of pulp and paperboard									..	
22 PUBLISHING, PRINTING & REPRODUCTION	10	228	95	38	3,906	0	0	1,075,000	..	8,147
221 Publishing									..	
2211 Publishing of books & brochures									..	
2212 Publishing of newspapers and periodicals									..	
2213 Publishing of recorded media									..	
2219 Other publishing									..	
222 Printing and related service activities									..	
2221 Printing									..	
2222 Service activities related to printing									..	
223 Reproduction of recorded media									..	
23 COKE, REFINED PETROLEUM PRODUCTS	24,292	3,154	1,178	38,000	218,128	0	0	5,500,000	..	304,552
231 Coke oven products									..	
232 Refined petroleum products	24,292	3,154	1,178	38,000	218,128	0	0	5,500,000	..	304,552
233 Processing of nuclear fuel									..	
24 CHEMICALS & CHEMICAL PRODUCTS	0	29	342	10,108	111,501	0	18,790	19,288,889	..	210,210
241 Basic chemicals	0	29	295	4,874	50,067	0	5,810	14,766,667	..	114,235
2411 Basic chemicals, exc. fertilizers & nitrogen compounds	0	0	276	3,011	15,660	0	5,730	12,469,445	..	69,567
2412 Fertilizers and nitrogen compounds	0	0	0	181	24,426	0	80	1,152,778	..	28,837
2413 Plastics in primary forms and synthetic rubber	0	29	19	1,682	9,981	0	0	1,144,444	..	15,831
242 Other chemical products	0	0	47	5,234	61,434	0	12,980	4,522,222	..	95,975
2421 Pesticides and other agro-chemical products									..	
2422 Paints, varnishes and similar coatings									..	
2423 Pharmaceuticals, medicinal chem. & botanical prod.									..	
2424 Soap and detergents, perfumes etc.									..	
2429 Other chemical products, nec									..	
243 Man-made fibres									..	
25 RUBBER AND PLASTICS PRODUCTS	10	257	181	1,843	8,388	0	0	2,508,333	..	19,709
251 Rubber products	0	124	48	1,729	3,420	0	0	825,000	..	8,291
2511 Rubber tyres and tubes									..	
2519 Other rubber products									..	
252 Plastic products	10	133	133	114	4,968	0	0	1,683,333	..	11,418
26 OTHER NON-METALLIC MINERAL PRODUCTS	36,699	200	342	3,477	30,861	0	0	2,313,889	..	79,909
261 Glass and glass products	0	190	0	332	9,441	0	0	377,778	..	11,323
269 Non-metallic mineral products, nec	36,699	10	342	3,145	21,420	0	0	1,936,111	..	68,586
2691 Pottery, china and earthenware									..	
2692 Refractory ceramic products									..	
2693 Structural non-refractory clay & ceramic prod.									..	
2694 Cement, lime and plaster	36,699	10	342	3,145	21,420	0	0	1,936,111	..	68,586
2695 Articles of concrete, cement and plaster									..	
2696 Cutting, shaping and finishing of stone									..	
2699 Other non-metallic mineral products, nec									..	

ISIC Revision 3 Industry Sector		Solid TJ	LPG TJ	Distiloil TJ	RFO TJ	Gas TJ	Biomass TJ	Steam TJ	Electr MWh	Own Use MWh	TOTAL TJ
27	**BASIC METALS**	128,022	0	1,767	26,163	71,127	0	0	44,155,555	..	386,039
271	Basic iron and steel	114,874	0	1,282	14,183	49,149	0	0	7,908,333	..	207,958
272	Basic precious and non-ferrous metals	13,148	0	485	11,980	21,978	0	0	36,247,222	..	178,081
273	Casting of metals	..	..	..	..	..	..	..	..	..	..
2731	Casting of iron and steel	..	..	..	..	..	..	..	..	..	..
2732	Casting non-ferrous metals	..	..	..	..	..	..	..	..	..	..
28	**FABRICATED METAL PRODUCTS**	0	789	295	428	16,263	0	0	2,166,667	..	25,575
281	Str. metal prod., tanks, reservoirs, steam generators	..	..	..	..	..	..	..	..	..	..
2811	Structural metal products	..	..	..	..	..	..	..	..	..	..
2812	Tanks, reservoirs and containers of metal	..	..	..	..	..	..	..	..	..	..
2813	Steam generators, exc. central heating hot water boilers	..	..	..	..	..	..	..	..	..	..
289	Other fabricated metal products	..	..	..	..	..	..	..	..	..	..
2891	Forging, pressing, stamping & roll-forming of metal	..	..	..	..	..	..	..	..	..	..
2892	Treatment and coating of metals	..	..	..	..	..	..	..	..	..	..
2893	Cutlery, hand tools and general hardware	..	..	..	..	..	..	..	..	..	..
2899	Other fabricated metal products, nec	..	..	..	..	..	..	..	..	..	..
29	**MACHINERY AND EQUIPMENT, NEC**	19	589	333	314	16,416	0	0	3,150,000	..	29,011
291	General purpose machinery	..	..	..	..	..	..	..	..	..	..
2911	Engines and turbines	..	..	..	..	..	..	..	..	..	..
2912	Pumps, compressors, taps and valves	..	..	..	..	..	..	..	..	..	..
2913	Bearings, gears, gearing and driving elements	..	..	..	..	..	..	..	..	..	..
2914	Ovens, furnaces and furnace burners	..	..	..	..	..	..	..	..	..	..
2915	Lifting and handling equipment	..	..	..	..	..	..	..	..	..	..
2919	Other general purpose machinery	..	..	..	..	..	..	..	..	..	..
292	Special purpose machinery	..	..	..	..	..	..	..	..	..	..
2921	Agricultural and forestry machinery	..	..	..	..	..	..	..	..	..	..
2922	Machine-tools	..	..	..	..	..	..	..	..	..	..
2923	Machinery for metallurgy	..	..	..	..	..	..	..	..	..	..
2924	Machinery for mining, quarrying and construction	..	..	..	..	..	..	..	..	..	..
2925	Machinery for food, beverage & tobacco processing	..	..	..	..	..	..	..	..	..	..
2926	Machinery for textile, apparel & leather production	..	..	..	..	..	..	..	..	..	..
2927	Machinery for weapons and ammunition	..	..	..	..	..	..	..	..	..	..
2929	Other special purpose machinery	..	..	..	..	..	..	..	..	..	..
293	Domestic appliances, nec	..	..	..	..	..	..	..	..	..	..
30	**OFFICE, ACCOUNTING & COMPUTING MACHINERY**	..	..	..	..	..	..	..	..	..	..
31	**ELECTRICAL MACHINERY & APPARATUS, NEC**	..	..	..	..	..	..	..	..	..	..
311	Electric motors, generators and transformers	..	..	..	..	..	..	..	..	..	..
312	Electricity distribution and control apparatus	..	..	..	..	..	..	..	..	..	..
313	Insulated wire and cable	..	..	..	..	..	..	..	..	..	..
314	Accumulators, primary cells & primary batteries	..	..	..	..	..	..	..	..	..	..
315	Electric lamps and lighting equipment	..	..	..	..	..	..	..	..	..	..
319	Other electrical equipment, nec	..	..	..	..	..	..	..	..	..	..
32	**RADIO, TV & COMMUNICATION EQUIP. & APP.**	..	..	..	..	..	..	..	..	..	..
321	Electronic valves, tubes, other electronic components	..	..	..	..	..	..	..	..	..	..
322	TV & radio transmitters, apparatus for line telephony	..	..	..	..	..	..	..	..	..	..
323	TV & radio receivers, recording apparatus	..	..	..	..	..	..	..	..	..	..
33	**MEDICAL PRECISION & OPTICAL INSTRUMENTS**	..	..	..	..	..	..	..	..	..	..
331	Medical appliances and instruments	..	..	..	..	..	..	..	..	..	..
3311	Medical, surgical equipment & orthopaedic app.	..	..	..	..	..	..	..	..	..	..
3312	Instruments & appliances for measuring, checking etc	..	..	..	..	..	..	..	..	..	..
3313	Industrial process control equipment	..	..	..	..	..	..	..	..	..	..
332	Optical instruments and photographic equipment	..	..	..	..	..	..	..	..	..	..
333	Watches and clocks	..	..	..	..	..	..	..	..	..	..
34	**MOTOR VEHICLES, TRAILERS & SEMI-TRAILERS**	1,188	960	589	1,017	27,270	0	0	4,994,444	..	49,004
341	Motor vehicles	1,026	162	171	105	13,833	0	0	1,819,444	..	21,847
342	Bodies (coachwork) for motor vehicles	..	..	..	..	..	..	..	..	..	..
343	Parts, accessories for motor vehicles & their engines	162	798	418	912	13,437	0	0	3,175,000	..	27,157
35	**OTHER TRANSPORT EQUIPMENT**	..	..	..	..	..	..	..	..	..	..
351	Building and repairing of ships and boats	..	..	..	..	..	..	..	..	..	..
3511	Building and repairing of ships	..	..	..	..	..	..	..	..	..	..
3512	Building, repairing of pleasure & sporting boats	..	..	..	..	..	..	..	..	..	..
352	Railway, tramway locomotives & rolling stock	..	..	..	..	..	..	..	..	..	..
353	Aircraft and spacecraft	..	..	..	..	..	..	..	..	..	..
359	Transport equipment, nec	..	..	..	..	..	..	..	..	..	..
3591	Motorcycles	..	..	..	..	..	..	..	..	..	..
3592	Bicycles and invalid carriages	..	..	..	..	..	..	..	..	..	..
3599	Other transport equipment, nec	..	..	..	..	..	..	..	..	..	..
36	**FURNITURE; MANUFACTURING, NEC**	4,053	8,841	9,555	26	151,056	0	560	3,077,778	..	185,171
361	Furniture	..	..	..	..	..	..	..	..	..	..
369	Manufacturing, nec	..	..	..	..	..	..	..	..	..	..
3691	Jewellery and related articles	..	..	..	..	..	..	..	..	..	..
3692	Musical instruments	..	..	..	..	..	..	..	..	..	..
3693	Sports goods	..	..	..	..	..	..	..	..	..	..
3694	Games and toys	..	..	..	..	..	..	..	..	..	..
3699	Other manufacturing, nec	..	..	..	..	..	..	..	..	..	..
37	**RECYCLING**	..	..	..	..	..	..	..	..	..	..
371	Recycling of metal waste and scrap	..	..	..	..	..	..	..	..	..	..
372	Recycling of non-metal waste and scrap	..	..	..	..	..	..	..	..	..	..
CERR	Non-specified industry	..	..	..	..	..	..	..	..	..	..
15-37	**TOTAL MANUFACTURING**	198,322	18,041	28,092	167,523	836,500	267,440	20,800	151,694,444	..	2,082,818

ISIC Revision 3 Industry Sector	Solid TJ	LPG TJ	Distiloil TJ	RFO TJ	Gas TJ	Biomass TJ	Steam TJ	Electr MWh	Own Use MWh	TOTAL TJ
15 FOOD PRODUCTS AND BEVERAGES	48	1,606	1,729	5,425	53,082	0	0	7,330,556	..	88,280
151 Production, processing and preserving (PPP)	0	408	351	3,829	13,050	0	0	1,602,778	..	23,408
1511 PPP of meat and meat products	..	..	..	..	..	..	..	..	..	..
1512 Processing and preserving of fish products	..	..	..	..	..	..	..	..	..	..
1513 Processing, preserving of fruit & vegetables	..	..	..	..	..	..	..	..	..	..
1514 Vegetable and animal oils and fats	..	..	..	..	..	..	..	..	..	..
152 Dairy products	38	399	266	893	5,958	0	0	880,556	..	10,724
153 Grain mill prod., starches & prepared animal feeds	..	..	..	..	..	..	..	..	..	..
1531 Grain mill products	..	..	..	..	..	..	..	..	..	..
1532 Starches and starch products	..	..	..	..	..	..	..	..	..	..
1533 Prepared animal feeds	..	..	..	..	..	..	..	..	..	..
154 Other food products	10	656	1,026	513	24,264	0	0	4,197,222	..	41,579
1541 Bakery products	..	..	..	..	..	..	..	..	..	..
1542 Sugar	..	..	..	..	..	..	..	..	..	..
1543 Cocoa, chocolate and sugar confectionery	..	..	..	..	..	..	..	..	..	..
1544 Macaroni, noodles, couscous & similar farinaceous prod.	..	..	..	..	..	..	..	..	..	..
1549 Other food products, nec	..	..	..	..	..	..	..	..	..	..
155 Beverages	0	143	86	190	9,810	0	0	650,000	..	12,569
1551 Distilling, rectifying and blending of spirits	..	..	..	..	..	..	..	..	..	..
1552 Wines	..	..	..	..	..	..	..	..	..	..
1553 Malt liquors and malt	..	..	..	..	..	..	..	..	..	..
1554 Soft drinks; production of mineral waters	..	..	..	..	..	..	..	..	..	..
16 TOBACCO PRODUCTS	..	..	..	..	..	..	..	..	..	..
17 TEXTILES	0	257	257	779	14,607	0	0	1,750,000	..	22,200
171 Spinning, weaving and finishing of textiles	0	76	190	389	9,495	0	0	1,147,222	..	14,280
1711 Preparation and spinning of textile fibres	..	..	..	..	..	..	..	..	..	..
1712 Finishing of textiles	..	..	..	..	..	..	..	..	..	..
172 Other textiles	..	..	..	..	..	..	..	..	..	..
1721 Made-up textile articles, except apparel	..	..	..	..	..	..	..	..	..	..
1722 Carpets and rugs	..	..	..	..	..	..	..	..	..	..
1723 Cordage, rope, twine and netting	..	..	..	..	..	..	..	..	..	..
1729 Other textiles, nec	..	..	..	..	..	..	..	..	..	..
173 Knitted and crocheted fabrics and articles	0	181	67	390	5,112	0	0	602,778	..	7,920
18 WEARING APPAREL, DRESSING & DYEING OF FUR	0	76	95	114	1,971	0	0	625,000	..	4,506
181 Wearing apparel, except fur apparel	..	..	..	..	..	..	..	..	..	..
182 Dressing and dyeing of fur; articles of fur	..	..	..	..	..	..	..	..	..	..
19 TANNING & DRESSING OF LEATHER, FOOTWEAR	0	10	48	0	576	0	0	119,444	..	1,064
191 Tanning and dressing of leather	..	..	..	..	..	..	..	..	..	..
1911 Tanning and dressing of leather	..	..	..	..	..	..	..	..	..	..
1912 Luggage, handbags, saddlery & harness	..	..	..	..	..	..	..	..	..	..
192 Footwear	..	..	..	..	..	..	..	..	..	..
20 WOOD AND WOOD PRODUCTS	38	722	656	1,017	13,536	0	0	4,566,667	..	32,409
201 Sawmilling and planing of wood	38	313	380	665	6,309	0	0	1,336,111	..	12,515
202 Products of wood, cork, straw & plaiting materials	0	409	276	352	7,227	0	0	3,230,556	..	19,894
2021 Veneer sheets	..	..	..	..	..	..	..	..	..	..
2022 Builders' carpentry and joinery	..	..	..	..	..	..	..	..	..	..
2023 Wooden containers	..	..	..	..	..	..	..	..	..	..
2029 Other products of wood	..	..	..	..	..	..	..	..	..	..
21 PAPER AND PAPER PRODUCTS	3,373	0	8,959	74,705	100,836	277,987	1,550	50,313,889	..	648,540
2101 Pulp, paper and paperboard	..	..	..	..	..	..	..	..	..	..
2102 Corrugated paper, paperboard and their containers	..	..	..	..	..	..	..	..	..	..
2109 Other articles of pulp and paperboard	..	..	..	..	..	..	..	..	..	..
22 PUBLISHING, PRINTING & REPRODUCTION	133	171	114	19	3,546	0	0	1,119,444	..	8,013
221 Publishing	..	..	..	..	..	..	..	..	..	..
2211 Publishing of books & brochures	..	..	..	..	..	..	..	..	..	..
2212 Publishing of newspapers and periodicals	..	..	..	..	..	..	..	..	..	..
2213 Publishing of recorded media	..	..	..	..	..	..	..	..	..	..
2219 Other publishing	..	..	..	..	..	..	..	..	..	..
222 Printing and related service activities	..	..	..	..	..	..	..	..	..	..
2221 Printing	..	..	..	..	..	..	..	..	..	..
2222 Service activities related to printing	..	..	..	..	..	..	..	..	..	..
223 Reproduction of recorded media	..	..	..	..	..	..	..	..	..	..
23 COKE, REFINED PETROLEUM PRODUCTS	24,406	1,435	1,007	40,366	207,901	0	0	5,636,111	..	295,405
231 Coke oven products	..	..	..	..	..	..	..	..	..	..
232 Refined petroleum products	24,406	1,435	1,007	40,366	207,901	0	0	5,636,111	..	295,405
233 Processing of nuclear fuel	..	..	..	..	..	..	..	..	..	..
24 CHEMICALS & CHEMICAL PRODUCTS	0	2,508	257	7,942	112,077	0	22,720	18,519,444	..	212,174
241 Basic chemicals	0	152	190	4,104	50,346	0	7,030	14,441,666	..	113,812
2411 Basic chemicals, exc. fertilizers & nitrogen compounds	0	85	180	2,821	17,289	0	6,920	12,113,889	..	70,905
2412 Fertilizers and nitrogen compounds	0	29	0	95	24,318	0	110	1,133,333	..	28,632
2413 Plastics in primary forms and synthetic rubber	0	38	10	1,188	8,739	0	0	1,194,444	..	14,275
242 Other chemical products	0	2,356	67	3,838	61,731	0	15,690	4,077,778	..	98,362
2421 Pesticides and other agro-chemical products	..	..	..	..	..	..	..	..	..	..
2422 Paints, varnishes and similar coatings	..	..	..	..	..	..	..	..	..	..
2423 Pharmaceuticals, medicinal chem. & botanical prod.	..	..	..	..	..	..	..	..	..	..
2424 Soap and detergents, perfumes etc.	..	..	..	..	..	..	..	..	..	..
2429 Other chemical products, nec	..	..	..	..	..	..	..	..	..	..
243 Man-made fibres	..	..	..	..	..	..	..	..	..	..
25 RUBBER AND PLASTICS PRODUCTS	0	228	171	1,330	7,794	0	0	2,502,778	..	18,533
251 Rubber products	0	76	47	1,263	3,474	0	0	827,778	..	7,840
2511 Rubber tyres and tubes	..	..	..	..	..	..	..	..	..	..
2519 Other rubber products	..	..	..	..	..	..	..	..	..	..
252 Plastic products	0	152	124	67	4,320	0	0	1,675,000	..	10,693
26 OTHER NON-METALLIC MINERAL PRODUCTS	32,690	143	371	3,088	29,556	0	0	2,125,000	..	73,498
261 Glass and glass products	0	124	10	238	8,469	0	0	397,222	..	10,271
269 Non-metallic mineral products, nec	32,690	19	361	2,850	21,087	0	0	1,727,778	..	63,227
2691 Pottery, china and earthenware	..	..	..	..	..	..	..	..	..	..
2692 Refractory ceramic products	..	..	..	..	..	..	..	..	..	..
2693 Structural non-refractory clay & ceramic prod.	..	..	..	..	..	..	..	..	..	..
2694 Cement, lime and plaster	32,690	19	361	2,850	21,087	0	0	1,727,778	..	63,227
2695 Articles of concrete, cement and plaster	..	..	..	..	..	..	..	..	..	..
2696 Cutting, shaping and finishing of stone	..	..	..	..	..	..	..	..	..	..
2699 Other non-metallic mineral products, nec	..	..	..	..	..	..	..	..	..	..

ISIC Revision 3 Industry Sector	Solid TJ	LPG TJ	Distiloil TJ	RFO TJ	Gas TJ	Biomass TJ	Steam TJ	Electr MWh	Own Use MWh	TOTAL TJ
27 **BASIC METALS**	138,539	0	1,454	17,281	79,227	0	0	48,283,333	..	410,321
271 Basic iron and steel	128,706	0	1,026	9,462	55,566	0	0	7,936,111	..	223,330
272 Basic precious and non-ferrous metals	9,833	0	428	7,819	23,661	0	0	40,347,222	..	186,991
273 Casting of metals	..	..	..	..	..	..	..	..	..	..
2731 Casting of iron and steel	..	..	..	..	..	..	..	..	..	..
2732 Casting non-ferrous metals	..	..	..	..	..	..	..	..	..	..
28 **FABRICATED METAL PRODUCTS**	10	770	266	171	14,436	0	0	2,105,556	..	23,233
281 Str. metal prod., tanks, reservoirs, steam generators	..	..	..	..	..	..	..	..	..	..
2811 Structural metal products	..	..	..	..	..	..	..	..	..	..
2812 Tanks, reservoirs and containers of metal	..	..	..	..	..	..	..	..	..	..
2813 Steam generators, exc. central heating hot water boilers	..	..	..	..	..	..	..	..	..	..
289 Other fabricated metal products	..	..	..	..	..	..	..	..	..	..
2891 Forging, pressing, stamping & roll-forming of metal	..	..	..	..	..	..	..	..	..	..
2892 Treatment and coating of metals	..	..	..	..	..	..	..	..	..	..
2893 Cutlery, hand tools and general hardware	..	..	..	..	..	..	..	..	..	..
2899 Other fabricated metal products, nec	..	..	..	..	..	..	..	..	..	..
29 **MACHINERY AND EQUIPMENT, NEC**	0	561	409	247	13,680	0	0	2,941,667	..	25,487
291 General purpose machinery	..	..	..	..	..	..	..	..	..	..
2911 Engines and turbines	..	..	..	..	..	..	..	..	..	..
2912 Pumps, compressors, taps and valves	..	..	..	..	..	..	..	..	..	..
2913 Bearings, gears, gearing and driving elements	..	..	..	..	..	..	..	..	..	..
2914 Ovens, furnaces and furnace burners	..	..	..	..	..	..	..	..	..	..
2915 Lifting and handling equipment	..	..	..	..	..	..	..	..	..	..
2919 Other general purpose machinery	..	..	..	..	..	..	..	..	..	..
292 Special purpose machinery	..	..	..	..	..	..	..	..	..	..
2921 Agricultural and forestry machinery	..	..	..	..	..	..	..	..	..	..
2922 Machine-tools	..	..	..	..	..	..	..	..	..	..
2923 Machinery for metallurgy	..	..	..	..	..	..	..	..	..	..
2924 Machinery for mining, quarrying and construction	..	..	..	..	..	..	..	..	..	..
2925 Machinery for food, beverage & tobacco processing	..	..	..	..	..	..	..	..	..	..
2926 Machinery for textile, apparel & leather production	..	..	..	..	..	..	..	..	..	..
2927 Machinery for weapons and ammunition	..	..	..	..	..	..	..	..	..	..
2929 Other special purpose machinery	..	..	..	..	..	..	..	..	..	..
293 Domestic appliances, nec	..	..	..	..	..	..	..	..	..	..
30 **OFFICE, ACCOUNTING & COMPUTING MACHINERY**	..	..	..	..	..	..	..	..	..	..
31 **ELECTRICAL MACHINERY & APPARATUS, NEC**	..	..	..	..	..	..	..	..	..	..
311 Electric motors, generators and transformers	..	..	..	..	..	..	..	..	..	..
312 Electricity distribution and control apparatus	..	..	..	..	..	..	..	..	..	..
313 Insulated wire and cable	..	..	..	..	..	..	..	..	..	..
314 Accumulators, primary cells & primary batteries	..	..	..	..	..	..	..	..	..	..
315 Electric lamps and lighting equipment	..	..	..	..	..	..	..	..	..	..
319 Other electrical equipment, nec	..	..	..	..	..	..	..	..	..	..
32 **RADIO, TV & COMMUNICATION EQUIP. & APP.**	..	..	..	..	..	..	..	..	..	..
321 Electronic valves, tubes, other electronic components	..	..	..	..	..	..	..	..	..	..
322 TV & radio transmitters, apparatus for line telephony	..	..	..	..	..	..	..	..	..	..
323 TV & radio receivers, recording apparatus	..	..	..	..	..	..	..	..	..	..
33 **MEDICAL PRECISION &OPTICAL INSTRUMENTS**	..	..	..	..	..	..	..	..	..	..
331 Medical appliances and instruments	..	..	..	..	..	..	..	..	..	..
3311 Medical, surgical equipment & orthopaedic app.	..	..	..	..	..	..	..	..	..	..
3312 Instruments & appliances for measuring, checking etc	..	..	..	..	..	..	..	..	..	..
3313 Industrial process control equipment	..	..	..	..	..	..	..	..	..	..
332 Optical instruments and photographic equipment	..	..	..	..	..	..	..	..	..	..
333 Watches and clocks	..	..	..	..	..	..	..	..	..	..
34 **MOTOR VEHICLES, TRAILERS & SEMI-TRAILERS**	1,007	836	371	808	26,721	0	0	4,802,778	..	47,033
341 Motor vehicles	817	133	38	57	14,229	0	0	1,877,778	..	22,034
342 Bodies (coachwork) for motor vehicles	..	..	..	..	..	..	..	..	..	..
343 Parts, accessories for motor vehicles & their engines	190	703	333	751	12,492	0	0	2,925,000	..	24,999
35 **OTHER TRANSPORT EQUIPMENT**	..	..	..	..	..	..	..	..	..	..
351 Building and repairing of ships and boats	..	..	..	..	..	..	..	..	..	..
3511 Building and repairing of ships	..	..	..	..	..	..	..	..	..	..
3512 Building, repairing of pleasure & sporting boats	..	..	..	..	..	..	..	..	..	..
352 Railway, tramway locomotives & rolling stock	..	..	..	..	..	..	..	..	..	..
353 Aircraft and spacecraft	..	..	..	..	..	..	..	..	..	..
359 Transport equipment, nec	..	..	..	..	..	..	..	..	..	..
3591 Motorcycles	..	..	..	..	..	..	..	..	..	..
3592 Bicycles and invalid carriages	..	..	..	..	..	..	..	..	..	..
3599 Other transport equipment, nec	..	..	..	..	..	..	..	..	..	..
36 **FURNITURE; MANUFACTURING, NEC**	2,866	9,915	8,432	10	129,231	0	310	3,136,111	..	162,054
361 Furniture	..	..	..	..	..	..	..	..	..	..
369 Manufacturing, nec	..	..	..	..	..	..	..	..	..	..
3691 Jewellery and related articles	..	..	..	..	..	..	..	..	..	..
3692 Musical instruments	..	..	..	..	..	..	..	..	..	..
3693 Sports goods	..	..	..	..	..	..	..	..	..	..
3694 Games and toys	..	..	..	..	..	..	..	..	..	..
3699 Other manufacturing, nec	..	..	..	..	..	..	..	..	..	..
37 **RECYCLING**	..	..	..	..	..	..	..	..	..	..
371 Recycling of metal waste and scrap	..	..	..	..	..	..	..	..	..	..
372 Recycling of non-metal waste and scrap	..	..	..	..	..	..	..	..	..	..
CERR **Non-specified industry**	..	..	..	..	..	..	..	..	..	..
15-37 **TOTAL MANUFACTURING**	203,110	19,238	24,596	153,302	808,777	277,987	24,580	155,877,778	..	2,072,750

ISIC Revision 3 Industry Sector		Solid TJ	LPG TJ	Distiloil TJ	RFO TJ	Gas TJ	Biomass TJ	Steam TJ	Electr MWh	Own Use MWh	TOTAL TJ
15	FOOD PRODUCTS AND BEVERAGES	0	1,425	2,185	4,684	51,066	0	0	6,050,000	..	81,140
151	Production, processing and preserving (PPP)	0	331	293	2,375	11,277	0	0	1,580,555	..	19,966
1511	PPP of meat and meat products	..	..	..	..	..	..	..	..	..	..
1512	Processing and preserving of fish products	..	..	..	..	..	..	..	..	..	..
1513	Processing, preserving of fruit & vegetables	..	..	..	..	..	..	..	..	..	..
1514	Vegetable and animal oils and fats	..	..	..	..	..	..	..	..	..	..
152	Dairy products	0	428	428	551	5,562	0	0	866,667	..	10,089
153	Grain mill prod., starches & prepared animal feeds	..	..	..	..	..	..	..	..	..	..
1531	Grain mill products	..	..	..	..	..	..	..	..	..	..
1532	Starches and starch products	..	..	..	..	..	..	..	..	..	..
1533	Prepared animal feeds	..	..	..	..	..	..	..	..	..	..
154	Other food products	0	561	1,302	1,406	26,154	0	0	2,922,222	..	39,943
1541	Bakery products	..	..	..	..	..	..	..	..	..	..
1542	Sugar	..	..	..	..	..	..	..	..	..	..
1543	Cocoa, chocolate and sugar confectionery	..	..	..	..	..	..	..	..	..	..
1544	Macaroni, noodles, couscous & similar farinaceous prod.	..	..	..	..	..	..	..	..	..	..
1549	Other food products, nec	..	..	..	..	..	..	..	..	..	..
155	Beverages	0	105	162	352	8,073	0	0	680,556	..	11,142
1551	Distilling, rectifying and blending of spirits	..	..	..	..	..	..	..	..	..	..
1552	Wines	..	..	..	..	..	..	..	..	..	..
1553	Malt liquors and malt	..	..	..	..	..	..	..	..	..	..
1554	Soft drinks; production of mineral waters	..	..	..	..	..	..	..	..	..	..
16	TOBACCO PRODUCTS	0	10	0	0	657	0	0	136,111	..	1,157
17	TEXTILES	0	276	200	646	14,391	0	0	1,955,556	..	22,553
171	Spinning, weaving and finishing of textiles	0	114	133	351	9,342	0	0	1,297,223	..	14,610
1711	Preparation and spinning of textile fibres	..	..	..	..	..	..	..	..	..	..
1712	Finishing of textiles	..	..	..	..	..	..	..	..	..	..
172	Other textiles	..	..	..	..	..	..	..	..	..	..
1721	Made-up textile articles, except apparel	..	..	..	..	..	..	..	..	..	..
1722	Carpets and rugs	..	..	..	..	..	..	..	..	..	..
1723	Cordage, rope, twine and netting	..	..	..	..	..	..	..	..	..	..
1729	Other textiles, nec	..	..	..	..	..	..	..	..	..	..
173	Knitted and crocheted fabrics and articles	0	162	67	295	5,049	0	0	658,333	..	7,943
18	WEARING APPAREL, DRESSING & DYEING OF FUR	0	57	86	162	2,259	0	0	741,667	..	5,234
181	Wearing apparel, except fur apparel	..	..	..	..	..	..	..	..	..	..
182	Dressing and dyeing of fur; articles of fur	..	..	..	..	..	..	..	..	..	..
19	TANNING & DRESSING OF LEATHER, FOOTWEAR	0	10	38	0	567	0	0	138,889	..	1,115
191	Tanning and dressing of leather	..	..	..	..	..	..	..	..	..	..
1911	Tanning and dressing of leather	..	..	..	..	..	..	..	..	..	..
1912	Luggage, handbags, saddlery & harness	..	..	..	..	..	..	..	..	..	..
192	Footwear	..	..	..	..	..	..	..	..	..	..
20	WOOD AND WOOD PRODUCTS	76	950	722	798	15,525	0	0	5,077,778	..	36,351
201	Sawmilling and planing of wood	0	380	437	190	6,795	0	0	1,763,889	..	14,152
202	Products of wood, cork, straw & plaiting materials	76	570	285	608	8,730	0	0	3,313,889	..	22,199
2021	Veneer sheets	..	..	..	..	..	..	..	..	..	..
2022	Builders' carpentry and joinery	..	..	..	..	..	..	..	..	..	..
2023	Wooden containers	..	..	..	..	..	..	..	..	..	..
2029	Other products of wood	..	..	..	..	..	..	..	..	..	..
21	PAPER AND PAPER PRODUCTS	1,777	0	5,833	66,878	102,528	287,137	1,300	49,505,556	..	643,673
2101	Pulp, paper and paperboard	..	..	..	..	..	..	..	..	..	..
2102	Corrugated paper, paperboard and their containers	..	..	..	..	..	..	..	..	..	..
2109	Other articles of pulp and paperboard	..	..	..	..	..	..	..	..	..	..
22	PUBLISHING, PRINTING & REPRODUCTION	10	181	95	38	4,122	0	0	1,436,111	..	9,616
221	Publishing	..	..	..	..	..	..	..	..	..	..
2211	Publishing of books & brochures	..	..	..	..	..	..	..	..	..	..
2212	Publishing of newspapers and periodicals	..	..	..	..	..	..	..	..	..	..
2213	Publishing of recorded media	..	..	..	..	..	..	..	..	..	..
2219	Other publishing	..	..	..	..	..	..	..	..	..	..
222	Printing and related service activities	..	..	..	..	..	..	..	..	..	..
2221	Printing	..	..	..	..	..	..	..	..	..	..
2222	Service activities related to printing	..	..	..	..	..	..	..	..	..	..
223	Reproduction of recorded media	..	..	..	..	..	..	..	..	..	..
23	COKE, REFINED PETROLEUM PRODUCTS	27,284	2,898	770	36,461	206,547	0	0	5,891,667	..	295,170
231	Coke oven products	..	..	..	..	..	..	..	..	..	..
232	Refined petroleum products	27,284	2,898	770	36,461	206,547	0	0	5,891,667	..	295,170
233	Processing of nuclear fuel	..	..	..	..	..	..	..	..	..	..
24	CHEMICALS & CHEMICAL PRODUCTS	0	2,499	238	6,612	109,494	0	15,320	18,191,667	..	199,653
241	Basic chemicals	0	228	228	2,689	49,635	0	4,740	14,083,333	..	108,220
2411	Basic chemicals, exc. fertilizers & nitrogen compounds	0	142	218	2,365	15,894	0	4,660	11,869,444	..	66,009
2412	Fertilizers and nitrogen compounds	0	57	0	124	25,065	0	80	1,011,111	..	28,966
2413	Plastics in primary forms and synthetic rubber	0	29	10	200	8,676	0	0	1,202,778	..	13,245
242	Other chemical products	0	2,271	10	3,923	59,859	0	10,580	4,108,334	..	91,433
2421	Pesticides and other agro-chemical products	..	..	..	..	..	..	..	..	..	..
2422	Paints, varnishes and similar coatings	..	..	..	..	..	..	..	..	..	..
2423	Pharmaceuticals, medicinal chem. & botanical prod.	..	..	..	..	..	..	..	..	..	..
2424	Soap and detergents, perfumes etc.	..	..	..	..	..	..	..	..	..	..
2429	Other chemical products, nec	..	..	..	..	..	..	..	..	..	..
243	Man-made fibres	..	..	..	..	..	..	..	..	..	..
25	RUBBER AND PLASTICS PRODUCTS	0	266	209	1,881	8,118	0	0	2,891,667	..	20,884
251	Rubber products	0	104	57	1,795	3,645	0	0	847,223	..	8,651
2511	Rubber tyres and tubes	..	..	..	..	..	..	..	..	..	..
2519	Other rubber products	..	..	..	..	..	..	..	..	..	..
252	Plastic products	0	162	152	86	4,473	0	0	2,044,444	..	12,233
26	OTHER NON-METALLIC MINERAL PRODUCTS	28,481	114	381	3,420	31,653	0	0	2,033,333	..	71,369
261	Glass and glass products	0	104	10	389	10,710	0	0	416,666	..	12,713
269	Non-metallic mineral products, nec	28,481	10	371	3,031	20,943	0	0	1,616,667	..	58,656
2691	Pottery, china and earthenware	..	..	..	..	..	..	..	..	..	..
2692	Refractory ceramic products	..	..	..	..	..	..	..	..	..	..
2693	Structural non-refractory clay & ceramic prod.	..	..	..	..	..	..	..	..	..	..
2694	Cement, lime and plaster	28,481	10	371	3,031	20,943	0	0	1,616,667	..	58,656
2695	Articles of concrete, cement and plaster	..	..	..	..	..	..	..	..	..	..
2696	Cutting, shaping and finishing of stone	..	..	..	..	..	..	..	..	..	..
2699	Other non-metallic mineral products, nec	..	..	..	..	..	..	..	..	..	..

ISIC Revision 3 Industry Sector		Solid TJ	LPG TJ	Distiloil TJ	RFO TJ	Gas TJ	Biomass TJ	Steam TJ	Electr MWh	Own Use MWh	TOTAL TJ
27	**BASIC METALS**	143,156	0	1,710	16,644	87,597	0	0	50,727,778	..	431,727
271	Basic iron and steel	130,388	0	1,064	8,949	62,019	0	0	8,294,444	..	232,280
272	Basic precious and non-ferrous metals	12,768	0	646	7,695	25,578	0	0	42,433,334	..	199,447
273	Casting of metals	..	..	..	..	..	..	..	..	..	..
2731	Casting of iron and steel	..	..	..	..	..	..	..	..	..	..
2732	Casting non-ferrous metals	..	..	..	..	..	..	..	..	..	..
28	**FABRICATED METAL PRODUCTS**	67	874	276	200	16,326	0	0	2,619,444	..	27,173
281	Str. metal prod., tanks, reservoirs, steam generators	..	..	..	..	..	..	..	..	..	..
2811	Structural metal products	..	..	..	..	..	..	..	..	..	..
2812	Tanks, reservoirs and containers of metal	..	..	..	..	..	..	..	..	..	..
2813	Steam generators, exc. central heating hot water boilers	..	..	..	..	..	..	..	..	..	..
289	Other fabricated metal products	..	..	..	..	..	..	..	..	..	..
2891	Forging, pressing, stamping & roll-forming of metal	..	..	..	..	..	..	..	..	..	..
2892	Treatment and coating of metals	..	..	..	..	..	..	..	..	..	..
2893	Cutlery, hand tools and general hardware	..	..	..	..	..	..	..	..	..	..
2899	Other fabricated metal products, nec	..	..	..	..	..	..	..	..	..	..
29	**MACHINERY AND EQUIPMENT, NEC**	0	608	361	238	13,689	0	0	4,213,889	..	30,066
291	General purpose machinery	..	..	..	..	..	..	..	..	..	..
2911	Engines and turbines	..	..	..	..	..	..	..	..	..	..
2912	Pumps, compressors, taps and valves	..	..	..	..	..	..	..	..	..	..
2913	Bearings, gears, gearing and driving elements	..	..	..	..	..	..	..	..	..	..
2914	Ovens, furnaces and furnace burners	..	..	..	..	..	..	..	..	..	..
2915	Lifting and handling equipment	..	..	..	..	..	..	..	..	..	..
2919	Other general purpose machinery	..	..	..	..	..	..	..	..	..	..
292	Special purpose machinery	..	..	..	..	..	..	..	..	..	..
2921	Agricultural and forestry machinery	..	..	..	..	..	..	..	..	..	..
2922	Machine-tools	..	..	..	..	..	..	..	..	..	..
2923	Machinery for metallurgy	..	..	..	..	..	..	..	..	..	..
2924	Machinery for mining, quarrying and construction	..	..	..	..	..	..	..	..	..	..
2925	Machinery for food, beverage & tobacco processing	..	..	..	..	..	..	..	..	..	..
2926	Machinery for textile, apparel & leather production	..	..	..	..	..	..	..	..	..	..
2927	Machinery for weapons and ammunition	..	..	..	..	..	..	..	..	..	..
2929	Other special purpose machinery	..	..	..	..	..	..	..	..	..	..
293	Domestic appliances, nec	..	..	..	..	..	..	..	..	..	..
30	**OFFICE, ACCOUNTING & COMPUTING MACHINERY**	..	..	..	..	..	..	..	..	..	..
31	**ELECTRICAL MACHINERY & APPARATUS, NEC**	..	..	..	..	..	..	..	..	..	..
311	Electric motors, generators and transformers	..	..	..	..	..	..	..	..	..	..
312	Electricity distribution and control apparatus	..	..	..	..	..	..	..	..	..	..
313	Insulated wire and cable	..	..	..	..	..	..	..	..	..	..
314	Accumulators, primary cells & primary batteries	..	..	..	..	..	..	..	..	..	..
315	Electric lamps and lighting equipment	..	..	..	..	..	..	..	..	..	..
319	Other electrical equipment, nec	..	..	..	..	..	..	..	..	..	..
32	**RADIO, TV & COMMUNICATION EQUIP. & APP.**	..	..	..	..	..	..	..	..	..	..
321	Electronic valves, tubes, other electronic components	..	..	..	..	..	..	..	..	..	..
322	TV & radio transmitters, apparatus for line telephony	..	..	..	..	..	..	..	..	..	..
323	TV & radio receivers, recording apparatus	..	..	..	..	..	..	..	..	..	..
33	**MEDICAL PRECISION &OPTICAL INSTRUMENTS**	..	..	..	..	..	..	..	..	..	..
331	Medical appliances and instruments	..	..	..	..	..	..	..	..	..	..
3311	Medical, surgical equipment & orthopaedic app.	..	..	..	..	..	..	..	..	..	..
3312	Instruments & appliances for measuring, checking etc	..	..	..	..	..	..	..	..	..	..
3313	Industrial process control equipment	..	..	..	..	..	..	..	..	..	..
332	Optical instruments and photographic equipment	..	..	..	..	..	..	..	..	..	..
333	Watches and clocks	..	..	..	..	..	..	..	..	..	..
34	**MOTOR VEHICLES, TRAILERS & SEMI-TRAILERS**	836	846	371	865	28,170	0	0	4,930,556	..	48,838
341	Motor vehicles	608	124	143	67	16,515	0	0	2,013,889	..	24,707
342	Bodies (coachwork) for motor vehicles	..	..	..	..	..	..	..	..	..	..
343	Parts, accessories for motor vehicles & their engines	228	722	228	798	11,655	0	0	2,916,667	..	24,131
35	**OTHER TRANSPORT EQUIPMENT**	..	..	..	..	..	..	..	..	..	..
351	Building and repairing of ships and boats	..	..	..	..	..	..	..	..	..	..
3511	Building and repairing of ships	..	..	..	..	..	..	..	..	..	..
3512	Building, repairing of pleasure & sporting boats	..	..	..	..	..	..	..	..	..	..
352	Railway, tramway locomotives & rolling stock	..	..	..	..	..	..	..	..	..	..
353	Aircraft and spacecraft	..	..	..	..	..	..	..	..	..	..
359	Transport equipment, nec	..	..	..	..	..	..	..	..	..	..
3591	Motorcycles	..	..	..	..	..	..	..	..	..	..
3592	Bicycles and invalid carriages	..	..	..	..	..	..	..	..	..	..
3599	Other transport equipment, nec	..	..	..	..	..	..	..	..	..	..
36	**FURNITURE; MANUFACTURING, NEC**	6,230	2,115	7,653	19	112,095	0	160	1,133,331	..	132,352
361	Furniture	..	..	..	..	..	..	..	..	..	..
369	Manufacturing, nec	..	..	..	..	..	..	..	..	..	..
3691	Jewellery and related articles	..	..	..	..	..	..	..	..	..	..
3692	Musical instruments	..	..	..	..	..	..	..	..	..	..
3693	Sports goods	..	..	..	..	..	..	..	..	..	..
3694	Games and toys	..	..	..	..	..	..	..	..	..	..
3699	Other manufacturing, nec	..	..	..	..	..	..	..	..	..	..
37	**RECYCLING**	..	..	..	..	..	..	..	..	..	..
371	Recycling of metal waste and scrap	..	..	..	..	..	..	..	..	..	..
372	Recycling of non-metal waste and scrap	..	..	..	..	..	..	..	..	..	..
CERR	Non-specified industry	..	..	..	..	..	..	..	..	..	..
15-37	**TOTAL MANUFACTURING**	207,917	13,129	21,128	139,546	804,804	287,137	16,780	157,675,000	..	2,058,071

ISIC Revision 3 Industry Sector	Solid TJ	LPG TJ	Distiloil TJ	RFO TJ	Gas TJ	Biomass TJ	Steam TJ	Electr MWh	Own Use MWh	TOTAL TJ
15　FOOD PRODUCTS AND BEVERAGES	0	1,587	2,480	4,237	55,683	0	0	6,252,778	..	86,497
151　Production, processing and preserving (PPP)	0	399	341	2,289	13,248	..	0	1,708,334	..	22,427
1511　PPP of meat and meat products	..	..	..	..	..	..	..	..	..	..
1512　Processing and preserving of fish products	..	..	..	..	..	..	..	..	..	..
1513　Processing, preserving of fruit & vegetables	..	..	..	..	..	..	..	..	..	..
1514　Vegetable and animal oils and fats	..	..	..	..	..	..	..	..	..	..
152　Dairy products	0	456	295	437	5,427	..	0	852,778	..	9,685
153　Grain mill prod., starches & prepared animal feeds	..	..	..	..	..	..	..	..	..	..
1531　Grain mill products	..	..	..	..	..	..	..	..	..	..
1532　Starches and starch products	..	..	..	..	..	..	..	..	..	..
1533　Prepared animal feeds	..	..	..	..	..	..	..	..	..	..
154　Other food products	0	684	1,112	1,311	28,332	0	0	2,969,444	..	42,129
1541　Bakery products	..	..	..	..	..	..	..	..	..	..
1542　Sugar	..	..	..	..	..	..	..	..	..	..
1543　Cocoa, chocolate and sugar confectionery	..	..	..	..	..	..	..	..	..	..
1544　Macaroni, noodles, couscous & similar farinaceous prod.	..	..	..	..	..	..	..	..	..	..
1549　Other food products, nec	..	..	..	..	..	..	..	..	..	..
155　Beverages	0	48	732	200	8,676	0	0	722,222	..	12,256
1551　Distilling, rectifying and blending of spirits	..	..	..	..	..	..	..	..	..	..
1552　Wines	..	..	..	..	..	..	..	..	..	..
1553　Malt liquors and malt	..	..	..	..	..	..	..	..	..	..
1554　Soft drinks; production of mineral waters	..	..	..	..	..	..	..	..	..	..
16　TOBACCO PRODUCTS	0	10	0	0	603	0	0	133,333	..	1,093
17　TEXTILES	0	314	247	760	13,914	0	0	1,844,444	..	21,875
171　Spinning, weaving and finishing of textiles	0	152	200	313	8,766	0	0	1,172,222	..	13,651
1711　Preparation and spinning of textile fibres	..	..	..	..	..	..	..	..	..	..
1712　Finishing of textiles	..	..	..	..	..	..	..	..	..	..
172　Other textiles	..	..	..	..	..	..	..	..	..	..
1721　Made-up textile articles, except apparel	..	..	..	..	..	..	..	..	..	..
1722　Carpets and rugs	..	..	..	..	..	..	..	..	..	..
1723　Cordage, rope, twine and netting	..	..	..	..	..	..	..	..	..	..
1729　Other textiles, nec	..	..	..	..	..	..	..	..	..	..
173　Knitted and crocheted fabrics and articles	0	162	47	447	5,148	0	0	672,222	..	8,224
18　WEARING APPAREL, DRESSING & DYEING OF FUR	0	86	95	114	2,169	0	0	733,333	..	5,104
181　Wearing apparel, except fur apparel	..	..	..	..	..	..	..	..	..	..
182　Dressing and dyeing of fur; articles of fur	..	..	..	..	..	..	..	..	..	..
19　TANNING & DRESSING OF LEATHER, FOOTWEAR	0	19	38	10	648	0	0	150,000	..	1,255
191　Tanning and dressing of leather	..	..	..	..	..	..	..	..	..	..
1911　Tanning and dressing of leather	..	..	..	..	..	..	..	..	..	..
1912　Luggage, handbags, saddlery & harness	..	..	..	..	..	..	..	..	..	..
192　Footwear	..	..	..	..	..	..	..	..	..	..
20　WOOD AND WOOD PRODUCTS	0	1,349	798	1,074	20,655	0	0	5,566,667	..	43,916
201　Sawmilling and planing of wood	0	484	408	152	9,360	0	0	1,994,445	..	17,584
202　Products of wood, cork, straw & plaiting materials	0	865	390	922	11,295	0	0	3,572,222	..	26,332
2021　Veneer sheets	..	..	..	..	..	..	..	..	..	..
2022　Builders' carpentry and joinery	..	..	..	..	..	..	..	..	..	..
2023　Wooden containers	..	..	..	..	..	..	..	..	..	..
2029　Other products of wood	..	..	..	..	..	..	..	..	..	..
21　PAPER AND PAPER PRODUCTS	1,995	0	5,216	64,904	105,120	285,222	1,090	51,277,778	..	648,147
2101　Pulp, paper and paperboard	..	..	..	..	..	..	..	..	..	..
2102　Corrugated paper, paperboard and their containers	..	..	..	..	..	..	..	..	..	..
2109　Other articles of pulp and paperboard	..	..	..	..	..	..	..	..	..	..
22　PUBLISHING, PRINTING & REPRODUCTION	19	181	86	29	4,536	0	0	1,408,333	..	9,921
221　Publishing	..	..	..	..	..	..	..	..	..	..
2211　Publishing of books & brochures	..	..	..	..	..	..	..	..	..	..
2212　Publishing of newspapers and periodicals	..	..	..	..	..	..	..	..	..	..
2213　Publishing of recorded media	..	..	..	..	..	..	..	..	..	..
2219　Other publishing	..	..	..	..	..	..	..	..	..	..
222　Printing and related service activities	..	..	..	..	..	..	..	..	..	..
2221　Printing	..	..	..	..	..	..	..	..	..	..
2222　Service activities related to printing	..	..	..	..	..	..	..	..	..	..
223　Reproduction of recorded media	..	..	..	..	..	..	..	..	..	..
23　COKE, REFINED PETROLEUM PRODUCTS	31,559	3,648	1,273	39,644	201,625	0	0	5,955,556	..	299,189
231　Coke oven products	..	..	..	..	..	..	..	..	..	..
232　Refined petroleum products	31,559	3,648	1,273	39,644	201,625	0	0	5,955,556	..	299,189
233　Processing of nuclear fuel	..	..	..	..	..	..	..	..	..	..
24　CHEMICALS & CHEMICAL PRODUCTS	0	19,989	171	10,631	121,113	0	9,330	18,019,444	..	226,104
241　Basic chemicals	0	4,836	171	3,230	71,136	0	2,890	14,808,333	..	135,573
2411　Basic chemicals, exc. fertilizers & nitrogen compounds	0	3,258	161	2,384	18,864	0	2,840	11,844,444	..	70,147
2412　Fertilizers and nitrogen compounds	0	1,549	0	105	34,605	0	50	1,316,667	..	41,049
2413　Plastics in primary forms and synthetic rubber	0	29	10	741	17,667	0	0	1,647,222	..	24,377
242　Other chemical products	0	15,153	0	7,401	49,977	0	6,440	3,211,111	..	90,531
2421　Pesticides and other agro-chemical products	..	..	..	..	..	..	..	..	..	..
2422　Paints, varnishes and similar coatings	..	..	..	..	..	..	..	..	..	..
2423　Pharmaceuticals, medicinal chem. & botanical prod.	..	..	..	..	..	..	..	..	..	..
2424　Soap and detergents, perfumes etc.	..	..	..	..	..	..	..	..	..	..
2429　Other chemical products, nec	..	..	..	..	..	..	..	..	..	..
243　Man-made fibres	..	..	..	..	..	..	..	..	..	..
25　RUBBER AND PLASTICS PRODUCTS	0	238	257	1,881	8,424	0	0	2,947,222	..	21,410
251　Rubber products	0	95	86	1,862	3,600	0	0	808,333	..	8,553
2511　Rubber tyres and tubes	..	..	..	..	..	..	..	..	..	..
2519　Other rubber products	..	..	..	..	..	..	..	..	..	..
252　Plastic products	0	143	171	19	4,824	0	0	2,138,889	..	12,857
26　OTHER NON-METALLIC MINERAL PRODUCTS	27,712	95	266	2,537	29,178	0	0	2,191,667	..	67,678
261　Glass and glass products	0	85	0	428	11,097	0	0	413,889	..	13,100
269　Non-metallic mineral products, nec	27,712	10	266	2,109	18,081	0	0	1,777,778	..	54,578
2691　Pottery, china and earthenware	..	..	..	..	..	..	..	..	..	..
2692　Refractory ceramic products	..	..	..	..	..	..	..	..	..	..
2693　Structural non-refractory clay & ceramic prod.	..	..	..	..	..	..	..	..	..	..
2694　Cement, lime and plaster	27,712	10	266	2,109	18,081	0	0	1,777,778	..	54,578
2695　Articles of concrete, cement and plaster	..	..	..	..	..	..	..	..	..	..
2696　Cutting, shaping and finishing of stone	..	..	..	..	..	..	..	..	..	..
2699　Other non-metallic mineral products, nec	..	..	..	..	..	..	..	..	..	..

ISIC Revision 3 Industry Sector		Solid TJ	LPG TJ	Distiloil TJ	RFO TJ	Gas TJ	Biomass TJ	Steam TJ	Electr MWh	Own Use MWh	TOTAL TJ
27	**BASIC METALS**	135,926	0	1,653	18,563	89,145	0	0	55,319,444	..	444,437
271	Basic iron and steel	125,324	0	950	9,462	64,656	0	0	8,447,222	..	230,802
272	Basic precious and non-ferrous metals	10,602	0	703	9,101	24,489	0	0	46,872,222	..	213,635
273	Casting of metals	..	..	..	..	..	..	..	..	..	..
2731	Casting of iron and steel	..	..	..	..	..	..	..	..	..	..
2732	Casting non-ferrous metals	..	..	..	..	..	..	..	..	..	..
28	**FABRICATED METAL PRODUCTS**	76	779	228	162	18,864	0	0	2,577,778	..	29,389
281	Str. metal prod., tanks, reservoirs, steam generators	..	..	..	..	..	..	..	..	..	..
2811	Structural metal products	..	..	..	..	..	..	..	..	..	..
2812	Tanks, reservoirs and containers of metal	..	..	..	..	..	..	..	..	..	..
2813	Steam generators, exc. central heating hot water boilers	..	..	..	..	..	..	..	..	..	..
289	Other fabricated metal products	..	..	..	..	..	..	..	..	..	..
2891	Forging, pressing, stamping & roll-forming of metal	..	..	..	..	..	..	..	..	..	..
2892	Treatment and coating of metals	..	..	..	..	..	..	..	..	..	..
2893	Cutlery, hand tools and general hardware	..	..	..	..	..	..	..	..	..	..
2899	Other fabricated metal products, nec	..	..	..	..	..	..	..	..	..	..
29	**MACHINERY AND EQUIPMENT, NEC**	38	589	162	247	14,499	0	0	3,188,889	..	27,015
291	General purpose machinery	..	..	..	..	..	..	..	..	..	..
2911	Engines and turbines	..	..	..	..	..	..	..	..	..	..
2912	Pumps, compressors, taps and valves	..	..	..	..	..	..	..	..	..	..
2913	Bearings, gears, gearing and driving elements	..	..	..	..	..	..	..	..	..	..
2914	Ovens, furnaces and furnace burners	..	..	..	..	..	..	..	..	..	..
2915	Lifting and handling equipment	..	..	..	..	..	..	..	..	..	..
2919	Other general purpose machinery	..	..	..	..	..	..	..	..	..	..
292	Special purpose machinery	..	..	..	..	..	..	..	..	..	..
2921	Agricultural and forestry machinery	..	..	..	..	..	..	..	..	..	..
2922	Machine-tools	..	..	..	..	..	..	..	..	..	..
2923	Machinery for metallurgy	..	..	..	..	..	..	..	..	..	..
2924	Machinery for mining, quarrying and construction	..	..	..	..	..	..	..	..	..	..
2925	Machinery for food, beverage & tobacco processing	..	..	..	..	..	..	..	..	..	..
2926	Machinery for textile, apparel & leather production	..	..	..	..	..	..	..	..	..	..
2927	Machinery for weapons and ammunition	..	..	..	..	..	..	..	..	..	..
2929	Other special purpose machinery	..	..	..	..	..	..	..	..	..	..
293	Domestic appliances, nec	..	..	..	..	..	..	..	..	..	..
30	**OFFICE, ACCOUNTING & COMPUTING MACHINERY**	..	..	..	..	..	..	..	..	..	..
31	**ELECTRICAL MACHINERY & APPARATUS, NEC**	..	..	..	..	..	..	..	..	..	..
311	Electric motors, generators and transformers	..	..	..	..	..	..	..	..	..	..
312	Electricity distribution and control apparatus	..	..	..	..	..	..	..	..	..	..
313	Insulated wire and cable	..	..	..	..	..	..	..	..	..	..
314	Accumulators, primary cells & primary batteries	..	..	..	..	..	..	..	..	..	..
315	Electric lamps and lighting equipment	..	..	..	..	..	..	..	..	..	..
319	Other electrical equipment, nec	..	..	..	..	..	..	..	..	..	..
32	**RADIO, TV & COMMUNICATION EQUIP. & APP.**	..	..	..	..	..	..	..	..	..	..
321	Electronic valves, tubes, other electronic components	..	..	..	..	..	..	..	..	..	..
322	TV & radio transmitters, apparatus for line telephony	..	..	..	..	..	..	..	..	..	..
323	TV & radio receivers, recording apparatus	..	..	..	..	..	..	..	..	..	..
33	**MEDICAL PRECISION &OPTICAL INSTRUMENTS**	..	..	..	..	..	..	..	..	..	..
331	Medical appliances and instruments	..	..	..	..	..	..	..	..	..	..
3311	Medical, surgical equipment & orthopaedic app.	..	..	..	..	..	..	..	..	..	..
3312	Instruments & appliances for measuring, checking etc	..	..	..	..	..	..	..	..	..	..
3313	Industrial process control equipment	..	..	..	..	..	..	..	..	..	..
332	Optical instruments and photographic equipment	..	..	..	..	..	..	..	..	..	..
333	Watches and clocks	..	..	..	..	..	..	..	..	..	..
34	**MOTOR VEHICLES, TRAILERS & SEMI-TRAILERS**	893	1,055	304	922	31,437	0	0	5,361,111	..	53,911
341	Motor vehicles	617	86	142	76	19,080	0	0	2,283,333	..	28,221
342	Bodies (coachwork) for motor vehicles	..	..	..	..	..	..	..	..	..	..
343	Parts, accessories for motor vehicles & their engines	276	969	162	846	12,357	0	0	3,077,778	..	25,690
35	**OTHER TRANSPORT EQUIPMENT**	..	..	..	..	..	..	..	..	..	..
351	Building and repairing of ships and boats	..	..	..	..	..	..	..	..	..	..
3511	Building and repairing of ships	..	..	..	..	..	..	..	..	..	..
3512	Building, repairing of pleasure & sporting boats	..	..	..	..	..	..	..	..	..	..
352	Railway, tramway locomotives & rolling stock	..	..	..	..	..	..	..	..	..	..
353	Aircraft and spacecraft	..	..	..	..	..	..	..	..	..	..
359	Transport equipment, nec	..	..	..	..	..	..	..	..	..	..
3591	Motorcycles	..	..	..	..	..	..	..	..	..	..
3592	Bicycles and invalid carriages	..	..	..	..	..	..	..	..	..	..
3599	Other transport equipment, nec	..	..	..	..	..	..	..	..	..	..
36	**FURNITURE; MANUFACTURING, NEC**	3,458	8,736	7,940	6,922	76,086	0	150	641,667	..	105,602
361	Furniture	..	..	..	..	..	..	..	..	..	..
369	Manufacturing, nec	..	..	..	..	..	..	..	..	..	..
3691	Jewellery and related articles	..	..	..	..	..	..	..	..	..	..
3692	Musical instruments	..	..	..	..	..	..	..	..	..	..
3693	Sports goods	..	..	..	..	..	..	..	..	..	..
3694	Games and toys	..	..	..	..	..	..	..	..	..	..
3699	Other manufacturing, nec	..	..	..	..	..	..	..	..	..	..
37	**RECYCLING**	..	..	..	..	..	..	..	..	..	..
371	Recycling of metal waste and scrap	..	..	..	..	..	..	..	..	..	..
372	Recycling of non-metal waste and scrap	..	..	..	..	..	..	..	..	..	..
CERR	**Non-specified industry**	..	..	..	..	..	..	..	..	..	..
15-37	**TOTAL MANUFACTURING**	201,676	38,675	21,214	152,637	793,699	285,222	10,570	163,569,444	..	2,092,543

ISIC Revision 3 Industry Sector		Solid TJ	LPG TJ	Distiloil TJ	RFO TJ	Gas TJ	Biomass TJ	Steam TJ	Electr MWh	Own Use MWh	TOTAL TJ
15	**FOOD PRODUCTS AND BEVERAGES**	10	1,596	2,081	5,881	67,356	0	0	6,575,000	..	100,594
151	Production, processing and preserving (PPP)	0	502	332	2,936	16,317	0	0	1,769,445		26,457
1511	PPP of meat and meat products	..	..	..	..	..		..	..		..
1512	Processing and preserving of fish products	..	..	..	..	..		..	..		..
1513	Processing, preserving of fruit & vegetables	..	..	..	..	..		..	..		..
1514	Vegetable and animal oils and fats	..	..	..	..	..		..	..		..
152	Dairy products	10	352	200	494	7,137	0	0	913,889		11,483
153	Grain mill prod., starches & prepared animal feeds	..	..	..	..	..		..	..		..
1531	Grain mill products	..	..	..	..	..		..	..		..
1532	Starches and starch products	..	..	..	..	..		..	..		..
1533	Prepared animal feeds	..	..	..	..	..		..	..		..
154	Other food products	0	675	1,055	1,919	34,983	0	0	3,183,333		50,092
1541	Bakery products	..	..	..	..	..		..	..		..
1542	Sugar	..	..	..	..	..		..	..		..
1543	Cocoa, chocolate and sugar confectionery	..	..	..	..	..		..	..		..
1544	Macaroni, noodles, couscous & similar farinaceous prod.	..	..	..	..	..		..	..		..
1549	Other food products, nec	..	..	..	..	..		..	..		..
155	Beverages	0	67	494	532	8,919	0	0	708,333		12,562
1551	Distilling, rectifying and blending of spirits	..	..	..	..	..		..	..		..
1552	Wines	..	..	..	..	..		..	..		..
1553	Malt liquors and malt	..	..	..	..	..		..	..		..
1554	Soft drinks; production of mineral waters	..	..	..	..	..		..	..		..
16	**TOBACCO PRODUCTS**	0	10	0	0	648	0	0	141,667	..	1,168
17	**TEXTILES**	0	399	285	1,007	20,934	0	0	1,930,556	..	29,575
171	Spinning, weaving and finishing of textiles	0	161	218	522	15,462	0	0	1,288,889		21,003
1711	Preparation and spinning of textile fibres	..	..	..	..	..		..	..		..
1712	Finishing of textiles	..	..	..	..	..		..	..		..
172	Other textiles	..	..	..	..	..		..	..		..
1721	Made-up textile articles, except apparel	..	..	..	..	..		..	..		..
1722	Carpets and rugs	..	..	..	..	..		..	..		..
1723	Cordage, rope, twine and netting	..	..	..	..	..		..	..		..
1729	Other textiles, nec	..	..	..	..	..		..	..		..
173	Knitted and crocheted fabrics and articles	0	238	67	485	5,472	0	0	641,667		8,572
18	**WEARING APPAREL, DRESSING & DYEING OF FUR**	0	76	86	114	2,394	0	0	722,222	..	5,270
181	Wearing apparel, except fur apparel	..	..	..	..	..		..	..		..
182	Dressing and dyeing of fur; articles of fur	..	..	..	..	..		..	..		..
19	**TANNING & DRESSING OF LEATHER, FOOTWEAR**	0	29	29	19	684	0	0	150,000	..	1,301
191	Tanning and dressing of leather	..	..	..	..	..		..	..		..
1911	Tanning and dressing of leather	..	..	..	..	..		..	..		..
1912	Luggage, handbags, saddlery & harness	..	..	..	..	..		..	..		..
192	Footwear	..	..	..	..	..		..	..		..
20	**WOOD AND WOOD PRODUCTS**	0	1,786	941	1,378	27,927	0	0	5,986,111	..	53,582
201	Sawmilling and planing of wood	0	551	456	133	12,699	0	0	2,097,222		21,389
202	Products of wood, cork, straw & plaiting materials	0	1,235	485	1,245	15,228	0	0	3,888,889		32,193
2021	Veneer sheets	..	..	..	..	..		..	..		..
2022	Builders' carpentry and joinery	..	..	..	..	..		..	..		..
2023	Wooden containers	..	..	..	..	..		..	..		..
2029	Other products of wood	..	..	..	..	..		..	..		..
21	**PAPER AND PAPER PRODUCTS**	2,337	0	4,912	59,024	117,504	317,758	3,160	53,141,667	..	696,005
2101	Pulp, paper and paperboard	..	..	..	..	..		..	..		..
2102	Corrugated paper, paperboard and their containers	..	..	..	..	..		..	..		..
2109	Other articles of pulp and paperboard	..	..	..	..	..		..	..		..
22	**PUBLISHING, PRINTING & REPRODUCTION**	10	181	86	10	5,121	0	0	1,472,222	..	10,708
221	Publishing	..	..	..	..	..		..	..		..
2211	Publishing of books & brochures	..	..	..	..	..		..	..		..
2212	Publishing of newspapers and periodicals	..	..	..	..	..		..	..		..
2213	Publishing of recorded media	..	..	..	..	..		..	..		..
2219	Other publishing	..	..	..	..	..		..	..		..
222	Printing and related service activities	..	..	..	..	..		..	..		..
2221	Printing	..	..	..	..	..		..	..		..
2222	Service activities related to printing	..	..	..	..	..		..	..		..
223	Reproduction of recorded media	..	..	..	..	..		..	..		..
23	**COKE, REFINED PETROLEUM PRODUCTS**	26,961	4,275	390	34,134	206,991	0	0	5,836,111	..	293,761
231	Coke oven products	..	..	..	..	..		..	..		..
232	Refined petroleum products	26,961	4,275	390	34,134	206,991	0	0	5,836,111		293,761
233	Processing of nuclear fuel	..	..	..	..	..		..	..		..
24	**CHEMICALS & CHEMICAL PRODUCTS**	0	24,805	285	11,847	131,481	0	7,020	18,969,444	..	243,728
241	Basic chemicals	0	3,971	275	5,444	56,502	0	2,170	16,155,555		126,522
2411	Basic chemicals, exc. fertilizers & nitrogen compounds	0	2,356	246	5,178	14,418	0	2,140	12,777,777		70,338
2412	Fertilizers and nitrogen compounds	0	1,577	0	57	34,173	0	30	1,511,111		41,277
2413	Plastics in primary forms and synthetic rubber	0	38	29	209	7,911	0	0	1,866,667		14,907
242	Other chemical products	0	20,834	10	6,403	74,979	0	4,850	2,813,889		117,206
2421	Pesticides and other agro-chemical products	..	..	..	..	..		..	..		..
2422	Paints, varnishes and similar coatings	..	..	..	..	..		..	..		..
2423	Pharmaceuticals, medicinal chem. & botanical prod.	..	..	..	..	..		..	..		..
2424	Soap and detergents, perfumes etc.	..	..	..	..	..		..	..		..
2429	Other chemical products, nec	..	..	..	..	..		..	..		..
243	Man-made fibres	..	..	..	..	..		..	..		..
25	**RUBBER AND PLASTICS PRODUCTS**	0	399	675	361	9,306	0	0	3,097,222	..	21,891
251	Rubber products	0	123	513	285	3,690	0	0	833,333	..	7,611
2511	Rubber tyres and tubes	..	..	..	..	..		..	..		..
2519	Other rubber products	..	..	..	..	..		..	..		..
252	Plastic products	0	276	162	76	5,616	0	0	2,263,889		14,280
26	**OTHER NON-METALLIC MINERAL PRODUCTS**	31,464	95	238	2,651	30,015	0	0	2,227,778	..	72,483
261	Glass and glass products	0	95	0	399	12,060	0	0	436,111		14,124
269	Non-metallic mineral products, nec	31,464	0	238	2,252	17,955	0	0	1,791,667		58,359
2691	Pottery, china and earthenware	..	..	..	..	..		..	..		..
2692	Refractory ceramic products	..	..	..	..	..		..	..		..
2693	Structural non-refractory clay & ceramic prod.	..	..	..	..	..		..	..		..
2694	Cement, lime and plaster	31,464	0	238	2,252	17,955	0	0	1,791,667		58,359
2695	Articles of concrete, cement and plaster	..	..	..	..	..		..	..		..
2696	Cutting, shaping and finishing of stone	..	..	..	..	..		..	..		..
2699	Other non-metallic mineral products, nec	..	..	..	..	..		..	..		..

ISIC Revision 3 Industry Sector		Solid TJ	LPG TJ	Distiloil TJ	RFO TJ	Gas TJ	Biomass TJ	Steam TJ	Electr MWh	Own Use MWh	TOTAL TJ
27	**BASIC METALS**	122,417	0	1,786	17,404	92,061	0	0	55,419,444	..	433,178
271	Basic iron and steel	111,207	0	1,083	8,854	68,706	0	0	8,683,333	..	221,110
272	Basic precious and non-ferrous metals	11,210	0	703	8,550	23,355	0	0	46,736,111	..	212,068
273	Casting of metals	..	..	..	..	..	..	..	..	..	..
2731	Casting of iron and steel	..	..	..	..	..	..	..	..	..	..
2732	Casting non-ferrous metals	..	..	..	..	..	..	..	..	..	..
28	**FABRICATED METAL PRODUCTS**	38	741	266	181	22,140	0	0	2,819,444	..	33,516
281	Str. metal prod., tanks, reservoirs, steam generators	..	..	..	..	..	..	..	..	..	..
2811	Structural metal products	..	..	..	..	..	..	..	..	..	..
2812	Tanks, reservoirs and containers of metal	..	..	..	..	..	..	..	..	..	..
2813	Steam generators, exc. central heating hot water boilers	..	..	..	..	..	..	..	..	..	..
289	Other fabricated metal products	..	..	..	..	..	..	..	..	..	..
2891	Forging, pressing, stamping & roll-forming of metal	..	..	..	..	..	..	..	..	..	..
2892	Treatment and coating of metals	..	..	..	..	..	..	..	..	..	..
2893	Cutlery, hand tools and general hardware	..	..	..	..	..	..	..	..	..	..
2899	Other fabricated metal products, nec	..	..	..	..	..	..	..	..	..	..
29	**MACHINERY AND EQUIPMENT, NEC**	0	675	209	162	15,093	0	0	3,263,889	..	27,889
291	General purpose machinery	..	..	..	..	..	..	..	..	..	..
2911	Engines and turbines	..	..	..	..	..	..	..	..	..	..
2912	Pumps, compressors, taps and valves	..	..	..	..	..	..	..	..	..	..
2913	Bearings, gears, gearing and driving elements	..	..	..	..	..	..	..	..	..	..
2914	Ovens, furnaces and furnace burners	..	..	..	..	..	..	..	..	..	..
2915	Lifting and handling equipment	..	..	..	..	..	..	..	..	..	..
2919	Other general purpose machinery	..	..	..	..	..	..	..	..	..	..
292	Special purpose machinery	..	..	..	..	..	..	..	..	..	..
2921	Agricultural and forestry machinery	..	..	..	..	..	..	..	..	..	..
2922	Machine-tools	..	..	..	..	..	..	..	..	..	..
2923	Machinery for metallurgy	..	..	..	..	..	..	..	..	..	..
2924	Machinery for mining, quarrying and construction	..	..	..	..	..	..	..	..	..	..
2925	Machinery for food, beverage & tobacco processing	..	..	..	..	..	..	..	..	..	..
2926	Machinery for textile, apparel & leather production	..	..	..	..	..	..	..	..	..	..
2927	Machinery for weapons and ammunition	..	..	..	..	..	..	..	..	..	..
2929	Other special purpose machinery	..	..	..	..	..	..	..	..	..	..
293	Domestic appliances, nec	..	..	..	..	..	..	..	..	..	..
30	**OFFICE, ACCOUNTING & COMPUTING MACHINERY**	..	..	..	..	..	..	..	..	..	..
31	**ELECTRICAL MACHINERY & APPARATUS, NEC**	..	..	..	..	..	..	..	..	..	..
311	Electric motors, generators and transformers	..	..	..	..	..	..	..	..	..	..
312	Electricity distribution and control apparatus	..	..	..	..	..	..	..	..	..	..
313	Insulated wire and cable	..	..	..	..	..	..	..	..	..	..
314	Accumulators, primary cells & primary batteries	..	..	..	..	..	..	..	..	..	..
315	Electric lamps and lighting equipment	..	..	..	..	..	..	..	..	..	..
319	Other electrical equipment, nec	..	..	..	..	..	..	..	..	..	..
32	**RADIO, TV & COMMUNICATION EQUIP. & APP.**	..	..	..	..	..	..	..	..	..	..
321	Electronic valves, tubes, other electronic components	..	..	..	..	..	..	..	..	..	..
322	TV & radio transmitters, apparatus for line telephony	..	..	..	..	..	..	..	..	..	..
323	TV & radio receivers, recording apparatus	..	..	..	..	..	..	..	..	..	..
33	**MEDICAL PRECISION &OPTICAL INSTRUMENTS**	..	..	..	..	..	..	..	..	..	..
331	Medical appliances and instruments	..	..	..	..	..	..	..	..	..	..
3311	Medical, surgical equipment & orthopaedic app.	..	..	..	..	..	..	..	..	..	..
3312	Instruments & appliances for measuring, checking etc	..	..	..	..	..	..	..	..	..	..
3313	Industrial process control equipment	..	..	..	..	..	..	..	..	..	..
332	Optical instruments and photographic equipment	..	..	..	..	..	..	..	..	..	..
333	Watches and clocks	..	..	..	..	..	..	..	..	..	..
34	**MOTOR VEHICLES, TRAILERS & SEMI-TRAILERS**	542	1,131	266	751	33,210	0	0	5,447,222	..	55,510
341	Motor vehicles	266	105	133	57	18,603	0	0	2,136,111	..	26,854
342	Bodies (coachwork) for motor vehicles	..	..	..	..	..	..	..	..	..	..
343	Parts, accessories for motor vehicles & their engines	276	1,026	133	694	14,607	0	0	3,311,111	..	28,656
35	**OTHER TRANSPORT EQUIPMENT**	..	..	..	..	..	..	..	..	..	..
351	Building and repairing of ships and boats	..	..	..	..	..	..	..	..	..	..
3511	Building and repairing of ships	..	..	..	..	..	..	..	..	..	..
3512	Building, repairing of pleasure & sporting boats	..	..	..	..	..	..	..	..	..	..
352	Railway, tramway locomotives & rolling stock	..	..	..	..	..	..	..	..	..	..
353	Aircraft and spacecraft	..	..	..	..	..	..	..	..	..	..
359	Transport equipment, nec	..	..	..	..	..	..	..	..	..	..
3591	Motorcycles	..	..	..	..	..	..	..	..	..	..
3592	Bicycles and invalid carriages	..	..	..	..	..	..	..	..	..	..
3599	Other transport equipment, nec	..	..	..	..	..	..	..	..	..	..
36	**FURNITURE; MANUFACTURING, NEC**	5,804	6,685	13,913	4,223	39,438	0	4,450	683,334	..	76,973
361	Furniture	..	..	..	..	..	..	..	..	..	..
369	Manufacturing, nec	..	..	..	..	..	..	..	..	..	..
3691	Jewellery and related articles	..	..	..	..	..	..	..	..	..	..
3692	Musical instruments	..	..	..	..	..	..	..	..	..	..
3693	Sports goods	..	..	..	..	..	..	..	..	..	..
3694	Games and toys	..	..	..	..	..	..	..	..	..	..
3699	Other manufacturing, nec	..	..	..	..	..	..	..	..	..	..
37	**RECYCLING**	..	..	..	..	..	..	..	..	..	..
371	Recycling of metal waste and scrap	..	..	..	..	..	..	..	..	..	..
372	Recycling of non-metal waste and scrap	..	..	..	..	..	..	..	..	..	..
CERR	**Non-specified industry**	..	..	..	..	..	..	..	..	..	..
15-37	**TOTAL MANUFACTURING**	189,583	42,883	26,448	139,147	822,303	317,758	14,630	167,883,333	..	2,157,132

ISIC Revision 3 Industry Sector	Solid TJ	LPG TJ	Distiloil TJ	RFO TJ	Gas TJ	Biomass TJ	Steam TJ	Electr MWh	Own Use MWh	TOTAL TJ
15 FOOD PRODUCTS AND BEVERAGES	48	589	884	4,465	58,671	0	730	6,588,889	..	89,107
151 Production, processing and preserving (PPP)	0	209	133	2,023	12,780	0	730	1,727,779	..	22,095
1511 PPP of meat and meat products	..	..	..	..	..	..	..	..	..	..
1512 Processing and preserving of fish products	..	..	..	..	..	..	..	..	..	..
1513 Processing, preserving of fruit & vegetables	..	..	..	..	..	..	..	..	..	..
1514 Vegetable and animal oils and fats	..	..	..	..	..	..	..	..	..	..
152 Dairy products	0	228	133	323	5,670	0	0	1,094,444	..	10,294
153 Grain mill prod., starches & prepared animal feeds	..	..	..	..	..	..	..	..	..	..
1531 Grain mill products	..	..	..	..	..	..	..	..	..	..
1532 Starches and starch products	..	..	..	..	..	..	..	..	..	..
1533 Prepared animal feeds	..	..	..	..	..	..	..	..	..	..
154 Other food products	48	133	333	1,805	32,391	0	0	3,133,333	..	45,990
1541 Bakery products	..	..	..	..	..	..	..	..	..	..
1542 Sugar	..	..	..	..	..	..	..	..	..	..
1543 Cocoa, chocolate and sugar confectionery	..	..	..	..	..	..	..	..	..	..
1544 Macaroni, noodles, couscous & similar farinaceous prod.	..	..	..	..	..	..	..	..	..	..
1549 Other food products, nec	..	..	..	..	..	..	..	..	..	..
155 Beverages	0	19	285	314	7,830	0	0	633,333	..	10,728
1551 Distilling, rectifying and blending of spirits	..	..	..	..	..	..	..	..	..	..
1552 Wines	..	..	..	..	..	..	..	..	..	..
1553 Malt liquors and malt	..	..	..	..	..	..	..	..	..	..
1554 Soft drinks; production of mineral waters	..	..	..	..	..	..	..	..	..	..
16 TOBACCO PRODUCTS	0	0	10	0	522	0	0	113,889	..	942
17 TEXTILES	0	57	95	675	16,920	0	710	1,994,444	..	25,637
171 Spinning, weaving and finishing of textiles	0	0	0	342	11,133	0	570	1,308,333	..	16,755
1711 Preparation and spinning of textile fibres	..	..	..	..	..	..	..	..	..	..
1712 Finishing of textiles	..	..	..	..	..	..	..	..	..	..
172 Other textiles	..	..	..	..	..	..	..	..	..	..
1721 Made-up textile articles, except apparel	..	..	..	..	..	..	..	..	..	..
1722 Carpets and rugs	..	..	..	..	..	..	..	..	..	..
1723 Cordage, rope, twine and netting	..	..	..	..	..	..	..	..	..	..
1729 Other textiles, nec	..	..	..	..	..	..	..	..	..	..
173 Knitted and crocheted fabrics and articles	0	57	95	333	5,787	0	140	686,111	..	8,882
18 WEARING APPAREL, DRESSING & DYEING OF FUR	0	29	38	67	2,565	0	0	672,222	..	5,119
181 Wearing apparel, except fur apparel	..	..	..	..	..	..	..	..	..	..
182 Dressing and dyeing of fur; articles of fur	..	..	..	..	..	..	..	..	..	..
19 TANNING & DRESSING OF LEATHER, FOOTWEAR	0	0	0	0	522	0	0	113,889	..	932
191 Tanning and dressing of leather	..	..	..	..	..	..	..	..	..	..
1911 Tanning and dressing of leather	..	..	..	..	..	..	..	..	..	..
1912 Luggage, handbags, saddlery & harness	..	..	..	..	..	..	..	..	..	..
192 Footwear	..	..	..	..	..	..	..	..	..	..
20 WOOD AND WOOD PRODUCTS	0	1,368	4,209	1,454	26,523	0	0	6,130,556	..	55,624
201 Sawmilling and planing of wood	0	237	1,368	285	11,241	0	0	2,247,221	..	21,221
202 Products of wood, cork, straw & plaiting materials	0	1,131	2,841	1,169	15,282	0	0	3,883,335	..	34,403
2021 Veneer sheets	..	..	..	..	..	..	..	..	..	..
2022 Builders' carpentry and joinery	..	..	..	..	..	..	..	..	..	..
2023 Wooden containers	..	..	..	..	..	..	..	..	..	..
2029 Other products of wood	..	..	..	..	..	..	..	..	..	..
21 PAPER AND PAPER PRODUCTS	1,406	0	6,204	52,640	118,116	355,256	3,390	55,969,444	..	738,502
2101 Pulp, paper and paperboard	..	..	..	..	..	..	..	..	..	..
2102 Corrugated paper, paperboard and their containers	..	..	..	..	..	..	..	..	..	..
2109 Other articles of pulp and paperboard	..	..	..	..	..	..	..	..	..	..
22 PUBLISHING, PRINTING & REPRODUCTION	0	48	19	0	4,878	0	0	1,480,556	..	10,275
221 Publishing	..	..	..	..	..	..	..	..	..	..
2211 Publishing of books & brochures	..	..	..	..	..	..	..	..	..	..
2212 Publishing of newspapers and periodicals	..	..	..	..	..	..	..	..	..	..
2213 Publishing of recorded media	..	..	..	..	..	..	..	..	..	..
2219 Other publishing	..	..	..	..	..	..	..	..	..	..
222 Printing and related service activities	..	..	..	..	..	..	..	..	..	..
2221 Printing	..	..	..	..	..	..	..	..	..	..
2222 Service activities related to printing	..	..	..	..	..	..	..	..	..	..
223 Reproduction of recorded media	..	..	..	..	..	..	..	..	..	..
23 COKE, REFINED PETROLEUM PRODUCTS	29,241	3,696	694	36,537	199,109	0	0	4,883,333	..	286,857
231 Coke oven products	..	..	..	..	..	..	..	..	..	..
232 Refined petroleum products	29,241	3,696	694	36,537	199,109	0	0	4,883,333	..	286,857
233 Processing of nuclear fuel	..	..	..	..	..	..	..	..	..	..
24 CHEMICALS & CHEMICAL PRODUCTS	0	31,749	827	4,104	138,879	0	6,180	21,433,333	..	258,899
241 Basic chemicals	0	27,816	323	1,311	77,409	0	4,130	18,775,000	..	178,579
2411 Basic chemicals, exc. fertilizers & nitrogen compounds	0	24,785	304	1,149	20,223	0	1,080	15,197,222	..	102,251
2412 Fertilizers and nitrogen compounds	0	3,021	0	29	39,879	0	20	1,491,667	..	48,319
2413 Plastics in primary forms and synthetic rubber	0	10	19	133	17,307	0	3,030	2,086,111	..	28,009
242 Other chemical products	0	3,933	504	2,793	61,470	0	2,050	2,658,333	..	80,320
2421 Pesticides and other agro-chemical products	..	..	..	..	..	..	..	..	..	..
2422 Paints, varnishes and similar coatings	..	..	..	..	..	..	..	..	..	..
2423 Pharmaceuticals, medicinal chem. & botanical prod.	..	..	..	..	..	..	..	..	..	..
2424 Soap and detergents, perfumes etc.	..	..	..	..	..	..	..	..	..	..
2429 Other chemical products, nec	..	..	..	..	..	..	..	..	..	..
243 Man-made fibres	..	..	..	..	..	..	..	..	..	..
25 RUBBER AND PLASTICS PRODUCTS	0	447	228	1,539	8,523	0	0	3,041,667	..	21,687
251 Rubber products	0	57	0	1,434	3,942	0	0	913,889	..	8,723
2511 Rubber tyres and tubes	..	..	..	..	..	..	..	..	..	..
2519 Other rubber products	..	..	..	..	..	..	..	..	..	..
252 Plastic products	0	390	228	105	4,581	0	0	2,127,778	..	12,964
26 OTHER NON-METALLIC MINERAL PRODUCTS	36,965	76	342	2,565	31,959	0	0	2,311,111	..	80,227
261 Glass and glass products	0	76	28	418	10,575	0	0	441,667	..	12,687
269 Non-metallic mineral products, nec	36,965	0	314	2,147	21,384	0	0	1,869,444	..	67,540
2691 Pottery, china and earthenware	..	..	..	..	..	..	..	..	..	..
2692 Refractory ceramic products	..	..	..	..	..	..	..	..	..	..
2693 Structural non-refractory clay & ceramic prod.	..	..	..	..	..	..	..	..	..	..
2694 Cement, lime and plaster	36,965	0	314	2,147	21,384	0	0	1,869,444	..	67,540
2695 Articles of concrete, cement and plaster	..	..	..	..	..	..	..	..	..	..
2696 Cutting, shaping and finishing of stone	..	..	..	..	..	..	..	..	..	..
2699 Other non-metallic mineral products, nec	..	..	..	..	..	..	..	..	..	..

ISIS Energy Data Programme (IEA/OECD)

ISIC Revision 3 Industry Sector	Solid TJ	LPG TJ	Distiloil TJ	RFO TJ	Gas TJ	Biomass TJ	Steam TJ	Electr MWh	Own Use MWh	TOTAL TJ
27	**BASIC METALS**	126,094	0	3,021	13,509	92,682	0	0	55,552,778	..	435,296
271	Basic iron and steel	113,611	0	1,206	8,255	69,534	0	0	8,655,556	..	223,766
272	Basic precious and non-ferrous metals	12,483	0	1,815	5,254	23,148	0	0	46,897,222	..	211,530
273	Casting of metals	..	..	..	..	..	..	..	..	..	..
2731	Casting of iron and steel	..	..	..	..	..	..	..	..	..	..
2732	Casting non-ferrous metals	..	..	..	..	..	..	..	..	..	..
28	**FABRICATED METAL PRODUCTS**	0	494	181	0	24,894	0	10	2,661,111	..	35,159
281	Str. metal prod., tanks, reservoirs, steam generators	..	..	..	..	..	..	..	..	..	..
2811	Structural metal products	..	..	..	..	..	..	..	..	..	..
2812	Tanks, reservoirs and containers of metal	..	..	..	..	..	..	..	..	..	..
2813	Steam generators, exc. central heating hot water boilers	..	..	..	..	..	..	..	..	..	..
289	Other fabricated metal products	..	..	..	..	..	..	..	..	..	..
2891	Forging, pressing, stamping & roll-forming of metal	..	..	..	..	..	..	..	..	..	..
2892	Treatment and coating of metals	..	..	..	..	..	..	..	..	..	..
2893	Cutlery, hand tools and general hardware	..	..	..	..	..	..	..	..	..	..
2899	Other fabricated metal products, nec	..	..	..	..	..	..	..	..	..	..
29	**MACHINERY AND EQUIPMENT, NEC**	0	295	86	0	14,004	0	0	3,244,444	..	26,065
291	General purpose machinery	..	..	..	..	..	..	..	..	..	..
2911	Engines and turbines	..	..	..	..	..	..	..	..	..	..
2912	Pumps, compressors, taps and valves	..	..	..	..	..	..	..	..	..	..
2913	Bearings, gears, gearing and driving elements	..	..	..	..	..	..	..	..	..	..
2914	Ovens, furnaces and furnace burners	..	..	..	..	..	..	..	..	..	..
2915	Lifting and handling equipment	..	..	..	..	..	..	..	..	..	..
2919	Other general purpose machinery	..	..	..	..	..	..	..	..	..	..
292	Special purpose machinery	..	..	..	..	..	..	..	..	..	..
2921	Agricultural and forestry machinery	..	..	..	..	..	..	..	..	..	..
2922	Machine-tools	..	..	..	..	..	..	..	..	..	..
2923	Machinery for metallurgy	..	..	..	..	..	..	..	..	..	..
2924	Machinery for mining, quarrying and construction	..	..	..	..	..	..	..	..	..	..
2925	Machinery for food, beverage & tobacco processing	..	..	..	..	..	..	..	..	..	..
2926	Machinery for textile, apparel & leather production	..	..	..	..	..	..	..	..	..	..
2927	Machinery for weapons and ammunition	..	..	..	..	..	..	..	..	..	..
2929	Other special purpose machinery	..	..	..	..	..	..	..	..	..	..
293	Domestic appliances, nec	..	..	..	..	..	..	..	..	..	..
30	**OFFICE, ACCOUNTING & COMPUTING MACHINERY**	..	..	..	..	..	..	..	..	..	..
31	**ELECTRICAL MACHINERY & APPARATUS, NEC**	..	..	..	..	..	..	..	..	..	..
311	Electric motors, generators and transformers	..	..	..	..	..	..	..	..	..	..
312	Electricity distribution and control apparatus	..	..	..	..	..	..	..	..	..	..
313	Insulated wire and cable	..	..	..	..	..	..	..	..	..	..
314	Accumulators, primary cells & primary batteries	..	..	..	..	..	..	..	..	..	..
315	Electric lamps and lighting equipment	..	..	..	..	..	..	..	..	..	..
319	Other electrical equipment, nec	..	..	..	..	..	..	..	..	..	..
32	**RADIO, TV & COMMUNICATION EQUIP. & APP.**	..	..	..	..	..	..	..	..	..	..
321	Electronic valves, tubes, other electronic components	..	..	..	..	..	..	..	..	..	..
322	TV & radio transmitters, apparatus for line telephony	..	..	..	..	..	..	..	..	..	..
323	TV & radio receivers, recording apparatus	..	..	..	..	..	..	..	..	..	..
33	**MEDICAL PRECISION &OPTICAL INSTRUMENTS**	..	..	..	..	..	..	..	..	..	..
331	Medical appliances and instruments	..	..	..	..	..	..	..	..	..	..
3311	Medical, surgical equipment & orthopaedic app.	..	..	..	..	..	..	..	..	..	..
3312	Instruments & appliances for measuring, checking etc	..	..	..	..	..	..	..	..	..	..
3313	Industrial process control equipment	..	..	..	..	..	..	..	..	..	..
332	Optical instruments and photographic equipment	..	..	..	..	..	..	..	..	..	..
333	Watches and clocks	..	..	..	..	..	..	..	..	..	..
34	**MOTOR VEHICLES, TRAILERS & SEMI-TRAILERS**	884	1,102	836	1,321	31,392	0	0	5,472,222	..	55,235
341	Motor vehicles	713	1,083	836	152	17,973	0	0	2,072,222	..	28,217
342	Bodies (coachwork) for motor vehicles	..	..	..	..	..	..	..	..	..	..
343	Parts, accessories for motor vehicles & their engines	171	19	0	1,169	13,419	0	0	3,400,000	..	27,018
35	**OTHER TRANSPORT EQUIPMENT**	..	..	..	..	..	..	..	..	..	..
351	Building and repairing of ships and boats	..	..	..	..	..	..	..	..	..	..
3511	Building and repairing of ships	..	..	..	..	..	..	..	..	..	..
3512	Building, repairing of pleasure & sporting boats	..	..	..	..	..	..	..	..	..	..
352	Railway, tramway locomotives & rolling stock	..	..	..	..	..	..	..	..	..	..
353	Aircraft and spacecraft	..	..	..	..	..	..	..	..	..	..
359	Transport equipment, nec	..	..	..	..	..	..	..	..	..	..
3591	Motorcycles	..	..	..	..	..	..	..	..	..	..
3592	Bicycles and invalid carriages	..	..	..	..	..	..	..	..	..	..
3599	Other transport equipment, nec	..	..	..	..	..	..	..	..	..	..
36	**FURNITURE; MANUFACTURING, NEC**	2,858	16,414	11,301	9,830	99,207	0	0	633,334	..	141,890
361	Furniture	..	..	..	..	..	..	..	..	..	..
369	Manufacturing, nec	..	..	..	..	..	..	..	..	..	..
3691	Jewellery and related articles	..	..	..	..	..	..	..	..	..	..
3692	Musical instruments	..	..	..	..	..	..	..	..	..	..
3693	Sports goods	..	..	..	..	..	..	..	..	..	..
3694	Games and toys	..	..	..	..	..	..	..	..	..	..
3699	Other manufacturing, nec	..	..	..	..	..	..	..	..	..	..
37	**RECYCLING**	..	..	..	..	..	..	..	..	..	..
371	Recycling of metal waste and scrap	..	..	..	..	..	..	..	..	..	..
372	Recycling of non-metal waste and scrap	..	..	..	..	..	..	..	..	..	..
CERR	**Non-specified industry**	..	..	..	..	..	..	..	..	..	..
15-37	**TOTAL MANUFACTURING**	197,496	56,364	28,975	128,706	869,366	355,256	11,020	172,297,222	..	2,267,453

ISIC Revision 3 Industry Sector	Solid TJ	LPG TJ	Distiloil TJ	RFO TJ	Gas TJ	Biomass TJ	Steam TJ	Electr MWh	Own Use MWh	TOTAL TJ
15 FOOD PRODUCTS AND BEVERAGES	38	808	1,368	5,073	57,843	0	1,080	6,627,778	..	90,070
151 Production, processing and preserving (PPP)	0	256	161	2,279	13,644	0	990	1,911,112		24,210
1511 PPP of meat and meat products	..	..	..	..	..		..	..		..
1512 Processing and preserving of fish products	..	..	..	..	..		..	..		..
1513 Processing, preserving of fruit & vegetables	..	..	..	..	..		..	..		..
1514 Vegetable and animal oils and fats	..	..	..	..	..		..	..		..
152 Dairy products	0	29	181	428	6,264	0	0	888,889		10,102
153 Grain mill prod., starches & prepared animal feeds	..	..	..	..	..		..	..		..
1531 Grain mill products	..	..	..	..	..		..	..		..
1532 Starches and starch products	..	..	..	..	..		..	..		..
1533 Prepared animal feeds	..	..	..	..	..		..	..		..
154 Other food products	38	504	627	1,834	29,808	0	90	3,244,444		44,581
1541 Bakery products	..	..	..	..	..		..	..		..
1542 Sugar	..	..	..	..	..		..	..		..
1543 Cocoa, chocolate and sugar confectionery	..	..	..	..	..		..	..		..
1544 Macaroni, noodles, couscous & similar farinaceous prod.	..	..	..	..	..		..	..		..
1549 Other food products, nec	..	..	..	..	..		..	..		..
155 Beverages	0	19	399	532	8,127	0	0	583,333		11,177
1551 Distilling, rectifying and blending of spirits	..	..	..	..	..		..	..		..
1552 Wines	..	..	..	..	..		..	..		..
1553 Malt liquors and malt	..	..	..	..	..		..	..		..
1554 Soft drinks; production of mineral waters	..	..	..	..	..		..	..		..
16 TOBACCO PRODUCTS	0	10	10	0	477	0	0	116,667		917
17 TEXTILES	0	48	86	1,321	15,966	0	680	1,941,667		25,091
171 Spinning, weaving and finishing of textiles	0	0	0	922	10,413	0	550	1,277,778		16,485
1711 Preparation and spinning of textile fibres	..	..	..	..	..		..	..		..
1712 Finishing of textiles	..	..	..	..	..		..	..		..
172 Other textiles	..	..	..	..	..		..	..		..
1721 Made-up textile articles, except apparel	..	..	..	..	..		..	..		..
1722 Carpets and rugs	..	..	..	..	..		..	..		..
1723 Cordage, rope, twine and netting	..	..	..	..	..		..	..		..
1729 Other textiles, nec	..	..	..	..	..		..	..		..
173 Knitted and crocheted fabrics and articles	0	48	86	399	5,553	0	130	663,889		8,606
18 WEARING APPAREL, DRESSING & DYEING OF FUR	0	19	29	67	2,610	0	0	666,667		5,125
181 Wearing apparel, except fur apparel	..	..	..	..	..		..	..		..
182 Dressing and dyeing of fur; articles of fur	..	..	..	..	..		..	..		..
19 TANNING & DRESSING OF LEATHER, FOOTWEAR	0	0	0	0	558	0	0	122,222		998
191 Tanning and dressing of leather	..	..	..	..	..		..	..		..
1911 Tanning and dressing of leather	..	..	..	..	..		..	..		..
1912 Luggage, handbags, saddlery & harness	..	..	..	..	..		..	..		..
192 Footwear	..	..	..	..	..		..	..		..
20 WOOD AND WOOD PRODUCTS	0	1,292	4,076	1,226	23,283	0	0	5,766,667		50,637
201 Sawmilling and planing of wood	0	275	1,454	86	11,655	0	0	2,388,889		22,070
202 Products of wood, cork, straw & plaiting materials	0	1,017	2,622	1,140	11,628	0	0	3,377,778		28,567
2021 Veneer sheets	..	..	..	..	..		..	..		..
2022 Builders' carpentry and joinery	..	..	..	..	..		..	..		..
2023 Wooden containers	..	..	..	..	..		..	..		..
2029 Other products of wood	..	..	..	..	..		..	..		..
21 PAPER AND PAPER PRODUCTS	2,005	0	7,135	57,333	111,069	331,436	5,680	54,838,889		712,078
2101 Pulp, paper and paperboard	..	..	..	..	..		..	..		..
2102 Corrugated paper, paperboard and their containers	..	..	..	..	..		..	..		..
2109 Other articles of pulp and paperboard	..	..	..	..	..		..	..		..
22 PUBLISHING, PRINTING & REPRODUCTION	0	29	19	0	4,293	0	0	1,344,444		9,181
221 Publishing	..	..	..	..	..		..	..		..
2211 Publishing of books & brochures	..	..	..	..	..		..	..		..
2212 Publishing of newspapers and periodicals	..	..	..	..	..		..	..		..
2213 Publishing of recorded media	..	..	..	..	..		..	..		..
2219 Other publishing	..	..	..	..	..		..	..		..
222 Printing and related service activities	..	..	..	..	..		..	..		..
2221 Printing	..	..	..	..	..		..	..		..
2222 Service activities related to printing	..	..	..	..	..		..	..		..
223 Reproduction of recorded media	..	..	..	..	..		..	..		..
23 COKE, REFINED PETROLEUM PRODUCTS	30,030	5,444	846	38,608	207,570	0	0	4,875,000		300,048
231 Coke oven products	..	..	..	..	..		..	..		..
232 Refined petroleum products	30,030	5,444	846	38,608	207,570	0	0	4,875,000		300,048
233 Processing of nuclear fuel	..	..	..	..	..		..	..		..
24 CHEMICALS & CHEMICAL PRODUCTS	0	11,714	903	4,304	136,107	0	14,010	20,955,556		242,478
241 Basic chemicals	0	11,714	475	2,822	73,872	0	6,290	18,294,444		161,033
2411 Basic chemicals, exc. fertilizers & nitrogen compounds	0	10,868	446	2,375	19,089	0	3,480	14,686,111		89,128
2412 Fertilizers and nitrogen compounds	0	836	0	19	38,007	0	20	1,458,333		44,132
2413 Plastics in primary forms and synthetic rubber	0	10	29	428	16,776	0	2,790	2,150,000		27,773
242 Other chemical products	0	0	428	1,482	62,235	0	7,720	2,661,112		81,445
2421 Pesticides and other agro-chemical products	..	..	..	..	..		..	..		..
2422 Paints, varnishes and similar coatings	..	..	..	..	..		..	..		..
2423 Pharmaceuticals, medicinal chem. & botanical prod.	..	..	..	..	..		..	..		..
2424 Soap and detergents, perfumes etc.	..	..	..	..	..		..	..		..
2429 Other chemical products, nec	..	..	..	..	..		..	..		..
243 Man-made fibres	..	..	..	..	..		..	..		..
25 RUBBER AND PLASTICS PRODUCTS	0	247	266	1,568	9,234	0	0	3,077,778		22,395
251 Rubber products	0	38	114	1,501	4,869	0	0	919,443		9,832
2511 Rubber tyres and tubes	..	..	..	..	..		..	..		..
2519 Other rubber products	..	..	..	..	..		..	..		..
252 Plastic products	0	209	152	67	4,365	0	0	2,158,335		12,563
26 OTHER NON-METALLIC MINERAL PRODUCTS	37,117	57	466	2,793	29,556	0	0	2,291,667		78,239
261 Glass and glass products	0	47	10	427	10,170	0	0	416,667		12,154
269 Non-metallic mineral products, nec	37,117	10	456	2,366	19,386	0	0	1,875,000		66,085
2691 Pottery, china and earthenware	..	..	..	..	..		..	..		..
2692 Refractory ceramic products	..	..	..	..	..		..	..		..
2693 Structural non-refractory clay & ceramic prod.	..	..	..	..	..		..	..		..
2694 Cement, lime and plaster	37,117	10	456	2,366	19,386	0	0	1,875,000		66,085
2695 Articles of concrete, cement and plaster	..	..	..	..	..		..	..		..
2696 Cutting, shaping and finishing of stone	..	..	..	..	..		..	..		..
2699 Other non-metallic mineral products, nec	..	..	..	..	..		..	..		..

ISIC Revision 3 Industry Sector	Solid TJ	LPG TJ	Distiloil TJ	RFO TJ	Gas TJ	Biomass TJ	Steam TJ	Electr MWh	Own Use MWh	TOTAL TJ
27 **BASIC METALS**	130,977	0	3,572	17,993	99,990	0	0	57,991,667	..	461,302
271 Basic iron and steel	117,259	0	1,149	8,369	75,339	0	0	8,888,889	..	234,116
272 Basic precious and non-ferrous metals	13,718	0	2,423	9,624	24,651	0	0	49,102,778	..	227,186
273 Casting of metals	..	..	..	..	..	..	..	..	..	..
2731 Casting of iron and steel	..	..	..	..	..	..	..	..	..	..
2732 Casting non-ferrous metals	..	..	..	..	..	..	..	..	..	..
28 **FABRICATED METAL PRODUCTS**	0	504	219	0	23,094	0	0	3,202,778	..	35,347
281 Str. metal prod., tanks, reservoirs, steam generators	..	..	..	..	..	..	..	..	..	..
2811 Structural metal products	..	..	..	..	..	..	..	..	..	..
2812 Tanks, reservoirs and containers of metal	..	..	..	..	..	..	..	..	..	..
2813 Steam generators, exc. central heating hot water boilers	..	..	..	..	..	..	..	..	..	..
289 Other fabricated metal products	..	..	..	..	..	..	..	..	..	..
2891 Forging, pressing, stamping & roll-forming of metal	..	..	..	..	..	..	..	..	..	..
2892 Treatment and coating of metals	..	..	..	..	..	..	..	..	..	..
2893 Cutlery, hand tools and general hardware	..	..	..	..	..	..	..	..	..	..
2899 Other fabricated metal products, nec	..	..	..	..	..	..	..	..	..	..
29 **MACHINERY AND EQUIPMENT, NEC**	0	238	76	86	14,022	0	0	3,138,889	..	25,722
291 General purpose machinery	..	..	..	..	..	..	..	..	..	..
2911 Engines and turbines	..	..	..	..	..	..	..	..	..	..
2912 Pumps, compressors, taps and valves	..	..	..	..	..	..	..	..	..	..
2913 Bearings, gears, gearing and driving elements	..	..	..	..	..	..	..	..	..	..
2914 Ovens, furnaces and furnace burners	..	..	..	..	..	..	..	..	..	..
2915 Lifting and handling equipment	..	..	..	..	..	..	..	..	..	..
2919 Other general purpose machinery	..	..	..	..	..	..	..	..	..	..
292 Special purpose machinery	..	..	..	..	..	..	..	..	..	..
2921 Agricultural and forestry machinery	..	..	..	..	..	..	..	..	..	..
2922 Machine-tools	..	..	..	..	..	..	..	..	..	..
2923 Machinery for metallurgy	..	..	..	..	..	..	..	..	..	..
2924 Machinery for mining, quarrying and construction	..	..	..	..	..	..	..	..	..	..
2925 Machinery for food, beverage & tobacco processing	..	..	..	..	..	..	..	..	..	..
2926 Machinery for textile, apparel & leather production	..	..	..	..	..	..	..	..	..	..
2927 Machinery for weapons and ammunition	..	..	..	..	..	..	..	..	..	..
2929 Other special purpose machinery	..	..	..	..	..	..	..	..	..	..
293 Domestic appliances, nec	..	..	..	..	..	..	..	..	..	..
30 **OFFICE, ACCOUNTING & COMPUTING MACHINERY**										
31 **ELECTRICAL MACHINERY & APPARATUS, NEC**	..	..	..	..	..	..	..	..	..	
311 Electric motors, generators and transformers	..	..	..	..	..	..	..	..	..	
312 Electricity distribution and control apparatus	..	..	..	..	..	..	..	..	..	
313 Insulated wire and cable	..	..	..	..	..	..	..	..	..	
314 Accumulators, primary cells & primary batteries	..	..	..	..	..	..	..	..	..	
315 Electric lamps and lighting equipment	..	..	..	..	..	..	..	..	..	
319 Other electrical equipment, nec	..	..	..	..	..	..	..	..	..	
32 **RADIO, TV & COMMUNICATION EQUIP. & APP.**	..	..	..	..	..	..	..	..	..	
321 Electronic valves, tubes, other electronic components	..	..	..	..	..	..	..	..	..	
322 TV & radio transmitters, apparatus for line telephony	..	..	..	..	..	..	..	..	..	
323 TV & radio receivers, recording apparatus	..	..	..	..	..	..	..	..	..	
33 **MEDICAL PRECISION &OPTICAL INSTRUMENTS**	..	..	..	..	..	..	..	..	..	
331 Medical appliances and instruments	..	..	..	..	..	..	..	..	..	
3311 Medical, surgical equipment & orthopaedic app.	..	..	..	..	..	..	..	..	..	
3312 Instruments & appliances for measuring, checking etc	..	..	..	..	..	..	..	..	..	
3313 Industrial process control equipment	..	..	..	..	..	..	..	..	..	
332 Optical instruments and photographic equipment	..	..	..	..	..	..	..	..	..	
333 Watches and clocks	..	..	..	..	..	..	..	..	..	
34 **MOTOR VEHICLES, TRAILERS & SEMI-TRAILERS**	874	1,188	884	1,273	34,326	0	0	5,508,333	..	58,375
341 Motor vehicles	703	1,159	333	342	19,629	0	0	2,075,000	..	29,636
342 Bodies (coachwork) for motor vehicles	..	..	..	..	..	..	..	..	..	..
343 Parts, accessories for motor vehicles & their engines	171	29	551	931	14,697	0	0	3,433,333	..	28,739
35 **OTHER TRANSPORT EQUIPMENT**	..	..	..	..	..	..	..	..	..	..
351 Building and repairing of ships and boats	..	..	..	..	..	..	..	..	..	..
3511 Building and repairing of ships	..	..	..	..	..	..	..	..	..	..
3512 Building, repairing of pleasure & sporting boats	..	..	..	..	..	..	..	..	..	..
352 Railway, tramway locomotives & rolling stock	..	..	..	..	..	..	..	..	..	..
353 Aircraft and spacecraft	..	..	..	..	..	..	..	..	..	..
359 Transport equipment, nec	..	..	..	..	..	..	..	..	..	..
3591 Motorcycles	..	..	..	..	..	..	..	..	..	..
3592 Bicycles and invalid carriages	..	..	..	..	..	..	..	..	..	..
3599 Other transport equipment, nec	..	..	..	..	..	..	..	..	..	..
36 **FURNITURE; MANUFACTURING, NEC**	3,874	262	11,908	3,626	101,232	0	0	597,220	..	123,052
361 Furniture	..	..	..	..	..	..	..	..	..	..
369 Manufacturing, nec	..	..	..	..	..	..	..	..	..	..
3691 Jewellery and related articles	..	..	..	..	..	..	..	..	..	..
3692 Musical instruments	..	..	..	..	..	..	..	..	..	..
3693 Sports goods	..	..	..	..	..	..	..	..	..	..
3694 Games and toys	..	..	..	..	..	..	..	..	..	..
3699 Other manufacturing, nec	..	..	..	..	..	..	..	..	..	..
37 **RECYCLING**	..	..	..	..	..	..	..	..	..	
371 Recycling of metal waste and scrap	..	..	..	..	..	..	..	..	..	
372 Recycling of non-metal waste and scrap	..	..	..	..	..	..	..	..	..	
CERR **Non-specified industry**										
15-37 **TOTAL MANUFACTURING**	204,915	21,860	31,863	135,271	871,230	331,436	21,450	173,063,889	..	2,241,055

ISIC Revision 3 Industry Sector	Solid TJ	LPG TJ	Distiloil TJ	RFO TJ	Gas TJ	Biomass TJ	Steam TJ	Electr MWh	Own Use MWh	TOTAL TJ
15 FOOD PRODUCTS AND BEVERAGES	14,596	13	4,530	3,570	12,210	..	20,047	1,234,845	..	59,411
151 Production, processing and preserving (PPP)	3,691	4	557	638	2,846	..	6,467	370,908	..	15,538
1511 PPP of meat and meat products	..	..	..	..	..	..	..	..	..	..
1512 Processing and preserving of fish products	..	..	..	..	..	..	..	..	..	..
1513 Processing, preserving of fruit & vegetables	..	..	..	..	..	..	..	..	..	..
1514 Vegetable and animal oils and fats	..	..	..	..	..	..	..	..	..	..
152 Dairy products	1,187	0	370	999	1,407	..	2,711	180,732	..	7,325
153 Grain mill prod., starches & prepared animal feeds	556	0	272	5	478	..	219	116,536	..	1,950
1531 Grain mill products	..	..	..	..	..	..	..	..	..	..
1532 Starches and starch products	..	..	..	..	..	..	..	..	..	..
1533 Prepared animal feeds	..	..	..	..	..	..	..	..	..	..
154 Other food products	7,578	2	3,124	1,661	3,862	..	6,261	315,008	..	23,622
1541 Bakery products	..	..	..	..	..	..	..	..	..	..
1542 Sugar	..	..	..	..	..	..	..	..	..	..
1543 Cocoa, chocolate and sugar confectionery	..	..	..	..	..	..	..	..	..	..
1544 Macaroni, noodles, couscous & similar farinaceous prod.	..	..	..	..	..	..	..	..	..	..
1549 Other food products, nec	..	..	..	..	..	..	..	..	..	..
155 Beverages	1,584	7	207	267	3,617	..	4,389	251,661	..	10,977
1551 Distilling, rectifying and blending of spirits	..	..	..	..	..	..	..	..	..	..
1552 Wines	..	..	..	..	..	..	..	..	..	..
1553 Malt liquors and malt	..	..	..	..	..	..	..	..	..	..
1554 Soft drinks; production of mineral waters	..	..	..	..	..	..	..	..	..	..
16 TOBACCO PRODUCTS	0	0	15	22	181	..	186	9,897	..	440
17 TEXTILES	11,184	3	1,045	2,625	2,134	..	12,601	967,590	..	33,075
171 Spinning, weaving and finishing of textiles	8,525	2	398	2,082	833	..	9,174	749,166	..	23,711
1711 Preparation and spinning of textile fibres	..	..	..	..	..	..	..	..	..	..
1712 Finishing of textiles	..	..	..	..	..	..	..	..	..	..
172 Other textiles	1,488	1	86	393	1,030	..	2,564	137,798	..	6,058
1721 Made-up textile articles, except apparel	..	..	..	..	..	..	..	..	..	..
1722 Carpets and rugs	..	..	..	..	..	..	..	..	..	..
1723 Cordage, rope, twine and netting	..	..	..	..	..	..	..	..	..	..
1729 Other textiles, nec	..	..	..	..	..	..	..	..	..	..
173 Knitted and crocheted fabrics and articles	1,171	0	561	150	271	..	863	80,626	..	3,306
18 WEARING APPAREL, DRESSING & DYEING OF FUR	368	0	57	50	699	..	711	64,258	..	2,116
181 Wearing apparel, except fur apparel	363	0	44	50	676	..	633	60,590	..	1,984
182 Dressing and dyeing of fur; articles of fur	5	0	13	0	23	..	78	3,668	..	132
19 TANNING & DRESSING OF LEATHER, FOOTWEAR	1,001	0	312	417	266	..	2,340	134,383	..	4,820
191 Tanning and dressing of leather	631	0	248	108	99	..	1,689	65,340	..	3,010
1911 Tanning and dressing of leather	..	..	..	..	..	..	..	..	..	..
1912 Luggage, handbags, saddlery & harness	..	..	..	..	..	..	..	..	..	..
192 Footwear	370	0	64	309	167	..	651	69,043	..	1,810
20 WOOD AND WOOD PRODUCTS	4,018	0	243	447	447	..	3,220	278,149	..	9,376
201 Sawmilling and planing of wood	1,264	0	201	169	3	..	1,003	93,830	..	2,978
202 Products of wood, cork, straw & plaiting materials	2,754	0	42	278	444	..	2,217	184,319	..	6,399
2021 Veneer sheets	..	..	..	..	..	..	..	..	..	..
2022 Builders' carpentry and joinery	..	..	..	..	..	..	..	..	..	..
2023 Wooden containers	..	..	..	..	..	..	..	..	..	..
2029 Other products of wood	..	..	..	..	..	..	..	..	..	..
21 PAPER AND PAPER PRODUCTS	23,010	2	7,436	3,247	1,793	..	18,047	1,295,825	..	58,200
2101 Pulp, paper and paperboard	..	..	..	..	..	..	..	..	..	..
2102 Corrugated paper, paperboard and their containers	..	..	..	..	..	..	..	..	..	..
2109 Other articles of pulp and paperboard	..	..	..	..	..	..	..	..	..	..
22 PUBLISHING, PRINTING & REPRODUCTION	103	0	97	0	291	..	268	767,116	..	3,521
221 Publishing	14	0	8	0	18	..	33	4,983	..	91
2211 Publishing of books & brochures	..	..	..	..	..	..	..	..	..	..
2212 Publishing of newspapers and periodicals	..	..	..	..	..	..	..	..	..	..
2213 Publishing of recorded media	..	..	..	..	..	..	..	..	..	..
2219 Other publishing	..	..	..	..	..	..	..	..	..	..
222 Printing and related service activities	89	0	88	0	220	..	182	753,316	..	3,291
2221 Printing	..	..	..	..	..	..	..	..	..	..
2222 Service activities related to printing	..	..	..	..	..	..	..	..	..	..
223 Reproduction of recorded media	0	0	1	0	53	..	53	8,817	..	139
23 COKE, REFINED PETROLEUM PRODUCTS	41,930	58	6,188	1,130	8,698	..	28,832	1,761,877	..	93,179
231 Coke oven products	0	0	6,188	0	0	..	0	0	..	6,188
232 Refined petroleum products	41,930	58	0	1,130	8,698	..	28,832	1,761,877	..	86,991
233 Processing of nuclear fuel	..	..	..	..	..	..	..	..	..	..
24 CHEMICALS & CHEMICAL PRODUCTS	18,899	35	223	799	5,826	..	17,343	1,772,561	..	49,506
241 Basic chemicals	16,032	28	148	753	3,903	..	14,665	1,574,707	..	41,198
2411 Basic chemicals, exc. fertilizers & nitrogen compounds	..	..	..	..	..	..	..	..	..	..
2412 Fertilizers and nitrogen compounds	..	..	..	..	..	..	..	..	..	..
2413 Plastics in primary forms and synthetic rubber	..	..	..	..	..	..	..	..	..	..
242 Other chemical products	370	7	49	46	1,923	..	1,825	133,888	..	4,702
2421 Pesticides and other agro-chemical products	..	..	..	..	..	..	..	..	..	..
2422 Paints, varnishes and similar coatings	..	..	..	..	..	..	..	..	..	..
2423 Pharmaceuticals, medicinal chem. & botanical prod.	..	..	..	..	..	..	..	..	..	..
2424 Soap and detergents, perfumes etc.	..	..	..	..	..	..	..	..	..	..
2429 Other chemical products, nec	..	..	..	..	..	..	..	..	..	..
243 Man-made fibres	2,497	0	26	0	0	..	853	63,966	..	3,606
25 RUBBER AND PLASTICS PRODUCTS	1,069	0	221	705	1,597	..	3,912	355,319	..	8,783
251 Rubber products	510	0	32	58	1,075	..	2,485	173,992	..	4,786
2511 Rubber tyres and tubes	..	..	..	..	..	..	..	..	..	..
2519 Other rubber products	..	..	..	..	..	..	..	..	..	..
252 Plastic products	559	0	189	647	522	..	1,427	181,327	..	3,997
26 OTHER NON-METALLIC MINERAL PRODUCTS	15,941	9	1,864	10,804	25,495	..	9,396	2,012,308	..	70,753
261 Glass and glass products	4,096	7	60	240	10,006	..	2,648	710,433	..	19,615
269 Non-metallic mineral products, nec	11,845	2	1,804	10,564	15,489	..	6,748	1,301,875	..	51,139
2691 Pottery, china and earthenware	..	..	..	..	..	..	..	..	..	..
2692 Refractory ceramic products	..	..	..	..	..	..	..	..	..	..
2693 Structural non-refractory clay & ceramic prod.	..	..	..	..	..	..	..	..	..	..
2694 Cement, lime and plaster	..	..	..	..	..	..	..	..	..	..
2695 Articles of concrete, cement and plaster	..	..	..	..	..	..	..	..	..	..
2696 Cutting, shaping and finishing of stone	..	..	..	..	..	..	..	..	..	..
2699 Other non-metallic mineral products, nec	..	..	..	..	..	..	..	..	..	..

ISIC Revision 3 Industry Sector		Solid TJ	LPG TJ	Distiloil TJ	RFO TJ	Gas TJ	Biomass TJ	Steam TJ	Electr MWh	Own Use MWh	TOTAL TJ
27	BASIC METALS	213,886	130	826	6,785	11,334	..	23,699	3,481,813	..	269,195
271	Basic iron and steel	211,762	58	272	6,577	9,701	..	21,806	3,064,860	..	261,209
272	Basic precious and non-ferrous metals	792	70	489	79	504	..	936	172,914	..	3,492
273	Casting of metals	1,332	2	65	129	1,129	..	957	244,039	..	4,493
2731	Casting of iron and steel	..	..	..	..	..	..	..	..	..	..
2732	Casting non-ferrous metals	..	..	..	..	..	..	..	..	..	..
28	FABRICATED METAL PRODUCTS	3,219	44	404	391	2,766	..	4,183	637,791	..	13,303
281	Str. metal prod., tanks, reservoirs, steam generators	1,039	1	198	249	896	..	1,136	216,481	..	4,298
2811	Structural metal products	..	..	..	..	..	..	..	..	..	..
2812	Tanks, reservoirs and containers of metal	..	..	..	..	..	..	..	..	..	..
2813	Steam generators, exc. central heating hot water boilers	..	..	..	..	..	..	..	..	..	..
289	Other fabricated metal products	2,180	43	206	142	1,870	..	3,047	421,310	..	9,005
2891	Forging, pressing, stamping & roll-forming of metal	..	..	..	..	..	..	..	..	..	..
2892	Treatment and coating of metals	..	..	..	..	..	..	..	..	..	..
2893	Cutlery, hand tools and general hardware	..	..	..	..	..	..	..	..	..	..
2899	Other fabricated metal products, nec	..	..	..	..	..	..	..	..	..	..
29	MACHINERY AND EQUIPMENT, NEC	11,948	36	801	1,521	7,984	..	12,306	1,154,660	..	38,753
291	General purpose machinery	3,194	0	186	243	2,841	..	4,039	403,281	..	11,955
2911	Engines and turbines	..	..	..	..	..	..	..	..	..	..
2912	Pumps, compressors, taps and valves	..	..	..	..	..	..	..	..	..	..
2913	Bearings, gears, gearing and driving elements	..	..	..	..	..	..	..	..	..	..
2914	Ovens, furnaces and furnace burners	..	..	..	..	..	..	..	..	..	..
2915	Lifting and handling equipment	..	..	..	..	..	..	..	..	..	..
2919	Other general purpose machinery	..	..	..	..	..	..	..	..	..	..
292	Special purpose machinery	8,002	36	613	1,278	5,016	..	8,140	698,962	..	25,601
2921	Agricultural and forestry machinery	..	..	..	..	..	..	..	..	..	..
2922	Machine-tools	..	..	..	..	..	..	..	..	..	..
2923	Machinery for metallurgy	..	..	..	..	..	..	..	..	..	..
2924	Machinery for mining, quarrying and construction	..	..	..	..	..	..	..	..	..	..
2925	Machinery for food, beverage & tobacco processing	..	..	..	..	..	..	..	..	..	..
2926	Machinery for textile, apparel & leather production	..	..	..	..	..	..	..	..	..	..
2927	Machinery for weapons and ammunition	..	..	..	..	..	..	..	..	..	..
2929	Other special purpose machinery	..	..	..	..	..	..	..	..	..	..
293	Domestic appliances, nec	752	0	2	0	127	..	127	52,417	..	1,197
30	OFFICE, ACCOUNTING & COMPUTING MACHINERY	190	0	1	14	11	..	174	12,754	..	436
31	ELECTRICAL MACHINERY & APPARATUS, NEC	935	13	51	75	2,297	..	2,927	311,685	..	7,420
311	Electric motors, generators and transformers	362	1	14	71	889	..	1,246	131,356	..	3,056
312	Electricity distribution and control apparatus	327	2	24	0	381	..	418	44,468	..	1,312
313	Insulated wire and cable	27	0	0	0	251	..	495	58,844	..	985
314	Accumulators, primary cells & primary batteries	70	10	2	0	86	..	77	20,819	..	320
315	Electric lamps and lighting equipment	15	0	1	0	104	..	66	15,129	..	240
319	Other electrical, nec	134	0	10	4	586	..	625	41,069	..	1,507
32	RADIO, TV & COMMUNICATION EQUIP. & APP.	465	3	54	83	255	..	583	77,629	..	1,722
321	Electronic valves, tubes, other electronic components	426	3	8	63	236	..	461	52,533	..	1,386
322	TV & radio transmitters, apparatus for line telephony	28	0	44	20	12	..	73	19,613	..	248
323	TV & radio receivers, recording apparatus	11	0	2	0	7	..	49	5,483	..	89
33	MEDICAL PRECISION &OPTICAL INSTRUMENTS	147	0	22	7	410	..	566	58,825	..	1,364
331	Medical appliances and instruments	143	0	21	0	332	..	371	45,294	..	1,030
3311	Medical, surgical equipment & orthopaedic app.	..	..	..	..	..	..	..	..	..	..
3312	Instruments & appliances for measuring, checking etc	..	..	..	..	..	..	..	..	..	..
3313	Industrial process control equipment	..	..	..	..	..	..	..	..	..	..
332	Optical instruments and photographic equipment	2	0	1	0	14	..	115	8,584	..	163
333	Watches and clocks	2	0	0	7	64	..	80	4,947	..	171
34	MOTOR VEHICLES, TRAILERS & SEMI-TRAILERS	6,920	2	186	545	4,877	..	7,813	627,466	..	22,602
341	Motor vehicles	5,250	0	101	501	4,169	..	5,666	448,401	..	17,301
342	Bodies (coachwork) for motor vehicles	352	0	6	0	218	..	362	17,710	..	1,002
343	Parts, accessories for motor vehicles & their engines	1,318	2	79	44	490	..	1,785	161,355	..	4,299
35	OTHER TRANSPORT EQUIPMENT	2,513	2	1,478	321	2,422	..	3,969	336,733	..	11,917
351	Building and repairing of ships and boats	22	0	2	0	39	..	45	3,926	..	122
3511	Building and repairing of ships	..	..	..	..	..	..	..	..	..	..
3512	Building, repairing of pleasure & sporting boats	..	..	..	..	..	..	..	..	..	..
352	Railway, tramway locomotives & rolling stock	1,586	0	1,420	313	1,670	..	2,426	220,441	..	8,209
353	Aircraft and spacecraft	619	0	7	8	582	..	1,074	62,748	..	2,516
359	Transport equipment, nec	286	2	49	0	131	..	424	49,618	..	1,071
3591	Motorcycles	..	..	..	..	..	..	..	..	..	..
3592	Bicycles and invalid carriages	..	..	..	..	..	..	..	..	..	..
3599	Other transport equipment, nec	..	..	..	..	..	..	..	..	..	..
36	FURNITURE; MANUFACTURING, NEC	3,607	3	459	235	2,057	..	3,007	252,469	..	10,277
361	Furniture	2,520	0	97	114	1,402	..	1,965	113,204	..	6,506
369	Manufacturing, nec	1,087	3	362	121	655	..	1,042	139,265	..	3,771
3691	Jewellery and related articles	..	..	..	..	..	..	..	..	..	..
3692	Musical instruments	..	..	..	..	..	..	..	..	..	..
3693	Sports goods	..	..	..	..	..	..	..	..	..	..
3694	Games and toys	..	..	..	..	..	..	..	..	..	..
3699	Other manufacturing, nec	..	..	..	..	..	..	..	..	..	..
37	RECYCLING	53	0	101	0	183	..	68	41,258	..	554
371	Recycling of metal waste and scrap	49	0	89	0	161	..	61	39,776	..	503
372	Recycling of non-metal waste and scrap	4	0	12	0	22	..	7	1,482	..	50
CERR	Non-specified industry	..	..	..	..	..	..	..	..	..	..
15-37	TOTAL MANUFACTURING	376,002	353	26,614	33,793	94,233	..	176,198	17,647,211	..	770,723

ISIC Revision 3 Industry Sector	Solid TJ	LPG TJ	Distiloil TJ	RFO TJ	Gas TJ	Biomass TJ	Steam TJ	Electr MWh	Own Use MWh	TOTAL TJ
15 FOOD PRODUCTS AND BEVERAGES	10,974	19	2,430	2,811	13,287	..	18,480	1,258,063	..	52,530
151 Production, processing and preserving (PPP)	3,227	6	561	392	2,955	..	6,002	402,428	..	14,592
1511 PPP of meat and meat products	..	..	..	..	..	..	..	..	..	..
1512 Processing and preserving of fish products	..	..	..	..	..	..	..	..	..	..
1513 Processing, preserving of fruit & vegetables	..	..	..	..	..	..	..	..	..	..
1514 Vegetable and animal oils and fats	..	..	..	..	..	..	..	..	..	..
152 Dairy products	787	0	293	830	1,835	..	2,505	179,031	..	6,895
153 Grain mill prod., starches & prepared animal feeds	338	0	335	29	772	..	210	131,884	..	2,159
1531 Grain mill products	..	..	..	..	..	..	..	..	..	..
1532 Starches and starch products	..	..	..	..	..	..	..	..	..	..
1533 Prepared animal feeds	..	..	..	..	..	..	..	..	..	..
154 Other food products	5,193	2	1,086	1,293	3,835	..	4,606	258,987	..	16,947
1541 Bakery products	..	..	..	..	..	..	..	..	..	..
1542 Sugar	..	..	..	..	..	..	..	..	..	..
1543 Cocoa, chocolate and sugar confectionery	..	..	..	..	..	..	..	..	..	..
1544 Macaroni, noodles, couscous & similar farinaceous prod.	..	..	..	..	..	..	..	..	..	..
1549 Other food products, nec	..	..	..	..	..	..	..	..	..	..
155 Beverages	1,429	11	155	267	3,890	..	5,157	285,733	..	11,938
1551 Distilling, rectifying and blending of spirits	..	..	..	..	..	..	..	..	..	..
1552 Wines	..	..	..	..	..	..	..	..	..	..
1553 Malt liquors and malt	..	..	..	..	..	..	..	..	..	..
1554 Soft drinks; production of mineral waters	..	..	..	..	..	..	..	..	..	..
16 TOBACCO PRODUCTS	0	0	1	0	174	..	0	14,088	..	226
17 TEXTILES	9,190	4	2,022	2,672	2,127	..	11,499	1,065,282	..	31,349
171 Spinning, weaving and finishing of textiles	6,872	3	207	2,092	948	..	8,418	845,100	..	21,582
1711 Preparation and spinning of textile fibres	..	..	..	..	..	..	..	..	..	..
1712 Finishing of textiles	..	..	..	..	..	..	..	..	..	..
172 Other textiles	1,485	1	1,778	497	952	..	2,115	154,903	..	7,386
1721 Made-up textile articles, except apparel	..	..	..	..	..	..	..	..	..	..
1722 Carpets and rugs	..	..	..	..	..	..	..	..	..	..
1723 Cordage, rope, twine and netting	..	..	..	..	..	..	..	..	..	..
1729 Other textiles, nec	..	..	..	..	..	..	..	..	..	..
173 Knitted and crocheted fabrics and articles	833	0	37	83	227	..	966	65,279	..	2,381
18 WEARING APPAREL, DRESSING & DYEING OF FUR	451	1	41	0	767	..	1,041	69,616	..	2,552
181 Wearing apparel, except fur apparel	448	1	34	0	697	..	965	66,603	..	2,385
182 Dressing and dyeing of fur; articles of fur	3	0	7	0	70	..	76	3,013	..	167
19 TANNING & DRESSING OF LEATHER, FOOTWEAR	655	0	109	351	370	..	1,498	110,685	..	3,381
191 Tanning and dressing of leather	372	0	50	48	237	..	804	42,446	..	1,664
1911 Tanning and dressing of leather	..	..	..	..	..	..	..	..	..	..
1912 Luggage, handbags, saddlery & harness	..	..	..	..	..	..	..	..	..	..
192 Footwear	283	0	59	303	133	..	694	68,239	..	1,718
20 WOOD AND WOOD PRODUCTS	3,568	0	211	777	367	..	2,790	299,260	..	8,790
201 Sawmilling and planing of wood	1,733	0	163	167	69	..	1,068	124,463	..	3,648
202 Products of wood, cork, straw & plaiting materials	1,835	0	48	610	298	..	1,722	174,797	..	5,142
2021 Veneer sheets	..	..	..	..	..	..	..	..	..	..
2022 Builders' carpentry and joinery	..	..	..	..	..	..	..	..	..	..
2023 Wooden containers	..	..	..	..	..	..	..	..	..	..
2029 Other products of wood	..	..	..	..	..	..	..	..	..	..
21 PAPER AND PAPER PRODUCTS	21,228	4	81	2,994	2,512	..	19,550	1,502,228	..	51,777
2101 Pulp, paper and paperboard	..	..	..	..	..	..	..	..	..	..
2102 Corrugated paper, paperboard and their containers	..	..	..	..	..	..	..	..	..	..
2109 Other articles of pulp and paperboard	..	..	..	..	..	..	..	..	..	..
22 PUBLISHING, PRINTING & REPRODUCTION	51	0	71	0	330	..	412	78,470	..	1,146
221 Publishing	4	0	7	0	40	..	40	10,249	..	128
2211 Publishing of books & brochures	..	..	..	..	..	..	..	..	..	..
2212 Publishing of newspapers and periodicals	..	..	..	..	..	..	..	..	..	..
2213 Publishing of recorded media	..	..	..	..	..	..	..	..	..	..
2219 Other publishing	..	..	..	..	..	..	..	..	..	..
222 Printing and related service activities	47	0	61	0	251	..	336	58,847	..	907
2221 Printing	..	..	..	..	..	..	..	..	..	..
2222 Service activities related to printing	..	..	..	..	..	..	..	..	..	..
223 Reproduction of recorded media	0	0	3	0	39	..	36	9,374	..	112
23 COKE, REFINED PETROLEUM PRODUCTS	99,518	2	3,470	4,262	6,741	..	30,843	1,820,627	..	151,390
231 Coke oven products	65,869	0	0	0	0	..	1,393	79,683	..	67,549
232 Refined petroleum products	33,649	2	3,470	4,262	6,741	..	29,450	1,740,944	..	83,841
233 Processing of nuclear fuel	..	..	..	..	..	..	..	..	..	..
24 CHEMICALS & CHEMICAL PRODUCTS	19,118	4	211	1,787	4,661	..	17,567	2,022,539	..	50,629
241 Basic chemicals	16,464	4	127	1,736	3,453	..	15,003	1,811,660	..	43,309
2411 Basic chemicals, exc. fertilizers & nitrogen compounds	..	..	..	..	..	..	..	..	..	..
2412 Fertilizers and nitrogen compounds	..	..	..	..	..	..	..	..	..	..
2413 Plastics in primary forms and synthetic rubber	..	..	..	..	..	..	..	..	..	..
242 Other chemical products	351	0	58	51	1,208	..	1,709	145,208	..	3,900
2421 Pesticides and other agro-chemical products	..	..	..	..	..	..	..	..	..	..
2422 Paints, varnishes and similar coatings	..	..	..	..	..	..	..	..	..	..
2423 Pharmaceuticals, medicinal chem. & botanical prod.	..	..	..	..	..	..	..	..	..	..
2424 Soap and detergents, perfumes etc.	..	..	..	..	..	..	..	..	..	..
2429 Other chemical products, nec	..	..	..	..	..	..	..	..	..	..
243 Man-made fibres	2,303	0	26	0	0	..	855	65,671	..	3,420
25 RUBBER AND PLASTICS PRODUCTS	1,010	4	118	564	1,653	..	3,583	487,832	..	8,688
251 Rubber products	424	0	60	7	1,098	..	2,024	180,006	..	4,261
2511 Rubber tyres and tubes	..	..	..	..	..	..	..	..	..	..
2519 Other rubber products	..	..	..	..	..	..	..	..	..	..
252 Plastic products	586	4	58	557	555	..	1,559	307,826	..	4,427
26 OTHER NON-METALLIC MINERAL PRODUCTS	12,301	10	694	11,518	18,443	..	10,856	1,706,882	..	59,967
261 Glass and glass products	1,074	7	122	186	5,318	..	1,947	443,404	..	10,250
269 Non-metallic mineral products, nec	11,227	3	572	11,332	13,125	..	8,909	1,263,478	..	49,717
2691 Pottery, china and earthenware	..	..	..	..	..	..	..	..	..	..
2692 Refractory ceramic products	..	..	..	..	..	..	..	..	..	..
2693 Structural non-refractory clay & ceramic prod.	..	..	..	..	..	..	..	..	..	..
2694 Cement, lime and plaster	..	..	..	..	..	..	..	..	..	..
2695 Articles of concrete, cement and plaster	..	..	..	..	..	..	..	..	..	..
2696 Cutting, shaping and finishing of stone	..	..	..	..	..	..	..	..	..	..
2699 Other non-metallic mineral products, nec	..	..	..	..	..	..	..	..	..	..

ISIC Revision 3 Industry Sector		Solid TJ	LPG TJ	Distiloil TJ	RFO TJ	Gas TJ	Biomass TJ	Steam TJ	Electr MWh	Own Use MWh	TOTAL TJ
27	**BASIC METALS**	244,263	3	2,487	6,513	12,885	..	15,807	3,333,894	..	293,960
271	Basic iron and steel	242,683	3	2,292	6,293	11,149	..	14,106	3,007,551	..	287,353
272	Basic precious and non-ferrous metals	369	0	148	91	808		782	103,187	..	2,569
273	Casting of metals	1,211	0	47	129	928		919	223,156		4,037
2731	Casting of iron and steel	..	..						..		
2732	Casting non-ferrous metals	..							..		
28	**FABRICATED METAL PRODUCTS**	2,760	31	876	330	2,362		4,721	566,354	..	13,119
281	Str. metal prod., tanks, reservoirs, steam generators	828	0	691	190	1,084		1,871	159,995		5,240
2811	Structural metal products	..		..		..			..		
2812	Tanks, reservoirs and containers of metal										
2813	Steam generators, exc. central heating hot water boilers	..		..		..		..	..		
289	Other fabricated metal products	1,932	31	185	140	1,278		2,850	406,359		7,879
2891	Forging, pressing, stamping & roll-forming of metal	..		..		..		..	..		
2892	Treatment and coating of metals	..							..		
2893	Cutlery, hand tools and general hardware	..				..			..		
2899	Other fabricated metal products, nec	..		..				..	..		
29	**MACHINERY AND EQUIPMENT, NEC**	8,586	33	1,123	1,402	6,959	..	12,134	1,200,319	..	34,558
291	General purpose machinery	2,781	0	411	274	2,185		4,204	373,936		11,201
2911	Engines and turbines	..									
2912	Pumps, compressors, taps and valves	..		..		..		..	..		
2913	Bearings, gears, gearing and driving elements	..		..		..		..	..		
2914	Ovens, furnaces and furnace burners	..		..		..		..	..		
2915	Lifting and handling equipment	..		..		..		..	..		
2919	Other general purpose machinery	..		..		..		..	..		
292	Special purpose machinery	5,195	33	711	1,128	4,594		7,846	775,153	..	22,298
2921	Agricultural and forestry machinery	..		..		..		..	..		
2922	Machine-tools	..		..		..		..	..		
2923	Machinery for metallurgy	..		..		..		..	..		
2924	Machinery for mining, quarrying and construction	..		..		..		..	..		
2925	Machinery for food, beverage & tobacco processing	..		..		..		..	..		
2926	Machinery for textile, apparel & leather production	..		..		..		..	..		
2927	Machinery for weapons and ammunition	..		..		..		..	..		
2929	Other special purpose machinery	..		..		..		..	..		
293	Domestic appliances, nec	610	0	1	0	180		84	51,230	..	1,059
30	**OFFICE, ACCOUNTING & COMPUTING MACHINERY**	143	0	1	0	1		91	9,248	..	269
31	**ELECTRICAL MACHINERY & APPARATUS, NEC**	887	3	70	1	1,229	..	2,548	316,268	..	5,877
311	Electric motors, generators and transformers	391	1	11	1	385		1,084	125,709	..	2,326
312	Electricity distribution and control apparatus	272	2	31	0	300		521	44,712	..	1,287
313	Insulated wire and cable	1	0	3	0	222		394	58,412		830
314	Accumulators, primary cells & primary batteries	46	0	5	0	91		115	20,313		330
315	Electric lamps and lighting equipment	22	0	10	0	92		89	18,926	..	281
319	Other electrical equipment, nec	155	0	10	0	139		345	48,196		823
32	**RADIO, TV & COMMUNICATION EQUIP. & APP.**	344	1	44	77	240	..	675	85,735	..	1,690
321	Electronic valves, tubes, other electronic components	306	1	4	63	223	..	503	57,652	..	1,308
322	TV & radio transmitters, apparatus for line telephony	24	0	38	14	14	..	127	19,842	..	288
323	TV & radio receivers, recording apparatus	14	0	2	0	3		45	8,241	..	94
33	**MEDICAL PRECISION &OPTICAL INSTRUMENTS**	82	0	22	3	444	..	651	61,104	..	1,422
331	Medical appliances and instruments	79	0	21	0	308		398	47,822	..	978
3311	Medical, surgical equipment & orthopaedic app.	..		..		..		..	..		..
3312	Instruments & appliances for measuring, checking etc	..		..		..		..	..		
3313	Industrial process control equipment	..		..		..		..	..		
332	Optical instruments and photographic equipment	1	0	1	0	11	..	123	7,330	..	162
333	Watches and clocks	2	0	0	3	125		130	5,952		281
34	**MOTOR VEHICLES, TRAILERS & SEMI-TRAILERS**	4,487	5	112	239	3,249	..	6,647	589,536	..	16,861
341	Motor vehicles	3,608	0	51	211	2,440		5,001	416,190	..	12,809
342	Bodies (coachwork) for motor vehicles	183	0	8	0	230		353	16,599		834
343	Parts, accessories for motor vehicles & their engines	696	5	53	28	579		1,293	156,747		3,218
35	**OTHER TRANSPORT EQUIPMENT**	1,943	5	87	84	1,357	..	3,320	187,409	..	7,471
351	Building and repairing of ships and boats	20	0	0	0	29		34	3,006		94
3511	Building and repairing of ships	..		..		..		..	..		
3512	Building, repairing of pleasure & sporting boats	..		..		..		..	..		
352	Railway, tramway locomotives & rolling stock	1,143	5	46	74	652		1,878	79,461	..	4,084
353	Aircraft and spacecraft	566	0	8	10	445		799	58,980	..	2,040
359	Transport equipment, nec	214	0	33	0	231		609	45,962	..	1,252
3591	Motorcycles	..		..		..		..	..		
3592	Bicycles and invalid carriages	..		..		..		..	..		
3599	Other transport equipment, nec	..		..		..		..	..		
36	**FURNITURE; MANUFACTURING, NEC**	2,588	2	165	158	1,763	..	2,554	281,338	..	8,243
361	Furniture	1,676	2	73	78	723		1,285	129,757	..	4,304
369	Manufacturing, nec	912	0	92	80	1,040		1,269	151,581	..	3,939
3691	Jewellery and related articles	..		..		..		..	..		
3692	Musical instruments	..		..		..		..	..		
3693	Sports goods	..		..		..		..	..		
3694	Games and toys	..		..		..		..	..		
3699	Other manufacturing, nec	..		..		..		..	..		
37	**RECYCLING**	45	0	1,168	0	228	..	130	43,541	..	1,728
371	Recycling of metal waste and scrap	34	0	1,144	0	182	..	67	40,346	..	1,572
372	Recycling of non-metal waste and scrap	11	0	24	0	46	..	63	3,195	..	156
CERR	**Non-specified industry**	..		..		..		..	..		
15-37	**TOTAL MANUFACTURING**	444,192	131	15,614	36,543	82,149	..	167,397	17,110,318	..	**807,623**

ISIC Revision 3 Industry Sector	Solid TJ	LPG TJ	Distiloil TJ	RFO TJ	Gas TJ	Biomass TJ	Steam TJ	Electr MWh	Own Use MWh	TOTAL TJ
15 FOOD PRODUCTS AND BEVERAGES	10,802	25	2,847	2,821	13,639	..	18,342	1,416,012	..	53,574
151 Production, processing and preserving (PPP)	3,054	6	892	265	2,971		5,879	389,469		14,469
1511 PPP of meat and meat products	..	..	..	..	..		..	..		..
1512 Processing and preserving of fish products	..	..	..	..	..		..	..		..
1513 Processing, preserving of fruit & vegetables	..	..	..	..	..		..	..		..
1514 Vegetable and animal oils and fats	..	..	..	..	..		..	..		..
152 Dairy products	423	0	475	520	1,943		2,366	268,391		6,693
153 Grain mill prod., starches & prepared animal feeds	349	0	549	186	797		265	159,948		2,722
1531 Grain mill products	..	..	..	..	..		..	..		..
1532 Starches and starch products	..	..	..	..	..		..	..		..
1533 Prepared animal feeds	..	..	..	..	..		..	..		..
154 Other food products	5,897	2	609	1,650	3,855		5,013	300,859		18,109
1541 Bakery products	..	..	..	..	..		..	..		..
1542 Sugar	..	..	..	..	..		..	..		..
1543 Cocoa, chocolate and sugar confectionery	..	..	..	..	..		..	..		..
1544 Macaroni, noodles, couscous & similar farinaceous prod.	..	..	..	..	..		..	..		..
1549 Other food products, nec	..	..	..	..	..		..	..		..
155 Beverages	1,079	17	322	200	4,073		4,819	297,345		11,580
1551 Distilling, rectifying and blending of spirits	..	..	..	..	..		..	..		..
1552 Wines	..	..	..	..	..		..	..		..
1553 Malt liquors and malt	..	..	..	..	..		..	..		..
1554 Soft drinks; production of mineral waters	..	..	..	..	..		..	..		..
16 TOBACCO PRODUCTS	0	0	22	0	195		157	19,863		446
17 TEXTILES	8,402	4	273	2,365	2,378	..	11,214	1,028,399	..	28,338
171 Spinning, weaving and finishing of textiles	6,343	3	143	1,899	1,179		8,379	771,069		20,722
1711 Preparation and spinning of textile fibres	..	..	..	..	..		..	..		..
1712 Finishing of textiles	..	..	..	..	..		..	..		..
172 Other textiles	1,083	1	66	459	887		1,923	188,292		5,097
1721 Made-up textile articles, except apparel	..	..	..	..	..		..	..		..
1722 Carpets and rugs	..	..	..	..	..		..	..		..
1723 Cordage, rope, twine and netting	..	..	..	..	..		..	..		..
1729 Other textiles, nec	..	..	..	..	..		..	..		..
173 Knitted and crocheted fabrics and articles	976	0	64	7	312		912	69,038		2,520
18 WEARING APPAREL, DRESSING & DYEING OF FUR	514	1	83	2	853	..	1,182	90,647	..	2,961
181 Wearing apparel, except fur apparel	510	1	76	2	822		1,102	87,033		2,826
182 Dressing and dyeing of fur; articles of fur	4	0	7	0	31		80	3,614		135
19 TANNING & DRESSING OF LEATHER, FOOTWEAR	607	0	106	386	417	..	1,181	81,080	..	2,989
191 Tanning and dressing of leather	391	0	46	70	256		792	41,520		1,704
1911 Tanning and dressing of leather	..	..	..	..	..		..	..		..
1912 Luggage, handbags, saddlery & harness	..	..	..	..	..		..	..		..
192 Footwear	216	0	60	316	161		389	39,560		1,284
20 WOOD AND WOOD PRODUCTS	3,157	1	383	846	523	..	3,649	332,255	..	9,755
201 Sawmilling and planing of wood	1,645	1	252	160	115		1,169	138,381		3,840
202 Products of wood, cork, straw & plaiting materials	1,512	0	131	686	408		2,480	193,874		5,915
2021 Veneer sheets	..	..	..	..	..		..	..		..
2022 Builders' carpentry and joinery	..	..	..	..	..		..	..		..
2023 Wooden containers	..	..	..	..	..		..	..		..
2029 Other products of wood	..	..	..	..	..		..	..		..
21 PAPER AND PAPER PRODUCTS	17,702	3	105	2,782	3,325	..	20,106	1,436,664	..	49,195
2101 Pulp, paper and paperboard	..	..	..	..	..		..	..		..
2102 Corrugated paper, paperboard and their containers	..	..	..	..	..		..	..		..
2109 Other articles of pulp and paperboard	..	..	..	..	..		..	..		..
22 PUBLISHING, PRINTING & REPRODUCTION	39	0	80	0	393	..	348	87,874	..	1,176
221 Publishing	3	0	20	0	79		32	18,519		201
2211 Publishing of books & brochures	..	..	..	..	..		..	..		..
2212 Publishing of newspapers and periodicals	..	..	..	..	..		..	..		..
2213 Publishing of recorded media	..	..	..	..	..		..	..		..
2219 Other publishing	..	..	..	..	..		..	..		..
222 Printing and related service activities	36	0	58	0	272		279	59,399		859
2221 Printing	..	..	..	..	..		..	..		..
2222 Service activities related to printing	..	..	..	..	..		..	..		..
223 Reproduction of recorded media	0	0	2	0	42		37	9,956		117
23 COKE, REFINED PETROLEUM PRODUCTS	103,154	2	6,880	4,412	6,672	..	29,621	1,757,331	..	157,067
231 Coke oven products	68,917	0	0	0	0		1,465	77,666		70,662
232 Refined petroleum products	34,237	2	6,880	4,412	6,672		28,156	1,679,665		86,406
233 Processing of nuclear fuel	..	..	..	..	..		..	..		..
24 CHEMICALS & CHEMICAL PRODUCTS	18,642	6	270	1,399	6,433	..	19,258	1,930,029	..	52,956
241 Basic chemicals	16,412	1	143	1,296	3,530		16,709	1,678,437		44,133
2411 Basic chemicals, exc. fertilizers & nitrogen compounds	..	..	..	..	..		..	..		..
2412 Fertilizers and nitrogen compounds	..	..	..	..	..		..	..		..
2413 Plastics in primary forms and synthetic rubber	..	..	..	..	..		..	..		..
242 Other chemical products	390	5	98	103	2,901		1,781	192,729		5,972
2421 Pesticides and other agro-chemical products	..	..	..	..	..		..	..		..
2422 Paints, varnishes and similar coatings	..	..	..	..	..		..	..		..
2423 Pharmaceuticals, medicinal chem. & botanical prod.	..	..	..	..	..		..	..		..
2424 Soap and detergents, perfumes etc.	..	..	..	..	..		..	..		..
2429 Other chemical products, nec	..	..	..	..	2		..	..		..
243 Man-made fibres	1,840	0	29	0	2		768	58,863		2,851
25 RUBBER AND PLASTICS PRODUCTS	915	6	126	504	1,937	..	4,021	470,303	..	9,202
251 Rubber products	382	2	37	2	1,338		2,439	208,598		4,951
2511 Rubber tyres and tubes	..	..	..	..	..		..	..		..
2519 Other rubber products	..	..	..	..	..		..	..		..
252 Plastic products	533	4	89	502	599		1,582	261,705		4,251
26 OTHER NON-METALLIC MINERAL PRODUCTS	11,698	23	2,423	9,592	27,524	..	11,148	2,043,230	..	69,764
261 Glass and glass products	1,117	15	188	184	12,141		2,780	755,913		19,146
269 Non-metallic mineral products, nec	10,581	8	2,235	9,408	15,383		8,368	1,287,317		50,617
2691 Pottery, china and earthenware	..	..	..	..	..		..	..		..
2692 Refractory ceramic products	..	..	..	..	..		..	..		..
2693 Structural non-refractory clay & ceramic prod.	..	..	..	..	..		..	..		..
2694 Cement, lime and plaster	..	..	..	..	..		..	..		..
2695 Articles of concrete, cement and plaster	..	..	..	..	..		..	..		..
2696 Cutting, shaping and finishing of stone	..	..	..	..	..		..	..		..
2699 Other non-metallic mineral products, nec	..	..	..	..	..		..	..		..

ISIC Revision 3 Industry Sector		Solid TJ	LPG TJ	Distiloil TJ	RFO TJ	Gas TJ	Biomass TJ	Steam TJ	Electr MWh	Own Use MWh	TOTAL TJ
27	BASIC METALS	236,096	7	771	10,401	15,333	..	20,441	3,653,535	..	296,202
271	Basic iron and steel	234,107	5	570	10,197	12,750	..	18,480	3,108,704		287,300
272	Basic precious and non-ferrous metals	246	0	127	78	1,308	..	596	176,412		2,990
273	Casting of metals	1,743	2	74	126	1,275	..	1,365	368,419		5,911
2731	Casting of iron and steel	..	..	..	..	..		..	..		..
2732	Casting non-ferrous metals	..	..	..	..	..		..	..		..
28	FABRICATED METAL PRODUCTS	2,957	55	572	330	3,499	..	4,432	656,977		14,210
281	Str. metal prod., tanks, reservoirs, steam generators	872	5	300	143	1,306	..	1,679	158,354		4,875
2811	Structural metal products	..	..	..	..	..		..	..		..
2812	Tanks, reservoirs and containers of metal	..	..	..	..	..		..	..		..
2813	Steam generators, exc. central heating hot water boilers	..	..	..	..	..		..	..		..
289	Other fabricated metal products	2,085	50	272	187	2,193	..	2,753	498,623		9,335
2891	Forging, pressing, stamping & roll-forming of metal	..	..	..	..	..		..	..		..
2892	Treatment and coating of metals	..	..	..	..	..		..	..		..
2893	Cutlery, hand tools and general hardware	..	..	..	..	..		..	..		..
2899	Other fabricated metal products, nec	..	..	..	..	..		..	..		..
29	MACHINERY AND EQUIPMENT, NEC	8,145	39	890	1,357	6,755	..	11,000	1,201,063	..	32,510
291	General purpose machinery	2,690	3	258	107	2,075	..	3,800	396,739		10,361
2911	Engines and turbines	..	..	..	..	..		..	..		..
2912	Pumps, compressors, taps and valves	..	..	..	..	..		..	..		..
2913	Bearings, gears, gearing and driving elements	..	..	..	..	..		..	..		..
2914	Ovens, furnaces and furnace burners	..	..	..	..	..		..	..		..
2915	Lifting and handling equipment	..	..	..	..	..		..	..		..
2919	Other general purpose machinery	..	..	..	..	..		..	..		..
292	Special purpose machinery	4,905	35	622	1,250	4,465	..	6,918	750,720		20,898
2921	Agricultural and forestry machinery	..	..	..	..	..		..	..		..
2922	Machine-tools	..	..	..	..	..		..	..		..
2923	Machinery for metallurgy	..	..	..	..	..		..	..		..
2924	Machinery for mining, quarrying and construction	..	..	..	..	..		..	..		..
2925	Machinery for food, beverage & tobacco processing	..	..	..	..	..		..	..		..
2926	Machinery for textile, apparel & leather production	..	..	..	..	..		..	..		..
2927	Machinery for weapons and ammunition	..	..	..	..	..		..	..		..
2929	Other special purpose machinery	..	..	..	..	..		..	..		..
293	Domestic appliances, nec	550	1	10	0	215	..	282	53,604		1,251
30	OFFICE, ACCOUNTING & COMPUTING MACHINERY	112	0	1	0	32	..	84	7,630	..	256
31	ELECTRICAL MACHINERY & APPARATUS, NEC	921	6	181	9	1,455	..	2,825	332,251	..	6,593
311	Electric motors, generators and transformers	329	3	43	0	415	..	1,254	127,664	..	2,504
312	Electricity distribution and control apparatus	202	1	37	0	371	..	527	43,642		1,295
313	Insulated wire and cable	14	2	4	0	288	..	394	62,004		925
314	Accumulators, primary cells & primary batteries	49	0	3	0	116	..	101	26,974		366
315	Electric lamps and lighting equipment	209	0	14	9	184	..	282	28,367		800
319	Other electrical equipment, nec	118	0	80	0	81	..	267	43,600		703
32	RADIO, TV & COMMUNICATION EQUIP. & APP.	336	0	53	59	262	..	683	84,858	..	1,698
321	Electronic valves, tubes, other electronic components	310	0	7	46	242	..	508	67,522	..	1,356
322	TV & radio transmitters, apparatus for line telephony	18	0	43	13	17	..	107	12,003		241
323	TV & radio receivers, recording apparatus	8	0	3	0	3	..	68	5,333		101
33	MEDICAL PRECISION &OPTICAL INSTRUMENTS	121	0	29	13	571	..	685	68,810	..	1,667
331	Medical appliances and instruments	119	0	27	0	431	..	421	54,131	..	1,193
3311	Medical, surgical equipment & orthopaedic app.	..	..	..		..	..	..	..		..
3312	Instruments & appliances for measuring, checking etc	..	..	..		..	..	..	..		..
3313	Industrial process control equipment	..	..	..		..	..	..	..		..
332	Optical instruments and photographic equipment	2	0	2	0	12	..	135	8,350		181
333	Watches and clocks	0	0	0	13	128	..	129	6,329		293
34	MOTOR VEHICLES, TRAILERS & SEMI-TRAILERS	3,222	2	803	109	2,360	..	5,455	550,836	..	13,934
341	Motor vehicles	2,457	0	114	89	1,151	..	3,625	341,693	..	8,666
342	Bodies (coachwork) for motor vehicles	243	0	17	0	219	..	342	17,486	..	884
343	Parts, accessories for motor vehicles & their engines	522	2	672	20	990	..	1,488	191,657	..	4,384
35	OTHER TRANSPORT EQUIPMENT	1,511	5	157	73	1,606	..	3,114	304,971	..	7,564
351	Building and repairing of ships and boats	11	0	1	0	27	..	30	106,757		453
3511	Building and repairing of ships	..		..		..	..	..	..		..
3512	Building, repairing of pleasure & sporting boats	..		..		..	..	..	..		..
352	Railway, tramway locomotives & rolling stock	1,086	5	77	73	846	..	1,887	93,068	..	4,309
353	Aircraft and spacecraft	210	0	27	0	435	..	689	57,897	..	1,569
359	Transport equipment, nec	204	0	52	0	298	..	508	47,249	..	1,232
3591	Motorcycles	..	..	..		..	..	..	..		..
3592	Bicycles and invalid carriages	..	..	..	..	..	..	..	..		..
3599	Other transport equipment, nec	..	..	..	..	..		..	..		..
36	FURNITURE; MANUFACTURING, NEC	2,798	43	274	184	1,875	..	3,015	260,947	..	9,128
361	Furniture	1,907	40	184	94	722	..	1,932	131,973	..	5,354
369	Manufacturing, nec	891	3	90	90	1,153	..	1,083	128,974	..	3,774
3691	Jewellery and related articles	..	..	..	..	..	..	..	..		..
3692	Musical instruments	..	..	..	..	..	..	..	..		..
3693	Sports goods	..	..	..	..	..	..	..	..		..
3694	Games and toys	..	..	..	..	..		..	..		..
3699	Other manufacturing, nec	..	..	..	..	..		..	..		..
37	RECYCLING	35	1	280	0	192	..	130	38,597		777
371	Recycling of metal waste and scrap	27	1	261	0	172	..	75	35,773	..	665
372	Recycling of non-metal waste and scrap	8	0	19	0	20	..	55	2,824	..	112
CERR	Non-specified industry	..	..	..	..	..	..	..	..		..
15-37	TOTAL MANUFACTURING	431,886	229	17,609	37,644	98,229	..	172,091	17,854,162		821,963

ISIC Revision 3 Industry Sector	Solid TJ	LPG TJ	Distiloil TJ	RFO TJ	Gas TJ	Biomass TJ	Steam TJ	Electr MWh	Own Use MWh	TOTAL TJ
15 FOOD PRODUCTS AND BEVERAGES	11,334	141	3,038	3,072	15,844	..	19,784	1,394,517	..	58,233
151 Production, processing and preserving (PPP)	3,004	10	845	237	3,473		6,091	395,622		15,084
1511 PPP of meat and meat products	..	..	..	..	..		..	..		..
1512 Processing and preserving of fish products	..	..	..	..	..		..	..		..
1513 Processing, preserving of fruit & vegetables	..	..	..	..	..		..	..		..
1514 Vegetable and animal oils and fats	..	..	..	..	..		..	..		..
152 Dairy products	316	0	572	717	2,354		3,001	185,041		7,626
153 Grain mill prod., starches & prepared animal feeds	363	92	660	179	1,030		258	174,868		3,212
1531 Grain mill products	..	..	..	..	..		..	..		..
1532 Starches and starch products	..	..	..	..	..		..	..		..
1533 Prepared animal feeds	..	..	..	..	..		..	..		..
154 Other food products	6,629	4	631	1,742	4,588		6,160	335,495		20,962
1541 Bakery products	..	..	..	..	..		..	..		..
1542 Sugar	..	..	..	..	..		..	..		..
1543 Cocoa, chocolate and sugar confectionery	..	..	..	..	..		..	..		..
1544 Macaroni, noodles, couscous & similar farinaceous prod.	..	..	..	..	..		..	..		..
1549 Other food products, nec	..	..	..	..	..		..	..		..
155 Beverages	1,022	35	330	197	4,399		4,274	303,491		11,350
1551 Distilling, rectifying and blending of spirits	..	..	..	..	..		..	..		..
1552 Wines	..	..	..	..	..		..	..		..
1553 Malt liquors and malt	..	..	..	..	..		..	..		..
1554 Soft drinks; production of mineral waters	..	..	..	..	..		..	..		..
16 TOBACCO PRODUCTS	0	0	26	11	195	..	204	20,190	..	509
17 TEXTILES	8,130	13	226	2,487	2,854	..	11,413	980,562	..	28,653
171 Spinning, weaving and finishing of textiles	6,214	1	136	2,010	1,459		8,495	781,333		21,128
1711 Preparation and spinning of textile fibres	..	..	..	..	..		..	..		..
1712 Finishing of textiles	..	..	..	..	..		..	..		..
172 Other textiles	1,024	12	51	448	1,121		1,965	136,248		5,111
1721 Made-up textile articles, except apparel	..	..	..	..	..		..	..		..
1722 Carpets and rugs	..	..	..	..	..		..	..		..
1723 Cordage, rope, twine and netting	..	..	..	..	..		..	..		..
1729 Other textiles, nec	..	..	..	..	..		..	..		..
173 Knitted and crocheted fabrics and articles	892	0	39	29	274		953	62,981		2,414
18 WEARING APPAREL, DRESSING & DYEING OF FUR	388	1	73	5	897	..	1,038	84,768	..	2,707
181 Wearing apparel, except fur apparel	382	1	70	5	891		983	82,206		2,628
182 Dressing and dyeing of fur; articles of fur	6	0	3	0	6		55	2,562		79
19 TANNING & DRESSING OF LEATHER, FOOTWEAR	429	0	67	405	468	..	737	72,299	..	2,366
191 Tanning and dressing of leather	207	0	46	66	251		495	33,073		1,184
1911 Tanning and dressing of leather	..	..	..	..	..		..	..		..
1912 Luggage, handbags, saddlery & harness	..	..	..	..	..		..	..		..
192 Footwear	222	0	21	339	217		242	39,226		1,182
20 WOOD AND WOOD PRODUCTS	3,565	0	380	1,162	668	..	3,381	345,988	..	10,402
201 Sawmilling and planing of wood	1,824	0	254	170	134		1,036	136,692		3,910
202 Products of wood, cork, straw & plaiting materials	1,741	0	126	992	534		2,345	209,296		6,491
2021 Veneer sheets	..	..	..	..	..		..	..		..
2022 Builders' carpentry and joinery	..	..	..	..	..		..	..		..
2023 Wooden containers	..	..	..	..	..		..	..		..
2029 Other products of wood	..	..	..	..	..		..	..		..
21 PAPER AND PAPER PRODUCTS	16,163	9	99	2,572	4,109	..	19,165	1,542,941	..	47,672
2101 Pulp, paper and paperboard	..	..	..	..	..		..	..		..
2102 Corrugated paper, paperboard and their containers	..	..	..	..	..		..	..		..
2109 Other articles of pulp and paperboard	..	..	..	..	..		..	..		..
22 PUBLISHING, PRINTING & REPRODUCTION	45	0	80	0	426	..	360	88,639	..	1,230
221 Publishing	3	0	19	0	88		43	18,320		219
2211 Publishing of books & brochures	..	..	..	..	..		..	..		..
2212 Publishing of newspapers and periodicals	..	..	..	..	..		..	..		..
2213 Publishing of recorded media	..	..	..	..	..		..	..		..
2219 Other publishing	..	..	..	..	..		..	..		..
222 Printing and related service activities	42	0	58	0	293		279	60,122		888
2221 Printing	..	..	..	..	..		..	..		..
2222 Service activities related to printing	..	..	..	..	..		..	..		..
223 Reproduction of recorded media	0	0	3	0	45		38	10,197		123
23 COKE, REFINED PETROLEUM PRODUCTS	67,591	0	548	2,194	5,225	..	8,239	350,420	..	85,059
231 Coke oven products	67,591	0	0	0	0		1,510	79,852		69,388
232 Refined petroleum products	0	0	548	2,194	5,225		6,729	270,568		15,670
233 Processing of nuclear fuel	..	..	..	..	..		..	..		..
24 CHEMICALS & CHEMICAL PRODUCTS	32,994	16	7,002	7,234	7,446	..	33,843	2,133,191	..	96,214
241 Basic chemicals	30,800	0	6,893	7,155	5,704		31,114	1,915,672		88,562
2411 Basic chemicals, exc. fertilizers & nitrogen compounds	..	..	..	..	..		..	..		..
2412 Fertilizers and nitrogen compounds	..	..	..	..	..		..	..		..
2413 Plastics in primary forms and synthetic rubber	..	..	..	..	..		..	..		..
242 Other chemical products	358	16	89	79	1,739		1,943	161,870		4,807
2421 Pesticides and other agro-chemical products	..	..	..	..	..		..	..		..
2422 Paints, varnishes and similar coatings	..	..	..	..	..		..	..		..
2423 Pharmaceuticals, medicinal chem. & botanical prod.	..	..	..	..	..		..	..		..
2424 Soap and detergents, perfumes etc.	..	..	..	..	..		..	..		..
2429 Other chemical products, nec	..	..	..	..	..		..	..		..
243 Man-made fibres	1,836	0	20	0	3		786	55,649		2,845
25 RUBBER AND PLASTICS PRODUCTS	1,143	13	139	355	2,135	..	4,142	513,481	..	9,776
251 Rubber products	382	6	47	20	1,483		2,450	213,663		5,157
2511 Rubber tyres and tubes	..	..	..	..	..		..	..		..
2519 Other rubber products	..	..	..	..	..		..	..		..
252 Plastic products	761	7	92	335	652		1,692	299,818		4,618
26 OTHER NON-METALLIC MINERAL PRODUCTS	10,901	17	1,194	9,297	28,929	..	11,770	1,976,295	..	69,223
261 Glass and glass products	496	7	117	14	12,494		3,050	671,841		18,597
269 Non-metallic mineral products, nec	10,405	10	1,077	9,283	16,435		8,720	1,304,454		50,626
2691 Pottery, china and earthenware	..	..	..	..	..		..	..		..
2692 Refractory ceramic products	..	..	..	..	..		..	..		..
2693 Structural non-refractory clay & ceramic prod.	..	..	..	..	..		..	..		..
2694 Cement, lime and plaster	..	..	..	..	..		..	..		..
2695 Articles of concrete, cement and plaster	..	..	..	..	..		..	..		..
2696 Cutting, shaping and finishing of stone	..	..	..	..	..		..	..		..
2699 Other non-metallic mineral products, nec	..	..	..	..	..		..	..		..

ISIC Revision 3 Industry Sector		Solid TJ	LPG TJ	Distiloil TJ	RFO TJ	Gas TJ	Biomass TJ	Steam TJ	Electr MWh	Own Use MWh	TOTAL TJ
27	**BASIC METALS**	226,691	56	676	8,621	15,829	..	16,353	3,379,198	..	280,391
271	Basic iron and steel	224,645	3	551	8,419	12,585	..	13,945	2,805,759	..	270,249
272	Basic precious and non-ferrous metals	227	45	68	35	1,815	..	870	186,463	..	3,731
273	Casting of metals	1,819	8	57	167	1,429	..	1,538	386,976	..	6,411
2731	Casting of iron and steel	..	..	..	..	..	..	..	..	..	..
2732	Casting non-ferrous metals	..	..	..	..	..	..	..	..	..	..
28	**FABRICATED METAL PRODUCTS**	1,896	74	623	195	5,559	..	4,635	708,102	..	15,531
281	Str. metal prod., tanks, reservoirs, steam generators	426	4	301	37	1,469	..	1,822	186,708	..	4,731
2811	Structural metal products	..	..	..	..	..	..	..	..	..	..
2812	Tanks, reservoirs and containers of metal	..	..	..	..	..	..	..	..	..	..
2813	Steam generators, exc. central heating hot water boilers	..	..	..	..	..	..	..	..	..	..
289	Other fabricated metal products	1,470	70	322	158	4,090	..	2,813	521,394	..	10,800
2891	Forging, pressing, stamping & roll-forming of metal	..	..	..	..	..	..	..	..	..	..
2892	Treatment and coating of metals	..	..	..	..	..	..	..	..	..	..
2893	Cutlery, hand tools and general hardware	..	..	..	..	..	..	..	..	..	..
2899	Other fabricated metal products, nec	..	..	..	..	..	..	..	..	..	..
29	**MACHINERY AND EQUIPMENT, NEC**	7,182	44	696	465	6,689	..	11,175	1,115,517	..	30,267
291	General purpose machinery	2,733	11	223	133	2,093	..	3,881	383,236	..	10,454
2911	Engines and turbines	..	..	..	..	..	..	..	..	..	..
2912	Pumps, compressors, taps and valves	..	..	..	..	..	..	..	..	..	..
2913	Bearings, gears, gearing and driving elements	..	..	..	..	..	..	..	..	..	..
2914	Ovens, furnaces and furnace burners	..	..	..	..	..	..	..	..	..	..
2915	Lifting and handling equipment	..	..	..	..	..	..	..	..	..	..
2919	Other general purpose machinery	..	..	..	..	..	..	..	..	..	..
292	Special purpose machinery	4,032	32	461	332	4,375	..	6,700	681,705	..	18,386
2921	Agricultural and forestry machinery	..	..	..	..	..	..	..	..	..	..
2922	Machine-tools	..	..	..	..	..	..	..	..	..	..
2923	Machinery for metallurgy	..	..	..	..	..	..	..	..	..	..
2924	Machinery for mining, quarrying and construction	..	..	..	..	..	..	..	..	..	..
2925	Machinery for food, beverage & tobacco processing	..	..	..	..	..	..	..	..	..	..
2926	Machinery for textile, apparel & leather production	..	..	..	..	..	..	..	..	..	..
2927	Machinery for weapons and ammunition	..	..	..	..	..	..	..	..	..	..
2929	Other special purpose machinery	..	..	..	..	..	..	..	..	..	..
293	Domestic appliances, nec	417	1	12	0	221	..	594	50,576	..	1,427
30	**OFFICE, ACCOUNTING & COMPUTING MACHINERY**	102	1	1	0	38	..	20	7,640	..	190
31	**ELECTRICAL MACHINERY & APPARATUS, NEC**	753	4	186	10	1,662	..	2,225	357,410	..	6,127
311	Electric motors, generators and transformers	236	1	24	0	420	..	844	132,417	..	2,002
312	Electricity distribution and control apparatus	116	1	39	0	330	..	431	41,548	..	1,067
313	Insulated wire and cable	17	0	6	0	348	..	317	65,270	..	923
314	Accumulators, primary cells & primary batteries	59	0	3	0	132	..	102	33,830	..	418
315	Electric lamps and lighting equipment	232	0	12	10	205	..	171	24,459	..	718
319	Other electrical equipment, nec	93	2	102	0	227	..	360	59,886	..	1,000
32	**RADIO, TV & COMMUNICATION EQUIP. & APP.**	284	0	61	43	372	..	760	77,189	..	1,798
321	Electronic valves, tubes, other electronic components	261	0	12	26	255	..	494	58,047	..	1,257
322	TV & radio transmitters, apparatus for line telephony	15	0	46	17	113	..	203	14,467	..	446
323	TV & radio receivers, recording apparatus	8	0	3	0	4	..	63	4,675	..	95
33	**MEDICAL PRECISION & OPTICAL INSTRUMENTS**	83	0	36	1	648	..	763	73,100	..	1,794
331	Medical appliances and instruments	81	0	31	0	473	..	492	55,677	..	1,277
3311	Medical, surgical equipment & orthopaedic app.	..	..	..	..	..	..	..	..	..	..
3312	Instruments & appliances for measuring, checking etc	..	..	..	..	..	..	..	..	..	..
3313	Industrial process control equipment	..	..	..	..	..	..	..	..	..	..
332	Optical instruments and photographic equipment	2	0	5	0	33	..	154	13,577	..	243
333	Watches and clocks	0	0	0	1	142	..	117	3,846	..	274
34	**MOTOR VEHICLES, TRAILERS & SEMI-TRAILERS**	976	4	174	66	2,725	..	4,933	577,534	..	10,957
341	Motor vehicles	525	0	100	45	1,350	..	3,489	368,390	..	6,835
342	Bodies (coachwork) for motor vehicles	242	0	17	0	271	..	255	16,450	..	844
343	Parts, accessories for motor vehicles & their engines	209	4	57	21	1,104	..	1,189	192,694	..	3,278
35	**OTHER TRANSPORT EQUIPMENT**	1,593	12	132	75	5,381	..	3,747	205,457	..	11,680
351	Building and repairing of ships and boats	9	0	0	0	29	..	27	2,373	..	74
3511	Building and repairing of ships	..	..	..	..	..	..	..	..	..	..
3512	Building, repairing of pleasure & sporting boats	..	..	..	..	..	..	..	..	..	..
352	Railway, tramway locomotives & rolling stock	1,066	5	69	75	4,166	..	1,990	88,741	..	7,690
353	Aircraft and spacecraft	283	0	6	0	590	..	928	57,763	..	2,015
359	Transport equipment, nec	235	7	57	0	596	..	802	56,580	..	1,901
3591	Motorcycles	..	..	..	..	..	..	..	..	..	..
3592	Bicycles and invalid carriages	..	..	..	..	..	..	..	..	..	..
3599	Other transport equipment, nec	..	..	..	..	..	..	..	..	..	..
36	**FURNITURE; MANUFACTURING, NEC**	2,805	45	246	208	1,962	..	2,306	363,186	..	8,879
361	Furniture	2,098	43	151	79	568	..	1,523	129,966	..	4,930
369	Manufacturing, nec	707	2	95	129	1,394	..	783	233,220	..	3,950
3691	Jewellery and related articles	..	..	..	..	..	..	..	..	..	..
3692	Musical instruments	..	..	..	..	..	..	..	..	..	..
3693	Sports goods	..	..	..	..	..	..	..	..	..	..
3694	Games and toys	..	..	..	..	..	..	..	..	..	..
3699	Other manufacturing, nec	..	..	..	..	..	..	..	..	..	..
37	**RECYCLING**	20	1	229	0	181	..	74	36,606	..	637
371	Recycling of metal waste and scrap	14	1	208	0	162	..	67	33,958	..	574
372	Recycling of non-metal waste and scrap	6	0	21	0	19	..	7	2,648	..	63
CERR	Non-specified industry	..	..	..	..	..	..	..	..	..	..
15-37	**TOTAL MANUFACTURING**	395,068	451	15,932	38,478	110,242	..	161,067	16,404,230	..	780,293

ISIC Revision 3 Industry Sector	Solid TJ	LPG TJ	Distiloil TJ	RFO TJ	Gas TJ	Biomass TJ	Steam TJ	Electr MWh	Own Use MWh	TOTAL TJ
15 FOOD PRODUCTS AND BEVERAGES	6,915	182	6,290	2,756	15,106	..	17,342	1,433,344	..	53,751
151 Production, processing and preserving (PPP)	2,568	11	1,091	124	3,017	..	4,973	417,897	..	13,288
1511 PPP of meat and meat products	..	..	..	..	..	..	..	..	..	..
1512 Processing and preserving of fish products	..	..	..	..	..	..	..	..	..	..
1513 Processing, preserving of fruit & vegetables	..	..	..	..	..	..	..	..	..	..
1514 Vegetable and animal oils and fats	..	..	..	..	..	..	..	..	..	..
152 Dairy products	196	0	1,049	540	1,993	..	2,734	169,994	..	7,124
153 Grain mill prod., starches & prepared animal feeds	200	2	1,148	256	1,301	..	603	210,149	..	4,267
1531 Grain mill products	..	..	..	..	..	..	..	..	..	..
1532 Starches and starch products	..	..	..	..	..	..	..	..	..	..
1533 Prepared animal feeds	..	..	..	..	..	..	..	..	..	..
154 Other food products	3,076	104	2,487	1,804	4,991	..	5,090	311,738	..	18,674
1541 Bakery products	..	..	..	..	..	..	..	..	..	..
1542 Sugar	..	..	..	..	..	..	..	..	..	..
1543 Cocoa, chocolate and sugar confectionery	..	..	..	..	..	..	..	..	..	..
1544 Macaroni, noodles, couscous & similar farinaceous prod.	..	..	..	..	..	..	..	..	..	..
1549 Other food products, nec	..	..	..	..	..	..	..	..	..	..
155 Beverages	875	65	515	32	3,804	..	3,942	323,566	..	10,398
1551 Distilling, rectifying and blending of spirits	..	..	..	..	..	..	..	..	..	..
1552 Wines	..	..	..	..	..	..	..	..	..	..
1553 Malt liquors and malt	..	..	..	..	..	..	..	..	..	..
1554 Soft drinks; production of mineral waters	..	..	..	..	..	..	..	..	..	..
16 TOBACCO PRODUCTS	0	1	76	0	180	..	153	21,401	..	487
17 TEXTILES	5,001	42	1,757	1,503	2,919	..	8,910	1,010,373	..	23,769
171 Spinning, weaving and finishing of textiles	3,903	5	1,391	1,229	2,017	..	7,008	786,587	..	18,385
1711 Preparation and spinning of textile fibres	..	..	..	..	..	..	..	..	..	..
1712 Finishing of textiles	..	..	..	..	..	..	..	..	..	..
172 Other textiles	465	37	327	257	766	..	1,050	165,902	..	3,499
1721 Made-up textile articles, except apparel	..	..	..	..	..	..	..	..	..	..
1722 Carpets and rugs	..	..	..	..	..	..	..	..	..	..
1723 Cordage, rope, twine and netting	..	..	..	..	..	..	..	..	..	..
1729 Other textiles, nec	..	..	..	..	..	..	..	..	..	..
173 Knitted and crocheted fabrics and articles	633	0	39	17	136	..	852	57,884	..	1,885
18 WEARING APPAREL, DRESSING & DYEING OF FUR	212	1	105	8	1,003	..	713	67,500	..	2,285
181 Wearing apparel, except fur apparel	208	1	102	8	1,000	..	713	67,006	..	2,273
182 Dressing and dyeing of fur; articles of fur	4	0	3	0	3	..	0	494	..	12
19 TANNING & DRESSING OF LEATHER, FOOTWEAR	272	0	286	230	243	..	810	70,816	..	2,096
191 Tanning and dressing of leather	130	0	57	19	107	..	340	26,825	..	750
1911 Tanning and dressing of leather	..	..	..	..	..	..	..	..	..	..
1912 Luggage, handbags, saddlery & harness	..	..	..	..	..	..	..	..	..	..
192 Footwear	142	0	229	211	136	..	470	43,991	..	1,346
20 WOOD AND WOOD PRODUCTS	3,570	2	933	574	1,122	..	3,691	314,107	..	11,023
201 Sawmilling and planing of wood	1,483	1	130	17	140	..	815	79,204	..	2,871
202 Products of wood, cork, straw & plaiting materials	2,087	1	803	557	982	..	2,876	234,903	..	8,152
2021 Veneer sheets	..	..	..	..	..	..	..	..	..	..
2022 Builders' carpentry and joinery	..	..	..	..	..	..	..	..	..	..
2023 Wooden containers	..	..	..	..	..	..	..	..	..	..
2029 Other products of wood	..	..	..	..	..	..	..	..	..	..
21 PAPER AND PAPER PRODUCTS	13,231	14	3,556	3,495	5,392	..	18,593	1,684,904	..	50,347
2101 Pulp, paper and paperboard	..	..	..	..	..	..	..	..	..	..
2102 Corrugated paper, paperboard and their containers	..	..	..	..	..	..	..	..	..	..
2109 Other articles of pulp and paperboard	..	..	..	..	..	..	..	..	..	..
22 PUBLISHING, PRINTING & REPRODUCTION	24	4	137	0	578	..	354	103,124	..	1,468
221 Publishing	16	2	66	0	291	..	266	58,760	..	853
2211 Publishing of books & brochures	..	..	..	..	..	..	..	..	..	..
2212 Publishing of newspapers and periodicals	..	..	..	..	..	..	..	..	..	..
2213 Publishing of recorded media	..	..	..	..	..	..	..	..	..	..
2219 Other publishing	..	..	..	..	..	..	..	..	..	..
222 Printing and related service activities	8	2	67	0	246	..	51	33,449	..	494
2221 Printing	..	..	..	..	..	..	..	..	..	..
2222 Service activities related to printing	..	..	..	..	..	..	..	..	..	..
223 Reproduction of recorded media	0	0	4	0	41	..	37	10,915	..	121
23 COKE, REFINED PETROLEUM PRODUCTS	49,716	0	2,145	1,588	5,439	..	10,811	341,001	..	70,927
231 Coke oven products	49,716	0	0	0	0	..	1,084	72,206	..	51,060
232 Refined petroleum products	0	0	2,145	1,588	5,439	..	9,727	268,795	..	19,867
233 Processing of nuclear fuel	..	..	..	..	..	..	..	..	..	..
24 CHEMICALS & CHEMICAL PRODUCTS	46,095	124	40,234	7,123	10,886	..	60,697	3,627,816	..	178,219
241 Basic chemicals	45,704	110	39,855	7,113	8,680	..	58,199	3,410,957	..	171,940
2411 Basic chemicals, exc. fertilizers & nitrogen compounds	..	..	..	..	..	..	..	..	..	..
2412 Fertilizers and nitrogen compounds	..	..	..	..	..	..	..	..	..	..
2413 Plastics in primary forms and synthetic rubber	..	..	..	..	..	..	..	..	..	..
242 Other chemical products	391	14	373	10	2,206	..	2,133	173,907	..	5,753
2421 Pesticides and other agro-chemical products	..	..	..	..	..	..	..	..	..	..
2422 Paints, varnishes and similar coatings	..	..	..	..	..	..	..	..	..	..
2423 Pharmaceuticals, medicinal chem. & botanical prod.	..	..	..	..	..	..	..	..	..	..
2424 Soap and detergents, perfumes etc.	..	..	..	..	..	..	..	..	..	..
2429 Other chemical products, nec	..	..	..	..	..	..	..	..	..	..
243 Man-made fibres	0	0	6	0	0	..	365	42,952	..	526
25 RUBBER AND PLASTICS PRODUCTS	729	7	267	2	2,275	..	3,248	585,768	..	8,637
251 Rubber products	140	1	73	0	1,461	..	2,214	230,504	..	4,719
2511 Rubber tyres and tubes	..	..	..	..	..	..	..	..	..	..
2519 Other rubber products	..	..	..	..	..	..	..	..	..	..
252 Plastic products	589	6	194	2	814	..	1,034	355,264	..	3,918
26 OTHER NON-METALLIC MINERAL PRODUCTS	9,565	46	9,446	5,891	29,163	..	6,744	2,083,334	..	68,355
261 Glass and glass products	197	17	103	3	13,718	..	1,927	847,586	..	19,016
269 Non-metallic mineral products, nec	9,368	29	9,343	5,888	15,445	..	4,817	1,235,748	..	49,339
2691 Pottery, china and earthenware	..	..	..	..	..	..	..	..	..	..
2692 Refractory ceramic products	..	..	..	..	..	..	..	..	..	..
2693 Structural non-refractory clay & ceramic prod.	..	..	..	..	..	..	..	..	..	..
2694 Cement, lime and plaster	..	..	..	..	..	..	..	..	..	..
2695 Articles of concrete, cement and plaster	..	..	..	..	..	..	..	..	..	..
2696 Cutting, shaping and finishing of stone	..	..	..	..	..	..	..	..	..	..
2699 Other non-metallic mineral products, nec	..	..	..	..	..	..	..	..	..	..

ISIC Revision 3 Industry Sector	Solid TJ	LPG TJ	Distiloil TJ	RFO TJ	Gas TJ	Biomass TJ	Steam TJ	Electr MWh	Own Use MWh	TOTAL TJ
27 **BASIC METALS**	208,328	16	9,463	8,940	14,990	..	28,537	3,308,317	..	282,184
271 Basic iron and steel	207,002	1	9,287	8,932	11,899	..	27,303	2,741,902	..	274,295
272 Basic precious and non-ferrous metals	138	0	64	0	1,692	..	372	182,818	..	2,924
273 Casting of metals	1,188	15	112	8	1,399	..	862	383,597	..	4,965
2731 Casting of iron and steel	..	..	..	..	..	..	..	..	..	..
2732 Casting non-ferrous metals	..	..	..	..	..	..	..	..	..	..
28 **FABRICATED METAL PRODUCTS**	1,020	91	796	31	3,598	..	3,074	680,098	..	11,058
281 Str. metal prod., tanks, reservoirs, steam generators	395	19	323	10	1,380	..	1,154	187,537	..	3,956
2811 Structural metal products	..	..	..	..	..	..	..	..	..	..
2812 Tanks, reservoirs and containers of metal	..	..	..	..	..	..	..	..	..	..
2813 Steam generators, exc. central heating hot water boilers	..	..	..	..	..	..	..	..	..	..
289 Other fabricated metal products	625	72	473	21	2,218	..	1,920	492,561	..	7,102
2891 Forging, pressing, stamping & roll-forming of metal	..	..	..	..	..	..	..	..	..	..
2892 Treatment and coating of metals	..	..	..	..	..	..	..	..	..	..
2893 Cutlery, hand tools and general hardware	..	..	..	..	..	..	..	..	..	..
2899 Other fabricated metal products, nec	..	..	..	..	..	..	..	..	..	..
29 **MACHINERY AND EQUIPMENT, NEC**	4,166	48	1,166	147	6,278	..	9,019	1,121,909	..	24,863
291 General purpose machinery	1,455	10	437	31	1,780	..	3,256	382,950	..	8,348
2911 Engines and turbines	..	..	..	..	..	..	..	..	..	..
2912 Pumps, compressors, taps and valves	..	..	..	..	..	..	..	..	..	..
2913 Bearings, gears, gearing and driving elements	..	..	..	..	..	..	..	..	..	..
2914 Ovens, furnaces and furnace burners	..	..	..	..	..	..	..	..	..	..
2915 Lifting and handling equipment	..	..	..	..	..	..	..	..	..	..
2919 Other general purpose machinery	..	..	..	..	..	..	..	..	..	..
292 Special purpose machinery	2,458	38	704	116	4,173	..	5,355	690,759	..	15,331
2921 Agricultural and forestry machinery	..	..	..	..	..	..	..	..	..	..
2922 Machine-tools	..	..	..	..	..	..	..	..	..	..
2923 Machinery for metallurgy	..	..	..	..	..	..	..	..	..	..
2924 Machinery for mining, quarrying and construction	..	..	..	..	..	..	..	..	..	..
2925 Machinery for food, beverage & tobacco processing	..	..	..	..	..	..	..	..	..	..
2926 Machinery for textile, apparel & leather production	..	..	..	..	..	..	..	..	..	..
2927 Machinery for weapons and ammunition	..	..	..	..	..	..	..	..	..	..
2929 Other special purpose machinery	..	..	..	..	..	..	..	..	..	..
293 Domestic appliances, nec	253	0	25	0	325	..	408	48,200	..	1,185
30 **OFFICE, ACCOUNTING & COMPUTING MACHINERY**	0	0	5	0	7	..	55	1,924	..	74
31 **ELECTRICAL MACHINERY & APPARATUS, NEC**	425	23	403	0	2,200	..	1,968	334,450	..	6,223
311 Electric motors, generators and transformers	178	1	52	0	435	..	589	95,878	..	1,600
312 Electricity distribution and control apparatus	47	16	149	0	416	..	408	52,413	..	1,225
313 Insulated wire and cable	6	2	14	0	264	..	268	72,025	..	813
314 Accumulators, primary cells & primary batteries	21	0	12	0	164	..	44	19,666	..	312
315 Electric lamps and lighting equipment	93	1	77	0	254	..	176	25,602	..	693
319 Other electrical equipment, nec	80	3	99	0	667	..	483	68,866	..	1,580
32 **RADIO, TV & COMMUNICATION EQUIP. & APP.**	186	3	100	28	359	..	525	93,967	..	1,539
321 Electronic valves, tubes, other electronic components	176	2	36	17	79	..	190	64,260	..	731
322 TV & radio transmitters, apparatus for line telephony	7	1	61	11	247	..	298	23,118	..	708
323 TV & radio receivers, recording apparatus	3	0	3	0	33	..	37	6,589	..	100
33 **MEDICAL PRECISION &OPTICAL INSTRUMENTS**	24	20	113	33	491	..	440	63,363	..	1,349
331 Medical appliances and instruments	22	19	107	33	468	..	356	50,175	..	1,186
3311 Medical, surgical equipment & orthopaedic app.	..	..	..	..	..	..	..	..	..	..
3312 Instruments & appliances for measuring, checking etc	..	..	..	..	..	..	..	..	..	..
3313 Industrial process control equipment	..	..	..	..	..	..	..	..	..	..
332 Optical instruments and photographic equipment	2	1	6	0	22	..	84	13,187	..	162
333 Watches and clocks	0	0	0	0	1	..	0	1	..	1
34 **MOTOR VEHICLES, TRAILERS & SEMI-TRAILERS**	891	8	415	17	2,360	..	4,615	646,750	..	10,634
341 Motor vehicles	555	1	290	6	1,093	..	3,319	405,933	..	6,725
342 Bodies (coachwork) for motor vehicles	8	0	20	0	196	..	95	17,217	..	381
343 Parts, accessories for motor vehicles & their engines	328	7	105	11	1,071	..	1,201	223,600	..	3,528
35 **OTHER TRANSPORT EQUIPMENT**	786	3	154	0	1,759	..	2,159	172,321	..	5,481
351 Building and repairing of ships and boats	6	0	2	0	29	..	29	2,125	..	74
3511 Building and repairing of ships	..	..	..	..	..	..	..	..	..	..
3512 Building, repairing of pleasure & sporting boats	..	..	..	..	..	..	..	..	..	..
352 Railway, tramway locomotives & rolling stock	471	1	75	0	1,128	..	1,445	102,660	..	3,490
353 Aircraft and spacecraft	262	1	58	0	431	..	525	45,144	..	1,440
359 Transport equipment, nec	47	1	19	0	171	..	160	22,392	..	479
3591 Motorcycles	..	..	..	..	..	..	..	..	..	..
3592 Bicycles and invalid carriages	..	..	..	..	..	..	..	..	..	..
3599 Other transport equipment, nec	..	..	..	..	..	..	..	..	..	..
36 **FURNITURE; MANUFACTURING, NEC**	1,776	61	428	121	834	..	2,105	221,709	..	6,123
361 Furniture	1,321	51	225	42	471	..	1,387	129,034	..	3,962
369 Manufacturing, nec	455	10	203	79	363	..	718	92,675	..	2,162
3691 Jewellery and related articles	..	..	..	..	..	..	..	..	..	..
3692 Musical instruments	..	..	..	..	..	..	..	..	..	..
3693 Sports goods	..	..	..	..	..	..	..	..	..	..
3694 Games and toys	..	..	..	..	..	..	..	..	..	..
3699 Other manufacturing, nec	..	..	..	..	..	..	..	..	..	..
37 **RECYCLING**	18	2	337	0	149	..	76	31,029	..	694
371 Recycling of metal waste and scrap	18	2	308	0	108	..	36	28,747	..	575
372 Recycling of non-metal waste and scrap	0	0	29	0	41	..	40	2,282	..	118
CERR **Non-specified industry**	..	..	..	..	..	..	..	..	..	..
15-37 **TOTAL MANUFACTURING**	352,950	698	78,612	32,487	107,331	..	184,639	18,019,325	..	821,587

ISIC Revision 3 Industry Sector		Solid TJ	LPG TJ	Distiloil TJ	RFO TJ	Gas TJ	Biomass TJ	Steam TJ	Electr MWh	Own Use MWh	TOTAL TJ
15	**FOOD PRODUCTS AND BEVERAGES**	5,138	683	3,428	8,430	8,174	58	522	1,820,524	183,734	32,325
151	Production, processing and preserving (PPP)	294	183	1,196	2,946	3,552	12	124	842,531	21,103	11,264
1511	PPP of meat and meat products	0	62	644	1,239	1,397	1	62	467,203	0	5,087
1512	Processing and preserving of fish products	294	19	306	530	1,431	9	55	180,695	0	3,295
1513	Processing, preserving of fruit & vegetables	0	3	147	91	243	0	3	44,086	60	645
1514	Vegetable and animal oils and fats	0	99	99	1,086	481	2	4	150,547	21,043	2,237
152	Dairy products	0	108	1,068	1,494	1,178	1	272	309,065	448	5,232
153	Grain mill prod., starches & prepared animal feeds	260	13	214	273	1,279	0	9	130,082	0	2,516
1531	Grain mill products	0	6	31	16	129	0	3	52,728	0	375
1532	Starches and starch products	0	0	37	0	101	0	3	18,274	0	207
1533	Prepared animal feeds	260	7	146	257	1,049	0	3	59,080	0	1,935
154	Other food products	4,103	148	703	1,892	1,265	45	95	341,392	119,779	9,049
1541	Bakery products	0	131	363	17	205	0	26	67,459	0	985
1542	Sugar	3,474	3	101	1,520	1	0	8	121,356	76,405	5,269
1543	Cocoa, chocolate and sugar confectionery	0	9	101	86	110	4	21	35,606	0	459
1544	Macaroni, noodles, couscous & similar farinaceous prod.	..	..	..	..	..	..	..	..	..	..
1549	Other food products, nec	629	5	138	269	949	41	40	116,971	43,374	2,336
155	Beverages	481	231	247	1,825	900	0	22	197,454	42,404	4,264
1551	Distilling, rectifying and blending of spirits	223	204	39	3	179	0	12	19,442	169	729
1552	Wines	..	..	..	..	..	..	..	..	..	..
1553	Malt liquors and malt	258	21	118	1,795	695	0	9	156,995	42,235	3,309
1554	Soft drinks; production of mineral waters	0	6	90	27	26	0	1	21,017	0	226
16	**TOBACCO PRODUCTS**	0	3	19	36	85	0	2	24,813	0	234
17	**TEXTILES**	0	79	152	424	1,113	3	39	146,650	30	2,338
171	Spinning, weaving and finishing of textiles	0	29	42	358	864	1	3	78,415	0	1,579
1711	Preparation and spinning of textile fibres	0	10	33	306	328	0	3	55,970	0	881
1712	Finishing of textiles	0	19	9	52	536	1	0	22,445	0	698
172	Other textiles	0	50	82	64	234	2	16	51,875	30	635
1721	Made-up textile articles, except apparel	0	2	45	9	42	0	3	9,560	0	135
1722	Carpets and rugs	0	46	15	48	114	2	9	19,550	30	304
1723	Cordage, rope, twine and netting	0	0	9	0	4	0	2	3,830	0	29
1729	Other textiles, nec	0	2	13	7	74	0	2	18,935	0	166
173	Knitted and crocheted fabrics and articles	0	0	28	2	15	0	20	16,360	0	124
18	**WEARING APPAREL, DRESSING & DYEING OF FUR**	0	0	99	6	10	0	27	21,211	628	216
181	Wearing apparel, except fur apparel	..	..	..	..	..	..	..	..	..	..
182	Dressing and dyeing of fur; articles of fur	..	..	..	..	..	..	..	..	..	..
19	**TANNING & DRESSING OF LEATHER, FOOTWEAR**	0	0	36	37	16	0	3	12,873	0	138
191	Tanning and dressing of leather	0	0	17	35	8	0	1	4,571	0	77
1911	Tanning and dressing of leather	0	0	14	35	7	0	0	3,721	0	69
1912	Luggage, handbags, saddlery & harness	0	0	3	0	1	0	1	850	0	8
192	Footwear	0	0	19	2	8	0	2	8,302	0	61
20	**WOOD AND WOOD PRODUCTS**	0	38	252	443	15	3,569	13	235,600	64,651	4,945
201	Sawmilling and planing of wood	0	2	61	21	6	449	0	28,551	0	642
202	Products of wood, cork, straw & plaiting materials	0	36	191	422	9	3,120	13	207,049	64,651	4,304
2021	Veneer sheets	0	6	27	289	0	827	0	58,795	8,350	1,331
2022	Builders' carpentry and joinery	0	28	143	129	3	2,021	11	132,853	56,301	2,611
2023	Wooden containers	0	1	12	0	1	41	2	4,893	0	75
2029	Other products of wood	0	1	9	4	5	231	0	10,508	0	288
21	**PAPER AND PAPER PRODUCTS**	1,625	53	208	989	2,068	16	27	404,484	61,564	6,221
2101	Pulp, paper and paperboard	1,625	25	22	862	1,236	0	1	277,376	61,564	4,548
2102	Corrugated paper, paperboard and their containers	0	23	157	122	119	12	21	67,772	0	698
2109	Other articles of pulp and paperboard	0	5	29	5	713	4	5	59,336	0	975
22	**PUBLISHING, PRINTING & REPRODUCTION**	0	12	192	25	148	47	204	154,713	0	1,185
221	Publishing	0	7	54	5	38	0	135	71,050	0	495
2211	Publishing of books & brochures	0	0	5	3	3	0	2	4,176	0	28
2212	Publishing of newspapers and periodicals	0	7	48	2	28	0	133	64,792	0	451
2213	Publishing of recorded media	..	..	..	..	..	..	..	..	..	..
2219	Other publishing	0	0	1	0	7	0	0	2,082	0	15
222	Printing and related service activities	0	5	138	20	110	47	69	83,663	0	690
2221	Printing	0	4	116	20	104	47	56	62,276	0	571
2222	Service activities related to printing	0	1	22	0	6	0	13	21,387	0	119
223	Reproduction of recorded media	..	..	..	..	..	..	..	..	..	..
23	**COKE, REFINED PETROLEUM PRODUCTS**	0	0	103	1,651	13,787	0	428	275,963	22,339	16,882
231	Coke oven products	..	..	..	..	..	..	..	..	..	..
232	Refined petroleum products	0	0	103	1,651	13,787	0	428	275,963	22,339	16,882
233	Processing of nuclear fuel	..	..	..	..	..	..	..	..	..	..
24	**CHEMICALS & CHEMICAL PRODUCTS**	84	78	579	2,449	1,944	11	874	830,303	52,709	8,818
241	Basic chemicals	0	33	255	1,235	840	10	506	419,810	30,229	4,281
2411	Basic chemicals, exc. fertilizers & nitrogen compounds	0	31	154	1,139	304	10	505	279,490	15	3,149
2412	Fertilizers and nitrogen compounds	0	2	23	85	506	0	0	129,316	30,214	973
2413	Plastics in primary forms and synthetic rubber	0	0	78	11	30	0	1	11,004	0	160
242	Other chemical products	84	45	324	1,214	1,104	1	368	410,493	22,480	4,537
2421	Pesticides and other agro-chemical products	..	..	..	..	..	..	..	..	..	..
2422	Paints, varnishes and similar coatings	0	2	62	27	44	1	48	27,291	0	282
2423	Pharmaceuticals, medicinal chem. & botanical prod.	0	5	156	949	360	0	305	191,218	22,480	2,382
2424	Soap and detergents, perfumes etc.	84	35	73	122	543	0	14	71,494	0	1,128
2429	Other chemical products, nec	0	3	33	116	157	0	1	120,490	0	744
243	Man-made fibres	..	..	..	..	..	..	..	..	..	..
25	**RUBBER AND PLASTICS PRODUCTS**	0	42	406	174	869	1	36	472,324	0	3,228
251	Rubber products	0	10	116	28	360	0	1	51,736	0	701
2511	Rubber tyres and tubes	..	..	..	..	..	..	..	..	..	..
2519	Other rubber products	0	10	116	28	360	0	1	51,736	0	701
252	Plastic products	0	32	290	146	509	1	35	420,588	0	2,527
26	**OTHER NON-METALLIC MINERAL PRODUCTS**	7,179	327	1,716	5,235	4,105	10	20	606,287	23,062	20,692
261	Glass and glass products	0	26	106	5	1,329	1	5	123,033	0	1,915
269	Non-metallic mineral products, nec	7,179	301	1,610	5,230	2,776	9	15	483,254	23,062	18,777
2691	Pottery, china and earthenware	0	7	27	76	151	0	1	12,797	0	308
2692	Refractory ceramic products	..	..	..	..	..	..	..	..	..	..
2693	Structural non-refractory clay & ceramic prod.	0	96	136	70	1,398	6	0	38,508	0	1,845
2694	Cement, lime and plaster	5,096	4	180	4,241	0	0	0	207,168	23,048	10,184
2695	Articles of concrete, cement and plaster	732	31	687	370	742	1	12	112,812	14	2,981
2696	Cutting, shaping and finishing of stone	0	1	6	0	0	0	0	1,448	0	12
2699	Other non-metallic mineral products, nec	1,351	162	574	473	485	2	2	110,521	0	3,447

ISIS Energy Data Programme (IEA/OECD)

ISIC Revision 3 Industry Sector	Solid TJ	LPG TJ	Distiloil TJ	RFO TJ	Gas TJ	Biomass TJ	Steam TJ	Electr MWh	Own Use MWh	TOTAL TJ
27	**BASIC METALS**	45	118	343	68	1,754	0	50	545,355	0	4,341
271	Basic iron and steel	45	85	211	38	1,675	0	29	489,630	0	3,846
272	Basic precious and non-ferrous metals	0	33	132	30	79	0	21	55,725	0	496
273	Casting of metals	..	..	..	..	..	..	..	..	..	..
2731	Casting of iron and steel	..	..	..	..	..	..	..	..	..	..
2732	Casting non-ferrous metals	..	..	..	..	..	..	..	..	..	..
28	**FABRICATED METAL PRODUCTS**	77	244	801	68	646	12	73	443,815	0	3,519
281	Str. metal prod., tanks, reservoirs, steam generators	2	48	360	17	85	7	25	82,364	0	841
2811	Structural metal products	1	23	285	14	28	1	19	47,745	0	543
2812	Tanks, reservoirs and containers of metal	1	23	55	3	55	6	3	25,390	0	237
2813	Steam generators, exc. central heating hot water boilers	0	2	20	0	2	0	3	9,229	0	60
289	Other fabricated metal products	75	196	441	51	561	5	48	361,451	0	2,678
2891	Forging, pressing, stamping & roll-forming of metal	..	..	..	..	..	..	..	..	..	..
2892	Treatment and coating of metals	0	17	67	3	53	0	5	35,072	0	271
2893	Cutlery, hand tools and general hardware	1	76	125	7	392	5	16	91,800	0	952
2899	Other fabricated metal products, nec	74	103	249	41	116	0	27	234,579	0	1,454
29	**MACHINERY AND EQUIPMENT, NEC**	44	176	1,211	725	457	11	157	497,133	13	4,571
291	General purpose machinery	0	90	613	401	227	1	105	297,465	0	2,508
2911	Engines and turbines	..	..	..	..	..	..	..	..	..	..
2912	Pumps, compressors, taps and valves	0	39	159	336	123	1	55	195,793	0	1,418
2913	Bearings, gears, gearing and driving elements	0	2	15	0	1	0	1	5,396	0	38
2914	Ovens, furnaces and furnace burners	0	0	5	0	0	0	0	635	0	7
2915	Lifting and handling equipment	0	42	234	23	45	0	15	44,543	0	519
2919	Other general purpose machinery	0	7	200	42	58	0	34	51,098	0	525
292	Special purpose machinery	10	56	495	159	202	0	32	131,955	13	1,429
2921	Agricultural and forestry machinery	10	46	186	30	139	0	3	41,281	13	563
2922	Machine-tools	0	4	99	20	9	0	10	18,792	0	210
2923	Machinery for metallurgy	..	..	..	..	..	..	..	..	..	..
2924	Machinery for mining, quarrying and construction	0	5	53	93	12	0	3	24,580	0	254
2925	Machinery for food, beverage & tobacco processing	0	1	68	9	11	0	10	16,018	0	157
2926	Machinery for textile, apparel & leather production	0	0	25	6	9	0	2	8,568	0	73
2927	Machinery for weapons and ammunition	..	..	..	..	..	..	..	..	..	..
2929	Other special purpose machinery	0	0	64	1	22	0	4	22,716	0	173
293	Domestic appliances, nec	34	30	103	165	28	10	20	67,713	0	634
30	**OFFICE, ACCOUNTING & COMPUTING MACHINERY**	0	0	31	0	16	0	15	18,130	0	127
31	**ELECTRICAL MACHINERY & APPARATUS, NEC**	0	13	196	206	46	1	81	112,994	0	950
311	Electric motors, generators and transformers	0	1	47	2	2	0	27	21,298	0	156
312	Electricity distribution and control apparatus	0	1	23	27	18	1	20	20,990	0	166
313	Insulated wire and cable	0	4	40	175	3	0	23	42,806	0	399
314	Accumulators, primary cells & primary batteries	0	2	7	0	1	0	0	8,821	0	42
315	Electric lamps and lighting equipment	0	5	62	2	18	0	8	12,616	0	140
319	Other electrical equipment, nec	0	0	17	0	4	0	3	6,463	0	47
32	**RADIO, TV & COMMUNICATION EQUIP. & APP.**	5	1	61	20	134	34	33	69,721	0	539
321	Electronic valves, tubes, other electronic components	3	0	15	7	23	0	7	21,350	0	132
322	TV & radio transmitters, apparatus for line telephony	0	0	18	0	35	0	19	7,643	0	100
323	TV & radio receivers, recording apparatus	2	1	28	13	76	34	7	40,728	0	308
33	**MEDICAL PRECISION &OPTICAL INSTRUMENTS**	0	2	105	22	54	2	57	62,719	— 0	468
331	Medical appliances and instruments	0	2	87	13	49	0	51	54,834	0	399
3311	Medical, surgical equipment & orthopaedic app.	0	2	49	10	18	0	22	32,392	0	218
3312	Instruments & appliances for measuring, checking etc	0	0	33	3	26	0	29	21,646	0	169
3313	Industrial process control equipment	0	0	5	0	5	0	0	796	0	13
332	Optical instruments and photographic equipment	0	0	18	9	5	2	6	7,885	0	68
333	Watches and clocks	..	..	..	..	..	..	..	..	..	..
34	**MOTOR VEHICLES, TRAILERS & SEMI-TRAILERS**	0	39	210	43	96	1	16	68,756	0	653
341	Motor vehicles	..	..	..	..	..	..	..	..	..	..
342	Bodies (coachwork) for motor vehicles	0	4	112	18	20	1	8	15,768	0	220
343	Parts, accessories for motor vehicles & their engines	0	35	98	25	76	0	8	52,988	0	433
35	**OTHER TRANSPORT EQUIPMENT**	4	20	219	116	38	0	98	142,056	0	1,006
351	Building and repairing of ships and boats	4	17	191	102	9	0	74	121,144	0	833
3511	Building and repairing of ships	4	17	174	100	8	0	74	119,097	0	806
3512	Building, repairing of pleasure & sporting boats	0	0	17	2	1	0	0	2,047	0	27
352	Railway, tramway locomotives & rolling stock	..	..	..	..	..	..	..	..	..	..
353	Aircraft and spacecraft	0	0	4	9	0	0	4	7,252	0	43
359	Transport equipment, nec	0	3	24	5	29	0	20	13,660	0	130
3591	Motorcycles	..	..	..	..	..	..	..	..	..	..
3592	Bicycles and invalid carriages	0	2	7	3	29	0	2	3,193	0	54
3599	Other transport equipment, nec	0	1	17	2	0	0	18	10,467	0	76
36	**FURNITURE; MANUFACTURING, NEC**	1	38	483	87	140	1,079	73	265,400	0	2,856
361	Furniture	1	35	406	74	113	1,060	31	194,286	0	2,419
369	Manufacturing, nec	0	3	77	13	27	19	42	71,114	0	437
3691	Jewellery and related articles	0	2	2	0	1	0	3	1,381	0	13
3692	Musical instruments	0	0	3	0	0	1	0	241	0	5
3693	Sports goods	..	..	..	..	..	..	..	..	..	..
3694	Games and toys	0	0	47	0	13	0	32	57,476	0	299
3699	Other manufacturing, nec	0	1	25	13	13	18	7	12,016	0	120
37	**RECYCLING**	..	..	..	..	..	..	..	..	..	..
371	Recycling of metal waste and scrap	..	..	..	..	..	..	..	..	..	..
372	Recycling of non-metal waste and scrap	..	..	..	..	..	..	..	..	..	..
CERR	**Non-specified industry**	..	..	..	..	..	..	..	..	..	..
15-37	**TOTAL MANUFACTURING**	14,202	1,966	10,850	21,254	35,715	4,855	2,848	7,231,824	408,730	116,253

ISIC Revision 3 Industry Sector	Solid TJ	LPG TJ	Distiloil TJ	RFO TJ	Gas TJ	Biomass TJ	Steam TJ	Electr MWh	Own Use MWh	TOTAL TJ
15	**FOOD PRODUCTS AND BEVERAGES**	4,069	193	2,521	7,716	11,067	28	801	1,867,620	117,906	32,694
151	Production, processing and preserving (PPP)	356	56	944	2,590	4,393	23	148	794,280	8,145	11,340
1511	PPP of meat and meat products	0	33	570	1,073	1,922	4	79	507,323	0	5,507
1512	Processing and preserving of fish products	356	19	203	364	1,950	12	48	176,728	0	3,588
1513	Processing, preserving of fruit & vegetables	0	4	132	93	291	6	6	63,497	60	760
1514	Vegetable and animal oils and fats	0	0	39	1,060	230	1	15	46,732	8,085	1,484
152	Dairy products	0	5	619	1,131	2,196	0	263	363,715	496	5,522
153	Grain mill prod., starches & prepared animal feeds	156	19	181	228	1,460	0	14	146,634	0	2,586
1531	Grain mill products	0	6	50	15	175	0	5	52,380	0	440
1532	Starches and starch products	0	0	37	0	262	0	4	34,973	0	429
1533	Prepared animal feeds	156	13	94	213	1,023	0	5	59,281	0	1,717
154	Other food products	2,984	81	510	2,050	1,811	5	349	361,022	67,487	8,847
1541	Bakery products	0	64	249	0	366	0	34	84,910	0	1,019
1542	Sugar	2,592	7	103	1,655	1	0	8	101,840	67,487	4,490
1543	Cocoa, chocolate and sugar confectionery	0	1	63	82	139	5	15	39,134	0	446
1544	Macaroni, noodles, couscous & similar farinaceous prod.	..	..	..	..	..		..	..	..	..
1549	Other food products, nec	392	9	95	313	1,305	0	292	135,138	0	2,892
155	Beverages	573	32	267	1,717	1,207	0	27	201,969	41,778	4,400
1551	Distilling, rectifying and blending of spirits	281	2	27	5	203	0	7	25,396	1,407	611
1552	Wines	..	..	..	..	..		..	..	..	..
1553	Malt liquors and malt	292	26	103	1,712	975	0	8	154,030	40,371	3,525
1554	Soft drinks; production of mineral waters	0	4	137	0	29	0	12	22,543	0	263
16	**TOBACCO PRODUCTS**	0	2	18	29	94	0	2	22,567	0	226
17	**TEXTILES**	0	48	140	187	1,189	1	35	144,489	25	2,120
171	Spinning, weaving and finishing of textiles	0	13	33	170	876	0	1	77,427	0	1,372
1711	Preparation and spinning of textile fibres	0	13	24	170	256	0	1	51,534	0	650
1712	Finishing of textiles	0	0	9	0	620	0	0	25,893	0	722
172	Other textiles	0	35	86	11	293	0	15	48,208	25	613
1721	Made-up textile articles, except apparel	0	1	50	0	57	0	4	10,982	0	152
1722	Carpets and rugs	0	27	10	0	154	0	8	13,001	25	246
1723	Cordage, rope, twine and netting	0	0	6	0	4	0	1	2,454	0	20
1729	Other textiles, nec	0	7	20	11	78	0	2	21,771	0	196
173	Knitted and crocheted fabrics and articles	0	0	21	6	20	1	19	18,854	0	135
18	**WEARING APPAREL, DRESSING & DYEING OF FUR**	0	0	84	7	23	0	31	21,209	23	221
181	Wearing apparel, except fur apparel	..	..	..	..	..		..	..	..	..
182	Dressing and dyeing of fur; articles of fur	..	..	..	..	..		..	..	..	..
19	**TANNING & DRESSING OF LEATHER, FOOTWEAR**	0	1	24	39	43	0	2	13,666	0	158
191	Tanning and dressing of leather	0	1	6	39	25	0	0	5,238	0	90
1911	Tanning and dressing of leather	0	1	1	39	24	0	0	4,276	0	80
1912	Luggage, handbags, saddlery & harness	0	0	5	0	1	0	0	962	0	9
192	Footwear	0	0	18	0	18	0	2	8,428	0	68
20	**WOOD AND WOOD PRODUCTS**	6	25	298	434	38	3,388	3	256,190	0	5,114
201	Sawmilling and planing of wood	0	2	40	41	5	376	0	27,221	0	562
202	Products of wood, cork, straw & plaiting materials	6	23	258	393	33	3,012	3	228,969	0	4,552
2021	Veneer sheets	0	5	37	185	1	781	0	48,133	0	1,182
2022	Builders' carpentry and joinery	6	16	193	207	23	2,049	3	163,732	0	3,086
2023	Wooden containers	0	1	20	0	1	24	0	5,447	0	66
2029	Other products of wood	0	1	8	1	8	158	0	11,657	0	218
21	**PAPER AND PAPER PRODUCTS**	796	41	142	795	1,294	13	1,006	361,405	44,102	5,229
2101	Pulp, paper and paperboard	796	14	16	54	1,019	0	974	225,448	44,102	3,526
2102	Corrugated paper, paperboard and their containers	0	23	94	166	174	11	27	75,140	0	766
2109	Other articles of pulp and paperboard	0	4	32	575	101	2	5	60,817	0	938
22	**PUBLISHING, PRINTING & REPRODUCTION**	0	9	153	7	161	50	190	165,994	0	1,168
221	Publishing	0	7	44	1	52	0	130	76,483	0	509
2211	Publishing of books & brochures	0	1	4	0	11	0	3	6,928	0	44
2212	Publishing of newspapers and periodicals	0	6	38	1	31	0	127	67,129	0	445
2213	Publishing of recorded media	..	..	..	..	..		..	..	..	..
2219	Other publishing	0	0	2	0	10	0	0	2,426	0	21
222	Printing and related service activities	0	2	109	6	109	50	60	89,511	0	658
2221	Printing	0	2	85	4	102	50	50	67,849	0	537
2222	Service activities related to printing	0	0	24	2	7	0	10	21,662	0	121
223	Reproduction of recorded media	..	..	..	..	..		..	..	..	..
23	**COKE, REFINED PETROLEUM PRODUCTS**	0	22	105	3,652	15,085	0	306	321,601	32,268	20,212
231	Coke oven products	..	..	..	..	..		..	..	..	..
232	Refined petroleum products	0	22	105	3,652	15,085	0	306	321,601	32,268	20,212
233	Processing of nuclear fuel	..	..	..	..	..		..	..	..	..
24	**CHEMICALS & CHEMICAL PRODUCTS**	764	102	513	2,113	2,321	23	1,043	975,720	47,616	10,220
241	Basic chemicals	0	72	254	1,144	1,433	23	626	566,689	13,729	5,543
2411	Basic chemicals, exc. fertilizers & nitrogen compounds	0	71	149	1,030	691	23	623	438,376	0	4,165
2412	Fertilizers and nitrogen compounds	0	1	46	110	721	0	0	109,399	13,729	1,222
2413	Plastics in primary forms and synthetic rubber	0	0	59	4	21	0	3	18,914	0	155
242	Other chemical products	764	30	259	969	888	0	417	409,031	33,887	4,678
2421	Pesticides and other agro-chemical products	..	..	..	..	..		..	..	..	..
2422	Paints, varnishes and similar coatings	0	3	46	18	87	0	41	24,702	0	284
2423	Pharmaceuticals, medicinal chem. & botanical prod.	0	3	123	848	451	0	354	210,764	18,883	2,470
2424	Soap and detergents, perfumes etc.	764	22	64	23	169	0	20	57,334	15,004	1,214
2429	Other chemical products, nec	0	2	26	80	181	0	2	116,231	0	709
243	Man-made fibres	..	..	..	..	..		..	..	..	..
25	**RUBBER AND PLASTICS PRODUCTS**	0	35	283	87	994	0	54	478,800	0	3,177
251	Rubber products	0	5	41	7	374	0	0	49,513	0	605
2511	Rubber tyres and tubes	..	..	..	..	..		..	..	..	..
2519	Other rubber products	0	5	41	7	374	0	0	49,513	0	605
252	Plastic products	0	30	242	80	620	0	54	429,287	0	2,571
26	**OTHER NON-METALLIC MINERAL PRODUCTS**	8,533	370	1,503	6,297	4,392	28	20	622,860	17	23,385
261	Glass and glass products	0	10	46	0	1,424	0	9	124,770	0	1,938
269	Non-metallic mineral products, nec	8,533	360	1,457	6,297	2,968	28	11	498,090	17	21,447
2691	Pottery, china and earthenware	0	12	19	72	91	0	2	10,562	0	234
2692	Refractory ceramic products	..	..	..	..	..		..	..	..	..
2693	Structural non-refractory clay & ceramic prod.	0	106	66	58	1,045	28	0	29,031	0	1,408
2694	Cement, lime and plaster	6,602	2	341	5,600	0	0	1	261,406	0	13,487
2695	Articles of concrete, cement and plaster	466	24	610	255	1,291	0	6	99,227	17	3,009
2696	Cutting, shaping and finishing of stone	0	1	6	0	0	0	0	1,438	0	12
2699	Other non-metallic mineral products, nec	1,465	215	415	312	541	0	2	96,426	0	3,297

ISIC Revision 3 Industry Sector	Solid TJ	LPG TJ	Distiloil TJ	RFO TJ	Gas TJ	Biomass TJ	Steam TJ	Electr MWh	Own Use MWh	TOTAL TJ
27	**BASIC METALS**	0	79	278	148	1,615	0	37	551,584	0	4,143
271	Basic iron and steel	0	65	178	42	1,506	0	21	484,262	0	3,555
272	Basic precious and non-ferrous metals	0	14	100	106	109	0	16	67,322	0	587
273	Casting of metals	..	..	..	..	..	..	..	..	..	..
2731	Casting of iron and steel	..	..	..	..	..	..	..	..	..	..
2732	Casting non-ferrous metals	..	..	..	..	..	..	..	..	..	..
28	**FABRICATED METAL PRODUCTS**	22	174	687	38	1,090	16	106	432,732	48	3,691
281	Str. metal prod., tanks, reservoirs, steam generators	2	35	316	19	192	10	49	100,054	0	983
2811	Structural metal products	2	18	263	6	75	1	25	69,402	0	640
2812	Tanks, reservoirs and containers of metal	0	17	49	7	83	9	2	22,934	0	250
2813	Steam generators, exc. central heating hot water boilers	0	0	4	6	34	0	22	7,718	0	94
289	Other fabricated metal products	20	139	371	19	898	6	57	332,678	48	2,707
2891	Forging, pressing, stamping & roll-forming of metal	..	..	..	..	..	..	..	..	..	..
2892	Treatment and coating of metals	0	9	49	5	176	0	7	29,305	48	351
2893	Cutlery, hand tools and general hardware	0	53	100	4	559	5	20	101,999	0	1,108
2899	Other fabricated metal products, nec	20	77	222	10	163	1	30	201,374	0	1,248
29	**MACHINERY AND EQUIPMENT, NEC**	15	158	1,038	575	943	10	234	512,879	0	4,819
291	General purpose machinery	0	96	587	363	500	0	134	321,613	0	2,838
2911	Engines and turbines	..	..	..	..	..	..	..	..	..	..
2912	Pumps, compressors, taps and valves	0	29	139	285	341	0	65	216,855	0	1,640
2913	Bearings, gears, gearing and driving elements	0	1	13	0	11	0	3	14,855	0	81
2914	Ovens, furnaces and furnace burners	0	0	51	0	3	0	0	724	0	57
2915	Lifting and handling equipment	0	30	190	40	60	0	31	40,624	0	497
2919	Other general purpose machinery	0	36	194	38	85	0	35	48,555	0	563
292	Special purpose machinery	0	53	379	172	300	0	65	127,254	0	1,427
2921	Agricultural and forestry machinery	0	43	139	24	165	0	8	31,574	0	493
2922	Machine-tools	0	0	68	2	5	0	11	9,889	0	122
2923	Machinery for metallurgy	..	..	..	..	..	..	..	..	..	..
2924	Machinery for mining, quarrying and construction	0	5	45	106	27	0	4	24,006	0	273
2925	Machinery for food, beverage & tobacco processing	0	4	54	13	21	0	30	22,216	0	202
2926	Machinery for textile, apparel & leather production	0	1	14	3	4	0	3	4,171	0	40
2927	Machinery for weapons and ammunition	..	..	..	..	..	..	..	..	..	..
2929	Other special purpose machinery	0	0	59	24	78	0	9	35,398	0	297
293	Domestic appliances, nec	15	9	72	40	143	10	35	64,012	0	554
30	**OFFICE, ACCOUNTING & COMPUTING MACHINERY**	0	1	9	0	19	0	14	8,945	0	75
31	**ELECTRICAL MACHINERY & APPARATUS, NEC**	0	19	151	35	96	0	136	99,464	0	795
311	Electric motors, generators and transformers	0	1	42	0	15	0	39	20,566	0	171
312	Electricity distribution and control apparatus	0	1	15	31	31	0	29	20,579	0	181
313	Insulated wire and cable	0	1	12	2	21	0	50	27,702	0	186
314	Accumulators, primary cells & primary batteries	0	5	5	0	3	0	0	9,187	0	46
315	Electric lamps and lighting equipment	0	11	40	2	22	0	14	16,233	0	147
319	Other electrical equipment, nec	0	0	37	0	4	0	4	5,197	0	64
32	**RADIO, TV & COMMUNICATION EQUIP. & APP.**	0	5	38	0	132	55	38	75,322	0	539
321	Electronic valves, tubes, other electronic components	0	0	12	0	32	0	11	29,223	0	160
322	TV & radio transmitters, apparatus for line telephony	0	0	4	0	3	0	20	5,920	0	48
323	TV & radio receivers, recording apparatus	0	5	22	0	97	55	7	40,179	0	331
33	**MEDICAL PRECISION &OPTICAL INSTRUMENTS**	2	3	75	14	94	0	64	66,274	0	491
331	Medical appliances and instruments	2	1	67	10	80	0	62	58,023	0	431
3311	Medical, surgical equipment & orthopaedic app.	2	1	31	8	39	0	16	32,002	0	212
3312	Instruments & appliances for measuring, checking etc	0	0	33	2	40	0	45	25,099	0	210
3313	Industrial process control equipment	0	0	3	0	1	0	1	922	0	8
332	Optical instruments and photographic equipment	0	2	8	4	14	0	2	8,251	0	60
333	Watches and clocks	..	..	..	..	..	..	..	..	..	..
34	**MOTOR VEHICLES, TRAILERS & SEMI-TRAILERS**	0	35	169	41	240	0	21	81,514	0	799
341	Motor vehicles	..	..	..	..	..	..	..	..	..	..
342	Bodies (coachwork) for motor vehicles	0	1	92	21	79	0	8	21,810	0	280
343	Parts, accessories for motor vehicles & their engines	0	34	77	20	161	0	13	59,704	0	520
35	**OTHER TRANSPORT EQUIPMENT**	4	21	176	106	96	0	99	146,639	0	1,030
351	Building and repairing of ships and boats	4	19	161	87	53	0	70	128,255	0	856
3511	Building and repairing of ships	4	19	145	86	49	0	70	125,648	0	825
3512	Building, repairing of pleasure & sporting boats	0	0	16	1	4	0	0	2,607	0	30
352	Railway, tramway locomotives & rolling stock	..	..	..	..	..	..	..	..	..	..
353	Aircraft and spacecraft	0	0	2	8	1	0	4	5,211	0	34
359	Transport equipment, nec	0	2	13	11	42	0	25	13,173	0	140
3591	Motorcycles	..	..	..	..	..	..	..	..	..	..
3592	Bicycles and invalid carriages	0	1	9	4	35	0	3	3,649	0	65
3599	Other transport equipment, nec	0	1	4	7	7	0	22	9,524	0	75
36	**FURNITURE; MANUFACTURING, NEC**	9	27	458	88	262	1,406	83	324,285	0	3,500
361	Furniture	9	26	393	78	202	1,401	31	244,830	0	3,021
369	Manufacturing, nec	0	1	65	10	60	5	52	79,455	0	479
3691	Jewellery and related articles	0	0	2	0	1	0	2	1,240	0	9
3692	Musical instruments	0	0	3	0	0	1	0	249	0	5
3693	Sports goods	..	..	..	..	..	..	..	..	..	..
3694	Games and toys	0	1	29	0	28	0	42	67,596	0	343
3699	Other manufacturing, nec	0	0	31	10	31	4	8	10,370	0	121
37	**RECYCLING**	..	..	..	..	..	..	..	..	..	..
371	Recycling of metal waste and scrap	..	..	..	..	..	..	..	..	..	..
372	Recycling of non-metal waste and scrap	..	..	..	..	..	..	..	..	..	..
CERR	**Non-specified industry**	..	..	..	..	..	..	..	..	..	..
15-37	**TOTAL MANUFACTURING**	14,220	1,370	8,863	22,408	41,288	5,018	4,325	7,551,759	242,005	123,807

ISIC Revision 3 Industry Sector	Solid TJ	LPG TJ	Distiloil TJ	RFO TJ	Gas TJ	Biomass TJ	Steam TJ	Electr MWh	Own Use MWh	TOTAL TJ
15 **FOOD PRODUCTS AND BEVERAGES**	3,561	246	2,791	7,104	14,029	59	1,043	2,009,530	99,705	35,708
151 Production, processing and preserving (PPP)	505	87	1,007	3,017	4,886	3	227	858,595	5,167	12,804
1511 PPP of meat and meat products	0	38	554	1,063	2,009	0	154	530,071	0	5,726
1512 Processing and preserving of fish products	505	32	261	747	2,415	1	60	211,440	0	4,782
1513 Processing, preserving of fruit & vegetables	0	16	172	40	342	2	9	73,237	0	845
1514 Vegetable and animal oils and fats	0	1	20	1,167	120	0	4	43,847	5,167	1,451
152 Dairy products	0	5	616	387	2,903	8	212	376,318	345	5,485
153 Grain mill prod., starches & prepared animal feeds	767	6	374	190	2,852	0	18	189,916	0	4,891
1531 Grain mill products	0	1	28	0	168	0	6	52,444	0	392
1532 Starches and starch products	0	0	34	0	63	0	4	10,558	0	139
1533 Prepared animal feeds	767	5	312	190	2,621	0	8	126,914	0	4,360
154 Other food products	2,015	111	529	1,976	2,532	48	497	398,464	64,786	8,909
1541 Bakery products	0	84	231	0	457	0	55	101,065	0	1,191
1542 Sugar	2,015	5	80	1,551	1	13	10	92,442	64,786	3,775
1543 Cocoa, chocolate and sugar confectionery	0	17	118	54	141	0	16	42,362	0	499
1544 Macaroni, noodles, couscous & similar farinaceous prod.	..	..	..	..	..	..	..	..	..	..
1549 Other food products, nec	0	5	100	371	1,933	35	416	162,595	0	3,445
155 Beverages	274	37	265	1,534	856	0	89	186,237	29,407	3,620
1551 Distilling, rectifying and blending of spirits	274	0	5	9	169	0	4	20,750	0	536
1552 Wines	..	..	..	..	..	..	..	..	..	..
1553 Malt liquors and malt	0	35	123	1,525	623	0	81	149,169	29,407	2,818
1554 Soft drinks; production of mineral waters	0	2	137	0	64	0	4	16,318	0	266
16 **TOBACCO PRODUCTS**	0	2	19	21	104	0	1	22,109	0	227
17 **TEXTILES**	0	40	176	157	977	1	38	139,430	0	1,891
171 Spinning, weaving and finishing of textiles	0	17	30	156	587	1	0	61,709	0	1,013
1711 Preparation and spinning of textile fibres	0	17	24	155	79	0	0	39,448	0	417
1712 Finishing of textiles	0	0	6	1	508	1	0	22,261	0	596
172 Other textiles	0	23	120	1	290	0	17	56,945	0	656
1721 Made-up textile articles, except apparel	0	2	62	0	81	0	2	16,028	0	205
1722 Carpets and rugs	0	6	15	0	115	0	8	14,625	0	197
1723 Cordage, rope, twine and netting	0	0	5	0	6	0	5	2,370	0	25
1729 Other textiles, nec	0	15	38	1	88	0	2	23,922	0	230
173 Knitted and crocheted fabrics and articles	0	0	26	0	100	0	21	20,776	0	222
18 **WEARING APPAREL, DRESSING & DYEING OF FUR**	0	0	94	1	24	0	31	19,535	37	220
181 Wearing apparel, except fur apparel	..	..	..	..	..	..	..	..	..	..
182 Dressing and dyeing of fur; articles of fur	..	..	..	..	..	..	..	..	..	..
19 **TANNING & DRESSING OF LEATHER, FOOTWEAR**	0	0	27	46	41	0	4	13,193	0	165
191 Tanning and dressing of leather	0	0	12	46	27	0	1	5,848	0	107
1911 Tanning and dressing of leather	0	0	5	46	27	0	0	4,367	0	94
1912 Luggage, handbags, saddlery & harness	0	0	7	0	0	0	1	1,481	0	13
192 Footwear	0	0	15	0	14	0	3	7,345	0	58
20 **WOOD AND WOOD PRODUCTS**	33	38	329	475	66	3,667	7	313,598	56,854	5,539
201 Sawmilling and planing of wood	33	1	51	0	5	457	0	39,979	0	691
202 Products of wood, cork, straw & plaiting materials	0	37	278	475	61	3,210	7	273,619	56,854	4,848
2021 Veneer sheets	0	4	14	248	4	542	0	57,003	0	1,017
2022 Builders' carpentry and joinery	0	31	237	227	50	2,378	7	197,875	56,854	3,438
2023 Wooden containers	0	2	14	0	2	114	0	5,522	0	152
2029 Other products of wood	0	0	13	0	5	176	0	13,219	0	242
21 **PAPER AND PAPER PRODUCTS**	60	52	165	598	2,084	4	1,035	378,611	45,320	5,198
2101 Pulp, paper and paperboard	60	7	16	0	1,584	0	1,007	228,773	45,320	3,334
2102 Corrugated paper, paperboard and their containers	0	36	128	148	182	1	26	88,329	0	839
2109 Other articles of pulp and paperboard	0	9	21	450	318	3	2	61,509	0	1,024
22 **PUBLISHING, PRINTING & REPRODUCTION**	0	9	125	23	349	53	215	184,395	0	1,438
221 Publishing	0	8	52	0	57	0	162	78,874	0	563
2211 Publishing of books & brochures	0	0	12	0	18	0	19	8,686	0	80
2212 Publishing of newspapers and periodicals	0	8	36	0	33	0	143	68,255	0	466
2213 Publishing of recorded media	..	..	..	..	..	..	..	..	..	..
2219 Other publishing	0	0	4	0	6	0	0	1,933	0	17
222 Printing and related service activities	0	1	73	23	292	53	53	105,521	0	875
2221 Printing	0	1	54	23	136	53	46	81,864	0	608
2222 Service activities related to printing	0	0	19	0	156	0	7	23,657	0	267
223 Reproduction of recorded media	..	..	..	..	..	..	..	..	..	..
23 **COKE, REFINED PETROLEUM PRODUCTS**	0	28	45	2,143	19,230	0	379	328,504	0	23,008
231 Coke oven products	..	..	..	..	..	..	..	..	..	..
232 Refined petroleum products	0	28	45	2,143	19,230	0	379	328,504	0	23,008
233 Processing of nuclear fuel	..	..	..	..	..	..	..	..	..	..
24 **CHEMICALS & CHEMICAL PRODUCTS**	728	127	408	2,094	2,496	18	1,082	1,029,299	48,760	10,483
241 Basic chemicals	0	80	143	1,113	1,604	18	593	585,178	14,720	5,605
2411 Basic chemicals, exc. fertilizers & nitrogen compounds	0	79	109	992	681	18	589	441,650	0	4,058
2412 Fertilizers and nitrogen compounds	0	1	21	121	906	0	0	126,665	14,720	1,452
2413 Plastics in primary forms and synthetic rubber	0	0	13	0	17	0	4	16,863	0	95
242 Other chemical products	728	47	265	981	892	0	489	444,121	34,040	4,878
2421 Pesticides and other agro-chemical products	..	..	..	..	..	..	..	..	..	..
2422 Paints, varnishes and similar coatings	1	4	32	2	100	0	47	30,684	0	296
2423 Pharmaceuticals, medicinal chem. & botanical prod.	0	10	168	847	354	0	410	214,404	20,663	2,486
2424 Soap and detergents, perfumes etc.	727	33	42	18	152	0	24	59,313	13,377	1,161
2429 Other chemical products, nec	0	0	23	114	286	0	8	139,720	0	934
243 Man-made fibres	..	..	..	..	..	..	..	..	..	..
25 **RUBBER AND PLASTICS PRODUCTS**	0	37	303	86	1,124	0	64	551,602	0	3,600
251 Rubber products	0	6	42	18	413	0	0	55,435	0	679
2511 Rubber tyres and tubes	..	..	..	..	..	..	..	..	..	..
2519 Other rubber products	0	6	42	18	413	0	0	55,435	0	679
252 Plastic products	0	31	261	68	711	0	64	496,167	0	2,921
26 **OTHER NON-METALLIC MINERAL PRODUCTS**	9,731	630	1,579	7,373	5,166	7	21	712,937	43	27,073
261 Glass and glass products	0	9	74	0	1,412	0	7	143,477	0	2,019
269 Non-metallic mineral products, nec	9,731	621	1,505	7,373	3,754	7	14	569,460	43	25,055
2691 Pottery, china and earthenware	0	14	24	62	97	0	3	11,126	0	240
2692 Refractory ceramic products	..	..	..	..	..	..	..	..	..	..
2693 Structural non-refractory clay & ceramic prod.	0	175	83	118	1,721	7	0	46,388	0	2,271
2694 Cement, lime and plaster	7,225	2	304	6,689	0	0	0	286,146	0	15,250
2695 Articles of concrete, cement and plaster	0	175	660	258	1,135	0	6	103,676	14	2,607
2696 Cutting, shaping and finishing of stone	0	0	1	0	0	0	0	0	0	1
2699 Other non-metallic mineral products, nec	2,506	255	433	246	801	0	5	122,124	29	4,686

ISIC Revision 3 Industry Sector		Solid TJ	LPG TJ	Distiloil TJ	RFO TJ	Gas TJ	Biomass TJ	Steam TJ	Electr MWh	Own Use MWh	TOTAL TJ
27	**BASIC METALS**	0	100	279	131	1,731	0	57	595,418	0	4,442
271	Basic iron and steel	0	95	187	9	1,638	0	38	523,366	0	3,851
272	Basic precious and non-ferrous metals	0	5	92	122	93	0	19	72,052	0	590
273	Casting of metals	..	..	..	..	..	..	..	..	..	..
2731	Casting of iron and steel	..	..	..	..	..	..	..	..	..	..
2732	Casting non-ferrous metals	..	..	..	..	..	..	..	..	..	..
28	**FABRICATED METAL PRODUCTS**	31	183	750	44	872	5	147	526,598	0	3,928
281	Str. metal prod., tanks, reservoirs, steam generators	0	39	335	22	259	0	73	123,635	0	1,173
2811	Structural metal products	0	25	294	6	164	0	48	87,756	0	853
2812	Tanks, reservoirs and containers of metal	0	14	32	0	76	0	1	24,004	0	209
2813	Steam generators, exc. central heating hot water boilers	0	0	9	16	19	0	24	11,875	0	111
289	Other fabricated metal products	31	144	415	22	613	5	74	402,963	0	2,755
2891	Forging, pressing, stamping & roll-forming of metal	..	..	..	..	..	..	..	..	..	..
2892	Treatment and coating of metals	0	12	83	5	112	0	13	43,054	0	380
2893	Cutlery, hand tools and general hardware	0	60	125	10	304	4	30	97,263	0	883
2899	Other fabricated metal products, nec	31	72	207	7	197	1	31	262,646	0	1,492
29	**MACHINERY AND EQUIPMENT, NEC**	0	209	1,046	192	1,450	1	401	614,378	0	5,511
291	General purpose machinery	0	113	631	93	794	0	224	412,864	0	3,341
2911	Engines and turbines	..	..	..	..	..	..	..	..	..	..
2912	Pumps, compressors, taps and valves	0	49	214	43	613	0	149	276,190	0	2,062
2913	Bearings, gears, gearing and driving elements	0	0	8	0	19	0	1	28,105	0	129
2914	Ovens, furnaces and furnace burners	0	0	5	0	5	0	2	1,125	0	16
2915	Lifting and handling equipment	0	44	231	5	49	0	43	50,775	0	555
2919	Other general purpose machinery	0	20	173	45	108	0	29	56,669	0	579
292	Special purpose machinery	0	91	368	86	375	1	128	140,350	0	1,554
2921	Agricultural and forestry machinery	0	84	112	44	225	0	33	37,658	0	634
2922	Machine-tools	0	0	59	2	5	1	18	10,798	0	124
2923	Machinery for metallurgy	..	..	..	..	..	..	..	..	..	..
2924	Machinery for mining, quarrying and construction	0	2	59	32	20	0	20	19,617	0	204
2925	Machinery for food, beverage & tobacco processing	0	2	41	1	43	0	37	26,775	0	220
2926	Machinery for textile, apparel & leather production	0	1	12	0	14	0	8	4,229	0	50
2927	Machinery for weapons and ammunition	..	..	..	..	..	..	..	..	..	..
2929	Other special purpose machinery	0	2	85	7	68	0	12	41,273	0	323
293	Domestic appliances, nec	0	5	47	13	281	0	49	61,164	0	615
30	**OFFICE, ACCOUNTING & COMPUTING MACHINERY**	0	0	20	0	29	0	13	10,462	0	100
31	**ELECTRICAL MACHINERY & APPARATUS, NEC**	0	22	303	20	186	0	120	125,016	0	1,101
311	Electric motors, generators and transformers	0	0	57	0	54	0	38	27,615	0	248
312	Electricity distribution and control apparatus	0	1	167	18	45	0	14	23,700	0	330
313	Insulated wire and cable	0	6	16	0	31	0	47	37,578	0	235
314	Accumulators, primary cells & primary batteries	0	4	2	1	8	0	0	13,010	0	62
315	Electric lamps and lighting equipment	0	11	34	1	36	0	9	15,083	0	145
319	Other electrical equipment, nec	0	0	27	0	12	0	12	8,030	0	80
32	**RADIO, TV & COMMUNICATION EQUIP. & APP.**	2	1	43	0	161	52	19	90,227	0	603
321	Electronic valves, tubes, other electronic components	0	0	9	0	28	0	10	33,130	0	166
322	TV & radio transmitters, apparatus for line telephony	0	0	4	0	16	0	6	11,190	0	66
323	TV & radio receivers, recording apparatus	2	1	30	0	117	52	3	45,907	0	370
33	MEDICAL PRECISION & OPTICAL INSTRUMENTS	0	6	96	1	79	0	79	74,179	0	528
331	Medical appliances and instruments	0	3	80	1	66	0	75	65,033	0	459
3311	Medical, surgical equipment & orthopaedic app.	0	3	40	1	42	0	25	35,550	0	239
3312	Instruments & appliances for measuring, checking etc	0	0	39	0	24	0	50	29,268	0	218
3313	Industrial process control equipment	0	0	1	0	0	0	0	215	0	2
332	Optical instruments and photographic equipment	0	3	16	0	13	0	4	9,146	0	69
333	Watches and clocks	..	..	..	..	..	..	..	..	..	..
34	**MOTOR VEHICLES, TRAILERS & SEMI-TRAILERS**	0	29	122	3	244	1	26	82,674	0	723
341	Motor vehicles	..	..	..	..	..	..	..	..	..	..
342	Bodies (coachwork) for motor vehicles	0	0	90	3	97	1	14	33,028	0	324
343	Parts, accessories for motor vehicles & their engines	0	29	32	0	147	0	12	49,646	0	399
35	**OTHER TRANSPORT EQUIPMENT**	10	7	247	65	297	0	111	185,674	0	1,405
351	Building and repairing of ships and boats	10	4	224	56	228	0	57	163,771	0	1,169
3511	Building and repairing of ships	10	4	213	56	224	0	56	161,736	0	1,145
3512	Building, repairing of pleasure & sporting boats	0	0	11	0	4	0	1	2,035	0	23
352	Railway, tramway locomotives & rolling stock	..	..	..	..	..	..	..	..	..	..
353	Aircraft and spacecraft	0	0	5	0	10	0	13	7,898	0	56
359	Transport equipment, nec	0	3	18	9	59	0	41	14,005	0	180
3591	Motorcycles	..	..	..	..	..	..	..	..	..	..
3592	Bicycles and invalid carriages	0	0	6	5	13	0	4	3,324	0	40
3599	Other transport equipment, nec	0	3	12	4	46	0	37	10,681	0	140
36	**FURNITURE; MANUFACTURING, NEC**	0	31	433	54	351	1,225	97	384,295	0	3,574
361	Furniture	0	30	353	54	252	1,220	39	298,887	0	3,024
369	Manufacturing, nec	0	1	80	0	99	5	58	85,408	0	550
3691	Jewellery and related articles	0	0	5	0	0	0	3	1,588	0	14
3692	Musical instruments	0	0	9	0	0	1	0	497	0	12
3693	Sports goods	..	..	..	..	..	..	..	..	..	..
3694	Games and toys	0	1	19	0	69	0	47	72,648	0	398
3699	Other manufacturing, nec	0	0	47	0	30	4	8	10,675	0	127
37	**RECYCLING**										
371	Recycling of metal waste and scrap	..	..	..	..	..	..	..	..	..	..
372	Recycling of non-metal waste and scrap	..	..	..	..	..	..	..	..	..	..
CERR	Non-specified industry	..	..	..	..	..	..	..	..	..	..
15-37	**TOTAL MANUFACTURING**	14,156	1,797	9,400	20,631	51,090	5,093	4,990	8,391,664	250,719	136,464

ISIC Revision 3 Industry Sector	Solid TJ	LPG TJ	Distiloil TJ	RFO TJ	Gas TJ	Biomass TJ	Steam TJ	Electr MWh	Own Use MWh	TOTAL TJ
15 **FOOD PRODUCTS AND BEVERAGES**	2,932	216	3,159	6,950	13,030	56	1,088	1,958,404	105,287	34,102
151 Production, processing and preserving (PPP)	467	70	825	3,272	4,488	4	234	845,226	21,131	12,327
1511 PPP of meat and meat products	0	20	427	1,129	1,967	0	157	521,501	0	5,577
1512 Processing and preserving of fish products	467	45	211	408	1,961	4	61	185,355	0	3,824
1513 Processing, preserving of fruit & vegetables	0	5	165	28	445	0	11	67,917	0	899
1514 Vegetable and animal oils and fats	0	0	22	1,707	115	0	5	70,453	21,131	2,027
152 Dairy products	0	1	1,268	405	3,413	6	231	375,551	59	6,676
153 Grain mill prod., starches & prepared animal feeds	121	22	189	92	1,583	0	24	144,944	0	2,553
1531 Grain mill products	0	17	32	9	141	0	8	56,002	0	409
1532 Starches and starch products	0	0	35	0	187	0	4	28,828	0	330
1533 Prepared animal feeds	121	5	122	83	1,255	0	12	60,114	0	1,814
154 Other food products	2,052	95	559	2,045	2,299	46	524	407,610	64,724	8,854
1541 Bakery products	0	87	255	0	491	0	38	95,556	0	1,215
1542 Sugar	2,052	4	86	1,600	0	13	13	92,443	64,724	3,868
1543 Cocoa, chocolate and sugar confectionery	0	1	112	58	108	0	24	36,699	0	435
1544 Macaroni, noodles, couscous & similar farinaceous prod.	..	..	..	..	..	..	..	..	..	..
1549 Other food products, nec	0	3	106	387	1,700	33	449	182,912	0	3,336
155 Beverages	292	28	318	1,136	1,247	0	75	185,073	19,373	3,693
1551 Distilling, rectifying and blending of spirits	292	0	34	8	182	0	4	25,935	0	613
1552 Wines	..	..	..	..	..	..	..	..	..	..
1553 Malt liquors and malt	0	17	150	1,128	1,015	0	59	144,176	19,373	2,818
1554 Soft drinks; production of mineral waters	0	11	134	0	50	0	12	14,962	0	261
16 **TOBACCO PRODUCTS**	0	3	23	1	104	0	12	20,487	0	217
17 **TEXTILES**	0	54	162	159	992	1	51	134,853	0	1,904
171 Spinning, weaving and finishing of textiles	0	17	32	158	639	1	3	59,111	0	1,063
1711 Preparation and spinning of textile fibres	0	17	16	158	103	0	2	35,287	0	423
1712 Finishing of textiles	0	0	16	0	536	1	1	23,824	0	640
172 Other textiles	0	37	107	0	330	0	22	56,921	0	701
1721 Made-up textile articles, except apparel	0	1	51	0	125	0	5	14,968	0	236
1722 Carpets and rugs	0	23	18	0	107	0	11	15,318	0	214
1723 Cordage, rope, twine and netting	0	0	5	0	4	0	4	2,499	0	22
1729 Other textiles, nec	0	13	33	0	94	0	2	24,136	0	229
173 Knitted and crocheted fabrics and articles	0	0	23	1	23	0	26	18,821	0	141
18 **WEARING APPAREL, DRESSING & DYEING OF FUR**	0	0	76	7	35	0	42	19,240	9	229
181 Wearing apparel, except fur apparel	..	..	..	..	..	..	..	..	..	..
182 Dressing and dyeing of fur; articles of fur	..	..	..	..	..	..	..	..	..	..
19 **TANNING & DRESSING OF LEATHER, FOOTWEAR**	0	0	28	28	47	0	4	12,137	0	151
191 Tanning and dressing of leather	0	0	10	28	33	0	2	4,560	0	89
1911 Tanning and dressing of leather	0	0	4	28	32	0	0	3,689	0	77
1912 Luggage, handbags, saddlery & harness	0	0	6	0	1	0	2	871	0	12
192 Footwear	0	0	18	0	14	0	2	7,577	0	61
20 **WOOD AND WOOD PRODUCTS**	0	36	299	560	85	3,499	8	315,203	53,941	5,428
201 Sawmilling and planing of wood	0	1	49	24	2	342	0	31,348	0	531
202 Products of wood, cork, straw & plaiting materials	0	35	250	536	83	3,157	8	283,855	53,941	4,897
2021 Veneer sheets	0	3	12	293	3	667	0	61,395	0	1,199
2022 Builders' carpentry and joinery	0	30	191	243	69	2,145	8	201,536	53,941	3,217
2023 Wooden containers	0	2	34	0	2	151	0	9,086	0	222
2029 Other products of wood	0	0	13	0	9	194	0	11,838	0	259
21 **PAPER AND PAPER PRODUCTS**	3	37	204	296	3,037	5	1,134	377,472	0	6,075
2101 Pulp, paper and paperboard	3	7	39	129	1,343	0	1,095	222,182	0	3,416
2102 Corrugated paper, paperboard and their containers	0	30	134	167	227	2	37	90,408	0	922
2109 Other articles of pulp and paperboard	0	0	31	0	1,467	3	2	64,882	0	1,737
22 **PUBLISHING, PRINTING & REPRODUCTION**	0	20	128	13	257	57	269	190,731	0	1,431
221 Publishing	0	19	51	0	60	0	200	82,677	0	628
2211 Publishing of books & brochures	0	0	9	0	16	0	24	7,615	0	76
2212 Publishing of newspapers and periodicals	0	19	40	0	39	0	176	71,441	0	531
2213 Publishing of recorded media	..	..	..	..	..	..	..	..	..	..
2219 Other publishing	0	0	2	0	5	0	0	3,621	0	20
222 Printing and related service activities	0	1	77	13	197	57	69	108,054	0	803
2221 Printing	0	1	61	12	187	57	60	85,669	0	686
2222 Service activities related to printing	0	0	16	1	10	0	9	22,385	0	117
223 Reproduction of recorded media	..	..	..	..	..	..	..	..	..	..
23 **COKE, REFINED PETROLEUM PRODUCTS**	0	0	54	2,408	21,420	0	314	368,121	0	25,521
231 Coke oven products	..	..	..	..	..	..	..	..	..	..
232 Refined petroleum products	0	0	54	2,408	21,420	0	314	368,121	0	25,521
233 Processing of nuclear fuel	..	..	..	..	..	..	..	..	..	..
24 **CHEMICALS & CHEMICAL PRODUCTS**	674	130	546	1,992	3,114	13	1,230	1,120,635	41,838	11,583
241 Basic chemicals	0	97	257	1,006	1,919	13	515	559,633	15,371	5,766
2411 Basic chemicals, exc. fertilizers & nitrogen compounds	0	96	151	942	730	13	510	422,856	0	3,964
2412 Fertilizers and nitrogen compounds	0	1	50	64	1,161	0	0	113,958	15,371	1,631
2413 Plastics in primary forms and synthetic rubber	0	0	56	0	28	0	5	22,819	0	171
242 Other chemical products	674	33	289	986	1,195	0	715	561,002	26,467	5,816
2421 Pesticides and other agro-chemical products	..	..	..	..	..	..	..	..	..	..
2422 Paints, varnishes and similar coatings	0	3	33	11	113	0	54	30,542	0	324
2423 Pharmaceuticals, medicinal chem. & botanical prod.	0	11	179	825	505	0	634	315,441	15,413	3,234
2424 Soap and detergents, perfumes etc.	674	17	50	22	162	0	25	57,349	11,054	1,117
2429 Other chemical products, nec	0	2	27	128	415	0	2	157,670	0	1,142
243 Man-made fibres	..	..	..	..	..	..	..	..	..	..
25 **RUBBER AND PLASTICS PRODUCTS**	0	41	348	74	1,166	2	72	545,396	8,635	3,635
251 Rubber products	0	1	55	22	424	0	1	54,928	8,635	670
2511 Rubber tyres and tubes	..	..	..	..	..	..	..	..	..	..
2519 Other rubber products	0	1	55	22	424	0	1	54,928	8,635	670
252 Plastic products	0	40	293	52	742	2	71	490,468	0	2,966
26 **OTHER NON-METALLIC MINERAL PRODUCTS**	9,787	631	2,321	7,679	4,977	7	35	729,628	180	28,063
261 Glass and glass products	0	13	56	20	1,394	0	16	140,768	0	2,006
269 Non-metallic mineral products, nec	9,787	618	2,265	7,659	3,583	7	19	588,860	180	26,057
2691 Pottery, china and earthenware	0	15	27	62	100	0	0	9,665	0	239
2692 Refractory ceramic products	..	..	..	..	..	..	..	..	..	..
2693 Structural non-refractory clay & ceramic prod.	0	153	93	99	1,604	6	0	44,497	0	2,115
2694 Cement, lime and plaster	7,524	2	1,025	7,017	0	0	0	303,685	0	16,661
2695 Articles of concrete, cement and plaster	0	196	645	253	1,102	0	11	107,786	0	2,595
2696 Cutting, shaping and finishing of stone	0	0	8	0	0	0	0	1,710	0	14
2699 Other non-metallic mineral products, nec	2,263	252	467	228	777	1	8	121,517	180	4,433

ISIC Revision 3 Industry Sector		Solid TJ	LPG TJ	Distiloil TJ	RFO TJ	Gas TJ	Biomass TJ	Steam TJ	Electr MWh	Own Use MWh	TOTAL TJ
27	BASIC METALS	0	94	251	72	1,776	0	77	632,745	0	4,548
271	Basic iron and steel	0	78	166	4	1,647	0	58	561,709	0	3,975
272	Basic precious and non-ferrous metals	0	16	85	68	129	0	19	71,036	0	573
273	Casting of metals	..	..	..	..	..	..	..	..	..	..
2731	Casting of iron and steel	..	..	..	..	..	..	..	..	..	..
2732	Casting non-ferrous metals	..	..	..	..	..	..	..	..	..	..
28	FABRICATED METAL PRODUCTS	26	188	822	39	1,129	21	160	519,836	0	4,256
281	Str. metal prod., tanks, reservoirs, steam generators	0	26	354	25	291	7	80	111,115	0	1,183
2811	Structural metal products	0	17	303	5	181	1	59	79,927	0	854
2812	Tanks, reservoirs and containers of metal	0	9	43	0	92	6	4	20,503	0	228
2813	Steam generators, exc. central heating hot water boilers	0	0	8	20	18	0	17	10,685	0	101
289	Other fabricated metal products	26	162	468	14	838	14	80	408,721	0	3,073
2891	Forging, pressing, stamping & roll-forming of metal	..	..	..	..	..	..	..	..	..	..
2892	Treatment and coating of metals	0	15	98	5	91	0	15	44,503	0	384
2893	Cutlery, hand tools and general hardware	0	68	125	1	548	7	27	114,246	0	1,187
2899	Other fabricated metal products, nec	26	79	245	8	199	7	38	249,972	0	1,502
29	MACHINERY AND EQUIPMENT, NEC	0	167	1,134	172	1,548	0	470	596,883	0	5,640
291	General purpose machinery	0	81	673	120	879	0	305	402,466	0	3,507
2911	Engines and turbines	..	..	..	..	..	..	..	..	..	..
2912	Pumps, compressors, taps and valves	0	36	235	67	665	0	218	276,606	0	2,217
2913	Bearings, gears, gearing and driving elements	0	0	11	0	16	0	8	19,357	0	105
2914	Ovens, furnaces and furnace burners	0	0	6	0	6	0	2	612	0	16
2915	Lifting and handling equipment	0	42	253	3	61	0	53	48,592	0	587
2919	Other general purpose machinery	0	3	168	50	131	0	24	57,299	0	582
292	Special purpose machinery	0	81	403	37	412	0	123	137,788	0	1,552
2921	Agricultural and forestry machinery	0	72	164	0	231	0	42	42,426	0	662
2922	Machine-tools	0	2	57	0	8	0	13	10,616	0	118
2923	Machinery for metallurgy	..	..	..	..	..	..	..	..	..	..
2924	Machinery for mining, quarrying and construction	0	4	45	37	30	0	4	18,590	0	187
2925	Machinery for food, beverage & tobacco processing	0	1	55	0	56	0	44	26,011	0	250
2926	Machinery for textile, apparel & leather production	0	1	9	0	4	0	9	3,514	0	36
2927	Machinery for weapons and ammunition	..	..	..	..	..	..	..	..	..	..
2929	Other special purpose machinery	0	1	73	0	83	0	11	36,631	0	300
293	Domestic appliances, nec	0	5	58	15	257	0	42	56,629	0	581
30	OFFICE, ACCOUNTING & COMPUTING MACHINERY	0	0	19	0	13	0	13	7,448	0	72
31	ELECTRICAL MACHINERY & APPARATUS, NEC	0	25	169	28	178	0	175	122,361	0	1,015
311	Electric motors, generators and transformers	0	4	63	0	41	0	61	32,544	0	286
312	Electricity distribution and control apparatus	0	1	25	27	40	0	20	24,335	0	201
313	Insulated wire and cable	0	6	14	0	42	0	59	33,852	0	243
314	Accumulators, primary cells & primary batteries	0	4	3	0	8	0	0	8,847	0	47
315	Electric lamps and lighting equipment	0	10	34	1	39	0	16	14,061	0	151
319	Other electrical equipment, nec	0	0	30	0	8	0	19	8,722	0	88
32	RADIO, TV & COMMUNICATION EQUIP. & APP.	0	0	70	0	239	59	30	88,025	0	715
321	Electronic valves, tubes, other electronic components	0	0	33	0	49	0	17	31,978	0	214
322	TV & radio transmitters, apparatus for line telephony	0	0	7	0	21	0	9	12,576	0	82
323	TV & radio receivers, recording apparatus	0	0	30	0	169	59	4	43,471	0	418
33	MEDICAL PRECISION &OPTICAL INSTRUMENTS	0	11	111	4	79	0	106	69,754	0	562
331	Medical appliances and instruments	0	8	96	4	60	0	99	60,195	0	484
3311	Medical, surgical equipment & orthopaedic app.	0	7	41	4	41	0	33	32,562	0	243
3312	Instruments & appliances for measuring, checking etc	0	1	54	0	19	0	65	27,417	0	238
3313	Industrial process control equipment	0	0	1	0	0	0	1	216	0	3
332	Optical instruments and photographic equipment	0	3	15	0	19	0	7	9,559	0	78
333	Watches and clocks	..	..	..	..	..	..	..	..	..	..
34	MOTOR VEHICLES, TRAILERS & SEMI-TRAILERS	0	35	143	0	273	0	31	89,628	0	805
341	Motor vehicles	..	..	..	..	..	..	..	..	..	..
342	Bodies (coachwork) for motor vehicles	0	6	111	0	111	0	19	40,126	0	391
343	Parts, accessories for motor vehicles & their engines	0	29	32	0	162	0	12	49,502	0	413
35	OTHER TRANSPORT EQUIPMENT	5	2	188	16	508	0	75	172,406	0	1,415
351	Building and repairing of ships and boats	5	1	159	16	435	0	36	150,436	0	1,194
3511	Building and repairing of ships	5	1	146	16	430	0	35	148,163	0	1,166
3512	Building, repairing of pleasure & sporting boats	0	0	13	0	5	0	1	2,273	0	27
352	Railway, tramway locomotives & rolling stock	..	..	..	..	..	..	..	..	..	..
353	Aircraft and spacecraft	0	0	10	0	22	0	0	10,372	0	69
359	Transport equipment, nec	0	1	19	0	51	0	39	11,598	0	152
3591	Motorcycles	..	..	..	..	..	..	..	..	..	..
3592	Bicycles and invalid carriages	0	0	9	0	38	0	0	3,336	0	59
3599	Other transport equipment, nec	0	1	10	0	13	0	39	8,262	0	93
36	FURNITURE; MANUFACTURING, NEC	0	28	507	41	386	1,070	134	365,046	0	3,480
361	Furniture	0	27	453	30	267	1,065	70	284,378	0	2,936
369	Manufacturing, nec	0	1	54	11	119	5	64	80,668	0	544
3691	Jewellery and related articles	0	0	2	10	12	0	3	2,212	0	35
3692	Musical instruments	0	0	4	0	0	1	1	274	0	7
3693	Sports goods	..	..	..	..	..	..	..	..	..	..
3694	Games and toys	0	1	11	0	77	0	49	68,118	0	383
3699	Other manufacturing, nec	0	0	37	1	30	4	11	10,064	0	119
37	RECYCLING	..	..	..	..	..	..	..	..	..	..
371	Recycling of metal waste and scrap	..	..	..	..	..	..	..	..	..	..
372	Recycling of non-metal waste and scrap	..	..	..	..	..	..	..	..	..	..
CERR	Non-specified industry	..	..	..	..	..	..	..	..	..	..
15-37	TOTAL MANUFACTURING	13,427	1,718	10,762	20,539	54,393	4,790	5,530	8,456,439	209,890	140,847

ISIC Revision 3 Industry Sector		Solid TJ	LPG TJ	Distiloil TJ	RFO TJ	Gas TJ	Biomass TJ	Steam TJ	Electr MWh	Own Use MWh	TOTAL TJ
15	**FOOD PRODUCTS AND BEVERAGES**	2,977	179	2,782	6,120	15,127	5	1,121	1,962,573	86,057	35,066
151	Production, processing and preserving (PPP)	443	75	762	2,803	5,213	5	188	846,953	4,197	12,523
1511	PPP of meat and meat products	0	30	436	1,013	2,165	2	122	534,546	0	5,692
1512	Processing and preserving of fish products	443	40	144	489	2,471	3	55	193,966	0	4,343
1513	Processing, preserving of fruit & vegetables	0	5	161	0	463	0	7	68,788	0	884
1514	Vegetable and animal oils and fats	0	0	21	1,301	114	0	4	49,653	4,197	1,604
152	Dairy products	0	0	1,056	302	3,893	0	220	382,492	62	6,848
153	Grain mill prod., starches & prepared animal feeds	73	6	169	8	1,845	0	19	144,275	0	2,639
1531	Grain mill products	0	0	26	8	192	0	8	52,781	0	424
1532	Starches and starch products	0	0	40	0	195	0	4	28,304	0	341
1533	Prepared animal feeds	73	6	103	0	1,458	0	7	63,190	0	1,874
154	Other food products	2,333	70	545	2,222	2,196	0	471	406,820	71,912	9,043
1541	Bakery products	0	65	240	0	485	0	37	98,245	0	1,181
1542	Sugar	2,332	0	96	1,867	1	0	8	94,210	71,912	4,384
1543	Cocoa, chocolate and sugar confectionery	0	1	122	55	109	0	25	38,067	0	449
1544	Macaroni, noodles, couscous & similar farinaceous prod.	..	..	..	..	..	..	..	..	..	..
1549	Other food products, nec	1	4	87	300	1,601	0	401	176,298	0	3,029
155	Beverages	128	28	250	785	1,980	0	223	182,033	9,886	4,014
1551	Distilling, rectifying and blending of spirits	128	0	8	6	226	0	151	26,290	0	614
1552	Wines	..	..	..	..	..	..	..	..	..	..
1553	Malt liquors and malt	0	17	142	779	1,399	0	53	139,823	9,886	2,858
1554	Soft drinks; production of mineral waters	0	11	100	0	355	0	19	15,920	0	542
16	**TOBACCO PRODUCTS**	0	2	20	0	105	0	11	19,937	0	210
17	**TEXTILES**	0	43	126	127	982	1	42	138,401	0	1,819
171	Spinning, weaving and finishing of textiles	0	18	12	125	679	1	4	56,797	0	1,043
1711	Preparation and spinning of textile fibres	0	18	9	125	104	0	3	32,269	0	375
1712	Finishing of textiles	0	0	3	0	575	1	1	24,528	0	668
172	Other textiles	0	25	91	1	282	0	20	64,824	0	652
1721	Made-up textile articles, except apparel	0	2	46	0	73	0	4	13,234	0	173
1722	Carpets and rugs	0	22	15	0	116	0	11	15,158	0	219
1723	Cordage, rope, twine and netting	0	0	5	0	4	0	3	2,492	0	21
1729	Other textiles, nec	0	1	25	1	89	0	2	33,940	0	240
173	Knitted and crocheted fabrics and articles	0	0	23	1	21	0	18	16,780	0	123
18	**WEARING APPAREL, DRESSING & DYEING OF FUR**	0	0	60	0	51	0	39	17,403	9	213
181	Wearing apparel, except fur apparel	..	..	..	..	..	..	..	..	..	..
182	Dressing and dyeing of fur; articles of fur	..	..	..	..	..	..	..	..	..	..
19	**TANNING & DRESSING OF LEATHER, FOOTWEAR**	0	0	23	0	43	0	4	10,618	0	108
191	Tanning and dressing of leather	0	0	8	0	32	0	1	3,482	0	54
1911	Tanning and dressing of leather	0	0	2	0	31	0	0	2,524	0	42
1912	Luggage, handbags, saddlery & harness	0	0	6	0	1	0	1	958	0	11
192	Footwear	0	0	15	0	11	0	3	7,136	0	55
20	**WOOD AND WOOD PRODUCTS**	0	36	294	669	91	3,530	8	328,792	55,310	5,613
201	Sawmilling and planing of wood	0	1	46	32	1	426	0	33,179	0	625
202	Products of wood, cork, straw & plaiting materials	0	35	248	637	90	3,104	8	295,613	55,310	4,987
2021	Veneer sheets	0	1	14	322	2	633	0	65,526	0	1,208
2022	Builders' carpentry and joinery	0	32	185	315	77	2,205	8	210,424	55,310	3,380
2023	Wooden containers	0	2	36	0	2	163	0	9,095	0	236
2029	Other products of wood	0	0	13	0	9	103	0	10,568	0	163
21	**PAPER AND PAPER PRODUCTS**	0	41	169	184	3,265	9	1,186	392,912	0	6,268
2101	Pulp, paper and paperboard	0	7	16	1	1,550	0	1,153	232,825	0	3,565
2102	Corrugated paper, paperboard and their containers	0	34	122	182	234	5	31	96,030	0	954
2109	Other articles of pulp and paperboard	0	0	31	1	1,481	4	2	64,057	0	1,750
22	**PUBLISHING, PRINTING & REPRODUCTION**	0	18	105	17	194	32	250	185,131	0	1,282
221	Publishing	0	17	50	1	59	0	178	83,672	0	606
2211	Publishing of books & brochures	0	0	8	0	15	0	23	8,455	0	76
2212	Publishing of newspapers and periodicals	0	17	39	1	39	0	155	72,937	0	514
2213	Publishing of recorded media	..	..	..	..	..	..	..	..	..	..
2219	Other publishing	0	0	3	0	5	0	0	2,280	0	16
222	Printing and related service activities	0	1	55	16	135	32	72	101,459	0	676
2221	Printing	0	1	40	15	126	32	62	79,070	0	561
2222	Service activities related to printing	0	0	15	1	9	0	10	22,389	0	116
223	Reproduction of recorded media	..	..	..	..	..	..	..	..	..	..
23	**COKE, REFINED PETROLEUM PRODUCTS**	0	0	75	1,318	18,382	0	257	293,010	0	21,087
231	Coke oven products	..	..	..	..	..	..	..	..	..	..
232	Refined petroleum products	0	0	75	1,318	18,382	0	257	293,010	0	21,087
233	Processing of nuclear fuel	..	..	..	..	..	..	..	..	..	..
24	**CHEMICALS & CHEMICAL PRODUCTS**	566	97	478	1,803	3,310	13	1,159	1,170,862	29,527	11,535
241	Basic chemicals	0	62	183	775	2,094	13	950	644,077	8,713	6,364
2411	Basic chemicals, exc. fertilizers & nitrogen compounds	0	54	95	703	720	13	945	508,981	0	4,362
2412	Fertilizers and nitrogen compounds	0	1	29	72	1,347	0	0	111,465	8,713	1,819
2413	Plastics in primary forms and synthetic rubber	0	7	59	0	27	0	5	23,631	0	183
242	Other chemical products	566	35	295	1,028	1,216	0	209	526,785	20,814	5,170
2421	Pesticides and other agro-chemical products	..	..	..	..	..	..	..	..	..	..
2422	Paints, varnishes and similar coatings	0	3	40	14	101	0	53	31,170	0	323
2423	Pharmaceuticals, medicinal chem. & botanical prod.	0	11	185	841	529	0	123	258,848	14,442	2,569
2424	Soap and detergents, perfumes etc.	566	18	46	27	164	0	28	62,708	6,372	1,052
2429	Other chemical products, nec	0	3	24	146	422	0	5	174,059	0	1,227
243	Man-made fibres	..	..	..	..	..	..	..	..	..	..
25	**RUBBER AND PLASTICS PRODUCTS**	0	39	286	81	1,319	3	67	582,397	0	3,892
251	Rubber products	0	1	23	33	572	0	1	64,244	0	861
2511	Rubber tyres and tubes	..	..	..	..	..	..	..	..	..	..
2519	Other rubber products	0	1	23	33	572	0	1	64,244	0	861
252	Plastic products	0	38	263	48	747	3	66	518,153	0	3,030
26	**OTHER NON-METALLIC MINERAL PRODUCTS**	9,250	616	3,124	7,941	5,501	6	37	752,826	708	29,183
261	Glass and glass products	0	11	58	18	1,347	0	18	143,684	163	1,969
269	Non-metallic mineral products, nec	9,250	605	3,066	7,923	4,154	6	19	609,142	545	27,214
2691	Pottery, china and earthenware	0	13	24	63	113	0	0	10,697	0	252
2692	Refractory ceramic products	..	..	..	..	..	..	..	..	..	..
2693	Structural non-refractory clay & ceramic prod.	0	66	74	120	1,754	6	0	46,037	0	2,186
2694	Cement, lime and plaster	7,320	3	1,816	7,439	0	0	0	311,992	0	17,701
2695	Articles of concrete, cement and plaster	0	291	685	130	1,225	0	11	114,378	545	2,752
2696	Cutting, shaping and finishing of stone	0	0	8	0	0	0	1	1,574	0	15
2699	Other non-metallic mineral products, nec	1,930	232	459	171	1,062	0	7	124,464	0	4,309

ISIC Revision 3 Industry Sector		Solid TJ	LPG TJ	Distiloil TJ	RFO TJ	Gas TJ	Biomass TJ	Steam TJ	Electr MWh	Own Use MWh	TOTAL TJ
27	**BASIC METALS**	0	81	215	82	1,873	0	82	659,242	0	4,706
271	Basic iron and steel	0	75	137	4	1,727	0	65	584,724	0	4,113
272	Basic precious and non-ferrous metals	0	6	78	78	146	0	17	74,518	0	593
273	Casting of metals	..	..	..	..	..	..	..	..	..	..
2731	Casting of iron and steel	..	..	..	..	..	..	..	..	..	..
2732	Casting non-ferrous metals	..	..	..	..	..	..	..	..	..	..
28	**FABRICATED METAL PRODUCTS**	24	167	757	33	1,120	18	172	531,368	0	4,204
281	Str. metal prod., tanks, reservoirs, steam generators	0	24	371	20	286	5	85	112,154	0	1,195
2811	Structural metal products	0	19	328	2	175	1	56	82,983	0	880
2812	Tanks, reservoirs and containers of metal	0	5	38	0	95	4	6	18,919	0	216
2813	Steam generators, exc. central heating hot water boilers	0	0	5	18	16	0	23	10,252	0	99
289	Other fabricated metal products	24	143	386	13	834	13	87	419,214	0	3,009
2891	Forging, pressing, stamping & roll-forming of metal	..	..	..	..	..	..	..	..	..	..
2892	Treatment and coating of metals	0	16	97	0	84	1	15	44,534	0	373
2893	Cutlery, hand tools and general hardware	0	54	111	1	544	8	30	119,274	0	1,177
2899	Other fabricated metal products, nec	24	73	178	12	206	4	42	255,406	0	1,458
29	**MACHINERY AND EQUIPMENT, NEC**	0	226	1,104	135	1,440	0	503	616,892	0	5,629
291	General purpose machinery	0	124	658	102	833	0	300	423,657	0	3,542
2911	Engines and turbines	..	..	..	..	..	..	..	..	..	..
2912	Pumps, compressors, taps and valves	0	41	237	9	610	0	201	288,370	0	2,136
2913	Bearings, gears, gearing and driving elements	0	0	10	0	16	0	16	18,476	0	109
2914	Ovens, furnaces and furnace burners	0	0	7	0	5	0	4	1,066	0	20
2915	Lifting and handling equipment	0	57	208	6	66	0	48	48,220	0	559
2919	Other general purpose machinery	0	26	196	87	136	0	31	67,525	0	719
292	Special purpose machinery	0	98	380	33	396	0	159	137,474	0	1,561
2921	Agricultural and forestry machinery	0	82	173	0	232	0	32	46,323	0	686
2922	Machine-tools	0	2	47	0	10	0	14	8,882	0	105
2923	Machinery for metallurgy	..	..	..	..	..	..	..	..	..	..
2924	Machinery for mining, quarrying and construction	0	3	43	31	26	0	39	18,129	0	207
2925	Machinery for food, beverage & tobacco processing	0	1	52	0	54	0	43	25,444	0	242
2926	Machinery for textile, apparel & leather production	0	1	8	0	0	0	7	3,097	0	27
2927	Machinery for weapons and ammunition	..	..	..	..	..	..	..	..	..	..
2929	Other special purpose machinery	0	9	57	2	74	0	24	35,599	0	294
293	Domestic appliances, nec	0	4	66	0	211	0	44	55,761	0	526
30	**OFFICE, ACCOUNTING & COMPUTING MACHINERY**	0	1	14	0	9	0	14	8,242	0	68
31	**ELECTRICAL MACHINERY & APPARATUS, NEC**	0	24	144	26	165	0	154	113,439	753	919
311	Electric motors, generators and transformers	0	2	51	0	37	0	57	36,476	753	276
312	Electricity distribution and control apparatus	0	0	26	25	41	0	15	22,337	0	187
313	Insulated wire and cable	0	6	14	0	32	0	55	27,289	0	205
314	Accumulators, primary cells & primary batteries	0	3	3	0	6	0	0	6,661	0	36
315	Electric lamps and lighting equipment	0	13	38	1	41	0	19	13,956	0	162
319	Other electrical equipment, nec	0	0	12	0	8	0	8	6,720	0	52
32	**RADIO, TV & COMMUNICATION EQUIP. & APP.**	0	2	71	0	218	52	27	91,259	0	699
321	Electronic valves, tubes, other electronic components	0	2	36	0	48	0	16	37,986	0	239
322	TV & radio transmitters, apparatus for line telephony	0	0	10	0	21	0	6	11,152	0	77
323	TV & radio receivers, recording apparatus	0	0	25	0	149	52	5	42,121	0	383
33	**MEDICAL PRECISION &OPTICAL INSTRUMENTS**	0	9	94	3	204	0	98	78,787	29	692
331	Medical appliances and instruments	0	5	79	3	179	0	89	64,529	29	587
3311	Medical, surgical equipment & orthopaedic app.	0	5	43	3	51	0	32	36,766	29	266
3312	Instruments & appliances for measuring, checking etc	0	0	35	0	128	0	57	27,457	0	319
3313	Industrial process control equipment	0	0	1	0	0	0	0	306	0	2
332	Optical instruments and photographic equipment	0	4	15	0	25	0	9	14,258	0	104
333	Watches and clocks	..	..	..	..	..	..	..	..	..	..
34	**MOTOR VEHICLES, TRAILERS & SEMI-TRAILERS**	0	34	135	0	244	0	31	86,679	0	756
341	Motor vehicles	..	..	..	..	..	..	..	..	..	..
342	Bodies (coachwork) for motor vehicles	0	5	118	0	104	0	18	39,228	0	386
343	Parts, accessories for motor vehicles & their engines	0	29	17	0	140	0	13	47,451	0	370
35	**OTHER TRANSPORT EQUIPMENT**	4	3	163	2	456	0	93	164,245	7	1,312
351	Building and repairing of ships and boats	4	2	126	2	412	0	49	146,802	7	1,123
3511	Building and repairing of ships	4	2	112	2	407	0	49	144,681	7	1,097
3512	Building, repairing of pleasure & sporting boats	0	0	14	0	5	0	0	2,121	0	27
352	Railway, tramway locomotives & rolling stock	..	..	..	..	..	..	..	..	..	..
353	Aircraft and spacecraft	0	0	14	0	19	0	13	7,121	0	72
359	Transport equipment, nec	0	1	23	0	25	0	31	10,322	0	117
3591	Motorcycles	..	..	..	..	..	..	..	..	..	..
3592	Bicycles and invalid carriages	0	0	11	0	16	0	0	2,625	0	36
3599	Other transport equipment, nec	0	1	12	0	9	0	31	7,697	0	81
36	**FURNITURE; MANUFACTURING, NEC**	7	25	434	31	366	1,032	120	377,077	0	3,372
361	Furniture	7	24	380	29	273	1,027	58	293,002	0	2,853
369	Manufacturing, nec	0	1	54	2	93	5	62	84,075	0	520
3691	Jewellery and related articles	0	0	1	0	0	0	2	1,350	0	8
3692	Musical instruments	0	0	4	0	0	1	0	272	0	6
3693	Sports goods	..	..	..	..	..	..	..	..	..	..
3694	Games and toys	0	0	12	1	63	0	46	72,358	0	382
3699	Other manufacturing, nec	0	1	37	1	30	4	14	10,095	0	123
37	**RECYCLING**	..	..	..	..	..	..	..	..	..	..
371	Recycling of metal waste and scrap	..	..	..	..	..	..	..	..	..	..
372	Recycling of non-metal waste and scrap	..	..	..	..	..	..	..	..	..	..
CERR	Non-specified industry	..	..	..	..	..	..	..	..	..	..
15-37	**TOTAL MANUFACTURING**	12,828	1,643	10,673	18,572	54,465	4,701	5,475	8,582,092	172,400	138,632

ISIC Revision 3 Industry Sector	Solid TJ	LPG TJ	Distiloil TJ	RFO TJ	Gas TJ	Biomass TJ	Steam TJ	Electr MWh	Own Use MWh	TOTAL TJ
15 FOOD PRODUCTS AND BEVERAGES	15,063	5,562	41,167	..	72,935	..	3,435	14,743,519	..	191,239
151 Production, processing and preserving (PPP)	35	2,243	8,882	..	11,583	..	1,193	3,780,215	..	37,545
1511 PPP of meat and meat products	..	..	..	..	..	..	..	..	..	..
1512 Processing and preserving of fish products	..	..	..	..	..	..	..	..	..	..
1513 Processing, preserving of fruit & vegetables	..	..	..	..	..	..	..	..	..	..
1514 Vegetable and animal oils and fats	..	..	..	..	..	..	..	..	..	..
152 Dairy products	399	919	10,673	..	12,800	..	473	2,858,448	..	35,554
153 Grain mill prod., starches & prepared animal feeds	4,634	1,043	4,524	..	16,988	..	346	3,447,692	..	39,947
1531 Grain mill products	..	..	..	..	..	..	..	..	..	..
1532 Starches and starch products	..	..	..	..	..	..	..	..	..	..
1533 Prepared animal feeds	..	..	..	..	..	..	..	..	..	..
154 Other food products	9,812	972	14,099	..	23,793	..	1,208	3,205,208	..	61,423
1541 Bakery products	..	..	..	..	..	..	..	..	..	..
1542 Sugar	..	..	..	..	..	..	..	..	..	..
1543 Cocoa, chocolate and sugar confectionery	..	..	..	..	..	..	..	..	..	..
1544 Macaroni, noodles, couscous & similar farinaceous prod.	..	..	..	..	..	..	..	..	..	..
1549 Other food products, nec	..	..	..	..	..	..	..	..	..	..
155 Beverages	183	385	2,989	..	7,771	..	215	1,451,956	..	16,770
1551 Distilling, rectifying and blending of spirits	..	..	..	..	..	..	..	..	..	..
1552 Wines	..	..	..	..	..	..	..	..	..	..
1553 Malt liquors and malt	..	..	..	..	..	..	..	..	..	..
1554 Soft drinks; production of mineral waters	..	..	..	..	..	..	..	..	..	..
16 TOBACCO PRODUCTS	..	..	..	..	..	..	..	..	..	..
17 TEXTILES	100	1,406	7,548	..	14,711	..	377	3,132,559	..	35,419
171 Spinning, weaving and finishing of textiles	79	1,151	5,912	..	10,806	..	312	2,318,356	..	26,606
1711 Preparation and spinning of textile fibres	..	..	..	..	..	..	..	..	..	..
1712 Finishing of textiles	..	..	..	..	..	..	..	..	..	..
172 Other textiles	0	178	1,313	..	2,772	..	28	565,375	..	6,326
1721 Made-up textile articles, except apparel	..	..	..	..	..	..	..	..	..	..
1722 Carpets and rugs	..	..	..	..	..	..	..	..	..	..
1723 Cordage, rope, twine and netting	..	..	..	..	..	..	..	..	..	..
1729 Other textiles, nec	..	..	..	..	..	..	..	..	..	..
173 Knitted and crocheted fabrics and articles	21	77	323	..	1,133	..	37	248,828	..	2,487
18 WEARING APPAREL, DRESSING & DYEING OF FUR	37	45	557	..	1,131	..	26	244,144	..	2,675
181 Wearing apparel, except fur apparel	37	37	478	..	1,127	..	26	241,797	..	2,575
182 Dressing and dyeing of fur; articles of fur	0	8	79	..	4	..	0	2,347	..	99
19 TANNING & DRESSING OF LEATHER, FOOTWEAR	40	177	717	..	313	..	95	280,954	..	2,353
191 Tanning and dressing of leather	40	102	435	..	159	..	94	114,467	..	1,242
1911 Tanning and dressing of leather	..	..	..	..	..	..	..	..	..	..
1912 Luggage, handbags, saddlery & harness	..	..	..	..	..	..	..	..	..	..
192 Footwear	0	75	282	..	154	..	1	166,487	..	1,111
20 WOOD AND WOOD PRODUCTS	91	262	1,972	..	2,564	..	0	1,545,464	..	10,453
201 Sawmilling and planing of wood	0	112	448	..	82	..	0	377,079	..	1,999
202 Products of wood, cork, straw & plaiting materials	91	150	1,524	..	2,482	..	0	1,168,385	..	8,453
2021 Veneer sheets	..	..	..	..	..	..	..	..	..	..
2022 Builders' carpentry and joinery	..	..	..	..	..	..	..	..	..	..
2023 Wooden containers	..	..	..	..	..	..	..	..	..	..
2029 Other products of wood	..	..	..	..	..	..	..	..	..	..
21 PAPER AND PAPER PRODUCTS	8,189	599	16,871	..	42,890	..	2,028	12,012,916	..	113,823
2101 Pulp, paper and paperboard	..	..	..	..	..	..	..	..	..	..
2102 Corrugated paper, paperboard and their containers	..	..	..	..	..	..	..	..	..	..
2109 Other articles of pulp and paperboard	..	..	..	..	..	..	..	..	..	..
22 PUBLISHING, PRINTING & REPRODUCTION	1	178	795	..	2,991	..	86	1,396,344	..	9,078
221 Publishing	1	4	92	..	258	..	23	194,286	..	1,077
2211 Publishing of books & brochures	..	..	..	..	..	..	..	..	..	..
2212 Publishing of newspapers and periodicals	..	..	..	..	..	..	..	..	..	..
2213 Publishing of recorded media	..	..	..	..	..	..	..	..	..	..
2219 Other publishing	..	..	..	..	..	..	..	..	..	..
222 Printing and related service activities	0	166	663	..	2,727	..	63	1,144,458	..	7,739
2221 Printing	..	..	..	..	..	..	..	..	..	..
2222 Service activities related to printing	..	..	..	..	..	..	..	..	..	..
223 Reproduction of recorded media	0	8	40	..	6	..	0	57,600	..	261
23 COKE, REFINED PETROLEUM PRODUCTS	..	..	..	..	..	..	..	..	..	..
231 Coke oven products	..	..	..	..	..	..	..	..	..	..
232 Refined petroleum products	..	..	..	..	..	..	..	..	..	..
233 Processing of nuclear fuel	..	..	..	..	..	..	..	..	..	..
24 CHEMICALS & CHEMICAL PRODUCTS	810	554	5,492	..	16,542	..	1,756	24,336,835	..	112,767
241 Basic chemicals	c	c	c	..	c	..	c	20,759,094	..	74,733
2411 Basic chemicals, exc. fertilizers & nitrogen compounds	..	..	..	..	..	..	..	..	..	..
2412 Fertilizers and nitrogen compounds	..	..	..	..	..	..	..	..	..	..
2413 Plastics in primary forms and synthetic rubber	..	..	..	..	..	..	..	..	..	..
242 Other chemical products	810	550	5,360	..	14,882	..	1,512	3,174,397	..	34,542
2421 Pesticides and other agro-chemical products	..	..	..	..	..	..	..	..	..	..
2422 Paints, varnishes and similar coatings	..	..	..	..	..	..	..	..	..	..
2423 Pharmaceuticals, medicinal chem. & botanical prod.	..	..	..	..	..	..	..	..	..	..
2424 Soap and detergents, perfumes etc.	..	..	..	..	..	..	..	..	..	..
2429 Other chemical products, nec	..	..	..	..	..	..	..	..	..	..
243 Man-made fibres	0	4	132	..	1,660	..	244	403,344	..	3,492
25 RUBBER AND PLASTICS PRODUCTS	513	732	4,453	..	11,046	..	992	5,907,562	..	39,003
251 Rubber products	510	63	2,104	..	5,796	..	844	1,671,774	..	15,335
2511 Rubber tyres and tubes	..	..	..	..	..	..	..	..	..	..
2519 Other rubber products	..	..	..	..	..	..	..	..	..	..
252 Plastic products	3	669	2,349	..	5,250	..	148	4,235,788	..	23,668
26 OTHER NON-METALLIC MINERAL PRODUCTS	15,396	2,085	67,815	..	57,796	..	0	7,437,562	..	169,867
261 Glass and glass products	0	716	22,378	..	21,352	..	0	2,807,863	..	54,554
269 Non-metallic mineral products, nec	15,396	1,369	45,437	..	36,444	..	0	4,629,699	..	115,313
2691 Pottery, china and earthenware	..	..	..	..	..	..	..	..	..	..
2692 Refractory ceramic products	..	..	..	..	..	..	..	..	..	..
2693 Structural non-refractory clay & ceramic prod.	..	..	..	..	..	..	..	..	..	..
2694 Cement, lime and plaster	..	..	..	..	..	..	..	..	..	..
2695 Articles of concrete, cement and plaster	..	..	..	..	..	..	..	..	..	..
2696 Cutting, shaping and finishing of stone	..	..	..	..	..	..	..	..	..	..
2699 Other non-metallic mineral products, nec	..	..	..	..	..	..	..	..	..	..

ISIC Revision 3 Industry Sector		Solid TJ	LPG TJ	Distiloil TJ	RFO TJ	Gas TJ	Biomass TJ	Steam TJ	Electr MWh	Own Use MWh	TOTAL TJ
27	**BASIC METALS**	65,190	975	4,811	..	37,388	..	209	23,881,540	..	194,547
271	Basic iron and steel	63,339	452	4,326	..	33,162	..	209	13,183,685	..	148,949
272	Basic precious and non-ferrous metals	c	c	c	..	c	..	c	8,826,188	..	31,774
273	Casting of metals	1,851	523	485	..	4,226	..	0	1,871,667	..	13,823
2731	Casting of iron and steel	..	..	..	..	..	..	..	..	..	..
2732	Casting non-ferrous metals	..	..	..	..	..	..	..	..	..	..
28	**FABRICATED METAL PRODUCTS**	539	1,971	2,508	..	12,832	..	36	3,882,247	..	31,862
281	Str. metal prod., tanks, reservoirs, steam generators	268	531	632	..	2,390	..	2	665,591	..	6,219
2811	Structural metal products	..	..	..	..	..	..	..	..	..	..
2812	Tanks, reservoirs and containers of metal	..	..	..	..	..	..	..	..	..	..
2813	Steam generators, exc. central heating hot water boilers	..	..	..	..	..	..	..	..	..	..
289	Other fabricated metal products	271	1,440	1,876	..	10,442	..	34	3,216,656	..	25,643
2891	Forging, pressing, stamping & roll-forming of metal	..	..	..	..	..	..	..	..	..	..
2892	Treatment and coating of metals	..	..	..	..	..	..	..	..	..	..
2893	Cutlery, hand tools and general hardware	..	..	..	..	..	..	..	..	..	..
2899	Other fabricated metal products, nec	..	..	..	..	..	..	..	..	..	..
29	**MACHINERY AND EQUIPMENT, NEC**	117	979	2,334	..	8,247	..	43	2,445,092	..	20,522
291	General purpose machinery	23	493	1,129	..	3,544	..	20	1,238,061	..	9,666
2911	Engines and turbines	..	..	..	..	..	..	..	..	..	..
2912	Pumps, compressors, taps and valves	..	..	..	..	..	..	..	..	..	..
2913	Bearings, gears, gearing and driving elements	..	..	..	..	..	..	..	..	..	..
2914	Ovens, furnaces and furnace burners	..	..	..	..	..	..	..	..	..	..
2915	Lifting and handling equipment	..	..	..	..	..	..	..	..	..	..
2919	Other general purpose machinery	..	..	..	..	..	..	..	..	..	..
292	Special purpose machinery	17	398	955	..	3,285	..	23	835,910	..	7,687
2921	Agricultural and forestry machinery	..	..	..	..	..	..	..	..	..	..
2922	Machine-tools	..	..	..	..	..	..	..	..	..	..
2923	Machinery for metallurgy	..	..	..	..	..	..	..	..	..	..
2924	Machinery for mining, quarrying and construction	..	..	..	..	..	..	..	..	..	..
2925	Machinery for food, beverage & tobacco processing	..	..	..	..	..	..	..	..	..	..
2926	Machinery for textile, apparel & leather production	..	..	..	..	..	..	..	..	..	..
2927	Machinery for weapons and ammunition	..	..	..	..	..	..	..	..	..	..
2929	Other special purpose machinery	..	..	..	..	..	..	..	..	..	..
293	Domestic appliances, nec	77	88	250	..	1,418	..	0	371,121	..	3,169
30	**OFFICE, ACCOUNTING & COMPUTING MACHINERY**	0	0	35	..	335	..	0	258,861	..	1,302
31	**ELECTRICAL MACHINERY & APPARATUS, NEC**	140	454	1,173	..	5,202	..	175	2,174,587	..	14,973
311	Electric motors, generators and transformers	3	48	164	..	898	..	0	236,134	..	1,963
312	Electricity distribution and control apparatus	41	51	275	..	1,032	..	151	592,240	..	3,682
313	Insulated wire and cable	0	58	281	..	855	..	6	501,357	..	3,005
314	Accumulators, primary cells & primary batteries	0	30	9	..	838	..	18	248,426	..	1,789
315	Electric lamps and lighting equipment	0	53	24	..	647	..	0	89,860	..	1,047
319	Other electrical equipment, nec	96	214	420	..	932	..	0	506,570	..	3,486
32	**RADIO, TV & COMMUNICATION EQUIP. & APP.**	0	47	606	..	2,060	..	54	1,947,535	..	9,778
321	Electronic valves, tubes, other electronic components	0	34	424	..	1,205	..	33	1,464,990	..	6,970
322	TV & radio transmitters, apparatus for line telephony	0	13	143	..	605	..	20	377,360	..	2,139
323	TV & radio receivers, recording apparatus	0	0	39	..	250	..	1	105,185	..	669
33	**MEDICAL PRECISION & OPTICAL INSTRUMENTS**	0	56	557	..	1,322	..	42	778,305	..	4,779
331	Medical appliances and instruments	0	35	425	..	1,187	..	42	641,556	..	3,999
3311	Medical, surgical equipment & orthopaedic app.	..	..	..	..	..	..	..	..	..	..
3312	Instruments & appliances for measuring, checking etc	..	..	..	..	..	..	..	..	..	..
3313	Industrial process control equipment	..	..	..	..	..	..	..	..	..	..
332	Optical instruments and photographic equipment	0	19	100	..	79	..	0	90,096	..	522
333	Watches and clocks	0	2	32	..	56	..	0	46,653	..	258
34	**MOTOR VEHICLES, TRAILERS & SEMI-TRAILERS**	1,659	519	4,630	..	13,603	..	275	5,201,594	..	39,412
341	Motor vehicles	1,388	109	3,436	..	8,830	..	7	3,070,099	..	24,822
342	Bodies (coachwork) for motor vehicles	4	93	236	..	572	..	0	139,482	..	1,407
343	Parts, accessories for motor vehicles & their engines	267	317	958	..	4,201	..	268	1,992,013	..	13,182
35	**OTHER TRANSPORT EQUIPMENT**	0	185	1,234	..	4,406	..	107	1,276,096	..	10,526
351	Building and repairing of ships and boats	0	44	135	..	206	..	0	108,470	..	775
3511	Building and repairing of ships	..	..	..	..	..	..	..	..	..	..
3512	Building, repairing of pleasure & sporting boats	..	..	..	..	..	..	..	..	..	..
352	Railway, tramway locomotives & rolling stock	0	49	197	..	897	..	0	96,309	..	1,490
353	Aircraft and spacecraft	0	47	780	..	3,065	..	9	986,409	..	7,452
359	Transport equipment, nec	0	45	122	..	238	..	98	84,908	..	809
3591	Motorcycles	..	..	..	..	..	..	..	..	..	..
3592	Bicycles and invalid carriages	..	..	..	..	..	..	..	..	..	..
3599	Other transport equipment, nec	..	..	..	..	..	..	..	..	..	..
36	**FURNITURE; MANUFACTURING, NEC**	20	662	1,253	..	2,730	..	7	1,223,662	..	9,077
361	Furniture	0	491	935	..	1,990	..	7	852,604	..	6,492
369	Manufacturing, nec	20	171	318	..	740	..	0	371,058	..	2,585
3691	Jewellery and related articles	..	..	..	..	..	..	..	..	..	..
3692	Musical instruments	..	..	..	..	..	..	..	..	..	..
3693	Sports goods	..	..	..	..	..	..	..	..	..	..
3694	Games and toys	..	..	..	..	..	..	..	..	..	..
3699	Other manufacturing, nec	..	..	..	..	..	..	..	..	..	..
37	**RECYCLING**	..	..	..	..	..	..	..	..	..	..
371	Recycling of metal waste and scrap	..	..	..	..	..	..	..	..	..	..
372	Recycling of non-metal waste and scrap	..	..	..	..	..	..	..	..	..	..
CERR	**Non-specified industry**	..	..	..	..	..	..	..	..	..	..
15-37	**TOTAL MANUFACTURING**	133,449	21,560	209,098	..	410,358	..	23,456	114,107,378	..	1,208,708

ISIC Revision 3 Industry Sector	Solid TJ	LPG TJ	Distiloil TJ	RFO TJ	Gas TJ	Biomass TJ	Steam TJ	Electr MWh	Own Use MWh	TOTAL TJ
15 FOOD PRODUCTS AND BEVERAGES	16,088	5,994	41,343	..	77,412	..	2,883	15,242,787	..	198,594
151 Production, processing and preserving (PPP)	8	2,564	9,581	..	12,052	..	898	3,904,532	..	39,159
1511 PPP of meat and meat products	..	..	..	..	..	..	..	..	..	..
1512 Processing and preserving of fish products	..	..	..	..	..	..	..	..	..	..
1513 Processing, preserving of fruit & vegetables	..	..	..	..	..	..	..	..	..	..
1514 Vegetable and animal oils and fats	..	..	..	..	..	..	..	..	..	..
152 Dairy products	717	1,104	11,438	..	13,990	..	517	3,083,783	..	38,868
153 Grain mill prod., starches & prepared animal feeds	4,903	1,060	4,381	..	17,279	..	231	3,520,351	..	40,527
1531 Grain mill products	..	..	..	..	..	..	..	..	..	..
1532 Starches and starch products	..	..	..	..	..	..	..	..	..	..
1533 Prepared animal feeds	..	..	..	..	..	..	..	..	..	..
154 Other food products	10,239	973	12,849	..	26,543	..	1,151	3,264,701	..	63,508
1541 Bakery products	..	..	..	..	..	..	..	..	..	..
1542 Sugar	..	..	..	..	..	..	..	..	..	..
1543 Cocoa, chocolate and sugar confectionery	..	..	..	..	..	..	..	..	..	..
1544 Macaroni, noodles, couscous & similar farinaceous prod.	..	..	..	..	..	..	..	..	..	..
1549 Other food products, nec	..	..	..	..	..	..	..	..	..	..
155 Beverages	221	293	3,094	..	7,548	..	86	1,469,420	..	16,532
1551 Distilling, rectifying and blending of spirits	..	..	..	..	..	..	..	..	..	..
1552 Wines	..	..	..	..	..	..	..	..	..	..
1553 Malt liquors and malt	..	..	..	..	..	..	..	..	..	..
1554 Soft drinks; production of mineral waters	..	..	..	..	..	..	..	..	..	..
16 TOBACCO PRODUCTS	..	..	..	..	..	..	..	..	..	..
17 TEXTILES	79	1,420	7,113	..	15,939	..	366	3,223,065	..	36,520
171 Spinning, weaving and finishing of textiles	57	1,243	5,527	..	11,804	..	311	2,334,446	..	27,346
1711 Preparation and spinning of textile fibres	..	..	..	..	..	..	..	..	..	..
1712 Finishing of textiles	..	..	..	..	..	..	..	..	..	..
172 Other textiles	0	118	1,297	..	2,970	..	17	642,254	..	6,714
1721 Made-up textile articles, except apparel	..	..	..	..	..	..	..	..	..	..
1722 Carpets and rugs	..	..	..	..	..	..	..	..	..	..
1723 Cordage, rope, twine and netting	..	..	..	..	..	..	..	..	..	..
1729 Other textiles, nec	..	..	..	..	..	..	..	..	..	..
173 Knitted and crocheted fabrics and articles	22	59	289	..	1,165	..	38	246,365	..	2,460
18 WEARING APPAREL, DRESSING & DYEING OF FUR	36	76	548	..	3,565	..	22	240,205	..	5,112
181 Wearing apparel, except fur apparel	36	70	470	..	3,565	..	22	236,457	..	5,014
182 Dressing and dyeing of fur; articles of fur	0	6	78	..	0	..	0	3,748	..	97
19 TANNING & DRESSING OF LEATHER, FOOTWEAR	34	104	661	..	458	..	96	271,529	..	2,331
191 Tanning and dressing of leather	34	49	437	..	289	..	95	103,890	..	1,278
1911 Tanning and dressing of leather	..	..	..	..	..	..	..	..	..	..
1912 Luggage, handbags, saddlery & harness	..	..	..	..	..	..	..	..	..	..
192 Footwear	0	55	224	..	169	..	1	167,639	..	1,053
20 WOOD AND WOOD PRODUCTS	96	274	1,961	..	3,062	..	0	1,646,413	..	11,320
201 Sawmilling and planing of wood	0	115	488	..	96	..	0	397,221	..	2,129
202 Products of wood, cork, straw & plaiting materials	96	159	1,473	..	2,966	..	0	1,249,192	..	9,191
2021 Veneer sheets	..	..	..	..	..	..	..	..	..	..
2022 Builders' carpentry and joinery	..	..	..	..	..	..	..	..	..	..
2023 Wooden containers	..	..	..	..	..	..	..	..	..	..
2029 Other products of wood	..	..	..	..	..	..	..	..	..	..
21 PAPER AND PAPER PRODUCTS	7,581	762	14,973	..	45,489	..	2,050	11,768,630	..	113,222
2101 Pulp, paper and paperboard	..	..	..	..	..	..	..	..	..	..
2102 Corrugated paper, paperboard and their containers	..	..	..	..	..	..	..	..	..	..
2109 Other articles of pulp and paperboard	..	..	..	..	..	..	..	..	..	..
22 PUBLISHING, PRINTING & REPRODUCTION	0	260	674	..	2,859	..	103	1,329,231	..	8,681
221 Publishing	0	9	121	..	149	..	11	157,777	..	858
2211 Publishing of books & brochures	..	..	..	..	..	..	..	..	..	..
2212 Publishing of newspapers and periodicals	..	..	..	..	..	..	..	..	..	..
2213 Publishing of recorded media	..	..	..	..	..	..	..	..	..	..
2219 Other publishing	..	..	..	..	..	..	..	..	..	..
222 Printing and related service activities	0	247	506	..	2,685	..	92	1,112,421	..	7,535
2221 Printing	..	..	..	..	..	..	..	..	..	..
2222 Service activities related to printing	..	..	..	..	..	..	..	..	..	..
223 Reproduction of recorded media	0	4	47	..	25	..	0	59,033	..	289
23 COKE, REFINED PETROLEUM PRODUCTS	..	..	..	..	..	..	..	..	..	..
231 Coke oven products	..	..	..	..	..	..	..	..	..	..
232 Refined petroleum products	..	..	..	..	..	..	..	..	..	..
233 Processing of nuclear fuel	..	..	..	..	..	..	..	..	..	..
24 CHEMICALS & CHEMICAL PRODUCTS	616	512	5,409	..	16,814	..	1,606	24,732,064	..	113,992
241 Basic chemicals	c	c	c	..	c	..	c	21,409,418	..	77,074
2411 Basic chemicals, exc. fertilizers & nitrogen compounds	..	..	..	..	..	..	..	..	..	..
2412 Fertilizers and nitrogen compounds	..	..	..	..	..	..	..	..	..	..
2413 Plastics in primary forms and synthetic rubber	..	..	..	..	..	..	..	..	..	..
242 Other chemical products	616	485	5,180	..	15,373	..	1,515	2,999,595	..	33,968
2421 Pesticides and other agro-chemical products	..	..	..	..	..	..	..	..	..	..
2422 Paints, varnishes and similar coatings	..	..	..	..	..	..	..	..	..	..
2423 Pharmaceuticals, medicinal chem. & botanical prod.	..	..	..	..	..	..	..	..	..	..
2424 Soap and detergents, perfumes etc.	..	..	..	..	..	..	..	..	..	..
2429 Other chemical products, nec	..	..	..	..	..	..	..	..	..	..
243 Man-made fibres	0	27	229	..	1,441	..	91	323,051	..	2,951
25 RUBBER AND PLASTICS PRODUCTS	428	754	4,755	..	12,127	..	1,030	6,117,035	..	41,115
251 Rubber products	427	128	2,343	..	6,933	..	899	1,713,184	..	16,897
2511 Rubber tyres and tubes	..	..	..	..	..	..	..	..	..	..
2519 Other rubber products	..	..	..	..	..	..	..	..	..	..
252 Plastic products	1	626	2,412	..	5,194	..	131	4,403,851	..	24,218
26 OTHER NON-METALLIC MINERAL PRODUCTS	11,606	2,005	67,265	..	63,260	..	0	7,763,821	..	172,086
261 Glass and glass products	0	706	20,901	..	24,489	..	0	2,986,441	..	56,847
269 Non-metallic mineral products, nec	11,606	1,299	46,364	..	38,771	..	0	4,777,380	..	115,239
2691 Pottery, china and earthenware	..	..	..	..	..	..	..	..	..	..
2692 Refractory ceramic products	..	..	..	..	..	..	..	..	..	..
2693 Structural non-refractory clay & ceramic prod.	..	..	..	..	..	..	..	..	..	..
2694 Cement, lime and plaster	..	..	..	..	..	..	..	..	..	..
2695 Articles of concrete, cement and plaster	..	..	..	..	..	..	..	..	..	..
2696 Cutting, shaping and finishing of stone	..	..	..	..	..	..	..	..	..	..
2699 Other non-metallic mineral products, nec	..	..	..	..	..	..	..	..	..	..

ISIC Revision 3 Industry Sector	Solid TJ	LPG TJ	Distiloil TJ	RFO TJ	Gas TJ	Biomass TJ	Steam TJ	Electr MWh	Own Use MWh	TOTAL TJ
27 BASIC METALS	4,225	1,059	5,526	..	39,541	..	247	24,117,240	..	137,420
271 Basic iron and steel	2,170	416	4,995	..	34,910	..	247	13,551,112	..	91,522
272 Basic precious and non-ferrous metals	c	c	c	..	c	..	c	8,612,700	..	31,006
273 Casting of metals	2,055	643	531	..	4,631	..	0	1,953,428	..	14,892
2731 Casting of iron and steel	..	..	..	..	..	..	..	..	..	..
2732 Casting non-ferrous metals	..	..	..	..	..	..	..	..	..	..
28 FABRICATED METAL PRODUCTS	508	2,010	2,475	..	13,123	..	67	4,094,243	..	32,922
281 Str. metal prod., tanks, reservoirs, steam generators	197	428	566	..	2,421	..	0	714,709	..	6,185
2811 Structural metal products	..	..	..	..	..	..	..	..	..	..
2812 Tanks, reservoirs and containers of metal	..	..	..	..	..	..	..	..	..	..
2813 Steam generators, exc. central heating hot water boilers	..	..	..	..	..	..	..	..	..	..
289 Other fabricated metal products	311	1,582	1,909	..	10,702	..	67	3,379,534	..	26,737
2891 Forging, pressing, stamping & roll-forming of metal	..	..	..	..	..	..	..	..	..	..
2892 Treatment and coating of metals	..	..	..	..	..	..	..	..	..	..
2893 Cutlery, hand tools and general hardware	..	..	..	..	..	..	..	..	..	..
2899 Other fabricated metal products, nec	..	..	..	..	..	..	..	..	..	..
29 MACHINERY AND EQUIPMENT, NEC	75	1,068	2,425	..	8,766	..	23	2,512,023	..	21,400
291 General purpose machinery	17	510	1,189	..	3,784	..	21	1,281,307	..	10,134
2911 Engines and turbines	..	..	..	..	..	..	..	..	..	..
2912 Pumps, compressors, taps and valves	..	..	..	..	..	..	..	..	..	..
2913 Bearings, gears, gearing and driving elements	..	..	..	..	..	..	..	..	..	..
2914 Ovens, furnaces and furnace burners	..	..	..	..	..	..	..	..	..	..
2915 Lifting and handling equipment	..	..	..	..	..	..	..	..	..	..
2919 Other general purpose machinery	..	..	..	..	..	..	..	..	..	..
292 Special purpose machinery	0	461	963	..	3,442	..	2	848,674	..	7,923
2921 Agricultural and forestry machinery	..	..	..	..	..	..	..	..	..	..
2922 Machine-tools	..	..	..	..	..	..	..	..	..	..
2923 Machinery for metallurgy	..	..	..	..	..	..	..	..	..	..
2924 Machinery for mining, quarrying and construction	..	..	..	..	..	..	..	..	..	..
2925 Machinery for food, beverage & tobacco processing	..	..	..	..	..	..	..	..	..	..
2926 Machinery for textile, apparel & leather production	..	..	..	..	..	..	..	..	..	..
2927 Machinery for weapons and ammunition	..	..	..	..	..	..	..	..	..	..
2929 Other special purpose machinery	..	..	..	..	..	..	..	..	..	..
293 Domestic appliances, nec	58	97	273	..	1,540	..	0	382,042	..	3,343
30 OFFICE, ACCOUNTING & COMPUTING MACHINERY	0	1	69	..	351	..	0	287,541	..	1,456
31 ELECTRICAL MACHINERY & APPARATUS, NEC	192	383	1,312	..	5,707	..	170	2,272,949	..	15,947
311 Electric motors, generators and transformers	1	49	168	..	936	..	147	269,188	..	2,270
312 Electricity distribution and control apparatus	0	52	270	..	1,140	..	7	571,258	..	3,526
313 Insulated wire and cable	0	64	329	..	885	..	6	494,738	..	3,065
314 Accumulators, primary cells & primary batteries	0	23	10	..	830	..	10	249,114	..	1,770
315 Electric lamps and lighting equipment	0	18	30	..	724	..	0	114,525	..	1,184
319 Other electrical equipment, nec	191	177	505	..	1,192	..	0	574,126	..	4,132
32 RADIO, TV & COMMUNICATION EQUIP. & APP.	0	143	691	..	2,190	..	42	1,998,796	..	10,262
321 Electronic valves, tubes, other electronic components	0	44	483	..	1,364	..	38	1,533,652	..	7,450
322 TV & radio transmitters, apparatus for line telephony	0	78	173	..	527	..	3	359,154	..	2,074
323 TV & radio receivers, recording apparatus	0	21	35	..	299	..	1	105,990	..	738
33 MEDICAL PRECISION &OPTICAL INSTRUMENTS	0	70	578	..	1,572	..	36	853,209	..	5,328
331 Medical appliances and instruments	0	48	437	..	1,432	..	36	712,514	..	4,518
3311 Medical, surgical equipment & orthopaedic app.	..	..	..	..	..	..	..	..	..	..
3312 Instruments & appliances for measuring, checking etc	..	..	..	..	..	..	..	..	..	..
3313 Industrial process control equipment	..	..	..	..	..	..	..	..	..	..
332 Optical instruments and photographic equipment	0	21	95	..	89	..	0	95,071	..	547
333 Watches and clocks	0	1	46	..	51	..	0	45,624	..	262
34 MOTOR VEHICLES, TRAILERS & SEMI-TRAILERS	1,356	568	5,508	..	15,434	..	279	5,312,799	..	42,271
341 Motor vehicles	1,012	122	3,946	..	10,092	..	8	2,944,114	..	25,779
342 Bodies (coachwork) for motor vehicles	5	104	249	..	687	..	0	136,965	..	1,538
343 Parts, accessories for motor vehicles & their engines	339	342	1,313	..	4,655	..	271	2,231,720	..	14,954
35 OTHER TRANSPORT EQUIPMENT	1	150	1,330	..	5,288	..	118	1,301,514	..	11,572
351 Building and repairing of ships and boats	0	40	165	..	213	..	0	103,956	..	792
3511 Building and repairing of ships	..	..	..	..	..	..	..	..	..	..
3512 Building, repairing of pleasure & sporting boats	..	..	..	..	..	..	..	..	..	..
352 Railway, tramway locomotives & rolling stock	1	30	127	..	965	..	0	97,681	..	1,475
353 Aircraft and spacecraft	0	42	908	..	3,848	..	2	1,015,911	..	8,457
359 Transport equipment, nec	0	38	130	..	262	..	116	83,966	..	848
3591 Motorcycles	..	..	..	..	..	..	..	..	..	..
3592 Bicycles and invalid carriages	..	..	..	..	..	..	..	..	..	..
3599 Other transport equipment, nec	..	..	..	..	..	..	..	..	..	..
36 FURNITURE; MANUFACTURING, NEC	19	619	1,108	..	2,797	..	8	1,237,717	..	9,007
361 Furniture	0	493	734	..	2,100	..	8	893,313	..	6,551
369 Manufacturing, nec	19	126	374	..	697	..	0	344,404	..	2,456
3691 Jewellery and related articles	..	..	..	..	..	..	..	..	..	..
3692 Musical instruments	..	..	..	..	..	..	..	..	..	..
3693 Sports goods	..	..	..	..	..	..	..	..	..	..
3694 Games and toys	..	..	..	..	..	..	..	..	..	..
3699 Other manufacturing, nec	..	..	..	..	..	..	..	..	..	..
37 RECYCLING	..	..	..	..	..	..	..	..	..	..
371 Recycling of metal waste and scrap	..	..	..	..	..	..	..	..	..	..
372 Recycling of non-metal waste and scrap	..	..	..	..	..	..	..	..	..	..
CERR Non-specified industry	..	..	..	..	..	..	..	..	..	..
15-37 TOTAL MANUFACTURING	71,260	20,898	209,939	..	442,359	..	24,613	116,322,811	..	1,187,831

ISIC Revision 3 Industry Sector	Solid TJ	LPG TJ	Distiloil TJ	RFO TJ	Gas TJ	Biomass TJ	Steam TJ	Electr MWh	Own Use MWh	TOTAL TJ
15 FOOD PRODUCTS AND BEVERAGES	14,062	5,940	38,641	..	80,165	..	2,873	15,274,910	..	196,671
151 Production, processing and preserving (PPP)	4	2,491	9,223	..	13,083	..	843	4,067,092	..	40,286
1511 PPP of meat and meat products	..	..	..	..	..	..	..	..	..	..
1512 Processing and preserving of fish products	..	..	..	..	..	..	..	..	..	..
1513 Processing, preserving of fruit & vegetables	..	..	..	..	..	..	..	..	..	..
1514 Vegetable and animal oils and fats	..	..	..	..	..	..	..	..	..	..
152 Dairy products	631	1,038	11,086	..	15,234	..	491	3,003,014	..	39,291
153 Grain mill prod., starches & prepared animal feeds	4,333	997	4,213	..	17,173	..	141	3,348,226	..	38,911
1531 Grain mill products	..	..	..	..	..	..	..	..	..	..
1532 Starches and starch products	..	..	..	..	..	..	..	..	..	..
1533 Prepared animal feeds	..	..	..	..	..	..	..	..	..	..
154 Other food products	8,868	1,134	11,216	..	27,937	..	1,314	3,377,322	..	62,627
1541 Bakery products	..	..	..	..	..	..	..	..	..	..
1542 Sugar	..	..	..	..	..	..	..	..	..	..
1543 Cocoa, chocolate and sugar confectionery	..	..	..	..	..	..	..	..	..	..
1544 Macaroni, noodles, couscous & similar farinaceous prod.	..	..	..	..	..	..	..	..	..	..
1549 Other food products, nec	..	..	..	..	..	..	..	..	..	..
155 Beverages	226	280	2,903	..	6,738	..	84	1,479,256	..	15,556
1551 Distilling, rectifying and blending of spirits	..	..	..	..	..	..	..	..	..	..
1552 Wines	..	..	..	..	..	..	..	..	..	..
1553 Malt liquors and malt	..	..	..	..	..	..	..	..	..	..
1554 Soft drinks; production of mineral waters	..	..	..	..	..	..	..	..	..	..
16 TOBACCO PRODUCTS	..	..	..	..	..	..	..	..	..	..
17 TEXTILES	38	1,473	6,706	..	15,355	..	435	3,167,194	..	35,409
171 Spinning, weaving and finishing of textiles	18	1,248	5,037	..	11,254	..	362	2,216,940	..	25,900
1711 Preparation and spinning of textile fibres	..	..	..	..	..	..	..	..	..	..
1712 Finishing of textiles	..	..	..	..	..	..	..	..	..	..
172 Other textiles	0	152	1,351	..	2,854	..	30	703,324	..	6,919
1721 Made-up textile articles, except apparel	..	..	..	..	..	..	..	..	..	..
1722 Carpets and rugs	..	..	..	..	..	..	..	..	..	..
1723 Cordage, rope, twine and netting	..	..	..	..	..	..	..	..	..	..
1729 Other textiles, nec	..	..	..	..	..	..	..	..	..	..
173 Knitted and crocheted fabrics and articles	20	73	318	..	1,247	..	43	246,930	..	2,590
18 WEARING APPAREL, DRESSING & DYEING OF FUR	27	117	524	..	1,088	..	45	218,546	..	2,588
181 Wearing apparel, except fur apparel	27	107	444	..	1,085	..	45	214,673	..	2,481
182 Dressing and dyeing of fur; articles of fur	0	10	80	..	3	..	0	3,873	..	107
19 TANNING & DRESSING OF LEATHER, FOOTWEAR	35	91	650	..	458	..	115	267,394	..	2,312
191 Tanning and dressing of leather	35	22	428	..	241	..	114	94,869	..	1,182
1911 Tanning and dressing of leather	..	..	..	..	..	..	..	..	..	..
1912 Luggage, handbags, saddlery & harness	..	..	..	..	..	..	..	..	..	..
192 Footwear	0	69	222	..	217	..	1	172,525	..	1,130
20 WOOD AND WOOD PRODUCTS	75	274	2,166	..	3,201	..	0	1,723,971	..	11,922
201 Sawmilling and planing of wood	0	169	600	..	111	..	0	435,442	..	2,448
202 Products of wood, cork, straw & plaiting materials	75	105	1,566	..	3,090	..	0	1,288,529	..	9,475
2021 Veneer sheets	..	..	..	..	..	..	..	..	..	..
2022 Builders' carpentry and joinery	..	..	..	..	..	..	..	..	..	..
2023 Wooden containers	..	..	..	..	..	..	..	..	..	..
2029 Other products of wood	..	..	..	..	..	..	..	..	..	..
21 PAPER AND PAPER PRODUCTS	8,109	777	14,553	..	48,957	..	2,296	11,202,669	..	115,022
2101 Pulp, paper and paperboard	..	..	..	..	..	..	..	..	..	..
2102 Corrugated paper, paperboard and their containers	..	..	..	..	..	..	..	..	..	..
2109 Other articles of pulp and paperboard	..	..	..	..	..	..	..	..	..	..
22 PUBLISHING, PRINTING & REPRODUCTION	0	372	738	..	3,446	..	80	1,539,727	..	10,179
221 Publishing	0	28	97	..	220	..	9	156,008	..	916
2211 Publishing of books & brochures	..	..	..	..	..	..	..	..	..	..
2212 Publishing of newspapers and periodicals	..	..	..	..	..	..	..	..	..	..
2213 Publishing of recorded media	..	..	..	..	..	..	..	..	..	..
2219 Other publishing	..	..	..	..	..	..	..	..	..	..
222 Printing and related service activities	0	336	599	..	3,200	..	71	1,315,571	..	8,942
2221 Printing	..	..	..	..	..	..	..	..	..	..
2222 Service activities related to printing	..	..	..	..	..	..	..	..	..	..
223 Reproduction of recorded media	0	8	42	..	26	..	0	68,148	..	321
23 COKE, REFINED PETROLEUM PRODUCTS	..	..	..	..	..	..	..	..	..	..
231 Coke oven products	..	..	..	..	..	..	..	..	..	..
232 Refined petroleum products	..	..	..	..	..	..	..	..	..	..
233 Processing of nuclear fuel	..	..	..	..	..	..	..	..	..	..
24 CHEMICALS & CHEMICAL PRODUCTS	450	507	5,604	..	17,990	..	1,323	25,532,253	..	117,790
241 Basic chemicals	c	c	c	..	c	..	c	22,000,464	..	79,202
2411 Basic chemicals, exc. fertilizers & nitrogen compounds	..	..	..	..	..	..	..	..	..	..
2412 Fertilizers and nitrogen compounds	..	..	..	..	..	..	..	..	..	..
2413 Plastics in primary forms and synthetic rubber	..	..	..	..	..	..	..	..	..	..
242 Other chemical products	450	459	5,394	..	16,166	..	1,232	3,129,929	..	34,969
2421 Pesticides and other agro-chemical products	..	..	..	..	..	..	..	..	..	..
2422 Paints, varnishes and similar coatings	..	..	..	..	..	..	..	..	..	..
2423 Pharmaceuticals, medicinal chem. & botanical prod.	..	..	..	..	..	..	..	..	..	..
2424 Soap and detergents, perfumes etc.	..	..	..	..	..	..	..	..	..	..
2429 Other chemical products, nec	..	..	..	..	..	..	..	..	..	..
243 Man-made fibres	0	48	210	..	1,824	..	91	401,860	..	3,620
25 RUBBER AND PLASTICS PRODUCTS	249	773	3,944	..	14,760	..	825	6,139,969	..	42,655
251 Rubber products	248	134	1,593	..	9,058	..	709	1,695,183	..	17,845
2511 Rubber tyres and tubes	..	..	..	..	..	..	..	..	..	..
2519 Other rubber products	..	..	..	..	..	..	..	..	..	..
252 Plastic products	1	639	2,351	..	5,702	..	116	4,444,786	..	24,810
26 OTHER NON-METALLIC MINERAL PRODUCTS	9,834	1,918	65,738	..	64,406	..	0	7,670,331	..	169,509
261 Glass and glass products	0	705	20,379	..	24,836	..	0	2,994,354	..	56,700
269 Non-metallic mineral products, nec	9,834	1,213	45,359	..	39,570	..	0	4,675,977	..	112,810
2691 Pottery, china and earthenware	..	..	..	..	..	..	..	..	..	..
2692 Refractory ceramic products	..	..	..	..	..	..	..	..	..	..
2693 Structural non-refractory clay & ceramic prod.	..	..	..	..	..	..	..	..	..	..
2694 Cement, lime and plaster	..	..	..	..	..	..	..	..	..	..
2695 Articles of concrete, cement and plaster	..	..	..	..	..	..	..	..	..	..
2696 Cutting, shaping and finishing of stone	..	..	..	..	..	..	..	..	..	..
2699 Other non-metallic mineral products, nec	..	..	..	..	..	..	..	..	..	..

ISIC Revision 3 Industry Sector	Solid TJ	LPG TJ	Distiloil TJ	RFO TJ	Gas TJ	Biomass TJ	Steam TJ	Electr MWh	Own Use MWh	TOTAL TJ
27 **BASIC METALS**	5,253	1,031	4,461	..	38,502	..	241	24,110,407	..	136,285
271 Basic iron and steel	3,192	382	4,056	..	33,689	..	241	13,421,491	..	89,877
272 Basic precious and non-ferrous metals	c	c	c	..	c	..	c	8,761,850	..	31,543
273 Casting of metals	2,061	649	405	..	4,813	..	0	1,927,066	..	14,865
2731 Casting of iron and steel	..	..	..	..	..	..	..	..	..	..
2732 Casting non-ferrous metals	..	..	..	..	..	..	..	..	..	..
28 **FABRICATED METAL PRODUCTS**	523	2,431	2,819	..	15,760	..	35	4,291,036	..	37,016
281 Str. metal prod., tanks, reservoirs, steam generators	163	577	670	..	3,113	..	0	757,460	..	7,250
2811 Structural metal products	..	..	..	..	..	..	..	..	..	..
2812 Tanks, reservoirs and containers of metal	..	..	..	..	..	..	..	..	..	..
2813 Steam generators, exc. central heating hot water boilers	..	..	..	..	..	..	..	..	..	..
289 Other fabricated metal products	360	1,854	2,149	..	12,647	..	35	3,533,576	..	29,766
2891 Forging, pressing, stamping & roll-forming of metal	..	..	..	..	..	..	..	..	..	..
2892 Treatment and coating of metals	..	..	..	..	..	..	..	..	..	..
2893 Cutlery, hand tools and general hardware	..	..	..	..	..	..	..	..	..	..
2899 Other fabricated metal products, nec	..	..	..	..	..	..	..	..	..	..
29 **MACHINERY AND EQUIPMENT, NEC**	63	1,003	2,628	..	9,653	..	31	2,502,193	..	22,386
291 General purpose machinery	9	558	1,272	..	3,944	..	24	1,281,371	..	10,420
2911 Engines and turbines	..	..	..	..	..	..	..	..	..	..
2912 Pumps, compressors, taps and valves	..	..	..	..	..	..	..	..	..	..
2913 Bearings, gears, gearing and driving elements	..	..	..	..	..	..	..	..	..	..
2914 Ovens, furnaces and furnace burners	..	..	..	..	..	..	..	..	..	..
2915 Lifting and handling equipment	..	..	..	..	..	..	..	..	..	..
2919 Other general purpose machinery	..	..	..	..	..	..	..	..	..	..
292 Special purpose machinery	0	359	1,078	..	4,043	..	7	839,985	..	8,511
2921 Agricultural and forestry machinery	..	..	..	..	..	..	..	..	..	..
2922 Machine-tools	..	..	..	..	..	..	..	..	..	..
2923 Machinery for metallurgy	..	..	..	..	..	..	..	..	..	..
2924 Machinery for mining, quarrying and construction	..	..	..	..	..	..	..	..	..	..
2925 Machinery for food, beverage & tobacco processing	..	..	..	..	..	..	..	..	..	..
2926 Machinery for textile, apparel & leather production	..	..	..	..	..	..	..	..	..	..
2927 Machinery for weapons and ammunition	..	..	..	..	..	..	..	..	..	..
2929 Other special purpose machinery	..	..	..	..	..	..	..	..	..	..
293 Domestic appliances, nec	54	86	278	..	1,666	..	0	380,837	..	3,455
30 **OFFICE, ACCOUNTING & COMPUTING MACHINERY**	0	2	44	..	388	..	17	307,380	..	1,558
31 **ELECTRICAL MACHINERY & APPARATUS, NEC**	169	439	1,348	..	6,533	..	171	2,387,985	..	17,257
311 Electric motors, generators and transformers	1	55	240	..	1,407	..	152	334,816	..	3,060
312 Electricity distribution and control apparatus	0	47	249	..	1,166	..	10	556,172	..	3,474
313 Insulated wire and cable	0	72	276	..	1,072	..	8	495,109	..	3,210
314 Accumulators, primary cells & primary batteries	0	18	11	..	858	..	1	246,298	..	1,775
315 Electric lamps and lighting equipment	0	65	48	..	664	..	0	121,581	..	1,215
319 Other electrical equipment, nec	168	182	524	..	1,366	..	0	634,009	..	4,522
32 **RADIO, TV & COMMUNICATION EQUIP. & APP.**	0	81	652	..	2,342	..	48	2,165,973	..	10,921
321 Electronic valves, tubes, other electronic components	0	50	448	..	1,452	..	42	1,668,999	..	8,000
322 TV & radio transmitters, apparatus for line telephony	0	18	169	..	583	..	5	387,970	..	2,172
323 TV & radio receivers, recording apparatus	0	13	35	..	307	..	1	109,004	..	748
33 **MEDICAL PRECISION &OPTICAL INSTRUMENTS**	0	76	507	..	1,649	..	42	874,279	..	5,421
331 Medical appliances and instruments	0	57	379	..	1,464	..	42	722,008	..	4,541
3311 Medical, surgical equipment & orthopaedic app.	..	..	..	..	..	..	..	..	..	..
3312 Instruments & appliances for measuring, checking etc	..	..	..	..	..	..	..	..	..	..
3313 Industrial process control equipment	..	..	..	..	..	..	..	..	..	..
332 Optical instruments and photographic equipment	0	18	87	..	148	..	0	97,511	..	604
333 Watches and clocks	0	1	41	..	37	..	0	54,760	..	276
34 **MOTOR VEHICLES, TRAILERS & SEMI-TRAILERS**	1,367	565	5,317	..	17,225	..	312	5,302,096	..	43,874
341 Motor vehicles	1,072	122	3,626	..	11,454	..	10	2,960,340	..	26,941
342 Bodies (coachwork) for motor vehicles	0	92	276	..	816	..	0	138,264	..	1,682
343 Parts, accessories for motor vehicles & their engines	295	351	1,415	..	4,955	..	302	2,203,492	..	15,251
35 **OTHER TRANSPORT EQUIPMENT**	1	203	1,459	..	5,716	..	120	1,338,083	..	12,316
351 Building and repairing of ships and boats	0	51	138	..	263	..	0	101,946	..	819
3511 Building and repairing of ships	..	..	..	..	..	..	..	..	..	..
3512 Building, repairing of pleasure & sporting boats	..	..	..	..	..	..	..	..	..	..
352 Railway, tramway locomotives & rolling stock	1	58	199	..	1,143	..	0	108,827	..	1,793
353 Aircraft and spacecraft	0	51	972	..	4,022	..	0	1,044,247	..	8,804
359 Transport equipment, nec	0	43	150	..	288	..	120	83,063	..	900
3591 Motorcycles	..	..	..	..	..	..	..	..	..	..
3592 Bicycles and invalid carriages	..	..	..	..	..	..	..	..	..	..
3599 Other transport equipment, nec	..	..	..	..	..	..	..	..	..	..
36 **FURNITURE; MANUFACTURING, NEC**	0	626	1,188	..	3,374	..	8	1,249,327	..	9,694
361 Furniture	0	426	868	..	2,587	..	8	894,039	..	7,108
369 Manufacturing, nec	0	200	320	..	787	..	0	355,288	..	2,586
3691 Jewellery and related articles	..	..	..	..	..	..	..	..	..	..
3692 Musical instruments	..	..	..	..	..	..	..	..	..	..
3693 Sports goods	..	..	..	..	..	..	..	..	..	..
3694 Games and toys	..	..	..	..	..	..	..	..	..	..
3699 Other manufacturing, nec	..	..	..	..	..	..	..	..	..	..
37 **RECYCLING**	..	..	..	..	..	..	..	..	..	..
371 Recycling of metal waste and scrap	..	..	..	..	..	..	..	..	..	..
372 Recycling of non-metal waste and scrap	..	..	..	..	..	..	..	..	..	..
CERR Non-specified industry	..	..	..	..	..	..	..	..	..	..
15-37 **TOTAL MANUFACTURING**	68,618	21,584	202,102	..	457,099	..	24,799	117,265,723	..	1,196,359

ISIC Revision 3 Industry Sector	Solid TJ	LPG TJ	Distiloil TJ	RFO TJ	Gas TJ	Biomass TJ	Steam TJ	Electr MWh	Own Use MWh	TOTAL TJ
15 **FOOD PRODUCTS AND BEVERAGES**	14,101	5,893	36,852	..	82,985	..	2,930	15,781,988	..	199,576
151 Production, processing and preserving (PPP)	2	2,563	8,438	..	12,613	..	830	4,170,954	..	39,461
1511 PPP of meat and meat products	..	..	..	..	..	..	..	..	..	..
1512 Processing and preserving of fish products	..	..	..	..	..	..	..	..	..	..
1513 Processing, preserving of fruit & vegetables	..	..	..	..	..	..	..	..	..	..
1514 Vegetable and animal oils and fats	..	..	..	..	..	..	..	..	..	..
152 Dairy products	786	982	10,250	..	15,505	..	389	3,000,364	..	38,713
153 Grain mill prod., starches & prepared animal feeds	4,944	960	4,592	..	17,366	..	323	3,657,118	..	41,351
1531 Grain mill products	..	..	..	..	..	..	..	..	..	..
1532 Starches and starch products	..	..	..	..	..	..	..	..	..	..
1533 Prepared animal feeds	..	..	..	..	..	..	..	..	..	..
154 Other food products	8,098	1,011	11,119	..	31,300	..	1,199	3,430,470	..	65,077
1541 Bakery products	..	..	..	..	..	..	..	..	..	..
1542 Sugar	..	..	..	..	..	..	..	..	..	..
1543 Cocoa, chocolate and sugar confectionery	..	..	..	..	..	..	..	..	..	..
1544 Macaroni, noodles, couscous & similar farinaceous prod.	..	..	..	..	..	..	..	..	..	..
1549 Other food products, nec	..	..	..	..	..	..	..	..	..	..
155 Beverages	271	377	2,453	..	6,201	..	189	1,523,082	..	14,974
1551 Distilling, rectifying and blending of spirits	..	..	..	..	..	..	..	..	..	..
1552 Wines	..	..	..	..	..	..	..	..	..	..
1553 Malt liquors and malt	..	..	..	..	..	..	..	..	..	..
1554 Soft drinks; production of mineral waters	..	..	..	..	..	..	..	..	..	..
16 **TOBACCO PRODUCTS**	..	..	..	..	..	..	..	..	..	..
17 **TEXTILES**	14	1,495	6,228	..	15,805	..	388	3,268,239	..	35,696
171 Spinning, weaving and finishing of textiles	4	1,176	4,868	..	11,556	..	315	2,321,633	..	26,277
1711 Preparation and spinning of textile fibres	..	..	..	..	..	..	..	..	..	..
1712 Finishing of textiles	..	..	..	..	..	..	..	..	..	..
172 Other textiles	0	255	1,124	..	3,068	..	28	713,920	..	7,045
1721 Made-up textile articles, except apparel	..	..	..	..	..	..	..	..	..	..
1722 Carpets and rugs	..	..	..	..	..	..	..	..	..	..
1723 Cordage, rope, twine and netting	..	..	..	..	..	..	..	..	..	..
1729 Other textiles, nec	..	..	..	..	..	..	..	..	..	..
173 Knitted and crocheted fabrics and articles	10	64	236	..	1,181	..	45	232,686	..	2,374
18 **WEARING APPAREL, DRESSING & DYEING OF FUR**	35	80	490	..	955	..	10	221,182	..	2,366
181 Wearing apparel, except fur apparel	35	63	426	..	954	..	10	217,327	..	2,270
182 Dressing and dyeing of fur; articles of fur	0	17	64	..	1	..	0	3,855	..	96
19 **TANNING & DRESSING OF LEATHER, FOOTWEAR**	40	64	652	..	463	..	101	258,387	..	2,250
191 Tanning and dressing of leather	40	19	466	..	275	..	100	98,471	..	1,254
1911 Tanning and dressing of leather	..	..	..	..	..	..	..	..	..	..
1912 Luggage, handbags, saddlery & harness	..	..	..	..	..	..	..	..	..	..
192 Footwear	0	45	186	..	188	..	1	159,916	..	996
20 **WOOD AND WOOD PRODUCTS**	54	465	1,972	..	3,927	..	0	1,811,611	..	12,940
201 Sawmilling and planing of wood	0	181	530	..	114	..	0	446,956	..	2,434
202 Products of wood, cork, straw & plaiting materials	54	284	1,442	..	3,813	..	0	1,364,655	..	10,506
2021 Veneer sheets	..	..	..	..	..	..	..	..	..	..
2022 Builders' carpentry and joinery	..	..	..	..	..	..	..	..	..	..
2023 Wooden containers	..	..	..	..	..	..	..	..	..	..
2029 Other products of wood	..	..	..	..	..	..	..	..	..	..
21 **PAPER AND PAPER PRODUCTS**	8,186	768	13,481	..	50,340	..	1,794	11,944,086	..	117,568
2101 Pulp, paper and paperboard	..	..	..	..	..	..	..	..	..	..
2102 Corrugated paper, paperboard and their containers	..	..	..	..	..	..	..	..	..	..
2109 Other articles of pulp and paperboard	..	..	..	..	..	..	..	..	..	..
22 **PUBLISHING, PRINTING & REPRODUCTION**	0	365	699	..	3,133	..	71	1,476,789	..	9,584
221 Publishing	0	4	83	..	208	..	7	178,041	..	943
2211 Publishing of books & brochures	..	..	..	..	..	..	..	..	..	..
2212 Publishing of newspapers and periodicals	..	..	..	..	..	..	..	..	..	..
2213 Publishing of recorded media	..	..	..	..	..	..	..	..	..	..
2219 Other publishing	..	..	..	..	..	..	..	..	..	..
222 Printing and related service activities	0	338	571	..	2,902	..	64	1,227,025	..	8,292
2221 Printing	..	..	..	..	..	..	..	..	..	..
2222 Service activities related to printing	..	..	..	..	..	..	..	..	..	..
223 Reproduction of recorded media	0	23	45	..	23	..	0	71,723	..	349
23 **COKE, REFINED PETROLEUM PRODUCTS**	..	..	..	..	..	..	..	..	..	..
231 Coke oven products	..	..	..	..	..	..	..	..	..	..
232 Refined petroleum products	..	..	..	..	..	..	..	..	..	..
233 Processing of nuclear fuel	..	..	..	..	..	..	..	..	..	..
24 **CHEMICALS & CHEMICAL PRODUCTS**	336	508	4,984	..	18,248	..	1,201	25,869,909	..	118,409
241 Basic chemicals	c	c	c	..	c	..	c	22,239,134	..	80,061
2411 Basic chemicals, exc. fertilizers & nitrogen compounds	..	..	..	..	..	..	..	..	..	..
2412 Fertilizers and nitrogen compounds	..	..	..	..	..	..	..	..	..	..
2413 Plastics in primary forms and synthetic rubber	..	..	..	..	..	..	..	..	..	..
242 Other chemical products	336	495	4,799	..	16,403	..	1,089	3,213,836	..	34,692
2421 Pesticides and other agro-chemical products	..	..	..	..	..	..	..	..	..	..
2422 Paints, varnishes and similar coatings	..	..	..	..	..	..	..	..	..	..
2423 Pharmaceuticals, medicinal chem. & botanical prod.	..	..	..	..	..	..	..	..	..	..
2424 Soap and detergents, perfumes etc.	..	..	..	..	..	..	..	..	..	..
2429 Other chemical products, nec	..	..	..	..	..	..	..	..	..	..
243 Man-made fibres	0	13	185	..	1,845	..	112	416,939	..	3,656
25 **RUBBER AND PLASTICS PRODUCTS**	171	746	3,262	..	16,257	..	818	6,678,663	..	45,297
251 Rubber products	169	135	1,099	..	10,419	..	692	1,802,380	..	19,003
2511 Rubber tyres and tubes	..	..	..	..	..	..	..	..	..	..
2519 Other rubber products	..	..	..	..	..	..	..	..	..	..
252 Plastic products	2	611	2,163	..	5,838	..	126	4,876,283	..	26,295
26 **OTHER NON-METALLIC MINERAL PRODUCTS**	8,924	2,135	65,188	..	67,377	..	0	7,732,828	..	171,462
261 Glass and glass products	0	762	20,192	..	27,016	..	0	3,085,768	..	59,079
269 Non-metallic mineral products, nec	8,924	1,373	44,996	..	40,361	..	0	4,647,060	..	112,383
2691 Pottery, china and earthenware	..	..	..	..	..	..	..	..	..	..
2692 Refractory ceramic products	..	..	..	..	..	..	..	..	..	..
2693 Structural non-refractory clay & ceramic prod.	..	..	..	..	..	..	..	..	..	..
2694 Cement, lime and plaster	..	..	..	..	..	..	..	..	..	..
2695 Articles of concrete, cement and plaster	..	..	..	..	..	..	..	..	..	..
2696 Cutting, shaping and finishing of stone	..	..	..	..	..	..	..	..	..	..
2699 Other non-metallic mineral products, nec	..	..	..	..	..	..	..	..	..	..

ISIC Revision 3 Industry Sector	Solid TJ	LPG TJ	Distiloil TJ	RFO TJ	Gas TJ	Biomass TJ	Steam TJ	Electr MWh	Own Use MWh	TOTAL TJ
27	**BASIC METALS**	**5,709**	**1,072**	**3,165**	..	**40,064**	..	**325**	**25,408,985**	..	**141,807**
271	Basic iron and steel	3,542	420	2,729	..	35,264	..	325	14,518,794	..	94,548
272	Basic precious and non-ferrous metals	c	c	c	..	c	..	c	8,907,188	..	32,066
273	Casting of metals	2,167	652	436	..	4,800	..	0	1,983,003	..	15,194
2731	Casting of iron and steel	..	..	..	..	..	..	..	..	..	..
2732	Casting non-ferrous metals	..	..	..	..	..	..	..	..	..	..
28	**FABRICATED METAL PRODUCTS**	**551**	**2,214**	**2,847**	..	**16,027**	..	**39**	**4,639,207**	..	**38,379**
281	Str. metal prod., tanks, reservoirs, steam generators	184	551	782	..	2,777	..	2	764,553	..	7,048
2811	Structural metal products	..	..	..	..	..	..	..	..	..	..
2812	Tanks, reservoirs and containers of metal	..	..	..	..	..	..	..	..	..	..
2813	Steam generators, exc. central heating hot water boilers	..	..	..	..	..	..	..	..	..	..
289	Other fabricated metal products	367	1,663	2,065	..	13,250	..	37	3,874,654	..	31,331
2891	Forging, pressing, stamping & roll-forming of metal	..	..	..	..	..	..	..	..	..	..
2892	Treatment and coating of metals	..	..	..	..	..	..	..	..	..	..
2893	Cutlery, hand tools and general hardware	..	..	..	..	..	..	..	..	..	..
2899	Other fabricated metal products, nec	..	..	..	..	..	..	..	..	..	..
29	**MACHINERY AND EQUIPMENT, NEC**	**69**	**967**	**2,449**	..	**9,167**	..	**46**	**2,603,860**	..	**22,072**
291	General purpose machinery	15	531	1,256	..	3,703	..	22	1,331,308	..	10,320
2911	Engines and turbines	..	..	..	..	..	..	..	..	..	..
2912	Pumps, compressors, taps and valves	..	..	..	..	..	..	..	..	..	..
2913	Bearings, gears, gearing and driving elements	..	..	..	..	..	..	..	..	..	..
2914	Ovens, furnaces and furnace burners	..	..	..	..	..	..	..	..	..	..
2915	Lifting and handling equipment	..	..	..	..	..	..	..	..	..	..
2919	Other general purpose machinery	..	..	..	..	..	..	..	..	..	..
292	Special purpose machinery	0	365	964	..	4,015	..	24	910,467	..	8,646
2921	Agricultural and forestry machinery	..	..	..	..	..	..	..	..	..	..
2922	Machine-tools	..	..	..	..	..	..	..	..	..	..
2923	Machinery for metallurgy	..	..	..	..	..	..	..	..	..	..
2924	Machinery for mining, quarrying and construction	..	..	..	..	..	..	..	..	..	..
2925	Machinery for food, beverage & tobacco processing	..	..	..	..	..	..	..	..	..	..
2926	Machinery for textile, apparel & leather production	..	..	..	..	..	..	..	..	..	..
2927	Machinery for weapons and ammunition	..	..	..	..	..	..	..	..	..	..
2929	Other special purpose machinery	..	..	..	..	..	..	..	..	..	..
293	Domestic appliances, nec	54	71	229	..	1,449	..	0	362,085	..	3,107
30	**OFFICE, ACCOUNTING & COMPUTING MACHINERY**	**0**	**1**	**80**	..	**374**	..	**4**	**265,579**	..	**1,415**
31	**ELECTRICAL MACHINERY & APPARATUS, NEC**	**137**	**387**	**1,053**	..	**6,489**	..	**134**	**2,440,824**	..	**16,987**
311	Electric motors, generators and transformers	0	35	152	..	1,430	..	117	339,466	..	2,956
312	Electricity distribution and control apparatus	0	42	224	..	1,096	..	8	574,702	..	3,439
313	Insulated wire and cable	0	77	201	..	1,135	..	9	527,504	..	3,321
314	Accumulators, primary cells & primary batteries	0	16	7	..	811	..	0	254,070	..	1,749
315	Electric lamps and lighting equipment	0	39	30	..	650	..	0	114,911	..	1,133
319	Other electrical equipment, nec	137	178	439	..	1,367	..	0	630,171	..	4,390
32	**RADIO, TV & COMMUNICATION EQUIP. & APP.**	**0**	**72**	**696**	..	**2,322**	..	**34**	**2,234,817**	..	**11,169**
321	Electronic valves, tubes, other electronic components	0	47	474	..	1,518	..	29	1,746,484	..	8,355
322	TV & radio transmitters, apparatus for line telephony	0	11	186	..	541	..	5	371,984	..	2,082
323	TV & radio receivers, recording apparatus	0	14	36	..	263	..	0	116,349	..	732
33	**MEDICAL PRECISION & OPTICAL INSTRUMENTS**	**0**	**68**	**510**	..	**1,599**	..	**16**	**904,479**	..	**5,449**
331	Medical appliances and instruments	0	50	399	..	1,406	..	16	744,536	..	4,551
3311	Medical, surgical equipment & orthopaedic app.	..	..	..	..	..	..	..	..	..	..
3312	Instruments & appliances for measuring, checking etc	..	..	..	..	..	..	..	..	..	..
3313	Industrial process control equipment	..	..	..	..	..	..	..	..	..	..
332	Optical instruments and photographic equipment	0	17	83	..	158	..	0	120,739	..	693
333	Watches and clocks	0	1	28	..	35	..	0	39,204	..	205
34	**MOTOR VEHICLES, TRAILERS & SEMI-TRAILERS**	**1,304**	**596**	**4,322**	..	**16,768**	..	**197**	**5,523,176**	..	**43,070**
341	Motor vehicles	1,011	112	2,747	..	11,433	..	9	3,049,185	..	26,289
342	Bodies (coachwork) for motor vehicles	0	151	217	..	664	..	0	142,140	..	1,544
343	Parts, accessories for motor vehicles & their engines	293	333	1,358	..	4,671	..	188	2,331,851	..	15,238
35	**OTHER TRANSPORT EQUIPMENT**	**0**	**166**	**1,794**	..	**5,616**	..	**126**	**1,392,891**	..	**12,716**
351	Building and repairing of ships and boats	0	39	140	..	259	..	0	93,627	..	775
3511	Building and repairing of ships	..	..	..	..	..	..	..	..	..	..
3512	Building, repairing of pleasure & sporting boats	..	..	..	..	..	..	..	..	..	..
352	Railway, tramway locomotives & rolling stock	0	51	145	..	914	..	0	108,737	..	1,501
353	Aircraft and spacecraft	0	34	1,375	..	4,188	..	0	1,107,715	..	9,585
359	Transport equipment, nec	0	42	134	..	255	..	126	82,812	..	855
3591	Motorcycles	..	..	..	..	..	..	..	..	..	..
3592	Bicycles and invalid carriages	..	..	..	..	..	..	..	..	..	..
3599	Other transport equipment, nec	..	..	..	..	..	..	..	..	..	..
36	**FURNITURE; MANUFACTURING, NEC**	**0**	**697**	**1,059**	..	**2,829**	..	**7**	**1,233,160**	..	**9,031**
361	Furniture	0	539	728	..	1,983	..	7	833,585	..	6,258
369	Manufacturing, nec	0	158	331	..	846	..	0	399,575	..	2,773
3691	Jewellery and related articles	..	..	..	..	..	..	..	..	..	..
3692	Musical instruments	..	..	..	..	..	..	..	..	..	..
3693	Sports goods	..	..	..	..	..	..	..	..	..	..
3694	Games and toys	..	..	..	..	..	..	..	..	..	..
3699	Other manufacturing, nec	..	..	..	..	..	..	..	..	..	..
37	**RECYCLING**	..	..	..	..	..	..	..	..	..	..
371	Recycling of metal waste and scrap	..	..	..	..	..	..	..	..	..	..
372	Recycling of non-metal waste and scrap	..	..	..	..	..	..	..	..	..	..
CERR	**Non-specified industry**	..	..	..	..	..	..	..	..	..	..
15-37	**TOTAL MANUFACTURING**	**66,226**	**20,621**	**188,585**	..	**469,744**	..	**22,617**	**121,690,660**	..	**1,205,879**

ISIC Revision 3 Industry Sector	Solid TJ	LPG TJ	Distiloil TJ	RFO TJ	Gas TJ	Biomass TJ	Steam TJ	Electr MWh	Own Use MWh	TOTAL TJ
15 **FOOD PRODUCTS AND BEVERAGES**	..	..	..	..	..	..	..	..	..	..
151 Production, processing and preserving (PPP)	..	..	..	..	..	..	..	..	..	..
1511 PPP of meat and meat products	..	..	..	..	..	..	..	..	..	..
1512 Processing and preserving of fish products	..	..	..	..	..	..	..	..	..	..
1513 Processing, preserving of fruit & vegetables	..	..	..	..	..	..	..	..	..	..
1514 Vegetable and animal oils and fats	..	..	..	..	..	..	..	..	..	..
152 Dairy products	..	..	..	..	..	..	..	..	..	..
153 Grain mill prod., starches & prepared animal feeds	..	..	..	..	..	..	..	..	..	..
1531 Grain mill products	..	..	..	..	..	..	..	..	..	..
1532 Starches and starch products	..	..	..	..	..	..	..	..	..	..
1533 Prepared animal feeds	..	..	..	..	..	..	..	..	..	..
154 Other food products	..	..	..	..	..	..	..	..	..	..
1541 Bakery products	..	..	..	..	..	..	..	..	..	..
1542 Sugar	..	..	..	..	..	..	..	..	..	..
1543 Cocoa, chocolate and sugar confectionery	..	..	..	..	..	..	..	..	..	..
1544 Macaroni, noodles, couscous & similar farinaceous prod.	..	..	..	..	..	..	..	..	..	..
1549 Other food products, nec	..	..	..	..	..	..	..	..	..	..
155 Beverages	..	..	..	..	..	..	..	..	..	..
1551 Distilling, rectifying and blending of spirits	..	..	..	..	..	..	..	..	..	..
1552 Wines	..	..	..	..	..	..	..	..	..	..
1553 Malt liquors and malt	..	..	..	..	..	..	..	..	..	..
1554 Soft drinks; production of mineral waters	..	..	..	..	..	..	..	..	..	..
16 **TOBACCO PRODUCTS**	..	..	..	..	..	..	..	..	..	..
17 **TEXTILES**	5	1,354	5,203	..	15,689	..	202	3,245,983	57,151	33,933
171 Spinning, weaving and finishing of textiles	1	1,142	3,924	..	11,425	..	137	2,286,487	46,982	24,691
1711 Preparation and spinning of textile fibres	..	..	..	..	..	..	..	..	..	..
1712 Finishing of textiles	..	..	..	..	..	..	..	..	..	..
172 Other textiles	0	161	1,043	..	3,061	..	26	709,258	9,972	6,808
1721 Made-up textile articles, except apparel	..	..	..	..	..	..	..	..	..	..
1722 Carpets and rugs	..	..	..	..	..	..	..	..	..	..
1723 Cordage, rope, twine and netting	..	..	..	..	..	..	..	..	..	..
1729 Other textiles, nec	..	..	..	..	..	..	..	..	..	..
173 Knitted and crocheted fabrics and articles	4	51	236	..	1,203	..	39	250,238	197	2,433
18 **WEARING APPAREL, DRESSING & DYEING OF FUR**	30	46	431	..	1,072	..	11	214,463	44	2,362
181 Wearing apparel, except fur apparel	30	32	375	..	1,072	..	11	210,747	44	2,279
182 Dressing and dyeing of fur; articles of fur	0	14	56	..	0	..	0	3,716	0	83
19 **TANNING & DRESSING OF LEATHER, FOOTWEAR**	44	110	380	..	457	..	108	236,827	640	1,949
191 Tanning and dressing of leather	44	70	190	..	265	..	108	88,673	200	996
1911 Tanning and dressing of leather	..	..	..	..	..	..	..	..	..	..
1912 Luggage, handbags, saddlery & harness	..	..	..	..	..	..	..	..	..	..
192 Footwear	0	40	190	..	192	..	0	148,154	440	954
20 **WOOD AND WOOD PRODUCTS**	9	247	1,442	..	3,543	..	0	1,391,713	15,228	10,196
201 Sawmilling and planing of wood	..	..	..	..	..	..	..	..	..	..
202 Products of wood, cork, straw & plaiting materials	9	247	1,442	..	3,543	..	0	1,391,713	15,228	10,196
2021 Veneer sheets	..	..	..	..	..	..	..	..	..	..
2022 Builders' carpentry and joinery	..	..	..	..	..	..	..	..	..	..
2023 Wooden containers	..	..	..	..	..	..	..	..	..	..
2029 Other products of wood	..	..	..	..	..	..	..	..	..	..
21 **PAPER AND PAPER PRODUCTS**	6,745	1,131	13,389	..	54,696	..	1,700	11,586,444	2,249,086	111,275
2101 Pulp, paper and paperboard	..	..	..	..	..	..	..	..	..	..
2102 Corrugated paper, paperboard and their containers	..	..	..	..	..	..	..	..	..	..
2109 Other articles of pulp and paperboard	..	..	..	..	..	..	..	..	..	..
22 **PUBLISHING, PRINTING & REPRODUCTION**	0	264	3,864	..	4,086	..	66	1,557,306	6,690	13,862
221 Publishing	0	21	100	..	333	..	7	173,139	1,126	1,080
2211 Publishing of books & brochures	..	..	..	..	..	..	..	..	..	..
2212 Publishing of newspapers and periodicals	..	..	..	..	..	..	..	..	..	..
2213 Publishing of recorded media	..	..	..	..	..	..	..	..	..	..
2219 Other publishing	..	..	..	..	..	..	..	..	..	..
222 Printing and related service activities	0	219	3,716	..	3,729	..	59	1,293,321	5,563	12,359
2221 Printing	..	..	..	..	..	..	..	..	..	..
2222 Service activities related to printing	..	..	..	..	..	..	..	..	..	..
223 Reproduction of recorded media	0	24	48	..	24	..	0	90,846	1	423
23 **COKE, REFINED PETROLEUM PRODUCTS**	..	..	..	..	..	..	..	..	..	..
231 Coke oven products	..	..	..	..	..	..	..	..	..	..
232 Refined petroleum products	..	..	..	..	..	..	..	..	..	..
233 Processing of nuclear fuel	..	..	..	..	..	..	..	..	..	..
24 **CHEMICALS & CHEMICAL PRODUCTS**	132	443	5,237	..	20,241	..	1,373	26,854,496	2,048,775	116,727
241 Basic chemicals	c	c	c	..	c	..	c	22,850,604	1,904,201	75,407
2411 Basic chemicals, exc. fertilizers & nitrogen compounds	..	..	..	..	..	..	..	..	..	..
2412 Fertilizers and nitrogen compounds	..	..	..	..	..	..	..	..	..	..
2413 Plastics in primary forms and synthetic rubber	..	..	..	..	..	..	..	..	..	..
242 Other chemical products	132	429	5,052	..	18,262	..	1,262	3,574,558	137,567	37,510
2421 Pesticides and other agro-chemical products	..	..	..	..	..	..	..	..	..	..
2422 Paints, varnishes and similar coatings	..	..	..	..	..	..	..	..	..	..
2423 Pharmaceuticals, medicinal chem. & botanical prod.	..	..	..	..	..	..	..	..	..	..
2424 Soap and detergents, perfumes etc.	..	..	..	..	..	..	..	..	..	..
2429 Other chemical products, nec	..	..	..	..	..	..	..	..	..	..
243 Man-made fibres	0	14	185	..	1,979	..	111	429,334	7,007	3,809
25 **RUBBER AND PLASTICS PRODUCTS**	137	758	3,719	..	17,978	..	1,043	7,295,183	214,827	49,124
251 Rubber products	137	143	1,133	..	11,390	..	889	2,053,766	152,967	20,535
2511 Rubber tyres and tubes	..	..	..	..	..	..	..	..	..	..
2519 Other rubber products	..	..	..	..	..	..	..	..	..	..
252 Plastic products	0	615	2,586	..	6,588	..	154	5,241,417	61,860	28,589
26 **OTHER NON-METALLIC MINERAL PRODUCTS**	8,580	2,225	64,673	..	68,833	..	0	8,074,258	42,283	173,226
261 Glass and glass products	0	680	19,792	..	28,260	..	0	3,217,541	14,031	60,265
269 Non-metallic mineral products, nec	8,580	1,545	44,881	..	40,573	..	0	4,856,717	28,252	112,961
2691 Pottery, china and earthenware	..	..	..	..	..	..	..	..	..	..
2692 Refractory ceramic products	..	..	..	..	..	..	..	..	..	..
2693 Structural non-refractory clay & ceramic prod.	..	..	..	..	..	..	..	..	..	..
2694 Cement, lime and plaster	..	..	..	..	..	..	..	..	..	..
2695 Articles of concrete, cement and plaster	..	..	..	..	..	..	..	..	..	..
2696 Cutting, shaping and finishing of stone	..	..	..	..	..	..	..	..	..	..
2699 Other non-metallic mineral products, nec	..	..	..	..	..	..	..	..	..	..

ISIC Revision 3 Industry Sector		Solid TJ	LPG TJ	Distiloil TJ	RFO TJ	Gas TJ	Biomass TJ	Steam TJ	Electr MWh	Own Use MWh	TOTAL TJ
27	BASIC METALS	6,096	1,051	7,433	..	42,713	..	298	26,203,211	1,426,157	146,788
271	Basic iron and steel	3,537	351	6,952	..	37,442	..	298	14,723,168	1,347,915	96,731
272	Basic precious and non-ferrous metals	c	c	c	..	c	..	c	9,294,910	34,894	33,336
273	Casting of metals	2,559	700	481	..	5,271	..	0	2,185,133	43,348	16,721
2731	Casting of iron and steel	..	..	..	..	..	..	..	..	..	..
2732	Casting non-ferrous metals	..	..	..	..	..	..	..	..	..	..
28	FABRICATED METAL PRODUCTS	631	2,395	2,770	..	17,364	..	52	5,018,096	31,113	41,165
281	Str. metal prod., tanks, reservoirs, steam generators	204	548	718	..	3,025	..	6	830,351	4,201	7,475
2811	Structural metal products	..	..	..	..	..	..	..	..	..	..
2812	Tanks, reservoirs and containers of metal	..	..	..	..	..	..	..	..	..	..
2813	Steam generators, exc. central heating hot water boilers	..	..	..	..	..	..	..	..	..	..
289	Other fabricated metal products	427	1,847	2,052	..	14,339	..	46	4,187,745	26,912	33,690
2891	Forging, pressing, stamping & roll-forming of metal	..	..	..	..	..	..	..	..	..	..
2892	Treatment and coating of metals	..	..	..	..	..	..	..	..	..	..
2893	Cutlery, hand tools and general hardware	..	..	..	..	..	..	..	..	..	..
2899	Other fabricated metal products, nec	..	..	..	..	..	..	..	..	..	..
29	MACHINERY AND EQUIPMENT, NEC	68	1,127	2,287	..	9,570	..	16	2,688,934	26,987	22,651
291	General purpose machinery	6	559	1,181	..	4,119	..	16	1,438,181	19,452	10,988
2911	Engines and turbines	..	..	..	..	..	..	..	..	..	..
2912	Pumps, compressors, taps and valves	..	..	..	..	..	..	..	..	..	..
2913	Bearings, gears, gearing and driving elements	..	..	..	..	..	..	..	..	..	..
2914	Ovens, furnaces and furnace burners	..	..	..	..	..	..	..	..	..	..
2915	Lifting and handling equipment	..	..	..	..	..	..	..	..	..	..
2919	Other general purpose machinery	..	..	..	..	..	..	..	..	..	..
292	Special purpose machinery	0	455	878	..	3,942	..	0	888,754	6,247	8,452
2921	Agricultural and forestry machinery	..	..	..	..	..	..	..	..	..	..
2922	Machine-tools	..	..	..	..	..	..	..	..	..	..
2923	Machinery for metallurgy	..	..	..	..	..	..	..	..	..	..
2924	Machinery for mining, quarrying and construction	..	..	..	..	..	..	..	..	..	..
2925	Machinery for food, beverage & tobacco processing	..	..	..	..	..	..	..	..	..	..
2926	Machinery for textile, apparel & leather production	..	..	..	..	..	..	..	..	..	..
2927	Machinery for weapons and ammunition	..	..	..	..	..	..	..	..	..	..
2929	Other special purpose machinery	..	..	..	..	..	..	..	..	..	..
293	Domestic appliances, nec	62	113	228	..	1,509	..	0	361,999	1,288	3,211
30	OFFICE, ACCOUNTING & COMPUTING MACHINERY	0	4	27	..	370	..	0	225,130	2,113	1,204
31	ELECTRICAL MACHINERY & APPARATUS, NEC	156	418	1,081	..	6,791	..	122	2,677,852	18,542	18,142
311	Electric motors, generators and transformers	0	51	129	..	1,542	..	103	347,675	2,252	3,069
312	Electricity distribution and control apparatus	0	47	213	..	1,091	..	11	572,453	1,953	3,416
313	Insulated wire and cable	0	73	236	..	1,113	..	8	571,431	10,252	3,450
314	Accumulators, primary cells & primary batteries	0	16	12	..	792	..	0	256,051	0	1,742
315	Electric lamps and lighting equipment	0	30	27	..	740	..	0	133,782	735	1,276
319	Other electrical equipment, nec	156	201	464	..	1,513	..	0	796,460	3,350	5,189
32	RADIO, TV & COMMUNICATION EQUIP. & APP.	0	120	618	..	2,617	..	43	2,357,292	38,220	11,747
321	Electronic valves, tubes, other electronic components	0	57	462	..	1,647	..	37	1,825,788	17,475	8,713
322	TV & radio transmitters, apparatus for line telephony	0	52	136	..	674	..	6	418,131	17,362	2,311
323	TV & radio receivers, recording apparatus	0	11	20	..	296	..	0	113,373	3,383	723
33	MEDICAL PRECISION &OPTICAL INSTRUMENTS	0	69	449	..	1,722	..	13	900,540	12,697	5,449
331	Medical appliances and instruments	0	44	327	..	1,548	..	11	759,467	12,640	4,619
3311	Medical, surgical equipment & orthopaedic app.	..	..	..	..	..	..	..	..	..	..
3312	Instruments & appliances for measuring, checking etc	..	..	..	..	..	..	..	..	..	..
3313	Industrial process control equipment	..	..	..	..	..	..	..	..	..	..
332	Optical instruments and photographic equipment	0	23	86	..	125	..	2	104,500	22	612
333	Watches and clocks	0	2	36	..	49	..	0	36,573	35	219
34	MOTOR VEHICLES, TRAILERS & SEMI-TRAILERS	1,372	556	4,869	..	17,382	..	159	5,824,534	126,491	44,851
341	Motor vehicles	1,082	36	3,282	..	11,655	..	8	3,173,054	81,161	27,194
342	Bodies (coachwork) for motor vehicles	0	135	263	..	799	..	0	161,990	0	1,780
343	Parts, accessories for motor vehicles & their engines	290	385	1,324	..	4,928	..	151	2,489,490	45,330	15,877
35	OTHER TRANSPORT EQUIPMENT	0	185	1,289	..	5,987	..	131	1,412,379	79,473	12,390
351	Building and repairing of ships and boats	0	54	136	..	223	..	0	93,033	3,069	737
3511	Building and repairing of ships	..	..	..	..	..	..	..	..	..	..
3512	Building, repairing of pleasure & sporting boats	..	..	..	..	..	..	..	..	..	..
352	Railway, tramway locomotives & rolling stock	0	48	133	..	929	..	0	106,631	0	1,494
353	Aircraft and spacecraft	0	33	909	..	4,549	..	0	1,127,786	76,404	9,276
359	Transport equipment, nec	0	50	111	..	286	..	131	84,929	0	884
3591	Motorcycles	..	..	..	..	..	..	..	..	..	..
3592	Bicycles and invalid carriages	..	..	..	..	..	..	..	..	..	..
3599	Other transport equipment, nec	..	..	..	..	..	..	..	..	..	..
36	FURNITURE; MANUFACTURING, NEC	0	635	922	..	2,635	..	7	1,277,086	14,730	8,743
361	Furniture	0	507	601	..	1,714	..	7	892,956	14,430	5,992
369	Manufacturing, nec	0	128	321	..	921	..	0	384,130	300	2,752
3691	Jewellery and related articles	..	..	..	..	..	..	..	..	..	..
3692	Musical instruments	..	..	..	..	..	..	..	..	..	..
3693	Sports goods	..	..	..	..	..	..	..	..	..	..
3694	Games and toys	..	..	..	..	..	..	..	..	..	..
3699	Other manufacturing, nec	..	..	..	..	..	..	..	..	..	..
37	RECYCLING	..	..	..	..	..	..	..	..	..	..
371	Recycling of metal waste and scrap	..	..	..	..	..	..	..	..	..	..
372	Recycling of non-metal waste and scrap	..	..	..	..	..	..	..	..	..	..
CERR	Non-specified industry	..	..	..	..	..	..	..	..	..	..
15-37	TOTAL MANUFACTURING	50,969	14,107	156,371	..	403,023	..	22,494	109,041,727	6,411,247	1,016,434

ISIC Revision 3 Industry Sector	Solid TJ	LPG TJ	Distiloil TJ	RFO TJ	Gas TJ	Biomass TJ	Steam TJ	Electr MWh	Own Use MWh	TOTAL TJ
15 FOOD PRODUCTS AND BEVERAGES	13,909	..	30,406	20,848	87,551	..	..	12,228,333	..	196,736
151 Production, processing and preserving (PPP)	..	..	..	..	..	..	..	..	..	..
1511 PPP of meat and meat products	..	..	..	..	..	..	..	..	..	..
1512 Processing and preserving of fish products	..	..	..	..	..	..	..	..	..	..
1513 Processing, preserving of fruit & vegetables	..	..	..	..	..	..	..	..	..	..
1514 Vegetable and animal oils and fats	..	..	..	..	..	..	..	..	..	..
152 Dairy products	..	..	..	..	..	..	..	..	..	..
153 Grain mill prod., starches & prepared animal feeds	..	..	..	..	..	..	..	..	..	..
1531 Grain mill products	..	..	..	..	..	..	..	..	..	..
1532 Starches and starch products	..	..	..	..	..	..	..	..	..	..
1533 Prepared animal feeds	..	..	..	..	..	..	..	..	..	..
154 Other food products	..	..	..	..	..	..	..	..	..	..
1541 Bakery products	..	..	..	..	..	..	..	..	..	..
1542 Sugar	..	..	..	..	..	..	..	..	..	..
1543 Cocoa, chocolate and sugar confectionery	..	..	..	..	..	..	..	..	..	..
1544 Macaroni, noodles, couscous & similar farinaceous prod.	..	..	..	..	..	..	..	..	..	..
1549 Other food products, nec	..	..	..	..	..	..	..	..	..	..
155 Beverages	..	..	..	..	..	..	..	..	..	..
1551 Distilling, rectifying and blending of spirits	..	..	..	..	..	..	..	..	..	..
1552 Wines	..	..	..	..	..	..	..	..	..	..
1553 Malt liquors and malt	..	..	..	..	..	..	..	..	..	..
1554 Soft drinks; production of mineral waters	..	..	..	..	..	..	..	..	..	..
16 TOBACCO PRODUCTS	164	..	957	353	677	..	..	245,000	..	3,033
17 TEXTILES	3,442	..	4,550	2,098	21,143	..	..	3,631,111	..	44,305
171 Spinning, weaving and finishing of textiles	..	..	..	..	..	..	..	..	..	..
1711 Preparation and spinning of textile fibres	..	..	..	..	..	..	..	..	..	..
1712 Finishing of textiles	..	..	..	..	..	..	..	..	..	..
172 Other textiles	..	..	..	..	..	..	..	..	..	..
1721 Made-up textile articles, except apparel	..	..	..	..	..	..	..	..	..	..
1722 Carpets and rugs	..	..	..	..	..	..	..	..	..	..
1723 Cordage, rope, twine and netting	..	..	..	..	..	..	..	..	..	..
1729 Other textiles, nec	..	..	..	..	..	..	..	..	..	..
173 Knitted and crocheted fabrics and articles	..	..	..	..	..	..	..	..	..	..
18 WEARING APPAREL, DRESSING & DYEING OF FUR	20	..	1,436	29	1,410	..	..	287,778	..	3,931
181 Wearing apparel, except fur apparel	..	..	..	..	..	..	..	..	..	..
182 Dressing and dyeing of fur; articles of fur	..	..	..	..	..	..	..	..	..	..
19 TANNING & DRESSING OF LEATHER, FOOTWEAR	260	..	711	157	666	..	..	212,222	..	2,558
191 Tanning and dressing of leather	..	..	..	..	..	..	..	..	..	..
1911 Tanning and dressing of leather	..	..	..	..	..	..	..	..	..	..
1912 Luggage, handbags, saddlery & harness	..	..	..	..	..	..	..	..	..	..
192 Footwear	..	..	..	..	..	..	..	..	..	..
20 WOOD AND WOOD PRODUCTS	423	..	3,896	3,305	4,849	..	..	2,894,444	..	22,893
201 Sawmilling and planing of wood	..	..	..	..	..	..	..	..	..	..
202 Products of wood, cork, straw & plaiting materials	..	..	..	..	..	..	..	..	..	..
2021 Veneer sheets	..	..	..	..	..	..	..	..	..	..
2022 Builders' carpentry and joinery	..	..	..	..	..	..	..	..	..	..
2023 Wooden containers	..	..	..	..	..	..	..	..	..	..
2029 Other products of wood	..	..	..	..	..	..	..	..	..	..
21 PAPER AND PAPER PRODUCTS	36,209	..	6,834	13,520	100,874	..	..	16,747,500	..	217,728
2101 Pulp, paper and paperboard	..	..	..	..	..	..	..	..	..	..
2102 Corrugated paper, paperboard and their containers	..	..	..	..	..	..	..	..	..	..
2109 Other articles of pulp and paperboard	..	..	..	..	..	..	..	..	..	..
22 PUBLISHING, PRINTING & REPRODUCTION	19	..	2,573	81	7,359	..	..	2,689,167	..	19,713
221 Publishing	..	..	..	..	..	..	..	..	..	..
2211 Publishing of books & brochures	..	..	..	..	..	..	..	..	..	..
2212 Publishing of newspapers and periodicals	..	..	..	..	..	..	..	..	..	..
2213 Publishing of recorded media	..	..	..	..	..	..	..	..	..	..
2219 Other publishing	..	..	..	..	..	..	..	..	..	..
222 Printing and related service activities	..	..	..	..	..	..	..	..	..	..
2221 Printing	..	..	..	..	..	..	..	..	..	..
2222 Service activities related to printing	..	..	..	..	..	..	..	..	..	..
223 Reproduction of recorded media	..	..	..	..	..	..	..	..	..	..
23 COKE, REFINED PETROLEUM PRODUCTS	40	..	2,525	48,305	32,643	..	..	6,113,056	..	105,520
231 Coke oven products	..	..	..	..	..	..	..	..	..	..
232 Refined petroleum products	..	..	..	..	..	..	..	..	..	..
233 Processing of nuclear fuel	..	..	..	..	..	..	..	..	..	..
24 CHEMICALS & CHEMICAL PRODUCTS	91,304	..	26,641	101,778	396,056	..	..	46,726,667	..	783,995
241 Basic chemicals	..	..	..	..	..	..	..	..	..	..
2411 Basic chemicals, exc. fertilizers & nitrogen compounds	..	..	..	..	..	..	..	..	..	..
2412 Fertilizers and nitrogen compounds	..	..	..	..	..	..	..	..	..	..
2413 Plastics in primary forms and synthetic rubber	..	..	..	..	..	..	..	..	..	..
242 Other chemical products	..	..	..	..	..	..	..	..	..	..
2421 Pesticides and other agro-chemical products	..	..	..	..	..	..	..	..	..	..
2422 Paints, varnishes and similar coatings	..	..	..	..	..	..	..	..	..	..
2423 Pharmaceuticals, medicinal chem. & botanical prod.	..	..	..	..	..	..	..	..	..	..
2424 Soap and detergents, perfumes etc.	..	..	..	..	..	..	..	..	..	..
2429 Other chemical products, nec	..	..	..	..	..	..	..	..	..	..
243 Man-made fibres	..	..	..	..	..	..	..	..	..	..
25 RUBBER AND PLASTICS PRODUCTS	1,273	..	6,494	1,789	27,041	..	..	9,833,056	..	71,996
251 Rubber products	..	..	..	..	..	..	..	..	..	..
2511 Rubber tyres and tubes	..	..	..	..	..	..	..	..	..	..
2519 Other rubber products	..	..	..	..	..	..	..	..	..	..
252 Plastic products	..	..	..	..	..	..	..	..	..	..
26 OTHER NON-METALLIC MINERAL PRODUCTS	103,842	..	18,247	27,423	125,280	..	..	13,193,889	..	322,290
261 Glass and glass products	..	..	..	..	..	..	..	..	..	..
269 Non-metallic mineral products, nec	..	..	..	..	..	..	..	..	..	..
2691 Pottery, china and earthenware	..	..	..	..	..	..	..	..	..	..
2692 Refractory ceramic products	..	..	..	..	..	..	..	..	..	..
2693 Structural non-refractory clay & ceramic prod.	..	..	..	..	..	..	..	..	..	..
2694 Cement, lime and plaster	..	..	..	..	..	..	..	..	..	..
2695 Articles of concrete, cement and plaster	..	..	..	..	..	..	..	..	..	..
2696 Cutting, shaping and finishing of stone	..	..	..	..	..	..	..	..	..	..
2699 Other non-metallic mineral products, nec	..	..	..	..	..	..	..	..	..	..

ISIS Energy Data Programme (IEA/OECD)

ISIC Revision 3 Industry Sector		Solid TJ	LPG TJ	Distiloil TJ	RFO TJ	Gas TJ	Biomass TJ	Steam TJ	Electr MWh	Own Use MWh	TOTAL TJ
27	**BASIC METALS**	438,081	..	5,436	63,143	194,608	..	..	39,367,222	..	842,990
271	Basic iron and steel	..	..	..	..	..	..	..	..	..	..
272	Basic precious and non-ferrous metals	..	..	..	..	..	..	..	..	..	..
273	Casting of metals	..	..	..	..	..	..	..	..	..	..
2731	Casting of iron and steel	..	..	..	..	..	..	..	..	..	..
2732	Casting non-ferrous metals	..	..	..	..	..	..	..	..	..	..
28	**FABRICATED METAL PRODUCTS**	765	..	10,215	292	34,078	..	..	7,428,333	..	72,092
281	Str. metal prod., tanks, reservoirs, steam generators	..	..	..	..	..	..	..	..	..	..
2811	Structural metal products	..	..	..	..	..	..	..	..	..	..
2812	Tanks, reservoirs and containers of metal	..	..	..	..	..	..	..	..	..	..
2813	Steam generators, exc. central heating hot water boilers	..	..	..	..	..	..	..	..	..	..
289	Other fabricated metal products	..	..	..	..	..	..	..	..	..	..
2891	Forging, pressing, stamping & roll-forming of metal	..	..	..	..	..	..	..	..	..	..
2892	Treatment and coating of metals	..	..	..	..	..	..	..	..	..	..
2893	Cutlery, hand tools and general hardware	..	..	..	..	..	..	..	..	..	..
2899	Other fabricated metal products, nec	..	..	..	..	..	..	..	..	..	..
29	**MACHINERY AND EQUIPMENT, NEC**	2,266	..	15,166	887	31,102	..	..	8,637,778	..	80,517
291	General purpose machinery	..	..	..	..	..	..	..	..	..	..
2911	Engines and turbines	..	..	..	..	..	..	..	..	..	..
2912	Pumps, compressors, taps and valves	..	..	..	..	..	..	..	..	..	..
2913	Bearings, gears, gearing and driving elements	..	..	..	..	..	..	..	..	..	..
2914	Ovens, furnaces and furnace burners	..	..	..	..	..	..	..	..	..	..
2915	Lifting and handling equipment	..	..	..	..	..	..	..	..	..	..
2919	Other general purpose machinery	..	..	..	..	..	..	..	..	..	..
292	Special purpose machinery	..	..	..	..	..	..	..	..	..	..
2921	Agricultural and forestry machinery	..	..	..	..	..	..	..	..	..	..
2922	Machine-tools	..	..	..	..	..	..	..	..	..	..
2923	Machinery for metallurgy	..	..	..	..	..	..	..	..	..	..
2924	Machinery for mining, quarrying and construction	..	..	..	..	..	..	..	..	..	..
2925	Machinery for food, beverage & tobacco processing	..	..	..	..	..	..	..	..	..	..
2926	Machinery for textile, apparel & leather production	..	..	..	..	..	..	..	..	..	..
2927	Machinery for weapons and ammunition	..	..	..	..	..	..	..	..	..	..
2929	Other special purpose machinery	..	..	..	..	..	..	..	..	..	..
293	Domestic appliances, nec	..	..	..	..	..	..	..	..	..	..
30	**OFFICE, ACCOUNTING & COMPUTING MACHINERY**	0	..	242	4	963	..	..	531,667	..	3,123
31	**ELECTRICAL MACHINERY & APPARATUS, NEC**	1,966	..	5,064	272	11,726	..	..	4,778,333	..	36,230
311	Electric motors, generators and transformers	..	..	..	..	..	..	..	..	..	..
312	Electricity distribution and control apparatus	..	..	..	..	..	..	..	..	..	..
313	Insulated wire and cable	..	..	..	..	..	..	..	..	..	..
314	Accumulators, primary cells & primary batteries	..	..	..	..	..	..	..	..	..	..
315	Electric lamps and lighting equipment	..	..	..	..	..	..	..	..	..	..
319	Other electrical equipment, nec	..	..	..	..	..	..	..	..	..	..
32	**RADIO, TV & COMMUNICATION EQUIP. & APP.**	45	..	1,096	1	3,008	..	..	2,130,556	..	11,820
321	Electronic valves, tubes, other electronic components	..	..	..	..	..	..	..	..	..	..
322	TV & radio transmitters, apparatus for line telephony	..	..	..	..	..	..	..	..	..	..
323	TV & radio receivers, recording apparatus	..	..	..	..	..	..	..	..	..	..
33	**MEDICAL PRECISION &OPTICAL INSTRUMENTS**	84	..	2,875	31	3,401	..	..	1,402,222	..	11,439
331	Medical appliances and instruments	..	..	..	..	..	..	..	..	..	..
3311	Medical, surgical equipment & orthopaedic app.	..	..	..	..	..	..	..	..	..	..
3312	Instruments & appliances for measuring, checking etc	..	..	..	..	..	..	..	..	..	..
3313	Industrial process control equipment	..	..	..	..	..	..	..	..	..	..
332	Optical instruments and photographic equipment	..	..	..	..	..	..	..	..	..	..
333	Watches and clocks	..	..	..	..	..	..	..	..	..	..
34	**MOTOR VEHICLES, TRAILERS & SEMI-TRAILERS**	3,651	..	4,795	652	37,642	..	..	13,187,222	..	94,214
341	Motor vehicles	..	..	..	..	..	..	..	..	..	..
342	Bodies (coachwork) for motor vehicles	..	..	..	..	..	..	..	..	..	..
343	Parts, accessories for motor vehicles & their engines	..	..	..	..	..	..	..	..	..	..
35	**OTHER TRANSPORT EQUIPMENT**	1,168	..	2,191	93	8,079	..	..	1,625,278	..	17,382
351	Building and repairing of ships and boats	..	..	..	..	..	..	..	..	..	..
3511	Building and repairing of ships	..	..	..	..	..	..	..	..	..	..
3512	Building, repairing of pleasure & sporting boats	..	..	..	..	..	..	..	..	..	..
352	Railway, tramway locomotives & rolling stock	..	..	..	..	..	..	..	..	..	..
353	Aircraft and spacecraft	..	..	..	..	..	..	..	..	..	..
359	Transport equipment, nec	..	..	..	..	..	..	..	..	..	..
3591	Motorcycles	..	..	..	..	..	..	..	..	..	..
3592	Bicycles and invalid carriages	..	..	..	..	..	..	..	..	..	..
3599	Other transport equipment, nec	..	..	..	..	..	..	..	..	..	..
36	**FURNITURE; MANUFACTURING, NEC**	166	..	4,014	703	3,983	..	..	2,039,167	..	16,207
361	Furniture	..	..	..	..	..	..	..	..	..	..
369	Manufacturing, nec	..	..	..	..	..	..	..	..	..	..
3691	Jewellery and related articles	..	..	..	..	..	..	..	..	..	..
3692	Musical instruments	..	..	..	..	..	..	..	..	..	..
3693	Sports goods	..	..	..	..	..	..	..	..	..	..
3694	Games and toys	..	..	..	..	..	..	..	..	..	..
3699	Other manufacturing, nec	..	..	..	..	..	..	..	..	..	..
37	**RECYCLING**	111	..	79	7	453	..	..	152,778	..	1,200
371	Recycling of metal waste and scrap	..	..	..	..	..	..	..	..	..	..
372	Recycling of non-metal waste and scrap	..	..	..	..	..	..	..	..	..	..
CERR	**Non-specified industry**	..	..	..	..	..	..	..	..	..	..
15-37	**TOTAL MANUFACTURING**	699,208	..	156,443	285,771	1,134,592	..	..	196,082,778	..	**2,981,912**

ISIC Revision 3 Industry Sector	Solid TJ	LPG TJ	Distiloil TJ	RFO TJ	Gas TJ	Biomass TJ	Steam TJ	Electr MWh	Own Use MWh	TOTAL TJ
15　FOOD PRODUCTS AND BEVERAGES	14,886	..	28,901	17,647	93,214	..		12,225,833	..	198,661
151　Production, processing and preserving (PPP)	..	..	..	..	..	..	..	..	..	..
1511　PPP of meat and meat products	..	..	..	..	..	..	..	..	..	..
1512　Processing and preserving of fish products	..	..	..	..	..	..	..	..	..	..
1513　Processing, preserving of fruit & vegetables	..	..	..	..	..	..	..	..	..	..
1514　Vegetable and animal oils and fats	..	..	..	..	..	..	..	..	..	..
152　Dairy products	..	..	..	..	..	..	..	..	..	..
153　Grain mill prod., starches & prepared animal feeds	..	..	..	..	..	..	..	..	..	..
1531　Grain mill products	..	..	..	..	..	..	..	..	..	..
1532　Starches and starch products	..	..	..	..	..	..	..	..	..	..
1533　Prepared animal feeds	..	..	..	..	..	..	..	..	..	..
154　Other food products	..	..	..	..	..	..	..	..	..	..
1541　Bakery products	..	..	..	..	..	..	..	..	..	..
1542　Sugar	..	..	..	..	..	..	..	..	..	..
1543　Cocoa, chocolate and sugar confectionery	..	..	..	..	..	..	..	..	..	..
1544　Macaroni, noodles, couscous & similar farinaceous prod.	..	..	..	..	..	..	..	..	..	..
1549　Other food products, nec	..	..	..	..	..	..	..	..	..	..
155　Beverages	..	..	..	..	..	..	..	..	..	..
1551　Distilling, rectifying and blending of spirits	..	..	..	..	..	..	..	..	..	..
1552　Wines	..	..	..	..	..	..	..	..	..	..
1553　Malt liquors and malt	..	..	..	..	..	..	..	..	..	..
1554　Soft drinks; production of mineral waters	..	..	..	..	..	..	..	..	..	..
16　TOBACCO PRODUCTS	142	..	756	1	819	..		237,222	..	2,572
17　TEXTILES	2,790	..	4,435	2,050	20,936	..		3,500,000	..	42,811
171　Spinning, weaving and finishing of textiles	..	..	..	..	..	..	..	..	..	..
1711　Preparation and spinning of textile fibres	..	..	..	..	..	..	..	..	..	..
1712　Finishing of textiles	..	..	..	..	..	..	..	..	..	..
172　Other textiles	..	..	..	..	..	..	..	..	..	..
1721　Made-up textile articles, except apparel	..	..	..	..	..	..	..	..	..	..
1722　Carpets and rugs	..	..	..	..	..	..	..	..	..	..
1723　Cordage, rope, twine and netting	..	..	..	..	..	..	..	..	..	..
1729　Other textiles, nec	..	..	..	..	..	..	..	..	..	..
173　Knitted and crocheted fabrics and articles	..	..	..	..	..	..	..	..	..	..
18　WEARING APPAREL, DRESSING & DYEING OF FUR	9	..	1,416	32	1,474	..		267,778	..	3,895
181　Wearing apparel, except fur apparel	..	..	..	..	..	..	..	..	..	..
182　Dressing and dyeing of fur; articles of fur	..	..	..	..	..	..	..	..	..	..
19　TANNING & DRESSING OF LEATHER, FOOTWEAR	184	..	768	125	655	..		210,278	..	2,489
191　Tanning and dressing of leather	..	..	..	..	..	..	..	..	..	..
1911　Tanning and dressing of leather	..	..	..	..	..	..	..	..	..	..
1912　Luggage, handbags, saddlery & harness	..	..	..	..	..	..	..	..	..	..
192　Footwear	..	..	..	..	..	..	..	..	..	..
20　WOOD AND WOOD PRODUCTS	401	..	4,065	2,596	5,522	..		2,976,944	..	23,301
201　Sawmilling and planing of wood	..	..	..	..	..	..	..	..	..	..
202　Products of wood, cork, straw & plaiting materials	..	..	..	..	..	..	..	..	..	..
2021　Veneer sheets	..	..	..	..	..	..	..	..	..	..
2022　Builders' carpentry and joinery	..	..	..	..	..	..	..	..	..	..
2023　Wooden containers	..	..	..	..	..	..	..	..	..	..
2029　Other products of wood	..	..	..	..	..	..	..	..	..	..
21　PAPER AND PAPER PRODUCTS	31,920	..	5,914	13,524	99,345	..		16,440,000	..	209,887
2101　Pulp, paper and paperboard	..	..	..	..	..	..	..	..	..	..
2102　Corrugated paper, paperboard and their containers	..	..	..	..	..	..	..	..	..	..
2109　Other articles of pulp and paperboard	..	..	..	..	..	..	..	..	..	..
22　PUBLISHING, PRINTING & REPRODUCTION	9	..	1,986	89	8,103	..		2,725,833	..	20,000
221　Publishing	..	..	..	..	..	..	..	..	..	..
2211　Publishing of books & brochures	..	..	..	..	..	..	..	..	..	..
2212　Publishing of newspapers and periodicals	..	..	..	..	..	..	..	..	..	..
2213　Publishing of recorded media	..	..	..	..	..	..	..	..	..	..
2219　Other publishing	..	..	..	..	..	..	..	..	..	..
222　Printing and related service activities	..	..	..	..	..	..	..	..	..	..
2221　Printing	..	..	..	..	..	..	..	..	..	..
2222　Service activities related to printing	..	..	..	..	..	..	..	..	..	..
223　Reproduction of recorded media	..	..	..	..	..	..	..	..	..	..
23　COKE, REFINED PETROLEUM PRODUCTS	3,681	..	2,287	45,112	32,902	..		6,083,333	..	105,882
231　Coke oven products	..	..	..	..	..	..	..	..	..	..
232　Refined petroleum products	..	..	..	..	..	..	..	..	..	..
233　Processing of nuclear fuel	..	..	..	..	..	..	..	..	..	..
24　CHEMICALS & CHEMICAL PRODUCTS	92,656	..	30,086	99,340	377,749	..		46,549,444	..	767,409
241　Basic chemicals	..	..	..	..	..	..	..	..	..	..
2411　Basic chemicals, exc. fertilizers & nitrogen compounds	..	..	..	..	..	..	..	..	..	..
2412　Fertilizers and nitrogen compounds	..	..	..	..	..	..	..	..	..	..
2413　Plastics in primary forms and synthetic rubber	..	..	..	..	..	..	..	..	..	..
242　Other chemical products	..	..	..	..	..	..	..	..	..	..
2421　Pesticides and other agro-chemical products	..	..	..	..	..	..	..	..	..	..
2422　Paints, varnishes and similar coatings	..	..	..	..	..	..	..	..	..	..
2423　Pharmaceuticals, medicinal chem. & botanical prod.	..	..	..	..	..	..	..	..	..	..
2424　Soap and detergents, perfumes etc.	..	..	..	..	..	..	..	..	..	..
2429　Other chemical products, nec	..	..	..	..	..	..	..	..	..	..
243　Man-made fibres	..	..	..	..	..	..	..	..	..	..
25　RUBBER AND PLASTICS PRODUCTS	1,022	..	6,892	1,777	26,084	..		9,631,944	..	70,450
251　Rubber products	..	..	..	..	..	..	..	..	..	..
2511　Rubber tyres and tubes	..	..	..	..	..	..	..	..	..	..
2519　Other rubber products	..	..	..	..	..	..	..	..	..	..
252　Plastic products	..	..	..	..	..	..	..	..	..	..
26　OTHER NON-METALLIC MINERAL PRODUCTS	97,674	..	15,179	25,181	126,185	..		13,209,444	..	311,773
261　Glass and glass products	..	..	..	..	..	..	..	..	..	..
269　Non-metallic mineral products, nec	..	..	..	..	..	..	..	..	..	..
2691　Pottery, china and earthenware	..	..	..	..	..	..	..	..	..	..
2692　Refractory ceramic products	..	..	..	..	..	..	..	..	..	..
2693　Structural non-refractory clay & ceramic prod.	..	..	..	..	..	..	..	..	..	..
2694　Cement, lime and plaster	..	..	..	..	..	..	..	..	..	..
2695　Articles of concrete, cement and plaster	..	..	..	..	..	..	..	..	..	..
2696　Cutting, shaping and finishing of stone	..	..	..	..	..	..	..	..	..	..
2699　Other non-metallic mineral products, nec	..	..	..	..	..	..	..	..	..	..

ISIC Revision 3 Industry Sector		Solid TJ	LPG TJ	Distiloil TJ	RFO TJ	Gas TJ	Biomass TJ	Steam TJ	Electr MWh	Own Use MWh	TOTAL TJ
27	**BASIC METALS**	413,745	..	5,933	57,411	190,272	..	..	38,416,111	..	805,659
271	Basic iron and steel	..	..	..	..	..	..	..	..	..	..
272	Basic precious and non-ferrous metals	..	..	..	..	..	..	..	..	..	..
273	Casting of metals	..	..	..	..	..	..	..	..	..	..
2731	Casting of iron and steel	..	..	..	..	..	..	..	..	..	..
2732	Casting non-ferrous metals	..	..	..	..	..	..	..	..	..	..
28	**FABRICATED METAL PRODUCTS**	650	..	10,317	281	34,652	..	..	7,208,056	..	71,849
281	Str. metal prod., tanks, reservoirs, steam generators	..	..	..	..	..	..	..	..	..	..
2811	Structural metal products	..	..	..	..	..	..	..	..	..	..
2812	Tanks, reservoirs and containers of metal	..	..	..	..	..	..	..	..	..	..
2813	Steam generators, exc. central heating hot water boilers	..	..	..	..	..	..	..	..	..	..
289	Other fabricated metal products	..	..	..	..	..	..	..	..	..	..
2891	Forging, pressing, stamping & roll-forming of metal	..	..	..	..	..	..	..	..	..	..
2892	Treatment and coating of metals	..	..	..	..	..	..	..	..	..	..
2893	Cutlery, hand tools and general hardware	..	..	..	..	..	..	..	..	..	..
2899	Other fabricated metal products, nec	..	..	..	..	..	..	..	..	..	..
29	**MACHINERY AND EQUIPMENT, NEC**	1,693	..	16,171	885	31,828	..	..	8,501,111	..	81,181
291	General purpose machinery	..	..	..	..	..	..	..	..	..	..
2911	Engines and turbines	..	..	..	..	..	..	..	..	..	..
2912	Pumps, compressors, taps and valves	..	..	..	..	..	..	..	..	..	..
2913	Bearings, gears, gearing and driving elements	..	..	..	..	..	..	..	..	..	..
2914	Ovens, furnaces and furnace burners	..	..	..	..	..	..	..	..	..	..
2915	Lifting and handling equipment	..	..	..	..	..	..	..	..	..	..
2919	Other general purpose machinery	..	..	..	..	..	..	..	..	..	..
292	Special purpose machinery	..	..	..	..	..	..	..	..	..	..
2921	Agricultural and forestry machinery	..	..	..	..	..	..	..	..	..	..
2922	Machine-tools	..	..	..	..	..	..	..	..	..	..
2923	Machinery for metallurgy	..	..	..	..	..	..	..	..	..	..
2924	Machinery for mining, quarrying and construction	..	..	..	..	..	..	..	..	..	..
2925	Machinery for food, beverage & tobacco processing	..	..	..	..	..	..	..	..	..	..
2926	Machinery for textile, apparel & leather production	..	..	..	..	..	..	..	..	..	..
2927	Machinery for weapons and ammunition	..	..	..	..	..	..	..	..	..	..
2929	Other special purpose machinery	..	..	..	..	..	..	..	..	..	..
293	Domestic appliances, nec	..	..	..	..	..	..	..	..	..	..
30	**OFFICE, ACCOUNTING & COMPUTING MACHINERY**	0	..	295	0	932	..	..	465,000	..	2,901
31	**ELECTRICAL MACHINERY & APPARATUS, NEC**	643	..	5,107	178	12,188	..	..	4,734,167	..	35,159
311	Electric motors, generators and transformers	..	..	..	..	..	..	..	..	..	..
312	Electricity distribution and control apparatus	..	..	..	..	..	..	..	..	..	..
313	Insulated wire and cable	..	..	..	..	..	..	..	..	..	..
314	Accumulators, primary cells & primary batteries	..	..	..	..	..	..	..	..	..	..
315	Electric lamps and lighting equipment	..	..	..	..	..	..	..	..	..	..
319	Other electrical equipment, nec	..	..	..	..	..	..	..	..	..	..
32	**RADIO, TV & COMMUNICATION EQUIP. & APP.**	49	..	1,159	1	3,447	..	..	2,285,278	..	12,883
321	Electronic valves, tubes, other electronic components	..	..	..	..	..	..	..	..	..	..
322	TV & radio transmitters, apparatus for line telephony	..	..	..	..	..	..	..	..	..	..
323	TV & radio receivers, recording apparatus	..	..	..	..	..	..	..	..	..	..
33	**MEDICAL PRECISION &OPTICAL INSTRUMENTS**	30	..	2,057	32	3,555	..	..	1,499,444	..	11,072
331	Medical appliances and instruments	..	..	..	..	..	..	..	..	..	..
3311	Medical, surgical equipment & orthopaedic app.	..	..	..	..	..	..	..	..	..	..
3312	Instruments & appliances for measuring, checking etc	..	..	..	..	..	..	..	..	..	..
3313	Industrial process control equipment	..	..	..	..	..	..	..	..	..	..
332	Optical instruments and photographic equipment	..	..	..	..	..	..	..	..	..	..
333	Watches and clocks	..	..	..	..	..	..	..	..	..	..
34	**MOTOR VEHICLES, TRAILERS & SEMI-TRAILERS**	3,554	..	6,335	672	40,461	..	..	12,678,333	..	96,664
341	Motor vehicles	..	..	..	..	..	..	..	..	..	..
342	Bodies (coachwork) for motor vehicles	..	..	..	..	..	..	..	..	..	..
343	Parts, accessories for motor vehicles & their engines	..	..	..	..	..	..	..	..	..	..
35	**OTHER TRANSPORT EQUIPMENT**	436	..	2,293	111	9,248	..	..	1,540,833	..	17,635
351	Building and repairing of ships and boats	..	..	..	..	..	..	..	..	..	..
3511	Building and repairing of ships	..	..	..	..	..	..	..	..	..	..
3512	Building, repairing of pleasure & sporting boats	..	..	..	..	..	..	..	..	..	..
352	Railway, tramway locomotives & rolling stock	..	..	..	..	..	..	..	..	..	..
353	Aircraft and spacecraft	..	..	..	..	..	..	..	..	..	..
359	Transport equipment, nec	..	..	..	..	..	..	..	..	..	..
3591	Motorcycles	..	..	..	..	..	..	..	..	..	..
3592	Bicycles and invalid carriages	..	..	..	..	..	..	..	..	..	..
3599	Other transport equipment, nec	..	..	..	..	..	..	..	..	..	..
36	**FURNITURE; MANUFACTURING, NEC**	86	..	4,316	761	4,009	..	..	2,027,500	..	16,471
361	Furniture	..	..	..	..	..	..	..	..	..	..
369	Manufacturing, nec	..	..	..	..	..	..	..	..	..	..
3691	Jewellery and related articles	..	..	..	..	..	..	..	..	..	..
3692	Musical instruments	..	..	..	..	..	..	..	..	..	..
3693	Sports goods	..	..	..	..	..	..	..	..	..	..
3694	Games and toys	..	..	..	..	..	..	..	..	..	..
3699	Other manufacturing, nec	..	..	..	..	..	..	..	..	..	..
37	**RECYCLING**	97	..	137	6	516	..	..	200,278	..	1,477
371	Recycling of metal waste and scrap	..	..	..	..	..	..	..	..	..	..
372	Recycling of non-metal waste and scrap	..	..	..	..	..	..	..	..	..	..
CERR	Non-specified industry		..				..	..		..	
15-37	**TOTAL MANUFACTURING**	666,357	..	156,805	267,812	1,124,096	..	..	193,614,167	..	2,912,081

ISIC Revision 3 Industry Sector	Solid TJ	LPG TJ	Distiloil TJ	RFO TJ	Gas TJ	Biomass TJ	Steam TJ	Electr MWh	Own Use MWh	TOTAL TJ
15	**FOOD PRODUCTS AND BEVERAGES**	14,145	..	24,823	17,361	95,247	..	..	12,377,778	..	196,136
151	Production, processing and preserving (PPP)	..	..	..	..	..	..	..	..	..	..
1511	PPP of meat and meat products	..	..	..	..	..	..	..	..	..	..
1512	Processing and preserving of fish products	..	..	..	..	..	..	..	..	..	..
1513	Processing, preserving of fruit & vegetables	..	..	..	..	..	..	..	..	..	..
1514	Vegetable and animal oils and fats	..	..	..	..	..	..	..	..	..	..
152	Dairy products	..	..	..	..	..	..	..	..	..	..
153	Grain mill prod., starches & prepared animal feeds	..	..	..	..	..	..	..	..	..	..
1531	Grain mill products	..	..	..	..	..	..	..	..	..	..
1532	Starches and starch products	..	..	..	..	..	..	..	..	..	..
1533	Prepared animal feeds	..	..	..	..	..	..	..	..	..	..
154	Other food products	..	..	..	..	..	..	..	..	..	..
1541	Bakery products	..	..	..	..	..	..	..	..	..	..
1542	Sugar	..	..	..	..	..	..	..	..	..	..
1543	Cocoa, chocolate and sugar confectionery	..	..	..	..	..	..	..	..	..	..
1544	Macaroni, noodles, couscous & similar farinaceous prod.	..	..	..	..	..	..	..	..	..	..
1549	Other food products, nec	..	..	..	..	..	..	..	..	..	..
155	Beverages	..	..	..	..	..	..	..	..	..	..
1551	Distilling, rectifying and blending of spirits	..	..	..	..	..	..	..	..	..	..
1552	Wines	..	..	..	..	..	..	..	..	..	..
1553	Malt liquors and malt	..	..	..	..	..	..	..	..	..	..
1554	Soft drinks; production of mineral waters	..	..	..	..	..	..	..	..	..	..
16	**TOBACCO PRODUCTS**	110	..	692	0	711	..	..	240,278	..	2,378
17	**TEXTILES**	2,734	..	3,898	1,831	19,827	..	..	3,514,722	..	40,943
171	Spinning, weaving and finishing of textiles	..	..	..	..	..	..	..	..	..	..
1711	Preparation and spinning of textile fibres	..	..	..	..	..	..	..	..	..	..
1712	Finishing of textiles	..	..	..	..	..	..	..	..	..	..
172	Other textiles	..	..	..	..	..	..	..	..	..	..
1721	Made-up textile articles, except apparel	..	..	..	..	..	..	..	..	..	..
1722	Carpets and rugs	..	..	..	..	..	..	..	..	..	..
1723	Cordage, rope, twine and netting	..	..	..	..	..	..	..	..	..	..
1729	Other textiles, nec	..	..	..	..	..	..	..	..	..	..
173	Knitted and crocheted fabrics and articles	..	..	..	..	..	..	..	..	..	..
18	**WEARING APPAREL, DRESSING & DYEING OF FUR**	7	..	1,500	15	1,329	..	..	246,667	..	3,739
181	Wearing apparel, except fur apparel	..	..	..	..	..	..	..	..	..	..
182	Dressing and dyeing of fur; articles of fur	..	..	..	..	..	..	..	..	..	..
19	**TANNING & DRESSING OF LEATHER, FOOTWEAR**	166	..	664	115	521	..	..	173,333	..	2,090
191	Tanning and dressing of leather	..	..	..	..	..	..	..	..	..	..
1911	Tanning and dressing of leather	..	..	..	..	..	..	..	..	..	..
1912	Luggage, handbags, saddlery & harness	..	..	..	..	..	..	..	..	..	..
192	Footwear	..	..	..	..	..	..	..	..	..	..
20	**WOOD AND WOOD PRODUCTS**	314	..	3,082	2,390	6,445	..	..	3,145,833	..	23,556
201	Sawmilling and planing of wood	..	..	..	..	..	..	..	..	..	..
202	Products of wood, cork, straw & plaiting materials	..	..	..	..	..	..	..	..	..	..
2021	Veneer sheets	..	..	..	..	..	..	..	..	..	..
2022	Builders' carpentry and joinery	..	..	..	..	..	..	..	..	..	..
2023	Wooden containers	..	..	..	..	..	..	..	..	..	..
2029	Other products of wood	..	..	..	..	..	..	..	..	..	..
21	**PAPER AND PAPER PRODUCTS**	30,293	..	4,370	11,444	108,406	..	..	16,997,778	..	215,705
2101	Pulp, paper and paperboard	..	..	..	..	..	..	..	..	..	..
2102	Corrugated paper, paperboard and their containers	..	..	..	..	..	..	..	..	..	..
2109	Other articles of pulp and paperboard	..	..	..	..	..	..	..	..	..	..
22	**PUBLISHING, PRINTING & REPRODUCTION**	3	..	1,751	44	8,707	..	..	2,811,111	..	20,625
221	Publishing	..	..	..	..	..	..	..	..	..	..
2211	Publishing of books & brochures	..	..	..	..	..	..	..	..	..	..
2212	Publishing of newspapers and periodicals	..	..	..	..	..	..	..	..	..	..
2213	Publishing of recorded media	..	..	..	..	..	..	..	..	..	..
2219	Other publishing	..	..	..	..	..	..	..	..	..	..
222	Printing and related service activities	..	..	..	..	..	..	..	..	..	..
2221	Printing	..	..	..	..	..	..	..	..	..	..
2222	Service activities related to printing	..	..	..	..	..	..	..	..	..	..
223	Reproduction of recorded media	..	..	..	..	..	..	..	..	..	..
23	**COKE, REFINED PETROLEUM PRODUCTS**	5,004	..	2,233	41,904	32,256	..	..	5,947,222	..	102,807
231	Coke oven products	..	..	..	..	..	..	..	..	..	..
232	Refined petroleum products	..	..	..	..	..	..	..	..	..	..
233	Processing of nuclear fuel	..	..	..	..	..	..	..	..	..	..
24	**CHEMICALS & CHEMICAL PRODUCTS**	95,482	..	29,334	103,489	391,902	..	..	49,351,667	..	797,873
241	Basic chemicals	..	..	..	..	..	..	..	..	..	..
2411	Basic chemicals, exc. fertilizers & nitrogen compounds	..	..	..	..	..	..	..	..	..	..
2412	Fertilizers and nitrogen compounds	..	..	..	..	..	..	..	..	..	..
2413	Plastics in primary forms and synthetic rubber	..	..	..	..	..	..	..	..	..	..
242	Other chemical products	..	..	..	..	..	..	..	..	..	..
2421	Pesticides and other agro-chemical products	..	..	..	..	..	..	..	..	..	..
2422	Paints, varnishes and similar coatings	..	..	..	..	..	..	..	..	..	..
2423	Pharmaceuticals, medicinal chem. & botanical prod.	..	..	..	..	..	..	..	..	..	..
2424	Soap and detergents, perfumes etc.	..	..	..	..	..	..	..	..	..	..
2429	Other chemical products, nec	..	..	..	..	..	..	..	..	..	..
243	Man-made fibres	..	..	..	..	..	..	..	..	..	..
25	**RUBBER AND PLASTICS PRODUCTS**	752	..	8,143	1,411	25,732	..	..	10,064,722	..	72,271
251	Rubber products	..	..	..	..	..	..	..	..	..	..
2511	Rubber tyres and tubes	..	..	..	..	..	..	..	..	..	..
2519	Other rubber products	..	..	..	..	..	..	..	..	..	..
252	Plastic products	..	..	..	..	..	..	..	..	..	..
26	**OTHER NON-METALLIC MINERAL PRODUCTS**	96,467	..	13,414	24,684	127,556	..	..	13,149,444	..	309,459
261	Glass and glass products	..	..	..	..	..	..	..	..	..	..
269	Non-metallic mineral products, nec	..	..	..	..	..	..	..	..	..	..
2691	Pottery, china and earthenware	..	..	..	..	..	..	..	..	..	..
2692	Refractory ceramic products	..	..	..	..	..	..	..	..	..	..
2693	Structural non-refractory clay & ceramic prod.	..	..	..	..	..	..	..	..	..	..
2694	Cement, lime and plaster	..	..	..	..	..	..	..	..	..	..
2695	Articles of concrete, cement and plaster	..	..	..	..	..	..	..	..	..	..
2696	Cutting, shaping and finishing of stone	..	..	..	..	..	..	..	..	..	..
2699	Other non-metallic mineral products, nec	..	..	..	..	..	..	..	..	..	..

ISIC Revision 3 Industry Sector		Solid TJ	LPG TJ	Distiloil TJ	RFO TJ	Gas TJ	Biomass TJ	Steam TJ	Electr MWh	Own Use MWh	TOTAL TJ
27	BASIC METALS	433,746	..	5,683	59,699	200,562	..	..	40,828,333	..	846,672
271	Basic iron and steel	..	..	..	..	..	..	..	..	..	..
272	Basic precious and non-ferrous metals	..	..	..	..	..	..	..	..	..	..
273	Casting of metals	..	..	..	..	..	..	..	..	..	..
2731	Casting of iron and steel	..	..	..	..	..	..	..	..	..	..
2732	Casting non-ferrous metals	..	..	..	..	..	..	..	..	..	..
28	FABRICATED METAL PRODUCTS	641	..	11,967	1,106	33,882	..	..	7,474,444	..	74,504
281	Str. metal prod., tanks, reservoirs, steam generators	..	..	..	..	..	..	..	..	..	..
2811	Structural metal products	..	..	..	..	..	..	..	..	..	..
2812	Tanks, reservoirs and containers of metal	..	..	..	..	..	..	..	..	..	..
2813	Steam generators, exc. central heating hot water boilers	..	..	..	..	..	..	..	..	..	..
289	Other fabricated metal products	..	..	..	..	..	..	..	..	..	..
2891	Forging, pressing, stamping & roll-forming of metal	..	..	..	..	..	..	..	..	..	..
2892	Treatment and coating of metals	..	..	..	..	..	..	..	..	..	..
2893	Cutlery, hand tools and general hardware	..	..	..	..	..	..	..	..	..	..
2899	Other fabricated metal products, nec	..	..	..	..	..	..	..	..	..	..
29	MACHINERY AND EQUIPMENT, NEC	1,232	..	12,174	2,651	29,990	..	..	8,487,222	..	76,601
291	General purpose machinery	..	..	..	..	..	..	..	..	..	..
2911	Engines and turbines	..	..	..	..	..	..	..	..	..	..
2912	Pumps, compressors, taps and valves	..	..	..	..	..	..	..	..	..	..
2913	Bearings, gears, gearing and driving elements	..	..	..	..	..	..	..	..	..	..
2914	Ovens, furnaces and furnace burners	..	..	..	..	..	..	..	..	..	..
2915	Lifting and handling equipment	..	..	..	..	..	..	..	..	..	..
2919	Other general purpose machinery	..	..	..	..	..	..	..	..	..	..
292	Special purpose machinery	..	..	..	..	..	..	..	..	..	..
2921	Agricultural and forestry machinery	..	..	..	..	..	..	..	..	..	..
2922	Machine-tools	..	..	..	..	..	..	..	..	..	..
2923	Machinery for metallurgy	..	..	..	..	..	..	..	..	..	..
2924	Machinery for mining, quarrying and construction	..	..	..	..	..	..	..	..	..	..
2925	Machinery for food, beverage & tobacco processing	..	..	..	..	..	..	..	..	..	..
2926	Machinery for textile, apparel & leather production	..	..	..	..	..	..	..	..	..	..
2927	Machinery for weapons and ammunition	..	..	..	..	..	..	..	..	..	..
2929	Other special purpose machinery	..	..	..	..	..	..	..	..	..	..
293	Domestic appliances, nec	..	..	..	..	..	..	..	..	..	..
30	OFFICE, ACCOUNTING & COMPUTING MACHINERY	0	..	251	0	989	..	..	462,778	..	2,906
31	ELECTRICAL MACHINERY & APPARATUS, NEC	577	..	3,684	163	11,484	..	..	4,756,944	..	33,033
311	Electric motors, generators and transformers	..	..	..	..	..	..	..	..	..	..
312	Electricity distribution and control apparatus	..	..	..	..	..	..	..	..	..	..
313	Insulated wire and cable	..	..	..	..	..	..	..	..	..	..
314	Accumulators, primary cells & primary batteries	..	..	..	..	..	..	..	..	..	..
315	Electric lamps and lighting equipment	..	..	..	..	..	..	..	..	..	..
319	Other electrical equipment, nec	..	..	..	..	..	..	..	..	..	..
32	RADIO, TV & COMMUNICATION EQUIP. & APP.	41	..	921	3	3,177	..	..	2,376,944	..	12,699
321	Electronic valves, tubes, other electronic components	..	..	..	..	..	..	..	..	..	..
322	TV & radio transmitters, apparatus for line telephony	..	..	..	..	..	..	..	..	..	..
323	TV & radio receivers, recording apparatus	..	..	..	..	..	..	..	..	..	..
33	MEDICAL PRECISION &OPTICAL INSTRUMENTS	40	..	1,650	29	3,282	..	..	1,433,056	..	10,160
331	Medical appliances and instruments	..	..	..	..	..	..	..	..	..	..
3311	Medical, surgical equipment & orthopaedic app.	..	..	..	..	..	..	..	..	..	..
3312	Instruments & appliances for measuring, checking etc	..	..	..	..	..	..	..	..	..	..
3313	Industrial process control equipment	..	..	..	..	..	..	..	..	..	..
332	Optical instruments and photographic equipment	..	..	..	..	..	..	..	..	..	..
333	Watches and clocks	..	..	..	..	..	..	..	..	..	..
34	MOTOR VEHICLES, TRAILERS & SEMI-TRAILERS	1,354	..	4,106	261	42,413	..	..	13,393,056	..	96,349
341	Motor vehicles	..	..	..	..	..	..	..	..	..	..
342	Bodies (coachwork) for motor vehicles	..	..	..	..	..	..	..	..	..	..
343	Parts, accessories for motor vehicles & their engines	..	..	..	..	..	..	..	..	..	..
35	OTHER TRANSPORT EQUIPMENT	121	..	2,070	192	8,319	..	..	1,527,778	..	16,202
351	Building and repairing of ships and boats	..	..	..	..	..	..	..	..	..	..
3511	Building and repairing of ships	..	..	..	..	..	..	..	..	..	..
3512	Building, repairing of pleasure & sporting boats	..	..	..	..	..	..	..	..	..	..
352	Railway, tramway locomotives & rolling stock	..	..	..	..	..	..	..	..	..	..
353	Aircraft and spacecraft	..	..	..	..	..	..	..	..	..	..
359	Transport equipment, nec	..	..	..	..	..	..	..	..	..	..
3591	Motorcycles	..	..	..	..	..	..	..	..	..	..
3592	Bicycles and invalid carriages	..	..	..	..	..	..	..	..	..	..
3599	Other transport equipment, nec	..	..	..	..	..	..	..	..	..	..
36	FURNITURE; MANUFACTURING, NEC	18	..	3,613	689	3,768	..	..	2,012,778	..	15,334
361	Furniture	..	..	..	..	..	..	..	..	..	..
369	Manufacturing, nec	..	..	..	..	..	..	..	..	..	..
3691	Jewellery and related articles	..	..	..	..	..	..	..	..	..	..
3692	Musical instruments	..	..	..	..	..	..	..	..	..	..
3693	Sports goods	..	..	..	..	..	..	..	..	..	..
3694	Games and toys	..	..	..	..	..	..	..	..	..	..
3699	Other manufacturing, nec	..	..	..	..	..	..	..	..	..	..
37	RECYCLING	114	..	205	8	521	..	..	264,167	..	1,799
371	Recycling of metal waste and scrap	..	..	..	..	..	..	..	..	..	..
372	Recycling of non-metal waste and scrap	..	..	..	..	..	..	..	..	..	..
CERR	Non-specified industry	..	..	..	..	..	..	..	..	..	..
15-37	TOTAL MANUFACTURING	683,361	..	140,228	269,489	1,157,026	..	..	201,038,056	..	2,973,841

ISIC Revision 3 Industry Sector		Solid TJ	LPG TJ	Distiloil TJ	RFO TJ	Gas TJ	Biomass TJ	Steam TJ	Electr MWh	Own Use MWh	TOTAL TJ
15	FOOD PRODUCTS AND BEVERAGES	4,591	17,936	9,133	106,911	18,648	..	..	15,325,823	..	212,392
151	Production, processing and preserving (PPP)	1,457	2,342	1,744	19,147	1,823			3,334,588	..	38,518
1511	PPP of meat and meat products	0	295	614	4,081	723			984,574	..	9,257
1512	Processing and preserving of fish products	1	1,759	769	4,826	438			1,451,762	..	13,019
1513	Processing, preserving of fruit & vegetables	57	214	265	3,777	20			353,958	..	5,607
1514	Vegetable and animal oils and fats	1,399	74	96	6,463	642			544,294	..	10,633
152	Dairy products	3	515	272	11,612	627			1,546,037	..	18,595
153	Grain mill prod., starches & prepared animal feeds	0	1,433	407	10,571	522			1,657,401	..	18,900
1531	Grain mill products	0	33	131	571	81			562,699	..	2,842
1532	Starches and starch products	0	1,399	205	9,121	438			756,447	..	13,886
1533	Prepared animal feeds	0	1	71	879	3			338,255	..	2,172
154	Other food products	2,786	11,170	4,692	43,031	10,066			6,533,566	..	95,266
1541	Bakery products	0	5,646	961	6,635	3,207			1,743,093	..	22,724
1542	Sugar	2,635	3	193	9,826	2,347			438,591	..	16,583
1543	Cocoa, chocolate and sugar confectionery	0	664	1,002	1,826	952			650,760	..	6,787
1544	Macaroni, noodles, couscous & similar farinaceous prod.	0	657	308	4,295	13			292,050	..	6,324
1549	Other food products, nec	151	4,200	2,228	20,449	3,547			3,409,072	..	42,848
155	Beverages	345	2,476	2,018	22,550	5,610			2,254,231	..	41,114
1551	Distilling, rectifying and blending of spirits	344	1,125	78	5,077	737			300,899	..	8,444
1552	Wines	1	22	307	2,142	110			229,105	..	3,407
1553	Malt liquors and malt	0	510	553	6,618	4,458			1,016,748	..	15,799
1554	Soft drinks; production of mineral waters	0	819	1,080	8,713	305			707,479	..	13,464
16	TOBACCO PRODUCTS	0	316	154	1,728	52			359,360	..	3,544
17	TEXTILES	1,930	6,745	1,813	59,270	2,960			8,362,794	..	102,824
171	Spinning, weaving and finishing of textiles	1,930	5,711	930	54,747	2,478			7,318,548	..	92,143
1711	Preparation and spinning of textile fibres	0	367	216	13,091	155			5,375,500	..	33,181
1712	Finishing of textiles	1,930	5,344	714	41,656	2,323			1,943,048	..	58,962
172	Other textiles	0	1,015	757	4,324	475			848,100	..	9,624
1721	Made-up textile articles, except apparel	0	153	155	529	14			186,635	..	1,523
1722	Carpets and rugs	0	172	145	466	158			117,467	..	1,364
1723	Cordage, rope, twine and netting	0	14	43	329	0			56,571	..	590
1729	Other textiles, nec	0	676	414	3,000	303			487,427	..	6,148
173	Knitted and crocheted fabrics and articles	0	19	126	199	7			196,146	..	1,057
18	WEARING APPAREL, DRESSING & DYEING OF FUR	0	115	1,046	1,701	62			486,405	..	4,675
181	Wearing apparel, except fur apparel	0	111	1,043	1,650	62			480,819	..	4,597
182	Dressing and dyeing of fur; articles of fur	0	4	3	51	0			5,586	..	78
19	TANNING & DRESSING OF LEATHER, FOOTWEAR	51	515	325	3,316	58			490,116	..	6,029
191	Tanning and dressing of leather	0	498	247	2,507	54			298,623	..	4,381
1911	Tanning and dressing of leather	0	495	219	2,503	46			280,572	..	4,273
1912	Luggage, handbags, saddlery & harness	0	3	28	4	8			18,051	..	108
192	Footwear	51	17	78	809	4			191,493	..	1,648
20	WOOD AND WOOD PRODUCTS	0	285	1,414	2,847	204			1,936,694	..	11,722
201	Sawmilling and planing of wood	0	6	608	458	0			426,970	..	2,609
202	Products of wood, cork, straw & plaiting materials	0	279	806	2,389	204			1,509,724	..	9,113
2021	Veneer sheets	0	265	556	2,229	204			1,264,680	..	7,807
2022	Builders' carpentry and joinery	0	8	127	117	0			158,204	..	822
2023	Wooden containers	0	3	51	11	0			20,082	..	137
2029	Other products of wood	0	3	72	32	0			66,758	..	347
21	PAPER AND PAPER PRODUCTS	91,625	4,404	3,006	207,325	4,396			33,672,227	..	431,976
2101	Pulp, paper and paperboard	91,625	3,202	1,667	196,310	4,201			31,762,972	..	411,352
2102	Corrugated paper, paperboard and their containers	0	308	808	6,972	137			964,886	..	11,699
2109	Other articles of pulp and paperboard	0	894	531	4,043	58			944,369	..	8,926
22	PUBLISHING, PRINTING & REPRODUCTION	0	2,505	1,136	2,111	3,530			3,096,268	..	20,429
221	Publishing	0	8	68	316	637			636,982	..	3,322
2211	Publishing of books & brochures	0	1	16	34	54			75,997	..	379
2212	Publishing of newspapers and periodicals	0	7	52	282	583			560,985	..	2,944
2213	Publishing of recorded media	..	..	..	..	..			..	..	..
2219	Other publishing										
222	Printing and related service activities	0	2,497	1,056	1,794	2,893			2,452,098	..	17,068
2221	Printing	0	2,462	1,003	1,669	2,804			2,227,004	..	15,955
2222	Service activities related to printing	0	35	53	125	89			225,094	..	1,112
223	Reproduction of recorded media	0	0	12	1	0			7,188	..	39
23	COKE, REFINED PETROLEUM PRODUCTS	572,242	26,007	147,312	111,664	43,064			8,429,763	..	930,636
231	Coke oven products	543,725	80	179	604	41,996			840,000	..	589,608
232	Refined petroleum products	28,517	25,639	147,090	110,708	1,068			7,409,336	..	339,696
233	Processing of nuclear fuel	0	288	43	352	0			180,424	..	1,333
24	CHEMICALS & CHEMICAL PRODUCTS	271,364	174,540	210,644	257,257	49,761			52,965,657	..	1,154,242
241	Basic chemicals	216,728	165,592	201,379	181,972	41,573			39,921,317	..	950,961
2411	Basic chemicals, exc. fertilizers & nitrogen compounds	126,240	89,791	125,423	120,474	32,424			25,085,739	..	584,661
2412	Fertilizers and nitrogen compounds	35,961	194	136	3,477	2,022			1,112,563	..	45,795
2413	Plastics in primary forms and synthetic rubber	54,527	75,607	75,820	58,021	7,127			13,723,015	..	320,505
242	Other chemical products	14,766	7,627	6,193	43,505	8,175			8,226,430	..	109,881
2421	Pesticides and other agro-chemical products	28	186	321	1,811	93			520,332	..	4,312
2422	Paints, varnishes and similar coatings	2	79	565	1,000	207			409,843	..	3,328
2423	Pharmaceuticals, medicinal chem. & botanical prod.	1,692	718	1,223	16,696	1,602			2,505,022	..	30,949
2424	Soap and detergents, perfumes etc.	0	207	1,435	6,388	1,923			726,452	..	12,568
2429	Other chemical products, nec	13,044	6,437	2,649	17,610	4,350			4,064,781	..	58,723
243	Man-made fibres	39,870	1,321	3,072	31,780	13			4,817,910	..	93,400
25	RUBBER AND PLASTICS PRODUCTS	3,867	3,095	3,097	37,528	2,501			13,954,111	..	100,323
251	Rubber products	1,593	834	1,117	14,095	1,384			3,709,123	..	32,376
2511	Rubber tyres and tubes	1,593	565	494	8,022	360			1,786,065	..	17,464
2519	Other rubber products	0	269	623	6,073	1,024			1,923,058	..	14,912
252	Plastic products	2,274	2,261	1,980	23,433	1,117			10,244,988	..	67,947
26	OTHER NON-METALLIC MINERAL PRODUCTS	288,707	30,759	21,956	119,160	15,969			21,769,061	..	554,920
261	Glass and glass products	0	9,393	2,018	46,607	4,580			4,813,599	..	79,927
269	Non-metallic mineral products, nec	288,707	21,366	19,938	72,553	11,389			16,955,462	..	474,993
2691	Pottery, china and earthenware	1	6,002	2,987	1,934	547			938,101	..	14,848
2692	Refractory ceramic products	2,950	5,842	3,704	10,563	140			888,280	..	26,397
2693	Structural non-refractory clay & ceramic prod.	1,592	3,006	3,019	4,187	69			265,707	..	12,830
2694	Cement, lime and plaster	268,127	207	2,182	22,940	5,139			9,385,918	..	332,384
2695	Articles of concrete, cement and plaster	3,854	1,945	4,768	17,771	532			1,479,133	..	34,195
2696	Cutting, shaping and finishing of stone	134	13	1,337	311	0			240,823	..	2,662
2699	Other non-metallic mineral products, nec	12,049	4,351	1,941	14,847	4,962			3,757,500	..	51,677

ISIC Revision 3 Industry Sector	Solid TJ	LPG TJ	Distiloil TJ	RFO TJ	Gas TJ	Biomass TJ	Steam TJ	Electr MWh	Own Use MWh	TOTAL TJ
27 BASIC METALS	2,711,557	56,963	37,898	127,981	496,429	..	..	87,788,393	..	3,746,866
271 Basic iron and steel	2,695,089	34,380	19,609	80,632	486,280	..	..	66,169,360	..	3,554,200
272 Basic precious and non-ferrous metals	16,101	7,578	8,718	31,081	4,012	..	..	11,088,450	..	107,408
273 Casting of metals	367	15,005	9,571	16,268	6,137	..	..	10,530,583	..	85,258
2731 Casting of iron and steel	35	1,561	1,737	5,620	1,148	..	..	1,746,135	..	16,387
2732 Casting non-ferrous metals	332	13,444	7,834	10,648	4,989	..	..	8,784,448	..	68,871
28 FABRICATED METAL PRODUCTS	2,044	28,229	18,698	26,639	12,643	..	..	23,592,049	..	173,184
281 Str. metal prod., tanks, reservoirs, steam generators	243	11,857	4,702	7,386	4,429	..	..	6,462,701	..	51,883
2811 Structural metal products	4	462	385	221	178	..	..	262,121	..	2,194
2812 Tanks, reservoirs and containers of metal	37	182	200	419	347	..	..	281,074	..	2,197
2813 Steam generators, exc. central heating hot water boilers	202	11,213	4,117	6,746	3,904	..	..	5,919,506	..	47,492
289 Other fabricated metal products	1,801	16,372	13,996	19,253	8,214	..	..	17,129,348	..	121,302
2891 Forging, pressing, stamping & roll-forming of metal	21	2,205	1,168	2,988	1,357	..	..	1,615,990	..	13,557
2892 Treatment and coating of metals	61	260	376	593	126	..	..	745,980	..	4,102
2893 Cutlery, hand tools and general hardware	114	6,784	2,071	2,360	2,291	..	..	2,603,525	..	22,993
2899 Other fabricated metal products, nec	1,605	7,123	10,381	13,312	4,440	..	..	12,163,853	..	80,651
29 MACHINERY AND EQUIPMENT, NEC	1,458	9,181	14,380	15,981	6,077	..	..	17,240,473	..	109,143
291 General purpose machinery	967	5,994	8,366	8,906	2,377	..	..	9,830,815	..	62,001
2911 Engines and turbines	111	1,338	574	916	198	..	..	868,463	..	6,263
2912 Pumps, compressors, taps and valves	60	1,644	1,280	1,378	282	..	..	2,177,953	..	12,485
2913 Bearings, gears, gearing and driving elements	0	78	31	12	191	..	..	67,127	..	554
2914 Ovens, furnaces and furnace burners	26	551	512	827	365	..	..	615,672	..	4,497
2915 Lifting and handling equipment	414	551	1,059	935	243	..	..	1,317,827	..	7,946
2919 Other general purpose machinery	356	1,832	4,910	4,838	1,098	..	..	4,783,773	..	30,256
292 Special purpose machinery	406	2,314	5,221	5,824	2,353	..	..	5,969,317	..	37,608
2921 Agricultural and forestry machinery	70	270	1,406	1,564	274	..	..	1,461,581	..	8,846
2922 Machine-tools	31	15	122	13	4	..	..	144,142	..	704
2923 Machinery for metallurgy	150	508	1,454	1,244	309	..	..	992,615	..	7,238
2924 Machinery for mining, quarrying and construction	4	15	84	33	4	..	..	36,098	..	270
2925 Machinery for food, beverage & tobacco processing	42	164	333	241	54	..	..	365,839	..	2,151
2926 Machinery for textile, apparel & leather production	0	5	60	202	45	..	..	87,759	..	628
2927 Machinery for weapons and ammunition	24	464	969	1,276	316	..	..	1,440,942	..	8,236
2929 Other special purpose machinery	85	873	793	1,251	1,347	..	..	1,440,341	..	9,534
293 Domestic appliances, nec	85	873	793	1,251	1,347	..	..	1,440,341	..	9,534
30 OFFICE, ACCOUNTING & COMPUTING MACHINERY	1	342	1,392	2,090	823	..	..	2,921,023	..	15,164
31 ELECTRICAL MACHINERY & APPARATUS, NEC	193	7,778	4,958	9,984	4,329	..	..	9,811,968	..	62,565
311 Electric motors, generators and transformers	188	456	716	1,085	186	..	..	1,123,952	..	6,677
312 Electricity distribution and control apparatus	4	327	804	1,244	343	..	..	1,161,138	..	6,902
313 Insulated wire and cable	1	4,739	1,345	4,727	1,297	..	..	2,804,973	..	22,207
314 Accumulators, primary cells & primary batteries	0	301	124	992	553	..	..	599,032	..	4,127
315 Electric lamps and lighting equipment	0	1,003	135	248	382	..	..	482,675	..	3,506
319 Other electrical equipment, nec	0	952	1,834	1,688	1,568	..	..	3,640,198	..	19,147
32 RADIO, TV & COMMUNICATION EQUIP. & APP.	46	4,725	3,851	17,479	5,292	..	..	16,471,119	..	90,689
321 Electronic valves, tubes, other electronic components	0	2,094	1,431	8,387	3,536	..	..	8,179,560	..	44,894
322 TV & radio transmitters, apparatus for line telephony	0	45	158	434	609	..	..	824,443	..	4,214
323 TV & radio receivers, recording apparatus	46	2,586	2,262	8,658	1,147	..	..	7,467,116	..	41,581
33 MEDICAL PRECISION &OPTICAL INSTRUMENTS	3	361	1,311	2,317	694	..	..	2,636,577	..	14,178
331 Medical appliances and instruments	1	269	770	1,111	384	..	..	1,386,872	..	7,528
3311 Medical, surgical equipment & orthopaedic app.	1	183	403	371	179	..	..	617,994	..	3,362
3312 Instruments & appliances for measuring, checking etc	0	5	68	45	38	..	..	82,878	..	454
3313 Industrial process control equipment	0	81	299	695	167	..	..	686,000	..	3,712
332 Optical instruments and photographic equipment	0	81	299	694	166	..	..	686,000	..	3,710
333 Watches and clocks	2	11	242	512	144	..	..	563,705	..	2,940
34 MOTOR VEHICLES, TRAILERS & SEMI-TRAILERS	11,354	16,039	16,909	25,090	12,770	..	..	21,567,336	..	159,804
341 Motor vehicles	7,916	5,799	7,173	10,388	6,725	..	..	6,594,670	..	61,742
342 Bodies (coachwork) for motor vehicles	0	1,450	1,198	1,621	1,513	..	..	1,092,800	..	9,716
343 Parts, accessories for motor vehicles & their engines	3,438	8,790	8,538	13,081	4,532	..	..	13,879,866	..	88,347
35 OTHER TRANSPORT EQUIPMENT	18	1,368	1,943	3,439	495	..	..	2,263,449	..	15,411
351 Building and repairing of ships and boats	7	69	129	155	11	..	..	101,949	..	738
3511 Building and repairing of ships	0	0	19	69	0	..	..	15,985	..	146
3512 Building, repairing of pleasure & sporting boats	7	69	110	86	11	..	..	85,964	..	592
352 Railway, tramway locomotives & rolling stock	7	69	110	86	11	..	..	85,964	..	592
353 Aircraft and spacecraft	0	18	356	713	125	..	..	298,900	..	2,288
359 Transport equipment, nec	4	1,212	1,348	2,485	348	..	..	1,776,636	..	11,793
3591 Motorcycles	..	..	..	..	..	..	..	..	..	..
3592 Bicycles and invalid carriages	0	18	29	8	2	..	..	13,986	..	107
3599 Other transport equipment, nec	4	1,194	1,319	2,477	346	..	..	1,762,650	..	11,686
36 FURNITURE; MANUFACTURING, NEC	164	1,304	1,543	2,486	357	..	..	1,834,470	..	12,458
361 Furniture	1	751	696	727	191	..	..	817,609	..	5,309
369 Manufacturing, nec	163	553	847	1,759	166	..	..	1,016,861	..	7,149
3691 Jewellery and related articles	0	28	31	318	24	..	..	197,400	..	1,112
3692 Musical instruments	1	72	68	167	4	..	..	110,282	..	709
3693 Sports goods	0	16	52	28	5	..	..	70,195	..	354
3694 Games and toys	2	317	460	1,176	99	..	..	525,381	..	3,945
3699 Other manufacturing, nec	160	120	236	70	34	..	..	113,603	..	1,029
37 RECYCLING	160	120	236	70	34	..	..	113,603	..	1,029
371 Recycling of metal waste and scrap	160	120	236	70	34	..	..	113,603	..	1,029
372 Recycling of non-metal waste and scrap	..	..	..	..	..	..	..	..	..	..
CERR Non-specified industry	..	..	..	..	..	..	..	..	..	..
15-37 TOTAL MANUFACTURING	3,961,375	393,632	504,155	1,144,374	681,148	..	..	347,088,739	..	7,934,203

JAPAN

ISIC Revision 3 Industry Sector		Solid TJ	LPG TJ	Distiloil TJ	RFO TJ	Gas TJ	Biomass TJ	Steam TJ	Electr MWh	Own Use MWh	TOTAL TJ
15	FOOD PRODUCTS AND BEVERAGES	4,755	18,621	9,363	111,319	20,841	..	..	15,913,395	..	222,187
151	Production, processing and preserving (PPP)	1,446	2,301	1,806	18,856	2,050	..	..	3,321,615	..	38,417
1511	PPP of meat and meat products	0	314	582	3,704	785	..	..	959,262	..	8,838
1512	Processing and preserving of fish products	1	1,713	845	4,875	493	..	..	1,439,756	..	13,110
1513	Processing, preserving of fruit & vegetables	55	187	282	3,757	83	..	..	372,584	..	5,705
1514	Vegetable and animal oils and fats	1,390	87	97	6,520	689	..	..	550,013	..	10,763
152	Dairy products	0	428	156	11,803	634	..	..	1,555,423	..	18,621
153	Grain mill prod., starches & prepared animal feeds	0	1,356	502	10,839	1,117	..	..	1,692,135	..	19,906
1531	Grain mill products	0	32	162	536	132	..	..	569,440	..	2,912
1532	Starches and starch products	0	1,323	269	9,396	982	..	..	783,278	..	14,790
1533	Prepared animal feeds	0	1	71	907	3	..	..	339,417	..	2,204
154	Other food products	2,921	12,267	4,815	43,629	11,228	..	..	6,922,314	..	99,780
1541	Bakery products	0	5,621	998	6,518	3,848	..	..	1,872,860	..	23,727
1542	Sugar	2,751	3	205	10,174	2,390	..	..	433,499	..	17,084
1543	Cocoa, chocolate and sugar confectionery	0	976	1,056	1,942	1,037	..	..	710,077	..	7,567
1544	Macaroni, noodles, couscous & similar farinaceous prod.	0	933	321	4,435	36	..	..	337,539	..	6,940
1549	Other food products, nec	170	4,734	2,235	20,560	3,917	..	..	3,568,339	..	44,462
155	Beverages	388	2,269	2,084	26,192	5,812	..	..	2,421,908	..	45,464
1551	Distilling, rectifying and blending of spirits	388	963	75	4,837	748	..	..	326,244	..	8,185
1552	Wines	0	22	293	2,003	138	..	..	228,634	..	3,279
1553	Malt liquors and malt	0	512	626	7,248	4,534	..	..	1,116,294	..	16,939
1554	Soft drinks; production of mineral waters	0	772	1,090	12,104	392	..	..	750,736	..	17,061
16	TOBACCO PRODUCTS	0	316	133	1,869	55	..	..	378,622	..	3,736
17	TEXTILES	1,871	7,186	1,713	60,524	3,438	..	..	8,361,505	..	104,833
171	Spinning, weaving and finishing of textiles	1,871	6,105	920	55,932	2,839	..	..	7,260,943	..	93,806
1711	Preparation and spinning of textile fibres	0	390	220	14,217	161	..	..	5,308,683	..	34,099
1712	Finishing of textiles	1,871	5,715	700	41,715	2,678	..	..	1,952,260	..	59,707
172	Other textiles	0	1,063	681	4,385	590	..	..	899,259	..	9,956
1721	Made-up textile articles, except apparel	0	192	150	622	20	..	..	215,161	..	1,759
1722	Carpets and rugs	0	166	101	436	234	..	..	119,730	..	1,368
1723	Cordage, rope, twine and netting	0	14	43	323	0	..	..	62,923	..	607
1729	Other textiles, nec	0	691	387	3,004	336	..	..	501,445	..	6,223
173	Knitted and crocheted fabrics and articles	0	18	112	207	9	..	..	201,303	..	1,071
18	WEARING APPAREL, DRESSING & DYEING OF FUR	0	140	1,155	1,701	64	..	..	539,577	..	5,002
181	Wearing apparel, except fur apparel	0	136	1,152	1,653	64	..	..	533,366	..	4,925
182	Dressing and dyeing of fur; articles of fur	0	4	3	48	0	..	..	6,211	..	77
19	TANNING & DRESSING OF LEATHER, FOOTWEAR	9	493	345	3,001	41	..	..	484,832	..	5,634
191	Tanning and dressing of leather	0	484	258	2,093	35	..	..	292,081	..	3,921
1911	Tanning and dressing of leather	0	481	229	2,090	28	..	..	272,770	..	3,810
1912	Luggage, handbags, saddlery & harness	0	3	29	3	7	..	..	19,311	..	112
192	Footwear	9	9	87	908	6	..	..	192,751	..	1,713
20	WOOD AND WOOD PRODUCTS	0	111	1,467	3,186	678	..	..	1,927,758	..	12,382
201	Sawmilling and planing of wood	0	0	559	491	0	..	..	399,070	..	2,487
202	Products of wood, cork, straw & plaiting materials	0	111	908	2,695	678	..	..	1,528,688	..	9,895
2021	Veneer sheets	0	91	651	2,479	678	..	..	1,253,033	..	8,410
2022	Builders' carpentry and joinery	0	9	137	171	0	..	..	181,812	..	972
2023	Wooden containers	0	4	37	13	0	..	..	22,134	..	134
2029	Other products of wood	0	7	83	32	0	..	..	71,709	..	380
21	PAPER AND PAPER PRODUCTS	93,339	4,640	3,074	201,844	8,992	..	..	34,236,944	..	435,142
2101	Pulp, paper and paperboard	93,339	3,617	1,681	191,398	8,229	..	..	32,158,744	..	414,035
2102	Corrugated paper, paperboard and their containers	0	300	811	6,897	708	..	..	1,174,000	..	12,942
2109	Other articles of pulp and paperboard	0	723	582	3,549	55	..	..	904,200	..	8,164
22	PUBLISHING, PRINTING & REPRODUCTION	0	2,392	1,149	2,453	3,432	..	..	3,386,767	..	21,618
221	Publishing	0	4	58	321	285	..	..	648,417	..	3,002
2211	Publishing of books & brochures	0	0	10	37	53	..	..	84,269	..	403
2212	Publishing of newspapers and periodicals	0	4	48	284	232	..	..	564,148	..	2,599
2213	Publishing of recorded media	..	..	..	..	..	..	..		..	
2219	Other publishing	..	..	..	..	..	..	..		..	
222	Printing and related service activities	0	2,388	1,074	2,129	3,147	..	..	2,730,610	..	18,568
2221	Printing	0	2,365	1,024	1,998	3,063	..	..	2,481,527	..	17,383
2222	Service activities related to printing	0	23	50	131	84	..	..	249,083	..	1,185
223	Reproduction of recorded media	0	0	17	3	0	..	..	7,740	..	48
23	COKE, REFINED PETROLEUM PRODUCTS	578,166	27,975	171,978	120,085	41,782	..	..	8,913,444	..	972,074
231	Coke oven products	551,794	140	179	465	40,671	..	..	841,382	..	596,278
232	Refined petroleum products	26,372	27,617	171,753	119,242	1,111	..	..	7,887,656	..	374,491
233	Processing of nuclear fuel	0	218	46	378	0	..	..	184,406	..	1,306
24	CHEMICALS & CHEMICAL PRODUCTS	276,827	180,740	235,452	253,361	49,292	..	..	55,114,885	..	1,187,086
241	Basic chemicals	227,501	163,976	225,589	171,817	40,665	..	..	41,971,531	..	980,646
2411	Basic chemicals, exc. fertilizers & nitrogen compounds	119,902	87,220	152,906	118,098	32,312	..	..	26,456,959	..	605,683
2412	Fertilizers and nitrogen compounds	37,563	180	154	3,703	1,835	..	..	1,153,758	..	47,589
2413	Plastics in primary forms and synthetic rubber	70,036	76,576	72,529	50,016	6,518	..	..	14,360,814	..	327,374
242	Other chemical products	14,317	8,330	6,864	45,807	8,612	..	..	8,269,764	..	113,701
2421	Pesticides and other agro-chemical products	24	188	330	2,395	97	..	..	560,105	..	5,050
2422	Paints, varnishes and similar coatings	3	69	663	1,072	186	..	..	450,511	..	3,615
2423	Pharmaceuticals, medicinal chem. & botanical prod.	1,429	698	1,267	17,058	1,878	..	..	2,549,612	..	31,509
2424	Soap and detergents, perfumes etc.	0	217	1,534	6,685	2,120	..	..	774,759	..	13,345
2429	Other chemical products, nec	12,861	7,158	3,070	18,597	4,331	..	..	3,934,777	..	60,182
243	Man-made fibres	35,009	1,434	2,999	35,737	15	..	..	4,873,590	..	92,739
25	RUBBER AND PLASTICS PRODUCTS	3,821	3,434	3,262	39,400	2,496	..	..	14,717,573	..	105,396
251	Rubber products	1,485	986	1,008	14,434	1,402	..	..	3,817,011	..	33,056
2511	Rubber tyres and tubes	1,485	668	475	8,202	379	..	..	1,858,256	..	17,899
2519	Other rubber products	0	318	533	6,232	1,023	..	..	1,958,755	..	15,158
252	Plastic products	2,336	2,448	2,254	24,966	1,094	..	..	10,900,562	..	72,340
26	OTHER NON-METALLIC MINERAL PRODUCTS	308,470	29,853	21,660	116,631	17,312	..	..	21,588,343	..	571,644
261	Glass and glass products	5	8,379	1,887	45,873	4,771	..	..	4,108,589	..	75,706
269	Non-metallic mineral products, nec	308,465	21,474	19,773	70,758	12,541	..	..	17,479,754	..	495,938
2691	Pottery, china and earthenware	1	5,966	2,735	1,784	511	..	..	921,654	..	14,315
2692	Refractory ceramic products	2,639	6,029	3,575	10,495	917	..	..	904,913	..	26,913
2693	Structural non-refractory clay & ceramic prod.	1,614	2,924	2,867	3,818	15	..	..	270,268	..	12,211
2694	Cement, lime and plaster	287,388	169	2,264	22,149	7,339	..	..	9,878,076	..	354,870
2695	Articles of concrete, cement and plaster	5,420	2,006	4,629	17,133	842	..	..	1,553,660	..	35,623
2696	Cutting, shaping and finishing of stone	97	14	1,567	334	0	..	..	201,348	..	2,737
2699	Other non-metallic mineral products, nec	11,306	4,366	2,136	15,045	2,917	..	..	3,749,835	..	49,269

ISIS Energy Data Programme (IEA/OECD)

ISIC Revision 3 Industry Sector	Solid TJ	LPG TJ	Distiloil TJ	RFO TJ	Gas TJ	Biomass TJ	Steam TJ	Electr MWh	Own Use MWh	TOTAL TJ
27 **BASIC METALS**	2,726,573	57,717	37,840	123,754	497,469	..	..	90,293,667	..	3,768,410
271 Basic iron and steel	2,709,380	34,429	19,367	76,859	487,162	..	..	67,395,254	..	3,569,820
272 Basic precious and non-ferrous metals	16,814	7,533	8,748	29,894	3,938	..	..	11,637,457	..	108,822
273 Casting of metals	379	15,755	9,725	17,001	6,369	..	..	11,260,956	..	89,768
2731 Casting of iron and steel	30	1,478	1,822	5,088	1,262	..	..	1,785,587	..	16,108
2732 Casting non-ferrous metals	349	14,277	7,903	11,913	5,107	..	..	9,475,369	..	73,660
28 **FABRICATED METAL PRODUCTS**	2,011	29,332	19,101	27,860	14,374	..	..	24,909,682	..	182,353
281 Str. metal prod., tanks, reservoirs, steam generators	239	12,749	4,892	8,516	4,442	..	..	7,078,265	..	56,320
2811 Structural metal products	23	695	378	259	221	..	..	288,362	..	2,614
2812 Tanks, reservoirs and containers of metal	40	180	289	740	196	..	..	408,124	..	2,914
2813 Steam generators, exc. central heating hot water boilers	176	11,874	4,225	7,517	4,025	..	..	6,381,779	..	50,791
289 Other fabricated metal products	1,772	16,583	14,209	19,344	9,932	..	..	17,831,417	..	126,033
2891 Forging, pressing, stamping & roll-forming of metal	7	2,298	1,240	3,123	1,189	..	..	1,682,524	..	13,914
2892 Treatment and coating of metals	56	213	367	588	126	..	..	757,661	..	4,078
2893 Cutlery, hand tools and general hardware	113	6,911	2,092	2,449	2,539	..	..	2,807,481	..	24,211
2899 Other fabricated metal products, nec	1,596	7,161	10,510	13,184	6,078	..	..	12,583,751	..	83,831
29 **MACHINERY AND EQUIPMENT, NEC**	1,865	9,610	15,030	18,031	8,165	..	..	18,383,585	..	118,882
291 General purpose machinery	1,390	5,882	8,448	10,940	4,271	..	..	10,480,135	..	68,659
2911 Engines and turbines	125	1,129	493	792	280	..	..	781,291	..	5,632
2912 Pumps, compressors, taps and valves	23	1,613	1,251	1,450	243	..	..	2,179,936	..	12,428
2913 Bearings, gears, gearing and driving elements	0	74	23	13	24	..	..	39,003	..	274
2914 Ovens, furnaces and furnace burners	22	510	462	735	725	..	..	654,067	..	4,809
2915 Lifting and handling equipment	868	486	1,376	2,979	1,578	..	..	1,776,598	..	13,683
2919 Other general purpose machinery	352	2,070	4,843	4,971	1,421	..	..	5,049,240	..	31,834
292 Special purpose machinery	397	2,710	5,502	5,885	2,600	..	..	6,362,153	..	39,998
2921 Agricultural and forestry machinery	46	329	1,386	1,536	559	..	..	1,587,649	..	9,572
2922 Machine-tools	24	14	121	12	4	..	..	146,987	..	704
2923 Machinery for metallurgy	178	621	1,428	1,327	295	..	..	1,025,709	..	7,542
2924 Machinery for mining, quarrying and construction	4	14	81	51	4	..	..	44,196	..	313
2925 Machinery for food, beverage & tobacco processing	38	172	330	268	64	..	..	376,066	..	2,226
2926 Machinery for textile, apparel & leather production	0	5	75	231	41	..	..	92,228	..	684
2927 Machinery for weapons and ammunition	29	538	1,002	1,255	339	..	..	1,548,021	..	8,736
2929 Other special purpose machinery	78	1,017	1,079	1,205	1,294	..	..	1,541,297	..	10,222
293 Domestic appliances, nec	78	1,018	1,080	1,206	1,294	..	..	1,541,297	..	10,225
30 **OFFICE, ACCOUNTING & COMPUTING MACHINERY**	0	358	1,448	2,433	950	..	..	3,192,561	..	16,682
31 **ELECTRICAL MACHINERY & APPARATUS, NEC**	212	8,625	5,334	10,351	4,033	..	..	10,524,872	..	66,445
311 Electric motors, generators and transformers	189	549	896	1,132	226	..	..	1,260,763	..	7,531
312 Electricity distribution and control apparatus	5	353	1,021	1,282	433	..	..	1,264,080	..	7,645
313 Insulated wire and cable	1	5,195	1,309	4,623	1,388	..	..	2,914,423	..	23,008
314 Accumulators, primary cells & primary batteries	17	323	116	1,041	621	..	..	627,810	..	4,378
315 Electric lamps and lighting equipment	0	1,093	166	244	382	..	..	516,793	..	3,745
319 Other electrical equipment, nec	0	1,112	1,826	2,029	983	..	..	3,941,003	..	20,138
32 **RADIO, TV & COMMUNICATION EQUIP. & APP.**	201	4,825	4,373	19,442	5,737	..	..	18,327,516	..	100,557
321 Electronic valves, tubes, other electronic components	0	2,102	1,619	9,831	3,955	..	..	9,330,905	..	51,098
322 TV & radio transmitters, apparatus for line telephony	0	48	168	400	364	..	..	756,708	..	3,704
323 TV & radio receivers, recording apparatus	201	2,675	2,586	9,211	1,418	..	..	8,239,903	..	45,755
33 **MEDICAL PRECISION &OPTICAL INSTRUMENTS**	3	417	1,351	2,520	875	..	..	2,812,678	..	15,292
331 Medical appliances and instruments	2	316	762	1,233	528	..	..	1,507,343	..	8,267
3311 Medical, surgical equipment & orthopaedic app.	2	222	411	380	250	..	..	659,942	..	3,641
3312 Instruments & appliances for measuring, checking etc	0	5	54	48	24	..	..	89,512	..	453
3313 Industrial process control equipment	0	89	297	805	254	..	..	757,889	..	4,173
332 Optical instruments and photographic equipment	0	89	297	805	254	..	..	757,889	..	4,173
333 Watches and clocks	1	12	292	482	93	..	..	547,446	..	2,851
34 **MOTOR VEHICLES, TRAILERS & SEMI-TRAILERS**	11,082	16,217	17,310	30,720	14,153	..	..	22,987,479	..	172,237
341 Motor vehicles	7,805	5,511	7,201	15,964	7,266	..	..	7,123,295	..	69,391
342 Bodies (coachwork) for motor vehicles	0	1,607	1,283	1,566	1,797	..	..	1,191,738	..	10,543
343 Parts, accessories for motor vehicles & their engines	3,277	9,099	8,826	13,190	5,090	..	..	14,672,446	..	92,303
35 **OTHER TRANSPORT EQUIPMENT**	939	1,985	2,080	3,341	449	..	..	2,312,737	..	17,120
351 Building and repairing of ships and boats	9	461	137	177	11	..	..	113,870	..	1,205
3511 Building and repairing of ships	0	373	21	79	1	..	..	19,021	..	542
3512 Building, repairing of pleasure & sporting boats	9	88	116	98	10	..	..	94,849	..	662
352 Railway, tramway locomotives & rolling stock	9	88	116	98	10	..	..	94,849	..	662
353 Aircraft and spacecraft	0	17	337	743	140	..	..	293,812	..	2,295
359 Transport equipment, nec	921	1,419	1,490	2,323	288	..	..	1,810,206	..	12,958
3591 Motorcycles	..	..	..	..	..	..	..		..	
3592 Bicycles and invalid carriages	0	20	48	5	4	..	..	14,253	..	128
3599 Other transport equipment, nec	921	1,399	1,442	2,318	284	..	..	1,795,953	..	12,829
36 **FURNITURE; MANUFACTURING, NEC**	1,151	1,507	1,643	2,347	291	..	..	1,887,856	..	13,735
361 Furniture	1	874	720	763	179	..	..	832,947	..	5,536
369 Manufacturing, nec	1,150	633	923	1,584	112	..	..	1,054,909	..	8,200
3691 Jewellery and related articles	0	31	40	302	7	..	..	190,309	..	1,065
3692 Musical instruments	0	89	68	208	5	..	..	124,339	..	818
3693 Sports goods	0	17	46	19	4	..	..	74,681	..	355
3694 Games and toys	919	374	554	1,000	62	..	..	525,446	..	4,801
3699 Other manufacturing, nec	231	122	215	55	34	..	..	140,134	..	1,161
37 **RECYCLING**	231	122	215	55	34	..	..	140,134	..	1,161
371 Recycling of metal waste and scrap	231	122	215	55	34	..	..	140,134	..	1,161
372 Recycling of non-metal waste and scrap										
CERR **Non-specified industry**	..	..	..	..	..	..	..		..	
15-37 **TOTAL MANUFACTURING**	4,011,526	399,616	556,476	1,156,228	694,953	..	..	361,336,412	..	8,119,610

ISIC Revision 3 Industry Sector	Solid TJ	LPG TJ	Distiloil TJ	RFO TJ	Gas TJ	Biomass TJ	Steam TJ	Electr MWh	Own Use MWh	TOTAL TJ
15 **FOOD PRODUCTS AND BEVERAGES**	4,803	19,224	9,321	108,669	23,749	..	..	16,301,336	..	224,451
151 Production, processing and preserving (PPP)	1,469	2,508	1,763	19,820	3,275	..	..	3,514,419	..	41,487
1511 PPP of meat and meat products	1	382	578	4,700	1,184	..	..	1,092,840	..	10,779
1512 Processing and preserving of fish products	1	1,836	859	4,907	560	..	..	1,454,572	..	13,399
1513 Processing, preserving of fruit & vegetables	56	196	257	3,657	205	..	..	399,502	..	5,809
1514 Vegetable and animal oils and fats	1,411	94	69	6,556	1,326	..	..	567,505	..	11,499
152 Dairy products	0	474	169	12,129	777	..	..	1,613,135	..	19,356
153 Grain mill prod., starches & prepared animal feeds	0	1,524	502	10,819	1,135	..	..	1,708,536	..	20,131
1531 Grain mill products	0	54	165	525	94	..	..	568,212	..	2,884
1532 Starches and starch products	0	1,469	263	9,371	1,038	..	..	796,086	..	15,007
1533 Prepared animal feeds	0	1	74	923	3	..	..	344,238	..	2,240
154 Other food products	2,950	12,851	4,743	43,198	12,366	..	..	7,083,239	..	101,608
1541 Bakery products	0	5,738	887	6,228	4,135	..	..	1,877,663	..	23,748
1542 Sugar	2,813	3	201	10,138	2,656	..	..	432,431	..	17,368
1543 Cocoa, chocolate and sugar confectionery	0	1,053	1,056	1,930	1,086	..	..	714,451	..	7,697
1544 Macaroni, noodles, couscous & similar farinaceous prod.	0	790	337	4,542	79	..	..	358,942	..	7,040
1549 Other food products, nec	137	5,267	2,262	20,360	4,410	..	..	3,699,752	..	45,755
155 Beverages	384	1,867	2,144	22,703	6,196	..	..	2,382,007	..	41,869
1551 Distilling, rectifying and blending of spirits	384	799	71	4,896	605	..	..	311,929	..	7,878
1552 Wines	0	20	268	1,968	227	..	..	239,222	..	3,344
1553 Malt liquors and malt	0	549	632	7,275	4,918	..	..	1,117,039	..	17,395
1554 Soft drinks; production of mineral waters	0	499	1,173	8,564	446	..	..	713,817	..	13,252
16 **TOBACCO PRODUCTS**	0	309	147	1,880	64	..	..	378,689	..	3,763
17 **TEXTILES**	2,734	7,321	1,550	57,303	3,695	..	..	8,106,203	..	101,785
171 Spinning, weaving and finishing of textiles	2,734	6,376	839	52,881	3,090	..	..	6,970,798	..	91,015
1711 Preparation and spinning of textile fibres	314	407	192	13,914	195	..	..	5,038,912	..	33,162
1712 Finishing of textiles	2,420	5,969	647	38,967	2,895	..	..	1,931,886	..	57,853
172 Other textiles	0	925	603	4,215	600	..	..	905,016	..	9,601
1721 Made-up textile articles, except apparel	0	168	137	582	10	..	..	239,468	..	1,759
1722 Carpets and rugs	0	130	92	329	304	..	..	108,480	..	1,246
1723 Cordage, rope, twine and netting	0	18	38	262	0	..	..	54,711	..	515
1729 Other textiles, nec	0	609	336	3,042	286	..	..	502,357	..	6,081
173 Knitted and crocheted fabrics and articles	0	20	108	207	5	..	..	230,389	..	1,169
18 **WEARING APPAREL, DRESSING & DYEING OF FUR**	0	120	1,111	1,639	80	..	..	550,294	..	4,931
181 Wearing apparel, except fur apparel	0	120	1,109	1,596	80	..	..	544,916	..	4,867
182 Dressing and dyeing of fur; articles of fur	0	0	2	43	0	..	..	5,378	..	64
19 **TANNING & DRESSING OF LEATHER, FOOTWEAR**	0	288	358	3,069	47	..	..	488,446	..	5,520
191 Tanning and dressing of leather	0	275	268	2,116	20	..	..	297,050	..	3,748
1911 Tanning and dressing of leather	0	253	214	2,114	9	..	..	277,567	..	3,589
1912 Luggage, handbags, saddlery & harness	0	22	54	2	11	..	..	19,483	..	159
192 Footwear	0	13	90	953	27	..	..	191,396	..	1,772
20 **WOOD AND WOOD PRODUCTS**	0	119	1,447	3,573	725	..	..	1,918,464	..	12,770
201 Sawmilling and planing of wood	0	0	577	455	1	..	..	413,234	..	2,521
202 Products of wood, cork, straw & plaiting materials	0	119	870	3,118	724	..	..	1,505,230	..	10,250
2021 Veneer sheets	0	102	604	2,895	708	..	..	1,216,776	..	8,689
2022 Builders' carpentry and joinery	0	11	147	153	0	..	..	198,877	..	1,027
2023 Wooden containers	0	4	41	15	0	..	..	20,992	..	136
2029 Other products of wood	0	2	78	55	16	..	..	68,585	..	398
21 **PAPER AND PAPER PRODUCTS**	91,858	4,627	3,408	198,168	17,043	..	..	33,865,288	..	437,019
2101 Pulp, paper and paperboard	91,858	3,526	1,936	187,476	16,145	..	..	31,792,295	..	415,393
2102 Corrugated paper, paperboard and their containers	0	324	989	6,911	790	..	..	1,136,606	..	13,106
2109 Other articles of pulp and paperboard	0	777	483	3,781	108	..	..	936,387	..	8,520
22 **PUBLISHING, PRINTING & REPRODUCTION**	0	2,552	1,179	2,609	3,944	..	..	3,554,542	..	23,080
221 Publishing	0	52	57	296	299	..	..	707,898	..	3,252
2211 Publishing of books & brochures	0	48	10	29	42	..	..	88,069	..	446
2212 Publishing of newspapers and periodicals	0	4	47	267	257	..	..	619,829	..	2,806
2213 Publishing of recorded media	..	..	..	..	..	..	..	..	..	..
2219 Other publishing	..	..	..	..	..	..	..	..	..	..
222 Printing and related service activities	0	2,494	1,105	2,312	3,645	..	..	2,818,827	..	19,704
2221 Printing	0	2,456	1,066	2,213	3,555	..	..	2,559,628	..	18,505
2222 Service activities related to printing	0	38	39	99	90	..	..	259,199	..	1,199
223 Reproduction of recorded media	0	6	17	1	0	..	..	27,817	..	124
23 **COKE, REFINED PETROLEUM PRODUCTS**	557,744	31,723	213,436	122,240	40,197	..	..	9,195,472	..	998,444
231 Coke oven products	528,479	148	2,670	332	39,138	..	..	856,777	..	573,851
232 Refined petroleum products	29,265	31,329	210,719	121,471	1,059	..	..	8,114,521	..	423,055
233 Processing of nuclear fuel	0	246	47	437	0	..	..	224,174	..	1,537
24 **CHEMICALS & CHEMICAL PRODUCTS**	267,871	180,664	240,932	254,664	49,380	..	..	55,616,060	..	1,193,729
241 Basic chemicals	213,015	170,925	231,462	168,556	40,409	..	..	40,919,616	..	971,678
2411 Basic chemicals, exc. fertilizers & nitrogen compounds	117,252	86,397	151,272	112,694	32,752	..	..	25,320,243	..	591,520
2412 Fertilizers and nitrogen compounds	37,554	201	158	3,634	1,750	..	..	1,060,470	..	47,115
2413 Plastics in primary forms and synthetic rubber	58,209	84,327	80,032	52,228	5,907	..	..	14,538,903	..	333,043
242 Other chemical products	14,185	8,460	6,233	48,251	8,958	..	..	9,324,411	..	119,655
2421 Pesticides and other agro-chemical products	22	46	292	1,865	129	..	..	487,894	..	4,110
2422 Paints, varnishes and similar coatings	2	59	584	1,071	215	..	..	451,726	..	3,557
2423 Pharmaceuticals, medicinal chem. & botanical prod.	1,328	670	1,395	17,703	2,212	..	..	2,639,028	..	32,809
2424 Soap and detergents, perfumes etc.	0	129	1,516	6,086	1,872	..	..	792,019	..	12,454
2429 Other chemical products, nec	12,833	7,556	2,446	21,526	4,530	..	..	4,953,744	..	66,724
243 Man-made fibres	40,671	1,279	3,237	37,857	13	..	..	5,372,033	..	102,396
25 **RUBBER AND PLASTICS PRODUCTS**	2,836	4,202	3,132	38,681	3,671	..	..	14,612,384	..	105,127
251 Rubber products	1,425	1,116	1,040	14,318	1,549	..	..	3,797,845	..	33,120
2511 Rubber tyres and tubes	1,415	663	508	8,295	449	..	..	1,860,104	..	18,026
2519 Other rubber products	10	453	532	6,023	1,100	..	..	1,937,741	..	15,094
252 Plastic products	1,411	3,086	2,092	24,363	2,122	..	..	10,814,539	..	72,006
26 **OTHER NON-METALLIC MINERAL PRODUCTS**	322,760	30,690	21,583	115,881	18,047	..	..	22,727,251	..	590,779
261 Glass and glass products	3	9,336	2,049	44,703	5,143	..	..	4,034,130	..	75,757
269 Non-metallic mineral products, nec	322,757	21,354	19,534	71,178	12,904	..	..	18,693,121	..	515,022
2691 Pottery, china and earthenware	1	5,958	2,626	2,017	1,026	..	..	909,439	..	14,902
2692 Refractory ceramic products	2,746	5,624	3,241	9,165	917	..	..	843,032	..	24,728
2693 Structural non-refractory clay & ceramic prod.	1,930	2,949	3,035	3,397	29	..	..	258,318	..	12,270
2694 Cement, lime and plaster	299,890	204	2,163	24,652	4,516	..	..	11,110,781	..	371,424
2695 Articles of concrete, cement and plaster	5,827	2,034	4,757	17,115	1,462	..	..	1,601,073	..	36,959
2696 Cutting, shaping and finishing of stone	260	17	1,657	320	1	..	..	221,532	..	3,053
2699 Other non-metallic mineral products, nec	12,103	4,568	2,055	14,512	4,953	..	..	3,748,946	..	51,687

ISIC Revision 3 Industry Sector	Solid TJ	LPG TJ	Distiloil TJ	RFO TJ	Gas TJ	Biomass TJ	Steam TJ	Electr MWh	Own Use MWh	TOTAL TJ
27 BASIC METALS	2,575,024	56,962	35,250	115,624	466,196	..	..	87,393,141	..	3,563,671
271 Basic iron and steel	2,556,580	33,487	17,674	70,935	455,096	..	..	64,822,503	..	3,367,133
272 Basic precious and non-ferrous metals	18,110	7,655	8,672	29,507	4,318	..	..	11,718,583	..	110,449
273 Casting of metals	334	15,820	8,904	15,182	6,782	..	..	10,852,055	..	86,089
2731 Casting of iron and steel	29	1,539	1,595	4,210	1,245	..	..	1,722,618	..	14,819
2732 Casting non-ferrous metals	305	14,281	7,309	10,972	5,537	..	..	9,129,437	..	71,270
28 FABRICATED METAL PRODUCTS	1,725	28,852	18,746	25,649	14,160	..	..	24,440,459	..	177,118
281 Str. metal prod., tanks, reservoirs, steam generators	189	12,685	4,917	7,999	4,819	..	..	7,114,387	..	56,221
2811 Structural metal products	2	536	357	250	174	..	..	288,109	..	2,356
2812 Tanks, reservoirs and containers of metal	38	170	398	760	234	..	..	410,819	..	3,079
2813 Steam generators, exc. central heating hot water boilers	149	11,979	4,162	6,989	4,411	..	..	6,415,459	..	50,786
289 Other fabricated metal products	1,536	16,167	13,829	17,650	9,341	..	..	17,326,072	..	120,897
2891 Forging, pressing, stamping & roll-forming of metal	27	2,316	1,261	2,991	1,308	..	..	1,726,164	..	14,117
2892 Treatment and coating of metals	30	216	318	507	122	..	..	714,082	..	3,764
2893 Cutlery, hand tools and general hardware	83	6,738	2,056	2,128	2,800	..	..	2,824,004	..	23,971
2899 Other fabricated metal products, nec	1,396	6,897	10,194	12,024	5,111	..	..	12,061,822	..	79,045
29 MACHINERY AND EQUIPMENT, NEC	1,118	8,989	14,368	14,185	6,813	..	..	17,236,017	..	107,523
291 General purpose machinery	754	5,754	8,027	8,005	3,043	..	..	9,818,946	..	60,931
2911 Engines and turbines	99	250	594	545	181	..	..	559,311	..	3,683
2912 Pumps, compressors, taps and valves	19	1,645	1,135	1,329	245	..	..	2,097,630	..	11,924
2913 Bearings, gears, gearing and driving elements	0	73	25	13	189	..	..	66,751	..	540
2914 Ovens, furnaces and furnace burners	12	369	425	744	157	..	..	557,363	..	3,714
2915 Lifting and handling equipment	350	1,660	1,168	1,240	1,034	..	..	1,829,922	..	12,040
2919 Other general purpose machinery	274	1,757	4,680	4,134	1,237	..	..	4,707,969	..	29,031
292 Special purpose machinery	307	2,327	5,306	5,021	2,438	..	..	5,947,140	..	36,809
2921 Agricultural and forestry machinery	14	218	1,238	1,127	356	..	..	1,419,309	..	8,063
2922 Machine-tools	18	15	108	14	4	..	..	126,166	..	613
2923 Machinery for metallurgy	156	537	1,187	1,017	316	..	..	924,438	..	6,541
2924 Machinery for mining, quarrying and construction	2	14	69	56	4	..	..	40,737	..	292
2925 Machinery for food, beverage & tobacco processing	38	197	337	266	58	..	..	391,002	..	2,304
2926 Machinery for textile, apparel & leather production	0	14	101	228	42	..	..	120,348	..	818
2927 Machinery for weapons and ammunition	22	423	1,231	1,153	326	..	..	1,455,209	..	8,394
2929 Other special purpose machinery	57	909	1,035	1,160	1,332	..	..	1,469,931	..	9,785
293 Domestic appliances, nec	57	908	1,035	1,159	1,332	..	..	1,469,931	..	9,783
30 OFFICE, ACCOUNTING & COMPUTING MACHINERY	0	331	1,377	2,274	995	..	..	3,182,326	..	16,433
31 ELECTRICAL MACHINERY & APPARATUS, NEC	166	8,290	5,166	10,129	3,792	..	..	10,275,091	..	64,533
311 Electric motors, generators and transformers	145	528	720	1,100	168	..	..	1,201,539	..	6,987
312 Electricity distribution and control apparatus	3	368	1,163	1,182	513	..	..	1,222,994	..	7,632
313 Insulated wire and cable	1	4,888	1,185	4,386	1,454	..	..	2,772,543	..	21,895
314 Accumulators, primary cells & primary batteries	17	357	112	997	626	..	..	647,531	..	4,440
315 Electric lamps and lighting equipment	0	1,083	165	238	256	..	..	521,497	..	3,619
319 Other electrical equipment, nec	0	1,066	1,821	2,226	775	..	..	3,908,987	..	19,960
32 RADIO, TV & COMMUNICATION EQUIP. & APP.	0	4,278	4,109	18,448	6,241	..	..	18,024,335	..	97,964
321 Electronic valves, tubes, other electronic components	0	1,791	1,539	9,715	4,385	..	..	9,489,237	..	51,591
322 TV & radio transmitters, apparatus for line telephony	0	36	155	385	444	..	..	724,900	..	3,630
323 TV & radio receivers, recording apparatus	0	2,451	2,415	8,348	1,412	..	..	7,810,198	..	42,743
33 MEDICAL PRECISION &OPTICAL INSTRUMENTS	5	497	1,219	2,322	819	..	..	2,642,820	..	14,376
331 Medical appliances and instruments	4	370	678	1,103	489	..	..	1,401,562	..	7,690
3311 Medical, surgical equipment & orthopaedic app.	4	252	377	298	244	..	..	588,293	..	3,293
3312 Instruments & appliances for measuring, checking etc	0	6	19	40	32	..	..	90,806	..	424
3313 Industrial process control equipment	0	112	282	765	213	..	..	722,463	..	3,973
332 Optical instruments and photographic equipment	0	112	281	765	213	..	..	722,463	..	3,972
333 Watches and clocks	1	15	260	454	117	..	..	518,795	..	2,715
34 MOTOR VEHICLES, TRAILERS & SEMI-TRAILERS	15,141	16,028	15,943	26,403	16,833	..	..	22,386,529	..	170,940
341 Motor vehicles	11,950	5,264	6,649	11,398	8,912	..	..	6,808,839	..	68,685
342 Bodies (coachwork) for motor vehicles	0	1,546	1,175	1,674	1,833	..	..	1,201,293	..	10,553
343 Parts, accessories for motor vehicles & their engines	3,191	9,218	8,119	13,331	6,088	..	..	14,376,397	..	91,702
35 OTHER TRANSPORT EQUIPMENT	64	1,512	2,827	4,095	584	..	..	3,118,902	..	20,310
351 Building and repairing of ships and boats	31	109	137	335	27	..	..	314,128	..	1,770
3511 Building and repairing of ships	0	28	18	70	1	..	..	18,861	..	185
3512 Building, repairing of pleasure & sporting boats	31	81	119	265	26	..	..	295,267	..	1,585
352 Railway, tramway locomotives & rolling stock	30	81	119	265	26	..	..	295,267	..	1,584
353 Aircraft and spacecraft	0	23	341	731	193	..	..	319,826	..	2,439
359 Transport equipment, nec	3	1,299	2,230	2,764	338	..	..	2,189,681	..	14,517
3591 Motorcycles	..	..	..	..	..	..	..	..	..	..
3592 Bicycles and invalid carriages	0	19	43	4	0	..	..	13,224	..	114
3599 Other transport equipment, nec	3	1,280	2,187	2,760	338	..	..	2,176,457	..	14,403
36 FURNITURE; MANUFACTURING, NEC	80	1,378	2,354	2,796	314	..	..	2,246,207	..	15,008
361 Furniture	0	751	721	721	232	..	..	820,864	..	5,380
369 Manufacturing, nec	80	627	1,633	2,075	82	..	..	1,425,343	..	9,628
3691 Jewellery and related articles	0	34	29	257	7	..	..	181,262	..	980
3692 Musical instruments	0	95	60	184	4	..	..	133,101	..	822
3693 Sports goods	0	19	57	87	6	..	..	83,457	..	469
3694 Games and toys	2	365	1,305	1,484	64	..	..	911,021	..	6,500
3699 Other manufacturing, nec	78	114	182	63	1	..	..	116,502	..	857
37 RECYCLING	77	114	182	63	1	..	..	116,502	..	856
371 Recycling of metal waste and scrap	77	114	182	63	1	..	..	116,502	..	856
372 Recycling of non-metal waste and scrap										
CERR Non-specified industry	..	..	..	..	..	..	..	..	..	..
15-37 TOTAL MANUFACTURING	3,844,006	409,070	599,145	1,130,364	677,390	..	..	358,376,758	..	7,950,131

ISIC Revision 3 Industry Sector	Solid TJ	LPG TJ	Distiloil TJ	RFO TJ	Gas TJ	Biomass TJ	Steam TJ	Electr MWh	Own Use MWh	TOTAL TJ
15 **FOOD PRODUCTS AND BEVERAGES**	4,293	20,816	9,704	107,397	25,985	..	..	16,699,643	..	228,314
151 Production, processing and preserving (PPP)	1,390	2,930	1,825	19,002	3,651	..	..	3,563,458	..	41,626
1511 PPP of meat and meat products	0	467	554	4,793	1,262	..	..	1,138,528	..	11,175
1512 Processing and preserving of fish products	1	1,967	795	4,770	575	..	..	1,454,509	..	13,344
1513 Processing, preserving of fruit & vegetables	0	387	426	3,624	32	..	..	413,773	..	5,959
1514 Vegetable and animal oils and fats	1,389	109	50	5,815	1,782	..	..	556,648	..	11,149
152 Dairy products	0	593	179	12,468	1,103	..	..	1,669,203	..	20,352
153 Grain mill prod., starches & prepared animal feeds	0	1,541	468	10,663	1,240	..	..	1,694,071	..	20,011
1531 Grain mill products	0	35	139	755	98	..	..	589,919	..	3,151
1532 Starches and starch products	0	1,505	265	9,007	1,138	..	..	765,852	..	14,672
1533 Prepared animal feeds	0	1	64	901	4	..	..	338,300	..	2,188
154 Other food products	2,528	13,744	4,916	42,967	13,152	..	..	7,390,717	..	103,914
1541 Bakery products	0	5,575	922	6,378	4,394	..	..	1,955,410	..	24,308
1542 Sugar	2,344	2	203	8,483	2,755	..	..	396,894	..	15,216
1543 Cocoa, chocolate and sugar confectionery	0	1,452	1,120	1,823	1,229	..	..	711,149	..	8,184
1544 Macaroni, noodles, couscous & similar farinaceous prod.	0	911	301	5,030	342	..	..	408,951	..	8,056
1549 Other food products, nec	184	5,804	2,370	21,253	4,432	..	..	3,918,313	..	48,149
155 Beverages	375	2,008	2,316	22,297	6,839	..	..	2,382,194	..	42,411
1551 Distilling, rectifying and blending of spirits	375	724	71	4,830	1,248	..	..	303,207	..	8,340
1552 Wines	0	24	280	1,967	251	..	..	255,325	..	3,441
1553 Malt liquors and malt	0	589	629	7,628	4,804	..	..	1,110,623	..	17,648
1554 Soft drinks; production of mineral waters	0	671	1,336	7,872	536	..	..	713,039	..	12,982
16 **TOBACCO PRODUCTS**		303	112	1,878	66	..	..	367,149	..	3,681
17 **TEXTILES**	3,075	7,047	1,471	52,377	4,013	..	..	7,237,524	..	94,038
171 Spinning, weaving and finishing of textiles	3,075	6,121	863	48,174	3,319	..	..	6,135,066	..	83,638
1711 Preparation and spinning of textile fibres	671	384	245	11,731	188	..	..	4,306,163	..	28,721
1712 Finishing of textiles	2,404	5,737	618	36,443	3,131	..	..	1,828,903	..	54,917
172 Other textiles	0	909	506	4,003	686	..	..	894,323	..	9,324
1721 Made-up textile articles, except apparel	0	257	131	625	32	..	..	245,190	..	1,928
1722 Carpets and rugs	0	140	87	392	283	..	..	111,330	..	1,303
1723 Cordage, rope, twine and netting	0	13	25	251	0	..	..	48,624	..	464
1729 Other textiles, nec	0	499	263	2,735	371	..	..	489,179	..	5,629
173 Knitted and crocheted fabrics and articles	0	17	102	200	8	..	..	208,135	..	1,076
18 **WEARING APPAREL, DRESSING & DYEING OF FUR**	0	126	1,105	1,564	71	..	..	515,940	..	4,723
181 Wearing apparel, except fur apparel	0	126	1,103	1,522	71	..	..	510,646	..	4,660
182 Dressing and dyeing of fur; articles of fur	0	0	2	42	0	..	..	5,294	..	63
19 **TANNING & DRESSING OF LEATHER, FOOTWEAR**	0	267	283	2,997	99	..	..	454,108	..	5,281
191 Tanning and dressing of leather	0	254	218	2,112	11	..	..	275,492	..	3,587
1911 Tanning and dressing of leather	0	234	192	2,109	10	..	..	256,936	..	3,470
1912 Luggage, handbags, saddlery & harness	0	20	26	3	1	..	..	18,556	..	117
192 Footwear	0	13	65	885	88	..	..	178,616	..	1,694
20 **WOOD AND WOOD PRODUCTS**	0	166	1,467	3,771	729	..	..	1,943,288	..	13,129
201 Sawmilling and planing of wood	0	0	560	400	7	..	..	384,391	..	2,351
202 Products of wood, cork, straw & plaiting materials	0	166	907	3,371	722	..	..	1,558,897	..	10,778
2021 Veneer sheets	0	146	671	3,191	701	..	..	1,254,215	..	9,224
2022 Builders' carpentry and joinery	0	13	137	125	0	..	..	218,445	..	1,061
2023 Wooden containers	0	3	38	8	0	..	..	22,078	..	128
2029 Other products of wood	0	4	61	47	21	..	..	64,159	..	364
21 **PAPER AND PAPER PRODUCTS**	97,991	4,691	4,593	197,452	19,125	..	..	33,343,192	..	443,887
2101 Pulp, paper and paperboard	97,991	3,529	3,186	187,090	18,517	..	..	31,251,925	..	422,820
2102 Corrugated paper, paperboard and their containers	0	356	990	6,953	419	..	..	1,115,216	..	12,733
2109 Other articles of pulp and paperboard	0	806	417	3,409	189	..	..	976,051	..	8,335
22 **PUBLISHING, PRINTING & REPRODUCTION**	0	2,566	1,125	2,359	4,572	..	..	3,693,658	..	23,919
221 Publishing	0	4	56	293	547	..	..	735,931	..	3,549
2211 Publishing of books & brochures	0	0	12	25	26	..	..	88,131	..	380
2212 Publishing of newspapers and periodicals	0	4	44	268	521	..	..	647,800	..	3,169
2213 Publishing of recorded media	..	..	..	..	..	..	..	..	..	..
2219 Other publishing	..	..	..	..	..	..	..	..	..	..
222 Printing and related service activities	0	2,560	1,057	2,064	4,024	..	..	2,929,399	..	20,251
2221 Printing	0	2,491	1,016	1,971	3,902	..	..	2,641,682	..	18,890
2222 Service activities related to printing	0	69	41	93	122	..	..	287,717	..	1,361
223 Reproduction of recorded media	0	2	12	2	1	..	..	28,328	..	119
23 **COKE, REFINED PETROLEUM PRODUCTS**	553,115	33,300	242,898	127,851	40,259	..	..	9,871,273	..	1,032,960
231 Coke oven products	521,712	133	3,942	368	38,835	..	..	944,118	..	568,389
232 Refined petroleum products	31,403	32,941	238,907	127,086	1,424	..	..	8,687,887	..	463,037
233 Processing of nuclear fuel	0	226	49	397	0	..	..	239,268	..	1,533
24 **CHEMICALS & CHEMICAL PRODUCTS**	268,845	173,346	239,138	260,370	52,013	..	..	56,174,230	..	1,195,939
241 Basic chemicals	215,561	162,322	228,868	172,779	42,001	..	..	42,613,859	..	974,941
2411 Basic chemicals, exc. fertilizers & nitrogen compounds	122,664	88,365	153,663	114,541	33,686	..	..	26,697,900	..	609,031
2412 Fertilizers and nitrogen compounds	34,685	121	166	3,473	1,686	..	..	1,129,456	..	44,197
2413 Plastics in primary forms and synthetic rubber	58,212	73,836	75,039	54,765	6,629	..	..	14,786,503	..	321,712
242 Other chemical products	12,781	9,718	6,834	47,608	9,999	..	..	8,372,898	..	117,082
2421 Pesticides and other agro-chemical products	18	856	644	1,924	51	..	..	492,109	..	5,265
2422 Paints, varnishes and similar coatings	2	99	512	1,000	243	..	..	465,822	..	3,533
2423 Pharmaceuticals, medicinal chem. & botanical prod.	1,489	650	1,700	17,804	2,542	..	..	2,692,066	..	33,876
2424 Soap and detergents, perfumes etc.	0	75	1,532	5,675	2,532	..	..	807,469	..	12,721
2429 Other chemical products, nec	11,272	8,038	2,446	21,205	4,631	..	..	3,915,432	..	61,688
243 Man-made fibres	40,503	1,306	3,436	39,983	13	..	..	5,187,473	..	103,916
25 **RUBBER AND PLASTICS PRODUCTS**	2,777	3,751	2,755	37,661	3,724	..	..	14,550,871	..	103,051
251 Rubber products	1,362	974	969	13,403	1,624	..	..	3,611,175	..	31,332
2511 Rubber tyres and tubes	1,351	612	490	7,308	383	..	..	1,678,396	..	16,186
2519 Other rubber products	11	362	479	6,095	1,241	..	..	1,932,779	..	15,146
252 Plastic products	1,415	2,777	1,786	24,258	2,100	..	..	10,939,696	..	71,719
26 **OTHER NON-METALLIC MINERAL PRODUCTS**	309,937	29,997	20,786	107,328	18,415	..	..	21,278,715	..	563,066
261 Glass and glass products	0	8,848	2,055	42,499	5,073	..	..	3,781,735	..	72,089
269 Non-metallic mineral products, nec	309,937	21,149	18,731	64,829	13,342	..	..	17,496,980	..	490,977
2691 Pottery, china and earthenware	1	5,555	2,350	1,954	1,096	..	..	900,079	..	14,196
2692 Refractory ceramic products	2,229	5,662	3,020	8,122	511	..	..	823,086	..	22,507
2693 Structural non-refractory clay & ceramic prod.	1,673	2,988	3,093	3,114	77	..	..	267,107	..	11,907
2694 Cement, lime and plaster	287,469	233	1,978	20,748	5,017	..	..	9,847,172	..	350,895
2695 Articles of concrete, cement and plaster	5,734	2,615	4,718	16,724	1,660	..	..	1,643,808	..	37,369
2696 Cutting, shaping and finishing of stone	175	16	1,652	346	1	..	..	230,964	..	3,021
2699 Other non-metallic mineral products, nec	12,656	4,080	1,920	13,821	4,980	..	..	3,784,764	..	51,082

ISIC Revision 3 Industry Sector	Solid TJ	LPG TJ	Distiloil TJ	RFO TJ	Gas TJ	Biomass TJ	Steam TJ	Electr MWh	Own Use MWh	TOTAL TJ
27	**BASIC METALS**	2,568,866	54,251	36,012	105,896	478,013	..	..	86,417,255	..	3,554,140
271	Basic iron and steel	2,551,205	31,300	18,012	63,120	466,934	..	..	64,175,751	..	3,361,604
272	Basic precious and non-ferrous metals	17,375	7,544	8,222	27,716	4,365	..	..	11,405,982	..	106,284
273	Casting of metals	286	15,407	9,778	15,060	6,714	..	..	10,835,522	..	86,253
2731	Casting of iron and steel	25	1,313	1,581	3,840	1,111	..	..	1,565,996	..	13,508
2732	Casting non-ferrous metals	261	14,094	8,197	11,220	5,603	..	..	9,269,526	..	72,745
28	**FABRICATED METAL PRODUCTS**	1,508	27,749	18,588	24,390	14,145	..	..	23,241,201	..	170,048
281	Str. metal prod., tanks, reservoirs, steam generators	178	12,150	5,093	7,696	4,869	..	..	6,836,986	..	54,599
2811	Structural metal products	26	527	413	242	209	..	..	297,213	..	2,487
2812	Tanks, reservoirs and containers of metal	36	136	532	589	348	..	..	360,034	..	2,937
2813	Steam generators, exc. central heating hot water boilers	116	11,487	4,148	6,865	4,312	..	..	6,179,739	..	49,175
289	Other fabricated metal products	1,330	15,599	13,495	16,694	9,276	..	..	16,404,215	..	115,449
2891	Forging, pressing, stamping & roll-forming of metal	22	2,337	1,259	2,822	1,442	..	..	1,682,057	..	13,937
2892	Treatment and coating of metals	22	198	404	507	129	..	..	661,835	..	3,643
2893	Cutlery, hand tools and general hardware	71	6,321	1,958	2,021	2,548	..	..	2,629,825	..	22,386
2899	Other fabricated metal products, nec	1,215	6,743	9,874	11,344	5,157	..	..	11,430,498	..	75,483
29	**MACHINERY AND EQUIPMENT, NEC**	916	8,789	13,385	13,367	7,069	..	..	16,290,945	..	102,173
291	General purpose machinery	575	5,683	7,678	7,737	3,066	..	..	9,261,377	..	58,080
2911	Engines and turbines	41	1,374	677	607	673	..	..	859,165	..	6,465
2912	Pumps, compressors, taps and valves	69	1,665	1,362	1,562	261	..	..	2,077,008	..	12,396
2913	Bearings, gears, gearing and driving elements	0	65	23	11	182	..	..	66,310	..	520
2914	Ovens, furnaces and furnace burners	15	405	373	878	144	..	..	549,837	..	3,794
2915	Lifting and handling equipment	200	533	1,006	968	384	..	..	1,297,022	..	7,760
2919	Other general purpose machinery	250	1,641	4,237	3,711	1,422	..	..	4,412,035	..	27,144
292	Special purpose machinery	295	2,198	4,776	4,547	2,640	..	..	5,609,930	..	34,652
2921	Agricultural and forestry machinery	24	219	1,182	1,056	321	..	..	1,255,745	..	7,323
2922	Machine-tools	18	16	107	15	4	..	..	127,469	..	619
2923	Machinery for metallurgy	105	475	1,084	929	345	..	..	903,393	..	6,190
2924	Machinery for mining, quarrying and construction	2	16	68	51	5	..	..	43,564	..	299
2925	Machinery for food, beverage & tobacco processing	36	134	266	184	154	..	..	305,064	..	1,872
2926	Machinery for textile, apparel & leather production	0	14	93	223	37	..	..	119,410	..	797
2927	Machinery for weapons and ammunition	65	417	1,045	1,007	411	..	..	1,435,647	..	8,113
2929	Other special purpose machinery	45	907	931	1,082	1,363	..	..	1,419,638	..	9,439
293	Domestic appliances, nec	46	908	931	1,083	1,363	..	..	1,419,638	..	9,442
30	**OFFICE, ACCOUNTING & COMPUTING MACHINERY**	0	280	1,443	2,246	984	..	..	3,138,969	..	16,253
31	**ELECTRICAL MACHINERY & APPARATUS, NEC**	146	8,103	5,138	9,927	4,149	..	..	10,716,547	..	66,043
311	Electric motors, generators and transformers	121	543	705	991	175	..	..	1,130,346	..	6,604
312	Electricity distribution and control apparatus	2	365	1,133	1,107	427	..	..	1,490,323	..	8,399
313	Insulated wire and cable	1	4,799	1,229	4,023	1,557	..	..	2,830,160	..	21,798
314	Accumulators, primary cells & primary batteries	17	296	78	981	667	..	..	623,425	..	4,283
315	Electric lamps and lighting equipment	0	1,138	124	273	250	..	..	528,721	..	3,688
319	Other electrical equipment, nec	5	962	1,869	2,552	1,073	..	..	4,113,572	..	21,270
32	**RADIO, TV & COMMUNICATION EQUIP. & APP.**	3	5,273	3,811	19,076	6,845	..	..	18,459,277	..	101,461
321	Electronic valves, tubes, other electronic components	0	2,017	1,366	10,065	4,422	..	..	9,943,941	..	53,668
322	TV & radio transmitters, apparatus for line telephony	0	432	161	399	652	..	..	859,027	..	4,736
323	TV & radio receivers, recording apparatus	3	2,824	2,284	8,612	1,771	..	..	7,656,309	..	43,057
33	**MEDICAL PRECISION & OPTICAL INSTRUMENTS**	5	542	1,206	2,282	830	..	..	2,535,275	..	13,992
331	Medical appliances and instruments	4	381	668	1,075	488	..	..	1,334,601	..	7,421
3311	Medical, surgical equipment & orthopaedic app.	4	267	365	309	224	..	..	568,049	..	3,214
3312	Instruments & appliances for measuring, checking etc	0	5	17	30	32	..	..	86,155	..	394
3313	Industrial process control equipment	0	109	286	736	232	..	..	680,397	..	3,812
332	Optical instruments and photographic equipment	0	110	285	735	231	..	..	680,397	..	3,810
333	Watches and clocks	1	51	253	472	111	..	..	520,277	..	2,761
34	**MOTOR VEHICLES, TRAILERS & SEMI-TRAILERS**	11,545	15,611	15,201	25,431	17,556	..	..	21,694,805	..	163,445
341	Motor vehicles	8,224	4,822	6,150	11,041	9,398	..	..	6,405,064	..	62,693
342	Bodies (coachwork) for motor vehicles	0	1,488	1,235	1,250	1,776	..	..	1,084,914	..	9,655
343	Parts, accessories for motor vehicles & their engines	3,321	9,301	7,816	13,140	6,382	..	..	14,204,827	..	91,097
35	**OTHER TRANSPORT EQUIPMENT**	516	1,462	2,058	3,963	633	..	..	2,377,117	..	17,190
351	Building and repairing of ships and boats	10	68	125	128	46	..	..	123,138	..	820
3511	Building and repairing of ships	1	2	21	54	1	..	..	13,770	..	129
3512	Building, repairing of pleasure & sporting boats	9	66	104	74	45	..	..	109,368	..	692
352	Railway, tramway locomotives & rolling stock	9	66	104	74	44	..	..	109,368	..	691
353	Aircraft and spacecraft	1	27	329	683	186	..	..	309,152	..	2,339
359	Transport equipment, nec	496	1,301	1,500	3,078	357	..	..	1,835,459	..	13,340
3591	Motorcycles	..	..	..	..	..	..	..	..	..	..
3592	Bicycles and invalid carriages	0	14	42	1	0	..	..	10,829	..	96
3599	Other transport equipment, nec	496	1,287	1,458	3,077	357	..	..	1,824,630	..	13,244
36	**FURNITURE; MANUFACTURING, NEC**	648	1,398	1,650	3,167	333	..	..	1,913,770	..	14,086
361	Furniture	0	809	655	740	236	..	..	818,035	..	5,385
369	Manufacturing, nec	648	589	995	2,427	97	..	..	1,095,735	..	8,701
3691	Jewellery and related articles	47	113	47	355	12	..	..	209,386	..	1,328
3692	Musical instruments	0	84	117	173	4	..	..	131,040	..	850
3693	Sports goods	0	8	42	17	5	..	..	66,022	..	310
3694	Games and toys	449	255	584	1,764	75	..	..	555,161	..	5,126
3699	Other manufacturing, nec	152	129	205	118	1	..	..	134,126	..	1,088
37	**RECYCLING**	152	129	205	118	1	..	..	134,126	..	1,088
371	Recycling of metal waste and scrap	152	129	205	118	1	..	..	134,126	..	1,088
372	Recycling of non-metal waste and scrap	..	..	..	..	..	..	..	..	..	..
CERR	**Non-specified industry**	..	..	..	..	..	..	..	..	..	..
15-37	**TOTAL MANUFACTURING**	3,824,338	399,963	624,134	1,112,868	699,629	..	..	353,048,878	..	7,931,908

ISIC Revision 3 Industry Sector	Solid TJ	LPG TJ	Distiloil TJ	RFO TJ	Gas TJ	Biomass TJ	Steam TJ	Electr MWh	Own Use MWh	TOTAL TJ
15 FOOD PRODUCTS AND BEVERAGES	4,326	20,731	10,326	108,325	26,336	..	..	17,232,014	..	232,079
151 Production, processing and preserving (PPP)	1,390	3,144	2,322	19,078	4,336	..	..	3,614,127	..	43,281
1511 PPP of meat and meat products	0	555	660	4,104	2,242	..	..	1,170,991	..	11,777
1512 Processing and preserving of fish products	0	2,089	1,301	6,132	542	..	..	1,488,309	..	15,422
1513 Processing, preserving of fruit & vegetables	0	382	287	3,412	25	..	..	417,583	..	5,609
1514 Vegetable and animal oils and fats	1,390	118	74	5,430	1,527	..	..	537,244	..	10,473
152 Dairy products	0	599	178	11,758	1,232	..	..	1,736,940	..	20,020
153 Grain mill prod., starches & prepared animal feeds	0	1,685	491	11,807	1,305	..	..	1,717,411	..	21,471
1531 Grain mill products	0	30	137	821	93	..	..	570,719	..	3,136
1532 Starches and starch products	0	1,385	227	9,275	1,142	..	..	766,547	..	14,789
1533 Prepared animal feeds	0	270	127	1,711	70	..	..	380,145	..	3,547
154 Other food products	2,517	13,252	5,038	43,765	13,293	..	..	7,654,252	..	105,420
1541 Bakery products	0	5,861	922	6,219	4,732	..	..	1,967,371	..	24,817
1542 Sugar	2,362	3	197	8,246	2,766	..	..	394,320	..	14,994
1543 Cocoa, chocolate and sugar confectionery	0	907	1,039	1,872	1,242	..	..	728,027	..	7,681
1544 Macaroni, noodles, couscous & similar farinaceous prod.	0	719	319	4,986	106	..	..	419,220	..	7,639
1549 Other food products, nec	155	5,762	2,561	22,442	4,447	..	..	4,145,313	..	50,290
155 Beverages	419	2,051	2,297	21,917	6,170	..	..	2,509,284	..	41,887
1551 Distilling, rectifying and blending of spirits	419	640	84	4,614	550	..	..	321,395	..	7,464
1552 Wines	0	26	266	1,782	326	..	..	239,829	..	3,263
1553 Malt liquors and malt	0	594	620	6,969	4,641	..	..	1,160,152	..	17,001
1554 Soft drinks; production of mineral waters	0	791	1,327	8,552	653	..	..	787,908	..	14,159
16 TOBACCO PRODUCTS	0	1	1	921	243	..	..	342,183	..	2,398
17 TEXTILES	3,024	7,291	1,685	52,563	4,026	..	..	7,445,874	..	95,394
171 Spinning, weaving and finishing of textiles	3,024	6,266	750	46,599	3,399	..	..	5,937,053	..	81,411
1711 Preparation and spinning of textile fibres	603	387	184	11,368	206	..	..	4,091,763	..	27,478
1712 Finishing of textiles	2,421	5,879	566	35,231	3,193	..	..	1,845,290	..	53,933
172 Other textiles	0	869	420	3,852	527	..	..	832,832	..	8,666
1721 Made-up textile articles, except apparel	0	162	123	504	39	..	..	201,404	..	1,553
1722 Carpets and rugs	0	130	74	317	235	..	..	99,534	..	1,114
1723 Cordage, rope, twine and netting	0	10	22	208	0	..	..	44,738	..	401
1729 Other textiles, nec	0	567	201	2,823	253	..	..	487,156	..	5,598
173 Knitted and crocheted fabrics and articles	0	156	515	2,112	100	..	..	675,989	..	5,317
18 WEARING APPAREL, DRESSING & DYEING OF FUR	0	123	1,150	1,683	99	..	..	620,967	..	5,290
181 Wearing apparel, except fur apparel	0	123	1,148	1,647	99	..	..	616,571	..	5,237
182 Dressing and dyeing of fur; articles of fur	0	0	2	36	0	..	..	4,396	..	54
19 TANNING & DRESSING OF LEATHER, FOOTWEAR	0	196	287	2,523	30	..	..	408,496	..	4,507
191 Tanning and dressing of leather	0	181	227	1,719	22	..	..	240,836	..	3,016
1911 Tanning and dressing of leather	0	177	205	1,714	22	..	..	222,334	..	2,918
1912 Luggage, handbags, saddlery & harness	0	4	22	5	0	..	..	18,502	..	98
192 Footwear	0	15	60	804	8	..	..	167,660	..	1,491
20 WOOD AND WOOD PRODUCTS	0	192	1,483	3,712	789	..	..	1,972,247	..	13,276
201 Sawmilling and planing of wood	0	0	565	417	3	..	..	387,420	..	2,380
202 Products of wood, cork, straw & plaiting materials	0	192	918	3,295	786	..	..	1,584,827	..	10,896
2021 Veneer sheets	0	171	644	3,073	773	..	..	1,247,106	..	9,151
2022 Builders' carpentry and joinery	0	14	161	153	0	..	..	239,237	..	1,189
2023 Wooden containers	0	4	55	20	0	..	..	21,693	..	157
2029 Other products of wood	0	3	58	49	13	..	..	76,791	..	399
21 PAPER AND PAPER PRODUCTS	103,558	4,862	5,947	198,375	20,685	..	..	33,917,969	..	455,532
2101 Pulp, paper and paperboard	103,558	3,756	4,433	188,609	19,683	..	..	31,835,084	..	434,645
2102 Corrugated paper, paperboard and their containers	0	413	978	6,940	487	..	..	1,139,740	..	12,921
2109 Other articles of pulp and paperboard	0	693	536	2,826	515	..	..	943,145	..	7,965
22 PUBLISHING, PRINTING & REPRODUCTION	0	2,698	1,227	2,500	5,148	..	..	3,990,770	..	25,940
221 Publishing	0	42	67	271	639	..	..	764,777	..	3,772
2211 Publishing of books & brochures	0	38	9	30	32	..	..	101,314	..	474
2212 Publishing of newspapers and periodicals	0	4	58	241	607	..	..	663,463	..	3,298
2213 Publishing of recorded media	..	..	..	..	..	..	..	..	..	..
2219 Other publishing	..	..	..	..	..	..	..	..	..	..
222 Printing and related service activities	0	2,655	1,066	1,903	4,478	..	..	3,039,936	..	21,046
2221 Printing	0	2,615	1,026	1,801	4,339	..	..	2,731,575	..	19,615
2222 Service activities related to printing	0	40	40	102	139	..	..	308,361	..	1,431
223 Reproduction of recorded media	0	1	94	326	31	..	..	186,057	..	1,122
23 COKE, REFINED PETROLEUM PRODUCTS	538,543	35,949	268,162	127,469	39,206	..	..	10,343,932	..	1,046,567
231 Coke oven products	501,187	172	4,643	355	37,831	..	..	923,976	..	547,514
232 Refined petroleum products	37,356	35,581	263,476	126,661	1,375	..	..	9,165,131	..	497,443
233 Processing of nuclear fuel	0	196	43	453	0	..	..	254,825	..	1,609
24 CHEMICALS & CHEMICAL PRODUCTS	244,219	162,054	200,266	269,637	50,252	..	..	54,083,190	..	1,121,127
241 Basic chemicals	197,246	150,083	189,808	170,882	39,086	..	..	39,443,881	..	889,103
2411 Basic chemicals, exc. fertilizers & nitrogen compounds	121,020	79,228	117,396	102,675	31,329	..	..	24,278,374	..	539,050
2412 Fertilizers and nitrogen compounds	32,757	111	164	3,429	1,630	..	..	1,062,094	..	41,915
2413 Plastics in primary forms and synthetic rubber	43,469	70,744	72,248	64,778	6,127	..	..	14,103,413	..	308,138
242 Other chemical products	13,368	10,670	6,901	55,966	11,153	..	..	9,553,599	..	132,451
2421 Pesticides and other agro-chemical products	22	984	654	2,397	141	..	..	536,781	..	6,130
2422 Paints, varnishes and similar coatings	2	91	512	909	263	..	..	476,111	..	3,491
2423 Pharmaceuticals, medicinal chem. & botanical prod.	1,394	710	1,788	17,270	2,241	..	..	2,697,085	..	33,113
2424 Soap and detergents, perfumes etc.	0	592	1,585	5,841	2,736	..	..	1,068,040	..	14,599
2429 Other chemical products, nec	11,950	8,293	2,362	29,549	5,772	..	..	4,775,582	..	75,118
243 Man-made fibres	33,605	1,301	3,557	42,789	13	..	..	5,085,710	..	99,574
25 RUBBER AND PLASTICS PRODUCTS	3,995	4,868	3,296	39,413	4,666	..	..	15,893,757	..	113,456
251 Rubber products	1,310	858	909	12,933	1,714	..	..	3,561,067	..	30,544
2511 Rubber tyres and tubes	1,310	569	485	7,278	541	..	..	1,783,732	..	16,604
2519 Other rubber products	0	289	424	5,655	1,173	..	..	1,777,335	..	13,939
252 Plastic products	2,685	4,010	2,387	26,480	2,952	..	..	12,332,690	..	82,912
26 OTHER NON-METALLIC MINERAL PRODUCTS	323,771	30,492	21,959	111,140	16,882	..	..	23,384,302	..	588,427
261 Glass and glass products	5	9,012	1,951	43,655	4,846	..	..	4,838,783	..	76,889
269 Non-metallic mineral products, nec	323,766	21,480	20,008	67,485	12,036	..	..	18,545,519	..	511,539
2691 Pottery, china and earthenware	1	5,532	2,312	1,813	1,084	..	..	949,309	..	14,160
2692 Refractory ceramic products	2,210	5,512	2,987	7,181	679	..	..	763,507	..	21,318
2693 Structural non-refractory clay & ceramic prod.	1,345	3,501	3,192	2,822	76	..	..	265,119	..	11,890
2694 Cement, lime and plaster	303,093	191	3,010	25,607	4,558	..	..	11,168,374	..	376,665
2695 Articles of concrete, cement and plaster	5,445	2,742	4,631	16,472	1,667	..	..	1,662,976	..	36,944
2696 Cutting, shaping and finishing of stone	64	15	1,610	177	0	..	..	213,200	..	2,634
2699 Other non-metallic mineral products, nec	11,608	3,987	2,266	13,413	3,972	..	..	3,523,034	..	47,929

ISIC Revision 3 Industry Sector	Solid TJ	LPG TJ	Distiloil TJ	RFO TJ	Gas TJ	Biomass TJ	Steam TJ	Electr MWh	Own Use MWh	TOTAL TJ
27 BASIC METALS	2,557,123	41,809	28,896	95,796	476,120	..	..	78,779,179	..	3,483,349
271 Basic iron and steel	2,532,162	30,020	18,904	63,169	469,989		..	63,684,999	..	3,343,510
272 Basic precious and non-ferrous metals	17,653	7,624	8,454	28,606	4,559		..	11,214,657	..	107,269
273 Casting of metals	7,308	4,165	1,538	4,021	1,572		..	3,879,523	..	32,570
2731 Casting of iron and steel	7,308	1,252	968	1,635	803		..	3,059,054	..	22,979
2732 Casting non-ferrous metals	0	2,913	570	2,386	769		..	820,469	..	9,592
28 FABRICATED METAL PRODUCTS	225	16,726	11,075	14,769	8,033		..	11,129,018	..	90,892
281 Str. metal prod., tanks, reservoirs, steam generators	134	2,714	4,973	4,285	1,240		..	3,191,296	..	24,835
2811 Structural metal products	80	2,113	2,975	3,349	813		..	2,511,444	..	18,371
2812 Tanks, reservoirs and containers of metal	15	464	473	226	282		..	301,497	..	2,545
2813 Steam generators, exc. central heating hot water boilers	39	137	1,525	710	145		..	378,355	..	3,918
289 Other fabricated metal products	91	14,012	6,102	10,484	6,793		..	7,937,722	..	66,058
2891 Forging, pressing, stamping & roll-forming of metal	26	4,353	1,912	4,960	1,478		..	2,744,525	..	22,609
2892 Treatment and coating of metals	1	3,093	1,751	3,089	2,544		..	1,843,121	..	17,113
2893 Cutlery, hand tools and general hardware	20	178	481	479	104		..	669,200	..	3,671
2899 Other fabricated metal products, nec	44	6,388	1,958	1,956	2,667		..	2,680,876	..	22,664
29 MACHINERY AND EQUIPMENT, NEC	1,179	6,914	9,564	11,228	5,665		..	12,087,227	..	78,064
291 General purpose machinery	933	4,324	4,770	6,681	3,074		..	6,155,045	..	41,940
2911 Engines and turbines	632	179	1,317	2,871	662		..	839,808	..	8,684
2912 Pumps, compressors, taps and valves	35	1,367	525	540	856		..	882,477	..	6,500
2913 Bearings, gears, gearing and driving elements	75	1,654	1,242	1,411	265		..	2,084,318	..	12,151
2914 Ovens, furnaces and furnace burners	0	35	22	2	168		..	60,753	..	446
2915 Lifting and handling equipment	9	367	353	744	159		..	529,120	..	3,537
2919 Other general purpose machinery	182	722	1,311	1,113	964		..	1,758,569	..	10,623
292 Special purpose machinery	227	1,646	3,917	3,634	1,189		..	4,439,810	..	26,596
2921 Agricultural and forestry machinery	0	339	347	225	89		..	209,336	..	1,754
2922 Machine-tools	25	224	1,020	902	314		..	1,189,832	..	6,768
2923 Machinery for metallurgy	20	16	107	11	3		..	125,160	..	608
2924 Machinery for mining, quarrying and construction	116	483	1,105	923	372		..	951,584	..	6,425
2925 Machinery for food, beverage & tobacco processing	0	17	72	51	2		..	44,271	..	301
2926 Machinery for textile, apparel & leather production	14	86	203	174	64		..	248,052	..	1,434
2927 Machinery for weapons and ammunition	0	6	75	214	30		..	90,637	..	651
2929 Other special purpose machinery	52	475	988	1,134	315		..	1,580,938	..	8,655
293 Domestic appliances, nec	19	944	877	913	1,402		..	1,492,372	..	9,528
30 OFFICE, ACCOUNTING & COMPUTING MACHINERY	1	283	1,364	2,285	844		..	3,184,444	..	16,241
31 ELECTRICAL MACHINERY & APPARATUS, NEC	118	5,293	4,673	6,816	3,617		..	9,455,821	..	54,558
311 Electric motors, generators and transformers	103	508	668	822	177		..	1,035,657	..	6,006
312 Electricity distribution and control apparatus	1	413	1,134	982	453		..	1,243,242	..	7,459
313 Insulated wire and cable	1	1,808	607	1,627	909		..	2,038,157	..	12,289
314 Accumulators, primary cells & primary batteries	13	336	340	955	767		..	680,485	..	4,861
315 Electric lamps and lighting equipment	0	1,166	143	255	241		..	535,514	..	3,733
319 Other electrical equipment, nec	0	1,062	1,781	3,175	1,070		..	3,922,766	..	20,210
32 RADIO, TV & COMMUNICATION EQUIP. & APP.	149	5,377	4,135	20,004	8,063		..	20,003,931	..	109,742
321 Electronic valves, tubes, other electronic components	53	2,836	2,122	13,269	6,128		..	13,452,080	..	72,835
322 TV & radio transmitters, apparatus for line telephony	0	39	183	348	659		..	846,103	..	4,275
323 TV & radio receivers, recording apparatus	96	2,502	1,830	6,387	1,276		..	5,705,748	..	32,632
33 MEDICAL PRECISION &OPTICAL INSTRUMENTS	6	836	966	2,239	759		..	2,323,913	..	13,172
331 Medical appliances and instruments	5	721	462	1,351	328		..	1,148,957	..	7,003
3311 Medical, surgical equipment & orthopaedic app.	2	465	100	1,008	45		..	458,586	..	3,271
3312 Instruments & appliances for measuring, checking etc	3	251	346	315	246		..	603,515	..	3,334
3313 Industrial process control equipment	0	5	16	28	37		..	86,856	..	399
332 Optical instruments and photographic equipment	0	106	266	613	250		..	740,071	..	3,899
333 Watches and clocks	1	9	238	275	181		..	434,885	..	2,270
34 MOTOR VEHICLES, TRAILERS & SEMI-TRAILERS	12,843	15,494	14,736	25,559	17,509		..	21,918,567	..	165,048
341 Motor vehicles	8,943	4,697	5,906	10,659	9,202		..	6,257,176	..	61,933
342 Bodies (coachwork) for motor vehicles	0	1,457	1,155	1,251	1,466		..	1,093,159	..	9,264
343 Parts, accessories for motor vehicles & their engines	3,900	9,340	7,675	13,649	6,841		..	14,568,232	..	93,851
35 OTHER TRANSPORT EQUIPMENT	10	647	1,052	2,383	407		..	1,438,489	..	9,678
351 Building and repairing of ships and boats	1	318	545	1,483	65		..	867,834	..	5,536
3511 Building and repairing of ships	0	317	533	1,433	64		..	852,582	..	5,416
3512 Building, repairing of pleasure & sporting boats	1	1	12	50	1		..	15,252	..	120
352 Railway, tramway locomotives & rolling stock	9	62	91	63	46		..	95,550	..	615
353 Aircraft and spacecraft	0	21	305	748	148		..	328,292	..	2,404
359 Transport equipment, nec	0	246	111	89	148		..	146,813	..	1,123
3591 Motorcycles	..	..	..	..			..		..	
3592 Bicycles and invalid carriages	0	225	77	62	148		..	128,248	..	974
3599 Other transport equipment, nec	0	21	34	27	0		..	18,565	..	149
36 FURNITURE; MANUFACTURING, NEC	451	1,575	1,714	2,560	423		..	2,059,408	..	14,137
361 Furniture	0	955	657	674	235		..	820,716	..	5,476
369 Manufacturing, nec	451	620	1,057	1,886	188		..	1,238,692	..	8,661
3691 Jewellery and related articles	0	18	10	28	20		..	46,546	..	244
3692 Musical instruments	50	105	48	451	12		..	230,272	..	1,495
3693 Sports goods	0	75	61	194	5		..	128,740	..	798
3694 Games and toys	0	14	27	13	6		..	54,508	..	256
3699 Other manufacturing, nec	401	408	911	1,200	145		..	778,626	..	5,868
37 RECYCLING	150	119	194	94	1		..	119,826	..	989
371 Recycling of metal waste and scrap	150	119	194	94	1	..	..	119,826	..	989
372 Recycling of non-metal waste and scrap	..	..	..	..		..	..		..	
CERR Non-specified industry	..	..	..	..			..		..	
15-37 TOTAL MANUFACTURING	3,793,691	364,530	594,158	1,101,994	689,803		..	332,135,524	..	7,739,864

ISIC Revision 3 Industry Sector	Solid TJ	LPG TJ	Distiloil TJ	RFO TJ	Gas TJ	Biomass TJ	Steam TJ	Electr MWh	Own Use MWh	TOTAL TJ
15 FOOD PRODUCTS AND BEVERAGES	4,567	22,110	11,162	112,179	29,119	..	..	17,978,662	..	243,860
151 Production, processing and preserving (PPP)	1,501	3,525	2,467	19,480	5,161	..	..	3,866,165	..	46,052
1511 PPP of meat and meat products	0	660	617	3,730	2,716	..	..	1,219,045	..	12,112
1512 Processing and preserving of fish products	0	2,375	1,333	6,742	578	..	..	1,610,409	..	16,825
1513 Processing, preserving of fruit & vegetables	0	387	441	3,285	50	..	..	445,484	..	5,767
1514 Vegetable and animal oils and fats	1,501	103	76	5,723	1,817	..	..	591,227	..	11,348
152 Dairy products	0	684	145	12,036	1,535	..	..	1,769,599	..	20,771
153 Grain mill prod., starches & prepared animal feeds	0	1,734	489	12,083	1,237	..	..	1,770,789	..	21,918
1531 Grain mill products	0	23	180	747	70	..	..	617,020	..	3,241
1532 Starches and starch products	0	1,455	210	9,642	1,105	..	..	794,182	..	15,271
1533 Prepared animal feeds	0	256	99	1,694	62	..	..	359,587	..	3,406
154 Other food products	2,644	13,777	5,680	46,172	15,040	..	..	8,027,530	..	112,212
1541 Bakery products	0	5,483	1,518	6,761	5,401	..	..	2,032,385	..	26,480
1542 Sugar	2,504	4	187	8,726	2,766	..	..	416,294	..	15,686
1543 Cocoa, chocolate and sugar confectionery	0	912	1,074	1,716	1,293	..	..	724,956	..	7,605
1544 Macaroni, noodles, couscous & similar farinaceous prod.	0	772	367	5,438	270	..	..	469,691	..	8,538
1549 Other food products, nec	140	6,606	2,534	23,531	5,310	..	..	4,384,204	..	53,904
155 Beverages	422	2,390	2,381	22,408	6,146	..	..	2,544,579	..	42,907
1551 Distilling, rectifying and blending of spirits	422	675	115	4,539	544	..	..	336,539	..	7,507
1552 Wines	0	25	243	1,776	365	..	..	268,954	..	3,377
1553 Malt liquors and malt	0	558	545	6,763	4,494	..	..	1,097,519	..	16,311
1554 Soft drinks; production of mineral waters	0	1,132	1,478	9,330	743	..	..	841,567	..	15,713
16 TOBACCO PRODUCTS	0	1	1	916	228	..	..	341,376	..	2,375
17 TEXTILES	2,978	7,715	1,517	49,014	4,460	..	..	6,888,674	..	90,483
171 Spinning, weaving and finishing of textiles	2,978	6,324	665	43,228	3,566	..	..	5,399,627	..	76,200
1711 Preparation and spinning of textile fibres	651	356	171	10,073	304	..	..	3,647,633	..	24,686
1712 Finishing of textiles	2,327	5,968	494	33,155	3,262	..	..	1,751,994	..	51,513
172 Other textiles	0	1,265	477	3,721	789	..	..	867,293	..	9,374
1721 Made-up textile articles, except apparel	0	152	125	540	83	..	..	230,187	..	1,729
1722 Carpets and rugs	0	140	83	324	259	..	..	103,977	..	1,180
1723 Cordage, rope, twine and netting	0	16	25	229	0	..	..	45,915	..	435
1729 Other textiles, nec	0	957	244	2,628	447	..	..	487,214	..	6,030
173 Knitted and crocheted fabrics and articles	0	126	375	2,065	105	..	..	621,754	..	4,909
18 WEARING APPAREL, DRESSING & DYEING OF FUR	0	132	1,019	1,624	86	..	..	578,559	..	4,944
181 Wearing apparel, except fur apparel	0	132	1,017	1,617	86	..	..	576,948	..	4,929
182 Dressing and dyeing of fur; articles of fur	0	0	2	7	0	..	..	1,611	..	15
19 TANNING & DRESSING OF LEATHER, FOOTWEAR	0	239	266	2,579	39	..	..	425,320	..	4,654
191 Tanning and dressing of leather	0	221	201	1,783	31	..	..	259,937	..	3,172
1911 Tanning and dressing of leather	0	217	181	1,777	29	..	..	239,129	..	3,065
1912 Luggage, handbags, saddlery & harness	0	4	20	6	2	..	..	20,808	..	107
192 Footwear	0	18	65	796	8	..	..	165,383	..	1,482
20 WOOD AND WOOD PRODUCTS	0	187	1,511	3,841	797	..	..	1,959,524	..	13,390
201 Sawmilling and planing of wood	0	4	627	451	3	..	..	389,536	..	2,487
202 Products of wood, cork, straw & plaiting materials	0	183	884	3,390	794	..	..	1,569,988	..	10,903
2021 Veneer sheets	0	144	612	3,102	789	..	..	1,151,232	..	8,791
2022 Builders' carpentry and joinery	0	31	186	238	4	..	..	330,253	..	1,648
2023 Wooden containers	0	5	40	3	0	..	..	20,200	..	121
2029 Other products of wood	0	3	46	47	1	..	..	68,303	..	343
21 PAPER AND PAPER PRODUCTS	114,955	8,627	6,385	202,084	21,587	..	..	34,593,093	..	478,173
2101 Pulp, paper and paperboard	114,955	7,542	4,964	192,300	20,421	..	..	32,437,912	..	456,958
2102 Corrugated paper, paperboard and their containers	0	486	983	7,004	577	..	..	1,206,537	..	13,394
2109 Other articles of pulp and paperboard	0	599	438	2,780	589	..	..	948,644	..	7,821
22 PUBLISHING, PRINTING & REPRODUCTION	0	3,167	1,466	2,563	6,150	..	..	4,348,662	..	29,001
221 Publishing	0	42	293	258	1,069	..	..	797,429	..	4,533
2211 Publishing of books & brochures	0	38	11	21	46	..	..	106,264	..	499
2212 Publishing of newspapers and periodicals	0	4	282	237	1,023	..	..	691,165	..	4,034
2213 Publishing of recorded media	..	..	..	..	..	..	..	..	..	..
2219 Other publishing	..	..	..	..	..	..	..	..	..	..
222 Printing and related service activities	0	3,124	1,104	2,003	5,054	..	..	3,373,584	..	23,430
2221 Printing	0	3,089	1,062	1,891	4,902	..	..	3,042,926	..	21,899
2222 Service activities related to printing	0	35	42	112	152	..	..	330,658	..	1,531
223 Reproduction of recorded media	0	1	69	302	27	..	..	177,649	..	1,039
23 COKE, REFINED PETROLEUM PRODUCTS	553,316	36,049	288,616	123,724	41,432	..	..	10,757,103	..	1,081,863
231 Coke oven products	515,995	103	5,863	295	40,169	..	..	905,562	..	565,685
232 Refined petroleum products	37,321	35,938	282,706	122,782	1,263	..	..	9,572,277	..	514,470
233 Processing of nuclear fuel	0	8	47	647	0	..	..	279,264	..	1,707
24 CHEMICALS & CHEMICAL PRODUCTS	232,651	148,969	245,021	286,388	49,663	..	..	57,624,969	..	1,170,142
241 Basic chemicals	190,115	135,129	234,336	182,785	37,188	..	..	41,926,796	..	930,489
2411 Basic chemicals, exc. fertilizers & nitrogen compounds	96,884	66,237	173,300	109,492	28,139	..	..	26,090,944	..	567,979
2412 Fertilizers and nitrogen compounds	31,295	112	153	3,558	1,459	..	..	1,115,805	..	40,594
2413 Plastics in primary forms and synthetic rubber	61,936	68,780	60,883	69,735	7,590	..	..	14,720,047	..	321,916
242 Other chemical products	12,762	12,742	6,667	57,759	12,460	..	..	10,707,225	..	140,936
2421 Pesticides and other agro-chemical products	22	891	735	2,434	195	..	..	557,659	..	6,285
2422 Paints, varnishes and similar coatings	2	88	447	948	453	..	..	498,921	..	3,734
2423 Pharmaceuticals, medicinal chem. & botanical prod.	256	1,122	1,882	18,236	2,754	..	..	2,758,731	..	34,181
2424 Soap and detergents, perfumes etc.	0	572	1,226	6,082	3,080	..	..	1,079,574	..	14,846
2429 Other chemical products, nec	12,482	10,069	2,377	30,059	5,978	..	..	5,812,340	..	81,889
243 Man-made fibres	29,774	1,098	4,018	45,844	15	..	..	4,990,948	..	98,716
25 RUBBER AND PLASTICS PRODUCTS	5,179	5,202	3,366	43,098	5,210	..	..	16,705,116	..	122,193
251 Rubber products	1,416	944	1,020	14,096	1,803	..	..	3,751,786	..	32,785
2511 Rubber tyres and tubes	1,416	601	495	7,966	851	..	..	1,888,394	..	18,127
2519 Other rubber products	0	343	525	6,130	952	..	..	1,863,392	..	14,658
252 Plastic products	3,763	4,258	2,346	29,002	3,407	..	..	12,953,330	..	89,408
26 OTHER NON-METALLIC MINERAL PRODUCTS	316,511	30,145	21,962	106,606	19,567	..	..	21,519,252	..	572,260
261 Glass and glass products	6	8,218	1,932	42,871	5,501	..	..	4,049,461	..	73,106
269 Non-metallic mineral products, nec	316,505	21,927	20,030	63,735	14,066	..	..	17,469,791	..	499,154
2691 Pottery, china and earthenware	0	5,580	2,387	1,479	1,378	..	..	1,012,669	..	14,470
2692 Refractory ceramic products	1,950	5,577	2,845	7,872	576	..	..	776,399	..	21,615
2693 Structural non-refractory clay & ceramic prod.	1,380	3,445	3,008	2,726	145	..	..	276,196	..	11,698
2694 Cement, lime and plaster	295,545	258	2,731	21,344	4,896	..	..	9,791,136	..	360,022
2695 Articles of concrete, cement and plaster	5,453	2,757	4,856	16,421	2,558	..	..	1,758,181	..	38,374
2696 Cutting, shaping and finishing of stone	61	13	1,580	286	0	..	..	204,870	..	2,678
2699 Other non-metallic mineral products, nec	12,116	4,297	2,623	13,607	4,513	..	..	3,650,340	..	50,297

ISIC Revision 3 Industry Sector	Solid TJ	LPG TJ	Distiloil TJ	RFO TJ	Gas TJ	Biomass TJ	Steam TJ	Electr MWh	Own Use MWh	TOTAL TJ
27 **BASIC METALS**	2,616,497	43,081	29,286	101,266	488,837	..	..	82,043,063	..	3,574,322
271 Basic iron and steel	2,591,165	30,779	18,748	67,663	482,183	..	..	66,145,380	..	3,428,661
272 Basic precious and non-ferrous metals	18,231	8,172	8,864	29,271	4,769	..	..	11,663,143	..	111,294
273 Casting of metals	7,101	4,130	1,674	4,332	1,885	..	..	4,234,540	..	34,366
2731 Casting of iron and steel	7,101	1,391	1,069	1,868	941	..	..	3,390,281	..	24,575
2732 Casting non-ferrous metals	0	2,739	605	2,464	944	..	..	844,259	..	9,791
28 **FABRICATED METAL PRODUCTS**	236	17,419	11,183	15,779	8,849	..	..	11,611,288	..	95,267
281 Str. metal prod., tanks, reservoirs, steam generators	123	2,923	5,303	4,447	1,377	..	..	3,258,261	..	25,903
2811 Structural metal products	78	2,286	3,048	3,342	776	..	..	2,528,399	..	18,632
2812 Tanks, reservoirs and containers of metal	16	497	405	276	229	..	..	306,789	..	2,527
2813 Steam generators, exc. central heating hot water boilers	29	140	1,850	829	372	..	..	423,073	..	4,743
289 Other fabricated metal products	113	14,496	5,880	11,332	7,472	..	..	8,353,027	..	69,364
2891 Forging, pressing, stamping & roll-forming of metal	44	4,558	2,008	5,854	1,671	..	..	2,978,717	..	24,858
2892 Treatment and coating of metals	2	3,043	1,848	3,428	2,851	..	..	1,981,772	..	18,306
2893 Cutlery, hand tools and general hardware	26	230	262	440	123	..	..	682,958	..	3,540
2899 Other fabricated metal products, nec	41	6,665	1,762	1,610	2,827	..	..	2,709,580	..	22,659
29 **MACHINERY AND EQUIPMENT, NEC**	1,219	6,994	9,442	11,434	6,243	..	..	12,536,291	..	80,463
291 General purpose machinery	971	4,401	4,670	6,261	3,405	..	..	6,255,094	..	42,226
2911 Engines and turbines	178	216	1,098	1,083	441	..	..	623,200	..	5,260
2912 Pumps, compressors, taps and valves	32	1,276	495	533	1,000	..	..	885,195	..	6,523
2913 Bearings, gears, gearing and driving elements	62	1,743	1,375	1,560	373	..	..	2,276,022	..	13,307
2914 Ovens, furnaces and furnace burners	0	46	26	8	180	..	..	72,354	..	520
2915 Lifting and handling equipment	14	327	390	557	186	..	..	450,415	..	3,095
2919 Other general purpose machinery	685	793	1,286	2,520	1,225	..	..	1,947,908	..	13,521
292 Special purpose machinery	228	1,710	3,885	4,343	1,354	..	..	4,773,395	..	28,704
2921 Agricultural and forestry machinery	0	367	369	231	111	..	..	232,539	..	1,915
2922 Machine-tools	21	280	1,144	1,259	379	..	..	1,381,219	..	8,055
2923 Machinery for metallurgy	4	8	13	4	1	..	..	21,136	..	106
2924 Machinery for mining, quarrying and construction	127	512	1,019	1,000	396	..	..	947,132	..	6,464
2925 Machinery for food, beverage & tobacco processing	0	18	60	60	2	..	..	46,874	..	309
2926 Machinery for textile, apparel & leather production	12	85	190	139	46	..	..	240,055	..	1,336
2927 Machinery for weapons and ammunition	0	5	71	208	40	..	..	90,744	..	651
2929 Other special purpose machinery	64	435	1,019	1,442	379	..	..	1,813,696	..	9,868
293 Domestic appliances, nec	20	883	887	830	1,484	..	..	1,507,802	..	9,532
30 **OFFICE, ACCOUNTING & COMPUTING MACHINERY**	0	322	1,449	2,358	762	..	..	3,172,873	..	16,313
31 **ELECTRICAL MACHINERY & APPARATUS, NEC**	114	5,928	4,732	6,471	4,607	..	..	9,749,817	..	56,951
311 Electric motors, generators and transformers	99	518	602	822	276	..	..	1,072,216	..	6,177
312 Electricity distribution and control apparatus	1	364	1,061	826	544	..	..	1,223,644	..	7,201
313 Insulated wire and cable	1	1,809	488	1,624	1,048	..	..	2,095,272	..	12,513
314 Accumulators, primary cells & primary batteries	11	378	527	1,151	719	..	..	715,495	..	5,362
315 Electric lamps and lighting equipment	0	1,180	145	251	230	..	..	582,741	..	3,904
319 Other electrical equipment, nec	2	1,679	1,909	1,797	1,790	..	..	4,060,449	..	21,795
32 **RADIO, TV & COMMUNICATION EQUIP. & APP.**	64	5,786	4,185	22,391	8,695	..	..	21,248,024	..	117,614
321 Electronic valves, tubes, other electronic components	64	3,316	2,166	14,754	6,052	..	..	14,156,732	..	77,316
322 TV & radio transmitters, apparatus for line telephony	0	30	181	353	662	..	..	856,482	..	4,309
323 TV & radio receivers, recording apparatus	0	2,440	1,838	7,284	1,981	..	..	6,234,810	..	35,988
33 **MEDICAL PRECISION &OPTICAL INSTRUMENTS**	4	861	976	1,962	892	..	..	2,283,724	..	12,916
331 Medical appliances and instruments	3	746	454	1,326	448	..	..	1,183,049	..	7,236
3311 Medical, surgical equipment & orthopaedic app.	0	479	97	1,013	178	..	..	469,109	..	3,456
3312 Instruments & appliances for measuring, checking etc	3	261	345	282	241	..	..	635,201	..	3,419
3313 Industrial process control equipment	0	6	12	31	29	..	..	78,739	..	361
332 Optical instruments and photographic equipment	0	106	280	429	256	..	..	672,277	..	3,491
333 Watches and clocks	1	9	242	207	188	..	..	428,398	..	2,189
34 **MOTOR VEHICLES, TRAILERS & SEMI-TRAILERS**	11,969	21,505	14,897	25,847	18,144	..	..	21,985,922	..	171,511
341 Motor vehicles	8,237	4,871	5,814	10,671	9,342	..	..	6,214,308	..	61,307
342 Bodies (coachwork) for motor vehicles	0	1,413	1,072	1,371	1,556	..	..	1,058,478	..	9,223
343 Parts, accessories for motor vehicles & their engines	3,732	15,221	8,011	13,805	7,246	..	..	14,713,136	..	100,982
35 **OTHER TRANSPORT EQUIPMENT**	7	566	1,067	2,386	416	..	..	1,433,990	..	9,604
351 Building and repairing of ships and boats	1	285	493	1,578	60	..	..	833,016	..	5,416
3511 Building and repairing of ships	0	283	476	1,514	59	..	..	818,948	..	5,280
3512 Building, repairing of pleasure & sporting boats	0	2	17	64	1	..	..	14,068	..	135
352 Railway, tramway locomotives & rolling stock	6	65	104	63	43	..	..	122,022	..	720
353 Aircraft and spacecraft	0	28	381	669	139	..	..	342,194	..	2,449
359 Transport equipment, nec	0	188	89	76	174	..	..	136,758	..	1,019
3591 Motorcycles	..	..	..	..	..	..	..	..	..	..
3592 Bicycles and invalid carriages	0	177	68	64	169	..	..	119,849	..	909
3599 Other transport equipment, nec	0	11	21	12	5	..	..	16,900	..	110
36 **FURNITURE; MANUFACTURING, NEC**	405	1,593	1,880	2,749	466	..	..	2,088,963	..	14,613
361 Furniture	0	937	803	695	277	..	..	849,400	..	5,770
369 Manufacturing, nec	405	656	1,077	2,054	189	..	..	1,239,563	..	8,843
3691 Jewellery and related articles	0	15	12	32	25	..	..	52,114	..	272
3692 Musical instruments	47	97	28	323	11	..	..	164,583	..	1,098
3693 Sports goods	0	83	57	203	16	..	..	126,649	..	815
3694 Games and toys	0	7	26	31	5	..	..	62,530	..	294
3699 Other manufacturing, nec	358	454	954	1,465	132	..	..	833,687	..	6,364
37 **RECYCLING**	179	122	225	104	1	..	..	137,591	..	1,126
371 Recycling of metal waste and scrap	179	122	225	104	1	..	..	137,591	..	1,126
372 Recycling of non-metal waste and scrap	..	..	..	..	..	..	..	..	..	..
CERR **Non-specified industry**	..	..	..	..	..	..	..	..	..	..
15-37 **TOTAL MANUFACTURING**	3,860,851	366,720	661,614	1,127,363	716,250	..	..	342,011,856	..	7,964,041

ISIC Revision 3 Industry Sector		Solid TJ	LPG TJ	Distiloil TJ	RFO TJ	Gas TJ	Biomass TJ	Steam TJ	Electr MWh	Own Use MWh	TOTAL TJ
15	FOOD PRODUCTS AND BEVERAGES	4,669	23,688	10,751	113,163	30,689		..	18,317,899	..	248,904
151	Production, processing and preserving (PPP)	1,475	3,302	2,270	19,993	4,790		..	3,902,055	..	45,877
1511	PPP of meat and meat products	0	705	672	3,981	2,136		..	1,208,442	..	11,844
1512	Processing and preserving of fish products	0	1,904	1,211	7,184	589		..	1,637,402	..	16,783
1513	Processing, preserving of fruit & vegetables	0	322	288	3,308	50		..	453,878	..	5,602
1514	Vegetable and animal oils and fats	1,475	371	99	5,520	2,015		..	602,333	..	11,648
152	Dairy products	0	1,448	139	12,807	1,627		..	1,800,747	..	22,504
153	Grain mill prod., starches & prepared animal feeds	0	1,723	486	12,495	1,326		..	1,782,563	..	22,447
1531	Grain mill products	0	23	138	817	23		..	589,495	..	3,123
1532	Starches and starch products	0	1,467	255	9,655	1,233		..	815,764	..	15,547
1533	Prepared animal feeds	0	233	93	2,023	70		..	377,304	..	3,777
154	Other food products	2,462	14,814	5,313	45,774	16,901		..	8,254,157	..	114,979
1541	Bakery products	0	6,452	896	6,941	5,874		..	2,106,590	..	27,747
1542	Sugar	2,357	4	161	8,141	3,339		..	417,817	..	15,506
1543	Cocoa, chocolate and sugar confectionery	0	1,007	1,022	1,763	1,296		..	749,025	..	7,784
1544	Macaroni, noodles, couscous & similar farinaceous prod.	0	799	376	5,562	361		..	465,314	..	8,773
1549	Other food products, nec	105	6,552	2,858	23,367	6,031		..	4,515,411	..	55,168
155	Beverages	732	2,401	2,543	22,094	6,045		..	2,578,377	..	43,097
1551	Distilling, rectifying and blending of spirits	732	657	144	4,512	519		..	338,070	..	7,781
1552	Wines	0	22	212	1,711	434		..	264,788	..	3,332
1553	Malt liquors and malt	0	564	494	6,887	4,289		..	1,104,706	..	16,211
1554	Soft drinks; production of mineral waters	0	1,158	1,693	8,984	803		..	870,813	..	15,773
16	TOBACCO PRODUCTS	0	1	1	892	251		..	331,875	..	2,340
17	TEXTILES	3,103	7,386	1,485	48,963	4,592		..	6,711,158	..	89,689
171	Spinning, weaving and finishing of textiles	3,103	6,321	632	43,202	3,697		..	5,093,314	..	75,291
1711	Preparation and spinning of textile fibres	737	249	179	9,247	460		..	3,324,475	..	22,840
1712	Finishing of textiles	2,366	6,072	453	33,955	3,237		..	1,768,839	..	52,451
172	Other textiles	0	931	451	3,735	804		..	997,117	..	9,511
1721	Made-up textile articles, except apparel	0	141	112	560	70		..	268,926	..	1,851
1722	Carpets and rugs	0	181	74	312	243		..	111,680	..	1,212
1723	Cordage, rope, twine and netting	0	17	23	211	0		..	43,475	..	408
1729	Other textiles, nec	0	592	242	2,652	491		..	573,036	..	6,040
173	Knitted and crocheted fabrics and articles	0	134	402	2,026	91		..	620,727	..	4,888
18	WEARING APPAREL, DRESSING & DYEING OF FUR	0	106	931	1,436	121		..	543,792	..	4,552
181	Wearing apparel, except fur apparel	0	106	929	1,398	121		..	538,950	..	4,494
182	Dressing and dyeing of fur; articles of fur	0	0	2	38	0		..	4,842	..	57
19	TANNING & DRESSING OF LEATHER, FOOTWEAR	2,641	241	310	2,934	47		..	477,280	..	7,891
191	Tanning and dressing of leather	2,641	229	243	2,117	41		..	318,383	..	6,417
1911	Tanning and dressing of leather	2,641	223	224	2,111	40		..	298,702	..	6,314
1912	Luggage, handbags, saddlery & harness	0	6	19	6	1		..	19,681	..	103
192	Footwear	0	12	67	817	6		..	158,897	..	1,474
20	WOOD AND WOOD PRODUCTS	0	171	1,521	3,899	886		..	2,040,271	..	13,822
201	Sawmilling and planing of wood	0	3	644	518	4		..	411,113	..	2,649
202	Products of wood, cork, straw & plaiting materials	0	168	877	3,381	882		..	1,629,158	..	11,173
2021	Veneer sheets	0	141	501	3,152	874		..	1,226,993	..	9,085
2022	Builders' carpentry and joinery	0	19	263	165	7		..	319,413	..	1,604
2023	Wooden containers	0	6	45	6	0		..	22,324	..	137
2029	Other products of wood	0	2	68	58	1		..	60,428	..	347
21	PAPER AND PAPER PRODUCTS	119,107	8,730	6,616	203,565	24,378		..	35,781,032	..	491,208
2101	Pulp, paper and paperboard	119,107	7,527	5,294	193,373	23,004		..	33,509,334	..	468,939
2102	Corrugated paper, paperboard and their containers	0	391	869	7,024	606		..	1,205,854	..	13,231
2109	Other articles of pulp and paperboard	0	812	453	3,168	768		..	1,065,844	..	9,038
22	PUBLISHING, PRINTING & REPRODUCTION	0	3,340	1,433	2,674	6,065		..	4,374,080	..	29,259
221	Publishing	0	46	347	265	831		..	762,354	..	4,233
2211	Publishing of books & brochures	0	42	66	41	86		..	120,330	..	668
2212	Publishing of newspapers and periodicals	0	4	281	224	745		..	642,024	..	3,565
2213	Publishing of recorded media	..	..	..	..	..		..		..	..
2219	Other publishing	..	..	..	..	..		..		..	..
222	Printing and related service activities	0	3,292	1,039	2,114	5,207		..	3,432,917	..	24,011
2221	Printing	0	3,229	989	1,983	4,985		..	3,082,016	..	22,281
2222	Service activities related to printing	0	63	50	131	222		..	350,901	..	1,729
223	Reproduction of recorded media	0	2	47	295	27		..	178,809	..	1,015
23	COKE, REFINED PETROLEUM PRODUCTS	546,360	38,796	287,480	126,356	40,891		..	11,519,072	..	1,081,352
231	Coke oven products	505,257	75	6,368	284	39,642		..	1,057,339	..	555,432
232	Refined petroleum products	41,103	38,718	281,065	125,346	1,249		..	10,160,375	..	524,058
233	Processing of nuclear fuel	0	3	47	726	0		..	301,358	..	1,861
24	CHEMICALS & CHEMICAL PRODUCTS	242,173	160,499	244,394	286,643	47,012		..	58,192,933	..	1,190,216
241	Basic chemicals	201,448	144,917	233,958	184,714	35,266		..	43,499,658	..	956,902
2411	Basic chemicals, exc. fertilizers & nitrogen compounds	105,173	92,196	192,726	122,592	26,022		..	28,511,599	..	641,351
2412	Fertilizers and nitrogen compounds	31,323	107	161	3,512	1,312		..	1,108,921	..	40,407
2413	Plastics in primary forms and synthetic rubber	64,952	52,614	41,071	58,610	7,932		..	13,879,138	..	275,144
242	Other chemical products	14,440	14,374	6,330	57,780	11,728		..	9,865,866	..	140,169
2421	Pesticides and other agro-chemical products	29	1,078	683	2,632	189		..	532,367	..	6,528
2422	Paints, varnishes and similar coatings	2	99	429	1,104	515		..	526,460	..	4,044
2423	Pharmaceuticals, medicinal chem. & botanical prod.	0	908	1,822	18,916	2,878		..	2,869,213	..	34,853
2424	Soap and detergents, perfumes etc.	0	1,178	1,300	5,794	2,699		..	1,055,368	..	14,770
2429	Other chemical products, nec	14,409	11,111	2,096	29,334	5,447		..	4,882,458	..	79,974
243	Man-made fibres	26,285	1,208	4,106	44,149	18		..	4,827,409	..	93,145
25	RUBBER AND PLASTICS PRODUCTS	2,870	4,771	4,240	43,113	5,931		..	16,700,538	..	121,047
251	Rubber products	1,398	1,025	999	14,537	2,752		..	3,887,957	..	34,708
2511	Rubber tyres and tubes	1,398	658	512	8,288	1,160		..	1,846,823	..	18,665
2519	Other rubber products	0	367	487	6,249	1,592		..	2,041,134	..	16,043
252	Plastic products	1,472	3,746	3,241	28,576	3,179		..	12,812,581	..	86,339
26	OTHER NON-METALLIC MINERAL PRODUCTS	318,270	30,114	22,315	107,119	19,679		..	21,851,989	..	576,164
261	Glass and glass products	6	8,192	1,850	41,773	5,139		..	4,154,041	..	71,915
269	Non-metallic mineral products, nec	318,264	21,922	20,465	65,346	14,540		..	17,697,948	..	504,250
2691	Pottery, china and earthenware	1	5,504	2,268	1,708	1,609		..	1,019,500	..	14,760
2692	Refractory ceramic products	2,251	5,507	2,750	8,096	611		..	788,823	..	22,055
2693	Structural non-refractory clay & ceramic prod.	1,225	3,395	2,842	2,806	128		..	274,330	..	11,384
2694	Cement, lime and plaster	297,730	280	3,763	21,575	4,354		..	9,934,441	..	363,466
2695	Articles of concrete, cement and plaster	5,651	2,857	4,959	16,662	3,248		..	1,775,669	..	39,769
2696	Cutting, shaping and finishing of stone	60	11	1,576	314	0		..	197,865	..	2,673
2699	Other non-metallic mineral products, nec	11,346	4,368	2,307	14,185	4,590		..	3,707,320	..	50,142

ISIS Energy Data Programme (IEA/OECD)

ISIC Revision 3 Industry Sector	Solid TJ	LPG TJ	Distiloil TJ	RFO TJ	Gas TJ	Biomass TJ	Steam TJ	Electr MWh	Own Use MWh	TOTAL TJ
27 BASIC METALS	2,619,815	42,946	29,031	100,452	482,312	..	..	81,746,052		3,568,842
271 Basic iron and steel	2,595,223	30,453	18,409	65,409	476,033	..	..	66,031,472	..	3,423,240
272 Basic precious and non-ferrous metals	17,620	8,262	8,926	30,592	4,603	..	..	11,456,286		111,246
273 Casting of metals	6,972	4,231	1,696	4,451	1,676	..	..	4,258,294		34,356
2731 Casting of iron and steel	6,972	1,461	1,142	2,123	800	..	..	3,430,185		24,847
2732 Casting non-ferrous metals	0	2,770	554	2,328	876			828,109	..	9,509
28 FABRICATED METAL PRODUCTS	234	18,278	11,211	16,531	8,784	..	..	12,097,497	..	98,589
281 Str. metal prod., tanks, reservoirs, steam generators	131	3,022	5,525	5,029	1,511			3,443,906		27,616
2811 Structural metal products	83	2,330	3,591	3,672	877			2,658,687		20,124
2812 Tanks, reservoirs and containers of metal	14	531	388	320	214			323,758		2,633
2813 Steam generators, exc. central heating hot water boilers	34	161	1,546	1,037	420			461,461		4,859
289 Other fabricated metal products	103	15,256	5,686	11,502	7,273			8,653,591		70,973
2891 Forging, pressing, stamping & roll-forming of metal	41	5,054	2,108	5,980	1,571			3,142,057		26,065
2892 Treatment and coating of metals	0	3,196	1,821	3,176	2,952			2,006,730		18,369
2893 Cutlery, hand tools and general hardware	24	215	268	491	110			712,685		3,674
2899 Other fabricated metal products, nec	38	6,791	1,489	1,855	2,640			2,792,119		22,865
29 MACHINERY AND EQUIPMENT, NEC	1,267	6,459	9,627	12,094	7,034	..	..	13,253,656	..	84,194
291 General purpose machinery	1,024	3,857	4,493	6,685	3,903	..		6,536,614		43,494
2911 Engines and turbines	184	197	1,155	1,096	517			656,307		5,512
2912 Pumps, compressors, taps and valves	32	567	513	579	415			917,925		5,411
2913 Bearings, gears, gearing and driving elements	81	1,788	1,172	1,937	353			2,280,935		13,542
2914 Ovens, furnaces and furnace burners	0	38	35	1	194			69,935		520
2915 Lifting and handling equipment	12	393	334	614	213			487,231		3,320
2919 Other general purpose machinery	715	874	1,284	2,458	2,211			2,124,281		15,189
292 Special purpose machinery	228	1,774	4,219	4,559	1,586			5,175,912		30,999
2921 Agricultural and forestry machinery	0	356	378	233	108			229,497		1,901
2922 Machine-tools	20	248	1,151	1,243	418			1,387,688		8,076
2923 Machinery for metallurgy	4	0	50	10	3			58,483		278
2924 Machinery for mining, quarrying and construction	127	591	1,159	1,055	356			997,914		6,880
2925 Machinery for food, beverage & tobacco processing	1	19	60	56	2			47,221		308
2926 Machinery for textile, apparel & leather production	8	83	195	142	31			248,488		1,354
2927 Machinery for weapons and ammunition	0	7	67	238	30			93,781		680
2929 Other special purpose machinery	68	470	1,159	1,582	638			2,112,840		11,523
293 Domestic appliances, nec	15	828	915	850	1,545			1,541,130		9,701
30 OFFICE, ACCOUNTING & COMPUTING MACHINERY	0	267	1,379	2,155	820			3,107,949		15,810
31 ELECTRICAL MACHINERY & APPARATUS, NEC	95	5,404	4,456	6,067	4,323	..	..	9,759,643	..	55,480
311 Electric motors, generators and transformers	92	602	559	821	277			1,038,417		6,089
312 Electricity distribution and control apparatus	0	366	1,088	869	626			1,280,127		7,557
313 Insulated wire and cable	0	1,872	630	1,530	1,108			2,207,972		13,089
314 Accumulators, primary cells & primary batteries	1	424	311	1,052	722			753,508		5,223
315 Electric lamps and lighting equipment	0	1,285	144	253	238			609,189		4,113
319 Other electrical equipment, nec	2	855	1,724	1,542	1,352			3,870,430		19,409
32 RADIO, TV & COMMUNICATION EQUIP. & APP.	2,840	6,603	4,794	23,600	9,656	..	..	22,669,132	..	129,102
321 Electronic valves, tubes, other electronic components	74	3,659	2,705	14,872	7,250			14,982,498		82,497
322 TV & radio transmitters, apparatus for line telephony	0	31	196	393	683			926,963		4,640
323 TV & radio receivers, recording apparatus	2,766	2,913	1,893	8,335	1,723			6,759,671		41,965
33 MEDICAL PRECISION &OPTICAL INSTRUMENTS	2	1,037	837	1,884	793	..	..	2,208,976	..	12,505
331 Medical appliances and instruments	1	892	332	1,283	404			1,157,003		7,077
3311 Medical, surgical equipment & orthopaedic app.	0	581	124	954	180			487,221		3,593
3312 Instruments & appliances for measuring, checking etc	1	300	189	300	191			579,119		3,066
3313 Industrial process control equipment	0	11	19	29	33			90,663		418
332 Optical instruments and photographic equipment	0	109	286	411	202			644,787		3,329
333 Watches and clocks	1	36	219	190	187			407,186		2,099
34 MOTOR VEHICLES, TRAILERS & SEMI-TRAILERS	12,892	15,930	14,317	25,389	19,839	..	..	21,781,706	..	166,781
341 Motor vehicles	8,945	4,818	5,821	10,720	10,630			6,251,868		63,441
342 Bodies (coachwork) for motor vehicles	97	1,394	1,120	1,133	1,710			1,125,099		9,504
343 Parts, accessories for motor vehicles & their engines	3,850	9,718	7,376	13,536	7,499			14,404,739		93,836
35 OTHER TRANSPORT EQUIPMENT	10	514	1,708	2,637	446	..	..	1,392,721	..	10,329
351 Building and repairing of ships and boats	1	258	599	1,795	88			825,626		5,713
3511 Building and repairing of ships	1	255	581	1,716	88			809,891		5,557
3512 Building, repairing of pleasure & sporting boats	0	3	18	79	0			15,735		157
352 Railway, tramway locomotives & rolling stock	9	38	100	49	71			99,520		625
353 Aircraft and spacecraft	0	17	941	740	121			343,669		3,056
359 Transport equipment, nec	0	201	68	53	166			123,906		934
3591 Motorcycles	..	..	..	..	..			..		
3592 Bicycles and invalid carriages	0	192	47	42	162			104,737		820
3599 Other transport equipment, nec	0	9	21	11	4			19,169		114
36 FURNITURE; MANUFACTURING, NEC	420	2,532	1,529	3,012	875	..	..	2,178,661	..	16,211
361 Furniture	0	921	608	959	685	..	..	920,569		6,487
369 Manufacturing, nec	420	1,611	921	2,053	190			1,258,092		9,724
3691 Jewellery and related articles	0	22	12	34	26			41,063		242
3692 Musical instruments	40	104	22	310	9			163,382		1,073
3693 Sports goods	0	83	60	240	5			145,826		913
3694 Games and toys	0	30	21	20	12			57,847		291
3699 Other manufacturing, nec	380	1,372	806	1,449	138			849,974		7,205
37 RECYCLING	178	122	200	73	1	..	..	136,611	..	1,066
371 Recycling of metal waste and scrap	178	122	200	73	1	..	..	136,611		1,066
372 Recycling of non-metal waste and scrap	..	..	..	..	..	..		..		
CERR Non-specified industry	..	..	..	..	..	..	..	..	..	..
15-37 TOTAL MANUFACTURING	3,876,946	377,935	660,566	1,134,651	715,425	..	..	347,174,523	..	8,015,351

ISIC Revision 3 Industry Sector	Solid TJ	LPG TJ	Distiloil TJ	RFO TJ	Gas TJ	Biomass TJ	Steam TJ	Electr MWh	Own Use MWh	TOTAL TJ
15 FOOD PRODUCTS AND BEVERAGES	4,632	25,327	10,192	113,945	34,405	..	..	348,747	..	189,756
151 Production, processing and preserving (PPP)	1,547	4,375	2,059	19,886	5,924	..	..	77,468	..	34,070
1511 PPP of meat and meat products	0	745	671	4,253	3,263	..	..	32,351	..	9,048
1512 Processing and preserving of fish products	2	2,432	1,034	6,460	719	..	..	42,988	..	10,802
1513 Processing, preserving of fruit & vegetables	0	406	282	3,520	46	..	..	2,129	..	4,262
1514 Vegetable and animal oils and fats	1,545	792	72	5,653	1,896	..	..	0	..	9,958
152 Dairy products	0	582	154	12,927	2,034	..	..	19,880	..	15,769
153 Grain mill prod., starches & prepared animal feeds	0	1,941	449	12,457	1,523	..	..	27,585	..	16,469
1531 Grain mill products	0	93	134	556	66	..	..	15,965	..	906
1532 Starches and starch products	0	1,580	233	9,964	1,368	..	..	4,069	..	13,160
1533 Prepared animal feeds	0	268	82	1,937	89	..	..	7,551	..	2,403
154 Other food products	2,403	15,779	4,951	46,553	18,215	..	..	191,629	..	88,591
1541 Bakery products	0	6,239	937	6,347	7,103	..	..	32,293	..	20,742
1542 Sugar	2,260	2	130	7,518	3,229	..	..	10,746	..	13,178
1543 Cocoa, chocolate and sugar confectionery	0	1,455	898	1,823	1,288	..	..	19,782	..	5,535
1544 Macaroni, noodles, couscous & similar farinaceous prod.	0	870	374	5,798	505	..	..	11,699	..	7,589
1549 Other food products, nec	143	7,213	2,612	25,067	6,090	..	..	117,109	..	41,547
155 Beverages	682	2,650	2,579	22,122	6,709	..	..	32,185	..	34,858
1551 Distilling, rectifying and blending of spirits	682	714	135	4,504	543	..	..	8,924	..	6,610
1552 Wines	0	23	186	1,594	383	..	..	396	..	2,187
1553 Malt liquors and malt	0	593	484	6,629	4,814	..	..	5,950	..	12,541
1554 Soft drinks; production of mineral waters	0	1,320	1,774	9,395	969	..	..	16,915	..	13,519
16 TOBACCO PRODUCTS	0	1	1	866	64	..	..	0	..	932
17 TEXTILES	3,926	8,045	1,381	46,755	5,139	..	..	248,348	..	66,140
171 Spinning, weaving and finishing of textiles	3,926	6,827	640	41,245	3,945	..	..	238,185	..	57,440
1711 Preparation and spinning of textile fibres	1,426	279	162	9,839	480	..	..	202,198	..	12,914
1712 Finishing of textiles	2,500	6,548	478	31,406	3,465	..	..	35,987	..	44,527
172 Other textiles	0	1,045	413	3,459	1,130	..	..	1,096	..	6,051
1721 Made-up textile articles, except apparel	0	148	78	506	74	..	..	403	..	807
1722 Carpets and rugs	0	188	75	279	248	..	..	0	..	790
1723 Cordage, rope, twine and netting	0	43	18	194	0	..	..	0	..	255
1729 Other textiles, nec	0	666	242	2,480	808	..	..	693	..	4,198
173 Knitted and crocheted fabrics and articles	0	173	328	2,051	64	..	..	9,067	..	2,649
18 WEARING APPAREL, DRESSING & DYEING OF FUR	0	91	1,173	1,547	135	..	..	1,736	..	2,952
181 Wearing apparel, except fur apparel	0	91	1,171	1,512	135	..	..	1,736	..	2,915
182 Dressing and dyeing of fur; articles of fur	0	0	2	35	0	..	..	0	..	37
19 TANNING & DRESSING OF LEATHER, FOOTWEAR	2,667	160	293	2,462	62	..	..	12,283	..	5,688
191 Tanning and dressing of leather	2,667	147	231	1,701	57	..	..	8,183	..	4,832
1911 Tanning and dressing of leather	2,667	146	212	1,696	56	..	..	8,183	..	4,806
1912 Luggage, handbags, saddlery & harness	0	1	19	5	1	..	..	0	..	26
192 Footwear	0	13	62	761	5	..	..	4,100	..	856
20 WOOD AND WOOD PRODUCTS	0	187	1,476	3,840	862	..	..	49,785	..	6,544
201 Sawmilling and planing of wood	0	3	596	519	42	..	..	5,034	..	1,178
202 Products of wood, cork, straw & plaiting materials	0	184	880	3,321	820	..	..	44,751	..	5,366
2021 Veneer sheets	0	125	485	2,884	817	..	..	39,142	..	4,452
2022 Builders' carpentry and joinery	0	47	286	322	0	..	..	5,133	..	673
2023 Wooden containers	0	12	29	6	0	..	..	476	..	49
2029 Other products of wood	0	0	80	109	3	..	..	0	..	192
21 PAPER AND PAPER PRODUCTS	119,010	4,922	7,306	201,371	25,865	..	..	781,610	..	361,288
2101 Pulp, paper and paperboard	119,010	3,960	5,720	191,554	24,230	..	..	769,189	..	347,243
2102 Corrugated paper, paperboard and their containers	0	372	1,080	6,831	549	..	..	4,841	..	8,849
2109 Other articles of pulp and paperboard	0	590	506	2,986	1,086	..	..	7,580	..	5,195
22 PUBLISHING, PRINTING & REPRODUCTION	0	3,586	1,421	2,559	9,362	..	..	27,832	..	17,028
221 Publishing	0	13	325	252	847	..	..	2,572	..	1,446
2211 Publishing of books & brochures	0	7	69	44	77	..	..	180	..	198
2212 Publishing of newspapers and periodicals	0	6	256	208	770	..	..	2,392	..	1,249
2213 Publishing of recorded media	..	..	..	..	..	..	..	..	..	..
2219 Other publishing	..	..	..	..	..	..	..	..	..	..
222 Printing and related service activities	0	3,570	1,055	1,964	8,488	..	..	25,260	..	15,168
2221 Printing	0	3,510	1,009	1,825	8,333	..	..	24,239	..	14,764
2222 Service activities related to printing	0	60	46	139	155	..	..	1,021	..	404
223 Reproduction of recorded media	0	3	41	343	27	..	..	0	..	414
23 COKE, REFINED PETROLEUM PRODUCTS	527,958	40,626	350,861	115,246	39,746	..	..	945,519	..	1,077,841
231 Coke oven products	497,829	96	7,097	263	38,641	..	..	0	..	543,926
232 Refined petroleum products	30,129	40,523	343,724	114,282	1,105	..	..	945,519	..	533,167
233 Processing of nuclear fuel	0	7	40	701	0	..	..	0	..	748
24 CHEMICALS & CHEMICAL PRODUCTS	246,640	169,588	218,197	292,304	55,095	..	..	1,735,465	..	988,072
241 Basic chemicals	206,078	154,564	207,240	191,212	37,151	..	..	1,398,465	..	801,279
2411 Basic chemicals, exc. fertilizers & nitrogen compounds	108,623	102,797	164,665	130,163	24,751	..	..	958,843	..	534,451
2412 Fertilizers and nitrogen compounds	29,511	108	140	3,144	1,403	..	..	96,848	..	34,655
2413 Plastics in primary forms and synthetic rubber	67,944	51,659	42,435	57,905	10,997	..	..	342,774	..	232,174
242 Other chemical products	12,987	14,036	6,619	56,782	17,646	..	..	225,994	..	108,884
2421 Pesticides and other agro-chemical products	33	1,061	636	2,753	191	..	..	0	..	4,674
2422 Paints, varnishes and similar coatings	2	103	421	1,028	523	..	..	1,351	..	2,082
2423 Pharmaceuticals, medicinal chem. & botanical prod.	0	945	2,204	18,746	3,327	..	..	108,697	..	25,613
2424 Soap and detergents, perfumes etc.	0	782	1,215	6,084	2,926	..	..	21,432	..	11,084
2429 Other chemical products, nec	12,952	11,145	2,143	28,171	10,679	..	..	94,514	..	65,430
243 Man-made fibres	27,575	988	4,338	44,310	298	..	..	111,006	..	77,909
25 RUBBER AND PLASTICS PRODUCTS	3,681	5,557	3,768	45,170	6,423	..	..	307,643	..	65,707
251 Rubber products	1,421	1,102	979	14,370	2,738	..	..	31,988	..	20,725
2511 Rubber tyres and tubes	1,421	690	475	8,270	1,183	..	..	5,648	..	12,059
2519 Other rubber products	0	412	504	6,100	1,555	..	..	26,340	..	8,666
252 Plastic products	2,260	4,455	2,789	30,800	3,685	..	..	275,655	..	44,981
26 OTHER NON-METALLIC MINERAL PRODUCTS	318,240	34,530	20,868	106,384	19,965	..	..	2,028,999	..	507,291
261 Glass and glass products	6	6,818	1,236	39,303	5,408	..	..	118,600	..	53,198
269 Non-metallic mineral products, nec	318,234	27,712	19,632	67,081	14,557	..	..	1,910,399	..	454,093
2691 Pottery, china and earthenware	1	5,394	1,913	1,890	1,577	..	..	9,194	..	10,808
2692 Refractory ceramic products	2,009	5,413	2,214	7,302	548	..	..	33,965	..	17,608
2693 Structural non-refractory clay & ceramic prod.	1,196	3,282	2,549	2,726	117	..	..	3,996	..	9,884
2694 Cement, lime and plaster	297,903	230	3,770	20,671	3,287	..	..	1,616,824	..	331,682
2695 Articles of concrete, cement and plaster	5,904	3,401	4,508	16,186	4,077	..	..	37,063	..	34,209
2696 Cutting, shaping and finishing of stone	53	8	1,927	344	0	..	..	18,161	..	2,397
2699 Other non-metallic mineral products, nec	11,168	9,984	2,751	17,962	4,951	..	..	191,196	..	47,504

ISIC Revision 3 Industry Sector		Solid TJ	LPG TJ	Distiloil TJ	RFO TJ	Gas TJ	Biomass TJ	Steam TJ	Electr MWh	Own Use MWh	TOTAL TJ
27	BASIC METALS	2,706,646	42,944	28,868	97,725	496,891	..	..	6,393,230	..	3,396,090
271	Basic iron and steel	2,681,470	30,372	18,080	66,233	490,127	..	..	5,804,681	..	3,307,179
272	Basic precious and non-ferrous metals	18,051	8,059	9,072	27,248	4,976	..	..	459,835	..	69,061
273	Casting of metals	7,125	4,513	1,716	4,244	1,788	..	..	128,714	..	19,849
2731	Casting of iron and steel	7,125	1,490	1,138	1,912	794	..	..	100,065	..	12,819
2732	Casting non-ferrous metals	0	3,023	578	2,332	994	..	..	28,649	..	7,030
28	FABRICATED METAL PRODUCTS	169	17,196	10,952	16,050	9,759	..	..	139,981	..	54,630
281	Str. metal prod., tanks, reservoirs, steam generators	109	2,823	5,089	4,924	1,627	..	..	48,316	..	14,746
2811	Structural metal products	76	2,178	3,435	3,907	951	..	..	35,655	..	10,675
2812	Tanks, reservoirs and containers of metal	1	512	354	283	231	..	..	5,360	..	1,400
2813	Steam generators, exc. central heating hot water boilers	32	133	1,300	734	445	..	..	7,301	..	2,670
289	Other fabricated metal products	60	14,373	5,863	11,126	8,132	..	..	91,665	..	39,884
2891	Forging, pressing, stamping & roll-forming of metal	36	5,096	2,064	5,937	1,584	..	..	31,314	..	14,830
2892	Treatment and coating of metals	0	3,121	1,823	3,159	3,320	..	..	21,730	..	11,501
2893	Cutlery, hand tools and general hardware	2	191	495	463	146	..	..	1,771	..	1,303
2899	Other fabricated metal products, nec	22	5,965	1,481	1,567	3,082	..	..	36,850	..	12,250
29	MACHINERY AND EQUIPMENT, NEC	1,177	6,452	10,226	11,825	9,185	..	..	318,913	..	40,013
291	General purpose machinery	1,045	3,727	4,671	7,046	6,120	..	..	273,765	..	23,595
2911	Engines and turbines	176	166	1,066	1,167	587	..	..	36,235	..	3,292
2912	Pumps, compressors, taps and valves	24	303	500	612	2,137	..	..	28,795	..	3,680
2913	Bearings, gears, gearing and driving elements	74	1,814	1,339	2,044	540	..	..	118,398	..	6,237
2914	Ovens, furnaces and furnace burners	0	15	26	1	198	..	..	0	..	240
2915	Lifting and handling equipment	11	399	339	604	249	..	..	184	..	1,603
2919	Other general purpose machinery	760	1,030	1,401	2,618	2,409	..	..	90,153	..	8,543
292	Special purpose machinery	117	1,841	4,678	4,024	1,689	..	..	25,854	..	12,442
2921	Agricultural and forestry machinery	0	364	778	208	121	..	..	2,867	..	1,481
2922	Machine-tools	25	271	1,099	1,207	511	..	..	6,877	..	3,138
2923	Machinery for metallurgy	12	5	23	4	2	..	..	1	..	46
2924	Machinery for mining, quarrying and construction	2	516	1,003	939	333	..	..	9,096	..	2,826
2925	Machinery for food, beverage & tobacco processing	1	24	56	58	3	..	..	1,724	..	148
2926	Machinery for textile, apparel & leather production	16	84	184	130	44	..	..	0	..	458
2927	Machinery for weapons and ammunition	0	7	62	190	40	..	..	0	..	299
2929	Other special purpose machinery	61	570	1,473	1,288	635	..	..	5,289	..	4,046
293	Domestic appliances, nec	15	884	877	755	1,376	..	..	19,294	..	3,976
30	OFFICE, ACCOUNTING & COMPUTING MACHINERY	0	271	1,442	2,432	918	..	..	4,720	..	5,080
31	ELECTRICAL MACHINERY & APPARATUS, NEC	101	5,246	4,542	5,940	5,850	..	..	133,667	..	22,160
311	Electric motors, generators and transformers	99	551	554	874	374	..	..	11,059	..	2,492
312	Electricity distribution and control apparatus	0	474	1,028	727	919	..	..	42,064	..	3,299
313	Insulated wire and cable	0	2,028	738	1,432	1,583	..	..	56,571	..	5,985
314	Accumulators, primary cells & primary batteries	0	323	85	1,196	723	..	..	2,828	..	2,337
315	Electric lamps and lighting equipment	0	935	108	229	382	..	..	3,160	..	1,665
319	Other electrical equipment, nec	2	935	2,029	1,482	1,869	..	..	17,985	..	6,382
32	RADIO, TV & COMMUNICATION EQUIP. & APP.	3	8,353	6,088	28,237	12,007	..	..	270,127	..	55,660
321	Electronic valves, tubes, other electronic components	3	4,385	3,917	15,486	9,128	..	..	170,790	..	33,534
322	TV & radio transmitters, apparatus for line telephony	0	57	206	377	472	..	..	845	..	1,115
323	TV & radio receivers, recording apparatus	0	3,911	1,965	12,374	2,407	..	..	98,492	..	21,012
33	MEDICAL PRECISION &OPTICAL INSTRUMENTS	1	1,346	1,254	1,816	893	..	..	21,645	..	5,388
331	Medical appliances and instruments	1	1,180	773	1,219	490	..	..	4,220	..	3,678
3311	Medical, surgical equipment & orthopaedic app.	0	830	121	900	262	..	..	1,050	..	2,117
3312	Instruments & appliances for measuring, checking etc	1	343	632	295	191	..	..	3,170	..	1,473
3313	Industrial process control equipment	0	7	20	24	37	..	..	0	..	88
332	Optical instruments and photographic equipment	0	145	281	429	209	..	..	16,722	..	1,124
333	Watches and clocks	0	21	200	168	194	..	..	703	..	586
34	MOTOR VEHICLES, TRAILERS & SEMI-TRAILERS	12,604	16,851	14,589	25,740	21,669	..	..	308,385	..	92,563
341	Motor vehicles	8,620	4,725	5,939	10,207	9,723	..	..	24,811	..	39,303
342	Bodies (coachwork) for motor vehicles	0	1,301	1,119	1,427	1,940	..	..	664	..	5,789
343	Parts, accessories for motor vehicles & their engines	3,984	10,825	7,531	14,106	10,006	..	..	282,910	..	47,470
35	OTHER TRANSPORT EQUIPMENT	3	519	1,721	2,982	1,082	..	..	55,570	..	6,507
351	Building and repairing of ships and boats	1	282	654	2,053	694	..	..	44,371	..	3,844
3511	Building and repairing of ships	0	280	645	1,981	694	..	..	44,371	..	3,760
3512	Building, repairing of pleasure & sporting boats	0	2	9	72	0	..	..	0	..	83
352	Railway, tramway locomotives & rolling stock	2	47	107	99	70	..	..	0	..	325
353	Aircraft and spacecraft	0	17	815	770	160	..	..	11,199	..	1,802
359	Transport equipment, nec	0	173	145	60	158	..	..	0	..	536
3591	Motorcycles	..	..	..	..	..	..	..	..	..	
3592	Bicycles and invalid carriages	0	162	113	50	158	..	..	0	..	483
3599	Other transport equipment, nec	0	11	32	10	0	..	..	0	..	53
36	FURNITURE; MANUFACTURING, NEC	345	3,006	1,470	2,870	763	..	..	34,518	..	8,578
361	Furniture	0	892	584	866	531	..	..	14,518	..	2,925
369	Manufacturing, nec	345	2,114	886	2,004	232	..	..	20,000	..	5,653
3691	Jewellery and related articles	0	12	12	13	22	..	..	0	..	59
3692	Musical instruments	45	108	19	299	10	..	..	11,407	..	522
3693	Sports goods	0	90	53	187	7	..	..	3,462	..	349
3694	Games and toys	0	8	15	5	30	..	..	0	..	58
3699	Other manufacturing, nec	300	1,896	787	1,500	163	..	..	5,131	..	4,664
37	RECYCLING	89	154	243	128	1	..	..	3,285	..	627
371	Recycling of metal waste and scrap	89	154	243	128	1	..	..	3,285	..	627
372	Recycling of non-metal waste and scrap	..	..	..	..	..	..	..	..	..	
CERR	Non-specified industry	..	..	..	..	..	..	..	..	..	
15-37	TOTAL MANUFACTURING	3,947,892	394,958	698,332	1,128,194	756,141	..	..	14,172,008	..	6,976,536

ISIC Revision 3 Industry Sector		Solid TJ	LPG TJ	Distiloil TJ	RFO TJ	Gas TJ	Biomass TJ	Steam TJ	Electr MWh	Own Use MWh	TOTAL TJ
15	**FOOD PRODUCTS AND BEVERAGES**	..	..	..	..	..	..	..	..	..	61,320
151	Production, processing and preserving (PPP)	..	..	..	..	..	..	..	..	..	..
1511	PPP of meat and meat products	..	..	..	..	..	..	..	..	..	..
1512	Processing and preserving of fish products	..	..	..	..	..	..	..	..	..	..
1513	Processing, preserving of fruit & vegetables	..	..	..	..	..	..	..	..	..	..
1514	Vegetable and animal oils and fats	..	..	..	..	..	..	..	..	..	..
152	Dairy products	..	..	..	..	..	..	..	..	..	..
153	Grain mill prod., starches & prepared animal feeds	..	..	..	..	..	..	..	..	..	..
1531	Grain mill products	..	..	..	..	..	..	..	..	..	..
1532	Starches and starch products	..	..	..	..	..	..	..	..	..	..
1533	Prepared animal feeds	..	..	..	..	..	..	..	..	..	..
154	Other food products	..	..	..	..	..	..	..	..	..	..
1541	Bakery products	..	..	..	..	..	..	..	..	..	..
1542	Sugar	..	..	..	..	..	..	..	..	..	..
1543	Cocoa, chocolate and sugar confectionery	..	..	..	..	..	..	..	..	..	..
1544	Macaroni, noodles, couscous & similar farinaceous prod.	..	..	..	..	..	..	..	..	..	..
1549	Other food products, nec	..	..	..	..	..	..	..	..	..	..
155	Beverages	..	..	..	..	..	..	..	..	..	..
1551	Distilling, rectifying and blending of spirits	..	..	..	..	..	..	..	..	..	..
1552	Wines	..	..	..	..	..	..	..	..	..	..
1553	Malt liquors and malt	..	..	..	..	..	..	..	..	..	..
1554	Soft drinks; production of mineral waters	..	..	..	..	..	..	..	..	..	..
16	**TOBACCO PRODUCTS**	..	..	..	..	..	..	..	..	..	..
17	**TEXTILES**	..	..	..	..	..	..	..	..	..	..
171	Spinning, weaving and finishing of textiles	..	..	..	..	..	..	..	..	..	..
1711	Preparation and spinning of textile fibres	..	..	..	..	..	..	..	..	..	..
1712	Finishing of textiles	..	..	..	..	..	..	..	..	..	..
172	Other textiles	..	..	..	..	..	..	..	..	..	..
1721	Made-up textile articles, except apparel	..	..	..	..	..	..	..	..	..	..
1722	Carpets and rugs	..	..	..	..	..	..	..	..	..	..
1723	Cordage, rope, twine and netting	..	..	..	..	..	..	..	..	..	..
1729	Other textiles, nec	..	..	..	..	..	..	..	..	..	..
173	Knitted and crocheted fabrics and articles	..	..	..	..	..	..	..	..	..	..
18	**WEARING APPAREL, DRESSING & DYEING OF FUR**	..	..	..	..	..	..	..	..	..	..
181	Wearing apparel, except fur apparel	..	..	..	..	..	..	..	..	..	..
182	Dressing and dyeing of fur; articles of fur	..	..	..	..	..	..	..	..	..	..
19	**TANNING & DRESSING OF LEATHER, FOOTWEAR**	..	..	..	..	..	..	..	..	..	..
191	Tanning and dressing of leather	..	..	..	..	..	..	..	..	..	..
1911	Tanning and dressing of leather	..	..	..	..	..	..	..	..	..	..
1912	Luggage, handbags, saddlery & harness	..	..	..	..	..	..	..	..	..	..
192	Footwear	..	..	..	..	..	..	..	..	..	..
20	**WOOD AND WOOD PRODUCTS**	..	..	..	..	..	..	..	..	..	..
201	Sawmilling and planing of wood	..	..	..	..	..	..	..	..	..	..
202	Products of wood, cork, straw & plaiting materials	..	..	..	..	..	..	..	..	..	..
2021	Veneer sheets	..	..	..	..	..	..	..	..	..	..
2022	Builders' carpentry and joinery	..	..	..	..	..	..	..	..	..	..
2023	Wooden containers	..	..	..	..	..	..	..	..	..	..
2029	Other products of wood	..	..	..	..	..	..	..	..	..	..
21	**PAPER AND PAPER PRODUCTS**	..	..	..	..	..	..	..	..	..	..
2101	Pulp, paper and paperboard	..	..	..	..	..	..	..	..	..	..
2102	Corrugated paper, paperboard and their containers	..	..	..	..	..	..	..	..	..	..
2109	Other articles of pulp and paperboard	..	..	..	..	..	..	..	..	..	..
22	**PUBLISHING, PRINTING & REPRODUCTION**	..	..	..	..	..	..	..	..	..	..
221	Publishing	..	..	..	..	..	..	..	..	..	..
2211	Publishing of books & brochures	..	..	..	..	..	..	..	..	..	..
2212	Publishing of newspapers and periodicals	..	..	..	..	..	..	..	..	..	..
2213	Publishing of recorded media	..	..	..	..	..	..	..	..	..	..
2219	Other publishing	..	..	..	..	..	..	..	..	..	..
222	Printing and related service activities	..	..	..	..	..	..	..	..	..	..
2221	Printing	..	..	..	..	..	..	..	..	..	..
2222	Service activities related to printing	..	..	..	..	..	..	..	..	..	..
223	Reproduction of recorded media	..	..	..	..	..	..	..	..	..	..
23	**COKE, REFINED PETROLEUM PRODUCTS**	..	..	..	..	..	..	..	..	..	..
231	Coke oven products	..	..	..	..	..	..	..	..	..	..
232	Refined petroleum products	..	..	..	..	..	..	..	..	..	..
233	Processing of nuclear fuel	..	..	..	..	..	..	..	..	..	..
24	**CHEMICALS & CHEMICAL PRODUCTS**	..	..	..	..	..	..	..	..	..	2,317
241	Basic chemicals	..	..	..	..	..	..	..	..	..	..
2411	Basic chemicals, exc. fertilizers & nitrogen compounds	..	..	..	..	..	..	..	..	..	..
2412	Fertilizers and nitrogen compounds	..	..	..	..	..	..	..	..	..	..
2413	Plastics in primary forms and synthetic rubber	..	..	..	..	..	..	..	..	..	..
242	Other chemical products	..	..	..	..	..	..	..	..	..	..
2421	Pesticides and other agro-chemical products	..	..	..	..	..	..	..	..	..	..
2422	Paints, varnishes and similar coatings	..	..	..	..	..	..	..	..	..	..
2423	Pharmaceuticals, medicinal chem. & botanical prod.	..	..	..	..	..	..	..	..	..	..
2424	Soap and detergents, perfumes etc.	..	..	..	..	..	..	..	..	..	..
2429	Other chemical products, nec	..	..	..	..	..	..	..	..	..	..
243	Man-made fibres	..	..	..	..	..	..	..	..	..	..
25	**RUBBER AND PLASTICS PRODUCTS**	..	..	..	..	..	..	..	..	..	..
251	Rubber products	..	..	..	..	..	..	..	..	..	..
2511	Rubber tyres and tubes	..	..	..	..	..	..	..	..	..	..
2519	Other rubber products	..	..	..	..	..	..	..	..	..	..
252	Plastic products	..	..	..	..	..	..	..	..	..	..
26	**OTHER NON-METALLIC MINERAL PRODUCTS**	..	..	..	..	..	..	..	..	..	5,074
261	Glass and glass products	..	..	..	..	..	..	..	..	..	..
269	Non-metallic mineral products, nec	..	..	..	..	..	..	..	..	..	..
2691	Pottery, china and earthenware	..	..	..	..	..	..	..	..	..	..
2692	Refractory ceramic products	..	..	..	..	..	..	..	..	..	..
2693	Structural non-refractory clay & ceramic prod.	..	..	..	..	..	..	..	..	..	..
2694	Cement, lime and plaster	..	..	..	..	..	..	..	..	..	..
2695	Articles of concrete, cement and plaster	..	..	..	..	..	..	..	..	..	..
2696	Cutting, shaping and finishing of stone	..	..	..	..	..	..	..	..	..	..
2699	Other non-metallic mineral products, nec	..	..	..	..	..	..	..	..	..	..

ISIC Revision 3 Industry Sector		Solid TJ	LPG TJ	Distiloil TJ	RFO TJ	Gas TJ	Biomass TJ	Steam TJ	Electr MWh	Own Use MWh	TOTAL TJ
27	BASIC METALS	..	..	..	..	..	..	..	..	..	7,858
271	Basic iron and steel	..	..	..	..	..	..	..	..	..	..
272	Basic precious and non-ferrous metals	..	..	..	..	..	..	..	..	..	..
273	Casting of metals	..	..	..	..	..	..	..	..	..	..
2731	Casting of iron and steel	..	..	..	..	..	..	..	..	..	..
2732	Casting non-ferrous metals	..	..	..	..	..	..	..	..	..	..
28	FABRICATED METAL PRODUCTS	..	..	..	..	..	..	..	..	..	286
281	Str. metal prod., tanks, reservoirs, steam generators	..	..	..	..	..	..	..	..	..	..
2811	Structural metal products	..	..	..	..	..	..	..	..	..	..
2812	Tanks, reservoirs and containers of metal	..	..	..	..	..	..	..	..	..	..
2813	Steam generators, exc. central heating hot water boilers	..	..	..	..	..	..	..	..	..	..
289	Other fabricated metal products	..	..	..	..	..	..	..	..	..	..
2891	Forging, pressing, stamping & roll-forming of metal	..	..	..	..	..	..	..	..	..	..
2892	Treatment and coating of metals	..	..	..	..	..	..	..	..	..	..
2893	Cutlery, hand tools and general hardware	..	..	..	..	..	..	..	..	..	..
2899	Other fabricated metal products, nec	..	..	..	..	..	..	..	..	..	..
29	MACHINERY AND EQUIPMENT, NEC	..	..	..	..	..	..	..	..	..	..
291	General purpose machinery	..	..	..	..	..	..	..	..	..	..
2911	Engines and turbines	..	..	..	..	..	..	..	..	..	..
2912	Pumps, compressors, taps and valves	..	..	..	..	..	..	..	..	..	..
2913	Bearings, gears, gearing and driving elements	..	..	..	..	..	..	..	..	..	..
2914	Ovens, furnaces and furnace burners	..	..	..	..	..	..	..	..	..	..
2915	Lifting and handling equipment	..	..	..	..	..	..	..	..	..	..
2919	Other general purpose machinery	..	..	..	..	..	..	..	..	..	..
292	Special purpose machinery	..	..	..	..	..	..	..	..	..	..
2921	Agricultural and forestry machinery	..	..	..	..	..	..	..	..	..	..
2922	Machine-tools	..	..	..	..	..	..	..	..	..	..
2923	Machinery for metallurgy	..	..	..	..	..	..	..	..	..	..
2924	Machinery for mining, quarrying and construction	..	..	..	..	..	..	..	..	..	..
2925	Machinery for food, beverage & tobacco processing	..	..	..	..	..	..	..	..	..	..
2926	Machinery for textile, apparel & leather production	..	..	..	..	..	..	..	..	..	..
2927	Machinery for weapons and ammunition	..	..	..	..	..	..	..	..	..	..
2929	Other special purpose machinery	..	..	..	..	..	..	..	..	..	..
293	Domestic appliances, nec	..	..	..	..	..	..	..	..	..	..
30	OFFICE, ACCOUNTING & COMPUTING MACHINERY	..	..	..	..	..	..	..	..	..	..
31	ELECTRICAL MACHINERY & APPARATUS, NEC	..	..	..	..	..	..	..	..	..	..
311	Electric motors, generators and transformers	..	..	..	..	..	..	..	..	..	..
312	Electricity distribution and control apparatus	..	..	..	..	..	..	..	..	..	..
313	Insulated wire and cable	..	..	..	..	..	..	..	..	..	..
314	Accumulators, primary cells & primary batteries	..	..	..	..	..	..	..	..	..	..
315	Electric lamps and lighting equipment	..	..	..	..	..	..	..	..	..	..
319	Other electrical equipment, nec	..	..	..	..	..	..	..	..	..	..
32	RADIO, TV & COMMUNICATION EQUIP. & APP.	..	..	..	..	..	..	..	..	..	..
321	Electronic valves, tubes, other electronic components	..	..	..	..	..	..	..	..	..	..
322	TV & radio transmitters, apparatus for line telephony	..	..	..	..	..	..	..	..	..	..
323	TV & radio receivers, recording apparatus	..	..	..	..	..	..	..	..	..	..
33	MEDICAL PRECISION &OPTICAL INSTRUMENTS	..	..	..	..	..	..	..	..	..	..
331	Medical appliances and instruments	..	..	..	..	..	..	..	..	..	..
3311	Medical, surgical equipment & orthopaedic app.	..	..	..	..	..	..	..	..	..	..
3312	Instruments & appliances for measuring, checking etc	..	..	..	..	..	..	..	..	..	..
3313	Industrial process control equipment	..	..	..	..	..	..	..	..	..	..
332	Optical instruments and photographic equipment	..	..	..	..	..	..	..	..	..	..
333	Watches and clocks	..	..	..	..	..	..	..	..	..	..
34	MOTOR VEHICLES, TRAILERS & SEMI-TRAILERS	..	..	..	..	..	..	..	..	..	..
341	Motor vehicles	..	..	..	..	..	..	..	..	..	..
342	Bodies (coachwork) for motor vehicles	..	..	..	..	..	..	..	..	..	..
343	Parts, accessories for motor vehicles & their engines	..	..	..	..	..	..	..	..	..	..
35	OTHER TRANSPORT EQUIPMENT	..	..	..	..	..	..	..	..	..	..
351	Building and repairing of ships and boats	..	..	..	..	..	..	..	..	..	..
3511	Building and repairing of ships	..	..	..	..	..	..	..	..	..	..
3512	Building, repairing of pleasure & sporting boats	..	..	..	..	..	..	..	..	..	..
352	Railway, tramway locomotives & rolling stock	..	..	..	..	..	..	..	..	..	..
353	Aircraft and spacecraft	..	..	..	..	..	..	..	..	..	..
359	Transport equipment, nec	..	..	..	..	..	..	..	..	..	..
3591	Motorcycles	..	..	..	..	..	..	..	..	..	..
3592	Bicycles and invalid carriages	..	..	..	..	..	..	..	..	..	..
3599	Other transport equipment, nec	..	..	..	..	..	..	..	..	..	..
36	FURNITURE; MANUFACTURING, NEC	..	..	..	..	..	..	..	..	..	..
361	Furniture	..	..	..	..	..	..	..	..	..	..
369	Manufacturing, nec	..	..	..	..	..	..	..	..	..	..
3691	Jewellery and related articles	..	..	..	..	..	..	..	..	..	..
3692	Musical instruments	..	..	..	..	..	..	..	..	..	..
3693	Sports goods	..	..	..	..	..	..	..	..	..	..
3694	Games and toys	..	..	..	..	..	..	..	..	..	..
3699	Other manufacturing, nec	..	..	..	..	..	..	..	..	..	..
37	RECYCLING	..	..	..	..	..	..	..	..	..	..
371	Recycling of metal waste and scrap	..	..	..	..	..	..	..	..	..	..
372	Recycling of non-metal waste and scrap	..	..	..	..	..	..	..	..	..	..
CERR	Non-specified industry	..	..	..	..	..	..	..	..	..	..
15-37	TOTAL MANUFACTURING	..	..	..	..	..	..	..	..	..	76,855

ISIC Revision 3 Industry Sector		Solid TJ	LPG TJ	Distiloil TJ	RFO TJ	Gas TJ	Biomass TJ	Steam TJ	Electr MWh	Own Use MWh	TOTAL TJ
15	FOOD PRODUCTS AND BEVERAGES	..	..	..	..	..	..	..	..	..	58,532
151	Production, processing and preserving (PPP)	..	..	..	..	..	..	..	..	..	..
1511	PPP of meat and meat products	..	..	..	..	..	..	..	..	..	..
1512	Processing and preserving of fish products	..	..	..	..	..	..	..	..	..	..
1513	Processing, preserving of fruit & vegetables	..	..	..	..	..	..	..	..	..	..
1514	Vegetable and animal oils and fats	..	..	..	..	..	..	..	..	..	..
152	Dairy products	..	..	..	..	..	..	..	..	..	..
153	Grain mill prod., starches & prepared animal feeds	..	..	..	..	..	..	..	..	..	..
1531	Grain mill products	..	..	..	..	..	..	..	..	..	..
1532	Starches and starch products	..	..	..	..	..	..	..	..	..	..
1533	Prepared animal feeds	..	..	..	..	..	..	..	..	..	..
154	Other food products	..	..	..	..	..	..	..	..	..	..
1541	Bakery products	..	..	..	..	..	..	..	..	..	..
1542	Sugar	..	..	..	..	..	..	..	..	..	..
1543	Cocoa, chocolate and sugar confectionery	..	..	..	..	..	..	..	..	..	..
1544	Macaroni, noodles, couscous & similar farinaceous prod.	..	..	..	..	..	..	..	..	..	..
1549	Other food products, nec	..	..	..	..	..	..	..	..	..	..
155	Beverages	..	..	..	..	..	..	..	..	..	..
1551	Distilling, rectifying and blending of spirits	..	..	..	..	..	..	..	..	..	..
1552	Wines	..	..	..	..	..	..	..	..	..	..
1553	Malt liquors and malt	..	..	..	..	..	..	..	..	..	..
1554	Soft drinks; production of mineral waters	..	..	..	..	..	..	..	..	..	..
16	TOBACCO PRODUCTS	..	..	..	..	..	..	..	..	..	..
17	TEXTILES	..	..	..	..	..	..	..	..	..	..
171	Spinning, weaving and finishing of textiles	..	..	..	..	..	..	..	..	..	..
1711	Preparation and spinning of textile fibres	..	..	..	..	..	..	..	..	..	..
1712	Finishing of textiles	..	..	..	..	..	..	..	..	..	..
172	Other textiles	..	..	..	..	..	..	..	..	..	..
1721	Made-up textile articles, except apparel	..	..	..	..	..	..	..	..	..	..
1722	Carpets and rugs	..	..	..	..	..	..	..	..	..	..
1723	Cordage, rope, twine and netting	..	..	..	..	..	..	..	..	..	..
1729	Other textiles, nec	..	..	..	..	..	..	..	..	..	..
173	Knitted and crocheted fabrics and articles	..	..	..	..	..	..	..	..	..	..
18	WEARING APPAREL, DRESSING & DYEING OF FUR	..	..	..	..	..	..	..	..	..	..
181	Wearing apparel, except fur apparel	..	..	..	..	..	..	..	..	..	..
182	Dressing and dyeing of fur; articles of fur	..	..	..	..	..	..	..	..	..	..
19	TANNING & DRESSING OF LEATHER, FOOTWEAR	..	..	..	..	..	..	..	..	..	..
191	Tanning and dressing of leather	..	..	..	..	..	..	..	..	..	..
1911	Tanning and dressing of leather	..	..	..	..	..	..	..	..	..	..
1912	Luggage, handbags, saddlery & harness	..	..	..	..	..	..	..	..	..	..
192	Footwear	..	..	..	..	..	..	..	..	..	..
20	WOOD AND WOOD PRODUCTS	..	..	..	..	..	..	..	..	..	..
201	Sawmilling and planing of wood	..	..	..	..	..	..	..	..	..	..
202	Products of wood, cork, straw & plaiting materials	..	..	..	..	..	..	..	..	..	..
2021	Veneer sheets	..	..	..	..	..	..	..	..	..	..
2022	Builders' carpentry and joinery	..	..	..	..	..	..	..	..	..	..
2023	Wooden containers	..	..	..	..	..	..	..	..	..	..
2029	Other products of wood	..	..	..	..	..	..	..	..	..	..
21	PAPER AND PAPER PRODUCTS	..	..	..	..	..	..	..	..	..	..
2101	Pulp, paper and paperboard	..	..	..	..	..	..	..	..	..	..
2102	Corrugated paper, paperboard and their containers	..	..	..	..	..	..	..	..	..	..
2109	Other articles of pulp and paperboard	..	..	..	..	..	..	..	..	..	..
22	PUBLISHING, PRINTING & REPRODUCTION	..	..	..	..	..	..	..	..	..	..
221	Publishing	..	..	..	..	..	..	..	..	..	..
2211	Publishing of books & brochures	..	..	..	..	..	..	..	..	..	..
2212	Publishing of newspapers and periodicals	..	..	..	..	..	..	..	..	..	..
2213	Publishing of recorded media	..	..	..	..	..	..	..	..	..	..
2219	Other publishing	..	..	..	..	..	..	..	..	..	..
222	Printing and related service activities	..	..	..	..	..	..	..	..	..	..
2221	Printing	..	..	..	..	..	..	..	..	..	..
2222	Service activities related to printing	..	..	..	..	..	..	..	..	..	..
223	Reproduction of recorded media	..	..	..	..	..	..	..	..	..	..
23	COKE, REFINED PETROLEUM PRODUCTS	..	..	..	..	..	..	..	..	..	..
231	Coke oven products	..	..	..	..	..	..	..	..	..	..
232	Refined petroleum products	..	..	..	..	..	..	..	..	..	..
233	Processing of nuclear fuel	..	..	..	..	..	..	..	..	..	..
24	CHEMICALS & CHEMICAL PRODUCTS	..	..	..	..	..	..	..	..	..	2,521
241	Basic chemicals	..	..	..	..	..	..	..	..	..	..
2411	Basic chemicals, exc. fertilizers & nitrogen compounds	..	..	..	..	..	..	..	..	..	..
2412	Fertilizers and nitrogen compounds	..	..	..	..	..	..	..	..	..	..
2413	Plastics in primary forms and synthetic rubber	..	..	..	..	..	..	..	..	..	..
242	Other chemical products	..	..	..	..	..	..	..	..	..	..
2421	Pesticides and other agro-chemical products	..	..	..	..	..	..	..	..	..	..
2422	Paints, varnishes and similar coatings	..	..	..	..	..	..	..	..	..	..
2423	Pharmaceuticals, medicinal chem. & botanical prod.	..	..	..	..	..	..	..	..	..	..
2424	Soap and detergents, perfumes etc.	..	..	..	..	..	..	..	..	..	..
2429	Other chemical products, nec	..	..	..	..	..	..	..	..	..	..
243	Man-made fibres	..	..	..	..	..	..	..	..	..	..
25	RUBBER AND PLASTICS PRODUCTS	..	..	..	..	..	..	..	..	..	..
251	Rubber products	..	..	..	..	..	..	..	..	..	..
2511	Rubber tyres and tubes	..	..	..	..	..	..	..	..	..	..
2519	Other rubber products	..	..	..	..	..	..	..	..	..	..
252	Plastic products	..	..	..	..	..	..	..	..	..	..
26	OTHER NON-METALLIC MINERAL PRODUCTS	..	..	..	..	..	..	..	..	..	5,334
261	Glass and glass products	..	..	..	..	..	..	..	..	..	..
269	Non-metallic mineral products, nec	..	..	..	..	..	..	..	..	..	..
2691	Pottery, china and earthenware	..	..	..	..	..	..	..	..	..	..
2692	Refractory ceramic products	..	..	..	..	..	..	..	..	..	..
2693	Structural non-refractory clay & ceramic prod.	..	..	..	..	..	..	..	..	..	..
2694	Cement, lime and plaster	..	..	..	..	..	..	..	..	..	..
2695	Articles of concrete, cement and plaster	..	..	..	..	..	..	..	..	..	..
2696	Cutting, shaping and finishing of stone	..	..	..	..	..	..	..	..	..	..
2699	Other non-metallic mineral products, nec	..	..	..	..	..	..	..	..	..	..

ISIC Revision 3 Industry Sector		Solid TJ	LPG TJ	Distiloil TJ	RFO TJ	Gas TJ	Biomass TJ	Steam TJ	Electr MWh	Own Use MWh	TOTAL TJ
27	**BASIC METALS**	..	..	..	..	..	..	..	..	..	7,297
271	Basic iron and steel	..	..	..	..	..	..	..	..	..	..
272	Basic precious and non-ferrous metals	..	..	..	..	..	..	..	..	..	..
273	Casting of metals	..	..	..	..	..	..	..	..	..	..
2731	Casting of iron and steel	..	..	..	..	..	..	..	..	..	..
2732	Casting non-ferrous metals	..	..	..	..	..	..	..	..	..	..
28	**FABRICATED METAL PRODUCTS**	..	..	..	..	..	..	..	..	..	281
281	Str. metal prod., tanks, reservoirs, steam generators	..	..	..	..	..	..	..	..	..	..
2811	Structural metal products	..	..	..	..	..	..	..	..	..	..
2812	Tanks, reservoirs and containers of metal	..	..	..	..	..	..	..	..	..	..
2813	Steam generators, exc. central heating hot water boilers	..	..	..	..	..	..	..	..	..	..
289	Other fabricated metal products	..	..	..	..	..	..	..	..	..	..
2891	Forging, pressing, stamping & roll-forming of metal	..	..	..	..	..	..	..	..	..	..
2892	Treatment and coating of metals	..	..	..	..	..	..	..	..	..	..
2893	Cutlery, hand tools and general hardware	..	..	..	..	..	..	..	..	..	..
2899	Other fabricated metal products, nec	..	..	..	..	..	..	..	..	..	..
29	**MACHINERY AND EQUIPMENT, NEC**	..	..	..	..	..	..	..	..	..	..
291	General purpose machinery	..	..	..	..	..	..	..	..	..	..
2911	Engines and turbines	..	..	..	..	..	..	..	..	..	..
2912	Pumps, compressors, taps and valves	..	..	..	..	..	..	..	..	..	..
2913	Bearings, gears, gearing and driving elements	..	..	..	..	..	..	..	..	..	..
2914	Ovens, furnaces and furnace burners	..	..	..	..	..	..	..	..	..	..
2915	Lifting and handling equipment	..	..	..	..	..	..	..	..	..	..
2919	Other general purpose machinery	..	..	..	..	..	..	..	..	..	..
292	Special purpose machinery	..	..	..	..	..	..	..	..	..	..
2921	Agricultural and forestry machinery	..	..	..	..	..	..	..	..	..	..
2922	Machine-tools	..	..	..	..	..	..	..	..	..	..
2923	Machinery for metallurgy	..	..	..	..	..	..	..	..	..	..
2924	Machinery for mining, quarrying and construction	..	..	..	..	..	..	..	..	..	..
2925	Machinery for food, beverage & tobacco processing	..	..	..	..	..	..	..	..	..	..
2926	Machinery for textile, apparel & leather production	..	..	..	..	..	..	..	..	..	..
2927	Machinery for weapons and ammunition	..	..	..	..	..	..	..	..	..	..
2929	Other special purpose machinery	..	..	..	..	..	..	..	..	..	..
293	Domestic appliances, nec	..	..	..	..	..	..	..	..	..	..
30	**OFFICE, ACCOUNTING & COMPUTING MACHINERY**	..	..	..	..	..	..	..	..	..	..
31	**ELECTRICAL MACHINERY & APPARATUS, NEC**	..	..	..	..	..	..	..	..	..	..
311	Electric motors, generators and transformers	..	..	..	..	..	..	..	..	..	..
312	Electricity distribution and control apparatus	..	..	..	..	..	..	..	..	..	..
313	Insulated wire and cable	..	..	..	..	..	..	..	..	..	..
314	Accumulators, primary cells & primary batteries	..	..	..	..	..	..	..	..	..	..
315	Electric lamps and lighting equipment	..	..	..	..	..	..	..	..	..	..
319	Other electrical equipment, nec	..	..	..	..	..	..	..	..	..	..
32	**RADIO, TV & COMMUNICATION EQUIP. & APP.**	..	..	..	..	..	..	..	..	..	..
321	Electronic valves, tubes, other electronic components	..	..	..	..	..	..	..	..	..	..
322	TV & radio transmitters, apparatus for line telephony	..	..	..	..	..	..	..	..	..	..
323	TV & radio receivers, recording apparatus	..	..	..	..	..	..	..	..	..	..
33	**MEDICAL PRECISION &OPTICAL INSTRUMENTS**	..	..	..	..	..	..	..	..	..	..
331	Medical appliances and instruments	..	..	..	..	..	..	..	..	..	..
3311	Medical, surgical equipment & orthopaedic app.	..	..	..	..	..	..	..	..	..	..
3312	Instruments & appliances for measuring, checking etc	..	..	..	..	..	..	..	..	..	..
3313	Industrial process control equipment	..	..	..	..	..	..	..	..	..	..
332	Optical instruments and photographic equipment	..	..	..	..	..	..	..	..	..	..
333	Watches and clocks	..	..	..	..	..	..	..	..	..	..
34	**MOTOR VEHICLES, TRAILERS & SEMI-TRAILERS**	..	..	..	..	..	..	..	..	..	..
341	Motor vehicles	..	..	..	..	..	..	..	..	..	..
342	Bodies (coachwork) for motor vehicles	..	..	..	..	..	..	..	..	..	..
343	Parts, accessories for motor vehicles & their engines	..	..	..	..	..	..	..	..	..	..
35	**OTHER TRANSPORT EQUIPMENT**	..	..	..	..	..	..	..	..	..	..
351	Building and repairing of ships and boats	..	..	..	..	..	..	..	..	..	..
3511	Building and repairing of ships	..	..	..	..	..	..	..	..	..	..
3512	Building, repairing of pleasure & sporting boats	..	..	..	..	..	..	..	..	..	..
352	Railway, tramway locomotives & rolling stock	..	..	..	..	..	..	..	..	..	..
353	Aircraft and spacecraft	..	..	..	..	..	..	..	..	..	..
359	Transport equipment, nec	..	..	..	..	..	..	..	..	..	..
3591	Motorcycles	..	..	..	..	..	..	..	..	..	..
3592	Bicycles and invalid carriages	..	..	..	..	..	..	..	..	..	..
3599	Other transport equipment, nec	..	..	..	..	..	..	..	..	..	..
36	**FURNITURE; MANUFACTURING, NEC**	..	..	..	..	..	..	..	..	..	..
361	Furniture	..	..	..	..	..	..	..	..	..	..
369	Manufacturing, nec	..	..	..	..	..	..	..	..	..	..
3691	Jewellery and related articles	..	..	..	..	..	..	..	..	..	..
3692	Musical instruments	..	..	..	..	..	..	..	..	..	..
3693	Sports goods	..	..	..	..	..	..	..	..	..	..
3694	Games and toys	..	..	..	..	..	..	..	..	..	..
3699	Other manufacturing, nec	..	..	..	..	..	..	..	..	..	..
37	**RECYCLING**	..	..	..	..	..	..	..	..	..	..
371	Recycling of metal waste and scrap	..	..	..	..	..	..	..	..	..	..
372	Recycling of non-metal waste and scrap	..	..	..	..	..	..	..	..	..	..
CERR	**Non-specified industry**	..	..	..	..	..	..	..	..	..	..
15-37	**TOTAL MANUFACTURING**	..	..	..	..	..	..	..	..	..	73,965

ISIC Revision 3 Industry Sector		Solid TJ	LPG TJ	Distiloil TJ	RFO TJ	Gas TJ	Biomass TJ	Steam TJ	Electr MWh	Own Use MWh	TOTAL TJ
15	**FOOD PRODUCTS AND BEVERAGES**	..	..	..	..	..	..	..	..	..	54,735
151	Production, processing and preserving (PPP)	..	..	..	..	..	..	..	..	..	..
1511	PPP of meat and meat products	..	..	..	..	..	..	..	..	..	..
1512	Processing and preserving of fish products	..	..	..	..	..	..	..	..	..	..
1513	Processing, preserving of fruit & vegetables	..	..	..	..	..	..	..	..	..	..
1514	Vegetable and animal oils and fats	..	..	..	..	..	..	..	..	..	..
152	Dairy products	..	..	..	..	..	..	..	..	..	..
153	Grain mill prod., starches & prepared animal feeds	..	..	..	..	..	..	..	..	..	..
1531	Grain mill products	..	..	..	..	..	..	..	..	..	..
1532	Starches and starch products	..	..	..	..	..	..	..	..	..	..
1533	Prepared animal feeds	..	..	..	..	..	..	..	..	..	..
154	Other food products	..	..	..	..	..	..	..	..	..	..
1541	Bakery products	..	..	..	..	..	..	..	..	..	..
1542	Sugar	..	..	..	..	..	..	..	..	..	..
1543	Cocoa, chocolate and sugar confectionery	..	..	..	..	..	..	..	..	..	..
1544	Macaroni, noodles, couscous & similar farinaceous prod.	..	..	..	..	..	..	..	..	..	..
1549	Other food products, nec	..	..	..	..	..	..	..	..	..	..
155	Beverages	..	..	..	..	..	..	..	..	..	..
1551	Distilling, rectifying and blending of spirits	..	..	..	..	..	..	..	..	..	..
1552	Wines	..	..	..	..	..	..	..	..	..	..
1553	Malt liquors and malt	..	..	..	..	..	..	..	..	..	..
1554	Soft drinks; production of mineral waters	..	..	..	..	..	..	..	..	..	..
16	**TOBACCO PRODUCTS**	..	..	..	..	..	..	..	..	..	..
17	**TEXTILES**	..	..	..	..	..	..	..	..	..	..
171	Spinning, weaving and finishing of textiles	..	..	..	..	..	..	..	..	..	..
1711	Preparation and spinning of textile fibres	..	..	..	..	..	..	..	..	..	..
1712	Finishing of textiles	..	..	..	..	..	..	..	..	..	..
172	Other textiles	..	..	..	..	..	..	..	..	..	..
1721	Made-up textile articles, except apparel	..	..	..	..	..	..	..	..	..	..
1722	Carpets and rugs	..	..	..	..	..	..	..	..	..	..
1723	Cordage, rope, twine and netting	..	..	..	..	..	..	..	..	..	..
1729	Other textiles, nec	..	..	..	..	..	..	..	..	..	..
173	Knitted and crocheted fabrics and articles	..	..	..	..	..	..	..	..	..	..
18	**WEARING APPAREL, DRESSING & DYEING OF FUR**	..	..	..	..	..	..	..	..	..	..
181	Wearing apparel, except fur apparel	..	..	..	..	..	..	..	..	..	..
182	Dressing and dyeing of fur; articles of fur	..	..	..	..	..	..	..	..	..	..
19	**TANNING & DRESSING OF LEATHER, FOOTWEAR**	..	..	..	..	..	..	..	..	..	..
191	Tanning and dressing of leather	..	..	..	..	..	..	..	..	..	..
1911	Tanning and dressing of leather	..	..	..	..	..	..	..	..	..	..
1912	Luggage, handbags, saddlery & harness	..	..	..	..	..	..	..	..	..	..
192	Footwear	..	..	..	..	..	..	..	..	..	..
20	**WOOD AND WOOD PRODUCTS**	..	..	..	..	..	..	..	..	..	..
201	Sawmilling and planing of wood	..	..	..	..	..	..	..	..	..	..
202	Products of wood, cork, straw & plaiting materials	..	..	..	..	..	..	..	..	..	..
2021	Veneer sheets	..	..	..	..	..	..	..	..	..	..
2022	Builders' carpentry and joinery	..	..	..	..	..	..	..	..	..	..
2023	Wooden containers	..	..	..	..	..	..	..	..	..	..
2029	Other products of wood	..	..	..	..	..	..	..	..	..	..
21	**PAPER AND PAPER PRODUCTS**	..	..	..	..	..	..	..	..	..	..
2101	Pulp, paper and paperboard	..	..	..	..	..	..	..	..	..	..
2102	Corrugated paper, paperboard and their containers	..	..	..	..	..	..	..	..	..	..
2109	Other articles of pulp and paperboard	..	..	..	..	..	..	..	..	..	..
22	**PUBLISHING, PRINTING & REPRODUCTION**	..	..	..	..	..	..	..	..	..	..
221	Publishing	..	..	..	..	..	..	..	..	..	..
2211	Publishing of books & brochures	..	..	..	..	..	..	..	..	..	..
2212	Publishing of newspapers and periodicals	..	..	..	..	..	..	..	..	..	..
2213	Publishing of recorded media	..	..	..	..	..	..	..	..	..	..
2219	Other publishing	..	..	..	..	..	..	..	..	..	..
222	Printing and related service activities	..	..	..	..	..	..	..	..	..	..
2221	Printing	..	..	..	..	..	..	..	..	..	..
2222	Service activities related to printing	..	..	..	..	..	..	..	..	..	..
223	Reproduction of recorded media	..	..	..	..	..	..	..	..	..	..
23	**COKE, REFINED PETROLEUM PRODUCTS**	..	..	..	..	..	..	..	..	..	..
231	Coke oven products	..	..	..	..	..	..	..	..	..	..
232	Refined petroleum products	..	..	..	..	..	..	..	..	..	..
233	Processing of nuclear fuel	..	..	..	..	..	..	..	..	..	..
24	**CHEMICALS & CHEMICAL PRODUCTS**	..	..	..	..	..	..	..	..	..	2,736
241	Basic chemicals	..	..	..	..	..	..	..	..	..	..
2411	Basic chemicals, exc. fertilizers & nitrogen compounds	..	..	..	..	..	..	..	..	..	..
2412	Fertilizers and nitrogen compounds	..	..	..	..	..	..	..	..	..	..
2413	Plastics in primary forms and synthetic rubber	..	..	..	..	..	..	..	..	..	..
242	Other chemical products	..	..	..	..	..	..	..	..	..	..
2421	Pesticides and other agro-chemical products	..	..	..	..	..	..	..	..	..	..
2422	Paints, varnishes and similar coatings	..	..	..	..	..	..	..	..	..	..
2423	Pharmaceuticals, medicinal chem. & botanical prod.	..	..	..	..	..	..	..	..	..	..
2424	Soap and detergents, perfumes etc.	..	..	..	..	..	..	..	..	..	..
2429	Other chemical products, nec	..	..	..	..	..	..	..	..	..	..
243	Man-made fibres	..	..	..	..	..	..	..	..	..	..
25	**RUBBER AND PLASTICS PRODUCTS**	..	..	..	..	..	..	..	..	..	..
251	Rubber products	..	..	..	..	..	..	..	..	..	..
2511	Rubber tyres and tubes	..	..	..	..	..	..	..	..	..	..
2519	Other rubber products	..	..	..	..	..	..	..	..	..	..
252	Plastic products	..	..	..	..	..	..	..	..	..	..
26	**OTHER NON-METALLIC MINERAL PRODUCTS**	..	..	..	..	..	..	..	..	..	5,802
261	Glass and glass products	..	..	..	..	..	..	..	..	..	..
269	Non-metallic mineral products, nec	..	..	..	..	..	..	..	..	..	..
2691	Pottery, china and earthenware	..	..	..	..	..	..	..	..	..	..
2692	Refractory ceramic products	..	..	..	..	..	..	..	..	..	..
2693	Structural non-refractory clay & ceramic prod.	..	..	..	..	..	..	..	..	..	..
2694	Cement, lime and plaster	..	..	..	..	..	..	..	..	..	..
2695	Articles of concrete, cement and plaster	..	..	..	..	..	..	..	..	..	..
2696	Cutting, shaping and finishing of stone	..	..	..	..	..	..	..	..	..	..
2699	Other non-metallic mineral products, nec	..	..	..	..	..	..	..	..	..	..

ISIC Revision 3 Industry Sector		Solid TJ	LPG TJ	Distiloil TJ	RFO TJ	Gas TJ	Biomass TJ	Steam TJ	Electr MWh	Own Use MWh	TOTAL TJ
27	**BASIC METALS**	..	..	..	..	..	..	..	..	..	7,744
271	Basic iron and steel	..	..	..	..	..	..	..	..	..	..
272	Basic precious and non-ferrous metals	..	..	..	..	..	..	..	..	..	..
273	Casting of metals	..	..	..	..	..	..	..	..	..	..
2731	Casting of iron and steel	..	..	..	..	..	..	..	..	..	..
2732	Casting non-ferrous metals	..	..	..	..	..	..	..	..	..	..
28	**FABRICATED METAL PRODUCTS**	..	..	..	..	..	..	..	..	..	271
281	Str. metal prod., tanks, reservoirs, steam generators	..	..	..	..	..	..	..	..	..	..
2811	Structural metal products	..	..	..	..	..	..	..	..	..	..
2812	Tanks, reservoirs and containers of metal	..	..	..	..	..	..	..	..	..	..
2813	Steam generators, exc. central heating hot water boilers	..	..	..	..	..	..	..	..	..	..
289	Other fabricated metal products	..	..	..	..	..	..	..	..	..	..
2891	Forging, pressing, stamping & roll-forming of metal	..	..	..	..	..	..	..	..	..	..
2892	Treatment and coating of metals	..	..	..	..	..	..	..	..	..	..
2893	Cutlery, hand tools and general hardware	..	..	..	..	..	..	..	..	..	..
2899	Other fabricated metal products, nec	..	..	..	..	..	..	..	..	..	..
29	**MACHINERY AND EQUIPMENT, NEC**	..	..	..	..	..	..	..	..	..	..
291	General purpose machinery	..	..	..	..	..	..	..	..	..	..
2911	Engines and turbines	..	..	..	..	..	..	..	..	..	..
2912	Pumps, compressors, taps and valves	..	..	..	..	..	..	..	..	..	..
2913	Bearings, gears, gearing and driving elements	..	..	..	..	..	..	..	..	..	..
2914	Ovens, furnaces and furnace burners	..	..	..	..	..	..	..	..	..	..
2915	Lifting and handling equipment	..	..	..	..	..	..	..	..	..	..
2919	Other general purpose machinery	..	..	..	..	..	..	..	..	..	..
292	Special purpose machinery	..	..	..	..	..	..	..	..	..	..
2921	Agricultural and forestry machinery	..	..	..	..	..	..	..	..	..	..
2922	Machine-tools	..	..	..	..	..	..	..	..	..	..
2923	Machinery for metallurgy	..	..	..	..	..	..	..	..	..	..
2924	Machinery for mining, quarrying and construction	..	..	..	..	..	..	..	..	..	..
2925	Machinery for food, beverage & tobacco processing	..	..	..	..	..	..	..	..	..	..
2926	Machinery for textile, apparel & leather production	..	..	..	..	..	..	..	..	..	..
2927	Machinery for weapons and ammunition	..	..	..	..	..	..	..	..	..	..
2929	Other special purpose machinery	..	..	..	..	..	..	..	..	..	..
293	Domestic appliances, nec	..	..	..	..	..	..	..	..	..	..
30	**OFFICE, ACCOUNTING & COMPUTING MACHINERY**	..	..	..	..	..	..	..	..	..	..
31	**ELECTRICAL MACHINERY & APPARATUS, NEC**	..	..	..	..	..	..	..	..	..	..
311	Electric motors, generators and transformers	..	..	..	..	..	..	..	..	..	..
312	Electricity distribution and control apparatus	..	..	..	..	..	..	..	..	..	..
313	Insulated wire and cable	..	..	..	..	..	..	..	..	..	..
314	Accumulators, primary cells & primary batteries	..	..	..	..	..	..	..	..	..	..
315	Electric lamps and lighting equipment	..	..	..	..	..	..	..	..	..	..
319	Other electrical equipment, nec	..	..	..	..	..	..	..	..	..	..
32	**RADIO, TV & COMMUNICATION EQUIP. & APP.**	..	..	..	..	..	..	..	..	..	..
321	Electronic valves, tubes, other electronic components	..	..	..	..	..	..	..	..	..	..
322	TV & radio transmitters, apparatus for line telephony	..	..	..	..	..	..	..	..	..	..
323	TV & radio receivers, recording apparatus	..	..	..	..	..	..	..	..	..	..
33	**MEDICAL PRECISION &OPTICAL INSTRUMENTS**	..	..	..	..	..	..	..	..	..	..
331	Medical appliances and instruments	..	..	..	..	..	..	..	..	..	..
3311	Medical, surgical equipment & orthopaedic app.	..	..	..	..	..	..	..	..	..	..
3312	Instruments & appliances for measuring, checking etc	..	..	..	..	..	..	..	..	..	..
3313	Industrial process control equipment	..	..	..	..	..	..	..	..	..	..
332	Optical instruments and photographic equipment	..	..	..	..	..	..	..	..	..	..
333	Watches and clocks	..	..	..	..	..	..	..	..	..	..
34	**MOTOR VEHICLES, TRAILERS & SEMI-TRAILERS**	..	..	..	..	..	..	..	..	..	..
341	Motor vehicles	..	..	..	..	..	..	..	..	..	..
342	Bodies (coachwork) for motor vehicles	..	..	..	..	..	..	..	..	..	..
343	Parts, accessories for motor vehicles & their engines	..	..	..	..	..	..	..	..	..	..
35	**OTHER TRANSPORT EQUIPMENT**	..	..	..	..	..	..	..	..	..	..
351	Building and repairing of ships and boats	..	..	..	..	..	..	..	..	..	..
3511	Building and repairing of ships	..	..	..	..	..	..	..	..	..	..
3512	Building, repairing of pleasure & sporting boats	..	..	..	..	..	..	..	..	..	..
352	Railway, tramway locomotives & rolling stock	..	..	..	..	..	..	..	..	..	..
353	Aircraft and spacecraft	..	..	..	..	..	..	..	..	..	..
359	Transport equipment, nec	..	..	..	..	..	..	..	..	..	..
3591	Motorcycles	..	..	..	..	..	..	..	..	..	..
3592	Bicycles and invalid carriages	..	..	..	..	..	..	..	..	..	..
3599	Other transport equipment, nec	..	..	..	..	..	..	..	..	..	..
36	**FURNITURE; MANUFACTURING, NEC**	..	..	..	..	..	..	..	..	..	..
361	Furniture	..	..	..	..	..	..	..	..	..	..
369	Manufacturing, nec	..	..	..	..	..	..	..	..	..	..
3691	Jewellery and related articles	..	..	..	..	..	..	..	..	..	..
3692	Musical instruments	..	..	..	..	..	..	..	..	..	..
3693	Sports goods	..	..	..	..	..	..	..	..	..	..
3694	Games and toys	..	..	..	..	..	..	..	..	..	..
3699	Other manufacturing, nec	..	..	..	..	..	..	..	..	..	..
37	**RECYCLING**	..	..	..	..	..	..	..	..	..	..
371	Recycling of metal waste and scrap	..	..	..	..	..	..	..	..	..	..
372	Recycling of non-metal waste and scrap	..	..	..	..	..	..	..	..	..	..
CERR	**Non-specified industry**	..	..	..	..	..	..	..	..	..	..
15-37	**TOTAL MANUFACTURING**	..	..	..	..	..	..	..	..	..	71,288

ISIC Revision 3 Industry Sector	Solid TJ	LPG TJ	Distiloil TJ	RFO TJ	Gas TJ	Biomass TJ	Steam TJ	Electr MWh	Own Use MWh	TOTAL TJ
15 FOOD PRODUCTS AND BEVERAGES	..	..	..	..	..	..	..	..	..	57,373
151 Production, processing and preserving (PPP)	..	..	..	..	..	..	..	..	..	..
1511 PPP of meat and meat products	..	..	..	..	..	..	..	..	..	..
1512 Processing and preserving of fish products	..	..	..	..	..	..	..	..	..	..
1513 Processing, preserving of fruit & vegetables	..	..	..	..	..	..	..	..	..	..
1514 Vegetable and animal oils and fats	..	..	..	..	..	..	..	..	..	..
152 Dairy products	..	..	..	..	..	..	..	..	..	..
153 Grain mill prod., starches & prepared animal feeds	..	..	..	..	..	..	..	..	..	..
1531 Grain mill products	..	..	..	..	..	..	..	..	..	..
1532 Starches and starch products	..	..	..	..	..	..	..	..	..	..
1533 Prepared animal feeds	..	..	..	..	..	..	..	..	..	..
154 Other food products	..	..	..	..	..	..	..	..	..	..
1541 Bakery products	..	..	..	..	..	..	..	..	..	..
1542 Sugar	..	..	..	..	..	..	..	..	..	..
1543 Cocoa, chocolate and sugar confectionery	..	..	..	..	..	..	..	..	..	..
1544 Macaroni, noodles, couscous & similar farinaceous prod.	..	..	..	..	..	..	..	..	..	..
1549 Other food products, nec	..	..	..	..	..	..	..	..	..	..
155 Beverages	..	..	..	..	..	..	..	..	..	..
1551 Distilling, rectifying and blending of spirits	..	..	..	..	..	..	..	..	..	..
1552 Wines	..	..	..	..	..	..	..	..	..	..
1553 Malt liquors and malt	..	..	..	..	..	..	..	..	..	..
1554 Soft drinks; production of mineral waters	..	..	..	..	..	..	..	..	..	..
16 TOBACCO PRODUCTS	..	..	..	..	..	..	..	..	..	..
17 TEXTILES	..	..	..	..	..	..	..	..	..	..
171 Spinning, weaving and finishing of textiles	..	..	..	..	..	..	..	..	..	..
1711 Preparation and spinning of textile fibres	..	..	..	..	..	..	..	..	..	..
1712 Finishing of textiles	..	..	..	..	..	..	..	..	..	..
172 Other textiles	..	..	..	..	..	..	..	..	..	..
1721 Made-up textile articles, except apparel	..	..	..	..	..	..	..	..	..	..
1722 Carpets and rugs	..	..	..	..	..	..	..	..	..	..
1723 Cordage, rope, twine and netting	..	..	..	..	..	..	..	..	..	..
1729 Other textiles, nec	..	..	..	..	..	..	..	..	..	..
173 Knitted and crocheted fabrics and articles	..	..	..	..	..	..	..	..	..	..
18 WEARING APPAREL, DRESSING & DYEING OF FUR	..	..	..	..	..	..	..	..	..	..
181 Wearing apparel, except fur apparel	..	..	..	..	..	..	..	..	..	..
182 Dressing and dyeing of fur; articles of fur	..	..	..	..	..	..	..	..	..	..
19 TANNING & DRESSING OF LEATHER, FOOTWEAR	..	..	..	..	..	..	..	..	..	..
191 Tanning and dressing of leather	..	..	..	..	..	..	..	..	..	..
1911 Tanning and dressing of leather	..	..	..	..	..	..	..	..	..	..
1912 Luggage, handbags, saddlery & harness	..	..	..	..	..	..	..	..	..	..
192 Footwear	..	..	..	..	..	..	..	..	..	..
20 WOOD AND WOOD PRODUCTS	..	..	..	..	..	..	..	..	..	..
201 Sawmilling and planing of wood	..	..	..	..	..	..	..	..	..	..
202 Products of wood, cork, straw & plaiting materials	..	..	..	..	..	..	..	..	..	..
2021 Veneer sheets	..	..	..	..	..	..	..	..	..	..
2022 Builders' carpentry and joinery	..	..	..	..	..	..	..	..	..	..
2023 Wooden containers	..	..	..	..	..	..	..	..	..	..
2029 Other products of wood	..	..	..	..	..	..	..	..	..	..
21 PAPER AND PAPER PRODUCTS	..	..	..	..	..	..	..	..	..	..
2101 Pulp, paper and paperboard	..	..	..	..	..	..	..	..	..	..
2102 Corrugated paper, paperboard and their containers	..	..	..	..	..	..	..	..	..	..
2109 Other articles of pulp and paperboard	..	..	..	..	..	..	..	..	..	..
22 PUBLISHING, PRINTING & REPRODUCTION	..	..	..	..	..	..	..	..	..	..
221 Publishing	..	..	..	..	..	..	..	..	..	..
2211 Publishing of books & brochures	..	..	..	..	..	..	..	..	..	..
2212 Publishing of newspapers and periodicals	..	..	..	..	..	..	..	..	..	..
2213 Publishing of recorded media	..	..	..	..	..	..	..	..	..	..
2219 Other publishing	..	..	..	..	..	..	..	..	..	..
222 Printing and related service activities	..	..	..	..	..	..	..	..	..	..
2221 Printing	..	..	..	..	..	..	..	..	..	..
2222 Service activities related to printing	..	..	..	..	..	..	..	..	..	..
223 Reproduction of recorded media	..	..	..	..	..	..	..	..	..	..
23 COKE, REFINED PETROLEUM PRODUCTS	..	..	..	..	..	..	..	..	..	..
231 Coke oven products	..	..	..	..	..	..	..	..	..	..
232 Refined petroleum products	..	..	..	..	..	..	..	..	..	..
233 Processing of nuclear fuel	..	..	..	..	..	..	..	..	..	2,737
24 CHEMICALS & CHEMICAL PRODUCTS	..	..	..	..	..	..	..	..	..	..
241 Basic chemicals	..	..	..	..	..	..	..	..	..	..
2411 Basic chemicals, exc. fertilizers & nitrogen compounds	..	..	..	..	..	..	..	..	..	..
2412 Fertilizers and nitrogen compounds	..	..	..	..	..	..	..	..	..	..
2413 Plastics in primary forms and synthetic rubber	..	..	..	..	..	..	..	..	..	..
242 Other chemical products	..	..	..	..	..	..	..	..	..	..
2421 Pesticides and other agro-chemical products	..	..	..	..	..	..	..	..	..	..
2422 Paints, varnishes and similar coatings	..	..	..	..	..	..	..	..	..	..
2423 Pharmaceuticals, medicinal chem. & botanical prod.	..	..	..	..	..	..	..	..	..	..
2424 Soap and detergents, perfumes etc.	..	..	..	..	..	..	..	..	..	..
2429 Other chemical products, nec	..	..	..	..	..	..	..	..	..	..
243 Man-made fibres	..	..	..	..	..	..	..	..	..	..
25 RUBBER AND PLASTICS PRODUCTS	..	..	..	..	..	..	..	..	..	..
251 Rubber products	..	..	..	..	..	..	..	..	..	..
2511 Rubber tyres and tubes	..	..	..	..	..	..	..	..	..	..
2519 Other rubber products	..	..	..	..	..	..	..	..	..	..
252 Plastic products	..	..	..	..	..	..	..	..	..	5,745
26 OTHER NON-METALLIC MINERAL PRODUCTS	..	..	..	..	..	..	..	..	..	..
261 Glass and glass products	..	..	..	..	..	..	..	..	..	..
269 Non-metallic mineral products, nec	..	..	..	..	..	..	..	..	..	..
2691 Pottery, china and earthenware	..	..	..	..	..	..	..	..	..	..
2692 Refractory ceramic products	..	..	..	..	..	..	..	..	..	..
2693 Structural non-refractory clay & ceramic prod.	..	..	..	..	..	..	..	..	..	..
2694 Cement, lime and plaster	..	..	..	..	..	..	..	..	..	..
2695 Articles of concrete, cement and plaster	..	..	..	..	..	..	..	..	..	..
2696 Cutting, shaping and finishing of stone	..	..	..	..	..	..	..	..	..	..
2699 Other non-metallic mineral products, nec	..	..	..	..	..	..	..	..	..	..

ISIS Energy Data Programme (IEA/OECD)

ISIC Revision 3 Industry Sector		Solid TJ	LPG TJ	Distiloil TJ	RFO TJ	Gas TJ	Biomass TJ	Steam TJ	Electr MWh	Own Use MWh	TOTAL TJ
27	**BASIC METALS**	..	..	..	..	..	..	..	..	..	7,254
271	Basic iron and steel	..	..	..	..	..	..	..	..	..	..
272	Basic precious and non-ferrous metals	..	..	..	..	..	..	..	..	..	..
273	Casting of metals	..	..	..	..	..	..	..	..	..	..
2731	Casting of iron and steel	..	..	..	..	..	..	..	..	..	..
2732	Casting non-ferrous metals	..	..	..	..	..	..	..	..	..	..
28	**FABRICATED METAL PRODUCTS**	..	..	..	..	..	..	..	..	..	272
281	Str. metal prod., tanks, reservoirs, steam generators	..	..	..	..	..	..	..	..	..	..
2811	Structural metal products	..	..	..	..	..	..	..	..	..	..
2812	Tanks, reservoirs and containers of metal	..	..	..	..	..	..	..	..	..	..
2813	Steam generators, exc. central heating hot water boilers	..	..	..	..	..	..	..	..	..	..
289	Other fabricated metal products	..	..	..	..	..	..	..	..	..	..
2891	Forging, pressing, stamping & roll-forming of metal	..	..	..	..	..	..	..	..	..	..
2892	Treatment and coating of metals	..	..	..	..	..	..	..	..	..	..
2893	Cutlery, hand tools and general hardware	..	..	..	..	..	..	..	..	..	..
2899	Other fabricated metal products, nec	..	..	..	..	..	..	..	..	..	..
29	**MACHINERY AND EQUIPMENT, NEC**	..	..	..	..	..	..	..	..	..	..
291	General purpose machinery	..	..	..	..	..	..	..	..	..	..
2911	Engines and turbines	..	..	..	..	..	..	..	..	..	..
2912	Pumps, compressors, taps and valves	..	..	..	..	..	..	..	..	..	..
2913	Bearings, gears, gearing and driving elements	..	..	..	..	..	..	..	..	..	..
2914	Ovens, furnaces and furnace burners	..	..	..	..	..	..	..	..	..	..
2915	Lifting and handling equipment	..	..	..	..	..	..	..	..	..	..
2919	Other general purpose machinery	..	..	..	..	..	..	..	..	..	..
292	Special purpose machinery	..	..	..	..	..	..	..	..	..	..
2921	Agricultural and forestry machinery	..	..	..	..	..	..	..	..	..	..
2922	Machine-tools	..	..	..	..	..	..	..	..	..	..
2923	Machinery for metallurgy	..	..	..	..	..	..	..	..	..	..
2924	Machinery for mining, quarrying and construction	..	..	..	..	..	..	..	..	..	..
2925	Machinery for food, beverage & tobacco processing	..	..	..	..	..	..	..	..	..	..
2926	Machinery for textile, apparel & leather production	..	..	..	..	..	..	..	..	..	..
2927	Machinery for weapons and ammunition	..	..	..	..	..	..	..	..	..	..
2929	Other special purpose machinery	..	..	..	..	..	..	..	..	..	..
293	Domestic appliances, nec	..	..	..	..	..	..	..	..	..	..
30	**OFFICE, ACCOUNTING & COMPUTING MACHINERY**	..	..	..	..	..	..	..	..	..	..
31	**ELECTRICAL MACHINERY & APPARATUS, NEC**	..	..	..	..	..	..	..	..	..	..
311	Electric motors, generators and transformers	..	..	..	..	..	..	..	..	..	..
312	Electricity distribution and control apparatus	..	..	..	..	..	..	..	..	..	..
313	Insulated wire and cable	..	..	..	..	..	..	..	..	..	..
314	Accumulators, primary cells & primary batteries	..	..	..	..	..	..	..	..	..	..
315	Electric lamps and lighting equipment	..	..	..	..	..	..	..	..	..	..
319	Other electrical equipment, nec	..	..	..	..	..	..	..	..	..	..
32	**RADIO, TV & COMMUNICATION EQUIP. & APP.**	..	..	..	..	..	..	..	..	..	..
321	Electronic valves, tubes, other electronic components	..	..	..	..	..	..	..	..	..	..
322	TV & radio transmitters, apparatus for line telephony	..	..	..	..	..	..	..	..	..	..
323	TV & radio receivers, recording apparatus	..	..	..	..	..	..	..	..	..	..
33	**MEDICAL PRECISION &OPTICAL INSTRUMENTS**	..	..	..	..	..	..	..	..	..	..
331	Medical appliances and instruments	..	..	..	..	..	..	..	..	..	..
3311	Medical, surgical equipment & orthopaedic app.	..	..	..	..	..	..	..	..	..	..
3312	Instruments & appliances for measuring, checking etc	..	..	..	..	..	..	..	..	..	..
3313	Industrial process control equipment	..	..	..	..	..	..	..	..	..	..
332	Optical instruments and photographic equipment	..	..	..	..	..	..	..	..	..	..
333	Watches and clocks	..	..	..	..	..	..	..	..	..	..
34	**MOTOR VEHICLES, TRAILERS & SEMI-TRAILERS**	..	..	..	..	..	..	..	..	..	..
341	Motor vehicles	..	..	..	..	..	..	..	..	..	..
342	Bodies (coachwork) for motor vehicles	..	..	..	..	..	..	..	..	..	..
343	Parts, accessories for motor vehicles & their engines	..	..	..	..	..	..	..	..	..	..
35	**OTHER TRANSPORT EQUIPMENT**	..	..	..	..	..	..	..	..	..	..
351	Building and repairing of ships and boats	..	..	..	..	..	..	..	..	..	..
3511	Building and repairing of ships	..	..	..	..	..	..	..	..	..	..
3512	Building, repairing of pleasure & sporting boats	..	..	..	..	..	..	..	..	..	..
352	Railway, tramway locomotives & rolling stock	..	..	..	..	..	..	..	..	..	..
353	Aircraft and spacecraft	..	..	..	..	..	..	..	..	..	..
359	Transport equipment, nec	..	..	..	..	..	..	..	..	..	..
3591	Motorcycles	..	..	..	..	..	..	..	..	..	..
3592	Bicycles and invalid carriages	..	..	..	..	..	..	..	..	..	..
3599	Other transport equipment, nec	..	..	..	..	..	..	..	..	..	..
36	**FURNITURE; MANUFACTURING, NEC**	..	..	..	..	..	..	..	..	..	..
361	Furniture	..	..	..	..	..	..	..	..	..	..
369	Manufacturing, nec	..	..	..	..	..	..	..	..	..	..
3691	Jewellery and related articles	..	..	..	..	..	..	..	..	..	..
3692	Musical instruments	..	..	..	..	..	..	..	..	..	..
3693	Sports goods	..	..	..	..	..	..	..	..	..	..
3694	Games and toys	..	..	..	..	..	..	..	..	..	..
3699	Other manufacturing, nec	..	..	..	..	..	..	..	..	..	..
37	**RECYCLING**	..	..	..	..	..	..	..	..	..	..
371	Recycling of metal waste and scrap	..	..	..	..	..	..	..	..	..	..
372	Recycling of non-metal waste and scrap	..	..	..	..	..	..	..	..	..	..
CERR	Non-specified industry	..	..	..	..	..	..	..	..	..	..
15-37	**TOTAL MANUFACTURING**	..	..	..	..	..	..	..	..	..	73,381

ISIC Revision 3 Industry Sector	Solid TJ	LPG TJ	Distiloil TJ	RFO TJ	Gas TJ	Biomass TJ	Steam TJ	Electr MWh	Own Use MWh	TOTAL TJ
15 FOOD PRODUCTS AND BEVERAGES	..	..	..	..	..	..	..	..	..	50,538
151 Production, processing and preserving (PPP)	..	..	..	..	..	..	..	..	..	..
1511 PPP of meat and meat products	..	..	..	..	..	..	..	..	..	..
1512 Processing and preserving of fish products	..	..	..	..	..	..	..	..	..	..
1513 Processing, preserving of fruit & vegetables	..	..	..	..	..	..	..	..	..	..
1514 Vegetable and animal oils and fats	..	..	..	..	..	..	..	..	..	..
152 Dairy products	..	..	..	..	..	..	..	..	..	..
153 Grain mill prod., starches & prepared animal feeds	..	..	..	..	..	..	..	..	..	..
1531 Grain mill products	..	..	..	..	..	..	..	..	..	..
1532 Starches and starch products	..	..	..	..	..	..	..	..	..	..
1533 Prepared animal feeds	..	..	..	..	..	..	..	..	..	..
154 Other food products	..	..	..	..	..	..	..	..	..	..
1541 Bakery products	..	..	..	..	..	..	..	..	..	..
1542 Sugar	..	..	..	..	..	..	..	..	..	..
1543 Cocoa, chocolate and sugar confectionery	..	..	..	..	..	..	..	..	..	..
1544 Macaroni, noodles, couscous & similar farinaceous prod.	..	..	..	..	..	..	..	..	..	..
1549 Other food products, nec	..	..	..	..	..	..	..	..	..	..
155 Beverages	..	..	..	..	..	..	..	..	..	..
1551 Distilling, rectifying and blending of spirits	..	..	..	..	..	..	..	..	..	..
1552 Wines	..	..	..	..	..	..	..	..	..	..
1553 Malt liquors and malt	..	..	..	..	..	..	..	..	..	..
1554 Soft drinks; production of mineral waters	..	..	..	..	..	..	..	..	..	..
16 TOBACCO PRODUCTS	..	..	..	..	..	..	..	..	..	..
17 TEXTILES	..	..	..	..	..	..	..	..	..	..
171 Spinning, weaving and finishing of textiles	..	..	..	..	..	..	..	..	..	..
1711 Preparation and spinning of textile fibres	..	..	..	..	..	..	..	..	..	..
1712 Finishing of textiles	..	..	..	..	..	..	..	..	..	..
172 Other textiles	..	..	..	..	..	..	..	..	..	..
1721 Made-up textile articles, except apparel	..	..	..	..	..	..	..	..	..	..
1722 Carpets and rugs	..	..	..	..	..	..	..	..	..	..
1723 Cordage, rope, twine and netting	..	..	..	..	..	..	..	..	..	..
1729 Other textiles, nec	..	..	..	..	..	..	..	..	..	..
173 Knitted and crocheted fabrics and articles	..	..	..	..	..	..	..	..	..	..
18 WEARING APPAREL, DRESSING & DYEING OF FUR	..	..	..	..	..	..	..	..	..	..
181 Wearing apparel, except fur apparel	..	..	..	..	..	..	..	..	..	..
182 Dressing and dyeing of fur; articles of fur	..	..	..	..	..	..	..	..	..	..
19 TANNING & DRESSING OF LEATHER, FOOTWEAR	..	..	..	..	..	..	..	..	..	..
191 Tanning and dressing of leather	..	..	..	..	..	..	..	..	..	..
1911 Tanning and dressing of leather	..	..	..	..	..	..	..	..	..	..
1912 Luggage, handbags, saddlery & harness	..	..	..	..	..	..	..	..	..	..
192 Footwear	..	..	..	..	..	..	..	..	..	..
20 WOOD AND WOOD PRODUCTS	..	..	..	..	..	..	..	..	..	..
201 Sawmilling and planing of wood	..	..	..	..	..	..	..	..	..	..
202 Products of wood, cork, straw & plaiting materials	..	..	..	..	..	..	..	..	..	..
2021 Veneer sheets	..	..	..	..	..	..	..	..	..	..
2022 Builders' carpentry and joinery	..	..	..	..	..	..	..	..	..	..
2023 Wooden containers	..	..	..	..	..	..	..	..	..	..
2029 Other products of wood	..	..	..	..	..	..	..	..	..	..
21 PAPER AND PAPER PRODUCTS	..	..	..	..	..	..	..	..	..	..
2101 Pulp, paper and paperboard	..	..	..	..	..	..	..	..	..	..
2102 Corrugated paper, paperboard and their containers	..	..	..	..	..	..	..	..	..	..
2109 Other articles of pulp and paperboard	..	..	..	..	..	..	..	..	..	..
22 PUBLISHING, PRINTING & REPRODUCTION	..	..	..	..	..	..	..	..	..	..
221 Publishing	..	..	..	..	..	..	..	..	..	..
2211 Publishing of books & brochures	..	..	..	..	..	..	..	..	..	..
2212 Publishing of newspapers and periodicals	..	..	..	..	..	..	..	..	..	..
2213 Publishing of recorded media	..	..	..	..	..	..	..	..	..	..
2219 Other publishing	..	..	..	..	..	..	..	..	..	..
222 Printing and related service activities	..	..	..	..	..	..	..	..	..	..
2221 Printing	..	..	..	..	..	..	..	..	..	..
2222 Service activities related to printing	..	..	..	..	..	..	..	..	..	..
223 Reproduction of recorded media	..	..	..	..	..	..	..	..	..	..
23 COKE, REFINED PETROLEUM PRODUCTS	..	..	..	..	..	..	..	..	..	..
231 Coke oven products	..	..	..	..	..	..	..	..	..	..
232 Refined petroleum products	..	..	..	..	..	..	..	..	..	..
233 Processing of nuclear fuel	..	..	..	..	..	..	..	..	..	2,903
24 CHEMICALS & CHEMICAL PRODUCTS	..	..	..	..	..	..	..	..	..	..
241 Basic chemicals	..	..	..	..	..	..	..	..	..	..
2411 Basic chemicals, exc. fertilizers & nitrogen compounds	..	..	..	..	..	..	..	..	..	..
2412 Fertilizers and nitrogen compounds	..	..	..	..	..	..	..	..	..	..
2413 Plastics in primary forms and synthetic rubber	..	..	..	..	..	..	..	..	..	..
242 Other chemical products	..	..	..	..	..	..	..	..	..	..
2421 Pesticides and other agro-chemical products	..	..	..	..	..	..	..	..	..	..
2422 Paints, varnishes and similar coatings	..	..	..	..	..	..	..	..	..	..
2423 Pharmaceuticals, medicinal chem. & botanical prod.	..	..	..	..	..	..	..	..	..	..
2424 Soap and detergents, perfumes etc.	..	..	..	..	..	..	..	..	..	..
2429 Other chemical products, nec	..	..	..	..	..	..	..	..	..	..
243 Man-made fibres	..	..	..	..	..	..	..	..	..	..
25 RUBBER AND PLASTICS PRODUCTS	..	..	..	..	..	..	..	..	..	..
251 Rubber products	..	..	..	..	..	..	..	..	..	..
2511 Rubber tyres and tubes	..	..	..	..	..	..	..	..	..	..
2519 Other rubber products	..	..	..	..	..	..	..	..	..	..
252 Plastic products	..	..	..	..	..	..	..	..	..	..
26 OTHER NON-METALLIC MINERAL PRODUCTS	..	..	..	..	..	..	..	..	..	6,117
261 Glass and glass products	..	..	..	..	..	..	..	..	..	..
269 Non-metallic mineral products, nec	..	..	..	..	..	..	..	..	..	..
2691 Pottery, china and earthenware	..	..	..	..	..	..	..	..	..	..
2692 Refractory ceramic products	..	..	..	..	..	..	..	..	..	..
2693 Structural non-refractory clay & ceramic prod.	..	..	..	..	..	..	..	..	..	..
2694 Cement, lime and plaster	..	..	..	..	..	..	..	..	..	..
2695 Articles of concrete, cement and plaster	..	..	..	..	..	..	..	..	..	..
2696 Cutting, shaping and finishing of stone	..	..	..	..	..	..	..	..	..	..
2699 Other non-metallic mineral products, nec	..	..	..	..	..	..	..	..	..	..

ISIC Revision 3 Industry Sector		Solid TJ	LPG TJ	Distiloil TJ	RFO TJ	Gas TJ	Biomass TJ	Steam TJ	Electr MWh	Own Use MWh	TOTAL TJ
27	**BASIC METALS**	..	..	..	..	..	..	..	..	..	7,963
271	Basic iron and steel	..	..	..	..	..	..	..	..	..	..
272	Basic precious and non-ferrous metals	..	..	..	..	..	..	..	..	..	..
273	Casting of metals	..	..	..	..	..	..	..	..	..	..
2731	Casting of iron and steel	..	..	..	..	..	..	..	..	..	..
2732	Casting non-ferrous metals	..	..	..	..	..	..	..	..	..	..
28	**FABRICATED METAL PRODUCTS**	..	..	..	..	..	..	..	..	..	268
281	Str. metal prod., tanks, reservoirs, steam generators	..	..	..	..	..	..	..	..	..	..
2811	Structural metal products	..	..	..	..	..	..	..	..	..	..
2812	Tanks, reservoirs and containers of metal	..	..	..	..	..	..	..	..	..	..
2813	Steam generators, exc. central heating hot water boilers	..	..	..	..	..	..	..	..	..	..
289	Other fabricated metal products	..	..	..	..	..	..	..	..	..	..
2891	Forging, pressing, stamping & roll-forming of metal	..	..	..	..	..	..	..	..	..	..
2892	Treatment and coating of metals	..	..	..	..	..	..	..	..	..	..
2893	Cutlery, hand tools and general hardware	..	..	..	..	..	..	..	..	..	..
2899	Other fabricated metal products, nec	..	..	..	..	..	..	..	..	..	..
29	**MACHINERY AND EQUIPMENT, NEC**	..	..	..	..	..	..	..	..	..	..
291	General purpose machinery	..	..	..	..	..	..	..	..	..	..
2911	Engines and turbines	..	..	..	..	..	..	..	..	..	..
2912	Pumps, compressors, taps and valves	..	..	..	..	..	..	..	..	..	..
2913	Bearings, gears, gearing and driving elements	..	..	..	..	..	..	..	..	..	..
2914	Ovens, furnaces and furnace burners	..	..	..	..	..	..	..	..	..	..
2915	Lifting and handling equipment	..	..	..	..	..	..	..	..	..	..
2919	Other general purpose machinery	..	..	..	..	..	..	..	..	..	..
292	Special purpose machinery	..	..	..	..	..	..	..	..	..	..
2921	Agricultural and forestry machinery	..	..	..	..	..	..	..	..	..	..
2922	Machine-tools	..	..	..	..	..	..	..	..	..	..
2923	Machinery for metallurgy	..	..	..	..	..	..	..	..	..	..
2924	Machinery for mining, quarrying and construction	..	..	..	..	..	..	..	..	..	..
2925	Machinery for food, beverage & tobacco processing	..	..	..	..	..	..	..	..	..	..
2926	Machinery for textile, apparel & leather production	..	..	..	..	..	..	..	..	..	..
2927	Machinery for weapons and ammunition	..	..	..	..	..	..	..	..	..	..
2929	Other special purpose machinery	..	..	..	..	..	..	..	..	..	..
293	Domestic appliances, nec	..	..	..	..	..	..	..	..	..	..
30	**OFFICE, ACCOUNTING & COMPUTING MACHINERY**	..	..	..	..	..	..	..	..	..	..
31	**ELECTRICAL MACHINERY & APPARATUS, NEC**	..	..	..	..	..	..	..	..	..	..
311	Electric motors, generators and transformers	..	..	..	..	..	..	..	..	..	..
312	Electricity distribution and control apparatus	..	..	..	..	..	..	..	..	..	..
313	Insulated wire and cable	..	..	..	..	..	..	..	..	..	..
314	Accumulators, primary cells & primary batteries	..	..	..	..	..	..	..	..	..	..
315	Electric lamps and lighting equipment	..	..	..	..	..	..	..	..	..	..
319	Other electrical equipment, nec	..	..	..	..	..	..	..	..	..	..
32	**RADIO, TV & COMMUNICATION EQUIP. & APP.**	..	..	..	..	..	..	..	..	..	..
321	Electronic valves, tubes, other electronic components	..	..	..	..	..	..	..	..	..	..
322	TV & radio transmitters, apparatus for line telephony	..	..	..	..	..	..	..	..	..	..
323	TV & radio receivers, recording apparatus	..	..	..	..	..	..	..	..	..	..
33	**MEDICAL PRECISION &OPTICAL INSTRUMENTS**	..	..	..	..	..	..	..	..	..	..
331	Medical appliances and instruments	..	..	..	..	..	..	..	..	..	..
3311	Medical, surgical equipment & orthopaedic app.	..	..	..	..	..	..	..	..	..	..
3312	Instruments & appliances for measuring, checking etc	..	..	..	..	..	..	..	..	..	..
3313	Industrial process control equipment	..	..	..	..	..	..	..	..	..	..
332	Optical instruments and photographic equipment	..	..	..	..	..	..	..	..	..	..
333	Watches and clocks	..	..	..	..	..	..	..	..	..	..
34	**MOTOR VEHICLES, TRAILERS & SEMI-TRAILERS**	..	..	..	..	..	..	..	..	..	..
341	Motor vehicles	..	..	..	..	..	..	..	..	..	..
342	Bodies (coachwork) for motor vehicles	..	..	..	..	..	..	..	..	..	..
343	Parts, accessories for motor vehicles & their engines	..	..	..	..	..	..	..	..	..	..
35	**OTHER TRANSPORT EQUIPMENT**	..	..	..	..	..	..	..	..	..	..
351	Building and repairing of ships and boats	..	..	..	..	..	..	..	..	..	..
3511	Building and repairing of ships	..	..	..	..	..	..	..	..	..	..
3512	Building, repairing of pleasure & sporting boats	..	..	..	..	..	..	..	..	..	..
352	Railway, tramway locomotives & rolling stock	..	..	..	..	..	..	..	..	..	..
353	Aircraft and spacecraft	..	..	..	..	..	..	..	..	..	..
359	Transport equipment, nec	..	..	..	..	..	..	..	..	..	..
3591	Motorcycles	..	..	..	..	..	..	..	..	..	..
3592	Bicycles and invalid carriages	..	..	..	..	..	..	..	..	..	..
3599	Other transport equipment, nec	..	..	..	..	..	..	..	..	..	..
36	**FURNITURE; MANUFACTURING, NEC**	..	..	..	..	..	..	..	..	..	..
361	Furniture	..	..	..	..	..	..	..	..	..	..
369	Manufacturing, nec	..	..	..	..	..	..	..	..	..	..
3691	Jewellery and related articles	..	..	..	..	..	..	..	..	..	..
3692	Musical instruments	..	..	..	..	..	..	..	..	..	..
3693	Sports goods	..	..	..	..	..	..	..	..	..	..
3694	Games and toys	..	..	..	..	..	..	..	..	..	..
3699	Other manufacturing, nec	..	..	..	..	..	..	..	..	..	..
37	**RECYCLING**	..	..	..	..	..	..	..	..	..	..
371	Recycling of metal waste and scrap	..	..	..	..	..	..	..	..	..	..
372	Recycling of non-metal waste and scrap	..	..	..	..	..	..	..	..	..	..
CERR	Non-specified industry	..	..	..	..	..	..	..	..	..	..
15-37	**TOTAL MANUFACTURING**	..	..	..	..	..	..	..	..	..	67,789

Year: 1995 **LUXEMBOURG**

ISIC Revision 3 Industry Sector	Solid TJ	LPG TJ	Distiloil TJ	RFO TJ	Gas TJ	Biomass TJ	Steam TJ	Electr MWh	Own Use MWh	TOTAL TJ
15 **FOOD PRODUCTS AND BEVERAGES**	..	..	..	..	..	..	..	..	..	37,302
151 Production, processing and preserving (PPP)	..	..	..	..	..	..	..	..	..	..
1511 PPP of meat and meat products	..	..	..	..	..	..	..	..	..	..
1512 Processing and preserving of fish products	..	..	..	..	..	..	..	..	..	..
1513 Processing, preserving of fruit & vegetables	..	..	..	..	..	..	..	..	..	..
1514 Vegetable and animal oils and fats	..	..	..	..	..	..	..	..	..	..
152 Dairy products	..	..	..	..	..	..	..	..	..	..
153 Grain mill prod., starches & prepared animal feeds	..	..	..	..	..	..	..	..	..	..
1531 Grain mill products	..	..	..	..	..	..	..	..	..	..
1532 Starches and starch products	..	..	..	..	..	..	..	..	..	..
1533 Prepared animal feeds	..	..	..	..	..	..	..	..	..	..
154 Other food products	..	..	..	..	..	..	..	..	..	..
1541 Bakery products	..	..	..	..	..	..	..	..	..	..
1542 Sugar	..	..	..	..	..	..	..	..	..	..
1543 Cocoa, chocolate and sugar confectionery	..	..	..	..	..	..	..	..	..	..
1544 Macaroni, noodles, couscous & similar farinaceous prod.	..	..	..	..	..	..	..	..	..	..
1549 Other food products, nec	..	..	..	..	..	..	..	..	..	..
155 Beverages	..	..	..	..	..	..	..	..	..	..
1551 Distilling, rectifying and blending of spirits	..	..	..	..	..	..	..	..	..	..
1552 Wines	..	..	..	..	..	..	..	..	..	..
1553 Malt liquors and malt	..	..	..	..	..	..	..	..	..	..
1554 Soft drinks; production of mineral waters	..	..	..	..	..	..	..	..	..	..
16 **TOBACCO PRODUCTS**	..	..	..	..	..	..	..	..	..	..
17 **TEXTILES**	..	..	..	..	..	..	..	..	..	..
171 Spinning, weaving and finishing of textiles	..	..	..	..	..	..	..	..	..	..
1711 Preparation and spinning of textile fibres	..	..	..	..	..	..	..	..	..	..
1712 Finishing of textiles	..	..	..	..	..	..	..	..	..	..
172 Other textiles	..	..	..	..	..	..	..	..	..	..
1721 Made-up textile articles, except apparel	..	..	..	..	..	..	..	..	..	..
1722 Carpets and rugs	..	..	..	..	..	..	..	..	..	..
1723 Cordage, rope, twine and netting	..	..	..	..	..	..	..	..	..	..
1729 Other textiles, nec	..	..	..	..	..	..	..	..	..	..
173 Knitted and crocheted fabrics and articles	..	..	..	..	..	..	..	..	..	..
18 **WEARING APPAREL, DRESSING & DYEING OF FUR**	..	..	..	..	..	..	..	..	..	..
181 Wearing apparel, except fur apparel	..	..	..	..	..	..	..	..	..	..
182 Dressing and dyeing of fur; articles of fur	..	..	..	..	..	..	..	..	..	..
19 **TANNING & DRESSING OF LEATHER, FOOTWEAR**	..	..	..	..	..	..	..	..	..	..
191 Tanning and dressing of leather	..	..	..	..	..	..	..	..	..	..
1911 Tanning and dressing of leather	..	..	..	..	..	..	..	..	..	..
1912 Luggage, handbags, saddlery & harness	..	..	..	..	..	..	..	..	..	..
192 Footwear	..	..	..	..	..	..	..	..	..	..
20 **WOOD AND WOOD PRODUCTS**	..	..	..	..	..	..	..	..	..	..
201 Sawmilling and planing of wood	..	..	..	..	..	..	..	..	..	..
202 Products of wood, cork, straw & plaiting materials	..	..	..	..	..	..	..	..	..	..
2021 Veneer sheets	..	..	..	..	..	..	..	..	..	..
2022 Builders' carpentry and joinery	..	..	..	..	..	..	..	..	..	..
2023 Wooden containers	..	..	..	..	..	..	..	..	..	..
2029 Other products of wood	..	..	..	..	..	..	..	..	..	..
21 **PAPER AND PAPER PRODUCTS**	..	..	..	..	..	..	..	..	..	..
2101 Pulp, paper and paperboard	..	..	..	..	..	..	..	..	..	..
2102 Corrugated paper, paperboard and their containers	..	..	..	..	..	..	..	..	..	..
2109 Other articles of pulp and paperboard	..	..	..	..	..	..	..	..	..	..
22 **PUBLISHING, PRINTING & REPRODUCTION**	..	..	..	..	..	..	..	..	..	..
221 Publishing	..	..	..	..	..	..	..	..	..	..
2211 Publishing of books & brochures	..	..	..	..	..	..	..	..	..	..
2212 Publishing of newspapers and periodicals	..	..	..	..	..	..	..	..	..	..
2213 Publishing of recorded media	..	..	..	..	..	..	..	..	..	..
2219 Other publishing	..	..	..	..	..	..	..	..	..	..
222 Printing and related service activities	..	..	..	..	..	..	..	..	..	..
2221 Printing	..	..	..	..	..	..	..	..	..	..
2222 Service activities related to printing	..	..	..	..	..	..	..	..	..	..
223 Reproduction of recorded media	..	..	..	..	..	..	..	..	..	..
23 **COKE, REFINED PETROLEUM PRODUCTS**	..	..	..	..	..	..	..	..	..	..
231 Coke oven products	..	..	..	..	..	..	..	..	..	..
232 Refined petroleum products	..	..	..	..	..	..	..	..	..	..
233 Processing of nuclear fuel	..	..	..	..	..	..	..	..	..	2,968
24 **CHEMICALS & CHEMICAL PRODUCTS**	..	..	..	..	..	..	..	..	..	..
241 Basic chemicals	..	..	..	..	..	..	..	..	..	..
2411 Basic chemicals, exc. fertilizers & nitrogen compounds	..	..	..	..	..	..	..	..	..	..
2412 Fertilizers and nitrogen compounds	..	..	..	..	..	..	..	..	..	..
2413 Plastics in primary forms and synthetic rubber	..	..	..	..	..	..	..	..	..	..
242 Other chemical products	..	..	..	..	..	..	..	..	..	..
2421 Pesticides and other agro-chemical products	..	..	..	..	..	..	..	..	..	..
2422 Paints, varnishes and similar coatings	..	..	..	..	..	..	..	..	..	..
2423 Pharmaceuticals, medicinal chem. & botanical prod.	..	..	..	..	..	..	..	..	..	..
2424 Soap and detergents, perfumes etc.	..	..	..	..	..	..	..	..	..	..
2429 Other chemical products, nec	..	..	..	..	..	..	..	..	..	..
243 Man-made fibres	..	..	..	..	..	..	..	..	..	..
25 **RUBBER AND PLASTICS PRODUCTS**	..	..	..	..	..	..	..	..	..	..
251 Rubber products	..	..	..	..	..	..	..	..	..	..
2511 Rubber tyres and tubes	..	..	..	..	..	..	..	..	..	..
2519 Other rubber products	..	..	..	..	..	..	..	..	..	..
252 Plastic products	..	..	..	..	..	..	..	..	..	5,843
26 **OTHER NON-METALLIC MINERAL PRODUCTS**	..	..	..	..	..	..	..	..	..	..
261 Glass and glass products	..	..	..	..	..	..	..	..	..	..
269 Non-metallic mineral products, nec	..	..	..	..	..	..	..	..	..	..
2691 Pottery, china and earthenware	..	..	..	..	..	..	..	..	..	..
2692 Refractory ceramic products	..	..	..	..	..	..	..	..	..	..
2693 Structural non-refractory clay & ceramic prod.	..	..	..	..	..	..	..	..	..	..
2694 Cement, lime and plaster	..	..	..	..	..	..	..	..	..	..
2695 Articles of concrete, cement and plaster	..	..	..	..	..	..	..	..	..	..
2696 Cutting, shaping and finishing of stone	..	..	..	..	..	..	..	..	..	..
2699 Other non-metallic mineral products, nec	..	..	..	..	..	..	..	..	..	..

ISIS Energy Data Programme (IEA/OECD)

ISIC Revision 3 Industry Sector		Solid TJ	LPG TJ	Distiloil TJ	RFO TJ	Gas TJ	Biomass TJ	Steam TJ	Electr MWh	Own Use MWh	TOTAL TJ
27	**BASIC METALS**	..	..	..	..	..	..	..	..	..	..
271	Basic iron and steel	..	..	..	..	..	..	..	..	..	7,766
272	Basic precious and non-ferrous metals	..	..	..	..	..	..	..	..	..	..
273	Casting of metals	..	..	..	..	..	..	..	..	..	..
2731	Casting of iron and steel	..	..	..	..	..	..	..	..	..	..
2732	Casting non-ferrous metals	..	..	..	..	..	..	..	..	..	..
28	**FABRICATED METAL PRODUCTS**	..	..	..	..	..	..	..	..	..	..
281	Str. metal prod., tanks, reservoirs, steam generators	..	..	..	..	..	..	..	..	..	276
2811	Structural metal products	..	..	..	..	..	..	..	..	..	..
2812	Tanks, reservoirs and containers of metal	..	..	..	..	..	..	..	..	..	..
2813	Steam generators, exc. central heating hot water boilers	..	..	..	..	..	..	..	..	..	..
289	Other fabricated metal products	..	..	..	..	..	..	..	..	..	..
2891	Forging, pressing, stamping & roll-forming of metal	..	..	..	..	..	..	..	..	..	..
2892	Treatment and coating of metals	..	..	..	..	..	..	..	..	..	..
2893	Cutlery, hand tools and general hardware	..	..	..	..	..	..	..	..	..	..
2899	Other fabricated metal products, nec	..	..	..	..	..	..	..	..	..	..
29	**MACHINERY AND EQUIPMENT, NEC**	..	..	..	..	..	..	..	..	..	..
291	General purpose machinery	..	..	..	..	..	..	..	..	..	..
2911	Engines and turbines	..	..	..	..	..	..	..	..	..	..
2912	Pumps, compressors, taps and valves	..	..	..	..	..	..	..	..	..	..
2913	Bearings, gears, gearing and driving elements	..	..	..	..	..	..	..	..	..	..
2914	Ovens, furnaces and furnace burners	..	..	..	..	..	..	..	..	..	..
2915	Lifting and handling equipment	..	..	..	..	..	..	..	..	..	..
2919	Other general purpose machinery	..	..	..	..	..	..	..	..	..	..
292	Special purpose machinery	..	..	..	..	..	..	..	..	..	..
2921	Agricultural and forestry machinery	..	..	..	..	..	..	..	..	..	..
2922	Machine-tools	..	..	..	..	..	..	..	..	..	..
2923	Machinery for metallurgy	..	..	..	..	..	..	..	..	..	..
2924	Machinery for mining, quarrying and construction	..	..	..	..	..	..	..	..	..	..
2925	Machinery for food, beverage & tobacco processing	..	..	..	..	..	..	..	..	..	..
2926	Machinery for textile, apparel & leather production	..	..	..	..	..	..	..	..	..	..
2927	Machinery for weapons and ammunition	..	..	..	..	..	..	..	..	..	..
2929	Other special purpose machinery	..	..	..	..	..	..	..	..	..	..
293	Domestic appliances, nec	..	..	..	..	..	..	..	..	..	..
30	**OFFICE, ACCOUNTING & COMPUTING MACHINERY**	..	..	..	..	..	..	..	..	..	..
31	**ELECTRICAL MACHINERY & APPARATUS, NEC**	..	..	..	..	..	..	..	..	..	..
311	Electric motors, generators and transformers	..	..	..	..	..	..	..	..	..	..
312	Electricity distribution and control apparatus	..	..	..	..	..	..	..	..	..	..
313	Insulated wire and cable	..	..	..	..	..	..	..	..	..	..
314	Accumulators, primary cells & primary batteries	..	..	..	..	..	..	..	..	..	..
315	Electric lamps and lighting equipment	..	..	..	..	..	..	..	..	..	..
319	Other electrical equipment, nec	..	..	..	..	..	..	..	..	..	..
32	**RADIO, TV & COMMUNICATION EQUIP. & APP.**	..	..	..	..	..	..	..	..	..	..
321	Electronic valves, tubes, other electronic components	..	..	..	..	..	..	..	..	..	..
322	TV & radio transmitters, apparatus for line telephony	..	..	..	..	..	..	..	..	..	..
323	TV & radio receivers, recording apparatus	..	..	..	..	..	..	..	..	..	..
33	**MEDICAL PRECISION & OPTICAL INSTRUMENTS**	..	..	..	..	..	..	..	..	..	..
331	Medical appliances and instruments	..	..	..	..	..	..	..	..	..	..
3311	Medical, surgical equipment & orthopaedic app.	..	..	..	..	..	..	..	..	..	..
3312	Instruments & appliances for measuring, checking etc	..	..	..	..	..	..	..	..	..	..
3313	Industrial process control equipment	..	..	..	..	..	..	..	..	..	..
332	Optical instruments and photographic equipment	..	..	..	..	..	..	..	..	..	..
333	Watches and clocks	..	..	..	..	..	..	..	..	..	..
34	**MOTOR VEHICLES, TRAILERS & SEMI-TRAILERS**	..	..	..	..	..	..	..	..	..	..
341	Motor vehicles	..	..	..	..	..	..	..	..	..	..
342	Bodies (coachwork) for motor vehicles	..	..	..	..	..	..	..	..	..	..
343	Parts, accessories for motor vehicles & their engines	..	..	..	..	..	..	..	..	..	..
35	**OTHER TRANSPORT EQUIPMENT**	..	..	..	..	..	..	..	..	..	..
351	Building and repairing of ships and boats	..	..	..	..	..	..	..	..	..	..
3511	Building and repairing of ships	..	..	..	..	..	..	..	..	..	..
3512	Building, repairing of pleasure & sporting boats	..	..	..	..	..	..	..	..	..	..
352	Railway, tramway locomotives & rolling stock	..	..	..	..	..	..	..	..	..	..
353	Aircraft and spacecraft	..	..	..	..	..	..	..	..	..	..
359	Transport equipment, nec	..	..	..	..	..	..	..	..	..	..
3591	Motorcycles	..	..	..	..	..	..	..	..	..	..
3592	Bicycles and invalid carriages	..	..	..	..	..	..	..	..	..	..
3599	Other transport equipment, nec	..	..	..	..	..	..	..	..	..	..
36	**FURNITURE; MANUFACTURING, NEC**	..	..	..	..	..	..	..	..	..	..
361	Furniture	..	..	..	..	..	..	..	..	..	..
369	Manufacturing, nec	..	..	..	..	..	..	..	..	..	..
3691	Jewellery and related articles	..	..	..	..	..	..	..	..	..	..
3692	Musical instruments	..	..	..	..	..	..	..	..	..	..
3693	Sports goods	..	..	..	..	..	..	..	..	..	..
3694	Games and toys	..	..	..	..	..	..	..	..	..	..
3699	Other manufacturing, nec	..	..	..	..	..	..	..	..	..	..
37	**RECYCLING**	..	..	..	..	..	..	..	..	..	..
371	Recycling of metal waste and scrap	..	..	..	..	..	..	..	..	..	..
372	Recycling of non-metal waste and scrap	..	..	..	..	..	..	..	..	..	..
CERR	**Non-specified industry**	..	..	..	..	..	..	..	..	..	..
15-37	**TOTAL MANUFACTURING**	..	..	..	..	..	..	..	..	..	54,155

ISIC Revision 3 Industry Sector	Solid TJ	LPG TJ	Distiloil TJ	RFO TJ	Gas TJ	Biomass TJ	Steam TJ	Electr MWh	Own Use MWh	TOTAL TJ
15 FOOD PRODUCTS AND BEVERAGES	..	..	..	..	..	..	..	..	..	34,980
151 Production, processing and preserving (PPP)	..	..	..	..	..	..	..	..	..	..
1511 PPP of meat and meat products	..	..	..	..	..	..	..	..	..	..
1512 Processing and preserving of fish products	..	..	..	..	..	..	..	..	..	..
1513 Processing, preserving of fruit & vegetables	..	..	..	..	..	..	..	..	..	..
1514 Vegetable and animal oils and fats	..	..	..	..	..	..	..	..	..	..
152 Dairy products	..	..	..	..	..	..	..	..	..	..
153 Grain mill prod., starches & prepared animal feeds	..	..	..	..	..	..	..	..	..	..
1531 Grain mill products	..	..	..	..	..	..	..	..	..	..
1532 Starches and starch products	..	..	..	..	..	..	..	..	..	..
1533 Prepared animal feeds	..	..	..	..	..	..	..	..	..	..
154 Other food products	..	..	..	..	..	..	..	..	..	..
1541 Bakery products	..	..	..	..	..	..	..	..	..	..
1542 Sugar	..	..	..	..	..	..	..	..	..	..
1543 Cocoa, chocolate and sugar confectionery	..	..	..	..	..	..	..	..	..	..
1544 Macaroni, noodles, couscous & similar farinaceous prod.	..	..	..	..	..	..	..	..	..	..
1549 Other food products, nec	..	..	..	..	..	..	..	..	..	..
155 Beverages	..	..	..	..	..	..	..	..	..	..
1551 Distilling, rectifying and blending of spirits	..	..	..	..	..	..	..	..	..	..
1552 Wines	..	..	..	..	..	..	..	..	..	..
1553 Malt liquors and malt	..	..	..	..	..	..	..	..	..	..
1554 Soft drinks; production of mineral waters	..	..	..	..	..	..	..	..	..	..
16 TOBACCO PRODUCTS	..	..	..	..	..	..	..	..	..	..
17 TEXTILES	..	..	..	..	..	..	..	..	..	..
171 Spinning, weaving and finishing of textiles	..	..	..	..	..	..	..	..	..	..
1711 Preparation and spinning of textile fibres	..	..	..	..	..	..	..	..	..	..
1712 Finishing of textiles	..	..	..	..	..	..	..	..	..	..
172 Other textiles	..	..	..	..	..	..	..	..	..	..
1721 Made-up textile articles, except apparel	..	..	..	..	..	..	..	..	..	..
1722 Carpets and rugs	..	..	..	..	..	..	..	..	..	..
1723 Cordage, rope, twine and netting	..	..	..	..	..	..	..	..	..	..
1729 Other textiles, nec	..	..	..	..	..	..	..	..	..	..
173 Knitted and crocheted fabrics and articles	..	..	..	..	..	..	..	..	..	..
18 WEARING APPAREL, DRESSING & DYEING OF FUR	..	..	..	..	..	..	..	..	..	..
181 Wearing apparel, except fur apparel	..	..	..	..	..	..	..	..	..	..
182 Dressing and dyeing of fur; articles of fur	..	..	..	..	..	..	..	..	..	..
19 TANNING & DRESSING OF LEATHER, FOOTWEAR	..	..	..	..	..	..	..	..	..	..
191 Tanning and dressing of leather	..	..	..	..	..	..	..	..	..	..
1911 Tanning and dressing of leather	..	..	..	..	..	..	..	..	..	..
1912 Luggage, handbags, saddlery & harness	..	..	..	..	..	..	..	..	..	..
192 Footwear	..	..	..	..	..	..	..	..	..	..
20 WOOD AND WOOD PRODUCTS	..	..	..	..	..	..	..	..	..	..
201 Sawmilling and planing of wood	..	..	..	..	..	..	..	..	..	..
202 Products of wood, cork, straw & plaiting materials	..	..	..	..	..	..	..	..	..	..
2021 Veneer sheets	..	..	..	..	..	..	..	..	..	..
2022 Builders' carpentry and joinery	..	..	..	..	..	..	..	..	..	..
2023 Wooden containers	..	..	..	..	..	..	..	..	..	..
2029 Other products of wood	..	..	..	..	..	..	..	..	..	..
21 PAPER AND PAPER PRODUCTS	..	..	..	..	..	..	..	..	..	..
2101 Pulp, paper and paperboard	..	..	..	..	..	..	..	..	..	..
2102 Corrugated paper, paperboard and their containers	..	..	..	..	..	..	..	..	..	..
2109 Other articles of pulp and paperboard	..	..	..	..	..	..	..	..	..	..
22 PUBLISHING, PRINTING & REPRODUCTION	..	..	..	..	..	..	..	..	..	..
221 Publishing	..	..	..	..	..	..	..	..	..	..
2211 Publishing of books & brochures	..	..	..	..	..	..	..	..	..	..
2212 Publishing of newspapers and periodicals	..	..	..	..	..	..	..	..	..	..
2213 Publishing of recorded media	..	..	..	..	..	..	..	..	..	..
2219 Other publishing	..	..	..	..	..	..	..	..	..	..
222 Printing and related service activities	..	..	..	..	..	..	..	..	..	..
2221 Printing	..	..	..	..	..	..	..	..	..	..
2222 Service activities related to printing	..	..	..	..	..	..	..	..	..	..
223 Reproduction of recorded media	..	..	..	..	..	..	..	..	..	..
23 COKE, REFINED PETROLEUM PRODUCTS	..	..	..	..	..	..	..	..	..	..
231 Coke oven products	..	..	..	..	..	..	..	..	..	..
232 Refined petroleum products	..	..	..	..	..	..	..	..	..	..
233 Processing of nuclear fuel	..	..	..	..	..	..	..	..	..	..
24 CHEMICALS & CHEMICAL PRODUCTS	..	..	..	..	..	..	..	..	..	2,885
241 Basic chemicals	..	..	..	..	..	..	..	..	..	..
2411 Basic chemicals, exc. fertilizers & nitrogen compounds	..	..	..	..	..	..	..	..	..	..
2412 Fertilizers and nitrogen compounds	..	..	..	..	..	..	..	..	..	..
2413 Plastics in primary forms and synthetic rubber	..	..	..	..	..	..	..	..	..	..
242 Other chemical products	..	..	..	..	..	..	..	..	..	..
2421 Pesticides and other agro-chemical products	..	..	..	..	..	..	..	..	..	..
2422 Paints, varnishes and similar coatings	..	..	..	..	..	..	..	..	..	..
2423 Pharmaceuticals, medicinal chem. & botanical prod.	..	..	..	..	..	..	..	..	..	..
2424 Soap and detergents, perfumes etc.	..	..	..	..	..	..	..	..	..	..
2429 Other chemical products, nec	..	..	..	..	..	..	..	..	..	..
243 Man-made fibres	..	..	..	..	..	..	..	..	..	..
25 RUBBER AND PLASTICS PRODUCTS	..	..	..	..	..	..	..	..	..	..
251 Rubber products	..	..	..	..	..	..	..	..	..	..
2511 Rubber tyres and tubes	..	..	..	..	..	..	..	..	..	..
2519 Other rubber products	..	..	..	..	..	..	..	..	..	..
252 Plastic products	..	..	..	..	..	..	..	..	..	..
26 OTHER NON-METALLIC MINERAL PRODUCTS	..	..	..	..	..	..	..	..	..	5,730
261 Glass and glass products	..	..	..	..	..	..	..	..	..	..
269 Non-metallic mineral products, nec	..	..	..	..	..	..	..	..	..	..
2691 Pottery, china and earthenware	..	..	..	..	..	..	..	..	..	..
2692 Refractory ceramic products	..	..	..	..	..	..	..	..	..	..
2693 Structural non-refractory clay & ceramic prod.	..	..	..	..	..	..	..	..	..	..
2694 Cement, lime and plaster	..	..	..	..	..	..	..	..	..	..
2695 Articles of concrete, cement and plaster	..	..	..	..	..	..	..	..	..	..
2696 Cutting, shaping and finishing of stone	..	..	..	..	..	..	..	..	..	..
2699 Other non-metallic mineral products, nec	..	..	..	..	..	..	..	..	..	..

ISIS Energy Data Programme (IEA/OECD)

ISIC Revision 3 Industry Sector		Solid TJ	LPG TJ	Distiloil TJ	RFO TJ	Gas TJ	Biomass TJ	Steam TJ	Electr MWh	Own Use MWh	TOTAL TJ
27	**BASIC METALS**	..	..	..	..	..	..	..	..	..	7,699
271	Basic iron and steel	..	..	..	..	..	..	..	..	..	..
272	Basic precious and non-ferrous metals	..	..	..	..	..	..	..	..	..	..
273	Casting of metals	..	..	..	..	..	..	..	..	..	..
2731	Casting of iron and steel	..	..	..	..	..	..	..	..	..	..
2732	Casting non-ferrous metals	..	..	..	..	..	..	..	..	..	..
28	**FABRICATED METAL PRODUCTS**	..	..	..	..	..	..	..	..	..	257
281	Str. metal prod., tanks, reservoirs, steam generators	..	..	..	..	..	..	..	..	..	..
2811	Structural metal products	..	..	..	..	..	..	..	..	..	..
2812	Tanks, reservoirs and containers of metal	..	..	..	..	..	..	..	..	..	..
2813	Steam generators, exc. central heating hot water boilers	..	..	..	..	..	..	..	..	..	..
289	Other fabricated metal products	..	..	..	..	..	..	..	..	..	..
2891	Forging, pressing, stamping & roll-forming of metal	..	..	..	..	..	..	..	..	..	..
2892	Treatment and coating of metals	..	..	..	..	..	..	..	..	..	..
2893	Cutlery, hand tools and general hardware	..	..	..	..	..	..	..	..	..	..
2899	Other fabricated metal products, nec	..	..	..	..	..	..	..	..	..	..
29	**MACHINERY AND EQUIPMENT, NEC**	..	..	..	..	..	..	..	..	..	..
291	General purpose machinery	..	..	..	..	..	..	..	..	..	..
2911	Engines and turbines	..	..	..	..	..	..	..	..	..	..
2912	Pumps, compressors, taps and valves	..	..	..	..	..	..	..	..	..	..
2913	Bearings, gears, gearing and driving elements	..	..	..	..	..	..	..	..	..	..
2914	Ovens, furnaces and furnace burners	..	..	..	..	..	..	..	..	..	..
2915	Lifting and handling equipment	..	..	..	..	..	..	..	..	..	..
2919	Other general purpose machinery	..	..	..	..	..	..	..	..	..	..
292	Special purpose machinery	..	..	..	..	..	..	..	..	..	..
2921	Agricultural and forestry machinery	..	..	..	..	..	..	..	..	..	..
2922	Machine-tools	..	..	..	..	..	..	..	..	..	..
2923	Machinery for metallurgy	..	..	..	..	..	..	..	..	..	..
2924	Machinery for mining, quarrying and construction	..	..	..	..	..	..	..	..	..	..
2925	Machinery for food, beverage & tobacco processing	..	..	..	..	..	..	..	..	..	..
2926	Machinery for textile, apparel & leather production	..	..	..	..	..	..	..	..	..	..
2927	Machinery for weapons and ammunition	..	..	..	..	..	..	..	..	..	..
2929	Other special purpose machinery	..	..	..	..	..	..	..	..	..	..
293	Domestic appliances, nec	..	..	..	..	..	..	..	..	..	..
30	**OFFICE, ACCOUNTING & COMPUTING MACHINERY**	..	..	..	..	..	..	..	..	..	..
31	**ELECTRICAL MACHINERY & APPARATUS, NEC**	..	..	..	..	..	..	..	..	..	..
311	Electric motors, generators and transformers	..	..	..	..	..	..	..	..	..	..
312	Electricity distribution and control apparatus	..	..	..	..	..	..	..	..	..	..
313	Insulated wire and cable	..	..	..	..	..	..	..	..	..	..
314	Accumulators, primary cells & primary batteries	..	..	..	..	..	..	..	..	..	..
315	Electric lamps and lighting equipment	..	..	..	..	..	..	..	..	..	..
319	Other electrical equipment, nec	..	..	..	..	..	..	..	..	..	..
32	**RADIO, TV & COMMUNICATION EQUIP. & APP.**	..	..	..	..	..	..	..	..	..	..
321	Electronic valves, tubes, other electronic components	..	..	..	..	..	..	..	..	..	..
322	TV & radio transmitters, apparatus for line telephony	..	..	..	..	..	..	..	..	..	..
323	TV & radio receivers, recording apparatus	..	..	..	..	..	..	..	..	..	..
33	**MEDICAL PRECISION &OPTICAL INSTRUMENTS**	..	..	..	..	..	..	..	..	..	..
331	Medical appliances and instruments	..	..	..	..	..	..	..	..	..	..
3311	Medical, surgical equipment & orthopaedic app.	..	..	..	..	..	..	..	..	..	..
3312	Instruments & appliances for measuring, checking etc	..	..	..	..	..	..	..	..	..	..
3313	Industrial process control equipment	..	..	..	..	..	..	..	..	..	..
332	Optical instruments and photographic equipment	..	..	..	..	..	..	..	..	..	..
333	Watches and clocks	..	..	..	..	..	..	..	..	..	..
34	**MOTOR VEHICLES, TRAILERS & SEMI-TRAILERS**	..	..	..	..	..	..	..	..	..	..
341	Motor vehicles	..	..	..	..	..	..	..	..	..	..
342	Bodies (coachwork) for motor vehicles	..	..	..	..	..	..	..	..	..	..
343	Parts, accessories for motor vehicles & their engines	..	..	..	..	..	..	..	..	..	..
35	**OTHER TRANSPORT EQUIPMENT**	..	..	..	..	..	..	..	..	..	..
351	Building and repairing of ships and boats	..	..	..	..	..	..	..	..	..	..
3511	Building and repairing of ships	..	..	..	..	..	..	..	..	..	..
3512	Building, repairing of pleasure & sporting boats	..	..	..	..	..	..	..	..	..	..
352	Railway, tramway locomotives & rolling stock	..	..	..	..	..	..	..	..	..	..
353	Aircraft and spacecraft	..	..	..	..	..	..	..	..	..	..
359	Transport equipment, nec	..	..	..	..	..	..	..	..	..	..
3591	Motorcycles	..	..	..	..	..	..	..	..	..	..
3592	Bicycles and invalid carriages	..	..	..	..	..	..	..	..	..	..
3599	Other transport equipment, nec	..	..	..	..	..	..	..	..	..	..
36	**FURNITURE; MANUFACTURING, NEC**	..	..	..	..	..	..	..	..	..	..
361	Furniture	..	..	..	..	..	..	..	..	..	..
369	Manufacturing, nec	..	..	..	..	..	..	..	..	..	..
3691	Jewellery and related articles	..	..	..	..	..	..	..	..	..	..
3692	Musical instruments	..	..	..	..	..	..	..	..	..	..
3693	Sports goods	..	..	..	..	..	..	..	..	..	..
3694	Games and toys	..	..	..	..	..	..	..	..	..	..
3699	Other manufacturing, nec	..	..	..	..	..	..	..	..	..	..
37	**RECYCLING**	..	..	..	..	..	..	..	..	..	..
371	Recycling of metal waste and scrap	..	..	..	..	..	..	..	..	..	..
372	Recycling of non-metal waste and scrap	..	..	..	..	..	..	..	..	..	..
CERR	Non-specified industry	..	..	..	..	..	..	..	..	..	..
15-37	**TOTAL MANUFACTURING**	..	..	..	..	..	..	..	..	..	51,551

Year: 1997 LUXEMBOURG

ISIC Revision 3 Industry Sector	Solid TJ	LPG TJ	Distiloil TJ	RFO TJ	Gas TJ	Biomass TJ	Steam TJ	Electr MWh	Own Use MWh	TOTAL TJ
15 FOOD PRODUCTS AND BEVERAGES	..	..	..	..	..	..	..	..	..	31,068
151 Production, processing and preserving (PPP)	..	..	..	..	..	..	..	..	..	..
1511 PPP of meat and meat products	..	..	..	..	..	..	..	..	..	..
1512 Processing and preserving of fish products	..	..	..	..	..	..	..	..	..	..
1513 Processing, preserving of fruit & vegetables	..	..	..	..	..	..	..	..	..	..
1514 Vegetable and animal oils and fats	..	..	..	..	..	..	..	..	..	..
152 Dairy products	..	..	..	..	..	..	..	..	..	..
153 Grain mill prod., starches & prepared animal feeds	..	..	..	..	..	..	..	..	..	..
1531 Grain mill products	..	..	..	..	..	..	..	..	..	..
1532 Starches and starch products	..	..	..	..	..	..	..	..	..	..
1533 Prepared animal feeds	..	..	..	..	..	..	..	..	..	..
154 Other food products	..	..	..	..	..	..	..	..	..	..
1541 Bakery products	..	..	..	..	..	..	..	..	..	..
1542 Sugar	..	..	..	..	..	..	..	..	..	..
1543 Cocoa, chocolate and sugar confectionery	..	..	..	..	..	..	..	..	..	..
1544 Macaroni, noodles, couscous & similar farinaceous prod.	..	..	..	..	..	..	..	..	..	..
1549 Other food products, nec	..	..	..	..	..	..	..	..	..	..
155 Beverages	..	..	..	..	..	..	..	..	..	..
1551 Distilling, rectifying and blending of spirits	..	..	..	..	..	..	..	..	..	..
1552 Wines	..	..	..	..	..	..	..	..	..	..
1553 Malt liquors and malt	..	..	..	..	..	..	..	..	..	..
1554 Soft drinks; production of mineral waters	..	..	..	..	..	..	..	..	..	..
16 TOBACCO PRODUCTS	..	..	..	..	..	..	..	..	..	..
17 TEXTILES	..	..	..	..	..	..	..	..	..	..
171 Spinning, weaving and finishing of textiles	..	..	..	..	..	..	..	..	..	..
1711 Preparation and spinning of textile fibres	..	..	..	..	..	..	..	..	..	..
1712 Finishing of textiles	..	..	..	..	..	..	..	..	..	..
172 Other textiles	..	..	..	..	..	..	..	..	..	..
1721 Made-up textile articles, except apparel	..	..	..	..	..	..	..	..	..	..
1722 Carpets and rugs	..	..	..	..	..	..	..	..	..	..
1723 Cordage, rope, twine and netting	..	..	..	..	..	..	..	..	..	..
1729 Other textiles, nec	..	..	..	..	..	..	..	..	..	..
173 Knitted and crocheted fabrics and articles	..	..	..	..	..	..	..	..	..	..
18 WEARING APPAREL, DRESSING & DYEING OF FUR	..	..	..	..	..	..	..	..	..	..
181 Wearing apparel, except fur apparel	..	..	..	..	..	..	..	..	..	..
182 Dressing and dyeing of fur; articles of fur	..	..	..	..	..	..	..	..	..	..
19 TANNING & DRESSING OF LEATHER, FOOTWEAR	..	..	..	..	..	..	..	..	..	..
191 Tanning and dressing of leather	..	..	..	..	..	..	..	..	..	..
1911 Tanning and dressing of leather	..	..	..	..	..	..	..	..	..	..
1912 Luggage, handbags, saddlery & harness	..	..	..	..	..	..	..	..	..	..
192 Footwear	..	..	..	..	..	..	..	..	..	..
20 WOOD AND WOOD PRODUCTS	..	..	..	..	..	..	..	..	..	..
201 Sawmilling and planing of wood	..	..	..	..	..	..	..	..	..	..
202 Products of wood, cork, straw & plaiting materials	..	..	..	..	..	..	..	..	..	..
2021 Veneer sheets	..	..	..	..	..	..	..	..	..	..
2022 Builders' carpentry and joinery	..	..	..	..	..	..	..	..	..	..
2023 Wooden containers	..	..	..	..	..	..	..	..	..	..
2029 Other products of wood	..	..	..	..	..	..	..	..	..	..
21 PAPER AND PAPER PRODUCTS	..	..	..	..	..	..	..	..	..	..
2101 Pulp, paper and paperboard	..	..	..	..	..	..	..	..	..	..
2102 Corrugated paper, paperboard and their containers	..	..	..	..	..	..	..	..	..	..
2109 Other articles of pulp and paperboard	..	..	..	..	..	..	..	..	..	..
22 PUBLISHING, PRINTING & REPRODUCTION	..	..	..	..	..	..	..	..	..	..
221 Publishing	..	..	..	..	..	..	..	..	..	..
2211 Publishing of books & brochures	..	..	..	..	..	..	..	..	..	..
2212 Publishing of newspapers and periodicals	..	..	..	..	..	..	..	..	..	..
2213 Publishing of recorded media	..	..	..	..	..	..	..	..	..	..
2219 Other publishing	..	..	..	..	..	..	..	..	..	..
222 Printing and related service activities	..	..	..	..	..	..	..	..	..	..
2221 Printing	..	..	..	..	..	..	..	..	..	..
2222 Service activities related to printing	..	..	..	..	..	..	..	..	..	..
223 Reproduction of recorded media	..	..	..	..	..	..	..	..	..	..
23 COKE, REFINED PETROLEUM PRODUCTS	..	..	..	..	..	..	..	..	..	..
231 Coke oven products	..	..	..	..	..	..	..	..	..	..
232 Refined petroleum products	..	..	..	..	..	..	..	..	..	..
233 Processing of nuclear fuel	..	..	..	..	..	..	..	..	..	3,046
24 CHEMICALS & CHEMICAL PRODUCTS	..	..	..	..	..	..	..	..	..	..
241 Basic chemicals	..	..	..	..	..	..	..	..	..	..
2411 Basic chemicals, exc. fertilizers & nitrogen compounds	..	..	..	..	..	..	..	..	..	..
2412 Fertilizers and nitrogen compounds	..	..	..	..	..	..	..	..	..	..
2413 Plastics in primary forms and synthetic rubber	..	..	..	..	..	..	..	..	..	..
242 Other chemical products	..	..	..	..	..	..	..	..	..	..
2421 Pesticides and other agro-chemical products	..	..	..	..	..	..	..	..	..	..
2422 Paints, varnishes and similar coatings	..	..	..	..	..	..	..	..	..	..
2423 Pharmaceuticals, medicinal chem. & botanical prod.	..	..	..	..	..	..	..	..	..	..
2424 Soap and detergents, perfumes etc.	..	..	..	..	..	..	..	..	..	..
2429 Other chemical products, nec	..	..	..	..	..	..	..	..	..	..
243 Man-made fibres	..	..	..	..	..	..	..	..	..	..
25 RUBBER AND PLASTICS PRODUCTS	..	..	..	..	..	..	..	..	..	..
251 Rubber products	..	..	..	..	..	..	..	..	..	..
2511 Rubber tyres and tubes	..	..	..	..	..	..	..	..	..	..
2519 Other rubber products	..	..	..	..	..	..	..	..	..	..
252 Plastic products	..	..	..	..	..	..	..	..	..	6,106
26 OTHER NON-METALLIC MINERAL PRODUCTS	..	..	..	..	..	..	..	..	..	..
261 Glass and glass products	..	..	..	..	..	..	..	..	..	..
269 Non-metallic mineral products, nec	..	..	..	..	..	..	..	..	..	..
2691 Pottery, china and earthenware	..	..	..	..	..	..	..	..	..	..
2692 Refractory ceramic products	..	..	..	..	..	..	..	..	..	..
2693 Structural non-refractory clay & ceramic prod.	..	..	..	..	..	..	..	..	..	..
2694 Cement, lime and plaster	..	..	..	..	..	..	..	..	..	..
2695 Articles of concrete, cement and plaster	..	..	..	..	..	..	..	..	..	..
2696 Cutting, shaping and finishing of stone	..	..	..	..	..	..	..	..	..	..
2699 Other non-metallic mineral products, nec	..	..	..	..	..	..	..	..	..	..

ISIS Energy Data Programme (IEA/OECD)

ISIC Revision 3 Industry Sector		Solid TJ	LPG TJ	Distiloil TJ	RFO TJ	Gas TJ	Biomass TJ	Steam TJ	Electr MWh	Own Use MWh	TOTAL TJ
27	BASIC METALS	..	..	..	..	..	..	..	..	..	7,517
271	Basic iron and steel	..	..	..	..	..	..	..	..	..	..
272	Basic precious and non-ferrous metals	..	..	..	..	..	..	..	..	..	..
273	Casting of metals	..	..	..	..	..	..	..	..	..	..
2731	Casting of iron and steel	..	..	..	..	..	..	..	..	..	..
2732	Casting non-ferrous metals	..	..	..	..	..	..	..	..	..	..
28	FABRICATED METAL PRODUCTS	..	..	..	..	..	..	..	..	..	263
281	Str. metal prod., tanks, reservoirs, steam generators	..	..	..	..	..	..	..	..	..	..
2811	Structural metal products	..	..	..	..	..	..	..	..	..	..
2812	Tanks, reservoirs and containers of metal	..	..	..	..	..	..	..	..	..	..
2813	Steam generators, exc. central heating hot water boilers	..	..	..	..	..	..	..	..	..	..
289	Other fabricated metal products	..	..	..	..	..	..	..	..	..	..
2891	Forging, pressing, stamping & roll-forming of metal	..	..	..	..	..	..	..	..	..	..
2892	Treatment and coating of metals	..	..	..	..	..	..	..	..	..	..
2893	Cutlery, hand tools and general hardware	..	..	..	..	..	..	..	..	..	..
2899	Other fabricated metal products, nec	..	..	..	..	..	..	..	..	..	..
29	MACHINERY AND EQUIPMENT, NEC	..	..	..	..	..	..	..	..	..	..
291	General purpose machinery	..	..	..	..	..	..	..	..	..	..
2911	Engines and turbines	..	..	..	..	..	..	..	..	..	..
2912	Pumps, compressors, taps and valves	..	..	..	..	..	..	..	..	..	..
2913	Bearings, gears, gearing and driving elements	..	..	..	..	..	..	..	..	..	..
2914	Ovens, furnaces and furnace burners	..	..	..	..	..	..	..	..	..	..
2915	Lifting and handling equipment	..	..	..	..	..	..	..	..	..	..
2919	Other general purpose machinery	..	..	..	..	..	..	..	..	..	..
292	Special purpose machinery	..	..	..	..	..	..	..	..	..	..
2921	Agricultural and forestry machinery	..	..	..	..	..	..	..	..	..	..
2922	Machine-tools	..	..	..	..	..	..	..	..	..	..
2923	Machinery for metallurgy	..	..	..	..	..	..	..	..	..	..
2924	Machinery for mining, quarrying and construction	..	..	..	..	..	..	..	..	..	..
2925	Machinery for food, beverage & tobacco processing	..	..	..	..	..	..	..	..	..	..
2926	Machinery for textile, apparel & leather production	..	..	..	..	..	..	..	..	..	..
2927	Machinery for weapons and ammunition	..	..	..	..	..	..	..	..	..	..
2929	Other special purpose machinery	..	..	..	..	..	..	..	..	..	..
293	Domestic appliances, nec	..	..	..	..	..	..	..	..	..	..
30	OFFICE, ACCOUNTING & COMPUTING MACHINERY	..	..	..	..	..	..	..	..	..	..
31	ELECTRICAL MACHINERY & APPARATUS, NEC	..	..	..	..	..	..	..	..	..	..
311	Electric motors, generators and transformers	..	..	..	..	..	..	..	..	..	..
312	Electricity distribution and control apparatus	..	..	..	..	..	..	..	..	..	..
313	Insulated wire and cable	..	..	..	..	..	..	..	..	..	..
314	Accumulators, primary cells & primary batteries	..	..	..	..	..	..	..	..	..	..
315	Electric lamps and lighting equipment	..	..	..	..	..	..	..	..	..	..
319	Other electrical equipment, nec	..	..	..	..	..	..	..	..	..	..
32	RADIO, TV & COMMUNICATION EQUIP. & APP.	..	..	..	..	..	..	..	..	..	..
321	Electronic valves, tubes, other electronic components	..	..	..	..	..	..	..	..	..	..
322	TV & radio transmitters, apparatus for line telephony	..	..	..	..	..	..	..	..	..	..
323	TV & radio receivers, recording apparatus	..	..	..	..	..	..	..	..	..	..
33	MEDICAL PRECISION &OPTICAL INSTRUMENTS	..	..	..	..	..	..	..	..	..	..
331	Medical appliances and instruments	..	..	..	..	..	..	..	..	..	..
3311	Medical, surgical equipment & orthopaedic app.	..	..	..	..	..	..	..	..	..	..
3312	Instruments & appliances for measuring, checking etc	..	..	..	..	..	..	..	..	..	..
3313	Industrial process control equipment	..	..	..	..	..	..	..	..	..	..
332	Optical instruments and photographic equipment	..	..	..	..	..	..	..	..	..	..
333	Watches and clocks	..	..	..	..	..	..	..	..	..	..
34	MOTOR VEHICLES, TRAILERS & SEMI-TRAILERS	..	..	..	..	..	..	..	..	..	..
341	Motor vehicles	..	..	..	..	..	..	..	..	..	..
342	Bodies (coachwork) for motor vehicles	..	..	..	..	..	..	..	..	..	..
343	Parts, accessories for motor vehicles & their engines	..	..	..	..	..	..	..	..	..	..
35	OTHER TRANSPORT EQUIPMENT	..	..	..	..	..	..	..	..	..	..
351	Building and repairing of ships and boats	..	..	..	..	..	..	..	..	..	..
3511	Building and repairing of ships	..	..	..	..	..	..	..	..	..	..
3512	Building, repairing of pleasure & sporting boats	..	..	..	..	..	..	..	..	..	..
352	Railway, tramway locomotives & rolling stock	..	..	..	..	..	..	..	..	..	..
353	Aircraft and spacecraft	..	..	..	..	..	..	..	..	..	..
359	Transport equipment, nec	..	..	..	..	..	..	..	..	..	..
3591	Motorcycles	..	..	..	..	..	..	..	..	..	..
3592	Bicycles and invalid carriages	..	..	..	..	..	..	..	..	..	..
3599	Other transport equipment, nec	..	..	..	..	..	..	..	..	..	..
36	FURNITURE; MANUFACTURING, NEC	..	..	..	..	..	..	..	..	..	..
361	Furniture	..	..	..	..	..	..	..	..	..	..
369	Manufacturing, nec	..	..	..	..	..	..	..	..	..	..
3691	Jewellery and related articles	..	..	..	..	..	..	..	..	..	..
3692	Musical instruments	..	..	..	..	..	..	..	..	..	..
3693	Sports goods	..	..	..	..	..	..	..	..	..	..
3694	Games and toys	..	..	..	..	..	..	..	..	..	..
3699	Other manufacturing, nec	..	..	..	..	..	..	..	..	..	..
37	RECYCLING	..	..	..	..	..	..	..	..	..	..
371	Recycling of metal waste and scrap	..	..	..	..	..	..	..	..	..	..
372	Recycling of non-metal waste and scrap	..	..	..	..	..	..	..	..	..	..
CERR	Non-specified industry	..	..	..	..	..	..	..	..	..	..
15-37	TOTAL MANUFACTURING	..	..	..	..	..	..	..	..	..	48,000

NETHERLANDS

ISIC Revision 3 Industry Sector		Solid TJ	LPG TJ	Distiloil TJ	RFO TJ	Gas TJ	Biomass TJ	Steam TJ	Electr MWh	Own Use MWh	TOTAL TJ
15	**FOOD PRODUCTS AND BEVERAGES**	2,350	230	830	870	46,610	..	18,890	6,241,667	1,297,222	87,580
151	Production, processing and preserving (PPP)	..	..	..	..	..	..	..	..	..	..
1511	PPP of meat and meat products	..	..	..	..	..	..	..	..	..	..
1512	Processing and preserving of fish products	..	..	..	..	..	..	..	..	..	..
1513	Processing, preserving of fruit & vegetables	..	..	..	..	..	..	..	..	..	..
1514	Vegetable and animal oils and fats	..	..	..	..	..	..	..	..	..	..
152	Dairy products	..	..	..	..	..	..	..	..	..	..
153	Grain mill prod., starches & prepared animal feeds	..	..	..	..	..	..	..	..	..	..
1531	Grain mill products	..	..	..	..	..	..	..	..	..	..
1532	Starches and starch products	..	..	..	..	..	..	..	..	..	..
1533	Prepared animal feeds	..	..	..	..	..	..	..	..	..	..
154	Other food products	..	..	..	..	..	..	..	..	..	..
1541	Bakery products	..	..	..	..	..	..	..	..	..	..
1542	Sugar	..	..	..	..	..	..	..	..	..	..
1543	Cocoa, chocolate and sugar confectionery	..	..	..	..	..	..	..	..	..	..
1544	Macaroni, noodles, couscous & similar farinaceous prod.	..	..	..	..	..	..	..	..	..	..
1549	Other food products, nec	..	..	..	..	..	..	..	..	..	..
155	Beverages	..	..	..	..	..	..	..	..	..	..
1551	Distilling, rectifying and blending of spirits	..	..	..	..	..	..	..	..	..	..
1552	Wines	..	..	..	..	..	..	..	..	..	..
1553	Malt liquors and malt	..	..	..	..	..	..	..	..	..	..
1554	Soft drinks; production of mineral waters	..	..	..	..	..	..	..	..	..	..
16	**TOBACCO PRODUCTS**	..	..	..	..	..	..	..	..	..	..
17	**TEXTILES**	40	0	20	20	5,700	..	590	530,556	30,556	8,170
171	Spinning, weaving and finishing of textiles	..	..	..	..	..	..	..	..	..	..
1711	Preparation and spinning of textile fibres	..	..	..	..	..	..	..	..	..	..
1712	Finishing of textiles	..	..	..	..	..	..	..	..	..	..
172	Other textiles	..	..	..	..	..	..	..	..	..	..
1721	Made-up textile articles, except apparel	..	..	..	..	..	..	..	..	..	..
1722	Carpets and rugs	..	..	..	..	..	..	..	..	..	..
1723	Cordage, rope, twine and netting	..	..	..	..	..	..	..	..	..	..
1729	Other textiles, nec	..	..	..	..	..	..	..	..	..	..
173	Knitted and crocheted fabrics and articles	..	..	..	..	..	..	..	..	..	..
18	**WEARING APPAREL, DRESSING & DYEING OF FUR**	..	..	..	..	..	..	..	..	..	..
181	Wearing apparel, except fur apparel	..	..	..	..	..	..	..	..	..	..
182	Dressing and dyeing of fur; articles of fur	..	..	..	..	..	..	..	..	..	..
19	**TANNING & DRESSING OF LEATHER, FOOTWEAR**	..	..	..	..	..	..	..	..	..	..
191	Tanning and dressing of leather	..	..	..	..	..	..	..	..	..	..
1911	Tanning and dressing of leather	..	..	..	..	..	..	..	..	..	..
1912	Luggage, handbags, saddlery & harness	..	..	..	..	..	..	..	..	..	..
192	Footwear	..	..	..	..	..	..	..	..	..	..
20	**WOOD AND WOOD PRODUCTS**	..	..	..	..	..	..	..	..	..	..
201	Sawmilling and planing of wood	..	..	..	..	..	..	..	..	..	..
202	Products of wood, cork, straw & plaiting materials	..	..	..	..	..	..	..	..	..	..
2021	Veneer sheets	..	..	..	..	..	..	..	..	..	..
2022	Builders' carpentry and joinery	..	..	..	..	..	..	..	..	..	..
2023	Wooden containers	..	..	..	..	..	..	..	..	..	..
2029	Other products of wood	..	..	..	..	..	..	..	..	..	..
21	**PAPER AND PAPER PRODUCTS**	0	30	10	60	10,410	..	14,100	3,127,778	1,227,778	31,450
2101	Pulp, paper and paperboard	..	..	..	..	..	..	..	..	..	..
2102	Corrugated paper, paperboard and their containers	..	..	..	..	..	..	..	..	..	..
2109	Other articles of pulp and paperboard	..	..	..	..	..	..	..	..	..	..
22	**PUBLISHING, PRINTING & REPRODUCTION**	..	..	..	..	..	..	..	..	..	..
221	Publishing	..	..	..	..	..	..	..	..	..	..
2211	Publishing of books & brochures	..	..	..	..	..	..	..	..	..	..
2212	Publishing of newspapers and periodicals	..	..	..	..	..	..	..	..	..	..
2213	Publishing of recorded media	..	..	..	..	..	..	..	..	..	..
2219	Other publishing	..	..	..	..	..	..	..	..	..	..
222	Printing and related service activities	..	..	..	..	..	..	..	..	..	..
2221	Printing	..	..	..	..	..	..	..	..	..	..
2222	Service activities related to printing	..	..	..	..	..	..	..	..	..	..
223	Reproduction of recorded media	..	..	..	..	..	..	..	..	..	..
23	**COKE, REFINED PETROLEUM PRODUCTS**	0	60	12,350	24,110	86,330	..	16,450	2,036,111	2,019,444	139,360
231	Coke oven products	..	..	..	..	..	..	..	..	..	..
232	Refined petroleum products	0	60	12,350	24,110	86,330	..	16,450	2,036,111	2,019,444	139,360
233	Processing of nuclear fuel	..	..	..	..	..	..	..	..	..	..
24	**CHEMICALS & CHEMICAL PRODUCTS**	210	290	61,550	730	86,730	..	77,950	8,230,556	4,005,556	242,670
241	Basic chemicals	130	130	61,390	710	78,870	..	67,090	6,525,000	2,963,889	221,140
2411	Basic chemicals, exc. fertilizers & nitrogen compounds	130	130	61,230	450	45,930	..	64,300	5,536,111	2,266,667	183,940
2412	Fertilizers and nitrogen compounds	0	0	160	260	32,940	..	2,790	988,889	697,222	37,200
2413	Plastics in primary forms and synthetic rubber	..	..	..	..	..	..	..	..	..	..
242	Other chemical products	80	160	160	20	7,860	..	10,860	1,705,556	1,041,667	21,530
2421	Pesticides and other agro-chemical products	..	..	..	..	..	..	..	..	..	..
2422	Paints, varnishes and similar coatings	..	..	..	..	..	..	..	..	..	..
2423	Pharmaceuticals, medicinal chem. & botanical prod.	..	..	..	..	..	..	..	..	..	..
2424	Soap and detergents, perfumes etc.	..	..	..	..	..	..	..	..	..	..
2429	Other chemical products, nec	..	..	..	..	..	..	..	..	..	..
243	Man-made fibres	..	..	..	..	..	..	..	..	..	..
25	**RUBBER AND PLASTICS PRODUCTS**	0	0	240	90	5,360	..	1,040	1,891,667	0	13,540
251	Rubber products	..	..	..	..	..	..	..	..	..	..
2511	Rubber tyres and tubes	..	..	..	..	..	..	..	..	..	..
2519	Other rubber products	..	..	..	..	..	..	..	..	..	..
252	Plastic products	..	..	..	..	..	..	..	..	..	..
26	**OTHER NON-METALLIC MINERAL PRODUCTS**	2,160	50	590	3,640	26,130	..	730	1,480,556	86,111	38,320
261	Glass and glass products	..	..	..	..	..	..	..	..	..	..
269	Non-metallic mineral products, nec	..	..	..	..	..	..	..	..	..	..
2691	Pottery, china and earthenware	..	..	..	..	..	..	..	..	..	..
2692	Refractory ceramic products	..	..	..	..	..	..	..	..	..	..
2693	Structural non-refractory clay & ceramic prod.	..	..	..	..	..	..	..	..	..	..
2694	Cement, lime and plaster	..	..	..	..	..	..	..	..	..	..
2695	Articles of concrete, cement and plaster	..	..	..	..	..	..	..	..	..	..
2696	Cutting, shaping and finishing of stone	..	..	..	..	..	..	..	..	..	..
2699	Other non-metallic mineral products, nec	..	..	..	..	..	..	..	..	..	..

ISIS Energy Data Programme (IEA/OECD)

ISIC Revision 3 Industry Sector		Solid TJ	LPG TJ	Distiloil TJ	RFO TJ	Gas TJ	Biomass TJ	Steam TJ	Electr MWh	Own Use MWh	TOTAL TJ
27	**BASIC METALS**	52,470	40	190	0	33,520	..	4,100	3,413,889	230,556	101,780
271	Basic iron and steel	52,470	40	140	0	29,960	..	2,700	2,088,889	213,889	92,060
272	Basic precious and non-ferrous metals	0	0	50	0	3,560	..	1,400	1,325,000	16,667	9,720
273	Casting of metals	..	..	..	..	..	..	..	..	..	..
2731	Casting of iron and steel	..	..	..	..	..	..	..	..	..	..
2732	Casting non-ferrous metals	..	..	..	..	..	..	..	..	..	..
28	**FABRICATED METAL PRODUCTS**	10	230	680	60	22,270	..	910	3,363,889	111,111	35,870
281	Str. metal prod., tanks, reservoirs, steam generators	..	..	..	..	..	..	..	..	..	..
2811	Structural metal products	..	..	..	..	..	..	..	..	..	..
2812	Tanks, reservoirs and containers of metal	..	..	..	..	..	..	..	..	..	..
2813	Steam generators, exc. central heating hot water boilers	..	..	..	..	..	..	..	..	..	..
289	Other fabricated metal products	..	..	..	..	..	..	..	..	..	..
2891	Forging, pressing, stamping & roll-forming of metal	..	..	..	..	..	..	..	..	..	..
2892	Treatment and coating of metals	..	..	..	..	..	..	..	..	..	..
2893	Cutlery, hand tools and general hardware	..	..	..	..	..	..	..	..	..	..
2899	Other fabricated metal products, nec	..	..	..	..	..	..	..	..	..	..
29	**MACHINERY AND EQUIPMENT, NEC**	..	..	..	..	..	..	..	..	..	..
291	General purpose machinery	..	..	..	..	..	..	..	..	..	..
2911	Engines and turbines	..	..	..	..	..	..	..	..	..	..
2912	Pumps, compressors, taps and valves	..	..	..	..	..	..	..	..	..	..
2913	Bearings, gears, gearing and driving elements	..	..	..	..	..	..	..	..	..	..
2914	Ovens, furnaces and furnace burners	..	..	..	..	..	..	..	..	..	..
2915	Lifting and handling equipment	..	..	..	..	..	..	..	..	..	..
2919	Other general purpose machinery	..	..	..	..	..	..	..	..	..	..
292	Special purpose machinery	..	..	..	..	..	..	..	..	..	..
2921	Agricultural and forestry machinery	..	..	..	..	..	..	..	..	..	..
2922	Machine-tools	..	..	..	..	..	..	..	..	..	..
2923	Machinery for metallurgy	..	..	..	..	..	..	..	..	..	..
2924	Machinery for mining, quarrying and construction	..	..	..	..	..	..	..	..	..	..
2925	Machinery for food, beverage & tobacco processing	..	..	..	..	..	..	..	..	..	..
2926	Machinery for textile, apparel & leather production	..	..	..	..	..	..	..	..	..	..
2927	Machinery for weapons and ammunition	..	..	..	..	..	..	..	..	..	..
2929	Other special purpose machinery	..	..	..	..	..	..	..	..	..	..
293	Domestic appliances, nec	..	..	..	..	..	..	..	..	..	..
30	**OFFICE, ACCOUNTING & COMPUTING MACHINERY**	..	..	..	..	..	..	..	..	..	..
31	**ELECTRICAL MACHINERY & APPARATUS, NEC**	..	..	..	..	..	..	..	..	..	..
311	Electric motors, generators and transformers	..	..	..	..	..	..	..	..	..	..
312	Electricity distribution and control apparatus	..	..	..	..	..	..	..	..	..	..
313	Insulated wire and cable	..	..	..	..	..	..	..	..	..	..
314	Accumulators, primary cells & primary batteries	..	..	..	..	..	..	..	..	..	..
315	Electric lamps and lighting equipment	..	..	..	..	..	..	..	..	..	..
319	Other electrical equipment, nec	..	..	..	..	..	..	..	..	..	..
32	**RADIO, TV & COMMUNICATION EQUIP. & APP.**	..	..	..	..	..	..	..	..	..	..
321	Electronic valves, tubes, other electronic components	..	..	..	..	..	..	..	..	..	..
322	TV & radio transmitters, apparatus for line telephony	..	..	..	..	..	..	..	..	..	..
323	TV & radio receivers, recording apparatus	..	..	..	..	..	..	..	..	..	..
33	**MEDICAL PRECISION &OPTICAL INSTRUMENTS**	..	..	..	..	..	..	..	..	..	..
331	Medical appliances and instruments	..	..	..	..	..	..	..	..	..	..
3311	Medical, surgical equipment & orthopaedic app.	..	..	..	..	..	..	..	..	..	..
3312	Instruments & appliances for measuring, checking etc	..	..	..	..	..	..	..	..	..	..
3313	Industrial process control equipment	..	..	..	..	..	..	..	..	..	..
332	Optical instruments and photographic equipment	..	..	..	..	..	..	..	..	..	..
333	Watches and clocks	..	..	..	..	..	..	..	..	..	..
34	**MOTOR VEHICLES, TRAILERS & SEMI-TRAILERS**	..	..	..	..	..	..	..	..	..	..
341	Motor vehicles	..	..	..	..	..	..	..	..	..	..
342	Bodies (coachwork) for motor vehicles	..	..	..	..	..	..	..	..	..	..
343	Parts, accessories for motor vehicles & their engines	..	..	..	..	..	..	..	..	..	..
35	**OTHER TRANSPORT EQUIPMENT**	..	..	..	..	..	..	..	..	..	..
351	Building and repairing of ships and boats	..	..	..	..	..	..	..	..	..	..
3511	Building and repairing of ships	..	..	..	..	..	..	..	..	..	..
3512	Building, repairing of pleasure & sporting boats	..	..	..	..	..	..	..	..	..	..
352	Railway, tramway locomotives & rolling stock	..	..	..	..	..	..	..	..	..	..
353	Aircraft and spacecraft	..	..	..	..	..	..	..	..	..	..
359	Transport equipment, nec	..	..	..	..	..	..	..	..	..	..
3591	Motorcycles	..	..	..	..	..	..	..	..	..	..
3592	Bicycles and invalid carriages	..	..	..	..	..	..	..	..	..	..
3599	Other transport equipment, nec	..	..	..	..	..	..	..	..	..	..
36	**FURNITURE; MANUFACTURING, NEC**	..	..	..	..	..	..	..	..	..	..
361	Furniture	..	..	..	..	..	..	..	..	..	..
369	Manufacturing, nec	..	..	..	..	..	..	..	..	..	..
3691	Jewellery and related articles	..	..	..	..	..	..	..	..	..	..
3692	Musical instruments	..	..	..	..	..	..	..	..	..	..
3693	Sports goods	..	..	..	..	..	..	..	..	..	..
3694	Games and toys	..	..	..	..	..	..	..	..	..	..
3699	Other manufacturing, nec	..	..	..	..	..	..	..	..	..	..
37	**RECYCLING**	..	..	..	..	..	..	..	..	..	..
371	Recycling of metal waste and scrap	..	..	..	..	..	..	..	..	..	..
372	Recycling of non-metal waste and scrap	..	..	..	..	..	..	..	..	..	..
CERR	**Non-specified industry**	..	..	..	..	..	..	..	..	..	..
15-37	**TOTAL MANUFACTURING**	57,240	930	76,460	29,580	323,060	..	134,760	30,316,669	9,008,334	**698,740**

ISIC Revision 3 Industry Sector	Solid TJ	LPG TJ	Distiloil TJ	RFO TJ	Gas TJ	Biomass TJ	Steam TJ	Electr MWh	Own Use MWh	TOTAL TJ
15 FOOD PRODUCTS AND BEVERAGES	1,470	140	940	720	45,270	..	19,390	5,752,778	1,422,222	83,520
151 Production, processing and preserving (PPP)	..	..	..	..	..	..	..	..	..	..
1511 PPP of meat and meat products	..	..	..	..	..	..	..	..	..	..
1512 Processing and preserving of fish products	..	..	..	..	..	..	..	..	..	..
1513 Processing, preserving of fruit & vegetables	..	..	..	..	..	..	..	..	..	..
1514 Vegetable and animal oils and fats	..	..	..	..	..	..	..	..	..	..
152 Dairy products	..	..	..	..	..	..	..	..	..	..
153 Grain mill prod., starches & prepared animal feeds	..	..	..	..	..	..	..	..	..	..
1531 Grain mill products	..	..	..	..	..	..	..	..	..	..
1532 Starches and starch products	..	..	..	..	..	..	..	..	..	..
1533 Prepared animal feeds	..	..	..	..	..	..	..	..	..	..
154 Other food products	..	..	..	..	..	..	..	..	..	..
1541 Bakery products	..	..	..	..	..	..	..	..	..	..
1542 Sugar	..	..	..	..	..	..	..	..	..	..
1543 Cocoa, chocolate and sugar confectionery	..	..	..	..	..	..	..	..	..	..
1544 Macaroni, noodles, couscous & similar farinaceous prod.	..	..	..	..	..	..	..	..	..	..
1549 Other food products, nec	..	..	..	..	..	..	..	..	..	..
155 Beverages	..	..	..	..	..	..	..	..	..	..
1551 Distilling, rectifying and blending of spirits	..	..	..	..	..	..	..	..	..	..
1552 Wines	..	..	..	..	..	..	..	..	..	..
1553 Malt liquors and malt	..	..	..	..	..	..	..	..	..	..
1554 Soft drinks; production of mineral waters	..	..	..	..	..	..	..	..	..	..
16 TOBACCO PRODUCTS	..	..	..	..	..	..	..	..	..	..
17 TEXTILES	20	0	20	10	5,640	..	420	577,778	22,222	8,110
171 Spinning, weaving and finishing of textiles	..	..	..	..	..	..	..	..	..	..
1711 Preparation and spinning of textile fibres	..	..	..	..	..	..	..	..	..	..
1712 Finishing of textiles	..	..	..	..	..	..	..	..	..	..
172 Other textiles	..	..	..	..	..	..	..	..	..	..
1721 Made-up textile articles, except apparel	..	..	..	..	..	..	..	..	..	..
1722 Carpets and rugs	..	..	..	..	..	..	..	..	..	..
1723 Cordage, rope, twine and netting	..	..	..	..	..	..	..	..	..	..
1729 Other textiles, nec	..	..	..	..	..	..	..	..	..	..
173 Knitted and crocheted fabrics and articles	..	..	..	..	..	..	..	..	..	..
18 WEARING APPAREL, DRESSING & DYEING OF FUR	..	..	..	..	..	..	..	..	..	..
181 Wearing apparel, except fur apparel	..	..	..	..	..	..	..	..	..	..
182 Dressing and dyeing of fur; articles of fur	..	..	..	..	..	..	..	..	..	..
19 TANNING & DRESSING OF LEATHER, FOOTWEAR	..	..	..	..	..	..	..	..	..	..
191 Tanning and dressing of leather	..	..	..	..	..	..	..	..	..	..
1911 Tanning and dressing of leather	..	..	..	..	..	..	..	..	..	..
1912 Luggage, handbags, saddlery & harness	..	..	..	..	..	..	..	..	..	..
192 Footwear	..	..	..	..	..	..	..	..	..	..
20 WOOD AND WOOD PRODUCTS	..	..	..	..	..	..	..	..	..	..
201 Sawmilling and planing of wood	..	..	..	..	..	..	..	..	..	..
202 Products of wood, cork, straw & plaiting materials	..	..	..	..	..	..	..	..	..	..
2021 Veneer sheets	..	..	..	..	..	..	..	..	..	..
2022 Builders' carpentry and joinery	..	..	..	..	..	..	..	..	..	..
2023 Wooden containers	..	..	..	..	..	..	..	..	..	..
2029 Other products of wood	..	..	..	..	..	..	..	..	..	..
21 PAPER AND PAPER PRODUCTS	10	40	60	80	8,940	..	14,720	3,172,222	1,313,889	30,540
2101 Pulp, paper and paperboard	..	..	..	..	..	..	..	..	..	..
2102 Corrugated paper, paperboard and their containers	..	..	..	..	..	..	..	..	..	..
2109 Other articles of pulp and paperboard	..	..	..	..	..	..	..	..	..	..
22 PUBLISHING, PRINTING & REPRODUCTION	..	..	..	..	..	..	..	..	..	..
221 Publishing	..	..	..	..	..	..	..	..	..	..
2211 Publishing of books & brochures	..	..	..	..	..	..	..	..	..	..
2212 Publishing of newspapers and periodicals	..	..	..	..	..	..	..	..	..	..
2213 Publishing of recorded media	..	..	..	..	..	..	..	..	..	..
2219 Other publishing	..	..	..	..	..	..	..	..	..	..
222 Printing and related service activities	..	..	..	..	..	..	..	..	..	..
2221 Printing	..	..	..	..	..	..	..	..	..	..
2222 Service activities related to printing	..	..	..	..	..	..	..	..	..	..
223 Reproduction of recorded media	..	..	..	..	..	..	..	..	..	..
23 COKE, REFINED PETROLEUM PRODUCTS	0	120	12,300	25,430	88,780	..	16,200	2,088,889	1,994,444	143,170
231 Coke oven products	..	..	..	..	..	..	..	..	..	..
232 Refined petroleum products	0	120	12,300	25,430	88,780	..	16,200	2,088,889	1,994,444	143,170
233 Processing of nuclear fuel	..	..	..	..	..	..	..	..	..	..
24 CHEMICALS & CHEMICAL PRODUCTS	330	450	64,050	280	77,570	..	86,250	8,627,777	3,697,222	246,680
241 Basic chemicals	150	320	64,050	280	67,720	..	74,200	6,619,444	2,597,222	221,200
2411 Basic chemicals, exc. fertilizers & nitrogen compounds	150	320	64,040	40	39,330	..	71,450	5,600,000	1,888,889	188,690
2412 Fertilizers and nitrogen compounds	0	0	10	240	28,390	..	2,750	1,019,444	708,333	32,510
2413 Plastics in primary forms and synthetic rubber	..	..	..	..	..	..	..	..	..	..
242 Other chemical products	180	130	0	0	9,850	..	12,050	2,008,333	1,100,000	25,480
2421 Pesticides and other agro-chemical products	..	..	..	..	..	..	..	..	..	..
2422 Paints, varnishes and similar coatings	..	..	..	..	..	..	..	..	..	..
2423 Pharmaceuticals, medicinal chem. & botanical prod.	..	..	..	..	..	..	..	..	..	..
2424 Soap and detergents, perfumes etc.	..	..	..	..	..	..	..	..	..	..
2429 Other chemical products, nec	..	..	..	..	..	..	..	..	..	..
243 Man-made fibres	..	..	..	..	..	..	..	..	..	..
25 RUBBER AND PLASTICS PRODUCTS	0	10	140	80	5,330	..	900	1,961,111	2,778	13,510
251 Rubber products	..	..	..	..	..	..	..	..	..	..
2511 Rubber tyres and tubes	..	..	..	..	..	..	..	..	..	..
2519 Other rubber products	..	..	..	..	..	..	..	..	..	..
252 Plastic products	..	..	..	..	..	..	..	..	..	..
26 OTHER NON-METALLIC MINERAL PRODUCTS	2,100	40	410	2,740	26,770	..	750	1,441,667	88,889	37,680
261 Glass and glass products	..	..	..	..	..	..	..	..	..	..
269 Non-metallic mineral products, nec	..	..	..	..	..	..	..	..	..	..
2691 Pottery, china and earthenware	..	..	..	..	..	..	..	..	..	..
2692 Refractory ceramic products	..	..	..	..	..	..	..	..	..	..
2693 Structural non-refractory clay & ceramic prod.	..	..	..	..	..	..	..	..	..	..
2694 Cement, lime and plaster	..	..	..	..	..	..	..	..	..	..
2695 Articles of concrete, cement and plaster	..	..	..	..	..	..	..	..	..	..
2696 Cutting, shaping and finishing of stone	..	..	..	..	..	..	..	..	..	..
2699 Other non-metallic mineral products, nec	..	..	..	..	..	..	..	..	..	..

ISIC Revision 3 Industry Sector	Solid TJ	LPG TJ	Distiloil TJ	RFO TJ	Gas TJ	Biomass TJ	Steam TJ	Electr MWh	Own Use MWh	TOTAL TJ
27 BASIC METALS	41,020	30	650	0	32,580	..	4,920	3,566,666	219,445	91,250
271 Basic iron and steel	41,020	30	150	0	28,720	..	3,500	2,208,333	205,556	80,630
272 Basic precious and non-ferrous metals	0	0	500	0	3,860	..	1,420	1,358,333	13,889	10,620
273 Casting of metals	..	..	..	..	..	..	..	..	..	..
2731 Casting of iron and steel	..	..	..	..	..	..	..	..	..	..
2732 Casting non-ferrous metals	..	..	..	..	..	..	..	..	..	..
28 FABRICATED METAL PRODUCTS	100	220	670	50	20,760	..	860	3,633,333	19,444	35,670
281 Str. metal prod., tanks, reservoirs, steam generators	..	..	..	..	..	..	..	..	..	..
2811 Structural metal products	..	..	..	..	..	..	..	..	..	..
2812 Tanks, reservoirs and containers of metal	..	..	..	..	..	..	..	..	..	..
2813 Steam generators, exc. central heating hot water boilers	..	..	..	..	..	..	..	..	..	..
289 Other fabricated metal products	..	..	..	..	..	..	..	..	..	..
2891 Forging, pressing, stamping & roll-forming of metal	..	..	..	..	..	..	..	..	..	..
2892 Treatment and coating of metals	..	..	..	..	..	..	..	..	..	..
2893 Cutlery, hand tools and general hardware	..	..	..	..	..	..	..	..	..	..
2899 Other fabricated metal products, nec	..	..	..	..	..	..	..	..	..	..
29 MACHINERY AND EQUIPMENT, NEC	..	..	..	..	..	..	..	..	..	..
291 General purpose machinery	..	..	..	..	..	..	..	..	..	..
2911 Engines and turbines	..	..	..	..	..	..	..	..	..	..
2912 Pumps, compressors, taps and valves	..	..	..	..	..	..	..	..	..	..
2913 Bearings, gears, gearing and driving elements	..	..	..	..	..	..	..	..	..	..
2914 Ovens, furnaces and furnace burners	..	..	..	..	..	..	..	..	..	..
2915 Lifting and handling equipment	..	..	..	..	..	..	..	..	..	..
2919 Other general purpose machinery	..	..	..	..	..	..	..	..	..	..
292 Special purpose machinery	..	..	..	..	..	..	..	..	..	..
2921 Agricultural and forestry machinery	..	..	..	..	..	..	..	..	..	..
2922 Machine-tools	..	..	..	..	..	..	..	..	..	..
2923 Machinery for metallurgy	..	..	..	..	..	..	..	..	..	..
2924 Machinery for mining, quarrying and construction	..	..	..	..	..	..	..	..	..	..
2925 Machinery for food, beverage & tobacco processing	..	..	..	..	..	..	..	..	..	..
2926 Machinery for textile, apparel & leather production	..	..	..	..	..	..	..	..	..	..
2927 Machinery for weapons and ammunition	..	..	..	..	..	..	..	..	..	..
2929 Other special purpose machinery	..	..	..	..	..	..	..	..	..	..
293 Domestic appliances, nec	..	..	..	..	..	..	..	..	..	..
30 OFFICE, ACCOUNTING & COMPUTING MACHINERY	..	..	..	..	..	..	..	..	..	..
31 ELECTRICAL MACHINERY & APPARATUS, NEC	..	..	..	..	..	..	..	..	..	..
311 Electric motors, generators and transformers	..	..	..	..	..	..	..	..	..	..
312 Electricity distribution and control apparatus	..	..	..	..	..	..	..	..	..	..
313 Insulated wire and cable	..	..	..	..	..	..	..	..	..	..
314 Accumulators, primary cells & primary batteries	..	..	..	..	..	..	..	..	..	..
315 Electric lamps and lighting equipment	..	..	..	..	..	..	..	..	..	..
319 Other electrical equipment, nec	..	..	..	..	..	..	..	..	..	..
32 RADIO, TV & COMMUNICATION EQUIP. & APP.	..	..	..	..	..	..	..	..	..	..
321 Electronic valves, tubes, other electronic components	..	..	..	..	..	..	..	..	..	..
322 TV & radio transmitters, apparatus for line telephony	..	..	..	..	..	..	..	..	..	..
323 TV & radio receivers, recording apparatus	..	..	..	..	..	..	..	..	..	..
33 MEDICAL PRECISION & OPTICAL INSTRUMENTS	..	..	..	..	..	..	..	..	..	..
331 Medical appliances and instruments	..	..	..	..	..	..	..	..	..	..
3311 Medical, surgical equipment & orthopaedic app.	..	..	..	..	..	..	..	..	..	..
3312 Instruments & appliances for measuring, checking etc	..	..	..	..	..	..	..	..	..	..
3313 Industrial process control equipment	..	..	..	..	..	..	..	..	..	..
332 Optical instruments and photographic equipment	..	..	..	..	..	..	..	..	..	..
333 Watches and clocks	..	..	..	..	..	..	..	..	..	..
34 MOTOR VEHICLES, TRAILERS & SEMI-TRAILERS	..	..	..	..	..	..	..	..	..	..
341 Motor vehicles	..	..	..	..	..	..	..	..	..	..
342 Bodies (coachwork) for motor vehicles	..	..	..	..	..	..	..	..	..	..
343 Parts, accessories for motor vehicles & their engines	..	..	..	..	..	..	..	..	..	..
35 OTHER TRANSPORT EQUIPMENT	..	..	..	..	..	..	..	..	..	..
351 Building and repairing of ships and boats	..	..	..	..	..	..	..	..	..	..
3511 Building and repairing of ships	..	..	..	..	..	..	..	..	..	..
3512 Building, repairing of pleasure & sporting boats	..	..	..	..	..	..	..	..	..	..
352 Railway, tramway locomotives & rolling stock	..	..	..	..	..	..	..	..	..	..
353 Aircraft and spacecraft	..	..	..	..	..	..	..	..	..	..
359 Transport equipment, nec	..	..	..	..	..	..	..	..	..	..
3591 Motorcycles	..	..	..	..	..	..	..	..	..	..
3592 Bicycles and invalid carriages	..	..	..	..	..	..	..	..	..	..
3599 Other transport equipment, nec	..	..	..	..	..	..	..	..	..	..
36 FURNITURE; MANUFACTURING, NEC	0	0	0	0	0	..	220	0	0	220
361 Furniture	..	..	..	..	..	..	..	..	..	..
369 Manufacturing, nec	0	0	0	0	0	..	220	0	0	220
3691 Jewellery and related articles	..	..	..	..	..	..	..	..	..	..
3692 Musical instruments	..	..	..	..	..	..	..	..	..	..
3693 Sports goods	..	..	..	..	..	..	..	..	..	..
3694 Games and toys	..	..	..	..	..	..	..	..	..	..
3699 Other manufacturing, nec	0	0	0	0	0	..	220	0	0	220
37 RECYCLING	..	..	..	..	..	..	..	..	..	..
371 Recycling of metal waste and scrap	..	..	..	..	..	..	..	..	..	..
372 Recycling of non-metal waste and scrap	..	..	..	..	..	..	..	..	..	..
CERR Non-specified industry	..	..	..	..	..	..	..	..	..	..
15-37 TOTAL MANUFACTURING	45,050	1,050	79,240	29,390	311,640	..	144,630	30,822,221	8,780,555	690,350

ISIC Revision 3 Industry Sector		Solid TJ	LPG TJ	Distiloil TJ	RFO TJ	Gas TJ	Biomass TJ	Steam TJ	Electr MWh	Own Use MWh	TOTAL TJ
15	**FOOD PRODUCTS AND BEVERAGES**	1,270	90	970	430	44,810		21,410	5,838,889	1,369,444	85,070
151	Production, processing and preserving (PPP)	..	..	..	..	..		..	..	..	..
1511	PPP of meat and meat products	..	..	..	..	..		..	..	..	..
1512	Processing and preserving of fish products	..	..	..	..	..		..	..	..	..
1513	Processing, preserving of fruit & vegetables	..	..	..	..	..		..	..	..	..
1514	Vegetable and animal oils and fats	..	..	..	..	..		..	..	..	..
152	Dairy products	..	..	..	..	..		..	..	..	..
153	Grain mill prod., starches & prepared animal feeds	..	..	..	..	..		..	..	..	..
1531	Grain mill products	..	..	..	..	..		..	..	..	..
1532	Starches and starch products	..	..	..	..	..		..	..	..	..
1533	Prepared animal feeds	..	..	..	..	..		..	..	..	..
154	Other food products	..	..	..	..	..		..	..	..	..
1541	Bakery products	..	..	..	..	..		..	..	..	..
1542	Sugar	..	..	..	..	..		..	..	..	..
1543	Cocoa, chocolate and sugar confectionery	..	..	..	..	..		..	..	..	..
1544	Macaroni, noodles, couscous & similar farinaceous prod.	..	..	..	..	..		..	..	..	..
1549	Other food products, nec	..	..	..	..	..		..	..	..	..
155	Beverages	..	..	..	..	..		..	..	..	..
1551	Distilling, rectifying and blending of spirits	..	..	..	..	..		..	..	..	..
1552	Wines	..	..	..	..	..		..	..	..	..
1553	Malt liquors and malt	..	..	..	..	..		..	..	..	..
1554	Soft drinks; production of mineral waters	..	..	..	..	..		..	..	..	..
16	**TOBACCO PRODUCTS**	..	..	..	..	..		..	..	..	..
17	**TEXTILES**	20	10	10	0	6,140		560	572,222	25,000	8,710
171	Spinning, weaving and finishing of textiles	..	..	..	..	..		..	..	..	..
1711	Preparation and spinning of textile fibres	..	..	..	..	..		..	..	..	..
1712	Finishing of textiles	..	..	..	..	..		..	..	..	..
172	Other textiles	..	..	..	..	..		..	..	..	..
1721	Made-up textile articles, except apparel	..	..	..	..	..		..	..	..	..
1722	Carpets and rugs	..	..	..	..	..		..	..	..	..
1723	Cordage, rope, twine and netting	..	..	..	..	..		..	..	..	..
1729	Other textiles, nec	..	..	..	..	..		..	..	..	..
173	Knitted and crocheted fabrics and articles	..	..	..	..	..		..	..	..	..
18	**WEARING APPAREL, DRESSING & DYEING OF FUR**	..	..	..	..	..		..	..	..	..
181	Wearing apparel, except fur apparel	..	..	..	..	..		..	..	..	..
182	Dressing and dyeing of fur; articles of fur	..	..	..	..	..		..	..	..	..
19	**TANNING & DRESSING OF LEATHER, FOOTWEAR**	..	..	..	..	..		..	..	..	..
191	Tanning and dressing of leather	..	..	..	..	..		..	..	..	..
1911	Tanning and dressing of leather	..	..	..	..	..		..	..	..	..
1912	Luggage, handbags, saddlery & harness	..	..	..	..	..		..	..	..	..
192	Footwear	..	..	..	..	..		..	..	..	..
20	**WOOD AND WOOD PRODUCTS**	..	..	..	..	..		..	..	..	..
201	Sawmilling and planing of wood	..	..	..	..	..		..	..	..	..
202	Products of wood, cork, straw & plaiting materials	..	..	..	..	..		..	..	..	..
2021	Veneer sheets	..	..	..	..	..		..	..	..	..
2022	Builders' carpentry and joinery	..	..	..	..	..		..	..	..	..
2023	Wooden containers	..	..	..	..	..		..	..	..	..
2029	Other products of wood	..	..	..	..	..		..	..	..	..
21	**PAPER AND PAPER PRODUCTS**	0	40	60	20	9,320		15,350	3,252,778	1,019,444	32,830
2101	Pulp, paper and paperboard	..	..	..	..	..		..	..	..	..
2102	Corrugated paper, paperboard and their containers	..	..	..	..	..		..	..	..	..
2109	Other articles of pulp and paperboard	..	..	..	..	..		..	..	..	..
22	**PUBLISHING, PRINTING & REPRODUCTION**	..	..	..	..	..		..	..	..	..
221	Publishing	..	..	..	..	..		..	..	..	..
2211	Publishing of books & brochures	..	..	..	..	..		..	..	..	..
2212	Publishing of newspapers and periodicals	..	..	..	..	..		..	..	..	..
2213	Publishing of recorded media	..	..	..	..	..		..	..	..	..
2219	Other publishing	..	..	..	..	..		..	..	..	..
222	Printing and related service activities	..	..	..	..	..		..	..	..	..
2221	Printing	..	..	..	..	..		..	..	..	..
2222	Service activities related to printing	..	..	..	..	..		..	..	..	..
223	Reproduction of recorded media	..	..	..	..	..		..	..	..	..
23	**COKE, REFINED PETROLEUM PRODUCTS**	0	60	13,410	25,020	83,380		8,000	4,661,111	4,661,111	129,870
231	Coke oven products	..	..	..	..	..		..	..	..	..
232	Refined petroleum products	0	60	13,410	25,020	83,380		8,000	4,661,111	4,661,111	129,870
233	Processing of nuclear fuel	..	..	..	..	..		..	..	..	..
24	**CHEMICALS & CHEMICAL PRODUCTS**	160	520	65,550	620	87,830		96,410	9,780,555	3,211,112	274,740
241	Basic chemicals	0	430	65,390	620	79,170		81,040	7,847,222	2,033,334	247,580
2411	Basic chemicals, exc. fertilizers & nitrogen compounds	0	430	65,380	410	51,620		77,710	6,822,222	1,341,667	215,280
2412	Fertilizers and nitrogen compounds	0	0	10	210	27,550		3,330	1,025,000	691,667	32,300
2413	Plastics in primary forms and synthetic rubber	..	..	..	..	..		..	..	..	..
242	Other chemical products	160	90	160	0	8,660		15,370	1,933,333	1,177,778	27,160
2421	Pesticides and other agro-chemical products	..	..	..	..	..		..	..	..	..
2422	Paints, varnishes and similar coatings	..	..	..	..	..		..	..	..	..
2423	Pharmaceuticals, medicinal chem. & botanical prod.	..	..	..	..	..		..	..	..	..
2424	Soap and detergents, perfumes etc.	..	..	..	..	..		..	..	..	..
2429	Other chemical products, nec	..	..	..	..	..		..	..	..	..
243	Man-made fibres	..	..	..	..	..		..	..	..	..
25	**RUBBER AND PLASTICS PRODUCTS**	0	10	80	100	5,600		910	1,838,889	2,778	13,310
251	Rubber products	..	..	..	..	..		..	..	..	..
2511	Rubber tyres and tubes	..	..	..	..	..		..	..	..	..
2519	Other rubber products	..	..	..	..	..		..	..	..	..
252	Plastic products	..	..	..	..	..		..	..	..	..
26	**OTHER NON-METALLIC MINERAL PRODUCTS**	2,110	50	460	2,670	22,790		550	1,372,222	58,333	33,360
261	Glass and glass products	..	..	..	..	..		..	..	..	..
269	Non-metallic mineral products, nec	..	..	..	..	..		..	..	..	..
2691	Pottery, china and earthenware	..	..	..	..	..		..	..	..	..
2692	Refractory ceramic products	..	..	..	..	..		..	..	..	..
2693	Structural non-refractory clay & ceramic prod.	..	..	..	..	..		..	..	..	..
2694	Cement, lime and plaster	..	..	..	..	..		..	..	..	..
2695	Articles of concrete, cement and plaster	..	..	..	..	..		..	..	..	..
2696	Cutting, shaping and finishing of stone	..	..	..	..	..		..	..	..	..
2699	Other non-metallic mineral products, nec	..	..	..	..	..		..	..	..	..

ISIC Revision 3 Industry Sector		Solid TJ	LPG TJ	Distiloil TJ	RFO TJ	Gas TJ	Biomass TJ	Steam TJ	Electr MWh	Own Use MWh	TOTAL TJ
27	**BASIC METALS**	56,400	0	140	0	34,160	..	4,890	2,969,444	238,889	105,420
271	Basic iron and steel	56,400	0	100	0	30,100	..	3,400	2,333,333	222,222	97,600
272	Basic precious and non-ferrous metals	0	0	40	0	4,060	..	1,490	636,111	16,667	7,820
273	Casting of metals	..	..	..	..	..	..	..	..	..	..
2731	Casting of iron and steel	..	..	..	..	..	..	..	..	..	..
2732	Casting non-ferrous metals	..	..	..	..	..	..	..	..	..	..
28	**FABRICATED METAL PRODUCTS**	0	260	710	0	22,690	..	730	4,125,000	30,556	39,130
281	Str. metal prod., tanks, reservoirs, steam generators	..	..	..	..	..	..	..	..	..	..
2811	Structural metal products	..	..	..	..	..	..	..	..	..	..
2812	Tanks, reservoirs and containers of metal	..	..	..	..	..	..	..	..	..	..
2813	Steam generators, exc. central heating hot water boilers	..	..	..	..	..	..	..	..	..	..
289	Other fabricated metal products	..	..	..	..	..	..	..	..	..	..
2891	Forging, pressing, stamping & roll-forming of metal	..	..	..	..	..	..	..	..	..	..
2892	Treatment and coating of metals	..	..	..	..	..	..	..	..	..	..
2893	Cutlery, hand tools and general hardware	..	..	..	..	..	..	..	..	..	..
2899	Other fabricated metal products, nec	..	..	..	..	..	..	..	..	..	..
29	**MACHINERY AND EQUIPMENT, NEC**	..	..	..	..	..	..	..	..	..	..
291	General purpose machinery	..	..	..	..	..	..	..	..	..	..
2911	Engines and turbines	..	..	..	..	..	..	..	..	..	..
2912	Pumps, compressors, taps and valves	..	..	..	..	..	..	..	..	..	..
2913	Bearings, gears, gearing and driving elements	..	..	..	..	..	..	..	..	..	..
2914	Ovens, furnaces and furnace burners	..	..	..	..	..	..	..	..	..	..
2915	Lifting and handling equipment	..	..	..	..	..	..	..	..	..	..
2919	Other general purpose machinery	..	..	..	..	..	..	..	..	..	..
292	Special purpose machinery	..	..	..	..	..	..	..	..	..	..
2921	Agricultural and forestry machinery	..	..	..	..	..	..	..	..	..	..
2922	Machine-tools	..	..	..	..	..	..	..	..	..	..
2923	Machinery for metallurgy	..	..	..	..	..	..	..	..	..	..
2924	Machinery for mining, quarrying and construction	..	..	..	..	..	..	..	..	..	..
2925	Machinery for food, beverage & tobacco processing	..	..	..	..	..	..	..	..	..	..
2926	Machinery for textile, apparel & leather production	..	..	..	..	..	..	..	..	..	..
2927	Machinery for weapons and ammunition	..	..	..	..	..	..	..	..	..	..
2929	Other special purpose machinery	..	..	..	..	..	..	..	..	..	..
293	Domestic appliances, nec	..	..	..	..	..	..	..	..	..	..
30	**OFFICE, ACCOUNTING & COMPUTING MACHINERY**	..	..	..	..	..	..	..	..	..	..
31	**ELECTRICAL MACHINERY & APPARATUS, NEC**	..	..	..	..	..	..	..	..	..	..
311	Electric motors, generators and transformers	..	..	..	..	..	..	..	..	..	..
312	Electricity distribution and control apparatus	..	..	..	..	..	..	..	..	..	..
313	Insulated wire and cable	..	..	..	..	..	..	..	..	..	..
314	Accumulators, primary cells & primary batteries	..	..	..	..	..	..	..	..	..	..
315	Electric lamps and lighting equipment	..	..	..	..	..	..	..	..	..	..
319	Other electrical equipment, nec	..	..	..	..	..	..	..	..	..	..
32	**RADIO, TV & COMMUNICATION EQUIP. & APP.**	..	..	..	..	..	..	..	..	..	..
321	Electronic valves, tubes, other electronic components	..	..	..	..	..	..	..	..	..	..
322	TV & radio transmitters, apparatus for line telephony	..	..	..	..	..	..	..	..	..	..
323	TV & radio receivers, recording apparatus	..	..	..	..	..	..	..	..	..	..
33	**MEDICAL PRECISION &OPTICAL INSTRUMENTS**	..	..	..	..	..	..	..	..	..	..
331	Medical appliances and instruments	..	..	..	..	..	..	..	..	..	..
3311	Medical, surgical equipment & orthopaedic app.	..	..	..	..	..	..	..	..	..	..
3312	Instruments & appliances for measuring, checking etc	..	..	..	..	..	..	..	..	..	..
3313	Industrial process control equipment	..	..	..	..	..	..	..	..	..	..
332	Optical instruments and photographic equipment	..	..	..	..	..	..	..	..	..	..
333	Watches and clocks	..	..	..	..	..	..	..	..	..	..
34	**MOTOR VEHICLES, TRAILERS & SEMI-TRAILERS**	..	..	..	..	..	..	..	..	..	..
341	Motor vehicles	..	..	..	..	..	..	..	..	..	..
342	Bodies (coachwork) for motor vehicles	..	..	..	..	..	..	..	..	..	..
343	Parts, accessories for motor vehicles & their engines	..	..	..	..	..	..	..	..	..	..
35	**OTHER TRANSPORT EQUIPMENT**	..	..	..	..	..	..	..	..	..	..
351	Building and repairing of ships and boats	..	..	..	..	..	..	..	..	..	..
3511	Building and repairing of ships	..	..	..	..	..	..	..	..	..	..
3512	Building, repairing of pleasure & sporting boats	..	..	..	..	..	..	..	..	..	..
352	Railway, tramway locomotives & rolling stock	..	..	..	..	..	..	..	..	..	..
353	Aircraft and spacecraft	..	..	..	..	..	..	..	..	..	..
359	Transport equipment, nec	..	..	..	..	..	..	..	..	..	..
3591	Motorcycles	..	..	..	..	..	..	..	..	..	..
3592	Bicycles and invalid carriages	..	..	..	..	..	..	..	..	..	..
3599	Other transport equipment, nec	..	..	..	..	..	..	..	..	..	..
36	**FURNITURE; MANUFACTURING, NEC**	0	0	0	0	0	..	260	0	0	260
361	Furniture	..	..	..	..	..	..	..	..	..	..
369	Manufacturing, nec	0	0	0	0	0	..	260	0	0	260
3691	Jewellery and related articles	..	..	..	..	..	..	..	..	..	..
3692	Musical instruments	..	..	..	..	..	..	..	..	..	..
3693	Sports goods	..	..	..	..	..	..	..	..	..	..
3694	Games and toys	..	..	..	..	..	..	..	..	..	..
3699	Other manufacturing, nec	0	0	0	0	0	..	260	0	0	260
37	**RECYCLING**	..	..	..	..	..	..	..	..	..	..
371	Recycling of metal waste and scrap	..	..	..	..	..	..	..	..	..	..
372	Recycling of non-metal waste and scrap	..	..	..	..	..	..	..	..	..	..
CERR	**Non-specified industry**	..	..	..	..	..	..	..	..	..	..
15-37	**TOTAL MANUFACTURING**	59,960	1,040	81,390	28,860	316,720	..	149,070	34,411,110	10,616,667	722,700

ISIC Revision 3 Industry Sector	Solid TJ	LPG TJ	Distiloil TJ	RFO TJ	Gas TJ	Biomass TJ	Steam TJ	Electr MWh	Own Use MWh	TOTAL TJ
15	**FOOD PRODUCTS AND BEVERAGES**	0	0	875	489	0	..	..	1,859,686	470	8,057
151	Production, processing and preserving (PPP)	..	..	..	..	..	..	..	..	..	..
1511	PPP of meat and meat products	..	..	..	..	..	..	..	..	..	..
1512	Processing and preserving of fish products	..	..	..	..	..	..	..	..	..	..
1513	Processing, preserving of fruit & vegetables	..	..	..	..	..	..	..	..	..	..
1514	Vegetable and animal oils and fats	..	..	..	..	..	..	..	..	..	..
152	Dairy products	..	..	..	..	..	..	..	..	..	..
153	Grain mill prod., starches & prepared animal feeds	..	..	..	..	..	..	..	..	..	..
1531	Grain mill products	..	..	..	..	..	..	..	..	..	..
1532	Starches and starch products	..	..	..	..	..	..	..	..	..	..
1533	Prepared animal feeds	..	..	..	..	..	..	..	..	..	..
154	Other food products	..	..	..	..	..	..	..	..	..	..
1541	Bakery products	..	..	..	..	..	..	..	..	..	..
1542	Sugar	..	..	..	..	..	..	..	..	..	..
1543	Cocoa, chocolate and sugar confectionery	..	..	..	..	..	..	..	..	..	..
1544	Macaroni, noodles, couscous & similar farinaceous prod.	..	..	..	..	..	..	..	..	..	..
1549	Other food products, nec	..	..	..	..	..	..	..	..	..	..
155	Beverages	..	..	..	..	..	..	..	..	..	..
1551	Distilling, rectifying and blending of spirits	..	..	..	..	..	..	..	..	..	..
1552	Wines	..	..	..	..	..	..	..	..	..	..
1553	Malt liquors and malt	..	..	..	..	..	..	..	..	..	..
1554	Soft drinks; production of mineral waters	..	..	..	..	..	..	..	..	..	..
16	**TOBACCO PRODUCTS**	..	..	..	..	..	..	..	..	..	..
17	**TEXTILES**	0	0	0	0	0	..	..	215,207	0	775
171	Spinning, weaving and finishing of textiles	..	..	..	..	..	..	..	..	..	..
1711	Preparation and spinning of textile fibres	..	..	..	..	..	..	..	..	..	..
1712	Finishing of textiles	..	..	..	..	..	..	..	..	..	..
172	Other textiles	..	..	..	..	..	..	..	..	..	..
1721	Made-up textile articles, except apparel	..	..	..	..	..	..	..	..	..	..
1722	Carpets and rugs	..	..	..	..	..	..	..	..	..	..
1723	Cordage, rope, twine and netting	..	..	..	..	..	..	..	..	..	..
1729	Other textiles, nec	..	..	..	..	..	..	..	..	..	..
173	Knitted and crocheted fabrics and articles	..	..	..	..	..	..	..	..	..	..
18	**WEARING APPAREL, DRESSING & DYEING OF FUR**	..	..	..	..	..	..	..	..	..	..
181	Wearing apparel, except fur apparel	..	..	..	..	..	..	..	..	..	..
182	Dressing and dyeing of fur; articles of fur	..	..	..	..	..	..	..	..	..	..
19	**TANNING & DRESSING OF LEATHER, FOOTWEAR**	..	..	..	..	..	..	..	..	..	..
191	Tanning and dressing of leather	..	..	..	..	..	..	..	..	..	..
1911	Tanning and dressing of leather	..	..	..	..	..	..	..	..	..	..
1912	Luggage, handbags, saddlery & harness	..	..	..	..	..	..	..	..	..	..
192	Footwear	..	..	..	..	..	..	..	..	..	..
20	**WOOD AND WOOD PRODUCTS**	0	0	0	0	0	..	..	1,907,418	23	6,867
201	Sawmilling and planing of wood	..	..	..	..	..	..	..	..	..	..
202	Products of wood, cork, straw & plaiting materials	..	..	..	..	..	..	..	..	..	..
2021	Veneer sheets	..	..	..	..	..	..	..	..	..	..
2022	Builders' carpentry and joinery	..	..	..	..	..	..	..	..	..	..
2023	Wooden containers	..	..	..	..	..	..	..	..	..	..
2029	Other products of wood	..	..	..	..	..	..	..	..	..	..
21	**PAPER AND PAPER PRODUCTS**	0	0	0	0	0	..	..	1,355,217	285	4,878
2101	Pulp, paper and paperboard	..	..	..	..	..	..	..	..	..	..
2102	Corrugated paper, paperboard and their containers	..	..	..	..	..	..	..	..	..	..
2109	Other articles of pulp and paperboard	..	..	..	..	..	..	..	..	..	..
22	**PUBLISHING, PRINTING & REPRODUCTION**	..	..	..	..	..	..	..	..	..	..
221	Publishing	..	..	..	..	..	..	..	..	..	..
2211	Publishing of books & brochures	..	..	..	..	..	..	..	..	..	..
2212	Publishing of newspapers and periodicals	..	..	..	..	..	..	..	..	..	..
2213	Publishing of recorded media	..	..	..	..	..	..	..	..	..	..
2219	Other publishing	..	..	..	..	..	..	..	..	..	..
222	Printing and related service activities	..	..	..	..	..	..	..	..	..	..
2221	Printing	..	..	..	..	..	..	..	..	..	..
2222	Service activities related to printing	..	..	..	..	..	..	..	..	..	..
223	Reproduction of recorded media	..	..	..	..	..	..	..	..	..	..
23	**COKE, REFINED PETROLEUM PRODUCTS**	0	0	16,663	0	26,861	..	..	0	0	43,524
231	Coke oven products	..	..	..	..	..	..	..	..	..	..
232	Refined petroleum products	0	0	16,663	0	26,861	..	..	0	0	43,524
233	Processing of nuclear fuel	..	..	..	..	..	..	..	..	..	..
24	**CHEMICALS & CHEMICAL PRODUCTS**	0	0	0	0	9,552	..	..	1,054,215	138	13,347
241	Basic chemicals	0	0	0	0	9,552	..	..	0	0	9,552
2411	Basic chemicals, exc. fertilizers & nitrogen compounds	..	..	..	..	..	..	..	..	..	..
2412	Fertilizers and nitrogen compounds	..	..	..	..	..	..	..	..	..	..
2413	Plastics in primary forms and synthetic rubber	..	..	..	..	..	..	..	..	..	..
242	Other chemical products	..	..	..	..	..	..	..	..	..	..
2421	Pesticides and other agro-chemical products	..	..	..	..	..	..	..	..	..	..
2422	Paints, varnishes and similar coatings	..	..	..	..	..	..	..	..	..	..
2423	Pharmaceuticals, medicinal chem. & botanical prod.	..	..	..	..	..	..	..	..	..	..
2424	Soap and detergents, perfumes etc.	..	..	..	..	..	..	..	..	..	..
2429	Other chemical products, nec	..	..	..	..	..	..	..	..	..	..
243	Man-made fibres	..	..	..	..	..	..	..	..	..	..
25	**RUBBER AND PLASTICS PRODUCTS**	..	..	..	..	..	..	..	..	..	..
251	Rubber products	..	..	..	..	..	..	..	..	..	..
2511	Rubber tyres and tubes	..	..	..	..	..	..	..	..	..	..
2519	Other rubber products	..	..	..	..	..	..	..	..	..	..
252	Plastic products	..	..	..	..	..	..	..	..	..	..
26	**OTHER NON-METALLIC MINERAL PRODUCTS**	0	0	0	0	0	..	..	238,643	0	859
261	Glass and glass products	..	..	..	..	..	..	..	..	..	..
269	Non-metallic mineral products, nec	..	..	..	..	..	..	..	..	..	..
2691	Pottery, china and earthenware	..	..	..	..	..	..	..	..	..	..
2692	Refractory ceramic products	..	..	..	..	..	..	..	..	..	..
2693	Structural non-refractory clay & ceramic prod.	..	..	..	..	..	..	..	..	..	..
2694	Cement, lime and plaster	..	..	..	..	..	..	..	..	..	..
2695	Articles of concrete, cement and plaster	..	..	..	..	..	..	..	..	..	..
2696	Cutting, shaping and finishing of stone	..	..	..	..	..	..	..	..	..	..
2699	Other non-metallic mineral products, nec	..	..	..	..	..	..	..	..	..	..

ISIC Revision 3 Industry Sector		Solid TJ	LPG TJ	Distiloil TJ	RFO TJ	Gas TJ	Biomass TJ	Steam TJ	Electr MWh	Own Use MWh	TOTAL TJ
27	**BASIC METALS**	13,715	0	312	88	0	..	..	5,483,221	226	33,854
271	Basic iron and steel	13,715	0	312	88	0	..	..	0	0	14,115
272	Basic precious and non-ferrous metals	0	0	0	0	0	..	..	5,483,221	226	19,739
273	Casting of metals	..	..	..	..	..	..	..	..	..	..
2731	Casting of iron and steel	..	..	..	..	..	..	..	..	..	..
2732	Casting non-ferrous metals	..	..	..	..	..	..	..	..	..	..
28	**FABRICATED METAL PRODUCTS**	0	0	0	0	0	..	..	411,470	0	1,481
281	Str. metal prod., tanks, reservoirs, steam generators	..	..	..	..	..	..	..	..	..	..
2811	Structural metal products	..	..	..	..	..	..	..	..	..	..
2812	Tanks, reservoirs and containers of metal	..	..	..	..	..	..	..	..	..	..
2813	Steam generators, exc. central heating hot water boilers	..	..	..	..	..	..	..	..	..	..
289	Other fabricated metal products	..	..	..	..	..	..	..	..	..	..
2891	Forging, pressing, stamping & roll-forming of metal	..	..	..	..	..	..	..	..	..	..
2892	Treatment and coating of metals	..	..	..	..	..	..	..	..	..	..
2893	Cutlery, hand tools and general hardware	..	..	..	..	..	..	..	..	..	..
2899	Other fabricated metal products, nec	..	..	..	..	..	..	..	..	..	..
29	**MACHINERY AND EQUIPMENT, NEC**	..	..	..	..	..	..	..	..	..	..
291	General purpose machinery	..	..	..	..	..	..	..	..	..	..
2911	Engines and turbines	..	..	..	..	..	..	..	..	..	..
2912	Pumps, compressors, taps and valves	..	..	..	..	..	..	..	..	..	..
2913	Bearings, gears, gearing and driving elements	..	..	..	..	..	..	..	..	..	..
2914	Ovens, furnaces and furnace burners	..	..	..	..	..	..	..	..	..	..
2915	Lifting and handling equipment	..	..	..	..	..	..	..	..	..	..
2919	Other general purpose machinery	..	..	..	..	..	..	..	..	..	..
292	Special purpose machinery	..	..	..	..	..	..	..	..	..	..
2921	Agricultural and forestry machinery	..	..	..	..	..	..	..	..	..	..
2922	Machine-tools	..	..	..	..	..	..	..	..	..	..
2923	Machinery for metallurgy	..	..	..	..	..	..	..	..	..	..
2924	Machinery for mining, quarrying and construction	..	..	..	..	..	..	..	..	..	..
2925	Machinery for food, beverage & tobacco processing	..	..	..	..	..	..	..	..	..	..
2926	Machinery for textile, apparel & leather production	..	..	..	..	..	..	..	..	..	..
2927	Machinery for weapons and ammunition	..	..	..	..	..	..	..	..	..	..
2929	Other special purpose machinery	..	..	..	..	..	..	..	..	..	..
293	Domestic appliances, nec	..	..	..	..	..	..	..	..	..	..
30	**OFFICE, ACCOUNTING & COMPUTING MACHINERY**	..	..	..	..	..	..	..	..	..	..
31	**ELECTRICAL MACHINERY & APPARATUS, NEC**	..	..	..	..	..	..	..	..	..	..
311	Electric motors, generators and transformers	..	..	..	..	..	..	..	..	..	..
312	Electricity distribution and control apparatus	..	..	..	..	..	..	..	..	..	..
313	Insulated wire and cable	..	..	..	..	..	..	..	..	..	..
314	Accumulators, primary cells & primary batteries	..	..	..	..	..	..	..	..	..	..
315	Electric lamps and lighting equipment	..	..	..	..	..	..	..	..	..	..
319	Other electrical equipment, nec	..	..	..	..	..	..	..	..	..	..
32	**RADIO, TV & COMMUNICATION EQUIP. & APP.**	..	..	..	..	..	..	..	..	..	..
321	Electronic valves, tubes, other electronic components	..	..	..	..	..	..	..	..	..	..
322	TV & radio transmitters, apparatus for line telephony	..	..	..	..	..	..	..	..	..	..
323	TV & radio receivers, recording apparatus	..	..	..	..	..	..	..	..	..	..
33	**MEDICAL PRECISION &OPTICAL INSTRUMENTS**	..	..	..	..	..	..	..	..	..	..
331	Medical appliances and instruments	..	..	..	..	..	..	..	..	..	..
3311	Medical, surgical equipment & orthopaedic app.	..	..	..	..	..	..	..	..	..	..
3312	Instruments & appliances for measuring, checking etc	..	..	..	..	..	..	..	..	..	..
3313	Industrial process control equipment	..	..	..	..	..	..	..	..	..	..
332	Optical instruments and photographic equipment	..	..	..	..	..	..	..	..	..	..
333	Watches and clocks	..	..	..	..	..	..	..	..	..	..
34	**MOTOR VEHICLES, TRAILERS & SEMI-TRAILERS**	..	..	..	..	..	..	..	..	..	..
341	Motor vehicles	..	..	..	..	..	..	..	..	..	..
342	Bodies (coachwork) for motor vehicles	..	..	..	..	..	..	..	..	..	..
343	Parts, accessories for motor vehicles & their engines	..	..	..	..	..	..	..	..	..	..
35	**OTHER TRANSPORT EQUIPMENT**	..	..	..	..	..	..	..	..	..	..
351	Building and repairing of ships and boats	..	..	..	..	..	..	..	..	..	..
3511	Building and repairing of ships	..	..	..	..	..	..	..	..	..	..
3512	Building, repairing of pleasure & sporting boats	..	..	..	..	..	..	..	..	..	..
352	Railway, tramway locomotives & rolling stock	..	..	..	..	..	..	..	..	..	..
353	Aircraft and spacecraft	..	..	..	..	..	..	..	..	..	..
359	Transport equipment, nec	..	..	..	..	..	..	..	..	..	..
3591	Motorcycles	..	..	..	..	..	..	..	..	..	..
3592	Bicycles and invalid carriages	..	..	..	..	..	..	..	..	..	..
3599	Other transport equipment, nec	..	..	..	..	..	..	..	..	..	..
36	**FURNITURE; MANUFACTURING, NEC**	0	0	0	0	0	..	..	46,265	0	167
361	Furniture	..	..	..	..	..	..	..	..	..	..
369	Manufacturing, nec	..	..	..	..	..	..	..	..	..	..
3691	Jewellery and related articles	..	..	..	..	..	..	..	..	..	..
3692	Musical instruments	..	..	..	..	..	..	..	..	..	..
3693	Sports goods	..	..	..	..	..	..	..	..	..	..
3694	Games and toys	..	..	..	..	..	..	..	..	..	..
3699	Other manufacturing, nec	..	..	..	..	..	..	..	..	..	..
37	**RECYCLING**	..	..	..	..	..	..	..	..	..	..
371	Recycling of metal waste and scrap	..	..	..	..	..	..	..	..	..	..
372	Recycling of non-metal waste and scrap	..	..	..	..	..	..	..	..	..	..
CERR	Non-specified industry	19,940	730	2,709	1,451	22,350	..	..	0	0	47,180
15-37	**TOTAL MANUFACTURING**	33,655	730	20,559	2,028	58,763	..	..	12,571,342	1,142	**160,988**

ISIC Revision 3 Industry Sector	Solid TJ	LPG TJ	Distiloil TJ	RFO TJ	Gas TJ	Biomass TJ	Steam TJ	Electr MWh	Own Use MWh	TOTAL TJ
15 FOOD PRODUCTS AND BEVERAGES	0	0	920	475	0		..	1,859,686	470	8,088
151 Production, processing and preserving (PPP)	..	..	..	..	..	..	..	..	..	..
1511 PPP of meat and meat products	..	..	..	..	..	..	..	..	..	..
1512 Processing and preserving of fish products	..	..	..	..	..	..	..	..	..	..
1513 Processing, preserving of fruit & vegetables	..	..	..	..	..	..	..	..	..	..
1514 Vegetable and animal oils and fats	..	..	..	..	..	..	..	..	..	..
152 Dairy products	..	..	..	..	..	..	..	..	..	..
153 Grain mill prod., starches & prepared animal feeds	..	..	..	..	..	..	..	..	..	..
1531 Grain mill products	..	..	..	..	..	..	..	..	..	..
1532 Starches and starch products	..	..	..	..	..	..	..	..	..	..
1533 Prepared animal feeds	..	..	..	..	..	..	..	..	..	..
154 Other food products	..	..	..	..	..	..	..	..	..	..
1541 Bakery products	..	..	..	..	..	..	..	..	..	..
1542 Sugar	..	..	..	..	..	..	..	..	..	..
1543 Cocoa, chocolate and sugar confectionery	..	..	..	..	..	..	..	..	..	..
1544 Macaroni, noodles, couscous & similar farinaceous prod.	..	..	..	..	..	..	..	..	..	..
1549 Other food products, nec	..	..	..	..	..	..	..	..	..	..
155 Beverages	..	..	..	..	..	..	..	..	..	..
1551 Distilling, rectifying and blending of spirits	..	..	..	..	..	..	..	..	..	..
1552 Wines	..	..	..	..	..	..	..	..	..	..
1553 Malt liquors and malt	..	..	..	..	..	..	..	..	..	..
1554 Soft drinks; production of mineral waters	..	..	..	..	..	..	..	..	..	..
16 TOBACCO PRODUCTS	..	..	..	..	..	..	..	215,207	0	775
17 TEXTILES	0	0	0	0	0		..			
171 Spinning, weaving and finishing of textiles	..	..	..	..	..	..	..	..	..	..
1711 Preparation and spinning of textile fibres	..	..	..	..	..	..	..	..	..	..
1712 Finishing of textiles	..	..	..	..	..	..	..	..	..	..
172 Other textiles	..	..	..	..	..	..	..	..	..	..
1721 Made-up textile articles, except apparel	..	..	..	..	..	..	..	..	..	..
1722 Carpets and rugs	..	..	..	..	..	..	..	..	..	..
1723 Cordage, rope, twine and netting	..	..	..	..	..	..	..	..	..	..
1729 Other textiles, nec	..	..	..	..	..	..	..	..	..	..
173 Knitted and crocheted fabrics and articles	..	..	..	..	..	..	..	..	..	..
18 WEARING APPAREL, DRESSING & DYEING OF FUR	..	..	..	..	..	..	..	..		
181 Wearing apparel, except fur apparel	..	..	..	..	..	..	..	..	..	..
182 Dressing and dyeing of fur; articles of fur	..	..	..	..	..	..	..	..	..	..
19 TANNING & DRESSING OF LEATHER, FOOTWEAR	..	..	..	..	..	..	..	..		
191 Tanning and dressing of leather	..	..	..	..	..	..	..	..	..	..
1911 Tanning and dressing of leather	..	..	..	..	..	..	..	..	..	..
1912 Luggage, handbags, saddlery & harness	..	..	..	..	..	..	..	..	..	..
192 Footwear	..	..	..	..	..	..	..	..	..	..
20 WOOD AND WOOD PRODUCTS	0	0	0	0	0		..	1,907,418	23	6,867
201 Sawmilling and planing of wood	..	..	..	..	..	..	..	..	..	..
202 Products of wood, cork, straw & plaiting materials	..	..	..	..	..	..	..	..	..	..
2021 Veneer sheets	..	..	..	..	..	..	..	..	..	..
2022 Builders' carpentry and joinery	..	..	..	..	..	..	..	..	..	..
2023 Wooden containers	..	..	..	..	..	..	..	..	..	..
2029 Other products of wood	..	..	..	..	..	..	..	..	..	..
21 PAPER AND PAPER PRODUCTS	0	0	0	0	0		..	1,355,217	285	4,878
2101 Pulp, paper and paperboard	..	..	..	..	..	..	..	..	..	..
2102 Corrugated paper, paperboard and their containers	..	..	..	..	..	..	..	..	..	..
2109 Other articles of pulp and paperboard	..	..	..	..	..	..	..	..	..	..
22 PUBLISHING, PRINTING & REPRODUCTION	..	..	..	..	..	..	..	..		
221 Publishing	..	..	..	..	..	..	..	..	..	..
2211 Publishing of books & brochures	..	..	..	..	..	..	..	..	..	..
2212 Publishing of newspapers and periodicals	..	..	..	..	..	..	..	..	..	..
2213 Publishing of recorded media	..	..	..	..	..	..	..	..	..	..
2219 Other publishing	..	..	..	..	..	..	..	..	..	..
222 Printing and related service activities	..	..	..	..	..	..	..	..	..	..
2221 Printing	..	..	..	..	..	..	..	..	..	..
2222 Service activities related to printing	..	..	..	..	..	..	..	..	..	..
223 Reproduction of recorded media	..	..	..	..	..	..	..	..	..	..
23 COKE, REFINED PETROLEUM PRODUCTS	0	0	17,230	0	22,194		..	0	0	39,424
231 Coke oven products	..	..	..	..	..	..	..	..	..	..
232 Refined petroleum products	0	0	17,230	0	22,194		..	0	0	39,424
233 Processing of nuclear fuel	..	..	..	..	..	..	..	..	..	..
24 CHEMICALS & CHEMICAL PRODUCTS	0	0	0	0	6,985		..	1,054,215	138	10,780
241 Basic chemicals	0	0	0	0	6,985		..	0	0	6,985
2411 Basic chemicals, exc. fertilizers & nitrogen compounds	..	..	..	..	..	..	..	..	..	..
2412 Fertilizers and nitrogen compounds	..	..	..	..	..	..	..	..	..	..
2413 Plastics in primary forms and synthetic rubber	..	..	..	..	..	..	..	..	..	..
242 Other chemical products	..	..	..	..	..	..	..	..	..	..
2421 Pesticides and other agro-chemical products	..	..	..	..	..	..	..	..	..	..
2422 Paints, varnishes and similar coatings	..	..	..	..	..	..	..	..	..	..
2423 Pharmaceuticals, medicinal chem. & botanical prod.	..	..	..	..	..	..	..	..	..	..
2424 Soap and detergents, perfumes etc.	..	..	..	..	..	..	..	..	..	..
2429 Other chemical products, nec	..	..	..	..	..	..	..	..	..	..
243 Man-made fibres	..	..	..	..	..	..	..	..	..	..
25 RUBBER AND PLASTICS PRODUCTS	..	..	..	..	..	..	..	..		
251 Rubber products	..	..	..	..	..	..	..	..	..	..
2511 Rubber tyres and tubes	..	..	..	..	..	..	..	..	..	..
2519 Other rubber products	..	..	..	..	..	..	..	..	..	..
252 Plastic products	..	..	..	..	..	..	..	..	..	..
26 OTHER NON-METALLIC MINERAL PRODUCTS	0	0	0	0	0		..	238,643	0	859
261 Glass and glass products	..	..	..	..	..	..	..	..	..	..
269 Non-metallic mineral products, nec	..	..	..	..	..	..	..	..	..	..
2691 Pottery, china and earthenware	..	..	..	..	..	..	..	..	..	..
2692 Refractory ceramic products	..	..	..	..	..	..	..	..	..	..
2693 Structural non-refractory clay & ceramic prod.	..	..	..	..	..	..	..	..	..	..
2694 Cement, lime and plaster	..	..	..	..	..	..	..	..	..	..
2695 Articles of concrete, cement and plaster	..	..	..	..	..	..	..	..	..	..
2696 Cutting, shaping and finishing of stone	..	..	..	..	..	..	..	..	..	..
2699 Other non-metallic mineral products, nec	..	..	..	..	..	..	..	..	..	..

245

ISIC Revision 3 Industry Sector	Solid TJ	LPG TJ	Distiloil TJ	RFO TJ	Gas TJ	Biomass TJ	Steam TJ	Electr MWh	Own Use MWh	TOTAL TJ
27	**BASIC METALS**	14,659	0	256	381	0	..	..	5,483,221	226	35,035
271	Basic iron and steel	14,659	0	256	381	0	..	..	0	0	15,296
272	Basic precious and non-ferrous metals	0	0	0	0	0	..	..	5,483,221	226	19,739
273	Casting of metals	..	..	..	..	..	..	..	..	..	..
2731	Casting of iron and steel	..	..	..	..	..	..	..	..	..	..
2732	Casting non-ferrous metals	..	..	..	..	..	..	..	..	..	..
28	**FABRICATED METAL PRODUCTS**	0	0	0	0	0	..	..	411,470	0	1,481
281	Str. metal prod., tanks, reservoirs, steam generators	..	..	..	..	..	..	..	..	..	..
2811	Structural metal products	..	..	..	..	..	..	..	..	..	..
2812	Tanks, reservoirs and containers of metal	..	..	..	..	..	..	..	..	..	..
2813	Steam generators, exc. central heating hot water boilers	..	..	..	..	..	..	..	..	..	..
289	Other fabricated metal products	..	..	..	..	..	..	..	..	..	..
2891	Forging, pressing, stamping & roll-forming of metal	..	..	..	..	..	..	..	..	..	..
2892	Treatment and coating of metals	..	..	..	..	..	..	..	..	..	..
2893	Cutlery, hand tools and general hardware	..	..	..	..	..	..	..	..	..	..
2899	Other fabricated metal products, nec	..	..	..	..	..	..	..	..	..	..
29	**MACHINERY AND EQUIPMENT, NEC**	..	..	..	..	..	..	..	..	..	..
291	General purpose machinery	..	..	..	..	..	..	..	..	..	..
2911	Engines and turbines	..	..	..	..	..	..	..	..	..	..
2912	Pumps, compressors, taps and valves	..	..	..	..	..	..	..	..	..	..
2913	Bearings, gears, gearing and driving elements	..	..	..	..	..	..	..	..	..	..
2914	Ovens, furnaces and furnace burners	..	..	..	..	..	..	..	..	..	..
2915	Lifting and handling equipment	..	..	..	..	..	..	..	..	..	..
2919	Other general purpose machinery	..	..	..	..	..	..	..	..	..	..
292	Special purpose machinery	..	..	..	..	..	..	..	..	..	..
2921	Agricultural and forestry machinery	..	..	..	..	..	..	..	..	..	..
2922	Machine-tools	..	..	..	..	..	..	..	..	..	..
2923	Machinery for metallurgy	..	..	..	..	..	..	..	..	..	..
2924	Machinery for mining, quarrying and construction	..	..	..	..	..	..	..	..	..	..
2925	Machinery for food, beverage & tobacco processing	..	..	..	..	..	..	..	..	..	..
2926	Machinery for textile, apparel & leather production	..	..	..	..	..	..	..	..	..	..
2927	Machinery for weapons and ammunition	..	..	..	..	..	..	..	..	..	..
2929	Other special purpose machinery	..	..	..	..	..	..	..	..	..	..
293	Domestic appliances, nec	..	..	..	..	..	..	..	..	..	..
30	**OFFICE, ACCOUNTING & COMPUTING MACHINERY**	..	..	..	..	..	..	..	..	..	..
31	**ELECTRICAL MACHINERY & APPARATUS, NEC**	..	..	..	..	..	..	..	..	..	..
311	Electric motors, generators and transformers	..	..	..	..	..	..	..	..	..	..
312	Electricity distribution and control apparatus	..	..	..	..	..	..	..	..	..	..
313	Insulated wire and cable	..	..	..	..	..	..	..	..	..	..
314	Accumulators, primary cells & primary batteries	..	..	..	..	..	..	..	..	..	..
315	Electric lamps and lighting equipment	..	..	..	..	..	..	..	..	..	..
319	Other electrical equipment, nec	..	..	..	..	..	..	..	..	..	..
32	**RADIO, TV & COMMUNICATION EQUIP. & APP.**	..	..	..	..	..	..	..	..	..	..
321	Electronic valves, tubes, other electronic components	..	..	..	..	..	..	..	..	..	..
322	TV & radio transmitters, apparatus for line telephony	..	..	..	..	..	..	..	..	..	..
323	TV & radio receivers, recording apparatus	..	..	..	..	..	..	..	..	..	..
33	**MEDICAL PRECISION &OPTICAL INSTRUMENTS**	..	..	..	..	..	..	..	..	..	..
331	Medical appliances and instruments	..	..	..	..	..	..	..	..	..	..
3311	Medical, surgical equipment & orthopaedic app.	..	..	..	..	..	..	..	..	..	..
3312	Instruments & appliances for measuring, checking etc	..	..	..	..	..	..	..	..	..	..
3313	Industrial process control equipment	..	..	..	..	..	..	..	..	..	..
332	Optical instruments and photographic equipment	..	..	..	..	..	..	..	..	..	..
333	Watches and clocks	..	..	..	..	..	..	..	..	..	..
34	**MOTOR VEHICLES, TRAILERS & SEMI-TRAILERS**	..	..	..	..	..	..	..	..	..	..
341	Motor vehicles	..	..	..	..	..	..	..	..	..	..
342	Bodies (coachwork) for motor vehicles	..	..	..	..	..	..	..	..	..	..
343	Parts, accessories for motor vehicles & their engines	..	..	..	..	..	..	..	..	..	..
35	**OTHER TRANSPORT EQUIPMENT**	..	..	..	..	..	..	..	..	..	..
351	Building and repairing of ships and boats	..	..	..	..	..	..	..	..	..	..
3511	Building and repairing of ships	..	..	..	..	..	..	..	..	..	..
3512	Building, repairing of pleasure & sporting boats	..	..	..	..	..	..	..	..	..	..
352	Railway, tramway locomotives & rolling stock	..	..	..	..	..	..	..	..	..	..
353	Aircraft and spacecraft	..	..	..	..	..	..	..	..	..	..
359	Transport equipment, nec	..	..	..	..	..	..	..	..	..	..
3591	Motorcycles	..	..	..	..	..	..	..	..	..	..
3592	Bicycles and invalid carriages	..	..	..	..	..	..	..	..	..	..
3599	Other transport equipment, nec	..	..	..	..	..	..	..	..	..	..
36	**FURNITURE; MANUFACTURING, NEC**	0	0	0	0	0	..	..	46,265	0	167
361	Furniture	..	..	..	..	..	..	..	..	..	..
369	Manufacturing, nec	..	..	..	..	..	..	..	..	..	..
3691	Jewellery and related articles	..	..	..	..	..	..	..	..	..	..
3692	Musical instruments	..	..	..	..	..	..	..	..	..	..
3693	Sports goods	..	..	..	..	..	..	..	..	..	..
3694	Games and toys	..	..	..	..	..	..	..	..	..	..
3699	Other manufacturing, nec	..	..	..	..	..	..	..	..	..	..
37	**RECYCLING**	..	..	..	..	..	..	..	..	..	..
371	Recycling of metal waste and scrap	..	..	..	..	..	..	..	..	..	..
372	Recycling of non-metal waste and scrap	..	..	..	..	..	..	..	..	..	..
CERR	**Non-specified industry**	20,375	724	2,467	1,228	22,967	..	..	0	0	47,761
15-37	**TOTAL MANUFACTURING**	35,034	724	20,873	2,084	52,146	..	..	12,571,342	1,142	**156,115**

ISIC Revision 3 Industry Sector	Solid TJ	LPG TJ	Distiloil TJ	RFO TJ	Gas TJ	Biomass TJ	Steam TJ	Electr MWh	Own Use MWh	TOTAL TJ
15 FOOD PRODUCTS AND BEVERAGES	0	0	760	297	0	..	..	1,859,686	470	7,750
151 Production, processing and preserving (PPP)	..	..	..	..	..	..	..	..	..	..
1511 PPP of meat and meat products	..	..	..	..	..	..	..	..	..	..
1512 Processing and preserving of fish products	..	..	..	..	..	..	..	..	..	..
1513 Processing, preserving of fruit & vegetables	..	..	..	..	..	..	..	..	..	..
1514 Vegetable and animal oils and fats	..	..	..	..	..	..	..	..	..	..
152 Dairy products	..	..	..	..	..	..	..	..	..	..
153 Grain mill prod., starches & prepared animal feeds	..	..	..	..	..	..	..	..	..	..
1531 Grain mill products	..	..	..	..	..	..	..	..	..	..
1532 Starches and starch products	..	..	..	..	..	..	..	..	..	..
1533 Prepared animal feeds	..	..	..	..	..	..	..	..	..	..
154 Other food products	..	..	..	..	..	..	..	..	..	..
1541 Bakery products	..	..	..	..	..	..	..	..	..	..
1542 Sugar	..	..	..	..	..	..	..	..	..	..
1543 Cocoa, chocolate and sugar confectionery	..	..	..	..	..	..	..	..	..	..
1544 Macaroni, noodles, couscous & similar farinaceous prod.	..	..	..	..	..	..	..	..	..	..
1549 Other food products, nec	..	..	..	..	..	..	..	..	..	..
155 Beverages	..	..	..	..	..	..	..	..	..	..
1551 Distilling, rectifying and blending of spirits	..	..	..	..	..	..	..	..	..	..
1552 Wines	..	..	..	..	..	..	..	..	..	..
1553 Malt liquors and malt	..	..	..	..	..	..	..	..	..	..
1554 Soft drinks; production of mineral waters	..	..	..	..	..	..	..	..	..	..
16 TOBACCO PRODUCTS	..	..	..	..	..	..	..	..	..	..
17 TEXTILES	0	0	0	0	0	..	..	215,207	0	775
171 Spinning, weaving and finishing of textiles	..	..	..	..	..	..	..	..	..	..
1711 Preparation and spinning of textile fibres	..	..	..	..	..	..	..	..	..	..
1712 Finishing of textiles	..	..	..	..	..	..	..	..	..	..
172 Other textiles	..	..	..	..	..	..	..	..	..	..
1721 Made-up textile articles, except apparel	..	..	..	..	..	..	..	..	..	..
1722 Carpets and rugs	..	..	..	..	..	..	..	..	..	..
1723 Cordage, rope, twine and netting	..	..	..	..	..	..	..	..	..	..
1729 Other textiles, nec	..	..	..	..	..	..	..	..	..	..
173 Knitted and crocheted fabrics and articles	..	..	..	..	..	..	..	..	..	..
18 WEARING APPAREL, DRESSING & DYEING OF FUR	..	..	..	..	..	..	..	..	..	..
181 Wearing apparel, except fur apparel	..	..	..	..	..	..	..	..	..	..
182 Dressing and dyeing of fur; articles of fur	..	..	..	..	..	..	..	..	..	..
19 TANNING & DRESSING OF LEATHER, FOOTWEAR	..	..	..	..	..	..	..	..	..	..
191 Tanning and dressing of leather	..	..	..	..	..	..	..	..	..	..
1911 Tanning and dressing of leather	..	..	..	..	..	..	..	..	..	..
1912 Luggage, handbags, saddlery & harness	..	..	..	..	..	..	..	..	..	..
192 Footwear	..	..	..	..	..	..	..	..	..	..
20 WOOD AND WOOD PRODUCTS	0	0	0	0	0	..	..	1,907,418	23	6,867
201 Sawmilling and planing of wood	..	..	..	..	..	..	..	..	..	..
202 Products of wood, cork, straw & plaiting materials	..	..	..	..	..	..	..	..	..	..
2021 Veneer sheets	..	..	..	..	..	..	..	..	..	..
2022 Builders' carpentry and joinery	..	..	..	..	..	..	..	..	..	..
2023 Wooden containers	..	..	..	..	..	..	..	..	..	..
2029 Other products of wood	..	..	..	..	..	..	..	..	..	..
21 PAPER AND PAPER PRODUCTS	0	0	0	0	0	..	..	1,355,217	285	4,878
2101 Pulp, paper and paperboard	..	..	..	..	..	..	..	..	..	..
2102 Corrugated paper, paperboard and their containers	..	..	..	..	..	..	..	..	..	..
2109 Other articles of pulp and paperboard	..	..	..	..	..	..	..	..	..	..
22 PUBLISHING, PRINTING & REPRODUCTION	..	..	..	..	..	..	..	..	..	..
221 Publishing	..	..	..	..	..	..	..	..	..	..
2211 Publishing of books & brochures	..	..	..	..	..	..	..	..	..	..
2212 Publishing of newspapers and periodicals	..	..	..	..	..	..	..	..	..	..
2213 Publishing of recorded media	..	..	..	..	..	..	..	..	..	..
2219 Other publishing	..	..	..	..	..	..	..	..	..	..
222 Printing and related service activities	..	..	..	..	..	..	..	..	..	..
2221 Printing	..	..	..	..	..	..	..	..	..	..
2222 Service activities related to printing	..	..	..	..	..	..	..	..	..	..
223 Reproduction of recorded media	..	..	..	..	..	..	..	..	..	..
23 COKE, REFINED PETROLEUM PRODUCTS	0	0	17,450	0	25,727	..	..	0	0	43,177
231 Coke oven products	..	..	..	..	..	..	..	..	..	..
232 Refined petroleum products	0	0	17,450	0	25,727	..	..	0	0	43,177
233 Processing of nuclear fuel	..	..	..	..	..	..	..	..	..	..
24 CHEMICALS & CHEMICAL PRODUCTS	0	0	0	0	8,446	..	..	1,054,215	138	12,241
241 Basic chemicals	0	0	0	0	8,446	..	..	0	0	8,446
2411 Basic chemicals, exc. fertilizers & nitrogen compounds	..	..	..	..	..	..	..	..	..	..
2412 Fertilizers and nitrogen compounds	..	..	..	..	..	..	..	..	..	..
2413 Plastics in primary forms and synthetic rubber	..	..	..	..	..	..	..	..	..	..
242 Other chemical products	..	..	..	..	..	..	..	..	..	..
2421 Pesticides and other agro-chemical products	..	..	..	..	..	..	..	..	..	..
2422 Paints, varnishes and similar coatings	..	..	..	..	..	..	..	..	..	..
2423 Pharmaceuticals, medicinal chem. & botanical prod.	..	..	..	..	..	..	..	..	..	..
2424 Soap and detergents, perfumes etc.	..	..	..	..	..	..	..	..	..	..
2429 Other chemical products, nec	..	..	..	..	..	..	..	..	..	..
243 Man-made fibres	..	..	..	..	..	..	..	..	..	..
25 RUBBER AND PLASTICS PRODUCTS	..	..	..	..	..	..	..	..	..	..
251 Rubber products	..	..	..	..	..	..	..	..	..	..
2511 Rubber tyres and tubes	..	..	..	..	..	..	..	..	..	..
2519 Other rubber products	..	..	..	..	..	..	..	..	..	..
252 Plastic products	..	..	..	..	..	..	..	..	..	..
26 OTHER NON-METALLIC MINERAL PRODUCTS	0	0	0	0	0	..	..	238,643	0	859
261 Glass and glass products	..	..	..	..	..	..	..	..	..	..
269 Non-metallic mineral products, nec	..	..	..	..	..	..	..	..	..	..
2691 Pottery, china and earthenware	..	..	..	..	..	..	..	..	..	..
2692 Refractory ceramic products	..	..	..	..	..	..	..	..	..	..
2693 Structural non-refractory clay & ceramic prod.	..	..	..	..	..	..	..	..	..	..
2694 Cement, lime and plaster	..	..	..	..	..	..	..	..	..	..
2695 Articles of concrete, cement and plaster	..	..	..	..	..	..	..	..	..	..
2696 Cutting, shaping and finishing of stone	..	..	..	..	..	..	..	..	..	..
2699 Other non-metallic mineral products, nec	..	..	..	..	..	..	..	..	..	..

ISIS Energy Data Programme (IEA/OECD)

ISIC Revision 3 Industry Sector	Solid TJ	LPG TJ	Distiloil TJ	RFO TJ	Gas TJ	Biomass TJ	Steam TJ	Electr MWh	Own Use MWh	TOTAL TJ
27 **BASIC METALS**	15,331	0	69	12	0	..	..	5,483,221	226	35,151
271 Basic iron and steel	15,331	0	69	12	0	..	..	0	0	15,412
272 Basic precious and non-ferrous metals	0	0	0	0	0	..	..	5,483,221	226	19,739
273 Casting of metals	..	..	..	..	..	..	..	..	..	..
2731 Casting of iron and steel	..	..	..	..	..	..	..	..	..	..
2732 Casting non-ferrous metals	..	..	..	..	..	..	..	..	..	..
28 **FABRICATED METAL PRODUCTS**	0	0	0	0	0	..	..	411,470	0	1,481
281 Str. metal prod., tanks, reservoirs, steam generators	..	..	..	..	..	..	..	..	..	..
2811 Structural metal products	..	..	..	..	..	..	..	..	..	..
2812 Tanks, reservoirs and containers of metal	..	..	..	..	..	..	..	..	..	..
2813 Steam generators, exc. central heating hot water boilers	..	..	..	..	..	..	..	..	..	..
289 Other fabricated metal products	..	..	..	..	..	..	..	..	..	..
2891 Forging, pressing, stamping & roll-forming of metal	..	..	..	..	..	..	..	..	..	..
2892 Treatment and coating of metals	..	..	..	..	..	..	..	..	..	..
2893 Cutlery, hand tools and general hardware	..	..	..	..	..	..	..	..	..	..
2899 Other fabricated metal products, nec	..	..	..	..	..	..	..	..	..	..
29 **MACHINERY AND EQUIPMENT, NEC**	..	..	..	..	..	..	..	..	..	..
291 General purpose machinery	..	..	..	..	..	..	..	..	..	..
2911 Engines and turbines	..	..	..	..	..	..	..	..	..	..
2912 Pumps, compressors, taps and valves	..	..	..	..	..	..	..	..	..	..
2913 Bearings, gears, gearing and driving elements	..	..	..	..	..	..	..	..	..	..
2914 Ovens, furnaces and furnace burners	..	..	..	..	..	..	..	..	..	..
2915 Lifting and handling equipment	..	..	..	..	..	..	..	..	..	..
2919 Other general purpose machinery	..	..	..	..	..	..	..	..	..	..
292 Special purpose machinery	..	..	..	..	..	..	..	..	..	..
2921 Agricultural and forestry machinery	..	..	..	..	..	..	..	..	..	..
2922 Machine-tools	..	..	..	..	..	..	..	..	..	..
2923 Machinery for metallurgy	..	..	..	..	..	..	..	..	..	..
2924 Machinery for mining, quarrying and construction	..	..	..	..	..	..	..	..	..	..
2925 Machinery for food, beverage & tobacco processing	..	..	..	..	..	..	..	..	..	..
2926 Machinery for textile, apparel & leather production	..	..	..	..	..	..	..	..	..	..
2927 Machinery for weapons and ammunition	..	..	..	..	..	..	..	..	..	..
2929 Other special purpose machinery	..	..	..	..	..	..	..	..	..	..
293 Domestic appliances, nec	..	..	..	..	..	..	..	..	..	..
30 **OFFICE, ACCOUNTING & COMPUTING MACHINERY**	..	..	..	..	..	..	..	..	..	..
31 **ELECTRICAL MACHINERY & APPARATUS, NEC**	..	..	..	..	..	..	..	..	..	..
311 Electric motors, generators and transformers	..	..	..	..	..	..	..	..	..	..
312 Electricity distribution and control apparatus	..	..	..	..	..	..	..	..	..	..
313 Insulated wire and cable	..	..	..	..	..	..	..	..	..	..
314 Accumulators, primary cells & primary batteries	..	..	..	..	..	..	..	..	..	..
315 Electric lamps and lighting equipment	..	..	..	..	..	..	..	..	..	..
319 Other electrical equipment, nec	..	..	..	..	..	..	..	..	..	..
32 **RADIO, TV & COMMUNICATION EQUIP. & APP.**	..	..	..	..	..	..	..	..	..	..
321 Electronic valves, tubes, other electronic components	..	..	..	..	..	..	..	..	..	..
322 TV & radio transmitters, apparatus for line telephony	..	..	..	..	..	..	..	..	..	..
323 TV & radio receivers, recording apparatus	..	..	..	..	..	..	..	..	..	..
33 **MEDICAL PRECISION & OPTICAL INSTRUMENTS**	..	..	..	..	..	..	..	..	..	..
331 Medical appliances and instruments	..	..	..	..	..	..	..	..	..	..
3311 Medical, surgical equipment & orthopaedic app.	..	..	..	..	..	..	..	..	..	..
3312 Instruments & appliances for measuring, checking etc	..	..	..	..	..	..	..	..	..	..
3313 Industrial process control equipment	..	..	..	..	..	..	..	..	..	..
332 Optical instruments and photographic equipment	..	..	..	..	..	..	..	..	..	..
333 Watches and clocks	..	..	..	..	..	..	..	..	..	..
34 **MOTOR VEHICLES, TRAILERS & SEMI-TRAILERS**	..	..	..	..	..	..	..	..	..	..
341 Motor vehicles	..	..	..	..	..	..	..	..	..	..
342 Bodies (coachwork) for motor vehicles	..	..	..	..	..	..	..	..	..	..
343 Parts, accessories for motor vehicles & their engines	..	..	..	..	..	..	..	..	..	..
35 **OTHER TRANSPORT EQUIPMENT**	..	..	..	..	..	..	..	..	..	..
351 Building and repairing of ships and boats	..	..	..	..	..	..	..	..	..	..
3511 Building and repairing of ships	..	..	..	..	..	..	..	..	..	..
3512 Building, repairing of pleasure & sporting boats	..	..	..	..	..	..	..	..	..	..
352 Railway, tramway locomotives & rolling stock	..	..	..	..	..	..	..	..	..	..
353 Aircraft and spacecraft	..	..	..	..	..	..	..	..	..	..
359 Transport equipment, nec	..	..	..	..	..	..	..	..	..	..
3591 Motorcycles	..	..	..	..	..	..	..	..	..	..
3592 Bicycles and invalid carriages	..	..	..	..	..	..	..	..	..	..
3599 Other transport equipment, nec	..	..	..	..	..	..	..	..	..	..
36 **FURNITURE; MANUFACTURING, NEC**	0	0	0	0	0	..	..	46,265	0	167
361 Furniture	..	..	..	..	..	..	..	..	..	..
369 Manufacturing, nec	..	..	..	..	..	..	..	..	..	..
3691 Jewellery and related articles	..	..	..	..	..	..	..	..	..	..
3692 Musical instruments	..	..	..	..	..	..	..	..	..	..
3693 Sports goods	..	..	..	..	..	..	..	..	..	..
3694 Games and toys	..	..	..	..	..	..	..	..	..	..
3699 Other manufacturing, nec	..	..	..	..	..	..	..	..	..	..
37 **RECYCLING**	..	..	..	..	..	..	..	..	..	..
371 Recycling of metal waste and scrap	..	..	..	..	..	..	..	..	..	..
372 Recycling of non-metal waste and scrap	..	..	..	..	..	..	..	..	..	..
CERR **Non-specified industry**	17,899	773	3,327	1,854	23,007	..	..	0	0	46,860
15-37 **TOTAL MANUFACTURING**	33,230	773	21,606	2,163	57,180	..	..	12,571,342	1,142	160,206

ISIC Revision 3 Industry Sector		Solid TJ	LPG TJ	Distiloil TJ	RFO TJ	Gas TJ	Biomass TJ	Steam TJ	Electr MWh	Own Use MWh	TOTAL TJ
15	FOOD PRODUCTS AND BEVERAGES	0	0	723	217	0		..	1,859,686	470	7,633
151	Production, processing and preserving (PPP)	..	..	..	..	..		..	..	..	..
1511	PPP of meat and meat products	..	..	..	..	..		..	..	..	..
1512	Processing and preserving of fish products	..	..	..	..	..		..	..	..	..
1513	Processing, preserving of fruit & vegetables	..	..	..	..	..		..	..	..	..
1514	Vegetable and animal oils and fats	..	..	..	..	..		..	..	..	..
152	Dairy products	..	..	..	..	..		..	..	..	..
153	Grain mill prod., starches & prepared animal feeds	..	..	..	..	..		..	..	..	..
1531	Grain mill products	..	..	..	..	..		..	..	..	..
1532	Starches and starch products	..	..	..	..	..		..	..	..	..
1533	Prepared animal feeds	..	..	..	..	..		..	..	..	..
154	Other food products	..	..	..	..	..		..	..	..	..
1541	Bakery products	..	..	..	..	..		..	..	..	..
1542	Sugar	..	..	..	..	..		..	..	..	..
1543	Cocoa, chocolate and sugar confectionery	..	..	..	..	..		..	..	..	..
1544	Macaroni, noodles, couscous & similar farinaceous prod.	..	..	..	..	..		..	..	..	..
1549	Other food products, nec	..	..	..	..	..		..	..	..	..
155	Beverages	..	..	..	..	..		..	..	..	..
1551	Distilling, rectifying and blending of spirits	..	..	..	..	..		..	..	..	..
1552	Wines	..	..	..	..	..		..	..	..	..
1553	Malt liquors and malt	..	..	..	..	..		..	..	..	..
1554	Soft drinks; production of mineral waters	..	..	..	..	..		..	..	..	..
16	TOBACCO PRODUCTS	..	..	..	..	..		..	215,207	0	775
17	TEXTILES	0	0	0	0	0		..			
171	Spinning, weaving and finishing of textiles	..	..	..	..	..		..	..	..	..
1711	Preparation and spinning of textile fibres	..	..	..	..	..		..	..	..	..
1712	Finishing of textiles	..	..	..	..	..		..	..	..	..
172	Other textiles	..	..	..	..	..		..	..	..	..
1721	Made-up textile articles, except apparel	..	..	..	..	..		..	..	..	..
1722	Carpets and rugs	..	..	..	..	..		..	..	..	..
1723	Cordage, rope, twine and netting	..	..	..	..	..		..	..	..	..
1729	Other textiles, nec	..	..	..	..	..		..	..	..	..
173	Knitted and crocheted fabrics and articles	..	..	..	..	..		..	..	..	..
18	WEARING APPAREL, DRESSING & DYEING OF FUR	..	..	..	..	..		..	..	..	..
181	Wearing apparel, except fur apparel	..	..	..	..	..		..	..	..	..
182	Dressing and dyeing of fur; articles of fur	..	..	..	..	..		..	..	..	..
19	TANNING & DRESSING OF LEATHER, FOOTWEAR	..	..	..	..	..		..	..	..	..
191	Tanning and dressing of leather	..	..	..	..	..		..	..	..	..
1911	Tanning and dressing of leather	..	..	..	..	..		..	..	..	..
1912	Luggage, handbags, saddlery & harness	..	..	..	..	..		..	..	..	..
192	Footwear	..	..	..	..	..		..	..	..	..
20	WOOD AND WOOD PRODUCTS	0	0	0	0	0		..	1,907,418	23	6,867
201	Sawmilling and planing of wood	..	..	..	..	..		..	..	..	..
202	Products of wood, cork, straw & plaiting materials	..	..	..	..	..		..	..	..	..
2021	Veneer sheets	..	..	..	..	..		..	..	..	..
2022	Builders' carpentry and joinery	..	..	..	..	..		..	..	..	..
2023	Wooden containers	..	..	..	..	..		..	..	..	..
2029	Other products of wood	..	..	..	..	..		..	..	..	..
21	PAPER AND PAPER PRODUCTS	0	0	0	0	0		..	1,355,217	285	4,878
2101	Pulp, paper and paperboard	..	..	..	..	..		..	..	..	..
2102	Corrugated paper, paperboard and their containers	..	..	..	..	..		..	..	..	..
2109	Other articles of pulp and paperboard	..	..	..	..	..		..	..	..	..
22	PUBLISHING, PRINTING & REPRODUCTION	..	..	..	..	..		..	..	..	..
221	Publishing	..	..	..	..	..		..	..	..	..
2211	Publishing of books & brochures	..	..	..	..	..		..	..	..	..
2212	Publishing of newspapers and periodicals	..	..	..	..	..		..	..	..	..
2213	Publishing of recorded media	..	..	..	..	..		..	..	..	..
2219	Other publishing	..	..	..	..	..		..	..	..	..
222	Printing and related service activities	..	..	..	..	..		..	..	..	..
2221	Printing	..	..	..	..	..		..	..	..	..
2222	Service activities related to printing	..	..	..	..	..		..	..	..	..
223	Reproduction of recorded media	..	..	..	..	..		..	..	..	..
23	COKE, REFINED PETROLEUM PRODUCTS	0	0	17,894	0	24,582		..	0	0	42,476
231	Coke oven products	..	..	..	..	..		..	..	..	..
232	Refined petroleum products	0	0	17,894	0	24,582		..	0	0	42,476
233	Processing of nuclear fuel	..	..	..	..	..		..	..	..	..
24	CHEMICALS & CHEMICAL PRODUCTS	0	0	0	0	8,017		..	1,054,215	138	11,812
241	Basic chemicals	0	0	0	0	8,017		..	0	0	8,017
2411	Basic chemicals, exc. fertilizers & nitrogen compounds	..	..	..	..	..		..	..	..	..
2412	Fertilizers and nitrogen compounds	..	..	..	..	..		..	..	..	..
2413	Plastics in primary forms and synthetic rubber	..	..	..	..	..		..	..	..	..
242	Other chemical products	..	..	..	..	..		..	..	..	..
2421	Pesticides and other agro-chemical products	..	..	..	..	..		..	..	..	..
2422	Paints, varnishes and similar coatings	..	..	..	..	..		..	..	..	..
2423	Pharmaceuticals, medicinal chem. & botanical prod.	..	..	..	..	..		..	..	..	..
2424	Soap and detergents, perfumes etc.	..	..	..	..	..		..	..	..	..
2429	Other chemical products, nec	..	..	..	..	..		..	..	..	..
243	Man-made fibres	..	..	..	..	..		..	..	..	..
25	RUBBER AND PLASTICS PRODUCTS	..	..	..	..	..		..	..	..	..
251	Rubber products	..	..	..	..	..		..	..	..	..
2511	Rubber tyres and tubes	..	..	..	..	..		..	..	..	..
2519	Other rubber products	..	..	..	..	..		..	..	..	..
252	Plastic products	..	..	..	..	..		..	..	..	..
26	OTHER NON-METALLIC MINERAL PRODUCTS	0	0	0	0	0		..	238,643	0	859
261	Glass and glass products	..	..	..	..	..		..	..	..	..
269	Non-metallic mineral products, nec	..	..	..	..	..		..	..	..	..
2691	Pottery, china and earthenware	..	..	..	..	..		..	..	..	..
2692	Refractory ceramic products	..	..	..	..	..		..	..	..	..
2693	Structural non-refractory clay & ceramic prod.	..	..	..	..	..		..	..	..	..
2694	Cement, lime and plaster	..	..	..	..	..		..	..	..	..
2695	Articles of concrete, cement and plaster	..	..	..	..	..		..	..	..	..
2696	Cutting, shaping and finishing of stone	..	..	..	..	..		..	..	..	..
2699	Other non-metallic mineral products, nec	..	..	..	..	..		..	..	..	..

ISIC Revision 3 Industry Sector		Solid TJ	LPG TJ	Distiloil TJ	RFO TJ	Gas TJ	Biomass TJ	Steam TJ	Electr MWh	Own Use MWh	TOTAL TJ
27	**BASIC METALS**	15,282	0	22	0	0	..		5,483,221	226	35,043
271	Basic iron and steel	15,282	0	22	0	0			0	0	15,304
272	Basic precious and non-ferrous metals	0	0	0	0	0		..	5,483,221	226	19,739
273	Casting of metals	..	..	..	..	..		..	..	..	..
2731	Casting of iron and steel	..	..	..	..	..		..	..	..	..
2732	Casting non-ferrous metals	..	..	..	..	..		..	..	..	..
28	**FABRICATED METAL PRODUCTS**	0	0	0	0	0			411,470	0	1,481
281	Str. metal prod., tanks, reservoirs, steam generators	..	..	..	..	..		..	..	..	..
2811	Structural metal products	..	..	..	..	..		..	..	..	..
2812	Tanks, reservoirs and containers of metal	..	..	..	..	..		..	..	..	..
2813	Steam generators, exc. central heating hot water boilers	..	..	..	..	..		..	..	..	..
289	Other fabricated metal products	..	..	..	..	..		..	..	..	..
2891	Forging, pressing, stamping & roll-forming of metal	..	..	..	..	..		..	..	..	..
2892	Treatment and coating of metals	..	..	..	..	..		..	..	..	..
2893	Cutlery, hand tools and general hardware	..	..	..	..	..		..	..	..	..
2899	Other fabricated metal products, nec	..	..	..	..	..		..	..	..	..
29	**MACHINERY AND EQUIPMENT, NEC**	..	..	..	..	..		..	..	..	..
291	General purpose machinery	..	..	..	..	..		..	..	..	..
2911	Engines and turbines	..	..	..	..	..		..	..	..	..
2912	Pumps, compressors, taps and valves	..	..	..	..	..		..	..	..	..
2913	Bearings, gears, gearing and driving elements	..	..	..	..	..		..	..	..	..
2914	Ovens, furnaces and furnace burners	..	..	..	..	..		..	..	..	..
2915	Lifting and handling equipment	..	..	..	..	..		..	..	..	..
2919	Other general purpose machinery	..	..	..	..	..		..	..	..	..
292	Special purpose machinery	..	..	..	..	..		..	..	..	..
2921	Agricultural and forestry machinery	..	..	..	..	..		..	..	..	..
2922	Machine-tools	..	..	..	..	..		..	..	..	..
2923	Machinery for metallurgy	..	..	..	..	..		..	..	..	..
2924	Machinery for mining, quarrying and construction	..	..	..	..	..		..	..	..	..
2925	Machinery for food, beverage & tobacco processing	..	..	..	..	..		..	..	..	..
2926	Machinery for textile, apparel & leather production	..	..	..	..	..		..	..	..	..
2927	Machinery for weapons and ammunition	..	..	..	..	..		..	..	..	..
2929	Other special purpose machinery	..	..	..	..	..		..	..	..	..
293	Domestic appliances, nec	..	..	..	..	..		..	..	..	..
30	**OFFICE, ACCOUNTING & COMPUTING MACHINERY**	..	..	..	..	..		..	..	..	..
31	**ELECTRICAL MACHINERY & APPARATUS, NEC**	..	..	..	..	..		..	..	..	..
311	Electric motors, generators and transformers	..	..	..	..	..		..	..	..	..
312	Electricity distribution and control apparatus	..	..	..	..	..		..	..	..	..
313	Insulated wire and cable	..	..	..	..	..		..	..	..	..
314	Accumulators, primary cells & primary batteries	..	..	..	..	..		..	..	..	..
315	Electric lamps and lighting equipment	..	..	..	..	..		..	..	..	..
319	Other electrical equipment, nec	..	..	..	..	..		..	..	..	..
32	**RADIO, TV & COMMUNICATION EQUIP. & APP.**	..	..	..	..	..		..	..	..	..
321	Electronic valves, tubes, other electronic components	..	..	..	..	..		..	..	..	..
322	TV & radio transmitters, apparatus for line telephony	..	..	..	..	..		..	..	..	..
323	TV & radio receivers, recording apparatus	..	..	..	..	..		..	..	..	..
33	**MEDICAL PRECISION &OPTICAL INSTRUMENTS**	..	..	..	..	..		..	..	..	..
331	Medical appliances and instruments	..	..	..	..	..		..	..	..	..
3311	Medical, surgical equipment & orthopaedic app.	..	..	..	..	..		..	..	..	..
3312	Instruments & appliances for measuring, checking etc	..	..	..	..	..		..	..	..	..
3313	Industrial process control equipment	..	..	..	..	..		..	..	..	..
332	Optical instruments and photographic equipment	..	..	..	..	..		..	..	..	..
333	Watches and clocks	..	..	..	..	..		..	..	..	..
34	**MOTOR VEHICLES, TRAILERS & SEMI-TRAILERS**	..	..	..	..	..		..	..	..	..
341	Motor vehicles	..	..	..	..	..		..	..	..	..
342	Bodies (coachwork) for motor vehicles	..	..	..	..	..		..	..	..	..
343	Parts, accessories for motor vehicles & their engines	..	..	..	..	..		..	..	..	..
35	**OTHER TRANSPORT EQUIPMENT**	..	..	..	..	..		..	..	..	..
351	Building and repairing of ships and boats	..	..	..	..	..		..	..	..	..
3511	Building and repairing of ships	..	..	..	..	..		..	..	..	..
3512	Building, repairing of pleasure & sporting boats	..	..	..	..	..		..	..	..	..
352	Railway, tramway locomotives & rolling stock	..	..	..	..	..		..	..	..	..
353	Aircraft and spacecraft	..	..	..	..	..		..	..	..	..
359	Transport equipment, nec	..	..	..	..	..		..	..	..	..
3591	Motorcycles	..	..	..	..	..		..	..	..	..
3592	Bicycles and invalid carriages	..	..	..	..	..		..	..	..	..
3599	Other transport equipment, nec	..	..	..	..	..		..	..	..	..
36	**FURNITURE; MANUFACTURING, NEC**	0	0	0	0	0			46,265	0	167
361	Furniture	..	..	..	..	..		..	..	..	..
369	Manufacturing, nec	..	..	..	..	..		..	..	..	..
3691	Jewellery and related articles	..	..	..	..	..		..	..	..	..
3692	Musical instruments	..	..	..	..	..		..	..	..	..
3693	Sports goods	..	..	..	..	..		..	..	..	..
3694	Games and toys	..	..	..	..	..		..	..	..	..
3699	Other manufacturing, nec	..	..	..	..	..		..	..	..	..
37	**RECYCLING**	..	..	..	..	..		..	..	..	..
371	Recycling of metal waste and scrap	..	..	..	..	..		..	..	..	..
372	Recycling of non-metal waste and scrap	..	..	..	..	..		..	..	..	..
CERR	**Non-specified industry**	20,070	1,006	2,034	1,460	24,217			0	0	48,787
15-37	**TOTAL MANUFACTURING**	35,352	1,006	20,673	1,677	56,816			12,571,342	1,142	**160,778**

ISIC Revision 3 Industry Sector	Solid TJ	LPG TJ	Distiloil TJ	RFO TJ	Gas TJ	Biomass TJ	Steam TJ	Electr MWh	Own Use MWh	TOTAL TJ
15 **FOOD PRODUCTS AND BEVERAGES**	0	0	714	266	0	..	..	1,859,686	470	7,673
151 Production, processing and preserving (PPP)	..	..	..	..	..	..	..	..	..	..
1511 PPP of meat and meat products	..	..	..	..	..	..	..	..	..	..
1512 Processing and preserving of fish products	..	..	..	..	..	..	..	..	..	..
1513 Processing, preserving of fruit & vegetables	..	..	..	..	..	..	..	..	..	..
1514 Vegetable and animal oils and fats	..	..	..	..	..	..	..	..	..	..
152 Dairy products	..	..	..	..	..	..	..	..	..	..
153 Grain mill prod., starches & prepared animal feeds	..	..	..	..	..	..	..	..	..	..
1531 Grain mill products	..	..	..	..	..	..	..	..	..	..
1532 Starches and starch products	..	..	..	..	..	..	..	..	..	..
1533 Prepared animal feeds	..	..	..	..	..	..	..	..	..	..
154 Other food products	..	..	..	..	..	..	..	..	..	..
1541 Bakery products	..	..	..	..	..	..	..	..	..	..
1542 Sugar	..	..	..	..	..	..	..	..	..	..
1543 Cocoa, chocolate and sugar confectionery	..	..	..	..	..	..	..	..	..	..
1544 Macaroni, noodles, couscous & similar farinaceous prod.	..	..	..	..	..	..	..	..	..	..
1549 Other food products, nec	..	..	..	..	..	..	..	..	..	..
155 Beverages	..	..	..	..	..	..	..	..	..	..
1551 Distilling, rectifying and blending of spirits	..	..	..	..	..	..	..	..	..	..
1552 Wines	..	..	..	..	..	..	..	..	..	..
1553 Malt liquors and malt	..	..	..	..	..	..	..	..	..	..
1554 Soft drinks; production of mineral waters	..	..	..	..	..	..	..	..	..	..
16 **TOBACCO PRODUCTS**	..	..	..	..	..	..	..	..	..	..
17 **TEXTILES**	0	0	0	0	0	..	..	215,207	0	775
171 Spinning, weaving and finishing of textiles	..	..	..	..	..	..	..	..	..	..
1711 Preparation and spinning of textile fibres	..	..	..	..	..	..	..	..	..	..
1712 Finishing of textiles	..	..	..	..	..	..	..	..	..	..
172 Other textiles	..	..	..	..	..	..	..	..	..	..
1721 Made-up textile articles, except apparel	..	..	..	..	..	..	..	..	..	..
1722 Carpets and rugs	..	..	..	..	..	..	..	..	..	..
1723 Cordage, rope, twine and netting	..	..	..	..	..	..	..	..	..	..
1729 Other textiles, nec	..	..	..	..	..	..	..	..	..	..
173 Knitted and crocheted fabrics and articles	..	..	..	..	..	..	..	..	..	..
18 **WEARING APPAREL, DRESSING & DYEING OF FUR**	..	..	..	..	..	..	..	..	..	..
181 Wearing apparel, except fur apparel	..	..	..	..	..	..	..	..	..	..
182 Dressing and dyeing of fur; articles of fur	..	..	..	..	..	..	..	..	..	..
19 **TANNING & DRESSING OF LEATHER, FOOTWEAR**	..	..	..	..	..	..	..	..	..	..
191 Tanning and dressing of leather	..	..	..	..	..	..	..	..	..	..
1911 Tanning and dressing of leather	..	..	..	..	..	..	..	..	..	..
1912 Luggage, handbags, saddlery & harness	..	..	..	..	..	..	..	..	..	..
192 Footwear	..	..	..	..	..	..	..	..	..	..
20 **WOOD AND WOOD PRODUCTS**	0	0	0	0	0	..	..	1,907,418	23	6,867
201 Sawmilling and planing of wood	..	..	..	..	..	..	..	..	..	..
202 Products of wood, cork, straw & plaiting materials	..	..	..	..	..	..	..	..	..	..
2021 Veneer sheets	..	..	..	..	..	..	..	..	..	..
2022 Builders' carpentry and joinery	..	..	..	..	..	..	..	..	..	..
2023 Wooden containers	..	..	..	..	..	..	..	..	..	..
2029 Other products of wood	..	..	..	..	..	..	..	..	..	..
21 **PAPER AND PAPER PRODUCTS**	0	0	0	0	0	..	..	1,355,217	285	4,878
2101 Pulp, paper and paperboard	..	..	..	..	..	..	..	..	..	..
2102 Corrugated paper, paperboard and their containers	..	..	..	..	..	..	..	..	..	..
2109 Other articles of pulp and paperboard	..	..	..	..	..	..	..	..	..	..
22 **PUBLISHING, PRINTING & REPRODUCTION**	..	..	..	..	..	..	..	..	..	..
221 Publishing	..	..	..	..	..	..	..	..	..	..
2211 Publishing of books & brochures	..	..	..	..	..	..	..	..	..	..
2212 Publishing of newspapers and periodicals	..	..	..	..	..	..	..	..	..	..
2213 Publishing of recorded media	..	..	..	..	..	..	..	..	..	..
2219 Other publishing	..	..	..	..	..	..	..	..	..	..
222 Printing and related service activities	..	..	..	..	..	..	..	..	..	..
2221 Printing	..	..	..	..	..	..	..	..	..	..
2222 Service activities related to printing	..	..	..	..	..	..	..	..	..	..
223 Reproduction of recorded media	..	..	..	..	..	..	..	..	..	..
23 **COKE, REFINED PETROLEUM PRODUCTS**	0	0	18,353	0	21,772	..	..	0	0	40,125
231 Coke oven products	..	..	..	..	..	..	..	..	..	..
232 Refined petroleum products	0	0	18,353	0	21,772	..	..	0	0	40,125
233 Processing of nuclear fuel	..	..	..	..	..	..	..	..	..	..
24 **CHEMICALS & CHEMICAL PRODUCTS**	0	0	0	0	13,688	..	..	1,054,215	138	17,483
241 Basic chemicals	0	0	0	0	13,688	..	..	0	0	13,688
2411 Basic chemicals, exc. fertilizers & nitrogen compounds	..	..	..	..	..	..	..	..	..	..
2412 Fertilizers and nitrogen compounds	..	..	..	..	..	..	..	..	..	..
2413 Plastics in primary forms and synthetic rubber	..	..	..	..	..	..	..	..	..	..
242 Other chemical products	..	..	..	..	..	..	..	..	..	..
2421 Pesticides and other agro-chemical products	..	..	..	..	..	..	..	..	..	..
2422 Paints, varnishes and similar coatings	..	..	..	..	..	..	..	..	..	..
2423 Pharmaceuticals, medicinal chem. & botanical prod.	..	..	..	..	..	..	..	..	..	..
2424 Soap and detergents, perfumes etc.	..	..	..	..	..	..	..	..	..	..
2429 Other chemical products, nec	..	..	..	..	..	..	..	..	..	..
243 Man-made fibres	..	..	..	..	..	..	..	..	..	..
25 **RUBBER AND PLASTICS PRODUCTS**	..	..	..	..	..	..	..	..	..	..
251 Rubber products	..	..	..	..	..	..	..	..	..	..
2511 Rubber tyres and tubes	..	..	..	..	..	..	..	..	..	..
2519 Other rubber products	..	..	..	..	..	..	..	..	..	..
252 Plastic products	..	..	..	..	..	..	..	..	..	..
26 **OTHER NON-METALLIC MINERAL PRODUCTS**	0	0	0	0	0	..	..	238,643	0	859
261 Glass and glass products	..	..	..	..	..	..	..	..	..	..
269 Non-metallic mineral products, nec	..	..	..	..	..	..	..	..	..	..
2691 Pottery, china and earthenware	..	..	..	..	..	..	..	..	..	..
2692 Refractory ceramic products	..	..	..	..	..	..	..	..	..	..
2693 Structural non-refractory clay & ceramic prod.	..	..	..	..	..	..	..	..	..	..
2694 Cement, lime and plaster	..	..	..	..	..	..	..	..	..	..
2695 Articles of concrete, cement and plaster	..	..	..	..	..	..	..	..	..	..
2696 Cutting, shaping and finishing of stone	..	..	..	..	..	..	..	..	..	..
2699 Other non-metallic mineral products, nec	..	..	..	..	..	..	..	..	..	..

ISIS Energy Data Programme (IEA/OECD)

ISIC Revision 3 Industry Sector		Solid TJ	LPG TJ	Distiloil TJ	RFO TJ	Gas TJ	Biomass TJ	Steam TJ	Electr MWh	Own Use MWh	TOTAL TJ
27	**BASIC METALS**	14,138	0	17	0	0		..	5,483,221	226	33,894
271	Basic iron and steel	14,138	0	17	0	0		..	0	0	14,155
272	Basic precious and non-ferrous metals	0	0	0	0	0		..	5,483,221	226	19,739
273	Casting of metals	..	..	..	..	..		..	..	..	..
2731	Casting of iron and steel	..	..	..	..	..		..	..	..	..
2732	Casting non-ferrous metals	..	..	..	..	..		..	..	..	..
28	**FABRICATED METAL PRODUCTS**	0	0	0	0	0		..	411,470	0	1,481
281	Str. metal prod., tanks, reservoirs, steam generators	..	..	..	..	..		..	..	..	..
2811	Structural metal products	..	..	..	..	..		..	..	..	..
2812	Tanks, reservoirs and containers of metal	..	..	..	..	..		..	..	..	..
2813	Steam generators, exc. central heating hot water boilers	..	..	..	..	..		..	..	..	..
289	Other fabricated metal products	..	..	..	..	..		..	..	..	..
2891	Forging, pressing, stamping & roll-forming of metal	..	..	..	..	..		..	..	..	..
2892	Treatment and coating of metals	..	..	..	..	..		..	..	..	..
2893	Cutlery, hand tools and general hardware	..	..	..	..	..		..	..	..	..
2899	Other fabricated metal products, nec	..	..	..	..	..		..	..	..	..
29	**MACHINERY AND EQUIPMENT, NEC**	..	..	..	..	..		..	..	..	..
291	General purpose machinery	..	..	..	..	..		..	..	..	..
2911	Engines and turbines	..	..	..	..	..		..	..	..	..
2912	Pumps, compressors, taps and valves	..	..	..	..	..		..	..	..	..
2913	Bearings, gears, gearing and driving elements	..	..	..	..	..		..	..	..	..
2914	Ovens, furnaces and furnace burners	..	..	..	..	..		..	..	..	..
2915	Lifting and handling equipment	..	..	..	..	..		..	..	..	..
2919	Other general purpose machinery	..	..	..	..	..		..	..	..	..
292	Special purpose machinery	..	..	..	..	..		..	..	..	..
2921	Agricultural and forestry machinery	..	..	..	..	..		..	..	..	..
2922	Machine-tools	..	..	..	..	..		..	..	..	..
2923	Machinery for metallurgy	..	..	..	..	..		..	..	..	..
2924	Machinery for mining, quarrying and construction	..	..	..	..	..		..	..	..	..
2925	Machinery for food, beverage & tobacco processing	..	..	..	..	..		..	..	..	..
2926	Machinery for textile, apparel & leather production	..	..	..	..	..		..	..	..	..
2927	Machinery for weapons and ammunition	..	..	..	..	..		..	..	..	..
2929	Other special purpose machinery	..	..	..	..	..		..	..	..	..
293	Domestic appliances, nec	..	..	..	..	..		..	..	..	..
30	**OFFICE, ACCOUNTING & COMPUTING MACHINERY**	..	..	..	..	..		..	..	..	..
31	**ELECTRICAL MACHINERY & APPARATUS, NEC**	..	..	..	..	..		..	..	..	..
311	Electric motors, generators and transformers	..	..	..	..	..		..	..	..	..
312	Electricity distribution and control apparatus	..	..	..	..	..		..	..	..	..
313	Insulated wire and cable	..	..	..	..	..		..	..	..	..
314	Accumulators, primary cells & primary batteries	..	..	..	..	..		..	..	..	..
315	Electric lamps and lighting equipment	..	..	..	..	..		..	..	..	..
319	Other electrical equipment, nec	..	..	..	..	..		..	..	..	..
32	**RADIO, TV & COMMUNICATION EQUIP. & APP.**	..	..	..	..	..		..	..	..	..
321	Electronic valves, tubes, other electronic components	..	..	..	..	..		..	..	..	..
322	TV & radio transmitters, apparatus for line telephony	..	..	..	..	..		..	..	..	..
323	TV & radio receivers, recording apparatus	..	..	..	..	..		..	..	..	..
33	**MEDICAL PRECISION &OPTICAL INSTRUMENTS**	..	..	..	..	..		..	..	..	..
331	Medical appliances and instruments	..	..	..	..	..		..	..	..	..
3311	Medical, surgical equipment & orthopaedic app.	..	..	..	..	..		..	..	..	..
3312	Instruments & appliances for measuring, checking etc	..	..	..	..	..		..	..	..	..
3313	Industrial process control equipment	..	..	..	..	..		..	..	..	..
332	Optical instruments and photographic equipment	..	..	..	..	..		..	..	..	..
333	Watches and clocks	..	..	..	..	..		..	..	..	..
34	**MOTOR VEHICLES, TRAILERS & SEMI-TRAILERS**	..	..	..	..	..		..	..	..	..
341	Motor vehicles	..	..	..	..	..		..	..	..	..
342	Bodies (coachwork) for motor vehicles	..	..	..	..	..		..	..	..	..
343	Parts, accessories for motor vehicles & their engines	..	..	..	..	..		..	..	..	..
35	**OTHER TRANSPORT EQUIPMENT**	..	..	..	..	..		..	..	..	..
351	Building and repairing of ships and boats	..	..	..	..	..		..	..	..	..
3511	Building and repairing of ships	..	..	..	..	..		..	..	..	..
3512	Building, repairing of pleasure & sporting boats	..	..	..	..	..		..	..	..	..
352	Railway, tramway locomotives & rolling stock	..	..	..	..	..		..	..	..	..
353	Aircraft and spacecraft	..	..	..	..	..		..	..	..	..
359	Transport equipment, nec	..	..	..	..	..		..	..	..	..
3591	Motorcycles	..	..	..	..	..		..	..	..	..
3592	Bicycles and invalid carriages	..	..	..	..	..		..	..	..	..
3599	Other transport equipment, nec	..	..	..	..	..		..	..	..	..
36	**FURNITURE; MANUFACTURING, NEC**	0	0	0	0	0		..	46,265	0	167
361	Furniture	..	..	..	..	..		..	..	..	..
369	Manufacturing, nec	..	..	..	..	..		..	..	..	..
3691	Jewellery and related articles	..	..	..	..	..		..	..	..	..
3692	Musical instruments	..	..	..	..	..		..	..	..	..
3693	Sports goods	..	..	..	..	..		..	..	..	..
3694	Games and toys	..	..	..	..	..		..	..	..	..
3699	Other manufacturing, nec	..	..	..	..	..		..	..	..	..
37	**RECYCLING**	..	..	..	..	..		..	..	..	..
371	Recycling of metal waste and scrap	..	..	..	..	..		..	..	..	..
372	Recycling of non-metal waste and scrap	..	..	..	..	..		..	..	..	..
CERR	Non-specified industry	16,567	1,007	2,150	1,934	25,307		..	0	0	46,965
15-37	**TOTAL MANUFACTURING**	30,705	1,007	21,234	2,200	60,767		..	12,571,342	1,142	**161,167**

ISIC Revision 3 Industry Sector	Solid TJ	LPG TJ	Distiloil TJ	RFO TJ	Gas TJ	Biomass TJ	Steam TJ	Electr MWh	Own Use MWh	TOTAL TJ
15 FOOD PRODUCTS AND BEVERAGES	0	0	686	329	0	..	..	1,694,209	186,832	6,442
151 Production, processing and preserving (PPP)	..	..	..	..	..	..	..	..	..	..
1511 PPP of meat and meat products	..	..	..	..	..	..	..	..	..	..
1512 Processing and preserving of fish products	..	..	..	..	..	..	..	..	..	..
1513 Processing, preserving of fruit & vegetables	..	..	..	..	..	..	..	..	..	..
1514 Vegetable and animal oils and fats	..	..	..	..	..	..	..	..	..	..
152 Dairy products	..	..	..	..	..	..	..	..	..	..
153 Grain mill prod., starches & prepared animal feeds	..	..	..	..	..	..	..	..	..	..
1531 Grain mill products	..	..	..	..	..	..	..	..	..	..
1532 Starches and starch products	..	..	..	..	..	..	..	..	..	..
1533 Prepared animal feeds	..	..	..	..	..	..	..	..	..	..
154 Other food products	..	..	..	..	..	..	..	..	..	..
1541 Bakery products	..	..	..	..	..	..	..	..	..	..
1542 Sugar	..	..	..	..	..	..	..	..	..	..
1543 Cocoa, chocolate and sugar confectionery	..	..	..	..	..	..	..	..	..	..
1544 Macaroni, noodles, couscous & similar farinaceous prod.	..	..	..	..	..	..	..	..	..	..
1549 Other food products, nec	..	..	..	..	..	..	..	..	..	..
155 Beverages	..	..	..	..	..	..	..	..	..	..
1551 Distilling, rectifying and blending of spirits	..	..	..	..	..	..	..	..	..	..
1552 Wines	..	..	..	..	..	..	..	..	..	..
1553 Malt liquors and malt	..	..	..	..	..	..	..	..	..	..
1554 Soft drinks; production of mineral waters	..	..	..	..	..	..	..	..	..	..
16 TOBACCO PRODUCTS	..	..	..	..	..	..	..	222,330	0	800
17 TEXTILES	0	0	0	0	0	..	..	..	..	..
171 Spinning, weaving and finishing of textiles	..	..	..	..	..	..	..	..	..	..
1711 Preparation and spinning of textile fibres	..	..	..	..	..	..	..	..	..	..
1712 Finishing of textiles	..	..	..	..	..	..	..	..	..	..
172 Other textiles	..	..	..	..	..	..	..	..	..	..
1721 Made-up textile articles, except apparel	..	..	..	..	..	..	..	..	..	..
1722 Carpets and rugs	..	..	..	..	..	..	..	..	..	..
1723 Cordage, rope, twine and netting	..	..	..	..	..	..	..	..	..	..
1729 Other textiles, nec	..	..	..	..	..	..	..	..	..	..
173 Knitted and crocheted fabrics and articles	..	..	..	..	..	..	..	..	..	..
18 WEARING APPAREL, DRESSING & DYEING OF FUR	..	..	..	..	..	..	..	..	..	..
181 Wearing apparel, except fur apparel	..	..	..	..	..	..	..	..	..	..
182 Dressing and dyeing of fur; articles of fur	..	..	..	..	..	..	..	..	..	..
19 TANNING & DRESSING OF LEATHER, FOOTWEAR	..	..	..	..	..	..	..	..	..	..
191 Tanning and dressing of leather	..	..	..	..	..	..	..	..	..	..
1911 Tanning and dressing of leather	..	..	..	..	..	..	..	..	..	..
1912 Luggage, handbags, saddlery & harness	..	..	..	..	..	..	..	..	..	..
192 Footwear	..	..	..	..	..	..	..	..	..	..
20 WOOD AND WOOD PRODUCTS	0	0	0	0	0	..	..	1,981,706	35,532	7,006
201 Sawmilling and planing of wood	..	..	..	..	..	..	..	..	..	..
202 Products of wood, cork, straw & plaiting materials	..	..	..	..	..	..	..	..	..	..
2021 Veneer sheets	..	..	..	..	..	..	..	..	..	..
2022 Builders' carpentry and joinery	..	..	..	..	..	..	..	..	..	..
2023 Wooden containers	..	..	..	..	..	..	..	..	..	..
2029 Other products of wood	..	..	..	..	..	..	..	..	..	..
21 PAPER AND PAPER PRODUCTS	0	0	0	0	0	..	..	1,222,781	273,048	3,419
2101 Pulp, paper and paperboard	..	..	..	..	..	..	..	..	..	..
2102 Corrugated paper, paperboard and their containers	..	..	..	..	..	..	..	..	..	..
2109 Other articles of pulp and paperboard	..	..	..	..	..	..	..	..	..	..
22 PUBLISHING, PRINTING & REPRODUCTION	..	..	..	..	..	..	..	..	..	..
221 Publishing	..	..	..	..	..	..	..	..	..	..
2211 Publishing of books & brochures	..	..	..	..	..	..	..	..	..	..
2212 Publishing of newspapers and periodicals	..	..	..	..	..	..	..	..	..	..
2213 Publishing of recorded media	..	..	..	..	..	..	..	..	..	..
2219 Other publishing	..	..	..	..	..	..	..	..	..	..
222 Printing and related service activities	..	..	..	..	..	..	..	..	..	..
2221 Printing	..	..	..	..	..	..	..	..	..	..
2222 Service activities related to printing	..	..	..	..	..	..	..	..	..	..
223 Reproduction of recorded media	..	..	..	..	..	..	..	..	..	..
23 COKE, REFINED PETROLEUM PRODUCTS	0	0	16,357	0	13,757	..	..	0	0	30,114
231 Coke oven products	..	..	..	..	..	..	..	..	..	..
232 Refined petroleum products	0	0	16,357	0	13,757	..	..	0	0	30,114
233 Processing of nuclear fuel	..	..	..	..	..	..	..	..	..	..
24 CHEMICALS & CHEMICAL PRODUCTS	0	0	0	0	19,763	..	..	1,004,936	79,221	23,096
241 Basic chemicals	0	0	0	0	19,763	..	..	0	0	19,763
2411 Basic chemicals, exc. fertilizers & nitrogen compounds	..	..	..	..	..	..	..	..	..	..
2412 Fertilizers and nitrogen compounds	..	..	..	..	..	..	..	..	..	..
2413 Plastics in primary forms and synthetic rubber	..	..	..	..	..	..	..	..	..	..
242 Other chemical products	..	..	..	..	..	..	..	..	..	..
2421 Pesticides and other agro-chemical products	..	..	..	..	..	..	..	..	..	..
2422 Paints, varnishes and similar coatings	..	..	..	..	..	..	..	..	..	..
2423 Pharmaceuticals, medicinal chem. & botanical prod.	..	..	..	..	..	..	..	..	..	..
2424 Soap and detergents, perfumes etc.	..	..	..	..	..	..	..	..	..	..
2429 Other chemical products, nec	..	..	..	..	..	..	..	..	..	..
243 Man-made fibres	..	..	..	..	..	..	..	..	..	..
25 RUBBER AND PLASTICS PRODUCTS	..	..	..	..	..	..	..	..	..	..
251 Rubber products	..	..	..	..	..	..	..	..	..	..
2511 Rubber tyres and tubes	..	..	..	..	..	..	..	..	..	..
2519 Other rubber products	..	..	..	..	..	..	..	..	..	..
252 Plastic products	..	..	..	..	..	..	..	..	..	..
26 OTHER NON-METALLIC MINERAL PRODUCTS	0	0	0	0	0	..	..	241,216	0	868
261 Glass and glass products	..	..	..	..	..	..	..	..	..	..
269 Non-metallic mineral products, nec	..	..	..	..	..	..	..	..	..	..
2691 Pottery, china and earthenware	..	..	..	..	..	..	..	..	..	..
2692 Refractory ceramic products	..	..	..	..	..	..	..	..	..	..
2693 Structural non-refractory clay & ceramic prod.	..	..	..	..	..	..	..	..	..	..
2694 Cement, lime and plaster	..	..	..	..	..	..	..	..	..	..
2695 Articles of concrete, cement and plaster	..	..	..	..	..	..	..	..	..	..
2696 Cutting, shaping and finishing of stone	..	..	..	..	..	..	..	..	..	..
2699 Other non-metallic mineral products, nec	..	..	..	..	..	..	..	..	..	..

ISIC Revision 3 Industry Sector		Solid TJ	LPG TJ	Distiloil TJ	RFO TJ	Gas TJ	Biomass TJ	Steam TJ	Electr MWh	Own Use MWh	TOTAL TJ
27	**BASIC METALS**	15,106	0	9	0	0	..	..	5,425,449	170,744	34,032
271	Basic iron and steel	15,106	0	9	0	0	..	..	0	0	15,115
272	Basic precious and non-ferrous metals	0	0	0	0	0	..	..	5,425,449	170,744	18,917
273	Casting of metals	..	..	..	..	..	..	..	..	..	..
2731	Casting of iron and steel	..	..	..	..	..	..	..	..	..	..
2732	Casting non-ferrous metals	..	..	..	..	..	..	..	..	..	..
28	**FABRICATED METAL PRODUCTS**	0	0	0	0	0	..	..	414,074	0	1,491
281	Str. metal prod., tanks, reservoirs, steam generators	..	..	..	..	..	..	..	..	..	..
2811	Structural metal products	..	..	..	..	..	..	..	..	..	..
2812	Tanks, reservoirs and containers of metal	..	..	..	..	..	..	..	..	..	..
2813	Steam generators, exc. central heating hot water boilers	..	..	..	..	..	..	..	..	..	..
289	Other fabricated metal products	..	..	..	..	..	..	..	..	..	..
2891	Forging, pressing, stamping & roll-forming of metal	..	..	..	..	..	..	..	..	..	..
2892	Treatment and coating of metals	..	..	..	..	..	..	..	..	..	..
2893	Cutlery, hand tools and general hardware	..	..	..	..	..	..	..	..	..	..
2899	Other fabricated metal products, nec	..	..	..	..	..	..	..	..	..	..
29	**MACHINERY AND EQUIPMENT, NEC**	..	..	..	..	..	..	..	..	..	..
291	General purpose machinery	..	..	..	..	..	..	..	..	..	..
2911	Engines and turbines	..	..	..	..	..	..	..	..	..	..
2912	Pumps, compressors, taps and valves	..	..	..	..	..	..	..	..	..	..
2913	Bearings, gears, gearing and driving elements	..	..	..	..	..	..	..	..	..	..
2914	Ovens, furnaces and furnace burners	..	..	..	..	..	..	..	..	..	..
2915	Lifting and handling equipment	..	..	..	..	..	..	..	..	..	..
2919	Other general purpose machinery	..	..	..	..	..	..	..	..	..	..
292	Special purpose machinery	..	..	..	..	..	..	..	..	..	..
2921	Agricultural and forestry machinery	..	..	..	..	..	..	..	..	..	..
2922	Machine-tools	..	..	..	..	..	..	..	..	..	..
2923	Machinery for metallurgy	..	..	..	..	..	..	..	..	..	..
2924	Machinery for mining, quarrying and construction	..	..	..	..	..	..	..	..	..	..
2925	Machinery for food, beverage & tobacco processing	..	..	..	..	..	..	..	..	..	..
2926	Machinery for textile, apparel & leather production	..	..	..	..	..	..	..	..	..	..
2927	Machinery for weapons and ammunition	..	..	..	..	..	..	..	..	..	..
2929	Other special purpose machinery	..	..	..	..	..	..	..	..	..	..
293	Domestic appliances, nec	..	..	..	..	..	..	..	..	..	..
30	**OFFICE, ACCOUNTING & COMPUTING MACHINERY**	..	..	..	..	..	..	..	..	..	..
31	**ELECTRICAL MACHINERY & APPARATUS, NEC**	..	..	..	..	..	..	..	..	..	..
311	Electric motors, generators and transformers	..	..	..	..	..	..	..	..	..	..
312	Electricity distribution and control apparatus	..	..	..	..	..	..	..	..	..	..
313	Insulated wire and cable	..	..	..	..	..	..	..	..	..	..
314	Accumulators, primary cells & primary batteries	..	..	..	..	..	..	..	..	..	..
315	Electric lamps and lighting equipment	..	..	..	..	..	..	..	..	..	..
319	Other electrical equipment, nec	..	..	..	..	..	..	..	..	..	..
32	**RADIO, TV & COMMUNICATION EQUIP. & APP.**	..	..	..	..	..	..	..	..	..	..
321	Electronic valves, tubes, other electronic components	..	..	..	..	..	..	..	..	..	..
322	TV & radio transmitters, apparatus for line telephony	..	..	..	..	..	..	..	..	..	..
323	TV & radio receivers, recording apparatus	..	..	..	..	..	..	..	..	..	..
33	**MEDICAL PRECISION &OPTICAL INSTRUMENTS**	..	..	..	..	..	..	..	..	..	..
331	Medical appliances and instruments	..	..	..	..	..	..	..	..	..	..
3311	Medical, surgical equipment & orthopaedic app.	..	..	..	..	..	..	..	..	..	..
3312	Instruments & appliances for measuring, checking etc	..	..	..	..	..	..	..	..	..	..
3313	Industrial process control equipment	..	..	..	..	..	..	..	..	..	..
332	Optical instruments and photographic equipment	..	..	..	..	..	..	..	..	..	..
333	Watches and clocks	..	..	..	..	..	..	..	..	..	..
34	**MOTOR VEHICLES, TRAILERS & SEMI-TRAILERS**	..	..	..	..	..	..	..	..	..	..
341	Motor vehicles	..	..	..	..	..	..	..	..	..	..
342	Bodies (coachwork) for motor vehicles	..	..	..	..	..	..	..	..	..	..
343	Parts, accessories for motor vehicles & their engines	..	..	..	..	..	..	..	..	..	..
35	**OTHER TRANSPORT EQUIPMENT**	..	..	..	..	..	..	..	..	..	..
351	Building and repairing of ships and boats	..	..	..	..	..	..	..	..	..	..
3511	Building and repairing of ships	..	..	..	..	..	..	..	..	..	..
3512	Building, repairing of pleasure & sporting boats	..	..	..	..	..	..	..	..	..	..
352	Railway, tramway locomotives & rolling stock	..	..	..	..	..	..	..	..	..	..
353	Aircraft and spacecraft	..	..	..	..	..	..	..	..	..	..
359	Transport equipment, nec	..	..	..	..	..	..	..	..	..	..
3591	Motorcycles	..	..	..	..	..	..	..	..	..	..
3592	Bicycles and invalid carriages	..	..	..	..	..	..	..	..	..	..
3599	Other transport equipment, nec	..	..	..	..	..	..	..	..	..	..
36	**FURNITURE; MANUFACTURING, NEC**	0	0	0	0	0	..	..	79,103	0	285
361	Furniture	..	..	..	..	..	..	..	..	..	..
369	Manufacturing, nec	..	..	..	..	..	..	..	..	..	..
3691	Jewellery and related articles	..	..	..	..	..	..	..	..	..	..
3692	Musical instruments	..	..	..	..	..	..	..	..	..	..
3693	Sports goods	..	..	..	..	..	..	..	..	..	..
3694	Games and toys	..	..	..	..	..	..	..	..	..	..
3699	Other manufacturing, nec	..	..	..	..	..	..	..	..	..	..
37	**RECYCLING**	..	..	..	..	..	..	..	..	..	..
371	Recycling of metal waste and scrap	..	..	..	..	..	..	..	..	..	..
372	Recycling of non-metal waste and scrap	..	..	..	..	..	..	..	..	..	..
CERR	**Non-specified industry**	15,061	1,005	2,536	1,693	22,352	..	..	0	0	42,647
15-37	**TOTAL MANUFACTURING**	30,167	1,005	19,588	2,022	55,872	..	..	12,285,804	745,377	150,200

ISIC Revision 3 Industry Sector	Solid TJ	LPG TJ	Distiloil TJ	RFO TJ	Gas TJ	Biomass TJ	Steam TJ	Electr MWh	Own Use MWh	TOTAL TJ
15 FOOD PRODUCTS AND BEVERAGES	0	0	711	406	0	..			1,859,686	470,651	6,118
151 Production, processing and preserving (PPP)	..	..	..	..	..	..			..	..	..
1511 PPP of meat and meat products	..		..		..				..		..
1512 Processing and preserving of fish products	..		..		..				..		..
1513 Processing, preserving of fruit & vegetables	..		..		..				..		..
1514 Vegetable and animal oils and fats	..		..		..				..		..
152 Dairy products	..		..		..				..		..
153 Grain mill prod., starches & prepared animal feeds	..	..	..		..				..		..
1531 Grain mill products	..		..		..				..		..
1532 Starches and starch products	..		..		..				..		..
1533 Prepared animal feeds	..		..		..				..		..
154 Other food products	..		..		..				..		..
1541 Bakery products	..		..		..				..		..
1542 Sugar	..		..		..				..		..
1543 Cocoa, chocolate and sugar confectionery	..		..		..				..		..
1544 Macaroni, noodles, couscous & similar farinaceous prod.	..		..		..				..		..
1549 Other food products, nec	..		..		..				..		..
155 Beverages	..		..		..				..		..
1551 Distilling, rectifying and blending of spirits	..		..		..				..		..
1552 Wines	..		..		..				..		..
1553 Malt liquors and malt	..		..		..				..		..
1554 Soft drinks; production of mineral waters	..		..		..				..		..
16 TOBACCO PRODUCTS	..		..		..				..		..
17 TEXTILES	0	0	0	0	0	..			215,207	0	775
171 Spinning, weaving and finishing of textiles	..		..		..				..		..
1711 Preparation and spinning of textile fibres	..		..		..				..		..
1712 Finishing of textiles	..		..		..				..		..
172 Other textiles	..		..		..				..		..
1721 Made-up textile articles, except apparel	..		..		..				..		..
1722 Carpets and rugs	..		..		..				..		..
1723 Cordage, rope, twine and netting	..		..		..				..		..
1729 Other textiles, nec	..		..		..				..		..
173 Knitted and crocheted fabrics and articles	..		..		..				..		..
18 WEARING APPAREL., DRESSING & DYEING OF FUR	..		..		..				..		..
181 Wearing apparel, except fur apparel	..		..		..				..		..
182 Dressing and dyeing of fur; articles of fur	..		..		..				..		..
19 TANNING & DRESSING OF LEATHER, FOOTWEAR	..		..		..				..		..
191 Tanning and dressing of leather	..		..		..				..		..
1911 Tanning and dressing of leather	..		..		..				..		..
1912 Luggage, handbags, saddlery & harness	..		..		..				..		..
192 Footwear	..		..		..				..		..
20 WOOD AND WOOD PRODUCTS	0	0	0	0	0	..			1,907,418	29,683	6,760
201 Sawmilling and planing of wood	..		..		..				..		..
202 Products of wood, cork, straw & plaiting materials	..		..		..				..		..
2021 Veneer sheets	..		..		..				..		..
2022 Builders' carpentry and joinery	..		..		..				..		..
2023 Wooden containers	..		..		..				..		..
2029 Other products of wood	..		..		..				..		..
21 PAPER AND PAPER PRODUCTS	0	0	0	0	0	..			1,355,217	267,000	3,918
2101 Pulp, paper and paperboard	..		..		..				..		..
2102 Corrugated paper, paperboard and their containers	..		..		..				..		..
2109 Other articles of pulp and paperboard	..		..		..				..		..
22 PUBLISHING, PRINTING & REPRODUCTION	..		..		..				..		..
221 Publishing	..		..		..				..		..
2211 Publishing of books & brochures	..		..		..				..		..
2212 Publishing of newspapers and periodicals	..		..		..				..		..
2213 Publishing of recorded media	..		..		..				..		..
2219 Other publishing	..		..		..				..		..
222 Printing and related service activities	..		..		..				..		..
2221 Printing	..		..		..				..		..
2222 Service activities related to printing	..		..		..				..		..
223 Reproduction of recorded media	..		..		..				..		..
23 COKE, REFINED PETROLEUM PRODUCTS	0	0	17,082	0	8,480	..			0	0	25,562
231 Coke oven products	..		..		..				..		..
232 Refined petroleum products	0	0	17,082	0	8,480				0	0	25,562
233 Processing of nuclear fuel	..		..		..				..		..
24 CHEMICALS & CHEMICAL PRODUCTS	0	0	0	0	26,850	..			1,054,215	132,185	30,169
241 Basic chemicals	0	0	0	0	26,850	..			0	0	26,850
2411 Basic chemicals, exc. fertilizers & nitrogen compounds	..		..		..				..		..
2412 Fertilizers and nitrogen compounds	..		..		..				..		..
2413 Plastics in primary forms and synthetic rubber	..		..		..				..		..
242 Other chemical products	..		..		..				..		..
2421 Pesticides and other agro-chemical products	..		..		..				..		..
2422 Paints, varnishes and similar coatings	..		..		..				..		..
2423 Pharmaceuticals, medicinal chem. & botanical prod.	..		..		..				..		..
2424 Soap and detergents, perfumes etc.	..		..		..				..		..
2429 Other chemical products, nec	..		..		..				..		..
243 Man-made fibres	..		..		..				..		..
25 RUBBER AND PLASTICS PRODUCTS	..		..		..				..		..
251 Rubber products	..		..		..				..		..
2511 Rubber tyres and tubes	..		..		..				..		..
2519 Other rubber products	..		..		..				..		..
252 Plastic products	..		..		..				..		..
26 OTHER NON-METALLIC MINERAL PRODUCTS	0	0	0	0	0	..			238,643	0	859
261 Glass and glass products	..		..		..				..		..
269 Non-metallic mineral products, nec	..		..		..				..		..
2691 Pottery, china and earthenware	..		..		..				..		..
2692 Refractory ceramic products	..		..		..				..		..
2693 Structural non-refractory clay & ceramic prod.	..		..		..				..		..
2694 Cement, lime and plaster	..		..		..				..		..
2695 Articles of concrete, cement and plaster	..		..		..				..		..
2696 Cutting, shaping and finishing of stone	..		..		..				..		..
2699 Other non-metallic mineral products, nec	..		..		..				..		..

ISIS Energy Data Programme (IEA/OECD)

ISIC Revision 3 Industry Sector	Solid TJ	LPG TJ	Distiloil TJ	RFO TJ	Gas TJ	Biomass TJ	Steam TJ	Electr MWh	Own Use MWh	TOTAL TJ
27　**BASIC METALS**	14,554	0	9	0	0		..	5,483,221	165,406	33,707
271　Basic iron and steel	14,554	0	9	0	0		..	0	0	14,563
272　Basic precious and non-ferrous metals	0	0	0	0	0		..	5,483,221	165,406	19,144
273　Casting of metals	..	..	..	..	..		..	..	..	..
2731　Casting of iron and steel	..	..	..	..	..		..	..	..	..
2732　Casting non-ferrous metals	..	..	..	..	..		..	..	..	..
28　**FABRICATED METAL PRODUCTS**	0	0	0	0	0		..	411,470	0	1,481
281　Str. metal prod., tanks, reservoirs, steam generators	..	..	..	..	..		..	..	..	..
2811　Structural metal products	..	..	..	..	..		..	..	..	..
2812　Tanks, reservoirs and containers of metal	..	..	..	..	..		..	..	..	..
2813　Steam generators, exc. central heating hot water boilers	..	..	..	..	..		..	..	..	..
289　Other fabricated metal products	..	..	..	..	..		..	..	..	..
2891　Forging, pressing, stamping & roll-forming of metal	..	..	..	..	..		..	..	..	..
2892　Treatment and coating of metals	..	..	..	..	..		..	..	..	..
2893　Cutlery, hand tools and general hardware	..	..	..	..	..		..	..	..	..
2899　Other fabricated metal products, nec	..	..	..	..	..		..	..	..	..
29　**MACHINERY AND EQUIPMENT, NEC**	..	..	..	..	..		..	..	..	..
291　General purpose machinery	..	..	..	..	..		..	..	..	..
2911　Engines and turbines	..	..	..	..	..		..	..	..	..
2912　Pumps, compressors, taps and valves	..	..	..	..	..		..	..	..	..
2913　Bearings, gears, gearing and driving elements	..	..	..	..	..		..	..	..	..
2914　Ovens, furnaces and furnace burners	..	..	..	..	..		..	..	..	..
2915　Lifting and handling equipment	..	..	..	..	..		..	..	..	..
2919　Other general purpose machinery	..	..	..	..	..		..	..	..	..
292　Special purpose machinery	..	..	..	..	..		..	..	..	..
2921　Agricultural and forestry machinery	..	..	..	..	..		..	..	..	..
2922　Machine-tools	..	..	..	..	..		..	..	..	..
2923　Machinery for metallurgy	..	..	..	..	..		..	..	..	..
2924　Machinery for mining, quarrying and construction	..	..	..	..	..		..	..	..	..
2925　Machinery for food, beverage & tobacco processing	..	..	..	..	..		..	..	..	..
2926　Machinery for textile, apparel & leather production	..	..	..	..	..		..	..	..	..
2927　Machinery for weapons and ammunition	..	..	..	..	..		..	..	..	..
2929　Other special purpose machinery	..	..	..	..	..		..	..	..	..
293　Domestic appliances, nec	..	..	..	..	..		..	..	..	..
30　**OFFICE, ACCOUNTING & COMPUTING MACHINERY**	..	..	..	..	..		..	..	..	..
31　**ELECTRICAL MACHINERY & APPARATUS, NEC**	..	..	..	..	..		..	..	..	..
311　Electric motors, generators and transformers	..	..	..	..	..		..	..	..	..
312　Electricity distribution and control apparatus	..	..	..	..	..		..	..	..	..
313　Insulated wire and cable	..	..	..	..	..		..	..	..	..
314　Accumulators, primary cells & primary batteries	..	..	..	..	..		..	..	..	..
315　Electric lamps and lighting equipment	..	..	..	..	..		..	..	..	..
319　Other electrical equipment, nec	..	..	..	..	..		..	..	..	..
32　**RADIO, TV & COMMUNICATION EQUIP. & APP.**	..	..	..	..	..		..	..	..	..
321　Electronic valves, tubes, other electronic components	..	..	..	..	..		..	..	..	..
322　TV & radio transmitters, apparatus for line telephony	..	..	..	..	..		..	..	..	..
323　TV & radio receivers, recording apparatus	..	..	..	..	..		..	..	..	..
33　**MEDICAL PRECISION &OPTICAL INSTRUMENTS**	..	..	..	..	..		..	..	..	..
331　Medical appliances and instruments	..	..	..	..	..		..	..	..	..
3311　Medical, surgical equipment & orthopaedic app.	..	..	..	..	..		..	..	..	..
3312　Instruments & appliances for measuring, checking etc	..	..	..	..	..		..	..	..	..
3313　Industrial process control equipment	..	..	..	..	..		..	..	..	..
332　Optical instruments and photographic equipment	..	..	..	..	..		..	..	..	..
333　Watches and clocks	..	..	..	..	..		..	..	..	..
34　**MOTOR VEHICLES, TRAILERS & SEMI-TRAILERS**	..	..	..	..	..		..	..	..	..
341　Motor vehicles	..	..	..	..	..		..	..	..	..
342　Bodies (coachwork) for motor vehicles	..	..	..	..	..		..	..	..	..
343　Parts, accessories for motor vehicles & their engines	..	..	..	..	..		..	..	..	..
35　**OTHER TRANSPORT EQUIPMENT**	..	..	..	..	..		..	..	..	..
351　Building and repairing of ships and boats	..	..	..	..	..		..	..	..	..
3511　Building and repairing of ships	..	..	..	..	..		..	..	..	..
3512　Building, repairing of pleasure & sporting boats	..	..	..	..	..		..	..	..	..
352　Railway, tramway locomotives & rolling stock	..	..	..	..	..		..	..	..	..
353　Aircraft and spacecraft	..	..	..	..	..		..	..	..	..
359　Transport equipment, nec	..	..	..	..	..		..	..	..	..
3591　Motorcycles	..	..	..	..	..		..	..	..	..
3592　Bicycles and invalid carriages	..	..	..	..	..		..	..	..	..
3599　Other transport equipment, nec	..	..	..	..	..		..	..	..	..
36　**FURNITURE; MANUFACTURING, NEC**	0	0	0	0	0		..	46,265	0	167
361　Furniture	..	..	..	..	..		..	..	..	..
369　Manufacturing, nec	..	..	..	..	..		..	..	..	..
3691　Jewellery and related articles	..	..	..	..	..		..	..	..	..
3692　Musical instruments	..	..	..	..	..		..	..	..	..
3693　Sports goods	..	..	..	..	..		..	..	..	..
3694　Games and toys	..	..	..	..	..		..	..	..	..
3699　Other manufacturing, nec	..	..	..	..	..		..	..	..	..
37　**RECYCLING**	..	..	..	..	..		..	..	..	..
371　Recycling of metal waste and scrap	..	..	..	..	..		..	..	..	..
372　Recycling of non-metal waste and scrap	..	..	..	..	..		..	..	..	..
CERR　Non-specified industry	11,683	1,010	2,679	2,169	21,314		..	0	0	38,855
15-37　**TOTAL MANUFACTURING**	26,237	1,010	20,481	2,575	56,644		..	12,571,342	1,064,925	148,371

ISIC Revision 3 Industry Sector	Solid TJ	LPG TJ	Distiloil TJ	RFO TJ	Gas TJ	Biomass TJ	Steam TJ	Electr MWh	Own Use MWh	TOTAL TJ
15 **FOOD PRODUCTS AND BEVERAGES**	0	0	810	544	0	..	..	2,109,610	490,058	7,184
151 Production, processing and preserving (PPP)	..	..	..	..	..	..	..	..	..	..
1511 PPP of meat and meat products	..	..	..	..	..	..	..	..	..	..
1512 Processing and preserving of fish products	..	..	..	..	..	..	..	..	..	..
1513 Processing, preserving of fruit & vegetables	..	..	..	..	..	..	..	..	..	..
1514 Vegetable and animal oils and fats	..	..	..	..	..	..	..	..	..	..
152 Dairy products	..	..	..	..	..	..	..	..	..	..
153 Grain mill prod., starches & prepared animal feeds	..	..	..	..	..	..	..	..	..	..
1531 Grain mill products	..	..	..	..	..	..	..	..	..	..
1532 Starches and starch products	..	..	..	..	..	..	..	..	..	..
1533 Prepared animal feeds	..	..	..	..	..	..	..	..	..	..
154 Other food products	..	..	..	..	..	..	..	..	..	..
1541 Bakery products	..	..	..	..	..	..	..	..	..	..
1542 Sugar	..	..	..	..	..	..	..	..	..	..
1543 Cocoa, chocolate and sugar confectionery	..	..	..	..	..	..	..	..	..	..
1544 Macaroni, noodles, couscous & similar farinaceous prod.	..	..	..	..	..	..	..	..	..	..
1549 Other food products, nec	..	..	..	..	..	..	..	..	..	..
155 Beverages	..	..	..	..	..	..	..	..	..	..
1551 Distilling, rectifying and blending of spirits	..	..	..	..	..	..	..	..	..	..
1552 Wines	..	..	..	..	..	..	..	..	..	..
1553 Malt liquors and malt	..	..	..	..	..	..	..	..	..	..
1554 Soft drinks; production of mineral waters	..	..	..	..	..	..	..	..	..	..
16 **TOBACCO PRODUCTS**	..	..	..	..	..	..	..	..	..	..
17 **TEXTILES**	0	0	0	0	0	..	..	199,958	0	720
171 Spinning, weaving and finishing of textiles	..	..	..	..	..	..	..	..	..	..
1711 Preparation and spinning of textile fibres	..	..	..	..	..	..	..	..	..	..
1712 Finishing of textiles	..	..	..	..	..	..	..	..	..	..
172 Other textiles	..	..	..	..	..	..	..	..	..	..
1721 Made-up textile articles, except apparel	..	..	..	..	..	..	..	..	..	..
1722 Carpets and rugs	..	..	..	..	..	..	..	..	..	..
1723 Cordage, rope, twine and netting	..	..	..	..	..	..	..	..	..	..
1729 Other textiles, nec	..	..	..	..	..	..	..	..	..	..
173 Knitted and crocheted fabrics and articles	..	..	..	..	..	..	..	..	..	..
18 **WEARING APPAREL, DRESSING & DYEING OF FUR**	..	..	..	..	..	..	..	..	..	..
181 Wearing apparel, except fur apparel	..	..	..	..	..	..	..	..	..	..
182 Dressing and dyeing of fur; articles of fur	..	..	..	..	..	..	..	..	..	..
19 **TANNING & DRESSING OF LEATHER, FOOTWEAR**	..	..	..	..	..	..	..	..	..	..
191 Tanning and dressing of leather	..	..	..	..	..	..	..	..	..	..
1911 Tanning and dressing of leather	..	..	..	..	..	..	..	..	..	..
1912 Luggage, handbags, saddlery & harness	..	..	..	..	..	..	..	..	..	..
192 Footwear	..	..	..	..	..	..	..	..	..	..
20 **WOOD AND WOOD PRODUCTS**	0	0	0	0	0	..	..	2,098,522	346,301	6,308
201 Sawmilling and planing of wood	..	..	..	..	..	..	..	..	..	..
202 Products of wood, cork, straw & plaiting materials	..	..	..	..	..	..	..	..	..	..
2021 Veneer sheets	..	..	..	..	..	..	..	..	..	..
2022 Builders' carpentry and joinery	..	..	..	..	..	..	..	..	..	..
2023 Wooden containers	..	..	..	..	..	..	..	..	..	..
2029 Other products of wood	..	..	..	..	..	..	..	..	..	..
21 **PAPER AND PAPER PRODUCTS**	0	0	0	0	0	..	..	1,358,723	243,000	4,017
2101 Pulp, paper and paperboard	..	..	..	..	..	..	..	..	..	..
2102 Corrugated paper, paperboard and their containers	..	..	..	..	..	..	..	..	..	..
2109 Other articles of pulp and paperboard	..	..	..	..	..	..	..	..	..	..
22 **PUBLISHING, PRINTING & REPRODUCTION**	..	..	..	..	..	..	..	..	..	..
221 Publishing	..	..	..	..	..	..	..	..	..	..
2211 Publishing of books & brochures	..	..	..	..	..	..	..	..	..	..
2212 Publishing of newspapers and periodicals	..	..	..	..	..	..	..	..	..	..
2213 Publishing of recorded media	..	..	..	..	..	..	..	..	..	..
2219 Other publishing	..	..	..	..	..	..	..	..	..	..
222 Printing and related service activities	..	..	..	..	..	..	..	..	..	..
2221 Printing	..	..	..	..	..	..	..	..	..	..
2222 Service activities related to printing	..	..	..	..	..	..	..	..	..	..
223 Reproduction of recorded media	..	..	..	..	..	..	..	..	..	..
23 **COKE, REFINED PETROLEUM PRODUCTS**	0	0	19,141	0	734	..	..	0	0	19,875
231 Coke oven products	..	..	..	..	..	..	..	..	..	..
232 Refined petroleum products	0	0	19,141	0	734	..	..	0	0	19,875
233 Processing of nuclear fuel	..	..	..	..	..	..	..	..	..	..
24 **CHEMICALS & CHEMICAL PRODUCTS**	0	0	0	0	32,508	..	..	950,024	91,783	35,598
241 Basic chemicals	0	0	0	0	32,508	..	..	0	0	32,508
2411 Basic chemicals, exc. fertilizers & nitrogen compounds	0	0	0	0	32,508	..	..	0	0	32,508
2412 Fertilizers and nitrogen compounds	..	..	..	..	..	..	..	..	..	..
2413 Plastics in primary forms and synthetic rubber	..	..	..	..	..	..	..	..	..	..
242 Other chemical products	..	..	..	..	..	..	..	..	..	..
2421 Pesticides and other agro-chemical products	..	..	..	..	..	..	..	..	..	..
2422 Paints, varnishes and similar coatings	..	..	..	..	..	..	..	..	..	..
2423 Pharmaceuticals, medicinal chem. & botanical prod.	..	..	..	..	..	..	..	..	..	..
2424 Soap and detergents, perfumes etc.	..	..	..	..	..	..	..	..	..	..
2429 Other chemical products, nec	..	..	..	..	..	..	..	..	..	..
243 Man-made fibres	..	..	..	..	..	..	..	..	..	..
25 **RUBBER AND PLASTICS PRODUCTS**	..	..	..	..	..	..	..	..	..	..
251 Rubber products	..	..	..	..	..	..	..	..	..	..
2511 Rubber tyres and tubes	..	..	..	..	..	..	..	..	..	..
2519 Other rubber products	..	..	..	..	..	..	..	..	..	..
252 Plastic products	..	..	..	..	..	..	..	..	..	..
26 **OTHER NON-METALLIC MINERAL PRODUCTS**	0	0	0	0	0	..	..	184,117	0	663
261 Glass and glass products	..	..	..	..	..	..	..	..	..	..
269 Non-metallic mineral products, nec	..	..	..	..	..	..	..	..	..	..
2691 Pottery, china and earthenware	..	..	..	..	..	..	..	..	..	..
2692 Refractory ceramic products	..	..	..	..	..	..	..	..	..	..
2693 Structural non-refractory clay & ceramic prod.	..	..	..	..	..	..	..	..	..	..
2694 Cement, lime and plaster	..	..	..	..	..	..	..	..	..	..
2695 Articles of concrete, cement and plaster	..	..	..	..	..	..	..	..	..	..
2696 Cutting, shaping and finishing of stone	..	..	..	..	..	..	..	..	..	..
2699 Other non-metallic mineral products, nec	..	..	..	..	..	..	..	..	..	..

ISIS Energy Data Programme (IEA/OECD)

ISIC Revision 3 Industry Sector	Solid TJ	LPG TJ	Distiloil TJ	RFO TJ	Gas TJ	Biomass TJ	Steam TJ	Electr MWh	Own Use MWh	TOTAL TJ
27 BASIC METALS	13,715	0	5	0	0	..	..	..	4,897,008	166,550	30,750
271 Basic iron and steel	13,715	0	5	0	0			..	0	0	13,720
272 Basic precious and non-ferrous metals	0	0	0	0	0			..	4,897,008	166,550	17,030
273 Casting of metals								..			
2731 Casting of iron and steel	..	..									
2732 Casting non-ferrous metals	..										
28 FABRICATED METAL PRODUCTS	0	0	0	0	0			..	401,235	0	1,444
281 Str. metal prod., tanks, reservoirs, steam generators	..	..	..	..	..			..	..	..	..
2811 Structural metal products	..	..						..			
2812 Tanks, reservoirs and containers of metal	..	..						..			
2813 Steam generators, exc. central heating hot water boilers	..	..						..			
289 Other fabricated metal products	..	..						..			
2891 Forging, pressing, stamping & roll-forming of metal	..	..						..			
2892 Treatment and coating of metals	..	..						..			
2893 Cutlery, hand tools and general hardware	..	..						..			
2899 Other fabricated metal products, nec	..	..						..			
29 MACHINERY AND EQUIPMENT, NEC	..	..	..	..	..			..	..	..	..
291 General purpose machinery	..	..						..			
2911 Engines and turbines	..	..						..			
2912 Pumps, compressors, taps and valves	..	..						..			
2913 Bearings, gears, gearing and driving elements	..	..						..			
2914 Ovens, furnaces and furnace burners	..	..						..			
2915 Lifting and handling equipment	..	..						..			
2919 Other general purpose machinery	..	..						..			
292 Special purpose machinery	..	..						..			
2921 Agricultural and forestry machinery	..	..						..			
2922 Machine-tools	..	..						..			
2923 Machinery for metallurgy	..	..						..			
2924 Machinery for mining, quarrying and construction	..	..						..			
2925 Machinery for food, beverage & tobacco processing	..	..						..			
2926 Machinery for textile, apparel & leather production	..	..						..			
2927 Machinery for weapons and ammunition	..	..						..			
2929 Other special purpose machinery	..	..						..			
293 Domestic appliances, nec	..	..						..			
30 OFFICE, ACCOUNTING & COMPUTING MACHINERY	..	..	..	..	..			..	..	..	..
31 ELECTRICAL MACHINERY & APPARATUS, NEC	..	..	..	..	..			..	..	..	..
311 Electric motors, generators and transformers	..	..						..			
312 Electricity distribution and control apparatus	..	..						..			
313 Insulated wire and cable	..	..						..			
314 Accumulators, primary cells & primary batteries	..	..						..			
315 Electric lamps and lighting equipment	..	..						..			
319 Other electrical equipment, nec	..	..						..			
32 RADIO, TV & COMMUNICATION EQUIP. & APP.	..	..	..	..	..			..	..	..	..
321 Electronic valves, tubes, other electronic components	..	..						..			
322 TV & radio transmitters, apparatus for line telephony	..	..						..			
323 TV & radio receivers, recording apparatus	..	..						..			
33 MEDICAL PRECISION &OPTICAL INSTRUMENTS	..	..	..	..	..			..	..	..	..
331 Medical appliances and instruments	..	..						..			
3311 Medical, surgical equipment & orthopaedic app.	..	..						..			
3312 Instruments & appliances for measuring, checking etc	..	..						..			
3313 Industrial process control equipment	..	..						..			
332 Optical instruments and photographic equipment	..	..						..			
333 Watches and clocks	..	..						..			
34 MOTOR VEHICLES, TRAILERS & SEMI-TRAILERS	..	..	..	..	..			..	..	..	..
341 Motor vehicles	..	..						..			
342 Bodies (coachwork) for motor vehicles	..	..						..			
343 Parts, accessories for motor vehicles & their engines	..	..						..			
35 OTHER TRANSPORT EQUIPMENT	..	..	..	..	..			..	..	..	..
351 Building and repairing of ships and boats	..	..						..			
3511 Building and repairing of ships	..	..						..			
3512 Building, repairing of pleasure & sporting boats	..	..						..			
352 Railway, tramway locomotives & rolling stock	..	..						..			
353 Aircraft and spacecraft	..	..						..			
359 Transport equipment, nec	..	..						..			
3591 Motorcycles	..	..						..			
3592 Bicycles and invalid carriages	..	..						..			
3599 Other transport equipment, nec	..	..						..			
36 FURNITURE; MANUFACTURING, NEC	0	0	0	0	0			..	167,902	0	604
361 Furniture	..	..		..	..			..	..	..	..
369 Manufacturing, nec	0	0	0	0	0			..	167,902	0	604
3691 Jewellery and related articles	..	..						..			
3692 Musical instruments	..	..						..			
3693 Sports goods	..	..						..			
3694 Games and toys	..	..						..			
3699 Other manufacturing, nec	..	..						..			
37 RECYCLING	..	..	..	..	..			..	..	..	..
371 Recycling of metal waste and scrap	..	..						..			
372 Recycling of non-metal waste and scrap	..	..						..			
CERR Non-specified industry	18,444	1,190	5,820	1,907	39,521			..	0	0	66,882
15-37 TOTAL MANUFACTURING	32,159	1,190	25,776	2,451	72,763			..	12,367,099	1,337,692	**174,045**

ISIC Revision 3 Industry Sector	Solid TJ	LPG TJ	Distiloil TJ	RFO TJ	Gas TJ	Biomass TJ	Steam TJ	Electr MWh	Own Use MWh	TOTAL TJ
15 **FOOD PRODUCTS AND BEVERAGES**	0	0	815	462	0	..		1,720,008	813,813	4,539
151 Production, processing and preserving (PPP)	0	0	0	0	0			613,680	0	2,209
1511 PPP of meat and meat products	..	..	..	..	..			..	..	..
1512 Processing and preserving of fish products	..	..	..	..	..			..	..	..
1513 Processing, preserving of fruit & vegetables	..	..	..	..	..			..	..	..
1514 Vegetable and animal oils and fats	..	..	..	..	..			..	..	..
152 Dairy products	0	0	0	0	0			476,704	0	1,716
153 Grain mill prod., starches & prepared animal feeds	..	..	..	..	..			..	..	..
1531 Grain mill products	..	..	..	..	..			..	..	..
1532 Starches and starch products	..	..	..	..	..			..	..	..
1533 Prepared animal feeds	..	..	..	..	..			..	..	..
154 Other food products	0	0	0	0	0			629,624	3,374	2,255
1541 Bakery products	..	..	..	..	..			..	..	..
1542 Sugar	..	..	..	..	..			..	..	..
1543 Cocoa, chocolate and sugar confectionery	..	..	..	..	..			..	..	..
1544 Macaroni, noodles, couscous & similar farinaceous prod.	..	..	..	..	..			..	..	..
1549 Other food products, nec	..	..	..	..	..			..	..	..
155 Beverages	..	..	..	..	..			..	..	..
1551 Distilling, rectifying and blending of spirits	..	..	..	..	..			..	..	..
1552 Wines	..	..	..	..	..			..	..	..
1553 Malt liquors and malt	..	..	..	..	..			..	..	..
1554 Soft drinks; production of mineral waters	..	..	..	..	..			..	..	..
16 **TOBACCO PRODUCTS**	..	..	..	..	..			..	..	..
17 **TEXTILES**	0	0	0	0	0		..	230,823	0	831
171 Spinning, weaving and finishing of textiles	..	..	..	..	..			..	..	..
1711 Preparation and spinning of textile fibres	..	..	..	..	..			..	..	..
1712 Finishing of textiles	..	..	..	..	..			..	..	..
172 Other textiles	..	..	..	..	..			..	..	..
1721 Made-up textile articles, except apparel	..	..	..	..	..			..	..	..
1722 Carpets and rugs	..	..	..	..	..			..	..	..
1723 Cordage, rope, twine and netting	..	..	..	..	..			..	..	..
1729 Other textiles, nec	..	..	..	..	..			..	..	..
173 Knitted and crocheted fabrics and articles	..	..	..	..	..			..	..	..
18 **WEARING APPAREL, DRESSING & DYEING OF FUR**	..	..	..	..	..			..	..	..
181 Wearing apparel, except fur apparel	..	..	..	..	..			..	..	..
182 Dressing and dyeing of fur; articles of fur	..	..	..	..	..			..	..	..
19 **TANNING & DRESSING OF LEATHER, FOOTWEAR**	..	..	..	..	..			..	..	..
191 Tanning and dressing of leather	..	..	..	..	..			..	..	..
1911 Tanning and dressing of leather	..	..	..	..	..			..	..	..
1912 Luggage, handbags, saddlery & harness	..	..	..	..	..			..	..	..
192 Footwear	..	..	..	..	..			..	..	..
20 **WOOD AND WOOD PRODUCTS**	0	0	0	0	0		..	1,861,496	1,010,521	3,064
201 Sawmilling and planing of wood	..	..	..	..	..			..	..	..
202 Products of wood, cork, straw & plaiting materials	..	..	..	..	..			..	..	..
2021 Veneer sheets	..	..	..	..	..			..	..	..
2022 Builders' carpentry and joinery	..	..	..	..	..			..	..	..
2023 Wooden containers	..	..	..	..	..			..	..	..
2029 Other products of wood	..	..	..	..	..			..	..	..
21 **PAPER AND PAPER PRODUCTS**	0	0	0	0	0		..	809,125	243,000	2,038
2101 Pulp, paper and paperboard	..	..	..	..	..			..	..	..
2102 Corrugated paper, paperboard and their containers	..	..	..	..	..			..	..	..
2109 Other articles of pulp and paperboard	..	..	..	..	..			..	..	..
22 **PUBLISHING, PRINTING & REPRODUCTION**	0	0	0	0	0		..	175,401	0	631
221 Publishing	..	..	..	..	..			..	..	..
2211 Publishing of books & brochures	..	..	..	..	..			..	..	..
2212 Publishing of newspapers and periodicals	..	..	..	..	..			..	..	..
2213 Publishing of recorded media	..	..	..	..	..			..	..	..
2219 Other publishing	..	..	..	..	..			..	..	..
222 Printing and related service activities	..	..	..	..	..			..	..	..
2221 Printing	..	..	..	..	..			..	..	..
2222 Service activities related to printing	..	..	..	..	..			..	..	..
223 Reproduction of recorded media	..	..	..	..	..			..	..	..
23 **COKE, REFINED PETROLEUM PRODUCTS**	0	0	19,761	0	0		..	241,367	0	20,630
231 Coke oven products	0	0	0	0	0			7,403	0	27
232 Refined petroleum products	0	0	19,761	0	0			233,964	0	20,603
233 Processing of nuclear fuel	..	..	..	..	..			..	..	..
24 **CHEMICALS & CHEMICAL PRODUCTS**	0	0	0	0	30,299		..	235,960	89,815	30,825
241 Basic chemicals	0	0	0	0	30,299			0	0	30,299
2411 Basic chemicals, exc. fertilizers & nitrogen compounds	0	0	0	0	30,299			0	0	30,299
2412 Fertilizers and nitrogen compounds	..	..	..	..	..			..	..	..
2413 Plastics in primary forms and synthetic rubber	..	..	..	..	..			..	..	..
242 Other chemical products	..	..	..	..	..			..	..	..
2421 Pesticides and other agro-chemical products	..	..	..	..	..			..	..	..
2422 Paints, varnishes and similar coatings	..	..	..	..	..			..	..	..
2423 Pharmaceuticals, medicinal chem. & botanical prod.	..	..	..	..	..			..	..	..
2424 Soap and detergents, perfumes etc.	..	..	..	..	..			..	..	..
2429 Other chemical products, nec	..	..	..	..	..			..	..	..
243 Man-made fibres	..	..	..	..	..			..	..	..
25 **RUBBER AND PLASTICS PRODUCTS**	0	0	0	0	0		..	256,759	0	924
251 Rubber products	..	..	..	..	..			..	..	..
2511 Rubber tyres and tubes	..	..	..	..	..			..	..	..
2519 Other rubber products	..	..	..	..	..			..	..	..
252 Plastic products	..	..	..	..	..			..	..	..
26 **OTHER NON-METALLIC MINERAL PRODUCTS**	0	0	0	0	0		..	240,611	0	866
261 Glass and glass products	..	..	..	..	..			..	..	..
269 Non-metallic mineral products, nec	..	..	..	..	..			..	..	..
2691 Pottery, china and earthenware	..	..	..	..	..			..	..	..
2692 Refractory ceramic products	..	..	..	..	..			..	..	..
2693 Structural non-refractory clay & ceramic prod.	..	..	..	..	..			..	..	..
2694 Cement, lime and plaster	..	..	..	..	..			..	..	..
2695 Articles of concrete, cement and plaster	..	..	..	..	..			..	..	..
2696 Cutting, shaping and finishing of stone	..	..	..	..	..			..	..	..
2699 Other non-metallic mineral products, nec	..	..	..	..	..			..	..	..

ISIS Energy Data Programme (IEA/OECD)

ISIC Revision 3 Industry Sector	Solid TJ	LPG TJ	Distiloil TJ	RFO TJ	Gas TJ	Biomass TJ	Steam TJ	Electr MWh	Own Use MWh	TOTAL TJ
27 BASIC METALS	14,788	0	3	0	0	..	..	5,013,632	303,685	31,747
271 Basic iron and steel	14,788	0	3	0	0	..	..	115,843	0	15,208
272 Basic precious and non-ferrous metals	0	0	0	0	0	..	..	4,897,789	0	17,632
273 Casting of metals	..	..	..	..	..	..	..	..	..	..
2731 Casting of iron and steel	..	..	..	..	..	..	..	..	..	..
2732 Casting non-ferrous metals	..	..	..	..	..	..	..	..	..	..
28 FABRICATED METAL PRODUCTS	0	0	0	0	0	..	..	142,045	0	511
281 Str. metal prod., tanks, reservoirs, steam generators	..	..	..	..	..	..	..	142,045	0	511
2811 Structural metal products	..	..	..	..	..	..	..	..	..	..
2812 Tanks, reservoirs and containers of metal	..	..	..	..	..	..	..	..	..	..
2813 Steam generators, exc. central heating hot water boilers	..	..	..	..	..	..	..	..	..	..
289 Other fabricated metal products	0	0	0	0	0	..	..	..	..	..
2891 Forging, pressing, stamping & roll-forming of metal	..	..	..	..	..	..	..	142,045	0	511
2892 Treatment and coating of metals	..	..	..	..	..	..	..	..	..	..
2893 Cutlery, hand tools and general hardware	..	..	..	..	..	..	..	..	..	..
2899 Other fabricated metal products, nec	..	..	..	..	..	..	..	..	..	..
29 MACHINERY AND EQUIPMENT, NEC	..	..	..	..	..	..	..	..	..	..
291 General purpose machinery	..	..	..	..	..	..	..	..	..	..
2911 Engines and turbines	..	..	..	..	..	..	..	..	..	..
2912 Pumps, compressors, taps and valves	..	..	..	..	..	..	..	..	..	..
2913 Bearings, gears, gearing and driving elements	..	..	..	..	..	..	..	..	..	..
2914 Ovens, furnaces and furnace burners	..	..	..	..	..	..	..	..	..	..
2915 Lifting and handling equipment	..	..	..	..	..	..	..	..	..	..
2919 Other general purpose machinery	..	..	..	..	..	..	..	..	..	..
292 Special purpose machinery	..	..	..	..	..	..	..	..	..	..
2921 Agricultural and forestry machinery	..	..	..	..	..	..	..	..	..	..
2922 Machine-tools	..	..	..	..	..	..	..	..	..	..
2923 Machinery for metallurgy	..	..	..	..	..	..	..	..	..	..
2924 Machinery for mining, quarrying and construction	..	..	..	..	..	..	..	..	..	..
2925 Machinery for food, beverage & tobacco processing	..	..	..	..	..	..	..	..	..	..
2926 Machinery for textile, apparel & leather production	..	..	..	..	..	..	..	..	..	..
2927 Machinery for weapons and ammunition	..	..	..	..	..	..	..	..	..	..
2929 Other special purpose machinery	..	..	..	..	..	..	..	..	..	..
293 Domestic appliances, nec	..	..	..	..	..	..	..	..	..	..
30 OFFICE, ACCOUNTING & COMPUTING MACHINERY	..	..	..	..	..	..	..	..	..	..
31 ELECTRICAL MACHINERY & APPARATUS, NEC	..	..	..	..	..	..	..	..	..	..
311 Electric motors, generators and transformers	..	..	..	..	..	..	..	..	..	..
312 Electricity distribution and control apparatus	..	..	..	..	..	..	..	..	..	..
313 Insulated wire and cable	..	..	..	..	..	..	..	..	..	..
314 Accumulators, primary cells & primary batteries	..	..	..	..	..	..	..	..	..	..
315 Electric lamps and lighting equipment	..	..	..	..	..	..	..	..	..	..
319 Other electrical equipment, nec	..	..	..	..	..	..	..	..	..	..
32 RADIO, TV & COMMUNICATION EQUIP. & APP.	..	..	..	..	..	..	..	..	..	..
321 Electronic valves, tubes, other electronic components	..	..	..	..	..	..	..	..	..	..
322 TV & radio transmitters, apparatus for line telephony	..	..	..	..	..	..	..	..	..	..
323 TV & radio receivers, recording apparatus	..	..	..	..	..	..	..	..	..	..
33 MEDICAL PRECISION &OPTICAL INSTRUMENTS	0	0	0	0	0	..	..	229,256	0	825
331 Medical appliances and instruments	0	0	0	0	0	..	..	229,256	0	825
3311 Medical, surgical equipment & orthopaedic app.	0	0	0	0	0	..	..	2,991	0	11
3312 Instruments & appliances for measuring, checking etc	..	..	..	..	..	..	..	..	..	..
3313 Industrial process control equipment	0	0	0	0	0	..	..	226,265	0	815
332 Optical instruments and photographic equipment	..	..	..	..	..	..	..	..	..	..
333 Watches and clocks	..	..	..	..	..	..	..	..	..	..
34 MOTOR VEHICLES, TRAILERS & SEMI-TRAILERS	..	..	..	..	..	..	..	..	..	..
341 Motor vehicles	..	..	..	..	..	..	..	..	..	..
342 Bodies (coachwork) for motor vehicles	..	..	..	..	..	..	..	..	..	..
343 Parts, accessories for motor vehicles & their engines	..	..	..	..	..	..	..	..	..	..
35 OTHER TRANSPORT EQUIPMENT	0	0	0	0	0	..	..	97,119	0	350
351 Building and repairing of ships and boats	..	..	..	..	..	..	..	97,119	0	350
3511 Building and repairing of ships	..	..	..	..	..	..	..	..	..	..
3512 Building, repairing of pleasure & sporting boats	..	..	..	..	..	..	..	..	..	..
352 Railway, tramway locomotives & rolling stock	..	..	..	..	..	..	..	..	..	..
353 Aircraft and spacecraft	..	..	..	..	..	..	..	..	..	..
359 Transport equipment, nec	..	..	..	..	..	..	..	..	..	..
3591 Motorcycles	..	..	..	..	..	..	..	..	..	..
3592 Bicycles and invalid carriages	..	..	..	..	..	..	..	..	..	..
3599 Other transport equipment, nec	..	..	..	..	..	..	..	..	..	..
36 FURNITURE; MANUFACTURING, NEC	0	0	0	0	0	..	..	83,163	0	299
361 Furniture	..	..	..	..	..	..	..	..	..	..
369 Manufacturing, nec	0	0	0	0	0	..	..	..	..	..
3691 Jewellery and related articles	..	..	..	..	..	..	..	83,163	0	299
3692 Musical instruments	..	..	..	..	..	..	..	..	..	..
3693 Sports goods	..	..	..	..	..	..	..	..	..	..
3694 Games and toys	..	..	..	..	..	..	..	..	..	..
3699 Other manufacturing, nec	..	..	..	..	..	..	..	..	..	..
37 RECYCLING	..	..	..	..	..	..	..	..	..	..
371 Recycling of metal waste and scrap	..	..	..	..	..	..	..	..	..	..
372 Recycling of non-metal waste and scrap	..	..	..	..	..	..	..	..	..	..
CERR Non-specified industry	18,461	1,123	5,073	1,312	46,075	..	..	0	0	72,044
15-37 TOTAL MANUFACTURING	33,249	1,123	25,652	1,774	76,374	..	..	11,336,765	2,460,834	170,124

ISIC Revision 3 Industry Sector	Solid TJ	LPG TJ	Distiloil TJ	RFO TJ	Gas TJ	Biomass TJ	Steam TJ	Electr MWh	Own Use MWh	TOTAL TJ
15 FOOD PRODUCTS AND BEVERAGES	0	445	4,391	2,854	0	1	..	3,043,197	..	18,647
151 Production, processing and preserving (PPP)	0	347	1,836	2,563	0	1	..	1,364,232	..	9,658
1511 PPP of meat and meat products	0	220	762	40	0	1	..	519,943	..	2,895
1512 Processing and preserving of fish products	0	100	814	2,395	0	0	..	588,140	..	5,426
1513 Processing, preserving of fruit & vegetables	0	4	168	75	0	0	..	134,513	..	731
1514 Vegetable and animal oils and fats	0	23	92	53	0	0	..	121,636	..	606
152 Dairy products	0	48	606	27	0	0	..	559,132	..	2,694
153 Grain mill prod., starches & prepared animal feeds	0	20	864	61	0	0	..	404,392	..	2,401
1531 Grain mill products	0	19	213	0	0	0	..	172,845	..	854
1532 Starches and starch products	0	0	0	61	0	0	..	4,028	..	76
1533 Prepared animal feeds	0	1	651	0	0	0	..	227,519	..	1,471
154 Other food products	0	28	618	42	0	0	..	481,844	..	2,423
1541 Bakery products	0	22	383	0	0	0	..	298,555	..	1,480
1542 Sugar	..	..	..	..	..	..	..	..	..	..
1543 Cocoa, chocolate and sugar confectionery	0	6	41	40	0	0	..	78,290	..	369
1544 Macaroni, noodles, couscous & similar farinaceous prod.	0	0	24	0	0	0	..	7,723	..	52
1549 Other food products, nec	0	0	170	2	0	0	..	97,276	..	522
155 Beverages	0	2	467	161	0	0	..	233,597	..	1,471
1551 Distilling, rectifying and blending of spirits	0	0	1	0	0	0	..	10,763	..	40
1552 Wines	0	0	2	0	0	0	..	403	..	3
1553 Malt liquors and malt	0	2	384	153	0	0	..	179,727	..	1,186
1554 Soft drinks; production of mineral waters	0	0	80	8	0	0	..	42,704	..	242
16 TOBACCO PRODUCTS	0	0	60	0	0	13	..	13,471	..	121
17 TEXTILES	0	6	213	118	0	0	..	170,009	..	949
171 Spinning, weaving and finishing of textiles	0	6	107	99	0	0	..	72,622	..	473
1711 Preparation and spinning of textile fibres	0	5	58	43	0	0	..	66,769	..	346
1712 Finishing of textiles	0	1	49	56	0	0	..	5,853	..	127
172 Other textiles	0	0	76	0	0	0	..	69,188	..	325
1721 Made-up textile articles, except apparel	0	0	8	0	0	0	..	15,354	..	63
1722 Carpets and rugs	0	0	3	0	0	0	..	1,163	..	7
1723 Cordage, rope, twine and netting	0	0	34	0	0	0	..	26,885	..	131
1729 Other textiles, nec	0	0	31	0	0	0	..	25,786	..	124
173 Knitted and crocheted fabrics and articles	0	0	30	19	0	0	..	28,199	..	151
18 WEARING APPAREL, DRESSING & DYEING OF FUR	0	1	19	0	0	0	..	18,459	..	86
181 Wearing apparel, except fur apparel	0	1	14	0	0	0	..	15,881	..	72
182 Dressing and dyeing of fur; articles of fur	0	0	5	0	0	0	..	2,578	..	14
19 TANNING & DRESSING OF LEATHER, FOOTWEAR	0	0	34	0	0	0	..	16,439	..	93
191 Tanning and dressing of leather	0	0	32	0	0	0	..	13,045	..	79
1911 Tanning and dressing of leather	0	0	31	0	0	0	..	10,989	..	71
1912 Luggage, handbags, saddlery & harness	0	0	1	0	0	0	..	2,056	..	8
192 Footwear	0	0	2	0	0	0	..	3,394	..	14
20 WOOD AND WOOD PRODUCTS	0	4	656	89	0	4,884	..	804,840	..	8,530
201 Sawmilling and planing of wood	0	2	411	0	0	4,457	..	311,698	..	5,992
202 Products of wood, cork, straw & plaiting materials	0	2	245	89	0	427	..	493,142	..	2,538
2021 Veneer sheets	0	2	134	89	0	419	..	322,310	..	1,804
2022 Builders' carpentry and joinery	0	0	66	0	0	8	..	152,560	..	623
2023 Wooden containers	0	0	41	0	0	0	..	9,440	..	75
2029 Other products of wood	0	0	4	0	0	0	..	8,832	..	36
21 PAPER AND PAPER PRODUCTS	207	14	162	2,956	0	12,448	..	7,067,593	..	41,230
2101 Pulp, paper and paperboard	207	13	77	2,913	0	12,448	..	6,812,811	..	40,184
2102 Corrugated paper, paperboard and their containers	0	0	83	30	0	0	..	150,728	..	656
2109 Other articles of pulp and paperboard	0	1	2	13	0	0	..	104,054	..	391
22 PUBLISHING, PRINTING & REPRODUCTION	0	317	144	1	0	9	..	450,243	..	2,092
221 Publishing	0	0	74	0	0	0	..	201,487	..	799
2211 Publishing of books & brochures	0	0	3	0	0	0	..	34,529	..	127
2212 Publishing of newspapers and periodicals	0	0	66	0	0	0	..	146,414	..	593
2213 Publishing of recorded media	0	0	0	0	0	0	..	12,192	..	44
2219 Other publishing	0	0	5	0	0	0	..	8,352	..	35
222 Printing and related service activities	0	317	70	1	0	9	..	235,384	..	1,244
2221 Printing	0	317	31	1	0	9	..	186,234	..	1,028
2222 Service activities related to printing	0	0	39	0	0	0	..	49,150	..	216
223 Reproduction of recorded media	0	0	0	0	0	0	..	13,372	..	48
23 COKE, REFINED PETROLEUM PRODUCTS	0	482	516	171	33,933	0	..	706,305	..	37,645
231 Coke oven products	..	..	..	..	..	..	..	..	..	..
232 Refined petroleum products	0	482	516	171	33,933	0	..	706,305	..	37,645
233 Processing of nuclear fuel	..	..	..	..	..	..	..	..	..	..
24 CHEMICALS & CHEMICAL PRODUCTS	0	379	1,062	1,500	13,779	0	..	6,123,780	..	38,766
241 Basic chemicals	0	263	768	1,353	13,779	0	..	5,840,645	..	37,189
2411 Basic chemicals, exc. fertilizers & nitrogen compounds	0	127	546	965	11,408	0	..	4,381,299	..	28,819
2412 Fertilizers and nitrogen compounds	0	0	6	283	452	0	..	1,018,216	..	4,407
2413 Plastics in primary forms and synthetic rubber	0	136	216	105	1,919	0	..	441,130	..	3,964
242 Other chemical products	0	116	294	147	0	0	..	283,135	..	1,576
2421 Pesticides and other agro-chemical products	..	..	..	..	..	..	..	..	..	..
2422 Paints, varnishes and similar coatings	0	2	95	0	0	0	..	48,579	..	272
2423 Pharmaceuticals, medicinal chem. & botanical prod.	0	0	89	124	0	0	..	142,560	..	726
2424 Soap and detergents, perfumes etc.	0	114	76	0	0	0	..	37,248	..	324
2429 Other chemical products, nec	0	0	34	23	0	0	..	54,748	..	254
243 Man-made fibres	..	..	..	..	..	..	..	..	..	..
25 RUBBER AND PLASTICS PRODUCTS	0	24	231	68	0	0	..	414,072	..	1,814
251 Rubber products	0	0	18	30	0	0	..	95,668	..	392
2511 Rubber tyres and tubes	0	0	12	0	0	0	..	77,208	..	290
2519 Other rubber products	0	0	6	30	0	0	..	18,460	..	102
252 Plastic products	0	24	213	38	0	0	..	318,404	..	1,421
26 OTHER NON-METALLIC MINERAL PRODUCTS	5,574	961	795	1,192	0	0	..	964,652	..	11,995
261 Glass and glass products	0	286	56	611	0	0	..	166,885	..	1,554
269 Non-metallic mineral products, nec	5,574	675	739	581	0	0	..	797,767	..	10,441
2691 Pottery, china and earthenware	0	49	26	0	0	0	..	36,282	..	206
2692 Refractory ceramic products	0	0	0	54	0	0	..	4,260	..	69
2693 Structural non-refractory clay & ceramic prod.	0	54	47	70	0	0	..	14,753	..	224
2694 Cement, lime and plaster	4,897	3	40	424	0	0	..	392,498	..	6,777
2695 Articles of concrete, cement and plaster	415	448	512	33	0	0	..	145,941	..	1,933
2696 Cutting, shaping and finishing of stone	0	0	41	0	0	0	..	69,152	..	290
2699 Other non-metallic mineral products, nec	262	121	73	0	0	0	..	134,881	..	942

ISIC Revision 3 Industry Sector	Solid TJ	LPG TJ	Distiloil TJ	RFO TJ	Gas TJ	Biomass TJ	Steam TJ	Electr MWh	Own Use MWh	TOTAL TJ
27 **BASIC METALS**	9	359	1,917	1,284	1,076	3	..	22,779,622	..	86,655
271 Basic iron and steel	7	4	240	452	344	3	..	5,981,571	..	22,584
272 Basic precious and non-ferrous metals	2	355	1,660	832	732	0	..	16,791,268	..	64,030
273 Casting of metals	0	0	17	0	0	0	..	6,783	..	41
2731 Casting of iron and steel	..	..	..	..	..	..	..		..	..
2732 Casting non-ferrous metals	0	0	17	0	0	0	..	6,783	..	41
28 **FABRICATED METAL PRODUCTS**	1	208	596	12	0	0	..	734,512	..	3,461
281 Str. metal prod., tanks, reservoirs, steam generators	1	164	244	12	0	0	..	257,154	..	1,347
2811 Structural metal products	1	161	217	12	0	0	..	244,195	..	1,270
2812 Tanks, reservoirs and containers of metal	0	3	22	0	0	0	..	11,297	..	66
2813 Steam generators, exc. central heating hot water boilers	0	0	5	0	0	0	..	1,662	..	11
289 Other fabricated metal products	0	44	352	0	0	0	..	477,358	..	2,114
2891 Forging, pressing, stamping & roll-forming of metal	..	..	..	..	..	..	..		..	..
2892 Treatment and coating of metals	0	8	114	0	0	0	..	151,689	..	668
2893 Cutlery, hand tools and general hardware	0	5	16	0	0	0	..	56,780	..	225
2899 Other fabricated metal products, nec	0	31	222	0	0	0	..	268,889	..	1,221
29 **MACHINERY AND EQUIPMENT, NEC**	0	56	619	0	0	0	..	601,321	..	2,840
291 General purpose machinery	0	19	310	0	0	0	..	277,280	..	1,327
2911 Engines and turbines	0	2	97	0	0	0	..	40,971	..	246
2912 Pumps, compressors, taps and valves	0	1	32	0	0	0	..	57,444	..	240
2913 Bearings, gears, gearing and driving elements	0	0	8	0	0	0	..	20,513	..	82
2914 Ovens, furnaces and furnace burners	0	0	12	0	0	0	..	7,600	..	39
2915 Lifting and handling equipment	0	3	108	0	0	0	..	83,390	..	411
2919 Other general purpose machinery	0	13	53	0	0	0	..	67,362	..	309
292 Special purpose machinery	0	35	272	0	0	0	..	265,135	..	1,261
2921 Agricultural and forestry machinery	0	30	162	0	0	0	..	80,690	..	482
2922 Machine-tools	0	0	6	0	0	0	..	10,214	..	43
2923 Machinery for metallurgy	0	0	0	0	0	0	..	5,725	..	21
2924 Machinery for mining, quarrying and construction	0	4	57	0	0	0	..	21,563	..	139
2925 Machinery for food, beverage & tobacco processing	0	0	3	0	0	0	..	6,809	..	28
2926 Machinery for textile, apparel & leather production	0	0	0	0	0	0	..	1,327	..	5
2927 Machinery for weapons and ammunition	0	0	15	0	0	0	..	108,595	..	406
2929 Other special purpose machinery	0	1	29	0	0	0	..	30,212	..	139
293 Domestic appliances, nec	0	2	37	0	0	0	..	58,906	..	251
30 **OFFICE, ACCOUNTING & COMPUTING MACHINERY**	0	0	10	0	0	0	..	8,332	..	40
31 **ELECTRICAL MACHINERY & APPARATUS, NEC**	0	20	203	399	0	0	..	188,831	..	1,302
311 Electric motors, generators and transformers	0	8	121	1	0	0	..	36,439	..	261
312 Electricity distribution and control apparatus	0	0	25	0	0	0	..	39,244	..	166
313 Insulated wire and cable	0	4	21	0	0	0	..	60,673	..	243
314 Accumulators, primary cells & primary batteries	0	0	3	0	0	0	..	5,807	..	24
315 Electric lamps and lighting equipment	0	8	24	0	0	0	..	30,301	..	141
319 Other electrical equipment, nec	0	0	9	398	0	0	..	16,367	..	466
32 **RADIO, TV & COMMUNICATION EQUIP. & APP.**	0	0	22	0	0	0	..	71,731	..	280
321 Electronic valves, tubes, other electronic components	0	0	0	0	0	0	..	11,815	..	43
322 TV & radio transmitters, apparatus for line telephony	0	0	14	0	0	0	..	38,544	..	153
323 TV & radio receivers, recording apparatus	0	0	8	0	0	0	..	21,372	..	85
33 **MEDICAL PRECISION &OPTICAL INSTRUMENTS**	0	1	7	0	0	0	..	74,618	..	277
331 Medical appliances and instruments	0	1	7	0	0	0	..	71,433	..	265
3311 Medical, surgical equipment & orthopaedic app.	0	1	1	0	0	0	..	18,885	..	70
3312 Instruments & appliances for measuring, checking etc	0	0	6	0	0	0	..	40,213	..	151
3313 Industrial process control equipment	0	0	0	0	0	0	..	12,335	..	44
332 Optical instruments and photographic equipment	0	0	0	0	0	0	..	3,185	..	11
333 Watches and clocks	..	..	..	..	..	..	..		..	..
34 **MOTOR VEHICLES, TRAILERS & SEMI-TRAILERS**	0	26	152	0	0	1	..	210,500	..	937
341 Motor vehicles	0	0	2	0	0	0	..	2,987	..	13
342 Bodies (coachwork) for motor vehicles	0	16	74	0	0	0	..	20,439	..	164
343 Parts, accessories for motor vehicles & their engines	0	10	76	0	0	1	..	187,074	..	760
35 **OTHER TRANSPORT EQUIPMENT**	0	73	632	101	0	3	..	572,786	..	2,871
351 Building and repairing of ships and boats	0	41	480	20	0	3	..	422,146	..	2,064
3511 Building and repairing of ships	0	41	452	20	0	3	..	403,898	..	1,970
3512 Building, repairing of pleasure & sporting boats	0	0	28	0	0	0	..	18,248	..	94
352 Railway, tramway locomotives & rolling stock	0	8	102	27	0	0	..	50,510	..	319
353 Aircraft and spacecraft	0	23	32	54	0	0	..	78,604	..	392
359 Transport equipment, nec	0	1	18	0	0	0	..	21,526	..	96
3591 Motorcycles	..	..	..	..	..	..	..		..	..
3592 Bicycles and invalid carriages	0	0	14	0	0	0	..	19,166	..	83
3599 Other transport equipment, nec	0	1	4	0	0	0	..	2,360	..	13
36 **FURNITURE; MANUFACTURING, NEC**	0	12	264	1	0	35	..	256,488	..	1,235
361 Furniture	0	1	202	0	0	27	..	197,458	..	941
369 Manufacturing, nec	0	11	62	1	0	8	..	59,030	..	295
3691 Jewellery and related articles	0	2	9	0	0	0	..	11,717	..	53
3692 Musical instruments	0	0	0	0	0	0	..	371	..	1
3693 Sports goods	0	0	22	0	0	0	..	13,617	..	71
3694 Games and toys	0	0	0	0	0	0	..	4,012	..	14
3699 Other manufacturing, nec	0	9	31	1	0	8	..	29,313	..	155
37 **RECYCLING**	0	0	0	0	0	0	..	14,908	..	54
371 Recycling of metal waste and scrap	0	0	0	0	0	0	..	3,859	..	14
372 Recycling of non-metal waste and scrap	0	0	0	0	0	0	..	11,049	..	40
CERR **Non-specified industry**	..	..	..	..	..	..	..		..	..
15-37 **TOTAL MANUFACTURING**	5,791	3,388	12,705	10,746	48,788	17,397	..	45,306,709	..	261,919

ISIC Revision 3 Industry Sector	Solid TJ	LPG TJ	Distiloil TJ	RFO TJ	Gas TJ	Biomass TJ	Steam TJ	Electr MWh	Own Use MWh	TOTAL TJ
15 FOOD PRODUCTS AND BEVERAGES	0	708	5,308	2,718	0	1		2,695,804	..	18,440
151 Production, processing and preserving (PPP)	0	462	2,123	2,349	0	1		1,253,225	..	9,447
1511 PPP of meat and meat products	0	169	962	19	0	1		466,253	..	2,830
1512 Processing and preserving of fish products	0	253	726	1,969	0	0		572,934	..	5,011
1513 Processing, preserving of fruit & vegetables	0	9	323	45	0	0		114,344	..	789
1514 Vegetable and animal oils and fats	0	31	112	316	0	0		99,694	..	818
152 Dairy products	0	197	1,296	39	0	0		453,036	..	3,163
153 Grain mill prod., starches & prepared animal feeds	0	19	542	121	0	0		344,610	..	1,923
1531 Grain mill products	0	19	119	0	0	0		132,632	..	615
1532 Starches and starch products	0	0	7	80	0	0		3,057	..	98
1533 Prepared animal feeds	0	0	416	41	0	0		208,921	..	1,209
154 Other food products	0	28	624	120	0	0		441,497	..	2,361
1541 Bakery products	0	21	348	0	0	0		275,167	..	1,360
1542 Sugar	..	..	..	..	..	..			..	..
1543 Cocoa, chocolate and sugar confectionery	0	6	51	73	0	0		82,651	..	428
1544 Macaroni, noodles, couscous & similar farinaceous prod.	0	0	21	0	0	0		7,869	..	49
1549 Other food products, nec	0	1	204	47	0	0		75,810	..	525
155 Beverages	0	2	723	89	0	0		203,436	..	1,546
1551 Distilling, rectifying and blending of spirits	0	0	1	0	0	0		10,125	..	37
1552 Wines	0	0	0	0	0	0		509	..	2
1553 Malt liquors and malt	0	2	427	89	0	0		112,558	..	923
1554 Soft drinks; production of mineral waters	0	0	295	0	0	0		80,244	..	584
16 TOBACCO PRODUCTS	0	0	57	0	0	13		10,982	..	110
17 TEXTILES	0	26	268	154	0	0		180,784	..	1,099
171 Spinning, weaving and finishing of textiles	0	26	139	129	0	0		97,984	..	647
1711 Preparation and spinning of textile fibres	0	24	98	64	0	0		91,892	..	517
1712 Finishing of textiles	0	2	41	65	0	0		6,092	..	130
172 Other textiles	0	0	97	0	0	0		60,497	..	315
1721 Made-up textile articles, except apparel	0	0	16	0	0	0		14,308	..	68
1722 Carpets and rugs	0	0	0	0	0	0		907	..	3
1723 Cordage, rope, twine and netting	0	0	47	0	0	0		24,355	..	135
1729 Other textiles, nec	0	0	34	0	0	0		20,927	..	109
173 Knitted and crocheted fabrics and articles	0	0	32	25	0	0		22,303	..	137
18 WEARING APPAREL, DRESSING & DYEING OF FUR	0	0	44	0	0	0		17,173	..	106
181 Wearing apparel, except fur apparel	0	0	17	0	0	0		15,785	..	74
182 Dressing and dyeing of fur; articles of fur	0	0	27	0	0	0		1,388	..	32
19 TANNING & DRESSING OF LEATHER, FOOTWEAR	0	0	55	0	0	0		14,257	..	106
191 Tanning and dressing of leather	0	0	51	0	0	0		11,619	..	93
1911 Tanning and dressing of leather	0	0	50	0	0	0		10,145	..	87
1912 Luggage, handbags, saddlery & harness	0	0	1	0	0	0		1,474	..	6
192 Footwear	0	0	4	0	0	0		2,638	..	13
20 WOOD AND WOOD PRODUCTS	0	4	717	147	0	4,957		875,857	..	8,978
201 Sawmilling and planing of wood	0	2	443	0	0	4,458		366,553	..	6,223
202 Products of wood, cork, straw & plaiting materials	0	2	274	147	0	499		509,304	..	2,755
2021 Veneer sheets	0	2	150	147	0	490		348,448	..	2,043
2022 Builders' carpentry and joinery	0	0	87	0	0	9		145,995	..	622
2023 Wooden containers	0	0	34	0	0	0		7,522	..	61
2029 Other products of wood	0	0	3	0	0	0		7,339	..	29
21 PAPER AND PAPER PRODUCTS	317	78	361	7,301	0	13,760		6,374,921	..	44,767
2101 Pulp, paper and paperboard	317	77	244	7,181	0	13,760		6,188,825	..	43,859
2102 Corrugated paper, paperboard and their containers	0	0	102	71	0	0		87,663	..	489
2109 Other articles of pulp and paperboard	0	1	15	49	0	0		98,433	..	419
22 PUBLISHING, PRINTING & REPRODUCTION	0	184	140	0	0	9		439,075	..	1,914
221 Publishing	0	0	87	0	0	0		206,668	..	831
2211 Publishing of books & brochures	0	0	2	0	0	0		32,878	..	120
2212 Publishing of newspapers and periodicals	0	0	79	0	0	0		152,478	..	628
2213 Publishing of recorded media	0	0	0	0	0	0		11,472	..	41
2219 Other publishing	0	0	6	0	0	0		9,840	..	41
222 Printing and related service activities	0	184	42	0	0	9		222,775	..	1,037
2221 Printing	0	184	29	0	0	9		171,457	..	839
2222 Service activities related to printing	0	0	13	0	0	0		51,318	..	198
223 Reproduction of recorded media	0	0	11	0	0	0		9,632	..	46
23 COKE, REFINED PETROLEUM PRODUCTS	0	851	349	148	33,195	0		711,696	..	37,105
231 Coke oven products	..	..	..	..	..	..			..	..
232 Refined petroleum products	0	851	349	148	33,195	0		711,696	..	37,105
233 Processing of nuclear fuel	..	..	..	..	..	..			..	..
24 CHEMICALS & CHEMICAL PRODUCTS	0	291	1,379	2,296	12,212	0		6,033,618	..	37,899
241 Basic chemicals	0	251	935	2,197	11,676	0		5,786,101	..	35,889
2411 Basic chemicals, exc. fertilizers & nitrogen compounds	0	101	585	1,511	10,041	0		4,471,790	..	28,336
2412 Fertilizers and nitrogen compounds	0	0	5	354	0	0		926,248	..	3,693
2413 Plastics in primary forms and synthetic rubber	0	150	345	332	1,635	0		388,063	..	3,859
242 Other chemical products	0	40	444	99	536	0		247,517	..	2,010
2421 Pesticides and other agro-chemical products	0	0	0	0	536	0		17	..	536
2422 Paints, varnishes and similar coatings	0	1	106	0	0	0		43,242	..	263
2423 Pharmaceuticals, medicinal chem. & botanical prod.	0	39	191	60	0	0		114,370	..	702
2424 Soap and detergents, perfumes etc.	0	0	119	0	0	0		27,261	..	217
2429 Other chemical products, nec	0	0	28	39	0	0		62,627	..	292
243 Man-made fibres	..	..	..	..	..	..			..	..
25 RUBBER AND PLASTICS PRODUCTS	0	26	228	130	0	0		433,523	..	1,945
251 Rubber products	0	0	8	51	0	0		97,987	..	412
2511 Rubber tyres and tubes	0	0	0	0	0	0		24,386	..	88
2519 Other rubber products	0	0	8	51	0	0		73,601	..	324
252 Plastic products	0	26	220	79	0	0		335,536	..	1,533
26 OTHER NON-METALLIC MINERAL PRODUCTS	6,057	1,112	872	1,053	0	0		1,073,755	..	12,960
261 Glass and glass products	0	337	63	689	0	0		195,353	..	1,792
269 Non-metallic mineral products, nec	6,057	775	809	364	0	0		878,402	..	11,167
2691 Pottery, china and earthenware	0	52	28	0	0	0		42,144	..	232
2692 Refractory ceramic products	0	0	0	57	0	0		4,008	..	71
2693 Structural non-refractory clay & ceramic prod.	0	66	57	102	0	0		14,569	..	277
2694 Cement, lime and plaster	5,220	0	38	205	0	0		401,263	..	6,908
2695 Articles of concrete, cement and plaster	486	541	530	0	0	0		130,603	..	2,027
2696 Cutting, shaping and finishing of stone	0	0	87	0	0	0		137,282	..	581
2699 Other non-metallic mineral products, nec	351	116	69	0	0	0		148,533	..	1,071

ISIS Energy Data Programme (IEA/OECD)

ISIC Revision 3 Industry Sector		Solid TJ	LPG TJ	Distiloil TJ	RFO TJ	Gas TJ	Biomass TJ	Steam TJ	Electr MWh	Own Use MWh	TOTAL TJ
27	**BASIC METALS**	0	371	2,137	1,181	1,192	0	..	24,107,536	..	91,668
271	Basic iron and steel	0	3	253	340	497	0	..	6,791,718	..	25,543
272	Basic precious and non-ferrous metals	0	355	1,766	841	695	0	..	17,113,028	..	65,264
273	Casting of metals	0	13	118	0	0	0	..	202,790	..	861
2731	Casting of iron and steel	0	5	106	0	0	0	..	170,917	..	726
2732	Casting non-ferrous metals	0	8	12	0	0	0	..	31,873	..	135
28	**FABRICATED METAL PRODUCTS**	1	264	531	13	0	0	..	508,209	..	2,639
281	Str. metal prod., tanks, reservoirs, steam generators	1	202	274	12	0	0	..	219,932	..	1,281
2811	Structural metal products	1	200	254	12	0	0	..	211,269	..	1,228
2812	Tanks, reservoirs and containers of metal	0	2	12	0	0	0	..	6,838	..	39
2813	Steam generators, exc. central heating hot water boilers	0	0	8	0	0	0	..	1,825	..	15
289	Other fabricated metal products	0	62	257	1	0	0	..	288,277	..	1,358
2891	Forging, pressing, stamping & roll-forming of metal	0	0	3	0	0	0	..	4,146	..	18
2892	Treatment and coating of metals	0	11	61	1	0	0	..	75,602	..	345
2893	Cutlery, hand tools and general hardware	0	6	36	0	0	0	..	56,375	..	245
2899	Other fabricated metal products, nec	0	45	157	0	0	0	..	152,154	..	750
29	**MACHINERY AND EQUIPMENT, NEC**	0	48	653	1	0	0	..	589,586	..	2,825
291	General purpose machinery	0	18	287	1	0	0	..	267,835	..	1,270
2911	Engines and turbines	0	13	87	1	0	0	..	64,953	..	335
2912	Pumps, compressors, taps and valves	0	1	41	0	0	0	..	51,727	..	228
2913	Bearings, gears, gearing and driving elements	0	0	5	0	0	0	..	14,819	..	58
2914	Ovens, furnaces and furnace burners	0	0	14	0	0	0	..	9,986	..	50
2915	Lifting and handling equipment	0	4	116	0	0	0	..	63,753	..	350
2919	Other general purpose machinery	0	0	24	0	0	0	..	62,597	..	249
292	Special purpose machinery	0	25	324	0	0	0	..	265,400	..	1,304
2921	Agricultural and forestry machinery	0	22	138	0	0	0	..	72,139	..	420
2922	Machine-tools	0	0	8	0	0	0	..	10,493	..	46
2923	Machinery for metallurgy	0	0	0	0	0	0	..	5,885	..	21
2924	Machinery for mining, quarrying and construction	0	3	82	0	0	0	..	27,181	..	183
2925	Machinery for food, beverage & tobacco processing	0	0	8	0	0	0	..	10,836	..	47
2926	Machinery for textile, apparel & leather production	0	0	0	0	0	0	..	1,571	..	6
2927	Machinery for weapons and ammunition	0	0	36	0	0	0	..	107,502	..	423
2929	Other special purpose machinery	0	0	52	0	0	0	..	29,793	..	159
293	Domestic appliances, nec	0	5	42	0	0	0	..	56,351	..	250
30	**OFFICE, ACCOUNTING & COMPUTING MACHINERY**	0	0	2	0	0	0	..	14,327	..	54
31	**ELECTRICAL MACHINERY & APPARATUS, NEC**	0	29	374	438	0	0	..	314,748	..	1,974
311	Electric motors, generators and transformers	0	4	147	0	0	0	..	122,519	..	592
312	Electricity distribution and control apparatus	0	0	21	0	0	0	..	44,848	..	182
313	Insulated wire and cable	0	15	27	0	0	0	..	53,896	..	236
314	Accumulators, primary cells & primary batteries	0	0	5	0	0	0	..	5,690	..	25
315	Electric lamps and lighting equipment	0	10	30	0	0	0	..	30,704	..	151
319	Other electrical equipment, nec	0	0	144	438	0	0	..	57,091	..	788
32	**RADIO, TV & COMMUNICATION EQUIP. & APP.**	0	0	26	0	0	0	..	60,061	..	242
321	Electronic valves, tubes, other electronic components	0	0	0	0	0	0	..	14,476	..	52
322	TV & radio transmitters, apparatus for line telephony	0	0	20	0	0	0	..	33,353	..	140
323	TV & radio receivers, recording apparatus	0	0	6	0	0	0	..	12,232	..	50
33	**MEDICAL PRECISION & OPTICAL INSTRUMENTS**	0	1	7	0	0	0	..	91,204	..	336
331	Medical appliances and instruments	0	1	6	0	0	0	..	86,555	..	319
3311	Medical, surgical equipment & orthopaedic app.	0	1	1	0	0	0	..	21,844	..	81
3312	Instruments & appliances for measuring, checking etc	0	0	5	0	0	0	..	49,490	..	183
3313	Industrial process control equipment	0	0	0	0	0	0	..	15,221	..	55
332	Optical instruments and photographic equipment	0	0	1	0	0	0	..	4,649	..	18
333	Watches and clocks	..	..	..	..	..	..	..	..	..	..
34	**MOTOR VEHICLES, TRAILERS & SEMI-TRAILERS**	0	37	178	0	0	0	..	165,023	..	809
341	Motor vehicles	0	0	7	0	0	0	..	3,379	..	19
342	Bodies (coachwork) for motor vehicles	0	23	61	0	0	0	..	20,833	..	159
343	Parts, accessories for motor vehicles & their engines	0	14	110	0	0	0	..	140,811	..	631
35	**OTHER TRANSPORT EQUIPMENT**	0	197	750	92	0	0	..	641,638	..	3,349
351	Building and repairing of ships and boats	0	156	561	20	0	0	..	505,199	..	2,556
3511	Building and repairing of ships	0	156	532	20	0	0	..	469,000	..	2,396
3512	Building, repairing of pleasure & sporting boats	0	0	29	0	0	0	..	36,199	..	159
352	Railway, tramway locomotives & rolling stock	0	6	119	5	0	0	..	44,852	..	291
353	Aircraft and spacecraft	0	33	43	67	0	0	..	79,615	..	430
359	Transport equipment, nec	0	2	27	0	0	0	..	11,972	..	72
3591	Motorcycles	..	..	..	..	..	..	..	..	..	..
3592	Bicycles and invalid carriages	0	0	23	0	0	0	..	9,694	..	58
3599	Other transport equipment, nec	0	2	4	0	0	0	..	2,278	..	14
36	**FURNITURE; MANUFACTURING, NEC**	0	13	273	0	0	39	..	273,641	..	1,310
361	Furniture	0	1	192	0	0	30	..	225,666	..	1,035
369	Manufacturing, nec	0	12	81	0	0	9	..	47,975	..	275
3691	Jewellery and related articles	0	1	12	0	0	0	..	9,276	..	46
3692	Musical instruments	0	0	0	0	0	0	..	360	..	1
3693	Sports goods	0	0	38	0	0	0	..	19,516	..	108
3694	Games and toys	0	0	0	0	0	0	..	573	..	2
3699	Other manufacturing, nec	0	11	31	0	0	9	..	18,250	..	117
37	**RECYCLING**	0	130	67	0	0	0	..	15,670	..	253
371	Recycling of metal waste and scrap	0	130	67	0	0	0	..	4,418	..	213
372	Recycling of non-metal waste and scrap	0	0	0	0	0	0	..	11,252	..	41
CERR	**Non-specified industry**	..	..	..	..	..	..	..	..	..	..
15-37	**TOTAL MANUFACTURING**	6,375	4,370	14,776	15,672	46,599	18,779	..	45,643,088	..	270,886

ISIC Revision 3 Industry Sector		Solid TJ	LPG TJ	Distilloil TJ	RFO TJ	Gas TJ	Biomass TJ	Steam TJ	Electr MWh	Own Use MWh	TOTAL TJ
15	**FOOD PRODUCTS AND BEVERAGES**	0	669	5,316	1,687	0	3	306	2,880,935	..	18,352
151	Production, processing and preserving (PPP)	0	506	2,794	1,404	0	3	306	1,318,853	..	9,761
1511	PPP of meat and meat products	0	177	771	3	0	1	0	504,926	..	2,770
1512	Processing and preserving of fish products	0	263	1,324	1,356	0	2	0	600,858	..	5,108
1513	Processing, preserving of fruit & vegetables	0	34	299	45	0	0	0	111,716	..	780
1514	Vegetable and animal oils and fats	0	32	400	0	0	0	306	101,353	..	1,103
152	Dairy products	0	113	683	10	0	0	0	498,075	..	2,599
153	Grain mill prod., starches & prepared animal feeds	0	8	595	91	0	0	0	363,263	..	2,002
1531	Grain mill products	0	8	128	0	0	0	0	103,852	..	510
1532	Starches and starch products	0	0	7	91	0	0	0	3,213	..	110
1533	Prepared animal feeds	0	0	460	0	0	0	0	256,198	..	1,382
154	Other food products	0	40	676	61	0	0	0	520,586	..	2,651
1541	Bakery products	0	35	419	0	0	0	0	354,221	..	1,729
1542	Sugar	..	..	..	..	..	..	..	..	..	..
1543	Cocoa, chocolate and sugar confectionery	0	5	42	61	0	0	0	77,981	..	389
1544	Macaroni, noodles, couscous & similar farinaceous prod.	0	0	30	0	0	0	0	5,908	..	51
1549	Other food products, nec	0	0	185	0	0	0	0	82,476	..	482
155	Beverages	0	2	568	121	0	0	0	180,158	..	1,340
1551	Distilling, rectifying and blending of spirits	0	0	8	0	0	0	0	9,987	..	44
1552	Wines	0	0	0	0	0	0	0	879	..	3
1553	Malt liquors and malt	0	2	364	121	0	0	0	130,081	..	955
1554	Soft drinks; production of mineral waters	0	0	196	0	0	0	0	39,211	..	337
16	**TOBACCO PRODUCTS**	0	0	48	0	0	54	0	12,722	..	148
17	**TEXTILES**	0	13	201	147	0	0	30	173,087	..	1,014
171	Spinning, weaving and finishing of textiles	0	12	98	108	0	0	30	91,126	..	576
1711	Preparation and spinning of textile fibres	0	12	52	44	0	0	30	85,282	..	445
1712	Finishing of textiles	0	0	46	64	0	0	0	5,844	..	131
172	Other textiles	0	1	85	13	0	0	0	60,350	..	316
1721	Made-up textile articles, except apparel	0	1	22	11	0	0	0	14,262	..	85
1722	Carpets and rugs	0	0	0	0	0	0	0	802	..	3
1723	Cordage, rope, twine and netting	0	0	39	0	0	0	0	30,281	..	148
1729	Other textiles, nec	0	0	24	2	0	0	0	15,005	..	80
173	Knitted and crocheted fabrics and articles	0	0	18	26	0	0	0	21,611	..	122
18	**WEARING APPAREL, DRESSING & DYEING OF FUR**	0	1	33	0	0	0	0	24,178	..	121
181	Wearing apparel, except fur apparel	0	1	18	0	0	0	0	23,082	..	102
182	Dressing and dyeing of fur; articles of fur	0	0	15	0	0	0	0	1,096	..	19
19	**TANNING & DRESSING OF LEATHER, FOOTWEAR**	0	0	53	0	0	0	0	17,444	..	116
191	Tanning and dressing of leather	0	0	44	0	0	0	0	12,096	..	88
1911	Tanning and dressing of leather	0	0	44	0	0	0	0	10,806	..	83
1912	Luggage, handbags, saddlery & harness	0	0	0	0	0	0	0	1,290	..	5
192	Footwear	0	0	9	0	0	0	0	5,348	..	28
20	**WOOD AND WOOD PRODUCTS**	0	5	743	237	0	5,055	70	828,552	..	9,093
201	Sawmilling and planing of wood	0	2	443	0	0	4,497	70	354,467	..	6,288
202	Products of wood, cork, straw & plaiting materials	0	3	300	237	0	558	0	474,085	..	2,805
2021	Veneer sheets	0	2	156	237	0	543	0	260,276	..	1,875
2022	Builders' carpentry and joinery	0	1	91	0	0	15	0	193,008	..	802
2023	Wooden containers	0	0	51	0	0	0	0	13,454	..	99
2029	Other products of wood	0	0	2	0	0	0	0	7,347	..	28
21	**PAPER AND PAPER PRODUCTS**	290	146	419	6,846	0	12,287	0	6,891,281	..	44,797
2101	Pulp, paper and paperboard	290	8	297	6,720	0	12,287	0	6,637,161	..	43,496
2102	Corrugated paper, paperboard and their containers	0	0	99	80	0	0	0	97,313	..	529
2109	Other articles of pulp and paperboard	0	138	23	46	0	0	0	156,807	..	772
22	**PUBLISHING, PRINTING & REPRODUCTION**	0	208	131	0	0	9	75	439,947	..	2,007
221	Publishing	0	0	101	0	0	0	36	208,602	..	888
2211	Publishing of books & brochures	0	0	4	0	0	0	0	34,887	..	130
2212	Publishing of newspapers and periodicals	0	0	81	0	0	0	36	149,823	..	656
2213	Publishing of recorded media	0	0	0	0	0	0	0	14,908	..	54
2219	Other publishing	0	0	16	0	0	0	0	8,984	..	48
222	Printing and related service activities	0	208	27	0	0	9	39	221,284	..	1,080
2221	Printing	0	208	23	0	0	9	39	172,169	..	899
2222	Service activities related to printing	0	0	4	0	0	0	0	49,115	..	181
223	Reproduction of recorded media	0	0	3	0	0	0	0	10,061	..	39
23	**COKE, REFINED PETROLEUM PRODUCTS**	0	352	508	27	28,431	0	0	647,944	..	31,651
231	Coke oven products	..	..	..	..	..	..	..	..	..	..
232	Refined petroleum products	0	352	508	27	28,431	0	0	647,944	..	31,651
233	Processing of nuclear fuel	..	..	..	..	..	..	..	..	..	..
24	**CHEMICALS & CHEMICAL PRODUCTS**	11,375	44	1,136	1,851	9,218	0	292	6,377,812	..	46,876
241	Basic chemicals	11,375	5	702	1,836	9,218	0	288	6,080,282	..	45,313
2411	Basic chemicals, exc. fertilizers & nitrogen compounds	11,375	5	580	1,310	8,743	0	288	4,593,434	..	38,837
2412	Fertilizers and nitrogen compounds	0	0	6	84	133	0	0	952,229	..	3,651
2413	Plastics in primary forms and synthetic rubber	0	0	116	442	342	0	0	534,619	..	2,825
242	Other chemical products	0	39	434	15	0	0	4	297,530	..	1,563
2421	Pesticides and other agro-chemical products	..	..	..	..	..	..	..	39	..	39
2422	Paints, varnishes and similar coatings	0	1	96	0	0	0	0	45,742	..	262
2423	Pharmaceuticals, medicinal chem. & botanical prod.	0	38	159	0	0	0	0	147,277	..	727
2424	Soap and detergents, perfumes etc.	0	0	141	0	0	0	0	38,080	..	278
2429	Other chemical products, nec	0	0	38	15	0	0	4	66,392	..	296
243	Man-made fibres	..	..	..	..	..	..	..	..	..	..
25	**RUBBER AND PLASTICS PRODUCTS**	0	14	249	96	0	0	74	422,722	..	1,955
251	Rubber products	0	0	9	29	0	0	0	57,682	..	246
2511	Rubber tyres and tubes	0	0	0	0	0	0	0	18,895	..	68
2519	Other rubber products	0	0	9	29	0	0	0	38,787	..	178
252	Plastic products	0	14	240	67	0	0	74	365,040	..	1,709
26	**OTHER NON-METALLIC MINERAL PRODUCTS**	9,358	1,219	1,003	1,234	0	0	179	1,030,826	..	16,704
261	Glass and glass products	0	363	44	665	0	0	34	194,418	..	1,806
269	Non-metallic mineral products, nec	9,358	856	959	569	0	0	145	836,408	..	14,898
2691	Pottery, china and earthenware	0	69	27	0	0	0	9	36,563	..	237
2692	Refractory ceramic products	0	65	2	53	0	0	0	14,084	..	171
2693	Structural non-refractory clay & ceramic prod.	0	0	1	150	0	0	0	6,110	..	173
2694	Cement, lime and plaster	5,233	0	45	366	0	0	79	455,911	..	7,364
2695	Articles of concrete, cement and plaster	653	626	731	0	0	0	0	128,130	..	2,471
2696	Cutting, shaping and finishing of stone	0	0	50	0	0	0	0	45,513	..	214
2699	Other non-metallic mineral products, nec	3,472	96	103	0	0	0	57	150,097	..	4,268

ISIS Energy Data Programme (IEA/OECD)

ISIC Revision 3 Industry Sector	Solid TJ	LPG TJ	Distiloil TJ	RFO TJ	Gas TJ	Biomass TJ	Steam TJ	Electr MWh	Own Use MWh	TOTAL TJ
27 BASIC METALS	32,017	335	1,856	1,110	1,756	1	3	24,655,490	..	125,838
271 Basic iron and steel	24,582	4	127	214	634	0	3	7,249,926	..	51,664
272 Basic precious and non-ferrous metals	7,435	320	1,618	896	1,122	0	0	17,195,390	..	73,294
273 Casting of metals	0	11	111	0	0	1	0	210,174	..	880
2731 Casting of iron and steel	0	1	75	0	0	1	0	176,299	..	712
2732 Casting non-ferrous metals	0	10	36	0	0	0	0	33,875	..	168
28 FABRICATED METAL PRODUCTS	5	264	526	11	0	7	111	639,020	..	3,224
281 Str. metal prod., tanks, reservoirs, steam generators	2	201	267	11	0	0	72	281,380	..	1,566
2811 Structural metal products	2	197	247	11	0	0	72	267,449	..	1,492
2812 Tanks, reservoirs and containers of metal	0	4	16	0	0	0	0	12,812	..	66
2813 Steam generators, exc. central heating hot water boilers	0	0	4	0	0	0	0	1,119	..	8
289 Other fabricated metal products	3	63	259	0	0	7	39	357,640	..	1,659
2891 Forging, pressing, stamping & roll-forming of metal	0	0	3	0	0	0	0	5,551	..	23
2892 Treatment and coating of metals	0	24	89	0	0	0	0	126,634	..	569
2893 Cutlery, hand tools and general hardware	0	5	17	0	0	0	0	49,470	..	209
2899 Other fabricated metal products, nec	3	34	150	0	0	7	30	175,985	..	858
29 MACHINERY AND EQUIPMENT, NEC	12	121	543	0	0	0	109	625,476	..	3,037
291 General purpose machinery	0	13	240	0	0	0	53	304,829	..	1,403
2911 Engines and turbines	0	4	60	0	0	0	0	45,373	..	227
2912 Pumps, compressors, taps and valves	0	1	32	0	0	0	53	94,192	..	425
2913 Bearings, gears, gearing and driving elements	0	0	4	0	0	0	0	28,226	..	106
2914 Ovens, furnaces and furnace burners	0	0	21	0	0	0	0	7,228	..	47
2915 Lifting and handling equipment	0	8	106	0	0	0	0	69,939	..	366
2919 Other general purpose machinery	0	0	17	0	0	0	0	59,871	..	233
292 Special purpose machinery	0	92	257	0	0	0	47	268,755	..	1,364
2921 Agricultural and forestry machinery	0	85	102	0	0	0	32	85,436	..	527
2922 Machine-tools	0	0	7	0	0	0	0	13,401	..	55
2923 Machinery for metallurgy	0	0	3	0	0	0	0	5,843	..	24
2924 Machinery for mining, quarrying and construction	0	7	59	0	0	0	0	27,725	..	166
2925 Machinery for food, beverage & tobacco processing	0	0	24	0	0	0	0	26,371	..	119
2926 Machinery for textile, apparel & leather production	0	0	0	0	0	0	0	1,632	..	6
2927 Machinery for weapons and ammunition	0	0	41	0	0	0	15	85,082	..	362
2929 Other special purpose machinery	0	0	21	0	0	0	0	23,265	..	105
293 Domestic appliances, nec	12	16	46	0	0	0	9	51,892	..	270
30 OFFICE, ACCOUNTING & COMPUTING MACHINERY	0	0	1	0	0	0	0	16,005	..	59
31 ELECTRICAL MACHINERY & APPARATUS, NEC	3,795	40	293	344	0	0	0	273,470	..	5,456
311 Electric motors, generators and transformers	0	0	115	0	0	0	0	69,074	..	364
312 Electricity distribution and control apparatus	0	0	15	0	0	0	0	52,978	..	206
313 Insulated wire and cable	0	29	30	0	0	0	0	86,439	..	370
314 Accumulators, primary cells & primary batteries	0	0	2	0	0	0	0	6,728	..	26
315 Electric lamps and lighting equipment	0	11	21	0	0	0	0	23,249	..	116
319 Other electrical equipment, nec	3,795	0	110	344	0	0	0	35,002	..	4,375
32 RADIO, TV & COMMUNICATION EQUIP. & APP.	0	0	9	0	0	0	10	60,376	..	236
321 Electronic valves, tubes, other electronic components	0	0	1	0	0	0	0	14,035	..	52
322 TV & radio transmitters, apparatus for line telephony	0	0	3	0	0	0	10	40,664	..	159
323 TV & radio receivers, recording apparatus	0	0	5	0	0	0	0	5,677	..	25
33 MEDICAL PRECISION &OPTICAL INSTRUMENTS	0	1	6	0	0	0	14	84,658	..	326
331 Medical appliances and instruments	0	1	6	0	0	0	14	81,419	..	314
3311 Medical, surgical equipment & orthopaedic app.	0	1	2	0	0	0	0	18,312	..	69
3312 Instruments & appliances for measuring, checking etc	0	0	4	0	0	0	14	53,986	..	212
3313 Industrial process control equipment	0	0	0	0	0	0	0	9,121	..	33
332 Optical instruments and photographic equipment	0	0	0	0	0	0	0	3,239	..	12
333 Watches and clocks	..	..	..	..	..	..	..	..	..	..
34 MOTOR VEHICLES, TRAILERS & SEMI-TRAILERS	0	148	174	0	0	0	28	161,823	..	933
341 Motor vehicles	0	0	12	0	0	0	0	3,653	..	25
342 Bodies (coachwork) for motor vehicles	0	26	82	0	0	0	0	16,982	..	169
343 Parts, accessories for motor vehicles & their engines	0	122	80	0	0	0	28	141,188	..	738
35 OTHER TRANSPORT EQUIPMENT	0	203	519	100	0	0	104	649,187	..	3,263
351 Building and repairing of ships and boats	0	151	406	21	0	0	92	526,276	..	2,565
3511 Building and repairing of ships	0	151	388	21	0	0	92	510,362	..	2,489
3512 Building, repairing of pleasure & sporting boats	0	0	18	0	0	0	0	15,914	..	75
352 Railway, tramway locomotives & rolling stock	0	19	66	0	0	0	0	40,572	..	231
353 Aircraft and spacecraft	0	33	26	79	0	0	12	70,509	..	404
359 Transport equipment, nec	0	0	21	0	0	0	0	11,830	..	64
3591 Motorcycles	..	..	..	..	..	..	..	..	..	..
3592 Bicycles and invalid carriages	0	0	21	0	0	0	0	9,888	..	57
3599 Other transport equipment, nec	0	0	0	0	0	0	0	1,942	..	7
36 FURNITURE; MANUFACTURING, NEC	0	9	276	0	0	42	45	269,383	..	1,342
361 Furniture	0	1	208	0	0	32	36	204,525	..	1,013
369 Manufacturing, nec	0	8	68	0	0	10	9	64,858	..	328
3691 Jewellery and related articles	0	1	11	0	0	0	0	12,510	..	57
3692 Musical instruments	0	0	0	0	0	0	0	272	..	1
3693 Sports goods	0	0	28	0	0	0	9	25,035	..	127
3694 Games and toys	0	0	0	0	0	0	0	2,449	..	9
3699 Other manufacturing, nec	0	7	29	0	0	10	0	24,592	..	135
37 RECYCLING	0	100	10	0	0	0	0	26,629	..	206
371 Recycling of metal waste and scrap	0	100	10	0	0	0	0	7,053	..	135
372 Recycling of non-metal waste and scrap	0	0	0	0	0	0	0	19,576	..	70
CERR Non-specified industry	..	..	..	..	..	..	..	..	..	..
15-37 TOTAL MANUFACTURING	56,852	3,892	14,053	13,690	39,405	17,458	1,450	47,208,967	..	316,752

Year: 1996 **NORWAY**

ISIC Revision 3 Industry Sector	Solid TJ	LPG TJ	Distiloil TJ	RFO TJ	Gas TJ	Biomass TJ	Steam TJ	Electr MWh	Own Use MWh	TOTAL TJ
15 FOOD PRODUCTS AND BEVERAGES	0	514	6,291	1,627	261	1	324	2,418,020	..	17,723
151 Production, processing and preserving (PPP)	0	423	2,815	1,348	248	1	324	1,127,638	..	9,218
1511 PPP of meat and meat products	0	139	942	0	0	0	0	407,457	..	2,548
1512 Processing and preserving of fish products	0	214	1,355	929	248	1	0	594,718	..	4,888
1513 Processing, preserving of fruit & vegetables	0	52	420	0	0	0	0	62,371	..	697
1514 Vegetable and animal oils and fats	0	18	98	419	0	0	324	63,092	..	1,086
152 Dairy products	0	38	1,234	30	1	0	0	314,039	..	2,434
153 Grain mill prod., starches & prepared animal feeds	0	12	768	19	12	0	0	381,416	..	2,184
1531 Grain mill products	0	12	76	0	0	0	0	110,364	..	485
1532 Starches and starch products	0	0	48	19	0	0	0	5,504	..	87
1533 Prepared animal feeds	0	0	644	0	12	0	0	265,548	..	1,612
154 Other food products	0	40	676	143	0	0	0	431,459	..	2,412
1541 Bakery products	0	22	446	0	0	0	0	293,108	..	1,523
1542 Sugar	..	..	..	..	..	..	..	..	..	..
1543 Cocoa, chocolate and sugar confectionery	0	4	78	75	0	0	0	69,373	..	407
1544 Macaroni, noodles, couscous & similar farinaceous prod.	0	0	18	0	0	0	0	8,333	..	48
1549 Other food products, nec	0	14	134	68	0	0	0	60,645	..	434
155 Beverages	0	1	798	87	0	0	0	163,468	..	1,474
1551 Distilling, rectifying and blending of spirits	0	0	55	0	0	0	0	20,406	..	128
1552 Wines	0	0	2	0	0	0	0	514	..	4
1553 Malt liquors and malt	0	1	424	87	0	0	0	82,351	..	808
1554 Soft drinks; production of mineral waters	0	0	317	0	0	0	0	60,197	..	534
16 TOBACCO PRODUCTS	0	0	56	0	0	7	0	9,323	..	97
17 TEXTILES	0	27	327	203	0	0	17	157,680	..	1,142
171 Spinning, weaving and finishing of textiles	0	26	146	142	0	0	17	71,274	..	588
1711 Preparation and spinning of textile fibres	0	25	88	91	0	0	17	61,392	..	442
1712 Finishing of textiles	0	1	58	51	0	0	0	9,882	..	146
172 Other textiles	0	1	132	61	0	0	0	67,757	..	438
1721 Made-up textile articles, except apparel	0	0	39	61	0	0	0	24,782	..	189
1722 Carpets and rugs	0	0	0	0	0	0	0	2,262	..	8
1723 Cordage, rope, twine and netting	0	0	58	0	0	0	0	29,794	..	165
1729 Other textiles, nec	0	0	35	0	0	0	0	10,919	..	74
173 Knitted and crocheted fabrics and articles	0	0	49	0	0	0	0	18,649	..	116
18 WEARING APPAREL, DRESSING & DYEING OF FUR	0	1	50	0	0	0	0	28,049	..	152
181 Wearing apparel, except fur apparel	0	1	35	0	0	0	0	26,684	..	132
182 Dressing and dyeing of fur; articles of fur	0	0	15	0	0	0	0	1,365	..	20
19 TANNING & DRESSING OF LEATHER, FOOTWEAR	0	0	56	0	0	0	0	13,937	..	106
191 Tanning and dressing of leather	0	0	50	0	0	0	0	9,225	..	83
1911 Tanning and dressing of leather	0	0	50	0	0	0	0	7,505	..	77
1912 Luggage, handbags, saddlery & harness	0	0	0	0	0	0	0	1,720	..	6
192 Footwear	0	0	6	0	0	0	0	4,712	..	23
20 WOOD AND WOOD PRODUCTS	0	4	749	412	0	5,039	68	685,930	..	8,741
201 Sawmilling and planing of wood	0	1	349	1	0	4,398	68	327,021	..	5,994
202 Products of wood, cork, straw & plaiting materials	0	3	400	411	0	641	0	358,909	..	2,747
2021 Veneer sheets	0	1	275	411	0	541	0	175,919	..	1,861
2022 Builders' carpentry and joinery	0	2	101	0	0	84	0	161,006	..	767
2023 Wooden containers	0	0	21	0	0	8	0	14,179	..	80
2029 Other products of wood	0	0	3	0	0	8	0	7,805	..	39
21 PAPER AND PAPER PRODUCTS	338	274	644	8,838	0	10,926	0	6,659,719	..	44,995
2101 Pulp, paper and paperboard	338	44	478	8,770	0	10,926	0	6,474,730	..	43,865
2102 Corrugated paper, paperboard and their containers	0	3	93	68	0	0	0	52,290	..	352
2109 Other articles of pulp and paperboard	0	227	73	0	0	0	0	132,699	..	778
22 PUBLISHING, PRINTING & REPRODUCTION	0	65	209	0	0	0	47	456,982	..	1,966
221 Publishing	0	0	136	0	0	0	22	209,315	..	912
2211 Publishing of books & brochures	0	0	8	0	0	0	0	28,964	..	112
2212 Publishing of newspapers and periodicals	0	0	109	0	0	0	22	155,480	..	691
2213 Publishing of recorded media	0	0	2	0	0	0	0	11,748	..	44
2219 Other publishing	0	0	17	0	0	0	0	13,123	..	64
222 Printing and related service activities	0	65	66	0	0	0	25	240,321	..	1,021
2221 Printing	0	65	56	0	0	0	12	199,032	..	850
2222 Service activities related to printing	0	0	10	0	0	0	13	41,289	..	172
223 Reproduction of recorded media	0	0	7	0	0	0	0	7,346	..	33
23 COKE, REFINED PETROLEUM PRODUCTS	0	0	72	30	33,431	0	0	532,108	..	35,449
231 Coke oven products	..	..	..	..	..	..	..	..	..	..
232 Refined petroleum products	0	0	72	30	33,431	0	0	532,108	..	35,449
233 Processing of nuclear fuel	..	..	..	..	..	..	..	..	..	..
24 CHEMICALS & CHEMICAL PRODUCTS	0	123	1,322	2,464	10,288	6	294	6,211,214	..	36,857
241 Basic chemicals	0	87	767	2,391	10,288	6	284	6,020,804	..	35,498
2411 Basic chemicals, exc. fertilizers & nitrogen compounds	0	87	623	1,688	9,751	0	284	4,653,215	..	29,185
2412 Fertilizers and nitrogen compounds	0	0	5	352	151	0	0	928,074	..	3,849
2413 Plastics in primary forms and synthetic rubber	0	0	139	351	386	6	0	439,515	..	2,464
242 Other chemical products	0	36	555	73	0	0	9	190,410	..	1,358
2421 Pesticides and other agro-chemical products	..	..	..	..	..	..	..	45	..	..
2422 Paints, varnishes and similar coatings	0	0	89	0	0	0	0	40,241	..	234
2423 Pharmaceuticals, medicinal chem. & botanical prod.	0	36	292	0	0	0	0	93,001	..	663
2424 Soap and detergents, perfumes etc.	0	0	140	0	0	0	0	33,209	..	260
2429 Other chemical products, nec	0	0	34	73	0	0	9	23,914	..	202
243 Man-made fibres	0	0	0	0	0	0	1	0	..	1
25 RUBBER AND PLASTICS PRODUCTS	0	6	298	89	0	0	37	349,201	..	1,687
251 Rubber products	0	0	19	65	0	0	18	42,198	..	254
2511 Rubber tyres and tubes	0	0	11	0	0	0	18	15,538	..	85
2519 Other rubber products	0	0	8	65	0	0	0	26,660	..	169
252 Plastic products	0	6	279	24	0	0	19	307,003	..	1,433
26 OTHER NON-METALLIC MINERAL PRODUCTS	6,140	1,309	1,962	655	77	1	115	1,103,318	..	14,231
261 Glass and glass products	0	348	167	465	0	0	20	197,010	..	1,709
269 Non-metallic mineral products, nec	6,140	961	1,795	190	77	1	95	906,308	..	12,522
2691 Pottery, china and earthenware	0	84	27	0	0	0	5	32,324	..	232
2692 Refractory ceramic products	0	70	4	46	0	0	0	12,813	..	166
2693 Structural non-refractory clay & ceramic prod.	0	64	2	144	0	0	0	3,616	..	223
2694 Cement, lime and plaster	5,214	0	79	0	77	0	51	491,772	..	7,191
2695 Articles of concrete, cement and plaster	511	677	513	0	0	1	39	138,052	..	2,238
2696 Cutting, shaping and finishing of stone	0	0	101	0	0	0	0	50,971	..	284
2699 Other non-metallic mineral products, nec	415	66	1,069	0	0	0	0	176,760	..	2,186

ISIC Revision 3 Industry Sector	Solid TJ	LPG TJ	Distiloil TJ	RFO TJ	Gas TJ	Biomass TJ	Steam TJ	Electr MWh	Own Use MWh	TOTAL TJ
27 BASIC METALS	12	547	1,921	1,287	1,758	1	3	22,738,388	..	87,387
271 Basic iron and steel	0	5	297	339	398	0	3	6,795,727	..	25,507
272 Basic precious and non-ferrous metals	12	532	1,494	948	1,360	0	0	15,750,876	..	61,049
273 Casting of metals	0	10	130	0	0	1	0	191,785	..	831
2731 Casting of iron and steel	0	5	91	0	0	0	0	160,405	..	673
2732 Casting non-ferrous metals	0	5	39	0	0	1	0	31,380	..	158
28 FABRICATED METAL PRODUCTS	1	98	670	1	0	0	52	495,162	..	2,605
281 Str. metal prod., tanks, reservoirs, steam generators	1	33	280	1	0	0	32	178,105	..	988
2811 Structural metal products	1	26	253	1	0	0	16	168,509	..	904
2812 Tanks, reservoirs and containers of metal	0	1	21	0	0	0	16	8,291	..	68
2813 Steam generators, exc. central heating hot water boilers	0	6	6	0	0	0	0	1,305	..	17
289 Other fabricated metal products	0	65	390	0	0	0	20	317,057	..	1,616
2891 Forging, pressing, stamping & roll-forming of metal	0	0	3	0	0	0	0	5,339	..	22
2892 Treatment and coating of metals	0	45	194	0	0	0	0	116,578	..	659
2893 Cutlery, hand tools and general hardware	0	0	31	0	0	0	4	41,667	..	185
2899 Other fabricated metal products, nec	0	20	162	0	0	0	16	153,473	..	751
29 MACHINERY AND EQUIPMENT, NEC	7	159	676	15	0	0	58	552,999	..	2,906
291 General purpose machinery	0	82	305	15	0	0	30	286,849	..	1,465
2911 Engines and turbines	0	0	81	0	0	0	15	43,063	..	251
2912 Pumps, compressors, taps and valves	0	57	50	3	0	0	15	71,104	..	381
2913 Bearings, gears, gearing and driving elements	0	0	7	0	0	0	0	17,935	..	72
2914 Ovens, furnaces and furnace burners	0	16	9	0	0	0	0	7,756	..	53
2915 Lifting and handling equipment	0	7	105	12	0	0	0	68,595	..	371
2919 Other general purpose machinery	0	2	53	0	0	0	0	78,396	..	337
292 Special purpose machinery	0	69	317	0	0	0	23	217,580	..	1,192
2921 Agricultural and forestry machinery	0	51	158	0	0	0	19	84,256	..	531
2922 Machine-tools	0	0	14	0	0	0	0	13,501	..	63
2923 Machinery for metallurgy	0	0	1	0	0	0	0	1,678	..	7
2924 Machinery for mining, quarrying and construction	0	5	59	0	0	0	0	29,763	..	171
2925 Machinery for food, beverage & tobacco processing	0	13	44	0	0	0	0	26,090	..	151
2926 Machinery for textile, apparel & leather production	0	0	2	0	0	0	0	1,213	..	6
2927 Machinery for weapons and ammunition	0	0	17	0	0	0	4	39,087	..	162
2929 Other special purpose machinery	0	0	22	0	0	0	0	21,992	..	101
293 Domestic appliances, nec	7	8	54	0	0	0	5	48,570	..	249
30 OFFICE, ACCOUNTING & COMPUTING MACHINERY	1	0	1	0	1	0	0	7,464	..	30
31 ELECTRICAL MACHINERY & APPARATUS, NEC	0	21	249	410	0	2	25	235,088	..	1,553
311 Electric motors, generators and transformers	0	6	68	0	0	0	6	55,942	..	281
312 Electricity distribution and control apparatus	0	0	9	0	0	0	5	41,060	..	162
313 Insulated wire and cable	0	8	21	0	0	2	6	61,475	..	258
314 Accumulators, primary cells & primary batteries	0	0	5	0	0	0	4	6,837	..	34
315 Electric lamps and lighting equipment	0	7	26	0	0	0	0	34,112	..	156
319 Other electrical equipment, nec	0	0	120	410	0	0	4	35,662	..	662
32 RADIO, TV & COMMUNICATION EQUIP. & APP.	0	0	11	0	0	0	5	53,785	..	210
321 Electronic valves, tubes, other electronic components	0	0	0	0	0	0	2	15,348	..	57
322 TV & radio transmitters, apparatus for line telephony	0	0	8	0	0	0	3	32,630	..	128
323 TV & radio receivers, recording apparatus	0	0	3	0	0	0	0	5,807	..	24
33 MEDICAL PRECISION &OPTICAL INSTRUMENTS	0	0	10	0	0	0	8	80,052	..	306
331 Medical appliances and instruments	0	0	10	0	0	0	8	75,284	..	289
3311 Medical, surgical equipment & orthopaedic app.	0	0	3	0	0	0	4	23,274	..	91
3312 Instruments & appliances for measuring, checking etc	0	0	5	0	0	0	4	40,851	..	156
3313 Industrial process control equipment	0	0	2	0	0	0	0	11,159	..	42
332 Optical instruments and photographic equipment	0	0	0	0	0	0	0	4,768	..	17
333 Watches and clocks	..	..	..	..	..	..	..		..	
34 MOTOR VEHICLES, TRAILERS & SEMI-TRAILERS	0	112	191	0	0	59	22	209,758	..	1,139
341 Motor vehicles	0	0	12	0	0	19	0	5,875	..	52
342 Bodies (coachwork) for motor vehicles	0	10	87	0	0	20	22	23,815	..	225
343 Parts, accessories for motor vehicles & their engines	0	102	92	0	0	20	0	180,068	..	862
35 OTHER TRANSPORT EQUIPMENT	0	106	542	82	0	0	65	640,803	..	3,102
351 Building and repairing of ships and boats	0	82	402	21	0	0	53	503,038	..	2,369
3511 Building and repairing of ships	0	82	370	21	0	0	53	484,236	..	2,269
3512 Building, repairing of pleasure & sporting boats	0	0	32	0	0	0	0	18,802	..	100
352 Railway, tramway locomotives & rolling stock	0	8	64	0	0	0	4	36,591	..	208
353 Aircraft and spacecraft	0	16	54	61	0	0	8	85,887	..	448
359 Transport equipment, nec	0	0	22	0	0	0	0	15,287	..	77
3591 Motorcycles	..	..	..	..	..	..	..		..	
3592 Bicycles and invalid carriages	0	0	21	0	0	0	0	13,573	..	70
3599 Other transport equipment, nec	0	0	1	0	0	0	0	1,714	..	7
36 FURNITURE; MANUFACTURING, NEC	0	1	342	0	0	349	27	285,201	..	1,746
361 Furniture	0	1	234	0	0	346	20	212,821	..	1,367
369 Manufacturing, nec	0	0	108	0	0	3	7	72,380	..	379
3691 Jewellery and related articles	0	0	11	0	0	0	0	9,699	..	46
3692 Musical instruments	0	0	0	0	0	0	0	592	..	2
3693 Sports goods	0	0	64	0	0	0	0	27,772	..	164
3694 Games and toys	0	0	0	0	0	0	0	1,690	..	6
3699 Other manufacturing, nec	0	0	33	0	0	3	7	32,627	..	160
37 RECYCLING	0	65	54	0	0	0	0	29,775	..	226
371 Recycling of metal waste and scrap	0	65	54	0	0	0	0	19,103	..	188
372 Recycling of non-metal waste and scrap	0	0	0	0	0	0	0	10,672	..	38
CERR Non-specified industry	..	..	..	..	..	..	..		..	
15-37 TOTAL MANUFACTURING	6,499	3,432	16,703	16,113	45,816	16,391	1,167	43,953,956		**264,355**

ISIC Revision 3 Industry Sector	Solid TJ	LPG TJ	Distiloil TJ	RFO TJ	Gas TJ	Biomass TJ	Steam TJ	Electr MWh	Own Use MWh	TOTAL TJ
15 **FOOD PRODUCTS AND BEVERAGES**	82,423	85	7,809	4,954	7,237	59	19,336	3,249,829	438,458	132,024
151 Production, processing and preserving (PPP)	17,523	77	2,642	1,303	2,252	39	2,314	1,228,716	0	30,573
1511 PPP of meat and meat products	..	..	..	..	..	..	..	..	..	..
1512 Processing and preserving of fish products	..	..	..	..	..	..	..	..	..	..
1513 Processing, preserving of fruit & vegetables	..	..	..	..	..	..	..	..	..	..
1514 Vegetable and animal oils and fats	..	..	..	..	..	..	..	..	..	..
152 Dairy products	17,209	0	1,869	184	301	0	36	562,295	0	21,623
153 Grain mill prod., starches & prepared animal feeds	1,087	1	363	571	2	0	32	138,204	0	2,554
1531 Grain mill products	..	..	..	..	..	..	..	..	..	..
1532 Starches and starch products	..	..	..	..	..	..	..	..	..	..
1533 Prepared animal feeds	..	..	..	..	..	..	..	..	..	..
154 Other food products	35,916	4	2,235	1,804	4,308	3	15,847	922,885	438,458	61,861
1541 Bakery products	..	..	..	..	..	..	..	..	..	..
1542 Sugar	..	..	..	..	..	..	..	..	..	..
1543 Cocoa, chocolate and sugar confectionery	..	..	..	..	..	..	..	..	..	..
1544 Macaroni, noodles, couscous & similar farinaceous prod.	..	..	..	..	..	..	..	..	..	..
1549 Other food products, nec	..	..	..	..	..	..	..	..	..	..
155 Beverages	10,688	3	700	1,092	374	17	1,107	397,729	0	15,413
1551 Distilling, rectifying and blending of spirits	..	..	..	..	..	..	..	..	..	..
1552 Wines	..	..	..	..	..	..	..	..	..	..
1553 Malt liquors and malt	..	..	..	..	..	..	..	..	..	..
1554 Soft drinks; production of mineral waters	..	..	..	..	..	..	..	..	..	..
16 **TOBACCO PRODUCTS**	752	0	64	18	42	0	266	62,541	0	1,367
17 **TEXTILES**	21,850	7	437	230	113	114	9,379	1,360,319	118,885	36,599
171 Spinning, weaving and finishing of textiles	17,995	7	258	230	76	60	8,546	985,548	118,885	30,292
1711 Preparation and spinning of textile fibres	..	..	..	..	..	..	..	..	..	..
1712 Finishing of textiles	..	..	..	..	..	..	..	..	..	..
172 Other textiles	1,643	0	89	0	33	7	291	217,064	0	2,844
1721 Made-up textile articles, except apparel	..	..	..	..	..	..	..	..	..	..
1722 Carpets and rugs	..	..	..	..	..	..	..	..	..	..
1723 Cordage, rope, twine and netting	..	..	..	..	..	..	..	..	..	..
1729 Other textiles, nec	..	..	..	..	..	..	..	..	..	..
173 Knitted and crocheted fabrics and articles	2,212	0	90	0	4	47	542	157,707	0	3,463
18 **WEARING APPAREL, DRESSING & DYEING OF FUR**	3,032	0	516	219	412	1	152	285,397	2,097	5,352
181 Wearing apparel, except fur apparel	3,032	0	516	219	412	1	152	285,397	2,097	5,352
182 Dressing and dyeing of fur; articles of fur	..	..	..	..	..	..	..	..	..	..
19 **TANNING & DRESSING OF LEATHER, FOOTWEAR**	2,396	0	403	6	14	20	623	187,171	0	4,136
191 Tanning and dressing of leather	1,866	0	207	6	8	20	295	56,232	0	2,604
1911 Tanning and dressing of leather	..	..	..	..	..	..	..	..	..	..
1912 Luggage, handbags, saddlery & harness	..	..	..	..	..	..	..	..	..	..
192 Footwear	530	0	196	0	6	0	328	130,939	0	1,531
20 **WOOD AND WOOD PRODUCTS**	10,748	0	1,411	1,660	147	3,583	3,541	953,186	355	24,520
201 Sawmilling and planing of wood	521	0	1,015	0	0	1,546	9	291,638	0	4,141
202 Products of wood, cork, straw & plaiting materials	10,227	0	396	1,660	147	2,037	3,532	661,548	355	20,379
2021 Veneer sheets	..	..	..	..	..	..	..	..	..	..
2022 Builders' carpentry and joinery	..	..	..	..	..	..	..	..	..	..
2023 Wooden containers	..	..	..	..	..	..	..	..	..	..
2029 Other products of wood	..	..	..	..	..	..	..	..	..	..
21 **PAPER AND PAPER PRODUCTS**	25,439	35	241	1,539	149	15,645	25,795	2,103,418	827,487	73,436
2101 Pulp, paper and paperboard	..	..	..	..	..	..	..	..	..	..
2102 Corrugated paper, paperboard and their containers	..	..	..	..	..	..	..	..	..	..
2109 Other articles of pulp and paperboard	..	..	..	..	..	..	..	..	..	..
22 **PUBLISHING, PRINTING & REPRODUCTION**	198	0	126	0	103	0	42	97,754	0	821
221 Publishing	..	..	..	..	..	..	..	..	..	..
2211 Publishing of books & brochures	..	..	..	..	..	..	..	..	..	..
2212 Publishing of newspapers and periodicals	..	..	..	..	..	..	..	..	..	..
2213 Publishing of recorded media	..	..	..	..	..	..	..	..	..	..
2219 Other publishing	..	..	..	..	..	..	..	..	..	..
222 Printing and related service activities	198	0	126	0	103	0	42	97,754	0	821
2221 Printing	..	..	..	..	..	..	..	..	..	..
2222 Service activities related to printing	..	..	..	..	..	..	..	..	..	..
223 Reproduction of recorded media	..	..	..	..	..	..	..	..	..	..
23 **COKE, REFINED PETROLEUM PRODUCTS**	342,879	38	610,959	48,156	51,179	11,512	47,399	2,255,862	1,493,652	1,114,866
231 Coke oven products	338,634	0	126	0	37,327	423	12,689	672,341	319,067	390,471
232 Refined petroleum products	4,245	38	610,833	48,156	13,852	11,089	34,710	1,583,521	1,174,585	724,395
233 Processing of nuclear fuel	..	..	..	..	..	..	..	..	..	..
24 **CHEMICALS & CHEMICAL PRODUCTS**	88,270	2	2,167	13,965	6,438	4,024	79,104	8,059,719	2,653,677	213,432
241 Basic chemicals	71,018	1	1,261	7,765	5,353	4,023	66,059	6,604,500	2,334,460	170,852
2411 Basic chemicals, exc. fertilizers & nitrogen compounds	..	..	..	..	..	..	..	..	..	..
2412 Fertilizers and nitrogen compounds	..	..	..	..	..	..	..	..	..	..
2413 Plastics in primary forms and synthetic rubber	..	..	..	..	..	..	..	..	..	..
242 Other chemical products	9,872	1	862	1,027	1,077	1	4,403	633,467	35,651	19,395
2421 Pesticides and other agro-chemical products	..	..	..	..	..	..	..	..	..	..
2422 Paints, varnishes and similar coatings	..	..	..	..	..	..	..	..	..	..
2423 Pharmaceuticals, medicinal chem. & botanical prod.	..	..	..	..	..	..	..	..	..	..
2424 Soap and detergents, perfumes etc.	..	..	..	..	..	..	..	..	..	..
2429 Other chemical products, nec	..	..	..	..	..	..	..	..	..	..
243 Man-made fibres	7,380	0	44	5,173	8	0	8,642	821,752	283,566	23,184
25 **RUBBER AND PLASTICS PRODUCTS**	13,011	0	838	115	35	0	4,520	716,708	150,803	20,556
251 Rubber products	7,590	0	98	32	1	0	2,662	352,657	116,447	11,233
2511 Rubber tyres and tubes	..	..	..	..	..	..	..	..	..	..
2519 Other rubber products	..	..	..	..	..	..	..	..	..	..
252 Plastic products	5,421	0	740	83	34	0	1,858	364,051	34,356	9,323
26 **OTHER NON-METALLIC MINERAL PRODUCTS**	101,966	76	2,364	4,408	24,261	150	1,450	3,112,831	3,972	145,867
261 Glass and glass products	4,173	16	302	17	18,069	0	528	478,823	0	24,829
269 Non-metallic mineral products, nec	97,793	60	2,062	4,391	6,192	150	922	2,634,008	3,972	121,038
2691 Pottery, china and earthenware	..	..	..	..	..	..	..	..	..	..
2692 Refractory ceramic products	..	..	..	..	..	..	..	..	..	..
2693 Structural non-refractory clay & ceramic prod.	..	..	..	..	..	..	..	..	..	..
2694 Cement, lime and plaster	..	..	..	..	..	..	..	..	..	..
2695 Articles of concrete, cement and plaster	..	..	..	..	..	..	..	..	..	..
2696 Cutting, shaping and finishing of stone	..	..	..	..	..	..	..	..	..	..
2699 Other non-metallic mineral products, nec	..	..	..	..	..	..	..	..	..	..

ISIC Revision 3 Industry Sector	Solid TJ	LPG TJ	Distiloil TJ	RFO TJ	Gas TJ	Biomass TJ	Steam TJ	Electr MWh	Own Use MWh	TOTAL TJ
27 **BASIC METALS**	297,054	3	1,896	3,731	112,714	8,450	39,839	10,465,004	1,525,878	495,868
271 Basic iron and steel	278,153	1	1,448	2,884	106,200	3,830	35,682	6,848,537	1,190,658	448,566
272 Basic precious and non-ferrous metals	12,728	0	245	751	4,835	4,606	3,413	2,928,118	319,068	35,971
273 Casting of metals	6,173	2	203	96	1,679	14	744	688,349	16,152	11,331
2731 Casting of iron and steel	..		..							..
2732 Casting non-ferrous metals	..		..		..					..
28 **FABRICATED METAL PRODUCTS**	7,686	12	1,186	138	2,886	0	728	770,941	0	15,411
281 Str. metal prod., tanks, reservoirs, steam generators	1,839	0	256	0	192	0	143	102,942	0	2,801
2811 Structural metal products	..		..		..		..			..
2812 Tanks, reservoirs and containers of metal	..	..	..		..					..
2813 Steam generators, exc. central heating hot water boilers	..		..		..					..
289 Other fabricated metal products	5,847	12	930	138	2,694	0	585	667,999	0	12,611
2891 Forging, pressing, stamping & roll-forming of metal	..		..		..		..			..
2892 Treatment and coating of metals	..		..		..		..			..
2893 Cutlery, hand tools and general hardware	..		..		..		..			..
2899 Other fabricated metal products, nec	..		..		..		..			..
29 **MACHINERY AND EQUIPMENT, NEC**	25,272	10	2,034	63	3,529	5	5,401	2,084,145	62,030	43,594
291 General purpose machinery	4,805	7	410	12	427	4	1,650	519,086	17,975	9,119
2911 Engines and turbines	..		..		..		..			..
2912 Pumps, compressors, taps and valves	..		..		..		..			..
2913 Bearings, gears, gearing and driving elements	..		..		..		..			..
2914 Ovens, furnaces and furnace burners	..		..		..		..			..
2915 Lifting and handling equipment	..		..		..		..			..
2919 Other general purpose machinery	..		..		..		..			..
292 Special purpose machinery	15,229	2	1,483	51	2,855	1	2,858	1,264,309	26,074	26,937
2921 Agricultural and forestry machinery	..		..		..		..			..
2922 Machine-tools	..		..		..		..			..
2923 Machinery for metallurgy	..		..		..		..			..
2924 Machinery for mining, quarrying and construction	..		..		..		..			..
2925 Machinery for food, beverage & tobacco processing	..		..		..		..			..
2926 Machinery for textile, apparel & leather production	..		..		..		..			..
2927 Machinery for weapons and ammunition	..		..		..		..			..
2929 Other special purpose machinery	..		..		..		..			..
293 Domestic appliances, nec	5,238	1	141	0	247	0	893	300,750	17,981	7,538
30 **OFFICE, ACCOUNTING & COMPUTING MACHINERY**	..	..	..	..	..		..	..		..
31 **ELECTRICAL MACHINERY & APPARATUS, NEC**	5,579	6	440	84	2,475	103	1,281	1,019,759	1,886	13,632
311 Electric motors, generators and transformers	857	0	32	0	53	0	337	237,908	0	2,135
312 Electricity distribution and control apparatus	2,682	0	205	0	313	0	101	301,697	0	4,387
313 Insulated wire and cable	586	0	52	16	25	17	90	125,442	0	1,238
314 Accumulators, primary cells & primary batteries	98	2	39	60	21	0	98	41,870	0	469
315 Electric lamps and lighting equipment	498	1	39	0	488	0	67	62,357	0	1,317
319 Other electrical equipment, nec	858	3	73	8	1,575	86	588	250,485	1,886	4,086
32 **RADIO, TV & COMMUNICATION EQUIP. & APP.**	3,019	0	115	292	695	0	273	301,702	0	5,480
321 Electronic valves, tubes, other electronic components	32	0	17	0	37	0	95	48,922	0	357
322 TV & radio transmitters, apparatus for line telephony	966	0	58	0	7	0	47	38,015	0	1,215
323 TV & radio receivers, recording apparatus	2,021	0	40	292	651	0	131	214,765	0	3,908
33 **MEDICAL PRECISION &OPTICAL INSTRUMENTS**	801	0	198	27	58	0	77	113,466	0	1,569
331 Medical appliances and instruments	687	0	192	27	51	0	56	95,999	0	1,359
3311 Medical, surgical equipment & orthopaedic app.	..	..	..		..		..			..
3312 Instruments & appliances for measuring, checking etc	..	..	..		..		..			..
3313 Industrial process control equipment	..		..		..		..			..
332 Optical instruments and photographic equipment	114	0	5	0	7	0	0	16,041	0	184
333 Watches and clocks	0	0	1	0	0	0	21	1,426	0	27
34 **MOTOR VEHICLES, TRAILERS & SEMI-TRAILERS**	9,574	14	751	87	1,573	8	2,261	942,348	51,915	17,474
341 Motor vehicles	7,013	10	495	4	1,126	6	2,110	635,097	51,915	12,863
342 Bodies (coachwork) for motor vehicles	125	0	43	16	0	2	88	20,162	0	347
343 Parts, accessories for motor vehicles & their engines	2,436	4	213	67	447	0	63	287,089	0	4,264
35 **OTHER TRANSPORT EQUIPMENT**	8,663	10	1,312	269	611	56	3,999	768,932	34,510	17,564
351 Building and repairing of ships and boats	1,449	2	735	257	265	50	2,030	370,089	0	6,120
3511 Building and repairing of ships	..		..		..		..			..
3512 Building, repairing of pleasure & sporting boats	..		..		..		..			..
352 Railway, tramway locomotives & rolling stock	4,387	0	244	12	254	0	873	173,957	3,819	6,382
353 Aircraft and spacecraft	2,436	2	297	0	76	0	991	176,981	30,691	4,329
359 Transport equipment, nec	391	6	36	0	16	6	105	47,905	0	732
3591 Motorcycles	..		..		..		..			..
3592 Bicycles and invalid carriages	..		..		..		..			..
3599 Other transport equipment, nec	..		..		..		..			..
36 **FURNITURE; MANUFACTURING, NEC**	6,316	1	758	80	186	1,614	756	511,526	14,344	11,501
361 Furniture	5,841	0	646	79	172	1,614	751	484,749	14,344	10,796
369 Manufacturing, nec	475	1	112	1	14	0	5	26,777	0	704
3691 Jewellery and related articles	..		..		..		..			..
3692 Musical instruments	..	..	..		..		..			..
3693 Sports goods	..	..	..		..		..			..
3694 Games and toys	..	..	..		..		..			..
3699 Other manufacturing, nec	..		..		..		..			..
37 **RECYCLING**	49	3	102	2	0	0	10	21,774	0	244
371 Recycling of metal waste and scrap	43	3	56	0	0	0	0	13,771	0	152
372 Recycling of non-metal waste and scrap	6	0	46	2	0	0	10	8,003	0	93
CERR Non-specified industry	..	..	..	..	..		..	..		..
15-37 **TOTAL MANUFACTURING**	1,056,977	302	636,127	80,043	214,857	45,344	246,232	39,444,332	7,379,949	**2,395,314**

ISIC Revision 3 Industry Sector		Solid TJ	LPG TJ	Distiloil TJ	RFO TJ	Gas TJ	Biomass TJ	Steam TJ	Electr MWh	Own Use MWh	TOTAL TJ
15	**FOOD PRODUCTS AND BEVERAGES**	84,690	132	8,777	6,556	3,964	82	5,606	3,792,535	479,465	121,734
151	Production, processing and preserving (PPP)	20,635	118	3,105	1,448	1,410	55	2,828	1,426,629	0	34,735
1511	PPP of meat and meat products	..	..	..	..	..	..	..	..	..	..
1512	Processing and preserving of fish products	..	..	..	..	..	..	..	..	..	..
1513	Processing, preserving of fruit & vegetables	..	..	..	..	..	..	..	..	..	..
1514	Vegetable and animal oils and fats	..	..	..	..	..	..	..	..	..	..
152	Dairy products	15,042	0	1,946	773	467	11	608	587,473	0	20,962
153	Grain mill prod., starches & prepared animal feeds	2,554	2	883	425	28	1	136	359,214	0	5,322
1531	Grain mill products	..	..	..	..	..	..	..	..	..	..
1532	Starches and starch products	..	..	..	..	..	..	..	..	..	..
1533	Prepared animal feeds	..	..	..	..	..	..	..	..	..	..
154	Other food products	37,524	6	2,127	2,496	1,574	0	890	1,067,142	479,465	46,733
1541	Bakery products	..	..	..	..	..	..	..	..	..	..
1542	Sugar	..	..	..	..	..	..	..	..	..	..
1543	Cocoa, chocolate and sugar confectionery	..	..	..	..	..	..	..	..	..	..
1544	Macaroni, noodles, couscous & similar farinaceous prod.	..	..	..	..	..	..	..	..	..	..
1549	Other food products, nec	..	..	..	..	..	..	..	..	..	..
155	Beverages	8,935	6	716	1,414	485	15	1,144	352,077	0	13,982
1551	Distilling, rectifying and blending of spirits	..	..	..	..	..	..	..	..	..	..
1552	Wines	..	..	..	..	..	..	..	..	..	..
1553	Malt liquors and malt	..	..	..	..	..	..	..	..	..	..
1554	Soft drinks; production of mineral waters	..	..	..	..	..	..	..	..	..	..
16	**TOBACCO PRODUCTS**	827	0	78	50	85	0	343	78,935	0	1,667
17	**TEXTILES**	20,505	7	498	335	172	109	5,132	1,387,266	102,465	31,383
171	Spinning, weaving and finishing of textiles	16,444	7	274	238	38	18	3,535	960,061	102,465	23,641
1711	Preparation and spinning of textile fibres	..	..	..	..	..	..	..	..	..	..
1712	Finishing of textiles	..	..	..	..	..	..	..	..	..	..
172	Other textiles	2,115	0	104	50	126	0	866	222,790	0	4,063
1721	Made-up textile articles, except apparel	..	..	..	..	..	..	..	..	..	..
1722	Carpets and rugs	..	..	..	..	..	..	..	..	..	..
1723	Cordage, rope, twine and netting	..	..	..	..	..	..	..	..	..	..
1729	Other textiles, nec	..	..	..	..	..	..	..	..	..	..
173	Knitted and crocheted fabrics and articles	1,946	0	120	47	8	91	731	204,415	0	3,679
18	**WEARING APPAREL, DRESSING & DYEING OF FUR**	2,440	1	644	316	143	2	231	313,529	1,613	4,900
181	Wearing apparel, except fur apparel	2,290	1	640	316	143	2	225	306,672	1,613	4,715
182	Dressing and dyeing of fur; articles of fur	150	0	4	0	0	0	6	6,857	0	185
19	**TANNING & DRESSING OF LEATHER, FOOTWEAR**	2,838	0	1,503	69	8	18	730	154,499	0	5,722
191	Tanning and dressing of leather	1,951	0	78	39	1	16	281	64,130	0	2,597
1911	Tanning and dressing of leather	..	..	..	..	..	..	..	..	..	..
1912	Luggage, handbags, saddlery & harness	..	..	..	..	..	..	..	..	..	..
192	Footwear	887	0	1,425	30	7	2	449	90,369	0	3,125
20	**WOOD AND WOOD PRODUCTS**	10,179	1	1,737	1,858	563	4,945	409	959,075	0	23,145
201	Sawmilling and planing of wood	1,091	0	854	1	0	1,875	30	240,778	0	4,718
202	Products of wood, cork, straw & plaiting materials	9,088	1	883	1,857	563	3,070	379	718,297	0	18,427
2021	Veneer sheets	..	..	..	..	..	..	..	..	..	..
2022	Builders' carpentry and joinery	..	..	..	..	..	..	..	..	..	..
2023	Wooden containers	..	..	..	..	..	..	..	..	..	..
2029	Other products of wood	..	..	..	..	..	..	..	..	..	..
21	**PAPER AND PAPER PRODUCTS**	27,468	41	382	2,404	193	16,831	2,415	2,331,417	858,383	55,037
2101	Pulp, paper and paperboard	..	..	..	..	..	..	..	..	..	..
2102	Corrugated paper, paperboard and their containers	..	..	..	..	..	..	..	..	..	..
2109	Other articles of pulp and paperboard	..	..	..	..	..	..	..	..	..	..
22	**PUBLISHING, PRINTING & REPRODUCTION**	298	0	225	7	40	0	290	90,450	0	1,186
221	Publishing	41	0	46	0	0	0	20	9,539	0	141
2211	Publishing of books & brochures	..	..	..	..	..	..	..	..	..	..
2212	Publishing of newspapers and periodicals	..	..	..	..	..	..	..	..	..	..
2213	Publishing of recorded media	..	..	..	..	..	..	..	..	..	..
2219	Other publishing	..	..	..	..	..	..	..	..	..	..
222	Printing and related service activities	257	0	179	7	40	0	270	80,911	0	1,044
2221	Printing	..	..	..	..	..	..	..	..	..	..
2222	Service activities related to printing	..	..	..	..	..	..	..	..	..	..
223	Reproduction of recorded media	..	..	..	..	..	..	..	..	..	..
23	**COKE, REFINED PETROLEUM PRODUCTS**	364,557	21	626,875	46,777	53,661	18,164	24,554	2,347,204	1,428,068	1,137,918
231	Coke oven products	360,466	0	126	0	38,692	336	12,525	708,765	371,882	413,358
232	Refined petroleum products	4,091	21	626,749	46,777	14,969	17,828	12,029	1,638,439	1,056,186	724,560
233	Processing of nuclear fuel	..	..	..	..	..	..	..	..	..	..
24	**CHEMICALS & CHEMICAL PRODUCTS**	89,453	4	2,733	10,879	8,237	3,600	26,963	8,508,144	2,672,007	162,879
241	Basic chemicals	71,722	2	1,268	6,139	7,685	3,592	22,729	7,057,325	2,431,307	129,791
2411	Basic chemicals, exc. fertilizers & nitrogen compounds	..	..	..	..	..	..	..	..	..	..
2412	Fertilizers and nitrogen compounds	..	..	..	..	..	..	..	..	..	..
2413	Plastics in primary forms and synthetic rubber	..	..	..	..	..	..	..	..	..	..
242	Other chemical products	10,806	2	1,427	380	547	8	2,490	635,367	32,764	17,829
2421	Pesticides and other agro-chemical products	..	..	..	..	..	..	..	..	..	..
2422	Paints, varnishes and similar coatings	..	..	..	..	..	..	..	..	..	..
2423	Pharmaceuticals, medicinal chem. & botanical prod.	..	..	..	..	..	..	..	..	..	..
2424	Soap and detergents, perfumes etc.	..	..	..	..	..	..	..	..	..	..
2429	Other chemical products, nec	..	..	..	..	..	..	..	..	..	..
243	Man-made fibres	6,925	0	38	4,360	5	0	1,744	815,452	207,936	15,259
25	**RUBBER AND PLASTICS PRODUCTS**	13,102	0	1,343	1,763	46	15	1,765	1,036,055	162,227	21,180
251	Rubber products	8,268	0	470	33	31	0	1,395	463,748	136,102	11,377
2511	Rubber tyres and tubes	..	..	..	..	..	..	..	..	..	..
2519	Other rubber products	..	..	..	..	..	..	..	..	..	..
252	Plastic products	4,834	0	873	1,730	15	15	370	572,307	26,125	9,803
26	**OTHER NON-METALLIC MINERAL PRODUCTS**	97,414	120	3,977	6,270	28,271	207	2,578	3,526,916	3,684	151,521
261	Glass and glass products	3,898	25	365	48	19,460	0	602	612,274	0	26,602
269	Non-metallic mineral products, nec	93,516	95	3,612	6,222	8,811	207	1,976	2,914,642	3,684	124,918
2691	Pottery, china and earthenware	..	..	..	..	..	..	..	..	..	..
2692	Refractory ceramic products	..	..	..	..	..	..	..	..	..	..
2693	Structural non-refractory clay & ceramic prod.	..	..	..	..	..	..	..	..	..	..
2694	Cement, lime and plaster	..	..	..	..	..	..	..	..	..	..
2695	Articles of concrete, cement and plaster	..	..	..	..	..	..	..	..	..	..
2696	Cutting, shaping and finishing of stone	..	..	..	..	..	..	..	..	..	..
2699	Other non-metallic mineral products, nec	..	..	..	..	..	..	..	..	..	..

ISIC Revision 3 Industry Sector	Solid TJ	LPG TJ	Distiloil TJ	RFO TJ	Gas TJ	Biomass TJ	Steam TJ	Electr MWh	Own Use MWh	TOTAL TJ
27 **BASIC METALS**	286,341	2	2,008	2,706	110,627	7,271	11,325	10,856,743	1,569,892	453,713
271 Basic iron and steel	267,770	0	1,541	1,906	104,275	2,951	9,828	7,077,195	1,223,241	409,345
272 Basic precious and non-ferrous metals	12,936	0	242	688	5,053	4,316	792	3,014,936	322,452	33,720
273 Casting of metals	5,635	2	225	112	1,299	4	705	764,612	24,199	10,647
2731 Casting of iron and steel	..	..	..	..	..		..	..	..	..
2732 Casting non-ferrous metals	..	..	..	..	..		..	..	..	..
28 **FABRICATED METAL PRODUCTS**	5,973	8	1,078	262	1,726	1	1,305	1,180,232	6	14,602
281 Str. metal prod., tanks, reservoirs, steam generators	2,387	4	374	26	293	0	489	171,681	0	4,191
2811 Structural metal products	..	..	..	..	..		..	..	..	..
2812 Tanks, reservoirs and containers of metal	..	..	..	..	..		..	..	..	..
2813 Steam generators, exc. central heating hot water boilers	..	..	..	..	..		..	..	..	..
289 Other fabricated metal products	3,586	4	704	236	1,433	1	816	1,008,551	6	10,411
2891 Forging, pressing, stamping & roll-forming of metal	..	..	..	..	..		..	..	..	..
2892 Treatment and coating of metals	..	..	..	..	..		..	..	..	..
2893 Cutlery, hand tools and general hardware	..	..	..	..	..		..	..	..	..
2899 Other fabricated metal products, nec	..	..	..	..	..		..	..	..	..
29 **MACHINERY AND EQUIPMENT, NEC**	22,383	16	1,833	147	3,619	11	5,915	2,370,834	64,625	42,226
291 General purpose machinery	6,409	8	629	62	501	5	2,086	686,149	23,238	12,086
2911 Engines and turbines	..	..	..	..	..		..	..	..	..
2912 Pumps, compressors, taps and valves	..	..	..	..	..		..	..	..	..
2913 Bearings, gears, gearing and driving elements	..	..	..	..	..		..	..	..	..
2914 Ovens, furnaces and furnace burners	..	..	..	..	..		..	..	..	..
2915 Lifting and handling equipment	..	..	..	..	..		..	..	..	..
2919 Other general purpose machinery	..	..	..	..	..		..	..	..	..
292 Special purpose machinery	12,875	7	1,077	85	2,960	6	3,705	1,462,904	27,571	25,882
2921 Agricultural and forestry machinery	..	..	..	..	..		..	..	..	..
2922 Machine-tools	..	..	..	..	..		..	..	..	..
2923 Machinery for metallurgy	..	..	..	..	..		..	..	..	..
2924 Machinery for mining, quarrying and construction	..	..	..	..	..		..	..	..	..
2925 Machinery for food, beverage & tobacco processing	..	..	..	..	..		..	..	..	..
2926 Machinery for textile, apparel & leather production	..	..	..	..	..		..	..	..	..
2927 Machinery for weapons and ammunition	..	..	..	..	..		..	..	..	..
2929 Other special purpose machinery	..	..	..	..	..		..	..	..	..
293 Domestic appliances, nec	3,099	1	127	0	158	0	124	221,781	13,816	4,258
30 **OFFICE, ACCOUNTING & COMPUTING MACHINERY**	59	0	5	0	2	0	86	7,932	0	181
31 **ELECTRICAL MACHINERY & APPARATUS, NEC**	4,736	9	370	74	1,447	81	1,583	762,115	0	11,044
311 Electric motors, generators and transformers	1,063	0	35	1	43	0	672	100,784	0	2,177
312 Electricity distribution and control apparatus	1,358	0	69	0	62	0	233	102,147	0	2,090
313 Insulated wire and cable	859	0	61	12	29	18	202	164,076	0	1,772
314 Accumulators, primary cells & primary batteries	42	4	22	36	47	0	76	39,109	0	368
315 Electric lamps and lighting equipment	515	1	64	0	436	0	64	77,561	0	1,359
319 Other electrical equipment	899	4	119	25	830	45	336	278,438	0	3,278
32 **RADIO, TV & COMMUNICATION EQUIP. & APP.**	1,321	0	216	130	885	0	533	329,893	0	4,273
321 Electronic valves, tubes, other electronic components	172	0	48	3	34	0	206	90,504	0	789
322 TV & radio transmitters, apparatus for line telephony	335	0	67	31	0	0	176	32,175	0	725
323 TV & radio receivers, recording apparatus	814	0	101	96	851	0	151	207,214	0	2,759
33 **MEDICAL PRECISION & OPTICAL INSTRUMENTS**	832	1	107	90	108	0	339	133,632	0	1,958
331 Medical appliances and instruments	832	1	97	90	104	0	315	120,986	0	1,875
3311 Medical, surgical equipment & orthopaedic app.	..	..	..	..	..		..	..	..	..
3312 Instruments & appliances for measuring, checking etc	..	..	..	..	..		..	..	..	..
3313 Industrial process control equipment	..	..	..	..	..		..	..	..	..
332 Optical instruments and photographic equipment	0	0	8	0	4	0	0	11,142	0	52
333 Watches and clocks	0	0	2	0	0	0	24	1,504	0	31
34 **MOTOR VEHICLES, TRAILERS & SEMI-TRAILERS**	9,935	15	799	104	1,099	6	1,958	927,081	57,715	17,046
341 Motor vehicles	6,994	10	477	2	906	6	1,743	597,925	57,715	12,083
342 Bodies (coachwork) for motor vehicles	433	0	38	22	1	0	97	23,059	0	674
343 Parts, accessories for motor vehicles & their engines	2,508	5	284	80	192	0	118	306,097	0	4,289
35 **OTHER TRANSPORT EQUIPMENT**	7,111	11	1,193	292	545	29	3,630	747,390	25,557	15,410
351 Building and repairing of ships and boats	1,031	2	774	272	144	29	2,151	340,982	0	5,631
3511 Building and repairing of ships	..	..	..	..	..		..	..	..	..
3512 Building, repairing of pleasure & sporting boats	..	..	..	..	..		..	..	..	..
352 Railway, tramway locomotives & rolling stock	3,817	0	238	20	300	0	509	183,711	3,429	5,533
353 Aircraft and spacecraft	1,890	2	154	0	92	0	869	175,351	22,128	3,559
359 Transport equipment, nec	373	7	27	0	9	0	101	47,346	0	687
3591 Motorcycles	..	..	..	..	..		..	..	..	..
3592 Bicycles and invalid carriages	..	..	..	..	..		..	..	..	..
3599 Other transport equipment, nec	..	..	..	..	..		..	..	..	..
36 **FURNITURE; MANUFACTURING, NEC**	5,813	2	1,137	188	194	2,383	500	588,940	13,288	12,289
361 Furniture	5,107	0	707	163	174	2,299	447	484,357	13,288	10,593
369 Manufacturing, nec	706	2	430	25	20	84	53	104,583	0	1,696
3691 Jewellery and related articles	..	..	..	..	..		..	..	..	..
3692 Musical instruments	..	..	..	..	..		..	..	..	..
3693 Sports goods	..	..	..	..	..		..	..	..	..
3694 Games and toys	..	..	..	..	..		..	..	..	..
3699 Other manufacturing, nec	..	..	..	..	..		..	..	..	..
37 **RECYCLING**	40	5	220	100	7	0	10	36,919	0	515
371 Recycling of metal waste and scrap	34	5	175	100	7	0	7	26,461	0	423
372 Recycling of non-metal waste and scrap	6	0	45	0	0	0	3	10,458	0	92
CERR **Non-specified industry**	..	..	..	..	..		..	..	..	..
15-37 **TOTAL MANUFACTURING**	1,058,315	396	657,738	81,377	215,642	53,755	98,200	42,467,736	7,438,995	**2,291,526**

ISIC Revision 3 Industry Sector	Solid TJ	LPG TJ	Distiloil TJ	RFO TJ	Gas TJ	Biomass TJ	Steam TJ	Electr MWh	Own Use MWh	TOTAL TJ
15 FOOD PRODUCTS AND BEVERAGES	104,026	178	10,841	8,988	15,090	95	5,431	4,916,014	565,147	160,312
151 Production, processing and preserving (PPP)	25,675	134	3,461	2,370	3,854	61	2,585	1,886,334	0	44,931
1511 PPP of meat and meat products	..	..	..	..	..	..	..	..	..	..
1512 Processing and preserving of fish products	..	..	..	..	..	..	..	..	..	..
1513 Processing, preserving of fruit & vegetables	..	..	..	..	..	..	..	..	..	..
1514 Vegetable and animal oils and fats	..	..	..	..	..	..	..	..	..	..
152 Dairy products	21,215	4	2,561	762	928	24	432	733,447	0	28,566
153 Grain mill prod., starches & prepared animal feeds	2,651	11	1,038	426	179	2	155	344,539	0	5,702
1531 Grain mill products	..	..	..	..	..	..	..	..	..	..
1532 Starches and starch products	..	..	..	..	..	..	..	..	..	..
1533 Prepared animal feeds	..	..	..	..	..	..	..	..	..	..
154 Other food products	42,926	11	2,457	2,876	9,126	8	1,157	1,410,417	565,147	61,604
1541 Bakery products	..	..	..	..	..	..	..	..	..	..
1542 Sugar	..	..	..	..	..	..	..	..	..	..
1543 Cocoa, chocolate and sugar confectionery	..	..	..	..	..	..	..	..	..	..
1544 Macaroni, noodles, couscous & similar farinaceous prod.	..	..	..	..	..	..	..	..	..	..
1549 Other food products, nec	..	..	..	..	..	..	..	..	..	..
155 Beverages	11,559	18	1,324	2,554	1,003	0	1,102	541,277	0	19,509
1551 Distilling, rectifying and blending of spirits	..	..	..	..	..	..	..	..	..	..
1552 Wines	..	..	..	..	..	..	..	..	..	..
1553 Malt liquors and malt	..	..	..	..	..	..	..	..	..	..
1554 Soft drinks; production of mineral waters	..	..	..	..	..	..	..	..	..	..
16 TOBACCO PRODUCTS	730	0	80	62	174	0	267	73,135	0	1,576
17 TEXTILES	24,851	9	604	454	277	99	4,935	1,325,033	98,673	35,644
171 Spinning, weaving and finishing of textiles	16,624	7	248	239	67	77	3,345	907,295	98,673	23,518
1711 Preparation and spinning of textile fibres	..	..	..	..	..	..	..	..	..	..
1712 Finishing of textiles	..	..	..	..	..	..	..	..	..	..
172 Other textiles	6,224	0	206	146	163	0	844	215,386	0	8,358
1721 Made-up textile articles, except apparel	..	..	..	..	..	..	..	..	..	..
1722 Carpets and rugs	..	..	..	..	..	..	..	..	..	..
1723 Cordage, rope, twine and netting	..	..	..	..	..	..	..	..	..	..
1729 Other textiles, nec	..	..	..	..	..	..	..	..	..	..
173 Knitted and crocheted fabrics and articles	2,003	2	150	69	47	22	746	202,352	0	3,767
18 WEARING APPAREL, DRESSING & DYEING OF FUR	2,509	2	543	292	1,090	1	259	306,663	738	5,797
181 Wearing apparel, except fur apparel	2,347	2	536	292	1,090	1	250	299,353	738	5,593
182 Dressing and dyeing of fur; articles of fur	162	0	7	0	0	0	9	7,310	0	204
19 TANNING & DRESSING OF LEATHER, FOOTWEAR	3,278	1	341	96	76	427	546	182,252	0	5,421
191 Tanning and dressing of leather	2,467	1	102	86	30	424	335	73,788	0	3,711
1911 Tanning and dressing of leather	..	..	..	..	..	..	..	..	..	..
1912 Luggage, handbags, saddlery & harness	..	..	..	..	..	..	..	..	..	..
192 Footwear	811	0	239	10	46	3	211	108,464	0	1,710
20 WOOD AND WOOD PRODUCTS	10,015	2	3,531	2,209	387	10,724	321	1,294,346	0	31,849
201 Sawmilling and planing of wood	424	0	1,249	6	2	1,497	5	243,782	0	4,061
202 Products of wood, cork, straw & plaiting materials	9,591	2	2,282	2,203	385	9,227	316	1,050,564	0	27,788
2021 Veneer sheets	..	..	..	..	..	..	..	..	..	..
2022 Builders' carpentry and joinery	..	..	..	..	..	..	..	..	..	..
2023 Wooden containers	..	..	..	..	..	..	..	..	..	..
2029 Other products of wood	..	..	..	..	..	..	..	..	..	..
21 PAPER AND PAPER PRODUCTS	26,653	52	584	1,497	297	17,496	2,037	2,353,022	872,655	53,945
2101 Pulp, paper and paperboard	..	..	..	..	..	..	..	..	..	..
2102 Corrugated paper, paperboard and their containers	..	..	..	..	..	..	..	..	..	..
2109 Other articles of pulp and paperboard	..	..	..	..	..	..	..	..	..	..
22 PUBLISHING, PRINTING & REPRODUCTION	199	0	252	30	163	4	318	619,196	0	3,195
221 Publishing	0	0	61	0	0	0	28	14,685	0	142
2211 Publishing of books & brochures	..	..	..	..	..	..	..	..	..	..
2212 Publishing of newspapers and periodicals	..	..	..	..	..	..	..	..	..	..
2213 Publishing of recorded media	..	..	..	..	..	..	..	..	..	..
2219 Other publishing	..	..	..	..	..	..	..	..	..	..
222 Printing and related service activities	199	0	191	30	163	4	290	604,511	0	3,053
2221 Printing	..	..	..	..	..	..	..	..	..	..
2222 Service activities related to printing	..	..	..	..	..	..	..	..	..	..
223 Reproduction of recorded media	..	..	..	..	..	..	..	..	..	..
23 COKE, REFINED PETROLEUM PRODUCTS	323,125	4,640	777,656	42,948	49,792	2,860	25,420	2,312,216	1,605,022	1,228,987
231 Coke oven products	318,799	0	122	0	32,198	185	12,622	658,349	351,496	365,031
232 Refined petroleum products	4,326	4,640	777,534	42,948	17,594	2,675	12,798	1,653,867	1,253,526	863,956
233 Processing of nuclear fuel	..	..	..	..	..	..	..	..	..	..
24 CHEMICALS & CHEMICAL PRODUCTS	85,686	18	2,251	12,745	8,743	3,526	22,356	8,658,877	2,606,571	157,113
241 Basic chemicals	70,540	4	1,226	7,782	7,721	3,457	18,824	7,052,720	2,385,821	126,355
2411 Basic chemicals, exc. fertilizers & nitrogen compounds	..	..	..	..	..	..	..	..	..	..
2412 Fertilizers and nitrogen compounds	..	..	..	..	..	..	..	..	..	..
2413 Plastics in primary forms and synthetic rubber	..	..	..	..	..	..	..	..	..	..
242 Other chemical products	8,645	14	994	436	1,018	69	2,254	821,286	34,287	16,263
2421 Pesticides and other agro-chemical products	..	..	..	..	..	..	..	..	..	..
2422 Paints, varnishes and similar coatings	..	..	..	..	..	..	..	..	..	..
2423 Pharmaceuticals, medicinal chem. & botanical prod.	..	..	..	..	..	..	..	..	..	..
2424 Soap and detergents, perfumes etc.	..	..	..	..	..	..	..	..	..	..
2429 Other chemical products, nec	..	..	..	..	..	..	..	..	..	..
243 Man-made fibres	6,501	0	31	4,527	4	0	1,278	784,871	186,463	14,495
25 RUBBER AND PLASTICS PRODUCTS	13,035	8	898	463	114	42	1,287	1,077,937	163,354	19,139
251 Rubber products	8,934	6	320	100	37	23	1,108	413,522	136,262	11,526
2511 Rubber tyres and tubes	..	..	..	..	..	..	..	..	..	..
2519 Other rubber products	..	..	..	..	..	..	..	..	..	..
252 Plastic products	4,101	2	578	363	77	19	179	664,415	27,092	7,613
26 OTHER NON-METALLIC MINERAL PRODUCTS	105,954	199	3,078	4,747	29,988	154	2,452	3,762,961	4,135	160,104
261 Glass and glass products	2,895	32	216	141	20,765	0	616	620,859	0	26,900
269 Non-metallic mineral products, nec	103,059	167	2,862	4,606	9,223	154	1,836	3,142,102	4,135	133,204
2691 Pottery, china and earthenware	..	..	..	..	..	..	..	..	..	..
2692 Refractory ceramic products	..	..	..	..	..	..	..	..	..	..
2693 Structural non-refractory clay & ceramic prod.	..	..	..	..	..	..	..	..	..	..
2694 Cement, lime and plaster	..	..	..	..	..	..	..	..	..	..
2695 Articles of concrete, cement and plaster	..	..	..	..	..	..	..	..	..	..
2696 Cutting, shaping and finishing of stone	..	..	..	..	..	..	..	..	..	..
2699 Other non-metallic mineral products, nec	..	..	..	..	..	..	..	..	..	..

ISIS Energy Data Programme (IEA/OECD)

ISIC Revision 3 Industry Sector		Solid TJ	LPG TJ	Distiloil TJ	RFO TJ	Gas TJ	Biomass TJ	Steam TJ	Electr MWh	Own Use MWh	TOTAL TJ
27	**BASIC METALS**	265,179	66	1,814	1,795	101,664	5,528	10,831	10,845,411	1,470,194	420,628
271	Basic iron and steel	245,506	2	1,399	952	95,275	498	9,217	7,061,659	1,102,874	374,301
272	Basic precious and non-ferrous metals	13,927	60	245	716	4,761	5,023	749	2,947,132	340,680	34,864
273	Casting of metals	5,746	4	170	127	1,628	7	865	836,620	26,640	11,463
2731	Casting of iron and steel	..		..	..	..		..		..	..
2732	Casting non-ferrous metals	..		..	..	..		..		..	..
28	**FABRICATED METAL PRODUCTS**	6,899	16	1,601	449	2,184	1	1,517	1,018,617	2	16,334
281	Str. metal prod., tanks, reservoirs, steam generators	3,644	7	693	173	389	0	607	309,604	0	6,628
2811	Structural metal products	..	..	..	..	..		..	..	..	..
2812	Tanks, reservoirs and containers of metal	..		..	..	..		..		..	..
2813	Steam generators, exc. central heating hot water boilers	..		..	..	..		..		..	..
289	Other fabricated metal products	3,255	9	908	276	1,795	0	910	709,013	2	9,705
2891	Forging, pressing, stamping & roll-forming of metal	..		..	..	..		..		..	..
2892	Treatment and coating of metals	..		..	..	..		..		..	..
2893	Cutlery, hand tools and general hardware	..		..	..	..		..		..	..
2899	Other fabricated metal products, nec	..		..	..	..		..		..	..
29	**MACHINERY AND EQUIPMENT, NEC**	26,094	28	5,015	176	2,267	12	4,930	2,201,552	57,751	46,240
291	General purpose machinery	7,270	11	818	55	754	3	1,742	722,245	22,830	13,171
2911	Engines and turbines	..	..	..	..	..		..	..	..	..
2912	Pumps, compressors, taps and valves	..	..	..	..	..		..	..	..	..
2913	Bearings, gears, gearing and driving elements	..	..	..	..	..		..	..	..	..
2914	Ovens, furnaces and furnace burners	..	..	..	..	..		..	..	..	..
2915	Lifting and handling equipment	..	..	..	..	..		..	..	..	..
2919	Other general purpose machinery	..	..	..	..	..		..	..	..	..
292	Special purpose machinery	16,142	15	4,103	117	1,325	9	3,019	1,285,717	23,714	29,273
2921	Agricultural and forestry machinery	..	..	..	..	..		..	..	..	..
2922	Machine-tools	..	..	..	..	..		..	..	..	..
2923	Machinery for metallurgy	..	..	..	..	..		..	..	..	..
2924	Machinery for mining, quarrying and construction	..	..	..	..	..		..	..	..	..
2925	Machinery for food, beverage & tobacco processing	..	..	..	..	..		..	..	..	..
2926	Machinery for textile, apparel & leather production	..	..	..	..	..		..	..	..	..
2927	Machinery for weapons and ammunition	..	..	..	..	..		..	..	..	..
2929	Other special purpose machinery	..	..	..	..	..		..	..	..	..
293	Domestic appliances, nec	2,682	2	94	4	188	0	169	193,590	11,207	3,796
30	**OFFICE, ACCOUNTING & COMPUTING MACHINERY**	97	0	7	0	2	0	74	8,195	0	210
31	**ELECTRICAL MACHINERY & APPARATUS, NEC**	4,845	8	444	114	1,630	87	2,137	855,065	0	12,343
311	Electric motors, generators and transformers	1,377	0	57	0	50	0	727	124,731	0	2,660
312	Electricity distribution and control apparatus	1,411	1	82	12	131	1	243	130,106	0	2,349
313	Insulated wire and cable	952	0	63	10	26	26	224	171,664	0	1,919
314	Accumulators, primary cells & primary batteries	19	2	27	25	69	0	103	47,597	0	416
315	Electric lamps and lighting equipment	490	1	100	31	491	1	39	98,447	0	1,507
319	Other electrical equipment, nec	596	4	115	36	863	59	801	282,520	0	3,491
32	**RADIO, TV & COMMUNICATION EQUIP. & APP.**	792	1	333	161	907	0	508	297,272	0	3,772
321	Electronic valves, tubes, other electronic components	162	0	68	2	32	0	159	66,745	0	663
322	TV & radio transmitters, apparatus for line telephony	294	0	181	37	3	0	167	47,806	0	854
323	TV & radio receivers, recording apparatus	336	1	84	122	872	0	182	182,721	0	2,255
33	**MEDICAL PRECISION &OPTICAL INSTRUMENTS**	962	1	150	132	152	1	388	177,673	0	2,426
331	Medical appliances and instruments	827	1	140	131	131	1	368	163,610	0	2,188
3311	Medical, surgical equipment & orthopaedic app.	..	..	..	..	..		..	..	..	..
3312	Instruments & appliances for measuring, checking etc	..	..	..	..	..		..	..	..	..
3313	Industrial process control equipment	..	..	..	..	..		..	..	..	..
332	Optical instruments and photographic equipment	135	0	8	1	21	0	0	12,439	0	210
333	Watches and clocks	0	0	2	0	0	0	20	1,624	0	28
34	**MOTOR VEHICLES, TRAILERS & SEMI-TRAILERS**	9,674	22	804	129	1,332	0	2,894	1,042,040	55,664	18,406
341	Motor vehicles	6,610	16	458	12	1,036	0	2,611	623,865	55,664	12,789
342	Bodies (coachwork) for motor vehicles	57	0	45	29	0	0	125	22,540	0	337
343	Parts, accessories for motor vehicles & their engines	3,007	6	301	88	296	0	158	395,635	0	5,280
35	**OTHER TRANSPORT EQUIPMENT**	6,258	15	1,199	260	933	55	4,233	732,393	4,633	15,573
351	Building and repairing of ships and boats	1,172	2	724	237	287	54	2,044	343,209	0	5,756
3511	Building and repairing of ships	..	..	..	..	..		..	..	..	..
3512	Building, repairing of pleasure & sporting boats	..	..	..	..	..		..	..	..	..
352	Railway, tramway locomotives & rolling stock	4,134	4	250	15	362	0	692	178,856	4,633	6,084
353	Aircraft and spacecraft	568	4	193	1	284	1	1,409	168,417	0	3,066
359	Transport equipment, nec	384	5	32	7	0	0	88	41,911	0	667
3591	Motorcycles	..	..	..	..	..		..	..	..	..
3592	Bicycles and invalid carriages	..	..	..	..	..		..	..	..	..
3599	Other transport equipment, nec	..	..	..	..	..		..	..	..	..
36	**FURNITURE; MANUFACTURING, NEC**	7,679	6	1,112	155	271	2,196	484	633,049	12,294	14,138
361	Furniture	6,556	4	909	148	227	2,128	414	543,085	12,294	12,297
369	Manufacturing, nec	1,123	2	203	7	44	68	70	89,964	0	1,841
3691	Jewellery and related articles	..	..	..	..	..		..	..	..	..
3692	Musical instruments	..	..	..	..	..		..	..	..	..
3693	Sports goods	..	..	..	..	..		..	..	..	..
3694	Games and toys	..	..	..	..	..		..	..	..	..
3699	Other manufacturing, nec	..	..	..	..	..		..	..	..	..
37	**RECYCLING**	51	5	287	98	12	0	15	119,304	0	897
371	Recycling of metal waste and scrap	34	5	255	96	12	0	8	44,672	0	571
372	Recycling of non-metal waste and scrap	17	0	32	2	0	0	7	74,632	0	327
CERR	Non-specified industry	..	..	..	..	..		..	..	..	..
15-37	**TOTAL MANUFACTURING**	1,028,591	5,277	813,425	78,000	217,545	43,308	93,640	44,812,223	7,516,833	2,414,049

ISIC Revision 3 Industry Sector	Solid TJ	LPG TJ	Distiloil TJ	RFO TJ	Gas TJ	Biomass TJ	Steam TJ	Electr MWh	Own Use MWh	TOTAL TJ
15 FOOD PRODUCTS AND BEVERAGES	91,312	222	9,916	6,816	12,858	262	5,080	5,083,137	555,087	142,767
151 Production, processing and preserving (PPP)	20,868	123	3,143	1,785	2,769	40	2,465	1,906,199	0	38,055
1511 PPP of meat and meat products	..	..	..	..	..	..	..	..	..	..
1512 Processing and preserving of fish products	..	..	..	..	..	..	..	..	..	..
1513 Processing, preserving of fruit & vegetables	..	..	..	..	..	..	..	..	..	..
1514 Vegetable and animal oils and fats	..	..	..	..	..	..	..	..	..	..
152 Dairy products	15,612	6	2,403	902	1,426	22	354	912,301	0	24,009
153 Grain mill prod., starches & prepared animal feeds	1,839	15	940	287	197	4	103	317,073	0	4,526
1531 Grain mill products	..	..	..	..	..	..	..	..	..	..
1532 Starches and starch products	..	..	..	..	..	..	..	..	..	..
1533 Prepared animal feeds	..	..	..	..	..	..	..	..	..	..
154 Other food products	43,456	52	2,424	2,192	6,839	17	1,159	1,311,305	555,087	58,861
1541 Bakery products	..	..	..	..	..	..	..	..	..	..
1542 Sugar	..	..	..	..	..	..	..	..	..	..
1543 Cocoa, chocolate and sugar confectionery	..	..	..	..	..	..	..	..	..	..
1544 Macaroni, noodles, couscous & similar farinaceous prod.	..	..	..	..	..	..	..	..	..	..
1549 Other food products, nec	..	..	..	..	..	..	..	..	..	..
155 Beverages	9,537	26	1,006	1,650	1,627	179	999	636,259	0	17,315
1551 Distilling, rectifying and blending of spirits	..	..	..	..	..	..	..	..	..	..
1552 Wines	..	..	..	..	..	..	..	..	..	..
1553 Malt liquors and malt	..	..	..	..	..	..	..	..	..	..
1554 Soft drinks; production of mineral waters	..	..	..	..	..	..	..	..	..	..
16 TOBACCO PRODUCTS	566	0	121	112	146	1	264	76,437	0	1,485
17 TEXTILES	16,474	50	565	309	1,866	112	4,191	1,472,131	60,127	28,650
171 Spinning, weaving and finishing of textiles	12,935	6	198	193	198	112	2,586	933,073	60,108	19,371
1711 Preparation and spinning of textile fibres	..	..	..	..	..	..	..	..	..	..
1712 Finishing of textiles	..	..	..	..	..	..	..	..	..	..
172 Other textiles	1,636	41	204	61	1,608	0	925	339,618	19	5,698
1721 Made-up textile articles, except apparel	..	..	..	..	..	..	..	..	..	..
1722 Carpets and rugs	..	..	..	..	..	..	..	..	..	..
1723 Cordage, rope, twine and netting	..	..	..	..	..	..	..	..	..	..
1729 Other textiles, nec	..	..	..	..	..	..	..	..	..	..
173 Knitted and crocheted fabrics and articles	1,903	3	163	55	60	0	680	199,440	0	3,582
18 WEARING APPAREL, DRESSING & DYEING OF FUR	1,806	3	659	149	373	1	216	531,464	0	5,120
181 Wearing apparel, except fur apparel	1,584	3	653	147	373	1	210	524,309	0	4,859
182 Dressing and dyeing of fur; articles of fur	222	0	6	2	0	0	6	7,155	0	262
19 TANNING & DRESSING OF LEATHER, FOOTWEAR	3,227	2	397	80	86	0	567	245,343	0	5,242
191 Tanning and dressing of leather	2,427	2	87	73	45	0	309	89,297	0	3,264
1911 Tanning and dressing of leather	..	..	..	..	..	..	..	..	..	..
1912 Luggage, handbags, saddlery & harness	..	..	..	..	..	..	..	..	..	..
192 Footwear	800	0	310	7	41	0	258	156,046	0	1,978
20 WOOD AND WOOD PRODUCTS	9,499	5	2,311	2,510	766	6,314	329	1,452,313	5,527	26,942
201 Sawmilling and planing of wood	281	1	1,141	7	150	1,274	1	337,570	0	4,070
202 Products of wood, cork, straw & plaiting materials	9,218	4	1,170	2,503	616	5,040	328	1,114,743	5,527	22,872
2021 Veneer sheets	..	..	..	..	..	..	..	..	..	..
2022 Builders' carpentry and joinery	..	..	..	..	..	..	..	..	..	..
2023 Wooden containers	..	..	..	..	..	..	..	..	..	..
2029 Other products of wood	..	..	..	..	..	..	..	..	..	..
21 PAPER AND PAPER PRODUCTS	28,670	82	518	1,492	686	17,948	2,203	2,579,836	891,594	57,677
2101 Pulp, paper and paperboard	..	..	..	..	..	..	..	..	..	..
2102 Corrugated paper, paperboard and their containers	..	..	..	..	..	..	..	..	..	..
2109 Other articles of pulp and paperboard	..	..	..	..	..	..	..	..	..	..
22 PUBLISHING, PRINTING & REPRODUCTION	168	1	1,928	6	410	4	251	136,633	0	3,260
221 Publishing	19	0	1,735	0	0	0	13	21,595	0	1,845
2211 Publishing of books & brochures	..	..	..	..	..	..	..	..	..	..
2212 Publishing of newspapers and periodicals	..	..	..	..	..	..	..	..	..	..
2213 Publishing of recorded media	..	..	..	..	..	..	..	..	..	..
2219 Other publishing	..	..	..	..	..	..	..	..	..	..
222 Printing and related service activities	149	1	193	6	410	4	238	115,038	0	1,415
2221 Printing	..	..	..	..	..	..	..	..	..	..
2222 Service activities related to printing	..	..	..	..	..	..	..	..	..	..
223 Reproduction of recorded media	..	..	..	..	..	..	..	..	..	..
23 COKE, REFINED PETROLEUM PRODUCTS	342,922	3,951	1,049,822	42,311	60,307	1,373	27,783	2,383,209	1,751,133	1,530,744
231 Coke oven products	338,852	0	370	0	36,378	0	14,119	677,940	442,227	390,568
232 Refined petroleum products	4,070	3,951	1,049,452	42,311	23,929	1,373	13,664	1,705,269	1,308,906	1,140,177
233 Processing of nuclear fuel	..	..	..	..	..	..	..	..	..	..
24 CHEMICALS & CHEMICAL PRODUCTS	74,982	17	1,969	12,883	11,977	7,192	23,743	8,273,524	2,400,020	153,908
241 Basic chemicals	62,994	7	1,092	7,903	10,785	7,163	20,630	6,890,924	2,194,492	127,481
2411 Basic chemicals, exc. fertilizers & nitrogen compounds	..	..	..	..	..	..	..	..	..	..
2412 Fertilizers and nitrogen compounds	..	..	..	..	..	..	..	..	..	..
2413 Plastics in primary forms and synthetic rubber	..	..	..	..	..	..	..	..	..	..
242 Other chemical products	7,491	10	837	237	1,188	29	1,976	597,018	34,118	13,794
2421 Pesticides and other agro-chemical products	..	..	..	..	..	..	..	..	..	..
2422 Paints, varnishes and similar coatings	..	..	..	..	..	..	..	..	..	..
2423 Pharmaceuticals, medicinal chem. & botanical prod.	..	..	..	..	..	..	..	..	..	..
2424 Soap and detergents, perfumes etc.	..	..	..	..	..	..	..	..	..	..
2429 Other chemical products, nec	..	..	..	..	..	..	..	..	..	..
243 Man-made fibres	4,497	0	40	4,743	4	0	1,137	785,582	171,410	12,632
25 RUBBER AND PLASTICS PRODUCTS	12,256	15	1,053	232	171	34	1,110	1,311,030	148,854	19,055
251 Rubber products	8,362	10	190	83	68	17	915	422,650	132,964	10,688
2511 Rubber tyres and tubes	..	..	..	..	..	..	..	..	..	..
2519 Other rubber products	..	..	..	..	..	..	..	..	..	..
252 Plastic products	3,894	5	863	149	103	17	195	888,380	15,890	8,367
26 OTHER NON-METALLIC MINERAL PRODUCTS	94,888	344	3,574	5,175	28,413	53	2,404	3,830,829	3,932	148,628
261 Glass and glass products	3,145	100	295	205	19,276	0	549	751,619	0	26,276
269 Non-metallic mineral products, nec	91,743	244	3,279	4,970	9,137	53	1,855	3,079,210	3,932	122,352
2691 Pottery, china and earthenware	..	..	..	..	..	..	..	..	..	..
2692 Refractory ceramic products	..	..	..	..	..	..	..	..	..	..
2693 Structural non-refractory clay & ceramic prod.	..	..	..	..	..	..	..	..	..	..
2694 Cement, lime and plaster	..	..	..	..	..	..	..	..	..	..
2695 Articles of concrete, cement and plaster	..	..	..	..	..	..	..	..	..	..
2696 Cutting, shaping and finishing of stone	..	..	..	..	..	..	..	..	..	..
2699 Other non-metallic mineral products, nec	..	..	..	..	..	..	..	..	..	..

ISIC Revision 3 Industry Sector		Solid TJ	LPG TJ	Distiloil TJ	RFO TJ	Gas TJ	Biomass TJ	Steam TJ	Electr MWh	Own Use MWh	TOTAL TJ
27	**BASIC METALS**	265,171	32	1,832	5,567	105,740	4,805	10,349	12,414,533	1,424,064	433,062
271	Basic iron and steel	247,587	27	1,319	4,778	99,078	0	8,596	7,403,870	1,096,849	384,090
272	Basic precious and non-ferrous metals	13,231	1	340	657	5,138	4,801	789	3,215,730	304,339	35,438
273	Casting of metals	4,353	4	173	132	1,524	4	964	1,794,933	22,876	13,533
2731	Casting of iron and steel	..	..	..	..	..	..	..	..	..	..
2732	Casting non-ferrous metals	..	..	..	..	..	..	..	..	..	..
28	**FABRICATED METAL PRODUCTS**	8,136	24	1,659	195	3,425	1	1,465	1,178,745	0	19,148
281	Str. metal prod., tanks, reservoirs, steam generators	3,742	10	787	25	553	0	664	365,440	0	7,097
2811	Structural metal products	..	..	..	..	..	..	..	..	..	..
2812	Tanks, reservoirs and containers of metal	..	..	..	..	..	..	..	..	..	..
2813	Steam generators, exc. central heating hot water boilers	..	..	..	..	..	..	..	..	..	..
289	Other fabricated metal products	4,394	14	872	170	2,872	0	801	813,305	0	12,051
2891	Forging, pressing, stamping & roll-forming of metal	..	..	..	..	..	..	..	..	..	..
2892	Treatment and coating of metals	..	..	..	..	..	..	..	..	..	..
2893	Cutlery, hand tools and general hardware	..	..	..	..	..	..	..	..	..	..
2899	Other fabricated metal products, nec	..	..	..	..	..	..	..	..	..	..
29	**MACHINERY AND EQUIPMENT, NEC**	21,305	65	2,145	134	5,163	5	4,286	2,223,746	43,316	40,953
291	General purpose machinery	5,902	17	702	18	3,209	2	1,540	754,710	10,871	14,068
2911	Engines and turbines	..	..	..	..	..	..	..	..	..	..
2912	Pumps, compressors, taps and valves	..	..	..	..	..	..	..	..	..	..
2913	Bearings, gears, gearing and driving elements	..	..	..	..	..	..	..	..	..	..
2914	Ovens, furnaces and furnace burners	..	..	..	..	..	..	..	..	..	..
2915	Lifting and handling equipment	..	..	..	..	..	..	..	..	..	..
2919	Other general purpose machinery	..	..	..	..	..	..	..	..	..	..
292	Special purpose machinery	12,442	45	1,354	116	1,636	3	2,652	1,288,419	22,238	22,806
2921	Agricultural and forestry machinery	..	..	..	..	..	..	..	..	..	..
2922	Machine-tools	..	..	..	..	..	..	..	..	..	..
2923	Machinery for metallurgy	..	..	..	..	..	..	..	..	..	..
2924	Machinery for mining, quarrying and construction	..	..	..	..	..	..	..	..	..	..
2925	Machinery for food, beverage & tobacco processing	..	..	..	..	..	..	..	..	..	..
2926	Machinery for textile, apparel & leather production	..	..	..	..	..	..	..	..	..	..
2927	Machinery for weapons and ammunition	..	..	..	..	..	..	..	..	..	..
2929	Other special purpose machinery	..	..	..	..	..	..	..	..	..	..
293	Domestic appliances, nec	2,961	3	89	0	318	0	94	180,617	10,207	4,078
30	**OFFICE, ACCOUNTING & COMPUTING MACHINERY**	71	0	11	0	2	0	46	8,702	0	161
31	**ELECTRICAL MACHINERY & APPARATUS, NEC**	3,726	10	686	169	1,888	68	1,962	918,112	351	11,813
311	Electric motors, generators and transformers	1,164	3	72	0	57	0	620	116,047	0	2,334
312	Electricity distribution and control apparatus	1,259	0	87	21	80	0	289	102,435	0	2,105
313	Insulated wire and cable	775	1	265	9	122	28	186	286,552	0	2,418
314	Accumulators, primary cells & primary batteries	8	1	53	16	90	0	103	46,269	0	438
315	Electric lamps and lighting equipment	191	1	101	100	650	0	87	85,535	0	1,438
319	Other electrical equipment, nec	329	4	108	23	889	40	677	281,274	351	3,081
32	**RADIO, TV & COMMUNICATION EQUIP. & APP.**	534	0	166	166	801	0	357	274,636	0	3,013
321	Electronic valves, tubes, other electronic components	87	0	37	0	61	0	114	57,330	0	505
322	TV & radio transmitters, apparatus for line telephony	279	0	79	38	3	0	106	26,661	0	601
323	TV & radio receivers, recording apparatus	168	0	50	128	737	0	137	190,645	0	1,906
33	**MEDICAL PRECISION &OPTICAL INSTRUMENTS**	931	1	164	107	187	1	295	261,243	0	2,626
331	Medical appliances and instruments	821	1	152	102	186	1	281	250,865	0	2,447
3311	Medical, surgical equipment & orthopaedic app.	..	..	..	..	..	..	..	..	..	..
3312	Instruments & appliances for measuring, checking etc	..	..	..	..	..	..	..	..	..	..
3313	Industrial process control equipment	..	..	..	..	..	..	..	..	..	..
332	Optical instruments and photographic equipment	110	0	11	5	1	0	0	8,890	0	159
333	Watches and clocks	0	0	1	0	0	0	14	1,488	0	20
34	**MOTOR VEHICLES, TRAILERS & SEMI-TRAILERS**	8,264	33	866	106	1,480	1	2,592	1,107,042	56,180	17,125
341	Motor vehicles	5,602	16	448	11	1,182	0	2,226	639,211	56,180	11,584
342	Bodies (coachwork) for motor vehicles	134	7	94	27	0	0	35	37,209	0	431
343	Parts, accessories for motor vehicles & their engines	2,528	10	324	68	298	0	331	430,622	0	5,109
35	**OTHER TRANSPORT EQUIPMENT**	5,395	16	1,025	148	960	4	3,361	711,519	2,893	13,460
351	Building and repairing of ships and boats	1,132	1	643	133	240	4	1,530	324,193	0	4,850
3511	Building and repairing of ships	..	..	..	..	..	..	..	..	..	..
3512	Building, repairing of pleasure & sporting boats	..	..	..	..	..	..	..	..	..	..
352	Railway, tramway locomotives & rolling stock	3,509	6	185	14	360	0	489	150,704	2,893	5,095
353	Aircraft and spacecraft	419	5	169	0	360	0	1,252	201,760	0	2,931
359	Transport equipment, nec	335	4	28	1	0	0	90	34,862	0	584
3591	Motorcycles	..	..	..	..	..	..	..	..	..	..
3592	Bicycles and invalid carriages	..	..	..	..	..	..	..	..	..	..
3599	Other transport equipment, nec	..	..	..	..	..	..	..	..	..	..
36	**FURNITURE; MANUFACTURING, NEC**	5,636	8	8,308	107	129	1,999	405	1,110,908	14,644	20,539
361	Furniture	4,525	3	1,457	91	62	1,914	356	540,574	14,644	10,301
369	Manufacturing, nec	1,111	5	6,851	16	67	85	49	570,334	0	10,237
3691	Jewellery and related articles	..	..	..	..	..	..	..	..	..	..
3692	Musical instruments	..	..	..	..	..	..	..	..	..	..
3693	Sports goods	..	..	..	..	..	..	..	..	..	..
3694	Games and toys	..	..	..	..	..	..	..	..	..	..
3699	Other manufacturing, nec	..	..	..	..	..	..	..	..	..	..
37	**RECYCLING**	60	4	210	20	7	0	16	91,255	0	646
371	Recycling of metal waste and scrap	39	4	163	13	7	0	6	21,419	0	309
372	Recycling of non-metal waste and scrap	21	0	47	7	0	0	10	69,836	0	336
CERR	**Non-specified industry**	..	..	..	..	..	..	..	..	..	..
15-37	**TOTAL MANUFACTURING**	995,999	4,885	1,089,905	78,794	237,841	40,178	93,275	47,676,327	7,357,722	2,686,024

ISIC Revision 3 Industry Sector	Solid TJ	LPG TJ	Distiloil TJ	RFO TJ	Gas TJ	Biomass TJ	Steam TJ	Electr MWh	Own Use MWh	TOTAL TJ
15 FOOD PRODUCTS AND BEVERAGES	2,690	2	8,179	2,289	8,415	6	..	513,145	..	23,428
151 Production, processing and preserving (PPP)	360	1	320	31	1,940	1		134,255		3,136
1511 PPP of meat and meat products	328	1	237	31	1,008	1		99,269		1,963
1512 Processing and preserving of fish products	..		..	..		..				
1513 Processing, preserving of fruit & vegetables	32	0	57	0	173	0		13,352		310
1514 Vegetable and animal oils and fats	..		..			..				
152 Dairy products	101	0	82	286	1,309	0		65,203		2,013
153 Grain mill prod., starches & prepared animal feeds	262	0	184	84	842	0		84,909		1,678
1531 Grain mill products	1	0	41	7	158	0		40,590		353
1532 Starches and starch products	244	0	6	0	266	0		12,140		560
1533 Prepared animal feeds	15	0	135	55	418	0		31,810		738
154 Other food products	1,851	0	195	1,824	1,414	5		131,110		5,761
1541 Bakery products	82	0	130	164	1,163	5		37,809		1,680
1542 Sugar	1,456	0	28	1,587	76	0		50,135		3,327
1543 Cocoa, chocolate and sugar confectionery	..		..			..				
1544 Macaroni, noodles, couscous & similar farinaceous prod.	..		..			..				
1549 Other food products, nec	85	0	32	41	110	0		15,281		323
155 Beverages	115	0	7,397	64	2,910	0		97,668		10,838
1551 Distilling, rectifying and blending of spirits	60	0	33	64	693	0		17,456		913
1552 Wines	2	0	7,261	0	96	0		5,749		7,380
1553 Malt liquors and malt	5	0	55	0	2,025	0		66,285		2,324
1554 Soft drinks; production of mineral waters	..		..			..				
16 TOBACCO PRODUCTS	..		..			..				
17 TEXTILES	1,075	0	61	3,224	1,817	19		240,545		7,062
171 Spinning, weaving and finishing of textiles	535	0	34	3,150	1,118	6		177,044		5,480
1711 Preparation and spinning of textile fibres	535	0	30	2,188	833	6		152,639		4,142
1712 Finishing of textiles	..		..			..				
172 Other textiles	30	0	14	64	185	13		21,427		383
1721 Made-up textile articles, except apparel	0	0	4	2	15	0		1,957		28
1722 Carpets and rugs	..		..			..				
1723 Cordage, rope, twine and netting	..		..			..				
1729 Other textiles, nec	15	0	3	0	11	0		7,859		57
173 Knitted and crocheted fabrics and articles	510	0	13	10	513	0		42,074		1,197
18 WEARING APPAREL, DRESSING & DYEING OF FUR	185	0	40	41	711	1		45,355		1,141
181 Wearing apparel, except fur apparel	185	0	39	41	702	1		44,984		1,130
182 Dressing and dyeing of fur; articles of fur	..		..			..				
19 TANNING & DRESSING OF LEATHER, FOOTWEAR	1,759	1	61	12	277	0		60,725		2,329
191 Tanning and dressing of leather	715	0	23	5	92	0		20,808		910
1911 Tanning and dressing of leather	712	0	20	1	76	0		19,380		879
1912 Luggage, handbags, saddlery & harness	3	0	3	4	16	0		1,428		31
192 Footwear	1,044	1	37	7	185	0		39,917		1,418
20 WOOD AND WOOD PRODUCTS	763	0	113	325	730	1,292		184,708		3,888
201 Sawmilling and planing of wood	160	0	48	209	467	455		81,818		1,634
202 Products of wood, cork, straw & plaiting materials	602	0	64	116	262	837		102,226		2,249
2021 Veneer sheets	168	0	29	116	246	617		68,586		1,423
2022 Builders' carpentry and joinery	431	0	26	0	5	220		26,949		779
2023 Wooden containers	..		..			..				
2029 Other products of wood	3	0	7	0	11	0		6,691		45
21 PAPER AND PAPER PRODUCTS	5,890	0	85	4,050	3,542	159		787,404		16,561
2101 Pulp, paper and paperboard	5,860	0	61	3,591	3,296	159		655,904		15,328
2102 Corrugated paper, paperboard and their containers	..		..			..				
2109 Other articles of pulp and paperboard	29	0	19	457	211	0		127,358		1,174
22 PUBLISHING, PRINTING & REPRODUCTION	0	0	34	14	191	0		33,466		359
221 Publishing	0	0	10	0	19	0		1,913		36
2211 Publishing of books & brochures	0	0	2	0	9	0		623		13
2212 Publishing of newspapers and periodicals	0	0	6	0	4	0		415		11
2213 Publishing of recorded media	..		..			..				
2219 Other publishing	0	0	2	0	5	0		859		10
222 Printing and related service activities	0	0	21	14	169	0		30,759		315
2221 Printing	0	0	12	14	134	0		21,113		236
2222 Service activities related to printing	0	0	6	0	22	0		6,886		53
223 Reproduction of recorded media	..		..			..				
23 COKE, REFINED PETROLEUM PRODUCTS	0	0	8,764	8,716	20,148	0		740,706		40,295
231 Coke oven products	..		..			..				
232 Refined petroleum products	..		..			..				
233 Processing of nuclear fuel	..		..			..				
24 CHEMICALS & CHEMICAL PRODUCTS	7,970	0	19,381	902	17,873	56		1,522,397		51,663
241 Basic chemicals	5,426	0	19,187	16	1,338	0		762,532		28,712
2411 Basic chemicals, exc. fertilizers & nitrogen compounds	306	0	5	0	157	0		66,772		708
2412 Fertilizers and nitrogen compounds	..		..			..				
2413 Plastics in primary forms and synthetic rubber	..		..			..				
242 Other chemical products	186	0	73	654	1,222	0		102,596		2,504
2421 Pesticides and other agro-chemical products	..		..			..				
2422 Paints, varnishes and similar coatings	..		..			..				
2423 Pharmaceuticals, medicinal chem. & botanical prod.	135	0	24	654	889	0		85,786		2,011
2424 Soap and detergents, perfumes etc.	5	0	42	0	108	0		7,200		181
2429 Other chemical products, nec	..		..			..				
243 Man-made fibres	..		..			..				
25 RUBBER AND PLASTICS PRODUCTS	151	0	69	865	4,174	0		266,852		6,220
251 Rubber products	77	0	37	753	2,011	0		133,744		3,359
2511 Rubber tyres and tubes	67	0	29	600	1,330	0		90,533		2,352
2519 Other rubber products	0	0	5	153	681	0		39,472		981
252 Plastic products	74	0	32	112	2,163	0		133,108		2,860
26 OTHER NON-METALLIC MINERAL PRODUCTS	4,310	6	415	5,359	11,143	979		817,128		25,154
261 Glass and glass products	340	6	46	4	3,791	0		192,640		4,881
269 Non-metallic mineral products, nec	3,970	0	367	5,355	7,351	979		624,488		20,270
2691 Pottery, china and earthenware	7	0	2	0	159	0		4,373		184
2692 Refractory ceramic products	14	0	55	658	1,682	0		103,143		2,780
2693 Structural non-refractory clay & ceramic prod.	70	0	59	106	2,112	9		65,651		2,592
2694 Cement, lime and plaster	3,566	0	146	4,587	3,232	970		420,460		14,015
2695 Articles of concrete, cement and plaster	309	0	97	4	160	0		29,007		674
2696 Cutting, shaping and finishing of stone	3	0	8	0	6	0		1,750		23
2699 Other non-metallic mineral products, nec	..		..			..				

ISIC Revision 3 Industry Sector	Solid TJ	LPG TJ	Distiloil TJ	RFO TJ	Gas TJ	Biomass TJ	Steam TJ	Electr MWh	Own Use MWh	TOTAL TJ
27 BASIC METALS	145,838	0	2,562	1,167	37,422	0	..	3,457,297	..	199,435
271 Basic iron and steel	142,505	0	2,517	1,166	33,555	0	..	2,687,073	..	189,416
272 Basic precious and non-ferrous metals	3,174	0	40	0	3,658	0	..	725,168	..	9,483
273 Casting of metals	159	0	4	0	209	0	..	45,056	..	534
2731 Casting of iron and steel	..		..		..			..		..
2732 Casting non-ferrous metals	..		..		..			..		..
28 FABRICATED METAL PRODUCTS	622	0	149	39	772	10	..	116,078	..	2,010
281 Str. metal prod., tanks, reservoirs, steam generators	444	0	81	19	417	9	..	63,235	..	1,198
2811 Structural metal products	82	0	59	19	136	9	..	22,423	..	386
2812 Tanks, reservoirs and containers of metal	68	0	21	0	81	0	..	20,618	..	244
2813 Steam generators, exc. central heating hot water boilers	..		..		..			..		..
289 Other fabricated metal products	173	0	65	20	355	1	..	51,917	..	801
2891 Forging, pressing, stamping & roll-forming of metal	..		..		..			..		..
2892 Treatment and coating of metals	14	0	8	1	5	0	..	3,699	..	41
2893 Cutlery, hand tools and general hardware	54	0	12	19	48	0	..	13,781	..	183
2899 Other fabricated metal products, nec	105	0	45	0	300	1	..	33,016	..	570
29 MACHINERY AND EQUIPMENT, NEC	2,348	6	3,069	1,152	3,290	90	..	450,278	..	11,576
291 General purpose machinery	1,075	1	2,793	24	795	90	..	175,664	..	5,410
2911 Engines and turbines	20	1	5	5	111	0	..	11,993	..	185
2912 Pumps, compressors, taps and valves	426	0	13	0	37	0	..	34,010	..	598
2913 Bearings, gears, gearing and driving elements	416	0	2,749	18	511	90	..	102,794	..	4,154
2914 Ovens, furnaces and furnace burners	..		..		..			..		..
2915 Lifting and handling equipment	174	0	11	1	101	0	..	11,191	..	327
2919 Other general purpose machinery	35	0	6	0	24	0	..	9,329	..	99
292 Special purpose machinery	1,030	0	253	1,112	2,182	0	..	223,066	..	5,380
2921 Agricultural and forestry machinery	351	0	87	2	109	0	..	22,976	..	632
2922 Machine-tools	63	0	20	286	230	0	..	38,137	..	736
2923 Machinery for metallurgy	..		..		..			..		..
2924 Machinery for mining, quarrying and construction	38	0	30	688	546	0	..	33,616	..	1,423
2925 Machinery for food, beverage & tobacco processing	0	0	8	0	45	0	..	3,479	..	66
2926 Machinery for textile, apparel & leather production	20	0	3	0	142	0	..	7,150	..	191
2927 Machinery for weapons and ammunition	..		..		..			..		..
2929 Other special purpose machinery	555	0	101	136	1,005	0	..	107,140	..	2,183
293 Domestic appliances, nec	242	5	12	15	229	0	..	48,827	..	679
30 OFFICE, ACCOUNTING & COMPUTING MACHINERY	0	0	5	0	90	0	..	16,108	..	153
31 ELECTRICAL MACHINERY & APPARATUS, NEC	314	0	60	144	17,580	0	..	75,253	..	18,369
311 Electric motors, generators and transformers	85	0	7	0	206	0	..	10,962	..	337
312 Electricity distribution and control apparatus	46	0	14	0	84	0	..	9,571	..	178
313 Insulated wire and cable	5	0	8	0	248	0	..	23,678	..	346
314 Accumulators, primary cells & primary batteries	..		..		..			..		..
315 Electric lamps and lighting equipment	158	0	8	138	16,869	0	..	14,505	..	17,225
319 Other electrical equipment, nec	19	0	18	5	151	0	..	13,814	..	243
32 RADIO, TV & COMMUNICATION EQUIP. & APP.	468	0	958	24	308	0	..	69,991	..	2,010
321 Electronic valves, tubes, other electronic components	1	0	940	0	152	0	..	47,473	..	1,264
322 TV & radio transmitters, apparatus for line telephony	13	0	9	0	155	0	..	11,696	..	219
323 TV & radio receivers, recording apparatus	454	0	8	24	1	0	..	10,822	..	526
33 MEDICAL PRECISION &OPTICAL INSTRUMENTS	124	0	26	0	494	0	..	29,013	..	748
331 Medical appliances and instruments	124	0	26	0	452	0	..	27,233	..	700
3311 Medical, surgical equipment & orthopaedic app.	20	0	13	0	314	0	..	15,210	..	402
3312 Instruments & appliances for measuring, checking etc	104	0	10	0	133	0	..	10,932	..	286
3313 Industrial process control equipment	0	0	3	0	5	0	..	1,091	..	12
332 Optical instruments and photographic equipment	..		..		..			..		..
333 Watches and clocks	..		..		..			..		..
34 MOTOR VEHICLES, TRAILERS & SEMI-TRAILERS	290	0	64	22	991	0	..	98,181	..	1,720
341 Motor vehicles	249	0	26	6	568	0	..	69,137	..	1,098
342 Bodies (coachwork) for motor vehicles	0	0	10	0	3	0	..	1,271	..	18
343 Parts, accessories for motor vehicles & their engines	41	0	28	16	420	0	..	27,773	..	605
35 OTHER TRANSPORT EQUIPMENT	2,105	0	25	100	670	0	..	106,380	..	3,283
351 Building and repairing of ships and boats	..		..		..			..		..
3511 Building and repairing of ships	..		..		..			..		..
3512 Building, repairing of pleasure & sporting boats	..		..		..			..		..
352 Railway, tramway locomotives & rolling stock	300	0	11	22	425	0	..	27,227	..	856
353 Aircraft and spacecraft	..		..		..			..		..
359 Transport equipment, nec	7	0	5	78	53	0	..	6,326	..	166
3591 Motorcycles	0	0	1	0	50	0	..	3,853	..	65
3592 Bicycles and invalid carriages	..		..		..			..		..
3599 Other transport equipment, nec	..		..		..			..		..
36 FURNITURE; MANUFACTURING, NEC	270	2	92	87	554	247	..	89,964	..	1,576
361 Furniture	248	0	75	54	504	247	..	80,430	..	1,418
369 Manufacturing, nec	21	1	17	32	50	0	..	9,124	..	154
3691 Jewellery and related articles	0	0	7	27	23	0	..	3,297	..	69
3692 Musical instruments	..		..		..			..		..
3693 Sports goods	..		..		..			..		..
3694 Games and toys	..		..		..			..		..
3699 Other manufacturing, nec	20	1	10	5	27	0	..	5,429	..	83
37 RECYCLING	10	0	68	1	33	0	..	7,641	..	140
371 Recycling of metal waste and scrap	0	0	33	0	26	0	..	4,432	..	75
372 Recycling of non-metal waste and scrap	10	0	35	1	7	0	..	3,209	..	65
CERR Non-specified industry	..		..		..			..		..
15-37 TOTAL MANUFACTURING	177,182	17	44,280	28,533	131,225	2,859	..	9,728,615	..	419,119

ISIC Revision 3 Industry Sector	Solid TJ	LPG TJ	Distiloil TJ	RFO TJ	Gas TJ	Biomass TJ	Steam TJ	Electr MWh	Own Use MWh	TOTAL TJ
15 FOOD PRODUCTS AND BEVERAGES	2,330	1	7,095	2,136	8,570	1	..	495,692		21,917
151 Production, processing and preserving (PPP)	199	0	317	83	1,867	1	..	132,160		2,943
1511 PPP of meat and meat products	..	..	..	..	..	..	..	..		
1512 Processing and preserving of fish products	..	..	..	..	..	..	..	..		
1513 Processing, preserving of fruit & vegetables	..	..	..	..	..	..	..	..		
1514 Vegetable and animal oils and fats	75	0	85	287	1,170	0	..	61,228		1,837
152 Dairy products	205	0	173	109	700	0	..	83,134		1,486
153 Grain mill prod., starches & prepared animal feeds	1	0	42	5	154	0	..	40,197		347
1531 Grain mill products	186	0	4	0	233	0	..	12,599		468
1532 Starches and starch products	15	0	123	82	313	0	..	29,732		640
1533 Prepared animal feeds	1,759	0	212	1,572	1,871	0	..	120,382		5,847
154 Other food products	46	0	143	112	1,394	0	..	38,843		1,835
1541 Bakery products	1,656	0	40	1,460	308	0	..	62,479		3,689
1542 Sugar	..	..	..	..	..	..	..	..		
1543 Cocoa, chocolate and sugar confectionery	..	..	..	..	..	..	..	..		
1544 Macaroni, noodles, couscous & similar farinaceous prod.	57	0	23	0	112	0	..	7,580		219
1549 Other food products, nec	54	0	31	84	604	0	..	16,944		834
155 Beverages	92	0	6,308	85	2,962	0	..	98,788		9,803
1551 Distilling, rectifying and blending of spirits	2	0	6,170	1	84	0	..	5,750		6,278
1552 Wines	4	0	65	0	2,173	0	..	69,587		2,493
1553 Malt liquors and malt	24	0	23	0	100	0	..	4,791		164
1554 Soft drinks; production of mineral waters	..	..	..	..	..	..	..	..		
16 TOBACCO PRODUCTS	..	..	..	..	..	..	..	..		
17 TEXTILES	872	0	47	3,310	1,550	12	..	215,295		6,566
171 Spinning, weaving and finishing of textiles	348	0	26	3,285	1,170	12	..	174,833		5,470
1711 Preparation and spinning of textile fibres	..	..	..	..	..	..	..	..		
1712 Finishing of textiles	12	0	8	2	60	0	..	8,996		114
172 Other textiles	0	0	4	2	19	0	..	1,926		32
1721 Made-up textile articles, except apparel	..	..	..	..	..	..	..	..		
1722 Carpets and rugs	..	..	..	..	..	..	..	..		
1723 Cordage, rope, twine and netting	8	0	4	0	10	0	..	5,406		41
1729 Other textiles, nec	512	0	13	23	319	0	..	31,466		980
173 Knitted and crocheted fabrics and articles	120	0	38	5	961	1	..	52,539		1,314
18 WEARING APPAREL, DRESSING & DYEING OF FUR	..	..	..	..	..	..	..	56,688		2,263
181 Wearing apparel, except fur apparel	..	..	..	..	..	..	..	..		
182 Dressing and dyeing of fur; articles of fur	..	..	..	..	..	..	..	..		
19 TANNING & DRESSING OF LEATHER, FOOTWEAR	1,707	2	54	14	282	0	..	21,021		3,980
191 Tanning and dressing of leather	697	0	25	2	102	0	..	19,005		902
1911 Tanning and dressing of leather	694	0	21	0	74	0	..	2,016		857
1912 Luggage, handbags, saddlery & harness	3	0	4	2	28	0	..	35,667		44
192 Footwear	1,010	2	29	12	180	0	..	185,347		1,361
20 WOOD AND WOOD PRODUCTS	703	0	182	319	899	1,210	..	105,024		2,124
201 Sawmilling and planing of wood	173	0	108	210	554	522	..	79,650		1,854
202 Products of wood, cork, straw & plaiting materials	530	0	74	109	345	688	..	..		
2021 Veneer sheets	..	..	..	..	..	..	..	..		
2022 Builders' carpentry and joinery	..	..	..	..	..	..	..	..		
2023 Wooden containers	..	..	..	..	..	..	..	..		
2029 Other products of wood	..	..	..	..	..	..	..	..		
21 PAPER AND PAPER PRODUCTS	5,501	1	95	5,411	2,741	169	..	655,980		16,280
2101 Pulp, paper and paperboard	..	..	..	..	..	..	..	..		
2102 Corrugated paper, paperboard and their containers	..	..	..	..	..	..	..	..		
2109 Other articles of pulp and paperboard	..	..	..	..	..	..	..	..		
22 PUBLISHING, PRINTING & REPRODUCTION	1	0	49	15	191	0	..	34,560		380
221 Publishing	..	..	..	..	..	..	..	..		
2211 Publishing of books & brochures	..	..	..	..	..	..	..	..		
2212 Publishing of newspapers and periodicals	..	..	..	..	..	..	..	..		
2213 Publishing of recorded media	..	..	..	..	..	..	..	..		
2219 Other publishing	..	..	..	..	..	..	..	..		
222 Printing and related service activities	..	..	..	..	..	..	..	..		
2221 Printing	..	..	..	..	..	..	..	..		
2222 Service activities related to printing	..	..	..	..	..	..	..	..		
223 Reproduction of recorded media	..	..	..	..	..	..	..	..		
23 COKE, REFINED PETROLEUM PRODUCTS	0	0	14,168	3,306	17,821		..	762,815		38,041
231 Coke oven products	..	..	..	..	..	..	..	..		
232 Refined petroleum products	..	..	..	..	..	..	..	..		
233 Processing of nuclear fuel	..	..	..	..	..	..	..	..		
24 CHEMICALS & CHEMICAL PRODUCTS	8,584	0	728	885	20,206	68	..	1,762,043		36,814
241 Basic chemicals	..	..	..	..	..	..	..	..		
2411 Basic chemicals, exc. fertilizers & nitrogen compounds	..	..	..	..	..	..	..	..		
2412 Fertilizers and nitrogen compounds	..	..	..	..	..	..	..	..		
2413 Plastics in primary forms and synthetic rubber	..	..	..	..	..	..	..	..		
242 Other chemical products	..	..	..	..	..	..	..	..		
2421 Pesticides and other agro-chemical products	..	..	..	..	..	..	..	..		
2422 Paints, varnishes and similar coatings	..	..	..	..	..	..	..	..		
2423 Pharmaceuticals, medicinal chem. & botanical prod.	..	..	..	..	..	..	..	..		
2424 Soap and detergents, perfumes etc.	..	..	..	..	..	..	..	..		
2429 Other chemical products, nec	..	..	..	..	..	..	..	..		
243 Man-made fibres	..	..	..	..	..	..	0	291,155		6,488
25 RUBBER AND PLASTICS PRODUCTS	105	0	76	1,201	4,058	0	..	151,033	..	3,441
251 Rubber products	64	0	48	1,195	1,590	0	..	104,748		2,485
2511 Rubber tyres and tubes	64	0	38	712	1,294	0	..	42,025		912
2519 Other rubber products	0	0	6	483	272	0	..	140,122		3,046
252 Plastic products	41	0	28	6	2,467	0	4	856,667	..	27,868
26 OTHER NON-METALLIC MINERAL PRODUCTS	8,174	12	402	5,601	10,591	4	..	205,270		4,913
261 Glass and glass products	398	12	46	3	3,715	0	4	651,397		22,953
269 Non-metallic mineral products, nec	7,774	0	356	5,598	6,876	4	..	..		
2691 Pottery, china and earthenware	..	..	..	..	..	..	..	..		
2692 Refractory ceramic products	..	..	..	..	..	..	..	..		
2693 Structural non-refractory clay & ceramic prod.	..	..	..	..	..	..	..	..		
2694 Cement, lime and plaster	..	..	..	..	..	..	..	..		
2695 Articles of concrete, cement and plaster	..	..	..	..	..	..	..	..		
2696 Cutting, shaping and finishing of stone	..	..	..	..	..	..	..	..		
2699 Other non-metallic mineral products, nec	..	..	..	..	..	..	..	..		

ISIS Energy Data Programme (IEA/OECD)

ISIC Revision 3 Industry Sector	Solid TJ	LPG TJ	Distiloil TJ	RFO TJ	Gas TJ	Biomass TJ	Steam TJ	Electr MWh	Own Use MWh	TOTAL TJ
27 BASIC METALS	142,350	0	1,392	1,176	39,359	0	..	4,113,730	..	199,086
271 Basic iron and steel	139,648	0	1,345	1,173	36,873	0	..	2,879,940	..	189,407
272 Basic precious and non-ferrous metals	2,324	0	42	3	2,269	0	..	1,188,516	..	8,917
273 Casting of metals	378	0	5	0	217	0	..	45,274	..	763
2731 Casting of iron and steel	..	..	..	..	..	..	..	..	..	..
2732 Casting non-ferrous metals	..	..	..	..	..	..	..	..	..	..
28 FABRICATED METAL PRODUCTS	560	2	170	80	818	3	..	131,682	..	2,107
281 Str. metal prod., tanks, reservoirs, steam generators	358	0	84	55	408	0	..	66,791	..	1,145
2811 Structural metal products	..	..	..	..	..	..	..	..	..	..
2812 Tanks, reservoirs and containers of metal	..	..	..	..	..	..	..	..	..	..
2813 Steam generators, exc. central heating hot water boilers	..	..	..	..	..	..	..	..	..	..
289 Other fabricated metal products	191	2	81	25	409	2	..	64,113	..	941
2891 Forging, pressing, stamping & roll-forming of metal	..	..	..	..	..	..	..	..	..	..
2892 Treatment and coating of metals	..	..	..	..	..	..	..	..	..	..
2893 Cutlery, hand tools and general hardware	..	..	..	..	..	..	..	..	..	..
2899 Other fabricated metal products, nec	..	..	..	..	..	..	..	..	..	..
29 MACHINERY AND EQUIPMENT, NEC	2,069	4	403	326	3,706	0	..	485,754	..	8,257
291 General purpose machinery	869	1	94	8	970	0	..	196,189	..	2,648
2911 Engines and turbines	4	1	7	6	133	0	..	11,621	..	193
2912 Pumps, compressors, taps and valves	370	0	14	0	59	0	..	23,110	..	526
2913 Bearings, gears, gearing and driving elements	358	0	35	1	650	0	..	132,853	..	1,522
2914 Ovens, furnaces and furnace burners	..	..	..	..	..	..	..	..	..	..
2915 Lifting and handling equipment	96	0	11	1	86	0	..	11,794	..	236
2919 Other general purpose machinery	37	0	11	0	28	0	..	10,259	..	113
292 Special purpose machinery	1,067	0	276	300	2,389	0	..	235,657	..	4,880
2921 Agricultural and forestry machinery	337	0	84	1	119	0	..	22,533	..	622
2922 Machine-tools	65	0	19	67	280	0	..	35,433	..	559
2923 Machinery for metallurgy	..	..	..	..	..	..	..	..	..	..
2924 Machinery for mining, quarrying and construction	31	0	29	109	519	0	..	36,992	..	821
2925 Machinery for food, beverage & tobacco processing	4	0	6	0	38	0	..	3,650	..	61
2926 Machinery for textile, apparel & leather production	30	0	6	0	137	0	..	8,216	..	203
2927 Machinery for weapons and ammunition	..	..	..	..	..	..	..	..	..	..
2929 Other special purpose machinery	596	0	126	123	1,259	0	..	116,632	..	2,524
293 Domestic appliances, nec	130	3	14	18	294	0	..	50,869	..	642
30 OFFICE, ACCOUNTING & COMPUTING MACHINERY	0	0	5	0	114	0	..	6,110	..	141
31 ELECTRICAL MACHINERY & APPARATUS, NEC	269	0	73	227	903	0	..	89,848	..	1,795
311 Electric motors, generators and transformers	..	..	..	..	..	..	..	..	..	..
312 Electricity distribution and control apparatus	..	..	..	..	..	..	..	..	..	..
313 Insulated wire and cable	..	..	..	..	..	..	..	..	..	..
314 Accumulators, primary cells & primary batteries	..	..	..	..	..	..	..	..	..	..
315 Electric lamps and lighting equipment	..	..	..	..	..	..	..	..	..	..
319 Other electrical equipment, nec	..	..	..	..	..	..	..	..	..	..
32 RADIO, TV & COMMUNICATION EQUIP. & APP.	455	0	24	42	295	0	..	41,814	..	967
321 Electronic valves, tubes, other electronic components	..	..	..	..	..	..	..	..	..	..
322 TV & radio transmitters, apparatus for line telephony	..	..	..	..	..	..	..	..	..	..
323 TV & radio receivers, recording apparatus	..	..	..	..	..	..	..	..	..	..
33 MEDICAL PRECISION &OPTICAL INSTRUMENTS	131	0	47	0	462	0	..	34,421	..	764
331 Medical appliances and instruments	131	0	30	0	410	0	..	31,847	..	686
3311 Medical, surgical equipment & orthopaedic app.	0	0	14	0	300	0	..	20,430	..	388
3312 Instruments & appliances for measuring, checking etc	131	0	9	0	102	0	..	10,421	..	280
3313 Industrial process control equipment	0	0	7	0	8	0	..	996	..	19
332 Optical instruments and photographic equipment	..	..	..	..	..	..	..	..	..	..
333 Watches and clocks	..	..	..	..	..	..	..	..	..	..
34 MOTOR VEHICLES, TRAILERS & SEMI-TRAILERS	264	0	81	17	1,205	0	..	113,507	..	1,976
341 Motor vehicles	..	..	..	..	..	..	..	..	..	..
342 Bodies (coachwork) for motor vehicles	..	..	..	..	..	..	..	..	..	..
343 Parts, accessories for motor vehicles & their engines	..	..	..	..	..	..	..	..	..	..
35 OTHER TRANSPORT EQUIPMENT	2,041	0	108	319	779	0	..	120,294	..	3,680
351 Building and repairing of ships and boats	..	..	..	..	..	..	..	..	..	..
3511 Building and repairing of ships	..	..	..	..	..	..	..	..	..	..
3512 Building, repairing of pleasure & sporting boats	..	..	..	..	..	..	..	..	..	..
352 Railway, tramway locomotives & rolling stock	79	0	35	59	563	0	..	38,769	..	876
353 Aircraft and spacecraft	..	..	..	..	..	..	..	..	..	..
359 Transport equipment, nec	5	0	11	260	27	0	..	9,531	..	337
3591 Motorcycles	..	..	..	..	..	..	..	..	..	..
3592 Bicycles and invalid carriages	..	..	..	..	..	..	..	..	..	..
3599 Other transport equipment, nec	..	..	..	..	..	..	..	..	..	..
36 FURNITURE; MANUFACTURING, NEC	237	1	84	98	479	356	..	97,154	..	1,605
361 Furniture	210	0	69	45	430	356	..	83,918	..	1,412
369 Manufacturing, nec	27	1	14	52	49	0	..	12,449	..	188
3691 Jewellery and related articles	0	0	5	36	23	0	..	3,005	..	75
3692 Musical instruments	..	..	..	..	..	..	..	..	..	..
3693 Sports goods	..	..	..	..	..	..	..	..	..	..
3694 Games and toys	2	0	0	0	2	0	..	1,225	..	8
3699 Other manufacturing, nec	25	1	9	16	24	0	..	8,219	..	105
37 RECYCLING	9	0	77	1	30	0	..	8,016	..	146
371 Recycling of metal waste and scrap	0	0	37	0	22	0	..	3,698	..	72
372 Recycling of non-metal waste and scrap	9	0	40	1	7	0	..	4,313	..	73
CERR Non-specified industry	..	..	..	..	..	..	..	..	..	..
15-37 TOTAL MANUFACTURING	176,482	22	25,398	24,489	116,020	1,824	..	10,611,111	..	382,435

SLOVAK REPUBLIC

ISIC Revision 3 Industry Sector		Solid TJ	LPG TJ	Distiloil TJ	RFO TJ	Gas TJ	Biomass TJ	Steam TJ	Electr MWh	Own Use MWh	TOTAL TJ
15	**FOOD PRODUCTS AND BEVERAGES**	2,378	29	11,195	2,778	9,223	0	..	546,942	..	27,572
151	Production, processing and preserving (PPP)	197	24	256	82	1,170	0	..	146,878	..	2,258
1511	PPP of meat and meat products	92	23	164	35	971	0	..	83,464	..	1,585
1512	Processing and preserving of fish products	0	0	19	0	12	0	..	2,893	..	41
1513	Processing, preserving of fruit & vegetables	19	0	42	47	153	0	..	19,821	..	332
1514	Vegetable and animal oils and fats	..	..	..	..	..	..	..	..	..	..
152	Dairy products	57	0	81	233	1,292	0	..	64,194	..	1,894
153	Grain mill prod., starches & prepared animal feeds	185	0	2,423	99	944	0	..	101,802	..	4,017
1531	Grain mill products	1	0	41	5	153	0	..	43,448	..	356
1532	Starches and starch products	167	0	4	0	334	0	..	15,513	..	561
1533	Prepared animal feeds	17	0	2,376	94	455	0	..	42,704	..	3,096
154	Other food products	1,807	5	234	2,154	2,192	0	..	131,052	..	6,864
1541	Bakery products	33	5	156	95	1,396	0	..	46,953	..	1,854
1542	Sugar	1,704	0	46	2,059	605	0	..	63,218	..	4,642
1543	Cocoa, chocolate and sugar confectionery	0	0	7	0	67	0	..	12,698	..	120
1544	Macaroni, noodles, couscous & similar farinaceous prod.	..	..	..	..	..	..	..	..	..	..
1549	Other food products, nec	70	0	23	0	124	0	..	8,183	..	246
155	Beverages	118	0	8,179	210	2,761	0	..	100,410	..	11,629
1551	Distilling, rectifying and blending of spirits	77	0	29	209	534	0	..	18,997	..	917
1552	Wines	1	0	8,032	0	118	0	..	5,062	..	8,169
1553	Malt liquors and malt	4	0	77	0	1,980	0	..	69,030	..	2,310
1554	Soft drinks; production of mineral waters	28	0	19	0	87	0	..	5,043	..	152
16	**TOBACCO PRODUCTS**	..	..	..	..	..	..	..	..	..	..
17	**TEXTILES**	755	0	61	2,506	2,436	1	..	219,610	..	6,550
171	Spinning, weaving and finishing of textiles	309	0	22	1,631	1,618	1	..	149,391	..	4,119
1711	Preparation and spinning of textile fibres	309	0	22	1,631	589	1	..	36,069	..	2,682
1712	Finishing of textiles	..	..	6	..	23	0	..	2,706	..	39
172	Other textiles	0	0	4	0	23	0	..	2,579	..	36
1721	Made-up textile articles, except apparel	..	..	..	..	..	..	..	..	..	..
1722	Carpets and rugs	..	..	..	..	..	..	..	..	..	..
1723	Cordage, rope, twine and netting	..	..	..	..	..	..	..	127	..	..
1729	Other textiles, nec	..	..	..	..	..	..	..	..	..	..
173	Knitted and crocheted fabrics and articles	446	0	30	17	260	0	..	39,099	..	894
18	**WEARING APPAREL, DRESSING & DYEING OF FUR**	77	0	38	10	923	1	..	54,920	..	1,247
181	Wearing apparel, except fur apparel	77	0	37	10	912	1	..	54,485	..	1,233
182	Dressing and dyeing of fur; articles of fur	..	..	..	..	..	..	..	..	..	..
19	**TANNING & DRESSING OF LEATHER, FOOTWEAR**	1,492	2	67	16	666	0	..	54,451	..	2,439
191	Tanning and dressing of leather	588	0	26	4	113	0	..	18,382	..	797
1911	Tanning and dressing of leather	587	0	22	0	88	0	..	16,604	..	757
1912	Luggage, handbags, saddlery & harness	1	0	4	4	25	0	..	1,778	..	40
192	Footwear	904	2	41	11	553	0	..	36,069	..	1,641
20	**WOOD AND WOOD PRODUCTS**	530	0	142	215	965	838	..	173,856	..	3,316
201	Sawmilling and planing of wood	63	0	62	0	537	401	..	66,452	..	1,302
202	Products of wood, cork, straw & plaiting materials	467	0	77	215	428	428	..	106,202	..	1,997
2021	Veneer sheets	86	0	48	5	396	302	..	66,796	..	1,077
2022	Builders' carpentry and joinery	377	0	14	1	13	120	..	25,337	..	616
2023	Wooden containers	0	0	3	3	0	0	..	2,031	..	13
2029	Other products of wood	3	0	12	206	19	6	..	12,038	..	289
21	**PAPER AND PAPER PRODUCTS**	5,926	2	3,760	4,235	2,700	177	..	886,632	..	19,992
2101	Pulp, paper and paperboard	5,901	2	3,735	3,803	2,357	177	..	728,014	..	18,596
2102	Corrugated paper, paperboard and their containers	1	0	2	0	39	0	..	5,635	..	62
2109	Other articles of pulp and paperboard	23	0	23	432	304	0	..	152,529	..	1,331
22	**PUBLISHING, PRINTING & REPRODUCTION**	1	0	43	16	211	0	..	37,551	..	406
221	Publishing	1	0	18	0	15	0	..	937	..	37
2211	Publishing of books & brochures	0	0	8	0	10	0	..	376	..	19
2212	Publishing of newspapers and periodicals	1	0	7	0	1	0	..	126	..	9
2213	Publishing of recorded media	..	..	..	..	..	..	..	..	..	..
2219	Other publishing	..	..	..	..	..	..	..	..	..	..
222	Printing and related service activities	0	0	24	16	192	0	..	35,982	..	362
2221	Printing	0	0	15	16	154	0	..	25,637	..	277
2222	Service activities related to printing	0	0	7	0	15	0	..	7,721	..	50
223	Reproduction of recorded media	..	..	..	..	..	..	..	..	..	..
23	**COKE, REFINED PETROLEUM PRODUCTS**	0	0	14,699	3,810	19,293	0	..	768,778	..	40,570
231	Coke oven products	..	..	..	..	..	..	..	..	..	..
232	Refined petroleum products	..	..	..	..	..	..	..	..	..	..
233	Processing of nuclear fuel	..	..	..	..	..	..	..	..	..	..
24	**CHEMICALS & CHEMICAL PRODUCTS**	8,389	0	843	794	21,805	49	..	1,762,250	..	38,224
241	Basic chemicals	5,888	0	601	3	2,452	0	..	956,053	..	12,386
2411	Basic chemicals, exc. fertilizers & nitrogen compounds	101	0	4	3	262	0	..	63,956	..	600
2412	Fertilizers and nitrogen compounds	..	..	..	..	..	..	..	..	..	..
2413	Plastics in primary forms and synthetic rubber	124	0	61	514	1,762	0	..	131,917	..	2,936
242	Other chemical products	..	..	..	..	..	..	..	..	..	..
2421	Pesticides and other agro-chemical products	10	0	10	0	167	0	..	7,355	..	213
2422	Paints, varnishes and similar coatings	114	0	34	514	1,468	0	..	117,193	..	2,552
2423	Pharmaceuticals, medicinal chem. & botanical prod.	0	0	17	0	127	0	..	7,369	..	171
2424	Soap and detergents, perfumes etc.	..	..	..	..	..	..	..	..	..	..
2429	Other chemical products, nec	..	..	..	..	..	..	..	..	..	..
243	Man-made fibres	..	..	..	..	..	..	..	..	..	..
25	**RUBBER AND PLASTICS PRODUCTS**	86	0	87	1,390	4,006	0	..	280,850	..	6,580
251	Rubber products	55	0	39	1,379	1,457	0	..	134,245	..	3,413
2511	Rubber tyres and tubes	55	0	29	868	1,169	0	..	89,462	..	2,443
2519	Other rubber products	0	0	7	511	262	0	..	41,336	..	929
252	Plastic products	31	0	47	11	2,548	0	..	146,376	..	3,164
26	**OTHER NON-METALLIC MINERAL PRODUCTS**	6,549	16	418	5,388	10,079	192	..	729,828	..	25,269
261	Glass and glass products	399	14	54	1	3,493	0	..	207,970	..	4,710
269	Non-metallic mineral products, nec	6,150	2	364	5,386	6,586	192	..	521,858	..	20,559
2691	Pottery, china and earthenware	6	0	2	0	197	0	..	6,061	..	227
2692	Refractory ceramic products	17	0	92	1,087	1,924	0	..	58,201	..	3,330
2693	Structural non-refractory clay & ceramic prod.	38	0	42	0	2,112	11	..	67,743	..	2,447
2694	Cement, lime and plaster	5,776	0	127	4,295	2,077	180	..	355,642	..	13,735
2695	Articles of concrete, cement and plaster	312	2	97	4	271	1	..	32,904	..	805
2696	Cutting, shaping and finishing of stone	1	0	3	0	4	0	..	956	..	11
2699	Other non-metallic mineral products, nec	..	..	..	..	..	..	..	..	..	..

ISIS Energy Data Programme (IEA/OECD)

ISIC Revision 3 Industry Sector	Solid TJ	LPG TJ	Distiloil TJ	RFO TJ	Gas TJ	Biomass TJ	Steam TJ	Electr MWh	Own Use MWh	TOTAL TJ
27 BASIC METALS	131,285	1	97	914	37,148	0	..	4,788,296	..	186,683
271 Basic iron and steel	128,497	0	38	910	34,466	0	..	2,910,045	..	174,387
272 Basic precious and non-ferrous metals	2,335	1	53	4	2,485	0	..	1,824,281	..	11,445
273 Casting of metals	453	0	6	0	189	0	..	53,886	..	842
2731 Casting of iron and steel	325	0	5	0	16	0	..	22,003	..	425
2732 Casting non-ferrous metals	..	..	..	..	..	..	..		..	..
28 FABRICATED METAL PRODUCTS	480	5	172	62	1,677	0	..	127,032	..	2,853
281 Str. metal prod., tanks, reservoirs, steam generators	325	0	85	53	458	0	..	35,604	..	1,049
2811 Structural metal products	78	0	67	17	171	0	..	20,530	..	407
2812 Tanks, reservoirs and containers of metal	28	0	17	36	101	0	..	15,074	..	236
2813 Steam generators, exc. central heating hot water boilers	..	..	..	..	..	..	..		..	..
289 Other fabricated metal products	139	5	79	9	471	0	..	65,099	..	937
2891 Forging, pressing, stamping & roll-forming of metal	..	..	..	..	..	..	..		..	..
2892 Treatment and coating of metals	11	1	11	1	18	0	..	5,407	..	61
2893 Cutlery, hand tools and general hardware	48	4	11	4	43	0	..	17,268	..	172
2899 Other fabricated metal products, nec	80	0	55	4	216	0	..	42,424	..	508
29 MACHINERY AND EQUIPMENT, NEC	1,794	3	465	337	4,148	0	..	601,859	..	8,914
291 General purpose machinery	796	1	117	18	1,232	0	..	225,678	..	2,976
2911 Engines and turbines	25	1	7	0	116	0	..	11,070	..	189
2912 Pumps, compressors, taps and valves	345	0	10	0	154	0	..	32,668	..	627
2913 Bearings, gears, gearing and driving elements	227	0	31	17	816	0	..	150,477	..	1,633
2914 Ovens, furnaces and furnace burners	..	..	..	..	..	..	..		..	..
2915 Lifting and handling equipment	148	0	11	1	27	0	..	11,179	..	227
2919 Other general purpose machinery	31	0	13	0	108	0	..	16,078	..	210
292 Special purpose machinery	889	0	322	300	2,604	0	..	318,010	..	5,260
2921 Agricultural and forestry machinery	233	0	187	0	93	0	..	15,810	..	570
2922 Machine-tools	101	0	20	216	431	0	..	117,889	..	1,192
2923 Machinery for metallurgy	..	..	..	..	..	..	..		..	..
2924 Machinery for mining, quarrying and construction	28	0	28	62	419	0	..	33,934	..	659
2925 Machinery for food, beverage & tobacco processing	0	0	5	0	39	0	..	3,707	..	57
2926 Machinery for textile, apparel & leather production	41	0	3	0	75	0	..	7,060	..	144
2927 Machinery for weapons and ammunition	..	..	..	..	..	..	..		..	..
2929 Other special purpose machinery	482	0	74	22	1,547	0	..	137,188	..	2,619
293 Domestic appliances, nec	97	2	14	19	266	0	..	56,073	..	600
30 OFFICE, ACCOUNTING & COMPUTING MACHINERY	0	0	7	0	23	0	..	2,738	..	40
31 ELECTRICAL MACHINERY & APPARATUS, NEC	104	0	60	375	841	0	..	91,608	..	1,710
311 Electric motors, generators and transformers	6	0	8	0	182	0	..	14,695	..	249
312 Electricity distribution and control apparatus	0	0	13	1	128	0	..	8,837	..	174
313 Insulated wire and cable	8	0	8	0	233	0	..	30,012	..	357
314 Accumulators, primary cells & primary batteries	..	..	..	..	..	..	..		..	..
315 Electric lamps and lighting equipment	89	0	5	273	32	0	..	13,572	..	448
319 Other electrical equipment, nec	1	0	21	0	239	0	..	20,728	..	336
32 RADIO, TV & COMMUNICATION EQUIP. & APP.	462	0	10	31	204	17	..	26,676	..	820
321 Electronic valves, tubes, other electronic components	0	0	6	14	114	0	..	7,992	..	163
322 TV & radio transmitters, apparatus for line telephony	1	0	4	0	89	0	..	9,063	..	127
323 TV & radio receivers, recording apparatus	..	..	..	..	..	..	..		..	..
33 MEDICAL PRECISION &OPTICAL INSTRUMENTS	107	0	31	0	466	0	..	32,271	..	720
331 Medical appliances and instruments	107	0	29	0	416	0	..	30,080	..	660
3311 Medical, surgical equipment & orthopaedic app.	0	0	15	0	299	0	..	19,353	..	384
3312 Instruments & appliances for measuring, checking etc	107	0	7	0	92	0	..	9,775	..	241
3313 Industrial process control equipment	0	0	7	0	25	0	..	952	..	35
332 Optical instruments and photographic equipment	..	..	..	..	..	..	..		..	..
333 Watches and clocks	..	..	..	..	..	..	..		..	..
34 MOTOR VEHICLES, TRAILERS & SEMI-TRAILERS	294	0	84	0	1,380	0	..	131,414	..	2,231
341 Motor vehicles	226	0	42	0	858	0	..	93,916	..	1,464
342 Bodies (coachwork) for motor vehicles	23	0	16	0	4	0	..	3,181	..	54
343 Parts, accessories for motor vehicles & their engines	45	0	26	0	518	0	..	34,317	..	713
35 OTHER TRANSPORT EQUIPMENT	2,073	0	73	275	1,167	3	..	124,461	..	4,039
351 Building and repairing of ships and boats	..	..	..	..	..	..	..		..	..
3511 Building and repairing of ships	..	..	..	..	..	..	..		..	..
3512 Building, repairing of pleasure & sporting boats	..	..	..	..	..	..	..		..	..
352 Railway, tramway locomotives & rolling stock	121	0	40	12	925	0	..	39,708	..	1,241
353 Aircraft and spacecraft	..	..	..	..	..	..	..		..	..
359 Transport equipment, nec	6	0	13	263	61	0	..	11,020	..	383
3591 Motorcycles	..	..	..	..	..	..	..		..	..
3592 Bicycles and invalid carriages	..	..	..	..	..	..	..		..	..
3599 Other transport equipment, nec	..	..	..	..	..	..	..		..	..
36 FURNITURE; MANUFACTURING, NEC	240	1	79	65	302	179	..	73,602	..	1,131
361 Furniture	221	0	62	20	244	179	..	64,511	..	958
369 Manufacturing, nec	19	1	5	45	56	0	..	9,002	..	158
3691 Jewellery and related articles	0	0	5	45	25	0	..	3,168	..	86
3692 Musical instruments	..	..	..	..	..	..	..		..	..
3693 Sports goods	..	..	..	..	..	..	..		..	..
3694 Games and toys	2	0	0	0	3	0	..	1,532	..	11
3699 Other manufacturing, nec	17	1	0	0	28	0	..	4,083	..	61
37 RECYCLING	7	1	74	1	31	0	..	7,331	..	140
371 Recycling of metal waste and scrap	0	0	42	0	20	0	..	3,590	..	75
372 Recycling of non-metal waste and scrap	7	1	32	1	11	0	..	3,741	..	65
CERR Non-specified industry	..	..	..	..	..	..	..		..	..
15-37 TOTAL MANUFACTURING	163,029	60	32,505	23,218	119,694	1,457	..	11,522,956	..	381,446

SLOVAK REPUBLIC

Year: **1997**

ISIC Revision 3 Industry Sector	Solid TJ	LPG TJ	Distiloil TJ	RFO TJ	Gas TJ	Biomass TJ	Steam TJ	Electr MWh	Own Use MWh	TOTAL TJ
15 FOOD PRODUCTS AND BEVERAGES	1,511	189	1,044	4,766	8,878	0	..	571,206	..	18,444
151 Production, processing and preserving (PPP)	151	9	287	2,220	1,756	0	..	119,188	..	4,852
1511 PPP of meat and meat products	133	9	216	34	865	0	..	99,783	..	1,616
1512 Processing and preserving of fish products	0	0	20	0	11	0	..	3,368	..	43
1513 Processing, preserving of fruit & vegetables	18	0	33	2,186	101	0	..	16,037	..	2,396
1514 Vegetable and animal oils and fats	35	0	108	223	1,270	0	..	64,689	..	1,869
152 Dairy products	38	4	201	53	1,057	0	..	99,511	..	1,711
153 Grain mill prod., starches & prepared animal feeds	1	0	55	1	122	0	..	37,714	..	315
1531 Grain mill products	28	4	3	0	455	0	..	22,394	..	567
1532 Starches and starch products	9	9	142	52	477	0	..	39,308	..	826
1533 Prepared animal feeds	1,169	9	255	2,085	2,380	0	..	164,003	..	6,488
154 Other food products	20	9	162	42	1,353	0	..	43,402	..	1,742
1541 Bakery products	1,077	0	50	2,043	845	0	..	97,287	..	4,365
1542 Sugar	0	0	7	0	65	0	..	13,960	..	122
1543 Cocoa, chocolate and sugar confectionery	0	0	..	0	..	0	..	..	..	..
1544 Macaroni, noodles, couscous & similar farinaceous prod.	..	0	..	0	..	0	..	..	..	..
1549 Other food products, nec	72	0	23	0	80	0	..	8,276	..	205
155 Beverages	117	167	193	185	2,415	0	..	93,445	..	3,413
1551 Distilling, rectifying and blending of spirits	82	161	20	183	499	0	..	10,227	..	982
1552 Wines	5	0	54	0	51	0	..	4,628	..	127
1553 Malt liquors and malt	2	3	72	0	1,729	0	..	57,925	..	2,015
1554 Soft drinks; production of mineral waters	26	0	26	0	68	0	..	6,091	..	142
16 TOBACCO PRODUCTS	..	..	..	..	..	..	..	..	..	..
17 TEXTILES	296	0	44	1,727	2,594	0	..	179,365	..	5,307
171 Spinning, weaving and finishing of textiles	110	0	22	1,721	2,382	0	..	152,793	..	4,785
1711 Preparation and spinning of textile fibres	103	0	11	1,721	2,382	0	..	151,915	..	4,764
1712 Finishing of textiles	..	..	7	0	24	0	..	2,169	..	39
172 Other textiles	0	0	6	0	18	0	..	1,862	..	31
1721 Made-up textile articles, except apparel	0	0	..	..	..	..	..	..	..	..
1722 Carpets and rugs	..	..	..	..	..	..	..	..	..	..
1723 Cordage, rope, twine and netting	..	..	..	..	0	0	..	160	..	2
1729 Other textiles, nec	0	0	1	0	0	0	..	16,720	..	444
173 Knitted and crocheted fabrics and articles	185	0	11	0	188	0	..	54,449	..	1,209
18 WEARING APPAREL, DRESSING & DYEING OF FUR	17	0	52	5	938	1	..	53,658	..	1,190
181 Wearing apparel, except fur apparel	17	0	49	5	925	1	..	53,167	..	1,176
182 Dressing and dyeing of fur; articles of fur	0	0	2	0	10	0	..	491	..	14
19 TANNING & DRESSING OF LEATHER, FOOTWEAR	1,058	1	35	11	250	0	..	46,290	..	1,522
191 Tanning and dressing of leather	375	0	13	0	138	0	..	14,069	..	577
1911 Tanning and dressing of leather	374	0	10	0	112	0	..	12,523	..	541
1912 Luggage, handbags, saddlery & harness	1	0	3	0	26	0	..	1,546	..	36
192 Footwear	683	1	22	11	112	0	..	32,221	..	945
20 WOOD AND WOOD PRODUCTS	456	0	133	125	1,053	510	..	166,410	..	2,876
201 Sawmilling and planing of wood	72	0	59	0	615	152	..	65,445	..	1,134
202 Products of wood, cork, straw & plaiting materials	384	0	71	125	438	343	..	99,779	..	1,720
2021 Veneer sheets	88	0	37	1	359	207	..	63,884	..	922
2022 Builders' carpentry and joinery	293	0	15	0	19	107	..	21,515	..	511
2023 Wooden containers	3	0	5	3	0	27	..	3,038	..	49
2029 Other products of wood	0	0	14	121	60	2	..	11,312	..	238
21 PAPER AND PAPER PRODUCTS	5,555	1	3,398	4,366	3,221	254	..	911,258	..	20,076
2101 Pulp, paper and paperboard	5,528	0	3,359	3,949	2,846	254	..	778,188	..	18,737
2102 Corrugated paper, paperboard and their containers	1	1	11	2	64	0	..	10,135	..	115
2109 Other articles of pulp and paperboard	25	0	27	415	311	0	..	122,875	..	1,220
22 PUBLISHING, PRINTING & REPRODUCTION	0	0	83	8	211	0	..	33,460	..	422
221 Publishing	0	0	15	0	10	0	..	1,118	..	29
2211 Publishing of books & brochures	0	0	2	0	6	0	..	474	..	10
2212 Publishing of newspapers and periodicals	0	0	10	0	4	0	..	392	..	15
2213 Publishing of recorded media	..	..	..	..	..	..	..	..	..	..
2219 Other publishing	..	..	..	..	..	..	..	..	..	..
222 Printing and related service activities	0	0	67	8	199	0	..	31,925	..	389
2221 Printing	0	0	50	8	179	0	..	23,774	..	323
2222 Service activities related to printing	0	0	13	0	17	0	..	6,754	..	54
223 Reproduction of recorded media	..	..	..	..	..	..	..	..	..	..
23 COKE, REFINED PETROLEUM PRODUCTS	0	0	12,961	3,144	20,645	0	..	781,104	..	39,562
231 Coke oven products	..	..	..	..	..	..	..	..	..	..
232 Refined petroleum products	..	..	..	..	..	..	..	..	..	..
233 Processing of nuclear fuel	..	..	..	..	..	..	..	..	..	..
24 CHEMICALS & CHEMICAL PRODUCTS	8,056	0	839	756	22,402	0	..	1,414,105	..	37,144
241 Basic chemicals	5,706	0	719	0	2,980	0	..	674,389	..	11,833
2411 Basic chemicals, exc. fertilizers & nitrogen compounds	26	0	7	0	257	0	..	105,478	..	670
2412 Fertilizers and nitrogen compounds	2	0	9	0	2	0	..	643	..	15
2413 Plastics in primary forms and synthetic rubber	47	0	67	433	2,030	0	..	146,313	..	3,104
242 Other chemical products	..	..	..	..	..	..	..	..	..	..
2421 Pesticides and other agro-chemical products	11	0	13	0	158	0	..	7,185	..	208
2422 Paints, varnishes and similar coatings	28	0	27	433	1,545	0	..	126,686	..	2,489
2423 Pharmaceuticals, medicinal chem. & botanical prod.	8	0	22	0	145	0	..	7,642	..	203
2424 Soap and detergents, perfumes etc.	0	0	4	0	182	0	..	4,800	..	203
2429 Other chemical products, nec	..	..	..	..	..	..	..	..	..	..
243 Man-made fibres	..	..	..	..	..	..	..	260,585	..	4,171
25 RUBBER AND PLASTICS PRODUCTS	83	0	104	1,071	1,975	0	..	152,026	..	3,375
251 Rubber products	69	0	45	1,071	1,643	0	..	103,625	..	2,219
2511 Rubber tyres and tubes	45	0	32	618	1,151	0	..	44,576	..	915
2519 Other rubber products	24	0	9	253	469	0	..	108,559	..	796
252 Plastic products	14	0	59	332	332	0	..	..	..	1,156
26 OTHER NON-METALLIC MINERAL PRODUCTS	7,068	14	446	5,264	11,418	22	..	877,444	..	27,391
261 Glass and glass products	15	13	56	0	3,124	14	..	187,028	..	3,895
269 Non-metallic mineral products, nec	7,053	1	390	5,264	8,294	8	..	690,416	..	23,495
2691 Pottery, china and earthenware	6	0	3	0	226	0	..	8,090	..	264
2692 Refractory ceramic products	11	0	75	914	3,452	0	..	109,324	..	4,846
2693 Structural non-refractory clay & ceramic prod.	38	0	53	0	2,112	7	..	67,381	..	2,453
2694 Cement, lime and plaster	6,368	0	135	4,346	2,076	1	..	450,585	..	14,547
2695 Articles of concrete, cement and plaster	277	1	103	4	195	1	..	32,426	..	698
2696 Cutting, shaping and finishing of stone	6	0	11	0	3	0	..	1,943	..	27
2699 Other non-metallic mineral products, nec	..	..	..	..	..	..	..	..	..	..

ISIS Energy Data Programme (IEA/OECD)

ISIC Revision 3 Industry Sector		Solid TJ	LPG TJ	Distiloil TJ	RFO TJ	Gas TJ	Biomass TJ	Steam TJ	Electr MWh	Own Use MWh	TOTAL TJ
27	BASIC METALS	133,748	3	107	1,117	37,945	633	..	4,610,574	..	190,151
271	Basic iron and steel	131,127	0	44	1,104	35,189	633	..	2,825,553	..	178,269
272	Basic precious and non-ferrous metals	2,411	3	56	5	2,553	0	..	1,731,212	..	11,260
273	Casting of metals	210	0	5	7	197	0	..	40,532	..	565
2731	Casting of iron and steel	103	0	2	7	21	0	..	8,416	..	163
2732	Casting non-ferrous metals	..	..	..	..	..	..	..	..	..	..
28	FABRICATED METAL PRODUCTS	302	15	258	75	687	3	..	125,504	..	1,792
281	Str. metal prod., tanks, reservoirs, steam generators	219	1	106	53	395	0	..	54,964	..	972
2811	Structural metal products	200	1	91	13	314	0	..	46,513	..	786
2812	Tanks, reservoirs and containers of metal	18	0	12	40	81	0	..	8,451	..	181
2813	Steam generators, exc. central heating hot water boilers	0	0	3	0	0	0	..	0	..	3
289	Other fabricated metal products	76	14	144	3	290	3	..	65,406	..	765
2891	Forging, pressing, stamping & roll-forming of metal	..	..	..	..	..	..	..	..	..	..
2892	Treatment and coating of metals	14	0	11	0	22	0	..	5,510	..	67
2893	Cutlery, hand tools and general hardware	39	7	17	0	46	0	..	23,565	..	194
2899	Other fabricated metal products, nec	23	6	116	3	211	3	..	33,085	..	481
29	MACHINERY AND EQUIPMENT, NEC	2,299	2	330	263	5,082	0	..	1,059,441	..	11,790
291	General purpose machinery	655	0	107	5	2,622	0	..	711,278	..	5,950
2911	Engines and turbines	4	0	8	0	74	0	..	8,505	..	117
2912	Pumps, compressors, taps and valves	359	0	19	0	96	0	..	30,307	..	583
2913	Bearings, gears, gearing and driving elements	144	0	25	0	798	0	..	141,304	..	1,476
2914	Ovens, furnaces and furnace burners	..	..	..	..	..	..	..	..	..	..
2915	Lifting and handling equipment	100	..	11	2	12	0	..	1,467	..	130
2919	Other general purpose machinery	23	0	17	3	1,604	0	..	523,634	..	3,532
292	Special purpose machinery	1,557	0	198	235	2,178	0	..	288,932	..	5,208
2921	Agricultural and forestry machinery	247	0	51	0	154	0	..	44,381	..	612
2922	Machine-tools	958	0	19	224	288	0	..	58,993	..	1,701
2923	Machinery for metallurgy	..	..	..	..	..	..	..	..	..	..
2924	Machinery for mining, quarrying and construction	15	..	26	0	314	0	..	25,278	..	446
2925	Machinery for food, beverage & tobacco processing	0	0	5	0	43	0	..	3,682	..	61
2926	Machinery for textile, apparel & leather production	31	0	4	0	138	0	..	9,597	..	208
2927	Machinery for weapons and ammunition	..	..	..	..	..	..	..	..	..	..
2929	Other special purpose machinery	302	0	85	11	1,216	0	..	141,693	..	2,124
293	Domestic appliances, nec	63	1	15	23	265	0	..	52,329	..	555
30	OFFICE, ACCOUNTING & COMPUTING MACHINERY	0	0	22	0	19	0	..	1,921	..	48
31	ELECTRICAL MACHINERY & APPARATUS, NEC	44	0	74	193	572	0	..	102,641	..	1,253
311	Electric motors, generators and transformers	7	0	13	93	156	0	..	18,676	..	336
312	Electricity distribution and control apparatus	0	0	12	0	98	0	..	9,812	..	145
313	Insulated wire and cable	3	0	11	0	22	0	..	22,077	..	115
314	Accumulators, primary cells & primary batteries	..	..	..	..	..	..	..	..	..	..
315	Electric lamps and lighting equipment	33	0	5	99	50	0	..	10,617	..	225
319	Other electrical equipment, nec	1	0	32	1	229	0	..	39,486	..	405
32	RADIO, TV & COMMUNICATION EQUIP. & APP.	461	0	20	18	175	0	..	33,913	..	796
321	Electronic valves, tubes, other electronic components	0	0	11	9	119	0	..	14,513	..	191
322	TV & radio transmitters, apparatus for line telephony	1	0	6	0	56	0	..	9,703	..	98
323	TV & radio receivers, recording apparatus	439	0	3	9	0	0	..	8,874	..	483
33	MEDICAL PRECISION & OPTICAL INSTRUMENTS	0	0	38	0	434	0	..	31,349	..	585
331	Medical appliances and instruments	0	0	36	0	389	0	..	29,496	..	531
3311	Medical, surgical equipment & orthopaedic app.	0	0	15	0	279	0	..	16,437	..	353
3312	Instruments & appliances for measuring, checking etc	0	0	13	0	96	0	..	10,126	..	145
3313	Industrial process control equipment	0	0	8	0	14	0	..	2,633	..	31
332	Optical instruments and photographic equipment	..	..	..	..	..	..	..	..	..	..
333	Watches and clocks	..	..	..	..	..	..	..	..	..	..
34	MOTOR VEHICLES, TRAILERS & SEMI-TRAILERS	178	0	93	1	1,410	0	..	140,868	..	2,189
341	Motor vehicles	75	0	36	0	197	0	..	105,214	..	687
342	Bodies (coachwork) for motor vehicles	37	0	20	1	5	0	..	4,162	..	78
343	Parts, accessories for motor vehicles & their engines	66	0	37	0	408	0	..	31,492	..	624
35	OTHER TRANSPORT EQUIPMENT	1,722	0	64	168	1,236	0	..	111,516	..	3,591
351	Building and repairing of ships and boats	46	0	26	0	102	0	..	15,226	..	229
3511	Building and repairing of ships	46	0	26	0	102	0	..	15,226	..	229
3512	Building, repairing of pleasure & sporting boats	..	..	..	..	..	..	..	..	..	..
352	Railway, tramway locomotives & rolling stock	70	0	29	14	865	0	..	40,382	..	1,123
353	Aircraft and spacecraft	..	..	..	..	..	..	..	..	..	..
359	Transport equipment, nec	4	0	7	154	114	0	..	8,214	..	309
3591	Motorcycles	..	..	..	..	..	..	..	..	..	..
3592	Bicycles and invalid carriages	..	..	..	..	..	..	..	..	..	..
3599	Other transport equipment, nec	..	..	..	..	..	..	..	..	..	..
36	FURNITURE; MANUFACTURING, NEC	181	0	214	41	1,119	464	..	91,896	..	2,350
361	Furniture	161	0	196	11	1,069	411	..	81,413	..	2,141
369	Manufacturing, nec	13	0	15	30	49	53	..	10,483	..	198
3691	Jewellery and related articles	0	0	3	30	22	0	..	2,592	..	64
3692	Musical instruments	..	..	..	..	..	..	..	..	..	..
3693	Sports goods	..	..	..	..	..	..	..	..	..	..
3694	Games and toys	0	0	2	0	7	0	..	1,039	..	13
3699	Other manufacturing, nec	13	0	10	0	20	0	..	3,895	..	57
37	RECYCLING	6	0	127	1	15	0	..	4,941	..	167
371	Recycling of metal waste and scrap	1	0	99	0	5	0	..	1,027	..	109
372	Recycling of non-metal waste and scrap	5	0	27	1	6	0	..	2,752	..	49
CERR	Non-specified industry	..	..	..	..	..	..	..	..	..	..
15-37	TOTAL MANUFACTURING	163,041	225	20,486	23,120	122,279	1,887	..	11,610,240	..	372,835

Year: 1994 SWEDEN

ISIC Revision 3 Industry Sector	Solid TJ	LPG TJ	Distiloil TJ	RFO TJ	Gas TJ	Biomass TJ	Steam TJ	Electr MWh	Own Use MWh	TOTAL TJ
15 FOOD PRODUCTS AND BEVERAGES	909	1,907	3,090	3,670	3,929	76	1,054	2,485,364	..	23,582
151 Production, processing and preserving (PPP)	0	1,098	986	547	775	62	579	794,963	..	6,909
1511 PPP of meat and meat products	0	198	715	213	247	62	283	489,340	..	3,480
1512 Processing and preserving of fish products	0	92	113	33	0	0	7	45,513	..	409
1513 Processing, preserving of fruit & vegetables	0	77	150	209	402	0	2	145,806	..	1,365
1514 Vegetable and animal oils and fats	0	731	8	92	126	0	287	114,304	..	1,655
152 Dairy products	4	98	465	1,285	262	9	80	454,014	..	3,837
153 Grain mill prod., starches & prepared animal feeds	0	158	144	153	241	3	127	220,550	..	1,620
1531 Grain mill products	0	4	85	3	2	2	10	89,072	..	427
1532 Starches and starch products	0	78	3	84	59	0	0	22,491	..	305
1533 Prepared animal feeds	0	76	56	66	180	1	117	108,987	..	888
154 Other food products	905	248	1,313	1,196	2,069	2	263	784,958	..	8,822
1541 Bakery products	1	116	842	22	261	1	86	361,925	..	2,632
1542 Sugar	904	0	110	852	1,320	0	0	119,935	..	3,618
1543 Cocoa, chocolate and sugar confectionery	0	20	66	224	56	0	113	141,699	..	989
1544 Macaroni, noodles, couscous & similar farinaceous prod.	0	0	70	0	0	0	0	16,757	..	130
1549 Other food products, nec	0	112	225	98	432	1	64	144,642	..	1,453
155 Beverages	0	305	182	489	582	0	5	230,879	..	2,394
1551 Distilling, rectifying and blending of spirits	0	2	33	202	41	0	4	42,082	..	433
1552 Wines	0	295	0	98	0	0	0	34,835	..	518
1553 Malt liquors and malt	0	0	102	102	106	0	0	73,463	..	574
1554 Soft drinks; production of mineral waters	0	8	47	87	435	0	1	80,499	..	868
16 TOBACCO PRODUCTS	0	0	0	35	29	0	31	24,122	..	182
17 TEXTILES	0	373	618	375	115	1	73	248,226	..	2,449
171 Spinning, weaving and finishing of textiles	0	142	413	309	68	0	10	72,656	..	1,204
1711 Preparation and spinning of textile fibres	0	115	302	160	68	0	7	52,234	..	840
1712 Finishing of textiles	0	27	111	149	0	0	3	20,422	..	364
172 Other textiles	0	229	164	54	47	1	62	155,213	..	1,116
1721 Made-up textile articles, except apparel	0	4	93	24	0	1	3	21,136	..	201
1722 Carpets and rugs	0	0	2	0	0	0	1	813	..	6
1723 Cordage, rope, twine and netting	0	0	1	0	0	0	2	2,665	..	13
1729 Other textiles, nec	0	225	68	30	47	0	56	130,599	..	896
173 Knitted and crocheted fabrics and articles	0	2	41	12	0	0	1	20,357	..	129
18 WEARING APPAREL, DRESSING & DYEING OF FUR	0	3	54	11	0	1	7	18,770	..	144
181 Wearing apparel, except fur apparel	0	3	54	11	0	1	7	18,770	..	144
182 Dressing and dyeing of fur; articles of fur	..	..	..	..	..	..	..	..	..	..
19 TANNING & DRESSING OF LEATHER, FOOTWEAR	0	1	40	6	0	7	87	22,359	..	221
191 Tanning and dressing of leather	0	0	34	1	0	6	79	15,702	..	177
1911 Tanning and dressing of leather	0	0	26	0	0	6	76	12,435	..	153
1912 Luggage, handbags, saddlery & harness	0	0	8	1	0	0	3	3,267	..	24
192 Footwear	0	1	6	5	0	1	8	6,657	..	45
20 WOOD AND WOOD PRODUCTS	29	122	1,691	813	0	29,942	866	1,907,738	..	40,331
201 Sawmilling and planing of wood	0	89	1,321	422	0	0	772	1,276,909	..	7,201
202 Products of wood, cork, straw & plaiting materials	29	33	370	391	0	0	94	630,829	..	3,188
2021 Veneer sheets	9	14	80	304	0	0	37	280,225	..	1,453
2022 Builders' carpentry and joinery	20	15	210	85	0	0	52	299,970	..	1,462
2023 Wooden containers	0	2	56	1	0	0	2	29,960	..	169
2029 Other products of wood	0	2	24	1	0	0	3	20,674	..	104
21 PAPER AND PAPER PRODUCTS	2,612	1,026	1,943	29,209	2,697	128,179	2,568	19,087,072	..	236,947
2101 Pulp, paper and paperboard	2,609	950	1,705	28,793	2,525	0	2,516	18,809,879	..	106,814
2102 Corrugated paper, paperboard and their containers	1	24	99	317	171	0	28	151,625	..	1,186
2109 Other articles of pulp and paperboard	2	52	139	99	1	0	24	125,568	..	769
22 PUBLISHING, PRINTING & REPRODUCTION	0	216	375	114	29	0	315	482,208	..	2,785
221 Publishing	0	10	152	5	0	0	156	145,240	..	846
2211 Publishing of books & brochures	0	2	25	2	0	0	7	14,945	..	90
2212 Publishing of newspapers and periodicals	0	5	113	3	0	0	142	110,410	..	660
2213 Publishing of recorded media	0	3	12	0	0	0	6	17,355	..	83
2219 Other publishing	0	0	2	0	0	0	1	2,530	..	12
222 Printing and related service activities	0	206	223	109	29	0	159	335,495	..	1,934
2221 Printing	0	205	180	108	28	0	149	294,954	..	1,732
2222 Service activities related to printing	0	1	43	1	1	0	10	40,541	..	202
223 Reproduction of recorded media	0	0	0	0	0	0	0	1,473	..	5
23 COKE, REFINED PETROLEUM PRODUCTS	0	9	44	367	0	0	68	618,083	..	2,713
231 Coke oven products	..	..	..	..	..	..	..	..	..	..
232 Refined petroleum products	0	9	38	367	0	0	28	599,041	..	2,599
233 Processing of nuclear fuel	0	0	6	0	0	0	40	19,042	..	115
24 CHEMICALS & CHEMICAL PRODUCTS	179	437	1,054	2,456	1,996	1	2,602	4,725,753	..	25,738
241 Basic chemicals	179	422	514	1,282	1,668	0	1,101	4,072,701	..	19,828
2411 Basic chemicals, exc. fertilizers & nitrogen compounds	171	420	378	828	1,410	0	781	2,872,082	..	14,327
2412 Fertilizers and nitrogen compounds	0	0	35	59	16	0	2	137,936	..	609
2413 Plastics in primary forms and synthetic rubber	8	2	101	395	242	0	318	1,062,683	..	4,892
242 Other chemical products	0	13	540	377	328	1	1,501	604,130	..	4,935
2421 Pesticides and other agro-chemical products	0	0	18	0	0	0	0	1,602	..	24
2422 Paints, varnishes and similar coatings	0	2	100	18	47	1	61	76,502	..	504
2423 Pharmaceuticals, medicinal chem. & botanical prod.	0	0	183	150	3	0	1,166	313,496	..	2,631
2424 Soap and detergents, perfumes etc.	0	0	71	77	10	0	10	46,137	..	334
2429 Other chemical products, nec	0	11	168	132	268	0	264	166,393	..	1,442
243 Man-made fibres	0	2	0	797	0	0	0	48,922	..	975
25 RUBBER AND PLASTICS PRODUCTS	127	58	482	487	62	636	265	943,714	..	5,514
251 Rubber products	0	1	125	111	20	0	125	223,475	..	1,187
2511 Rubber tyres and tubes	0	0	48	72	0	0	1	61,993	..	344
2519 Other rubber products	0	1	77	39	20	0	124	161,482	..	842
252 Plastic products	127	57	357	376	42	636	140	720,239	..	4,328
26 OTHER NON-METALLIC MINERAL PRODUCTS	7,719	3,315	1,772	2,057	419	3	65	1,209,809	..	19,705
261 Glass and glass products	112	2,026	346	694	2	1	1	441,733	..	4,772
269 Non-metallic mineral products, nec	7,607	1,289	1,426	1,363	417	2	64	768,076	..	14,933
2691 Pottery, china and earthenware	4	244	13	74	0	0	21	54,910	..	554
2692 Refractory ceramic products	1	1	9	0	62	0	8	15,349	..	136
2693 Structural non-refractory clay & ceramic prod.	0	193	94	48	62	0	0	20,023	..	469
2694 Cement, lime and plaster	6,153	369	41	609	0	0	0	312,531	..	8,297
2695 Articles of concrete, cement and plaster	277	355	621	435	73	2	14	189,880	..	2,461
2696 Cutting, shaping and finishing of stone	35	19	84	5	3	0	0	23,557	..	231
2699 Other non-metallic mineral products, nec	1,137	108	564	192	217	0	21	151,826	..	2,786

ISIS Energy Data Programme (IEA/OECD)

ISIC Revision 3 Industry Sector	Solid TJ	LPG TJ	Distiloil TJ	RFO TJ	Gas TJ	Biomass TJ	Steam TJ	Electr MWh	Own Use MWh	TOTAL TJ
27 **BASIC METALS**	40,870	7,449	2,271	4,800	4,152	10	724	7,306,688	..	86,580
271 Basic iron and steel	38,773	6,903	1,891	4,501	4,053	0	593	4,968,633	..	74,601
272 Basic precious and non-ferrous metals	2,095	545	371	297	99	10	129	2,329,211	..	11,931
273 Casting of metals	2	1	9	2	0	0	2	8,844	..	48
2731 Casting of iron and steel	1	0	2	0	0	0	0	1,234	..	7
2732 Casting non-ferrous metals	1	1	7	2	0	0	2	7,610	..	40
28 **FABRICATED METAL PRODUCTS**	148	312	1,537	293	175	15	409	1,446,161	..	8,095
281 Str. metal prod., tanks, reservoirs, steam generators	15	43	544	55	30	10	80	307,281	..	1,883
2811 Structural metal products	14	37	450	53	29	9	63	258,144	..	1,584
2812 Tanks, reservoirs and containers of metal	1	6	81	2	0	1	17	44,270	..	267
2813 Steam generators, exc. central heating hot water boilers	0	0	13	0	1	0	0	4,867	..	32
289 Other fabricated metal products	133	269	993	238	145	5	329	1,138,880	..	6,212
2891 Forging, pressing, stamping & roll-forming of metal	0	1	9	0	0	0	1	5,255	..	30
2892 Treatment and coating of metals	6	11	228	19	20	0	31	209,467	..	1,069
2893 Cutlery, hand tools and general hardware	2	7	160	104	7	0	179	246,067	..	1,345
2899 Other fabricated metal products, nec	125	250	596	115	118	5	118	678,091	..	3,768
29 **MACHINERY AND EQUIPMENT, NEC**	13	280	1,920	561	189	22	1,372	1,792,853	..	10,811
291 General purpose machinery	2	137	1,122	268	78	11	689	959,014	..	5,759
2911 Engines and turbines	0	0	132	76	0	1	22	54,549	..	427
2912 Pumps, compressors, taps and valves	2	29	178	54	0	0	57	191,564	..	1,010
2913 Bearings, gears, gearing and driving elements	0	23	19	69	7	0	315	193,019	..	1,128
2914 Ovens, furnaces and furnace burners	0	0	13	0	1	0	0	4,717	..	31
2915 Lifting and handling equipment	0	22	325	35	7	8	137	170,620	..	1,148
2919 Other general purpose machinery	0	63	455	34	63	2	158	344,545	..	2,015
292 Special purpose machinery	11	97	731	244	111	11	658	638,271	..	4,161
2921 Agricultural and forestry machinery	0	1	85	22	1	0	14	43,036	..	278
2922 Machine-tools	6	53	192	23	1	1	64	117,189	..	762
2923 Machinery for metallurgy	0	2	4	21	0	0	36	50,065	..	243
2924 Machinery for mining, quarrying and construction	5	8	174	102	103	10	60	169,408	..	1,072
2925 Machinery for food, beverage & tobacco processing	0	0	43	11	5	0	3	15,173	..	117
2926 Machinery for textile, apparel & leather production	0	18	10	0	0	0	7	17,759	..	99
2927 Machinery for weapons and ammunition	0	0	71	33	0	0	289	65,535	..	629
2929 Other special purpose machinery	0	15	152	32	1	0	185	160,106	..	961
293 Domestic appliances, nec	0	46	67	49	0	0	25	195,568	..	891
30 **OFFICE, ACCOUNTING & COMPUTING MACHINERY**	0	4	60	10	0	22	15	73,798	..	377
31 **ELECTRICAL MACHINERY & APPARATUS, NEC**	90	78	338	113	3	0	238	444,071	..	2,459
311 Electric motors, generators and transformers	0	7	25	0	0	0	17	62,493	..	274
312 Electricity distribution and control apparatus	0	28	85	25	1	0	74	81,930	..	508
313 Insulated wire and cable	0	11	69	3	0	0	38	144,813	..	642
314 Accumulators, primary cells & primary batteries	0	29	16	76	0	0	69	71,663	..	448
315 Electric lamps and lighting equipment	3	2	90	7	1	0	23	43,354	..	282
319 Other electrical equipment, nec	87	1	53	2	1	0	17	39,818	..	304
32 **RADIO, TV & COMMUNICATION EQUIP. & APP.**	0	2	129	272	0	0	345	301,516	..	1,833
321 Electronic valves, tubes, other electronic components	0	1	34	6	0	0	53	87,808	..	410
322 TV & radio transmitters, apparatus for line telephony	0	1	86	266	0	0	291	207,202	..	1,390
323 TV & radio receivers, recording apparatus	0	0	9	0	0	0	1	6,506	..	33
33 **MEDICAL PRECISION &OPTICAL INSTRUMENTS**	0	1	407	25	5	0	155	182,078	..	1,248
331 Medical appliances and instruments	0	1	392	25	5	0	136	170,650	..	1,173
3311 Medical, surgical equipment & orthopaedic app.	0	1	75	1	5	0	46	62,623	..	353
3312 Instruments & appliances for measuring, checking etc	0	0	316	24	0	0	90	89,144	..	751
3313 Industrial process control equipment	0	0	1	0	0	0	0	18,883	..	69
332 Optical instruments and photographic equipment	0	0	11	0	0	0	19	10,283	..	67
333 Watches and clocks	0	0	4	0	0	0	1	1,145	..	7
34 **MOTOR VEHICLES, TRAILERS & SEMI-TRAILERS**	406	595	1,603	1,234	404	0	1,021	1,699,816	..	11,382
341 Motor vehicles	406	234	1,097	1,000	371	0	775	1,059,495	..	7,697
342 Bodies (coachwork) for motor vehicles	0	214	187	69	1	0	35	179,095	..	1,151
343 Parts, accessories for motor vehicles & their engines	0	147	319	165	32	0	211	461,226	..	2,534
35 **OTHER TRANSPORT EQUIPMENT**	0	80	693	66	94	0	389	412,876	..	2,808
351 Building and repairing of ships and boats	0	2	96	42	77	0	14	84,367	..	535
3511 Building and repairing of ships	0	2	39	41	77	0	5	70,283	..	417
3512 Building, repairing of pleasure & sporting boats	0	0	57	1	0	0	9	14,084	..	118
352 Railway, tramway locomotives & rolling stock	0	76	177	0	4	0	103	83,964	..	662
353 Aircraft and spacecraft	0	0	398	24	0	0	266	228,402	..	1,510
359 Transport equipment, nec	0	2	22	0	13	0	6	16,143	..	101
3591 Motorcycles	0	0	3	0	0	0	0	965	..	6
3592 Bicycles and invalid carriages	0	0	13	0	13	0	3	8,610	..	60
3599 Other transport equipment, nec	0	2	6	0	0	0	3	6,568	..	35
36 **FURNITURE; MANUFACTURING, NEC**	3	47	470	88	2	156	145	373,985	..	2,257
361 Furniture	3	47	383	67	2	149	73	322,910	..	1,886
369 Manufacturing, nec	0	0	87	21	0	7	72	51,075	..	371
3691 Jewellery and related articles	0	0	6	0	0	0	9	5,848	..	36
3692 Musical instruments	0	0	0	0	0	0	0	204	..	1
3693 Sports goods	0	0	22	1	0	0	2	7,609	..	52
3694 Games and toys	0	0	10	0	0	0	0	7,201	..	36
3699 Other manufacturing, nec	0	0	49	20	0	7	61	30,213	..	246
37 **RECYCLING**	301	0	70	0	0	0	0	127,206	..	829
371 Recycling of metal waste and scrap	301	0	63	0	0	0	0	123,779	..	810
372 Recycling of non-metal waste and scrap	0	0	7	0	0	0	0	3,427	..	19
CERR **Non-specified industry**	..	..	..	..	..	..	..	..		
15-37 **TOTAL MANUFACTURING**	53,406	16,315	20,661	47,062	14,300	159,071	12,814	45,934,266		488,992

ISIC Revision 3 Industry Sector	Solid TJ	LPG TJ	Distiloil TJ	RFO TJ	Gas TJ	Biomass TJ	Steam TJ	Electr MWh	Own Use MWh	TOTAL TJ
15 FOOD PRODUCTS AND BEVERAGES	1,004	1,761	3,084	4,425	3,937	92	1,002	2,542,581	..	24,458
151 Production, processing and preserving (PPP)	8	1,017	1,086	680	823	85	511	877,760	..	7,370
1511 PPP of meat and meat products	2	146	727	260	253	71	214	499,313	..	3,471
1512 Processing and preserving of fish products	6	139	87	43	0	14	15	67,644	..	548
1513 Processing, preserving of fruit & vegetables	0	86	266	179	433	0	3	184,385	..	1,631
1514 Vegetable and animal oils and fats	0	646	6	198	137	0	279	126,418	..	1,721
152 Dairy products	0	93	417	1,364	279	0	90	448,684	..	3,858
153 Grain mill prod., starches & prepared animal feeds	0	143	138	247	262	4	96	234,642	..	1,735
1531 Grain mill products	0	3	71	5	5	3	18	87,432	..	420
1532 Starches and starch products	0	79	6	136	66	0	0	35,622	..	415
1533 Prepared animal feeds	0	61	61	106	191	1	78	111,588	..	900
154 Other food products	996	217	1,222	1,274	2,207	2	288	758,817	..	8,938
1541 Bakery products	0	110	862	16	292	1	97	355,728	..	2,659
1542 Sugar	996	0	47	912	1,404	0	0	119,135	..	3,788
1543 Cocoa, chocolate and sugar confectionery	0	7	56	250	73	0	122	142,558	..	1,021
1544 Macaroni, noodles, couscous & similar farinaceous prod.	0	0	74	1	1	1	12	15,616	..	145
1549 Other food products, nec	0	100	183	95	437	0	57	125,780	..	1,325
155 Beverages	0	291	221	860	366	1	17	222,678	..	2,558
1551 Distilling, rectifying and blending of spirits	0	1	28	180	46	0	3	37,749	..	394
1552 Wines	0	280	2	119	0	0	2	32,213	..	519
1553 Malt liquors and malt	0	4	124	255	137	1	8	77,051	..	806
1554 Soft drinks; production of mineral waters	0	6	67	306	183	0	4	75,665	..	838
16 TOBACCO PRODUCTS	0	0	0	27	49	0	20	26,425	..	191
17 TEXTILES	0	360	235	671	69	1	73	251,624	..	2,315
171 Spinning, weaving and finishing of textiles	0	137	112	598	28	0	15	74,786	..	1,159
1711 Preparation and spinning of textile fibres	0	128	46	449	28	0	11	56,421	..	865
1712 Finishing of textiles	0	9	66	149	0	0	4	18,365	..	294
172 Other textiles	0	222	88	62	41	1	57	156,767	..	1,035
1721 Made-up textile articles, except apparel	0	3	37	22	0	0	3	18,584	..	132
1722 Carpets and rugs	0	0	4	2	0	0	1	983	..	11
1723 Cordage, rope, twine and netting	0	0	2	0	0	0	2	2,560	..	13
1729 Other textiles, nec	0	219	45	38	41	0	51	134,640	..	879
173 Knitted and crocheted fabrics and articles	0	1	35	11	0	0	1	20,071	..	120
18 WEARING APPAREL, DRESSING & DYEING OF FUR	0	1	42	13	0	0	12	17,752	..	132
181 Wearing apparel, except fur apparel	0	1	42	13	0	0	12	17,752	..	132
182 Dressing and dyeing of fur; articles of fur	..	..	..	..	..	..	..	..	..	..
19 TANNING & DRESSING OF LEATHER, FOOTWEAR	0	1	34	6	0	8	86	22,110	..	215
191 Tanning and dressing of leather	0	0	30	1	0	7	80	16,290	..	177
1911 Tanning and dressing of leather	0	0	23	0	0	7	77	13,121	..	154
1912 Luggage, handbags, saddlery & harness	0	0	7	1	0	0	3	3,169	..	22
192 Footwear	0	0	4	5	0	1	6	5,820	..	37
20 WOOD AND WOOD PRODUCTS	8	173	1,693	1,259	2	30,319	1,188	2,051,183	..	42,026
201 Sawmilling and planing of wood	3	144	1,314	393	1	0	948	1,356,952	..	7,688
202 Products of wood, cork, straw & plaiting materials	5	29	379	866	1	0	240	694,231	..	4,019
2021 Veneer sheets	0	6	114	780	0	0	144	320,410	..	2,197
2022 Builders' carpentry and joinery	5	18	189	84	1	0	88	307,069	..	1,490
2023 Wooden containers	0	3	50	1	0	0	4	29,585	..	165
2029 Other products of wood	0	2	26	1	0	0	4	37,167	..	167
21 PAPER AND PAPER PRODUCTS	1,853	953	2,190	30,610	2,489	146,758	3,491	19,134,364	..	257,228
2101 Pulp, paper and paperboard	1,850	857	1,940	30,165	2,308	0	3,413	18,804,160	..	108,228
2102 Corrugated paper, paperboard and their containers	0	23	90	351	178	0	49	167,849	..	1,295
2109 Other articles of pulp and paperboard	3	73	160	94	3	0	29	162,355	..	946
22 PUBLISHING, PRINTING & REPRODUCTION	0	200	364	98	73	0	373	515,523	..	2,964
221 Publishing	0	12	139	3	0	0	151	157,593	..	872
2211 Publishing of books & brochures	0	2	23	0	0	0	5	15,834	..	87
2212 Publishing of newspapers and periodicals	0	8	105	3	0	0	140	121,939	..	695
2213 Publishing of recorded media	0	2	7	0	0	0	5	15,208	..	69
2219 Other publishing	0	0	4	0	0	0	1	4,612	..	22
222 Printing and related service activities	0	188	225	95	73	0	222	357,001	..	2,088
2221 Printing	0	187	185	94	72	0	211	313,578	..	1,878
2222 Service activities related to printing	0	1	40	1	1	0	11	43,423	..	210
223 Reproduction of recorded media	0	0	0	0	0	0	0	929	..	3
23 COKE, REFINED PETROLEUM PRODUCTS	0	6	152	341	0	0	73	673,998	..	2,998
231 Coke oven products	..	..	..	..	..	..	..	..	..	..
232 Refined petroleum products	0	5	141	341	0	0	32	650,549	..	2,861
233 Processing of nuclear fuel	0	1	11	0	0	0	41	23,449	..	137
24 CHEMICALS & CHEMICAL PRODUCTS	175	44	1,444	2,152	2,059	7	2,237	4,583,495	..	24,619
241 Basic chemicals	173	28	415	1,190	1,676	2	1,187	3,938,389	..	18,849
2411 Basic chemicals, exc. fertilizers & nitrogen compounds	171	25	268	808	1,429	0	826	2,413,906	..	12,217
2412 Fertilizers and nitrogen compounds	0	0	44	22	0	0	4	139,405	..	572
2413 Plastics in primary forms and synthetic rubber	2	3	103	360	247	2	357	1,385,078	..	6,060
242 Other chemical products	2	14	1,029	263	383	5	1,050	605,680	..	4,926
2421 Pesticides and other agro-chemical products	0	0	18	1	0	1	0	1,983	..	27
2422 Paints, varnishes and similar coatings	2	2	112	55	42	4	71	80,860	..	579
2423 Pharmaceuticals, medicinal chem. & botanical prod.	0	0	672	7	4	0	730	340,066	..	2,637
2424 Soap and detergents, perfumes etc.	0	0	53	77	42	0	12	50,380	..	365
2429 Other chemical products, nec	0	12	174	123	295	0	237	132,391	..	1,318
243 Man-made fibres	0	2	0	699	0	0	0	39,426	..	843
25 RUBBER AND PLASTICS PRODUCTS	115	52	518	495	69	1,090	246	962,027	..	6,048
251 Rubber products	0	0	121	110	21	0	121	231,221	..	1,205
2511 Rubber tyres and tubes	0	0	48	66	0	0	2	67,168	..	358
2519 Other rubber products	0	0	73	44	21	0	119	164,053	..	848
252 Plastic products	115	52	397	385	48	1,090	125	730,806	..	4,843
26 OTHER NON-METALLIC MINERAL PRODUCTS	9,231	3,673	1,811	1,964	452	8	74	1,287,195	..	21,847
261 Glass and glass products	17	2,233	376	674	112	2	6	398,885	..	4,856
269 Non-metallic mineral products, nec	9,214	1,440	1,435	1,290	340	6	68	888,310	..	16,991
2691 Pottery, china and earthenware	2	242	12	85	0	0	20	55,993	..	563
2692 Refractory ceramic products	0	0	7	0	13	0	7	12,942	..	74
2693 Structural non-refractory clay & ceramic prod.	0	198	106	35	31	0	0	18,919	..	438
2694 Cement, lime and plaster	7,414	439	84	528	0	0	0	357,019	..	9,750
2695 Articles of concrete, cement and plaster	313	313	696	483	94	6	18	206,011	..	2,665
2696 Cutting, shaping and finishing of stone	169	36	74	15	2	0	0	28,010	..	397
2699 Other non-metallic mineral products, nec	1,316	212	456	144	200	0	23	209,416	..	3,105

ISIC Revision 3 Industry Sector		Solid TJ	LPG TJ	Distiloil TJ	RFO TJ	Gas TJ	Biomass TJ	Steam TJ	Electr MWh	Own Use MWh	TOTAL TJ
27	**BASIC METALS**	39,433	12,203	2,202	5,078	4,081	2	847	7,904,927	..	92,304
271	Basic iron and steel	37,515	11,585	1,828	4,796	3,969	0	719	5,323,894	..	79,578
272	Basic precious and non-ferrous metals	1,915	617	365	280	112	2	126	2,569,850	..	12,668
273	Casting of metals	3	1	9	2	0	0	2	11,183	..	57
2731	Casting of iron and steel	2	0	5	0	0	0	1	6,104	..	30
2732	Casting non-ferrous metals	1	1	4	2	0	0	1	5,079	..	27
28	**FABRICATED METAL PRODUCTS**	111	388	1,617	328	207	19	424	1,661,236	..	9,074
281	Str. metal prod., tanks, reservoirs, steam generators	17	57	529	54	29	14	76	313,463	..	1,904
2811	Structural metal products	13	48	432	51	29	12	58	259,316	..	1,577
2812	Tanks, reservoirs and containers of metal	4	9	78	3	0	2	18	51,483	..	299
2813	Steam generators, exc. central heating hot water boilers	0	0	19	0	0	0	0	2,664	..	29
289	Other fabricated metal products	94	331	1,088	274	178	5	348	1,347,773	..	7,170
2891	Forging, pressing, stamping & roll-forming of metal	0	1	13	0	0	0	2	8,357	..	46
2892	Treatment and coating of metals	8	20	274	6	28	2	46	307,568	..	1,491
2893	Cutlery, hand tools and general hardware	2	8	172	141	8	0	180	259,427	..	1,445
2899	Other fabricated metal products, nec	84	302	629	127	142	3	120	772,421	..	4,188
29	**MACHINERY AND EQUIPMENT, NEC**	7	272	1,998	492	219	11	1,438	1,957,148	..	11,483
291	General purpose machinery	0	153	1,214	171	98	8	1,027	1,076,334	..	6,546
2911	Engines and turbines	0	0	179	14	0	1	21	55,777	..	416
2912	Pumps, compressors, taps and valves	0	13	219	19	0	0	59	202,499	..	1,039
2913	Bearings, gears, gearing and driving elements	0	28	21	75	27	0	332	229,794	..	1,310
2914	Ovens, furnaces and furnace burners	0	0	11	0	1	0	1	4,451	..	29
2915	Lifting and handling equipment	0	25	354	22	7	6	155	194,520	..	1,269
2919	Other general purpose machinery	0	87	430	41	63	1	459	389,293	..	2,482
292	Special purpose machinery	7	76	724	239	121	3	387	692,495	..	4,050
2921	Agricultural and forestry machinery	1	10	104	29	1	2	35	57,613	..	389
2922	Machine-tools	6	19	178	23	1	0	63	124,951	..	740
2923	Machinery for metallurgy	0	1	28	14	0	0	26	75,658	..	341
2924	Machinery for mining, quarrying and construction	0	16	126	106	113	1	80	168,627	..	1,049
2925	Machinery for food, beverage & tobacco processing	0	0	41	13	5	0	4	15,021	..	117
2926	Machinery for textile, apparel & leather production	0	3	6	0	0	0	7	19,721	..	87
2927	Machinery for weapons and ammunition	0	5	75	26	0	0	32	61,252	..	359
2929	Other special purpose machinery	0	22	166	28	1	0	140	169,652	..	968
293	Domestic appliances, nec	0	43	60	82	0	0	24	188,319	..	887
30	**OFFICE, ACCOUNTING & COMPUTING MACHINERY**	0	1	41	1	0	27	18	35,188	..	215
31	**ELECTRICAL MACHINERY & APPARATUS, NEC**	7	90	330	115	3	13	249	467,867	..	2,491
311	Electric motors, generators and transformers	0	9	36	0	0	0	24	75,094	..	339
312	Electricity distribution and control apparatus	0	22	81	22	1	1	95	81,728	..	516
313	Insulated wire and cable	0	17	61	2	0	0	32	162,304	..	696
314	Accumulators, primary cells & primary batteries	7	28	28	84	0	12	60	79,749	..	506
315	Electric lamps and lighting equipment	0	14	78	7	1	0	22	43,500	..	279
319	Other electrical equipment, nec	0	0	46	0	1	0	16	25,492	..	155
32	**RADIO, TV & COMMUNICATION EQUIP. & APP.**	0	9	187	81	0	0	266	408,658	..	2,014
321	Electronic valves, tubes, other electronic components	0	1	56	5	0	0	75	102,769	..	507
322	TV & radio transmitters, apparatus for line telephony	0	7	119	76	0	0	190	297,571	..	1,463
323	TV & radio receivers, recording apparatus	0	1	12	0	0	0	1	8,318	..	44
33	**MEDICAL PRECISION &OPTICAL INSTRUMENTS**	0	4	210	33	4	0	143	193,775	..	1,092
331	Medical appliances and instruments	0	4	197	33	4	0	127	181,713	..	1,019
3311	Medical, surgical equipment & orthopaedic app.	0	4	87	7	1	0	84	84,041	..	486
3312	Instruments & appliances for measuring, checking etc	0	0	107	26	3	0	43	75,288	..	450
3313	Industrial process control equipment	0	0	3	0	0	0	0	22,384	..	84
332	Optical instruments and photographic equipment	0	0	9	0	0	0	16	11,093	..	65
333	Watches and clocks	0	0	4	0	0	0	0	969	..	7
34	**MOTOR VEHICLES, TRAILERS & SEMI-TRAILERS**	482	719	1,696	1,030	521	2	1,469	1,770,817	..	12,294
341	Motor vehicles	469	262	1,172	779	479	0	1,184	1,069,155	..	8,194
342	Bodies (coachwork) for motor vehicles	1	280	191	78	1	0	41	200,029	..	1,312
343	Parts, accessories for motor vehicles & their engines	12	177	333	173	41	2	244	501,633	..	2,788
35	**OTHER TRANSPORT EQUIPMENT**	0	72	669	31	68	10	427	396,577	..	2,705
351	Building and repairing of ships and boats	0	1	116	3	47	10	53	87,105	..	544
3511	Building and repairing of ships	0	1	63	2	47	1	43	74,184	..	424
3512	Building, repairing of pleasure & sporting boats	0	0	53	1	0	9	10	12,921	..	120
352	Railway, tramway locomotives & rolling stock	0	70	103	0	2	0	95	77,332	..	548
353	Aircraft and spacecraft	0	0	425	27	0	0	274	215,517	..	1,502
359	Transport equipment, nec	0	1	25	1	19	0	5	16,623	..	111
3591	Motorcycles	0	0	4	1	0	0	0	1,523	..	10
3592	Bicycles and invalid carriages	0	0	14	0	19	0	4	8,433	..	67
3599	Other transport equipment, nec	0	1	7	0	0	0	1	6,667	..	33
36	**FURNITURE; MANUFACTURING, NEC**	10	71	486	91	2	129	189	419,556	..	2,488
361	Furniture	3	70	399	44	2	115	95	349,801	..	1,987
369	Manufacturing, nec	7	1	87	47	0	14	94	69,755	..	501
3691	Jewellery and related articles	0	0	5	1	0	0	8	5,956	..	35
3692	Musical instruments	0	0	0	0	0	0	0	331	..	1
3693	Sports goods	0	0	17	11	0	0	0	8,430	..	58
3694	Games and toys	7	1	17	13	0	14	6	26,232	..	152
3699	Other manufacturing, nec	0	0	48	22	0	0	80	28,806	..	254
37	**RECYCLING**	324	0	103	0	0	0	0	145,666	..	951
371	Recycling of metal waste and scrap	324	0	99	0	0	0	0	144,089	..	942
372	Recycling of non-metal waste and scrap	0	0	4	0	0	0	0	1,577	..	10
CERR	**Non-specified industry**	..	..	..	..	..	..	..	..	..	..
15-37	**TOTAL MANUFACTURING**	52,760	21,053	21,106	49,341	14,304	178,496	14,345	47,429,692	..	522,152

ISIC Revision 3 Industry Sector	Solid TJ	LPG TJ	Distiloil TJ	RFO TJ	Gas TJ	Biomass TJ	Steam TJ	Electr MWh	Own Use MWh	TOTAL TJ
15 FOOD PRODUCTS AND BEVERAGES	665	1,681	3,518	4,232	3,825	188	857	2,359,794	..	23,461
151 Production, processing and preserving (PPP)	0	762	1,099	713	2,067	141	344	850,646	..	8,188
1511 PPP of meat and meat products	0	85	639	402	1,450	140	291	412,612	..	4,492
1512 Processing and preserving of fish products	0	2	96	94	1	1	1	72,905	..	457
1513 Processing, preserving of fruit & vegetables	0	102	247	217	455	0	2	182,502	..	1,680
1514 Vegetable and animal oils and fats	0	573	117	0	161	0	50	182,627	..	1,558
152 Dairy products	0	95	554	1,213	277	7	121	360,650	..	3,565
153 Grain mill prod., starches & prepared animal feeds	0	376	538	127	171	11	103	290,403	..	2,371
1531 Grain mill products	0	5	167	0	11	0	28	102,785	..	581
1532 Starches and starch products	0	298	129	87	79	9	0	85,700	..	911
1533 Prepared animal feeds	0	73	242	40	81	2	75	101,918	..	880
154 Other food products	665	260	981	1,505	935	29	135	680,371	..	6,959
1541 Bakery products	0	125	715	262	320	27	6	330,748	..	2,646
1542 Sugar	665	0	1	849	51	0	0	66,945	..	1,807
1543 Cocoa, chocolate and sugar confectionery	0	16	97	315	127	0	82	142,829	..	1,151
1544 Macaroni, noodles, couscous & similar farinaceous prod.	0	0	2	0	0	0	0	0	..	2
1549 Other food products, nec	0	119	166	79	437	2	47	139,849	..	1,353
155 Beverages	0	188	346	674	375	0	154	177,724	..	2,377
1551 Distilling, rectifying and blending of spirits	0	187	90	14	0	0	0	25,919	..	384
1552 Wines	0	0	7	0	53	0	0	11,365	..	101
1553 Malt liquors and malt	0	0	102	58	109	0	0	33,144	..	388
1554 Soft drinks; production of mineral waters	0	1	147	602	213	0	154	107,296	..	1,503
16 TOBACCO PRODUCTS	0	0	0	0	61	0	17	21,886	..	157
17 TEXTILES	0	474	268	523	128	0	61	307,629	..	2,561
171 Spinning, weaving and finishing of textiles	0	183	151	475	89	0	25	100,860	..	1,286
1711 Preparation and spinning of textile fibres	0	113	84	134	89	0	1	73,339	..	685
1712 Finishing of textiles	0	70	67	341	0	0	24	27,521	..	601
172 Other textiles	0	291	86	46	39	0	36	194,225	..	1,197
1721 Made-up textile articles, except apparel	0	65	42	36	7	0	8	26,787	..	254
1722 Carpets and rugs	0	0	5	0	0	0	0	1,030	..	9
1723 Cordage, rope, twine and netting	0	0	1	3	0	0	2	4,493	..	22
1729 Other textiles, nec	0	226	38	7	32	0	26	161,915	..	912
173 Knitted and crocheted fabrics and articles	0	0	31	2	0	0	0	12,544	..	78
18 WEARING APPAREL, DRESSING & DYEING OF FUR	0	1	103	8	0	0	55	65,081	..	401
181 Wearing apparel, except fur apparel	0	1	103	8	0	0	55	65,081	..	401
182 Dressing and dyeing of fur; articles of fur	..	..	..	..	..	..	..	..	..	..
19 TANNING & DRESSING OF LEATHER, FOOTWEAR	0	1	11	26	0	0	101	20,733	..	214
191 Tanning and dressing of leather	0	1	7	9	0	0	95	14,659	..	165
1911 Tanning and dressing of leather	0	1	3	8	0	0	95	12,828	..	153
1912 Luggage, handbags, saddlery & harness	0	0	4	1	0	0	0	1,831	..	12
192 Footwear	0	0	4	17	0	0	6	6,074	..	49
20 WOOD AND WOOD PRODUCTS	13	197	1,916	1,653	1	27,482	1,706	2,423,357	..	41,692
201 Sawmilling and planing of wood	0	184	1,308	759	1	22,744	1,222	1,682,399	..	32,275
202 Products of wood, cork, straw & plaiting materials	13	13	608	894	0	4,738	484	740,958	..	9,417
2021 Veneer sheets	13	8	232	855	0	3,437	363	411,618	..	6,390
2022 Builders' carpentry and joinery	0	5	306	36	0	1,214	54	269,089	..	2,584
2023 Wooden containers	0	0	36	1	0	28	0	21,315	..	142
2029 Other products of wood	0	0	34	2	0	59	67	38,936	..	302
21 PAPER AND PAPER PRODUCTS	1,262	2,026	3,291	27,479	1,965	104,202	2,931	20,131,814	..	215,631
2101 Pulp, paper and paperboard	1,262	1,360	1,590	26,394	1,777	104,004	2,654	19,485,434	..	209,189
2102 Corrugated paper, paperboard and their containers	0	390	1,573	579	186	0	110	346,022	..	4,084
2109 Other articles of pulp and paperboard	0	276	128	506	2	198	167	300,358	..	2,358
22 PUBLISHING, PRINTING & REPRODUCTION	20	207	287	35	204	0	424	551,965	..	3,164
221 Publishing	0	0	113	5	95	0	175	163,949	..	978
2211 Publishing of books & brochures	0	0	23	0	43	0	8	17,526	..	137
2212 Publishing of newspapers and periodicals	0	0	87	5	52	0	162	137,556	..	801
2213 Publishing of recorded media	0	0	0	0	0	0	5	8,867	..	37
2219 Other publishing	0	0	3	0	0	0	0	0	..	3
222 Printing and related service activities	20	207	174	30	109	0	249	386,869	..	2,182
2221 Printing	0	207	142	28	108	0	215	314,440	..	1,832
2222 Service activities related to printing	20	0	32	2	1	0	34	72,429	..	350
223 Reproduction of recorded media	0	0	0	0	0	0	0	1,147	..	4
23 COKE, REFINED PETROLEUM PRODUCTS	1,856	30	105	2,064	0	0	55	803,455	..	7,002
231 Coke oven products	1,856	30	91	2,064	0	0	22	775,002	..	6,853
232 Refined petroleum products	0	0	14	0	0	0	33	28,453	..	149
233 Processing of nuclear fuel	..	..	..	..	..	..	..	..	..	..
24 CHEMICALS & CHEMICAL PRODUCTS	174	897	1,528	3,440	1,572	953	1,746	4,334,964	..	25,916
241 Basic chemicals	173	838	929	2,112	823	910	630	3,842,940	..	20,250
2411 Basic chemicals, exc. fertilizers & nitrogen compounds	173	500	796	1,916	536	910	406	2,822,635	..	15,398
2412 Fertilizers and nitrogen compounds	0	0	26	0	0	0	4	157,847	..	598
2413 Plastics in primary forms and synthetic rubber	0	338	107	196	287	0	220	862,458	..	4,253
242 Other chemical products	1	59	599	640	749	43	1,116	450,795	..	4,830
2421 Pesticides and other agro-chemical products	0	0	15	0	1	0	44	1,349	..	65
2422 Paints, varnishes and similar coatings	1	0	87	34	133	0	74	75,784	..	602
2423 Pharmaceuticals, medicinal chem. & botanical prod.	0	59	288	436	256	0	718	235,273	..	2,604
2424 Soap and detergents, perfumes etc.	0	0	53	39	3	43	8	31,610	..	260
2429 Other chemical products, nec	0	0	156	131	356	0	272	106,779	..	1,299
243 Man-made fibres	0	0	0	688	0	0	0	41,229	..	836
25 RUBBER AND PLASTICS PRODUCTS	0	84	779	200	107	147	254	1,042,355	..	5,323
251 Rubber products	0	8	192	117	51	8	160	230,340	..	1,365
2511 Rubber tyres and tubes	0	0	63	0	0	0	8	39,962	..	215
2519 Other rubber products	0	8	129	117	51	8	152	190,378	..	1,150
252 Plastic products	0	76	587	83	56	139	94	812,015	..	3,958
26 OTHER NON-METALLIC MINERAL PRODUCTS	10,494	3,141	1,697	2,070	902	127	125	1,148,307	..	22,690
261 Glass and glass products	121	1,958	371	680	593	5	8	351,260	..	5,001
269 Non-metallic mineral products, nec	10,373	1,183	1,326	1,390	309	122	117	797,047	..	17,689
2691 Pottery, china and earthenware	0	282	9	175	0	1	78	76,408	..	820
2692 Refractory ceramic products	0	0	27	0	61	0	0	14,143	..	139
2693 Structural non-refractory clay & ceramic prod.	0	225	140	36	40	0	0	20,652	..	515
2694 Cement, lime and plaster	8,465	97	98	689	3	0	0	342,334	..	10,584
2695 Articles of concrete, cement and plaster	212	371	438	420	80	8	12	127,535	..	2,000
2696 Cutting, shaping and finishing of stone	0	0	64	0	0	0	0	13,405	..	112
2699 Other non-metallic mineral products, nec	1,696	208	550	70	125	113	27	202,570	..	3,518

ISIS Energy Data Programme (IEA/OECD)

ISIC Revision 3 Industry Sector	Solid TJ	LPG TJ	Distiloil TJ	RFO TJ	Gas TJ	Biomass TJ	Steam TJ	Electr MWh	Own Use MWh	TOTAL TJ
27 BASIC METALS	10,089	8,163	1,722	4,943	1,407	24	909	7,873,444	..	55,601
271 Basic iron and steel	7,649	7,856	1,284	4,576	1,171	23	786	5,177,209	..	41,983
272 Basic precious and non-ferrous metals	2,439	287	421	365	236	1	122	2,680,543	..	13,521
273 Casting of metals	1	20	17	2	0	0	1	15,692	..	97
2731 Casting of iron and steel	0	0	11	0	0	0	0	5,104	..	29
2732 Casting non-ferrous metals	1	20	6	2	0	0	1	10,588	..	68
28 FABRICATED METAL PRODUCTS	50	582	2,219	518	199	93	569	1,688,472	..	10,308
281 Str. metal prod., tanks, reservoirs, steam generators	12	127	426	129	24	67	91	277,142	..	1,874
2811 Structural metal products	7	118	364	99	20	67	61	241,775	..	1,606
2812 Tanks, reservoirs and containers of metal	5	9	44	26	4	0	21	28,121	..	210
2813 Steam generators, exc. central heating hot water boilers	0	0	18	4	0	0	9	7,246	..	57
289 Other fabricated metal products	38	455	1,793	389	175	26	478	1,411,330	..	8,435
2891 Forging, pressing, stamping & roll-forming of metal	0	0	747	95	7	1	9	50,108	..	1,039
2892 Treatment and coating of metals	9	102	359	46	16	12	107	412,397	..	2,136
2893 Cutlery, hand tools and general hardware	1	12	164	124	4	4	172	247,186	..	1,371
2899 Other fabricated metal products, nec	28	341	523	124	148	9	190	701,639	..	3,889
29 MACHINERY AND EQUIPMENT, NEC	3	297	2,126	898	351	55	1,760	2,030,910	..	12,801
291 General purpose machinery	0	148	889	337	137	16	1,045	977,421	..	6,091
2911 Engines and turbines	0	0	91	100	11	0	37	49,149	..	416
2912 Pumps, compressors, taps and valves	0	26	150	19	1	8	97	186,884	..	974
2913 Bearings, gears, gearing and driving elements	0	28	20	87	22	0	275	217,932	..	1,217
2914 Ovens, furnaces and furnace burners	0	0	7	1	0	0	0	3,396	..	20
2915 Lifting and handling equipment	0	27	297	88	4	7	214	187,418	..	1,312
2919 Other general purpose machinery	0	67	324	42	99	1	422	332,642	..	2,153
292 Special purpose machinery	3	113	1,192	379	208	11	695	872,485	..	5,742
2921 Agricultural and forestry machinery	0	15	91	38	11	5	17	60,315	..	394
2922 Machine-tools	3	17	432	52	79	0	184	161,549	..	1,349
2923 Machinery for metallurgy	0	1	33	0	0	0	26	75,191	..	331
2924 Machinery for mining, quarrying and construction	0	35	193	105	118	0	70	175,910	..	1,154
2925 Machinery for food, beverage & tobacco processing	0	4	23	33	0	0	6	44,587	..	227
2926 Machinery for textile, apparel & leather production	0	3	14	0	0	0	9	19,727	..	97
2927 Machinery for weapons and ammunition	0	0	220	29	0	3	136	77,548	..	667
2929 Other special purpose machinery	0	38	186	122	0	3	247	257,658	..	1,524
293 Domestic appliances, nec	0	36	45	182	6	28	20	181,004	..	969
30 OFFICE, ACCOUNTING & COMPUTING MACHINERY	0	0	22	0	12	36	0	40,834	..	217
31 ELECTRICAL MACHINERY & APPARATUS, NEC	1	82	270	181	1	0	854	541,603	..	3,339
311 Electric motors, generators and transformers	0	1	69	88	0	0	71	92,247	..	561
312 Electricity distribution and control apparatus	1	30	80	0	0	0	396	142,857	..	1,021
313 Insulated wire and cable	0	10	35	1	0	0	192	148,111	..	771
314 Accumulators, primary cells & primary batteries	0	31	11	80	0	0	154	74,079	..	543
315 Electric lamps and lighting equipment	0	10	41	1	1	0	31	66,048	..	322
319 Other electrical equipment, nec	0	0	34	11	0	0	10	18,261	..	121
32 RADIO, TV & COMMUNICATION EQUIP. & APP.	0	10	201	97	12	0	276	452,536	..	2,225
321 Electronic valves, tubes, other electronic components	0	9	37	2	5	0	74	130,278	..	596
322 TV & radio transmitters, apparatus for line telephony	0	1	148	72	7	0	180	291,489	..	1,457
323 TV & radio receivers, recording apparatus	0	0	16	23	0	0	22	30,769	..	172
33 MEDICAL PRECISION &OPTICAL INSTRUMENTS	0	0	78	48	13	1	255	209,774	..	1,150
331 Medical appliances and instruments	0	0	72	43	13	0	230	185,858	..	1,027
3311 Medical, surgical equipment & orthopaedic app.	0	0	28	5	10	0	77	69,627	..	371
3312 Instruments & appliances for measuring, checking etc	0	0	43	38	3	0	138	108,492	..	613
3313 Industrial process control equipment	0	0	1	0	0	0	15	7,739	..	44
332 Optical instruments and photographic equipment	0	0	4	0	0	0	24	21,871	..	107
333 Watches and clocks	0	0	2	5	0	1	1	2,045	..	16
34 MOTOR VEHICLES, TRAILERS & SEMI-TRAILERS	437	682	1,790	1,090	497	0	1,113	1,867,798	..	12,333
341 Motor vehicles	427	163	1,143	828	437	0	890	1,122,623	..	7,929
342 Bodies (coachwork) for motor vehicles	1	331	126	119	0	0	40	198,030	..	1,330
343 Parts, accessories for motor vehicles & their engines	9	188	521	143	60	0	183	547,145	..	3,074
35 OTHER TRANSPORT EQUIPMENT	34	67	473	76	83	11	683	457,819	..	3,075
351 Building and repairing of ships and boats	0	0	62	17	1	10	70	54,048	..	355
3511 Building and repairing of ships	0	0	44	1	1	0	63	43,394	..	265
3512 Building, repairing of pleasure & sporting boats	0	0	18	16	0	10	7	10,654	..	89
352 Railway, tramway locomotives & rolling stock	34	64	137	14	62	0	233	128,092	..	1,005
353 Aircraft and spacecraft	0	3	252	44	0	0	379	258,063	..	1,607
359 Transport equipment, nec	0	0	22	1	20	1	1	17,616	..	108
3591 Motorcycles	0	0	4	0	0	0	0	213	..	5
3592 Bicycles and invalid carriages	0	0	16	1	20	0	1	15,238	..	93
3599 Other transport equipment, nec	0	0	2	0	0	1	0	2,165	..	11
36 FURNITURE; MANUFACTURING, NEC	0	60	425	167	51	462	109	399,036	..	2,711
361 Furniture	0	60	370	113	0	445	70	283,659	..	2,079
369 Manufacturing, nec	0	0	55	54	51	17	39	115,377	..	631
3691 Jewellery and related articles	0	0	22	0	51	0	16	55,383	..	288
3692 Musical instruments	0	0	10	0	0	0	0	1,330	..	15
3693 Sports goods	0	0	1	0	0	5	6	3,967	..	26
3694 Games and toys	0	0	3	0	0	10	2	6,607	..	39
3699 Other manufacturing, nec	0	0	19	54	0	2	15	48,090	..	263
37 RECYCLING	294	0	69	0	0	0	1	140,321	..	869
371 Recycling of metal waste and scrap	294	0	64	0	0	0	1	137,981	..	856
372 Recycling of non-metal waste and scrap	0	0	5	0	0	0	0	2,340	..	13
CERR Non-specified industry	..	..	..	..	..	..	..	..	..	..
15-37 TOTAL MANUFACTURING	25,392	18,682	22,898	49,748	11,391	133,781	14,861	48,913,887	..	452,843

ISIC Revision 3 Industry Sector		Solid TJ	LPG TJ	Distiloil TJ	RFO TJ	Gas TJ	Biomass TJ	Steam TJ	Electr MWh	Own Use MWh	TOTAL TJ
15	**FOOD PRODUCTS AND BEVERAGES**	0	8	1,936	86	2,947	191	178	52,194	0	5,534
151	Production, processing and preserving (PPP)	0	0	160	0	379	0	0	3,389	0	551
1511	PPP of meat and meat products	..	..	..	..	..	..	..	..	..	..
1512	Processing and preserving of fish products	..	..	..	..	..	..	..	..	..	..
1513	Processing, preserving of fruit & vegetables	0	0	160	0	379	0	0	3,389	0	551
1514	Vegetable and animal oils and fats	..	..	..	..	..	..	..	..	..	..
152	Dairy products	0	0	63	0	8	0	0	1,917	0	78
153	Grain mill prod., starches & prepared animal feeds	0	0	63	0	8	0	0	1,917	0	78
1531	Grain mill products	..	..	..	..	..	..	..	..	..	..
1532	Starches and starch products	..	..	..	..	..	..	..	..	..	..
1533	Prepared animal feeds	..	..	..	..	..	..	..	..	..	..
154	Other food products	0	8	1,576	86	2,153	175	178	41,194	0	4,324
1541	Bakery products	..	..	..	..	..	..	..	..	..	..
1542	Sugar	..	..	..	..	..	..	..	..	..	..
1543	Cocoa, chocolate and sugar confectionery	0	1	130		149	6	77	10,111	0	399
1544	Macaroni, noodles, couscous & similar farinaceous prod.	..	..	..	..	..	..	..	..	..	..
1549	Other food products, nec	0	7	1,446	86	2,004	169	101	31,083	0	3,925
155	Beverages	0	0	137	0	407	16	0	5,694	0	580
1551	Distilling, rectifying and blending of spirits	..	..	..	..	..	..	..	..	..	..
1552	Wines	..	..	..	..	..	..	..	..	..	..
1553	Malt liquors and malt	0	0	137	0	407	16	0	5,694	0	580
1554	Soft drinks; production of mineral waters	..	..	..	..	..	..	..	..	..	..
16	**TOBACCO PRODUCTS**	..	..	..	..	..	..	..	..	..	..
17	**TEXTILES**	0	0	863	0	1,221	52	60	52,000	0	2,383
171	Spinning, weaving and finishing of textiles	..	..	..	..	..	..	..	..	..	..
1711	Preparation and spinning of textile fibres	..	..	..	..	..	..	..	..	..	..
1712	Finishing of textiles	..	..	..	..	..	..	..	..	..	..
172	Other textiles	0	0	863	0	1,221	52	60	52,000	0	2,383
1721	Made-up textile articles, except apparel	..	..	..	..	..	..	..	..	..	..
1722	Carpets and rugs	..	..	..	..	..	..	..	..	..	..
1723	Cordage, rope, twine and netting	..	..	..	..	..	..	..	..	..	..
1729	Other textiles, nec	0	0	863	0	1,221	52	60	52,000	0	2,383
173	Knitted and crocheted fabrics and articles	..	..	..	..	..	..	..	..	..	..
18	**WEARING APPAREL, DRESSING & DYEING OF FUR**	..	..	..	..	..	..	..	..	..	..
181	Wearing apparel, except fur apparel	..	..	..	..	..	..	..	..	..	..
182	Dressing and dyeing of fur; articles of fur	..	..	..	..	..	..	..	..	..	..
19	**TANNING & DRESSING OF LEATHER, FOOTWEAR**	..	..	..	..	..	..	..	..	..	..
191	Tanning and dressing of leather	..	..	..	..	..	..	..	..	..	..
1911	Tanning and dressing of leather	..	..	..	..	..	..	..	..	..	..
1912	Luggage, handbags, saddlery & harness	..	..	..	..	..	..	..	..	..	..
192	Footwear	..	..	..	..	..	..	..	..	..	..
20	**WOOD AND WOOD PRODUCTS**	..	..	..	..	..	..	..	..	..	..
201	Sawmilling and planing of wood	..	..	..	..	..	..	..	..	..	..
202	Products of wood, cork, straw & plaiting materials	..	..	..	..	..	..	..	..	..	..
2021	Veneer sheets	..	..	..	..	..	..	..	..	..	..
2022	Builders' carpentry and joinery	..	..	..	..	..	..	..	..	..	..
2023	Wooden containers	..	..	..	..	..	..	..	..	..	..
2029	Other products of wood	..	..	..	..	..	..	..	..	..	..
21	**PAPER AND PAPER PRODUCTS**	0	41	924	3,051	7,142	2,509	1,845	111,806	5,694	15,894
2101	Pulp, paper and paperboard	0	41	924	3,051	7,142	2,509	1,845	111,806	5,694	15,894
2102	Corrugated paper, paperboard and their containers	..	..	..	..	..	..	..	..	..	..
2109	Other articles of pulp and paperboard	..	..	..	..	..	..	..	..	..	..
22	**PUBLISHING, PRINTING & REPRODUCTION**	..	..	..	..	..	..	..	..	..	..
221	Publishing	..	..	..	..	..	..	..	..	..	..
2211	Publishing of books & brochures	..	..	..	..	..	..	..	..	..	..
2212	Publishing of newspapers and periodicals	..	..	..	..	..	..	..	..	..	..
2213	Publishing of recorded media	..	..	..	..	..	..	..	..	..	..
2219	Other publishing	..	..	..	..	..	..	..	..	..	..
222	Printing and related service activities	..	..	..	..	..	..	..	..	..	..
2221	Printing	..	..	..	..	..	..	..	..	..	..
2222	Service activities related to printing	..	..	..	..	..	..	..	..	..	..
223	Reproduction of recorded media	..	..	..	..	..	..	..	..	..	..
23	**COKE, REFINED PETROLEUM PRODUCTS**	..	..	..	..	..	..	..	..	..	..
231	Coke oven products	..	..	..	..	..	..	..	..	..	..
232	Refined petroleum products	..	..	..	..	..	..	..	..	..	..
233	Processing of nuclear fuel	..	..	..	..	..	..	..	..	..	..
24	**CHEMICALS & CHEMICAL PRODUCTS**	0	2	3,808	418	9,348	4,915	814	228,917	61,111	19,909
241	Basic chemicals	..	..	..	..	..	..	..	..	..	..
2411	Basic chemicals, exc. fertilizers & nitrogen compounds	..	..	..	..	..	..	..	..	..	..
2412	Fertilizers and nitrogen compounds	..	..	..	..	..	..	..	..	..	..
2413	Plastics in primary forms and synthetic rubber	0	2	3,808	418	9,348	4,915	814	228,917	61,111	19,909
242	Other chemical products	..	..	..	..	..	..	..	..	..	..
2421	Pesticides and other agro-chemical products	0	0	171	0	9	0	0	1,556	0	186
2422	Paints, varnishes and similar coatings	..	..	..	..	..	..	..	..	..	..
2423	Pharmaceuticals, medicinal chem. & botanical prod.	0	0	117	0	35	0	11	1,250	0	168
2424	Soap and detergents, perfumes etc.	0	2	3,520	418	9,304	4,915	803	226,111	61,111	19,556
2429	Other chemical products, nec	..	..	..	..	..	..	..	..	..	..
243	Man-made fibres	..	..	..	..	..	..	..	..	..	..
25	**RUBBER AND PLASTICS PRODUCTS**	..	..	..	..	..	..	..	..	..	..
251	Rubber products	..	..	..	..	..	..	..	..	..	..
2511	Rubber tyres and tubes	..	..	..	..	..	..	..	..	..	..
2519	Other rubber products	..	..	..	..	..	..	..	..	..	..
252	Plastic products	..	..	..	..	..	..	..	..	..	..
26	**OTHER NON-METALLIC MINERAL PRODUCTS**	4,126	0	530	4,451	1,679	3,622	3	44,778	3,083	14,561
261	Glass and glass products	..	..	..	..	..	..	..	..	..	..
269	Non-metallic mineral products, nec	4,126	0	530	4,451	1,679	3,622	3	44,778	3,083	14,561
2691	Pottery, china and earthenware	..	..	..	..	..	..	..	..	..	..
2692	Refractory ceramic products	..	..	..	..	..	..	..	..	..	..
2693	Structural non-refractory clay & ceramic prod.	0	0	266	1,096	1,642	90	0	7,417	0	3,121
2694	Cement, lime and plaster	4,126	0	264	3,355	37	3,532	3	37,361	3,083	11,440
2695	Articles of concrete, cement and plaster	..	..	..	..	..	..	..	..	..	..
2696	Cutting, shaping and finishing of stone	..	..	..	..	..	..	..	..	..	..
2699	Other non-metallic mineral products, nec	..	..	..	..	..	..	..	..	..	..

ISIC Revision 3 Industry Sector		Solid TJ	LPG TJ	Distiloil TJ	RFO TJ	Gas TJ	Biomass TJ	Steam TJ	Electr MWh	Own Use MWh	TOTAL TJ
27	BASIC METALS	0	106	270	17	1,110	55	3	74,722	58,778	1,618
271	Basic iron and steel	..	..	..	..	..	..	..	..	..	..
272	Basic precious and non-ferrous metals	..	..	..	..	..	..	..	..	..	..
273	Casting of metals	0	106	270	17	1,110	55	3	74,722	58,778	1,618
2731	Casting of iron and steel	..	..	..	..	..	..	..	..	..	..
2732	Casting non-ferrous metals	0	106	270	17	1,110	55	3	74,722	58,778	1,618
28	FABRICATED METAL PRODUCTS	0	71	77	0	70	0	0	4,444	0	234
281	Str. metal prod., tanks, reservoirs, steam generators	..	..	..	..	..	..	..	..	..	..
2811	Structural metal products	..	..	..	..	..	..	..	..	..	..
2812	Tanks, reservoirs and containers of metal	..	..	..	..	..	..	..	..	..	..
2813	Steam generators, exc. central heating hot water boilers	..	..	..	..	..	..	..	..	..	..
289	Other fabricated metal products	0	71	77	0	70	0	0	4,444	0	234
2891	Forging, pressing, stamping & roll-forming of metal	..	..	..	..	..	..	..	..	..	..
2892	Treatment and coating of metals	..	..	..	..	..	..	..	..	..	..
2893	Cutlery, hand tools and general hardware	..	..	..	..	..	..	..	..	..	..
2899	Other fabricated metal products, nec	0	71	77	0	70	0	0	4,444	0	234
29	MACHINERY AND EQUIPMENT, NEC	647	91	3,940	243	4,917	83	448	255,389	6,611	11,265
291	General purpose machinery	647	91	3,940	243	4,917	83	448	255,389	6,611	11,265
2911	Engines and turbines	..	..	..	..	..	..	..	..	..	..
2912	Pumps, compressors, taps and valves	..	..	..	..	..	..	..	..	..	..
2913	Bearings, gears, gearing and driving elements	..	..	..	..	..	..	..	..	..	..
2914	Ovens, furnaces and furnace burners	..	..	..	..	..	..	..	..	..	..
2915	Lifting and handling equipment	..	..	..	..	..	..	..	..	..	..
2919	Other general purpose machinery	647	91	3,940	243	4,917	83	448	255,389	6,611	11,265
292	Special purpose machinery	..	..	..	..	..	..	..	..	..	..
2921	Agricultural and forestry machinery	..	..	..	..	..	..	..	..	..	..
2922	Machine-tools	..	..	..	..	..	..	..	..	..	..
2923	Machinery for metallurgy	..	..	..	..	..	..	..	..	..	..
2924	Machinery for mining, quarrying and construction	..	..	..	..	..	..	..	..	..	..
2925	Machinery for food, beverage & tobacco processing	..	..	..	..	..	..	..	..	..	..
2926	Machinery for textile, apparel & leather production	..	..	..	..	..	..	..	..	..	..
2927	Machinery for weapons and ammunition	..	..	..	..	..	..	..	..	..	..
2929	Other special purpose machinery	..	..	..	..	..	..	..	..	..	..
293	Domestic appliances, nec	..	..	..	..	..	..	..	..	..	..
30	OFFICE, ACCOUNTING & COMPUTING MACHINERY	..	..	..	..	..	..	..	..	..	..
31	ELECTRICAL MACHINERY & APPARATUS, NEC	..	..	..	..	..	..	..	..	..	..
311	Electric motors, generators and transformers	..	..	..	..	..	..	..	..	..	..
312	Electricity distribution and control apparatus	..	..	..	..	..	..	..	..	..	..
313	Insulated wire and cable	..	..	..	..	..	..	..	..	..	..
314	Accumulators, primary cells & primary batteries	..	..	..	..	..	..	..	..	..	..
315	Electric lamps and lighting equipment	..	..	..	..	..	..	..	..	..	..
319	Other electrical equipment, nec	..	..	..	..	..	..	..	..	..	..
32	RADIO, TV & COMMUNICATION EQUIP. & APP.	..	..	..	..	..	..	..	..	..	..
321	Electronic valves, tubes, other electronic components	..	..	..	..	..	..	..	..	..	..
322	TV & radio transmitters, apparatus for line telephony	..	..	..	..	..	..	..	..	..	..
323	TV & radio receivers, recording apparatus	..	..	..	..	..	..	..	..	..	..
33	MEDICAL PRECISION &OPTICAL INSTRUMENTS	..	..	..	..	..	..	..	..	..	..
331	Medical appliances and instruments	..	..	..	..	..	..	..	..	..	..
3311	Medical, surgical equipment & orthopaedic app.	..	..	..	..	..	..	..	..	..	..
3312	Instruments & appliances for measuring, checking etc	..	..	..	..	..	..	..	..	..	..
3313	Industrial process control equipment	..	..	..	..	..	..	..	..	..	..
332	Optical instruments and photographic equipment	..	..	..	..	..	..	..	..	..	..
333	Watches and clocks	..	..	..	..	..	..	..	..	..	..
34	MOTOR VEHICLES, TRAILERS & SEMI-TRAILERS	..	..	..	..	..	..	..	..	..	..
341	Motor vehicles	..	..	..	..	..	..	..	..	..	..
342	Bodies (coachwork) for motor vehicles	..	..	..	..	..	..	..	..	..	..
343	Parts, accessories for motor vehicles & their engines	..	..	..	..	..	..	..	..	..	..
35	OTHER TRANSPORT EQUIPMENT	..	..	..	..	..	..	..	..	..	..
351	Building and repairing of ships and boats	..	..	..	..	..	..	..	..	..	..
3511	Building and repairing of ships	..	..	..	..	..	..	..	..	..	..
3512	Building, repairing of pleasure & sporting boats	..	..	..	..	..	..	..	..	..	..
352	Railway, tramway locomotives & rolling stock	..	..	..	..	..	..	..	..	..	..
353	Aircraft and spacecraft	..	..	..	..	..	..	..	..	..	..
359	Transport equipment, nec	..	..	..	..	..	..	..	..	..	..
3591	Motorcycles	..	..	..	..	..	..	..	..	..	..
3592	Bicycles and invalid carriages	..	..	..	..	..	..	..	..	..	..
3599	Other transport equipment, nec	..	..	..	..	..	..	..	..	..	..
36	FURNITURE; MANUFACTURING, NEC	..	..	..	..	..	..	..	..	..	..
361	Furniture	..	..	..	..	..	..	..	..	..	..
369	Manufacturing, nec	..	..	..	..	..	..	..	..	..	..
3691	Jewellery and related articles	..	..	..	..	..	..	..	..	..	..
3692	Musical instruments	..	..	..	..	..	..	..	..	..	..
3693	Sports goods	..	..	..	..	..	..	..	..	..	..
3694	Games and toys	..	..	..	..	..	..	..	..	..	..
3699	Other manufacturing, nec	..	..	..	..	..	..	..	..	..	..
37	RECYCLING	..	..	..	..	..	..	..	..	..	..
371	Recycling of metal waste and scrap	..	..	..	..	..	..	..	..	..	..
372	Recycling of non-metal waste and scrap	..	..	..	..	..	..	..	..	..	..
CERR	Non-specified industry	..	..	..	..	..	..	..	..	..	..
15-37	TOTAL MANUFACTURING	4,773	319	12,348	8,266	28,434	11,427	3,351	824,250	135,277	71,398

ISIC Revision 3 Industry Sector	Solid TJ	LPG TJ	Distiloil TJ	RFO TJ	Gas TJ	Biomass TJ	Steam TJ	Electr MWh	Own Use MWh	TOTAL TJ
15 FOOD PRODUCTS AND BEVERAGES	13,444	4,861	31,862	31,420	63,883	0	182	10,766,181	579,000	182,326
151 Production, processing and preserving (PPP)	..	..	..	..	..	..	..	..	..	..
1511 PPP of meat and meat products	..	..	..	..	..	..	..	..	..	..
1512 Processing and preserving of fish products	..	..	..	..	..	..	..	..	..	..
1513 Processing, preserving of fruit & vegetables	..	..	..	..	..	..	..	..	..	..
1514 Vegetable and animal oils and fats	..	..	..	..	..	..	..	..	..	..
152 Dairy products	..	..	..	..	..	..	..	..	..	..
153 Grain mill prod., starches & prepared animal feeds	..	..	..	..	..	..	..	..	..	..
1531 Grain mill products	..	..	..	..	..	..	..	..	..	..
1532 Starches and starch products	..	..	..	..	..	..	..	..	..	..
1533 Prepared animal feeds	..	..	..	..	..	..	..	..	..	..
154 Other food products	..	..	..	..	..	..	..	..	..	..
1541 Bakery products	..	..	..	..	..	..	..	..	..	..
1542 Sugar	..	..	..	..	..	..	..	..	..	..
1543 Cocoa, chocolate and sugar confectionery	..	..	..	..	..	..	..	..	..	..
1544 Macaroni, noodles, couscous & similar farinaceous prod.	..	..	..	..	..	..	..	..	..	..
1549 Other food products, nec	..	..	..	..	..	..	..	..	..	..
155 Beverages	..	..	..	..	..	..	..	..	..	..
1551 Distilling, rectifying and blending of spirits	..	..	..	..	..	..	..	..	..	..
1552 Wines	..	..	..	..	..	..	..	..	..	..
1553 Malt liquors and malt	..	..	..	..	..	..	..	..	..	..
1554 Soft drinks; production of mineral waters	..	..	..	..	..	..	..	..	..	..
16 TOBACCO PRODUCTS	0	97	268	321	555	0	0	177,649	0	1,881
17 TEXTILES	6,503	1,107	2,506	4,790	14,623	0	20	2,322,251	0	37,909
171 Spinning, weaving and finishing of textiles	..	..	..	..	..	..	..	..	..	..
1711 Preparation and spinning of textile fibres	..	..	..	..	..	..	..	..	..	..
1712 Finishing of textiles	..	..	..	..	..	..	..	..	..	..
172 Other textiles	..	..	..	..	..	..	..	..	..	..
1721 Made-up textile articles, except apparel	..	..	..	..	..	..	..	..	..	..
1722 Carpets and rugs	..	..	..	..	..	..	..	..	..	..
1723 Cordage, rope, twine and netting	..	..	..	..	..	..	..	..	..	..
1729 Other textiles, nec	..	..	..	..	..	..	..	..	..	..
173 Knitted and crocheted fabrics and articles	..	..	..	..	..	..	..	..	..	..
18 WEARING APPAREL, DRESSING & DYEING OF FUR	9	51	1,161	2,842	2,952	0	63	487,794	0	8,834
181 Wearing apparel, except fur apparel	..	..	..	..	..	..	..	..	..	..
182 Dressing and dyeing of fur; articles of fur	..	..	..	..	..	..	..	..	..	..
19 TANNING & DRESSING OF LEATHER, FOOTWEAR	250	59	282	244	1,116	0	0	225,385	0	2,762
191 Tanning and dressing of leather	..	..	..	..	..	..	..	..	..	..
1911 Tanning and dressing of leather	..	..	..	..	..	..	..	..	..	..
1912 Luggage, handbags, saddlery & harness	..	..	..	..	..	..	..	..	..	..
192 Footwear	..	..	..	..	..	..	..	..	..	..
20 WOOD AND WOOD PRODUCTS	1,239	1,012	3,460	403	3,372	0	21	1,286,752	0	14,139
201 Sawmilling and planing of wood	..	..	..	..	..	..	..	..	..	..
202 Products of wood, cork, straw & plaiting materials	..	..	..	..	..	..	..	..	..	..
2021 Veneer sheets	..	..	..	..	..	..	..	..	..	..
2022 Builders' carpentry and joinery	..	..	..	..	..	..	..	..	..	..
2023 Wooden containers	..	..	..	..	..	..	..	..	..	..
2029 Other products of wood	..	..	..	..	..	..	..	..	..	..
21 PAPER AND PAPER PRODUCTS	18,194	1,157	2,974	7,593	30,388	0	11	5,645,070	839,000	77,619
2101 Pulp, paper and paperboard	..	..	..	..	..	..	..	..	..	..
2102 Corrugated paper, paperboard and their containers	..	..	..	..	..	..	..	..	..	..
2109 Other articles of pulp and paperboard	..	..	..	..	..	..	..	..	..	..
22 PUBLISHING, PRINTING & REPRODUCTION	24	83	1,051	1,118	7,521	0	0	2,333,110	0	18,196
221 Publishing	..	..	..	..	..	..	..	..	..	..
2211 Publishing of books & brochures	..	..	..	..	..	..	..	..	..	..
2212 Publishing of newspapers and periodicals	..	..	..	..	..	..	..	..	..	..
2213 Publishing of recorded media	..	..	..	..	..	..	..	..	..	..
2219 Other publishing	..	..	..	..	..	..	..	..	..	..
222 Printing and related service activities	..	..	..	..	..	..	..	..	..	..
2221 Printing	..	..	..	..	..	..	..	..	..	..
2222 Service activities related to printing	..	..	..	..	..	..	..	..	..	..
223 Reproduction of recorded media	..	..	..	..	..	..	..	..	..	..
23 COKE, REFINED PETROLEUM PRODUCTS	47,851	0	168,039	83,081	2,543	0	0	8,450,000	2,357,000	323,449
231 Coke oven products	..	..	..	..	..	..	..	..	..	..
232 Refined petroleum products	..	..	..	..	..	..	..	..	..	..
233 Processing of nuclear fuel	..	..	..	..	..	..	..	..	..	..
24 CHEMICALS & CHEMICAL PRODUCTS	31,901	4,912	14,244	30,985	88,555	0	155	18,200,950	3,638,000	223,179
241 Basic chemicals	17,729	4,340	9,814	21,149	69,722	0	71	13,684,367	0	172,089
2411 Basic chemicals, exc. fertilizers & nitrogen compounds	16,619	4,315	8,053	18,984	55,442	0	50	10,055,716	0	139,664
2412 Fertilizers and nitrogen compounds	0	0	442	385	4,388	0	0	1,466,962	0	10,496
2413 Plastics in primary forms and synthetic rubber	1,110	25	1,319	1,780	9,892	0	21	2,161,689	0	21,929
242 Other chemical products	6,749	571	4,041	4,264	15,863	0	73	3,816,745	0	45,301
2421 Pesticides and other agro-chemical products	0	0	1,005	178	866	0	22	182,524	0	2,728
2422 Paints, varnishes and similar coatings	3	3	997	627	1,820	0	0	490,769	0	5,217
2423 Pharmaceuticals, medicinal chem. & botanical prod.	3,913	5	1,034	1,403	6,826	0	0	1,720,122	0	19,373
2424 Soap and detergents, perfumes etc.	1	68	126	1,666	2,679	0	0	708,957	0	7,092
2429 Other chemical products, nec	2,832	495	879	390	3,672	0	51	714,373	0	10,891
243 Man-made fibres	7,423	1	389	5,572	2,970	0	11	699,838	0	18,885
25 RUBBER AND PLASTICS PRODUCTS	25,489	2,555	6,514	7,205	21,587	0	65	7,222,372	0	89,416
251 Rubber products	2,845	61	2,771	1,700	7,216	0	21	1,674,869	0	20,644
2511 Rubber tyres and tubes	..	..	..	..	..	..	..	..	..	..
2519 Other rubber products	..	..	..	..	..	..	..	..	..	..
252 Plastic products	22,644	2,494	3,743	5,505	14,371	0	44	5,547,503	0	68,772
26 OTHER NON-METALLIC MINERAL PRODUCTS	62,449	2,456	9,155	11,603	49,378	0	278	5,458,156	0	154,968
261 Glass and glass products	78	0	1,028	9,114	9,347	0	66	830,472	0	22,623
269 Non-metallic mineral products, nec	62,371	2,456	8,127	2,489	40,031	0	212	4,627,684	0	132,346
2691 Pottery, china and earthenware	327	0	2,283	1,437	7,965	0	163	346,426	0	13,422
2692 Refractory ceramic products	1,637	1,616	1,591	708	10,999	0	0	377,271	0	17,909
2693 Structural non-refractory clay & ceramic prod.	..	..	..	..	..	..	..	..	..	..
2694 Cement, lime and plaster	59,125	0	979	31	12,350	0	10	1,955,167	0	79,534
2695 Articles of concrete, cement and plaster	1,282	840	3,274	313	8,717	0	39	1,948,820	0	21,481
2696 Cutting, shaping and finishing of stone	..	..	..	..	..	..	..	..	..	..
2699 Other non-metallic mineral products, nec	..	..	..	..	..	..	..	..	..	..

ISIC Revision 3 Industry Sector	Solid TJ	LPG TJ	Distiloil TJ	RFO TJ	Gas TJ	Biomass TJ	Steam TJ	Electr MWh	Own Use MWh	TOTAL TJ
27 BASIC METALS	155,645	1,476	8,849	23,309	82,186	0	49	15,781,910	1,208,000	323,980
271 Basic iron and steel	148,693	603	6,676	21,968	52,646	0	0	9,088,195	1,208,000	258,955
272 Basic precious and non-ferrous metals	3,712	301	1,453	1,233	17,543	0	11	4,320,340	0	39,806
273 Casting of metals	3,240	572	720	108	11,997	0	38	2,373,375	0	25,219
2731 Casting of iron and steel	..	..	..	..	..	..	..	..	..	..
2732 Casting non-ferrous metals	..	..	..	..	..	..	..	..	..	..
28 FABRICATED METAL PRODUCTS	165	48	2,715	552	20,062	0	52	3,378,090	0	35,755
281 Str. metal prod., tanks, reservoirs, steam generators	..	..	..	..	..	..	..	..	..	..
2811 Structural metal products	..	..	..	..	..	..	..	..	..	..
2812 Tanks, reservoirs and containers of metal	..	..	..	..	..	..	..	..	..	..
2813 Steam generators, exc. central heating hot water boilers	..	..	..	..	..	..	..	..	..	..
289 Other fabricated metal products	..	..	..	..	..	..	..	..	..	..
2891 Forging, pressing, stamping & roll-forming of metal	..	..	..	..	..	..	..	..	..	..
2892 Treatment and coating of metals	..	..	..	..	..	..	..	..	..	..
2893 Cutlery, hand tools and general hardware	..	..	..	..	..	..	..	..	..	..
2899 Other fabricated metal products, nec	..	..	..	..	..	..	..	..	..	..
29 MACHINERY AND EQUIPMENT, NEC	1,506	1,847	11,939	5,494	21,540	0	32	6,042,210	0	64,110
291 General purpose machinery	..	..	..	..	..	..	..	..	..	..
2911 Engines and turbines	..	..	..	..	..	..	..	..	..	..
2912 Pumps, compressors, taps and valves	..	..	..	..	..	..	..	..	..	..
2913 Bearings, gears, gearing and driving elements	..	..	..	..	..	..	..	..	..	..
2914 Ovens, furnaces and furnace burners	..	..	..	..	..	..	..	..	..	..
2915 Lifting and handling equipment	..	..	..	..	..	..	..	..	..	..
2919 Other general purpose machinery	..	..	..	..	..	..	..	..	..	..
292 Special purpose machinery	..	..	..	..	..	..	..	..	..	..
2921 Agricultural and forestry machinery	..	..	..	..	..	..	..	..	..	..
2922 Machine-tools	..	..	..	..	..	..	..	..	..	..
2923 Machinery for metallurgy	..	..	..	..	..	..	..	..	..	..
2924 Machinery for mining, quarrying and construction	..	..	..	..	..	..	..	..	..	..
2925 Machinery for food, beverage & tobacco processing	..	..	..	..	..	..	..	..	..	..
2926 Machinery for textile, apparel & leather production	..	..	..	..	..	..	..	..	..	..
2927 Machinery for weapons and ammunition	..	..	..	..	..	..	..	..	..	..
2929 Other special purpose machinery	..	..	..	..	..	..	..	..	..	..
293 Domestic appliances, nec	..	..	..	..	..	..	..	..	..	..
30 OFFICE, ACCOUNTING & COMPUTING MACHINERY	0	22	27	648	1,003	0	22	499,540	0	3,520
31 ELECTRICAL MACHINERY & APPARATUS, NEC	436	716	1,488	2,430	7,327	0	0	1,791,574	0	18,847
311 Electric motors, generators and transformers	..	..	..	..	..	..	..	..	..	..
312 Electricity distribution and control apparatus	..	..	..	..	..	..	..	..	..	..
313 Insulated wire and cable	..	..	..	..	..	..	..	..	..	..
314 Accumulators, primary cells & primary batteries	..	..	..	..	..	..	..	..	..	..
315 Electric lamps and lighting equipment	..	..	..	..	..	..	..	..	..	..
319 Other electrical equipment, nec	..	..	..	..	..	..	..	..	..	..
32 RADIO, TV & COMMUNICATION EQUIP. & APP.	240	127	1,730	1,129	7,334	0	10	2,335,363	0	18,977
321 Electronic valves, tubes, other electronic components	..	..	..	..	..	..	..	..	..	..
322 TV & radio transmitters, apparatus for line telephony	..	..	..	..	..	..	..	..	..	..
323 TV & radio receivers, recording apparatus	..	..	..	..	..	..	..	..	..	..
33 MEDICAL PRECISION &OPTICAL INSTRUMENTS	0	0	1,167	1,044	2,274	0	42	618,653	0	6,754
331 Medical appliances and instruments	..	..	..	..	..	..	..	..	..	..
3311 Medical, surgical equipment & orthopaedic app.	..	..	..	..	..	..	..	..	..	..
3312 Instruments & appliances for measuring, checking etc	..	..	..	..	..	..	..	..	..	..
3313 Industrial process control equipment	..	..	..	..	..	..	..	..	..	..
332 Optical instruments and photographic equipment	..	..	..	..	..	..	..	..	..	..
333 Watches and clocks	..	..	..	..	..	..	..	..	..	..
34 MOTOR VEHICLES, TRAILERS & SEMI-TRAILERS	2,831	96	4,661	2,408	17,156	0	30	3,358,627	0	39,273
341 Motor vehicles	..	..	..	..	..	..	..	..	..	..
342 Bodies (coachwork) for motor vehicles	..	..	..	..	..	..	..	..	..	..
343 Parts, accessories for motor vehicles & their engines	..	..	..	..	..	..	..	..	..	..
35 OTHER TRANSPORT EQUIPMENT	3,254	1,072	2,807	2,286	12,350	0	32	2,875,053	0	32,151
351 Building and repairing of ships and boats	..	..	..	..	..	..	..	..	..	..
3511 Building and repairing of ships	..	..	..	..	..	..	..	..	..	..
3512 Building, repairing of pleasure & sporting boats	..	..	..	..	..	..	..	..	..	..
352 Railway, tramway locomotives & rolling stock	..	..	..	..	..	..	..	..	..	..
353 Aircraft and spacecraft	..	..	..	..	..	..	..	..	..	..
359 Transport equipment, nec	..	..	..	..	..	..	..	..	..	..
3591 Motorcycles	..	..	..	..	..	..	..	..	..	..
3592 Bicycles and invalid carriages	..	..	..	..	..	..	..	..	..	..
3599 Other transport equipment, nec	..	..	..	..	..	..	..	..	..	..
36 FURNITURE; MANUFACTURING, NEC	1,010	347	17,038	7,786	4,471	0	11	1,420,758	0	35,778
361 Furniture	..	..	..	..	..	..	..	..	..	..
369 Manufacturing, nec	..	..	..	..	..	..	..	..	..	..
3691 Jewellery and related articles	..	..	..	..	..	..	..	..	..	..
3692 Musical instruments	..	..	..	..	..	..	..	..	..	..
3693 Sports goods	..	..	..	..	..	..	..	..	..	..
3694 Games and toys	..	..	..	..	..	..	..	..	..	..
3699 Other manufacturing, nec	..	..	..	..	..	..	..	..	..	..
37 RECYCLING										
371 Recycling of metal waste and scrap	..	..	..	..	..	..	..	..	..	..
372 Recycling of non-metal waste and scrap	..	..	..	..	..	..	..	..	..	..
CERR Non-specified industry	0	0	0	0	0	0	0	0	3,480,000	-12,528
15-37 TOTAL MANUFACTURING	372,440	24,101	293,937	228,691	462,176	14,802	1,075	100,677,448	12,101,000	1,716,097

ISIC Revision 3 Industry Sector	Solid TJ	LPG TJ	Distiloil TJ	RFO TJ	Gas TJ	Biomass TJ	Steam TJ	Electr MWh	Own Use MWh	TOTAL TJ
15 FOOD PRODUCTS AND BEVERAGES	16,348	4,643	27,461	31,659	60,675	0	181	10,674,048	601,000	177,230
151 Production, processing and preserving (PPP)	..	..	..	..	..	..	..	..	..	..
1511 PPP of meat and meat products	..	..	..	..	..	..	..	..	..	..
1512 Processing and preserving of fish products	..	..	..	..	..	..	..	..	..	..
1513 Processing, preserving of fruit & vegetables	..	..	..	..	..	..	..	..	..	..
1514 Vegetable and animal oils and fats	..	..	..	..	..	..	..	..	..	..
152 Dairy products	..	..	..	..	..	..	..	..	..	..
153 Grain mill prod., starches & prepared animal feeds	..	..	..	..	..	..	..	..	..	..
1531 Grain mill products	..	..	..	..	..	..	..	..	..	..
1532 Starches and starch products	..	..	..	..	..	..	..	..	..	..
1533 Prepared animal feeds	..	..	..	..	..	..	..	..	..	..
154 Other food products	..	..	..	..	..	..	..	..	..	..
1541 Bakery products	..	..	..	..	..	..	..	..	..	..
1542 Sugar	..	..	..	..	..	..	..	..	..	..
1543 Cocoa, chocolate and sugar confectionery	..	..	..	..	..	..	..	..	..	..
1544 Macaroni, noodles, couscous & similar farinaceous prod.	..	..	..	..	..	..	..	..	..	..
1549 Other food products, nec	..	..	..	..	..	..	..	..	..	..
155 Beverages	..	..	..	..	..	..	..	..	..	..
1551 Distilling, rectifying and blending of spirits	..	..	..	..	..	..	..	..	..	..
1552 Wines	..	..	..	..	..	..	..	..	..	..
1553 Malt liquors and malt	..	..	..	..	..	..	..	..	..	..
1554 Soft drinks; production of mineral waters	0	93	223	401	673	0	0	165,112	0	1,984
16 TOBACCO PRODUCTS	7,322	1,039	2,492	4,610	14,444	0	18	2,089,642	0	37,448
17 TEXTILES	7,322	1,039	2,492	4,610	14,444	0	18	2,089,642	0	37,448
171 Spinning, weaving and finishing of textiles	..	..	..	..	..	..	..	..	..	..
1711 Preparation and spinning of textile fibres	..	..	..	..	..	..	..	..	..	..
1712 Finishing of textiles	..	..	..	..	..	..	..	..	..	..
172 Other textiles	..	..	..	..	..	..	..	..	..	..
1721 Made-up textile articles, except apparel	..	..	..	..	..	..	..	..	..	..
1722 Carpets and rugs	..	..	..	..	..	..	..	..	..	..
1723 Cordage, rope, twine and netting	..	..	..	..	..	..	..	..	..	..
1729 Other textiles, nec	..	..	..	..	..	..	..	..	..	..
173 Knitted and crocheted fabrics and articles	..	..	..	..	..	..	..	..	..	..
18 WEARING APPAREL, DRESSING & DYEING OF FUR	15	74	1,126	3,182	2,725	0	57	422,396	0	8,700
181 Wearing apparel, except fur apparel	..	..	..	..	..	..	..	..	..	..
182 Dressing and dyeing of fur; articles of fur	..	..	..	..	..	..	..	..	..	..
19 TANNING & DRESSING OF LEATHER, FOOTWEAR	420	86	490	163	1,032	0	0	186,122	0	2,861
191 Tanning and dressing of leather	..	..	..	..	..	..	..	..	..	..
1911 Tanning and dressing of leather	..	..	..	..	..	..	..	..	..	..
1912 Luggage, handbags, saddlery & harness	..	..	..	..	..	..	..	..	..	..
192 Footwear	..	..	..	..	..	..	..	..	..	..
20 WOOD AND WOOD PRODUCTS	586	1,120	3,507	245	4,585	0	19	1,570,989	0	15,718
201 Sawmilling and planing of wood	..	..	..	..	..	..	..	..	..	..
202 Products of wood, cork, straw & plaiting materials	..	..	..	..	..	..	..	..	..	..
2021 Veneer sheets	..	..	..	..	..	..	..	..	..	..
2022 Builders' carpentry and joinery	..	..	..	..	..	..	..	..	..	..
2023 Wooden containers	..	..	..	..	..	..	..	..	..	..
2029 Other products of wood	..	..	..	..	..	..	..	..	..	..
21 PAPER AND PAPER PRODUCTS	21,751	1,324	3,106	9,613	29,970	0	11	5,555,751	872,000	82,637
2101 Pulp, paper and paperboard	..	..	..	..	..	..	..	..	..	..
2102 Corrugated paper, paperboard and their containers	..	..	..	..	..	..	..	..	..	..
2109 Other articles of pulp and paperboard	..	..	..	..	..	..	..	..	..	..
22 PUBLISHING, PRINTING & REPRODUCTION	47	140	1,059	1,008	8,090	0	0	2,457,319	0	19,190
221 Publishing	..	..	..	..	..	..	..	..	..	..
2211 Publishing of books & brochures	..	..	..	..	..	..	..	..	..	..
2212 Publishing of newspapers and periodicals	..	..	..	..	..	..	..	..	..	..
2213 Publishing of recorded media	..	..	..	..	..	..	..	..	..	..
2219 Other publishing	..	..	..	..	..	..	..	..	..	..
222 Printing and related service activities	..	..	..	..	..	..	..	..	..	..
2221 Printing	..	..	..	..	..	..	..	..	..	..
2222 Service activities related to printing	..	..	..	..	..	..	..	..	..	..
223 Reproduction of recorded media	..	..	..	..	..	..	..	..	..	..
23 COKE, REFINED PETROLEUM PRODUCTS	49,084	0	163,152	91,183	2,712	0	0	8,274,000	2,145,000	328,195
231 Coke oven products	..	..	..	..	..	..	..	..	..	..
232 Refined petroleum products	..	..	..	..	..	..	..	..	..	..
233 Processing of nuclear fuel	..	..	..	..	..	..	..	..	..	..
24 CHEMICALS & CHEMICAL PRODUCTS	35,003	5,386	13,843	37,469	58,785	0	158	17,351,960	3,639,000	200,011
241 Basic chemicals	20,770	5,017	9,458	25,583	45,609	0	74	12,816,907	0	152,652
2411 Basic chemicals, exc. fertilizers & nitrogen compounds	19,273	4,983	8,272	22,591	37,762	0	54	9,931,056	0	128,687
2412 Fertilizers and nitrogen compounds	0	0	276	380	2,622	0	0	995,995	0	6,864
2413 Plastics in primary forms and synthetic rubber	1,497	34	910	2,612	5,225	0	20	1,889,856	0	17,101
242 Other chemical products	7,057	368	4,038	4,901	11,564	0	73	3,839,230	0	41,822
2421 Pesticides and other agro-chemical products	0	0	1,251	310	704	0	21	203,310	0	3,018
2422 Paints, varnishes and similar coatings	7	2	1,136	900	1,051	0	0	458,478	0	4,747
2423 Pharmaceuticals, medicinal chem. & botanical prod.	3,771	6	933	1,831	4,816	0	0	1,717,694	0	17,541
2424 Soap and detergents, perfumes etc.	1	73	109	1,313	1,799	0	0	664,389	0	5,687
2429 Other chemical products, nec	3,278	287	609	547	3,194	0	52	795,359	0	10,830
243 Man-made fibres	7,176	1	347	6,985	1,612	0	11	695,823	0	18,637
25 RUBBER AND PLASTICS PRODUCTS	24,088	4,091	6,031	7,526	23,816	0	61	7,407,093	0	92,279
251 Rubber products	3,425	102	2,482	1,268	7,189	0	20	1,637,328	0	20,380
2511 Rubber tyres and tubes	..	..	..	..	..	..	..	..	..	..
2519 Other rubber products	..	..	..	..	..	..	..	..	..	..
252 Plastic products	20,663	3,989	3,549	6,258	16,627	0	41	5,769,765	0	71,898
26 OTHER NON-METALLIC MINERAL PRODUCTS	50,476	2,398	7,863	12,457	41,001	0	259	5,285,909	0	133,483
261 Glass and glass products	78	0	1,132	10,342	8,366	0	63	918,005	0	23,286
269 Non-metallic mineral products, nec	50,398	2,398	6,731	2,115	32,635	0	196	4,367,904	0	110,197
2691 Pottery, china and earthenware	327	0	2,024	1,242	7,231	0	153	388,489	0	12,376
2692 Refractory ceramic products	1,217	1,627	968	442	9,213	0	0	375,185	0	14,818
2693 Structural non-refractory clay & ceramic prod.	..	..	..	..	..	..	..	..	..	..
2694 Cement, lime and plaster	47,972	0	818	45	9,116	0	9	1,782,794	0	64,378
2695 Articles of concrete, cement and plaster	882	771	2,921	386	7,075	0	34	1,821,436	0	18,626
2696 Cutting, shaping and finishing of stone	..	..	..	..	..	..	..	..	..	..
2699 Other non-metallic mineral products, nec	..	..	..	..	..	..	..	..	..	..

ISIS Energy Data Programme (IEA/OECD)

ISIC Revision 3 Industry Sector	Solid TJ	LPG TJ	Distiloil TJ	RFO TJ	Gas TJ	Biomass TJ	Steam TJ	Electr MWh	Own Use MWh	TOTAL TJ
27 BASIC METALS	151,111	1,559	8,018	24,463	76,798	0	45	15,805,170	1,367,000	313,971
271 Basic iron and steel	144,388	606	6,147	22,320	49,572	0	0	9,355,532	1,367,000	251,792
272 Basic precious and non-ferrous metals	4,043	447	1,279	2,108	16,801	0	10	4,193,123	0	39,783
273 Casting of metals	2,680	506	592	35	10,425	0	35	2,256,515	0	22,396
2731 Casting of iron and steel	..	..	..	..	..	..	..	..	..	..
2732 Casting non-ferrous metals	..	..	..	..	..	..	..	..	..	..
28 FABRICATED METAL PRODUCTS	142	37	2,560	426	19,694	0	47	2,903,185	0	33,357
281 Str. metal prod., tanks, reservoirs, steam generators	..	..	..	..	..	..	..	..	..	..
2811 Structural metal products	..	..	..	..	..	..	..	..	..	..
2812 Tanks, reservoirs and containers of metal	..	..	..	..	..	..	..	..	..	..
2813 Steam generators, exc. central heating hot water boilers	..	..	..	..	..	..	..	..	..	..
289 Other fabricated metal products	..	..	..	..	..	..	..	..	..	..
2891 Forging, pressing, stamping & roll-forming of metal	..	..	..	..	..	..	..	..	..	..
2892 Treatment and coating of metals	..	..	..	..	..	..	..	..	..	..
2893 Cutlery, hand tools and general hardware	..	..	..	..	..	..	..	..	..	..
2899 Other fabricated metal products, nec	..	..	..	..	..	..	..	..	..	..
29 MACHINERY AND EQUIPMENT, NEC	1,409	1,610	10,409	4,665	21,644	0	29	5,295,965	0	58,831
291 General purpose machinery	..	..	..	..	..	..	..	..	..	..
2911 Engines and turbines	..	..	..	..	..	..	..	..	..	..
2912 Pumps, compressors, taps and valves	..	..	..	..	..	..	..	..	..	..
2913 Bearings, gears, gearing and driving elements	..	..	..	..	..	..	..	..	..	..
2914 Ovens, furnaces and furnace burners	..	..	..	..	..	..	..	..	..	..
2915 Lifting and handling equipment	..	..	..	..	..	..	..	..	..	..
2919 Other general purpose machinery	..	..	..	..	..	..	..	..	..	..
292 Special purpose machinery	..	..	..	..	..	..	..	..	..	..
2921 Agricultural and forestry machinery	..	..	..	..	..	..	..	..	..	..
2922 Machine-tools	..	..	..	..	..	..	..	..	..	..
2923 Machinery for metallurgy	..	..	..	..	..	..	..	..	..	..
2924 Machinery for mining, quarrying and construction	..	..	..	..	..	..	..	..	..	..
2925 Machinery for food, beverage & tobacco processing	..	..	..	..	..	..	..	..	..	..
2926 Machinery for textile, apparel & leather production	..	..	..	..	..	..	..	..	..	..
2927 Machinery for weapons and ammunition	..	..	..	..	..	..	..	..	..	..
2929 Other special purpose machinery	..	..	..	..	..	..	..	..	..	..
293 Domestic appliances, nec	..	..	..	..	..	..	..	..	..	..
30 OFFICE, ACCOUNTING & COMPUTING MACHINERY	0	36	44	634	1,192	0	23	562,065	0	3,952
31 ELECTRICAL MACHINERY & APPARATUS, NEC	433	732	1,544	2,659	7,785	0	0	1,977,447	0	20,272
311 Electric motors, generators and transformers	..	..	..	..	..	..	..	..	..	..
312 Electricity distribution and control apparatus	..	..	..	..	..	..	..	..	..	..
313 Insulated wire and cable	..	..	..	..	..	..	..	..	..	..
314 Accumulators, primary cells & primary batteries	..	..	..	..	..	..	..	..	..	..
315 Electric lamps and lighting equipment	..	..	..	..	..	..	..	..	..	..
319 Other electrical equipment, nec	..	..	..	..	..	..	..	..	..	..
32 RADIO, TV & COMMUNICATION EQUIP. & APP.	243	86	928	1,407	7,217	0	10	2,444,277	0	18,690
321 Electronic valves, tubes, other electronic components	..	..	..	..	..	..	..	..	..	..
322 TV & radio transmitters, apparatus for line telephony	..	..	..	..	..	..	..	..	..	..
323 TV & radio receivers, recording apparatus	..	..	..	..	..	..	..	..	..	..
33 MEDICAL PRECISION &OPTICAL INSTRUMENTS	0	0	856	829	2,421	0	40	668,391	0	6,552
331 Medical appliances and instruments	..	..	..	..	..	..	..	..	..	..
3311 Medical, surgical equipment & orthopaedic app.	..	..	..	..	..	..	..	..	..	..
3312 Instruments & appliances for measuring, checking etc	..	..	..	..	..	..	..	..	..	..
3313 Industrial process control equipment	..	..	..	..	..	..	..	..	..	..
332 Optical instruments and photographic equipment	..	..	..	..	..	..	..	..	..	..
333 Watches and clocks	..	..	..	..	..	..	..	..	..	..
34 MOTOR VEHICLES, TRAILERS & SEMI-TRAILERS	3,379	138	3,501	1,808	17,615	0	28	3,251,547	0	38,175
341 Motor vehicles	..	..	..	..	..	..	..	..	..	..
342 Bodies (coachwork) for motor vehicles	..	..	..	..	..	..	..	..	..	..
343 Parts, accessories for motor vehicles & their engines	..	..	..	..	..	..	..	..	..	..
35 OTHER TRANSPORT EQUIPMENT	3,702	959	2,556	3,164	11,552	0	31	2,819,313	0	32,114
351 Building and repairing of ships and boats	..	..	..	..	..	..	..	..	..	..
3511 Building and repairing of ships	..	..	..	..	..	..	..	..	..	..
3512 Building, repairing of pleasure & sporting boats	..	..	..	..	..	..	..	..	..	..
352 Railway, tramway locomotives & rolling stock	..	..	..	..	..	..	..	..	..	..
353 Aircraft and spacecraft	..	..	..	..	..	..	..	..	..	..
359 Transport equipment, nec	..	..	..	..	..	..	..	..	..	..
3591 Motorcycles	..	..	..	..	..	..	..	..	..	..
3592 Bicycles and invalid carriages	..	..	..	..	..	..	..	..	..	..
3599 Other transport equipment, nec	..	..	..	..	..	..	..	..	..	..
36 FURNITURE; MANUFACTURING, NEC	546	255	15,556	12,435	4,895	0	9	1,419,828	0	38,807
361 Furniture	..	..	..	..	..	..	..	..	..	..
369 Manufacturing, nec	..	..	..	..	..	..	..	..	..	..
3691 Jewellery and related articles	..	..	..	..	..	..	..	..	..	..
3692 Musical instruments	..	..	..	..	..	..	..	..	..	..
3693 Sports goods	..	..	..	..	..	..	..	..	..	..
3694 Games and toys	..	..	..	..	..	..	..	..	..	..
3699 Other manufacturing, nec	..	..	..	..	..	..	..	..	..	..
37 RECYCLING	..	..	..	..	..	..	..	..	..	..
371 Recycling of metal waste and scrap	..	..	..	..	..	..	..	..	..	..
372 Recycling of non-metal waste and scrap	..	..	..	..	..	..	..	..	..	..
CERR Non-specified industry	0	0	0	0	0	0	0	0	3,532,000	-12,715
15-37 TOTAL MANUFACTURING	366,105	25,806	276,325	252,006	419,321	15,048	1,026	98,587,529	12,156,000	**1,666,791**

ISIC Revision 3 Industry Sector	Solid TJ	LPG TJ	Distiloil TJ	RFO TJ	Gas TJ	Biomass TJ	Steam TJ	Electr MWh	Own Use MWh	TOTAL TJ
15 FOOD PRODUCTS AND BEVERAGES	17,064	4,774	26,218	31,740	63,779	0	184	10,557,554	574,000	179,700
151 Production, processing and preserving (PPP)	..	..	..	..	..	..	..	..	..	..
1511 PPP of meat and meat products	..	..	..	..	..	..	..	..	..	..
1512 Processing and preserving of fish products	..	..	..	..	..	..	..	..	..	..
1513 Processing, preserving of fruit & vegetables	..	..	..	..	..	..	..	..	..	..
1514 Vegetable and animal oils and fats	..	..	..	..	..	..	..	..	..	..
152 Dairy products	..	..	..	..	..	..	..	..	..	..
153 Grain mill prod., starches & prepared animal feeds	..	..	..	..	..	..	..	..	..	..
1531 Grain mill products	..	..	..	..	..	..	..	..	..	..
1532 Starches and starch products	..	..	..	..	..	..	..	..	..	..
1533 Prepared animal feeds	..	..	..	..	..	..	..	..	..	..
154 Other food products	..	..	..	..	..	..	..	..	..	..
1541 Bakery products	..	..	..	..	..	..	..	..	..	..
1542 Sugar	..	..	..	..	..	..	..	..	..	..
1543 Cocoa, chocolate and sugar confectionery	..	..	..	..	..	..	..	..	..	..
1544 Macaroni, noodles, couscous & similar farinaceous prod.	..	..	..	..	..	..	..	..	..	..
1549 Other food products, nec	..	..	..	..	..	..	..	..	..	..
155 Beverages	..	..	..	..	..	..	..	..	..	..
1551 Distilling, rectifying and blending of spirits	..	..	..	..	..	..	..	..	..	..
1552 Wines	..	..	..	..	..	..	..	..	..	..
1553 Malt liquors and malt	..	..	..	..	..	..	..	..	..	..
1554 Soft drinks; production of mineral waters	..	..	..	..	..	..	..	..	..	..
16 TOBACCO PRODUCTS	0	95	150	322	847	0	0	153,676	0	1,967
17 TEXTILES	7,926	978	2,304	4,104	12,874	0	18	2,058,694	0	35,615
171 Spinning, weaving and finishing of textiles	..	..	..	..	..	..	..	..	..	..
1711 Preparation and spinning of textile fibres	..	..	..	..	..	..	..	..	..	..
1712 Finishing of textiles	..	..	..	..	..	..	..	..	..	..
172 Other textiles	..	..	..	..	..	..	..	..	..	..
1721 Made-up textile articles, except apparel	..	..	..	..	..	..	..	..	..	..
1722 Carpets and rugs	..	..	..	..	..	..	..	..	..	..
1723 Cordage, rope, twine and netting	..	..	..	..	..	..	..	..	..	..
1729 Other textiles, nec	..	..	..	..	..	..	..	..	..	..
173 Knitted and crocheted fabrics and articles	..	..	..	..	..	..	..	..	..	..
18 WEARING APPAREL, DRESSING & DYEING OF FUR	19	83	1,082	3,210	2,327	0	59	407,458	0	8,247
181 Wearing apparel, except fur apparel	..	..	..	..	..	..	..	..	..	..
182 Dressing and dyeing of fur; articles of fur	..	..	..	..	..	..	..	..	..	..
19 TANNING & DRESSING OF LEATHER, FOOTWEAR	528	96	303	203	889	0	0	173,857	0	2,645
191 Tanning and dressing of leather	..	..	..	..	..	..	..	..	..	..
1911 Tanning and dressing of leather	..	..	..	..	..	..	..	..	..	..
1912 Luggage, handbags, saddlery & harness	..	..	..	..	..	..	..	..	..	..
192 Footwear	..	..	..	..	..	..	..	..	..	..
20 WOOD AND WOOD PRODUCTS	597	1,194	3,443	173	2,032	0	18	1,167,717	0	11,661
201 Sawmilling and planing of wood	..	..	..	..	..	..	..	..	..	..
202 Products of wood, cork, straw & plaiting materials	..	..	..	..	..	..	..	..	..	..
2021 Veneer sheets	..	..	..	..	..	..	..	..	..	..
2022 Builders' carpentry and joinery	..	..	..	..	..	..	..	..	..	..
2023 Wooden containers	..	..	..	..	..	..	..	..	..	..
2029 Other products of wood	..	..	..	..	..	..	..	..	..	..
21 PAPER AND PAPER PRODUCTS	24,428	1,456	3,358	11,237	28,551	0	11	5,053,404	832,000	84,238
2101 Pulp, paper and paperboard	..	..	..	..	..	..	..	..	..	..
2102 Corrugated paper, paperboard and their containers	..	..	..	..	..	..	..	..	..	..
2109 Other articles of pulp and paperboard	..	..	..	..	..	..	..	..	..	..
22 PUBLISHING, PRINTING & REPRODUCTION	74	212	1,120	457	8,529	0	0	2,436,316	0	19,163
221 Publishing	..	..	..	..	..	..	..	..	..	..
2211 Publishing of books & brochures	..	..	..	..	..	..	..	..	..	..
2212 Publishing of newspapers and periodicals	..	..	..	..	..	..	..	..	..	..
2213 Publishing of recorded media	..	..	..	..	..	..	..	..	..	..
2219 Other publishing	..	..	..	..	..	..	..	..	..	..
222 Printing and related service activities	..	..	..	..	..	..	..	..	..	..
2221 Printing	..	..	..	..	..	..	..	..	..	..
2222 Service activities related to printing	..	..	..	..	..	..	..	..	..	..
223 Reproduction of recorded media	..	..	..	..	..	..	..	..	..	..
23 COKE, REFINED PETROLEUM PRODUCTS	50,523	0	131,920	74,032	8,281	0	0	7,792,000	2,320,000	284,455
231 Coke oven products	..	..	..	..	..	..	..	..	..	..
232 Refined petroleum products	..	..	..	..	..	..	..	..	..	..
233 Processing of nuclear fuel	..	..	..	..	..	..	..	..	..	..
24 CHEMICALS & CHEMICAL PRODUCTS	33,969	4,959	12,740	32,657	69,374	0	159	17,572,930	3,473,000	204,618
241 Basic chemicals	22,439	4,733	9,025	21,925	52,507	0	74	12,616,155	0	156,121
2411 Basic chemicals, exc. fertilizers & nitrogen compounds	20,767	4,698	8,294	18,678	44,743	0	54	10,235,599	0	134,082
2412 Fertilizers and nitrogen compounds	0	0	178	255	2,766	0	0	571,529	0	5,257
2413 Plastics in primary forms and synthetic rubber	1,672	35	553	2,992	4,998	0	20	1,809,027	0	16,782
242 Other chemical products	6,074	225	3,421	4,722	15,314	0	74	4,213,479	0	44,999
2421 Pesticides and other agro-chemical products	0	0	1,175	396	1,000	0	20	240,629	0	3,457
2422 Paints, varnishes and similar coatings	11	1	825	1,028	1,158	0	0	480,875	0	4,754
2423 Pharmaceuticals, medicinal chem. & botanical prod.	2,858	7	787	1,873	6,129	0	0	1,839,324	0	18,276
2424 Soap and detergents, perfumes etc.	1	68	116	821	2,306	0	54	940,860	0	12,637
2429 Other chemical products, nec	3,204	149	518	604	4,721	0	11	743,296	0	16,001
243 Man-made fibres	5,456	1	294	6,010	1,553	0	62	6,236,652	0	78,036
25 RUBBER AND PLASTICS PRODUCTS	27,437	5,034	5,181	6,359	11,511	0	19	1,281,958	0	15,521
251 Rubber products	4,646	128	2,089	968	3,056	0	..	..	..	..
2511 Rubber tyres and tubes	..	..	..	..	..	..	..	..	..	..
2519 Other rubber products	22,791	4,906	3,092	5,391	8,455	0	43	4,954,694	0	62,515
252 Plastic products	..	..	..	..	..	0	249	5,080,867	0	125,386
26 OTHER NON-METALLIC MINERAL PRODUCTS	48,527	2,379	6,768	11,539	37,633	0	63	915,849	0	22,005
261 Glass and glass products	82	0	1,090	9,691	7,782	0	186	4,165,018	0	103,381
269 Non-metallic mineral products, nec	48,445	2,379	5,678	1,848	29,851	0	146	388,210	0	11,705
2691 Pottery, china and earthenware	338	0	1,989	1,119	6,715	0	0	380,009	0	13,677
2692 Refractory ceramic products	1,174	1,828	627	201	8,479	0	8	1,664,931	0	60,650
2693 Structural non-refractory clay & ceramic prod.	..	..	..	..	..	..	..	..	..	..
2694 Cement, lime and plaster	46,215	0	443	54	7,936	0	32	1,731,868	0	17,350
2695 Articles of concrete, cement and plaster	718	551	2,619	474	6,721	0	..	..	..	..
2696 Cutting, shaping and finishing of stone	..	..	..	..	..	..	..	..	..	..
2699 Other non-metallic mineral products, nec	..	..	..	..	..	..	..	..	..	..

ISIS Energy Data Programme (IEA/OECD)

ISIC Revision 3 Industry Sector		Solid TJ	LPG TJ	Distiloil TJ	RFO TJ	Gas TJ	Biomass TJ	Steam TJ	Electr MWh	Own Use MWh	TOTAL TJ
27	**BASIC METALS**	146,576	1,573	8,366	23,826	77,439	0	44	15,084,110	1,419,000	307,018
271	Basic iron and steel	138,669	590	6,488	21,564	50,303	0	0	9,145,022	1,419,000	245,428
272	Basic precious and non-ferrous metals	4,775	545	1,457	2,227	17,395	0	10	3,906,873	0	40,474
273	Casting of metals	3,132	438	421	35	9,741	0	34	2,032,215	0	21,117
2731	Casting of iron and steel	..	..	..	..	..	..	..	..	..	..
2732	Casting non-ferrous metals	..	..	..	..	..	..	..	..	..	..
28	**FABRICATED METAL PRODUCTS**	119	35	2,878	449	16,132	0	45	2,855,500	0	29,938
281	Str. metal prod., tanks, reservoirs, steam generators	..	..	..	..	..	..	..	..	..	..
2811	Structural metal products	..	..	..	..	..	..	..	..	..	..
2812	Tanks, reservoirs and containers of metal	..	..	..	..	..	..	..	..	..	..
2813	Steam generators, exc. central heating hot water boilers	..	..	..	..	..	..	..	..	..	..
289	Other fabricated metal products	..	..	..	..	..	..	..	..	..	..
2891	Forging, pressing, stamping & roll-forming of metal	..	..	..	..	..	..	..	..	..	..
2892	Treatment and coating of metals	..	..	..	..	..	..	..	..	..	..
2893	Cutlery, hand tools and general hardware	..	..	..	..	..	..	..	..	..	..
2899	Other fabricated metal products, nec	..	..	..	..	..	..	..	..	..	..
29	**MACHINERY AND EQUIPMENT, NEC**	1,313	1,788	9,657	5,518	18,687	0	28	5,471,580	0	56,689
291	General purpose machinery	..	..	..	..	..	..	..	..	..	..
2911	Engines and turbines	..	..	..	..	..	..	..	..	..	..
2912	Pumps, compressors, taps and valves	..	..	..	..	..	..	..	..	..	..
2913	Bearings, gears, gearing and driving elements	..	..	..	..	..	..	..	..	..	..
2914	Ovens, furnaces and furnace burners	..	..	..	..	..	..	..	..	..	..
2915	Lifting and handling equipment	..	..	..	..	..	..	..	..	..	..
2919	Other general purpose machinery	..	..	..	..	..	..	..	..	..	..
292	Special purpose machinery	..	..	..	..	..	..	..	..	..	..
2921	Agricultural and forestry machinery	..	..	..	..	..	..	..	..	..	..
2922	Machine-tools	..	..	..	..	..	..	..	..	..	..
2923	Machinery for metallurgy	..	..	..	..	..	..	..	..	..	..
2924	Machinery for mining, quarrying and construction	..	..	..	..	..	..	..	..	..	..
2925	Machinery for food, beverage & tobacco processing	..	..	..	..	..	..	..	..	..	..
2926	Machinery for textile, apparel & leather production	..	..	..	..	..	..	..	..	..	..
2927	Machinery for weapons and ammunition	..	..	..	..	..	..	..	..	..	..
2929	Other special purpose machinery	..	..	..	..	..	..	..	..	..	..
293	Domestic appliances, nec	..	..	..	..	..	..	..	..	..	..
30	**OFFICE, ACCOUNTING & COMPUTING MACHINERY**	0	51	60	618	1,187	0	26	524,225	0	3,829
31	**ELECTRICAL MACHINERY & APPARATUS, NEC**	374	859	1,153	3,033	7,195	0	0	1,857,952	0	19,303
311	Electric motors, generators and transformers	..	..	..	..	..	..	..	..	..	..
312	Electricity distribution and control apparatus	..	..	..	..	..	..	..	..	..	..
313	Insulated wire and cable	..	..	..	..	..	..	..	..	..	..
314	Accumulators, primary cells & primary batteries	..	..	..	..	..	..	..	..	..	..
315	Electric lamps and lighting equipment	..	..	..	..	..	..	..	..	..	..
319	Other electrical equipment, nec	..	..	..	..	..	..	..	..	..	..
32	**RADIO, TV & COMMUNICATION EQUIP. & APP.**	143	69	985	1,773	6,083	0	9	2,156,384	0	16,825
321	Electronic valves, tubes, other electronic components	..	..	..	..	..	..	..	..	..	..
322	TV & radio transmitters, apparatus for line telephony	..	..	..	..	..	..	..	..	..	..
323	TV & radio receivers, recording apparatus	..	..	..	..	..	..	..	..	..	..
33	**MEDICAL PRECISION &OPTICAL INSTRUMENTS**	0	0	1,106	780	2,191	0	40	601,899	0	6,284
331	Medical appliances and instruments	..	..	..	..	..	..	..	..	..	..
3311	Medical, surgical equipment & orthopaedic app.	..	..	..	..	..	..	..	..	..	..
3312	Instruments & appliances for measuring, checking etc	..	..	..	..	..	..	..	..	..	..
3313	Industrial process control equipment	..	..	..	..	..	..	..	..	..	..
332	Optical instruments and photographic equipment	..	..	..	..	..	..	..	..	..	..
333	Watches and clocks	..	..	..	..	..	..	..	..	..	..
34	**MOTOR VEHICLES, TRAILERS & SEMI-TRAILERS**	4,018	200	3,906	1,417	20,439	0	28	3,386,717	0	42,200
341	Motor vehicles	..	..	..	..	..	..	..	..	..	..
342	Bodies (coachwork) for motor vehicles	..	..	..	..	..	..	..	..	..	..
343	Parts, accessories for motor vehicles & their engines	..	..	..	..	..	..	..	..	..	..
35	**OTHER TRANSPORT EQUIPMENT**	3,977	868	2,434	3,117	11,630	0	29	2,812,073	0	32,178
351	Building and repairing of ships and boats	..	..	..	..	..	..	..	..	..	..
3511	Building and repairing of ships	..	..	..	..	..	..	..	..	..	..
3512	Building, repairing of pleasure & sporting boats	..	..	..	..	..	..	..	..	..	..
352	Railway, tramway locomotives & rolling stock	..	..	..	..	..	..	..	..	..	..
353	Aircraft and spacecraft	..	..	..	..	..	..	..	..	..	..
359	Transport equipment, nec	..	..	..	..	..	..	..	..	..	..
3591	Motorcycles	..	..	..	..	..	..	..	..	..	..
3592	Bicycles and invalid carriages	..	..	..	..	..	..	..	..	..	..
3599	Other transport equipment, nec	..	..	..	..	..	..	..	..	..	..
36	**FURNITURE; MANUFACTURING, NEC**	546	212	14,542	9,538	1,885	0	9	927,485	0	30,071
361	Furniture	..	..	..	..	..	..	..	..	..	..
369	Manufacturing, nec	..	..	..	..	..	..	..	..	..	..
3691	Jewellery and related articles	..	..	..	..	..	..	..	..	..	..
3692	Musical instruments	..	..	..	..	..	..	..	..	..	..
3693	Sports goods	..	..	..	..	..	..	..	..	..	..
3694	Games and toys	..	..	..	..	..	..	..	..	..	..
3699	Other manufacturing, nec	..	..	..	..	..	..	..	..	..	..
37	**RECYCLING**	..	..	..	..	..	..	..	..	..	..
371	Recycling of metal waste and scrap	..	..	..	..	..	..	..	..	..	..
372	Recycling of non-metal waste and scrap	..	..	..	..	..	..	..	..	..	..
CERR	**Non-specified industry**	0	0	0	0	0	0	0	0	3,371,000	-12,136
15-37	**TOTAL MANUFACTURING**	368,158	26,915	239,674	226,102	409,495	16,055	1,018	94,369,050	11,989,000	**1,583,985**

ISIC Revision 3 Industry Sector	Solid TJ	LPG TJ	Distiloil TJ	RFO TJ	Gas TJ	Biomass TJ	Steam TJ	Electr MWh	Own Use MWh	TOTAL TJ
15 FOOD PRODUCTS AND BEVERAGES	17,024	4,645	24,194	28,448	57,537	0	185	11,123,644	696,000	169,573
151 Production, processing and preserving (PPP)	..	..	..	..	..	..	..	..	..	..
1511 PPP of meat and meat products	..	..	..	..	..	..	..	..	..	..
1512 Processing and preserving of fish products	..	..	..	..	..	..	..	..	..	..
1513 Processing, preserving of fruit & vegetables	..	..	..	..	..	..	..	..	..	..
1514 Vegetable and animal oils and fats	..	..	..	..	..	..	..	..	..	..
152 Dairy products	..	..	..	..	..	..	..	..	..	..
153 Grain mill prod., starches & prepared animal feeds	..	..	..	..	..	..	..	..	..	..
1531 Grain mill products	..	..	..	..	..	..	..	..	..	..
1532 Starches and starch products	..	..	..	..	..	..	..	..	..	..
1533 Prepared animal feeds	..	..	..	..	..	..	..	..	..	..
154 Other food products	..	..	..	..	..	..	..	..	..	..
1541 Bakery products	..	..	..	..	..	..	..	..	..	..
1542 Sugar	..	..	..	..	..	..	..	..	..	..
1543 Cocoa, chocolate and sugar confectionery	..	..	..	..	..	..	..	..	..	..
1544 Macaroni, noodles, couscous & similar farinaceous prod.	..	..	..	..	..	..	..	..	..	..
1549 Other food products, nec	..	..	..	..	..	..	..	..	..	..
155 Beverages	..	..	..	..	..	..	..	..	..	..
1551 Distilling, rectifying and blending of spirits	..	..	..	..	..	..	..	..	..	..
1552 Wines	..	..	..	..	..	..	..	..	..	..
1553 Malt liquors and malt	..	..	..	..	..	..	..	..	..	..
1554 Soft drinks; production of mineral waters	..	..	..	..	..	..	..	..	..	..
16 TOBACCO PRODUCTS	0	89	102	271	871	0	0	145,826	0	1,858
17 TEXTILES	8,584	1,093	2,576	4,828	16,921	0	18	2,419,179	0	42,729
171 Spinning, weaving and finishing of textiles	..	..	..	..	..	..	..	..	..	..
1711 Preparation and spinning of textile fibres	..	..	..	..	..	..	..	..	..	..
1712 Finishing of textiles	..	..	..	..	..	..	..	..	..	..
172 Other textiles	..	..	..	..	..	..	..	..	..	..
1721 Made-up textile articles, except apparel	..	..	..	..	..	..	..	..	..	..
1722 Carpets and rugs	..	..	..	..	..	..	..	..	..	..
1723 Cordage, rope, twine and netting	..	..	..	..	..	..	..	..	..	..
1729 Other textiles, nec	..	..	..	..	..	..	..	..	..	..
173 Knitted and crocheted fabrics and articles	..	..	..	..	..	..	..	..	..	..
18 WEARING APPAREL, DRESSING & DYEING OF FUR	22	102	1,022	3,083	2,913	0	59	463,938	0	8,871
181 Wearing apparel, except fur apparel	..	..	..	..	..	..	..	..	..	..
182 Dressing and dyeing of fur; articles of fur	..	..	..	..	..	..	..	..	..	..
19 TANNING & DRESSING OF LEATHER, FOOTWEAR	623	118	281	203	1,155	0	0	198,833	0	3,096
191 Tanning and dressing of leather	..	..	..	..	..	..	..	..	..	..
1911 Tanning and dressing of leather	..	..	..	..	..	..	..	..	..	..
1912 Luggage, handbags, saddlery & harness	..	..	..	..	..	..	..	..	..	..
192 Footwear	..	..	..	..	..	..	..	..	..	..
20 WOOD AND WOOD PRODUCTS	403	1,416	3,218	163	4,175	0	19	1,254,014	0	13,908
201 Sawmilling and planing of wood	..	..	..	..	..	..	..	..	..	..
202 Products of wood, cork, straw & plaiting materials	..	..	..	..	..	..	..	..	..	..
2021 Veneer sheets	..	..	..	..	..	..	..	..	..	..
2022 Builders' carpentry and joinery	..	..	..	..	..	..	..	..	..	..
2023 Wooden containers	..	..	..	..	..	..	..	..	..	..
2029 Other products of wood	..	..	..	..	..	..	..	..	..	..
21 PAPER AND PAPER PRODUCTS	24,365	1,330	3,903	9,528	25,294	0	11	5,747,809	1,200,000	80,803
2101 Pulp, paper and paperboard	..	..	..	..	..	..	..	..	..	..
2102 Corrugated paper, paperboard and their containers	..	..	..	..	..	..	..	..	..	..
2109 Other articles of pulp and paperboard	..	..	..	..	..	..	..	..	..	..
22 PUBLISHING, PRINTING & REPRODUCTION	97	256	1,116	536	8,282	0	0	2,997,951	0	21,080
221 Publishing	..	..	..	..	..	..	..	..	..	..
2211 Publishing of books & brochures	..	..	..	..	..	..	..	..	..	..
2212 Publishing of newspapers and periodicals	..	..	..	..	..	..	..	..	..	..
2213 Publishing of recorded media	..	..	..	..	..	..	..	..	..	..
2219 Other publishing	..	..	..	..	..	..	..	..	..	..
222 Printing and related service activities	..	..	..	..	..	..	..	..	..	..
2221 Printing	..	..	..	..	..	..	..	..	..	..
2222 Service activities related to printing	..	..	..	..	..	..	..	..	..	..
223 Reproduction of recorded media	..	..	..	..	..	..	..	..	..	..
23 COKE, REFINED PETROLEUM PRODUCTS	44,908	0	178,232	96,900	10,384	0	0	7,638,000	2,667,000	348,320
231 Coke oven products	..	..	..	..	..	..	..	..	..	..
232 Refined petroleum products	..	..	..	..	..	..	..	..	..	..
233 Processing of nuclear fuel	..	..	..	..	..	..	..	..	..	..
24 CHEMICALS & CHEMICAL PRODUCTS	29,395	5,414	13,825	33,054	76,007	0	161	18,526,590	3,599,000	211,595
241 Basic chemicals	21,629	5,247	9,170	21,655	56,236	0	74	12,964,407	0	160,683
2411 Basic chemicals, exc. fertilizers & nitrogen compounds	19,975	5,206	8,724	17,323	48,764	0	54	10,793,780	0	138,904
2412 Fertilizers and nitrogen compounds	0	0	110	234	3,076	0	0	338,894	0	4,640
2413 Plastics in primary forms and synthetic rubber	1,654	41	336	4,098	4,396	0	20	1,831,733	0	17,139
242 Other chemical products	4,457	166	4,271	5,942	18,446	0	75	4,740,359	0	50,422
2421 Pesticides and other agro-chemical products	0	0	1,592	586	1,296	0	20	296,573	0	4,562
2422 Paints, varnishes and similar coatings	12	1	1,092	1,401	1,198	0	0	528,857	0	5,608
2423 Pharmaceuticals, medicinal chem. & botanical prod.	1,717	8	1,004	2,270	7,115	0	0	2,015,578	0	19,370
2424 Soap and detergents, perfumes etc.	0	75	87	897	2,678	0	0	773,415	0	6,521
2429 Other chemical products, nec	2,728	82	496	788	6,159	0	55	1,125,936	0	14,361
243 Man-made fibres	3,309	1	384	5,457	1,325	0	12	821,824	0	13,447
25 RUBBER AND PLASTICS PRODUCTS	21,681	6,459	5,811	6,860	21,398	0	65	6,115,159	0	84,289
251 Rubber products	4,437	166	2,122	1,184	5,146	0	18	1,209,019	0	17,425
2511 Rubber tyres and tubes	..	..	..	..	..	..	..	..	..	..
2519 Other rubber products	..	..	..	..	..	..	..	..	..	..
252 Plastic products	17,244	6,293	3,689	5,676	16,252	0	47	4,906,140	0	66,863
26 OTHER NON-METALLIC MINERAL PRODUCTS	26,968	2,809	8,019	13,213	32,591	0	256	4,789,350	0	101,098
261 Glass and glass products	53	0	1,053	11,023	6,283	0	69	862,299	0	21,585
269 Non-metallic mineral products, nec	26,915	2,809	6,966	2,190	26,308	0	187	3,927,051	0	79,512
2691 Pottery, china and earthenware	218	0	2,330	1,444	5,586	0	145	376,759	0	11,079
2692 Refractory ceramic products	763	2,221	867	293	6,652	0	0	336,960	0	12,009
2693 Structural non-refractory clay & ceramic prod.	..	..	..	..	..	..	..	..	..	..
2694 Cement, lime and plaster	25,481	0	716	0	8,314	0	8	1,591,938	0	40,250
2695 Articles of concrete, cement and plaster	453	588	3,053	453	5,756	0	34	1,621,394	0	16,174
2696 Cutting, shaping and finishing of stone	..	..	..	..	..	..	..	..	..	..
2699 Other non-metallic mineral products, nec	..	..	..	..	..	..	..	..	..	..

ISIS Energy Data Programme (IEA/OECD)

ISIC Revision 3 Industry Sector	Solid TJ	LPG TJ	Distiloil TJ	RFO TJ	Gas TJ	Biomass TJ	Steam TJ	Electr MWh	Own Use MWh	TOTAL TJ
27 BASIC METALS	148,804	1,512	7,788	31,224	84,937	0	42	14,339,790	1,393,000	320,915
271 Basic iron and steel	139,225	559	6,486	29,714	56,171	0	0	9,003,355	1,393,000	259,552
272 Basic precious and non-ferrous metals	5,930	600	1,204	1,472	19,344	0	10	3,602,434	0	41,529
273 Casting of metals	3,649	353	98	38	9,422	0	32	1,734,001	0	19,834
2731 Casting of iron and steel	..	..	..	..	..	..	..	..	..	..
2732 Casting non-ferrous metals	..	..	..	..	..	..	..	..	..	..
28 FABRICATED METAL PRODUCTS	129	33	3,015	419	15,471	0	44	2,718,888	0	28,899
281 Str. metal prod., tanks, reservoirs, steam generators	..	..	..	..	..	..	..	..	..	..
2811 Structural metal products	..	..	..	..	..	..	..	..	..	..
2812 Tanks, reservoirs and containers of metal	..	..	..	..	..	..	..	..	..	..
2813 Steam generators, exc. central heating hot water boilers	..	..	..	..	..	..	..	..	..	..
289 Other fabricated metal products	..	..	..	..	..	..	..	..	..	..
2891 Forging, pressing, stamping & roll-forming of metal	..	..	..	..	..	..	..	..	..	..
2892 Treatment and coating of metals	..	..	..	..	..	..	..	..	..	..
2893 Cutlery, hand tools and general hardware	..	..	..	..	..	..	..	..	..	..
2899 Other fabricated metal products, nec	..	..	..	..	..	..	..	..	..	..
29 MACHINERY AND EQUIPMENT, NEC	1,581	1,870	9,692	5,667	18,670	0	28	5,410,482	0	56,986
291 General purpose machinery	..	..	..	..	..	..	..	..	..	..
2911 Engines and turbines	..	..	..	..	..	..	..	..	..	..
2912 Pumps, compressors, taps and valves	..	..	..	..	..	..	..	..	..	..
2913 Bearings, gears, gearing and driving elements	..	..	..	..	..	..	..	..	..	..
2914 Ovens, furnaces and furnace burners	..	..	..	..	..	..	..	..	..	..
2915 Lifting and handling equipment	..	..	..	..	..	..	..	..	..	..
2919 Other general purpose machinery	..	..	..	..	..	..	..	..	..	..
292 Special purpose machinery	..	..	..	..	..	..	..	..	..	..
2921 Agricultural and forestry machinery	..	..	..	..	..	..	..	..	..	..
2922 Machine-tools	..	..	..	..	..	..	..	..	..	..
2923 Machinery for metallurgy	..	..	..	..	..	..	..	..	..	..
2924 Machinery for mining, quarrying and construction	..	..	..	..	..	..	..	..	..	..
2925 Machinery for food, beverage & tobacco processing	..	..	..	..	..	..	..	..	..	..
2926 Machinery for textile, apparel & leather production	..	..	..	..	..	..	..	..	..	..
2927 Machinery for weapons and ammunition	..	..	..	..	..	..	..	..	..	..
2929 Other special purpose machinery	..	..	..	..	..	..	..	..	..	..
293 Domestic appliances, nec	..	..	..	..	..	..	..	..	..	..
30 OFFICE, ACCOUNTING & COMPUTING MACHINERY	0	65	62	479	1,005	0	29	488,897	0	3,400
31 ELECTRICAL MACHINERY & APPARATUS, NEC	501	988	1,103	3,394	6,158	0	0	1,920,536	0	19,058
311 Electric motors, generators and transformers	..	..	..	..	..	..	..	..	..	..
312 Electricity distribution and control apparatus	..	..	..	..	..	..	..	..	..	..
313 Insulated wire and cable	..	..	..	..	..	..	..	..	..	..
314 Accumulators, primary cells & primary batteries	..	..	..	..	..	..	..	..	..	..
315 Electric lamps and lighting equipment	..	..	..	..	..	..	..	..	..	..
319 Other electrical equipment, nec	..	..	..	..	..	..	..	..	..	..
32 RADIO, TV & COMMUNICATION EQUIP. & APP.	136	57	1,082	2,173	4,834	0	10	2,133,105	0	15,971
321 Electronic valves, tubes, other electronic components	..	..	..	..	..	..	..	..	..	..
322 TV & radio transmitters, apparatus for line telephony	..	..	..	..	..	..	..	..	..	..
323 TV & radio receivers, recording apparatus	..	..	..	..	..	..	..	..	..	..
33 MEDICAL PRECISION &OPTICAL INSTRUMENTS	0	0	999	756	1,908	0	42	621,181	0	5,941
331 Medical appliances and instruments	..	..	..	..	..	..	..	..	..	..
3311 Medical, surgical equipment & orthopaedic app.	..	..	..	..	..	..	..	..	..	..
3312 Instruments & appliances for measuring, checking etc	..	..	..	..	..	..	..	..	..	..
3313 Industrial process control equipment	..	..	..	..	..	..	..	..	..	..
332 Optical instruments and photographic equipment	..	..	..	..	..	..	..	..	..	..
333 Watches and clocks	..	..	..	..	..	..	..	..	..	..
34 MOTOR VEHICLES, TRAILERS & SEMI-TRAILERS	3,447	296	3,420	1,539	19,780	0	28	3,392,428	0	40,723
341 Motor vehicles	..	..	..	..	..	..	..	..	..	..
342 Bodies (coachwork) for motor vehicles	..	..	..	..	..	..	..	..	..	..
343 Parts, accessories for motor vehicles & their engines	..	..	..	..	..	..	..	..	..	..
35 OTHER TRANSPORT EQUIPMENT	3,236	836	2,665	2,956	10,140	0	28	2,783,102	0	29,880
351 Building and repairing of ships and boats	..	..	..	..	..	..	..	..	..	..
3511 Building and repairing of ships	..	..	..	..	..	..	..	..	..	..
3512 Building, repairing of pleasure & sporting boats	..	..	..	..	..	..	..	..	..	..
352 Railway, tramway locomotives & rolling stock	..	..	..	..	..	..	..	..	..	..
353 Aircraft and spacecraft	..	..	..	..	..	..	..	..	..	..
359 Transport equipment, nec	..	..	..	..	..	..	..	..	..	..
3591 Motorcycles	..	..	..	..	..	..	..	..	..	..
3592 Bicycles and invalid carriages	..	..	..	..	..	..	..	..	..	..
3599 Other transport equipment, nec	..	..	..	..	..	..	..	..	..	..
36 FURNITURE; MANUFACTURING, NEC	391	209	15,818	8,370	3,483	0	9	902,993	0	31,531
361 Furniture	..	..	..	..	..	..	..	..	..	..
369 Manufacturing, nec	..	..	..	..	..	..	..	..	..	..
3691 Jewellery and related articles	..	..	..	..	..	..	..	..	..	..
3692 Musical instruments	..	..	..	..	..	..	..	..	..	..
3693 Sports goods	..	..	..	..	..	..	..	..	..	..
3694 Games and toys	..	..	..	..	..	..	..	..	..	..
3699 Other manufacturing, nec	..	..	..	..	..	..	..	..	..	..
37 RECYCLING										
371 Recycling of metal waste and scrap	..	..	..	..	..	..	..	..	..	..
372 Recycling of non-metal waste and scrap	..	..	..	..	..	..	..	..	..	..
CERR Non-specified industry	0	0	0	0	0	0	0	0	2,225,000	-8,010
15-37 TOTAL MANUFACTURING	332,295	29,597	287,943	254,064	423,914	15,481	1,034	96,131,695	11,780,000	1,647,994

ISIC Revision 3 Industry Sector	Solid TJ	LPG TJ	Distiloil TJ	RFO TJ	Gas TJ	Biomass TJ	Steam TJ	Electr MWh	Own Use MWh	TOTAL TJ
15 FOOD PRODUCTS AND BEVERAGES	16,541	4,639	28,065	25,260	68,624	73	189	10,191,466	685,000	177,614
151 Production, processing and preserving (PPP)	..	..	..	..	..	..	..	..	..	..
1511 PPP of meat and meat products	..	..	..	..	..	..	..	..	..	..
1512 Processing and preserving of fish products	..	..	..	..	..	..	..	..	..	..
1513 Processing, preserving of fruit & vegetables	..	..	..	..	..	..	..	..	..	..
1514 Vegetable and animal oils and fats	..	..	..	..	..	..	..	..	..	..
152 Dairy products	..	..	..	..	..	..	..	..	..	..
153 Grain mill prod., starches & prepared animal feeds	..	..	..	..	..	..	..	..	..	..
1531 Grain mill products	..	..	..	..	..	..	..	..	..	..
1532 Starches and starch products	..	..	..	..	..	..	..	..	..	..
1533 Prepared animal feeds	..	..	..	..	..	..	..	..	..	..
154 Other food products	..	..	..	..	..	..	..	..	..	..
1541 Bakery products	..	..	..	..	..	..	..	..	..	..
1542 Sugar	..	..	..	..	..	..	..	..	..	..
1543 Cocoa, chocolate and sugar confectionery	..	..	..	..	..	..	..	..	..	..
1544 Macaroni, noodles, couscous & similar farinaceous prod.	..	..	..	..	..	..	..	..	..	..
1549 Other food products, nec	..	..	..	..	..	..	..	..	..	..
155 Beverages	..	..	..	..	..	..	..	..	..	..
1551 Distilling, rectifying and blending of spirits	..	..	..	..	..	..	..	..	..	..
1552 Wines	..	..	..	..	..	..	..	..	..	..
1553 Malt liquors and malt	..	..	..	..	..	..	..	..	..	..
1554 Soft drinks; production of mineral waters	..	..	..	..	..	..	..	..	..	..
16 TOBACCO PRODUCTS	0	88	221	237	1,117	0	0	124,344	0	2,111
17 TEXTILES	6,781	1,152	2,450	4,670	13,368	0	18	2,959,263	0	39,092
171 Spinning, weaving and finishing of textiles	..	..	..	..	..	..	..	..	..	..
1711 Preparation and spinning of textile fibres	..	..	..	..	..	..	..	..	..	..
1712 Finishing of textiles	..	..	..	..	..	..	..	..	..	..
172 Other textiles	..	..	..	..	..	..	..	..	..	..
1721 Made-up textile articles, except apparel	..	..	..	..	..	..	..	..	..	..
1722 Carpets and rugs	..	..	..	..	..	..	..	..	..	..
1723 Cordage, rope, twine and netting	..	..	..	..	..	..	..	..	..	..
1729 Other textiles, nec	..	..	..	..	..	..	..	..	..	..
173 Knitted and crocheted fabrics and articles	..	..	..	..	..	..	..	..	..	..
18 WEARING APPAREL, DRESSING & DYEING OF FUR	18	115	1,054	2,964	2,205	0	61	551,225	0	8,401
181 Wearing apparel, except fur apparel	..	..	..	..	..	..	..	..	..	..
182 Dressing and dyeing of fur; articles of fur	..	..	..	..	..	..	..	..	..	..
19 TANNING & DRESSING OF LEATHER, FOOTWEAR	519	132	373	162	895	0	0	234,372	0	2,925
191 Tanning and dressing of leather	..	..	..	..	..	..	..	..	..	..
1911 Tanning and dressing of leather	..	..	..	..	..	..	..	..	..	..
1912 Luggage, handbags, saddlery & harness	..	..	..	..	..	..	..	..	..	..
192 Footwear	..	..	..	..	..	..	..	..	..	..
20 WOOD AND WOOD PRODUCTS	907	1,600	3,679	186	6,011	0	20	1,367,820	0	17,327
201 Sawmilling and planing of wood	..	..	..	..	..	..	..	..	..	..
202 Products of wood, cork, straw & plaiting materials	..	..	..	..	..	..	..	..	..	..
2021 Veneer sheets	..	..	..	..	..	..	..	..	..	..
2022 Builders' carpentry and joinery	..	..	..	..	..	..	..	..	..	..
2023 Wooden containers	..	..	..	..	..	..	..	..	..	..
2029 Other products of wood	..	..	..	..	..	..	..	..	..	..
21 PAPER AND PAPER PRODUCTS	21,899	1,376	4,309	9,611	34,309	0	12	5,652,871	1,214,000	87,496
2101 Pulp, paper and paperboard	..	..	..	..	..	..	..	..	..	..
2102 Corrugated paper, paperboard and their containers	..	..	..	..	..	..	..	..	..	..
2109 Other articles of pulp and paperboard	..	..	..	..	..	..	..	..	..	..
22 PUBLISHING, PRINTING & REPRODUCTION	105	346	830	452	11,069	0	0	3,174,299	0	24,229
221 Publishing	..	..	..	..	..	..	..	..	..	..
2211 Publishing of books & brochures	..	..	..	..	..	..	..	..	..	..
2212 Publishing of newspapers and periodicals	..	..	..	..	..	..	..	..	..	..
2213 Publishing of recorded media	..	..	..	..	..	..	..	..	..	..
2219 Other publishing	..	..	..	..	..	..	..	..	..	..
222 Printing and related service activities	..	..	..	..	..	..	..	..	..	..
2221 Printing	..	..	..	..	..	..	..	..	..	..
2222 Service activities related to printing	..	..	..	..	..	..	..	..	..	..
223 Reproduction of recorded media	..	..	..	..	..	..	..	..	..	..
23 COKE, REFINED PETROLEUM PRODUCTS	40,970	0	178,856	100,684	8,048	0	0	5,913,000	2,704,000	340,110
231 Coke oven products	..	..	..	..	..	..	..	..	..	..
232 Refined petroleum products	..	..	..	..	..	..	..	..	..	..
233 Processing of nuclear fuel	..	..	..	..	..	..	..	..	..	..
24 CHEMICALS & CHEMICAL PRODUCTS	34,576	7,574	18,585	44,072	91,325	695	164	17,479,890	2,875,000	249,569
241 Basic chemicals	21,954	7,418	11,171	27,175	63,505	0	81	12,017,066	0	174,565
2411 Basic chemicals, exc. fertilizers & nitrogen compounds	20,014	7,358	10,891	19,676	55,817	0	60	10,179,357	0	150,462
2412 Fertilizers and nitrogen compounds	0	0	92	325	3,835	0	0	188,908	0	4,932
2413 Plastics in primary forms and synthetic rubber	1,940	60	188	7,174	3,853	0	21	1,648,801	0	19,172
242 Other chemical products	4,654	154	3,889	9,418	22,701	0	72	4,663,018	0	57,675
2421 Pesticides and other agro-chemical products	0	0	1,063	1,085	1,701	0	18	318,419	0	5,013
2422 Paints, varnishes and similar coatings	16	1	1,324	2,417	1,257	0	0	507,191	0	6,841
2423 Pharmaceuticals, medicinal chem. & botanical prod.	1,873	12	872	3,471	8,414	0	0	1,915,004	0	21,536
2424 Soap and detergents, perfumes etc.	0	105	65	1,126	3,217	0	0	743,857	0	7,191
2429 Other chemical products, nec	2,765	36	565	1,319	8,112	0	54	1,178,547	0	17,094
243 Man-made fibres	1,832	2	345	7,479	1,068	0	11	799,806	0	13,616
25 RUBBER AND PLASTICS PRODUCTS	20,830	7,692	5,958	6,660	28,648	0	72	6,232,828	0	92,298
251 Rubber products	5,281	198	2,143	1,304	6,384	0	20	1,215,608	0	19,706
2511 Rubber tyres and tubes	..	..	..	..	..	..	..	..	..	..
2519 Other rubber products	..	..	..	..	..	..	..	..	..	..
252 Plastic products	15,549	7,494	3,815	5,356	22,264	0	52	5,017,220	0	72,592
26 OTHER NON-METALLIC MINERAL PRODUCTS	31,065	3,213	8,084	14,533	33,303	0	254	4,748,742	0	107,547
261 Glass and glass products	290	0	1,276	12,201	8,133	0	64	1,098,722	0	25,919
269 Non-metallic mineral products, nec	30,775	3,213	6,808	2,332	25,170	0	190	3,650,020	0	81,628
2691 Pottery, china and earthenware	62	0	1,565	963	6,226	0	145	630,414	0	11,230
2692 Refractory ceramic products	837	2,869	1,381	747	6,791	0	0	337,722	0	13,841
2693 Structural non-refractory clay & ceramic prod.	..	..	..	..	..	..	..	..	..	..
2694 Cement, lime and plaster	29,616	0	701	0	7,143	0	8	1,395,917	0	42,493
2695 Articles of concrete, cement and plaster	260	344	3,161	622	5,010	0	37	1,285,967	0	14,063
2696 Cutting, shaping and finishing of stone	..	..	..	..	..	..	..	..	..	..
2699 Other non-metallic mineral products, nec	..	..	..	..	..	..	..	..	..	..

ISIC Revision 3 Industry Sector		Solid TJ	LPG TJ	Distiloil TJ	RFO TJ	Gas TJ	Biomass TJ	Steam TJ	Electr MWh	Own Use MWh	TOTAL TJ
27	**BASIC METALS**	170,432	1,631	8,319	31,901	109,585	1,114	42	14,572,390	1,346,000	370,639
271	Basic iron and steel	141,897	604	5,972	29,793	68,334	0	0	9,327,266	1,346,000	275,333
272	Basic precious and non-ferrous metals	7,901	687	1,390	2,068	23,722	0	10	3,566,615	0	48,618
273	Casting of metals	5,290	340	594	40	10,705	0	32	1,678,509	0	23,044
2731	Casting of iron and steel	..	..	..	..	..	..	..	..	..	..
2732	Casting non-ferrous metals	..	..	..	..	..	..	..	..	..	..
28	**FABRICATED METAL PRODUCTS**	79	32	2,227	350	11,598	0	45	2,704,731	0	24,068
281	Str. metal prod., tanks, reservoirs, steam generators	..	..	..	..	..	..	..	..	..	..
2811	Structural metal products	..	..	..	..	..	..	..	..	..	..
2812	Tanks, reservoirs and containers of metal	..	..	..	..	..	..	..	..	..	..
2813	Steam generators, exc. central heating hot water boilers	..	..	..	..	..	..	..	..	..	..
289	Other fabricated metal products	..	..	..	..	..	..	..	..	..	..
2891	Forging, pressing, stamping & roll-forming of metal	..	..	..	..	..	..	..	..	..	..
2892	Treatment and coating of metals	..	..	..	..	..	..	..	..	..	..
2893	Cutlery, hand tools and general hardware	..	..	..	..	..	..	..	..	..	..
2899	Other fabricated metal products, nec	..	..	..	..	..	..	..	..	..	..
29	**MACHINERY AND EQUIPMENT, NEC**	1,034	1,896	9,419	4,940	13,876	0	29	5,319,969	0	50,346
291	General purpose machinery	..	..	..	..	..	..	..	..	..	..
2911	Engines and turbines	..	..	..	..	..	..	..	..	..	..
2912	Pumps, compressors, taps and valves	..	..	..	..	..	..	..	..	..	..
2913	Bearings, gears, gearing and driving elements	..	..	..	..	..	..	..	..	..	..
2914	Ovens, furnaces and furnace burners	..	..	..	..	..	..	..	..	..	..
2915	Lifting and handling equipment	..	..	..	..	..	..	..	..	..	..
2919	Other general purpose machinery	..	..	..	..	..	..	..	..	..	..
292	Special purpose machinery	..	..	..	..	..	..	..	..	..	..
2921	Agricultural and forestry machinery	..	..	..	..	..	..	..	..	..	..
2922	Machine-tools	..	..	..	..	..	..	..	..	..	..
2923	Machinery for metallurgy	..	..	..	..	..	..	..	..	..	..
2924	Machinery for mining, quarrying and construction	..	..	..	..	..	..	..	..	..	..
2925	Machinery for food, beverage & tobacco processing	..	..	..	..	..	..	..	..	..	..
2926	Machinery for textile, apparel & leather production	..	..	..	..	..	..	..	..	..	..
2927	Machinery for weapons and ammunition	..	..	..	..	..	..	..	..	..	..
2929	Other special purpose machinery	..	..	..	..	..	..	..	..	..	..
293	Domestic appliances, nec	..	..	..	..	..	..	..	..	..	..
30	**OFFICE, ACCOUNTING & COMPUTING MACHINERY**	0	71	76	210	635	0	37	424,115	0	2,556
31	**ELECTRICAL MACHINERY & APPARATUS, NEC**	119	1,013	814	3,249	4,115	0	0	1,965,687	0	16,386
311	Electric motors, generators and transformers	..	..	..	..	..	..	..	..	..	..
312	Electricity distribution and control apparatus	..	..	..	..	..	..	..	..	..	..
313	Insulated wire and cable	..	..	..	..	..	..	..	..	..	..
314	Accumulators, primary cells & primary batteries	..	..	..	..	..	..	..	..	..	..
315	Electric lamps and lighting equipment	..	..	..	..	..	..	..	..	..	..
319	Other electrical equipment, nec	..	..	..	..	..	..	..	..	..	..
32	**RADIO, TV & COMMUNICATION EQUIP. & APP.**	0	40	1,144	2,211	2,867	0	12	2,022,775	0	13,556
321	Electronic valves, tubes, other electronic components	..	..	..	..	..	..	..	..	..	..
322	TV & radio transmitters, apparatus for line telephony	..	..	..	..	..	..	..	..	..	..
323	TV & radio receivers, recording apparatus	..	..	..	..	..	..	..	..	..	..
33	**MEDICAL PRECISION &OPTICAL INSTRUMENTS**	0	0	962	614	1,200	0	41	588,323	0	4,935
331	Medical appliances and instruments	..	..	..	..	..	..	..	..	..	..
3311	Medical, surgical equipment & orthopaedic app.	..	..	..	..	..	..	..	..	..	..
3312	Instruments & appliances for measuring, checking etc	..	..	..	..	..	..	..	..	..	..
3313	Industrial process control equipment	..	..	..	..	..	..	..	..	..	..
332	Optical instruments and photographic equipment	..	..	..	..	..	..	..	..	..	..
333	Watches and clocks	..	..	..	..	..	..	..	..	..	..
34	**MOTOR VEHICLES, TRAILERS & SEMI-TRAILERS**	3,869	414	3,400	1,558	17,189	0	31	3,079,067	0	37,546
341	Motor vehicles	..	..	..	..	..	..	..	..	..	..
342	Bodies (coachwork) for motor vehicles	..	..	..	..	..	..	..	..	..	..
343	Parts, accessories for motor vehicles & their engines	..	..	..	..	..	..	..	..	..	..
35	**OTHER TRANSPORT EQUIPMENT**	3,061	756	2,472	2,419	8,293	0	27	2,526,593	0	26,124
351	Building and repairing of ships and boats	..	..	..	..	..	..	..	..	..	..
3511	Building and repairing of ships	..	..	..	..	..	..	..	..	..	..
3512	Building, repairing of pleasure & sporting boats	..	..	..	..	..	..	..	..	..	..
352	Railway, tramway locomotives & rolling stock	..	..	..	..	..	..	..	..	..	..
353	Aircraft and spacecraft	..	..	..	..	..	..	..	..	..	..
359	Transport equipment, nec	..	..	..	..	..	..	..	..	..	..
3591	Motorcycles	..	..	..	..	..	..	..	..	..	..
3592	Bicycles and invalid carriages	..	..	..	..	..	..	..	..	..	..
3599	Other transport equipment, nec	..	..	..	..	..	..	..	..	..	..
36	**FURNITURE; MANUFACTURING, NEC**	770	206	16,703	7,405	4,641	0	10	917,861	0	33,039
361	Furniture	..	..	..	..	..	..	..	..	..	..
369	Manufacturing, nec	..	..	..	..	..	..	..	..	..	..
3691	Jewellery and related articles	..	..	..	..	..	..	..	..	..	..
3692	Musical instruments	..	..	..	..	..	..	..	..	..	..
3693	Sports goods	..	..	..	..	..	..	..	..	..	..
3694	Games and toys	..	..	..	..	..	..	..	..	..	..
3699	Other manufacturing, nec	..	..	..	..	..	..	..	..	..	..
37	**RECYCLING**	..	..	..	..	..	..	..	..	..	..
371	Recycling of metal waste and scrap	..	..	..	..	..	..	..	..	..	..
372	Recycling of non-metal waste and scrap	..	..	..	..	..	..	..	..	..	..
CERR	**Non-specified industry**	414	0	222	0	1,190	25,974	0	0	1,950,000	20,780
15-37	**TOTAL MANUFACTURING**	353,989	33,986	298,222	264,348	474,111	27,856	1,064	92,751,631	10,774,000	**1,748,695**

Year: 1995

ISIC Revision 3 Industry Sector	Solid TJ	LPG TJ	Distiloil TJ	RFO TJ	Gas TJ	Biomass TJ	Steam TJ	Electr MWh	Own Use MWh	TOTAL TJ
15 FOOD PRODUCTS AND BEVERAGES	14,832	4,850	27,423	19,214	82,048	174	193	10,291,151	797,000	182,913
151 Production, processing and preserving (PPP)	..	..	..	..	..	..	..	..	..	..
1511 PPP of meat and meat products	..	..	..	..	..	..	..	..	..	..
1512 Processing and preserving of fish products	..	..	..	..	..	..	..	..	..	..
1513 Processing, preserving of fruit & vegetables	..	..	..	..	..	..	..	..	..	..
1514 Vegetable and animal oils and fats	..	..	..	..	..	..	..	..	..	..
152 Dairy products	..	..	..	..	..	..	..	..	..	..
153 Grain mill prod., starches & prepared animal feeds	..	..	..	..	..	..	..	..	..	..
1531 Grain mill products	..	..	..	..	..	..	..	..	..	..
1532 Starches and starch products	..	..	..	..	..	..	..	..	..	..
1533 Prepared animal feeds	..	..	..	..	..	..	..	..	..	..
154 Other food products	..	..	..	..	..	..	..	..	..	..
1541 Bakery products	..	..	..	..	..	..	..	..	..	..
1542 Sugar	..	..	..	..	..	..	..	..	..	..
1543 Cocoa, chocolate and sugar confectionery	..	..	..	..	..	..	..	..	..	..
1544 Macaroni, noodles, couscous & similar farinaceous prod.	..	..	..	..	..	..	..	..	..	..
1549 Other food products, nec	..	..	..	..	..	..	..	..	..	..
155 Beverages	..	..	..	..	..	..	..	..	..	..
1551 Distilling, rectifying and blending of spirits	..	..	..	..	..	..	..	..	..	..
1552 Wines	..	..	..	..	..	..	..	..	..	..
1553 Malt liquors and malt	..	..	..	..	..	..	..	..	..	..
1554 Soft drinks; production of mineral waters	..	..	..	..	..	..	..	..	..	..
16 TOBACCO PRODUCTS	0	87	222	196	1,244	0	0	117,699	0	2,173
17 TEXTILES	5,949	1,080	2,470	3,256	15,065	0	18	2,619,146	0	37,267
171 Spinning, weaving and finishing of textiles	..	..	..	..	..	..	..	..	..	..
1711 Preparation and spinning of textile fibres	..	..	..	..	..	..	..	..	..	..
1712 Finishing of textiles	..	..	..	..	..	..	..	..	..	..
172 Other textiles	..	..	..	..	..	..	..	..	..	..
1721 Made-up textile articles, except apparel	..	..	..	..	..	..	..	..	..	..
1722 Carpets and rugs	..	..	..	..	..	..	..	..	..	..
1723 Cordage, rope, twine and netting	..	..	..	..	..	..	..	..	..	..
1729 Other textiles, nec	..	..	..	..	..	..	..	..	..	..
173 Knitted and crocheted fabrics and articles	..	..	..	..	..	..	..	..	..	..
18 WEARING APPAREL, DRESSING & DYEING OF FUR	16	108	968	1,832	2,492	0	60	489,271	0	7,237
181 Wearing apparel, except fur apparel	..	..	..	..	..	..	..	..	..	..
182 Dressing and dyeing of fur; articles of fur	..	..	..	..	..	..	..	..	..	..
19 TANNING & DRESSING OF LEATHER, FOOTWEAR	478	130	402	163	1,059	0	0	217,764	0	3,016
191 Tanning and dressing of leather	..	..	..	..	..	..	..	..	..	..
1911 Tanning and dressing of leather	..	..	..	..	..	..	..	..	..	..
1912 Luggage, handbags, saddlery & harness	..	..	..	..	..	..	..	..	..	..
192 Footwear	..	..	..	..	..	..	..	..	..	..
20 WOOD AND WOOD PRODUCTS	716	1,823	4,007	226	6,103	0	19	1,501,281	0	18,299
201 Sawmilling and planing of wood	..	..	..	..	..	..	..	..	..	..
202 Products of wood, cork, straw & plaiting materials	..	..	..	..	..	..	..	..	..	..
2021 Veneer sheets	..	..	..	..	..	..	..	..	..	..
2022 Builders' carpentry and joinery	..	..	..	..	..	..	..	..	..	..
2023 Wooden containers	..	..	..	..	..	..	..	..	..	..
2029 Other products of wood	..	..	..	..	..	..	..	..	..	..
21 PAPER AND PAPER PRODUCTS	20,414	1,338	3,755	6,476	42,231	0	12	6,027,090	1,494,000	90,545
2101 Pulp, paper and paperboard	..	..	..	..	..	..	..	..	..	..
2102 Corrugated paper, paperboard and their containers	..	..	..	..	..	..	..	..	..	..
2109 Other articles of pulp and paperboard	..	..	..	..	..	..	..	..	..	..
22 PUBLISHING, PRINTING & REPRODUCTION	96	337	980	724	12,389	0	0	3,323,430	0	26,490
221 Publishing	..	..	..	..	..	..	..	..	..	..
2211 Publishing of books & brochures	..	..	..	..	..	..	..	..	..	..
2212 Publishing of newspapers and periodicals	..	..	..	..	..	..	..	..	..	..
2213 Publishing of recorded media	..	..	..	..	..	..	..	..	..	..
2219 Other publishing	..	..	..	..	..	..	..	..	..	..
222 Printing and related service activities	..	..	..	..	..	..	..	..	..	..
2221 Printing	..	..	..	..	..	..	..	..	..	..
2222 Service activities related to printing	..	..	..	..	..	..	..	..	..	..
223 Reproduction of recorded media	..	..	..	..	..	..	..	..	..	..
23 COKE, REFINED PETROLEUM PRODUCTS	47,692	0	186,646	95,337	10,984	0	0	6,513,000	2,942,000	353,515
231 Coke oven products	..	..	..	..	..	..	..	..	..	..
232 Refined petroleum products	..	..	..	..	..	..	..	..	..	..
233 Processing of nuclear fuel	..	..	..	..	..	..	..	..	..	..
24 CHEMICALS & CHEMICAL PRODUCTS	30,608	6,679	16,668	28,560	78,741	608	167	18,596,370	3,802,000	215,291
241 Basic chemicals	18,804	6,255	9,441	17,787	51,154	0	83	12,734,274	0	149,367
2411 Basic chemicals, exc. fertilizers & nitrogen compounds	16,805	5,958	8,187	13,758	42,627	0	60	10,518,371	0	125,261
2412 Fertilizers and nitrogen compounds	0	0	64	836	3,550	0	0	385,126	0	5,836
2413 Plastics in primary forms and synthetic rubber	1,999	297	1,190	3,193	4,977	0	23	1,830,777	0	18,270
242 Other chemical products	4,737	423	4,317	6,663	18,556	0	73	4,922,816	0	52,491
2421 Pesticides and other agro-chemical products	0	0	1,192	350	1,062	0	21	165,565	0	3,221
2422 Paints, varnishes and similar coatings	15	1	1,359	1,861	1,108	0	0	549,233	0	6,321
2423 Pharmaceuticals, medicinal chem. & botanical prod.	2,230	5	870	2,710	7,619	0	0	2,131,823	0	21,109
2424 Soap and detergents, perfumes etc.	84	401	467	765	1,881	0	0	847,309	0	6,648
2429 Other chemical products, nec	2,408	16	429	977	6,886	0	52	1,228,886	0	15,192
243 Man-made fibres	1,797	1	288	4,110	1,021	0	11	939,280		10,609
25 RUBBER AND PLASTICS PRODUCTS	17,348	9,574	7,333	5,382	29,552	0	73	6,952,088	0	94,290
251 Rubber products	4,459	251	2,810	1,357	6,680	0	20	1,376,067	..	20,531
2511 Rubber tyres and tubes	..	..	..	..	..	..	..	..	..	..
2519 Other rubber products	..	..	..	..	..	..	..	..	..	..
252 Plastic products	12,889	9,323	4,523	4,025	22,872	0	53	5,576,021	0	73,759
26 OTHER NON-METALLIC MINERAL PRODUCTS	27,963	3,083	7,655	9,472	37,881	0	256	5,403,361	0	105,762
261 Glass and glass products	253	0	1,322	7,437	9,757	0	66	1,265,897	0	23,392
269 Non-metallic mineral products, nec	27,710	3,083	6,333	2,035	28,124	0	190	4,137,464	0	82,370
2691 Pottery, china and earthenware	59	0	1,586	891	6,957	0	146	727,826	0	12,259
2692 Refractory ceramic products	776	2,773	1,255	713	7,958	0	0	405,503	0	14,935
2693 Structural non-refractory clay & ceramic prod.	..	..	..	..	..	..	..	..	..	..
2694 Cement, lime and plaster	26,651	0	734	0	7,719	0	8	1,559,627	0	40,727
2695 Articles of concrete, cement and plaster	224	310	2,758	431	5,490	0	36	1,444,508	0	14,449
2696 Cutting, shaping and finishing of stone	..	..	..	..	..	..	..	..	..	..
2699 Other non-metallic mineral products, nec	..	..	..	..	..	..	..	..	..	..

ISIS Energy Data Programme (IEA/OECD)

ISIC Revision 3 Industry Sector

ISIC	Industry Sector	Solid TJ	LPG TJ	Distiloil TJ	RFO TJ	Gas TJ	Biomass TJ	Steam TJ	Electr MWh	Own Use MWh	TOTAL TJ
27	**BASIC METALS**	170,837	1,664	7,996	31,304	111,262	1,914	44	14,909,660	1,746,000	372,410
271	Basic iron and steel	147,438	615	5,925	29,673	68,247	0	0	9,537,703	1,746,000	279,948
272	Basic precious and non-ferrous metals	4,335	698	1,230	1,591	23,622	0	10	3,635,725	0	44,575
273	Casting of metals	2,122	351	548	40	10,817	0	34	1,736,232	0	20,162
2731	Casting of iron and steel	..	..	..	..	..	..	..	..	..	..
2732	Casting non-ferrous metals	..	..	..	..	..	..	..	..	..	..
28	**FABRICATED METAL PRODUCTS**	72	35	2,562	239	12,599	0	46	2,755,339	0	25,472
281	Str. metal prod., tanks, reservoirs, steam generators	..	..	..	..	..	..	..	..	..	..
2811	Structural metal products	..	..	..	..	..	..	..	..	..	..
2812	Tanks, reservoirs and containers of metal	..	..	..	..	..	..	..	..	..	..
2813	Steam generators, exc. central heating hot water boilers	..	..	..	..	..	..	..	..	..	..
289	Other fabricated metal products	..	..	..	..	..	..	..	..	..	..
2891	Forging, pressing, stamping & roll-forming of metal	..	..	..	..	..	..	..	..	..	..
2892	Treatment and coating of metals	..	..	..	..	..	..	..	..	..	..
2893	Cutlery, hand tools and general hardware	..	..	..	..	..	..	..	..	..	..
2899	Other fabricated metal products, nec	..	..	..	..	..	..	..	..	..	..
29	**MACHINERY AND EQUIPMENT, NEC**	882	2,065	9,579	3,380	14,948	0	29	5,397,291	0	50,313
291	General purpose machinery	..	..	..	..	..	..	..	..	..	..
2911	Engines and turbines	..	..	..	..	..	..	..	..	..	..
2912	Pumps, compressors, taps and valves	..	..	..	..	..	..	..	..	..	..
2913	Bearings, gears, gearing and driving elements	..	..	..	..	..	..	..	..	..	..
2914	Ovens, furnaces and furnace burners	..	..	..	..	..	..	..	..	..	..
2915	Lifting and handling equipment	..	..	..	..	..	..	..	..	..	..
2919	Other general purpose machinery	..	..	..	..	..	..	..	..	..	..
292	Special purpose machinery	..	..	..	..	..	..	..	..	..	..
2921	Agricultural and forestry machinery	..	..	..	..	..	..	..	..	..	..
2922	Machine-tools	..	..	..	..	..	..	..	..	..	..
2923	Machinery for metallurgy	..	..	..	..	..	..	..	..	..	..
2924	Machinery for mining, quarrying and construction	..	..	..	..	..	..	..	..	..	..
2925	Machinery for food, beverage & tobacco processing	..	..	..	..	..	..	..	..	..	..
2926	Machinery for textile, apparel & leather production	..	..	..	..	..	..	..	..	..	..
2927	Machinery for weapons and ammunition	..	..	..	..	..	..	..	..	..	..
2929	Other special purpose machinery	..	..	..	..	..	..	..	..	..	..
293	Domestic appliances, nec	..	..	..	..	..	..	..	..	..	..
30	**OFFICE, ACCOUNTING & COMPUTING MACHINERY**	0	84	114	164	689	0	41	470,698	0	2,787
31	**ELECTRICAL MACHINERY & APPARATUS, NEC**	80	1,055	1,363	2,255	4,051	0	0	1,957,834	0	15,852
311	Electric motors, generators and transformers	..	..	..	..	..	..	..	..	..	..
312	Electricity distribution and control apparatus	..	..	..	..	..	..	..	..	..	..
313	Insulated wire and cable	..	..	..	..	..	..	..	..	..	..
314	Accumulators, primary cells & primary batteries	..	..	..	..	..	..	..	..	..	..
315	Electric lamps and lighting equipment	..	..	..	..	..	..	..	..	..	..
319	Other electrical equipment, nec	..	..	..	..	..	..	..	..	..	..
32	**RADIO, TV & COMMUNICATION EQUIP. & APP.**	0	45	1,314	1,651	3,037	0	13	2,167,570	0	13,863
321	Electronic valves, tubes, other electronic components	..	..	..	..	..	..	..	..	..	..
322	TV & radio transmitters, apparatus for line telephony	..	..	..	..	..	..	..	..	..	..
323	TV & radio receivers, recording apparatus	..	..	..	..	..	..	..	..	..	..
33	**MEDICAL PRECISION &OPTICAL INSTRUMENTS**	0	0	970	464	1,192	0	41	590,877	0	4,794
331	Medical appliances and instruments	..	..	..	..	..	..	..	..	..	..
3311	Medical, surgical equipment & orthopaedic app.	..	..	..	..	..	..	..	..	..	..
3312	Instruments & appliances for measuring, checking etc	..	..	..	..	..	..	..	..	..	..
3313	Industrial process control equipment	..	..	..	..	..	..	..	..	..	..
332	Optical instruments and photographic equipment	..	..	..	..	..	..	..	..	..	..
333	Watches and clocks	..	..	..	..	..	..	..	..	..	..
34	**MOTOR VEHICLES, TRAILERS & SEMI-TRAILERS**	3,630	492	3,451	1,130	19,403	0	32	3,546,808	0	40,907
341	Motor vehicles	..	..	..	..	..	..	..	..	..	..
342	Bodies (coachwork) for motor vehicles	..	..	..	..	..	..	..	..	..	..
343	Parts, accessories for motor vehicles & their engines	..	..	..	..	..	..	..	..	..	..
35	**OTHER TRANSPORT EQUIPMENT**	2,446	804	2,658	1,694	8,260	0	25	2,605,462	0	25,267
351	Building and repairing of ships and boats	..	..	..	..	..	..	..	..	..	..
3511	Building and repairing of ships	..	..	..	..	..	..	..	..	..	..
3512	Building, repairing of pleasure & sporting boats	..	..	..	..	..	..	..	..	..	..
352	Railway, tramway locomotives & rolling stock	..	..	..	..	..	..	..	..	..	..
353	Aircraft and spacecraft	..	..	..	..	..	..	..	..	..	..
359	Transport equipment, nec	..	..	..	..	..	..	..	..	..	..
3591	Motorcycles	..	..	..	..	..	..	..	..	..	..
3592	Bicycles and invalid carriages	..	..	..	..	..	..	..	..	..	..
3599	Other transport equipment, nec	..	..	..	..	..	..	..	..	..	..
36	**FURNITURE; MANUFACTURING, NEC**	593	227	16,669	5,291	4,577	0	9	978,467	0	30,888
361	Furniture	..	..	..	..	..	..	..	..	..	..
369	Manufacturing, nec	..	..	..	..	..	..	..	..	..	..
3691	Jewellery and related articles	..	..	..	..	..	..	..	..	..	..
3692	Musical instruments	..	..	..	..	..	..	..	..	..	..
3693	Sports goods	..	..	..	..	..	..	..	..	..	..
3694	Games and toys	..	..	..	..	..	..	..	..	..	..
3699	Other manufacturing, nec	..	..	..	..	..	..	..	..	..	..
37	**RECYCLING**	..	..	..	..	..	..	..	..	..	..
371	Recycling of metal waste and scrap	..	..	..	..	..	..	..	..	..	..
372	Recycling of non-metal waste and scrap	..	..	..	..	..	..	..	..	..	..
CERR	**Non-specified industry**	304	0	326	0	1,812	32,156	0	0	1,976,000	27,484
15-37	**TOTAL MANUFACTURING**	344,956	35,560	305,531	218,410	501,619	34,852	1,078	97,431,657	12,757,000	1,746,835

ISIC Revision 3 Industry Sector		Solid TJ	LPG TJ	Distiloil TJ	RFO TJ	Gas TJ	Biomass TJ	Steam TJ	Electr MWh	Own Use MWh	TOTAL TJ
15	**FOOD PRODUCTS AND BEVERAGES**	11,165	5,373	26,158	17,162	90,542	190	195	11,195,832	799,000	188,214
151	Production, processing and preserving (PPP)	..	..	..	..	..	..	..	..	..	..
1511	PPP of meat and meat products	..	..	..	..	..	..	..	..	..	..
1512	Processing and preserving of fish products	..	..	..	..	..	..	..	..	..	..
1513	Processing, preserving of fruit & vegetables	..	..	..	..	..	..	..	..	..	..
1514	Vegetable and animal oils and fats	..	..	..	..	..	..	..	..	..	..
152	Dairy products	..	..	..	..	..	..	..	..	..	..
153	Grain mill prod., starches & prepared animal feeds	..	..	..	..	..	..	..	..	..	..
1531	Grain mill products	..	..	..	..	..	..	..	..	..	..
1532	Starches and starch products	..	..	..	..	..	..	..	..	..	..
1533	Prepared animal feeds	..	..	..	..	..	..	..	..	..	..
154	Other food products	..	..	..	..	..	..	..	..	..	..
1541	Bakery products	..	..	..	..	..	..	..	..	..	..
1542	Sugar	..	..	..	..	..	..	..	..	..	..
1543	Cocoa, chocolate and sugar confectionery	..	..	..	..	..	..	..	..	..	..
1544	Macaroni, noodles, couscous & similar farinaceous prod.	..	..	..	..	..	..	..	..	..	..
1549	Other food products, nec	..	..	..	..	..	..	..	..	..	..
155	Beverages	..	..	..	..	..	..	..	..	..	..
1551	Distilling, rectifying and blending of spirits	..	..	..	..	..	..	..	..	..	..
1552	Wines	..	..	..	..	..	..	..	..	..	..
1553	Malt liquors and malt	..	..	..	..	..	..	..	..	..	..
1554	Soft drinks; production of mineral waters	..	..	..	..	..	..	..	..	..	..
16	**TOBACCO PRODUCTS**	0	99	80	276	1,427	0	0	131,788	0	2,356
17	**TEXTILES**	3,334	1,003	2,471	2,339	17,935	0	17	2,468,149	0	35,984
171	Spinning, weaving and finishing of textiles	..	..	..	..	..	..	..	..	..	..
1711	Preparation and spinning of textile fibres	..	..	..	..	..	..	..	..	..	..
1712	Finishing of textiles	..	..	..	..	..	..	..	..	..	..
172	Other textiles	..	..	..	..	..	..	..	..	..	..
1721	Made-up textile articles, except apparel	..	..	..	..	..	..	..	..	..	..
1722	Carpets and rugs	..	..	..	..	..	..	..	..	..	..
1723	Cordage, rope, twine and netting	..	..	..	..	..	..	..	..	..	..
1729	Other textiles, nec	..	..	..	..	..	..	..	..	..	..
173	Knitted and crocheted fabrics and articles	..	..	..	..	..	..	..	..	..	..
18	**WEARING APPAREL, DRESSING & DYEING OF FUR**	9	105	977	1,395	3,112	0	61	483,737	0	7,400
181	Wearing apparel, except fur apparel	..	..	..	..	..	..	..	..	..	..
182	Dressing and dyeing of fur; articles of fur	..	..	..	..	..	..	..	..	..	..
19	**TANNING & DRESSING OF LEATHER, FOOTWEAR**	276	125	311	164	1,299	0	0	211,474	0	2,936
191	Tanning and dressing of leather	..	..	..	..	..	..	..	..	..	..
1911	Tanning and dressing of leather	..	..	..	..	..	..	..	..	..	..
1912	Luggage, handbags, saddlery & harness	..	..	..	..	..	..	..	..	..	..
192	Footwear	..	..	..	..	..	..	..	..	..	..
20	**WOOD AND WOOD PRODUCTS**	437	2,244	4,341	165	6,433	0	19	1,702,473	0	19,768
201	Sawmilling and planing of wood	..	..	..	..	..	..	..	..	..	..
202	Products of wood, cork, straw & plaiting materials	..	..	..	..	..	..	..	..	..	..
2021	Veneer sheets	..	..	..	..	..	..	..	..	..	..
2022	Builders' carpentry and joinery	..	..	..	..	..	..	..	..	..	..
2023	Wooden containers	..	..	..	..	..	..	..	..	..	..
2029	Other products of wood	..	..	..	..	..	..	..	..	..	..
21	**PAPER AND PAPER PRODUCTS**	7,917	1,217	4,330	5,074	44,797	0	12	6,126,509	1,837,000	78,789
2101	Pulp, paper and paperboard	..	..	..	..	..	..	..	..	..	..
2102	Corrugated paper, paperboard and their containers	..	..	..	..	..	..	..	..	..	..
2109	Other articles of pulp and paperboard	..	..	..	..	..	..	..	..	..	..
22	**PUBLISHING, PRINTING & REPRODUCTION**	35	304	950	179	13,473	0	0	3,421,721	0	27,259
221	Publishing	..	..	..	..	..	..	..	..	..	..
2211	Publishing of books & brochures	..	..	..	..	..	..	..	..	..	..
2212	Publishing of newspapers and periodicals	..	..	..	..	..	..	..	..	..	..
2213	Publishing of recorded media	..	..	..	..	..	..	..	..	..	..
2219	Other publishing	..	..	..	..	..	..	..	..	..	..
222	Printing and related service activities	..	..	..	..	..	..	..	..	..	..
2221	Printing	..	..	..	..	..	..	..	..	..	..
2222	Service activities related to printing	..	..	..	..	..	..	..	..	..	..
223	Reproduction of recorded media	..	..	..	..	..	..	..	..	..	..
23	**COKE, REFINED PETROLEUM PRODUCTS**	45,584	0	195,836	94,597	12,934	0	0	6,687,000	3,017,000	362,163
231	Coke oven products	..	..	..	..	..	..	..	..	..	..
232	Refined petroleum products	..	..	..	..	..	..	..	..	..	..
233	Processing of nuclear fuel	..	..	..	..	..	..	..	..	..	..
24	**CHEMICALS & CHEMICAL PRODUCTS**	26,919	6,706	18,170	23,020	157,024	688	172	19,026,680	4,097,000	286,446
241	Basic chemicals	18,332	6,562	11,040	13,827	104,669	0	83	12,671,125	0	200,129
2411	Basic chemicals, exc. fertilizers & nitrogen compounds	16,452	6,503	10,751	9,652	92,421	0	59	10,527,877	0	173,738
2412	Fertilizers and nitrogen compounds	0	0	82	130	5,058	0	0	160,529	0	5,848
2413	Plastics in primary forms and synthetic rubber	1,880	59	207	4,045	7,190	0	24	1,982,719	0	20,543
242	Other chemical products	3,512	142	3,780	5,075	40,438	0	77	5,368,539	0	72,351
2421	Pesticides and other agro-chemical products	0	0	883	660	3,426	0	22	413,091	0	6,478
2422	Paints, varnishes and similar coatings	15	1	1,059	1,304	2,244	0	0	583,572	0	6,724
2423	Pharmaceuticals, medicinal chem. & botanical prod.	1,025	11	1,107	1,890	15,166	0	0	2,224,146	0	27,206
2424	Soap and detergents, perfumes etc.	0	97	163	535	5,643	0	55	840,835	0	9,465
2429	Other chemical products, nec	2,472	33	568	686	13,959	0	0	1,306,895	0	22,478
243	Man-made fibres	1,823	2	287	4,118	2,045	0	12	987,016	0	11,840
25	**RUBBER AND PLASTICS PRODUCTS**	12,833	12,045	6,631	5,002	30,535	0	72	7,729,156	0	94,943
251	Rubber products	3,144	297	2,189	992	6,566	0	19	1,452,786	..	18,437
2511	Rubber tyres and tubes	..	..	..	..	..	..	..	..	..	..
2519	Other rubber products	..	..	..	..	..	..	..	..	..	..
252	Plastic products	9,689	11,748	4,442	4,010	23,969	0	53	6,276,370	0	76,506
26	**OTHER NON-METALLIC MINERAL PRODUCTS**	31,026	2,810	7,727	5,518	39,785	0	248	5,300,006	0	106,194
261	Glass and glass products	293	0	1,337	4,362	10,080	0	65	1,294,284	0	20,796
269	Non-metallic mineral products, nec	30,733	2,810	6,390	1,156	29,705	0	183	4,005,722	0	85,398
2691	Pottery, china and earthenware	62	0	2,006	567	7,717	0	141	742,623	0	13,166
2692	Refractory ceramic products	602	2,497	1,287	440	8,417	0	0	397,833	0	14,675
2693	Structural non-refractory clay & ceramic prod.	..	..	..	..	..	..	..	..	..	..
2694	Cement, lime and plaster	29,861	0	667	0	8,854	0	8	1,644,377	0	45,310
2695	Articles of concrete, cement and plaster	208	313	2,430	149	4,717	0	34	1,220,889	0	12,246
2696	Cutting, shaping and finishing of stone	..	..	..	..	..	..	..	..	..	..
2699	Other non-metallic mineral products, nec	..	..	..	..	..	..	..	..	..	..

ISIC Revision 3 Industry Sector	Solid TJ	LPG TJ	Distiloil TJ	RFO TJ	Gas TJ	Biomass TJ	Steam TJ	Electr MWh	Own Use MWh	TOTAL TJ
27 BASIC METALS	180,864	1,433	8,475	24,785	124,911	1,117	44	15,735,390	1,863,000	391,570
271 Basic iron and steel	156,525	542	6,240	23,719	74,493	0	0	10,171,063	1,863,000	291,428
272 Basic precious and non-ferrous metals	2,761	581	1,220	1,026	24,835	0	10	3,707,873	0	43,781
273 Casting of metals	894	310	544	40	11,837	0	34	1,856,454	0	20,342
2731 Casting of iron and steel	..	..	..	..	..	..	..	..	..	..
2732 Casting non-ferrous metals	..	..	..	..	..	..	..	..	..	..
28 FABRICATED METAL PRODUCTS	73	41	2,634	215	14,864	0	45	2,641,967	0	27,383
281 Str. metal prod., tanks, reservoirs, steam generators	..	..	..	..	..	..	..	..	..	..
2811 Structural metal products	..	..	..	..	..	..	..	..	..	..
2812 Tanks, reservoirs and containers of metal	..	..	..	..	..	..	..	..	..	..
2813 Steam generators, exc. central heating hot water boilers	..	..	..	..	..	..	..	..	..	..
289 Other fabricated metal products	..	..	..	..	..	..	..	..	..	..
2891 Forging, pressing, stamping & roll-forming of metal	..	..	..	..	..	..	..	..	..	..
2892 Treatment and coating of metals	..	..	..	..	..	..	..	..	..	..
2893 Cutlery, hand tools and general hardware	..	..	..	..	..	..	..	..	..	..
2899 Other fabricated metal products, nec	..	..	..	..	..	..	..	..	..	..
29 MACHINERY AND EQUIPMENT, NEC	762	2,444	9,645	2,986	17,505	0	29	5,115,243	0	51,786
291 General purpose machinery	..	..	..	..	..	..	..	..	..	..
2911 Engines and turbines	..	..	..	..	..	..	..	..	..	..
2912 Pumps, compressors, taps and valves	..	..	..	..	..	..	..	..	..	..
2913 Bearings, gears, gearing and driving elements	..	..	..	..	..	..	..	..	..	..
2914 Ovens, furnaces and furnace burners	..	..	..	..	..	..	..	..	..	..
2915 Lifting and handling equipment	..	..	..	..	..	..	..	..	..	..
2919 Other general purpose machinery	..	..	..	..	..	..	..	..	..	..
292 Special purpose machinery	..	..	..	..	..	..	..	..	..	..
2921 Agricultural and forestry machinery	..	..	..	..	..	..	..	..	..	..
2922 Machine-tools	..	..	..	..	..	..	..	..	..	..
2923 Machinery for metallurgy	..	..	..	..	..	..	..	..	..	..
2924 Machinery for mining, quarrying and construction	..	..	..	..	..	..	..	..	..	..
2925 Machinery for food, beverage & tobacco processing	..	..	..	..	..	..	..	..	..	..
2926 Machinery for textile, apparel & leather production	..	..	..	..	..	..	..	..	..	..
2927 Machinery for weapons and ammunition	..	..	..	..	..	..	..	..	..	..
2929 Other special purpose machinery	..	..	..	..	..	..	..	..	..	..
293 Domestic appliances, nec	..	..	..	..	..	..	..	..	..	..
30 OFFICE, ACCOUNTING & COMPUTING MACHINERY	0	96	176	166	753	0	42	512,266	0	3,077
31 ELECTRICAL MACHINERY & APPARATUS, NEC	40	1,206	928	2,267	4,300	0	0	2,094,061	0	16,280
311 Electric motors, generators and transformers	..	..	..	..	..	..	..	..	..	..
312 Electricity distribution and control apparatus	..	..	..	..	..	..	..	..	..	..
313 Insulated wire and cable	..	..	..	..	..	..	..	..	..	..
314 Accumulators, primary cells & primary batteries	..	..	..	..	..	..	..	..	..	..
315 Electric lamps and lighting equipment	..	..	..	..	..	..	..	..	..	..
319 Other electrical equipment, nec	..	..	..	..	..	..	..	..	..	..
32 RADIO, TV & COMMUNICATION EQUIP. & APP.	0	51	1,774	1,657	3,219	0	13	2,315,244	0	15,049
321 Electronic valves, tubes, other electronic components	..	..	..	..	..	..	..	..	..	..
322 TV & radio transmitters, apparatus for line telephony	..	..	..	..	..	..	..	..	..	..
323 TV & radio receivers, recording apparatus	..	..	..	..	..	..	..	..	..	..
33 MEDICAL PRECISION & OPTICAL INSTRUMENTS	0	0	1,078	341	1,299	0	43	649,198	0	5,098
331 Medical appliances and instruments	..	..	..	..	..	..	..	..	..	..
3311 Medical, surgical equipment & orthopaedic app.	..	..	..	..	..	..	..	..	..	..
3312 Instruments & appliances for measuring, checking etc	..	..	..	..	..	..	..	..	..	..
3313 Industrial process control equipment	..	..	..	..	..	..	..	..	..	..
332 Optical instruments and photographic equipment	..	..	..	..	..	..	..	..	..	..
333 Watches and clocks	..	..	..	..	..	..	..	..	..	..
34 MOTOR VEHICLES, TRAILERS & SEMI-TRAILERS	1,955	591	4,089	1,149	21,431	0	33	3,679,608	0	42,495
341 Motor vehicles	..	..	..	..	..	..	..	..	..	..
342 Bodies (coachwork) for motor vehicles	..	..	..	..	..	..	..	..	..	..
343 Parts, accessories for motor vehicles & their engines	..	..	..	..	..	..	..	..	..	..
35 OTHER TRANSPORT EQUIPMENT	1,402	1,004	3,039	1,518	9,524	0	27	2,809,932	0	26,630
351 Building and repairing of ships and boats	..	..	..	..	..	..	..	..	..	..
3511 Building and repairing of ships	..	..	..	..	..	..	..	..	..	..
3512 Building, repairing of pleasure & sporting boats	..	..	..	..	..	..	..	..	..	..
352 Railway, tramway locomotives & rolling stock	..	..	..	..	..	..	..	..	..	..
353 Aircraft and spacecraft	..	..	..	..	..	..	..	..	..	..
359 Transport equipment, nec	..	..	..	..	..	..	..	..	..	..
3591 Motorcycles	..	..	..	..	..	..	..	..	..	..
3592 Bicycles and invalid carriages	..	..	..	..	..	..	..	..	..	..
3599 Other transport equipment, nec	..	..	..	..	..	..	..	..	..	..
36 FURNITURE; MANUFACTURING, NEC	378	287	15,658	5,829	4,921	0	9	1,131,936	0	31,157
361 Furniture	..	..	..	..	..	..	..	..	..	..
369 Manufacturing, nec	..	..	..	..	..	..	..	..	..	..
3691 Jewellery and related articles	..	..	..	..	..	..	..	..	..	..
3692 Musical instruments	..	..	..	..	..	..	..	..	..	..
3693 Sports goods	..	..	..	..	..	..	..	..	..	..
3694 Games and toys	..	..	..	..	..	..	..	..	..	..
3699 Other manufacturing, nec	..	..	..	..	..	..	..	..	..	..
37 RECYCLING	..	..	..	..	..	..	..	..	..	..
371 Recycling of metal waste and scrap	..	..	..	..	..	..	..	..	..	..
372 Recycling of non-metal waste and scrap	..	..	..	..	..	..	..	..	..	..
CERR Non-specified industry	208	0	356	0	1,707	24,973	0	0	2,048,000	19,871
15-37 TOTAL MANUFACTURING	325,217	39,184	315,834	195,804	623,730	26,973	1,081	101,169,370	13,661,000	1,842,853

Annex I

DEFINITIONS OF
ISIC REVISION 2
MANUFACTURING INDUSTRY SECTORS

INTERNATIONAL STANDARD INDUSTRIAL CLASSIFICATION (ISIC) MANUFACTURING INDUSTRY CODES (Revision 2)

Code	Industry
3100	**FOOD, BEVERAGES AND TOBACCO**
311/2	**FOOD**
3111	Slaughtering, preparing and preserving meat
3112	Dairy products
3113	Canning, preserving of fruits and vegetables
3114	Canning, preserving and processing of fish
3115	Vegetable and animal oils and fats
3116	Grain mill products
3117	Bakery products
3118	Sugar factories and refineries
3119	Cocoa, chocolate and sugar confectionery
3121	Other food products
3122	Prepared animal feeds
3130	**BEVERAGES**
3131	Distilling, rectifying and blending of spirits
3132	Wine industries
3133	Malt liquors and malts
3134	Soft drinks
3140	**TOBACCO**
3200	**TEXTILES, APPAREL AND LEATHER**
3210	**TEXTILES**
3211	Spinning weaving and finishing textiles
3212	Made-up goods excluding wearing apparel
3213	Knitting mills
3214	Carpets and rugs
3215	Cordage, rope and twine
3219	Other textiles
3220	**WEARING APPAREL, EXCEPT FOOTWEAR**
3230	**LEATHER AND FUR PRODUCTS**
3231	Tanneries and leather finishing
3232	Fur dressing and dyeing industries
3233	Leather prods. ex. footwear and wearing apparel
3240	**FOOTWEAR, EX. RUBBER AND PLASTIC**
3300	**WOOD PRODUCTS AND FURNITURE**
3310	**WOOD PRODUCTS, EXCEPT FURNITURE**
3311	Sawmills, planing and other wood mills
3312	Wooden and cane containers
3319	Other wood and cork products
3320	**FURNITURE, FIXTURES, EXCL. METALLIC**
3400	**PAPER, PUBLISHING AND PRINTING**
3410	**PAPER AND PRODUCTS**
3411	Pulp, paper and paperboard articles
3412	Containers of paper and paperboard
3419	Other pulp, paper and paperboard articles
3420	**PRINTING AND PUBLISHING**

ISIS Energy Data Programme (IEA/OECD)

Code	Industry
3500	**CHEMICAL PRODUCTS**
3510	**INDUSTRIAL CHEMICALS**
3511	Basic industrial chemicals excl. fertilizers
3512	Fertilizers and pesticides
3513	Synthetic resins and plastic materials
3520	**OTHER CHEMICALS**
3521	Paints, varnishes and lacquers
3522	Drugs and medicines
3523	Soap, cleaning preparations, perfumes, cosmetics
3529	Other chemical products
3530	**PETROLEUM REFINERIES**
3540	**MISC. PETROLEUM AND COAL PRODUCTS**
3550	**RUBBER PRODUCTS**
3551	Tyres and tubes
3559	Other rubber products
3560	**PLASTIC PRODUCTS**
3600	**NON-METALLIC MINERAL PRODUCTS**
3610	**POTTERY, CHINA, EARTHENWARE**
3620	**GLASS AND PRODUCTS**
3690	**OTHER NON-METAL. MINERAL PRODUCTS**
3691	Structural clay products
3692	Cement, lime and plaster
3699	Other non-metallic mineral products
3700	**BASIC METAL INDUSTRIES**
3710	**IRON AND STEEL**
3720	**NON-FERROUS METALS**
3800	**METAL PRODUCTS, MACHINERY, EQUIP.**
3810	**METAL PRODUCTS**
3811	Cutlery, hand tools and general hardware
3812	Furniture and fixtures primarily of metal
3813	Structural metal products
3819	Other fabricated metal products
3820	**NON-ELECTRICAL MACHINERY**
3821	Engines and turbines
3822	Agricultural machinery and equipment
3823	Metal and wood working machinery
3824	Special industrial machinery
3825	Office, computing and accounting machinery
3829	Other non-electrical machinery and equipment
3830	**ELECTRICAL MACHINERY**
3831	Electrical industrial machinery
3832	Radio, TV and communications equipment
3833	Electrical appliances and housewares
3839	Other electrical apparatus and supplies
3840	**TRANSPORT EQUIPMENT**
3841	Shipbuilding
3842	Railroad equipment
3843	Motor vehicles
3844	Motorcycles and bicycles
3845	Aircraft
3849	Other transport equipment

Code	Industry
3850	**PROFESSIONAL AND SCIENTIFIC EQUIPMENT**
3851	Professional equipment
3852	Photographic and optical goods
3853	Watches and clocks
3900	**OTHER MANUFACTURING INDUSTRIES**
3901	Jewellery and related articles
3902	Musical instruments
3903	Sporting and athletic goods
3909	Other manufactures
SERR	**Unallocated industry**
3000	**TOTAL MANUFACTURING**

ISIS Energy Data Programme (IEA/OECD)

Annex II

DEFINITIONS OF
ISIC REVISION 3
MANUFACTURING INDUSTRY SECTORS

ISIS Energy Data Programme (IEA/OECD)

INTERNATIONAL STANDARD INDUSTRIAL CLASSIFICATION (ISIC) MANUFACTURING INDUSTRY CODES (Revision 3)

Code	Industry
1500	**MANUFACTURE OF FOOD PRODUCTS AND BEVERAGES**
1510	Production, processing and preservation of meat, fish, fruit, vegetables, oils and fats
1511	Production, processing and preserving of meat and meat products
1512	Processing and preserving of fish and fish products
1513	Processing and preserving of fruit and vegetables
1514	Manufacture of vegetable and animal oils and fats
1520	Manufacture of dairy products
1530	Manufacture of grain mill products, starches and starch products, and prepared animal feeds
1531	Manufacture of grain mill products
1532	Manufacture of starches and starch products
1533	Manufacture of prepared animal feeds
1540	Manufacture of other food products, n.e.c.
1541	Manufacture of bakery products
1542	Manufacture of sugar
1543	Manufacture of cocoa, chocolate and sugar confectionery
1544	Manufacture of macaroni, noodles, couscous and similar farinaceous products
1549	Manufacture of other food products, n.e.c.
1550	Manufacture of beverages
1551	Distilling, rectifying and blending of spirits; ethyl alcohol production from fermented materials
1552	Manufacture of wines
1553	Manufacture of malt liquors and malt
1554	Manufacture of soft drinks; production of mineral waters
1600	**MANUFACTURE OF TOBACCO PRODUCTS**
1700	**MANUFACTURE OF TEXTILES**
1710	Spinning weaving and finishing of textiles
1711	Preparation and spinning of textile fibres; weaving of textiles
1712	Finishing of textiles
1720	Manufacture of other textiles
1721	Manufacture of made-up textile articles, except apparel
1722	Manufacture of carpets and rugs
1723	Manufacture of cordage, rope, twine and netting
1729	Manufacture of other textiles, nec
1730	Manufacture of knitted and crocheted fabrics and articles
1800	**MANUFACTURE OF WEARING APPAREL, DRESSING AND DYEING OF FUR**
1810	Manufacture of wearing apparel, except fur apparel
1820	Dressing and dyeing of fur; manufacture of articles of fur
1900	**TANNING AND DRESSING OF LEATHER; MANUFACTURE OF LUGGAGE, HANDBAGS, SADDLERY, HARNESS AND FOOTWEAR**
1910	Tanning and dressing of leather; manufacture of luggage, handbags, saddlery, harness and footwear
1911	Tanning and dressing of leather
1912	Luggage, handbags and the like, saddlery and harness
1920	Manufacture of footwear

ISIS Energy Data Programme (IEA/OECD)

Code	Industry
2000	**MANUFACTURE OF WOOD AND OF PRODUCTS OF WOOD AND CORK, EXCEPT FURNITURE; MANUFACTURE OF ARTICLES OF STRAW AND PLAITING MATERIALS**
2010	Sawmilling and planing of wood
2020	Manufacture of products of wood, cork, straw and plaiting materials
2021	Manufacture of veneer sheets; manufacture of plywood, laminboard, particle board and other panels and boards
2022	Manufacture of builders' carpentry and joinery
2023	Manufacture of wooden containers
2029	Manufacture of other products of wood; articles of cork, straw and plaiting materials
2100	**MANUFACTURE OF PAPER AND PAPER PRODUCTS**
2101	Manufacture of pulp, paper and paperboard
2102	Manufacture of corrugated paper and paperboard and containers of paper and paperboard
2109	Manufacture of other articles of pulp and paperboard
2200	**PUBLISHING, PRINTING AND REPRODUCTION OF RECORDED MEDIA**
2210	Publishing
2211	Publishing of books, brochures, musical books and other publications
2212	Publishing of newspapers, journals and periodicals
2213	Publishing of recorded media
2219	Other publishing
2220	Printing and service activities related to printing
2221	Printing
2222	Service activities related to printing
2230	Reproduction of recorded media
2300	**MANUFACTURE OF COKE, REFINED PETROLEUM PRODUCTS AND NUCLEAR FUEL**
2310	Manufacture of coke oven products
2320	Manufacture of refined petroleum products
2330	Processing of nuclear fuel
2400	**MANUFACTURE OF CHEMICALS AND CHEMICAL PRODUCTS**
2410	Manufacture of basic chemicals
2411	Manufacture of basic chemicals, except fertilizers and nitrogen compounds
2412	Manufacture of fertilizers and nitrogen compounds
2413	Manufacture of plastics in primary forms and synthetic rubber
2420	Manufacture of other chemical products
2421	Manufacture of pesticides and other agro-chemical products
2422	Manufacture of paints, varnishes and similar coatings, printing ink and mastics
2423	Manufacture of pharmaceuticals, medicinal chemicals and botanical products
2424	Manufacture of soap and detergents, cleaning and polishing preparations, perfumes and toilet preparations
2429	Manufacture of other chemical products, n.e.c.
2430	Manufacture of man-made fibres
2500	**MANUFACTURE OF RUBBER AND PLASTICS PRODUCTS**
2510	Manufacture of rubber products
2511	Manufacture of rubber tyres and tubes; retreading and rebuilding of rubber tyres
2519	Manufacture of other rubber products
2520	Manufacture of plastic products
2600	**MANUFACTURE OF OTHER NON-METALLIC MINERAL PRODUCTS**
2610	Manufacture of glass and glass products
2690	Manufacture of non-metallic mineral products, n.e.c.

Code	Industry
2691	Manufacture of non-structural non-refractory ceramic ware ("Pottery, china and earthenware")
2692	Manufacture of refractory ceramic products
2693	Manufacture of structural non-refractory clay and ceramic products
2694	Manufacture of cement, lime and plaster
2695	Manufacture of articles of concrete, cement and plaster
2696	Cutting, shaping and finishing of stone
2699	Manufacture of other non-metallic mineral products, nec
2700	**MANUFACTURE OF BASIC METALS**
2710	Manufacture of basic iron and steel
2720	Manufacture of basic precious and non-ferrous metals
2730	Casting of metals
2731	Casting of iron and steel
2732	Casting non-ferrous metals
2800	**MANUFACTURE OF FABRICATED METAL PRODUCTS, EXCEPT MACHINERY AND EQUIPMENT**
2810	Manufacture of structural metal products, tanks, reservoirs and steam generators
2811	Manufacture of structural metal products
2812	Manufacture of tanks, reservoirs and containers of metal
2813	Manufacture of steam generators, except central heating hot water boilers
2890	Manufacture of other fabricated metal products; metal working service activities
2891	Forging, pressing, stamping and roll-forming of metal; powder metallurgy
2892	Treatment and coating of metals; general mechanical engineering on a fee or contract basis
2893	Manufacture of cutlery, hand tools and general hardware
2899	Manufacture of other fabricated metal products, nec
2900	**MANUFACTURE OF MACHINERY AND EQUIPMENT, N.E.C.**
2910	Manufacture of general purpose machinery
2911	Manufacture of engines and turbines, except aircraft, vehicle and cycle engines
2912	Manufacture of pumps, compressors, taps and valves
2913	Manufacture of bearings, gears, gearing and driving elements
2914	Manufacture of ovens, furnaces and furnace burners
2915	Manufacture of lifting and handling equipment
2919	Manufacture of other general purpose machinery
2920	Manufacture of special purpose machinery
2921	Manufacture of agricultural and forestry machinery
2922	Manufacture of machine-tools
2923	Manufacture of machinery for metallurgy
2924	Manufacture of machinery for mining, quarrying and construction
2925	Machinery for food, beverage and tobacco processing
2926	Manufacture of machinery for textile, apparel and leather production
2927	Manufacture of machinery for weapons and ammunition
2929	Manufacture of other special purpose machinery
2930	Manufacture of domestic appliances, n.e.c.
3000	**MANUFACTURE OF OFFICE, ACCOUNTING AND COMPUTING MACHINERY**
3100	**MANUFACTURE OF ELECTRICAL MACHINERY AND APPARATUS, NEC**
3110	Manufacture of electric motors, generators and transformers
3120	Manufacture of electricity distribution and control apparatus
3130	Manufacture of insulated wire and cable
3140	Manufacture of accumulators, primary cells and primary batteries
3150	Manufacture of electric lamps and lighting equipment
3190	Manufacture of other electrical equipment, n.e.c.

ISIS Energy Data Programme (IEA/OECD)

Code	Industry
3200	**MANUFACTURE OF RADIO, TELEVISION AND COMMUNICATION EQUIPMENT AND APPARATUS**
3210	Manufacture of Electronic valves and tubes and other electronic components
3220	Manufacture of television and radio transmitters and apparatus for line telephony and line telegraphy
3230	Manufacture of television and radio receivers, sound or video recording or reproducing apparatus, and associated goods
3300	**MANUFACTURE OF MEDICAL, PRECISION AND OPTICAL INSTRUMENTS, WATCHES AND CLOCKS**
3310	Manufacture of medical appliances and instruments and appliances for measuring, checking, testing, navigating and other purposes, except optical instruments
3311	Manufacture of medical and surgical equipment and orthopaedic appliances
3312	Manufacture of instruments and appliances for measuring, checking, testing, navigating and other purposes, except industrial process control equipment
3313	Manufacture of industrial process control equipment
3320	Manufacture of optical instruments and photographic equipment
3330	Manufacture of watches and clocks
3400	**MANUFACTURE OF MOTOR VEHICLES, TRAILERS AND SEMI-TRAILERS**
3410	Manufacture of motor vehicles
3420	Manufacture of bodies (coachwork) for motor vehicles; manufacture of trailers and semi-trailers
3430	Manufacture of parts and accessories for motor vehicles and their engines
3500	**MANUFACTURE OF OTHER TRANSPORT EQUIPMENT**
3510	Building and repairing of ships and boats
3511	Building and repairing of ships
3512	Building and repairing of pleasure and sporting boats
3520	Manufacture of railway and tramway locomotives and rolling stock
3530	Manufacture of aircraft and spacecraft
3590	Manufacture of transport equipment, n.e.c.
3591	Manufacture of motorcycles
3592	Manufacture of bicycles and invalid carriages
3599	Manufacture of other transport equipment, nec
3600	**MANUFACTURE OF FURNITURE; MANUFACTURING, N.E.C.**
3610	Manufacture of furniture
3690	Manufacturing, n.e.c.
3691	Manufacture of jewellery and related articles
3692	Manufacture of musical instruments
3693	Manufacture of sports goods
3694	Manufacture of games and toys
3699	Other manufacturing, n.e.c.
3700	**RECYCLING**
3710	Recycling of metal waste and scrap
3720	Recycling of non-metal waste and scrap
CERR	**Unallocated Industry**
15_37	**TOTAL MANUFACTURING**

Annex III

GENERAL CONVERSION FACTORS

ISIS Energy Data Programme (IEA/OECD)

General Conversion Factors for Energy

To:	TJ	Gcal	Mtoe	MBtu	GWh
From:	multiply by:				
TJ	1	238.8	2.388×10^{-5}	947.8	0.2778
Gcal	4.1868×10^{-3}	1	10^{-7}	3.968	1.163×10^{-3}
Mtoe	4.1868×10^{4}	10^{7}	1	3.968×10^{7}	11630
MBtu	1.0551×10^{-3}	0.252	2.52×10^{-8}	1	2.931×10^{-4}
GWh	3.6	860	8.6×10^{-5}	3412	1

Decimal Prefixes

10^{1}	deca (da)	10^{-1}	deci (d)
10^{2}	hecto (h)	10^{-2}	centi (c)
10^{3}	kilo (k)	10^{-3}	milli (m)
10^{6}	mega (M)	10^{-6}	micro (μ)
10^{9}	giga (G)	10^{-9}	nano (n)
10^{12}	tera (T)	10^{-12}	pico (p)
10^{15}	peta (P)	10^{-15}	femto (f)
10^{18}	exa (E)	10^{-18}	atto (a)

ISIS Energy Data Programme (IEA/OECD)

Did you Know?

This publication is available in electronic form

Many OECD publications and data sets are now available in electronic form to suit your needs at affordable prices.

For our statistical publications we use powerful software platforms (Ivation's Beyond 20/20 or STATWISE) that allow you to get the maximum value from the data. Other publications are available using the simple Acrobat/PDF presentation. **Delivery platforms** range from magnetic tape through CD-Rom and diskettes to online via internet. **Stand alone and network** versions are offered for many titles.

For more information about electronic editions of this publication, or to ask for a catalogue of all our electronic publications, contact your nearest OECD Centre (see overleaf).

Le saviez-vous ?

La version électronique de cette publication est disponible !

Désormais, afin de mieux répondre à vos besoins, un grand nombre de publications et de données de l'OCDE sont disponibles sous forme électronique à des prix très abordables.

Nos études statistiques sont présentées sur des logiciels puissants (Beyond 20/20 ou Statwise) permettant d'optimiser les données au maximum. Certaines publications sont également disponibles sur Acrobat/PDF.

Par ailleurs, **un éventail très large de supports** vous est proposé : bande magnétique, Cédérom, disquette et interrogation en ligne via Internet. De nombreux titres sont également proposés en **versions monoposte et réseau.**

Pour de plus amples informations sur les versions électroniques de cette publication ou pour obtenir le catalogue de nos éditions électroniques, n'hésitez pas à contacter le Centre OCDE le plus proche (voir verso).

OECD-OCDE

A POTENT INSTRUMENT OF GLOBAL CHANGE
UN INSTRUMENT PUISSANT DE CHANGEMENT ET DE REFORME DANS LE MONDE

Where to send your request:

Où envoyer votre demande:

In Austria, Germany and Switzerland / En Allemagne, en Autriche et en Suisse

OECD Centre Bonn / Centre OCDE de Bonn
August-Bebel-Allee 6,
D-53175 Bonn
Tel.: (49-228) 959 1215
Fax: (49-228) 959 1218
E-mail: bonn.contact@oecd.org
Internet: www.oecd.org/bonn

In Latin America / En Amérique latine

OECD Centre Mexico / Centre OCDE de Mexico
Edificio INFOTEC
Av. San Fernando No. 37
Col. Toriello Guerra
Tlalpan C.P. 14050,
Mexico D.F.
Tel.: (525) 528 10 38
Fax: (525) 606 13 07
E-mail: mexico.contact@oecd.org
Internet: rtn.net.mx/ocde/

In the United States / Aux États-Unis

OECD Center Washington / Centre OCDE de Washington
2001 L Street N.W., Suite 650
Washington, DC 20036-4922
Tel.: (202) 785 6323
Toll free / Numéro vert : (800) 456-6323
Fax: (202) 785 0350
E-mail: washington.contact@oecd.org
Internet: www.oecdwash.org

In Asia / En Asie

OECD Centre Tokyo / Centre OCDE de Tokyo
Landic Akasaka Bldg.
2-3-4 Akasaka, Minato-ku,
Tokyo 107-0052
Tel.: (81-3) 3586 2016
Fax: (81-3) 3584 7929
E-mail : center@oecdtokyo.org
Internet: www.oecdtokyo.org

In the rest of the world / Dans le reste du monde
OECD Paris Centre / Centre OCDE de Paris
2 rue André-Pascal, 75775 Paris Cedex 16, France
Orders / Commandes : Fax: 33 (0)1 49 10 42 76

Enquiries / Renseignements : Tel: 33 (0)1 45 24 81 22 Fax: 33 (0) 1 45 24 19 50
E-mail : sales@oecd.org

Online Ordering: www.oecd.org/publications *(secure payment with credit card)*
Commande en ligne : www.oecd.org/publications *(paiement sécurisé par carte de crédit)*

OECD Main Switchboard / Standard OCDE : 33 (0) 1 45 24 82 00
Internet: www.oecd.org

OECD PUBLICATIONS, 2, rue André-Pascal, 75775 PARIS CEDEX 16
PRINTED IN FRANCE
(30 2000 01 3 P) ISBN 92-64-05887-7 – No. 51138 2000